THE PHYSICS OF REALITY

Space, Time, Matter, Cosmos

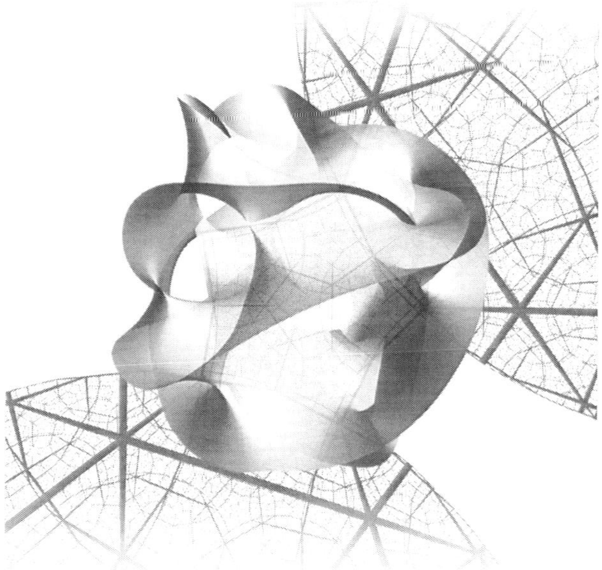

THE PHYSICS OF REALITY
Space, Time, Matter, Cosmos

Proceedings of the 8th Symposium Honoring Mathematical Physicist Jean-Pierre Vigier
Covent Garden, London, UK, 15 – 18 August 2012

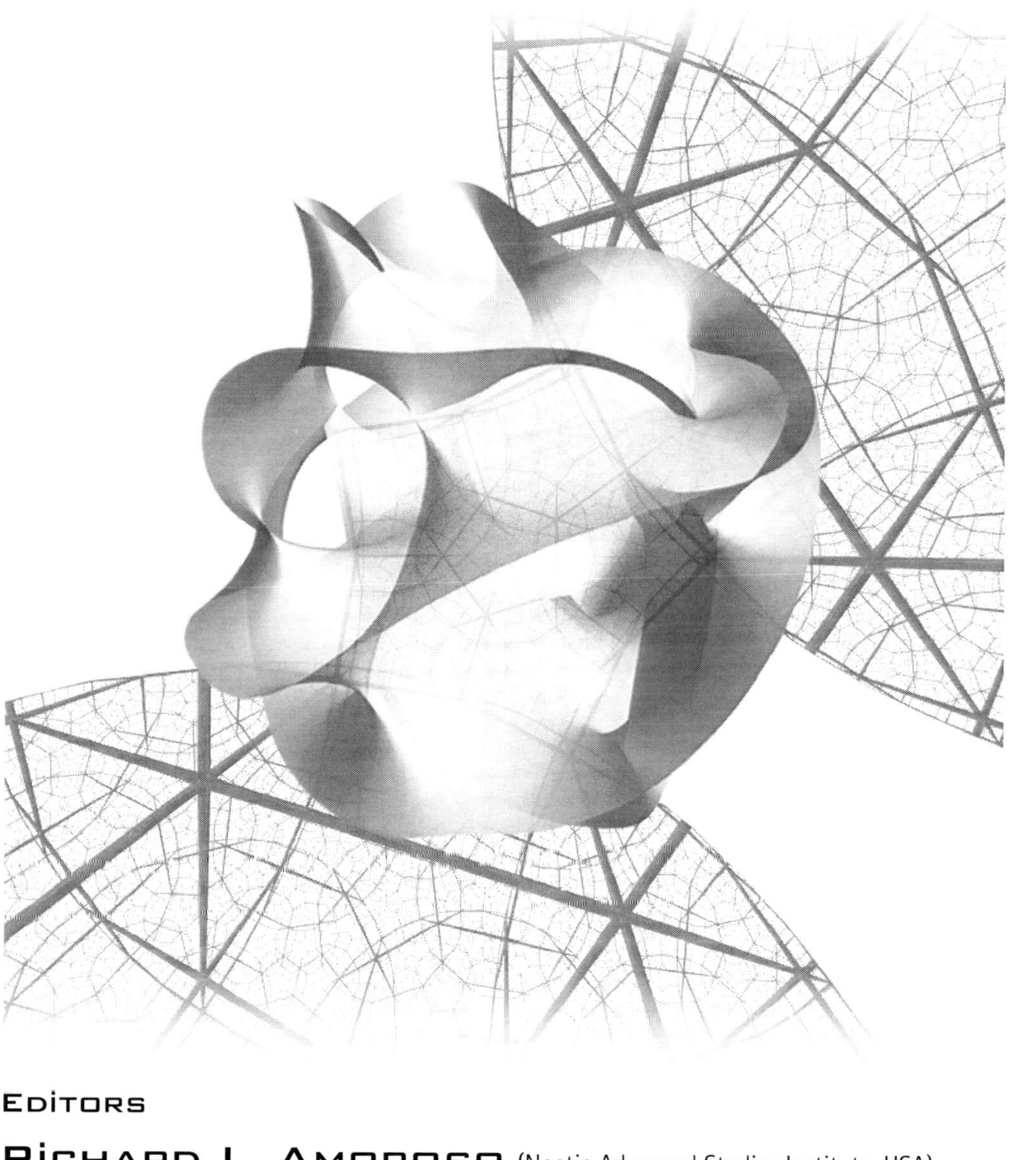

EDITORS

RICHARD L. AMOROSO (Noetic Advanced Studies Institute, USA)
LOUIS H. KAUFFMAN (University of Illinois at Chicago, USA)
PETER ROWLANDS (University of Liverpool, UK)

World Scientific

NEW JERSEY • LONDON • SINGAPORE • BEIJING • SHANGHAI • HONG KONG • TAIPEI • CHENNAI

Published by

World Scientific Publishing Co. Pte. Ltd.
5 Toh Tuck Link, Singapore 596224
USA office: 27 Warren Street, Suite 401-402, Hackensack, NJ 07601
UK office: 57 Shelton Street, Covent Garden, London WC2H 9HE

British Library Cataloguing-in-Publication Data
A catalogue record for this book is available from the British Library.

THE PHYSICS OF REALITY
Space, Time, Matter, Cosmos
Proceedings of the 8th Symposium Honoring Mathematical Physicist Jean-Pierre Vigier

ISBN 978-981-4504-77-5

Printed and bound in Great Britain by
Marston Book Services Limited, Oxfordshire

Jean-Pierre Vigier

L'hérétique de la physique!

Foreword

Biographical Reflections on an 'Emperor's New Clothes'

I fully agree with you about the significance and educational value of methodology as well as history and philosophy of science. So many people today—and even professional scientists—seem to me like somebody who has seen thousands of trees but has never seen a forest. A knowledge of the historic and philosophical background gives that kind of independence from prejudices of his generation from which most scientists are suffering. This independence created by philosophical insight is—in my opinion—the mark of distinction between a mere artisan or specialist and a real seeker after truth [1].—Einstein

How does it happen that a properly endowed natural scientist comes to concern himself with epistemology? Is there no more valuable work in his specialty? I hear many of my colleagues saying, and I sense it from many more, that they feel this way. I cannot share this sentiment.... Concepts that have proven useful in ordering things easily achieve such an authority over us that we forget their earthly origins and accept them as unalterable givens. Thus they come to be stamped as 'necessities of thought,' 'a priori givens.' The path of scientific advance is often made impassable for a long time through such errors. For that reason, it is by no means an idle game if we become practiced in analyzing the long-commonplace concepts and exhibiting those circumstances upon which their justification and usefulness depend, how they have grown up, individually, out of the givens of experience. By this means, their all-too-great authority will be broken [2].—Einstein

A conference proceedings provides an opportunity to explore new ideas, accommodated by a less stringent review process, and, supposedly, a setting with colleagues in an optimally tolerant frame of mind, as an aid in the dissemination of those new theoretical or empirical ideas that are, perhaps, still colored as a little 'outside the box.' I mention this umbrella although it is generally known, because in addition to the usual battery of interesting forefront theories and research reports, a surprising seminal foray into a new form of research method, possibly completing the tools of epistemology, occurred in this particular Vigier symposium. I feel some discussions on this matter is appropriate in the Vigier tradition, as Vigier himself was branded *L'hérétique de la physique*!

The last great age of discovery happened about 100 years ago with the advent of the quantum theory. It is this editor's opinion that the next age is immanent and that this age of discovery will also bring us from the current 'modern age' to a new age of mind or consciousness, additionally completing the tools of human epistemology—which began in the dark ages of myth and superstition, then entered the age of reason and logic until logic failed, for example, in terms of demonstrating whether a heavier object falls faster under the influence of a gravitational field. If one were to read the ancient texts as to why a heavier object should fall faster, the logic is so appealing and elegant that even knowing it to be otherwise today, one is inclined to believe it. Galileo is credited with instituting the modern age of empiricism at that time. But now, empiricism has also become troubled. Some experiments are said to require an accelerator the size of the Universe or a computation that takes the age of the Universe to complete. Most pertinently, a Nobel Prize was given for the discovery of the Big Bang cosmology. Critics said this award was given 20 years too early. Hubble discovered redshift, not a Doppler expansion of the Universe. The issue here is not bad experimental data, but erroneous or biased interpretation of that data as a 'proof' of the Big Bang.

The ancient Greek philosopher Plato said, 'no matter how great one's intelligence is, or how vast the breadth of one's wisdom is, noetic insight is beyond the self; it is transcendent—coming from the cosmos.' Transcendence is the third and final tool completing human epistemology. It has been shown that transcendence can be used as a tool in scientific theory formation [3]. This is not something to be taken without the usual scientific insistence on rigor. The usual entry ticket to the profession of science is a PhD, which takes 8 to 10 years of university study with often another two years in post-doc research. The utility of transcendence does not come automatically to the natural man and, thus, seems to require a minimum of about five years of intensive study in a 'spiritual' discipline—a semblance of the effort required to achieve entry-level virtuosity as a violinist or pianist, for example. Regarding the philosophy of physics, centuries ago, the study of causality and the fundamental nature of space, time, matter or the Universe (cosmology) were part of metaphysics. Until recently, it was said, 'first comes speculation, then speculation squared, then cosmology.' Now, cosmology is a hot arena of research.

I will repeat this statement once more later on in this foreword: *Before Einstein came along, relativity was considered metaphysical and outside the domain of the natural science. Man is a spiritual being. We anticipate with the immanent discovery of the mind (the basis of an observer)* [3–5] *that transcendence is immanently reaching the end of its classification as metaphysics, as a form of 'empirical metaphysics' will arise.*

The BCS-V8 symposium is assuredly the first in history to utilize transcendence as part of an international physics symposium, making this volume a truly Galilean- or Copernican-class document! With this introduction regarding this matter, I will first set the stage by relating to you an unusual incident regarding Jean-Pierre Vigier's past because, although the incident itself is transcendent, it elucidates the context for describing the transcendent path to the new physics that follows eight years later and is included as a seminal part of the development of a chapter in this proceedings [6–8].

I consider myself Vigier's final student and colleague; during the last decade of his life, I went to *Pierre et Marie Curie Universite* in Paris to collaborate with him two to six times a year. Perhaps, because he was older or, perhaps, because he was indelibly cemented in his role as *L'hérétique de la science,* not so many 'bothered' him; but the large box of papers and books he received weekly from all over the world suggested otherwise. Vigier was a powerful soul; he reminded me of a blast furnace: his light was so bright. Even in his eighties, most scientists half his age did not seem to have half his intellectual strength and dynamic erudition. I had mentioned to Vigier that a main interest of mine was consciousness or the role of the observer in the physical science. He emphatically told me he had absolutely no interest in the subject. About a year later, I brought a copy of my first book on the subject to show him [9], not entirely because it was a book on the mind–body problem, but because it was an attempt to show him that I was actually doing something of substance, as he had made the comment to me once 'that it was not necessarily easy to tell if a person was intelligent or not by sitting down with them in a discussion.' Much to my surprise, we spent much of the remainder of the day discussing consciousness and the work of all the natural philosophers of any merit throughout history. When we finished, I commented, 'I thought you weren't interested in consciousness/metaphysics.' He said unequivocally, '*that doesn't mean I don't know about it….*'

On a related note, I had told Jean-Pierre that I was a High Priest in the LDS (Mormon) church. As is the case in much of Europe and probably more so in France, Vigier typically informed me he had little interest in organized religion. Jean-Pierre told me he grew up as a Calvinist, which according to Vigier was a sect that believed God put dinosaur bones in the ground to trick mankind into believing that the Earth was older than 7,000 years as literal biblical interpretation suggests. He mentioned his visit to the Natural History Museum at five years old and that, at the next Sunday school, he exclaimed loudly, 'I saw the bones. I saw the bones.' Apparently, he made such a disturbance that the school authorities or his parents kept him from attending Calvinism from then on, …, and he smiled broadly. I loved that the child was still very much alive in Vigier.

Three or four days after Jean-Pierre passed away on May 4, 2004, as a result of hitting his head on a brass cornice at his bank in Paris the year before, I was house sitting for some friends in the Oakland hills, CA. I had dropped my friends off at the airport by 6 AM. I had brought tons of work, but as I often do on the first day in that environment, I turned on the Sci-Fi channel and watched it, junk or not, continuously until 10 PM that night, when I finally turned it off in disgust at myself for wasting a precious day. I was sitting on a couch in their idyllic family room, where often I could see deer amble through the canopy of fir trees. I closed my eyes and meditatively queried, 'should I go to bed, should I turn the TV back on, or should I finally start working?' At that moment, I had a vision that lasted no longer than it takes to say, 'He will help you.' The voice was clear but the faces were dim and appeared smaller than the normal size. I immediately recognized Vigier. The speaker was one Steve White, both a physicist at Lawrence Livermore National Laboratory and my former stake president (regional LDS leader). I only recognized him because of the unique way he jerked his head—which I will explain later—as he had done when I visited him on his death bed in the hospital some years earlier.

I was quite surprised, as one might imagine, at this unusual occurrence. I knew with some assurance that Vigier knew no other Mormons besides me and that, in France, there were very few LDS people or religious people at all for that matter. As I pondered this, I realized it suggested something about the nature of the 'spirit world' that, in three days, Vigier, who died in Paris, not only was able to find my deceased stake president and make arrangements to come and visit me across the world in California, but also have a 'transcendent experience' arranged for me. Jean-Pierre and, I had been in the middle of several projects and, in my initial arrogance, I assumed he was somehow going to help me move this work forward. I had gotten to the point in life of developing my transcendent abilities, with sufficient decades of hard work (an LDS charge), where I could upon occasion, mostly in the sacred Mormon temples, sense the presence of a deceased spirit. But this usually took the form of merely a simple emotion like gratitude during a performance of a proxy ordinance. That was all I was capable of discerning. Mental content or words were much more challenging.

During those initial days and weeks, I felt Jean-Pierre's presence almost constantly. I tried to listen. Eventually, I even waved my arms around, exclaiming, 'OK, I'm ready,' all to no avail. Then, I sardonically thought, perhaps, a Ouija board could work, scoffing in frustration. This was three or four months later. Finally, embarrassed in retrospect at my myopic and arrogant perception, I finally have the 'ah ha' moment came to me: Jean-Pierre didn't appear in order to offer me help; he wanted me to help him! It is a Christian doctrine that proxy ordinances done for deceased friends or family members lead to security and privileges in the spirit world [10]. Since de Broglie never married and had no other family, I decided to perform his proxy work also, in case it gave Vigier and his friend de Broglie some special association in the spirit world.

Vigier continued to be around from time to time with diminishing returns. I relate two additional pertinent anecdotes to embellish this history and for what is to come (and the commonality of such occurrences in some circles), as I am not sure when such an opportunity will come again for me to promulgate such experiences in an open arena.

The first incident relates to the late husband of my close friend and colleague Elizabeth Rauscher. After William Van Bise passed away, I told her the story about Vigier and asked permission to perform Bill's proxy ordinances. I didn't sense his presence during the ordinances as I did during Vigier's; but three or four months later, I was awoken by a bright red-orange ball of light at about 4 AM. It was a visitation from William, and I was bowled over by the force of his gratitude. To me, this was another lesson in the nature of the spirit world—the difference between when a proxy ordinance is boldly requested and when an emissary is sent afterwards to 'sell' or proselytize an ordinance to an unsuspecting recipient.

The other incident occurred recently in Budapest, Hungary, near the end of a visit to an old building with my colleagues. The old building was both an institute and a home to its director. I was packing a small duffel bag in the large room (with a about-20-foot ceiling) where I slept during my visit. His young daughter had a day earlier made a present of a small plastic bag of seeds, saying she thought I'd enjoy them. They were so good I kept the bag in hopes that, on a future visit, I could find more. I pressed shut a small zippered pouch on the side of the duffel. At that moment, the bag flew about 10 feet straight up in the air. I paused frozen. I am a physicist; I thought that is impossible. There couldn't have been sufficient air in the plastic bag; in fact, chances are there was virtually no air in it. If it did somehow fly out, it certainly couldn't have gone so high. As I continued to think, looking up to gauge the distance again, I heard two guys laughing. Peering into the space, I immediately recognized one of the guys who 'punked' me as my father, who had been quite a tease in life and to whom I probably owe much of my sense of humor. The other person kept recognition from me for two more days. It turned out to be Vigier! Vigier also had quite a sense of humor, but more than that, I was delighted and nonplussed that he had made friends with my father in the spirit world. I wondered also if, due to the nature of the 'veil' between the life on Earth and the life in the spirit world, it have to take the two of them to perform such a prank…

Now, we come to the incident at the core of and the reason for the delineation in this foreword. I attest to the veracity of these anecdotes; I have no reason to relate them other than the advancement of human knowledge, especially in the face of embarrassment to myself or my colleagues. This and the other incidents I have related are foreign to most people, and those who have them hold them sacred and keep quiet about them, and in fact, this is generally what the authorities of Mormonism earnestly suggest. But for me, it is different; it has become a part of my profession to 'open this door.' Finally, true transcendent abilities are no more difficult to acquire than the virtuosity in a musical instrument, i.e., at least five years of very hard work is needed. Man is a spiritual being.

It was early May 2012. I had just returned from a lecture tour in Gramado, Brazil, a few days before. Vigier was around, but I didn't pay much attention to him as circumstances had given me a troubled mind, along with the weariness of 40 hours of travel from door to door, from California to remote Brazil. My excuse also was I had never been able to go beyond recognizing his presence, in any case, and had no idea what he was able to read from my mind. And I had an 18-hour drive (ahead of me in my old Honda Civic), which did not improve my mood. Typically, when I receive these visitations from the dead, they occur as lit spheres about one-third the size of a normal head in an inner space, I assume, near mine, in a proximity such that I can sense some information as I mentioned, like who they are and an emotion. That is how it is for me, for another variance occurs. The apparitions always glow to greater or lesser degrees. This time, Vigier's image turned dark after a while, starting just before I began driving. In my tired funk, I still ignored him. After finally arriving in the Escalante Desert, I rested, sitting at my desk in my small fifth-wheel trailer. I was finally relaxed enough to realize Vigier was ticked at me. I would have, at this point, finally tried to listen; but in the interim, probably frustrated, he had enlisted the aid of Steve White again. At that moment, I had one of the most profound transcendent experiences of my career.

I was very, very surprised; Vigier wanted me to write a paper for this conference on 'tight bound states (TBS) in the hydrogen atom.' I knew the quote from the LDS prophet Brigham Young, who had said, 'all scientific discovery comes as revelation from God,' and I felt this is true like the poetic muse, but most scientists denigrated spirituality and wouldn't experience this directly, although there are a few anecdotes in the history of science. However, I did not think that direct interventions of this type occurred, and if one is not aware of the 'open faucet' of transcendence, as it seems to be the case for the typical scientist, the transcendent aspects of theory formation would be subconscious in any case. I pride myself as having a very open mind; but I had absolutely no interest in TBS because I myopically denied their possibility out of prejudice (more Bohr orbits below the ground-state orbit?)—firstly, as Vigier tried to use them to explain cold fusion, and secondly, during the 2000 symposium at UC Berkeley (the first symposium I organized and the third symposium in his honor), he gave his keynote address on TBS instead of a paper on CMB and Redshift that we had been preparing together for months. So I felt slighted and was puerilely miffed at him in this regard. I had given absolutely no thought to TBS in the intervening 12 years.

This vision was different from any I had ever had; instead of a glowing word, paragraph or graphic image, or dream, or voice—something like a pillar of light was put over me, and I suddenly 'knew stuff'—I was infused with knowledge! Much new physics was infused into my mind by Vigier in this regard with the aid of Steve White or by other methods invisible to me during my stay in the high desert. Later, I was able to find a copy of Vigier's seminal TBS paper [6] on the internet. During the time between these two events, I soon realized TBS is profoundly significant not only to developing and possibly experimentally proving a basis for my work on multiverse cosmology, but also as a boon to M-Theory. (See the TBS chapters in this volume.)

I do not believe in random coincidence, but the transcendent 'pillar of light' I just mentioned was a marked event like the planting of a flag on the Moon; thus, the following is obtuse enough to include. On the day I arrived in the desert, there was an eclipse of the Sun near sunset; a couple weeks later since the day I left, there was an eclipse of the Moon on the opposite horizon near sunrise. Why mention it then? The reason is because of its apparent importance. There are now 14 experimental protocols that could demonstrate HAM cosmology. TBS was the 12th; it is by far the simplest and least expensive. So? Demonstrating TBS will also 'prove' the existence of large-scale additional dimensions (LSXD) and, more importantly, open the door to the next age of discovery, i.e., classical mechanics, quantum mechanics and, now, unified field mechanics!

I have decided not to include Steve White as a coauthor of the paper, but I shall credit him here. This is where his unusual head jerk comes in. When I visited Steve on his death bed (cancer of the spleen, which is rarely detected until it is too late) with a woman friend Pauline, we did not know beforehand that he was going to die. When he told Pauline he was, she immediately began to weep. He had been looking at her. Then, he turned his head very quickly with a surprising jerk and gazed into my eyes. I was startled inside, but as a man, I held firm, deciding to show no emotion other than a note of sadness on my countenance. Steve was a big (overweight) man, but he also had a presence larger than life. He had no reason to care anything about me overtly; I was one of thousands in his 'flock', but I loved him because he was the type of person who exuded warmth and made one feel like one were a best friend.

Well, here's the rub: a couple of months later (not after his death but after the Escalante Desert TBS infusion), back at my home then in Oakland, CA, early one morning, Steve appeared to me and did a few of those (now-famous-to-me) head turns in a row. I had been pondering a putative protocol for demonstrating TBS (see the papers in this volume); this head-jerk scenario led to what I call a working nomenclature, a 'couple punch.' Vigier often talked of spin–spin coupling or spin–orbit coupling, etc. This so-called couple punch has become a part of the experimental strategy: to set up the resonant coupling in conjunction with a kick at various phase nodes in the continuous-state cycle [4], as an aid to exploring the possible relation of TBS to additional-dimensional (LSXD) parameters in HAM's unique version of oscillating Poincaré-dodecahedral-AdS_5-space M-Theoretic Calabi –Yau mirror symmetry.

There you have it. My mind has been taken to an empyreal realm. I attest to the veracity of this experience in hopes that it provides an indicium to help usher in a new age of epistemology. There is nothing to gain in wasting effort if this were a fabrication of fiction. There are not overly many similar incidents in the history of science. Descartes's 'revelation from God,' as he called it, on the distinction between *res cogitans* and *res extensa* is a prominent case. Another is Friedrich August Kekulé von Stradonitz's dream on the structure of benzene in the shape of a ring with alternating single and double carbon bonds as a snake eating its tail. There are different levels of inspiration. There is a saying among those trying to develop transcendent abilities concerning a threshold experience, 'how do you know the insight wasn't just the warped action of the pepperoni pizza you ate recently?' Where to draw the line between imagination and transcendence is a difficult challenge.

Plato talked of 'noetic insight' as beyond the individual. Transcendence as a tool in scientific theory formation has the ability to save billions and enormous amounts of time. Einstein said he owed a portion of his success to 'remembering the future.' It is not to take the place of empiricism, only to facilitate it!

A salient purpose in recording these events is to bring a strong scintilla of evidence of reintegrating science and theology, not as mutually exclusive but as the opposite ends of the long spectrum of human epistemology. We give a preliminary formula for developing transcendent abilities here [3]. One of the symposium delegates following my commentary on the TBS transcendent scenario politely brought me a poster he made during my discourse time; it said 'Mysphyt: Mystic Physicist' (short for MYStic PHYsicisT). I smiled in acknowledgment, and he said he was serious. What is your opinion? Perhaps, at this stage, such a position can only be strengthened in retrospect... Note: when I mentioned to him I would include this in the foreword, he said, 'the first time I used this was in a class with Fritjof Capra, who said I should learn to spell.'

We have made an initial foray into completing the tools of epistemology in relation to the TBS scenario. We have stated that the seed of comprehension began in the dark ages of myth and superstition. The ancient Greeks are credited with introducing logic. Analysis and reason brought about the birth of science and a golden age. It is curious that Aristotle was strongly opposed to experiment. When logic was found to fail in certain circumstances, those like Galileo instituted the current age of empiricism. But now, there are problems with empiricism, such as the misinterpretation of data and the suggestion by some that a supercollider the size of the Universe or a computation that takes the age of the Universe is required to obtain some results. What remains to complete the tools of epistemology [3]? It may come as a surprise to some and be met with resistance by others; but it was the ancient Greek philosopher Plato who hinted at the answer. Plato suggested that no matter how great one's intelligence or how vast one's wisdom is, there is what he called a 'noetic insight' that is beyond the self and comes from the cosmos. This, of course, is transcendence. What will make it unpopular for a while is that it is based on anthropic principles, that the Universe is designed for intelligence and that some form of unified-field–regime teleological-action principle exists that proffers guidance and insight. The proof, of course, will be the rabbits pulled out of a hat—breakthroughs that can't be easily understood or developed without a causative transcendent action involved. We have given one example here in relation to TBS in the hydrogen atom. We realize the difficulty of using that as a proof of transcendent intervention in scientific theory formation. How the reader accepts a testament of someone saying they were incapable of making such and such a discovery on their own efforts when, for all practical purposes, to an observer, the evidence (the fact of having the discovery) suggests that that can't be true because here it is and they have done it...

'*Every great advance in science has issued from a new audacity of imagination.*'—John Dewey

Writing the two TBS papers [7,8] required and led to, in an interesting manner, an understanding of the geometric topology of LSXD, which I can't resist relating narratively: If you are traveling an uncommon path—considering today's population of 7,051,664,961 souls, at the time of writing—heading North from the ancient homeland of the Kawaiisu people, now called Tehachapi (Kawaiisu language for 'difficult path'), towards Mojave on Hwy 58, CA, and have just past Love♥s Travel Stop (curious because my surname means Love), you might want to put your vehicle into neutral at the crest of the hill and coast down the off-ramp off the highway at exit 156, Sand Cyn Rd., and pull off the road at the bottom of the hill, coming to a stop in the dirt area on the right. That's what I did on September 9, 2012, when the 'timing belt' on the old Honda Civic I purchased for US$50 suddenly broke.

When you look up, out the front windshield, you will see hundreds of wind generators on the Mojave mountain range a few miles ahead—unusual but getting less and less uncommon in today's climate of the burgeoning proliferation of wind power; but bearing straight ahead, there is an obvious group of generators in a row. Counting to the sixth generator, one finds a pair of generators very close together, one a little behind the other. The trefoil-propeller rotations seem chaotic; but periodically, there is an apparent momentary coupling of four of the blades, giving the appearance of the face of a cube (see Figs. in [7,8]).

I had already been working on a dual-quaternion representation (*i, j, k* vertex or singularity that could be visually represented as a rotating trefoil propeller, in its cyclic permutations) of HD Calabi–Yau mirror symmetry. I work best when I can visualize geometric structures, and this wind-generator incident solidified my rumination. So now, I have an HD nilpotent [4,11] potentia in my cosmology [4] that resulted in a 3D Copenhagen quantum-state correspondence. Carrying the model all the way back to the U_F, the Calabi–Yau symmetry has a second doubling or complement to its duality such that this nilpotent totality of still higher-dimensional potentia has a resultant that is now a fully realized 3-space Euclidean cube (see [7,8]).

Could I have derived this essential symmetry another way? Perhaps, but I tell you outright I don't think I was capable of telekinetically snapping the timing belt in the vehicle and be stranded in the middle of nowhere, which is the last thing I consciously would have wanted. Many troubles and expenses came with this incident, but I will not relate them here. In general, I have pledged allegiance to and given myself over to the 'yoke' of this particular '*Zeitgeist*,' which leaves me to wonder if this incident wasn't by design! Curious coincidence, Love♥s Travel Stop, 'difficult path' or design—I'll let the reader decide. I'm just extremely happy to have the indelible required insight … 'to open this door' at long last.

Since much of this forward will not be perceived as 'politically correct,' I will first undergo the folly of putative justification. Let's begin with an editorial in the 12 October 2012 issue of Science. Noted physicist James S. Langer, a past president of the American Physical Society with numerous other exceptional achievements, says—while speaking about the special section in that issue focusing on physics in developmental biology (biophysics)—in his essay 'Enabling Scientific Innovation' [12]: '*… our ability to take advantage of … opportunities has been eroded in an environment where innovation and interdisciplinary research are systematically discouraged.*'

My first precession, not a crime, is under the umbrella of biophysics. Before we go further, let's review what is meant by a science of 'physics' and see if there is any wiggle room to do something not necessarily popular but perfectly legal. The etymology of the term physics comes from the Greek word φύσις ('*physis*'), meaning nature, which in the broadest sense is what the discipline of physics charges itself to do—to analyze matter, motion, space, time, energy and force in order to determine how the multiverse behaves by understanding natural phenomena. Physics is the oldest discipline, and because it is also the most fundamental discipline, most other sciences are based on and constrained by its principles.

Before Einstein, relativity was considered metaphysical. If the TBS experiment and any of the other 13 actually end up demonstrating LSXD or teleological U_F effects, as divined for this noetic researcher, through the utility of transcendence as a tool in scientific theory formation [3], then what has been claimed here will no longer be considered metaphysical, and a new regime (as has happened many times before) will be added to the natural science called physics!

References and Notes

[1] Einstein, A. (1944) letter to Robert A. Thornton, 7 December, in Hilliard III, A. G. (1983). In memoriam, 1982: A tribute to Dr. Robert A. Thornton, Journal of Negro History, 142–143.

[2] Einstein, A. (1916) Memorial notice for equalize Ernst Mach, *Physikalische Zeitschrift* 17: 101–102.

[3] Amoroso, R.L. (2010) Completing epistemology: the utility of Transcendence as a tool in scientific theory formation, in R.L. Amoroso (ed.) Complementarity of Mind and Body: Realizing the Dream of Descartes, Einstein, and Eccles, New York: Nova Science.

[4] Amoroso, R.L. & Rauscher, E.A. (2009) The Holographic Anthropic Multiverse: Formalizing the Ultimate Geometry of Reality, Singapore: World Scientific.

[5] Amoroso, R.L. (2014) Through the Looking Glass: Discovery of the Mind, Hackensack: World Scientific.

[6] Dragié, A., Marić, Z. & Vigier, J-P (2000) New quantum mechanical tight bound states and 'cold fusion' experiments, Physics Letters A, 265: 163–167.

[7] Amoroso, R.L. & Vigier, J-P (2013) Evidencing 'tight bound states' in the hydrogen atom: Empirical manipulation of large-scale XD in violation of QED, in R.L. Amoroso, L.H. Kauffman & P. Rowlands (eds.) The Physics of Reality, Space, Time, Matter, Cosmos, Hackensack: World Scientific.

[8] Amoroso, R.L. (2013) "Shut The Front Door!": Obviating the Challenge of Large-Scale Extra Dimensions and Psycho-physical Bridging, in R.L. Amoroso, L.H. Kauffman & P. Rowlands (eds.) The Physics of Reality, Space, Time, Matter, Cosmos, Hackensack: World Scientific.

[9] Amoroso, R.L. et al. (eds.) (2000) Science and the Primacy Consciousness: Intimation of a 21st Century Revolution Orinda: Noetic Press.

[10] Holy Bible, New Testament: 1 Corinthians 15:29, *Else what shall they do which are baptized for the dead, if the dead rise not at all? Why are they then baptized for the dead?*

[11] Rowlands, P. (2013) Space and antispace, in R.L. Amoroso, L.H. Kauffman & P. Rowlands (eds.) The Physics of Reality, Space, Time, Matter, Cosmos, Hackensack: World Scientific.

[12] Langer, J.S. (2012) Enabling scientific innovation, Science, 338: 6104, 171.

Richard L. Amoroso
Noetic Advanced Studies Institute
Escalante Desert, UT
July 2013

Preface

The eighth Vigier symposium held as a joint meeting with the British Computer Society - Cybernetics Machine Group (BCS-V8) at its national headquarters in Covent Garden, London, was very successful with about 50 delegates coming from dozens of countries around the world. We gratefully acknowledge the support and service of the BCS. It may not have been possible to hold the conference without its efficient and pleasant assistance.

As the 52 chapters will reflect, this symposium was a delightful and erudite cornucopia of ideas. Rather than encumbering this preface with a précis on each of the 52 chapters, we have decided for the sake of brevity to let the published abstract of each paper substitute for that undertaking.

The volume covers a broad spectrum of theoretical and mathematical physics by researchers from over 20 nations from four continents. Like Vigier himself, the Vigier symposia are noted for addressing *avant-garde,* cutting-edge topics in contemporary physics. Of the six proceedings honoring J.-P. Vigier, this is perhaps the most exciting as several important breakthroughs are introduced for the first time even to the extent of one bold 'noeticist' daring to introduce a method for completing the tools of epistemology. The most interesting breakthrough—in view of the recent violations of QED—is a continuation of the pioneering work by Vigier on tight bound states (TBS) in hydrogen (energy levels below the lowest Bohr orbit). The new experimental protocol described not only promising empirical proof of large-scale extra dimensions (LSXD) but also the birth of the field of unified field mechanics, ushering in a new age of discovery.

Work on quantum computing redefines the qubit in a manner that the uncertainty principle may be routinely violated; other breakthroughs occur in the utility of quaternion algebra in extending our understanding of the nature of the fermionic singularity or point particle. There are several other discoveries of equal magnitude making this volume a must-have acquisition for the library of any serious, forward-looking researcher.

This volume features prominent authors such as mathematician L. H. Kauffman, an internationally renowned knot theorist, and Walter Schempp, a grandson of Kepler and an NMR pioneer.

For nearly 20 years, Vigier symposia proceedings have been uniquely known internationally for their unbridled forays into cutting-edge topics that other physicists have no want to address. Advancement in science does not occur by 'hugging the status quo.'

This volume differs from others especially in its unique treatment of historical dilemmas such as the nature of the singularity or point particle, its introduction of a relativistic qubit (r-qubit) beyond the usual Bloch 2-sphere model, its studies of complex space and new cosmological modeling—to cite a few salient examples.

This volume is very important to the advancement of many fields in physics because in an age of conservatism relating to the Standard Model, Vigier symposium attendees have been noted for nearly 20 years to 'color outside the box' in areas such as extended electromagnetic theory, Dirac polarized vacuum, m_γ, thermodynamics, alternatives to Big Bang cosmology, and quantum computing by the violation of the uncertainty principle. The recent experimental violation of QED at NIST, for example, has been predicted by Vigier attendees for over a decade.

Peter Rowlands, Liverpool, UK
Peter J. Marcer, Var, France
24 April 2013

Contents

Section II
Special and General Relativity

Laws of Form, Majorana Fermions, and Discrete Physics

LOUIS H. KAUFFMAN

Department of Mathematics, Statistics & Computer Science
University of Illinois at Chicago, Chicago, IL, 60607-704 USA
Kauffman@uic.edu

The present paper is a discussion of how the idea of a distinction gives rise to fundamental physics. We discuss everything and nothing, and in the end we see that they are not distinct.

Keywords: G. Spencer-Brown, Laws of form, Majorana Fermions, Mark of distinction

1. Introduction

The present paper is a discussion of how the idea of a distinction gives rise to fundamental physics. We will contact the physics and the mathematics in the course of developing our theme. In this introduction we give a brief tour of the terrain of the paper. To begin with, we recall the calculus of indications of G. Spencer-Brown [3]. This is a formal system composed of a single symbol, the 'mark'.

This symbol is seen to make a distinction in the plane upon which it is written and hence it is possible to inscribe further marks either inside a given mark or outside that mark. Below we illustrate an internal inscription.

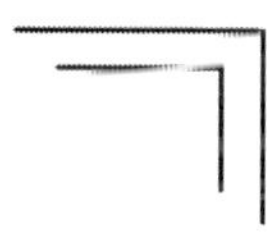

And further below we indicate an adjacent inscription.

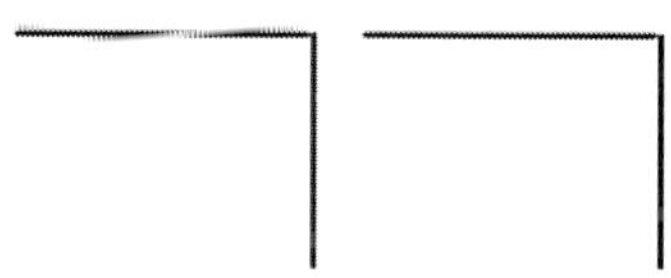

With the capacity for inscriptions of marks both inside and outside other marks, that base capacity of this formalism is the same as a typographical system of parentheses. What is represented is a possibility of systematic arrangements of distinctions so that one can say of any given distinction whether it is inside or outside of any other distinction. This entirely general consideration for distinctions is imaged or modeled in the expressions of the 'calculus of indications'.

There are two rules for the expansion or simplification of expressions in this calculus, as formulated by Spencer-Brown. One can condense two adjacent empty marks to a single mark as shown below.

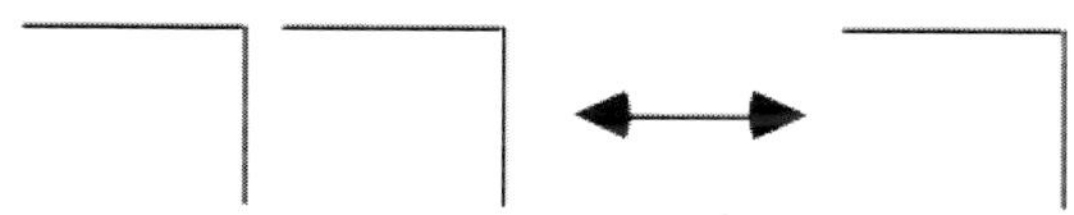

One can cancel or create two marks that are nested, so long as the innermost mark is empty.

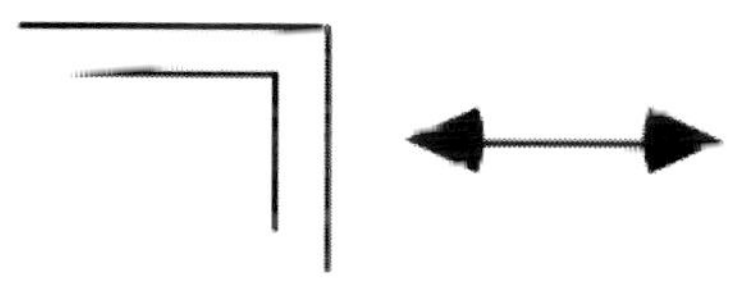

In this case, the double arrow points to nothing but an empty plane. The two nested marks arise from an empty plane or return to an empty plane. These two interactions of a distinction with itself are fundamental. Using a physical metaphor, we see that the mark is a logical particle that is its own anti-particle. It can interact with itself to either produce itself (as in the adjacent marks) or to annihilate itself (as in the nested marks). *Thus from fundamental considerations, we have arrived at a formalism that naturally contains the fusion algebra for a Majorana Fermion, a particle that is its own anti-particle.*

In the framework of physics there are at least two algebras that are associated with a particle or collection of particles. One is the fusion algebra that tells us how the particles can interact with one another. One usually works with a collection of particles that is closed under the interactions being studied, so that the possible interaction of any two particles can be written as a sum of possibilities. In the case of a Majorana fermion **P**, we have the fusion rule $\mathbf{P^2 = P + 1}$ that indicates that **P** can interact with itself to produce either itself or annihilate itself (the product of the annihilation is represented by the numeral **1**). See [18].

Along with fusion algebra, there is also an algebra of creation and annihilation operators for a particle. In the case of Majorana fermions, that creation and annihilation operators **a** and **b** for a pair of Majorana fermions behave as follows: $\mathbf{a^2 = b^2 = 1}$ **and** $\mathbf{ab = -ba}$. In other words, Majorana fermions generate a Clifford algebra with the above relations. If there is a set of Majorana particles $\{c_1, c_2, ..., c_n\}$ then we have

$$c_i^2 = 1 \text{ for all i, and } c_i c_j = - c_j c_i \text{ when i is } \neq \text{ to j.}$$

As we shall see, this algebra gives insight into the structure of fermions in general, and to the relationships of fermions with topology. At the foundation of this paper is the question of the relationship of the fermions with the properties of a distinction.

It is worth mentioning at once that if we take three Majorana fermion operators a, **b** and **c**. Then we can form $\mathbf{I = ba}$, $\mathbf{J = cb}$ and $\mathbf{K = ac}$ and it follows immediately that $\mathbf{I^2 = J^2 = K^2 = IJK = -1}$. Thus Hamilton's quaternions emerge at once from the ground of these operators. This is not all. If we define $\mathbf{S = (1 + ba)}$ and $\mathbf{T = (1 + cb)}$, then it follows that $\mathbf{STS = TST}$. This is the fundamental braiding relation illustrated below. For more information about these relationships with topology, see [18].

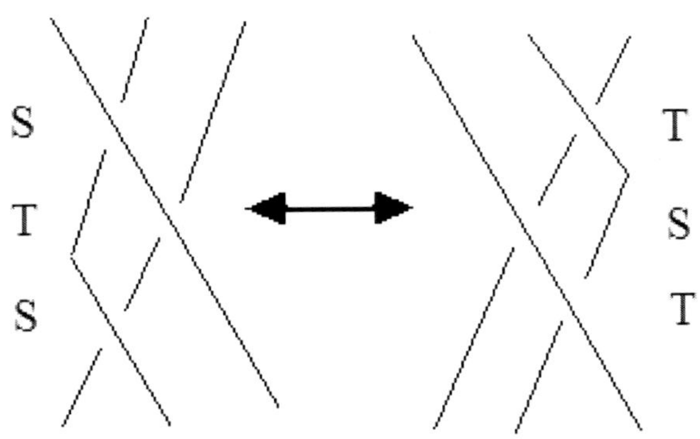

Figure 1. - The Braid Relation STS = TST.

Is there a natural way in which the Clifford algebra for the Majorana fermions is related to the act of distinction? In fact, it is the self-referential nature of the distinction and the resulting temporal dynamics that gives rise to exactly this algebraic structure. We now explain how this comes about.

The fact that the mark annihilates itself allows us to define a fundamental recursion in the form.

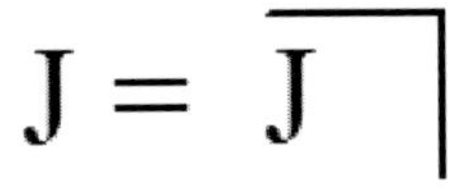

The entity J will be referred to as the reentering mark. If there were no relations then the associated recursion would take the form of building up a sequence of nested symbols as shown below.

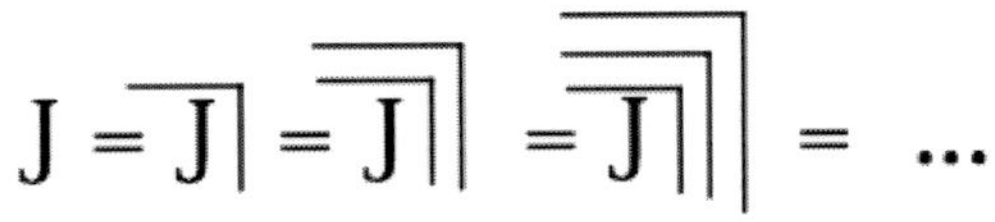

If we give **J** a marked starting value then we get the following pattern of nests and their reductions (written below each nest). The reductions oscillate between marked and unmarked states. This oscillation is the simple dynamics associated with the reentering mark.

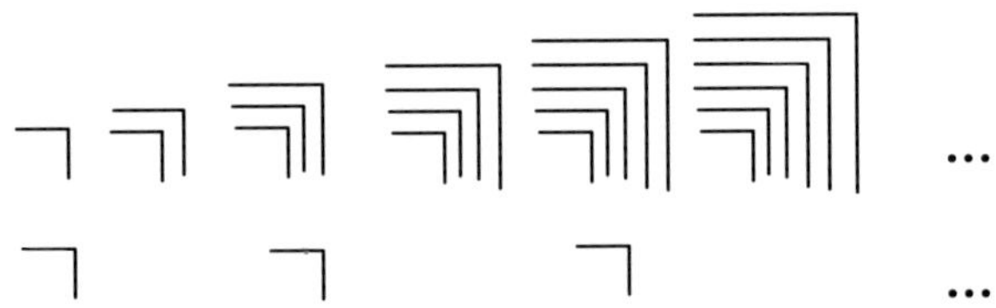

If we do not know the starting point of the oscillation, but simply look in the middle of the process then we see an unending alternation with no beginning or end.

Here we have indicated the unmarked state by two nested marks. When the oscillation is seen without beginning or end, it can be regarded in two distinct ways: as an oscillation from marked to unmarked, or as an oscillation from unmarked to marked. In this way the reentering marked becomes the recursive process generated by the equation and J is an algebraic or imaginary fixed point for the recursion.

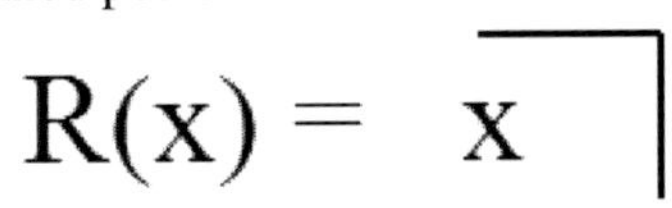

The analog with the complex numbers is sufficiently strong that we now shift to the complex numbers themselves as a specific instantiation of this recursive process. In the body of the paper we will return to the more rarified discussion leading from the distinction itself. We begin with recursive processes that work with the real numbers **+1** and **-1**. Think of the oscillatory process generated by **R(x) = -1/x.**

The fixed point is **i** with $\mathbf{i}^2 = \mathbf{-1}$, and the processes generated over the real numbers must be directly related to the idealized **i**. There are two *iterant views*: **[+1,-1]** and **[-1,+1]**.

These are seen as points of view of an alternating process that engender the square roots of negative unity.

... +1 -1 +1 -1 +1 -1 +1 -1 +1 -1 +1 -1 +1 -1 +1 -1 +1 -1 ...

To see this, we introduce a *temporal shift operator,* η such that

$$[a,b]\eta = \eta\,[b,a] \text{ and } \eta\,\eta = 1$$

so that concatenated observations can include a time step of one-half period of the process **...abababab...** .

We combine iterant views term-by-term as in **[a,b][c,d] = [ac,bd]**. Then with $\mathbf{I} = [1,-1]\eta$ (**I** is a combined view/operator), $\mathbf{I}\,\mathbf{I} = [1,-1]h\eta[1,-1]\eta = [1,-1][-1,1]\eta\,\eta = [-1,-1] = -1.$

This is a process-oriented construction of the square root of minus one, and hence of the complex numbers. The key to rethinking the complex numbers in these terms is to understand that **I** represents a discrete oscillating temporal process. *We take as a matter of principle that the usual real variable **t** for time is better represented as **It** so that time is seen to be a process, an observation and a magnitude all at once.* This principle of the *temporal nexus* is supported by our analysis of **I** as an eigenform, an imaginary fixed-point of an oscillatory recursive process.

The reader should take careful note of this structure. Let **e = [1,-1]** and **-e = [-1,1].** Note that $\mathbf{e}^2 = \mathbf{ee} = [1,-1][1,-1] = [1,1] = 1$, $\eta^2 = 1$ and $\mathbf{e}\,\eta = -\,\eta\,\mathbf{e}.$

Thus **e** and η form a Majorana Fermion pair and the formalism of $\mathbf{e}\,\eta$ as a representation of the square root of minus one is part of the general formalism for the Clifford algebra associated with such a fermion. By following the line by which this Clifford algebra comes into being we have discovered that this pair can be interpreted in terms of elementary or primordial space (**e**) and time (η). It is customary at the usual mathematical level to take a commuting square root of minus one (one that commutes with all of the other algebraic elements) and denote it by **i**. We shall follow this convention and use the lower case **i** for the central square root of negative unity. Then we can form a Fermion Operator $f = \mathbf{e} + \mathbf{i}\,\eta$ that combines these two elemental Majorana Fermions.

Note that

$$ff = (\mathbf{e} + \mathbf{i}\,\eta)(\mathbf{e} + \mathbf{i}\,\eta) = \mathbf{e}^2 - \eta^2 + \mathbf{i}\,(\mathbf{e}\,\eta + \eta\,\mathbf{e}) = 0.$$

In this way, the algebra of Majorana Fermions, and the algebra of Fermions arises from the recursive properties of a single distinction. Putting this distinction in the framework of the calculus of indications creates a unified context for the both the fusion algebra and the creation/annihilation algebra for these fermions and aligns them with space-time structure. It is the purpose of this essay to unpack further these ideas and structures that live at the ground of the mathematical articulation of physics.

Nilpotent Universality

It is worth pointing out in this introduction that the context of Laws of Form and Distinction leads to a suggestion for the form of a universal equation. We let **U** stand for the universe. **U** operates on **y**, the state of the universe. The universal equation should therefore be **Uy = 0**. Ah! The Universe herself is a state of the Universe and so we must have **UU = 0**. We do not have to look far for solutions. In the rarefied air of Laws of Form we may define

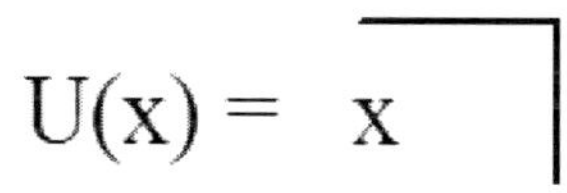

and then we have, for

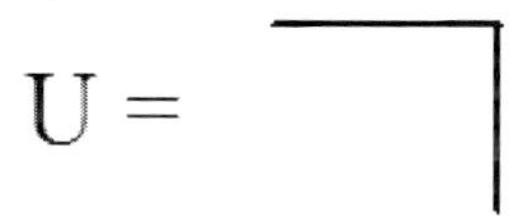

That

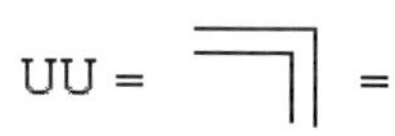

where the right most equality terminates in the void. The void is the Universal Player for the role of zero. The reader may think that this point of view is just philosophical, but in fact it is a beautiful entry point into the Dirac equation once one goes through the iterants of the mark into the Majorana fermion operators. See Section **VI**. This is the most significant point of the present paper. By examining the way we make and imagine distinctions, the deep patterns of physics come into view.

2. The Eigenform Model

We have seen how the concept of an object has evolved to make what we call objects (and the objective world) processes that are interdependent with the actions of observers. The notion of a fixed object has become a notion of a process that produces the apparent stability of the object. This process can be simplified in a model to become a recursive process where a rule or rules are applied time and time again. The resulting object of such a process is the *eigenform* of the process, and the process itself is the *eigenbehaviour*. See [4-13,16,17].

In this way we have a model for thinking about object as token for eigenbehaviour. This model examines the result of a simple recursive process carried to its limit. For example, suppose that

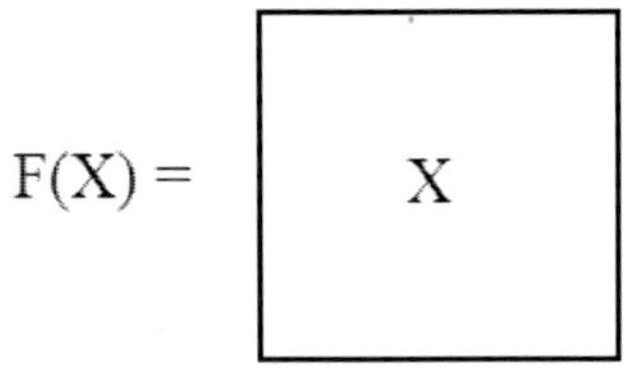

Figure 2.

That is, each step in the process encloses the results of the previous step within a box. Here is an illustration of the first few steps of the process applied to an empty box X:

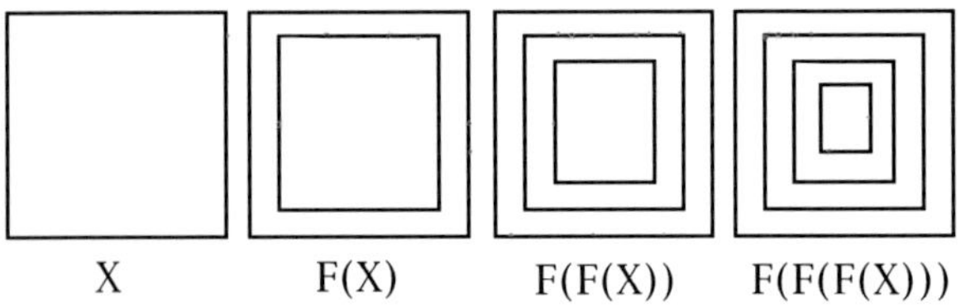

Figure 3.

If we continue this process, then successive nests of boxes resemble one another, and in the limit of infinitely many boxes, we find that the infinite nest of boxes is invariant under the addition of one more surrounding box. Hence this infinite nest of boxes is a fixed point for the recursion. In other words, if X denotes the infinite nest of boxes, then, **X = F(X).**

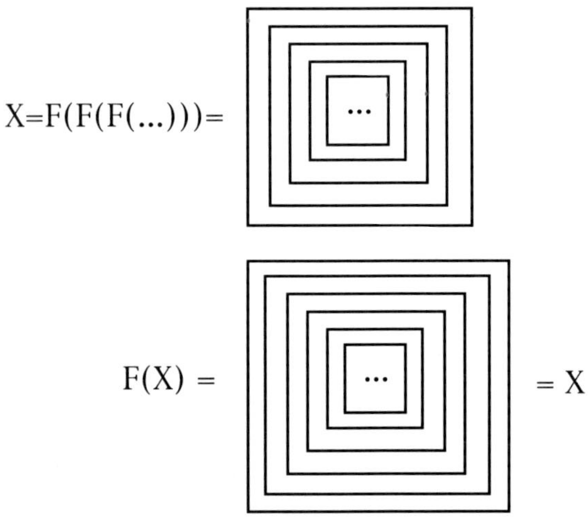

Figure 4.

This equation is a description of a state of affairs. The form of an infinite nest of boxes is invariant under the operation of adding one more surrounding box. The infinite nest of boxes is one of the simplest eigenforms. In the process of observation, we interact with ourselves and with the world to produce stabilities that become the objects of our perception. These objects, like the infinite nest of boxes, may go beyond the specific properties of the world in which we operate. They attain their stability through the limiting process that goes outside the immediate world of individual actions. We make an imaginative leap to complete such objects to become tokens for eigenbehaviours. It is impossible to make an infinite nest of boxes. We do not make it. We *imagine* it. And in imagining that infinite nest of boxes, we arrive at the eigenform.

The leap of imagination to the infinite eigenform is a model of the human ability to create signs and symbols. In the case of the eigenform **X** with **X = F(X), X** can be regarded as the name of the process itself or as the name of the limit process. Note that if you are told that **X = F(X)**, then substituting **F(X)** for **X**, you can write **X = F(F(X)).** Substituting again and again, you have

X = F(F(F(X))) = F(F(F(F(X)))) = F(F(F(F(F(X))))) = ...

The process arises from the symbolic expression of its eigenform. In this view *the eigenform is an implicate order for the process that generates it.* Sometimes one stylizes the structure by indicating where the eigenform **X** reenters its own indicational space by an arrow or other graphical device. See the picture below for the case of the nested boxes.

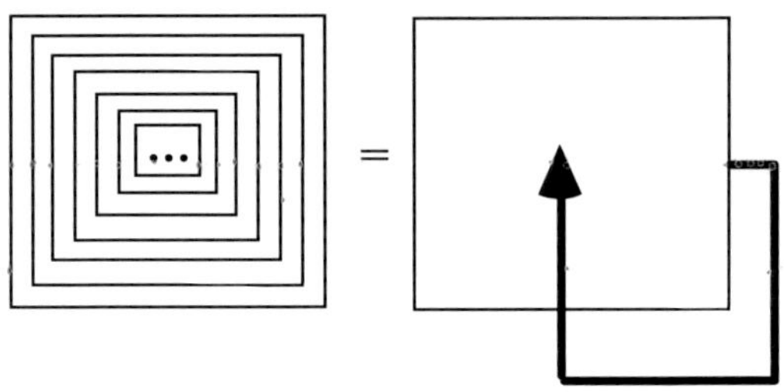

Figure 5.

Does the infinite nest of boxes exist? Certainly it does not exist in this page or anywhere in the physical world with which we are familiar. The infinite nest of boxes exists in the imagination. It is a symbolic entity. Eigenform is the imagined boundary in the reciprocal relationship of the object (the It) and the process leading to the object (the process leading to It). In the diagram below we have indicated these relationships with respect to the eigenform of nested boxes. Note that the It is illustrated as a finite approximation (to the infinite limit) that is sufficient to allow an observer to infer/perceive the generating process that underlies it.

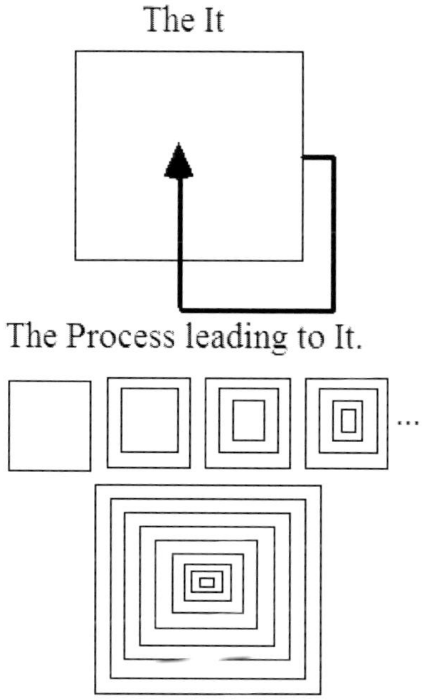

Figure 6.

Just so, an object in the world (cognitive, physical, ideal,...) provides a conceptual center for the exploration of a skein of relationships related to its context and to the processes that generate it. An object can have varying degrees of reality just as does an eigenform. If we take the suggestion to heart that objects are tokens for eigenbehaviors, then an object in itself is an entity, participating in a network of interactions, taking on its apparent solidity and stability from these interactions.

An object is an amphibian between the symbolic and imaginary world of the mind and the complex world of personal experience. The object, when viewed as process, is a dialogue between these worlds. The object when seen as a sign for itself, or in and of itself, is imaginary. Why are objects apparently solid? Of course you cannot walk through a brick wall even if you think about it differently. I do not mean apparent in the sense of thought alone. I mean apparent in the sense of appearance. The wall appears solid to me because of the actions that I can perform. The wall is quite transparent to a neutrino, and will not even be an eigenform for that neutrino. This example shows quite sharply how the nature of an object is entailed in the properties of its observer.

The eigenform model can be expressed in quite abstract and general terms. Suppose that we are given a recursion (not necessarily numerical) with the equation **X(t+1) = F(X(t)).** Here **X(t)** denotes the condition of observation at time **t. X(t)** could be as simple as a set of nested boxes, or as complex as the entire configuration of your body in relation to the known universe at time **t.** Then **F(X(t))** denotes the result of applying the operations symbolized by **F** to the condition at time **t**. You could, for simplicity, assume that **F** is independent of time. Time independence of the recursion **F** will give us simple answers and we can later discuss what will happen if the actions depend upon the time. In the time independent case we can write

J = F(F(F(...)))

the infinite concatenation of F upon itself. Then

F(J) = J

since adding one more **F** to the concatenation changes nothing. Thus **J,** the infinite concatenation of the operation upon itself leads to a fixed point for **F**. **J** is said to be the eigenform for the recursion **F**. We see that every recursion has an eigenform. Every recursion has an (imaginary) fixed point.

Lambda Calculus Incantation

There is a way to the fixed point that does not invoke infinity.

Let **Gx = F(xx)**. You can postulate that the **x**'s act on themselves or you can regard **G** as an operator that reproduces a copy of **x** and places the two copies adjacent to one another surrounded by the distinction of **F( )**. But then **GG = F(GG)** and by sheer magic, **F** has a fixed point! See [1,9,11,12,16,17] for more about this point of view and its relationship with the Lambda Calculus of Church and Curry.

The Koch Fractal

We end this section with one more example. This is the eigenform of the Koch fractal [13]. In this case one can write symbolically the eigenform equation

K = K { K K } K

to indicate that the Koch Fractal reenters its own indicational space four times (that is, it is made up of four copies of itself, each one-third the size of the original. The curly brackets in the center of this equation refer to the fact that the two middle copies within the fractal are inclined with respect to one another and with respect to the two outer copies. In the figure below (Fig. 7) we show the geometric configuration of the reentry.

In the geometric recursion, each line segment at a given stage is replaced by four line segments of one third its length, arranged according to the pattern of reentry as shown in the figure above. The recursion corresponding to the Koch eigenform is illustrated in the next figure (Fig. 8). Here we see the sequence of approximations leading to the infinite self-reflecting eigenform that is known as the Koch snowflake fractal.

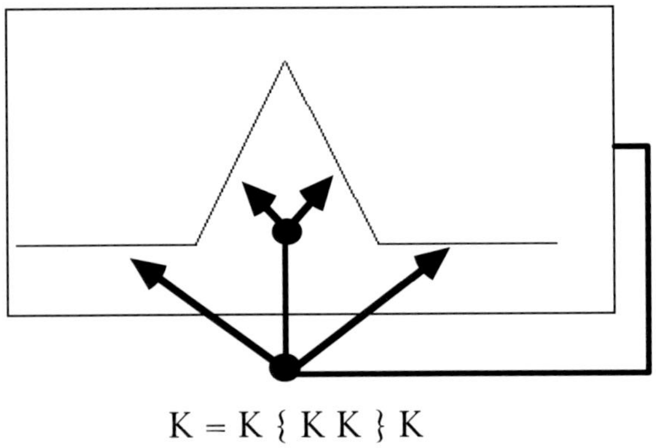

Figure 7.

Figure 8.

Five stages of recursion are shown. To the eye, the last stage vividly illustrates how the ideal fractal form contains four copies of itself, each one-third the size of the whole. The abstract schema

K = K { K K } K

for this fractal can itself be iterated to produce a skeleton of the geometric recursion:

K = K { K K } K

= K { K K } K { K { K K } K K { K K } K } K { K K } K

= ...

We have only performed one line of this skeletal recursion. There are sixteen **K**'s in this second expression just as there are sixteen line segments in the second stage of the geometric recursion. Comparison with this symbolic recursion shows how geometry aids the intuition. The interaction of eigenforms with the geometry of physical, mental, symbolic and spiritual landscapes is an entire subject that is in need of deep exploration.

It is usually thought that the recognition of an object arises in some simple way from the assumed existence of the object and the action of our perceiving systems. This is a fine tuning to the point where the action of the perceiver and the perception of the object are indistinguishable. Such tuning requires an intermixing of the perceiver and the perceived that goes beyond description. Yet in the mathematical levels, such as number or fractal pattern, part of the process is slowed down to the point where we can begin to apprehend the process. There is a stability in the comparison, in the correspondence that is a process happening at once in the present time. The closed loop of perception occurs in the eternity of present individual time. Each such process depends upon linked and ongoing eigenbehaviors and yet is seen as simple by the perceiving mind. The perceiving mind is itself an eigenform.

3. The Square Root of Minus One and the Algebra of Iterants

The purpose of this section is to place **i**, square root of minus one, and its algebra in a process context. We then see that the square root of minus one is intimately related to time and the mathematical expression of time. The main point is that **i** can be interpreted as a primitive dynamical system that undergoes oscillation and nevertheless has a significant algebraic form associated with this iteration.

Traditionally **i**, the square root of minus one, arises from the fact that it is suggested by the solutions to quadratic equations. The simplest instance is the equation $x^2 + 1 = 0$. A solution to this equation must have square equal to **-1**, but there are no real numbers whose square is **-1**. The square of a negative number is positive and the square of a positive number is positive and the square of zero is zero. Thus mathematicians (reluctantly at first) introduced an ideal number **i** such that $i^2 = -1$. This number can be used to find solutions to other quadratic equations. For example **1+i** and **1-i** are the roots of $x^2 - 2x + 2 = 0$.

The use of such complex numbers became indispensable when it was realized that they could be used in the solution of cubic equations in such a way that the real solution of a cubic was expressed naturally as a combination of two complex numbers such that the imaginary parts canceled out and all that was left was the real solution. Mathematicians began to seriously use complex numbers in the 1500's. It was not until around 1800 that Gauss and Argand discovered a geometric interpretation that opened up the subject to genuinely deep explorations.

The geometric interpretation of the complex numbers is well-known and we shall not repeat it here except to remark that **i** seen as a vector of unit length in the plane, and perpendicular to the horizontal axis. The horizontal axis is taken to be the real numbers. Multiplication by **i** is interpreted geometrically as rotation by **90** degrees. Thus **ii** is rotation by **180** degrees. Indeed the **180** degree rotate of **+1** is **-1,** and so **ii = -1**.

We now describe a process point of view for complex numbers. Think of the oscillatory process generated by

$$R(x) = -1/x.$$

The fixed point would satisfy **i = -1/I** and multiplying, we get that **ii = -1.** On the other hand the iteration of **R** on **1** yields **+1, R(1) = -1, R(R(1)) = + 1, -1, +1, ...**

Thus there must be a linkage between this ideal number **i** whose square is **-1** and the recursion that leads to an oscillation. The square root of minus one is a perfect example of an eigenform that occurs in a new and wider domain than the original context in which its recursive process arose. The process has no fixed point in the original domain.

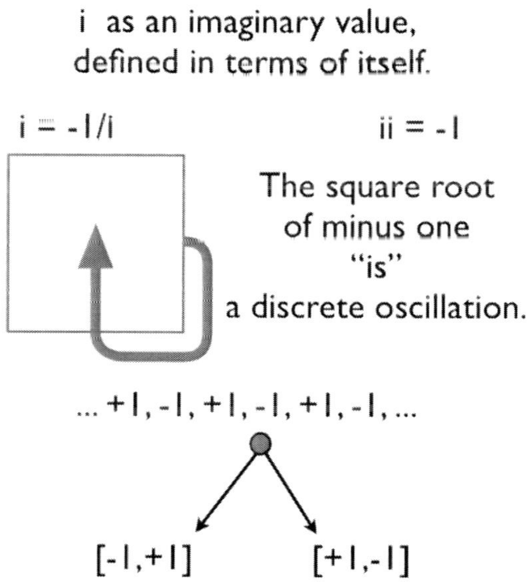

Figure 9.

Looking at the oscillation between **+1** and **-1**, we see that there are naturally two viewpoints that we denote by **[+1,-1]** and **[-1,+1]**. These viewpoints correspond to whether one regards the oscillation at time zero as starting with **+1** or with **-1**. We shall let **I{+1,-1}** stand for an undisclosed alternation or ambiguity between **+1** and **-1** and call **I{+1,-1}** an *iterant*. There are two *iterant views*: **[+1,-1]** and **[-1,+1]**.

Given an iterant **[a,b]**, we can think of **[b,a]** as the same process with a shift of one time step. These two iterant views, seen as points of view of an alternating process, will become the square roots of negative unity, **i** and **-i.** We introduce a *temporal shift operator* η such that

$$[a,b]\eta = \eta\,[b,a] \text{ and } \eta\,\eta = 1$$

for any iterant **[a,b]**, so that concatenated observations can include a time step of one-half period of the process **...abababab...** .

We combine iterant views term-by-term as in

$$[a,b][c,d] = [ac,bd].$$

We now define i by the equation, $i = [1,-1]\eta$.

This makes **i** both a view of the iterant process and an operator that takes into account a step in time.

We calculate

$$ii = [1,-1]\eta\ [1,-1]\eta = [1,-1][-1,1]\eta\,\eta = [-1,-1] = -1.$$

Thus we have constructed the square root of minus one by using an iterant viewpoint. The key to rethinking the complex numbers in these terms is to understand that **i** represents a discrete oscillating temporal process and it is an eigenform participating in the algebraic structure of the complex numbers.

The Temporal Nexus

We take as a matter of principle that the usual real variable **t** *for time is better represented as* **it** *so that time is seen to be a process, an observation and a magnitude all at once.* This principle of imaginary time is justified by the eigenform approach to the structure of time and the structure of the square root of minus one. An example of the use of the Temporal Nexus, consider the expression $x^2 + y^2 + z^2 + t^2$, the square of the Euclidean distance of a point **(x,y,z,t)** from the origin in Euclidean four-dimensional space. Now replace **t** by **it**, and find

$$x^2 + y^2 + z^2 + (it)^2 = x^2 + y^2 + z^2 - t^2,$$

the squared distance in hyperbolic metric for special relativity. By replacing **t** by its process operator value **it** we make the transition to the physical mathematics of special relativity. We will discuss this theme further and relate the Temporal Nexus, the role of complex numbers in quantum mechanics and the role of temporal shift operators in discrete physics.

4. Quantum Physics, Eigenvalue, and Eigenform

Quantum is inextricably related to the complex numbers and in particular to the square root of minus one. We wish to show that the process view of the square root of minus one and the Temporal Nexus inform the epistemology of quantum theory. The reader should recall the Temporal Nexus from the previous section. *We take as a matter of principle that the usual real variable **t** for time is better represented as **it** so that time is seen to be a process, an observation and a magnitude all at once.* This principle of imaginary time is justified by our approach to the structure of time and the structure of the square root of minus one.

In quantum physics [11], the state of a physical system is modeled by a vector in a high-dimensional space, called a Hilbert space. As time goes on the vector rotates in this high dimensional space. Observable quantities correspond to Hermitian operators H on this vector space. Linearity is usually a simplifying assumption in mathematical models, but it is an essential feature of quantum mechanics. *In quantum mechanics observation is founded on the production of eigenvectors, **v** with **Gv = v** where **v** is a vector in a Hilbert space and **G** is a linear operator on that space.*

Many of the strange and fascinating properties of quantum mechanics emanate directly from this model of observation. In order to observe a quantum state, its vector is projected into an eigenvector for that particular mode of observation. By projecting the vector into that mode and not another, one manages to make the observation, but at the cost of losing information about the other possibilities inherent in the vector. This is the source, in the mathematical model, of the complementarities that allow exact determination of the position of a particle at the expense of nearly complete uncertainty about its momentum (or vice versa the determination of momentum at the expense of knowledge of the position).

Observation and quantum evolution (the determinate rotation of the state vector in the high dimensional Hilbert space) are interlocked. Each observation discontinuously projects the state vector to an eigenvector. The intervals between observations allow the continuous evolution of the state vector. This tapestry of interaction of the continuous and the discrete is the basis for the quantum mechanical description of the world. Is the quantum model, in its details, a consequence of general principles about systems? This is an exploration that needs to be made. We can only ask the question here. But the mysteries of the interpretation of quantum mechanics all hinge on an assumption of a world external to the quantum language. Thinking in terms of eigenform we can begin to look at how the physics of objects emerges from the model itself.

Where are the eigenforms in quantum physics? They are in the mathematics itself. For example, we have the simplest wave-function

$$\psi(x,t) = e^{i(kx - wt)}.$$

Since we know that the function $\mathbf{E(x) = e^x}$ is an eigenform for operation of differentiation with respect to $\mathbf{x}$, $\psi(\mathbf{x,t})$ is a special multiple eigenform from which the energy can be extracted by temporal differentiation, and the momentum can be extracted by spatial differentiation. We see in $\psi(\mathbf{x,t})$ the complexity of an individual who presents many possible sides to the world. $\psi(\mathbf{x,t})$ is an eigenform for more than one operator and it is a composition of $\mathbf{e^x}$ and **i**, the square root on minus one, a primordial eigenvalue related to time. It is this internal complexity that is mirrored in the uncertainty relations of Heisenberg and the complementarity of Bohr. The eigenforms themselves, as wavefunctions, are *inside* the mathematical model, on the *other side* of that which can be observed by the physicist.

We have seen eigenforms as the constructs of the observer, and in that sense they are on the side of the observer, even if the process that generates them is outside the realm of his perception. This suggests that we think again about the nature of the wave function in quantum mechanics. Is it also a construct of the observer? To see quantum mechanics and the world in terms of eigenforms requires a turning around, a shift of perception where indeed we shall find that the distinction between model and reality has disappeared into the world of appearance.

This is a reversal of epistemology, a complete turning of the world upside down. Eigenform has tricked us into considering the world of our experience and we find that it is our world, generated by our actions. The world has become objective through the self-generated stabilities of those actions.

A Quick Review of Quantum Mechanics

De Broglie hypothesized two fundamental relationships: between energy and frequency, and between momentum and wave number. These relationships are summarized in the equations

$$E = \hbar\omega,$$

$$p = \hbar k,$$

where **E** denotes the energy associated with a wave and **p** denotes the momentum associated with the wave. Here $\hbar = h/2\pi$ where **h** is Planck's constant. Schrödinger answered the question: *Where is the wave equation for De Broglie's waves?* Writing an elementary wave in complex form

$$\psi = \psi(x,t) = \exp(i(kx - wt)),$$

we see that we can extract De Broglie's energy and momentum by differentiating:

$$i\hbar\partial\psi/\partial t = E\psi \quad \text{and} \quad -i\hbar\partial\psi/\partial x = p\psi.$$

This led Schrödinger to postulate *the identification of dynamical variables with operators* so that the first equation, $i\hbar\partial\psi/\partial t = E\psi$, is promoted to the status of an equation of motion while the second equation becomes the definition of momentum as an operator:

$$p = -i\hbar\partial/\partial x.$$

Once **p** is identified as an operator, the numerical value of momentum is associated with an eigenvalue of this operator, just as in the example above. In our example $p\psi = \hbar k\psi$. In this formulation, the position operator is just multiplication by **x** itself. Once we have fixed specific operators for position and momentum, the operators for other physical quantities can be expressed in terms of them. We obtain the energy operator by substitution of the momentum operator in the classical formula for the energy:

$$E = (1/2)mv^2 + V$$
$$E = p^2/2m + V$$
$$E = -(\hbar^2/2m)\partial^2/\partial x^2 + V.$$

Here **V** is the potential energy, and its corresponding operator depends upon the details of the application. With this operator identification for **E**, Schrödinger's equation

$$i\hbar\partial\psi/\partial t = -(\hbar^2/2m)\partial^2\psi/\partial x^2 + V\psi$$

or equivalently,

$$\partial\psi/\partial t = i(\hbar/2m)\partial^2\psi/\partial x^2 - (i/\hbar)V\psi$$

is an equation in the first derivatives of time and in second derivatives of space. In this form of the theory one considers general solutions to the differential equation and this in turn leads to excellent results in a myriad of applications. It is useful to point out that Schrödinger's equation without the potential term has the form

$$\partial\psi/\partial t = i(\hbar/2m)\partial^2\psi/\partial x^2.$$

It is in this form that we shall compare it with discrete processes in the next section. There we shall use the Temporal Nexus to obtain new insight into the role of **i** in this equation.

In quantum theory, observation is modeled by the concept of eigenvalues for corresponding operators. *The quantum model of an observation is a projection of the wave function into an eigenstate.* An energy spectrum, $\{E_k\}$ corresponds to wave functions, ψ satisfying the Schrödinger equation, such that there are constants E_k with $E\psi = E_k\psi$. An *observable* (such as energy) **E** is a Hermitian operator on a Hilbert space of wavefunctions. Since Hermitian operators have real eigenvalues, this provides the link with measurement for the quantum theory.

It is important to notice that there is no mechanism postulated in this theory for how a wave function is sent into an eigenstate by an observable. Just as mathematical logic need not demand causality behind an implication between propositions, the logic of quantum mechanics does not demand a specified cause behind an observation. This absence of an assumption of causality in logic does not obviate the possibility of causality in the world. Similarly, the absence of causality in quantum observation does not obviate causality in the physical world. Nevertheless, the debate over the interpretation of quantum theory has often led its participants into asserting that causality has been demolished in physics.

Note that the operators for position and momentum satisfy the equation $xp - px = \hbar i$. This corresponds directly to the equation obtained by Heisenberg, on other grounds, that dynamical variables can no longer necessarily commute with one another. In this way, the points of view of De Broglie, Schrödinger and Heisenberg came together, and quantum mechanics was born. In the course of this development, interpretations varied widely. Eventually, physicists came to regard the wave function not as a generalized wave packet, but as a carrier of information about possible observations. In this way of thinking $\psi^*\psi$ (ψ^* denotes the complex conjugate of ψ) represents the probability of finding the particle (A particle is an observable with local spatial characteristics.) at a given point in spacetime. Strictly speaking, it is the spatial integral of $\psi^*\psi$ that is interpreted as a total probability with $\psi^*\psi$ the probability density. This way of thinking is supported by the fact that the total spatial integral is time-invariant as a consequence of Schrödinger's equation!

5. Iterants, Complex Numbers, and Quantum Mechanics

We have seen that there are indeed eigenforms in quantum mechanics. *The eigenforms in quantum mechanics are the mathematical functions such as* e^x *that are invariant under operators such as* ***D = d/dx.***

But we wish to examine the relationship between recursion, reflexive spaces and the properties of the quantum world. The hint we have received from quantum theory is that we should begin with mathematics which is replete with eigenforms. In fact, this hint seems very rich when we consider that **i**, the square root of minus one, is a key eigenform in our panoply of eigenforms and it is a key ingredient in quantum mechanics intimately related to the role of time.

Revisiting the Temporal Nexus

If we define

$$R(A) = -1/A$$

then **R(i) = i** since $i^2 = -1$ is equivalent to **i = -1/i.** Using the infinite recursion we would then write

$$i = -1/-1/-1/-1/-1/\ldots,$$

making **i** an infinite reentry form for the operator **R.** We will write

i = [-1/*]

where * denotes the reentry of the whole form into that place in the right-hand part of the expression. Similarly, if **F(x) = 1 + 1/x,** then the eigenform would be **[1 + 1/*]** and we could write **(1+√5)/2 = [1 + 1/*].** With this in place we can now consider wave functions in quantum mechanics such as

y(x,t) = exp(i(kx - wt)) = exp([-1/*] (kx - wt))

and we can consider classical formulas in mathematics such as *Euler's formula*

exp([-1/*]j) = cos(j) + [-1/*] sin(j)

in this light. We start here with Euler's formula, for this formula is the key relation between complex numbers, **i**, waves and periodicity.

We return to the finite nature of **[-1/*]**. This eigenform is an oscillator between **-1** and **+1**. It is only **i** in its idealization. In its appropriate synchronization it has the property that **i = -1/i**. As a real oscillator, the equation **R(i) = -1/i** tells us that when **i** is **1**, then i is transformed to **-1** and when **i** is **-1** then **i** is transformed to **+1**. There is no fixed point for **R** in the real domain. The eigenform is achieved by leaving the real domain for a new and larger domain.

We know that this larger domain can be conceptualized as the plane with Euclidean rotational geometry, but here we explore the larger domain in terms of eigenforms. *We find that* ***i*** *itself is a fundamental discrete process, and it is in the microworld of such discrete physical processes that not only quantum mechanics, but also classical mechanics is born.*

Iterants and Iterant Views

In order to think about i, consider an infinite oscillation between **+1** and **-1**:

... -1,+1,-1,+1,-1,+1,-1,+1,...

This oscillation can be seen in two distinct ways. It can be seen as **[-1,+1]** (a repetition in this order) or as **[+1,-1]** (a repetition in the opposite order). This suggests regarding an infinite alternation such as

... a,b,a,b,a,b,a,b,a,b,a,b,a,b,...

as an entity that can be seen in two possible ways, indicated by the ordered pairs **[a,b]** and **[c,d]**. We shall call the infinite alternation of **a** and **b** the *iterant* of **a** and **b** and denote it by **I{a,b}**. Just as with a set **{a,b}**, the iterant is independent of the order of **a** and **b**.

We have **I{a,b} = I{b,a}**, but there are two distinct views of any iterant and these are denoted by **[a,b]** and **[b,a]**. The key to iterants is that two representatives of an iterant can *by themselves* appear identical, but *taken together* are seen to be different. For example, consider

... a,b,a,b,a,b,a,b,a,b,a,b,a,b,...

and also consider

...b,a,b,a,b,a,b,a,b,a,b,a,b,a,...

There is no way to tell the difference between these two iterants except by a direct comparison as shown below

... a,b,a,b,a,b,a,b,a,b,a,b,a,b,...
... b,a,b,a,b,a,b,a,b,a,b,a,b,a,...

In the direct comparison we see that if one of them is **[a,b]**, then the other one should be **[b,a]**. Still, there is no reason to assign one of them to be **[a,b]** and the other **[b,a]**. It is a strictly relative matter. The two iterants are entangled (to borrow a term from quantum mechanics) and if one of them is observed to be **[a,b]**, then the other is necessarily observed to be **[b,a]**.

Let's go back to the square root of minus one as an oscillatory eigenform.

... -1,+1,-1,+1,-1,+1,-1,+1,...

What is the operation **R(x) = -1/x** in this case? We usually think of a starting value and then the new operation shifts everything by one value with **R(+1) = -1** and **R(-1) = +1**. Thus would suggest that **R(... -1,+1,-1,+1,-1,+1,-1,...) = ... +1,-1,+1,-1,+1, -1,+1,...** and these sequences will be different when we compare, them even though they are identical as individual iterants.

... -1,+1,-1,+1,-1,+1,-1,+1,...

... +1,-1,+1,-1,+1,-1,+1,-1,...

However, we would like to take the eigenform/iterant concept and make a more finite algebraic model by using the iterant views, **[-1,+1]** and **[+1,-1]**.

Certainly we should consider the transform **P[a,b] = [b,a]** and we take

-[a,b] = [-a, -b],

so that

-P[a,b] = [-b,-a].

Then

-P[1,-1] = [1,-1].

In this sense the operation **-P** has eigenforms **[1,-1]** and **[-1,1]**. You can think of **P** as the shift by one-half of a period in the process **...abababab....** . Then **[-1,1]** is an eigenform for the operator that combines negation and shift. We will take a shorthand for the operator **P** via

P[a,b] = [a,b]' = [b,a].

If **x = [a,b]**

then **x' = [b,a]**.

We can add and multiply iterant views by the combinations

[a,b][c,d] = [ac,bd],
[a,b] + [c,d] = [a+c, b+ d],
k[a,b] = [ka,k,b] when k is a number**.**

We take **1 = [1,1]** and **-1 = [-1,-1].** This is a natural algebra of iterant views, but note that **[-1,+1][-1,+1] = [1,1] = 1**, so we do not yet have the square root of minus one. Consider **[a,b]** as representative of a process of observation of the iterant **I{a,b}. [a,b]** is an *iterant view*. We wish to combine **[a,b]** and **[c,d]** as *processes of observation.*

Suppose that observing **I{a,b}** requires a step in time. That being the case, **[a,b]** will have shifted to **[b,a]** in the course of the single time step. We need an algebraic structure to handle the temporality. To this end, we introduce an operator, **h** with the property that

$$[a,b]\eta = \eta[b,a] \text{ with } \eta^2 = \eta\eta = 1$$

where **1** means the identity operator. You can think of η as a *temporal shift operator* that can act on a sequence of individual observations. The algebra generated by iterant views and the operator η is taken to be associative. Here the interpretation is that **XY** denotes first observe **X**, then observe **Y**. Thus **X**η**Y**η = **XY'**$\eta\eta$ = **XY'** and we see that **Y** has been shifted by the presence of the operator η, just in accord with our temporal interpretation above.

We can now have a theory where **i** and its conjugate **-i** correspond to the two views of the iterant **I{-1,+1}.** Let **i = [1,-1]**η and **-i = [-1,1]**η . We get a square root of minus one:

$$ii = [1,-1]\eta[1,-1]\eta = [1,-1][-1,1]\eta\eta = [-1,-1] = -[1,1] = -1.$$

The square roots of minus one are iterant views coupled with temporal shift operators. Not so simple, but not so complex either! If **e = [1,-1]** then **e' = [-1,1] = -e** and **ee = [1,1] = 1** with **ee' = -1.**

$$i = e\eta$$
$$ii = e\eta e\eta = ee'\eta\eta = ee' = -1$$

With this definition of **i**, *we have an algebraic interpretation of complex numbers that allows one to think of them as observations of discrete processes.* This algebra contains more than just the complex numbers.

With **x = [a,b]** and **y = [c,d]**, consider the products **(x**η**)(y**η**)** and **(y**η**)(x**η**):**

$$(x\eta)(y\eta) = [a,b]\eta[c,d]\eta = [a,b][d,c] = [ad,bc]$$

$$(y\eta)(x\eta) = [c,d]\eta[a,b]\eta = [c,d][b,a] = [cb,da].$$

Thus

$$(x\eta)(y\eta) - (y\eta)(x\eta)$$
$$= [ad-bc, -(ad-bc)]$$
$$= (ad - bc)[1,-1].$$

Thus

$$x\eta y\eta - y\eta x\eta = (ad - bc)i\eta .$$

We see that, with temporal shifts, the algebra of observations is non-commutative. Note that for these processes, represented by vectors [a,b], the commutator **xhy**η - ψ **hx**η = **(ad -bc)ih** is given by the determinant of the matrix corresponding to two process vectors, and hence will be non-zero whenever the two process vectors are non-zero and represent different spatial rays in the plane.

There is more. *The full algebra of iterant views can be taken to be generated by elements of the form*

$$[a,b] + [c,d]\eta$$

and it is not hard to see that this is isomorphic with 2 x 2 matrix algebra with the correspondence given by the diagram below.

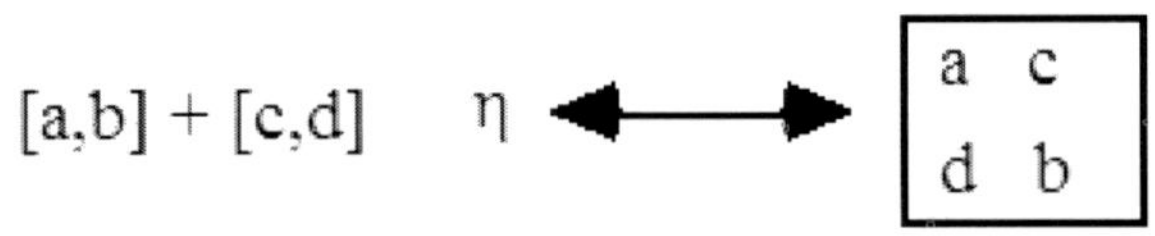

Figure 10.

We see from this excursion that there is a full interpretation for the complex numbers (and indeed matrix algebra) as an observational system taking into account time shifts for underlying iterant processes. Let **A** = **[a,b]** and **B** = **[c,d]** and let **C** = **[r,s], D** = **[t,u].** With **A'** = **[b,a]**, we have

$$(A + Bh)(C + D\eta) = (AC + BD') + (AD + BC')\eta .$$

This writes 2 x 2 matrix algebra in the form of a hypercomplex number system. From the point of view of iterants, the sum **[a,b]** + **[b,c]**η can be regarded as a superposition of two types of observation of the iterants **I{a,b}** and **I{c,d}.** The operator-view **[c,d]**η includes the shift that will move the viewpoint from **[c,d]** to **[d,c]**, while **[a,b]** does not contain this shift. Thus a shift of viewpoint on **[c,d]** in this superposition does not affect the values of **[a,b].** One can think of the corresponding process as having the form shown below.

... a a a a a a a a a a a a a a a ...
... c d c d c d c d c d c d c d ...
... b b b b b b b b b b b b b b b ...

The snapshot **[c,d]** changes to **[d,c]** in the horizontal time-shift while the vertical snapshot **[a,b]** remains invariant under the shift. It is interesting to note that in the spatial explication of the process we can imagine the horizontal oscillation corresponding to **[c,d]**η as making a boundary (like a frieze pattern), while the vertical iterant parts a and b mark the two sides of that boundary.

Returning to Quantum Mechanics

You can regard ψ **(x,t) =exp(i(kx - wt))** as containing a micro-oscillatory system with the special synchronizations of the iterant view **i = [+1,-1]**η . It is these synchronizations that make the big eigenform of the exponential ψ **(x,t)** work correctly with respect to differentiation, allowing it to create the appearance of rotational behavior, wave behavior and the semblance of the continuum. Note that **exp(ij) = cos(j) + i sin(j)** in this way of thinking is an infinite series involving powers of **i**. The exponential is synchronized via **i** to separate out its classical trigonometric parts. In the parts we have **cos(j) + i sin(j) = [cos(j), cos(j)] + [sin(j), -sin(j)]J,** a superposition of the constant cosine iterant and the oscillating sine iterant. Euler's formula is the result of a synchronization of iterant processes.

One can blend the classical geometrical view of the complex numbers with the iterant view by thinking of a point that orbits the origin of the complex plane, intersecting the real axis periodically and producing, in the real axis, a periodic oscillation in relation to its orbital movement in the higher dimensional space.

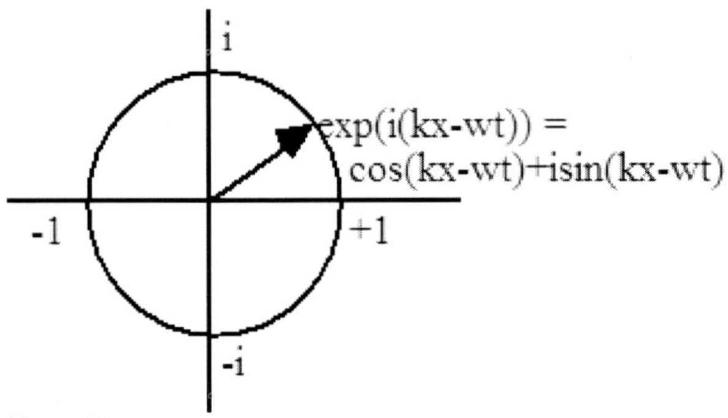

Figure 11.

The diagram above is the familiar depiction of a vector in the complex plane that represents the phase of a wave-function. I hope that the reader can now look at this picture in a new way, seeing **i = [+1,-1]h** as a discrete oscillation with built-in time shift and the exponential as a process oscillating between **cos(kx-wt) + sin(kx-wt)** and **cos(kx-wt)-sin(kx-wt)**. The exponential function takes the simple oscillation between **+(kx-wt)** and **-(kx-wt)** and converts it by a complex of observations of this discrete process to the trigonometric wave-forms. All this goes on beneath the surface of the Schrödinger equation. This is the production of the eigenforms from which may be extracted the energy, position and momentum.

Higher Orders of Iterant Structure

What works for 2 x 2 matrices generalizes to n x n matrix algebra, but then the operations on a vector $[x_1,x_2,...,x_n]$ constitute all permutations of n objects. A generating element of iterant algebra is now of the form $x\,s = [x_1,x_2,...,x_n]s$ where s is an element of the symmetric group S_n. The iterant algebra is the linear span of all elements $x\,s$, and we take the rule of multiplication as $x\,s\,y\,t = xy^s\,st$ where y^s denotes the vector obtained from y by permuting its coordinates via s; xy is the vector whose k-th coordinate is the product of the k-th coordinate of x and the k-th coordinate of y ; st is the composition of the two permutations s and t .

Hamilton's Quaternions

Here is an example. Hamilton's Quaternions are generated by the iterant views

$$I = [+1,-1,-1,+1]s$$
$$J = [+1,+1,-1,-1]l$$
$$K = [+1,-1,+1,-1]t$$

where

$$s = (12)(34)$$
$$l = (13)(24)$$
$$t = (14)(23).$$

Here we represent the permutations as products of transpositions (ij). The transposition (ij) interchanges i and j, leaving all other elements of $\{1,2,...,n\}$ fixed. One can verify that $I^2 = J^2 = K^2 = IJK = -1.$ For example,

$$\begin{aligned} I^2 &= [+1,-1,-1,+1]s\,[+1,-1,-1,+1]s \\ &= [+1,-1,-1,+1][-1,+1,+1,-1]s\,s \\ &= [-1,-1,-1,-1] \\ &= -1. \end{aligned}$$

and

$$\begin{aligned} IJ &= [+1,-1,-1,+1]s\,[+1,+1,-1,-1]l \\ &= [+1,-1,-1,+1][+1,+1,-1,-1]\,s\,l \\ &= [+1,-1,+1,-1]\,(12)(34)(13)(24) \\ &= [+1,-1,+1,-1]\,(14)(23) \\ &= [+1,-1,+1,-1]\,t. \end{aligned}$$

In a sequel to this paper, we will investigate this iterant approach to the quaternions and other algebras related to fundamental physics. For now it suffices to point out that the quaternions of the form $a + bI + cJ + dK$ with $a^2 + b^2 + c^2 + d^2 = 1$ (a,b,c,d real numbers) constitute the group $SU(2)$, ubiquitous in physics and fundamental to quantum theory. Thus the formal structure of all processes in quantum mechanics can be represented as actions of iterant viewpoints. Nevertheless, we must note that making an iterant interpretation of an entity like $I = [+1,-1,-1,+1]s$ is a conceptually natural departure from our original period two iterant notion. Now we are considering iterants such as $I\{+1,-1,-1,+1\}$ where the iterant is a multi-set and the permutation group acts to produce all possible orderings of that multi-set. The iterant itself is not an oscillation. It represents an implicate form that can be seen in any of its possible orders. Once seen, these orders are subject to permutations that produce the possible views of the iterant. Algebraic structures such as the quaternions appear in the explication of such implicate forms.

The reader will also note that we have moved into a different conceptual domain from the original emphasis in this paper on eigenform in relation to to recursion. Indeed, each generating quaternion is an eigenform for the transformation $R(x) = -1/x$. The richness of the quaternions arises from the closed algebra that arises with its infinity of eigenforms that satisfy this equation, all of the form $U = aI + bJ + cK$ where $a^2 + b^2 + c^2 = 1$. This kind of significant extra structure in the eigenforms comes from paying attention to specific aspects of implicate and explicate structure, relationships with geometry and ideas and inputs from the perceptual, conceptual and physical worlds. Just as with our earlier examples (with cellular automata) of phenomena arising in the course of the recursion, we see the same phenomena here in the evolution of mathematical and theoretical physical structures in the course of the recursion that constitutes scientific conversation.

Quaternions and SU(2) Using Complex-Number Iterants

Since complex numbers commute with one another, we could consider iterants whose values are in the complex numbers. This is just like considering matrices whose entries are complex numbers. For this purpose we shall allow given a version of i that commutes with the iterant shift operator η. Let this commuting i be denoted by ι (iota). Then we are assuming that

$$\iota^2 = -1$$
$$\eta\,\eta\,\iota = \iota\,\eta$$
$$\eta^2 = +1.$$

We then consider iterant views of the form $[a + b\iota, c + d\iota]$ and $[a + b\iota\,, c + d\iota\,]\eta = \eta\,[c + d\iota\,, a + b\iota\,]$. In particular, we have $e = [1,-1]$, and $i = e\eta$ is quite distinct from ι . Note, as before, that $e\eta = -\eta\,e$ and that $e^2 = 1$. Now let

$$I = \iota e$$
$$J = e\eta$$

$$\mathbf{K} = \mathbf{i}\eta .$$

We have used the commuting version of the square root of minus one in these definitions, and indeed we find the quaternions once more.

$$\mathbf{I^2 = ie\ ie = i\ i\ e\ e = (-1)(+1) = -1,}$$
$$\mathbf{J^2 = e\eta\ \ e\eta = e\,(-e)\ \eta\,\eta\ = -1,}$$
$$\mathbf{K^2 = i\eta\ \ i\eta = i\ i\ \eta\,\eta = -1,}$$
$$\mathbf{IJK = ie\ e\eta\ \ i\eta\ = i\ 1\ i\ \eta\ \ \eta = i\ i = -1.}$$

Thus

$$\mathbf{I^2 = J^2 = K^2 = IJK = -1.}$$

This must look a bit cryptic at first glance, but the construction shows how the structure of the quaternions comes directly from the non-commutative structure of our period two iterants. In other, words, quaternions can be represented by 2 x 2 matrices. This is way it has been presented in standard language. The group **SU(2)** of 2 x 2 unitary matrices of determinant one is isomorphic to the quaternions of length one.

$$\begin{bmatrix} z & w \\ -\bar{w} & \bar{z} \end{bmatrix} = [z, \bar{z}] + [w, -\bar{w}]\eta$$

Figure 12.

In the equation above, we indicate the matrix form of an element of **SU(2)** and its corresponding complex valued iterant. You can easily verify that

1: z=1, w=0,
I: z=i, w =0,
J: z=0, w = 1,
K: z=0, w = i.

This gives the generators of the quaternions as we have indicated them above and also as generators of **SU(2)**. Similarly, **H = [a,b] + [c + di, c-di]h** represents a Hermitian 2 x 2 matrix and hence an observable for quantum processes mediated by **SU(2)**. Hermitian matrices have real eigenvalues. It is curious how certain key iterant combinations turn out to be essential for the relations with quantum observation.

6. Relativity and the Dirac Equation

In this section we shall take the convention that Planck's constant is equal to 1, and that the speed of light is equal to 1. This will simplify the algebra and the forms of the equations.

Starting with the algebra structure of **e** and η and adding a commuting square root of **-1**, **i**, we have constructed fermion algebra and quaternion algebra. We can now go further and construct the Dirac equation. This may sound circular, in that the fermions arise from solving the Dirac equation, but in fact the algebra underlying this equation has the same properties as the creation and annihilation algebra for fermions, so it is by way of this algebra that we will come to the Dirac equation.

If the speed of light is equal to **1** (by convention), then energy **E,** momentum **p** and mass **m** are related by the (Einstein) equation $\mathbf{E^2 = p^2 + m^2}$. Dirac constructed his equation by looking for an algebraic square root of $\mathbf{p^2 + m^2}$ so that he could have a linear operator for **E** that would take the same role as the Hamiltonian in the Schrödinger equation. We will get to this operator by first taking the case where **p** is a scalar (we use one dimension of space and one dimension of time).

Let **E = ap + bm** where **a** and **b** are elements of a possibly non-commutative, associative algebra. Then

$$\mathbf{E^2 = a^2\,p2 + b^2\,m^2 + pm(ab + ba).}$$

Hence we will satisfy, $\mathbf{E^2 = p^2 + m^2}$ if $\mathbf{a^2 = b^2 = 1}$ and **ab + ba = 0.** This is our familiar Clifford algebra pattern and we can use the iterant algebra generated by **e** and **h** if we wish.

Then, because the quantum operator for momentum is $\mathbf{-i\ \partial/\partial x}$ and the operator for energy is $\mathbf{i\partial/\partial t}$, we have the Dirac equation

$$\mathbf{i\partial\psi/\partial t = -ia\partial\psi/\partial x + bm\psi\,.}$$

Let $\mathbf{O = i\partial/\partial t + ia\ \partial/\partial x - m}$ so that the Dirac equation takes the form **Oy(x,t) = 0.** Now note that

$$\mathbf{Oe^{i(px\,-\,Et)} = (E - a\,p + bm)\,e^{i(px\,-\,Et)}}$$

and that if

$$\mathbf{U = (E - ap + bm)ba = baE + bp + am,}$$

then $\mathbf{U^2 = -E^2 + p^2 + m^2 = 0}$, from which it follows that $\mathbf{y = U\,e^{i(px\,-\,Et)}}$ is a (plane wave) solution to the Dirac equation.

In fact, this calculation suggests that we should multiply the operator **O** by **ba** on the right, obtaining the operator

$$D = O\,ba = i\,ba\,\partial/\partial t + i\,b\partial/\partial x + am,$$

and the equivalent Dirac equation $D\psi = 0$. In fact for the specific ψ above we will now have

$$D (U e^{i(px - Et)}) = U^2\ e^{i(px - Et)} = 0.$$

This way of reconfiguring the Dirac equation in relation to nilpotent algebra elements **U** is due to Peter Rowlands [14]. We will explore this relationship with the Rowlands formulation in a separate paper.

Majorana Solutions

Return now to the original version of the Dirac equation.

$$i\partial\psi/\partial t = -ia\partial\psi/\partial x + bm\psi.$$

We can rewrite this as

$$\partial y/\partial t = a\partial\psi/\partial x + ibm\psi.$$

We see that if **ib** is real, then we can write a fully real version of the Dirac equation. For example, we can take the equation $\partial y/\partial t = e\partial\psi/\partial x + ehm\psi$. where we represent

$$e = \begin{matrix} -1 & 0 \\ 0 & 1 \end{matrix}$$

and

$$h = \begin{matrix} 0 & 1 \\ 1 & 0 \end{matrix}$$

as matrix versions of the iterants associated with the reentering mark. For the case of one dimension of space and one dimension of time, this is the Majorana representation for the Dirac equation (compare with N1). Since the equation can have real solutions, these are their own complex conjugates and correspond to particles that are their own anti-particles. As the reader can check, the corresponding Rowland nilpotent **U** is given by the formula

$$U = -i\,h\,E + i\,e\,h\,p + e\,m.$$

Nilpotency

For effective application to the topics in this paper, one needs to use three dimensions of space and one dimension of time. This will be explored in another paper, but we will begin the story here by recounting Rowland's approach to the Dirac equation in our algebraic terms. In this form, let **i** denote a commuting square root of minus one. Let **I, J** and **K** be any operators such that $I^2 = J^2 = K^2 = -1$ and $IJ = -JI$, $IK = -KI$, $JK = -KJ$. For example, **I**, **J** and **K** can be the generators of the quaternions with the extra relation $IJK = -1$. As we have seen, these can be constructed from Majorana fermion operators and from iterants, starting with elementary discrete waveforms. We then note that

$$(iIE + Jp + Km)^2$$
$$= I^2E^2 + J^2p^2 + K^2m^2 + (IJ+JI)Ep + (IK+KI)Em + (JK+KJ)pm = E^2 - p^2 - m^2 = 0.$$

Thus we see that there is a natural construction for a nilpotent algebraic operator $U = iIE + Jp + Km$. We define the differential operator $D = (I\,\partial/\partial t + i\,J\,\partial/\partial x + Km)$ and note that

$$D\,e^{i(Et - px)} = (iIE + Jp + Km)\,e^{i(Et - px)} = U\,e^{i(Et - px)}.$$

Thus $D(Ue^{i(Et - px)}) = U^2 = 0$ and we have the quaternionic version of the Dirac equation as $D\psi = 0$. It is now a small step to import another copy of the quaternions to support the three directions for the full momentum p and with the aid of this doubled complex quaternion algebra, accomplish the full Dirac equation. We will leave the analysis of the Majorana fermions in the sense of real solutions to the three dimensional Dirac equation to a sequel to the present paper.

In the present paper we have given a picture of how, starting with the mark as a logical and recursive particle, one can tell a story that reaches the Dirac equation and its algebra.

7. Time Series and Discrete Physics

In this section we shall use the convention (outside of iterants) that successive observations, first A and then B will be denoted BA rather than AB. This is to follow previous conventions that we have used. We continue to interpret iterant observation sequences in the opposite order as in the previous section. This section is based on our work in [15] *but takes a different interpretation of the meaning of the diffusion equation in relation to quantum mechanics.*

We have just reformulated the complex numbers and expanded the context of matrix algebra to an interpretation of **i** as an oscillatory process and matrix elements as combined spatial and temporal oscillatory processes (in the sense that **[a,b]** is not affected in its order by a time step, while **[a,b]h** includes the time dynamic in its interactive capability, and 2 x 2 matrix algebra is the algebra of iterant views **[a,b] + [c,d]h**).

We now consider elementary discrete physics in one dimension. Consider a time series of positions **x(t), t = 0, Dt, 2Dt, 3Dt, ...** . We can define the velocity **v(t)** by the formula **v(t) = (v(t+ D) - v(t))/Dt = Dx(t)** where **D** denotes this discrete derivative. In order to obtain **v(t)** we need at least one tick **Dt** of the discrete clock. Just as in the iterant algebra, we need a time-shift operator to handle the fact that once we have observed **v(t)**, the time has moved up by one tick.

Thus we shall add an operator **J** that in this context accomplishes the time shift: **x(t)J = Jx(t+Dt).** We then *redefine* the derivative to include this shift: **Dx(t) = J(x(t+ D) - x(t))/Dt .** The result of this definition is that a successive observation of the form **x(Dx)** is distinct from an observation of the form **(Dxn)x**. In the first case, we observe the velocity and then **x** is measured at **t + Dt** . In the second case, we measure **x** at **t** and then measure the velocity. Here are the two calculations:

$$\begin{aligned} x(Dx) &= x(t)\,(J(x(t+D) - x(t))/Dt) \\ &= (J/D)(x(t+Dt))(x(t+Dt) - x(t)) \\ &= (J/Dt)(x(t+Dt)^2 - x(t+Dt)x(t)). \end{aligned}$$

$$\begin{aligned} (Dx)x &= (J(x(t+Dt) - x(t))/Dt)x(t) \\ &= (J/Dt)(x(t+Dt)x(t) - x(t)^2). \end{aligned}$$

We measure the difference between these two results by taking a commutator **[A,B] = AB - BA** and we get the following formula where we write **Dx = x(t+ Dt) - x(t).**

$$\begin{aligned} [x,(Dx)] &= x(Dx) - (Dx)x \\ &= (J/Dt)(x(t+Dt) - x(t))^2 \\ &= J\,(Dx)^2/Dt \end{aligned}$$

This final result is worth marking:

$$[x,(Dx)] = J\,(Dx)^2/Dt.$$

From this result we see that the commutator of **x** and **Dx** will be constant if $(Dx)^2/Dt = K$ is a constant. For a given time-step, this means that $(Dx)^2 = K\,Dt$ so that **Dx = + √(K Dt)** or **- √(K Dt)**. In other words, **x(t + Dt) = x(t) + √(K Dt)** or **x(t) - √(K Dt).** This is a Brownian process with diffusion constant equal to **K.**

Digression on Brownian Processes and the Diffusion Equation

Assume, for the purpose of discussion that in the above process, at each next time, it is equally likely to have + or - in the formulas **x(t + Dt) = x(t) + √(K Dt)** or **x(t) - √(K Dt).** Let **P(x,t)** denote the probability of the particle being at the location x at time t in this process. Then we have

$$P(x, t + Dt) = (1/2)(P(x - Dx) + P(x + Dx)).$$

Hence

$$\begin{aligned} &(P(x, t + Dt) - P(x,t))/Dt) \\ &= ((Dx)^2/2Dt)\,(P(x - Dx) - 2P(x, t) + P(x + Dx,t))/(Dx)^2) \\ &= (K/2)(P(x - Dx) - 2P(x,t) + P(x + Dx,t))/(Dx)^2). \end{aligned}$$

Thus we see that **P(x,t)** satisfies the a discretization of the diffusion equation

$$\partial P/\partial t = (K/2)\partial^2 P/\partial x^2.$$

Compare the diffusion equation with the Schrödinger equation with zero potential shown below

$$i\hbar\partial\psi/\partial t = -(\hbar^2/2m)\partial^2\psi/\partial x^2.$$

In the Schrödinger equation we see that we can rewrite it in the form $\partial\psi/\partial t = i(\hbar/2m)\partial^2\psi/\partial x^2$. Thus, if we were to make a literal comparison with the diffusion equation we would take $K = i(\hbar/m)$ *and we would identify*

$$(Dx)^2/Dt = i(\hbar/m).$$

Whence

$$Dx = ((1+i)/\sqrt{2})\sqrt{[(\hbar/m)Dt]}$$

and the corresponding Brownian process is

$$x(t + Dt) = x + Dx \text{ or } x - Dx.$$

The process is a step-process along a diagonal line in the complex plane. We are looking at a Brownian process with complex values! What can this possibly mean? Note that if we take this point of view, then **x** is a complex variable and the partial derivative with respect to **x** is taken with respect to this complex variable.

In this view of a complexified version of the Schrödinger equation, the solutions for **Dx** as above are real probabilities. We shall have to move the **x** variation to real **x** to get the usual Schrödinger equation, and this will result in complex valued wave functions in its solutions.

In our context, the complex numbers are themselves oscillating and synchronized processes. We have **i = [1,-1]h** where **h** is a shifter satisfying the rules of the last section, and **[1,-1]** is a view of the iterant that oscillates between plus and minus one. Thus we are now observing that solutions to the Schrödinger equation can be construed as Brownian paths in a more complicated discrete space that is populated by both probabilistic and synchronized oscillations. This demands further discussion, which we now undertake.

The first comment that needs to be made is that since in the iterant context **Dx** is an oscillatory quantity it does make sense to calculate the partial derivatives using the limits as **Dx** and **Dt** approach zero, but this means that the interpretation of the Schrödinger equation as a diffusion equation and the wave function as a probability is dependent on this generalization of the derivative. *If we take* ***Dx*** *to be real, then we will get complex solutions to Schrödinger's equation.* In fact we can write $\psi(x, t + Dt) = (1-i)\psi(x, t) + (i/2)\psi(x - Dx) + (i/2)\psi(x + Dx)$ and then we will have, in the limit, $\partial y/\partial t = i(\hbar/2m)\partial^2\psi/\partial x^2$ if we take $(Dx)^2/Dt = (\hbar/m)$. It is interesting to compare these two choices. In one case we took

$$(Dx)^2/Dt = i(\hbar/m)$$

and obtained *a Brownian process with imaginary steps.* In the other case we took $(Dx)^2/Dt = (\hbar/m)$ and obtained *a real valued process with imaginary probability weights.* These are complementary points of view about the same structure.

With $(Dx)^2/Dt = (\hbar/m)$, $\psi(x, t)$ is no longer the classical probability for a simple Brownian process. We can imagine that the coefficients **(1-i)** and **(i/2)** in the expansion of $\psi(x, t + Dt)$ are somehow analogous to probability weights, and that these weights would correspond to the generalized Brownian process where the real-valued particle can move left or right by **Dx** or just stay put. Note that we have $(1-i) + (i/2) + (i/2) = 1$, signaling a direct analogy with probability where the probability values are imaginary. But this must be explored in the iterant epistemology!

Note that **1-i = [1,1] -[1,-1]h** and so at any given time represents either **[1,1] - [1,-1] = [0,2]** or **[1,1] - [-1,1] = [2,0]**. It is very peculiar to try to conceptualize this in terms of probability or amplitudes. Yet we know that in the standard interpretations of quantum mechanics one derives probability from the products of complex numbers and their conjugates. To this end it is worth seeing how the product of **a+bi** and **a-bi** works out:

$$(a + bi)(a-bi) = aa + bia + a(-bi) + (bi)(-bi)$$
$$= aa + abi - abi - bbii$$
$$= aa - bb(-1)$$
$$= aa + bb.$$

It is really the rotational nature of **exp(it)** that comes in and makes this work. **exp(it)exp(-it) = exp(it - it) = exp(0) = 1**. The structure is in the exponent. The additive combinatory properties of the complex numbers are all under the wing of the rotation group.

A fundamental symmetry is at work, and that symmetry is a property of the synchronization of the periodicities of underlying process. The fundamental iterant process of **i** disappears in the multiplication of a complex number by its conjugate. In its place is a pattern of apparent actuality. It is actual just to the extent that one regards **i** as only possibility. On making a reality of **i** itself we have removed the boundary between mathematics and the reality that it is supposed to describe. There is no such boundary.

8. Epilogue and Simplicity

Finally, we arrive at the simplest place. Time and the square root of minus one are inseparable in the temporal nexus. The square root of minus one is a symbol and algebraic operator for the simplest oscillatory process. As a symbolic form, **i** is an eigenform satisfying the equation **i = -1/i.** One does not have an increment of time all alone as in classical **Dt.** One has **iDt**, a combination of an interval and the elemental dynamic that is time. With this understanding, we can return to the commutator for a discrete process and use **iDt** for the temporal increment.

We found that discrete observation led to the commutator equation

$$[x,Dx] = J\,(Dx)^2/Dt$$

which we will simplify to

$$[q, p/m] = (Dx)^2/Dt.$$

taking **q** for the position **x** and **p/m** for velocity, the time derivative of position.

Understanding that **Dt** should be replaced by **iDt**, and that, by comparison with the diffusion equation,

$$(Dx)^2/Dt = \hbar/m,$$

we have

$$[q, p/m] = (Dx)^2/i\,Dt = -i\,\hbar/m,$$

whence

$$[p,q] = i\hbar,$$

and we have arrived at Heisenberg's fundamental relationship between position and momentum. This mode of arrival is predicated on the recognition that only **i Dt** represents a true interval of time.

In the notion of time there is an inherent clock or an inherent shift of phase that is making a synchrony in our ability to observe, a precise dynamic beneath the apparent dynamic of the observed process. Once this substitution is made, *once the correct imaginary value is placed in the temporal circuit, the patterns of quantum mechanics appear.* With the help of the Majorana fermion operators associated with the simplest discrete process the oscillation of the primordial clock behind the square root of minus one, we described a a path to the Dirac equation that is harmonious with the nilpotent approach of Peter Rowlands.

The problem that we have examined in this paper is the problem to understand the nature of quantum mechanics. In fact, we hope that the problem is seen to disappear the more we enter into the present viewpoint. A viewpoint is only on the periphery. The iterant from which the viewpoint emerges is in a superposition of indistinguishables, and can only be approached by varying the viewpoint until one is released from the particularities that a point of view contains.

It is not just the eigenvalues of Hermitian operators that are the structures of the observation, but rather a multiplicity of eigenforms that populate mathematics at all levels. These forms are the indicators of process. Mathematics comes alive as an interrelated orchestration of processes. It is these processes that become the exemplary operators and elements of the mathematics that are put together to form the physical theory. We hope that the reader will be unable, ever again, to look at Schrödinger's equation Heisenberg's commutator or the Dirac equation the same way, after reading this argument.

References

[1] Barendregt, H. P. (1985) The Lambda Calculus - Its Syntax and Semantics, North Holland Pub.

[2] Bortoft, H. (1971) The Whole - Counterfeit and Authentic, Systematics , vol. 9, No. 2, pp. 43-73.

[3] Spencer-Brown, G. (1969) Laws of Form, George Allen and Unwin Ltd.

[4] von Foerster, H. (1981) Observing Systems, The Systems Inquiry Series, Intersystems Publications.

[5] von Foerster, H. (1981) Objects: Tokens for (eigen-) behaviors, in Observing Systems, The Systems Inquiry Series, Intersystems Publications, pp. 274 - 285.

[6] von Foerster, H. (1981) Notes on an epistemology for living things, in Observing Systems, The Systems Inquiry Series, Intersystems Publications, pp. 258 - 271.

[7] von Foerster, H. (1981) On constructing a reality, in Observing Systems, The Systems Inquiry Series, Intersystems Publications, pp. 288 - 309.

[8] Kauffman, L.H. (1987) Self-reference and recursive forms, Journal of Social and Biological Structures, 53-72.

[9] Kauffman, L.H. (1995) Knot logic, in Knots and Applications, ed. by L. H. Kauffman, World Scientific Pub. Co., pp. 1-110.

[10] Kauffman, L.H. (2001) The mathematics of Charles Sanders Peirce, in Cybernetics and Human Knowing, Volume 8, No. 1-2, pp. 79-110.

[11] Kauffman, L.H. (2005) Eigenform, Kybernetes - The Intl J. of Systems and Cybernetics, Vol. 34, No. 1/2, Emerald Group Publishing Ltd, p. 129-150.

[12] Kauffman, L.H. (2003) Eigenforms - Objects as Tokens for Eigenbehaviors, Cybernetics and Human Knowing, Vol. 10, No. 3-4, pp. 73-90.

[13] Mandelbrot, B.B (1982) The Fractal Geometry of Nature, W. H. Freeman and Company.

[14] Rowlands, P. (2007) Zero to Infinity - The Foundations of Physics, Series on Knots and Everything - Vol. 41, World Scientific Publishing Company, Singapore.

[15] Kauffman, L.H. (2004) Non-Commutative Worlds, New Journal of Physics, Vol. 6, 173 (47 pages).

[16] Louis H. Kauffman, L.H. (2009) Reflexivity and Eigenform - The Shape of Process, Constructivist Foundations, Vol. 4, No. 3, pp. 121-137.

[17] Kauffman, L.H. (2010) Reflexivity and foundations of physics, in R.L. Amoroso, P. Rowlands (eds.) Search for Fundamental Theory - The VIIth Intenational Symposium Honoring French Mathematical Physicist Jean-Pierre Vigier, Imperial College, London, UK, 12-14 July, New York: AIP - American Institute of Physics Pub, pp.48-89.

[18] Kauffman, L.H. Knot logic and topological quantum computing with Majorana fermions, To appear in Logic and Algebraic Structures in Quantum Computing and Information, Lecture Notes in Logic, J. Chubb, J. Chubb, Ali Eskandarian, & V. Harizano (eds.), Cambridge University Press.

The Tie That Binds: A Fundamental Unit of 'Change' in Space and Time

JAMES E. BEICHLER

Research Institute for Paraphysics, Retired
P.O. Box 624, Belpre, Ohio 45714 USA
Jebco1st@aol.com

Why, despite all efforts to the contrary, have attempts at unification based on the supposedly more fundamental quantum theory failed miserably? The truth is that the essential idea or concept of the quantum itself has never been fully understood. What is the quantum, or rather, what is its ultimate nature? Science may be able to work adequately with the quantum; in a sense science is quite articulate in the language of the quantum, i.e., its mathematical interpretation of the quantum mechanics, but science has no idea of the true physical nature of the quantum. Scientists and philosophers have wasted energy and efforts on irrelevant issues such as the debate over determinism and indeterminism instead of carefully analyzing the physical source of the quantum. Only with a true understanding of the physical nature of the quantum will the unification of the quantum and relativity ever become a reality.

Keywords: Quantum, Change, Unification, Spacetime, Relativity, Hidden variables, Five-dimensional, Einstein, Kaluza, Clifford

1. Introduction

Many quantum scientists have admitted over the years that they know everything about the quantum but have no idea what the quantum is. In fact, the physical origin of the quantum is one of the greatest unspoken mysteries in all of science. After many decades of successes the quantum theory is still beset by important philosophical problems that will not go away. Recently, the physicist N. David Mermin stated that:

> Quantum mechanics is the most useful and powerful theory physicists have ever devised. Yet today, nearly 90 years after its formulation, disagreement about the meaning of the theory is stronger than ever. New interpretations appear every year. None ever disappear [1].

Mermin argued that the problem exists solely because physicists do not really understand probability theory, but he missed the point completely. The misunderstanding is actually a product of the physical interpretation of the mathematics involved rather than a misunderstanding of the mathematics itself.

During the late eighteenth and nineteenth centuries, mathematics was rigorized by purging mathematics of any references and connections to its physical origins in the natural world. The abstract purified mathematics that evolved from this process was later used to describe the quantum world. The concept of non-commuting variables and variable pairs that was later used in the quantum theory is wholly a mathematical abstraction that has no real counterpart in the physical world. The individual quantities involved certainly have real world counterparts, but the non-commuting relationship between the real quantities does not exist physically. Yet the real physical quantities represented in the Heisenberg uncertainty principle that are interpreted probabilistically are only connected by their non-commutability. Understanding this fact is the key to understanding the main problems of the quantum theory. Unfortunately, the unification of the quantum and relativity can never be accomplished, in spite all of the efforts to do so, until these issues are resolved or otherwise eliminated.

In reality, an important unexplored fundamental relationship exists between physical space, time and the quantum that is essential for physical existence itself. This relationship has the potential to solve the problems of quantum mechanics as well as lead to the unification of the quantum and relativity without upsetting or challenging the past successes of either theoretical framework. Simply put, the mathematical formulas usually associated with special relativity can be derived by simple algebraic means from the Heisenberg uncertainty formulas by suppressing Planck's constant and identifying natural limits on the uncertainties involved. Doing so not only yields the true meaning of the quantum but yields the identity of the 'hidden variables' in the Heisenberg uncertainty formulas for the first time. The logical process involved leads to a completely new understanding of the quantum and how it relates to the fundamental reality of our physical universe. The application of this knowledge to the fundamental interaction between space and time, which we commonly sense and interpret as 'change', forms the geometrical framework of our physical and material world and can lead directly to the unification of quantum and relativity.

2. The Challenge

The central problem with quantum theory that few are willing to admit exists let alone address and solve can be simply stated – what is the quantum? The quantum theory, as presented by the Heisenberg uncertainty principle, is an example of mathematics that has been dissociated from the real physical world coming back to haunt physicists who are unable to properly interpret the mathematical implications within a real physical setting. The uncertainty principle simply states that any attempt to establish physical reality by measuring specific pairs of non-commuting variables is severely limited by the quantum and appears in the forms

$$\Delta x \Delta p \geq \hbar / 2 \tag{1}$$

And

$$\Delta E \Delta t \geq \hbar / 2. \tag{2}$$

The variables of uncertainty in each formula are non-commuting conjugate pairs. The concept of non-commuting conjugate pairs is physically ambiguous even though it has a precise mathematical meaning, yet quantum mechanics is built upon the physics of such pairs while the Copenhagen Interpretation of the uncertainty principle declares that it is impossible to determine physical reality beyond the pairs because it cannot be measured until consciousness collapses the wave packet representing the pairs. Yet that 'fact' is not necessarily true, it is an interpretation based on the mathematical 'certainty' and 'absolute' nature of the uncertainty principle itself. Uncertainty is inviolable and sacrosanct to all quantum scientists.

However, time and space are not simultaneously represented in either formula even though the formulas supposedly represent a quantum action or event occurring in space-time. While space is represented by the uncertainty of position in the first equation, time is subdued as merely implicit in the uncertainty in momentum, which in reality is a material change in position. Such a pairing of physical quantities may be acceptable in mathematics (which by admission does not deal directly with physical reality), but it is implied by nature. The mathematical viability of non-commuting variables does not necessarily render a viable pair of quantities for distinguishing physical reality because an event must be localizable in both space and time simultaneously before its action can be determined through either observation or measurement. Once again, the general concept of space is only implied by the uncertainty in energy in the second equation, while time is directly represented as an uncertainly. Space is only indirectly represented by the uncertainty in energy since energy results from a change in position (kinetic). In both formulas, the fact that simultaneous changes in space and time are only implied in situations where the other is directly quantified does not necessarily mean that a true physical variation has taken place, even when a measurement that should 'change' the system in some way has been made.

Noting these discrepancies between the mathematics and the physical reality that scientists claim they represent changes the interpretation of the physics involved. To truly represent a 'complete' physical reality the two formulas must be brought back together, such that

$$\Delta x \Delta p \geq \hbar / 2 \leq \Delta E \Delta t, \tag{3}$$

$$\Delta x \Delta p = \Delta E \Delta t. \tag{4}$$

and finally

$$\frac{\Delta x}{\Delta t} \Delta p = \Delta E. \tag{5}$$

Bringing time and space back together within one simple relationship is clearly tantamount to suppressing Planck's constant and returning the implied measurement to a combined space-time framework. At this point the physics starts to get interesting.

Now consider two possibilities – either the ratio $\Delta x/\Delta t \leq c$ or $\Delta x/\Delta t > c$ where "c" is just a physical constant. In this case "c" is just an arbitrary number (as in $E = mc^2$) and these equivalences should not yet be interpreted as meaning that the quantity $\Delta x/\Delta t$ is a velocity. The physical constant "c" is used merely because it is the only universal physical constant used in quantum theory that has the proper units of distance over time. Furthermore, there is no logical reason why a limit to the ratio cannot be established in a purely mathematical manner without reference to a physical quantity such as speed.

For simplicity sake, the relationship $\Delta x/\Delta t \leq c$ can be called the 'Einstein certainty ratio' and $\Delta x/\Delta t > c$ the 'Heisenberg uncertainty ratio'. It is also convenient to introduce a new concept called 'relative uncertainty', by which the complementary quantities of quantum mechanical uncertainty vary in such a way that an application of the Einstein certainty ratio yields different results for the uncertainty principle than would the application of the Heisenberg uncertainty ratio, which would yield an 'absolute uncertainty'. By a careful choice of the uncertainties involved, all of the formulations that normally represent special relativity – the Lorentz-Fitzgerald contraction, time dilation, increase of mass with increasing speed and matter-energy equivalence – can be simply derived.

Since the mathematical formulas expressed in special relativity can be derived from the Heisenberg uncertainty equations under these specific physical conditions – a restriction of change or motion to the constant c – the uncertainty in the Heisenberg picture of the quantum can only come from the attempted mathematical isolation of space and time from each other. The uncertainty is inherent in relating only the specific pairs of non-commuting variables ΔxΔp and ΔEΔt together, which paints an unnatural and unrealistic yet purely rigorous mathematical picture of reality. To determine a true physical picture of physical reality, it is necessary to consider the static position in both space and time simultaneously and not isolate them from one another by substituting time-like (Δp) and space-like (ΔE) mathematical surrogates for the real physical quantities. Stated in another way, time (duration) is the 'hidden variable' in the relationship $\Delta x\Delta p \geq h$ while space (distance) is the 'hidden variable' in the relationship $\Delta E\Delta t \geq h$.

Even if it is mathematically feasible or possibly even desirable for some convoluted reason, it is physically impossible with any certainty to consider only space or spatial position (Δx) with a corresponding material change in spatial position (Δp) without reference to time or consequently a temporal position (Δt) with a corresponding material change in temporal position (ΔE) without direct reference to space. Hence, the uncertainty many scientists and philosophers have thought exists in nature is actually the result of a completely unnatural (nonphysical and completely mathematical) attempt to treat time and space as separate and physically distinct from one another. Uncertainty is a mathematical artifact rather than a physical reality. Quantum mechanics does offer a new and valuable way of looking at nature by establishing a limit to how greatly scientists can corrupt the space-time continuum and still gain measurable observations of reality, but quantum mechanics itself does not give an accurate sustainable picture of normal reality. Quantum theory merely gives science new information about the limits of the physical world under very specific circumstances, but it does not itself represent physical reality in either part or as a whole.

This notion can be better visualized using a common Minkowski space-time diagram. The Einstein certainty ratio of Δx/Δt that is less than c corresponds to local events inside the light cone, the Einstein certainty ratio of Δx/Δt that is equal to c corresponds to the defining light cone itself and the Heisenberg uncertainty ratio of Δx/Δt that is greater than c corresponds to events outside of the light cone in the area of space-time that Hermann Minkowski designated the "absolute elsewhere". The light cone is similar to uncertainty in that the origin of the space-time axes represents some form of physical event such as an observation or measurement. Physically valid events outside the light cone could still be connected to events within the light cone by quantum entanglement, which has never really been identified or defined but could be interpreted as a byproduct of the overall space-time geometry. Quantum entanglement could thus be reduced to a form of quantum relativity within a space-time structure defined by general relativity.

Further physical insight and information can be gained from these mathematical manipulations by suppressing the Einstein certainty ratio when modifying the values of uncertainty and suppressing the quantum. Again, we start from the hybrid relationship

$$\Delta x\Delta p = \Delta E\Delta t. \tag{6}$$

When no reference is made to the speed of light, which further delimits or generalizes the physics involved, Newton's second law of motion can be derived. According to the energy-work theorem, the quantity ΔE should be equivalent to a force applied through some 'uncertain' distance, yielding

$$\Delta x\Delta p = F\Delta x\Delta t. \tag{7}$$

and

$$F = \frac{\Delta p}{\Delta t} \tag{8}$$

which reduces to

$$F = \frac{dp}{dt} \tag{9}$$

in the limit where the uncertainties approach the natural limits of physical reality near zero according to the basic theorems of calculus.

This last equation is none other than the second law of motion as originally stated by Newton. It seems logical that suppressing the speed of light as a necessary factor should yield something Newtonian rather than something Einsteinian. Furthermore, this derivation reinforces the basic idea that the Heisenberg uncertainty relationships do not individually give a complete picture of physical reality. Each of the basic uncertainty relationships that form the fundamental principles of quantum mechanics only gives a partial and thus incomplete picture of physical reality. It is only when they are brought together that physical reality emerges from Heisenberg's mathematical statements. Space and time cannot be treated separately as independent variables when addressing physical reality, especially when the changes that occur over the forward movement of time approach their smallest physical measurements in the 'quantum' realm of the microscopic world.

These new relationships imply quite strongly if not outright require that Planck's constant must be identified as a binding or coupling constant for space and time. This interpretation would mean that the quantum is just the smallest measurable unit of pure 'change' or physical interaction that is possible within the material world represented by the space-time continuum. Scientists have consistently misinterpreted and misrepresented the quantum since the inception of quantum mechanics in spite of the fact, or perhaps even because of the fact, that the concept of physical 'change' at any given instant cannot be isolated to absolute point positions in either space or time alone without reference to the other. On the other hand, quantum mechanics evokes the concept of a material particle that is confined to exist at a dimensionless point in the space-time continuum even though the concept of a dimensionless point in geometry or anywhere else is actually a purely mathematical concept. This occurs even though quantum theory is notoriously non-geometrical while relativity is notoriously geometrical. These two competing and mutually exclusive worldviews of material existence, which reflect the conceptual differences between the discrete and continuous, can thus be reduced to a geometrical problem between the fundamental concepts of point- and extension-based geometries.

Within this context, the difference between reality as it really exists, how it is consciously perceived and/or mathematically interpreted – especially in those cases where modern scientists normally refer to as 'uncertain' – comes from the difference between material measured values and the expected values only when non-commutable and thus logically derived variables are used. Our scientific knowledge of material reality proceeds from our experience or simultaneous measurement of the three commutable quantities of space (position), time (duration) and matter (mass), a fact that is not evident in the uncertainty relationships. The simple 'changes' in these quantities by which we determine reality can only occur when a change in time accompanies a constant or varying position in space. Since some variation in time is a minimal requirement for 'change' to occur, the concept of 'change' is more readily associated with time than either space or matter, but 'change' cannot be reduced to a variation in time alone. In the case of the Heisenberg uncertainty principle, material reality is only 'implied' by the pairs of non-commuting variables but cannot be fully determined by either of the twinned pairs of $\Delta x \Delta p$ and $\Delta E \Delta t$ alone.

3. The Geometrical Nature of the Problem

The calculus and methods of calculation used in those cases where the Heisenberg uncertainty principle is applied ultimately depend upon a rigorous mathematical definition of instantaneous velocity or speed. This definition depends on two fundamental ideas: (1) the idea of a moment or instant that is all but a zero measure of time. In calculus this notion is made rigorous by the limit as the time interval approaches zero because a zero interval of time would yield an infinite velocity or speed; And, (2) a concept of continuity whereby any unbroken line element or spatial extension would consist of an infinite number of connected dimensionless points. On the other hand, the concept of speed in physical reality only approaches the quantum limit within the context of either the uncertainty in momentum or the uncertainty in energy. Nor can the real physical speed of a material object exceed the speed of light, posing yet another non-mathematical limit on physical reality. The mathematical situation presented by these realities creates a logical paradox that has gone completely unrecognized in science – mathematics is accepted completely and wholly when applied to physical situations without question or limitations even though real limitations exist. Yet precedents to this logical contradiction also occur within both geometry and general relativity and noting that fact can lead to a method to unify physics.

So there is a gross conceptual divide between the mathematical system of calculus and Heisenberg's interpretation of reality, which is completely non-geometrical. Quantum theory seems to depend upon a physical concept of a dimensionless point in space and time that is physically impossible to achieve let alone measure. Calculus also depends on the concept of a dimensionless point, but as a limit to what can be conceptually justified as physical, while it guarantees the existence of dimensionless points through the inclusion of continuity theorems. On the other hand, space-time theories in general are based upon extension-geometries. General relativity is based on the concept of a metric extension as defined in Riemannian geometry. Put another way, while both geometry and calculus use diminishing extensions to visualize the concept of a point and explain space itself, there exists no method or logical argument by which points in space could generate extensions let alone the extension-based geometry necessary to represent the concept of space in physics. The only method that comes close to forming an extended space from ideal mathematical points is the quantum method of perturbation which merely smears a dimensionless point over an already existing three-dimensional space. This method gives rise to such misconceptions as a fuzzy point, quantum foams and quantum fluctuations to explain continuity at the lowest possible imaginable level of physical reality.

It thus seems that a mathematical method of generating a space or spaces of as many dimensions as deemed necessary from dimensionless points should have been developed since the inverse logical argument is a necessary requirement for mathematical rigor in both geometry and calculus. But this method has never been developed nor even questioned by mathematicians. This oversight creates a gaping hole in mathematical logic, to put it lightly, especially in cases where mathematical logic applies to physical realities such as our commonly experienced three-dimensional space. This discrepancy has never been discussed because of the inherent problems associated with it. For example, to form a continuous extension, two points must at least be contiguous, i.e., making contact or touching, but neighboring dimensionless points cannot make contact with each other. To solve this fundamental conceptual problem, it would be necessary to derive and establish the legitimacy of a mathematical theorem that guarantees the reality of a point-generated physical space. Such a theorem-based mathematical system would depend upon the ability to generate extensions as well as real spaces from the ideal dimensionless points that constitute the space, i.e., the dimensions of space must be shown to depend upon the dimensionless points in that space. This theorem could be described as a 'reality' or 'existence' theorem.

The major obstacle to solving this problem would be how to define continuity relative to contiguity, so a conceptual definition of contiguous dimensionless points must be established first. Two dimensionless points, A and B, could never be contiguous through contact because contact would render them 'overlapping' or coincident. However, two different independent dimensionless points could be considered contiguous without actually depending on contact between them if and only if they were so close that no other dimensionless point could be placed between them to separate them. This situation is hard to imagine, but the concept is mathematically valid. Given this concept, take two dimensionless points, A and B, in close proximity to each other. In order to generate a one-dimensional continuous extension from them, these two points must be placed in positions contiguous to each other. But this cannot be accomplished within the space they are generating. There are literally an infinite (indefinable) number of directions either point could go to position itself contiguous to the other point. However, this problem can also be overcome. The points are only restricted to be dimensionless in the dimension(s) of space they share, which would constitute real one-, two- or three-dimensional spaces. So what then would guarantee the reality of a three-dimensional space given this situation and conditions?

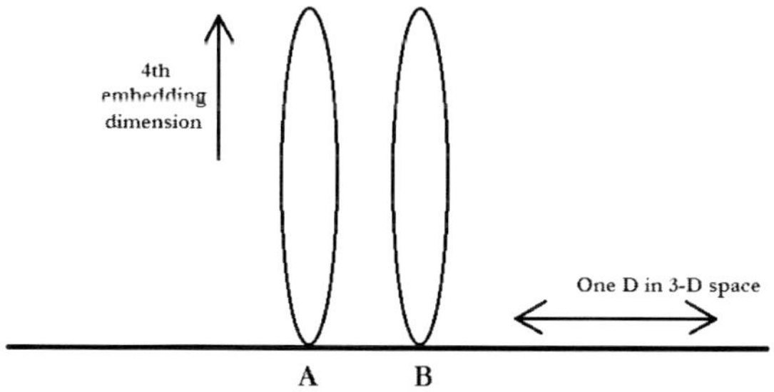

According to Gödel's theorem, only the internal logical consistency of a mathematical system can be proven within that system. The validity (reality) of that system, based on the primary foundational theorem from which the system is generated, can only be determined or proven logically true from outside of the system. So, all that present mathematics or physics can determine – prove of verify in either case – is the logical consistency of the system based upon their theorems and/or theories of the three-dimensionality of space. A reality theorem in mathematics would thus depend on dimensionless points in three-dimensional space utilizing a higher four-dimensional embedding space to guarantee that the three-dimensioned space could be generated from dimensionless points. This solution to the problem is already implied in Bernhard Riemann's original concept of space curvature whereby an n-dimensional space is embedded in an $n+1$-dimensional manifold. Einstein and others have either interpreted or physically abstracted this notion as an internally closed three-dimensional space coupled to a fourth dimension of time, which does not 'necessarily' represent the true physical situation.

In the next step of the 'proof', draw perpendicular lines in the fourth direction from both points A and B that are an infinitesimal distance apart in three-dimensional space.

These lines would normally remain parallel in the fourth direction and equidistant apart no matter how far they are extended, which does nothing to verify the reality of the three-dimensional space. However, if the three-dimensional space is curved, then the lines drawn from the points will draw closer three-dimensionally the further they are extended in the fourth direction. Once they have moved at least as far as the infinitesimal distance between them they would meet.

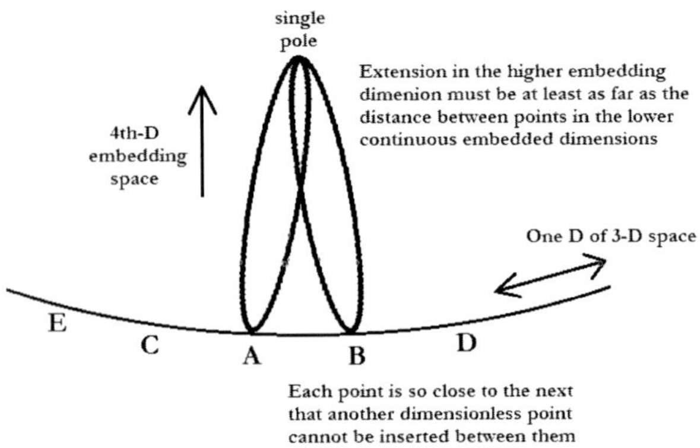

The extension in the fourth direction would then turn back to the other side of the three-dimensional space and return to the points from which they originated, maintaining continuity of the dimensionless points in three-dimensional space.

Now take another point, C, at the same infinitesimal distance from A, but in the opposite direction in three-dimensional space and repeat the procedure. C and A will coincide at one point in the fourth direction that is at least equal to or greater than the infinitesimal distance between them in three-dimensional space. In fact, A, B and C will all coincide at a single point in the fourth direction. Add two more points to either side of B and C – D and E – following the same procedure these points will also coincide at the same point in the fourth direction as A, B and C. Eventually an infinite number of points to either side of B and C would converge and form a closed space in one of the three-dimensions of three-dimensional space. All of these points would coincide at the same point in the higher fourth dimension of space.

When the same procedure is conducted for the other two dimensions of three-dimensional space, it would be found that all extensions in the fourth direction of space would coincide at a single point that is at least as far from three-dimensional space in the fourth direction as the sum of an infinite number of infinitesimal distances that separate the infinite number of points that make up the closed space in any one dimension of the three-dimensional space. The real three-dimensional space that is formed by this logical procedure would be internally double polar elliptical (spherical), but externally single polar elliptical in the fourth direction. The embedding higher-dimensional space is single polar elliptical (spherical) and at least as large as all of the dimensions in the real closed three-dimensional space. This space is exactly that type or form of physical hyperspace proposed by William Kingdon Clifford [2,3] and envisioned by other mathematicians in the late nineteenth century after they were first introduced to Riemann's geometry.

This structure, however, does not yet completely answer the question of how the individual dimensionless points in three-dimensional space can be contiguous to each other and thus form continuity in any given direction within three-dimensional space. The answer is implied by the geometrical structure and can be easily explained by reversing this process. If instead we start from the single polar point in the fourth dimension where all points in the three-dimensional subspace coincide and move away from that point along the fourth direction toward the full three-dimensional spherical space the order by which the individual dimensionless points of three-dimensional space separate would determine their order and placement in the three-dimensional space. Therefore the reality and existence of three-dimensional space built of dimensionless points is guaranteed. The interesting part of this procedure and the ensuing geometrical structure comes in the physical attributes that emerge for our own physical space of experience.

The mathematical conditions placed on this higher physical space that are required to render three-dimensional space real are quite straightforward. They are four in number. The first two are easily recognizable: (1) a one-dimensional line extending in the fourth direction from a dimensionless point in three-dimensional space must complete a circuit and return to that same point from the opposite direction; and (2) all lines extending from all dimensionless points in three-dimensional space must be of equal length. These are just the mathematical conditions that Kaluza set on his five-dimensional extension of Einstein's four-dimensional space-time continuum [4]. The third and fourth conditions are not so easily recognized: (3) the one-dimensional lines extending into the higher dimension from three-dimensional space must all be at least as long as a circumference line that completely encircles the embedded three-dimensional space. That is because (4) all of the one-dimensional lines that extend from the dimensionless points in three-dimensional space must pass through a single common point in their circuit before returning to the three-dimensional space in the opposite direction.

These last two conditions can only be fulfilled if

the higher embedding space is macroscopically sized, just as Einstein and his colleagues proved in the 1930s. It must also assume a single-polar elliptical Riemannian shape. In laymen's terms, each of the one-dimensional lines extended in the fourth direction of space works like a Möbius strip. This means that each point in three-dimensional space would have an inherent half-twist to it. These last two conditions impose the same overall structure as the four-dimensional space that Clifford envisioned in the 1870s, but the physical consequences of these conditions go beyond what Clifford imagined [2,3]. They also rule out the possibility that the superstring theories that depend on infinitesimally small and compacted higher dimensions could possibly represent physical reality.

The half-twist in each dimensionless geometrical point of our real three-dimensional space means that rotations of extended lines around a central point are possible in our common experiential space. In other words, three-dimensional space can be characterized by its support of either translational or rotational motions, which just happens to be observationally and experimentally true. Furthermore, this twist to the points in three-dimensional space accounts for the half-spin of elementary material particles and, in fact, establishes a geometrical requirement that all real stable material particles have half-spin. Material particles can only be stable and real if they meet the geometrical condition of a half-spin in the higher fourth dimension. Only protons, neutrons, electrons (muons and tauons) and neutrinos are real material particles, meaning, of course, that all of the geometrical points within the space they occupy are constrained by the half-spin of the same extended fourth dimension of space as opposed to other possible artificially constructed physical configurations.

Einstein adopted the geometrical structure of a double polar Riemannian sphere to model gravity as four-dimensional space-time curvature, but it is really space alone that is curved in the higher dimension. However, Einstein's positivistic leanings got the better of him when he interpreted the curvature as an intrinsic property of the four-dimensional spacetime continuum. On the other hand, the Riemannian geometry that he used was only a metrical geometry and could not directly account for the individual points in space or the physics that depends on the geometry of the individual dimensionless points. Classical electro-magnetic theory implies both an extension- and a point-geometry because electricity has a three-dimensional scalar potential structure and magnetism has a four-dimensional vector potential structure. Both are associated with points in three-dimensional space, but the magnetic vector potential must also have magnitude and direction even though it exists within a dimensionless three-dimensional point, which implies a higher dimensional embedding space [7].

Common Newtonian gravity only implies the need for an extension-geometry, so Einstein was unable to completely unify gravity and electromagnetism within the Riemannian space-time framework let alone incorporate the quantum into his ultimate interpretation of the world geometry.

This, however, does not mean that the task is impossible. Einstein was on the right track when he adopted Kaluza's five-dimensional space-time framework in the late 1930s as well as when he adopted Cartan geometry in 1929 and the symmetric/anti-symmetric tensor calculus after 1945. All of these geometric systems offer some limited form of a combined point- and extension-geometric structure. However, Einstein failed to realize two things while pursuing these particular solutions to unification. The combination of these two geometries through a Kaluza five-dimensional structure could lead to a single field theory, but Einstein only attempted his unifications of physics with these geometric structures individually. And secondly, he did not realize that the quantum could never emerge from an over-restriction of the mathematics of his geometrical structure as he hoped, but rather that the quantum is a fundamental property of space-time itself [4-6,8].

The quantum is the smallest possible physical unit of change that can be measured, while Planck's constant is the binding constant for space and time whether space is three or four dimensional, curved or non-curved. Einstein's attempted unification geom-etries only represented space curvature in a higher spatial dimension independent of any possible combination or connection to time – Einstein's curvature was a non-time based structure which he used dynamically to show how positions could change over time corresponding to gravitational attractions in three-dimensional space. He did not completely consider secondary effects of mutual changes in curvature. Therefore, the various space-time geometries that he tried could never provide a source from which the quantum could emerge logically from the geometrical structure.

In more general terms, the quantum is all about function (physical processes) and relativity is all about structures (form). Together they form a duality in which certain areas of physical reality overlap, but neither one can completely replace the other within our commonly experienced four-dimensional space-time continuum. For their part, modern quantum theorists do no better because they dismiss geometry altogether and in so doing they either base reality on incomplete pictures of quantized processes or develop statistical excuses as a substitute (alternate mapping) for the geometric structure of space-time. Both sides of the debate – quantum theory and relativity – need simply look at reality and analyze what they are doing in a serious critical manner to find the theoretical keys to unification. Then, and only then, will the method of unification become obvious. In reality there is no fundamental difference between the physics of Newton, Einstein, Bohr, Heisenberg, De Broglie or Schrödinger that cannot be overcome as indicated in this analysis and without inventing artificial mathematical and physical gimmicks such as virtual particles, particles whose existence can never be verified, supersymmetric particles, strings, superstrings, branes and many more dimensions of space than are necessary. Scientists have failed to realize why the difference between the quantum and relativity exists and have therefore missed the solution of how to unify them.

4. The Spacetime Quantum

Heisenberg's uncertainty is only guaranteed if the momentum p and the energy E are truly fundamental quantities such that they represent physical changes without direct reference to changes in either time or space. This interpretation raises a new question: is it possible to have a change in position, represented by the uncertainty in position, while there is no corresponding change in either time or position to be represented within the corresponding uncertainty in the change in momentum? This question is physically nonsensical yet it is directly implied by the Heisenberg uncertainty principle, hence Einstein and others' criticisms of quantum mechanics. The only truly fundamental uncertainties are spatial and temporal positions, which must be measured simultaneously to create a true perception of physical reality. There is no reason, whether scientific or mathematical, to conclude that momentum and energy are completely fundamental quantities at the subatomic realm of physical reality other than a faith in that *a priori* assumption by quantum scientists and philosophers.

This point of contention between worldviews can be better illustrated by a comparison between space-time and the dimensionless point or 'moment' of time that lies at the origin in a space-time diagram. This point corresponds to a quantum 'event' which occurs at a particular location in three-dimensional space. Relative time and relative space imply the existence of a 'point' in space-time where they meet as does calculus in mathematics when the limit of the time interval grows small enough that it approaches zero time during the measurement of an instantaneous speed. The same implication is found in Newtonian science, but for Newton those 'points' where the Euclidean three-dimensional space and time came together and collectively formed what he termed 'absolute space' and 'absolute time'.

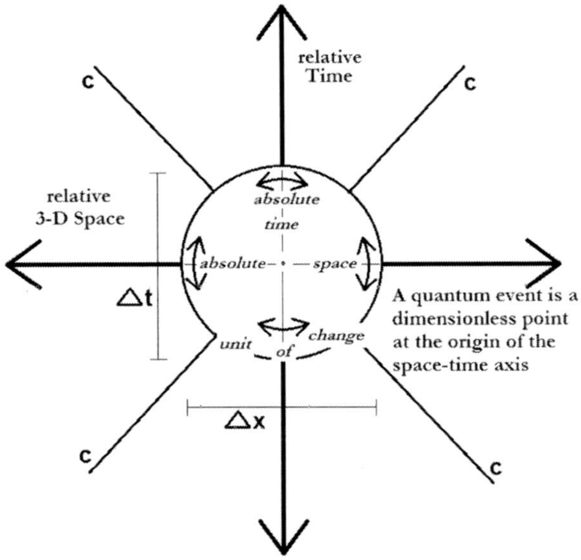

In Newton's theory a relative time interval could never go to zero, it could only approach absolute time. At this level of physical reality, there is no fundamental discrepancy between worldviews, the differences in these worldviews comes at another level of reality defined by high speeds approaching the speed of light (forming the light cone in the diagram) and extremely massive objects.

All of the different paradigms interpret the 'zone' bounded by the basic 'unit of change' differently, hence the fundamental problems and discrepancies between the quantum, relativistic and Newtonian worldviews. Relativity is a continuity theory whereby the idea of a dimensionless point is no stranger although the concept has never been defined beyond the purely mathematical necessity for continuity to complete the abstract pictures of geometry and calculus. In other words, the physical necessity for dimensionless points of space and time has been largely ignored by physical scientists, possibly because it sounds too much like absolute space and time or perhaps because it raises the likelihood of mathematical infinities (singularities), and has been assumed *a priori* without question.

Quantum theory is different from all other theories by having overtly adopted an actual physical limit of measurement that will not allow intervals of either space or time to go to zero, even as abstracted physical quantities. But quantum theory is based on a probabilistic point-geometry insofar as it mimics or maps out a pseudo-geometric picture of physical reality. This presents a paradox that adds to the confusion over the quantum and the only solution to this physical conundrum would be found in developing the geometry of space-time that it directly implies. It is in this sense then that a quantum point event can be pictured as a small sphere of uncertainty surrounding the abstract dimensionless point in space-time where the two axes of the space-time diagram come together. This sphere corresponds to the quantum of measurement. All that quantum theory can say about reality exists within this sphere that surrounds a point 'event', so quantum reality is strictly limited to an existence within this sphere alone since the quantum theory is neither mathematically geometrical nor physically relative to the rest of space-time.

When the quantum event – a material interaction, measurement or observation – occurs at a moment in time, the uncertainty would lie along the vertical axis invoking $\Delta E \Delta t \geq h$ and when it occurs at a point in space along the horizontal axis it would invoke the relationship $\Delta x \Delta p \geq h$. The uncertainty in position and time together form the axes of the interior of the sphere or 'fuzzy zone' of uncertainty surrounding the zero point of the space-time diagram representing the event. This 'fuzzy zone' of uncertainty would amount to what some scientists have called a 'fuzzy point', 'quantum fluctuation' or even a 'quantum foam' when the uncertainty is wrongly projected over the whole of space and time. David Bohm more accurately called the projection of such points over all of space the quantum potential field, while others mistakenly populate this lowest level of reality, the substrate of space-time, with invented unreal virtual particles and virtual photons.

Whatever the point location and thus the quantum event is called, the concept has created a conceptual conundrum (mess) that has seldom if ever been analyzed properly by scientists, mathematicians or philosophers. It would be far more accurate to call this 'fuzzy zone' the smallest possible 'unit of change' that can be measured or observed during any subatomic physical interaction. Quantum theory only deals with unrelated and physically unconnected point events (unless entanglement is taken into account), so the empty space and time between different independent events cannot be characterized by any one individual event that is represented by an infinitely extended wave function. Even though the quantum event still corresponds to an individual dimensionless point at the origin of the space-time axis, it must still conform to the dimensional structure established by the relative space and time axes outside of the 'fuzzy zone' before it can be considered real. However, changing from a dimensionless existence to a dimensioned experience poses a problem when that is normally described as the collapse of the wave function.

Put another way, the dimensionless point at the location where the space and time axes come together – the independent space-time 'now' of the observed event – can occupy only one of an infinite number of possible locations or orthogonal directions (thus the ↕ and ↔ symbols used in the diagram) corresponding to the x, y, z, and t axes of normal space-time. Each individual quantum 'event' must conform to the already established direction of relative time because time goes on, or moves forward, regardless of any one event external to the 'fuzzy zone' surrounding the event. All point events have a shared time outside of their individual 'fuzzy zones'. When the time-axis of the single quantum event collapses from all of its infinite possible orientations within the 'fuzzy zone' and aligns with the flow of time surrounding it, its spatial axes automatically align with the external spatial directions x, y, and z that define the common external three-dimensional space. This alignment corresponds to what is normally called 'collapsing the wave function'.

This process utilizes a Hilbert space of infinite dimensions whereby each point-located event establishes its own independent relative four-dimensional space-time framework, but there are an infinite number of such point-locations in this singular universe created by any one quantum event and all are pegged to the one central point of space-time origin. So the problem of 'uncertainty' inside the 'fuzzy zone' bounded by the quantum or 'unit of change' reduces to the singling out of one of the infinite number of possible axes orientations within the 'fuzzy zone' rather than a 'collapse of the wave function' that extends throughout all of space. This alignment occurs when the wave packet collapses to create 'reality' relative to the context of the normal space-time continuum. This quantum structure is intimately related to the mathematical concept of a Hilbert space. (See for example Brody and Hughston [9].) A Hilbert space is a purely mathematical projective space of rays that is non-linear with curves and described by Riemannian metrics. A precedent for this interpretation of the quantum already exists in general relativity since David Hilbert used such a construction to develop his own general relativistic structure of space-time independent of Einstein's metric development of general relativity.

In other words, the first question asked in quantum theory should be – what is the probability that the individual random four-dimensional space-time framework of a measurement beginning from a random dimensionless point in space-time would conform to the external physical and geometrical restrictions (requirements) of the relative space-time established by other material bodies in the universe? This probabilistic interpretation of the Heisenberg uncertainty principle and quantum mechanics is a far cry from assuming that the point-position in space and the point-moment in time corresponding to a quantum event are smeared out as a probability density cloud over an infinite number of possible locations corresponding to all of space-time before collapse of the wave packet, as presently assumed in the standard quantum interpretation.

Logically speaking, all of the probability is wound up inside the 'fuzzy zone' and the universe is still left deterministic outside of the 'fuzzy zone' since the collapse of the wave packet brings the event into alignment with external deterministic reality as well as Newtonian physics. Moreover, this model of the quantum yields an interpretation of quantum theory that corresponds well to the model of reality posed by both special and general relativity. Both the indeterminism of the quantum inside the 'fuzzy zone' and the determinism of relativity outside of the 'fuzzy zone' coexist to create physical reality. The width of the 'fuzzy zone' in both space and time is defined by the quantum, so the quantum is and can be nothing other than the binding constant for space-time. Scientists 'create' the uncertainty that they normally attribute to nature by their attempts to mimic mathematician's ideal mathematical abstractions on the stage of physical reality. The smaller we attempt to make the 'unit of change' size limit, either by abstraction or experimental measurement, the greater the radical 'change' of orientation of the event point's axes and the harder it becomes to align the time axis of the point to the external passage of time and thus the externally determined space-time geometry. This situation creates a greater uncertainty in the 'quantum measurement' and, in fact, is the source of the so-called measurement problem in quantum theory.

5. Conclusion

Debates regarding the deterministic and indeterm-inistic nature of reality do nothing but obfuscate the true physical meaning of the quantum. In the end, all questions about determinism and indeterminism are irrelevant to physics. They unnecessarily widen the philosophical gap between the quantum and relativity. Newtonian physics and relativity theory have never been as deterministic as quantum theorists have claimed and the absolute indeterminism of quantum physics is nothing more than a popular myth. The true difference between the quantum and relativity is geometrical even though the difference is normally couched as a difference between the discrete and continuous natures of reality.

Quantum theory is ideally point-geometric and relativity is physically an extension-based geometry. Both the quantum and relativity depend on the concept of a field (or fields), but the nature of the field differs from one model to the other. However, the single field that Einstein sought to discover in his theoretical research to develop a unified field theory was and still remains the best hope for the unification of the quantum and relativity. Within this geometrical structure of space-time, Planck's constant is the binding constant of four-dimensional space to time and the quantum is reduced to the smallest possible (meaningful) measure of physical 'change' no matter how that 'change' is brought about.

References

[1] Mermin, P. (2012) Physics Today 65, 8-10.

[2] Clifford, W.K. (1870) Trans Camb Phil Soc 2, 157-158.

[3] Clifford, W.K. (1873) Proc Lond Math Soc (12 June), 381-395.

[4] Kaluza, Th. (19210 Sitzsber Preuss Akad Wissen 54, 966-972.

[5] Einstein, A. et al. (1938) Ann Math 39, 3: 693-701.

[6] Einstein, A. et al. (1941) Theodor von Karman Anniv Vol, 212-225.

[7] Beichler, J (2007) Scient Expl 21: 523-542.

[8] Einstein, A. (1929) Sitzsber Preuss Akad Wissen, 2-7.

[9] Brody, D. et al., arxiv.org/pdf/quant-ph/9906086v2.pdf.

Space and Antispace

PETER ROWLANDS

Physics Department, University of Liverpool, Oliver Lodge Laboratory, Oxford St, Liverpool, L69 7ZE, UK,
p.rowlands@liverpool.ac.uk

Physics at the fundamental level can be effectively reduced to an explanation of the structures and interactions of fermions. Fermions appear to be singularities rather than extended objects, but there is no obvious way of creating such structures within the 3-dimensional space of observation. However, the algebra associated with the Dirac equation appears to suggest that the fermion requires a double, rather than a single, vector space, and this would seem to be confirmed by the double rotation required by spin ½ objects, and the associated effects of *zitterberwegung* and Berry phase shift. Further investigation of the second 'space' reveals that it is, in effect, an 'antispace', which contains the same information as real space but in a less accessible form. The two spaces effectively cancel to produce a norm 0 (nilpotent) object which has exactly the mathematical structure required to be a fermionic singularity.

Keywords: Berry phase shift, Fermion, Quaternions, Singularity, *Zitterberwegung*

1. Fermion Singularities and Double Spaces

1.1. *Fermions as singularities*

Physics at the fundamental level is about fermions and their interactions (which brings in the bosons). Ultimately, if we understand fermions, we will understand fundamental physics. Experimental evidence suggests that fermions have no size. They are singularities in space. Physics assumes that we can define dynamic physical objects ('particles' or fermions) as existing 'in space'. It is not at all obvious what this actually means, for, though the concept of space is inconceivable without matter, 3-D space has no mechanism within itself for constructing the physical singularities that make up material particles. So, how are such singularities constructed within ordinary space?

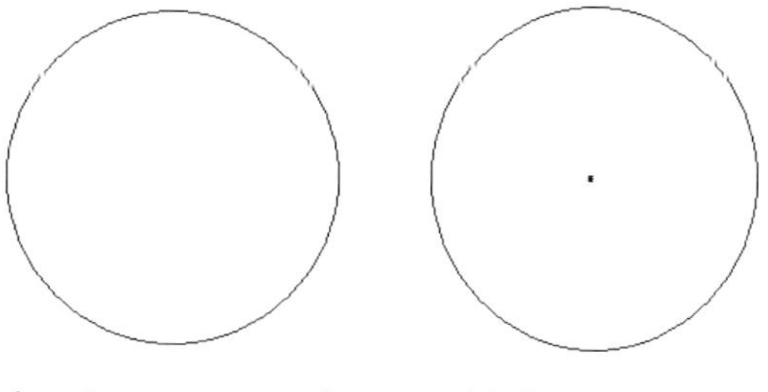

Here, we can use a reverse argument, where we first look at the effect of defining a space containing singularities – a multiply-connected space.

If we imagine parallel transporting a vector round a complete circuit of a simply-connected space (one with no singularities), then we would expect it to be pointing in the same direction at the end as it was at the beginning. However, if we should perform the same operation on a multiply-connected space (one enclosing a singularity), then the vector would end up pointing in the opposite direction at the end to that at the start. There would be a Berry phase effect, a phase shift of π or 180°. For the vector to return to the original starting position, and point in the same direction, we would have to perform the circuit twice.

The double circuit provides us with a direct analogy with the behaviour of a fermion with spin ½. The fermion is a singularity and, in effect, exists in its own multiply-connected space, so requiring a double circuit to return to its starting position. As is also well known, the fermion also undergoes the quantum motion called *zitterbewegung*, in which it continually switches between real space and vacuum, as though this were a second 'space'. It is as though the double circuit in real space is required because the fermion is only in this space for half its existence. It seems to be no coincidence, then, that the fermion algebra (the gamma matrices) requires a commutative combination of two vector spaces for its full mathematical representation. We may well suspect, therefore, that to construct a physical 'singularity', for example, a fundamental particle, we require two spaces.

1.2. *Quaternion and Clifford algebras*

At this point, we need to describe a number of significant algebras. The four quaternion units, $\boldsymbol{i}, \boldsymbol{j}, \boldsymbol{k}$, 1, follow the well-known multiplication rules:

$$\begin{aligned} \boldsymbol{i}^2 = \boldsymbol{j}^2 = \boldsymbol{k}^2 = \boldsymbol{ijk} &= -1 \\ \boldsymbol{ij} = -\boldsymbol{ji} &= \boldsymbol{k} \\ \boldsymbol{jk} = -\boldsymbol{kj} &= \boldsymbol{i} \\ \boldsymbol{ki} = -\boldsymbol{ik} &= \boldsymbol{j}. \end{aligned}$$

The nonscalar units are cyclic, norm –1, and anticommutative, and will be represented by bold italic symbols.

Multivariate vector units, also known as the units of the geometrical or Clifford algebra of 3D space, $\mathbf{i}$, $\mathbf{j}$, $\mathbf{k}$, i, are effectively complexified quaternions $(i\boldsymbol{i}) = \mathbf{i}$, $(i\boldsymbol{j}) = \mathbf{j}$, $(i\boldsymbol{k}) = \mathbf{k}$, $(i1) = i$, where i is the regular (pseudoscalar) square root of –1, and follow the multiplication rules:

$$\begin{aligned} \mathbf{i}^2 = \mathbf{j}^2 = \mathbf{k}^2 = -i\mathbf{ijk} &= 1 \\ \mathbf{ij} = -\mathbf{ji} &= i\mathbf{k} \\ \mathbf{jk} = -\mathbf{kj} &= i\mathbf{i} \\ \mathbf{ki} = -\mathbf{ik} &= i\mathbf{j}\,. \end{aligned}$$

They are isomorphic to Pauli matrices. If we complexify *this* algebra, we revert to quaternions, so $(i\mathbf{i}) = \boldsymbol{i}$, $(i\mathbf{j}) = \boldsymbol{j}$, $(i\mathbf{k}) = \boldsymbol{k}$, etc. The vectors in this algebra have a full (algebraic) product:

$$\mathbf{ab} = \mathbf{i.b} + i\, \mathbf{a} \times \mathbf{b}$$

from which all the rules concerning unit vector multiplication may be derived. The vector units $\mathbf{i}$, $\mathbf{j}$, $\mathbf{k}$ will be represented by bold symbols. Terms like $i\mathbf{i}$, $i\mathbf{j}$, $i\mathbf{k}$, generated by the product of two orthogonal vector units, are pseudovectors (e.g. area, angular momentum) or bivectors, and are algebraically equivalent to quaternions. Terms like i, generated by the product of three orthogonal vector units, are pseudoscalars (e.g. volume) or trivectors.

The Clifford or geometrical algebra algebra generated by a multivariate vector space of 3 dimensions thus requires 16 units for its full description, which are + and – versions of:

$\mathbf{i}$ $\mathbf{j}$ $\mathbf{k}$	vector		
$i\mathbf{i}$ $i\mathbf{j}$ $i\mathbf{k}$	bivector	pseudovector	quaternion
i	trivector	pseudoscalar	
1	scalar		

Let us now take a dual vector space which is commutative to the first, using bold upper-case symbols for the units:

$\mathbf{I}$ $\mathbf{J}$ $\mathbf{K}$	vector		
$i\mathbf{I}$ $i\mathbf{J}$ $i\mathbf{K}$	bivector	pseudovector	quaternion
i	trivector	pseudoscalar	
1	scalar		

If we take a commutative combination of the two spaces (a tensor product), allowing for + and – signs, we obtain an algebra structured on 64 units:

$\mathbf{i}$	$\mathbf{j}$	$\mathbf{k}$	$i\mathbf{i}$	$i\mathbf{j}$	$i\mathbf{k}$	i 1
$\mathbf{I}$	$\mathbf{J}$	$\mathbf{K}$	$i\mathbf{I}$	$i\mathbf{J}$	$i\mathbf{K}$	
$\mathbf{iI}$	$\mathbf{iJ}$	$\mathbf{iK}$	$i\mathbf{iI}$	$i\mathbf{iJ}$	$i\mathbf{iK}$	
$\mathbf{jI}$	$\mathbf{jJ}$	$\mathbf{jK}$	$i\mathbf{jI}$	$i\mathbf{jJ}$	$i\mathbf{jK}$	
$\mathbf{kI}$	$\mathbf{kJ}$	$\mathbf{kK}$	$i\mathbf{kI}$	$i\mathbf{kJ}$	$i\mathbf{kK}$	

Alternatively, we can take the algebraic product of the four quaternion units, 1, $\boldsymbol{i}, \boldsymbol{j}, \boldsymbol{k}$, and the four vector units i, $\mathbf{i}$, $\mathbf{j}$, $\mathbf{k}$, to obtain + and – versions of:

$\boldsymbol{i}$	$\boldsymbol{j}$	$\boldsymbol{k}$	$i\boldsymbol{i}$	$i\boldsymbol{j}$	$i\boldsymbol{k}$	i 1
$\mathbf{i}$	$\mathbf{j}$	$\mathbf{k}$	$i\mathbf{i}$	$i\mathbf{j}$	$i\mathbf{k}$	i 1
$\mathbf{i}\boldsymbol{i}$	$\mathbf{i}\boldsymbol{j}$	$\mathbf{i}\boldsymbol{k}$	$i\mathbf{i}\boldsymbol{i}$	$i\mathbf{i}\boldsymbol{j}$	$i\mathbf{i}\boldsymbol{k}$	
$\mathbf{j}\boldsymbol{i}$	$\mathbf{j}\boldsymbol{j}$	$\mathbf{j}\boldsymbol{k}$	$i\mathbf{j}\boldsymbol{i}$	$i\mathbf{j}\boldsymbol{j}$	$i\mathbf{j}\boldsymbol{k}$	
$\mathbf{k}\boldsymbol{i}$	$\mathbf{k}\boldsymbol{j}$	$\mathbf{k}\boldsymbol{k}$	$i\mathbf{k}\boldsymbol{i}$	$i\mathbf{k}\boldsymbol{j}$	$i\mathbf{k}\boldsymbol{k}$	

This is exactly isomorphic to the previous algebra and can be described as a vector quaternion algebra.

A third version of the same algebra could be obtained by complexifiying the algebraic product of two commutative sets of quaternion units 1, $\boldsymbol{i}, \boldsymbol{j}, \boldsymbol{k}$, and 1, $\boldsymbol{I}, \boldsymbol{J}, \boldsymbol{K}$. This algebra has + and – versions of:

$\boldsymbol{i}$	$\boldsymbol{j}$	$\boldsymbol{k}$	$i\boldsymbol{i}$	$i\boldsymbol{j}$	$i\boldsymbol{k}$	i 1
$\boldsymbol{I}$	$\boldsymbol{J}$	$\boldsymbol{K}$	$i\boldsymbol{I}$	$i\boldsymbol{J}$	$i\boldsymbol{K}$	
$\boldsymbol{Ii}$	$\boldsymbol{Ij}$	$\boldsymbol{Ik}$	$i\boldsymbol{Ii}$	$i\boldsymbol{Ij}$	$i\boldsymbol{Ik}$	
$\boldsymbol{Ji}$	$\boldsymbol{Jj}$	$\boldsymbol{Jk}$	$i\boldsymbol{Ji}$	$i\boldsymbol{Jj}$	$i\boldsymbol{Jk}$	
$\boldsymbol{Ki}$	$\boldsymbol{Kj}$	$\boldsymbol{Kk}$	$i\boldsymbol{Ki}$	$i\boldsymbol{Kj}$	$i\boldsymbol{Kk}$	

This can be described as a complexified double quaternion algebra.

1.3. *The Dirac algebra*

The 64-unit algebra is recognisably the algebra of the Dirac equation, or the γ matrices, just as the algebra of a single vector space is recognisably that of the Pauli or σ matrices. In fact, all possible versions of the γ matrices can be derived from a commutative combination of two sets of σ matrices, say σ_1, σ_2, σ_3 and Σ_1, Σ_2, Σ_3 [1-3].

The units of the Dirac algebra can, like the γ matrix products, be represented as a group of order 64, requiring only a *pentad* of 5 generators. Many sets of 5 generators can be derived, for example, the asterisked units in the vector-quaternion set:

i	j*	k	i**i**	i**j**	i**k***	i	1
i	**j**	**k**	i**i**	i**j**	i**k**	i	1
ii*	**i**j	**i**k	i**i**i	i**i**j	i**i**k		
ji*	**j**j	**j**k	i**j**i	i**j**j	i**j**k		
ki*	**k**j	**k**k	i**k**i	i**k**j	i**k**k.		

All the pentads have the same overall structure. Typically, the pentad is something like (in vector quaternion form):

$$i\mathbf{k} \qquad \mathbf{i}i \qquad \mathbf{j}i \qquad \mathbf{k}i \quad j$$

or, using double vector spaces

$$\mathbf{K} \qquad i\mathbf{iI} \qquad i\mathbf{jI} \qquad i\mathbf{kI} \quad i\mathbf{J}$$

All other ways of representing a generating pentad have the same structure, with the rotation symmetry of one vector space preserved (here, the lower case symbols) and with that of the other broken (here, the upper case symbols) in precisely the same way.

2. The Nilpotent Dirac Operator

2.1. *Nilpotency*

Physically, what results from defining the 5 generators is that the algebraic symbols describe particular physical quantities, to which we may attach scalar coefficients (E, p_x, p_y, p_z, m):

energy	momentum	rest mass
$i\mathbf{k}$	$\mathbf{i}\,\mathbf{i}\;\mathbf{i}\,\mathbf{j}\;i\mathbf{k}$	j
E	$p_x\;p_y\;p_z$	m

The characteristics of the physical terms are determined by the algebraic operators, alone, not the coefficients, which are merely scalar multiples.
Physically, it would appear that the combined object now described, is a nilpotent, or square root of zero, an algebraic object with norm 0. That is,

$$(i\mathbf{k}E + \mathbf{i}\mathbf{i}p_x + i\mathbf{j}p_y + i\mathbf{k}p_z + jm)^2 = 0$$

or, collecting the three momentum terms:

$$(i\mathbf{k}E + i\mathbf{p} + jm)\,(i\mathbf{k}E + i\mathbf{p} + jm) = 0.$$

Multiplying this out gives us Einstein's energy-momentum equation:

$$E^2 - p^2 - m^2 = 0. \tag{1}$$

The duality between the two constituent spaces thus comes from the fact that physical singularities constructed from units of the form

$$i\mathbf{k} \qquad \mathbf{i}i \qquad \mathbf{j}i \qquad \mathbf{k}i \quad j$$

are nilpotent in this way or square to zero (norm 0). This is the case with the fundamental singularity of physics, the fermion, and the nilpotent structure can be interpreted either classically or quantum mechanically to produce the Einstein or Dirac equations.

2.2. *The nilpotent Dirac equation*

A quite simple way to derive this expression is to take the Einstein relation (1) and factorise to give

$$(\pm\, i\mathbf{k}E \pm i\mathbf{p} + jm)\;(\pm\, i\mathbf{k}E \pm i\mathbf{p} + jm) = 0$$

Here, the four terms in first bracket could be considered as arranged as a row vector and the four terms in the second bracket as a column vector. If we apply a canonical quantization procedure to the first bracket in the squared expressions, to replace the terms E and $\mathbf{p}$ by the operators $E \to i\partial/\partial t$, $\mathbf{p} \to -i\nabla$ (with units where $\hbar = 1$), and assume that the operators act on the phase factor for a free fermion, $e^{-i(Et - \mathbf{p.r})}$, we obtain the nilpotent Dirac equation for a free fermion:

$$\left(\mp \boldsymbol{k}\frac{\partial}{\partial t} \mp i\boldsymbol{i}\nabla + \boldsymbol{j}m\right)(\pm i\boldsymbol{k}E \pm i\mathbf{p} + \boldsymbol{j}m)e^{-i(Et-\mathbf{p.r})} = 0$$

We can also use the convention that E and $\mathbf{p}$ represent operators as well as amplitudes to express it as

$$(\mp i\boldsymbol{k}E \pm i\mathbf{p} + \boldsymbol{j}m)(\pm i\boldsymbol{k}E \pm i\mathbf{p} + \boldsymbol{j}m)e^{-i(Et-\mathbf{p.r})} = 0$$

The significant fact now is that the amplitude is always nilpotent.

Four simultaneous solutions are required for the wavefunction: 2 for fermion / antifermion × 2 for spin up / spin down. Rather than a 4 × 4 matrix differential operator and a column vector wavefunction, we use a row vector operator and a column vector wavefunction, each of which may be represented in abbreviated form by $(\pm\, i\mathbf{k}E \pm i\mathbf{p} + jm)$.

In the nilpotent formalism, the four solutions can be represented as, say:

$(i\mathbf{k}E + i\mathbf{p} + jm)$ fermion spin up

$(\boldsymbol{k}E - \mathbf{ip} + \boldsymbol{j}m)$	fermion spin down
$(-\boldsymbol{k}E + \mathbf{ip} + \boldsymbol{j}m)$	antifermion spin down
$(-\boldsymbol{k}E - \mathbf{ip} + \boldsymbol{j}m)$	antifermion spin up

The fermion need not be free. We can incorporate field terms or covariant derivatives into the operator, with, for example, $E \rightarrow i\partial / \partial t + e\phi + \ldots$, and $\mathbf{p} \rightarrow -i\nabla + e\mathbf{A} + \ldots$. We can still represent the operator as $(\boldsymbol{ik}E + \boldsymbol{i}\mathbf{p} + \boldsymbol{j}m)$, but the phase term will no longer be $e^{-i(Et - \mathbf{p.r})}$. It will be whatever is needed to create an amplitude of the general form $(\boldsymbol{ik}E + \boldsymbol{i}\mathbf{p} + \boldsymbol{j}m)$, which squares to zero, with the eigenvalues E and $\mathbf{p}$ representing more complicated expressions that will result from the presence of the field terms.

In principle, this means that we can do relativistic quantum mechanics for a fermion in any state, subject to any number of interactions, simply by defining an operator of the form

$$(\pm \boldsymbol{ik}E \pm \boldsymbol{i}\mathbf{p} + \boldsymbol{j}m).$$

This will then uniquely determine the phase factor which makes the amplitude nilpotent. There is no need to define any equation at all. We simply require the condition that:

$$(\text{operator acting on phase factor})^2 = \text{amplitude}^2 = 0$$

2.3. *Pauli exclusion, vacuum, and nonlocality*

The nilpotent structure immediately gives us a *formal* way of separating the local from the nonlocal. The bracketed term representing the fermion creation operator or wavefunction determines how conservation of energy applies to that fermion, as squaring the wavefunction and equating to zero gives us back the energy-momentum equation, and, of course, it is local, as the required Lorentzian structure is intrinsic.

However, the addition and multiplication of nilpotent wavefunctions construct the nonlocal processes of superposition and combination, and these processes do not require a Lorentzian structure. In effect, anything inside the fermion bracket is local and anything outside it is nonlocal.

A particle with a nilpotent wavefunction will be automatically Pauli exclusive, as the combination state with an identical particle $\psi_1\psi_1$ will be zero. Such wavefunctions are also Pauli exclusive in the conventional sense of being automatically antisymmetric. So, if we take a combination of the form $(\psi_1\psi_2 - \psi_2\psi_1)$, for nilpotent wavefunctions $\psi_1 = (\pm \boldsymbol{ik}E_1 \pm \boldsymbol{i}\mathbf{p}_1 + \boldsymbol{j}m_1)$ and $\psi_2 = (\pm \boldsymbol{ik}E_2 \pm \boldsymbol{i}\mathbf{p}_2 + \boldsymbol{j}m_2)$, we will obtain:

$$(\pm \boldsymbol{ik}E_1 \pm \boldsymbol{i}\mathbf{p}_1 + \boldsymbol{j}m_1)(\pm \boldsymbol{ik}E_2 \pm \boldsymbol{i}\mathbf{p}_2 + \boldsymbol{j}m_2)$$
$$-(\pm \boldsymbol{ik}E_2 \pm \boldsymbol{i}\mathbf{p}_2 + \boldsymbol{j}m_2)(\pm \boldsymbol{ik}E_1 \pm \boldsymbol{i}\mathbf{p}_1 + \boldsymbol{j}m_1)$$
$$= 4\mathbf{p}_1\mathbf{p}_2 - 4\mathbf{p}_2\mathbf{p}_1 = 8\,i\,\mathbf{p}_1 \times \mathbf{p}_2 = -8\,i\,\mathbf{p}_2 \times \mathbf{p}_1$$

and we see immediately that

$$(\psi_1\psi_2 - \psi_2\psi_1) = -(\psi_2\psi_1 - \psi_1\psi_2).$$

The result is actually quite remarkable. It implies that, instantaneously, any nilpotent wavefunction must have a $\mathbf{p}$ vector in spin space (a kind of spin 'phase') at a *different orientation* to any other. The wavefunctions of all nilpotent fermions might then instantaneously correlate because the planes of their $\mathbf{p}$ vector directions (which we can define in terms of projective geometry) must all intersect, and the intersections actually create the *meaning* of Euclidean space, with an intrinsic spherical symmetry generated by the fermions themselves.

Here, we see the nilpotent condition defined in two different spaces. In the case of the vacuum space we might imagine nilpotency as creating a unique direction on a set of axes $\boldsymbol{k}, \boldsymbol{i}, \boldsymbol{j}$ defined by the unique values of E, $\mathbf{p}$ and m, while the real space condition is defined by the axes $\mathbf{i}, \mathbf{j}, \mathbf{k}$. At the same time, in the real space, the $\mathbf{p}$ vector carries *all the information* available to a fermionic state, its direction also determining its E and $\mathbf{p}$ values uniquely.

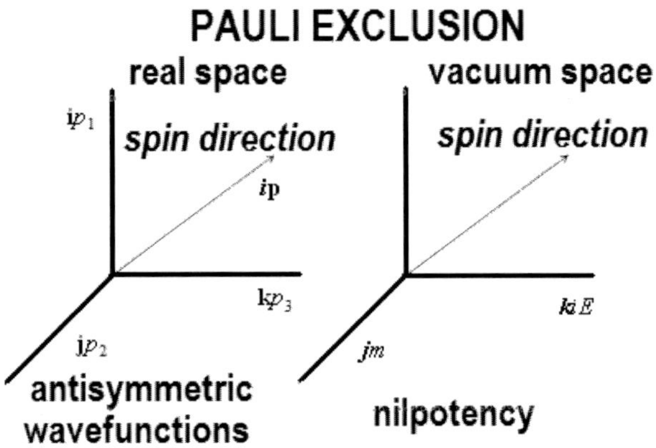

In the vacuum representation, half of the possibilities on one axis (those with $-m$) would be eliminated automatically (as being in the same direction as those with m), as would all those with zero m (since the directions would all be along the line $E = p$); such hypothetical massless particles would be impossible, in addition, for fermions and antifermions with the same helicity, as E, p has the same direction as $-E, -p$.

Pauli exclusion, a fundamentally nonlocal phenomenon, can also be seen as an immediate consequence of defining the total structure of the universe to be exactly zero. Imagine creating a fermion wavefunction of the form $\psi_f = (\boldsymbol{ik}E + \boldsymbol{i}\mathbf{p} + \boldsymbol{j}m)$ from *absolutely nothing*; then we must simultaneously create the dual term, $\psi_v = -(ikE + \mathbf{i}\mathbf{p} + jm)$, which negates it both in superposition and combination:

$$\psi_f + \psi_v = (ikE + \mathbf{i}\mathbf{p} + jm) - (ikE + \mathbf{i}\mathbf{p} + jm) = 0$$

$$\psi_f \psi_v = -(ikE + \mathbf{i}\mathbf{p} + jm)(ikE + \mathbf{i}\mathbf{p} + jm) = 0.$$

If the totality of the universe is zero, then that the act of creating a fermion (with all its special energy conditions, etc.) as a singularity simultaneously creates a kind of 'hole in nothing', which we call vacuum, or the rest of the universe. We create source and sink simultaneously, one point-like and localised, and the other delocalised. Only the space related to the localised state can be observed. Pauli exclusion tells us that no two fermions can have the same quantum numbers because the combination state would be zero. It also implies that no two fermions can share the same vacuum.

Vacuum is intrinsically nonlocal. Because the fermion is localized, then the rest of the universe is necessarily nonlocalized. If the fermion is a point, as experiments suggests that it may be, then the rest of the universe is defined as everything outside that point. So the nonlocal connection which makes Pauli exclusion possible can be thought to occur through the vacua for each fermion.

The object described by the Dirac equation, the fermion, is a singularity, and here the norm zero condition seemingly creates the 'singularity' state out of a combination of two vector spaces, and *no other information*. If the resulting condition is nilpotent or norm 0, then the information contained in the two spaces is not independent, but each depends on the other. The second space, in creating a zero totality, acts as a kind of 'antispace'.

Negative energy represents vacuum rather than the local quantized state, and the apparent disparity between matter and antimatter in the universe is really a result of the fact that, as in the four-component Dirac wavefunction, one set of states exists in observable real space, and the other in an unobservable 'vacuum space', as required by zero totality.

The dual nature of $\pm i$ is also a factor in creating the existence of two states of helicity. Once we have decided on a sign convention for **p**, the spin state of the particle (or, more conventionally, the helicity or handedness $\boldsymbol{\sigma}.\mathbf{p}$) is determined by the ratio of the signs of E and **p**. So $\mathbf{i}\mathbf{p} / ikE$ has the same helicity as $(-\mathbf{i}\mathbf{p}) / (-ikE)$, but the opposite helicity to $\mathbf{i}\mathbf{p} / (-ikE)$.

2.4. *Bosons and baryons*

In the 4-component spinor structure, the observed particle state is the first in the column, while the others are the accompanying vacuum states, or states into which the observed particle could transform by respective P, T and C transformations:

P	$i\,(ikE + \mathbf{i}\mathbf{p} + jm)\,i = (ikE - \mathbf{i}\mathbf{p} + jm)$
T	$k\,(ikE + \mathbf{i}\mathbf{p} + jm)\,k = (-ikE + \mathbf{i}\mathbf{p} + jm)$
C	$-j\,(ikE + \mathbf{i}\mathbf{p} + jm)\,j = (-ikE - \mathbf{i}\mathbf{p} + jm)$

Replacing the observed fermion state spin up with any of the others would simultaneously transform all four states by P, T or C.

Among the most important results which are seemingly unique to Nilpotent quantum mechanics are the descriptions of three different boson-type states, which are combinations of the fermion state which any of the P, T or C transformed ones, the result being a scalar wavefunction.

$(\pm ikE \pm \mathbf{i}\mathbf{p} + jm)\,(\mp ikE \pm \mathbf{i}\mathbf{p} + jm)$	spin 1 boson
$(\pm ikE \pm \mathbf{i}\mathbf{p} + jm)\,(\mp ikE \mp \mathbf{i}\mathbf{p} + jm)$	spin 0 boson
$(\pm ikE \pm \mathbf{i}\mathbf{p} + jm)\,(\pm ikE \mp \mathbf{i}\mathbf{p} + jm)$	fermion-fermion

One of the most significant aspects of this formalization is that a spin 1 boson can be massless, but a spin 0 boson cannot, as then $(\pm ikE \pm \mathbf{i}\mathbf{p})\,(\mp ikE \mp \mathbf{i}\mathbf{p})$ would immediately zero: hence Goldstone bosons must become Higgs bosons in the Higgs mechanism.

Another way of looking at this is to say that the fermion and antifermion cannot both be purely left-handed (or both purely right-handed) – or massless – and act via a weak interaction to produce a bosonic state. That is,

$(ikE + \mathbf{i}\mathbf{p})$	LH fermion
$(-ikE - \mathbf{i}\mathbf{p})$	LH antifermion

cannot be combined together in a single state (via a weak interaction) unless a nonzero mass term is introduced. This chirality is a direct consequence of the structure of the Dirac equation even in the conventional formalism, but is an immediate consequence of the nilpotent version.

We can also devise structures for baryons, by separating out three components of **p** in the form

$$(ikE \pm \mathbf{i}\mathbf{i}p_x + jm)\,(ikE \pm \mathbf{i}\mathbf{j}p_y + jm)\,(ikE \pm \mathbf{i}\mathbf{k}p_z + jm)$$

then we have a nonzero combination. Spin is defined in only one direction at a time, so, at any given instant, the wavefunction will reduce (after normalization) to

$$(ikE \pm \mathbf{i}\mathbf{i}p_x + jm)\,(ikE + jm)\,(ikE + jm) \rightarrow (ikE \pm \mathbf{i}\mathbf{i}p_x + jm)$$

$$(\boldsymbol{i}\mathbf{k}E + \boldsymbol{j}m)\ (\boldsymbol{i}\mathbf{k}E \pm \boldsymbol{i}\mathbf{j}p_y + \boldsymbol{j}m)\ (\boldsymbol{i}\mathbf{k}E + \boldsymbol{j}m) \to (\boldsymbol{i}\mathbf{k}E \mp \boldsymbol{i}\mathbf{j}p_y + \boldsymbol{j}m)$$

$$(\boldsymbol{i}\mathbf{k}E + \boldsymbol{j}m)\ (\boldsymbol{i}\mathbf{k}E + \boldsymbol{j}m)\ (\boldsymbol{i}\mathbf{k}E \pm \boldsymbol{i}\mathbf{k}p_z + \boldsymbol{j}m) \to (\boldsymbol{i}\mathbf{k}E \pm \boldsymbol{i}\mathbf{k}p_z + \boldsymbol{j}m). \qquad (2)$$

The change of sign in the second case is significant. This leads to baryonic structure. Since we need to maintain the symmetry between the three directions of momentum, there will be six possible outcomes, using both + and – values of momentum terms, that is, a superposition of six combination states:

$$(\boldsymbol{i}\mathbf{k}E + \boldsymbol{i}\mathbf{i}p_x + \boldsymbol{j}m)\ (\boldsymbol{i}\mathbf{k}E + \boldsymbol{j}m)\ (\boldsymbol{i}\mathbf{k}E + \boldsymbol{j}m)$$
$$(\boldsymbol{i}\mathbf{k}E - \boldsymbol{i}\mathbf{i}p_x + \boldsymbol{j}m)\ (\boldsymbol{i}\mathbf{k}E + \boldsymbol{j}m)\ (\boldsymbol{i}\mathbf{k}E + \boldsymbol{j}m)$$
$$(\boldsymbol{i}\mathbf{k}E + \boldsymbol{j}m)\ (\boldsymbol{i}\mathbf{k}E + \boldsymbol{i}\mathbf{j}p_y + \boldsymbol{j}m)\ (\boldsymbol{i}\mathbf{k}E + \boldsymbol{j}m)$$
$$(\boldsymbol{i}\mathbf{k}E + \boldsymbol{j}m)\ (\boldsymbol{i}\mathbf{k}E - \boldsymbol{i}\mathbf{j}p_y + \boldsymbol{j}m)\ (\boldsymbol{i}\mathbf{k}E + \boldsymbol{j}m)$$
$$(\boldsymbol{i}\mathbf{k}E + \boldsymbol{j}m)\ (\boldsymbol{i}\mathbf{k}E + \boldsymbol{j}m)\ (\boldsymbol{i}\mathbf{k}E + \boldsymbol{i}\mathbf{k}p_z + \boldsymbol{j}m)$$
$$(\boldsymbol{i}\mathbf{k}E + \boldsymbol{j}m)\ (\boldsymbol{i}\mathbf{k}E + \boldsymbol{j}m)\ (\boldsymbol{i}\mathbf{k}E - \boldsymbol{i}\mathbf{k}p_z + \boldsymbol{j}m)$$

which, on the application of (2), become

$$(\boldsymbol{i}\mathbf{k}E + \boldsymbol{i}\mathbf{i}p_x + \boldsymbol{j}m)$$
$$(\boldsymbol{i}\mathbf{k}E - \boldsymbol{i}\mathbf{i}p_x + \boldsymbol{j}m)$$
$$(\boldsymbol{i}\mathbf{k}E - \boldsymbol{i}\mathbf{j}p_y + \boldsymbol{j}m)$$
$$(\boldsymbol{i}\mathbf{k}E + \boldsymbol{i}\mathbf{j}p_y + \boldsymbol{j}m)$$
$$(\boldsymbol{i}\mathbf{k}E + \boldsymbol{i}\mathbf{k}p_z + \boldsymbol{j}m)$$
$$(\boldsymbol{i}\mathbf{k}E + \boldsymbol{i}\mathbf{k}p_z + \boldsymbol{j}m).$$

Significantly, the symmetry is only possible if the baryon incorporates both left-handed and right-handed components. *It must therefore have mass.* In principle, also, this mass comes from the same process as the Higgs mechanism, even though the switching mechanism between the different **p** components is provided by massless gluons. A demonstration that the proton must have mass, despite the strong interaction being conveyed by massless gluons, is a significant part of one of the prize challenge problems defined by the Clay Institute.

3. Vacuum and Antispace

3.1. *Partitioning the vacuum*

In the nilpotent formalism, the vacuum can be structured, directly reflecting the structure of matter. If we take $(\pm\, \boldsymbol{i}\mathbf{k}E \pm \boldsymbol{i}\mathbf{p} + \boldsymbol{j}m)$ and post-multiply it by the idempotent $\boldsymbol{k}(\pm\, \boldsymbol{i}\mathbf{k}E \pm \boldsymbol{i}\mathbf{p} + \boldsymbol{j}m)$ any number of times, the only effect is to introduce a scalar multiple, which can be normalized away.

$$(\pm\, \boldsymbol{i}\mathbf{k}E \pm \boldsymbol{i}\mathbf{p} + \boldsymbol{j}m)\boldsymbol{k}(\pm\, \boldsymbol{i}\mathbf{k}E \pm \boldsymbol{i}\mathbf{p} + \boldsymbol{j}m)\boldsymbol{k}(\pm\, \boldsymbol{i}\mathbf{k}E \pm \boldsymbol{i}\mathbf{p} + \boldsymbol{j}m) \ldots \to (\pm\, \boldsymbol{i}\mathbf{k}E \pm \boldsymbol{i}\mathbf{p} + \boldsymbol{j}m)$$

Similarly with $\boldsymbol{j}(\pm\, \boldsymbol{i}\mathbf{k}E \pm \boldsymbol{i}\mathbf{p} + \boldsymbol{j}m)$ or $\boldsymbol{i}(\pm\, \boldsymbol{i}\mathbf{k}E \pm \boldsymbol{i}\mathbf{p} + \boldsymbol{j}m)$ (though in the case of $\boldsymbol{i}(\pm\, \boldsymbol{i}\mathbf{k}E \pm \boldsymbol{i}\mathbf{p} + \boldsymbol{j}m)$, there will be additional vector multiples which can again be normalized away).

The three idempotent terms have the mathematical characteristics of vacuum operators. They also correspond to the respective transformations of alternate brackets by T, C and P, or to the production of the spin 1, spin 0 and fermion-fermion bosonic states. The character of the original fermionic state remains unchanged, while vacuum versions of the three bosonic states are created.

In principle, the quaternionic operators $\boldsymbol{k}, \boldsymbol{j}, \boldsymbol{i}$, like T, C and P, which are associated respectively with E, m and $\mathbf{p}$, split the continuous vacuum into discrete units, and we can to some extent regard these units as the respective 'charges' or sources of the weak, electric and strong interactions, acting to create the vacuum necessary for these forces to act.

In effect, the operators, $\boldsymbol{k}$, $\boldsymbol{i}$ and $\boldsymbol{j}$ perform the functions of weak, strong and electric charges, acting to partition the *continuous* (gravitational) vacuum represented by $-(\boldsymbol{i}\mathbf{k}E + \boldsymbol{i}\mathbf{p} + \boldsymbol{j}m)$, and responsible for zero-point energy, into discrete components, whose special characteristics are determined by the respective pseudoscalar, vector and scalar natures of their associated terms, iE, $\mathbf{p}$ and m. In this sense, they are related to 'real' weak, strong and electric localized charges, though they are delocalized.

One of the most important aspects of the nilpotent structure – with its pseudoscalar, vector and scalar components – is that it *already incorporates the fundamental interactions.* Simply defining a nilpotent fermion by this mathematical formalism means that it is *necessarily* acting according to some or all of these interactions. They arise solely from its internal structure. Coulomb terms, for example, are simply the result of spherical symmetry of point sources.

In principle, the interactions intrinsic to a fermion are a product of the three types of quantity (pseudoscalar, multivariate vector and scalar) which the nilpotent representation contains, and which are attached to the respective coefficients quaternion $\boldsymbol{k}, \boldsymbol{i}, \boldsymbol{j}$. Ultimately, and more subtly, these are consequences of the need for a discrete (point) source to preserve spherical symmetry and hence to conserve angular momentum. We can, in fact, identify the 'interactions' and their associated symmetries as being connected with the *three separately conserved aspects* of angular momentum:

Magnitude (scalar, $U(1)$), spherical symmetry does not depend on the length of the radius vector) ($\boldsymbol{j}$);

Direction (vector, $SU(3)$), spherical symmetry does not depend on the choice of axes) ($\boldsymbol{i}$);

And *handedness* (pseudoscalar, $SU(2)$), spherical symmetry does not depend on whether the rotation is left- or right-handed) ($\boldsymbol{k}$).

We can also see how the 3 bosonic states are related to vacua produced by the *3 quaternionic operators*:

The action of the *weak vacuum operator*

$(\boldsymbol{k}E + \boldsymbol{i}\mathbf{p} + \boldsymbol{j}m)\ \boldsymbol{k}\ (\boldsymbol{k}E + \boldsymbol{i}\mathbf{p} + \boldsymbol{j}m)\ \boldsymbol{k}\ (\boldsymbol{ik}E + \boldsymbol{i}\mathbf{p} + \boldsymbol{j}m)\ \boldsymbol{k}\ (\boldsymbol{ik}E + \boldsymbol{i}\mathbf{p} + \boldsymbol{j}m) \ldots$

is equivalent to

$(\boldsymbol{ik}E + \boldsymbol{i}\mathbf{p} + \boldsymbol{j}m)\ (-\boldsymbol{ik}E + \boldsymbol{i}\mathbf{p} + \boldsymbol{j}m)\ (\boldsymbol{ik}E + \boldsymbol{i}\mathbf{p} + \boldsymbol{j}m)\ (-\boldsymbol{ik}E + \boldsymbol{i}\mathbf{p} + \boldsymbol{j}m) \ldots$

or spin 1 boson creation.

The action of the *electric vacuum operator*

$(\boldsymbol{ik}E + \boldsymbol{i}\mathbf{p} + \boldsymbol{j}m)\, \boldsymbol{j}\ (\boldsymbol{ik}E + \boldsymbol{i}\mathbf{p} + \boldsymbol{j}m)\, \boldsymbol{j}\ (\boldsymbol{ik}E + \boldsymbol{i}\mathbf{p} + \boldsymbol{j}m)\, \boldsymbol{j}\ (\boldsymbol{ik}E + \boldsymbol{i}\mathbf{p} + \boldsymbol{j}m) \ldots$

is equivalent to

$(\boldsymbol{ik}E + \boldsymbol{i}\mathbf{p} + \boldsymbol{j}m)\ (-\boldsymbol{ik}E - \boldsymbol{i}\mathbf{p} + \boldsymbol{j}m)\ (\boldsymbol{ik}E + \boldsymbol{i}\mathbf{p} + \boldsymbol{j}m)\ (-\boldsymbol{ik}E - \boldsymbol{i}\mathbf{p} + \boldsymbol{j}m) \ldots$

or spin 0 boson creation.

The action of the *strong vacuum operator*

$(\boldsymbol{ik}E + \boldsymbol{i}\mathbf{p} + \boldsymbol{j}m)\, \boldsymbol{i}\ (\boldsymbol{ik}E + \boldsymbol{i}\mathbf{p} + \boldsymbol{j}m)\, \boldsymbol{i}\ (\boldsymbol{ik}E + \boldsymbol{i}\mathbf{p} + \boldsymbol{j}m)\, \boldsymbol{i}\ (\boldsymbol{ik}E + \boldsymbol{i}\mathbf{p} + \boldsymbol{j}m) \ldots$

is equivalent to

$(\boldsymbol{ik}E + \boldsymbol{i}\mathbf{p} + \boldsymbol{j}m)\ (\boldsymbol{ik}E - \boldsymbol{i}\mathbf{p} + \boldsymbol{j}m)\ (\boldsymbol{ik}E + \boldsymbol{i}\mathbf{p} + \boldsymbol{j}m)\ (\boldsymbol{ik}E - \boldsymbol{i}\mathbf{p} + \boldsymbol{j}m) \ldots$

or the creation of the paired fermion bosonic state.

We can thus describe the vacuum partitions as strong, weak and electric, and assign to them particular roles within existing physics:

$\boldsymbol{k}(\boldsymbol{ik}E + \boldsymbol{i}\mathbf{p} + \boldsymbol{j}m)$	weak	fermion creation
$\boldsymbol{i}(\boldsymbol{ik}E + \boldsymbol{i}\mathbf{p} + \boldsymbol{j}m)$	strong	gluon plasma
$\boldsymbol{j}(\boldsymbol{ik}E + \boldsymbol{i}\mathbf{p} + \boldsymbol{j}m)$	electric	isospin / hypercharge.

It is notable how this idea connects with the gravity-gauge theory correspondence now appearing in string theory, but always a fundamental component of the present theory and its earlier manifestations. (The electric vacuum partition – empty or filled – can be seen as responsible for the transition between weak isospin up and down states.)

3.2. *Antispace*

The term which is dual to space in the nilpotent formalism can be characterised in a number of distinct ways. 'Vacuum space' refers to the fact that the 3 axes represent a partitioning of the vacuum. 'Charge space' refers to the fact that it separates out the 3 components of charge – electric, strong and weak, and through its combination with ordinary space gives them the $U(1)$, $SU(3)$ and $SU(2)$ characteristics – their character as symmetries of 3-dimensionality are evidence that they cannot be explained by enlarging space to higher dimensions. 'Antispace' refers to the fact that it is the dual to ordinary space, effectively zeroing it in fermion creation.

The two 'spaces' – real space (with units **i**, **j**, **k**) and 'antispace' (with units, **I**, **J**, **K**) – though seemingly different, are truly dual, each containing the same information, and the duality manifests itself directly in many physical forms.

Pauli exclusion by antisymmetric wavefunctions

uses **i**, **j**, **k**

but by nilpotency uses **I**, **J**, **K**.

Both sets of coordinates yield information about the same physical quantity: angular momentum.

spin ½ from anticommuting aspects of **p** components

uses **i**, **j**, **k**

but spin ½ using Thomas precession (relativity)

uses **I**, **J**, **K**

The velocity addition law from 2 D of space

uses **i**, **j**, **k**

but the same using relativistic space-time

uses **I**, **J**, **K**

The holographic principle with the bounding 'area' defined by two spatial coordinates

uses **i**, **j**, **k**

but with one unit of space and one of time

uses **I**, **J**, **K**

The reason that we can only observe one space is that the other is not a single physical quantity, and is only a 'space' in a mathematical sense. That is, it has the mathematical structure of observed space, but is constructed from three different physical quantities representing the three subalgebras – scalar (mass), pseudoscalar (time) and bivector (charge). These represent the total physical information that is not spatial, but this combination cannot be realised in a single physical form [1-5].

Space is realised immediately as a vector quantity:

i	**j**	**k**	SPACE

But we can derive a vector quantity from a combination of:

*i***I**	*i***J**	*i***K**	CHARGE
i			+ TIME
1			+ MASS
I	**J**	**K**	= ANTISPACE

When the two spaces are combined in a norm 0 object (singularity or nilpotent), the symmetry of one of the two spaces is broken. Essentially, the space with the unbroken symmetry (lower case symbols) is that of space as we know it, the space of observation. The one with the broken symmetry (upper case symbols) is a dual space, which is unobservable, but which determines what happens in the observed space. The key fact is that the act of creating a singularity using these two spaces determines that they are precisely dual and that each contains the same information as the other, though in a different form as regards observation.

4. Conclusion

At the most fundamental level we have so far reached, physics seems to be concerned with point-like fermions and their interactions, the interactions generating bosons, the only other particles known to exist. An argument based purely on topology suggests that some of the properties fermions are known to possess (in particular, spin ½) could be explained by imagining a fermion to exist simultaneously in two spaces, one of which is recognizable as 'real space', the space which is the only known source of all physical observation, and the other can be seen as a 'vacuum space', which remains unobservable.

To express this in algebraic form, we require a double Clifford algebra of 3-dimensional space, which can also be expressed as a double multivariate vector algebra, a vector quaternion algebra, a complexified double quaternion algebra, or by using a tensor product of two sets of Pauli matrices. This is recognizable as the 64-part algebra of the Dirac equation for the fermion. It is also the algebra created by the four symmetrical parameters mass, time, charge and space, which have been identified as the only truly fundamental parameters required by physics [1,5]. A key fact is that the minimum number of generators for this structure (5) requires the symmetry of one of the two spaces to be broken, and this becomes the origin of the broken symmetry ($SU(3) \times SU(2) \times U(1)$), forming the Standard Model of particle physics.

Once the true origin and structure of the algebra is realized, as a combination of two vector spaces, an especially powerful version of relativistic quantum mechanics can be created in a purely algebraic way, directly from the 5 generators of the algebra. This identifies the fermion wavefunction as a nilpotent or square root of zero. Zero totality is an immediate consequence of the way the four components in the wavefunction are structured, with equal amounts positive and negative energy, momentum, angular, momentum, space and time, matter and antimatter.

These, however, are distributed between the real and vacuum spaces in such a way that we can only observe positive energy and positive time and that the universe, as observed, can be formed predominantly from matter, rather than antimatter. In addition causality will always be *observed* as backward ($+t$), though, in vacuum, it will be forward ($-t$), meaning that the vacuum – everything outside of a fermion – holds the key to its future. (Of course, this does not mean that 'determinism' can be applied as the future can only be fixed to the same extent that the fermion's parameters can be fixed by observation.)

Even the seemingly 'physical' properties of the parameters can be seen to be fundamentally algebraic. Space and mass are 'real' (norm 1), charge and time 'imaginary' (norm –1). Mass and time have a commutative (nondimensional) algebra, space and charge a noncommutative (dimensional) one. Mass and charge are conserved, space and time nonconserved, meaning that the last two, unlike the first two, incorporate an incomplete quaternion algebra [1,5]. Ultimately, physics is indeed a version of mathematics. The vacuum space can be seen as a kind of 'antispace', which contains the same information as the real space, but with opposite effect, so that the combination creates the zero normed or nilpotent state that constitutes the fermion as a point singularity. The opposite nature of the information from the two spaces is compatible with the idea that the totality of information in the entire universe is zero, while the broken symmetry at the fundamental level ensures that observation in space is not immediately cancelled by the unobserved counter-information in antispace.

The zero totality can be further explained in terms of a universal rewrite system [1,5,6] in which a zero-totality system continually rewrites itself in a unique and non-repeatable sequence, which cannot be restarted and can only progress from the previous point, and which also generates its own mathematics. If the universe is such a system, then from the point of view of an inside observer it will always appear asymmetric, even though the observer can perceive that zero totality will necessarily hold at an unobservable level.

References

[1] Rowlands, P. (2007) *Zero to Infinity*, World Scientific, Singapore, Hackensack, NJ, and London.

[2] Rowlands, P. Physical interpretations of nilpotent quantum mechanics, arXiv: 1004.1523.

[3] Rowlands, P. What is vacuum?, arXiv:0810.0224.

[4] Rowlands, P. (2010) Dual vector spaces and physical singularities, *AIP Conference Proceedings*, 1316, 102-111.

[5] Rowlands, P. Are there alternatives to our present theories of physical reality? arXiv:0912.3433.

[6] Rowlands, P. & Diaz, B. A universal alphabet and rewrite system, arXiv:cs.OH/0209026.

[7] Diaz, B. & Rowlands, P. (2005) International Journal of Computing Anticipatory Systems, 16, 203-18.

Zero-Totality in Action–Reaction Space: A Generalization of Newton's Third Law?

SABAH E. KARAM

Morgan State University, 1700 E. Cold Spring Ln.
Planning & Information Technology, Baltimore, MD, 21251
Sabah.Karam@morgan.edu

In order to present the universe as a zero-totality the key concepts of nothingness and duality are required. Diaz and Rowlands introduce processes of conjugation, complexification, and dimensionalization using a universal alphabet and rewrite system to describe a physical universe composed of nilpotents. This paper will apply the concept of conjugation to the Newtonian duality action-reaction by introducing associated dual spaces called action space-reaction space. An attempt to generalize Newton's third law of motion, utilizing the concept of dual spaces, will follow in a manner suggestive of the zero-totality fermion-vacuum relationship.

Keywords: Zero-totality, Conjugation, Duality, Newton's laws of motion, Action-space, Reaction-space.

1. Introduction

Diaz and Rowlands [1,2] have argued convincingly that the four fundamental parameters of physics are space, time, mass-energy, and charge. Superimposing charge on the other three parameters they identify a conserved and quantized Dirac state from which they derive the Dirac equation. The nilpotent version of the Dirac equation is derived from first principles and replaces the traditional matrix representation by an algebraic one. The Dirac algebra is applied to the baryon wavefunction, the strong interaction potential, and electroweak mixing allowing for a comprehensive treatment of the quantum field in terms of a quaternion-vector model. The 'wavefunctions' in the model are quaternion state vectors which are mathematical objects having a broader range of application than standard wavefunctions. The physical origin of quaternion wavefunctions lead to applications in particle physics not available in the conventional model [3].

The Dirac algebra incorporates three mathematical procedures called conjugation, complexification, and dimensionalization beginning in a zero-state and iterating to infinity. This paper will expand on conjugation, duality, and zero-state concepts used in the Dirac algebra in a non-standard attempt to generalize Newton's third law of motion. Conjugation and the zero-state depend on the context in which they are used. In an arithmetic setting numbers are conjugated by assigning alternating signs + and –, so that any number created is automatically paired with its opposite. Combining the two numbers under the operation of addition recovers the numeric 0. In a physical setting conjugation requires creation and annihilation operators. For any given elementary particle, say an electron, an antiparticle, the positron, is automatically created annihilating the created particle. Conjugation is then a process of bringing any element introduced into the system back to an original pre-defined zero-state of nothing. *Nothing* is not a scientific term even when referring to a vacuum, a region of space containing no matter, which still may contain gravitational fields. Zero is the starting point for a set of mathematical and physical rules used in constructing objects that preserve the original structures. Conjugation is a process that conserves a zero state. A negative charge, such as an electron, can only be part of the structure if a positive one is also part of the structure. A collision between the particle-antiparticle pair satisfies several conservation laws including the conservation law of electric charge in which the net charge before and after the collision is zero.

The concept of a zero-state appears in many areas of physics. It is a generally accepted theory in physics that the universe has zero (constant) total energy. The laws of thermodynamics (TD) can also be viewed as zero-state conditions ensuring that negative energy occurs in negative time and that any state satisfying TD laws are positive. Energy in isolated thermodynamic systems always remains constant. According to the 'big bang' and inflationary theory of the universe, matter and antimatter, as well as photons, are produced by the energy of the vacuum that was released following the phase transition. Matter consists of positive energy but is exactly balanced by gravitational energy thereby maintaining zero-energy of the universe.

The first claims concerning zero-total energy of an inflationary universe were made in the early 1960s. In the following years an idea that the universe may be a large-scale quantum-mechanical vacuum fluctuation emerged. It is theorized that positive mass-energy may be balanced by negative gravitational potential energy. By the turn of the twentieth century, researchers determined why the energy of a gravitational field is critical to the idea of a zero-energy universe. The gravitational property was used to show that the energy of a gravity field is truly negative. There are other cosmological arguments that matter-energy considered to be positive because it is increasing is exactly balanced by negative gravitational-energy resulting in a total-energy that is zero. An expanding universe extracts energy from the gravitational field, whereby more matter is created and hence more inertia [4-6]. Chemical reactions exhibit entangled zero-state properties as observed in redox reactions which involve a simultaneous transfer of electrons. Oxidation is a loss of electrons and reduction is a gain of electrons. Oxidation-reduction reactions are similar to acid-base reactions in which molecular, atomic, and ionic reactions are balanced with respect to electric charge. This also includes covalent bonding in which a stable balance of conjugate, attractive and repulsive, forces occurs.

A balancing act similar to that played out in gravity-matter interactions and redox reactions is evident in many conservation laws such as momentum, energy, mass, charge, and spin. Many mathematical equations describing physical events can be rewritten satisfying a zero-state condition that allows for new kinds of analysis and interpretations. Newton's laws of motion will be re-examined in light of Rowlands and Diaz's constraint that the physical universe "conserves" a Dirac state of space, time, mass and charge. Grouping the four parameters into a single mathematical expression creates a consistent and complete framework required needed to extend a physical reality from zero to infinity. In an attempt to generalize Newton's third law, "to every action there is an equal and opposite reaction," a key feature of the Dirac state will be employed, viz. the sum of all creation and annihilation operators maintains a zero-state. For fermionic fields this means that the positive frequency operators add a positive energy particle while the negative frequency operators annihilate a positive energy particle thereby maintaining a nothingness-state by simultaneously raising and lowering the energy. Inflationary universes and chemical balances conserve a zero state in a similar fashion.

2. Zero-Totality

Zero-totality conditions are generally satisfied in closed systems. Consider the conservation law of momentum where an interaction takes place between particles having masses m_1 and m_2 and traveling with velocities u_1 and u_2. If the final velocities are v_1 and v_2 then Newton's third law applies and we can write $m_1u_1 + m_2u_2 = m_1v_1 + m_2 v_2$. We rewrite this last equality in a form which underscores a zero state condition $m_1 (u_1 - v_1) + m_2 (u_2 - v_2) = 0$. If we let M represent the mass-vector (m_1, m_2) of the system, V the resulting velocity-vector $(u_1 - v_1, u_2 - v_2)$, and use dot product notation the last equation can now be written in the form $M \cdot V = 0$ allowing us to interpret momentum conservation in two ways. First, the mass-vector is orthogonal to the velocity vector and second the zero-state is conserved. It appears that momentum plays the same role in classical mechanics that the *mean* plays in statistics and probability theory. In a statistical environment zero is conserved by the mean; the sum of the deviations of observations from the expectation value is 0. Squaring and summing deviations from the mean produces a positive value allowing us to create a new statistic called the standard deviation. The continuous Gaussian distribution can then be uniquely defined using two parameters: mean and standard deviation.

The probability density function of a quantum harmonic oscillator is an exact Gaussian distribution with second-degree terms. Squared terms enter into a description of wavefunctions and include an orthogonality condition. The nilpotent Dirac wavefunction incorporates several fundamental parameters including space, time, mass and charge, energy and momentum. Real and virtual components in the nilpotent state vectors also include parameters of mass and charge and *zitterbewegung* is understood to be a switching between charge and mass providing state vectors a supersymmetric status. The oscillation is interpreted as the quantum equivalent of real *action* followed by virtual *reaction*. A new kinematical model of the *zitterbewegung* phenomenon, describing the inertial property of varying electron mass as being a relation between momentum and velocity, suggests that mass is not only a "measure of self-interaction" but also a measure of the "coupling strength of the electron to its own field" [7]. *Zitterbewegung* may be caused by interference between positive and negative energy wave components. If this is the case, then it is a good example of action-reaction conserving the zero-state. We will return to this idea later in the paper. The motivation for developing a kinematical model is to explain the magnetic dipole field of the electron. The dipole field is viewed as an "*average*" over a *zitterbewegung* period and actual field is theorized to contain a high-frequency component oscillating with the *zitterbewegung* frequency and may be responsible for electron diffraction and Pauli exclusion [8].

The idea that *zitterbewegung* is a kind of averaging can be mathematically demonstrated using a Foldy-Wouthuysen transformation. *Zitterbewegung* terms vanish upon taking expectation values for wave-packets that are made up only of positive or exclusively negative energy waves. These observations highlight the importance of squared terms in physics but are not the main topic of this paper. Chapter 16 of *Zero to Infinity* (*ZtI*) [9] calls attention to how duality generated conjugations account for squared terms and why the factor 2 is found in so many, seemingly unrelated, physical phenomena. The techniques developed in *ZtI* provide mechanisms that permit the mathematical manipulation of zero and also eliminate the problems found in averages and orthogonality conditions. The hierarchy problem also vanishes. In fact, no infinities arise in the Dirac states as they do in the statistical calculations of quantum electrodynamics (QED) field theory, and renormalization is not required. The nilpotent vector representation of the Dirac equation is second quantized and self-dual by virtue of the operator and state vectors being identical (*ZtI*, p. 291) [9].

There are other approaches to resolving complexity issues related to particle physics. The term dualism is used to characterize a general property of matter and has been used to explain one of the long-standing puzzles in the hierarchy of masses of the fundamental fermions [10]. The history of particle physics reveals that if a regular pattern is observed in the properties of matter, then it can be explained by invoking some underlying structures. An explanation for patterns of the fermion masses is based on a space-time dualism structure. Two reciprocal manifestations of space unified through an "inversion" region are used to address the complexity problem. The notion of dualism may be related to the ideas under discussion but will not be examined in this paper.

3. Zero-Totalities are Governed by Dualities

Many have suggested that the "discovery of fundamental dualities is the result of all attempts at scientific investigation" (*ZtI*, p. 64). Maintaining a zero-totality universe relies on a duality concept that is dependent on a process called conjugation. The dependence that physics has on dualities has been observed by others and has been called thema-antithema. A historical study of theme-based dualities identified as *discreetness-continuum*, *projection-retrojection*, and *evolution-devolution* reveals that scientists typically have an allegiance to one of the themes (thema) while ignoring the conjugate theme (anti-thema) [11]. The original theme may be dominant but, failing to explain experimentally observed facts, requires an alternative theory called the antithema. The competition between "conjugate notions" of reality is not restricted to the field of physics but is found in many other disciplines as well. We will investigate disciplines with conjugated variables and duality models in the next section.

Dual-pairs, which negate each other, are required in creating something from nothing, the empty state, or the zero-state. The requirement is what is meant by conjugation. We will investigate this process in greater detail as we focus on the interdependence of duality and conjugation. In number theory the zero-state is a numeric 0. In set theory we start with the null-set as the empty state. We then create opposite pairs {+1, –1} and dual sets{S, ~S}. Under the operations of addition and union, respectively, the zero-state and empty state are conserved. Zero conservation can be also inferred from the law of conservation of momentum. Consider the physical system of an arrow in a stretched bow. The momentum of the system before releasing the arrow is zero. The momentum after releasing the arrow is zero. Both the arrow and bow acquire equal momentum but in opposite directions. The dual pair consists of equal momentum but in opposite directions, i.e. velocities of opposite values. Not only is Newton's third law of motion satisfied but zero is also conserved. The conjugation process in both the mathematical and physical examples is what is meant by duality. Duality preserves the total zero state, the 0 value, the null set, and the momentum. Starting with the idea of zero, or nothing, derived from dualities is a clue how to produce something from nothing. Physicists believe that the universe is composed entirely of fermion-antifermion interactions. The nilpotent Dirac equation mathematically describes the interaction and is itself based on a duality concept leading to an observation that "duality is not merely a 'component' of physics but an expression of the fundamental nature of physics itself" (*ZtI*, p. 443).

The Dirac equation is a relativistic generalization of the Schrödinger equation which can be written in a zero-state form satisfying conjugation and duality conditions. The novel mathematical formulation provides an insight into the nature of the Dirac equation and unifies a broad range of hypotheses about fermions and their interactions. Rewriting and factoring the relativistic equation for energy $E^2 = p^2 + m^2$ (by convention, c = 1) a zero-totality condition

$$(\pm ikE \pm i\mathbf{p} + \mathrm{l}jm)\,(\pm ikE \pm i\mathbf{p} + \mathrm{l}jm) = E^2 - p^2 - m^2 = 0$$

emerges. In the context of Dirac's equation this is a nilpotent which squares to zero. The conjugate or dual state is the negative of the Dirac state $-(\pm ikE \pm i\mathbf{p} + \mathrm{l}jm)$ which is interpreted as being the rest of the universe or 'vacuum state' which is created simultaneously. Interactions between discrete charged objects occur because of the continuous vacuum state and agree with Mach's principle of inertial mass. It is extracted from a zero-totality superposition since it results in $(\pm ikE \pm i\mathbf{p} + \mathrm{l}jm) - (\pm ikE \pm i\mathbf{p} + \mathrm{l}jm) = 0$. The physical picture is a quantum mechanical universe created from nilpotent states within an infinite-dimensional Hilbert space. The 'universe' can be described as an entanglement of all possible nilpotent states and because it is zero-state balance between positive and negative charges it is electrically neutral. The Grassman algebra, used to represent the nilpotent states, is equivalent to the complex Hilbert space of modern quantum theory and represents a nonlocal superposition of fermionic (Dirac) states. The unique states also satisfy Pauli exclusion since fermions (electrons, neutrinos, quarks) with half-integer spin cannot occupy the same quantum state simultaneously. The universe is seen as being composed entirely of nilpotent fermionic or antifermionic wavefunctions. Boson wavefunctions are simply combinations of the two. Key insights into the physical world can be gained from purely mathematical models. The equations of Maxwell provided evidence that light is an electromagnetic phenomenon. The equation of Dirac implied the existence of a new form of matter called antimatter. Duality is central to a hypothesized inflationary universe in which energy is extracted from the gravitational field and produces matter.

A Dirac algebra is one in which the individual nilpotents $\Psi_n = (\pm ikE_n \pm \mathbf{ip}_n + jm_n)$ are wavefunctions satisfying the antisymmetric property $\Psi_n{}^\wedge \Psi_m = - \Psi_m{}^\wedge \Psi_n$, $n \neq m$. We will rewrite this expression and reinterpret it the context of Newton's third law. We then use it in a heuristic argument ultimately connecting it to zero-totality. First let us rewrite the last property of nilpotents in the form $\Psi_n{}^\wedge \Psi_m + \Psi_m{}^\wedge \Psi_n = 0$ to call attention to zero conservation. The left side is understood as being a superposition of two fermion states having opposite spin and whose combined *zitterbewegung* can be viewed as a fermionic quantum dance or, euphemistically speaking, a virtual *jitterbug*. This description is consistent with *zitterbewegung* described as "*a new kind of bound state where the helical world line of one lepton is intertwined with the helical world line of another*" [7].

In the case n = m, we find a one-particle self-interaction in the Dirac state. The last equation can be rewritten in the form $2\ \Psi_n{}^\wedge \Psi_n = 0$. This equation can be interpreted as Pauli exclusion; each nilpotent wave function is unique, and the product of identical ones would zero the entire set of fermion states. An alternative way of looking at this last equation is to disregard the purely algebraic nature of the equation. We will replace the 0 on the right side of the equation with a Z in order to emphasize a physical zero-state condition. The new relationship is not an algebraic one, meaning that factor 2 does not cancel. The new formulation $2\ \Psi_n{}^\wedge \Psi_n = Z$ can be interpreted as electromagnetic self-interaction with the fermion bound to an oscillating electromagnetic field. This can be viewed as a de Broglie's pilot wave and is consistent with the observation that "*electron mass and spin can be identified with the energy and angular momentum of electromagnetic self-interaction*" [7]. A zero-totality condition is also satisfied.

We will now define action space as the space in which all discrete self-interacting fermions conserve the zero state, i.e., $2\ \Psi_1{}^\wedge \Psi_1 + 2\ \Psi_2{}^\wedge \Psi_2 + 2\ \Psi_3{}^\wedge \Psi_3 + \ldots = Z$. The continuous vacuum is produced by the gravitational component and generates, for any given fermion, a state vector that is equivalent to $-1(ikE + \mathbf{ip} + jm)$ with negative energy. As a result any combination of a fermion and total vacuum produces a zero-state. The infinite superposition of all Dirac fermion states acts as the discrete Action Space for which a Reaction Space is generated, in this case the continuous gravitational vacuum. This transition from a discrete space to a continuous space, and vice versa, is the distinguishing feature of the Dirac algebra. The transition from discreteness to continuity is evident in many scientific theories and has been the source of seemingly contradictory notions of reality. Fermion states are incomplete without a vacuum and a supersymmetric partner. The discrete fermionic action space of kinetic energy requires a continuous reaction space vacuum of steady-state potential energy. The real fermion and the set of dual vacuum conjugate images combine to produce a conserved physical system. The system is conserved by a "simultaneously incorporating of both action space and reaction space of Newton's third law of motion" (*ZtI*, p. 149). We symbolically write Newton's third law in the form action + reaction = 0 in order to emphasize a zero-totality condition.

ZtI has presented novel mathematical techniques to analyze the Dirac equation having a coherent and integrated physical interpretation. The group symmetry for the Dirac algebra is almost identical to that for the Standard Model. Transitions in black body radiation from half to integral values are explained in terms of radiation reaction in which a transition is made between a discrete space and continuous space. The distinction between rest and relativistic masses becomes clear. Rest mass defines an isolated object having continuous kinetic energy. Relativistic mass, on the other hand, includes the effects of the environment. A photon, having no rest mass, acts as a classical potential energy term with discrete potential energy mc^2. Radiation pressure is produced by photon gas and the justification for action-reaction effects can be seen in the doubling of the value of the energy term that comes from the doubling of momentum when photons rebound from barriers, as in an absorption-emission process. The same thing happens in radiation reaction, a doubling of energy.

The mathematical constructs in *ZtI* treat discrete and continuous phenomenon simultaneously. In describing Dirac quantum states both continuous (fields, waves) and discrete (charges, particles) properties are incorporated. The results of Young's double-slit experiment are illustrative of the difficulties encountered in explaining quantum mechanics and originate in the discreteness-continuity dilemma. Reframing wave-particle duality in a nilpotent setting resolves the apparent dilemma.

4. Conjugation and Duality Concepts in Other Disciplines

Conjugation and dualities are indispensable in the Dirac algebra and can be found in almost every field of study including the natural sciences, philosophy, the humanities, social sciences, linguistics, and the medical sciences. Some paired terms are obviously dualistic, and identifying them is not difficult. These include *supply-demand, static-dynamic, thesis-antithesis, biotic-abiotic, thema-antithema, chaos-order*, and *intentional-unintentional.* There are dual pairs that, for the uninitiated, could never be recognized as properties that are in an oppositional relationship. The dual pairs *activation-inhibition, agency-structure, anabolic-catabolic, analog-digital, empirical-rational, glucagon-insulin, holistic-atomistic, ideal-real, immune-endocrine, inductive-deductive, integration-differentiation, anions-cations, osteoblast-osteoclast, nature-nurture, niche-biotope, oxidation-reduction, permittivity-permeability, reflective-reflexive, semantic-syntactic, systolic-diastolic* are part of this growing list.

It is a celebrated fact that duality is the central theme of quantum mechanics. Light exhibits both wave (field) and particle (matter) properties. The three hundred year old debate about the nature of light, dating back to the late 1600s when competing theories of light were proposed by Huygens and Newton, was finally resolved with the recognition that light has both wave and particle characteristics. Other disciplines have a similar history of evolving theories and witnessed the emergence of competing models [12]-[25]. The resulting standoff left no clear winner and there was final recognition that dual models were required to explain experimental results. A sampling of duality models from various disciplines sends a strong signal that duality and zero-totality may be as fundamental to other fields of study as they are to physics as the following examples will show.

A reflective-impulsive model has been introduced by social psychologists to explain cognition and behavior as a function of two interconnected mental faculties. Each faculty operates according to a different principle but interact in dualistic manner. The advantages of this dual-system model are "(a) their integrative power, (b) their foundation in well-established constructs of cognition and neuroscience; and (c) the ease with which they can explain the interplay of judgments and nonjudgmental processes" [26]. Dual cognitive science models identify paired voluntary-involuntary actions as conjugated variables.

The duality theory of production is a dual model that imposes a number of simplifying assumptions regarding economic production technologies, including dual concavity-convexity assumptions. Convexity is important in recovering technology information from economic models. Concave cost functions and convex profit functions are convex act as the conjugate functions in economic theories.

In optimization theory a dual problem is secondary problem with the property that its objective function is always a bound on the original, primal, mathematical problem. The same data are used in constructing the constraints in both the primary problem and the dual one. The coefficients in both problems are interpreted in a complementary manner, but the objective function reverses. If the objective function in the primary problem is to maximize a function, then the objective in the dual problem is to minimize the function. If the constraints in the primary model are constraints from above, then in the dual model, the constraints are constraints from below. Moving the conjugate constraints in the dual space corresponds to minimizing the slack between the original vector and the optimal vector.

An important property of set algebra, a principle of duality for sets, asserts that for any true statement about sets, the dual statement obtained by interchanging unions (U) and intersections ($\cap$) results in a true statement. A statement is said to be self-dual if it is equal to its own dual. DeMorgan's laws are transformation rules in which the Boolean operation of negation transforms the valid rule of conjunctions and disjunctions, into a complementary equally valid rule.

In electromagnetic theory and electrical engineering many properties are paired dualistically. These include *electric fields-magnetic fields, permittivity-permeability, piezoelectric-magnetostrictive, ferroelectric-ferro-magnetic materials, electrets-permanent magnets, Faraday effect-Kerr effect, voltage-current, parallel-serial (circuits), impedance-admittance, reactance-susceptance, short circuit-open circuit, time domain-frequency domain,* and *conductance-resistance.*

The earliest use of a duality principle occurred in 1825 in projective geometry: Given any theorem in plane projective geometry, exchanging the terms "point" and "line" everywhere results in a new, equally valid theorem. Other examples include: duality in order theory, dual polyhedron and geometric duals.

In biology, dualism is the theory that blood cells

have two origins, one from the lymphatic system and one from the bone marrow. The functions of chlorophyll and hemoglobin, inhaling and exhaling, muscle contraction and extension, anabolic and catabolic metabolism, osteoblasts and osteoclasts (bone creation-bone annihilation), lymphatic and myeloid elements are known to have opposite functions. The methods of category theory have been successfully applied to the investigation of what is referred to as the "basic dualism of biology" the phenotype and genotype of a given organism.

The wave-particle nature of light is a central concept of quantum mechanics. A less orthodox interpretation is the *duality condition* which is described by an inequality allowing wave and particle attributes to exist simultaneously. It is postulated that a stronger appearance of one of the attributes leads to a weaker appearance of the other meaning that the wave nature and particle nature can co-exist. A more recent experiment to be discussed in the last section of this paper demonstrates that wave and particle features can exist simultaneously.

In genetics, complementarity is the correspondence of DNA molecular components (nitrogenous bases) in the double helix, such that adenine in one strand is opposite to thymine in the other strand, and cytosine in one strand is opposite guanine in the other. This one-to-one relationship of the bases is called Chargaff's rule of base-pairing. Complementarity also means that one strand of nucleic acid (DNA or RNA) can pair with and serve as a template for its complementary strand.

In economics, two goods are considered to be complementary if the cross elasticity of demand is negative, as the price of one good increases the demand for the other decreases. Duality is the very foundation of a double entry book keeping system. Every transaction has a double (or dual) effect on the business as recorded in the accounts. For example, when an asset is bought another asset, cash (or bank), is simultaneously decreased or a liability such as creditors is simultaneously increased.

In neoclassical microeconomics duality refers to the existence, given appropriate regularity conditions, of indirect, dual, functions that embody the same essential information on preferences or technology as the more familiar direct, primal, functions such as production and utility functions. Dual functions contain information about both the optimal behavior and the structure of the underlying technology, or preferences, while the primal functions describe only the latter.

In molecular biology, code-duality refers to the fact that living systems always form a unity of two coded and interacting messages: the analog coded message of the organism itself and its re-description in the digital code of DNA. As analog codes, the organisms recognize and interact with each other in the ecological space, giving rise to a horizontal semiotic system (the ecological hierarchy), while as digital codes they (after eventual recombination through meiosis and fertilization in sexually reproducing species) are passively carried forward in time between generations. This, of course, is the process responsible for nature's vertical semiotic system, the genealogical hierarchy. Similar dichotomies are made in pathology, physiology, and in the diagnosis and treatment of medical problems. This philosophy centers on the ideas of the importance of maintaining the dynamic balance of these opposites for proper health. Investigations into the dynamic duality relationships between Symbiosis and Pathogenesis of microbial infections, which range along a continuum from conflict to cooperation, are considered a priority in the health community.

In cognitive psychology a duality arises naturally from the distinct roles played by a referent and a probe in comparative judgment. Dual-system models explain social cognition and behavior as a joint function of two interconnected mental faculties, each operating according to different principles. Dual terms such as: *reflective-impulsive, reflective-reflexive, propositional-associative, systematic-heuristic,* and *intentional-unintentional* are found primarily in social cognitive studies.

In learning theory, structure and function form a dualism. In the social sciences a dualism exists between human action and social structure. This issue has continued to divide sociologists. Symbolic interactionism stresses the active-creative components of human behavior and functionalism-structuralism the constraining nature of social influences on individual actions.

Opposing views of evolution-development (the evo-devo debate) exist over the roles and responsibilities assigned to such pairs as *structure-function, genes-environment, random-directed variations, innate-acquired characteristics, instructive-selective information, and self-organization-natural selection.* This debate is a result of self-referencing formulated in the ontic (internal-external) and epistemic (individual-local, population global) considerations that are necessary to integrate developmental and evolutionary theories.

In the computational sciences, programming techniques are based on the concept of an "object" which is a data structure encapsulated with a set of routines that operate on the data. In object-oriented paradigms an object has dual characteristics: state and behavior. The following list depicts some of the dualistic categories used in software engineering: *digital-analog, synchronous-asynchronous, hierarchical-flat, static-dynamic, distributed-central processsing, sequential-parallel, top down-bottom up, goal oriented-process oriented, explicit-implicit, active-passive,* and *predictable-statistical.*

The dualities between syntax and semantics and Fourier duality in algebraic theories are part of the same family of constructs. Abstract-concrete representations and theories are used to model the relationship between syntax and semantics.

Duality concepts can be appealed to when classifying academic disciplines. Some fields of study utilize variables that are continuous, while others require variables that are discrete. In the following list the first of each pair utilizes variables that take on continuous values or transformations and the second requires discrete ones: *calculus-statistics, analog computers-digital computers, ontology-epistemology, category theory-set theory, relativity theory-quantum mechanics, field theory-particle physics, and topology-geometry.*

Many engineering applications are characterized by combined discrete-continuous requirements and produce functions that are piecewise continuous. This is due to changes in modes of state variables or behavior. In chemical processes, for example, a discontinuity is associated with the beginning of a reaction. In materials engineering the phase change of elastic-plastic deformation represents a discrete change in the nature of the response of a mechanical system, even though the response is continuous.

5. The Well-Defined Meaning of Duality in *ZtI*

The Dirac nilpotent algebra is built upon clearly formulated duality concepts and emerges as a foundational approach to physics in a single structure. The mathematical structure created is a group of 64 conjugate variables composed of real-complex combinations of vectors and quaternions from which the nilpotent form of Dirac equation is derived. Conserved and non-conserved elements such as momentum-space are treated as conjugate pairs. The novel structure can be applied to classical, relativistic, and quantum systems. For example, the real-complex conjugate number systems allow transformations between the space-time of relativity theory and the electro-magnetic setting of Maxwell's equations. The equations of special relativity can be derived directly from the Dirac nilpotent. The two approaches to quantum mechanics are the discrete Heisenberg and the continuous Schrödinger. Each theory incorporates the missing feature in the other property when measurements in a real physical system are made. The wave-particle duality is a classic case of the continuous-discrete phenomenon. Making a distinction between the discreteness of matter and charge and the continuity of the vacuum allows for a re-examination of the entire domain of physical reality from a duality perspective. The Dirac algebra is one in which a conjugated nilpotent structure is preserved in a zero-totality dualistic environment. The classification of dualities in *ZtI* has two unique features not found in other disciplines. First, *ZtI* dualities can be uniquely categorized as being either discrete or continuous. Second, the variables relevant to each of the paired categories play a role in the environment of the other providing a mathematical and physical view from a dual perspective. The interaction (transformation) between a discrete space and a continuous space looks much like a harmonic oscillator in which the sum of the kinetic and potential energy is constant. The key feature of the discrete *action space*-continuous *reaction space* environment is to maintain zero-totality. The weak charge is a property unique to fermionic matter and the dipolar weak interaction yields a harmonic oscillator solution of the Dirac equation which is also typical of the components found in gaseous and condensed matter.

We note that the term *interaction* can be replaced with *transformation* without any loss of meaning. In mathematics the term 'transformation' means a rule or procedure describing a mapping (morphism) of one structure into another related structure. In set theory, mappings are functions; in algebra, they are linear transformations; in group theory, they are called are homomorphisms; and in topology, they are continuous functions. Transformations in this paper have been from an action space (e.g. Dirac fermion states) to a reaction space (e.g. vacuum state). The symmetry trans-formations found in the standard model of particle physics and central to the Dirac states in *ZtI* are: charge-symmetry (antiparticles replace particles), parity-symmetry (spatial variables are replaced by conjugates), and time-symmetry (direction of time is reversed).

The harmonic oscillator is a classic system such that when it is displaced from an equilibrium position experiences a restoring force. The solution of the Dirac equation depends on creation-annihilation operators for the production fermion-antifermion states which emerge from and disappear into the vacuum. The operators are essentially the same ones used in Quantum Field Theory. *Zitterbewegung* fluctuations in the vacuum are seen as the creation-annihilation of a weak dipolar fermion-antifermion pair via a harmonic oscillator creation-annihilation mechanism. The same fluctuations are responsible for the Casimir or Van der Waals force. The action of the weak dipole moment is seen as the cause of the creation of fermion-antifermion combinations from the vacuum and also as the cause of mutual annihilation. *ZtI* emphasizes that physical variables and phenomena are at their roots conjugated and dualistic and discrete-continuous properties are superimposed in a Dirac nilpotent algebra. The first term in each pair of the following examples from *ZtI* is associated with entities having a discrete nature, the second continuous: *particles-waves, Heisenberg-Schrödinger equations, quantum mechanics-wave mechanics, QED theory-SED theory, inertial mass-gravitational mass, matter-gravitational field, discrete matter-filled vacuum, potential energy-kinetic energy, charged particles-mass-energy, space dimension-time dimension, Leibnizian differentials-Newton fluxions, countable numbers-uncountable numbers,* and *fermion state-vacuum state.* Mathematically formulated nilpotent states in *ZtI* accomplish a seemingly impossible task of simultaneously treating discrete and continuous variables. This novel mathematical treatment allows for a reinterpretation of physical reality from a dualistic perspective. In the new framework gravity is a considered to be a continuous source which is responsible for the linkage of infinitely many discrete Dirac nilpotent states of which the universe is composed.

Inertia can be understood in this new framework as an interaction between matter in discrete space and the gravitational vacuum in a continuous space. Continuity implies *instantaneous interaction* whereas discreteness requires a *quantum interface*. This can consider a new approach to uniting general relativity (GR) and quantum mechanics (QM). The key difference between older quantum gravity (QG) theories is that in *ZtI* quantum-gravity is replaced with quantum-inertia. Gravitational mass and inertial mass are equivalent but the former, being continuous, and the latter, being discrete, means that it is inertia rather than gravity that should be quantized. Treating gravity as a continuous instantaneous force eliminates the need for a quantized attractive force rather than a fictitious repulsive force.

The dichotomization of variables, equations, states, physical properties, and mathematical operations offers a new approach to overcoming dilemmas of measurement and observation and has implications beyond the realm of physics. The implications for the foundations of physics are profound when we consider that a region of space, whether continuous or discrete, is the domain of interaction of material objects as well as fields. Newton's laws are concerned with objects and forces and the rules governing their interaction. In the context of the fermion space-vacuum space interaction Newton's third law of motion is one that connects discrete and continuous forces. Newton's first law of motion, on the other hand, applies to bodies with zero velocity and bodies moving with constant non-zero velocity. The first law can also be stated more generally in terms of force. If the net force, i.e., the sum of all vector forces, is zero then the velocity of the object is constant. The concept of zero conservation (of forces) is relevant here. The symbol *Z* was introduced in Section 3 to symbolize a numeric zero. We will now expand the meaning of zero-totality, represented by Z, to signify not only an algebraically manipulated numerical value, but also an object at rest, an object moving with constant velocity, an object experiencing zero total net force, an equilibrium point, a stability condition, an adiabatic condition, a steady state condition, a symbiosis, a balance, a symmetry condition, homeostasis, etc.

Appealing to the definition given to Z we will generalize Newton's first law of motion and call it the **Law of Zero-Totality**. This law applies to any field of study built on a duality model and conjugated variables. We now state a generalization of Newton's third law of motion as **Action + Reaction = Z.** The physical environment in which action and reaction occurs will be called Action Space-Reaction Space. The action space-reaction spaces in *ZtI*, for example, are the Dirac states of discrete fermions and the continuous vacuum. The predominant function of the fermion action space- vacuum reaction space is to accommodate the harmonic oscillation solutions of the Dirac equation and their *zitterbewegung* fluctuations.

6. Action Space–Reaction Space

The notion of *space* is not new in physics. Utilizing Fourier transforms quantum mechanics can be formulated using momentum space wavefunctions or particle space, sometimes called real space, wavefunctions. In this section we will construct phenomenological spaces in which to view Newton's third law. Newtonian methods have been used effectively for over two hundred years in generating formulas needed for space travel and a rationale for wearing seat belts. We know that the laws of conservation of energy, momentum, and angular momentum have more general validity than Newton's laws. The laws apply to both light and matter as well as to classical and quantum physics. Until recently [27] it has been impossible to demonstrate in a single experiment both a continuous and discrete property of light. That is why the double-slit experiment is such a paradox. Properties of light appear to be a particle in one experimental setting and a wave in another. *ZtI* resolves many paradoxes by focusing on the continuity-discreteness duality.

Lorentz invariance arises solely from the mathematical description of the parameters in terms of real or imaginary numbers but the result of this forced union is that, *for the purpose of measurement*, either a quantity which is continuous must become discrete, or a quantity which is discrete must become continuous. The first is the particle option (time and mass become discrete), while the second leaves us with waves (space and charge become continuous) (*ZtI*, p. 223).

Many paradoxes materialize when making a transition from continuous states to discrete states or vice versa. Newtonian fluxions are the mathematical precursor of the differential calculus. Newton was concerned with velocities and acceleration and he defined velocity as "*the ultimate ratio of evanescent quantities*." These evanescent quantities appear in the definition of velocity as an instantaneous rate of change resulting in a ratio that appears as the undefined ratio 0/0 originating in the limit $y'(x_0) = \lim \Delta y/\Delta x$. Newton's definition of velocity is a mathematical procedure linking discreteness and continuity. The definition is valid for well-defined and differentiable functions. A continuous limiting process over real numbers, resulting in a discrete number, resolves many of Zeno's paradoxes but is nevertheless puzzling and contrary to common sense.

The methods of the differential calculus were introduced by Newton to describe unseen earthly and astronomical forces, falling objects and orbiting planets. They appeared at the time when two divergent views about the nature of light were being debated. The interplay between continuous and discrete objects and forces amplified the paradoxes. *ZtI* has revealed the true nature of the dilemma and has been the motivation for defining action spaces-reaction spaces where interactions between the continuous and discrete can be studied. The key paradigms in the theory of light, special and general relativity, and quantum mechanics may be nothing more than an issue related to continuous-discrete interactions.

> Continuous vacuum energy, such as we would expect from a 'filled' vacuum, is what we mean by nonlocality. It is the continuing connectedness, through the vacuum, of apparently discrete fermionic states, and it is required to maintain Pauli exclusion. Rest mass is a localization and therefore discretization of the continuous total vacuum energy (*ZtI*, p. 510).

The interactions typical of vacuum-fermion are the primordial spaces in which Newton's third law functions. The continuous space-discrete space duality is the most logical environment for Newton's third law of motion to operate. The links between the continuous and discrete physical domains, between the changing and the fixed, between the real and imaginary, and between the orderable and nonorderable can now be described. The continuous-discrete duality is apparent in the distinction between potential and kinetic energies, which is also a distinction between conserved-nonconserved quantities, or fixed-changing conditions. The duality may also be expressed in terms of the distinction made between space like theories of Heisenberg approach and the time like theories of Schrödinger. In the quantum mechanics-stochastic electrodynamics duality, we see not only a discrete-continuous dichotomy but also a real-imaginary one.

We can represent symbolically the action space-reaction space zero-totality conditions of many of the examples discussed above using a new symbol (+) to represent the interaction between continuous spaces and discrete spaces.

potential energy (+) kinetic energy = Z,
inertia (+) gravity = Z,
relativistic push (+) newtonian pull = Z, and
fermions (+) vacuum = Z.

The interaction between fermion space and vacuum space means that the discrete fermionic self-interaction (action space) makes a field contribution force to the continuous vacuum (reaction space). The weak van der Walls force causing the fermion-vacuum interaction and the oscillation between the two spaces, in which zero-totality is maintained, is the generalized setting of Newton's third law. Aggregation of matter, referred to as *complexity* in *ZtI*, originates in the harmonic oscillation of interacting fermion-vacuum spaces. This *zitterbwegung* activity produces subatomic, atomic, chemical, molecular, organic, non-organic, planetary, galactic, and ultimately cosmological structures and is responsible for complexity (aggregation).

The quantum delayed-choice experiment referred to earlier in which 'strong nonlocal correlations were observed" showed that the photon behaved simultaneously as a wave and a particle [27]. If conservation laws are also satisfied then we can unambiguously write

particle (+) wave = Z.

There have been other attempts at refuting Bohr's principle of complementarity that both wave and particle behaviors can be exhibited in the same experiment. A double-slit experiment initially performed at Harvard University in 2001 claimed that the Englert–Greenberger duality relation was violated but that linear momentum was conserved [28].

7. Conclusion

Competing wave-particle theories about the dual nature of light took centuries to resolve with an ultimate realization that both theories are correct. The theories describe dualistic or, as some say, complementary features of the same phenomenon. No foundational principles, however, have ever been proposed that address the resulting paradoxes of the dual theories. That is why the relativity theory-quantum mechanics debate is now reaching its one-hundredth year. The roots of the debate go back to the 16th century but under a new name. There have been attempts, by string theorists and others, to unify GR-QM into a theory of everything. M-theory and quantum gravity, however, are nothing more than mathematical appendages on the original theories and they introduce unyielding singularities and infinities. *ZtI* has not only uncovered the roots of the debate but has provided the mathematical machinery for accommodating conflicting methodologies and interpretations. We have a description of elementary particles that is consistent with the Dirac equation, special relativity, and quantum mechanics. This is precisely the kind of foundational approach needed to understand and resolve issues presented in the original accounts. A full accounting of relativity in the context of quantum mechanics can be found in *ZtI*.

Galilean inertia was codified in Newton's first law of motion. Both laws find their ultimate expression in Einstein's special theory of relativity. Newton defined inertia to mean "the innate force possessed by an object which resists changes in motion." Inert objects are resistant to change especially if they are inanimate. Laws that apply to orbiting planets are not very easily transported into the life sciences. We may be able to apply, however, the generalized form of Newton's first and third laws to other fields in which duality models exist without compromising the theoretical framework in which living matter and forces operate. The following relationships are constructed from the examples in Section 4. They suggest a way of viewing dual-model theories in other disciplines in a generalized Newtonian framework.

As a first example we look at the reflexive-compulsive model in social psychology which can be written in the form

Voluntary (+) Involuntary action = Z, or
Reflective (+) Reflexive thought = Z

The symbol (+) represents the interaction between the conjugate dual modes of thought. The vacillating (oscillatory) nature of reflective-reflexive thinking is well known. According to the dual system model the reflective system depends on symbolic thought and is the basis of conscious experiences. Conscious experiences can only be communicated in a discrete symbol dependent manner whereas reflexive thought is a continuous process.

The Sym-Pat focal point duality which models the dynamic continuum from cooperation to conflict can also be reformulated in the formalism of zero-totality as

Symbiosis (+) Pathogenesis = Z.

For the conjugate variables in evolutionary theory we may write the relationship that describes the interaction between genes and the environment

Nature (+) Nurture = Z.

In molecular biology there is a fundamental distinction made the inheritance of traits and evolution. Biologists are interested in knowing to what extent an organism's phenotype affect its genotype. The duality models that have been proposed appear to satisfy a zero-totality condition

Genotype (+) Phenotype = Z.

In an examination of the epistemological and logical terminology problem found in language [29], educational philosophers introduced a new term to describe concepts related to dualisms and dualities. The new term, *distinctions*, implies a dualistic interaction, and there are over one hundred pairs revealed in the treatise. It is argued that self-action, interaction, and transactions are responsible for the loss of precision and efficiency in the use of words. Conjugate terms are introduced having precise meanings to remedy the terminological crises prevalent in language today.

Rowlands and co-collaborators (Diaz, Cullerne, Koberlein, Marcer, Mitchell, Schempp, Hill) have made a strong case for the foundational role of dualities in physics. Other theories incorporating a *duality principle* may find it appropriate to apply the generalizations of Newton's laws presented in this paper. Biological systems create order and destroy order by extracting, storing, and transmitting information in a continuous fashion. There are two fundamental, yet opposing, forces that compete with each other in all bio-systems. They are comparable to the creation and annihilation operators of fermion states. One force leads to uniformity over ecological time and the other leads to diversity over evolutionary time. Chapter 19 of *ZtI*, co-authored with biologist Vanessa Hill, demonstrates that the geometric and algebraic structures required in particle physics also apply equally to replicating biological systems. Translation, transcription, replication, the genetic code and the grouping of amino acids are known to be vitalized by the same fundamental processes that drive the elementary particles. DNA, RNA, and the genetic code efficiently process information and can be represented by mathematical structures, both algebraic and geometric. The surprising development is that these same mathematical nilpotent structures are used to describe nature at atomic and sub-atomic scales!

There is a plethora of seemingly divergent *principles of duality* in the literature. Complementarity, if no longer true about the wave-particle phenomenon, is thriving in the academic world where disciplines coexist and produce terminological tsunamis after each earth-shaking discovery. The author, in the spirit of the times, has contributed a new term to the wave of terminological confusion by coining a new term called the *Dualogy Principle*. The new principle attempts to accommodate theories in which duality models have clearly defined conjugated variables and/or transformations [30]. Restating the principle in the language of the generalized forms of Newton's first and third laws we write:

> Dualogy Principle: In order to describe a physical or a non-physical event co-dependent dual conjugated variables defining the state and/or transformations that define the behavior of the phenomenon are required. The state variables satisfy the Law of Zero-Totality and the transformations satisfy Action + Reaction = Z.

It should be noted that Descartes dualism and Bohr complementarity do not satisfy the Dualogy Principle; in each the conjugated states are irreconcilable or independent.

The original statement of the dualogy principle had relevance in a recent study on the living cell. Experimental results have led to a proposal by Sungchul Ji that the Gibbs free energy levels of enzymes in living cells are quantized with discrete units of information and energy [31]. Conversion (transformation) of information to energy is possible. Maxwell's demon, a being that violates the second law of thermodynamics, is replaced by a being that "receives energy and dissipates heat into his environment while performing his molecular work" then Maxwell's angel, can carry out the assigned work without violating the second law. The dual responsibility, *taking in energy-dissipating heat* simultaneously, is one performed by molecular machines. The two fundamental entities (conjugate variables) which operate in all molecular machines have been identified as information and energy. A dual model of the living cell is presented along with dual theories called *the conformon theory of molecular machines* and the *cell language theory of biopolymer interactions*. Conformons store mechanical energy in the form of Gibbs free energy and genetic information and are responsible for goal-directed molecular actions driving the cells. The new theory about the molecular properties of cells treats the interaction between energy and information and may be amenable to treatment within a zero-totality framework. The fields of thermodynamics and information theory appear to contradict the theory of evolution. If biologists look at the contradictory notions of order and disorder as a dualistic interaction of action-reaction spaces then it may be reasonable to consider that their interaction also satisfies zero-totality and can written in the form

$$\text{Entropy } (+) \text{ Evolution} = Z.$$

The action space in which evolution occurs has been identified as space-time [32]. It is reasonable to call the reaction space of entropy temperature-space.

The primary purpose of this paper is to derive a generalized form of Newton's third law of motion from the concept of zero-totality. It quickly became clear that such a generalization only made sense if Newton's first law was viewed not as a law of motion or inertia but rather as a law regarding forces. Many of the arguments that have been presented in this paper have admittedly been heuristic, but the profound nature of the foundational issues raised and answered in *ZtI* justifies this heuristic approach. An attempt has been made to apply generalized Newtonian laws to fields of study outside of physics. Duality models in other disciplines have paradoxical characteristics reminiscent of the wave-particle description of light. In *ZtI* the three mechanisms of conjugation, complexification, and dimensionalization are used to describe fundamental mathematical and physical processes in general and to derive the Dirac equation in particular. The processes of conjugation, complexification, and dimensionalization are all considered to be nothing more than "alternative forms to duality" (*ZtI*, p. 13). The nilpotent form of the Dirac equation can be represented in a compact 5-vector form with mass as the fifth dimension on equal mathematical footing with the four dimensions of spacetime. An extra fifth dimension also appears in a variant of the Kaluza-Klein model unifying gravity and electromagnetism called the induced matter theory. The form includes all of the conjugated variables found in classical and quantum physics. State information about one of the conjugate pairs can be exchanged for state information about the other. The nilpotent equation permits statements about conservation in action space to be transformed into equivalent statements about nonconservation in reaction space, and vice versa. The Dirac nilpotent is self-dual and being the square root of zero produces two solutions simultaneously. Boson states are generated from fermions and, in general, the nilpotent algebra produces a doubling effect and the reversal of properties. Discreteness is produced from continuity; a conserved state is created from a nonconserved state; and, in general, a conjugated state is paired with a nonconjugted one. *ZtI* provides a unique foundational approach to physics providing a rigorous mathematical framework which can handle incompatible concepts like discreteness and continuity simultaneously. The "*mathematics of duality*" maps onto the mathematics of reality and is a serious contender for a theory of everything. The second main purpose of this paper is to argue that zero-totality and duality apply to many other academic disciplines and have relevance far beyond the scope of particle physics itself.

References

[1] Rowlands, P. (1983) The fundamental parameters of physics, Speculat. Sci Tech., 6, 69-80.
[2] Diaz, B. & Rowlands, P. (2002) A universal alphabet and rewrite system, arXiv:cs/0209026.
[3] Rowlands, P. (20030 The nilpotent Dirac equation and its applications in particle physics, arXiv:quant-ph/0301071.
[4] Feynman, R.P. (1962) Lectures on Gravitation, Reading: Addison-Wesley.
[5] Guth, A.H. (1988) The Inflationary Universe: The Quest for a New Theory of Cosmic Origins, Cambridge: Persus.
[6] Hawking, S. (2001) The Universe in a Nutshell, New York: Bantam.
[7] Hestenes, D. (1985) Quantum mechanics from self-interaction, Found. Physics, 15:1; 63-87.
[8] Hestenes, D. (1993) *Zitterbewegung* modeling, Found. Physics, 25: 3; 53-87.
[9] Rowlands, P. (2007) Zero to Infinity: The Foundations of Physics, Singapore: World Scientific.
[10] Yershov, V. (2005) Dualism of spacetime as the origin of the fermion mass hierarchy, Apeiron, 12 (1), 1-25.
[11] Holton, G. (1988) Thematic Origins of Scientific Thought: Kepler to Einstein, Cambridge: Harvard Univ. Press.
[12] Andrade, E. (2002) The Interrelations between Genotype / Phenotype / Environment: A Semiotic Contribution to the Evo-Devo Debate, http://www.library.utoronto.ca /see/SEED/Vol5-2/Andrade.pdf
[13] Birkett, S. & Roe, P.H. (1989) The mathematical foundations of bond graphs. II: Duality, J. Franklin Inst. 326 (5), 691-708.
[14] Deutsch, R. (2006) Duality models in social psychology: From dual processes to interacting systems, Psychological Inquiry, 17 (3), 166-172.
[15] Peruzzo, A., Shadbolt, P., Brunner, N., Popescu, S. & O'Brien, J.L. (2012) A quantum-delayed choice experiment. Science, 338 (6107), 634-637.
[16] Afshar, S. (2007) Violation of the principle of complementarity and its implications, arXiv:quant-ph/0301071.
[17] Dewey, J. & Bentley, A. F. (1949) Knowing and the Known, Boston: Beacon Press.
[18] Karam, S.E. (2007) Dualogy: A category-theoretic approach to duality, 2nd Intl. Conf. on Complementarity, Duality, and Global Optimization in Sci. and Eng. http://dualityscience.com/yahoo_site_admin/assets/docs/cdgo 2007_PDF.21362207.pdf.
[19] Ji, S. (2012) Molecular Theory of the Living Cell: Concepts, Molecular Mechanisms, and Biomedical Applications, New York: Springer, p. 389.
[20] Ho, M.W. (1997) Towards a theory of the organism, Integrative Physiological Behavioral Science, 32, 343-363.
[21] Birkett, S. & Roe, P.H. (1990) The mathematical foundations of bond graphs. III: Matroid theory, J. Franklin Inst. 327 (1) 87-108.
[22] Cheon, T. & Shigehara, T. (1999) Fermion-Boson duality of one-dimensional quantum particles with generalized contact interactions, Phys. Rev. Lett. 82, 2536-2539.
[23] Drossos, C.A. (2005) Structures, points, and levels of reality, in G. Sica (Ed.) Essays on the Foundations of Mathematics and Logic, Monza: Polimetrica International Scientific, p. 83-114.
[24] Eco, U. (1984) Semiotics and the Philosophy of Language, Bloomington: Indiana University Press.
[25] Englert, B. (1996) Visibility and which-way information: An inequality, Phys. Rev. Lett. 77 (11), 2154-2157.
[26] Macnamara, J. & Reyes, G.E. (1994) The Logical Foundation of Cognition, London: Oxford Univ. Press.
[27] Melhado, E.M. & Berzelius, J. (1984) The Emergence of his Chemical System. Madison; The University of Wisconsin Press.
[28] Popper, K.R. (1974) Studies in the Philosophy of Biology, Berkeley: University of California Press.
[29] Rosen, R. (1991) Life Itself, New York: Columbia University Press.
[30] Rosen, R. (1958) The representation of biological systems from the standpoint of the theory of categories, Bull. Mathematical Biophysics, 20, 317-341.
[31] Sfard, A. (1991) On the dual nature of mathematical conceptions: Reflections on processes and objects as different sides of the same coin, Educational Studies in Mathematics, 22 (1), 1-36.
[32] Huang, S.-H., Zhou, W. & Jong, A. (2008) Focal point theory models for dissecting dynamic duality problems of microbial infections, Journal of Biomedical Biotechnology. doi: 10.1155/2008/856314.

A Computational Unification of Scientific Law: Spelling out a Universal Semantics for Physical Reality

PETER J. MARCER

55 rue Jean Jaures, 83600 Frejus, Var, France
peter.marcer@orange.fr

PETER ROWLANDS

Physics Department, University of Liverpool, Oliver Lodge Laboratory,
Oxford St, Liverpool, L69 7ZE, UK
p.rowlands@liverpool.ac.uk

The principal criteria Cn (n = 1 to 23) and grammatical production rules are set out of a universal computational rewrite language spelling out a semantic description of an emergent, self-organizing architecture for the cosmos. These language productions already predicate: (1) Einstein's conservation law of energy, momentum and mass and, subsequently, (2) with respect to gauge invariant relativistic space time (both Lorentz special & Einstein general); (3) Standard Model elementary particle physics; (4) the periodic table of the elements & chemical valence; and (5) the molecular biological basis of the DNA / RNA genetic code; so enabling the Cybernetic Machine specialist Groups Mission Statement premise;** (6) that natural semantic language thinking at the higher level of the self-organized emergent chemical molecular complexity of the human brain (only surpassed by that of the cosmos itself!) would be realized (7) by this same universal semantic language via (8) an architecture of a conscious human brain/mind and self which, it predicates consists of its neural / glia and microtubule substrates respectively, so as to endow it with; (9) the intelligent semantic capability to be able to specify, symbolize, spell out and understand the cosmos that conceived it; and (10) provide a quantum physical explanation of consciousness and of how (11) the dichotomy between first person subjectivity and third person objectivity or 'hard problem' is resolved.

Keywords: Cybernetics, DNA, Self-organization, Semantics

1. Introduction

In *Zero to Infinity* [1], Rowlands constructs a self-organized emergent semantic language of universal computational rewrite productions/rules using the mathematical concept of nilpotent such that there exists an infinite alphabet of unique operators $X \neq 0$, defining its universal grammar where $X^2 = 0$, X is nilpotent-'squares to zero', and each unique X is identifiable with a fermion as Pauli's exclusion principle requires. This provides a self-similar fractal language construct in accord with Wittgenstein's semantic principle [2], that there is necessarily only one proposition, i.e. X, to each fact / fermion that answers to it and the sense of the proposition cannot be expressed except by repeating it.

The open ended $0 \rightarrow 0$ indeterministic nilpotent logistics of this universal alphabetic semantic language construct of the fermions is a good basis from which to begin a unification, since in quantum physics, there are only fermions and their boson interactions – the latter being part of the former. According to the Cybernetic Machine Group's mission statement premise,** *'In Science, Nature sets the rules, but it must never be forgotten, that it is only because life has exploited these rules successfully for billions of years to our evolutionary advantage, that human brains are able to understand them. The mission, at the foundations of Computing / Information Processing, if one accepts the premise, is therefore to identify how these rules were exploited to achieve this end.'* In line with this, we postulate:-

(1) that this universal computational rewrite grammar should predicate all the semantic truths of a nontrivial emergent fermion cosmology in the form of X, a single generic nilpotent algebraic Dirac operator (**2.1** below),

the quantum mechanical trace $\sum X^2$ of which is always a zero totality, even though the description of each emergent fermion is nontrivial, i.e. $X \neq 0$!;
(2) so that its infinite emergent semantic alphabet, each symbol of which is unique, is isomorphic to the complete permutation / symmetry group, known as the Galois group, by means of which all the finite groups can be realised,
(3) where if one identifies the permutations with bijections on the symbol set (which it turns out are the creation, annihilation operations on the set 0 / vacuum state), this remains true of all groups including the continuous and Lie groups [3].

That is to say, the physics of the emergent process concerns the symmetry breaking of the Galois group, as a description of the cosmos's degenerate ground state into its nondegenerate component fermion parts, where quantum normalisation is satisfied since by definition for *any* group a unit element exists, and each group element has an inverse. Furthermore since any permutation is an example of an automorphism, the cosmic emergence, Conway tells us [5], is isomorphic to the unique birthorder field automorphism. These facts are in accord with the experimental findings of elementary particle physics, where the symmetries $U(1)$, $SU(2)$, $SU(3)$ form a Lie group, the gauge group of the Standard Model, the Lie algebra of which correctly defines the quantizations of the scalar boson, the 3 vector bosons and 8 gluons [1].

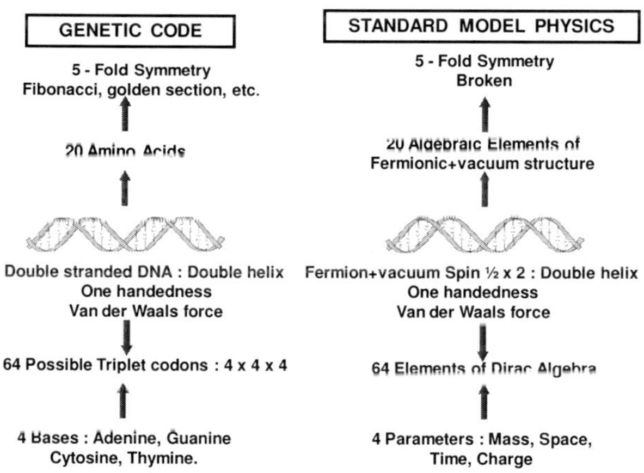

Figure 1. Self-similarity of SM physics and the genetic code (after V. Hill and P. Rowlands).

And as the evidence in *Zero to Infinity* [1] and later publications show, the cosmology is not only self organising in line with Rowlands' construction of its emergent description (where each newly emergent unique anti-commutative construct X must differ from but be compatible with all the Xs that precede it), it predicates the well known Thermodynamic Laws of Physics (**2.1** and **4**), the Standard Model of elementary particle quantum physics [1, Chapter 5,6,10], the periodic table of the elements and chemical valence [6]; and the chemical molecular basis of the RNA / DNA genetic code [1, Chapter 19] as productions of this self similar universal self-organised semantic language. Such Standard Model matter can thus be thought of as the 'construct's machine code' which is subsequently rewritten (in later computational productions) at higher levels of structural complexity – and in particular the molecular biological and neurophysiological [1, Chapters 19, 20 respectively; see Hill and Rowlands' figure 1 of the self-similarity of the Standard Model physics and the Genetic Code, which shows they exhibit the same group symmetries and rules; and 4].

Remarkably in this fixed past, irreversible emergent cosmology (see **3.5**), Heisenberg uncertainty is a duality, defined by the 3-dimensional Heisenberg Lie group G and its Lie algebra g as spelt out in **3.4 /** C6, such that signal recovery follows decay (figure 2), and is not an obstacle to quantum measurement / computation but an actual mechanism – the focus of an holographic lens – which allows exponential complexity, often inherent in Nature, to be optimally controlled C7; and in **4,** the Second Law – idealised as the experimentally-validated Quantum Carnot Engine [7] is able to describe a single heat bath cosmology, in terms of quantum phase coherence such that entropy is not simply a measure of disorder but also a metric of physical complexity determining the creation of entirely novel quantum ensemble states / properties of matter, which may never have existed before, so that a novel neuron ensemble state property could be that of human consciousness.

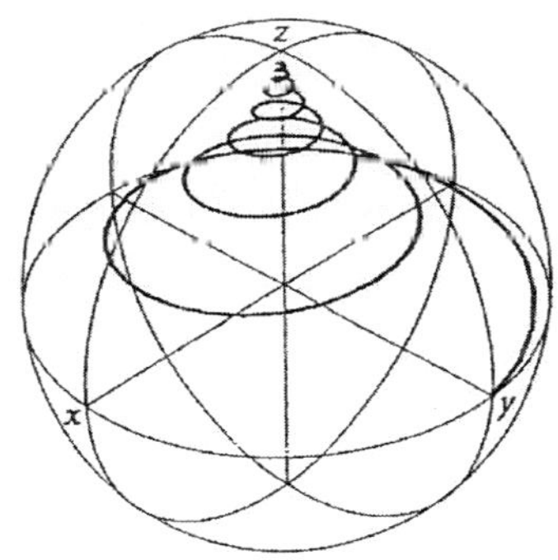

Figure 2. Quantum wave collapse and re-expansion based on the 3D Heisenberg Lie group and algebra: courtesy of Walter Schempp.

1.1. *Preliminary Summary Based On Existing Publications*

The architecture of this self-organizing emergent computational quantum cosmology specified in terms of the basic properties of energy, momentum and mass in **2.1**, is thus:- (1) made manifest in terms of matter fundamentally composed of the Standard Model elementary particles with (2) electromagnetic, weak and strong force charge quantizations and spin [1, Chapter 10]; leading, firstly (3) to the largely stable universal structure of all the known elements of the periodic table which, as Weyl shows [6], is embodied in the fundamental quantum proton (ion) / electron dual structure of the hydrogen atom and its infinite energy level quantizations. An inseparable quantum cosmos based on Pauli exclusion, which (4) can be mathematically described through quantum phase θ, measure the unit disk, onto which there is a natural projection from the infinite unit vector basis of the Hilbert space, appropriate to quantum physics; (5) manifesting itself through the process of phase conjugate adaptive resonance C7, whenever the Wittgenstein proposition 'the 3D object image of a dynamic 3D object is coincident with object itself' holds and; (6) as a consequence of the genetic code and human genome, humans are, it postulates, the intelligent offspring / agents / secondary sources of this self-organising emergent cosmos as the source of all energy, momentum and mass (**2.1**).

2. The Criteria Cn

Here, we discuss the minimum well-established criteria Cn for quantum physics consistent with Rowlands' self-organising nontrivial computational irreversible cosmology grounded on the mathematical property of nilpotency.

2.1. *The Algebraic Equation for the Conservation of Energy, Momentum, and Mass*

In *Zero to Infinity*, Chapter 5, Rowlands' 0 → 0 logistic symbolic language construct predicates: a single algebraic generic Dirac (fermion) operator

$$\mathrm{X} = ikE + \mathbf{ip} + jmc$$

where i is $\sqrt{-1}$; $\boldsymbol{k}, \boldsymbol{j}, \boldsymbol{i}$ are quaternion units and c is taken as 1, which satisfies Einstein's well known conservation law for energy E, momentum $\mathbf{p}$ and rest mass m,

$$E^2 - \mathbf{p}^2 - m^2 = 0 = \mathrm{X}^2$$
$$= (\boldsymbol{ik}E + \mathbf{ip} + \boldsymbol{j}mc)\,(\boldsymbol{ik}E + \mathbf{ip} + \boldsymbol{j}mc)$$

Thus, $(\boldsymbol{ik}E + \mathbf{ip} + \boldsymbol{j}mc)$ squares to zero, so that all the solutions X are nilpotent, self-similar and of fractal dimension 2, see the universal computational attractor Fig. 3, first mooted by Wolfram [8].

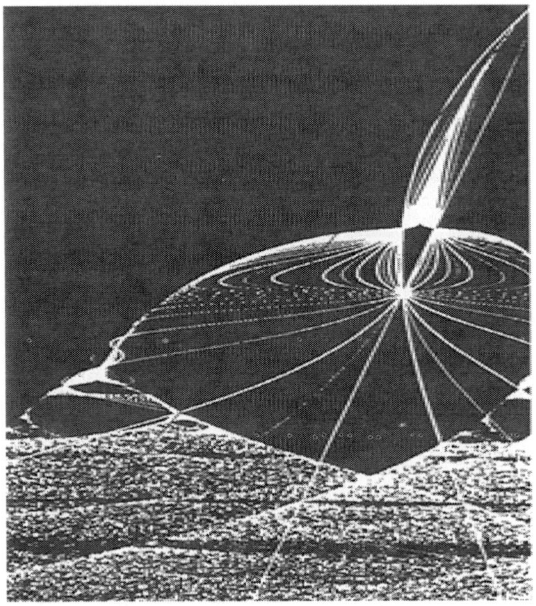

Figure 3. Graph of the canonical equation of chaos showing the universal fractal attractor of dimension 2.

This single operator delivers the entire universal computational semantic / grammatical language apparatus of quantum physical normalised amplitude and phase θ appropriate to all the solutions of the conservation law, where each such individual solution X conforms to the usual Dirac fermion properties, including spin [1]. That is, it specifies an entirely quantum cosmos.

Further this universal quantum computational semantics/grammar based on *a* 0 → 0 logistics includes that of universal digital computation based on 0 ↔ 1 (the universal binary syntax) as in Deutsch's initial 1985 Royal Society paper [9]; and Pauli's exclusion principle provides the essential computational canonical labelling and the distinction between quantum and classical automata [10, D. Z. Albert, 1983]. For as Feynman [11] shows: the universal computational primitive NAND of binary computation 0 ↔ 1 must be supplemented by the further primitives of signal exchange; fan out / in; and the universal unit wire, to be able to describe the quantum physical workings of the digital computer; an explanation which predicates a stored programme NAND 'Von Neumann" signal architecture for it / the universal digital computer [12] and of Von Neumann's discovery of the theory of universal computer construction [13].

The last postulates that these semantics include the cosmology's universal geometric (space-time) construction where, for example, the universal signal unit wire is realised in the form of the universal chemical bond at higher rewrite levels of molecular chemical complexity beyond those of the elementary particles (a line of force) and the atomic elements of the periodic table (orbit / valence / spectral line).

3. Gauge Invariance and Other Well-Established Criteria Cn

3.1. *Gauge Invariance*

It is implicit in the generic way the conservation law / Dirac operator above is formulated, independent of any particular 'space-time' coordinate system, that it is gauge invariant. Thus following Weyl [6, 1928, Dover edition, page 219, stanza 6, 'Energy and Momentum. Remarks on the Interchange of Past and Future'; Chapter IV, 'Application of the theory of groups to Quantum Mechanics'] in independent confirmation of Rowlands [14] as it describes this 'Nilpotent Universal Computational Rewrite System (NUCRS)' emergent construct:-

C1) if the conservation law is to hold everywhere, i.e. at each point P of 3+1 space time, it is essential at each P, that a (normal) set of co-ordinate axes consisting of four mutually perpendicular directions, be erected, in order that there exists a metric at P to describe the quantum wavefield quantity ψ in terms of its components, where these components at P are obtainable from each other by means of local Lorentz space time transformations which leave P invariant (as is shown to be the case in [14]), and also as Rowlands and Diaz show [15] this criteria applies ad infinitum since the NUCRS alphabet is isomorphic to the infinite square roots of –1, and therefore can be projected onto the infinite dimensional Hilbert space as quantum field physics requires;

C2) from the viewpoint of a physics action integral formulation, there should exist a differential form of this single Dirac operator, which must be gauge invariant (as in the case of the established theories of general relativity and electromagnetism) under arbitrary infinitesimal transformations of the (3+1) space time coordinates: so that a differential form of the nilpotent generalization of Dirac's famous equation can exist and be written [1] as

$$(+\boldsymbol{k}\partial/\partial t + i\nabla + \boldsymbol{j}m)(\pm i\boldsymbol{k}E \pm i\mathbf{p} + \boldsymbol{j}m)\,e^{-i(Et - \mathbf{p}.\mathbf{r})} = 0 \quad (1)$$

where E, $\mathbf{p}$, m, t and $\mathbf{r}$ are respectively energy, momentum, mass, time, space and the symbols ± 1, $\pm i$, $\pm \boldsymbol{i}$, $\pm \boldsymbol{j}$, $\pm \boldsymbol{k}$, $\pm \mathbf{i}$, $\pm \mathbf{j}$, $\pm \mathbf{k}$, are used to represent the units required by the scalar, pseudo-scalar, quaternion and multivariate vector groups respectively;

C3) the quantities ψ at P, however, are not simply just locally bound together as in special relativity, as gauge invariance is only to be understood from this standpoint, i.e. the rotations of these local directions at the points P, referred to in C1, can be performed quite independently. This can only be the case, because the transformation laws of the wave quantity ψ with respect to its four components are (as is well known) only determined up to a common fixed phase factor exp $i(\theta = \theta_0)$ where the phase θ depends arbitrarily (but gauge invariantly) on the position of P in space time.

3.2. *Leading Questions and the Phase Architecture*

The importance of the quantum phase θ is spelt out below in C5, but the question of how in the language of a normalised quantum mechanics/field theory, the infinity of possible solutions of Einstein's conservation equation above can be realised / delivered in relation to the single semantic phase parameter θ (normalized quantum metric, the unit disk) as gauge invariant phases of the cosmological wavefunction ψ and to how the problem of Heisenberg uncertainty is resolved with respect to measurement / computation, need to be addressed.

3.3. *The Multi-Valued Riemann Surface*

C4) An appropriate mathematical means is that of a many valued function of a Riemann surface, whereby the complex function $\theta \rightarrow \theta$ (measure the unit disk) each branch of which is analytic, is converted into an analytic function on a more general surface by associating each branch with a separate plane or sheet interconnected in a consistent fashion (as per the symmetries $U(1)$, $SU(2)$ and $SU(3)$ in the Standard Model for example). That is, each gauge invariant phase θ_j corresponds to a separate plane / sheet j, so that all are interconnected in uniquely consistent fashion (defining the quantum inseparability / phase coherence / entanglement θ (**4**), where each newly emergent associated more general surface j is realised uniquely in terms of a solution j of the conservation of energy, momentum and mass equation above as specified by the single Dirac operator solution j in question. That is, each newly emergent surface/plane j and associated unique parameter θ_j corresponds to each new unique Dirac operator solution j where [1] this describes a unique fermion (spin ½) and the logistic mappings $\theta \rightarrow \theta$ / 0 → 0 are indeterminate / multivalued and fractal so that spin ½ can be taken to be the 'permutative' indicator of their linear statistical independence, i.e. the probability of heads and tail of an unbiased coin. See figure 3 of universal computational attractor.

3.4. *Quantum Phase Computational Architecture*

C5) the criterion C4 permits the identification of each newly emergent sheet θ_j with an hologram plane j described in terms of the 3D Heisenberg Lie group G ÷

by its one dimensional centre C_G, such that

$$\Psi(t')\, dt' \otimes \varphi(t)\, dt$$

($\otimes$ tensor multiplication) is specified by the sesquilinear holographic transform of complex-valued writing-wavelet-packet amplitude densities

$$H_1(\psi, \varphi; x, y).dx \wedge dy$$

so as to define the holographic lattices that form 2D pixel arrays inside the hologram plane (x, y). That is to say, an object image bearing beam ψ may be recorded in the hologram plane by mixing it with an unfocused linearly polarized coherent non object-bearing beam φ such that coherent quantum wavelet mixing provides the probability of detecting say a photon within the unit area attached to the point (x, y) of the hologram plane $R \oplus R$ (where $\oplus$ signifies tensor addition) and the necessary holographic trace transform of $\psi(t)dt$ specified by $H_1(\psi; x, y).dx \wedge dy$ (which describes the self-interference of the photons) extends the complex Hilbert space mapping $L^2(R)$ into $L^2(R \otimes R)$, such that $H_1(\psi; x, y).dx \wedge dy = H_1(\varphi; x, y).dx \wedge dy$ implies $\psi(t)dt = c\varphi(t)dt$ where $c \varepsilon T$ denotes a constant phase factor.

The free choice of the phase factor c reflects the fact that only phase difference can be measured. That is, quantum holographic image encoding exhibits the mixing of a coherent reference signal beam by linear Mach-Zehnder interferometry, needed to incrementally record the phase of an object image containing signal beam in the hologram plane $R \otimes R$ and the hidden symmetries of the hologram transform $\psi(t')dt' \otimes \varphi(t)dt$ are such that they can be expressed by the group of automorphisms of the Heisenberg Lie group G, keeping its centre C_G pointwise fixed. These automorphisms are those of the metaplectic group Mp (1, R) which forms a twofold cover of the symplectic group Sp (1, R) = SL (2, R). That is to say its natural action on the hologram plane preserves the centre frequencies $\nu \neq 0$, and in relation to quantum holographic pattern recognition processing in terms of the 3D Heisenberg Lie Group G and Lie algebra g, the holographic phase transforms $H_{\nu,}$ frequency ν, describe the phase conjugate adaptive resonant coupled frequency / phase (ν, θ) memory organisation of four wavelet mixing of the quantum wave amplitudes ψ, φ, ψ', φ' to produce a hologram spectrum which constitutes the 'pages' of an associative read/write filter bank memory where as the identity below shows there is little or no cross talk between the pages

$$< H_\nu(\psi,\varphi;.,.)|H_{\nu'}(\psi',\varphi';.,.)> = <\psi\otimes\varphi|\psi'\otimes\varphi'> \; (\nu = \nu',\ \nu \neq 0)$$
$$= 0 \quad (0 \neq \nu \neq \nu' \neq 0)$$

so that this (θ, ν) organised memory of very narrow frequency spectral windows/pages is able to carry quantum holograms on the hologram planes. (The authors are indebted to Schempp's papers [16, 17] where the above mathematical descriptive is more fully laid out and referenced.)

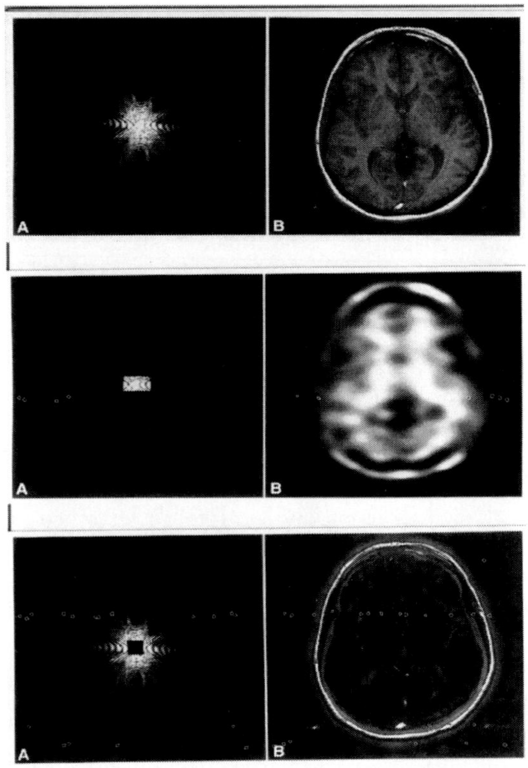

Figure 4. Three views of a single quantum wave diffraction pattern / hologram, each showing the whole brain slice image respectively, resulting, firstly, from the whole pattern; then, with all but its centre removed; and finally from just its centre. Notice the difference from a classical hologram pattern which is degenerate and not nondegenerate as here.

C6) This is computationally possible because, as was discovered by Weyl in 1928, the 3D Heisenberg Lie group G in the Hilbert space L^2 (R) defines Heisenberg uncertainty in terms of the Robertson relations [16] and has a dual(ity) in the form the Lie algebra g of G so that there exists a natural Lie exponential diffeomorphism, i.e. a differential mapping with a differential inverse such that exp g maps onto G and log G maps onto g (as on a slide rule), allowing the usual uncertainty limitations to be overcome, i.e. so that the quantum holographic encoding / decoding required above is possible. This duality implies that essentially unlimited compression of holographic data is possible with regard to signal transmission and memory, because at the most primitive level – that of memory and signal transmission – these are one and the same; and

C7) where, for example see figures 2 / 4, in Magnetic (nuclear spin) Resonance Imaging described by means of quantum holography (schema as outlined above) constitutes a process of phase conjugate adaptive

resonance, so that the 3D MRI image of the MRI brain slice selected, is 3-dimensionally coincident with that of the slice's selected nuclear spins in order to produce the desired resonant MRI signal imagery of the slice. That is, to say, from the two fold cover above, that the brain slice selected and its phase conjugate image are automorphisms.
C8) Furthermore this duality of a Lie Group and its Lie algebra is a description fundamental to the empirically validated Standard Model elementary particle quantum physics. And so by rewrite extrapolation to the higher level of the molecular complexity of the human neurophysiology of the nervous system and the neural brain, their dualities can be postulated to be those of the human self and the human mind respectively in the form of the quantum mechanisms of their microtubule and glial substrates realised as further consequence of Nature's fundamental chemical electron / ion duality, see Noboli papers [18, 19].

3.5. *The Initial Phase, Θ_0, as The Universal Cosmological Internal Measurement Standard*

A physical interpretation of θ (as it applies to gauge invariant phases of the cosmological wave function) is that it defines the initial totally degenerate physical state of 'matter' (see also **4)** of the cosmology θ_0 (quantum mechanically unknown but fixed arbitrarily) from which each new nondegenerate fermion state 'constituting the cosmology's individual spectral lines θ_j' are ultimately emergent.

An actual physical system where Nature implements such a Riemann surface is, it can be postulated, that of the geometric helical structures $\theta \rightarrow \theta$ of RNA (nilpotent X {A = U} ≠ 0) and DNA (nilpotent X^2 {A = T} = 0) where the planes in which the base pairings take place are the sheets / (hologram) planes j in question, and each unique gauge invariant θ_j, corresponds to an associated gauge invariant relativistic 3+1 space time geometry j:-
C9) this nilpotent Dirac theory leads in a most natural way to the usual expressions s^t and $c^s = \mathbf{c}(s^x, s^y, s^z)$, known as the 'probability' current and density respectively which depend on the wave field quantity ψ, so as to define the number of particles which will on average pass through a surface element in unit time, where the total number of particles present is the product of the area and the normal component of $\mathbf{c}^s$, C10), as is observed, for example, in the linearly independent synaptic action of a neuron firing as predicated by Ramon Y Cajal [20]; followed on the neural firing by the probabilistic release of a single vesicle from the hexagonal synaptic vesicular grid to provide the synaptic action across the synaptic junction as a flux / probability current of some 10^4 various neurotransmitter (vesicular) molecules as observed by Sir John Eccles [21] – both Nobel Laureates; figure 5.

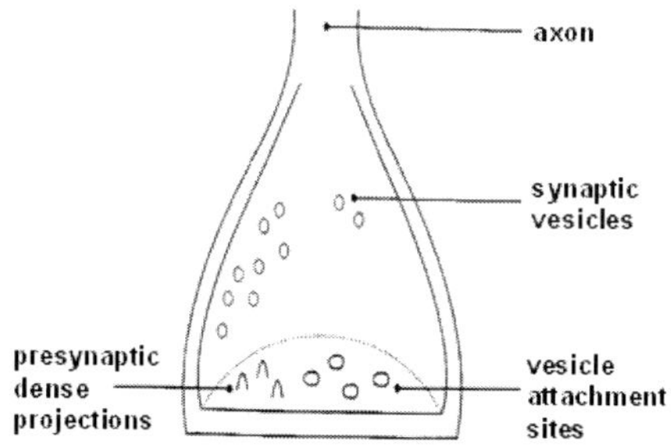

Figure 5. Diagrammatic morphology of the human synapse as observed by Sir John Eccles [21].

C11) the initial fixed phase factor exp $i(\theta = \theta_0)$ thus acts as the internal computational measurement standard, so that the quantum cosmology is necessarily self-referential and self-organising and exp $i(\theta=\theta_0)$ is the arbitrary fixed measure of its initial entanglement, which cannot itself be measured since by implication, there remains no measurement standard against which to measure it. And as is well known, the quantum field / system vacuum state plays such a role, i.e. it cannot itself be measured even though it produces measurable phenomena such as the Lamb shift of atomic spectra and condenser Casimir effect, which can be;
C12) these all lead to the postulate that there must exist a unique single semantic function of θ in the complex plane that describes each newly emergent entanglement θ_j relative to the initial arbitrary measurement standard θ_0 so that each such newly emergent gauge invariant θ_j must lie on some line in the complex plane characterised by the common physical parameter fermion spin = ½. That is, a 'gauge invariant' Riemann surface θ (measure the unit disk) must exist independently of any particular space time coordinate system where Pauli's fermion exclusion principle continues to apply, such that each gauge invariant phase θ_j is unique, single valued and linearly independent; a criterion which it can be postulated is that of the Riemann Zeta function which describes the entire functionality in the complex plane (x, y); a proven quantum system description [22] which Berry postulates has time reversal asymmetry [23] as has the NUCRS in line with its fermion $0 \rightarrow 0$ logistical fractal construction.

4. The Fundamental Role of Phase, θ, Resulting in Entirely New States of Matter

This is described by the discovery of the Quantum Carnot Engine, which extends the Second Law of Thermodynamics, and is experimentally validated by Scully *et al.* in the quantum mechanical field domain, as implicit in the helical structure of RNA / DNA genetic code molecules.

C13) In classical thermodynamics, irreversibility is ideally described by defining the Carnot engine (whose cycle consists of alternate isothermal or constant temperature, and adiabatic or constant entropy, phases) as the one that is theoretically most efficient; and one of the many statements of the second law of thermodynamics is that no engine can be more efficient than a Carnot engine. However to ensure the incorporation of thermodynamics with quantum dynamical laws, as is proposed in the nilpotent cosmology described here, the cosmos must be treated as a *quantum* Carnot engine (QCE) consisting of a single heat bath in which the ensembles of elementary particles retain a small amount of quantum coherence parameter phase θ, so as to constitute new states / properties of matter, called by Scully *et al* [7] a 'phaseonium', the very first of which is the ensemble of dark / indistinguishable / totally degenerate matter. This explanation of how this initial phaseonium fuels the natural structure of the universe, is then in complete accord with Nilpotent Universal Computational Rewrite System (NUCRS) description above as a quantum physical explanation of the origin of new emergent states of nondegenerate fermion matter;

C14) For although in the NUCRS, the Second Law of Thermodynamics continues to hold since quantum coherent systems must be quantum physically prepared for measurement, quantum coherence fundamentally alters Carnot heat engine operation (i.e. the Carnot efficiency η for converting heat into work) from what is thermodynamically possible classically,

$$\eta = 1 - T_c / T_h \quad \text{to} \quad \eta_\theta = \eta - \theta \cos\theta \qquad (2)$$

where T_c and T_h are the temperatures of the low temperature entropy sink and the high temperature source respectively as in the classical Carnot engine, and θ is the quantum coherent phase, such that by the proper choice of θ, work is obtained even when $T_h = T_c$, i.e. even when there is only one thermal bath, as is the case in a self organising quantum cosmology with its own internal measurement standard θ_0 and subsequent relative measurement standards, in the form of the gauge invariant θj of the emergent new states / properties of matter;

C15) there is now much recent experiment l evidence, that the usual classical wavelength limitations of measurement are overcome by quantum physically coherent means. For example as set out in Walter Schempp's paper [24], where the latest telescopes employing laser technology allow a 'tuning' of the incoming light so as give stable and clear imagery of the star configurations under investigation and where if, say, the existence of a black hole is suspected at the centre of the galaxy, the same mathematics/physics, as explains how such tuning clarifies the imagery, is also able to explain the very unusual star orbits observed (figure 6), attributable to the single virtual orbit focus of the suspected black hole in question,

C16) these quantum coherent wavefield phenomena contradict Bohr's assertion that the quantum wave field mathematics used in these examples has no corresponding physical meaning / interpretation.

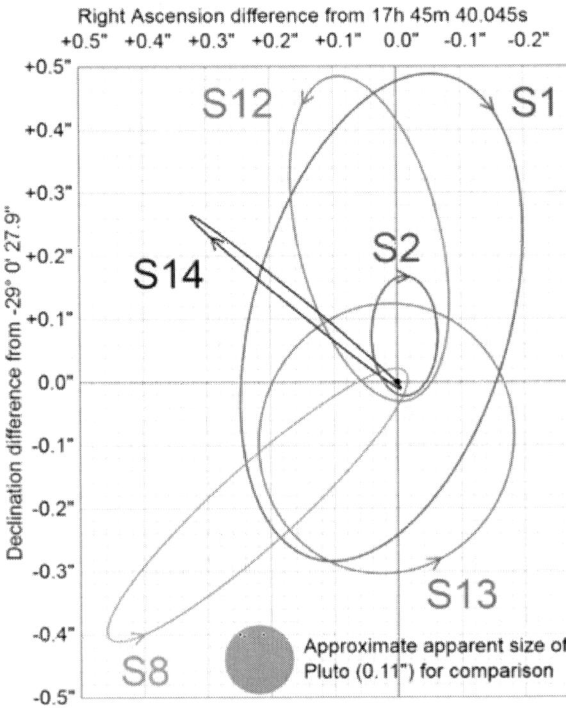

Figure 6. Star orbits detected at galactic centre, predicating a unseen black hole at their common elliptic focus.

4.1. *The Significance of Phase in the Nilpotent Representation of Quantum Mechanics*

Phase plays a significant role in the nilpotent representation of a fermion. The expression ($\pm ikE \pm i\mathbf{p} + jm$) applies either to operator or amplitude. If we use it to mean operator, and convert the E and $\mathbf{p}$ terms to differentials or differentials with attached potentials, it selects the *unique* phase factor required to generate ($\pm ikE \pm i\mathbf{p} + jm$) as a nilpotent amplitude. The phase is an expression of gauge invariance or the degree of variability of the space and time coordinates within the constraints determined by the conserved quantities in the system. In effect, the operator *codes* the space and time variation, the phase factor *decodes* it.

This variation is our entire knowledge of the system, and we can regard the amplitude without the phase factor, which is expressed in terms of conserved quantities, as the local manifestation of the fermion, and the phase factor as the vacuum or nonlocal manifestation, incorporating all that will determine the subsequent behaviour of the fermion – in effect, the rest of the universe as observed by that fermion.

It is possible to show that the entire information about a fermion and its vacuum is encoded in the instantaneous direction of the **p** operator, or, alternatively, in the values of E, p and m (or, in effect, E and p, as, for a nilpotent, the third term becomes redundant). Each fermion has a unique instantaneous direction for **p**, which can be regarded as a unique phase, though only the relative direction can be known. We can use the principles of projective geometry to regard the directional vectors as generating perpendicular planes, which in the case of fermions, must all intersect. This becomes a manifestation of Pauli exclusion.

In the nilpotent structure, phase and amplitude are dual in that each contains the same information, differently organized. In addition, we can regard E, p and m as incorporating holographic information as phase, amplitude and reference phase. The analogy with holography comes from the fact that, to reconstruct the complete structure we only need two of these, as nilpotency makes the third redundant. The origin of the holographic principle also becomes apparent, when we extend the nilpotency condition to all self-organizing systems, as we only need the equivalent of two orthogonal components to reconstruct the entire system. The principle is also clearly connected with gravity as a unifying principle for all the interactions (gravity-gauge theory correspondence), as the entire nilpotent structure represents the gravitational vacuum, while the component parts represent the vacuum partitions generated by the other forces.

5. Black-Hole Physics

It follows from 4 that galactic black hole physics concerns gravitational holographic focii as described in 3.4 C6 by means of the 3D Heisenberg Lie group G and its Lie algebraic dual g. Each black hole thus acts in 3D space as the focus of a gravitational lens, from which the matter absorbed returns to the Universe as a whole as Standard Model quantizations of elementary particle physics by the diffeomorphic diffusion process whereby exp g = G and log G = g; the asymmetry of which therefore breaks that of the Galois group see **1**, end of page. This says the emergence of the cosmos is a process of phase conjugate adaptive resonance whereby no information is lost and the role of the black holes is to recycle matter as 'garbage collectors' for reuse in its further evolutionary emergent self organisation. This postulates that the Universe as a whole acts as a White Hole source for the Lie transformation of matter in 3D space, where galactic black holes are its secondary anti-sources such that by Huygens principle of secondary sources, the White Whole together the black holes always constitute an almost zero totality, and that more generally Lie commutator relations constitute Lie Group 'germs' for their group Lie algebras, as for example researched extensively, by Hoffman [25] which reference to his many published papers in relation to his postulate that the human neurons function as Lie group germs enabling him to work out the various types of geometries and behaviours as actually seen in the neurons of the human cerebral cortex.

6. Consciousness and the Hard Problem

1) The solution to the 'hard problem of consciousness' of how objective third party person to person communication can be achieved is astonishingly simple, in spite of the seemingly contradictory fact that in the postulated self-organised cosmos, each individual human observer has / knows something so unique about themselves [9], that it is impossible to communicate it to any other individual:-

2) C17) if any kind of wave is incident on an object, the 3D geometric object image is spontaneously encoded in the wavefront by means of local changes of amplitude and phase. This spontaneous encoding happens totally independently of any observer; a fact demonstrable by the holographic capture of the wavefront incident on any 3D object by means of a reference wave to form a hologram, from which the subsequent removal of the reference wave, reveals the 3D holographic image of the object;

3) C18) this holographic capture of the light incident on the retina, is, we postulate, what happens in the human eye. For this would enable each of us to see any object where it actually is in 3D space by means of phase conjugate 3D dynamic object imagery, with the retina acting as an active phase conjugate mirror, Figure 4. And it would apply to all our senses, allowing us to see, hear and feel, etc. any 3D object exactly where it is located outside the body in 3D space! Although of course, the interference pattern of the incident light on the retina with the reference wave the retina provides is sent as a physical hologram to the brain, for the purposes of memory and cognition (in the form of a pixel signal array as specified in 3.4). Each retinal / sensory cell is therefore a secondary source of the reference wave the brain provides. It makes the holographic capture of the illumination incident on the retina possible, and Figure 7 illustrates how the captured signal can in principle be transferred onward into the brain itself. It shows how a 3D acoustic image produced in water can be converted into a 3D visual image above the water (with the water surface pattern

acting as a hologram!). A process, we postulate, happens at each human sensory apparatus where each consists of a set of secondary sources for the reference wave needed for the holographic capture of the incoming illumination, be it visual, acoustic, etc.

4) C19) Let's try this out! If you snap your fingers where do you hear the snap? Why, exactly where the snap happened outside your head. And if you pick up a 3D object where do you see it? Why, of course, exactly where the object is. And if you reach out to touch a 3D object, where do you find it? Why, where you see it and exactly where it is located in 3D space. And that's the whole point! *For although our whole experience of any 3D object is subjective, we can all communicate that experience to one another, simply by pointing to the object in question, or an image of it, or by saying its name & our description of it in a language we all share.* For the object itself quite literally defines its own 'objectivity' because no two objects on our level of scale can be in the same place at the same time!

5) And the NUCRS has this objectivity at the quantum level too because it concerns only fermions which by Pauli exclusion are unique;

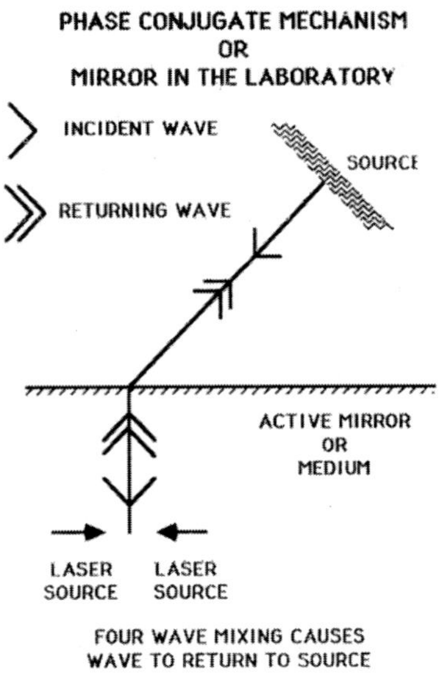

Figure 7. The phase conjugate mirror above can also be regarded as the focusing of a holographic lens.

6) C20) It would be absurd for our senses not to use all the information about 3D objects incident on each sensory apparatus in the form of both amplitude and phase, for this information allows the immediate real time holographic reconstruction / perception of all the incident 3D object imagery in a single computational step without the need of any prior knowledge! And when the new 3D imagery is then transferred to the brain (in the form of a holographic memory bank and filter of all the previous holographic inputs, 3.4), the holographic filtering provides the brain with the cognition required in the most efficient way possible, as well as alerting the brain to anything novel that the memory has never before experienced;

7) C21) the above criteria also explain why our sensory perception of 'reality' always takes place in terms of 3D objects within a relativistic 3+1 dimensional space-time, so that a) what we experience can be immediately added to the associative holographic memory bank of the brain for rapid recall and filtering by means of all that is already stored there about the particular objects in question and their properties acquired as the result of our past sensory experience, and also b) of what we have learnt from others about such objects as stored symbolically in the other half of our brain by means of our separate natural semantic language capabilities, including our own natural language thinking, which today includes our mathematical, logical and conventional computational ones: where these two halves of the brain are integrated in a single linked operable whole by means of the *corpus callosum*, by means of a signal path / unit wire from the object image to its language name;

8) C22) as the brain organisation above makes clear, it makes good sense that while all the processes described above proceed in a totally automatic, largely unconscious manner, a minority of them need to be brought, because of their importance to our survival, to the attention of our conscious awareness for conscious thought, decision taking and possible immediate exceptional action, as when we are learning something new, requiring, for example, the conscious control of say the steering wheel when we start to learn to drive, but which in response to normal driving circumstances once learnt, can then be carried out largely unconsciously, in accord with the repertoire of our driving experience as stored in the brain, or as when one 'thinks aloud' while consciously trying to solve a problem never before encountered;

9) C23) indeed it follows logically from the above argument, that, at the molecular complexity of each individual human being, each of us acts a) as a secondary source many times removed of the original unified field source that brought the self-organised cosmos into existence in the very first place and so we function b) as a potential agency for the cosmos's future evolutionary emergence with the possibility ultimately to bring into existence something that never before existed, if only of our own offspring, and which gives each of us the amazing potentiality not only to advance our understanding of cosmic evolution but to enable the cosmos to understand itself. This expresses the beginning of philosophical and scientific enquiry, as, for example, in the last verse x. 129 of the *Rig Veda* (written down at least two thousand years ago) where the poet of the Vedas consciously saw that for the progress of the mind and thought, mankind requires

doubt and faith, for the poet asks 'who knows therefore whence comes this creation?'And concludes in the spirit of his / her time 'only that god who sees in the highest heaven: he only knows whence comes this universe and whether it was made or uncreated. He only knows, or perhaps he knows not?'

References and Notes

[1] Rowlands, P. (2007) Zero to Infinity: The Foundations of Physics, Singapore and Hackensack, N.J., World Scientific.

[2] Wittgenstein, L. (1975) Philosophical Remarks, Oxford University Press.

[3] Marcer, P. & Rowlands, P. (2010) AIP Conference Proceedings, 1316, 90-101.

[4] Marcer, P. & Rowlands, P. (2008) How intelligence evolved?, Technical Report SS-07-08, AAAI Press, pp 46-51.

[5] Conway, J.H. (1976) On Numbers and Games, London: Academic Press.

[6] Weyl, H. (1931) The Theory of Groups and Quantum Mechanics, 1928, H. P. Robertson (Trans.) Dover Publications.

[7] Scully, M.O. et al, (2003) Science, 299, 862-864; Collini, E. et al, (2010) Nature, 463, 644-647; Miller, K. D. (2010) Science, 330, 1059-60; Kaschube, M. et al. (2010) Science, 330, 113-16, where three distantly related mammals are shown to share a self organized visual cortex brain architecture characterized by density π. The result is explained in P. Marcer and P. Rowlands, unpublished letter to Science, 6 December 2010: 'The appearance of π in the density of the squared 'map period' of the neurons in the visual cortex, strongly suggests the presence of a geometrical phase term in the spatial structure of the system. If we regard the 'pinwheel' as the equivalent of a singularity in the space, then a double circuit through the cycle of orientations or 'map period' would be needed to return to the original phase state. So the singularity would correspond to a double map period (2λ) in any given direction of the two-dimensional cortical sheet, and each pinwheel singularity would situate itself in a circle with radius length λ in this two-dimensional space, creating a pinwheel density $/\lambda^2$ of π. We have previously proposed that a characteristic structure for the space of self-organizing systems at all levels of complexity results from a dual vector system, or equivalent, for which a geometrical phase of π becomes an identifying feature.'

[8] Wolfram, S. (1984) Physica, IOP, Preface.

[9] Deutsch, D. (1985) Proc. Roy. Soc., A 400, 97-117.

[10] Albert, A. Z. (1983) Phys. Lett., 98 A, no. 5/6, 249-8.

[11] Feynman, R. P. (1986) Found. Phys., 16, 6, 507-531.

[12] von Neumann, J. (1945) First draft report on the EDVAC.

[13] von Neumann, J. (1966) Theory of Automata: construction, reproduction, homogeneity. Part II of A. W. Burks, The Theory of Self-reproducing Automata, Urbana: University of Illinois Press.

[14] Rowlands, P. Physical interpretations of nilpotent quantum mechanics, arXiv: 1004.1523

[15] Diaz, D. & Rowlands, P. (2006) International Journal of Computing Anticipatory Systems, 19, 229-235.

[16] Schempp, W. (1998) Journal of Mathematical Imaging and Vision, 2, 279-326; Magnetic Resonance Imaging (1992) New York, Wiley, p. 124, figure 22.

[17] Schempp, W. (1993) Magnetic resonance imaging and solitonic nanotechnology (1993) Nanobiology, B, 109-164.

[18] Noboli, R. (1987) Phys. Rev., A 35(4), 1902-1922.

[19] Noboli, R. (1985) Phys. Rev., A 32(6), 3618-3626.

[20] Ramon Y Cajal, S. (1911) *Histologie du Systeme Neuveux de l'Homme et des Vertebres*, Vol II, *Maloine*, Paris.

[21] Eccles, J. C. (1986) Proc. Roy. Soc., B 227, 411-428 (1986); How the Self Controls its Brain, New York: Springer-Verlag.

[22] Riemann Zeta Function as a description of a quantum system – Proof of Author

[23] Berry, M. V. (1986) Riemann's Zeta function: a model of quantum chaos?, in T. H. Seligman & H. Nisioka (Eds.), Quantum chaos and statistical nuclear physics, Springer lecture notes no. 263, Springer Berlin, pp. 1-17.

[24] Schempp, W. (2011) presented at CASYS Conference, Liège, Belgium; paper in press.

[25] Hoffman, W. C. (1989) Applied Mathematics & Computation, 32, 137-167.

Eliminating the Gravitational Constant, *G*, and Improving SI Units, Using Two New Planck-Unit Frameworks With All Parameters as Ratios of Only *H*, *C*, and *a*

MICHAEL LAWRENCE

Maldwyn Theoretical Physics, Cranfield Park, Burstall, Suffolk, IP8 3DT, United Kingdom
mike.lawrence@maldwynphysics.org

This paper presents a tool kit of new ways of understanding fundamental physics and the relationships between parameters. The novel insights and predictions include: A self-contained consistent new Planck unit set of maximal-sized parameters from which all observed values can be compared and easily combined in equations. A self-contained consistent new Planck unit set of electron-charge-based parameters, some of which are directly observable in experiments. The interpretation of the gravitational constant, G as a dimensionless ratio eliminates the need to test the equivalence of gravitational and inertial mass. That all parameters can be displayed in terms of only h and c for the Planck maximal-parameter set and additionally in a for the electron-charge-based set, previously considered impossible. A new dimensional analysis employed to describe parameter dimensionality and to uncover any law of nature or universal constant. That all electron-charge-based Planck parameters can be described solely in terms of ratios of R_k and K_j and so will benefit from the precision of measurement of these two parameters. That most electromagnetic parameters can be reinterpreted in terms of mechanical parameters. By adjusting currently misaligned SI units to be self-consistent and consistent with these Double-adjusted-Planck units, greater clarity will ensue.

Keywords: Gravitational constant; Planck units; SI units; Dimensionality; Parameters.
PACS numbers: 13.40.Em, 06.20.Jr, 06.30.Ka, 12.20.Ds, 12.10.-g

1. Background and Methodology

The discovery that the von Klitzing constant R_k [1] and the Josephson constant K_j [2] could be measured directly, although they are composed of Planck and near-Planck sized parameters h and q_e, has improved the precision of measurement of h and some SI electromagnetic units [3]. It is unfortunate that the misalignment of SI units between mechanical and electromagnetic parameters has not been addressed before. This paper looks at how the SI system should be adjusted to be self-consistent and consistent with the most fundamental set of Planck units that are hypothetically possible. It then reinterprets the meanings of, and relationships between, those fundamental parameters.

The paper starts by defining the most fundamental set of Planck units that are hypothetically possible, and then a second linked set based around the maximum values that are actually observed – the difference being that the former is built on a hypothetical Planck charge, Q_*, defined later, the latter on the electron charge q_e.

As part of the process in defining the new Planck units, it is shown that the gravitational constant G is only a dimensionless ratio, and can be subsumed within the new mass and distance units. What emerges from the q_e set are values in the new fundamental units for R_k and K_j. These two constants are members of the set of q_e units, as should be expected, although K_j appears inversely and twice the anticipated size.

The dimensional analysis used to subsume G is employed to show that R_k can be considered as equivalent to a velocity, and that many of the electromagnetic parameters can similarly be considered equivalent to mechanical parameters. This invites a reinterpretation of not just R_k and K_j, but of all electromagnetic parameters.

By using the new fundamental units as a framework, it is shown what adjustments are required to some of the electromagnetic SI units, so that the units are both self-consistent and consistent with the fundamental framework. This reintroduces the ratio $\sqrt{G}$ as a replacement for permeability, which is also shown to be dimensionless.

The measured value of R_k is shown to equate to a speed greater than light speed. Although it is not clear whether this increased maximum velocity applies to either physical objects, the media through which the physical objects travel or patterns created by subluminal physical objects, this can be experimentally tested.

The final output is to display all the Q_* parameter set as ratios of only h and c, and all the q_e parameter set as ratios of only h, c and $d = \sqrt{\alpha/2\pi}$. This highlights how the parameters are linked and shows that the laws of nature would be constructed in the same way regardless of the relative sizes of h, c and α. The dimensional analysis enables new laws to be constructed and new constants of nature to be uncovered, although it is not clear that there are any of the latter needed since h and c are all that are required to generate all the Q_* fundamental parameter set. Examples are given of simplifications of existing formulae using the new framework.

2. Significance and Objectives

The usefulness of the paper is partly in providing a toolkit for looking at the relationships between fundamental parameters and for displaying equations stripped of their usual SI based constants and ratios so that the underlying physics can be better understood.

Although to an extent this is already done [4], the new units introduced means that there need no longer be odd conversion factors used when considering electromagnetic and mechanical units together.

A prediction is made on the maximum faster than light velocity that can be obtained. The observed value of R_k implies that there is a limit above light speed, probably in terms of electromagnetic field pattern motion, and that limit can be tested in the laboratory.

The most fundamental set of units possible is produced and the set of SI units compared to it, with two adjustments proposed to ensure the self-consistency of SI units and of their consistency with the fundamental units framework. Also shown is how increased precision in the accuracy of other parameters can be constructed out of ratios of R_k and K_j. The gravitational constant G is eliminated from gravitational equations, confirming the equivalence of inertial and gravitational mass in a novel manner there being only the former remaining. The ratio $\sqrt{G}$ is reintroduced to make the second adjustment to SI units and is thereafter a measure of the permeability or interaction within materials, rather than between masses.

This paper sets out with a number of objectives. The starting point is a discussion of units and the lack of consistency between current SI units and Planck units. It is shown below that the current set of SI units does not have a consistent relationship with the most fundamental set of Planck units, described as Double-adjusted Planck units (DAPU units), and is not internally self-consistent.

By showing the changes needed to SI units in order to gain consistency, it becomes possible to better understand the relationships between parameters such as mass, length, charge, magnetic flux, resistance and time.

This paper is emphatically not an exercise in numerology. Although, by necessity, it manipulates numbers and relationships to produce self-consistent values for all the parameters considered in terms of the adjusted new SI units, that is not a major objective. What this paper provides is a toolkit for investigating the deeper relationships between parameters, which are currently hidden by their partially-aligned SI values and less than optimal Planck units.

3. Units

The paper by Mohr et al. [5] explains the current state, where SI units are being bought more into the quantum measurement realm. The excellent paper by M. J. Duff et al [6] includes a broad and varied introduction to the problems of fundamental units and also covers their relationship with SI units. The issue is not new [7]. To paraphrase Dr Okun [8] "The use of fundamental units h and c in SI has introduced greater accuracy in some of the units, but some electromagnetic units are based on pre-relativistic classical electrodynamics and so their measurement is not as accurate as other units. The use of permeability and permittivity spoils the perfection of the special relativistic spirit and, whilst this is useful for engineers, it results in the four physical parameters D, H and E, B having four different dimensions". It is only by starting at the most basic and simple physical maximal sized set of Planck type units - and maintaining the integrity of the relationships within that set by not stretching the parameter space unequally - that it is possible to see that the electromagnetic and mechanical parameters are misaligned and that the current value of permeability results in a further misalignment. A new form of dimensional analysis underpins this and allows both mechanical and electromagnetic parameters to be treated on an identical basis.

4. The Foundations

The most basic two formulae for defining a Planck unit system are the force equation $F = GM^2/L^2 = Q^2c^2/L^2$ and the angular momentum equation, $h = McL$. The normal usage of the latter is to define a Planck mass M_p and Planck Length L_p such that $\hbar = M_p c L_p$ and $M_p = \sqrt{\hbar c/G}$. Unfortunately this introduces the 2π factor in many equations, where it serves only to confuse. The preferred definition to be used here as a starting point is to define the system without the 2π

factor. There are two stages to the change, the first is the adjusted Planck unit (APU) and the second the double-adjusted Planck unit (DAPU). Initially the APU mass M_o and APU length L_o are related by $h = M_o c L_o$ and M_o is defined to be $M_o = \sqrt{hc/G}$.

However, to achieve the right relationship between M and L in parameter space, as described below, requires looking at the force equation at the same time. Rearranging to give $FL^2 = GM^2 = (Qc)^2$ provides the simple relationship that the APU mass M_o and APU charge Q_o are related such that $M_o\sqrt{G} = Q_o c$. Since the latter equation does not include L_o it is not immediately apparent that compared with the Planck parameters M_p and L_p there is a need to adjust both by the factor $\sqrt{G}$ in addition to the $\sqrt{2\pi}$ factor, so that $h = M_o c L_o = 2\pi(M_p\sqrt{2\pi G})c(L_p/\sqrt{2\pi G})$ if the latter factor is distributed in the same way as $\sqrt{G}$. This stretches parameter space equally along the mass and length parameters, rather than just the mass parameter as is usually done when trying to eliminate G [9].

Now the second stage of the change can be made into DAPU units. The DAPU mass M_* is defined to be $M_* = M_o\sqrt{G} = Q_o c = Q_* c$ where Q_* is the DAPU charge.

Following the angular momentum equation, M_* and DAPU length L_* are related by $h = M_* c L_*$ with $L_* = L_o/\sqrt{G}$.

The result is the foundation of a DAPU parameter set and units based on

$$h = M_* c L_* \tag{1}$$

and

$$F_* = M_*^2/L_*^2 = Q_*^2 c^2/L_*^2 \tag{2}$$

which was an initial objective, in that the formulae exclude G. The dimensionality of G will be shown to be zero later.

This is the most basic set of Planck parameters that can be devised because only the two universal constants h and c are used, and this is the minimum number of constants required to establish relationships between the parameters.

The subsuming of G within the mass and distance units also eliminates the difference between gravitational and inertial masses, since there is no longer any gravitational mass. This is not equivalent to making $G = 1$, as will be shown below, because the effect of subsuming G into M_* and L_* is to stretch current parameter space into the more symmetric DAPU parameters space, which does not occur when simply setting $G = 1$.

The base parameter space consists of M, L, c, h and Q. Since Q can be related to M and c only, the minimum parameter space is just M, L, c and h. Because h and c are the two basic universal constants, to maintain the topology and symmetry of the base parameter space requires that the other two parameters M and L are stretched proportionately together. Provided Q is treated in the same way as M, it will stay symmetric. Any non-symmetric stretching results in an asymmetric set of parameters and will require the use of factors such as $\sqrt{\alpha/2\pi}$ in the relationships between the stretched parameters.

5. SI Units and DAPU

The above two relationships hold in the new DAPU system in DAPU units, but unfortunately in SI units there is the first misalignment that becomes apparent. To align the charge and mass side of the equation in SI units requires that the base unit size Planck charge is decreased by the factor $\sqrt{1 \times 10^{-7}}$ relative to the mass side since $GM_o^2/L_o^2 = Q_o^2 c^2 (10^{-7})/L_o^2$ in SI units. To identify this difference, each equation in future may, where it might otherwise confuse, be identified either as being in DAPU or SI units, so that

$$Q_* = M_*/c(DAPU) = M_*\sqrt{1 \times 10^7}/c(SI) \tag{3}$$

It is useful for display purposes, as will be used liberally later, to define a factor

$$d = \sqrt{\alpha/2\pi} \tag{4}$$

that represents the ratio $d = q_{e*}/Q_*$, where q_{e*} is the DAPU size of the electronic charge.

The second SI misalignment appears when comparing electromagnetic and mechanical SI units that have material content requiring permeability or permittivity. The use of permeability u_* as $4\pi \times 10^{-7}$ causes the factor $4\pi \times 10^{-7}/\sqrt{G} = 6.501$ to appear in some parameters when compared with what their DAPU based value should be. This arises from some parameters whose SI units may mix electromagnetic and mechanical parameters within their definition, such as the Farad. So the second SI re-alignment is to define u_* to be equal to the ratio $\sqrt{G}$ rather than the usual $4\pi \times 10^{-7}$, which relegates G from gravitational to permeability use, so that it represents a measure of the strength of interactions within materials not between masses.

The result is that the proposed new adjusted-SI units (NSI) which should be used are either the same as the normal SI units or are different to normal SI units by a power of either the ratio $\sqrt{G}$, the $\sqrt{1 \times 10^{-7}}$ factor, the 6.501 factor or a combination of these. Wherever there is a factor q_{e*} used, the same power of $\sqrt{1 \times 10^{-7}}$ is used. Where there is no q_{e*} or u_* factor, the NSI and SI values are the same. Where the current SI unit is adjusted by a power of the $\sqrt{1 \times 10^{-7}}$ factor, the parameter has a cedilla above it. So the SI unit Watts, W becomes $\tilde{W}$ in NSI where $\tilde{W} = \sqrt{1 \times 10^{-7}}W$.

Because most of the parameter examples used in this paper do not have any specific material dependence, as would be the case for the magnetic field H, there is no use of permeability u_* or permittivity ε_* within most of the parameter examples given, except to show that H and magnetic inductance B have the same underlying units. For the examples used here, there are no complications of additional 6.501 usage or identification of double adjusted SI units, other than in the permittivity ε_* and capacitance C_*, where the SI unit the Farad F is adjusted by that factor to be $F^{\#}$ in NSI with $F^{\#} = F/6.501$.

So the adjustment of SI units to make them self-consistent across both mechanical and electromagnetic parameters, and to ensure that they have the same overall shape in parameter space as the underlying DAPU units allows the direct comparison of all parameters in either DAPU or NSI units, with the only difference being the actual number value in each set of units. For the Q_* set of parameters, in DAPU the maximal values are always one. For the q_{e*} set of parameters, the maximal values are always powers of d. For both these sets, the NSI values are shown in Tables 1 and 2, with translation factors between units in Table 3.

6. Dimensionality of *G*

The subsuming of G with the APU mass M_o to produce the DAPU mass M_*, and the APU length L_o to produce the DAPU length L_*, would seem to ignore the units of G, effectively treating G as unity and without units. But this is not the case. The units of G are $m^3kg^{-1}s^{-2}$. A consideration of the standard laws of nature and the fundamental constants through a form of dimensional analysis shows that if each parameter is assigned an appropriate dimensionality, every fundamental constant, other than c, will have a total dimensionality of zero.

The dimensional analysis consists of solving for a basis vector in vector parameter space which produces zeroes of dimension for four important constants of nature, h, G, Permeability (u) and Boltsmann's constant, k_B. Using G in the analysis may appear circular, but the analysis supports its use. It also shows that Boltzmann's constant, like G, is simply a factor that can be discarded in the correct units and that there may exist other parameters, as yet unrecognized, that correspond to missing dimensionalities.

The dimensionalities of the main SI, NSI, APU or DAPU parameters in terms of a hypothetical dimension Y that emerge from the consideration are:

Mass $M_* = Y^{+1}$ Velocity $c = Y^{+2}$
Length $L_* = Y^{-3}$ Energy $E_* = Y^{+5}$
Charge $Q_* = Y^{-1}$ Time $T_* = Y^{-5}$
$h = Y^0$ $G = Y^0$

The units of G are $m^3kg^{-1}s^{-2} = Y^{-9}Y^{-1}Y^{+10} = Y^0$ dimensionality and h has units $kgm^2s^{-1} = Y^{+1}Y^{-6}Y^{+5} = Y^0$ dimensionality. So the units of both h and G are actually irrelevant because they represent fundamental constants with zero dimensionality.

Thus adjusting the APU mass to the DAPU mass, and APU length to DAPU length, involves only multiplying or dividing by the ratio $\sqrt{G}$ as a dimensionless number, and does not affect the dimensionality of the units of mass or length, other than changing the sizes of the base Planck mass and distance units. This stretches the current parameter space into the more symmetric DAPU parameter space, which is different to treating G as equal to one, which does not affect the current parameter space topology.

The same dimensional analysis can be done for permeability

$$u_* = NA^{-2} = kgm^{-1}s^{-2}(\sqrt{kgms^{-1}})^{-2} = Y^0$$

which shows that the replacement of u_* by $\sqrt{G}$ does not affect the units used because they are both dimensionless.

This hypothetical dimensionality tool can be used to produce any law of nature by creating equations where the dimensionalities are equal on both sides. One example from the tables would be $F = Ma$, where force is Y^8 and is equal to the product of mass Y^1 and acceleration Y^7, so that both sides have Y^8 dimensionality. Another example would be the product of volume and viscosity which produces Y^0 on one side and could represent a new constant of nature on the other. To produce a constant of nature, aside from c, the minimum that is required is that it has Y^0 dimensionality. In this instance, there is no need for a new constant since the product of volume and viscosity is equal to Planck's constant, through $V_*\eta_* = h$.

However, producing laws of nature through the dimensional analysis does not provide the exact relationship between the non-maximal parameters, because these depend on the specific context in which the parameters are being considered. An example would be the kinetic energy of a particle in motion $E_{ke} = (\gamma_v - 1)mc^2 \simeq 0.5mv_e^2$ compared with the rest mass energy of the same particle, $E_{rm} = mc^2$. Dimensionally these exhibit the same relationships between mass, energy and velocity but they describe different specific aspects of that relationship.

7. Values of the Q_* Set of Parameters

Table 1 provides a list of the main Q_* parameter set and their NSI values at their maximal Planck sizes. The column headed 'NSI units' means that where the current electromagnetic SI units appear they have been adjusted by a power of the factor $\sqrt{1 \times 10^7}$ mentioned earlier and their use is denoted by a cedilla above the unit or $F^{\#}$ describes the SI unit F adjusted by the 6.501 factor. Note that the factor d does not appear in Table 1 because these values are all based on the DAPU charge, Q_*.

Table 1. Values of the Q_* Set of Parameters

Parameter (X_*)	Q_*DAPU set's NSI Value	NSI Units	DAPU equivalent	As Constants
Gravitational Constant(G)	1	$\mathrm{m}^3 kg^{-1} s^{-2}$	$none$	$none$
Permeability(u_*)	$\sqrt{6.67428 \times 10^{-11}}$	NA^{-2}	$none$	$\sqrt{G}$
Boltzmann's Constant (k_B)	1	$J\breve{K}^{-1}$	$none$	$none$
Angular Momentum(h)	$6.62606896 \times 10^{-34}$	Js	$kg\mathrm{m}^2 s^{-1}$	h
Mass(m_*)	$4.45695580 \times 10^{-13}$	kg	kg	$\sqrt{hc}$
Magnetic Flux(ϕ_*)	$4.45695580 \times 10^{-13}$	$\widetilde{W}$	$\sqrt{kg\mathrm{m}}\mathrm{m}s^{-1}$	$\sqrt{hc}$
Charge-mass$(q_* c)$	$4.45695580 \times 10^{-13}$	$\tilde{C}\mathrm{m}s^{-1}$	$\sqrt{kg\mathrm{m}}\mathrm{m}s^{-1}$	$\sqrt{hc}$
Velocity(v_*)	2.99792458×10^{8}	$\mathrm{m}s^{-1}$	$\mathrm{m}s^{-1}$	c
Resistance(R_*)	2.99792458×10^{8}	$\widetilde{\Omega}$	$\mathrm{m}s^{-1}$	c
Momentum$(m_* v_*)$	$1.33616173 \times 10^{-4}$	$kg\mathrm{m}s^{-1}$	$kg\mathrm{m}s^{-1}$	$c\sqrt{hc}$
Current(ι_*)	$8.98755179 \times 10^{16}$	$\tilde{A}$	$\sqrt{kg\mathrm{m}}s^{-1}$	c^2
Action(m_*/r_*)	$8.98755179 \times 10^{16}$	$kg\mathrm{m}^{-1}$	$kg\mathrm{m}^{-1}$	c^2
Angular Frequency(w_*)	$6.04538246 \times 10^{37}$	Hz	s^{-1}	$c^2\sqrt{c/h}$
Frequency(f_*)	$6.04538246 \times 10^{37}$	Hz	s^{-1}	$c^2\sqrt{c/h}$
Energy(E_*)	4.00571211×10^{4}	J	$kg\mathrm{m}^2 s^{-2}$	$c^2\sqrt{hc}$
Temperature (K_*)	4.00571211×10^{4}	$\breve{K}$	$\breve{K}$	$c^2\sqrt{hc}$
Potential Difference(V_*)	$2.69440024 \times 10^{25}$	$\tilde{V}$	$\sqrt{kg\mathrm{m}}\mathrm{m}s^{-2}$	c^3
Acceleration(a_*)	$1.81236007 \times 10^{46}$	$\mathrm{m}s^{-2}$	$\mathrm{m}s^{-2}$	$c^3\sqrt{c/h}$
Magnetic Inductance(B_*)	$1.81236007 \times 10^{46}$	$\tilde{A}\mathrm{m}^{-1}$	$\mathrm{m}s^{-2}$	$c^3\sqrt{c/h}$
Magnetic Field (H_*)	$2.21841235 \times 10^{51}$	$\tilde{A}\mathrm{m}^{-1}$	$\mathrm{m}s^{-2}$	$c^3\sqrt{c/hG}$
Force(F_*)	$8.07760871 \times 10^{33}$	N	$kg\mathrm{m}s^{-2}$	c^4
Electric Field(ξ_*)	$5.43331879 \times 10^{54}$	$\tilde{V}\mathrm{m}^{-1}$	$\sqrt{kg\mathrm{m}}\mathrm{m}^{-2}s^{-2}$	$c^4\sqrt{c/h}$
Viscosity (η_*)	$5.43331879 \times 10^{54}$	$P_a s$	$kg\mathrm{m}^{-1}s^{-1}$	$c^4\sqrt{c/h}$
Mass Density(ρ_*)	$3.65466491 \times 10^{75}$	$kg\mathrm{m}^{-3}$	$kg\mathrm{m}^{-3}$	c^5/h
Current Density(J_*)	$3.65466491 \times 10^{75}$	$\tilde{A}\mathrm{m}^{-2}$	$\sqrt{kg\mathrm{m}}\mathrm{m}^{-2}s^{-1}$	c^5/h
Power(P_*)	$2.42160617 \times 10^{42}$	Js^{-1}	$kg\mathrm{m}^2 s^{-3}$	c^5
Pressure(p_*)	$3.28464901 \times 10^{92}$	$N\mathrm{m}^{-2}$	$kg\mathrm{m}^{-1}s^{-2}$	c^7/h
Energy Density(ϵ_*)	$3.28464901 \times 10^{92}$	$J\mathrm{m}^{-3}$	$kg\mathrm{m}^{-1}s^{-2}$	c^7/h
Charge(q_*)	$1.48668043 \times 10^{-21}$	$\tilde{C}$	$\sqrt{kg\mathrm{m}}$	$\sqrt{h/c}$
Conductance(ς_*)	$3.33564095 \times 10^{-9}$	$\widetilde{\Omega}^{-1}$	$\mathrm{m}^{-1}s$	c^{-1}
Moment$(m_* r_*)$	$2.21021870 \times 10^{-42}$	$kg\mathrm{m}$	$kg\mathrm{m}$	h/c
Distance(L_*)	$4.95903212 \times 10^{-30}$	m	m	$c^{-1}\sqrt{h/c}$
Inductance$(\mathcal{L}_*)$	$4.95903212 \times 10^{-30}$	$\widetilde{H}$	$\sqrt{kg\mathrm{m}}\mathrm{m}^{-1}s^{-1}$	$c^{-1}\sqrt{h/c}$
Permittivity(ε_*)	$1.36193501 \times 10^{-12}$	$F^{\#}\mathrm{m}^{-1}$	$\mathrm{m}^{-2}s^2$	$c^{-2}/\sqrt{G}$
Time(T_*)	$1.65415506 \times 10^{-38}$	s	s	$c^{-2}\sqrt{h/c}$
Area(A_*)	$2.45919996 \times 10^{-59}$	m^2	m^2	h/c^3
Volume(V_*)	$1.21952516 \times 10^{-88}$	m^3	m^3	$h\sqrt{h/c}/c^4$

8. Why This is Not Numerology

What is important here is that the relationships between the parameters in both tables are easily displayed in terms of only h and c for the Q_* set in Table 1 and in terms of only h, c and d for the q_{e*} set in Table 2 (other than permeability, permittivity and H which have G content). So each parameter has a simple relationship to each other one. The actual NSI values of these parameters bear out these relationships numerically, but they are only a confirmation of what the fundamental constants already show.

9. R_k and K_j — Members of the q_{e*} Parameter Set Whose Values can be Measured Directly

Within the q_{e*} set in Table 2 are two parameters that deserve further consideration, R_k and K_j. The maximal value for Resistance R_{e*} is equal to the von Klitzing constant R_k,

$$R_{e*} = R_k(DAPU) \quad (5)$$

and the value of the Magnetic Flux ϕ_{e*} is equal to twice the inverse of the Josephson constant K_j,

$$\phi_{e*} = (2/K_j)(DAPU) \quad (6)$$

Table 3 shows that the NSI values of R_k and K_j are identical to R_{e*} and $2/\phi_{e*}$ when translated into DAPU units by multiplying by the factor 1×10^{-7} for R_j ($2.58128076 \times 10^4\Omega$(SI)) and $\sqrt{1 \times 10^{-7}}$ for K_j ($4.835870 \times 10^{14} HV^{-1}$(SI)). To ensure clarity, new parameters will be defined

$$R''_k = R_{e*} = R_k(DAPU) \quad (7)$$

and

$$K''_j = 2/\phi_{e*} = K_j(DAPU) \quad (8)$$

where R''_k and K''_j denote the DAPU interpretations of R_k and K_j which can be more easily compared with other constants in DAPU units. This will be explored further later.

10. Values of the q_{e*} Set of Parameters

In DAPU the value of each parameter in Table 1 is one. To arrive at the maximal real values that can be found experimentally, the list needs to be adjusted to use q_e instead of Q_* since we do not observe Q_* charges usually. The maximal values in NSI units of some parameters under this limitation are listed in Table 2. Note that the factor d is inversely proportional to the dimensionality of every parameter.

Table 2. Values of the q_{e*} Set of Parameters

Parameter (X_{e*})	q_e DAPU set's NSI Value	NSI Units	DAPU equivalent	As Constants
Permeability(u_{e*})	$\sqrt{6.67428 \times 10^{-11}}$	NA^{-2}	$none$	$\sqrt{G}$
Angular Momentum(h)	$6.62606896 \times 10^{-34}$	Js	$kg\mathrm{m}^2 s^{-1}$	h
Boltzmann (k_B)	1	$J\breve{K}^{-1}$	$none$	$none$
Mass(m_{e*})	$1.30781284 \times 10^{-11}$	kg	kg	$d^{-1}\sqrt{hc}$
Magnetic Flux(ϕ_{e*})	$1.30781284 \times 10^{-11}$	$\widetilde{W}$	$\sqrt{kg\mathrm{m}}\mathrm{m}s^{-1}$	$d^{-1}\sqrt{hc}$
Charge-mass($q_{e*}c$)	$1.30781284 \times 10^{-11}$	$\tilde{C}\mathrm{m}s^{-1}$	$\sqrt{kg\mathrm{m}}\mathrm{m}s^{-1}$	$d^{-1}\sqrt{hc}$
Velocity(v_{e*})	$2.58128076 \times 10^{11}$	$\mathrm{m}s^{-1}$	$\mathrm{m}s^{-1}$	$d^{-2}c$
Resistance(R_{e*})	$2.58128076 \times 10^{11}$	$\widetilde{\Omega}$	$\mathrm{m}s^{-1}$	$d^{-2}c$
Momentum($m_{e*}v_{e*}$)	$3.37583212 \times 10^{00}$	$kg\mathrm{m}s^{-1}$	$kg\mathrm{m}s^{-1}$	$d^{-3}c\sqrt{hc}$
Current(ι_{e*})	$6.66301034 \times 10^{22}$	$\tilde{A}$	$\sqrt{kg\mathrm{m}}s^{-1}$	$d^{-4}c^2$
Action(m_{e*}/r_{e*})	$6.66301034 \times 10^{22}$	$kg\mathrm{m}^{-1}$	$kg\mathrm{m}^{-1}$	$d^{-4}c^2$
Angular Frequency(w_{e*})	$1.31510410 \times 10^{45}$	Hz	s^{-1}	$d^{-5}c^2\sqrt{c/h}$
Frequency(f_{e*})	$1.31510410 \times 10^{45}$	Hz	s^{-1}	$d^{-5}c^2\sqrt{c/h}$
Energy(E_{e*})	$8.71397049 \times 10^{11}$	J	$kg\mathrm{m}^2 s^{-2}$	$d^{-5}c^2\sqrt{hc}$
Temperature (K_{e*})	$8.71397049 \times 10^{11}$	$\breve{K}$	$\breve{K}$	$d^{-5}c^2\sqrt{hc}$
Potential Difference(V_{e*})	$1.71991004 \times 10^{34}$	$\tilde{V}$	$\sqrt{kg\mathrm{m}}\mathrm{m}s^{-2}$	$d^{-6}c^3$
Acceleration(a_{e*})	$3.39465292 \times 10^{56}$	$\mathrm{m}s^{-2}$	$\mathrm{m}s^{-2}$	$d^{-7}c^3\sqrt{c/h}$
Magnetic Inductance(B_{e*})	$3.39465292 \times 10^{56}$	$\tilde{A}\mathrm{m}^{-1}$	$\mathrm{m}s^{-2}$	$d^{-7}c^3\sqrt{c/h}$
Magnetic Field (H_{e*})	$4.15521180 \times 10^{61}$	$\tilde{A}\mathrm{m}^{-1}$	$\mathrm{m}s^{-2}$	$d^{-7}c^3\sqrt{c/hG}$
Force(F_{e*})	$4.43957068 \times 10^{45}$	N	$kg\mathrm{m}s^{-2}$	$d^{-8}c^4$
Electric Field(ξ_{e*})	$8.76255225 \times 10^{67}$	$V\mathrm{m}^{-1}$	$\sqrt{kg\mathrm{m}}\mathrm{m}^{-2}s^{-2}$	$d^{-9}c^4\sqrt{c/h}$
Viscosity (η_{e*})	$8.76255225 \times 10^{67}$	$P_a s$	$kg\mathrm{m}^{-1}s^{-1}$	$d^{-9}c^4\sqrt{c/h}$
Mass Density(ρ_{e*})	$1.72949881 \times 10^{90}$	$kg\mathrm{m}^{-3}$	$kg\mathrm{m}^{-3}$	$d^{-10}c^5/h$
Current Density(J_{e*})	$1.72949881 \times 10^{90}$	$\tilde{A}\mathrm{m}^{-2}$	$\sqrt{kg\mathrm{m}}\mathrm{m}^{-2}s^{-1}$	$d^{-10}c^5/h$
Power(P_{e*})	$1.14597784 \times 10^{57}$	Js^{-1}	$kg\mathrm{m}^2 s^{-3}$	$d^{-10}c^5$
Pressure(p_{e*})	$1.15236684 \times 10^{113}$	$N\mathrm{m}^{-2}$	$kg\mathrm{m}^{-1}s^{-2}$	$d^{-14}c^7/h$
Energy Density(ϵ_{e*})	$1.15236684 \times 10^{113}$	$J\mathrm{m}^{-3}$	$kg\mathrm{m}^{-1}s^{-2}$	$d^{-14}c^7/h$
Charge(q_{e*})	$5.06652691 \times 10^{-23}$	$\tilde{C}$	$\sqrt{kg\mathrm{m}}$	$d\sqrt{h/c}$
Conductance(ς_{e*})	$3.87404585 \times 10^{-12}$	$\widetilde{\Omega}^{-1}$	$\mathrm{m}^{-1}s$	d^2c^{-1}
Moment($m_{e*}r_{e*}$)	$2.56696950 \times 10^{-45}$	$kg\mathrm{m}$	$kg\mathrm{m}$	d^2h/c
Distance(L_{e*})	$1.96279576 \times 10^{-34}$	m	m	$d^3c^{-1}\sqrt{h/c}$
Inductance($\mathcal{L}_{e*}$)	$1.96279576 \times 10^{-34}$	$\widetilde{H}$	$\sqrt{kg\mathrm{m}}\mathrm{m}^{-1}s^{-1}$	$d^3c^{-1}\sqrt{h/c}$
Permittivity(ε_{e*})	$1.83707675 \times 10^{-18}$	$F^{\#}\mathrm{m}^{-1}$	$\mathrm{m}^{-2}s^2$	$d^4c^{-2}/\sqrt{G}$
Time(T_{e*})	$7.60396075 \times 10^{-46}$	s	s	$d^5c^{-2}\sqrt{h/c}$
Area(A_{e*})	$3.85256718 \times 10^{-68}$	m^2	m^2	d^6h/c^3
Volume(V_{e*})	$7.56180251 \times 10^{-102}$	m^3	m^3	$d^9h\sqrt{h/c}/c^4$
Capacitance(C_{e*})	$2.94580926 \times 10^{-57}$	$F^{\#}$	$\mathrm{m}^{-1}s^2$	$d^7c^{-3}\sqrt{h/c}$

11. Parameters, Physical Constants, and Laws of Nature

Some of the q_{e*} set of parameters, such as velocity v_{e*}, appear to be larger than their Q_* set versions. As will be shown below, it is possible to interpret R_k as equivalent to a velocity and, if so, this suggests either that faster than light travel by physical objects, or patterns produced by subluminal physical objects, through media may be a possibility or that in order to pass through such maximal media an unachievable speed faster than light is required. Which is the case should be investigated. Parameters where $X_{e*} > X_*$, will be considered further below.

All the parameters above have been produced using standard relationships and formulae. It is interesting to observe that some parameters on the mechanical side have identical size and dimension partners on the electromagnetic side, for example mass M_* and magnetic flux ϕ_*. One interpretation could be that magnetic flux is the equivalent of the mass in an electromagnetic system, and that resistance R_{e*} is the equivalent of velocity, v_{e*}. Dimensional analysis supports this and the appropriateness of this interpretation will be considered later.

To ensure that the above values can be understood properly, the following series of relationships at the Q_* level can be culled from the standard laws and the results confirmed to be correct using their NSI values in Table 1 as:

$$\begin{aligned} F_* = (M_*/L_*)^2 &= (\phi_*/L_*)^2 = (Q_*c/L_*)^2 \\ &= M_*a_* = \phi_*B_* = Q_*cB_* \\ &= Q_*\xi_* = V_*c = i_*^2 \\ &= hc/L_*^2 \end{aligned} \tag{9}$$

It is also possible to use the same relationships at the q_{e*} level, using the parameter values from Table 2 thus:

$$\begin{aligned} F_{e*} = (M_{e*}/L_{e*})^2 &= (\phi_{e*}/L_{e*})^2 \\ &= (Q_{e*}c/L_{e*})^2 = M_{e*}a_{e*} \\ &= \phi_{e*}B_{e*} = Q_{e*}cB_{e*} \\ &= Q_{e*}\xi_{e*} = V_{e*}c = i_{e*}^2 \\ &= hc/L_{e*}^2 \end{aligned} \tag{10}$$

Since the values of some electromagnetic parameters are identical to the values of some mechanical parameters, it suggests that mechanical formulae could be used with electromagnetic parameters substituted instead, and vice versa. One example would be the simple $L_{e*} = v_{e*}T = \mathcal{L}_{e*}$ which suggests that in some way electromagnetic inductance is equivalent to a mechanical distance. Were this done in SI units, the mix of mechanical and electromagnetic parameters would not show that the parameters were interchangeable because of the misalignment of those two types of parameter in the SI units system.

The tables show that most electromagnetic parameters can be reinterpreted in terms of mechanical parameters. It requires a complete reinterpretation of what is understood by the terms magnetic inductance (acceleration), magnetic flux (mass), inductance (distance), current density (mass density) and other electromagnetic parameters.

12. Describing All Parameters Using Only Two From the Set of Parameters

Now it is possible to reinterpret the only two fundamental constants left, aside from the factor d which defines the electron charge-based system that we experience because of the relative size of the charge on the electron q_{e*} versus the DAPU Planck charge Q_*, in term of the two base parameters which have only dimension $Y^{\pm 1}$ which are charge Q_* and mass M_*.

$$M_*Q_* = h \tag{11}$$

$$M_*/Q_* = c \tag{12}$$

So the two constants h and c represent the only two possible ratios of the DAPU mass and DAPU charge, each used once. That ought to infer something fundamental about any hypothetical underlying structure of matter.

Also important is that the same reinterpretation can be done for h and c using R_k and K_j. However, for consistency, the DAPU constants $R"_k$ and $K"_j$ will be used, but the same relationships remain.

$$R"_k(K"_j/2)^2 = h^{-1} \tag{13}$$

$$d^2R"_k = c \tag{14}$$

The comparison with M_* and Q_* is not identical. $K"_j$ is an inverse magnetic flux and so equivalent to a mass of size $2/K"_j$. However, $R"_k$ is not a charge, but is a resistance or velocity. But by using the same relationship between equivalent mass and charge, with the factor of two, the value of h can be recovered as:

- Equivalent mass x electron charge $= (2/K"_j)\ q_{e*} = h$

And the same can be achieved with the ratio of equivalent mass to charge to recover the velocity c/d^2 or $R"_k$ as:

- Equivalent mass / electron charge $= (2/K"_j)/q_{e*} = h/q_{e*}^2 = Q_*^2c/[d^2Q_*^2] = c/d^2 = R"_k = 2\pi c/\alpha$

This shows that if inverse magnetic flux can be considered as equivalent to a mass, then resistance can be considered as equivalent to a velocity.

13. Parameters as Ratios of R''_k and $K''_j/2$

It is possible to generate examples of the usual constants of nature or DAPU parameters, other than G which was subsumed into M_* and L_*, using just R''_k and $K''_j/2$ (or with powers of d included for the Q_* set) as follows:

$M_* = d/(K''_j/2)$ $\quad Q_* = 1/[R''_k(K''_j/2)d]$

$L_* = 1/[R''^2_k(K''_j/2)d^3]$ $\quad c = R''_k d^2$

$M_{e*} = 1/(K''_j/2) = \phi_{e*}$ $\quad \iota_{e*} = R''^2_k$

$q_{e*} = 1/[R''_k(K''_j/2)]$ $\quad V_{e*} = R''^3_k$

$L_{e*} = 1/[R''^2_k(K''_j/2)]$ $\quad h = 1/[R''_k(K''_j/2)^2]$

$T_* = 1/[R''^3_k(K''_j/2)d^5]$ $\quad E_* = R''^2_k d^5/(K''_j/2)$

These relationships can be checked by using the following standard law formulae, in either Q_{e*} or q_{e*} form, and the DAPU values of the parameters in Table 1 or 2:

$V_{e*} = \iota_{e*} \times R''_k$ or $B_{e*} = \phi_{e*}/A_{e*}$ or $E_{e*} = m_{e*} \times v^2_{e*}$ or $F_{e*} = p_{e*} \times A_{e*}$ or $q_{e*} = \iota_{e*} \times T_{e*}$ or $E_{e*} \times T_{e*} = h$

Many parameters have been left out of the list for brevity, including those based on materials which require the permeability factor u_* and would mean the inclusion of the ratio $\sqrt{G}$ in the formula of constants producing those parameters. Of particular interest is the value of $\iota_{e*} = R''^2_k$ which suggests that the SI unit of Ampere could be defined using the DAPU value of R''^2_k as its sole reference point.

It may be possible to improve the accuracy of measurement of some of the constants by using the new relationships uncovered between R''_k and K''_j. It is not only h that can be made more precisely from ratios of R_k and K_j. There are many more composites of R''_k and K''_j that produce other parameters which may not have been measured to as great an accuracy as R_k and K_j have been.

14. Translating Between SI and APU/DAPU Units

Table 3 shows the relative factors required to translate between DAPU/APU/SI units. The SI values should be multiplied by the factors in the appropriate column to produce the DAPU or APU values of that parameter.

Table 3. How to Translate Between SI and APU/DAPU Units

Parameter NSI	Parameter	DAPU value	DAPU factor X_*	APU factor X_o	SI value for Planck unit	SI Name
$6.62606896 \times 10^{-34}$	h	h	2π	2π	$1.0545716 \times 10^{-34}$	$\hbar$
$4.45695580 \times 10^{-13}$	M_*	$\sqrt{hc}$	$\sqrt{2\pi G}$	$\sqrt{2\pi}$	2.1764374×10^{-8}	M_{Planck}
$1.48668043 \times 10^{-21}$	Q_*	$\sqrt{h/c}$	$\sqrt{1x10^{-7}}$	$\sqrt{1x10^{-7}}$	$4.7012963 \times 10^{-18}$	Q_{Planck}
$5.06652691 \times 10^{-23}$	q_{e*}	$\sqrt{\alpha/2\pi}\sqrt{h/c}$	$\sqrt{1x10^{-7}}$	$\sqrt{1x10^{-7}}$	$1.6021765 \times 10^{-19}$	e
$4.95903212 \times 10^{-30}$	L_*	$\sqrt{h/c^3}$	$\sqrt{2\pi/G}$	$\sqrt{2\pi}$	$1.6162525 \times 10^{-35}$	L_{Planck}
none	G	*none*	*none*	1	6.67428×10^{-11}	G
2.99792458×10^{8}	c	c	1	1	2.99792458×10^{8}	c
$2.58128076 \times 10^{11}$	R_{e*}	$2\pi c/\alpha$	$1x10^{+7}$	$1x10^{+7}$	2.58128076×10^{4}	R_k
$1.52927081 \times 10^{11}$	$2/\phi_{e*}$	$2\sqrt{\alpha/(2\pi hc)}$	$\sqrt{1x10^{-7}}$	$\sqrt{1x10^{-7}}$	$4.83597891 \times 10^{14}$	K_j

15. Faster Than Light Speed?

The parameters in Table 2, based on q_{e*}, that have $X_{e*} > X_*$, have sizes greater than their Q_* DAPU set values in Table 1. This leads to parameters like $v_{e*} = c/d^2 = 2\pi c/\alpha$ which is greater than light speed. It is the d factor, the ratio q_{e*}/Q_*, that alters the parameter values in Table 2. Where the parameter has d^{+x} the parameter will be smaller than its Planck parent and where the parameter has d^{-x} it will be larger - the whole q_{e*} parameter space has been stretched out of symmetry when compared with the Q_* parameter space topology, even though the same laws and relationships still apply.

For all physical objects at the maximal values for each q_{e*} parameter possessed, the actual value of d is immaterial. Such objects obey the same laws regardless of the relative size of the electron charge q_{e*} to DAPU charge Q_*. It is only at the lower levels, below maximal values, that the ratio of d to, for example, the masses of the particles will produce varying sizes of physical effect, such as differing electron energy levels in atoms, dependent on the mass of the electron and its orbital velocity.

Whether the maximal values $X_{e*} > X_*$ can actually be attained is a question for experimental verification or rebuttal. That R_k and K_J have been measured to be the sizes that they are [10] makes it certain that some of them can, since $R_k = c/d^2 = 2\pi c/\alpha$. So although equivalent to a velocity, it is not clear if $R_k > c$ means that q_{e*} based physical objects can exceed c in velocity. The interpretation preferred here is that the factor v/d^2 represents a limitation on the minimum velocity required for electrons to pass across the media.

16. Simplifying Expressions

An example of the use of simplification enabled by the use of DAPU units would be to compare the standard expression for the principal energy levels of the one-electron atom with the same in DAPU units, where each observed parameter is displayed as a fraction of its maximal DAPU value. It is not suggested that this is the only way to arrive at the simplification, but it shows how thinking in DAPU units enabled simplicity to emerge.

The simplest example of a standard equation for the allowed energy levels, to use as an example, is the Bohr SI equation which provides that

$$E_{nmass} = -1/n^2(k^2me^4/(2\hbar^2)) = -R/n^2 \quad (15)$$

where R is the Rydberg constant and the other parameters are as usually described. The factor k has the value $k = 8.988 \times 10^9 Nm^2C^{-2}$ (SI) and can be converted into DAPU units by dividing by $(\sqrt{1 \times 10^{-7}})^2 = 1 \times 10^{-7}$ to produce the value $k = 8.988 \times 10^{16}(DAPU) = c^2$. The formula can be manipulated using

$$F = e^2c^2/r_e^2 = mv_e^2/r_e = \hbar v_e/r_e^2 \quad (16)$$

so that in DAPU terms this becomes a simpler to understand kinetic energy-like ratio

$$E_{nmass}/E_* = -0.5mv_e^2/n^2 \quad (17)$$

Whilst the DAPU equation is admittedly not so easy to measure, it does bring out that what is being measured is simply different variations in the kinetic energy of the electron. Had the Bohr example started with the potential energy separated out, then so would the DAPU presentation.

Another simplification providing greater clarity might be based on the expression for the magnetic moment of an orbiting electron, usually described in SI units as

$$\mu_e = eh/2m \quad (18)$$

but which can be recast using the same equations as previously to become

$$n\mu_e/\mu_{e*} = 0.5ev_er_e \quad (19)$$

so that

$$E_{ncharge}/E_{e*} = \mu_e c\omega_e/n^2 = 0.5[ec]v_e^2/n^2 \quad (20)$$

which implies that in the same way that moving mass occurs in units of orbital angular momentum in $n\hbar = mv_er_e$, so does charge although in units of orbital magnetic momentum as $n\mu_e = 0.5ev_er_e$ in DAPU, using the same form as the $\hbar$ angular momentum equation. And that the form of the equation for energy of motion of charge is comparable with that for the kinetic energy of the mass, but with ec replacing the mass m in the kinetic energy equation.

17. Equivalence of Electromagnetic and Mechanical Parameters in Experiments

The new law of nature mentioned earlier, producing Planck's constant h as the product of DAPU volume V_* and viscosity, η_* together with the equivalence in DAPU units of viscosity η_* and electric field ξ_*, provide two interesting possibilities, one already experimentally hinted at.

Firstly, that any fundamental physical framework based on a single particle volume size would have constant viscosity acting on the motion of every such particle. This would be equivalent to the action of air resistance on a skydiver, providing a terminal velocity.

The same type of action on such fundamental particles could be the underlying reason for the terminal velocity that we describe as light speed, the irreversible arrow of time and could also provide an additional redshift factor to the passage of photons almost completely directly related to their distance travelled, reducing the size and expansion rate of the universe.

Secondly, and having potential experimental justification, is that viscosity η_* and electric field ξ_* could be the same parameter in different disguises. A recent paper[11] mentioned that the 'stickiness' of spiders' silk could be turned on and off through the application of an electric field. If such stickiness and viscosity are related, then this would show directly how viscosity is related to electric field and vice versa. This effect would not the same as the creation of magnetorheological fluids [12] with dual fluids, but would be describing a deeper level of equivalence.

18. Conclusions

This paper presents new ways of understanding the relationships between parameters. This is a toolkit for providing a better understanding of the fundamentals of physics. The novel insights and predictions include:

i. A self-contained and consistent new Planck unit set of maximal Q_* based parameters from which all observed values can be compared and easily combined in equations.
ii. A self-contained and consistent new Planck unit set of electron charge-size based parameters, some of which are directly observable in experiments.
iii. The interpretation of the gravitational constant G as a dimensionless ratio and its relegation from gravitational to permeability use, so that it represents a measure of the strength of interactions within materials not between masses, and the elimination of the need to test the equivalence of gravitational and inertial masses.
iv. That all parameters can be displayed in terms of only h and c for the Q_* parameter set and in terms of only h, c and d for the q_{e*} set (other than permeability, permittivity and H which have G content), which was previously considered impossible.
v. There exists a new hypothetical dimensionality analysis that can be used to describe parameter dimensions and to uncover any law of nature or any universal constants. All that is required to produce a law of nature is to create an equation where the dimensionalities are equal on both sides. To produce a constant of nature, aside from c, the minimum that is required is that it has Y^0 dimensionality.
vi. That most Q_* and q_{e*} parameters can be described solely in terms of ratios of the R_k and K_j (and d for the Q_* set) and so will benefit from the precision of measurement of these two parameters.
vii. That the experimentally observed value of R_k implies either that the velocity of a current within certain electromagnetic materials could be in excess of light speed, the patterns produced by subluminal physical objects could have a maximum velocity of c/d^2 or that such a minimum velocity is required in order to pass through those material. This is open to further experimental work to confirm which is the case.
viii. That most electromagnetic parameters can be reinterpreted in terms of mechanical parameters. It requires a complete reinterpretation of what is understood by the terms magnetic inductance (acceleration), magnetic flux (mass), inductance (distance), current density (mass density) and other electromagnetic parameters. One possible experimental verification exists in equating viscosity and electric field.
ix. That the reinterpretation of R_k and $K_j/2$ with their current excellent precision of measurement, should enable increased accuracy in the estimation of the values of other parameters and fundamental constants identified as novel composite functions of $R"_k$ and $K"_j/2$.
x. A universal method of discovering laws of nature that applies regardless of any stretching of parameter space. A unit with $q_e/Q_* \neq \sqrt{\alpha/2\pi}$ would still have the same relationships between parameters although the numerical values of the results would be different.
xi. Physics can be better understood when stripped to its bare essentials and without the use of a system of SI units that are currently misaligned across the electromagnetic and mechanical parameters. By adjusting SI units to be self-consistent and consistent with DAPU units, greater clarity will ensue.

The two adjustments necessary to align and make SI units self-consistent and also consistent with the simplicity of DAPU units have been proposed, producing a system of New SI units.

References

[1] von Klitzing, K.; Dorda, G.; Pepper, M. (1980) New method for high-accuracy determination of the fine-structure constant based on quantized Hall resistance Physical Review Letters, vol.45, no.6, p. 494-7.
[2] Josephson, B. D. (1974) The discovery of tunnelling supercurrents. Rev. Mod. Phys.; 46(2): 251-254.
[3] Mohr, P. J.; Taylor, B. N.; Newell, D. B. (2008) CODATA Recommended Values of the Fundamental Physical Constants: Rev. Mod. Phys. 80: 633–730. doi:10.1103/RevModPhys.80.633. http://physics.nist.gov/cuu/Constants/codata.pdf.
[4] Duff, M.J. Okun, L.B. & Veneziano, G. (2002) Comments: Trialogue on the number of fundamental constants JHEP 0203; 023 arXiv:physics/0110060
[5] Mohr, P. (2010) Recent progress in fundamental constants and the International System of Units 3rd workshop on Precision Physics and Fundamental Physical Constants.
[6] Duff, M.J. Okun, L.B. & Veneziano, G. (2002) Comments: Trialogue on the number of fundamental

constants JHEP 0203 (2002) 023 arXiv:physics/0110060

[7] Fiorentini, G.(1), Okun, L.B.(2) & Vysotsky, M. (1,2) (2006) INFN, *Sezione di Ferrara*, Ferrara, Italy, (2) ITEP, Moscow, Russia) Is G a conversion factor or a fundamental unit? arXiv:physics/0112006.

[8] L. B. Okun, L.B. (2003) Comments: Fundamental units: physics and metrology Talk at the Seminar Astrophysics, Clocks and Fundamental Constants (302. WE-Heraues-Seminar, Bad Honnef, Germany; arXiv:physics/0310069

[9] Fiorentini, G.(1), Okun, L.B.(2) & Vysotsky, M. (1,2) (2006) INFN, *Sezione di Ferrara*, Ferrara, Italy, (2) ITEP, Moscow, Russia) Is G a conversion factor or a fundamental unit? arXiv:physics/0112006.

[10] Mohr, P. J.; Taylor, B. N.; Newell, D. B. (2008) CODATA Recommended Values of the Fundamental Physical Constants: Rev. Mod. Phys. 80: 633–730. doi:10.1103/RevModPhys.80.633; http://physics.nist.gov/cuu/Constants/codata.pdf.

[11] Tao, H., Brenckle, M.A., Yang, M. Zhang, J., Liu, M., Siebert, S.M., Averitt, R.D., Mannoor, M.S., McAlpine, M.C., Rogers, J.A., Kaplan, D.L. & Omenetto, F.G. (2012) Silk-based conformal, adhesive, edible food sensors, in Advanced Materials, Volume 24, Issue 8, pages 1067–1072.

[12] Andrade, E. N., Da, C. & Dodd, C., (2002) The effect of an electric field on the viscosity of liquids, in Proceedings of the Royal Society of London. Series A, Mathematical and Physical Sciences, Volume 187, Issue 1010, pp. 296-337.

A Programmable Cellular-Automata Polarized Dirac Vacuum

DRAHCIR S. OSOROMA

Noetic Advanced Studies Institute,
Escalante Desert, Beryl, UT84714-5104 USA
Osoroma@noeticadvancedstudies.us, www.noeticadvancedstudies.us

We explore properties of a 'Least Cosmological Unit' (LCU) as an inherent spacetime raster tiling or tessellating the unique backcloth of Holographic Anthropic Multiverse (HAM) cosmology as an array of programmable cellular automata. The HAM vacuum is a scale-invariant HD extension of a covariant polarized Dirac vacuum with 'bumps' and 'holes' typically described by extended electromagnetic theory corresponding to an Einstein energy-dependent spacetime metric admitting a periodic photon mass. The new cosmology incorporates a unique form of M-Theoretic Calabi-Yau-Poincaré Dodecadedral-AdS_5-DS_5space (PDS) with mirror symmetry best described by an HD extension of Cramer's Transactional Interpretation when integrated also with an HD extension of the de Broglie-Bohm-Vigier causal interpretation of quantum theory. We incorporate a unique form of large-scale additional dimensionality (LSXD) bearing some similarity to that conceived by Randall and Sundrum; and extend the fundamental basis of our model to the Unified Field, U_F. A Sagnac Effect rf-pulsed incursive resonance hierarchy is utilized to manipulate and ballistically program the geometric-topological properties of this putative LSXD space-spacetime network. The model is empirically testable; and it is proposed that a variety of new technologies will arise from ballistic programming of tessellated LCU vacuum cellular automata.

Keywords: Ballistic programming, Cellular automata, Dirac vacuum, Quantum computing, Quantum theory, Spacetime, Unified field

1. Introduction — A Unique String (M-Theoretic) Nilpotent Vacuum

We postulate that space is tiled or tessellated with a space-antispace nilpotent array (The two spaces effectively cancel to produce a norm or square root of 0 nilpotent object) [1,2] of 12D 'least cosmological units' (LCU) [3,4] (like unit cells building up crystal lattice structure) forming a continuous-state spacetime raster [4]. The localized surface of this raster is a programmable Dirac vacuum cellular automata array [5] with Large Scale Higher Dimensions (LSXD) consistent with mirror symmetric Calabi-Yau brane topology [6,7]. The LSXD topology appears to have holographic properties the fundamental curvature of which fluctuates between 0, ± 1 in contrast to Big Bang cosmological topology which suggests an original fixed curvature. This pertains to the observed Hubble Universe, H_R which is closed and finite in time (but open and infinite in the HAM LSXD multiverse). This means that observationally our 3D Euclidean virtual reality is a flat Euclidean torus [8] during the zero curvature moment of fluctuation. But the LSXD oscillating curvature fluctuates from hyperbolic < 1 in one extrema to Poincaré Dodecahedral > 1 on the other. See [9] this volume for details.

We also note that some aspects of the original Hadronic form of string theory are better suited to HAM cosmology such as variable string tension, T_s, no superpartners and the admission of tachyons [10] especially what we see as an essential part of the continuous-state backcloth for ballistic computing. Pioneering Hadronic string work of Veneziano discovered that for the scattering amplitude to describe the scattering of a particle that appears in the theory, an obvious self-consistency condition, the lightest particle must be a tachyon [11].

The vacuum is thought to contain infinite energy but none of it until now has been considered accessible beyond virtual quanta for a duration of the Planck time. It seems if this were not so the fabric of the cosmos and our perception of reality would unravel; so that we may not really 'dip a ladle in' and draw any soup out. How then may we utilize or engineer vacuum energy for various purposes? The 'vacuum' is everything and nothing (nilpotent), everywhere and nowhere; it has infinite potentia. In the midst of this lies our virtual Euclidean/Minkowski 3D/4D holographic regime of spacetime. Relative to us as observers the surface geometric topology of spacetime is stochastic quantum foam of virtual quanta called the zero point field (ZPF) said to be governed by the Copenhagen uncertainty principle. To get at the vacuum we must reach beyond or pass through this virtual surface barrier or domain wall of stochastic uncertainty.

Typically a string theory originates in a Minkowski space of 9 + 1 Dimensions, M_{9+1}, then replacing M_{9+1} by M_{3+1} x M where the new M is a 6D complex Calabi-Yau manifold - because the string theory program is essentially the study of compactification on Calabi-Yau spaces. Calabi-Yau manifolds are complex HD analogues of K3 surfaces which are smooth regular minimal surfaces often defined as compact Kähler manifolds which are three mutually compatible structures - a complex, a Riemannian, and a symplectic structure [6].

The XD of stringy spacetime has been conjectured to take the form of a 6D Calabi-Yau manifold, which initially led to the idea of mirror symmetry because Calabi-Yau 3-folds tend to come in mirror pairs, M,W [6]. Essentially, Calabi-Yau manifolds are topologies satisfying the requirement of string theoretic hyperspace with six "unseen" spatial dimensions. Most string theorists believe the XD are Planck scale, but in our model we utilize a LSXD M-Theoretic brane-world model of 11D [4] to which we add an additional XD to embed action of the U_F for a total of 12D. The LSXD are still 'unseen' in HAM cosmology, but not because they are curled up at the Planck scale. They are unseen because of a form of 'subtractive interfere-ometry' as part of the nature of reality itself. See [12] this volume. However in HAM cosmology the 6D Calabi-Yau mirror symmetric manifold is only the intermediary between Euclidean space and the upper limit Poincaré oscillating Dodecadedral-AdS_5-DS_5 involute space [8,9].

2. Ballistic Cellular-Automata Programming

Recently Margolus claimed: "*The maximum computational density allowed by the laws of physics is available only in a format that mimics the basic spatial locality of physical law*" [5]. We suggest this 'law' is a limit of the localized Copenhagen vacuum and propose a putative virtually unlimited basis for ballistic programming for the U_F continuous-state properties of HAM cosmology [4,9]. As yet scalable quantum computing has been theorized but not realized. We have come to the conclusion that quantum computing in terms of the usual consideration of the qubit is not feasible [9]. At the time of writing all QC developers have been utilizing a qubit as a 2-level quantum system written as a superposition of the basis vectors represented geometrically on a Bloch 2-sphere, $\Psi^2 = |0\rangle + |1\rangle = \langle \psi^* | \psi \rangle = 1$. This is very nice for algebraically keeping track of quantum states but useless in reality for programming or performing bulk QC! Technically one could say there will be no such thing as 'quantum computing' because scalable bulk universal QC will require U_F Mechanics; but since U_F computers will use quantum states the term QC will stick. Programming, operating and data readout of a QC requires U_F Mechanics because it is necessary to surmount the quantum uncertainty principle which is a trivial process for U_F mechanics [4,13] but invisible to quantum mechanical techniques. In summary this occurs not only because of the unique properties of both the HAM LCU array and the associated LSXD regime interfacing the U_F with quantum and classical mechanics able to surmount the quantum uncertainty principle [4], but mainly because of the unfettered utility the coherence of the U_F supplies to the nonlocal ballistic processing of the mean-free path (MFP) in terms of this inherent U_F coherence's governance of the continuous-state LCU periodic gating mechanism.

MFP ballistic transport (Fig. 5) or ballistic programming as we will call it here has been achieved to a degree in superconducting semiconductors and carbon nanotubes like graphene [14-21]. Spacetime cellular automata programming has similarities to the Meissner effect in that there are transition effects and regimes. The current understanding of quantum theory is illustrated in Tbl. 1 and the additions introduced by HAM U_F theory in Tbl. 2.

TABLE 1 - USUAL LAYERS OF VIRTUAL REALITY

1. 3D Classical - Observer Macroscopy
2. 4D Quantum - Uncertainty
3. Planck Scale ZPF Stochastic Barrier
4. Heizenberg Potentia - Basement of Reality

TABLE 2 - HAM COSMOLOGY U_F ADDITIONS

1. Changes Basis of Potentia (4) in Table 1
A. Continuous-State Synchronization Backbone
B. Calabi-Yau Dual Mirror 3-Form - Midlevel 1
C. Triune Dodecahedral AdS_5 - DS_5 Involute - Level 2
2. LSXD 12D Sea of U_F Nilpotent Infinite Coherence

Ballistic vacuum cellular automata programming is not theoretically possible within the Copenhagen Interpretation of quantum theory because the uncertainty principle by definition blocks the complete utility of the MFP. Because of the available coherence inherent in the 'synchronization backbone' of the U_F, an 'infinite' ballistic MFP is accessible at certain periodic nodes of the continuous-state cycle. This property of the continuous-state process 'synchronization backbone' [22] allows putative vacuum program to be theoretically much simpler to achieve than otherwise would be expected by current physics. The salient reason is this property essentially provides a built in mechanism for surmounting quantum uncertainty [4] dramatically simplifying vacuum programming. It is like getting half the device for free.

The HAM cosmology model of ballistic vacuum programming is based essentially on:

- An LSXD extension of the covariant polarized Dirac vacuum.

- Extended form of De Broglie-Bohm piloted ontological quantum theory to LSXD.
- Extending Cramer's standing-wave future-past Transactional Interpretation to LSXD.
- LCU (Close-packed Calabi-Yau mirror symmetric 3-tori) cellular automata spacetime tiling/tessellation.
- It should be emphasized that the dual Calabi-Yau mirror symmetry is an intermediate Continuous-State topology. The highest or deepest level of the nilpotent structure is a triune Poincaré dodecahedral involute of AdS_5 - DS_5 topology with oscillating curvature [8,9] (facilitates HD Continuous-state gating).

In general a MFP will exhibit ballistic transport when the length, L of the active part of a device (or in our case vacuum LCU cellular automata array LSXD diagonals) is $L \leq \lambda_{MFP}$ which in this form is simply the scattering length for the carrier, i.e. electrons or as we hope to demonstrate ontological 'topological switching' of the Unified Field, U_F brane topology.

3. Overview of Theoretical Physical Parameters for the Cellular-Automata Vacuum

- Observed reality is like a virtual HD standing-wave of future-past advanced-retarded parameters (derived by extending Cramer's Transactional Interpretation of quantum theory).
- All current thinking confined to the limits of any/all standard models, i.e. particle physics, cosmology or quantum theory is insufficient and we therefore can safely emphasize that 'infinite' ballistic vacuum programming within these confines cannot be achieved.
- Most particularly regarding matter in the extended de Broglie-Bohm-Vigier causal stochastic interpretation of quantum theory, the wave function is physically real as are both 'wave and particle' which may exist simultaneously. The properties of the Dirac equation is extended from the original concept of matter to include both spacetime and domain walls of the reality of the observer, all of which are created-annihilated and recreated in a covariant continuous state scale-invariant process (wave-particle duality is elevated to a principle of cosmology).
- The Dirac Polarized vacuum (inherent bumps and holes) is a programmable 'ocean' of infinite nilpotent potentia, part of which we treat as a tessellated backcloth of 'close-packed' least cosmological units (LCU) with Riemann sphere zero to infinity rotational properties like a 'spin glass' cellular automata.
- In this general context the key to vacuum programming is simply to ballistically program the MFP of this LSXD spacetime array, not in the usually considered linear path but for all coordinates simultaneously in a minimum of 6 spatial dimensions (6D). 6D may not turn out to be adequate; the three temporal and three unitary (for quantum potential or piloting) may also need to be addressed. We 'guess' three may drop out and just a 9D spatial matrix will be required.
- Some form of Noetic Transformation (not given here) is required (the next set of transformations beyond the Galilean-Lorentz-Poincaré utilizing U_F Mechanics) in understanding the programming.

Like any principle in physics that at first seems incomprehensible, once it is understood it becomes simple. In this U_F Mechanics cosmology ballistic computing is all about the Continuous-State structure of the LCU array tessellating the backcloth. The quantum regime is not 'the basement of reality'. The quantum uncertainty principle / exclusion principle is like a 'set of handcuffs' limiting access to the 3rd regime of U_F Mechanics. Because of the nature of reality / time (see [12] this volume) access is periodic or cyclical; this is where the continuous-state dynamics enters in. The coordinates or metric fixing of Euclidean reality contains a 'leapfrog' cycle of coordinate fixing and unfixing that is also like a relay race where a baton is passed (coupling) at a certain phase in the cycle. This is inherent in the continuous-state Cramer-like standing-wave of 3D virtual reality. The fact that the Dirac covariant polarized vacuum is a form of Einstein energy-dependent spacetime metric, $\hat{M}_4$ is part of what allows it to be programmable (accessible/able to be manipulated).

Surmounting uncertainty is a 'gentle' approach combining the usual (making correspondence) phenomenology of the standard model of quantum theory with a new ontological energyless (no exchange quanta) methodology that by application of a coherently controlled incursive resonance hierarchy, timed to the correct phase nodes, enables access to the coherent elements of the U_F. You see the governance of reality (by a super quantum potential, i.e. the U_F) is in itself a form of ballistic quantum computing of spacetime or reality itself; our trick is merely to understand how to access it, and manipulate it in order to utilize it in new technologies [4,9].

This is key for resonant manipulation; not the usual 4D context of the required Causal Stochastic Interpretation of quantum theory, but its extended-completed LSXD version as inherent in the LCU tessellated backcloth of HAM cosmology. Once we are conceptually over that critical hurdle (difficult for those adhering to the 4D status quo) it is a straight forward matter to utilize the protocols of [4] Chaps. 9 & 11 to ballistically program the vacuum in terms of the additional features introduced and extended here for amorphous Ising model lattice-gas cellular automata following the seminal work of Toffoli and from other parameters found in the literature such as non-Newtonian fluid mechanics and various nanotechnology techniques [14-21].

Einstein postulated that Planck's quantization rule applied to an atom oscillating about its equilibrium point in a solid [23]. For vacuum programming we must extend this principle to spherical 'standing de Broglie matter-waves' in spacetime the internal motion of which obeys the Lorentz transformation. Simultaneous points produced by the wave are wavelet centers according to Huygens' Principle [24] that reinforce their common envelope (of the main wave) because if the waves are parallel they summate. Another important consideration in this regard is that all of the energy of the particle is focused at one point for all observers [25]. This is aligned with M-theory where all elementary particles are not discrete points but comprised of fundamental resonance modes of vibrating strings of a fixed tension [26]. It is beyond the scope of this chapter but the LCU of HAM cosmology has a variable string tension which is essential to the gating mechanism for surmounting uncertainty. Suffice it to say that the fixed T_S of standard M-Theory is like a wavefunction collapse or resultant moment, whereas in the HAM version of M-Theory the variable T_S is operating 'behind the scene' as an inherent component of the continuous-state process *UF* dynamics.

In considering the $360^\circ - 720^\circ$ Dirac spherical rotation of the electron spinor suggesting HD topological components for the fundamental structure of matter the idea is not as difficult to conceptually explore. The intuitive reader will realize this additional topology is almost a proof, as shown by the representation of physical quantum states in abstract *mathematical spaces* (for ease of calculation - not physically real, although most physicists might disagree because they 'believe' the physics is real) like Hilbert space, phase space or configuration space, of the incompleteness of quantum theory which when complete will entail an actual LSXD physically real brane topology space for the manipulation of observable properties of operators with mirror symmetry. The probability amplitude is not considered real, but in Cramer's transactional interpretation [27] an event is considered to be a physically real standing wave with all off diagonal elements also real so that an extended or completed form of quantum theory would be manifest in a physically real conformal scale invariant space. The great import of this situation is that the phase elements of quantum mechanics are physically real which has far reaching consequences and is required for our development of automata technology.

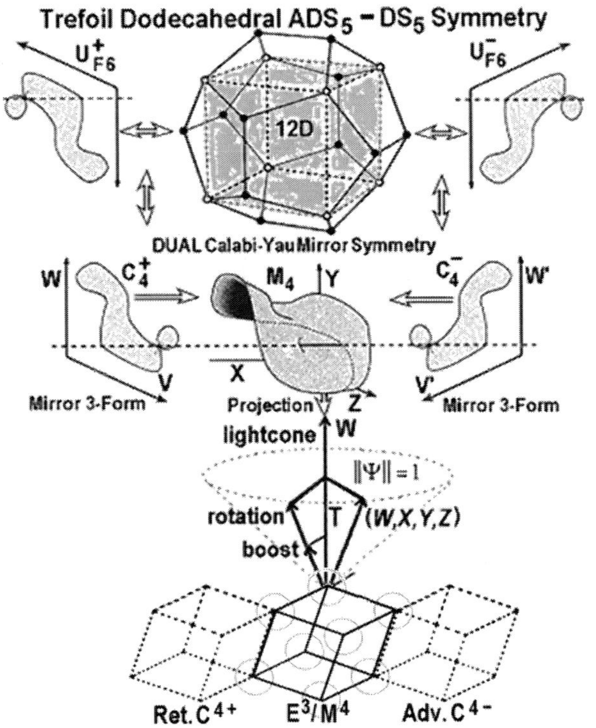

Figure 1. Complete local / nonlocal conceptualization LSXD regime of HAM cosmology illustrating the hierarchy of its geometric topology including Poincaré Dodecahedral involute properties (top), as well as the continuous-state exciplex 'hysteresis loop' of noeon injection (not shown) of U_F into every spacetime point (and thus atom) in tessellated LCU spacetime (bottom). It is suggested that this continuous-state mechanism is how our Euclidean-Minkowski virtual holographic reality is produced in a similar fashion by Cramer future-past standing-wave transaction parameters of spacetime (bottom). Calabi-Yau mirror symmetric resultant (middle).

It is essential to program spacetime itself [28], so as Copenhagen has ignored the physicality of this HD topology; we must in a sense take the perspective of an HD observer and ignore the 'particle-in-a-box' resultant and deal with the HD topological structural-phenomenology, which we suppose is a roundabout way of saying we must move well beyond non-relativistic quantum mechanics to an extended or dualistic form of relativistic quantum field theory (perhaps topological quantum field theory). By dualistic we mean taking our fundamental basis not from the resultant particulate matter as a stage for sequencing the locus of the evolution of quantum states for present events but from the LSXD mirror symmetric standing-wave elements from which the resultant particle arises (in Euclidean / Minkowski space with inherent uncertainty) which can be thought of as two Calabi-Yau 3-forms (Fig. 1). This is an HD extension of Cramer's transactional interpretation [27], a standing-wave model requiring a pair (actually a doubled pair to go all the way to causal separation) of Dirac equations, one for R_1 (Ret. C^{4+}) and one for R_2 (Adv, C^{4-}) – the future-past advanced-retarded components of the virtual discretized present instant forming our perceived virtual reality (Fig. 1).

Physicists have generally accepted the empirical reality of de Broglie matter waves, $\lambda = h/\rho$ for accelerated particles; but with little utility other than to demonstrate their existence through a variety of diffraction experiments applied to elementary particles, atoms and molecules in order to confirm the wave-particle duality of all matter [29-37]. De Broglie waves are physically real matter-waves associated with any moving particle traveling in a surrounding material medium [38,39].

$$h\nu_0 = m_0 c^2 . \tag{1}$$

In reading De Broglie's original paper where he mentions that "...the quantum principle suggests associating this internal energy with a simple periodic phenomenon of frequency, ν_0 such that, $h\nu_0 = m_0 c^2$" [36]. But it wasn't so much the reminder of the inherent energy, it was the manner we were struck by de Broglie's statements about the *periodicity of the internal energy*, a notion left out of most textbooks and is a key element of the Vigier causal interpretation of quantum theory [40,41] that we felt could be applied to cellular automata. Toffoli gives the metaphor of a band leader conducting the various instrument players in a band; this is the usual Turing machine programming of a CPU. Then he relates that the programming of cellular automata is internal for each processing unit allowing nonlinear ballistic computation [14,15].

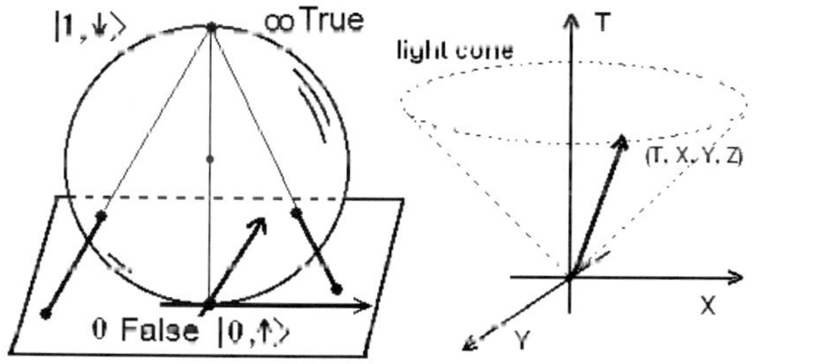

Figure 2. a) Block sphere rendition of a qubit. b) Relativistic qubit or r-qubit with more degrees of freedom as an aid in surmounting quantum uncertainty to facilitate ballistic ordering of the MFP.

Simplistically if one tears out a tiny segment of a holographic film the whole image may still be reproduced only much less bright and with much less resolution. Continuing the metaphor, then if one considers the tiny piece the usual energy associated with a coordinate point in the vacuum zero-point field, one must now utilize or set up a back-reaction potential that asymptotically approaches infinity as the MFP ballistic coherence length increases [42,43]. In reverse (the zero-point becomes the infinity point), by a coherent control cumulative interaction process, the wave structure of matter may undergo a power-factored constructive interference when set up in a Mirror symmetry-T-duality spacetime resonance hierarchy. The tricky part at the limit of our current comprehension is a manipulation of the arrow of time [12] essential to creating a focused array.

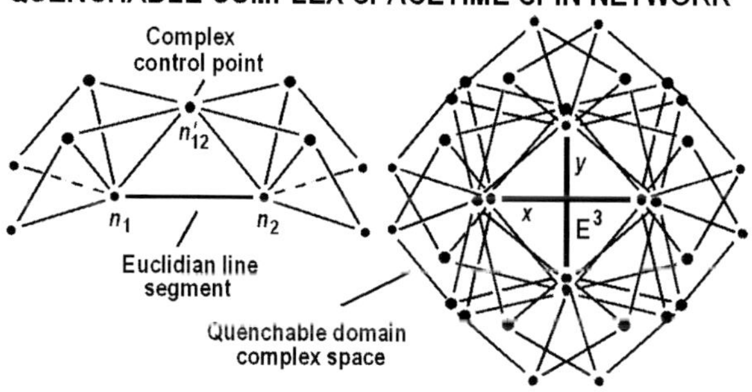

Figure 3. Spin networks of the programmable matter array with periodic control points for setting up the ordering of spacetime as programmable cellular automata. Forming these substructures enables MFP energy cascades in conjunction with ballistic programming. a) attempts to illustrate how a U_F control point can govern or coherently control a Euclidean standing-wave (continuous-state virtual) line segment. b) A Euclidean representation of a quantum 'particle in a box' and the 64 control points required for ballistic programming (12 for z-axis and 12 for time not shown).

To achieve this result a new basis for the standard model of particle theory is required, one cast in an 12D LSXD M-Theoretic Holographic Anthropic Multiverse (HAM) with a commensurate form of Relativistic Quantum Field Theory utilizing Dirac spherical rotation parameters and de Broglie matter-waves. The effectiveness of ballistic programming is directly proportional to the amount of unfettered MFP vacuum energy available to it. This vacuum energy is not utilized directly. What is needed instead is to move from the current Copenhagen phenomenological viewpoint to the extended (completed) de Broglie-Bohm 'ontological' point of view as the basis for programming the vacuum. What we have failed to realize from the Copenhagen view is that we are the vacuum and to operate our devices from the perspective that it also is the vacuum rather than from the local perspective where so-called collapse of the wave-function and the uncertainty principle rule the day. What we are trying to delineate is that nothing new needs to be created or built; it already exists. We need merely to uncouple the perspective of operation and recouple it periodically (like a holophote or lighthouse) to the required resonant phase modality in the continuous-state hierarchy.

As stated we are required to utilize a concept of static de Broglie matter-waves rather than the customary point of view of an associated propagating wave. The associated material medium for our 'static' de Broglie matter-waves [44] is the vacuum of spacetime itself which we will look at as a programmable tessellation of Ising model lattice gas cellular automata arising from the continuous-state parameters of the close-packed cosmological LCUs inherent to HAM cosmology which tile the spacetime backcloth. Conformally correlated with this vacuum regime we couple resonantly and program a Nanoscale kinematic matrix substrate of amorphous program-able matter to facilitate an ontological phase cascade.

If the close-packed cosmological least-units tiling the spacetime backcloth are considered to have properties like an Ising model lattice-gas Bloch sphere cellular automata array then space becomes programmable as has been suggested [28,45]. A Bloch sphere is a form of Riemann sphere here purported to fill the spacetime raster as Calabi-Yau dual 3-forms where relativistic, r-qubits become physically real rotatable transformable Ising lattice-gas Riemann spheres to which when perfect rolling motion resonance techniques are applied cascade transformations for ballistic computing can be set up. See Fig. 2(a) and 4; Fig. 2(b) is the relativistic qubit (r-qubit).

The 64 elements shown in Fig. 3(b) arise from the geometric algebra of quaternion duality. The quaternions are the cyclic noncommutative operators with elements 1,-1,*i*, -*i*, *j*,-*j*, *k*,-*k* of order 8. Quaternion complexification may continue with further dualing to order 16, order 32 and finally order 64. Additional dualing is possible using the same basis but a group of 64 elements is sufficient to establish all possible combinations of complexified dualing to infinity [2]! Thus clearly efficiently optimizes the U_F's ability to produce ballistic programming.

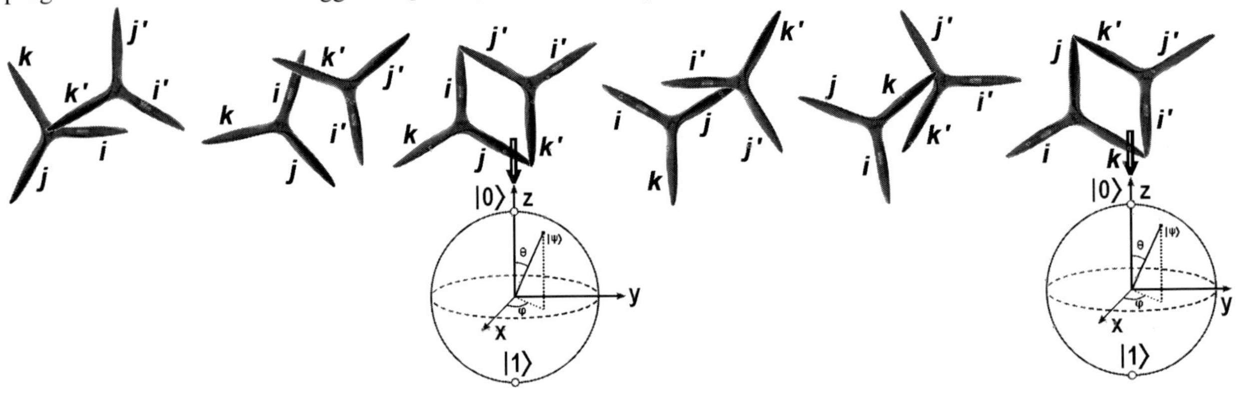

Figure 4. Conceptual emergence of a Standard Model Riemann Bloch 2-sphere qubit as a spacetime resultant as part of the locus of virtual points forming reality from LSXD continuous-state dual-quaternion nilpotent propagators.

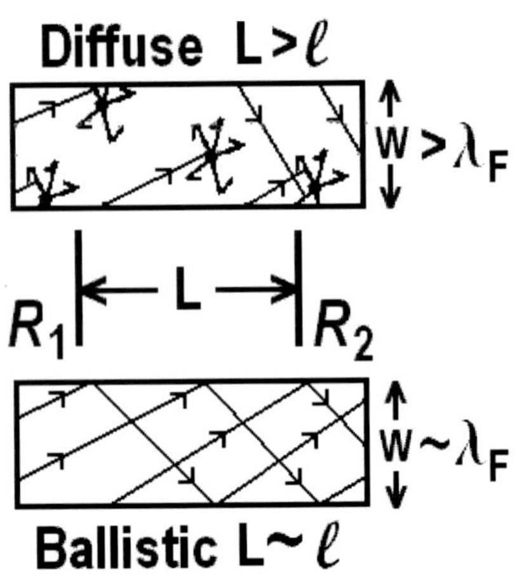

Figure 5. By resonant phase coherence the basis for a ballistic transport avalanche may be programmed into the spacetime topology using amorphous nanotech materials that simulate or map to the structural-phenomenology of spacetime. L is the coherence length of the resonator, and *l* the mean free path (MFP).

In the HAM cosmology as stated the spacetime raster is self-organized and thus has all the properties of complex self-organized systems such as incursion and evolution controlled by an external action principle [46,47] (coherence of the U_F). When a 'point' on an automata grid is accessed, 2D *x,y* for simplicity, the LCUs points parameters are updated. If the MFP is below threshold, i.e. quenchable, the neighboring elements of the array help to maintain equilibrium; but for a $z(x,y) \rightarrow z(x,y)+1 >$ threshold an avalanche occurs [48-50]. This has an associated asymptotic power law that fractally propagates with a domino effect of varied stepwise levels and thresholds of quenchable and infinite ballistic transport (Fig. 5) parameters mediated by the ability of the algorithm to program the amorphous nanoscale material for coherent control of the spacetime hierarchy with a continuous fractal-like incursion through the LSXD hierarchy that occurs when ballistic computing of the spacetime topology is achieved.

We are not sure if fractality bears any relation to our cellular automata model but the metaphor is illustrative. The Mandelbrot fractal set can be mathematically produced by the Feigenbaum fractal generator, $F_C(x) = x^2 + C$ which produces an iteration fulcrum with a period-doubling bifurcation cascade by repeated iteration of, F_C which is a family of complex polynomials from the critical point, x_0

$$\begin{aligned} x_0 &= x \\ x_1 &= F_C(x) = x^2 + C \\ x_2 &= F_C(F_C(x)) = (x^2 + C)^2 + C \\ x_3 &= F_C^3(x) \\ &\vdots \\ x_n &F_C n(x) \end{aligned} \tag{2}$$

where C is a complex number and for $C = i$ the sequence is 0,i,(-1+i),-i,(-1+i),-1... The map may escape to infinity or stay within the Mandelbrot set of a disk with infinite radii [51]. In contrast the Mandelbrot fractal set generator, $F_C(Z) = Z^2 + C$ where $Z = x - iy$ and $c = c_1 + ic_2$ maps a subset of the complex plane for values of c whose orbits don't escape to infinity by

$$\begin{aligned} Z^2 + C = (x + iy)^2 + (c_1 + ic_2) = \\ x^2 + y^2 + c_1 + (2xy + c_2) \end{aligned} \tag{3}$$

The scalar equation in spherical coordinates of wave motion in spacetime which has spherical symmetry [52,53]

$$\nabla^2\Phi - \frac{1}{c^2}\partial^2\frac{\Phi}{\partial t^2} = 0 \tag{4}$$

where Φ is the wave amplitude. The equation has two solutions

$$\begin{aligned} \Phi_{out} &= \frac{1}{r}\Phi_{\max}\exp(i\omega t - ikr) \\ \Phi_{in} &= \frac{1}{r}\Phi_{\max}\exp(i\omega t + ikr) \end{aligned} \tag{5}$$

which for the programming of spacetime can be applied to the propagation of Cramer's advanced-retarded waves from an emission locus at x,t = 0,0 by Eqs. (5) & (6).

$$\begin{aligned} F_{1-Ret} &= F_0 e^{-ikx} e^{-2\pi ift}, \quad F_{2-Ret} = F_0 e^{ikx} e^{-2\pi ift} \\ F_{3-Adv} &= F_0 e^{-ikx} e^{2\pi ift}, \quad F_{4-Adv} = F_0 e^{ikx} e^{2\pi ift} \end{aligned} \tag{6}$$

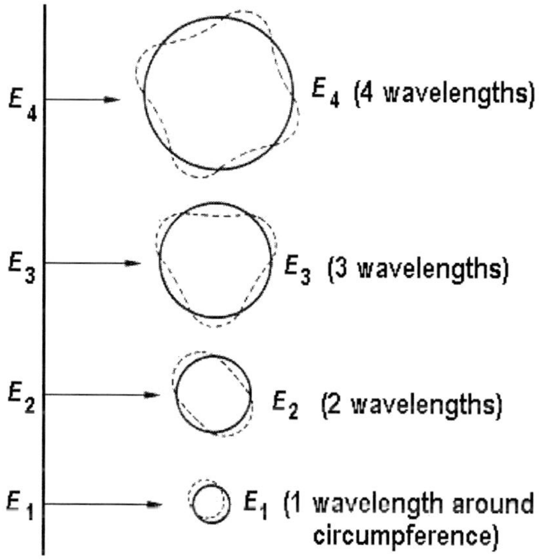

Figure 6. A Ring may vibrate with n standing wavelengths depending on the relationship of the circumference to the multiple number of whole wavelengths. Simplified here, it is suggested that the topology of spacetime and matter vibrate on and as hyperspherical surfaces.

Traditionally electron standing-waves oscillate about the atomic nucleus (Fig. 6). Here we attempt to expand the wave nature of matter itself as static waves centered on the locus of least spacetime units as it is annihilated and recreated in the arrow of time relative to the observer. This requires a conversion of the de Broglie wave equation, $mvr = n(h/2\pi)$ to a static form amenable to the parameters of continuous-state cosmology [25,44]. For Hyperspherical Representation the magnitudes of the radial coordinates of a two-state wavefunction, $\psi(\vec{r}_1, \vec{r}_2)$ in hyperspherical representation are replaced by the hyper-spherical radius, R and the hyperspherical angle, α such that

$$R \equiv \left(r_1^2 + r_2^2\right)^{1/2} \text{ and } \alpha \equiv \arctan\frac{r_2}{r_1} \tag{7}$$

in order that the symmetries may be more clearly shown. The hyperspherical radius, R represents the size of the two-state system and the hyperspherical angle, α is a measure of the radial correlation of the two-state system [54]. It is critical to note that when $\alpha = \pi/4$, $r_1 = r_2$; and when $\alpha = 0$ or $\pi/2$ one of the states is at a greater distance from the least-unit vertex than the other.

4. Spacetime Programming Utilizing LCU Cellular Automata

Programmable matter is defined as a material that locally adjusts its response to external inputs through programmed control. Amorphous Ising model lattice-gas cellular automata can be used for programming spacetime if designed to mirror the spacetime structure utilized. Each independent computational element in the amorphous or stochastic (accepting all) medium is identically programmed on a topological surface which in this case conforms to the least-unit tori of spacetime.

There are too many units to program individually so programming is achieved by neighbor connectedness. Toffoli formed a metaphor to describe this neighbor model [14,15]. Usually a marching band has a leader, this will not work for cellular automata where local self-assembly is internalized for each individual unit which acts as its own agent. This is a fundamental requirement for a massive ballistic response. The automata nanostructure must contain a computing substrate that is composed of fine-grained computing nodes distributed throughout space which communicate using only this nearest neighbor type of interactions [14,15,18,19,55]. According to Drexler [55] the closely packed computational units may be constructed to simulate a fractal system that for us would mean has the required continuous-state incursive properties.

5. Properties of de Broglie Matter-Waves

De Broglie by considering a material moving object of restmass, m_0 for a stationery observer suggested that a phase wave, or 'pilot' wave, accompanies a particle because the principle of inertia said it should possess an internal energy equal to m_0c^2 [36]. This phase wave arises as an inevitable consequence of de Broglie's assumption of the internal periodic phenomenon of the particle and the Lorentz transformation laws of the special theory of relativity

$$h\nu_0 - m_0c^2, \tag{8}$$

with $v - \beta c,\ (\beta < 1)$ for total energy $\nu - m_0c^2 / h\sqrt{1-\beta^2}$. De Broglie's result arose from a combination of the principle of Einstein's special relativity and the quantum relationship for the observer which he initially applied to a photon of nonzero restmass, $m_\gamma\ (<10^{-50}g)$ which because of its associated internal motion he associated with a piloting phase wave of frequency, ν at each point in space.

MacKinnon [25,44,56] described the de Broglie wave packet for stationery states and nondispersive wave packets of a free particle. He states that the nondispersive wave packet, ψ is a solution of

$$\Box\psi = 0 \tag{9}$$

where

$$\Box = \nabla^2 - \frac{1}{c^2}\frac{\partial^2}{\partial t^2}. \tag{10}$$

From this MacKinnon shows that the nondispersive wave packet for a particle relative to the observer has the form

$$\psi = \sin(kr/kr)\exp\left[i(\omega t - k_0x)\right] \tag{11}$$

where

$$k = m_0c/\hbar,\quad r = \left\{\frac{(x - vt^2)}{1-(v^2/c^2)} + y^2 + z^2\right\}^{1/2}, \tag{12}$$
$$\omega = mc^2/\hbar \quad k_0 = mv/\hbar.$$

Equation (12) is a spherically symmetric solution to Eq. (11) after being subjected to the Lorentz transform as initially obtained by de Broglie.

Of critical interest to us is MacKinnon's work to set up a de Broglie wave packet for a stationery state. Although we are interested in relativistic waves it is not for de Broglie waves for the usual particles in coordinate motion but for de Broglie waves for stationery matter with internal 'continuous-state' relativistic effects.

Consider two identical particles moving in opposite directions relative to an observer at x^* and t^*

$$\psi_1^* = A\cos(\omega t^* - kx^*),$$
$$\psi_2^* = A\cos(\omega t^* + kx^*) \tag{13}$$

which represent standing waves when solved by the Schrödinger equation for a particle in a box and cannot depend on the reference frame [44]. MacKinnon concludes that these stationery states are static and for which Bohm postulated a quantum potential to account for it. MacKinnon carries this point further [25] to suggest that:

> The motion of a particle in spacetime does not depend on the motion relative to it of any observer or any frame of reference [and] if the particle has an internal vibration of the type hypothesized by de Broglie, the phase of that vibration at any point in spacetime must appear to be the same for all observers...Each observer or reference frame will have its own de Broglie wave for the particle. The phase of the particle's vibration must, by definition, be the same as that for all possible de Broglie

waves at the point where the particle is. By superimposing all these possible de Broglie waves, a [nondispersive] wave packet is formed centered in space on the particle.

In his original work de Broglie was not able to properly form a wave packet that could localize the particle; MacKinnon was able to construct a wave packet from de Broglie's original wave phenomena that is also nondispersive [25].

Surmounting uncertainty by Coherent Control of Standing Matter-Waves entails a conformal scale-invariant continuous-state wave-particle seesaw leapfrog dynamics inherent in the topology of spacetime as a template within a brane topology hierarchy amenable to application of incursive resonance. This nanoscale programmable automata substrate also the leading edge or wave envelope would not merely be a Huygens wave front but be programmed with an annihilation-creation vector spin structure which would asymptotically increase the effectiveness of ballistic programming by optimizing the periodicity of the MFP. Also with sufficiently versatile programming this 'surface' would not create a percussive back-reaction but annihilate or damp the phases of the incoming shock waves to attenuation by destructive rather than constructive interference techniques.

References

[1] Rowlands, P. (2013) Space and Antispace, in R.L. Amoroso, P. Rowlands & L.H. Kauffman (eds.) The Physics of Reality: Space, Time, Matter, Cosmos, Hackensack: World Scientific.

[2] Rowlands, P. (2007) Zero to Infinity: The Foundations of Physics, Singapore: World Scientific.

[3] Stevens, H.H. (1989) Size of a least unit, in M. Kafatos (ed.) Bell's Theorem, Quantum Theory and Conceptions of the Universe, Dordrecht: Kluwer Academic.

[4] Amoroso, R.L. & Rauscher, E.A. (2009) The Holographic Anthropic Multiverse, Singapore: World Scientific.

[5] Margolus, N. (1995) CAM-8: A computer architecture based on cellular Automata, arXiv:comp-gas/9509001v1.

[6] Hübsch, T (1994) Calabi Yau Manifolds: A Bestiary for Physicists, Singapore: World Scientific.

[7] Randall, L. (2006) Warped Passages: Unraveling the Mysteries of the Universe's Hidden Dimensions, New York: Harper-Collins.

[8] Luminet, J-P (2008). The Wraparound Universe, Wellesley: AK Peters.

[9] Amoroso, RL. Kauffman, L.H. & Giandinoto, S. (2013) Universal Quantum Computing; 3rd Gen Prototyping Utilizing Relativistic 'Trivector' Qubit Modeling Surmounting Uncertainty, in R.L. Amoroso, P. Rowlands, L.H. Kauffman (eds.) The Physics of Reality: Space, Time, Matter, Cosmos, Hackensack: World Scientific.

[10] Amoroso, RL. & Rauscher, E.A. (2010) Empirical protocol for measuring virtual tachyon / tardon interactions in a Dirac vacuum, in R.L. Amoroso, P. Rowlands, S. Jeffers (eds.) Search for Fundamental Theory, AIP Conference Proceedings Vol. 1316, p. 199.

[11] Veneziano, G. (1968) Construction of a crossing-symmetric, Regge-behaved amplitude for linearly rising trajectories, Nuovo Cim. 57A 190.

[12] Amoroso, R.L. (2013) Time?, in R.L. Amoroso, P. Rowlands, L.H. Kauffman (eds.) The Physics of Reality: Space, Time, Matter, Cosmos, Hackensack: World Scientific.

[13] Amoroso, R.L., Kauffman, L.H., Rowlands, P. & Rauscher, E.A. Introduction to Unified Field Mechanics, in process.

[14] Toffoli, T. Programmable Matter: An introduction, Cambridge: MIT Univ. Press (in preparation).

[15] Toffoli, T. & Margolus, N. (1987) Cellular Automata Machines; A New Environment for Modelling, Cambridge: MIT Univ. Press.

[16] Ionicioiu, R., Amaratunga, G. & Udrea, F. (2000) Quantum computation with ballistic electrons, arXiv:quant-ph/0011051v1.

[17] Viswanathan, G.M. & Viswanathan, T.M. (2008) Spontaneous symmetry breaking and finite time singularities in D-dimensional incompressible flow with fractional dissipation, arXiv:0807.1563v1 [physics.flu-dyn].

[18] Kodama, T. & Koide, T. (2008) Memory effects and transport coefficients for non-Newtonian fluids, arXiv:0812.4138v1.

[19] Abelson, H. et al. (2000) Amorphous computing, Communications of the ACM, 43:74-82.

[20] Wolfram, S. (1994) Cellular Automata and Complexity, Reading: Addison-Wesley.

[21] Du, X., Skachko, I., Barker, A., & Andrei, E. Y. (2008) Approaching ballistic transport in suspended graphene. Nature Nanotechnology, 3(8), 491-495.

[22] Accardi, L. (1988) Foundations of quantum mechanics: A quantum probabilistic approach, in G. Tarozzi & A. van der Merwe (eds.) The Nature of Quantum Paradoxes, Dordrecht: Kluwer Academic.

[23] Peebles, P.J.E. (1992) Quantum Mechanics, Princeton: Princeton Univ. Press.

[24] Huygens, C. (1690) *Traite de la Lumiere*, (Treatise on Light).

[25] McKinnon, L. (1978) A nondispersive de Broglie wave packet,
Foundations of Physics, 8:3-4; 157-176.

[26] Kaku, M. (1998) Introduction to Superstrings & M-Theory, 2nd ed., New York: Springer.

[27] Cramer, J.G. (1986) The Transactional Interpretation of Quantum Mechanics, Rev. Mod. Phys 58, 647-687.

[28] Zizzi, P. (2005) Spacetime at the Planck scale. The quantum computer view, in R.L Amoroso, B. Lehnert & J-P Vigier (eds.), Beyond the Standard Model: Searching for Unity in Physics, Oakland: Noetic Press.

[29] Kulhánek, J. (1965) Propagation of de Broglie waves in space-time, Il Nuovo Cimento, 38:3; 1955-1965.

[30] Croca, J.R., Garuccio, A., Lepore, V.L. & Moreira, R.N. (1990) Quantum-optical predictions for an experiment on the de Broglie waves detection, Foundations Physics Letters, 3:6; 557-564.

[31] Feoli, A. & Scarpetta, G. (1998) De Broglie Matter Waves from the Linearized Einstein Field Equations, Foundations Phys L, 11:4; 395-403.
[32] Lepore, V.L. (1992) Homodyne detection of de Broglie waves, Foundations of Physics Letters, 5:5; 469-478
[33] Croca, J.R.., Ferrero, M., Garuccio, A. & Lepore, V.L. (1997) An experiment to test the reality of de Broglie waves, Foundations of Physics Letters, 10:5; 441-447.
[34] Barut, A.O. (1990) Quantum theory of single events: Localized de Broglie wavelets, Schrödinger waves, and classical trajectories, Foundations of Physics, 20: 10.
[35] Rauch, H. (2008) Test of quantum mechanics by neutron interferometry, European Physical J - Special Topics, V.159, No.1.
[36] de Broglie, L. (1923) Radiation, waves and quanta, *Comptes Rendus*, Vol. 177. 507-510.
[37] Davisson, C.J.& Germer, L.H. (1929) A test for polarization of electron waves by reflection, Phys Rev 33: 5; 760-772.
[38] Dewdney, C., Kypeianidis, A., Vigier, J-P., Garuccio, A. &. Gueret, Ph. (1984) Time-dependent neutron interferometry: Evidence in favour of de Broglie waves, *Lettere Al Nuovo Cimento*, 40:16; 1971-1985.
[39] Croca, R. (1987) Neutron interferometry can prove (or refute) the existence of de Broglie's waves, Foundations Phys, Vol. 17, No. 10.
[40] Rybakov, Y.P. (1974) On the causal interpretation of quantum mechanics, Found. Physics, Vol. 4, No. 2.
[41] Vigier, J-P (2000) selected papers, in S. Jeffers, B. Lehnert, N. Abramson, & L. Chebotarev (eds.) Jean-Pierre Vigier and the Stochastic Interpretation of Quantum Mechanics, Montreal: Aperion.
[42] Yang, S., Hammack, A.T., Fogler, M.M. & Butov, L.V. (2006) Coherence length of cold exciton gases in coupled quantum wells, arXiv:cond-mat/0606683v1.
[43] Kleinert, H. (2005) Vortex origin of tricritical point in Ginzburg-Landau theory, arXiv:cond-mat/0509430v1.
[44] McKinnon, L. (1979) The de Broglie wave packet for a simple stationery state, Found. Phys., 9:9-10; 787-791.
[45] Lloyd, S. (2006) Programming the Universe, New York: Knoph.
[46] Amoroso, R.L. (ed.) The Complementarity of Mind and Body: Realizing the Dream of Descartes, Einstein and Eccles, New York: Nova Science Publishers.
[47] Amoroso, R.L & Amoroso, P.J. (2004) The fundamental limit and origin of complexity in biological systems: A new model for the origin of life, in D. Dubois, (ed.) Computing Anticipatory Systems, Proceedings of CASYS04, 6th Int, Conf., Liege, Belgium, 11-16 Aug, 2003, AIP Proceedings Vol. 718, pp. 144-159.
[48] Jensen, H.J. (1998) Self-Organized Criticality - Emergent Complex Behavior in Physical and Biological Systems, Cambridge Univ. Press.
[49] Krink, T. & Tomsen, R. (2001) Self-organized criticality and mass extinction in evolutionary algorithms, Proc. IEEE int. Conf, on Evolutionary Computing, pp. 1155-1161.
[50] Bak, P., Tang, C. & Wiesenfeld, T.(1987) Self-organized criticality: an explanation of 1/f noise, Phys Review Letters. Vol. 59, No. 4.
[51] Feigenbaum, M.J. (1980) Universal behavior in nonlinear systems, Los Alamos Science v1, p4-27.
[52] Wolff, M. (2002) Cosmology, the quantum universe and electron spin, in R.L. Amoroso, G. Hunter, M. Kafatos & J-P Vigier (eds.) Gravitation and Cosmology: From the Hubble Radius to the Planck Scale, pp. 517-524, Dordrecht: Kluwer.
[53] Wolff, M. (2008) Schrödinger's Universe and the Origin of the Natural Laws, Parker: Outskirts Press.
[54] Starace, A.F. (1988) Hyperspherical description of two-electron systems, in Briggs, J.S., Kleinpoppen, H. & Lutz, H.O. (eds.) Fundamental Processes of Atomic Dynamics, pp. 235-258, NY: Plenum.
[55] Drexler, K.E. (1992) Nanosystems : Molecular Machinery, Manufacturing and Computation, New York : Wiley & Sons.
[56] MacKinnon, L. (1981) A fundamental equation in quantum mechanics? *Let Al Nuovo Cimento*, 32:10; 311-316.

Exploring Novel Cyclic Extensions of Hamilton's Dual-Quaternion Algebra

RICHARD L. AMOROSO

Physics Lab, Noetic Advanced Studies Institute
902 W 5400 N, Beryl, UT 84714-5104 USA
amoroso@noeticadvancedstudies.us

PETER ROWLANDS

Department of Physics
University of Liverpool, Liverpool, L69 7ZE UK
p.rowlands@liverpool.ac.uk

LOUIS H. KAUFFMAN

Department of Mathematics, Statistics & Computer Science
University of Illinois at Chicago, Chicago, IL, 60607-704 USA
Kauffman@uic.edu

We make a preliminary exploratory study of higher dimensional (HD) orthogonal forms of the quaternion algebra in order to explore putative novel Nilpotent/Idempotent/Dirac symmetry properties. Stage-1 transforms the dual quaternion algebra in a manner that extends the standard anticommutative 3-form, *i, j, k* into a 5D/6D triplet. Each is a copy of the others and each is self-commutative and believed to represent spin or different orientations of a 3-cube. The triplet represents a copy of the original that contains no new information other than rotational perspective and maps back to the original quaternion vertex or to a second point in a line element. In Stage-2 we attempt to break the inherent quaternionic property of algebraic closure by stereographic projection of the Argand plane onto rotating Riemann 4-spheres. Finally, we explore the properties of various topological symmetries in order to study anticommutative - commutative cycles in the periodic rotational motions of the quaternion algebra in additional HD dualities.

Keywords: Algebraic closure, Antispace, Dirac algebra, Dual quaternions, Idempotent, Nilpotent, Quaternions

There is a world of strange beauty to be explored here... John C. Baez [1].

1. Introduction — Brief Quaternion Review

Since quaternions are not ubiquitously appreciated by physicists we review them briefly here. For many years prior to discovering the quaternions William Rowan Hamilton tried unsuccessfully to extend the common system of complex numbers, $a + bi$ where the real numbers, a are represented by the x-axis and the imaginary numbers, bi are represented orthogonally on the y-axis. Hamilton considered this a 2-dimensional (2D) space and desired to extend the system to 3D by utilizing an additional set of orthogonal complex numbers, cj. This attempt failed because he found that this system of 'triplets', $a + bi + cj$ did not exhibit the necessary algebraic property called 'closure' that would complete the algebra, i.e. the product ij had no meaning within the algebra and therefore had to represent something else. In 1843 Hamilton finally realized he needed a 3rd system of imaginary numbers, dk to define an algebraic system properly exhibiting the needed property of closure [2-4] as in Figs. 1 and 8.

$$i^2 = j^2 = k^2 = ijk = -1 \tag{1}$$

with additional anticommutative multiplication rules

$$ij = -ji = k, \quad jk = -kj = i, \quad ki = -ik = j\,. \tag{2}$$

which can be represented graphically as

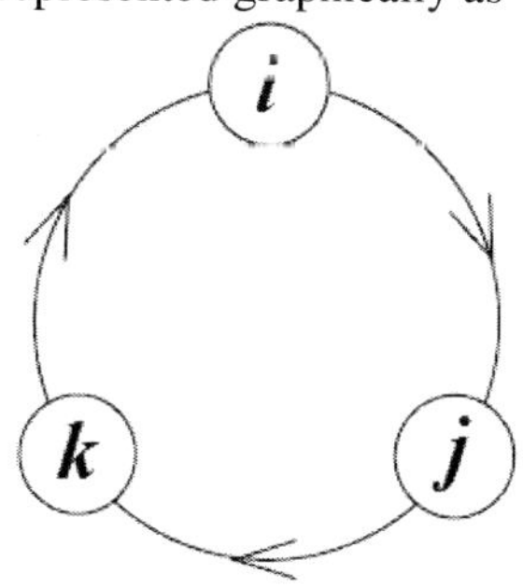

Figure 1. Graphic summary of the quaternion cyclic permutation rules. Adapted from Baez. [1].

Or represented in tabular form the dual quaternions, $\pm 1, \pm i, \pm j, \pm k$ form a non-Abelian group of order 8.

TABLE 1. Quaternion Cyclic Multiplication

	1	*i*	*j*	*k*
1	1	*i*	*j*	*k*
i	*i*	-1	*k*	-*j*
j	*j*	-*k*	-1	*i*
k	*k*	*j*	-*i*	-1

But by introducing the property of closure Hamilton sacrificed the principle of commutivity, $ij = -ji = k$. Frobenius proved in 1878 that no other system of complex algebra was possible with the same algebraic rules. This convinced him that he found a true explanation of the 3-dimensionality of space; with the 4th or real component in this case representing time as the real component (the 1 in quaternions, Tbl. 1). The Quaternion, Q_4 and Octonion, O_8 algebras arise naturally from restrictions inherent in the properties producing the discreteness of 3-dimensionality, i.e. the property of anticommutivity as realized by Hamilton as the key requirement for achieving the property of 'closure' when he discovered the *i, j, k* basis for the quaternion algebra or that in complex representation, $H = a \cdot 1 + bi + cj + dk$ is not commutative.

2. Background and Utility of New Quaternion Extensions

Our problem is similar to the one originally facing Hamilton. We also wish to add additional complex dimensionality but in a manner that also reintroduces commutivity in a periodic manner. Our procedure in this exercise is to manipulate the inherent cyclicality of the normative quaternion algebra (anticommutative) in a manner that the property of closure is broken periodically during a transformation through a Large Scale Additional Dimensional (LSXD) 'continuous-state' mirror symmetric space evolving through periodic anticommutative and commutative modes [5,6]. This requires a new form of transformation of the line element. The usual representation of the line element, ds in an n-dimensional metric space is $ds^2 = d\mathrm{q} \cdot d\mathrm{q} = g(d\mathrm{q}, d\mathrm{q})$ where g is the metric tensor, and $d\mathrm{q}$ an infinitesimal displacement. In n-dimensional coordinates q = $(q^1, q^2, q^3 ... q^n)$ and the square of arc length is $ds^2 = \sum g_{ij} dq^i dq^j$. Indices i and j take values 1, 2, 3 ... n. The metric chosen determines the line element. We will continue this development in an ensuing section in terms of stereographically projected Riemann manifolds where a Riemann surface does not entail an inherent particular metric.

According to Rowlands [4,7-10] quaternion cyclicity reduces the number of operators by a factor of 2 and *prevents the possibility* of defining further complex terms. To overcome this generalization of term limitations we explore utilizing shadow dualities employing additional copies of the quaternion algebra mapped into a 5D and 6D space to transform a new dualed 'trivector' quaternion algebra. As far as we know this can only occur under the auspices of a continuous-state cosmology within a dual metric where one coordinate system is fixed and then the other is fixed in a sort of leap-frog baton passing fashion [5]. We will explain this in more detail later with as much rigor currently at our disposal and conceptually in terms of a wonderfully illustrative metaphor in the 'Walking of the Moai' [11,12]. Correspondence is made to the usual 3D quaternion algebra as the resultant or re-closure that occurs at the end of the cycle repeats over and over continuously.

According to Rowlands' the nilpotent duality or a zero totality universe model requires an anti-commutative system that has this property of closure, whereas a commutative system remains open and degenerate to infinity [4]. This duality is fundamental to the continuous-state (like an internalized gravitational free-fall) postulate of Amoroso's Holographic Anthropic Multiverse cosmology [5] which is compatible with Calabi-Yau mirror symmetry properties of M-Theory suggesting a putative relationship to physicality. We suspect also that in the future correspondences will be found between the continuous-state hypothesis and the Rowlands' nilpotent universal rewrite system [4,5].

3. A New Concept in Quaternion Algebra — The First Triplet

We start with a standard quaternion in italicized notation *i,j,k* which physically represents a fermionic vertex in space and therefore has a copy as a vacuum anti-space *Zitterberwegung* partner [10], ***i,j,k*** (bold italic notation) which together can form a nilpotent [4,6-10]. Multiplying the two (space-antispace) nilpotent components together we get 3 copies (3 sets of three) of the original quaternion (uppercase notation) at some new level of postulated higher space; but still with no new information such that each of the three copies commute with themselves internally. Our initial consideration is that these three copies are spin-like objects that conceptually represent 3 different views of a cube such as: **I**) planar face, **J**) edge and **K**) corner (as if collapsed to a hexagon). See Fig. 2.

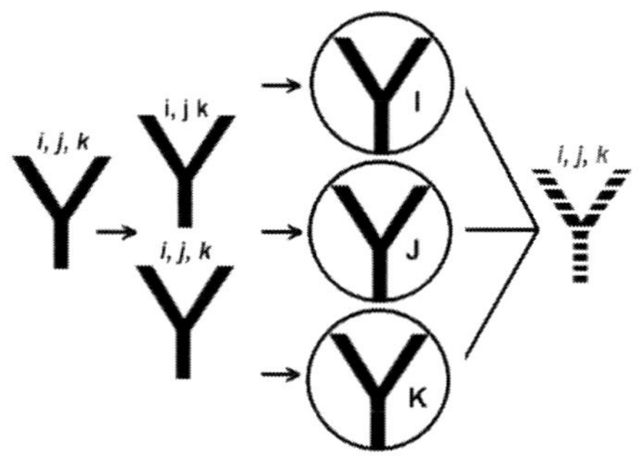

Figure 2. This figure and next (Fig. 3) represent different aspects of the usual quaternion algebra containing no new information beyond the original fermionic vertex. We postulate they represent conceptually 3 different ways (orientations) of viewing a cube. The first, $\mathbf{Y_I}$ a usual planar face, second, $\mathbf{Y_J}$ perhaps like view onto an edge and the third, $\mathbf{Y_K}$ from a corner which would make the cube look like it was collapsed to a 2D hexagon. In summary: Fig. 2(a) Usual *i,j,k* quaternion. (b) Space-antispace dual quaternion, *i,j,k and* ***i,j,k***. (c) The new **IJK** nilpotent triplet. (d) Reduced back to the original quaternion in 2(a).

We pass through a stage of three commutative objects, each of which is constructed from *ijk* (original quaternion) and ***ijk*** (nilpotent vacuum copy) which are thus a set of dual quaternions. Knowing one of these gives us the other two automatically.

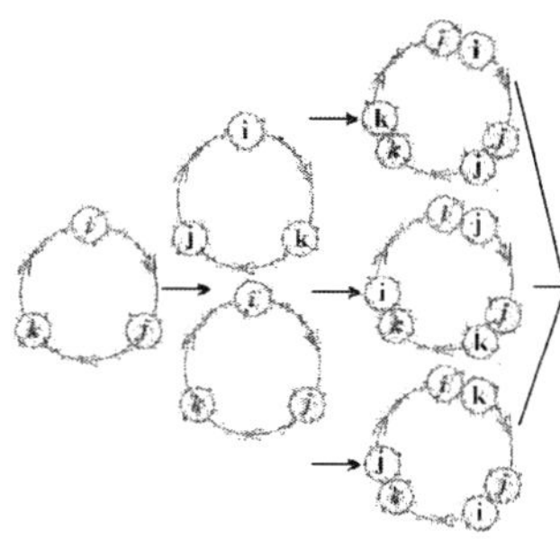

Figure 3. Varied view of Fig. 2 but highlighting the cyclicality of the algebraic elements. Reduces to original *i,j,k* quaternion fermionic vertex. Also in summary (same as Fig. 2): Fig. 3(a) Usual *i,j,k* quaternion. (b) Space-antispace dual quaternion, *i,j,k and* ***i,j,k*** (c) The new nilpotent triplet. (d) Reduced back to the original quaternion in 3(a).

Here is the algebraic derivation of the geometric objects in Figs. 2 & 3. In the quaternion algebra below, the usual italicized anticommutative quaternion, *i,j,k* acquires a dual vacuum (bold italicized) quaternion, ***i,j,k***. The product of these two quaternions becomes three commutative algebras (Figs. 2 & 3).

$$\begin{pmatrix} i\mathrm{i} \\ j\mathrm{j} \\ k\mathrm{k} \end{pmatrix}\begin{pmatrix} i\mathrm{j} \\ j\mathrm{k} \\ k\mathrm{i} \end{pmatrix}\begin{pmatrix} i\mathrm{k} \\ j\mathrm{i} \\ k\mathrm{j} \end{pmatrix} \qquad (3)$$

And their product becomes the new quaternion, **I J K** (Fig. 2(c)). Each of these algebras is a copy of the others and each is commutative on itself. Selecting out the three versions makes an anticommutative set. In Eq. (3) Mathtype would not allow bold font so we used regular and italic to illustrate the dualing.

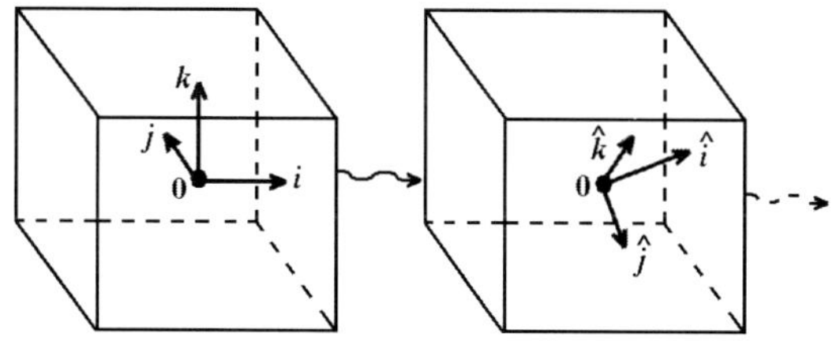

Figure 4. Different complex orientations of the Euclidean 3-cube projected from a quaternion space-antispace fermionic vertex or singularity, *i,j,k*.

4. Vectors, Scalars, Quaternions, and Commutivity

In this section notation is as follows. Vector are in regular and bold font. Quaternions are in regular italic and bold italic font. For the trivector we add the half quote mark (').

i j k	***i*i *i*j *i*k**	*i'*	1
i j k	*i*i *i*j *i*k	*i'*	1
vector	*bivector*	*trivector*	*scalar*
	quaternions		

Vectors

i j k anticommute
i j k anticommute
ii jj kk commute

Quaternions

i j k *anticommute*
i j k *anticommute*

ii ij ik ***j*** *i'**k** nilpotent*
ii ij ik ***j*** - *nilpotent*
(*5th term automatic*)

ii' jj kk *commute*

ai bj ck 1d
Ai Bj Ck 1D

ai bj ck 1d
ai bj ck 1D

ii jj kk
ij jk ki
ik ji kj

ijk anticommute SPACE
ijk *anticommute* VACUUM ANTISPACE
TOTAL 6 Degrees of freedom

This is Fig. 2 (b).

ii jj kk commute SPINOR SPACE I
ij jk ki commute SPINOR SPACE J
ik ji kj commute SPINOR SPACE K
TOTAL 9 Degrees of freedom

The above (**I J K**) is Fig. 2 (c).

IJ (*ii jj kk*) (*ij jk ki*) → (*i j k –i –j –k –ik –ji –kj*)
JK (*ij jk ki*) (*ik ji kj*) → (*i j k –i –j –k –ii –jj –kk*)
KI (*ik ji kj*) (*ii jj kk*) → (*i j k –i –j –k – ij –jk –ki*)

Reverse **IJ = – JI, JK = – KJ, KI = – IK. I, J, K** anticommutative, becomes the new *i j k* at the next level in the cycle. Because of nilpotency, *i j k* and ***i j k*** are dual and contain the same information. Also, **I, J, K** is made up only of these terms and contains the same information again so we can map **I, J, K** directly onto *i j k* and then pull out its dual by nilpotency and go round the cycle again continuously. The **I, J, K** are a **new** set of quaternions with the same structure as *i j k* but with only the same infomration as as contained in the usual *i j k* (real) and its dual ***i j k*** (vacuum).

We assume **I, J, K** are Dirac phase rotations of hypercube B. At phase K by stereographic projection (topological switching) a copy of A is boosted to a Riemann sphere "above" K. We assume the Riemann sphere by conformal scale-invariance inherent to the cosmology. This gives us + *i,j,k* and - *i,j,k* on opposite sides of the boosted Riemann sphere. The **I, J, K**s all rotate so they take turns cyclically being in position to be boosted.

5. Toward Completing the Hypercube — The Second Triplet

In addition to boosting into the higher space we wish to turn the nilpotent into an idempotent that transforms into the vacuum in a similar manner. By multiplying we also get three copies that are three sets of three in the same manner as in the nilpotent boost into the higher space above that are also spin-like. And that will likewise by stereographic projection boost new copies into Riemann spheres above K.

It is suggested that the orientation of the Cs in Fig. 5 is periodic and that all three take turns boosting orientation information to the hypercube in the higher space. Through more layers of boosting into the higher space, in conjunction with additional symmetric idempotent copies in the vacuum space we wish to end up with a structure that is like a bag of close-packed ping pong balls or rotating Riemann spheres where the original quaternion sphere always remains at the center, but through a rewrite system (where the nilpotents are annihilation and creation 'vectors') that continuously boost copies of the original orientation information from the center to the surface spheres and back again. It is possible to describe these cubes as component hyperspheres or dodecahedral polyhedrons.

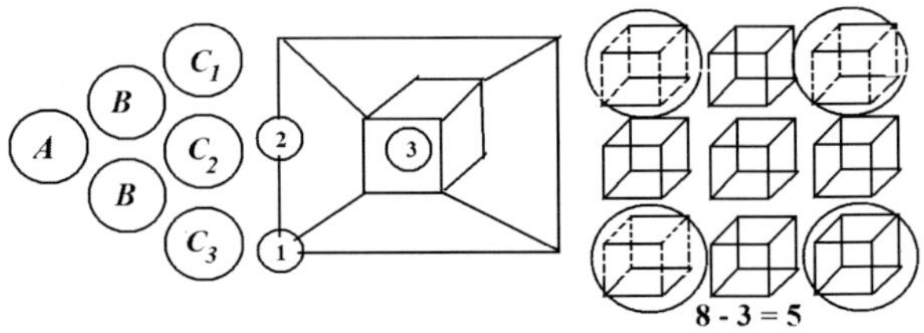

Figure 5. The first part cycles A, B, C (as in Figs. 1 & 3) is the algebra progression as in Figs. 2 & 3. The center hypercube represents the orientation provided by the three phases of the cube from C1, C2 & C3. The 3rd part represents the 8 cubes exploded off a hypercube, suggesting that there are many orientations to explore with the new algebraic triplets.

5.1 *Higher Doublings — Planar to Riemann Sphere*

Quaternions

i j k anticommute SPACE (Real) ⇄
Complex Argand Plane *i*

i j k *anticommute* VACUUM (+ C3) ⇄
Complex Argand Plane ***i***

-i -j-k *anticommute* VACUUM (- C3) ⇄
Stereographic Riemann Sphere ***I***

-i -j-k *anticommute* VACUUM (± Virtual) ⇄
Stereographic Riemann Sphere **I**

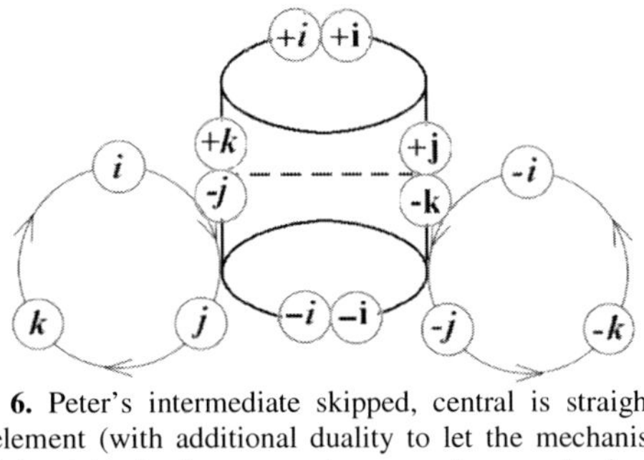

Figure. 6. Peter's intermediate skipped, central is straight to the spinor element (with additional duality to let the mechanism work cyclically), right is the symmetry reversing mechanism (when coupled at certain phase of the central spinor). To achieve causal separation it is postulated this structure is doubled again?

These are not planar (as usual Q) but stereographic projections onto Riemann sphere. In the central portion closure (when it appears) subtracts the extra degrees of freedom in the extra duality. When not subtracted the

Riemann sphere may "flip" the symmetry to achieve the basis for the commutative mode node. Doubled again, perhaps something like that shown in Fig. 6 to make it hyperspherical and causally separated.

In the London Underground tunnel near Imperial College there is an entrance to The Victoria and Albert Museum. In front of the entrance is a display sign encased in a 3 or 4 meter long clear plastic tube about a meter in diameter. Inside the tube components of a large (V & A) rotate such that the A becomes the V and vice versa (with the help of the ampersand) [19,20]. These rotations are insufficient because they are planar and preserve the symmetry and closure of the original quaternion algebra [2,3]. With an additional copy on the Riemann sphere (mirror symmetries) 2 rotations occur simultaneously; but closure is not yet broken. This requires a 3rd doubling in order to acquire properties of Dirac spherical rotation. With sufficient degrees of freedom one preserves the usual anticommutative parameters of closure. The other by 'boosting' through a Necker-like topological switching reverses the symmetry and passes cyclically through a commutative mode. And then back down through a reverse Necker n-spinor switch to return to the origin or 2nd point in the line element reducing back to the usual quaternion form. This is also beautifully illustrated in the 'Walking of the Moai on Rapa Nui' [11,12].

5.2. *Nilpotnet Idempotent Vacuum Doublings*

The nilpotent when multiplied by $\underset{\sim}{k}$ becomes idempotent.

$$\left(i\underset{\sim}{k}E+if+\underset{\sim}{j}m\right)\left(i\underset{\sim}{k}E+if+\underset{\sim}{j}m\right)$$
$$\left(i\underset{\sim}{k}E+if+\underset{\sim}{j}m\right)\left(i\underset{\sim}{k}E+if+\underset{\sim}{j}m\right)$$
$$\left(i\underset{\sim}{k}E+if+\underset{\sim}{j}m\right)\left(i\underset{\sim}{k}E+if+\underset{\sim}{j}m\right)$$
$$=0$$

5.3. *As Generalized Equation*

The anticommutative

Anticommutative

$(i_1+j_1+k_1)(i_2+j_2+k_2)=$

Commutative Commutative Commutative

$(i_1i_2+j_1j_2+k_1k_2)(i_1j_2+j_1k_2+k_1i_2)\ (i_1k_2+j_1i_2+k_1j_2)=$

I + J + K

a new iteration of the original i, j, k

5.4. *As Tensor Trasformation*

Transform (X Y Z)

$$\begin{pmatrix} i_1j_1 & i_2j_2 & i_3j_3 \\ j_1k_1 & j_2k_2 & j_3k_3 \\ k_1i_1 & k_2i_2 & k_3i_3 \end{pmatrix} = \begin{pmatrix} \mathrm{I} \\ \mathrm{J} \\ \mathrm{K} \end{pmatrix}$$

6. Quaternion Mirrorhouses

Utilizing hypersets, to represent two parallel mirrors, a simple 2-mirror mirrorhouse may be represented as: X = {Y}, Y = {X} (ignoring the inversion effect of mirroring). In constructing a 3-mirror mirrorhouse interestingly the hyper-structure turns out to be precisely the structure of the quaternion imaginaries. Let i, j and k be hypersets representing three facing mirrors [13]. We then have

$$i=\{j,k\},\ j=\{k,i\} \text{ and } k=\{i,j\}$$

where the notation $i = \{j,k\}$ means, e.g. that mirror i reflects mirrors j and k in that order.

With three mirrors ordering is vital because mirroring inverts left/right-handedness. If we denote the mirror inversion operation by "-" we have i = {j,k} = -{k,j}, j = {k,i} = -{i,k} and k = {i,j} = -{j,i} which is the structure of the quaternion triple of imaginaries: i = j*k = -k*j, j = k*i = -i*k, k = i*j = -j*i.

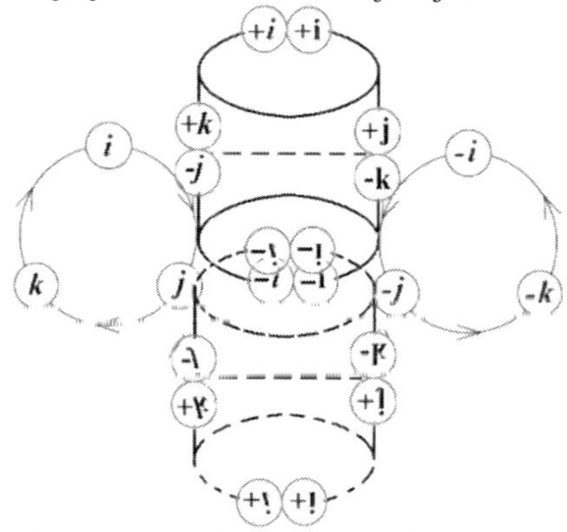

Figure 7. Possible new duality doublings in higher space.

The quaternion algebra therefore is the precise model of the holographic hyper-structure of three facing mirrors, where we see mirror inversion as the quaternionic anticommutation. The two versions of the quaternion multiplication table correspond to the two possible ways of arranging three mirrors into a triangular mirrorhouse.

Following Goertzel's [13] construction of a house of three mirrors; he demonstrates that the construction has the exact structure of the quaternion imaginaries. He lets i, j and k be hypersets representing the three facing mirrors such that i = {j,k}, j = {k,i} and k = {i,j} The notation i={j,k} means, e.g. that mirror i reflects mirrors j and k in that order. Goertzel points out that mirroring inverts left/right-handedness. Denoting mirror inversion by the minus sign "-" yields i = {j,k} = -{k,j}, j = {k,i} = -{i,k} and k = {i,j} = -{j,i} which is the exact structure of the quaternion imaginary triplets: i = j*k = -k*j, j = k*i = -i*k, k = i*j = -j*i. Goertzel then claims the quaternion algebra is therefore the precise model of the holographic hyper-structure of three facing mirrors, where mirror inversion is the quaternionic anticommutation.

The two versions of the quaternion multiplication table correspond to the two possible ways of arranging three mirrors into a triangular mirrorhouse.

i*j = k	j*i = - k
j*k = i	k*j = - i
k*i = j	i*k = - j

Which with dual counterpropagating mirror reflections is like a standing-wave.

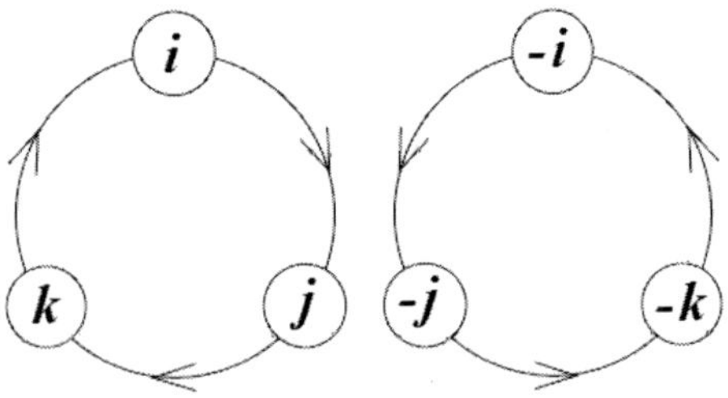

Figure 8. (a) the usual quaternion series cycle, (b) The dual counterpropagating standing-wave mirror image.

6.1. *Observation as Mirroring*

To map the elements inside the mirrorhouse/algebraic framework noted previously, it suffices to interpret the X = {Y}, Y = {X} as "X observes Y", "Y observes X" (e.g. we may have X = primary subset, Y = inner virtual other), and the above i = {j,k}, j = {k,i}, k = {i,j} as "i observes {j observing k}", "j observes {k observing i}" and "k observes {i observing j}". Then we can define the - observation as an inverter of observer and observed, so that e.g. {j,k} = -{k,j}. We then obtain the quaternions

$$i = j*k = -k*j,\ j = k*i = -i*k,\ k = i*j = -j*i$$

where multiplication is observation and negation is reversal of the order of observation. Three inter-observers = quaternions.

This is where the standard mathematics of quataternion algebra has remained. In an actual mirrorhouse the reflections, like Bancos ghost extend into infinity. We wish to build an additional dual trivector mirrorhouse on top of the quaternion mirrorhouse that reduce back to the usual quaternion mirrorhouse. The idea is that in the additional mirrorhouses the cyclic reflections would reverse anticommutative effects to commutative effects.

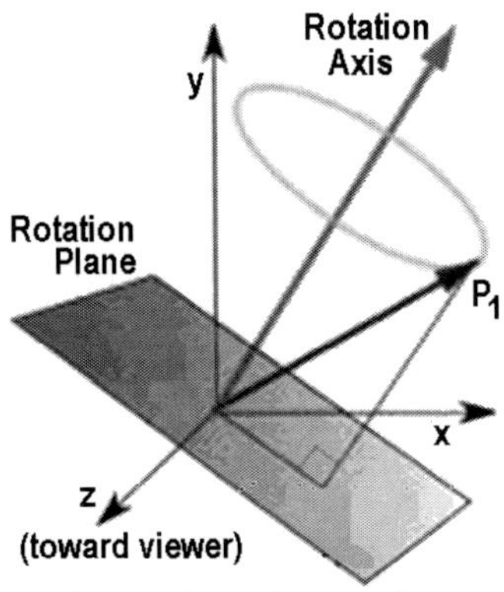

Figure 9. A quaternion can be used to transform one 3D vector into another.

The quaternions can be physically modeled by three mirrors facing each other in a triangle; and the octonions modeled by four mirrors facing each other in a tetrahedron (a cube is a tetrahedron), or more complex packing structures related to tetrahedra. Using these facts, we wish to explore the structure of spacetime as that of a quaternionic. Buckminster Fuller (1982) for example viewed the tetrahedron as an essential structure for internal and external reality. In addition in nature the structure of HD space is an interacting tiling of topologically switching adjacent quaternionic or octonionic mirrorhouses that break Hamilton's closure cyclicality with an inherent natural periodicity that oscillates from Non-Abelian to Abelian continuously. A 3D quaternion transform is shown in Fig. 9.

6.2. *Mirror-Symmetry Experiment*

One of us (RLA) curious to experience firsthand the symmetries of a quaternion mirrorhouse got permission on 30 Jan 2012, 1:30 - 2 PM to use Sealing room 6, in the Oakland, CA LDS Temple for an experiment.

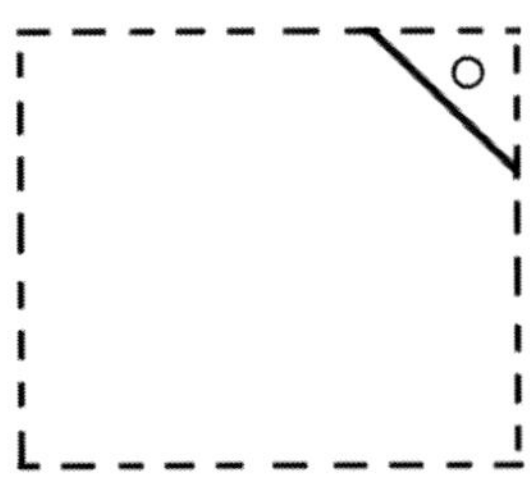

Figure 10. Quaternion mirror house experiment. The dashed line-sided square represents the four mirrored walls of a large room. The solid line in the upper right corner a 6 foot mirror supported on a small table, and the circle a chair inside the mirror house corner.

Getting inside the mirror house 4 images appeared. I initially thought it might be 3 images because I had made a triangle, but the 2 corners reflected off the mirror behind my head create 4 proximal images. Not counting the observer it's like viewing an infinite series of quaternion vertices doubling *ad infinitum* with each mirror having the 4-fold quaternion vertex (Fig. 10).

If I closed my right eye, the face in the left vertex mirror closes its left eye (eye on the left). Then in the 2nd level image it is the right eye in right vertex image that closes. This is standard mirror symmetry imagery, but interesting to experience and verify 1st hand because it helps pondering how to reflect, rotate and invert more complex image components in order to create a new cyclical quaternion algebra that will bot commute and anticommute.

7. Calabi–Yau Manifolds — Brief Review

In superstring theory, the extra dimensions (XD) of spacetime are sometimes conjectured to take the form of a 6-dimensional Calabi-Yau manifold which led to the idea of mirror symmetry. The classical formulation of mirror symmetry relates 2 Calabi-Yau 3-folds M & W whose Hodge numbers three folds M $h^{1,1}$ and $h^{1,2}$ are swapped. Mirror symmetry is a special example of T-duality: the Calabi-Yau manifold may be written as a fiber bundle whose fiber is a three-dimensional torus. The simultaneous action of T-duality on all three dimensions of this torus is equivalent to mirror symmetry. Mirror symmetry allowed the physicists to calculate many quantities that seemed virtually incalculable before, by invoking the "mirror" description of a given physical situation, which can be often much easier.

There are two different, but closely related, string theory statements of mirror symmetry.

1. Type IIA string theory on a Calabi-Yau M is mirror dual to Type IIB on W.

2. Type IIB string theory on a Calabi-Yau M is mirror dual to Type IIA on W.
This follows from the fact that Calabi-Yau hodge numbers satisfy $h^{1,1} \geq 1$ but $h^{2,1} \geq 0$. If the Hodge numbers of M are such that $h^{2,1} = 0$ then by definition its mirror dual W is not Calabi-Yau. As a result mirror symmetry allows for the definition of an extended space of compact spaces, which are defined by the W of the above two mirror symmetries.

Essentially, Calabi–Yau manifolds are shapes that satisfy the requirement of space for the six "unseen" spatial dimensions of string theory, which are currently considered to be smaller than our currently observable lengths as they have not yet been detected. An unpopular alternative that we embrace and have designed experiments to test for [14] is known as large extra dimensions (LSXD). This often occurs in braneworld models - that the Calabi–Yau brane topology is large but we are confined to a small subset on which it intersects a D-brane.

8. Search for a Commutative–Anticommutative Cyclical Algebra

We view the Hubble Universe, H_R as a nilpotent state of zero totality arising from a Higher Dimensional (HD) absolute space of infinite potentia. The world of apparent nonzero states that we observe, R must be associated with a zero creating conjugate, R^* such that $(R)(R) \Rightarrow (R)(R^*)$ in order to maintain the fundamental nilpotent condition of reality.

Since the quaternion set is isomorphic to the Pauli matrices we believe the Quaternion and Octonion sets are not merely a form of pure mathematics but indicative that the physical geometry of the universe is a fundamental generator of these algebras. Baez [1] stated that the properties of these algebras have: "so far…just begun to be exploited". Our purpose in this present work is to discover a new type of dualing in the cyclic dimensionality that alternates between the principle of closure and openness, i.e. anticommutivity and commutivity.

For our purposes we wish to apply this to an extended version of Cramer's Transactional Interpretation (TI) of quantum theory where according to Cramer the present instant of a quantum state is like a standing-wave of the future-past [5]. Cramer never elaborates on what he means by "standing-wave". In our extension the standing-wave transactional present is a metaphorically more like a hyperspherical standing wave with HD Calabi-Yau mirror house symmetry because the richer topology of brane transformations is equipped to handle the duality of closure and openness essential to our derivation of a new algebra we hope to use as the empirical foundation of Unified Field Mechanics, U_F [14].

We symbolize closure and openness as a form of wave-particle duality, which in our model we elevate to a principle of cosmology. So that rather than having the usual fixed closure built in to the quaternion algebra, we manipulate the standing-wave mirror house cycles so that closure is periodic instead. To do this we might keep a background doubling as a synchronization backbone rather than the 'reduction by a factor of 2' which as Rowlands states:

> [4] …describing a set of operators such as *i, j, k* as 'cyclic' means reducing the amount of independent

information they contain by a factor of 2, because *k* for example arises purely from the product of *jk*. It could even be argued that the necessity of maintaining the equivalence of Q_4 and C_2 x C_2 x C_2 representations is the determining factor in making the quaternion operators cyclic.

Rowlands further discusses (pg. 15) the dualing of Q_4:

[4] by complexifying it to the multivatiate vector group, 1, -1, *i*, -*i*, **i**, -**i**, **j**, -**j**, **k**, -**k**, *i***i**, -*i***i**, *i***j**, -*i***j**, *i***k**, -*i***k** of order 16, which has a related C_2 x C_2 x C_2 x C_2 formalism and which may also be written, 1, -1, *i*, -*i*, *i***i**, -*i***i**, *i***j**, -*i***j**, *i***k**, -*i***k**, **i**, -**i**, **j**, -**j**, **k**, -**k**, where a complex quaternion such as, *i***i** becomes the equivalent of the multivariate vector **i**.

There are two options

1) $(AB)(AB) \rightarrow$ (R^*) (Anticommutative)
2) $(AB)(AB) \rightarrow$ (R) (Commutative)

Option 1 was chosen by Hamilton when he discovered that the 3-dimensionality of the quaternion algebra required closure. The closed cycle of the anti-commuting option (zero creating) with distinguish-able terms; but the commutative option has a series of terms that are completely indistinguishable (ontological) because the anticommuting partner must be unique. There are a number of forms of these open and closed dualities.

Planes through the origin of this 3D vector space give subalgebras of O isomorphic to the quaternions, lines through the origin give subalgebras isomorphic to the complex numbers, and the origin itself gives a subalgebra isomorphic to the real numbers.

What we really have here is reminiscent of a description of octonions as a 'twisted group algebra'. Given any group G, the group algebra $\mathbb{R}[G]$ consists of all finite formal linear combinations of elements of G with real coefficients. This is an associative algebra with the product coming from that of G. We can use any function

$$\alpha : G^2 \rightarrow \{\pm 1\}$$

to `twist' this product, defining a new product

$$*\mathbb{R}[G]\times\mathbb{R}[G]\rightarrow\mathbb{R}[G]$$

by: $g * h = \alpha(g,h)gh$ Where $g,h \in G \subset \mathbb{R}[G]$. One can figure out an equation involving α that guarantees this new product will be associative. In this case we call α a '2-cocycle'. If α satisfies a certain extra equation, the product will also be commutative, and we call α a 'stable 2-cocycle'. For example, the group algebra $\mathbb{R}[\mathbb{Z}_2]$ is isomorphic to a product of 2 copies of $\mathbb{R}$, but we can twist it by a stable 2-cocyle to obtain the complex numbers. The group algebra $\mathbb{R}[\mathbb{Z}_2^2]$ is isomorphic to a product of 4 copies of $\mathbb{R}$, but we can twist it by a 2-cocycle to obtain the quaternions. Similarly, the group algebra $\mathbb{R}[\mathbb{Z}_2^3]$ is a product of 8 copies of $\mathbb{R}$, and what we have really done in this section is describe a function α that allows us to twist this group algebra to obtain the octonions. Since the octonions are nonassociative, this function is not a 2-cocycle. However, its coboundary is a 'stable 3-cocycle', which allows one to define a new associator and braiding for the category of $\mathbb{Z}_2^3$-graded vector spaces, making it into a symmetric monoidal category [15]. In this symmetric monoidal category, the octonions are a commutative monoid object. In less technical terms: this category provides a context in which the octonions *are* commutative and associative [1].

We do not believe an Octonion approach is correct for developing our new algebra and only provides a *gedanken* medium for exploring dualities and doublings. As in the Fano plane (Fig. 11) an ordinary 3-cube has 8 possible quaternionic vertices. A hypercube as in Fig. 5 explodes into eight 3-cubes for a total of 64 quaternionic vertices (see Fig. 4). Without yet having finished formalizing the new algebra we believe 64 symmetry components are required for its invention. See [4] for a discussion pertaining to 64 quaternionic elements.

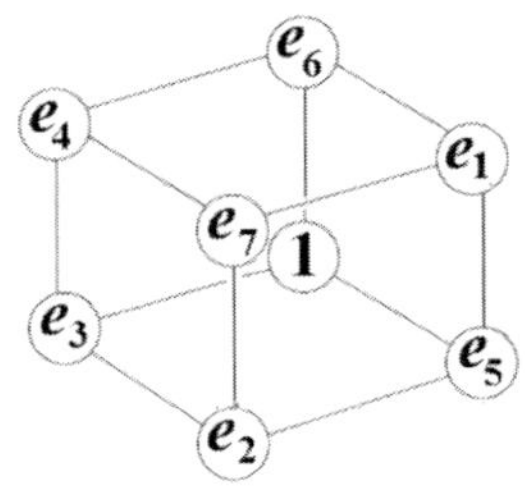

Figure 11. The Fano plane is the projective plane over the 2-element field $\mathbb{Z}_2$ consisting of lines through the origin in the vector space $\mathbb{Z}_2^3$. Since every such line contains a single nonzero element, we can also think of the Fano plane as consisting of the seven nonzero elements of $\mathbb{Z}_2^3$. If we think of the origin in $\mathbb{Z}_2^3$ as corresponding to $1 \in O$, we get this picture of the octonions.

The quaternions are a 4-dimensional algebra with basis 1, *i*, *j*, *k*. To describe the product we could give a multiplication table like Tbl. 1, but it is easier to remember that:

- 1 is the multiplicative identity,
- *i*, *j* and *k* are square roots of -1,

- $ij = k$, $ji = -k$, and all identities obtained from these by cyclic permutations of (i,j,k).

We can summarize the quaternion rule in a diagram, Fig. 8. When we multiply two elements going clockwise around the circle we get the next element: for example, $i,j = k$. But when we multiply two going around counterclockwise, we get *minus* of the next element: for example, $j,i = -k$ [1].

This cyclicality is key to our continuous-state commutative anticommutative leap-frogging of the metric. Einstein believed 'the photon provides its own medium and no ether is required'. This is true in general but the medium must be imbedded in a cosmology. Therefore is the Rowland's concept that the fermion vertex is fundamental similar, i.e. needing a cosmological embedding. Correspondence to HAM cosmology could provide the proper embedding. This is a primary postulate herein.

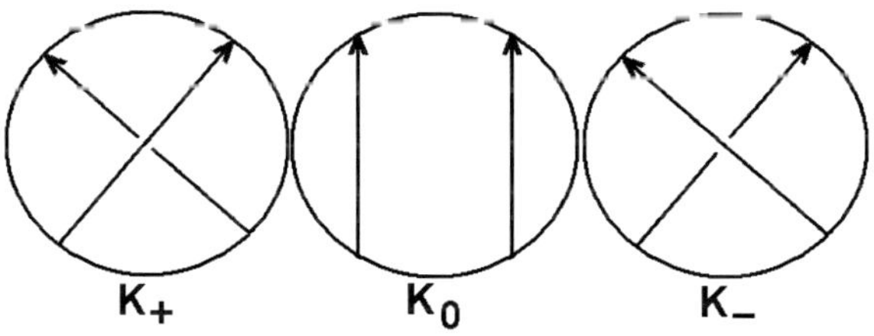

Figure 12. Crossing change triple. Imagine embedded in Figs. 6-8.

In Fig. 12 the ordered set (K_+, K_-, K_0) is called a crossing-change-triple. Let (K_+, K_-, K_0) be a twist-move-triple See [16-18] for a variations and in depth discussion.

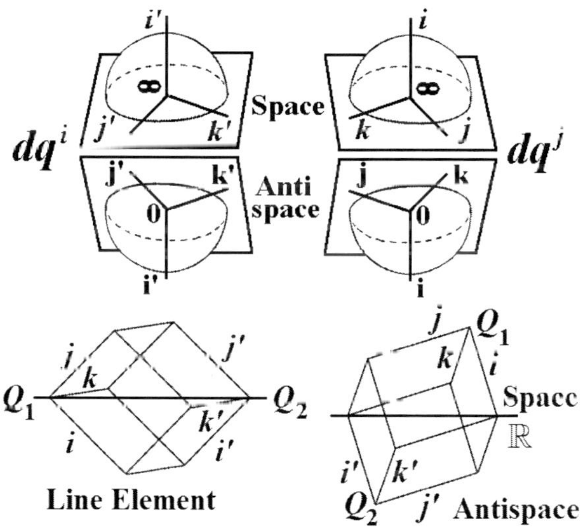

Figure 13. Riemann sphere quaternionic line elements.

Riemann sphere manifolds are key to algebraic topology. The Riemann sphere is usually represented as the unit sphere $x^2 + y^2 + z^2 = 1$ in real 3D space, R^3 (x,y,z) where two stereographic projections from the unit sphere are made onto the complex plane by $\zeta = x + iy$ and $\xi = x - iy$, written as

$$\varsigma = \frac{x+iy}{1-z} = \cot\left(\tfrac{1}{2}\phi\right)e^{i\theta}$$

and

$$\xi = \frac{x-iy}{1+z} = \tan\left(\tfrac{1}{2}\phi\right)e^{-i\theta}.$$

In order to cover the whole unit sphere two complex planes are required because each stereographic projection alone is missing one point, either the point at zero or infinity. The extended complex numbers consist of the complex numbers C together with ∞. The extended complex numbers are written as $\hat{\mathbb{C}} = \mathbb{C} \cup \{\infty\}$. Geometrically, the set of extended complex numbers is referred to as the Riemann sphere (or extended complex plane) seen as glued back-to-back at $z = 0$. Note that the two complex planes are identified differently with the plane $z = 0$. An orientation-reversal is necessary to maintain consistent orientation on the sphere, and in particular complex conjugation causes the transition maps to be holomorphic.

The transition maps between ζ-coordinates and ξ-coordinates are obtained by composing one projection with the inverse of the other. They turn out to be $\zeta = 1/\xi$ and $\xi = 1/\zeta$, as described above. Thus the unit sphere is diffeomorphic to the Riemann sphere. Under this diffeomorphism, the unit circle in the ζ-chart, the unit circle in the ξ-chart, and the equator of the unit sphere are all identified. The unit disk $|\zeta| < 1$ is identified with the southern hemisphere $z < 0$, while the unit disk $|\xi| < 1$ is identified with the northern hemisphere $z > 0$. A Riemann surface does not have a unique Riemannian metric which means we may develop a relevant metric for the noetic transformation.

This implies a dual or mirror metricity where one is fixed in reality or 3-space and the 'mirror' confined to complex space where it then cyclically 'topologically switches' into real space and the real space metric takes its place in complex space. A component of the metric fixing-unfixing when applied to the 'Walking of the Moai on Rapa Nui' relates to recent work of Kauffman:

> When the oscillation (**HAM continuous-state**) is seen without beginning or end, it can be regarded in two distinct ways: as an oscillation from marked to unmarked, or as an oscillation from unmarked to marked. In this way the re-entering marked becomes the recursive process generated by the equation and J is an algebraic or imaginary fixed point for the recursion.

$$R(x) = \overline{\;x\;}|$$

The analog with the complex numbers is sufficiently strong that we now shift to the complex numbers themselves as a specific instantiation of this recursive process. In the body of the paper we will return to the more rarified discussion leading from the distinction itself. We begin with recursive processes that work with the real numbers **+1** and **-1**. Think of the oscillatory process generated by **R(x) = -1/x** [17].

The fixed point is **i** with $\mathbf{i^2 = -1}$, and the processes generated over the real numbers must be directly related to the idealized **i**. There are two i*terant views*: **[+1,-1]** and **[-1,+1]** (Fig. 14).

These are seen as points of view of an alternating process that engender the square roots of negative unity.

... +1 -1 +1 -1 +1 -1 +1 -1 +1 -1 +1 -1 +1 -1 +1 -1...

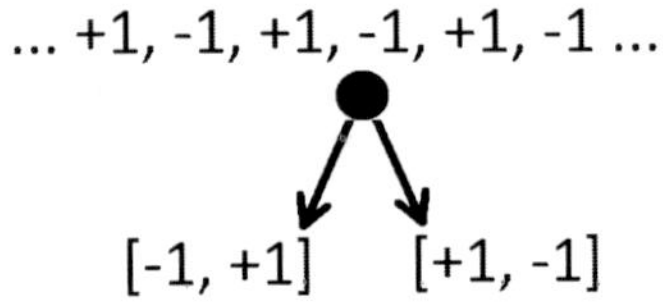

Figure 14. Quantizing the Iterant views.

To see this, we introduce a *temporal shift operator,* **h** such that **[a,b]h = h [b,a]** and **h h = 1** so that concatenated observations can include a time step of one-half period of the process **...abababab...** (Fig. 16) [17].

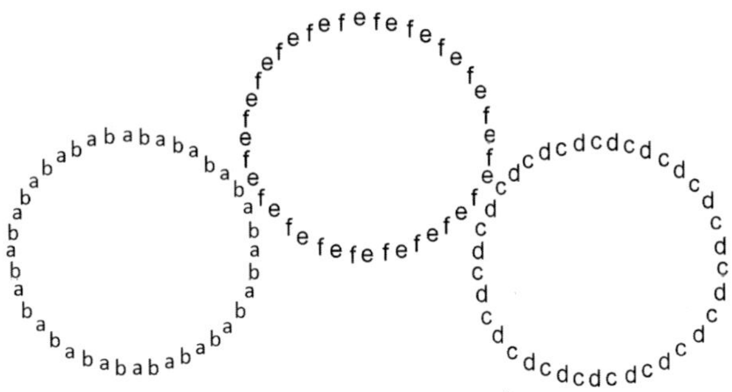

Figure 15. Triune temporal shift operator for 1st triad dualing.

Figure 15 reminds us of the 'walking of the Moai on Rapa Nui' [11,12] because if for example the left foot is the ababab iterator, when it is lifted the cdcdcd iterator represents the 'fixed metric' and the efefef acts as the shift operator enabling a 'complex crossing change' from commutivity to anticommutivity or vice versa. However as we already stated several more triune sets of additional mirror house doublings or dualings are required.

The 3rd complex metric is involved in making an evolution from dual quaternions to a 3rd quaternion we choose to name a trivector that acts as a baton passing mechanism between the space-antispace or dual quaternion vector space. The trivector facilitates a 'leap-frogging' between anti-commutative and commutative modes of HD space. This inaugurates a Mobius or transformation between the Riemann dual stereographic projection complex planes.

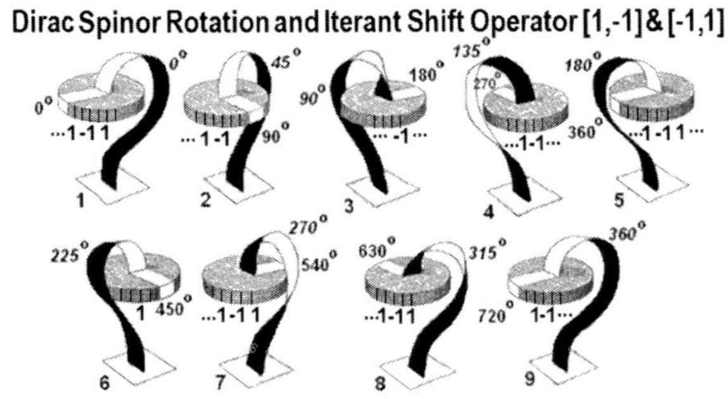

Figure 16. Temporal *i* clock shift operator with addition of Dirac spherical rotation in preparation for Mobius transformation crossing changes. See Fig. 12.

In the midst of the HAM continuous-state dimensional reduction compactification process (a U_F process) there is a central (focused around the singularity or LCU) coordinate 'fixing-unfixing'. In prior work we [5] have related this to the integration of G-em coordinates, arbitrarily fixing one metric. That work proceeded discovery of the continuous-state process wherein now an inherent cyclicality occurs in first fixing one metric and then fixing the other. This is an essential aspect of the gating mechanism for the U_F dynamics to surmount quantum uncertainty.

Geometrically, a standard Möbius transformation can be obtained by first performing stereographic projection from the plane to the unit 2-sphere, rotating and moving the sphere to a new location and orientation in space, and then performing stereographic projection (from the new position of the sphere) to the plane. These transformations preserve angles, map every straight line to a line or circle, and map every circle to a line or circle. Möbius transformations are defined on the extended complex plane (i.e. the complex plane augmented by the point at infinity): $\hat{\mathbb{C}} = \mathbb{C} \cup \{\infty\}$.

This extended complex plane can be thought of as a sphere, the Riemann sphere, or as the complex projective line. Every Möbius transformation is a bijective conformal map of the Riemann sphere to itself. Every such map is by necessity a Möbius transformation. Geometrically this map is the Riemann stereographic projection of a rotation by 90° around $\pm i$ with period 4, which takes the continuous cycle $0 \to 1 \to \infty \to -1 \to 0$.

An LSXD point particle representation of a fermionic singularity [10]. The 8 + 8 or 16 ($\mathbb{C}^{\pm 4}$ complex space) 2-spheres with future-past retarded-advanced contours are representations of HD components of a Cramer 'standing-wave' transaction. This can be considered in terms of Figs. X,Y,Z as Calabi-Yau dual mirror symmetries. To produce the quaternion trivector representation (Figs. 2 & 3) a 3rd singularity-contour map is required which is then also dualed, i.e. resulting in 6 singularity/contour maps. This many are required to oscillate from anticommutivity to commutivity to provide the *cyclic* opportunity to violate 4D quantum uncertainty [6]!

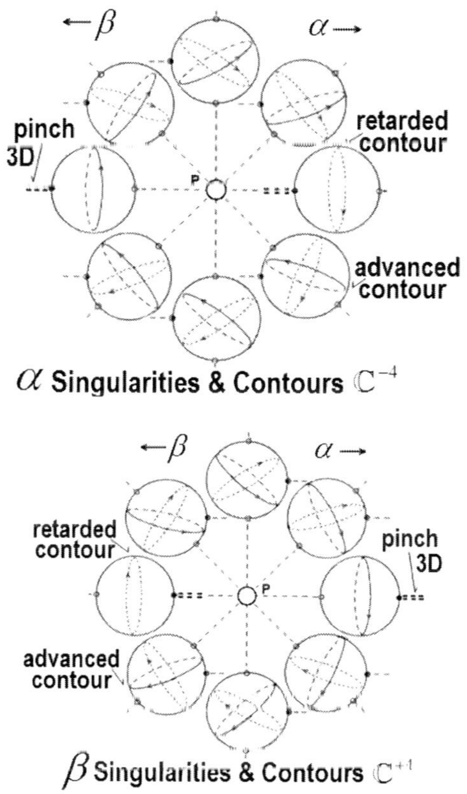

Figure 17. Two of six HD Nilpotent trivector r-qubit symmetries.

Figure 17 begins to illustrate the mirror symmetries involved in the more complex dualities. If the central points P represent mirrored 3-cubes each of which already has 8 quaternionic vertices, then each sphere in Fig. 17 represents one of eight 3-cubes exploded off a hypercube which is each mirrored in the α and β singularities and contours. We believe one more mirror house doubling is required beyond that scenario to complete a commutative-anticommutative iterative shift operator!

References and Notes

[1] John C. Baez (2005) On Quaternions and Octonions: Their Geometry, Arithmetic, and Symmetry by John H. Conway and Derek A. Smith, Bull. Amer. Math. Soc. 42 (2005), 229-243.

[2] Hamilton, W. R. (1853) Lectures on Quaternions: Containing a Systematic Statement of a New Mathematical Method. Dublin: Hodges and Smith.

[3] Hamilton, W. R. (1866) Elements of Quaternions. London: Longmans, Green.

[4] Rowlands, P. (2007) Zero to Infinity: The Foundations of Physics, Singapore: World Scientific.

[5] Amoroso, R.L. & Rauscher, E.A. (2009) The Holographic Anthropic Multiverse, Singapore: World Scientific.

[6] Kauffman, L.H. (2004) Non-commutative worlds. New Journal of Physics. 6, 2-46.

[7] Rowlands, P. (1994) An algebra combining vectors and quaternions: A comment on James D. Edmonds' paper. Speculat. Sci. Tech., 17, 279-282.

[8] P. Rowlands, Some interpretations of the Dirac algebra. Speculat. Sci. Tech., 19, 243-51, 1996.

[9] P. Rowlands, The physical consequences of a new version of the Dirac equation, in G. Hunter, S. Jeffers, and J-P. Vigier (eds.), Causality and Locality in Modern Physics and Astronomy: Open Questions and Possible Solutions (Fundamental Theories of Physics, vol. 97, Kluwer Academic Publishers, Dordrecht, 1998, 397-402.

[10] Rowlands, P. (2013) Space antispace v8

[11]Lipo, C.P., Hunt, T.L. & Haoa S.R. (2012) The 'walking' megalithic statues (moai) of Easter Island, J Archaeological Science, 1-8, in press

[12] Amazing Video, Walking of the Moai on Rapa Nui (Easter Island) http://www.youtube.com/watch?v=yvvES47OdmY.

[13] Goertzel, B., Aam, O., Smith, F.T. & Palmer, K. (2007) Mirror Neurons, Mirrorhouses, and the Algebraic Structure of the Self, Google online http://www site.

[14] See Chaps. 30 & 52 this volume

[15] Conway J.H. & Smith D.A. (2003) On Quaternions and Octonions: Their Geometry and Symmetry, Natic: A.K. Peters Ltd.

[16] Kauffman, L.H. (2013) Laws of form, Majorana fermions and discrete physics, pp. 1-18 V8

[17] Kauffman, L.H. (2013) i clock.

[18] Kauffman, L.H. & Ogasa, E. (2012) Local moves on knots and products of knots, http://arxiv.org/abs/1210.4667v1.

[19] London V & A Museum sign, side view http://www.youtube.com/watch?v=aat6Oic8biU Rotation in Calabi-Yau Mirror Symmetry like the surface of a brane

[20] London V & A Museum sign, end view http://www.youtube.com/watch?v=i0QRRcc03wE

[21] Amoroso RL, Kauffmann LH, Rauscher EA, Rowlands PR & Vigier J-P (2010) "Hidden" Parameters Describing Internal Motion Within Extended Particle Elements, in Search for Fundamental Theory, The VIIth International Symposium Honoring French Mathematical Physicist Jean-Pierre Vigier (ed. RL Amoroso et al. AIP Conference Proceedings. 1316, 1-27.

Explicit and Implicit Uncertainties and the Uncertainty Principle in the Special Theory of Relativity

OLEG V. MATVEJEV

Joint-Stock Company "Sinerta"
Vilnius, LT-03150, Lithuania

VADIM N. MATVEEV

Joint-Stock Company "Sinerta"
Vilnius, LT-03150, Lithuania

A macroscopic object equipped with synchronized clocks is examined. General physical relations are directly derived from Lorentz transformations for the case of one-dimensional motion (along the X axis) – the uncertainty relation of the object's x coordinate and the projection of its impulse along the X axis, p_x, and the uncertainty relation of the object's observation time, t, and its energy, E. The uncertainty relations take the form $dp_xdx > H$ and $dEdt > H$. The H value in the relation has action dimensions and is dependent upon the precision of the rod's clocks and its mass. It is shown that if the macroscopic object in and of itself performs the function of an ideal physical clock, the uncertainty relations derived in the limiting case then take the usual form of $dp_xdx \geq h$ and $dEdt \geq h$, where h is the Planck constant.

Keywords: Lorentz transformation, Special relativity, Uncertainty principle

1. Introduction

The uncertainty relation of a microparticle coordinate and the projection of its impulse along the coordinate axis, as well as the uncertainty relation of microparticle time and energy, falls into the ranks of the most important quantum relations that express the known uncertainty principle in a mathematical form. Quantum relations are not observed in the macrocosm in the sense that they are vanishingly small with respect to macrobodies and do not show themselves in practice. According to our data, no general physical relations have been reported in extant literature that extend to the macrocosm and that limit the simultaneous specification of the impulse and the coordinate of physical bodies. In light of existing notions concerning the properties of macroscopic bodies, the absence of such general physical uncertainty relations seems natural, since uncertainty relations have not been connected to measuring instrument errors, while it is possible to talk about macrobody coordinate, impulse, and energy inaccuracies, as they seem just as obvious today as measurement errors.

The objective of the work at hand was to demonstrate the existence of general physical uncertainty relations that extend to macrobodies. This objective was achieved over the course of solving a problem that consisted of determining the velocity and the velocity-related physical values of objects based on the degree of desynchronization of moving clocks.

The relations $\Delta p_x \Delta x \geq H$ and $\Delta E \Delta t \geq H$ were directly derived from Lorentz transformations. The first relation connects the uncertainty of the projection of the impulse, p_x, of the body under consideration, Δp_x, to the uncertainty of the x coordinate, Δx, while the second relation connects the uncertainty of the energy, E, of the object, ΔE, to the observation time, t, uncertainty, Δt. The H value in the relations has an action dimension and is dependent upon the precision of the rod's clocks and its mass.

Taking into account the fact that in Russia the recommendations formulated by metrologists for replacing the terms "error" with the term "uncertainty" (explicit uncertainty) are not compulsory in nature, and bearing in mind that these recommendations are still in the discussional stage [1-2], we used both terms in the work at hand.

We will take error to mean the inaccuracy of measurement results that is due to purely metrological causes. Increasing an instrument's measurement accuracy can reduce the error. The absolute error of the distance between two points can serve as an example of an error.

We will call the inaccuracy of measurement results that cannot be eliminated by means of increasing measurement instrument accuracy, and which may be due to terminological, conceptual, or linguistic causes, the implicit uncertainty of physical values. The uncertainty of the distance between two spheres that are located close to one another can serve as an example of this implicit uncertainty. This distance remains uncertain with an accuracy of up to the dimensions of the spheres, even in the presence of ideal measurement accuracy, while the thing that remains unclear is what is meant by the sought distance – the distance between the centers of mass of the spheres, the distance between their geometric centers, the distance between the closest points of the spheres, or something else. Construing uncertainty in this manner is mentioned in extant literature, albeit in general terms. In this vein, for example, taking into account the fact that it is only possible to specify the location of a spatially extended body by determining the position of a single solitary point that belongs to it with some degree of uncertainty, the uncertainty of the position of a sphere determined by the position of its center, which equals the radius of this sphere, is written about in reference [3].

The uncertainty of specifying the moment in time of a short-term process that does not occur instantaneously, but rather occupies a certain finite, let's say, very short time interval, can serve as another example of an uncertainty of this type. This uncertainty can be regarded as equal to half the duration of the process, if the moment in time of its transit is called the moment into which the middle of the process falls. These coordinate and time uncertainties, Δx and Δt, are precisely the ones that play a part in the uncertainty relations we derived.

2. Determining Velocity of a Rod by Reading Appurtenant Synchronized Clocks

We will visualize a thin rod of proper length, L, at two points, a and b, on which synchronously running clocks, A and B, are installed at a distance of d from one another.

Let's say that clocks A and B, like the clocks appurtenant to the K^0 reference system, where the rod is at rest, show this system's time; i.e., the readings of clocks A and B are always in agreement with the readings of the K^0 system's clocks. The length, d, of section ab, which is located between points a and b, may be equal to or less than the rod length, L; i.e., the condition $L \geq d$ holds true in the general case. If $L > d$, then the rod will look something like this:

----------*A*--------------------*B*-----------

Here, the A and B characters conditionally designate clock A and clock B, while the broken line shows the body of the rod.

If the distance, d, is equal to the rod length, L – i.e., if $L = d$, clocks A and B are then found at the ends of the rod.

We will call the arrangement that consists of this rod and the two running clocks, A and B, situated on it rod, R; i.e., clocks A and B will be regarded as integral parts of rod R, and we will treat the readings of clocks A and B as characteristic attributes of rod R.

Let's say that rod R, which is positioned parallel to the X^0 axis of inertial system K^0, is at rest relative to this system and moves at a constant velocity, V', along the X' axis of another reference system, K', remaining parallel to the X' axis and its direction of movement (the X^0 and X' axes of the K^0 and K' systems slide along one another over the course of their relative motion). We will call this rod motion longitudinal motion (with respect to its orientation in space), and it alone will be referred to in the future.

Pursuant to inverse Lorentz transformations, at a moment in time of t' of the K' system, the readings, $\tau_{A,t'}$ and $\tau_{B,t'}$, of clocks A and B, which are in agreement with the readings of the K^0 system's clocks, are determined by the relations

$$\tau_{A,t'} = \frac{t' + \frac{V'}{c^2} x'_{A,t'}}{\sqrt{1 - (\frac{V'}{c})^2}} \quad \text{and} \quad \tau_{B,t'} = \frac{t' + \frac{V'}{c^2} x'_{B,t'}}{\sqrt{1 - (\frac{V'}{c})^2}},$$

where $x'_{A,t'}$ and $x'_{B,t'}$ – the coordinates of clocks A and B of rod R within system K' at a moment in tie of t', while c – the speed of light in a vacuum.

As follows from the relations presented above, the difference in the readings of clocks A and B of rod R at a moment in time of t' for system K' equals

$$\tau_{B,t'} - \tau_{A,t'} = \frac{(x'_{B,t'} - x'_{A,t'})V'}{c^2 \sqrt{1 - (\frac{V'}{c})^2}}. \qquad (1)$$

Introducing the following notation for the purpose of saving writing space

$$U' = \frac{V'}{\sqrt{1 - (\frac{V'}{c})^2}}, \qquad (2)$$

then from formula (1), with allowance for notation (2), we obtain:

$$U' = \frac{c^2(\tau_{B,t'} - \tau_{A,t'})}{x'_{B,t'} - x'_{A,t'}}. \qquad (3)$$

Thus, having the data on the coordinates and

readings of clocks A and B at a moment in time of t', formula (3) can be used to find the value of U', and this value can in turn be used to find the velocity, V', of rod R in system K'. We will call the data characterizing the object's spatially distributed elements, but that relate to one and the same moment in time, t', single-time data.

In addition to the possibility of determining the rod's velocity using single-time data, the possibility also exists of determining the velocity, V', of rod R in system K' using single-coordinate data. We will call the data successively recorded at different moments in time, but at one and the same point (with one and the same coordinate), which characterize the elements of a spatially extended object at the times that they are located at this (or near this) point, single-coordinate data. Data of this type include, for example, the $\tau_{A,x'}$ and $\tau_{B,x'}$ readings of clocks A and B of rod R that are successively recorded at moments in time of $t'_{A,x'}$ and $t'_{B,x'}$ at a point in system K' with a coordinate of x', past which it is moving within this reference system.

According to inverse Lorentz transformations, the single-coordinate $\tau_{A,x'}$ and $\tau_{B,x'}$ readings of clocks A and B are connected to the moments in time, $t'_{A,x'}$ and $t'_{B,x'}$, during which clocks A and B, at a point with a coordinate of x', show the relations

$$\tau_{A,x'} = \frac{t'_{A,x'} + \frac{V'}{c^2}x'}{\sqrt{1-(\frac{V'}{c})^2}} \text{ and } \tau_{B,x'} = \frac{t'_{B,x'} + \frac{V'}{c^2}x'}{\sqrt{1-(\frac{V'}{c})^2}}.$$

As follows from the relations presented above, the difference in the readings of clocks A and B of rod R at a point with a K' system coordinate of x' equals

$$\tau_{B,x'} - \tau_{A,x'} = \frac{t'_{B,x'} - t'_{A,x'}}{\sqrt{1-(\frac{V'}{c})^2}}. \tag{4}$$

Introducing the notation $\Gamma' = 1/\sqrt{1-(V'/c)^2}$, then from formula (4), we obtain

$$\Gamma' = \frac{\tau_{B,x'} - \tau_{A,x'}}{t'_{B,x'} - t'_{A,x'}}, \tag{5}$$

from which the velocity, V', of rod R can be learned as necessary.

The determination of velocity using single-coordinate data is interesting in that it does not require distance measurements, but rather is based on the measurement of the $\tau_{B,x'} - \tau_{A,x'}$ and $t'_{B,x'} - t'_{A,x'}$ time intervals.

3. Relation of Rod Velocity and Coordinate Uncertainties Calculated by Single-Time Data

In talking about the determination of the velocity of rod R using the difference in the readings of clocks A and B, we tacitly proceeded on the basis of the fact that clocks A and B of rod R run ideally; i.e., the readings of clocks A and B are in absolutely precise agreement with the ideally accurate readings of the K^0 system clocks. We speculate that a certain maximum absolute error exists in the time readings of each of clocks A and B, which are appurtenant to rod R. We will designate the absolute error of these clocks as $\Delta\tau$. Here, adhering to the generally accepted assumption, we will presume that all the clocks appurtenant to any inertial reference system, including the K^0 system, run with ideal accuracy.

The existence of an absolute error, $\Delta\tau$, in the readings of each of clocks A and B means that, at a given moment in time, t^0, within the K^0 system, the readings of each of these clocks of rod R can differ from the readings of the K^0 system clocks located close to them and running with ideal accuracy by a value that does not exceed $\Delta\tau$. In this regard, remarks made above concerning agreement of the reading of the K^0 system clocks with those of clocks A and B must be interpreted with allowance for the finite accuracy of the latter.

If the rate of the K' system clocks, the measured $x'_{B,t'} - x'_{A,t'}$ value, and the speed of light c value are as accurate as desired, the error in the U' value calculated using formula (3) will then be solely due to the existence of an absolute error, $\Delta(\tau_{B,t'} - \tau_{A,t'})$, in the difference of the $\tau_{B,t'} - \tau_{A,t'}$ readings of clocks A and B.

In this instance, the $\Delta U'$ error, with allowance for formula (3), is expressed by the equality

$$\Delta U' = \frac{c^2 \Delta(\tau_{B,t'} - \tau_{A,t'})}{x'_{B,t'} - x'_{A,t'}} \tag{6}$$

In instances when the maximum absolute error of the difference in the readings of clocks A and B consists of the $\Delta\tau$ errors of each of these clocks – i.e., when $\Delta(\tau_{B,t'} - \tau_{A,t'}) = 2\Delta\tau$, it follows from equality (6) that:

$$(x'_{B,t'} - x'_{A,t'})\Delta U' = 2c^2\Delta\tau. \tag{7}$$

We note that the $\Delta\tau$ error is not dependent upon reference system selection, since it consists of the maximum possible difference between the readings of each of clocks A and B and the readings of the K^0 system clocks located close to them, where rod R is at rest. It is clear that this difference is not dependent upon the reference system within which it is picked up.

We will now imagine that, in addition to the condition of the single-time nature of the readings of clocks A and B within system K', the requirement of the single-coordinate nature of the specification of the location where rod R is situated at a moment in time of t' is satisfied. Let's say the essence of this requirement consists of using a single solitary x'_R coordinate to specify the location of the projection of rod R on the X' axis. This requirement can be satisfied if the rod length is ignored and it is regarded as a point. But if, due to the necessity of taking the property of the rod's spatial extent into account, it is impossible to ignore its length, the requirement of the single-coordinate nature of the specification of the position of rod R can only be satisfied in part. For example, the coordinate of any point appurtenant to rod R can be specified as its x'_R coordinate and a reference to its uncertainty can accompany the specification of this coordinate. In particular, the coordinate of the geometric center of rod R, or the coordinate of its center of mass, can serve as the coordinate of this point. In such cases, the distance from the point with a coordinate of x'_R to the point of the rod's projection on the X' axis farthest away from it can be regarded as the uncertainty, $\Delta x'_R$, of the x'_R coordinate.

When the position of rod R is specified in this manner, the x'_R coordinate indicates the location of one of a set of points of the projection of rod R that lies on the X' axis. If we give this point preference for one reason or another, the $\Delta x'_R$ uncertainty will then determine a range of point coordinates that, in the presence of other considerations, could also be regarded at the point coordinates of rod R.

For example, if the coordinate, x'_{ab}, of the geometric center (the midpoint) of the ab section of rod R parallel to the X' axis is selected as this section's coordinate, the uncertainty, $\Delta x'_{ab}$, of the x'_{ab} coordinate of the rod's ab section can by definition be regarded as a value equal to half the length of the section of the moving (within the K' system) rod.

Since the length, $d' = d\sqrt{1-(V'/c)^2}$, of the rod's moving ab section within the K' system equals $x'_{B,t'} - x'_{A,t'}$, the uncertainty, $\Delta x'_{ab}$, of the x'_{ab} coordinate of section ab of rod R then equals $\frac{1}{2}(x'_{B,t'} - x'_{A,t'})$. Thus, formula (7) can be presented in the form

$$\Delta U'\Delta x'_{ab} = c^2\Delta\tau \qquad (8)$$

Within the framework we have provided when introducing the determination of uncertainty, the $\Delta x'_{ab}$ value is absolutely an uncertainty and not an error, since it is dependent upon the length of the rod's ab section and cannot be reduced by means of increasing measurement accuracy.

Because $L \geq d$ in the general case, the $\Delta x'$ uncertainty of the x' coordinate of rod R within the arbitrary positioning of clocks A and B thereon ($\Delta x' = \frac{1}{2}L'$) can then generally both equal and exceed the $\Delta x'_{ab}$ value. And since the relation $\Delta x' \geq \Delta x'_{ab}$ holds true, then in the general case of the arbitrary positioning of the clocks on the rod, formula (8) takes the form:

$$\Delta U'\Delta x' \geq c^2\Delta\tau . \qquad (9)$$

Relation (9) pertains to the general case of the arbitrary positioning of the clocks on the rod and makes the transition to the equality $\Delta U'\Delta x' = c^2\Delta\tau$ in the special and most favorable case (as far as the determination of the U' value using single-time data) of clock positioning on the ends of the rod ($\Delta x' = \Delta x'_{ab}$).

The product of the error, $\Delta U'$, of the U' value and the uncertainty, $\Delta x'$, of the x' coordinate of rod R during an instantaneous observation will only be dependent upon the error in the readings of clocks A and B of rod R. Therefore, relation (9) remains unchanged in any inertial reference system and can be written for an arbitrary system in the form:

$$\Delta U_x\Delta x \geq c^2\Delta\tau \qquad (10)$$

If the mass of rod R with clocks A and B is known without a doubt and equals M_R (here and further on, the concepts of Lorentz-invariant mass [4] will be used), relation (10) can be transformed into the relation $M_R\Delta U_x\Delta x \geq M_R c^2\Delta\tau$ by multiplying its left-hand and right-hand members times M_R, whence, taking into account the fact that $M_R U_x = p_x$, we obtain:

$$\Delta p_x\Delta x \geq M_R c^2\Delta\tau . \qquad (11)$$

Introducing the notation

$$H = M_R c^2\Delta\tau , \qquad (12)$$

then from formula (11), we obtain

$$\Delta p_x\Delta x \geq H . \qquad (13)$$

The $\Delta\tau$ error of clocks A and B of rod R is one of the internal parameters of rod R and is not a part of the parameters of the measuring devices that are external relative to rod R. Therefore, the ΔU_x and Δp_x errors in relations (10) and (13) can tentatively be referred to as external (relative to rod R) uncertainties. The internal properties of rod R itself (a salient feature of clocks A and B appurtenant to it) determine the external uncertainties ΔU_x and Δp_x, and in the presence of a specific Δx uncertainty, they cannot be eliminated by means of increasing the accuracy of the measuring instruments located beyond rod R.

4. Relation of Rod Energy and Time Uncertainties Calculated Using Single-Coordinate Data

The error, $\Delta\Gamma'$, of the Γ' value can be derived from formula (5). Only the $\tau_{B,x'} - \tau_{A,x'}$ value error due to the $\Delta\tau$ error in the readings of clocks A and B determines the $\Delta\Gamma'$ error of the Γ' value, since the $t'_{B,x'} - t'_{A,x'}$ time interval at point x', according to our initial assumption, can be measured by the K' system clocks with an ideal accuracy. Therefore, from formula (5), we obtain:

$$\Delta\Gamma' = \frac{\Delta(\tau_{B,x'} - \tau_{A,x'})}{t'_{B,x'} - t'_{A,x'}},$$

or with allowance for the fact that $\Delta(\tau_{B,x'} - \tau_{A,x'}) = 2\Delta\tau$,

$$\Delta\Gamma' = \frac{2\Delta\tau}{t'_{B,x'} - t'_{A,x'}},$$

whence it follows that:

$$\Delta\Gamma'(t'_{B,x'} - t'_{A,x'}) = 2\Delta\tau . \quad (14)$$

We will now assume that, in addition to the condition of the single-coordinate nature of the observation clock readings, the requirement of the single-time nature of the specification of the observation time of the ab section of rod R, moving past point x', is imposed. Let's say the requirement consists of using a single solitary moment in time, $t'_{ab,x'}$, to specify the section ab observation time at point x'.

Since the observation at a point with a coordinate of x' is carried out over a time interval of $t'_{B,x'} - t'_{A,x'}$, it is then only possible to approximately specify the observation time by indicating, for example, the moment in time, $t'_{ab,x'}$, of the middle of the time interval, $t'_{B,x'} - t'_{A,x'}$, that goes toward observation. Here, the uncertainty, $\Delta t'_{ab,x'}$, of the moment in the observation time that equals half the $t'_{B,x'} - t'_{A,x'}$ observation time can be specified; i.e., it can be assumed that $\Delta t'_{ab,x'} = \frac{1}{2}(t'_{B,x'} - t'_{A,x'})$. Then formula (14) can be presented in the form:

$$\Delta\Gamma'\Delta t'_{ab,x'} = \Delta\tau . \quad (15)$$

For the arbitrary positioning of the clocks on the rod, the $\Delta t'_{x'}$ observation time for the entire rod, R, with a length of L proves to be greater than the observation time, $\Delta t'_{ab,x'}$, for the ab rod section; i.e., the condition $\Delta t'_{x'} \geq \Delta t'_{ab,x'}$ is satisfied, as a result of which it follows from formula (15) that

$$\Delta\Gamma'\Delta t'_{x'} \geq \Delta\tau . \quad (16)$$

During an accurate observation, the product of the uncertainty, $\Delta t'_{x'}$, of a moment in time, $t'_{x'}$, of the observation of rod R and the $\Delta\Gamma'$ error of the Γ' value will only be dependent upon the $\Delta\tau$ error of the readings of clocks A and B of rod R. Therefore, relation (16) remains unchanged within any inertial reference system and can be written for an arbitrary system in the form:

$$\Delta\Gamma\Delta t_x \geq \Delta\tau . \quad (17)$$

Multiplying the left-hand and right-hand members of relation (17) times M_Rc^2, bearing in mind that $\Delta\Gamma M_Rc^2 = \Delta E$, and using notation (12), we obtain

$$\Delta E\Delta t_x \geq H . \quad (18)$$

If the $\Delta\tau$ error is regarded as a rod parameter that is not a part of the measuring equipment parameters, the $\Delta\Gamma$ and ΔE values can then be referred to as uncertainties.

5. Clock Relation by Impulse and Coordinate Uncertainties of Spatially Extended Body

We will assume that each of clocks A and B discretely changes its reading with frequency, v. The maximum absolute $\Delta\tau$ error in the readings of the moments in time of each of the clocks will equal $1/v$. We will also assume that this change in the clock readings occurs synchronously; i.e., not only are the clock readings at a given moment in time synchronized, but also the times of the change in readings.

In this instance, the maximum error of the difference in the readings of clocks A and B will equal not the sum of the two $\Delta\tau$ errors, but rather a single $\Delta\tau$ error. Then relations (13) and (18), deduced from the condition of the equality of the clock A and B difference error to the two $\Delta\tau$ errors, take the form:

$$\Delta p_x\Delta x \geq H/2 \quad (19)$$

and

$$\Delta E\Delta t_x \geq H/2 . \quad (20)$$

Let's imagine that each of clocks A and B consists of two components, one of which, being artificial ("manmade"), performs the function of a display and provides discrete time readings, changing them in a keeping with external signals, while the other one – we will call it a physical clock – generates these signals in a natural, easy manner and controls the change in the display's readings.

For the sake of clarity, we will visualize the

second part of the physical clock as a piece of radioactive material with a long half-life (the material's radiant power can be regarded as constant for a sufficiently long time frame).

If the display reacts to a specific portion, ε, of the physical clock material's absorbed gamma-radiation energy by changing its readings, then in the presence of a material radiant power than equals P, the frequency, ν, of the change in the display readings will equal P/ε.

We will call the ε portion of the absorbed energy that leads to a change in the display's reading the energy of the perceived (by the display) physical clock signals.

We will assume that, regardless of the quantity of the physical clock material, the display absorbs all of the energy radiated, and that each ε portion of the energy absorbed performs the function of a reading change signal that is perceived by the display. The frequency, ν, of the physical clock materials signals perceived by the display, and accordingly the frequency of the change in the physical clock's readings, will then be proportional to the quantity of the physical clock's material. This means that if the frequency of the perceived signals equals ν_0 in the presence of a physical clock material unit mass of m_0, it will then equal $M_0\nu_0/m_0$ in the presence of a mass of M_0; i.e.,

$$\nu = M_0 \frac{\nu_0}{m_0}, \tag{21}$$

Since the maximum absolute $\Delta\tau$ error of the readings of each of the clocks equals $1/\nu$, the following expression can then be derived from equality (21):

$$\mathrm{m}\,\Delta\tau = \frac{m_0}{M_0 \nu_0}. \tag{22}$$

Taking the mass of the rod without the clocks to be negligible as compared to the mass of the physical clock material concentrated in clocks A and B, or assuming that the rod R consists entirely of the physical clock's material, the $2M_0$ value (since the mass of the material in both clocks equals $2M_0$) can be set to equal the mass, M_R, of rod R with clocks A and B. Taking this into account, it follows from formulas (12) and (22) that

$$H = \frac{2c^2 m_0}{\nu_0} \tag{23}$$

The H value is dependent upon physical clock type and display sensitivity, and is not a physical constant. The H becomes different when the display's sensitivity changes, or when the physical clock's radioactive material is replaced with a radioactive material that has a different radiant power.

In the context of the physical clock, relations (19) and (20) pertain to the case of the arbitrary distribution of the physical clock's material along the rod and make the transition to equalities in the special case when the physical clock's material is concentrated at the ends of rod R.

At first glance, the relations derived only hold true for the techniques of instantaneous and point observations of an object, and the uncertainties going into these relations consist exclusively of the uncertainties inherent in these techniques. In actuality, however, it is impossible to measure even the constant velocity of this rod R, equipped with clocks A and B, with absolute accuracy using conventional methods (based on the path traversed and the time), if the word combination "rod R with clocks A and B" is taken to mean a specific object, the complement of characteristic traits of which includes the difference in the readings of clocks A and B (or the events that characterize this object). In such cases, the velocity and the values derived from it (the impulse and energy) prove to be tied not to an extraneous reference system, but rather to this object's characteristics traits, for example, to time or event marks [5].

In order to grasp the essence of the latter comment, we will address the concepts of relativity and uncertainty.

It was shown in reference [5] that many of the questions arising in the quagmire of physical relativism are erased if attention is directed to the presence of uncertainty in physical relativity. Rather than talking about the relativity of the velocity of a body and about the fact that the velocity of a body without specifying a reference system makes no physical sense, it is proposed in reference [5] that the term "uncertain velocity" be used with respect to the irrelative velocity of a body. In this case, the velocity of a body that is not tied to a specific reference system should be referred to as uncertain velocity. The uncertainty of the irrelative velocity of a body equals the c constant. It can be said of the irrelative velocity of this body, for example, that it equals zero, while its uncertainty equals c. This specification of velocity does not differ qualitatively in any way from the conventional specification of a velocity value supplemented by the specification of its error.

We draw attention to two methods that make it possible to avoid the uncertainty of the velocity of a body and to give it certainty.

The first generally accepted method consists of specifying the reference system within which the velocity of a body is examined. After the reference system is specified, the velocity of this body becomes certain.

The second method, despite its obviousness, is scarcely mentioned at all in physics. This method consists of individualizing an object, which in turn consists of describing it in greater detail, and under certain conditions, it is capable of replacing the selection of a reference system.

In the example we examined, the object with respect to which relations (19) and (20) hold true is not a rod, R, equipped with clocks A and B, but rather a rod, R, equipped with clocks A and B, and possessing a predetermined difference, $\tau_B - \tau_A$, in the readings, τ_A and τ_B, of clocks A and B.

If one and the same rod, R, equipped with clocks A and B, is by definition regarded as a set of more concrete subobjects, each of which has an inherent difference, $\tau_B - \tau_A$, in the readings of clocks A and B, then these different objects will possess different velocities. The possibility of thus dividing an object into more concrete subobjects eludes relativists, since it puts an end to the objective nature of physical relativism. Relativists are incapable of understanding something that was clear to Heracles, who saw the difference between one and the same river and particular concrete "subobjects" of this river that differed from one another. At the same time, relativists themselves note that one and the same object differs in different reference systems due to the relativity of its single-time nature; i.e., it breaks down into subobjects of sorts. It is strange that they don't comprehend the fact that, having described a subobject, its velocity can often be determined based on "external appearance" without specifying a reference system.

For example, in the case we examined, a subobject such as a rod, R, that has a specific difference, $\tau_{B,t'} - \tau_{A,t'}$, in the readings, $\tau_{A,t'}$ and $\tau_{B,t'}$, of clocks A and B, moves at a specific longitudinal velocity, V_x. When the clocks run with ideal accuracy, the V_x velocity value is functionally dependent upon the $\tau_{B,t'} - \tau_{A,t'}$ difference value. A subobject such as a rod, R, with a difference, $\tau_{B,t'} - \tau_{A,t'}$, in the readings, $\tau_{A,t'}$ and $\tau_{B,t'}$, of clocks A and B that equals zero is at rest. Its longitudinal velocity, V_x, equals zero and cannot differ from zero in any reference system, since subobjects with a difference, $\tau_{B,t'} - \tau_{A,t'}$, in the readings of clocks A and B that equals zero do not exist in any reference system where this rod moves longitudinally. This rod, R, with clocks A and B, has different longitudinal velocities in different reference systems, but the subject rod, R, with a difference, $\tau_{B,t'} - \tau_{A,t'}$, in the readings of clocks A and B that equals zero, being a concrete subobject, only has a longitudinal velocity that equals zero. This fact is entirely independent of the manner in which the velocity is measured – based on the clock readings or using the path and distance traversed.

In addition to this, if clocks A and B provide discrete readings, the $\tau_{B,t'} - \tau_{A,t'}$ difference will also have a certain discreteness. Here, the longitudinal velocity value, V_x, will have an uncertainty of ΔV_x. In this case, a rod, R, with a given difference in the readings of clocks A and B, being a subobject, may be found in a certain set of reference systems in a state of motion at various longitudinal velocities that differ from one another by a magnitude that does not exceed a certain value, which is the ΔV_x uncertainty. This uncertainty in accuracy equals the uncertainty that occurs during the single-time determination of the V_x velocity of a given subobject.

If clocks A and B do not run, but rather stand still, continuously indicating an identical time, then the ΔV_x uncertainty of the longitudinal velocity of rod, R, with a difference of $\tau_{B,t'} - \tau_{A,t'}$ in the readings of clocks A and B that equals zero, will equal the speed of light, c. Such a rod, R, with a difference of $\tau_{B,t'} - \tau_{A,t'}$ in the readings of clocks A and B that equals zero, can be found in all reference systems and possesses any velocity over a range of zero to the speed of light, c.

6. Conclusion

The purpose of the work at hand consisted not of finding areas of common interest between Lorentzian physics and quantum mechanics, but rather of demonstrating the existence of general physical uncertainty relations in the physics of the macrocosm. The general physical relations derived are externally reminiscent of the known uncertainty relations of quantum mechanics; however, the physical essence of the values that go into the relations and that contain the relations themselves are different than those in quantum mechanics. In relation (13), the Δx uncertainty is taken to mean the uncertainty in specifying the position of a projection with a length of Δx of a spatially extended object using a single solitary coordinate, x, while in the Heisenberg relation, Δx is a probabilistic characteristic of the position of a microparticle described by the root-mean-square deviation from the mean value.

The Δp_x value in relation (13) is the p_x impulse error, which in the presence of a predetermined Δx uncertainty, generally speaking, can be reduced by using another type of physical clock, while Δp_x in the Heisenberg relation is an uncertainty that it is fundamentally impossible to reduce in the presence of a predetermined Δx value.

Different meanings are placed on the H and h values.

The H value in relations (13), (18), (19), and (20) is dependent upon display sensitivity and physical clock type. If the sensitivity of the display is changed or the physical clock's radioactive material is replaced with a radioactive material that has a different mass radiation frequency, the H value will be different. However, the fundamental Planck constant, h, is unrelated to the physical properties of any specific material. The fact that these relations prove to be connected to the Heisenberg relation, not only externally, but also internally, seems especially strange.

First, there is the question of the minimum value, H_{min}, of an H parameter with an action dimension. Can one assume that this value may be as small as desired in the macrocosm and may correspond , for example, to the condition $H_{min} << h$?

And second, relations (19) and (20) are formally transformed into the relations

$$\Delta p_x \Delta x \geq h/2 \text{ and } \Delta E \Delta t_x \geq h/2$$

via the simple substitution of the energy, $h\nu_0$, of a photon with a frequency of ν_0 in place of the unit energy, m_0c^2, of the physical clock in formula (23); i.e., by taking the unit mass of a photon, the ν_0 frequency of which numerically equals the ν_0 frequency of the signals of a hypothetical change in the clock readings as the physical clock. It would be more correct to proceed on the basis of the concept of Lorentz-invariant mass and the equality of photon mass to zero, then a pair of photons coming from opposite directions with equal energies of $E = h\nu$ and with a resultant impulse, P, that equals zero, should be regarded as a physical clock of a unit mass of m_0. Since the unit mass, m_0, of this clock equals

$$\sqrt{\left(\tfrac{2E}{c^2}\right)^2 - \left(\tfrac{P}{c}\right)^2} = 2E/c^2,$$

while the ν_0 summation frequency of the electromagnetic oscillations equals 2ν, then taking the ν_0 frequency of the unit mass physical clock to equal the 2ν summation frequency of the photon pair, it follows from formula (23) that $H = H_{min} = h$.

Another approach is also possible.

In considering formula (23), if it is assumed that a certain H_{min} value exists that is common to all types of physical clocks, then $2m_0c^2/\nu_0 = H_{min}$. The latter equality only occurs when $m_0c^2 = ½\ H_{min}\nu_0$. If the lowest energy of a hypothetic signal of a change in clock readings is taken as m_0c^2, then H_{min} must be equal to the Planck constant h, since the $½h\nu_0$ value is the minimum possible energy of the zero-point oscillations of an oscillator with a frequency of ν_0.

The relations expressed by formula (23), if we move to them from relations (19) and (20), do not reflect the statistical nature of the generation of the clock reading change signals; thus, when using this approach, reference can only be made to the order of the parameter present in the right-hand member of the relations and not its precise value.

References

[1] Tarbeyev, Y.V., Slayev, V.A. & Chunovkina, A.G. (1997) Problems with using the international guide to the expression of uncertainty in measurement in Russia, *Izmeritelnaya Tekhnika* [Measurement Techniques] 1.

[2] Mendeleyev, D.I. (Trans.) (2002) European Society for Analytical Chemistry/International Cooperation on Traceability in Analytical Chemistry (EURACHEM/CITAC) Guide Entitled: A Quantitative Description of Uncertainty in Analytical Measurement, (2nd Edition – Translated From English, St. Petersburg: Russian National Metrology Research Institute (RNMRI).

[3] Sukhanov, A.D., & Golubeva, O.N. (2006) Lectures on Quantum Physics, Moscow: *Vysshaya Shkola* [Higher School] Publishing House, 54.

[4] Okun, L.B. (2008) UFN (Advances in the Physical Sciences). 178, 541.

[5] Matveev, V.N. (2000) Entering the Third Millennium Without Physical Relativity, Moscow: *CheRo* Publishing House.

Simulations of Relativistic Effects, Relativistic Time, and the Constancy of Light Velocity

VADIM N. MATVEEV & OLEG V. MATVEJEV

Joint-Stock Company "Sinerta", Address
Vilnius, LT-03150, Lithuania

Based on pre-Einstein classical mechanics, a theoretical model is constructed that describes the behavior of objects in a liquid environment that conduct themselves in accordance with the formal laws of the special theory of relativity. This model reproduces Lorentz contraction, time dilation, the relativity of simultaneity, the Doppler effect in its symmetrical relativistic form, the twin paradox effects, Bell effect, the relativistic addition of velocities. The model makes it possible to obtain Lorentz transforms and to simulate Minkowski four-dimensional space-time.

Keywords: Clocks, Relativity, Speed of light, Time

1. Introduction

In Soviet scientific literature, the question of synchronizing clocks is mentioned in general terms, if at all. In popular articles, as well as in special literature, almost no attention is given to this central problem in Einstein's entire special theory of relativity (STR).

It should be noted that the problem of the speed of light in one direction is not of topical interest in practice, since the speed of light is actually measured using one solitary clock and a mirror. During this solitary clock method, the time interval between light pulse dispatch to the mirror and the reception of the pulse returned to the initial point after being reflected from the mirror is measured. Velocity is calculated for the doubled distance between the clock and the mirror and the light travel time in the back and forth directions. Strictly speaking, velocity measured in this manner constitutes the average speed in the back and forth directions – this is because the speed there may not equal the speed back. The equality of this average speed of the c constant is an experimental fact.

It is frequently said that Rømer measured the speed of light in one direction. It may seem strange, but Rømer velocity is also the velocity obtained under the tacit assumption of the equality of the speeds of light in opposite directions. The fact of the matter is that Rømer and Cassini were speculating about the movement of Jupiter's satellites, automatically assuming that the observers' space was isotropic. The Australian physicist Karlov [1] showed that Rømer actually measured the speed of light by implicitly making the assumption of the equality of the speeds of light back and forth.

Poincaré examined the proposition of the equality of the speed of light from A to B and the speed from B to A, and this proposition in particular became the principal postulate of Einstein's 1905 work [2], although it was not presented in the form of a postulate, but rather in the form of a "definition", which preceded two Einsteinian principals that are often called postulates. In a later work [3], Einstein called this "definition" an assumption, during which he noted that it pertains not only to the speed of light, but also to velocity in general. In this work, Einstein wrote: "But if velocity at the speed of light is fundamentally impossible to measure without arbitrary assumptions, we then also have the right to make arbitrary assumptions about the speed of light. We will now assume that the speed of light propagation in a vacuum from point A to point B is equal to the speed of light passing from B to A". In truth, unlike Poincare, who adhered to the conventionalist point of view, Einstein, by alluding to the impossibility of measuring velocity in one direction without arbitrary assumptions, was inclined to regard the arbitrary assumption of the inequality of the speed of light in opposite directions as unnatural and "highly improbable" [4].

Why nature does not permit us to measure the speed of light in one direction without arbitrary assumptions! Is this randomness or something deeper?

The developers of the alternative theories did not answer this question.

One of the authors of this booklet tried to answer these questions. By 2000, he had written the book "Entering the Third Millennium Without Physical Relativity?", which the *CheRo* Publishing House released that same year [5]. In the book, based on the principle of the equal status of assumptions of the equality and inequality of the speed of light in opposite directions, a means was proposed for solving the synchronization problem and the related problem of the dependence of the magnitude of the physical quantities of a body that are inherent in the body itself upon the reference systems.

The problem of relativistic quantities was solved by means of refining the concept of an "object" and viewing an object as a set of subobjects (objects with a higher degree of concretization), each of which has not relative, but rather absolute dimensions. The relativity of simultaneity is responsible for the existence of these subobjects.

The refinements in the concept of a "physical object" proved to be sufficient to do away with the relativity of the magnitude of physical quantities without the involvement of a dedicated reference system or a global environment. For this reason, the author felt that the matter was resolved (at least as far as he was concerned) and that dealing with a global environment was superfluous. The solution at which we arrived over the course of our joint work on developing the approach described in the book "Entering the Third Millennium Without Physical Relativity?" was even more unexpected. We discovered the possibility of simulating relativistic effects using the simplest techniques of pre-Einsteinian classical physics based on the example of the movement of objects in a material medium. Here, in order to facilitate simulation, it was not necessary for us to examine movement at velocities commensurate with the speed of light. Within the model, effects are manifested in an explicit form at usual "terrestrial" velocities dealt with in everyday life. The possibility of simulating STR effects with the involvement of an environment and the absence of such models in other versions made it necessary to take a new look at the old and the apparently once and for all resolved problem of the existence of a world environment.

In the article a theoretical STR model is described, which we also call an STR simulation. Einstein's STR is simulated based on examples of barges, shuttles, and boats traveling at usual aquatic environment speeds. Simulation did not require anything from us other than the most elementary rules of classical physics.

2. The Essence of the Simulation

The behavior of objects that, being slow-moving, nonetheless act in accordance with laws similar to those of the special theory of relativity are examined in the article.

Individual barges and groups of barges located on the surface of a flat-bottomed water body with a depth of h, filled with slack water, will serve as the objects of our conceptual observation. The barges are equipped with hardware components that perform metrological operations. The hardware components have at their disposal speedboats that whisk over the water surface between the barges and high-speed underwater shuttles that travel between the barges and the bottom. The velocity of the speedboats and the shuttles equals V and is unapproachable for the other watercraft, i.e., the velocity of the barges, v, which do not fall into the category of high-speed watercraft, correspond to the inequality $v < V$.

2.1. *Simulation of Time and the Lorentz Contraction*

Each barge is equipped with a clock, during which a high-speed shuttle that continuously moves along a plumb line (relative to this barge) between the barge and the bottom performs the function of the pendulum. Each shuttle trip to the bottom and back requires a time of $\Delta t(1) = 2h/V_Z$, where V_Z – the speed of submersion and surfacing of the underwater shuttle, and is accompanied by the replacement of the clock readings with a standard unit quantity that is uniform for the barge. This standard quantity for the barges both at rest and in motion equals $2h/V$. The shuttle clock "mechanism" controls not only the clock's hands, but also all the barge hardware components, thereby ensuring the proportionality of their working speed to the clock's working speed. We assumed that the t time scale for barges at rest relative to the water equaled the time scale of our conventional "terrestrial" clock, as well as that the reading replacement rate on the barges at rest and our clock was identical.

During phase one, we examined a group of barges at rest. Here, we assumed that the clock readings on the this group's various barges were not synchronized, i.e., when the working speeds of the clocks on each barge in the group are identical, their readings at a given moment in time might be different.

Assuming that the barges might change their position due to certain external causes (for example, because of the wind), we entrusted the hardware components with the function of maintaining the distance between this group's barges by means of interaction between the barges using the speedboats.

The procedure for maintaining distance consists of the following.

A speedboat is dispatched from each barge to the neighboring barge, and upon reaching it, the boat goes back again. The barge's hardware components, using their clocks, measure the boat's time of movement to the neighboring barge and back, then approach or move away from the neighboring barge as necessary in order to maintain this time and the immutability of the "radar" distance. This method for maintaining the "radar" distance between barges does not require the synchronization of the readings on the different barges and makes it possible to track the distance of the neighboring barges from each of the barges independently, without resorting to the measurement of the boat movement time from one barge to the another using synchronously running clocks on these barges.

We then examined a group of barges located at the points of intersection of an imaginary coordinate grid of the K' coordinate system. The group is initially at rest on the water body surface, after which it is transformed, together with the K' coordinate system that belongs to it, from a state of rest to a state of motion at a velocity of v in the direction of the X' axis (this axis lies on the water surface). As the group of barges accelerates to a velocity of v, the speed at which the clocks tick and the response time of the hardware components on the barges decreases. This occurs due to the fact that as the barges move at a velocity of v, the V_Z submersion and surfacing speed of a shuttle moving through the water between a barge and the bottom along the hypotenuses of right-angled triangles is equal to $V\sqrt{1-(v/V)^2}$. Time on the moving barges, which we named simulated time, t', also flows more slowly than our time, t, by $1/\sqrt{1-(v/V)^2}$ times.

The lateral dimensions of the group are retained in this instance.

Indeed, let us assume that a boat is sent out from and returns to a barge, $r'_{o'}$, that is moving within the R' group and that is located at the origin of coordinates of the K' system along the Y' axis at a point with a coordinate of y' relative to the neighboring barge, $r'_{y'}$, in this same group. If the Y' axis is located on the water surface perpendicular to the X' axis, the boat then moves over the water surface along the hypotenuses of right-angled triangles at a velocity of V. This corresponds to the movement of a boat along the Y' axis at a velocity of V_Y in our time scales and with a length equaling $V\sqrt{1-(v/V)^2}$. Since the t' time equals $t\sqrt{1-(v/V)^2}$, the simulated time of movement of the boat from barge $r'_{o'}$ to barge $r'_{y'}$ and back, $\Delta t'$, is independent of the speed of movement of the R' group, as well as the distance between barges $r'_{o'}$ and $r'_{y'}$, and the hardware components perceive the group as unchanged when the velocity changes.

However, the longitudinal dimensions (in the direction of the X' axis) of the barge group in motion are contracted for the following reason.

In negotiating the distance, $l_{o'x'}$, between barge $r'_{o'}$, which is located at the origin of coordinates, O', and barge $r'_{x'}$, which is located on the X' axis at a point with a coordinate of x', the boat needs a Δt_1 time that equals $l_{o'x'}/(V-v)$ in order to move from barge $r'_{o'}$ to barge $r'_{x'}$, and a Δt_2 time that equals $l_{o'x'}/(V+v)$ for the trip back. The total time of movement, $\Delta t_1 + \Delta t_2$, from barge $r'_{o'}$ to barge $r'_{x'}$ and back comes to

$$2l_{o'x'}V/(V^2-v^2).$$

According to the barge's slow clocks, the $\Delta t'_1 + \Delta t'_2$ time is $1/\sqrt{1-(v/V)^2}$ times shorter and comes to

$$2l_{o'x'}/V\sqrt{1-(v/V)^2}.$$

If the hardware components do not maintain the distance between barges, the instruments on barge $r'_{o'}$ would then perceive this as an increase in the distance between the barges in the direction of the X' axis by

$$1/\sqrt{1-(v/V)^2}$$

times. But the instruments, in tracking the distance between the barges using the radar technique, keep the radar distance unchanged, which we perceive as the contraction of $l_{o'x'}$ by

$$1/\sqrt{1-(v/V)^2}$$

times. We called the physical quantities expressed through simulated distances and times the simulated quantities.

We then made the transition to an examination of the synchronization of the clocks of two groups of barges – group R and group R' – and the related coordinate systems, K and K'. The R group and the K system are at rest on the water, while the R' group and the K' system are in motion on the water and relative to the R group at a velocity of v.

2.2. *Synchronization of the R and R' Group Clocks*

Imagine that at a certain moment in time when the origin of coordinates and the axes of the K and K' coordinate systems coincide, the readings on all the barges of the barge groups at rest and in motion were reset to zero. From that moment in time forward, the synchronous change in the readings on all the barges of the barge group in motion occurs more slowly than the synchronous change in the readings on the barges of the group at rest.

If the hardware components on the barges of the group at rest, R, track the clock of barge r' of the group in motion, R', which is moving past them, they then record the slowness of the clock rate on moving barge r'. If the hardware components on the barges of the group in motion, R', track the clock on barge r of the R group, which moving past them, but is at rest relative to water, they then record the fastness of the clock rate on barge r. There is no symmetry of any kind. What we have here is the asymmetry of the clock movement speeds on the barges at rest and in motion. The simulated time readings of the group in motion are linked to the time readings of the group at rest by the transformations $t' = t\sqrt{1-(v/V)^2}$ and $t = t'/\sqrt{1-(v/V)^2}$. The coordinate transformations take the form of

$$x' = (x - vt)/\sqrt{1-(v/V)^2} ,$$
$$x = (x' + v't')\sqrt{1-(v'/V')^2} ,$$
$$y' = y,$$

where the primed quantities are expressed in the distance and time scales of the barge group in motion.

It is clear that if the hardware components of the group in motion, R', now measure the speed of movement of a boat from one of the barges in their group to another barge in this same group using the synchronously running clocks on these barges, they will then find that the boat's speed of movement in the barge group's direction of movement, which we see from the outside, and opposite its direction of movement, are different.

2.3. *Simulation of the Doppler Effect Using Simulated Time*

Let's say that of two inertial barges, r and r', one is at rest, r, while the other, r', is moving away from it at a velocity of v. We will presume that speedboats carrying information are dispatched from each barge to the other barge. Expressed through t time for barge r and through t' simulated time for barge r', the frequencies of speedboat dispatches, f_r and $f'_{r'}$, are numerically equal to one another. Each such boat can be viewed as a "signal" that carries a message.

It is clear that if the simulated f_r and $f'_{r'}$ frequencies, i.e., respectively expressed in t and t' times, are identical, then the f_r frequency we record for the signals (the boat dispatches) sent from the barge at rest is $1/\sqrt{1-(v/V)^2}$ times higher than the conventional $f_{r'}$ frequency we record (an expressed through our conventional time) for the signals (the boat dispatches) sent from the barge in motion.

The distance between the boats moving away from barge r at a velocity of V equals V/f_r. From our point of view, being the point of view of outside observers, the speed of movement of the boats headed toward receding barge r', being equal to V relative to the barge at rest and to the water, equals $V - v$ relative to barge r', moving at a velocity of v. Dividing the $V - v$ velocity by the distance between boats, which equals V/f_r, we obtain the frequency of information receipt (boat arrival) from barge r to receding barge r', $f_{r'\leftarrow r}$, which equals $f_r(V-v)/V$, i.e.,

$$f_{r'\leftarrow r} = f_r(V-v)/V . \qquad (1)$$

Taking into account the fact that the conventional frequency, $f_{r'\leftarrow r}$, is $1/\sqrt{1-(v/V)^2}$ times smaller than the simulated frequency, $f'_{r-r'}$, expressed through simulated time, t', i.e., with allowance for the fact that $f_{r'\leftarrow r} = f'_{r'\leftarrow r}\sqrt{1-(v/V)^2}$, then from formula (1) we obtain

$$f'_{r'\leftarrow r} = f_r \frac{\sqrt{1-v/V}}{\sqrt{1+v/V}} . \qquad (2)$$

In the event that the approach of barge r' to a barge other than barge r is true, which it is not hard to demonstrate, the equality is

$$f'_{r'\leftarrow r} = f_r \frac{\sqrt{1+v/V}}{\sqrt{1-v/V}} . \qquad (3)$$

But what happens to the boats that carry the information from the barge in motion, r', to the barge at rest, r?

From our point of view, the speed of movement of the boats relative to barge r' equals $V + v$. Therefore, the distance between the boats departing from barge r' at a velocity of $V + v$ equals $(V + v)/f_{r'}$. Dividing the V velocity at which the boats move toward the barge at rest, r, by the distance between the boats, which equals $(V + v)/f_{r'}$, we obtain the frequency of information receipt (boat arrival) from barge r' at the barge at rest, r, $f_{r\leftarrow r'}$:

$$f_{r\leftarrow r'} = \frac{f_{r'}}{1+v/V} . \qquad (4)$$

The frequency of the signals (the boat dispatches) that we record, $f_{r'}$, equals $f'_{r'}\sqrt{1-(v/V)^2}$, where $f'_{r'}$

is the simulated frequency of the signals (the boat dispatches) from barge r'. Taking into account the fact that the $f_{r\leftarrow r'}$ frequency equals the simulated frequency, which has the same notation, $f_{r\leftarrow r'}$, and substituting $f_{r'} = f'_{r'}\sqrt{1-(v/V)^2}$, we obtain

$$f_{r\leftarrow r'} = f'_{r'}\frac{\sqrt{1-v/V}}{\sqrt{1+v/V}}. \qquad (5)$$

In the event that the convergence of barges r' and r is true, which it is not hard to demonstrate, the equality is

$$f_{r\leftarrow r'} = f'_{r'}\frac{\sqrt{1+v/V}}{\sqrt{1-v/V}}. \qquad (6)$$

Formulas (2) and (5), as well as formulas (3) and (6), similar to the formulas of the special theory of relativity, demonstrate the symmetry of the frequencies of the information reaching them that the hardware components on barges r and r' record in simulated times.

We arbitrarily selected barges r and r'. The conclusion concerning the impossibility of observing asymmetrical processes extends to all the pairs of barges that are located on the surface of a water body. The symmetry of the processes accessible for recording on the barges at rest and in motion demonstrates that, using documented information alone without resorting to contact with the water, automatic devices cannot, independently of one another, solve a problem that consists of determining which among them is located on the barge in motion and which is located on the barge at rest.

2.4. *Simulation of the Symmetry of Relativistic Effects*

We then assumed that the hardware components on the R and R' group barges are not in contact with the water and have no information concerning their motion or rest relative to the water. Not finding a basis for synchronization during which the velocity of a boat back and forth is assumed to be different, the hardware components resynchronize the clocks within the group of barges in motion, R', so that the boat's speed of movement there becomes identical to the boat's speed of movement back. In this instance, the time after resynchronization, t'' – we named this t'' time the double simulated time – is linked to the simulated time, t', by the correlation $t''=t'-x'v/V^2$, while the coordinates and the clock readings are linked by the transformations

$$x'=(x-vt)\Big/\sqrt{1-(v/V)^2},$$
$$y'=y, \qquad (7)$$
$$t''=(t-xv/V^2)\Big/\sqrt{1-(v/V)^2},$$

as well as

$$x=(x'+v''t'')\Big/\sqrt{1-(v''/V'')^2},$$
$$y=y', \qquad (8)$$
$$t=(t''+x'v''/V''^2)\Big/\sqrt{1-(v''/V'')^2}.$$

Here, the quantities with double primes are expressed through the double simulated time. The transformations obtained are consistent with the direct and inverse Lorentz transformations to notational accuracy. In particular, this results in the fact that by tracking the clock rate of barge r at rest in the water, which is motionless relative to the water, but moves past a barge of the group in motion, the hardware components on the group of barges in motion, R', detect the slowness of the clock rate on barge r. The results of the measurements made by the hardware components on the barges of the groups in motion and at rest become symmetrical. The same thing is true of the distances.

2.5. *Addition of Velocities*

The result of the addition of the simulated velocities is obvious, since it follows from the transformations in formulas (7) and (8), which are indistinguishable from the Lorentz transformations. If, for example, a certain watercraft moves at a velocity of u within the R group Σ coordinate system in the direction of the X axis and the equation $x = ut_R$ describes the dependence of its x coordinate upon time, then by substituting ut_R in place of x in the first transformation in formula (2), we obtain

$$x'=(u-v)t_R\Big/\sqrt{1-(v/V)^2}.$$

By substituting $x = ut_R$ in the last transformation in formula (2), we obtain

$$t''_{R'}=(1-uv/V^2)t_R)\Big/\sqrt{1-(v/V)^2}.$$

Dividing the x' value of the watercraft we are examining by $t''_{R'}$, we obtain the double simulated velocity, u'', which can be expressed by the formula

$$u''=\frac{u-v}{1-uv/V^2}.$$

If the watercraft move in accordance with the law $x = -ut_R$, i.e., in the direction opposite that of the X axis, the velocity addition formula then takes the form

$$u''=\frac{u+v}{1+uv/V^2}.$$

It is apparent from the latter formula that the watercraft's double simulated velocity cannot exceed the V velocity, and when u and v velocities approach V, it is not $2V$ that is approached, but rather V.

2.6. *Simulation of the Simplest Effects of Noninertial Bodies*

The simulation model we are proposing also makes it possible to examine the simplest effects on noninertial barges and within noninertial barge groups.

Let's say that each individual barge group simulates a solid physical body. The immutability of the distances between the component parts of the solid body, fixed in its own coordinate system, simulate the maintenance of the radar distance between the barges in each group.

We will examine two barge groups at rest, which, located a great distance from one another, simultaneously get under way and accelerate in accordance with identical programs along the line on which they are located according to our clock and according to the synchronized clocks of the barge groups at rest. In this case, according to our observations, the longitudinal distances between the barges in each of the two groups moving one after the other begin to shorten after a certain time. Here, the instruments on the barges will maintain the immutability of the radar distances within each group. However, according to our observation results, the distance between the barge groups remains immutable, first, due to the sameness of the acceleration programs, and second, since the tracking and maintenance of the distances between the barges are only performed within each group. I.e., from our point of view as outside observers, each group is contracted in the direction of movement, while the distance between the groups remains the same as before. After acceleration comes to an end, if the group instruments measure the distance between the groups using boats that make trips back and forth, or using lengths standards, they will then find that the groups have drifted apart and that the distance between them has increased. This effect is presently called the Bell paradox, although it was described before Bell, in particular, by D. V. Skobeltsyn [6].

The increase in the distance between the groups recorded by the instruments is due to the contraction of the groups, as well as the length standard that pertains to them, in the direction of movement and has a purely metrological relative nature.

After a certain time, if the barge groups begin to perform a reverse operation and simultaneously begin to decelerate, the result of this deceleration will then be dependent upon group clock synchronization.

After acceleration comes to an end and before deceleration begins, if the instruments within the groups do not resynchronize the clocks and simultaneously begin to decelerate in our time, t, and in that same time, t_R (as well as simultaneously in single stimulated time, $t'_{R'}$), the reverse process will occur in such a manner that the groups and the length standards that pertain to them, following the completion of deceleration and standstill, will again be expanded, and the groups will return to the original state. Due to the expansion of the length standards, the instruments will record a decrease in distance to the initial (starting) value. This will occur at an immutable actual distance between the groups.

If, however, the instruments of a pair of groups do resynchronize the clocks "according to Einstein" and the groups simultaneously begin to decelerate in double simulated time, $t''_{R'}$, then according to our clock, they begin to decelerate at different times (the rear group of the pair begins to decelerate earlier that the front group). In this instance, despite the expansion of the groups and the standards, after the groups of barges stop in the water, the instruments record a sequent increase in the distance between the groups, since, as it is easy to demonstrate, the increase in the distance between the groups due to the actual difference in the times that they begin to decelerate exceeds the expansion of the groups and the length standards by

$$1/\sqrt{1-(v/V)^2}$$

times.

2.7. *Simulated "Space-Time"*

In deriving the transformations in formulas (7) and (8), we did, as a matter of fact, also simulate pseudo-Euclidean space-time, since these transformations ensure the invariance of the space-time interval in the Σ and Σ' systems, as well as in any other systems associated with groups of barges in motion at different velocities. It is clear that this "space-time" has nothing at all to do with enigmatic multidimensional worlds and is an elementary mathematical construct that pertains more to the formalization of the measurement errors caused by the failure to take the presence of an aquatic environment into account than to the behavioral features of the barges on the water surface.

3. Conclusion

The special theory of relativity is closely linked to philosophical suppositions, and many of the problems that arise over the course of interpreting its physical content are philosophical in nature. Hermann Weyl probably foresaw something similar when he wrote: "Despite the disheartening leapfrog of philosophical systems, we cannot refuse to address it unless we want knowledge to be transformed into senseless chaos" [7].

What is the essence of kinematic phenomena? Is Lorentz contraction a real or apparent phenomenon? Is four-dimensional space-time an objective reality or a mathematical formalism? There are not so much answers to these and other questions as answers that mutually exclude one another.

A clear illustration of the confusion that exists in the theory of relativity consists the campaign to modify the concept of mass.

The concept of relativistic, i.e., velocity-dependent, mass has enjoyed the broadest possible use since the time that Einstein's special theory of relativity came into being, and it is still used in many articles and books today. Einstein, Born, Pauli, Møller, Gamow, Feynman, and other eminent physicists of the twentieth century wrote about the dependence of mass upon velocity. The fact of the experimental confirmation of the dependence of the mass of elementary particles upon velocity in particle accelerators has continuously been cited in learning aids and bibliographic literature. In addition, after academician L. B. Okun published his article in the physics journal [8], a campaign was launched in the Soviet Union, then in Russia to replace statement concerning the dependence of mass upon velocity with a statement concerning its independence upon the latter. This campaign culminated in the "revocation" of the physical statement concerning the dependence of mass upon velocity and the adjustment of the world-renowned formula "E equals mc square" by means of the seemingly insignificant replacement therein of the letter E with E_0 (with a zero subscript). One result of this campaign consisted of the fact that the obsolete concept of mass dependent upon velocity was stripped from the State Physics Program in Russia in 2006 and "Recommendations for Describing the STR With Allowance for a Standard" were formulated.

In the article by L. B. Okun [8], the following situation that American physicist C. Adler experienced is described.

"Does mass really depend on velocity, dad?", C. Adler was asked by his son.

"Well, yes…" "Actually, no, but don't tell your teacher", C. Adler answered.

The next day – Adler writes – his son dropped physics.

In addition, the special theory of relativity is very simple and does not involve any problems other than the problems of its interpretation. It can be described in the simplest language and using the simplest examples from our everyday life.

The simulation presented in this article demonstrates the simplicity of the special theory of relativity and its "earthiness". It is not difficult to see that the possibility of using a "four-dimensional formalism" that does not differ from Minkowski formalism in this model issues from a simulation that yields Lorentz transformations.

References

[1] Karlov. L. (1970) Australian Journal of Physics. 23, 243.
[2] Einstein, A. (1966) Collected Scientific Works, *Nauka* [Science] Moscow: Publishing House, 1, 9.
[3] Einstein, A. (1966) Collected Scientific Works, *Nauka* [Science] Moscow: Publishing House, 1, 181.
[4] Einstein, A. (1966) Collected Scientific Works, *Nauka* [Science] Moscow: Publishing House, 1, 539.
[5] Matveev, V.N. (2000) Entering the Third Millennium Without Physical Relativity, Moscow: *CheRo* Publishing House.
[6] Skobeltsyn, D.V. (1996) The Twin Paradox in the Theory of Relativity, Moscow: *Nauka* Publishing House.
[7] Weyl. H. (2004) Space, Time, and Matter, Moscow: Union of Soviet Socialist Republics (URSS) Publishing House.
[8] Okun, L.B. (1989) UFN (Advances in the Physical Sciences) 158, 3, 511.

On the Maximum Speed of Matter

DIONYSIOS G. RAFTOPOULOS

National Technical University of Athens (NTUA)
7 Korai St., 19013 Anavissos, Attica - Greece
www.dgraftopoulos.eu, draft@otenet.gr

In this paper we examine the analytical production of the Lorentz Transformation regarding its fundamental conclusion i.e. that the speed of Light in vacuum is the uppermost limit for the speed of matter, hence superluminal speeds are unattainable. This examination covers the four most prominent relevant sources of bibliography: Albert Einstein's historic paper (1905) titled: *"On the Electrodynamics of moving Bodies"* on which his Special Relativity Theory is founded. His famous textbook titled: *"Relativity, The Special and General Theory"*, A. P. French's textbook titled *"Special Relativity"*, Wolfgang Rindler's textbook titled. *"Essential Relativity"*. Special emphasis is placed on the critical analysis of Einstein's *gedanken experiment* as it is presented in his original paper, where he considers a moving, straight, rigid rod at the ends of which there are two clocks, whose synchronization is checked according to his own definition as given in part 1 of his paper. By applying the second fundamental hypothesis (principle) of SRT, we arrive at the conclusion that this noetic experiment can be concluded *only* if the rod's speed V with regards the stationary system and measured from it, is less than the speed of light C also with regards the stationary system and measured from it. In the opposite case, said noetic experiment would be meaningless as it could never be concluded for the Observer of the stationary system, at least in the Euclidean Space. Finally, we show that in all four cases under examination the relationship $v < c$ is *not a conclusion* of the Lorentz Transformation, but rather a *hidden pre-condition* of its analytical production, in other words, the LT can *only* be valid for $v < c$ Consequently, there doesn't exist a definite and rigid law of Physics forbidding matter to travel with superluminal velocity in vacuum.

Keywords: Special relativity theory, Lorentz transformation, Superluminal velocity of matter.

1. Introduction

It is well known that Special Relativity Theory forbids matter to travel with superluminal velocity, i.e. with a velocity v greater than that of light in vacuum c. Indeed Albert Einstein in p.48 of his book titled *"Relativity, The Special and General Theory"* [1] writes: *"It therefore follows that the length of a rigid metre-rod moving in the direction of its length with a velocity* v *is* $\sqrt{1-\frac{v^2}{c^2}}$ *of a metre. The rigid rod is thus shorter when in motion than when at rest, and the more quickly it is moving, the shorter is the rod. For the velocity* $v=c$ *we should have* $\sqrt{1-\frac{v^2}{c^2}}=0$, *and for still greater velocities the square root becomes imaginary. From this we conclude that in the theory of relativity the velocity* c *plays the part of a limiting velocity, which can neither be reached nor exceeded by any real body. Of course this feature of the velocity* c *as a limiting velocity also clearly follows from the equations of the Lorentz transformation, for these become meaningless if we choose values of* v *greater than* c "

So, how did Einstein's conclusion, which has since acquired the power of a physical law, arise? To properly answer this, it is necessary to examine first the analytical production of the Lorentz Transformation, from which the aforementioned conclusion is derived, by studying some well-known cases in the bibliography.

2. Examination of the Analytical Production of the Lorentz Transformation

2.1. From Einstein's historic paper (1905) titled: *"On the Electrodynamics of moving Bodies"*[2] where the author, having first defined the concept of the synchronization of two clocks at relative rest, formulates the two principles (hypotheses) of his theory:

"1. The laws by which the states of physical systems undergo change are not affected, whether these changes of state be referred to the one or the other of two systems of coordinates in uniform translational motion.
2. Any ray of light moves in the "stationary" system of co-ordinates with the determined velocity c, whether the ray be emitted by a stationary or by a moving body.

Hence: $velocity = \frac{light\ path}{time\ interval}$ *where time interval is to be taken in the sense of the definition in §1."*

Following that, the author constructs a noetic experiment:

He considers a rigid rod AB moving at a speed v, in a linear and uniform fashion, on the X axis of a stationary system of coordinates. Furthermore, he imagines that at the very tips of the rod's two ends there are two clocks which, by hypothesis, are synchronized with the clocks located at the relative positions of the stationary system. He also imagines that at the clocks' positions on the rod, there are two observers moving with the rod. Einstein considers that the two moving observers monitor whether those two clocks are synchronous to one another, in their own moving system.

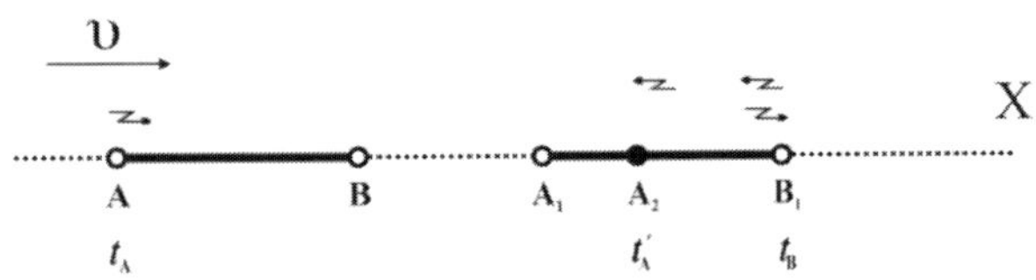

Figure 1. Einstein's noetic (*gedanken*) experiment.

Said monitoring is achieved via light. Einstein considers that a light signal starts at the rod's endpoint A, reaches its leading point B, where it is reflected and returns back to endpoint A. The author maintains that the two observers, moving with the rod, will ascertain that those two clocks are not synchronous, as the two light paths AB and BA are *not* covered by light in equal times.

In our paper titled *"On Synchronized Clocks at the Ends of a Moving Rod"* [3], we have shown that Einstein's argument is invalid in the moving system. Here, however, we shall try to answer a different question: What is the necessary condition for Einstein's noetic (gedanken) "experiment" to be concluded? The simple answer is that Light emitted from the rod's endpoint A, *must* be observed reaching leading point B, *from both systems (moving and stationary)* as, otherwise, comparing the measurements from both systems would be entirely meaningless.

The "experiment" is indeed concluded in the moving system. However, for it to be concluded in the stationary system as well, the rod ought to move at a speed, as measured from the stationary system, which is lower than the speed of light, i.e., in any case, $v < c$. That is because, according to A. Einstein's second principle (hypothesis), the speed of light relative to the stationary system, is c, *regardless of the speed of its source.* Therefore, for light emitted from the rod's endpoint A to reach its leading point B, observed from the stationary system, the speed of point B, hence of the whole (rigid) rod, ought to be subluminal ($v < c$) relative to the stationary system in Euclidean space.

Fundamental Remark:

The relationship $v < c$ is definitely not a conclusion (as it appears to be in A. Einstein's paper), but rather a (hidden) precondition. If that were not so, i.e. for $v \geq c$, the light signal emitted from endpoint A, considered from the stationary system, would never reach the moving rod's leading point B, at least not in the Euclidean space. Thus the "experiment" would have never been concluded and, consequently, A. Einstein's paper would never have been written" [3].

Further on, Einstein, in §3 of his article, formulates the mathematical equations that describe the events on the moving rod, i.e. the forth and back travel of light, and, under the condition that the two clocks on the moving rod must be synchronous, derives the well-known Lorentz Transformation. Hence:

Conclusion 2.1.:

The Lorentz Transformation, as derived in Einstein's original article [2], is indeed valid only for $v < c$, which is, therefore, a *necessary pre-condition* for its creation and not, as it seems, its conclusion.

2.2. From Einstein's book [1], written 15 years later, and in Appendix 1 (p.147). The author presents a simple derivation of the Lorentz transformation, writing:

"A light-signal, which is proceeding along the positive axis of x, is transmitted according to the equation

$$x = ct \quad or \quad x - ct = 0 \tag{1}$$

Since the same light-signal has to be transmitted relative to K′ with velocity c, the propagation relative to the system K′ will be represented by the analogous formula

$$x' - ct' = 0 \tag{2}$$

Of course the initial conditions are: For $t = t' = 0$, then $x = x' = 0$ and the system K′ is moving to the positive side of the axis x of the system K.

I note that the crucial point is that equations (1) and (2) refer to the same light signal. I am afraid that the simultaneous validity of equations (1) and (2) is not always true. I believe that the analytic functions of the Cartesian coordinates have no natural meaning, without previously establishing their geometrical equivalent which, however, is correlated with the appropriate natural hypotheses. So we are obliged to investigate the issue geometrically.

Investigation:

2.2.1 Let velocity v of frame K′ relative to frame K be less than the velocity of light c, which is assumed to be the same for both frames. Then in fact a light signal moving towards the positive end of x in frame K and obviously, relative to K, obeying equation (1), by moving faster than the zero of K′(start of x' axis of K′) relative to K, it follows that it leaves behind (by the Euclidean meaning) the zero of K′, in other words the light signal in question is moving towards the positive end of x' in K′.

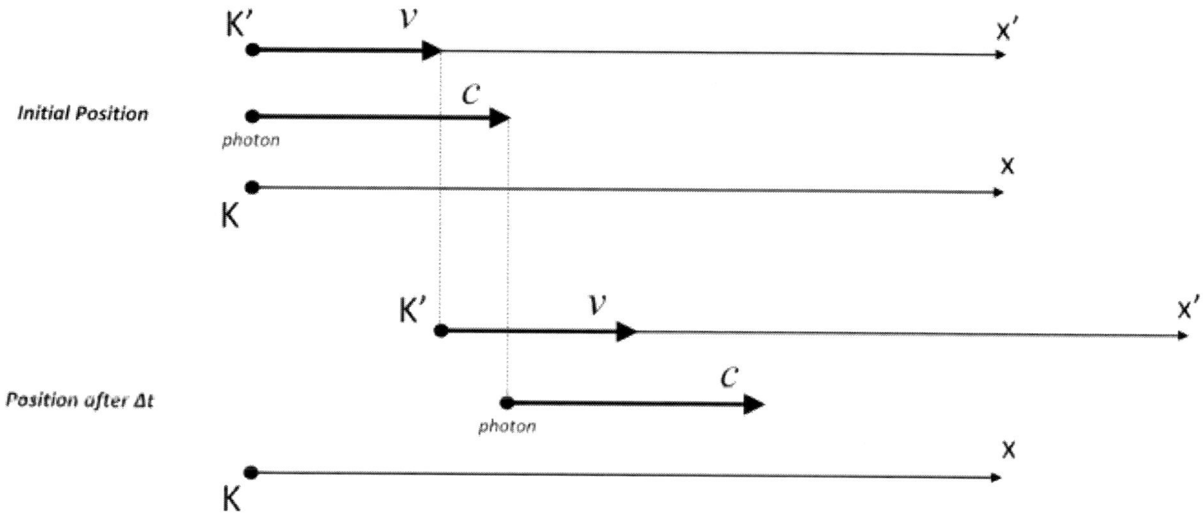

Figure 2. The relative position of the photon and of the two relatively moving reference systems, for $v < c$.

Therefore, respecting Albert Einstein's 2nd fundamental principle (hypothesis), we are obliged to write:

$$x'=ct' \quad or \quad x'-ct'=0 \qquad (2)$$

because in the case of $v < c$, x' is positive.

2.2.2. Let the velocity v of the frame K′, relative to K, be greater than the velocity of light c, which is the same for both frames by assumption.

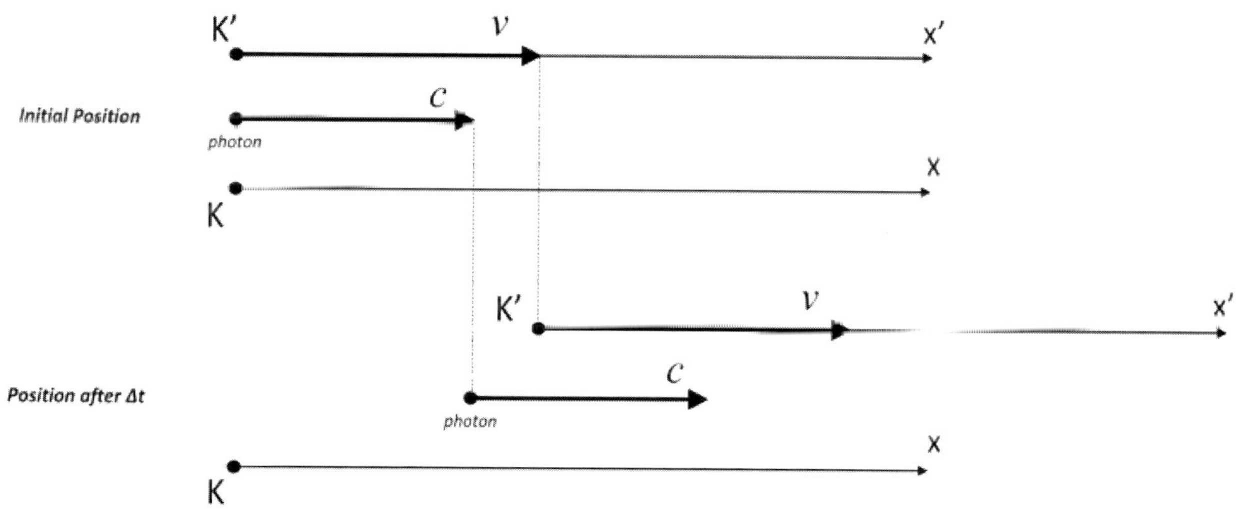

Figure 3. The relative position of the photon and of the two relatively moving reference systems, for $v > c$.

Then a light signal moving towards the positive end of the x in the frame K, obeying equation (1), as it is now moving slower than the zero of K′, relative to K, it follows that it trails behind (with the Euclidean meaning) the zero of K′, in other words the light signal in question moves towards the negative end of x' in K′.

Therefore, in this case, respecting also Einstein's 2nd hypothesis, we are obliged to write:

$$x'=-ct' \quad or \quad x'+ct'=0 \qquad (2a)$$

because in the case of $v>c$, x' is negative.

And here it should be stressed that we are definitely obliged to place the minus (-) sign in the equation describing the movement of light, as t' is *always positive*. And why t' should always be positive? Basically, for two fundamental reasons:

A. Because, at a theoretical level, no one has ever succeeded in giving a logically coherent definition of *Time flowing backwards.*

B. Because, at the empirical level, it can be safely argued that no one has ever seen clocks running backwards on their own. That is to say, that in the Human practical experience, i.e. in the Human perceptible world, Time definitely has an *"Arrow"*, (as the renowned British astronomer Sir A. Eddington argued), which means that clocks have one and only one direction of rotation.

When I am referring to *"clocks"*, on a deeper level, in reality I refer to natural events, as clocks manifest their operation solely through these events. Thus no one has ever witnessed a clock running backwards on its own, just as no one has ever witnessed a plate, having fallen from a table and shattered to pieces, reassembling itself and magically returning onto the table, safe and sound. Thus, by combining equations (1) & (2), which are valid only for $v<c$ with those written by Einstein in Appendix 1 of his book, we can formulate the Lorentz Transformation as we know it. However, by combining equations (1) & (2a), which are valid only for $v>c$, with those written by Einstein in Appendix 1 of his book we cannot formulate the Lorentz Transformation.

Conclusion 2.2.:

The Lorentz Transformation, as derived in Einstein's book [2], is valid *only* for $v<c$. In other words this is an *essential pre-condition* for the analytic derivation of the Lorentz Transformation and *not* its conclusion.

2.3. From A. P. French's textbook titled: *"Special Relativity"* [4]. The author in chapter 3 and under the subtitle *"The Lorentz-Einstein transformations"* (p.p.76-81), derives the transformation, considering that both equations ($x=ct$, $x'=ct'$) are simultaneously valid *for the same light signal* [equations (3-10) p.78].

Thus everything I have mentioned so far (and have previously analyzed in great detail in my book titled *"The Theory of the Harmonicity of the Field of Light"* (5)), applies also in this case.

2.4. From Wolfgang Rindler's textbook titled: *"Essential Relativity"* [6]. The author in § 2.6, p.32 of the book titled: *"The Lorentz Transformation"*, writes: *"Now, by the "second postulate" (constancy of the speed of light), we know that $x = ct$ must imply $x'= ct'$, and vice versa"*. Thus, our aforementioned arguments also apply here. It should be noted that, in everything presented so far, we have neither evoked the existence of *Tachyons* nor disputed the causality principle.

3. Final Conclusion

The relationship $v<c$, is definitely *not a conclusion* , as widely believed, of the Lorentz Transformation, but instead a *hidden pre-condition* of its analytical production. This means that the Lorentz Transformation is *only* valid for $v<c$.

Consequently, a law of Physics that forbids matter to travel with superluminal velocity *does not exist*.

References

[1] Einstein, A. (2005) Relativity, The Special and General Theory, (New Edition), New York: Pi Press.

[2] Einstein, A. (1952) in H. A. Lorentz, A. Einstein, H. Minkowski & H. Weyl, The Principle of relativity, New York: Dover Publications.

[3] Raftopoulos, D.G. (2007) On synchronized clocks at the ends of a moving rod, Physics Essays, Vol.20, issue 2.

[4] French, A.P. (1966) Special Relativity, New York: W.W. Norton & Co.

[5] Raftopoulos, D.G. (2011) The theory of the harmonicity of the field of light,Vol. A (Published by the author, in Greek, Athens 2004, and English, Athens: Angelakis Publ.

[6] Rindler, W. (1977) Essential Relativity, Second Edition, Springer-Verlag.

Spacetime and Quantum Propagation From Digital Clocks

GARNET. N. ORD

Department of Mathematics, Ryerson University
Toronto, Ont. Canada
gord@ryerson.ca

Minkowski spacetime predates quantum mechanics and is frequently regarded as an extension of the classical paradigm of Newtonian physics, rather than a harbinger of quantum mechanics. By inspecting how discrete clocks operate in a relativistic world we show that this view is misleading. Discrete relativistic clocks implicate classical spacetime provided a continuum limit is taken in such a way that successive ticks of the clock yield a smooth worldline. The classical picture emerges but does so by confining unitary propagation into spacetime regions between ticks that have zero area in the continuum limit. Clocks allowed a continuum limit that does not force inter-event intervals to zero, satisfy the Dirac equation. This strongly suggests that the origin of quantum propagation is to be found in the shift from Newton's absolute time to Minkowski's frame dependent time and is ultimately relativistic in origin.

Keywords: Dirac equation, Quantum theory, Relativity, Spacetime.

1. Introduction

Since the beginning of the 20th century we have had two seemingly fundamental theories that suggest very different pictures of the physical world. On one hand, relativity with its emphasis on 'events', 'worldlines' and reference frames, speaks of a physical world of objects, and as a theory is known to be very precise on large space and time scales.

On the other hand, quantum mechanics has trouble interpreting classical concepts like events, particles and worldlines since these involve extremes of the uncertainty principle. The precision of quantum mechanics arises not from its description of objects, but from its description of propagation through time. In all interpretations of quantum mechanics, the precise mechanism by which energy levels are found can be related to a form of wave propagation that allows solutions of a wave equation to detect boundary conditions. Whether the waves are just a mathematical device, or have a physical analog depends on the interpretation placed on the mathematics. So far there have been no definitive experiments to distinguish between the large number of interpretations.

While both theories have experimental results suggesting similar precision within their domains of relevance, relativity appears to pose less problems in terms of interpretation. After all, relativity is a mathematical expression of a physical picture while quantum mechanics is a collection of rules found by trial and error. This being the case, it seems reasonable to work from relativity using a classical picture to see if we can better understand how time is treated in Minkowski spacetime, and to see if this lends some insight into quantum propagation.

To this end we shall investigate a simple model of a digital clock. (This clock has been explored in previous work from several perspectives. See [1-5].) We shall see that in the digital to analog conversion of this clock, the concept of spacetime arises naturally. By examining just how this happens, we are able to see that the continuum limit, while very plausibly relevant on large scales, on small scales leaves questions about whether a natural uncertainty principle can be safely ignored. By changing the continuum limit to acknowledge the uncertainty principle we discover that our clock satisfies the Dirac equation.

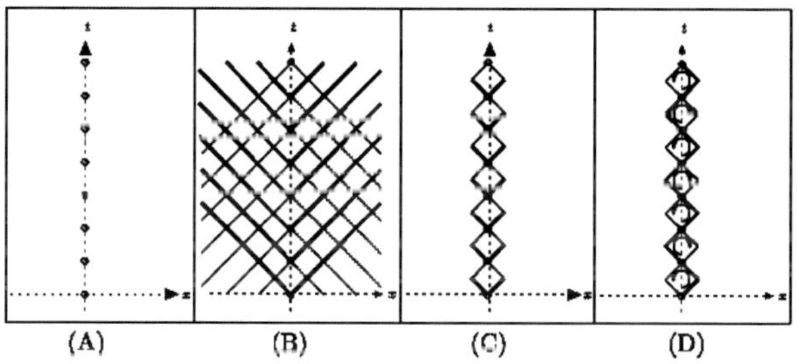

Figure 1. (A) Periodic events in a two dimensional spacetime. (B) Past and future light-cones for the events. Units are chosen so that $c = 1$. (C) The causal areas between events. (D) The area boundary is chosen to provide a single traversal. This results in an alternating orientation of causal areas.

The result of this investigation very strongly suggests that quantum propagation and special relativity, rather than being separate paradigms that intersect at the Dirac equation, are actually different manifestations of Einstein's second postulate that the speed of light is constant in all inertial frames. Thus if we are to look for a physical origin for quantum mechanics, we should make sure we start the search in the relativistic domain.

2. Minkowski's Clock

2.1. *The Clock and the Continuum Limit*

To begin with, consider a two dimensional space-time in which to build a simple digital clock [1, 2]. The clock will be assumed periodic and for reasons that shall appear later, will be based on a period four process. We call this model Minkowski's clock as it illustrates the emergence of Minkowski's spacetime.

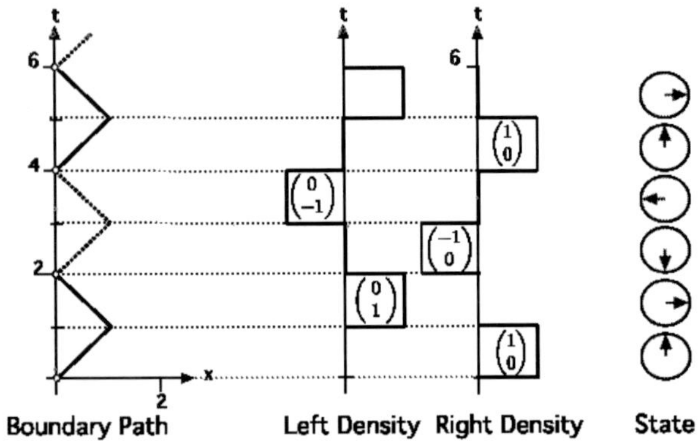

Figure 2. The right boundary of the chain of causal areas can be in any one of four states. The boundary densities and column vector description are illustrated. On-worldline events occur at the even integers. The continuum limit (7) suggest that the left and right densities taken together can be replaced by a 'clock face' that interpolates between the four states.

Figure 1(A) represents a periodic sequence of events at the origin in an inertial frame. (B) shows the past and future light cones of these events and (C) the causal areas that lie between the sequential events. Here the scale is such that the speed of light is one and the 'causal' areas are simply those areas between events that are in the future of the earlier event and the past of the later event. The significance of these areas is that regardless of the actual clock mechanism, the areas represent the maximum time-like domain of influence that successive ticks have in common. In (D) we distinguish the area boundaries to show that we can imagine the boundaries to be drawn as a single continuous curve from $t = 0$ to the latest event and thence back to the origin. Drawing the boundary this way gives the causal areas an orientation and places the events at intersections of smooth segments of the two curves. If we think of the boundary curves as particle paths, there are two kinds of events. The outside corners represent impulsive accelerations while the path-crossing points themselves are simply points where the curves intersect. We call the latter intersections on-worldline events and the former off-worldline events. From a path perspective, our clock is equivalent to a couple of featureless photons confined between two walls that generate a 'tick' at crossing points.

The task at this point is to use the events to provide a digital clock that can subsequently be used as a model for an analog clock. Notice that in terms of paths, the second on-worldline event has 2 accelerations in the two paths to the event, the third on-worldline event has four such accelerations, etc. We can think of the digital clock as counting unit impulses in the area boundaries. For simplicity, we can split the boundary in half and just use one side of it, as in Fig. 2. In this figure, a two-component 'state vector' records projections onto the left and right light cones and because the area boundaries are themselves null segments, successive states are orthogonal. The states themselves yield a period 4 sequence

$$s_k \in S_0 = \left\{ \begin{pmatrix} 1 \\ 0 \end{pmatrix}, \begin{pmatrix} 0 \\ 1 \end{pmatrix}, \begin{pmatrix} -1 \\ 0 \end{pmatrix}, \begin{pmatrix} 0 \\ -1 \end{pmatrix} \right\} \tag{1}$$

and the actual sequence can be inverted to establish the time so, for example, if the sequence so far is

$$\left\{ \begin{pmatrix} 1 \\ 0 \end{pmatrix}, \begin{pmatrix} 0 \\ 1 \end{pmatrix}, \begin{pmatrix} -1 \\ 0 \end{pmatrix} \right\} \tag{2}$$

the time is between $t = 2$ and $t = 3$.

If we think of the column vectors as the active display of the digital clock, algebraically the clock mechanism corresponds to left multiplication by the transfer matrix

$$T_D = \begin{pmatrix} 0 & -1 \\ 1 & 0 \end{pmatrix} \tag{3}$$

where we note that $T_D^2 = -I_2$. The 'tick' matrix is a square root of minus the identity and a fourth root of the identity matrix. In terms of the Pauli matrices $T_D = -i\sigma_y$. If the clock is in state s_k at time k then the subsequent state, after the next event is

$$s_{k+1} = T_D s_k \,. \tag{4}$$

Since

$$s_k = T_D^k s_0 \tag{5}$$

the power of the transfer matrix corresponds to the discrete displacement in time.

To make this clock analog, we have to construct a process that agrees with the clock at the integer events where the clock ticks, but provides extra events between ticks. The simplest way to do this is to construct a clock, similar to the original, but running at a higher frequency. Fig. 3 shows an example of the original clock with a similar clock running at three times the original frequency. As the clock frequency is increased to increase the density of on-worldline events, the causal areas between individual events decreases, and approach zero in the continuum limit. The dimension of the causal chain of areas collapses to one from its value of two for the discrete clock.

In terms of the transfer matrix T_D, a clock running at n times the frequency, with n an integer, will have a transfer matrix T_n such that $T_n{}^n = T_D$ so that we can take T_n as the n-th root of T_D via an eigenvalue expansion:

$$T_n = \begin{pmatrix} \cos\left(\frac{\pi}{2n}\right) & -\sin\left(\frac{\pi}{2n}\right) \\ \sin\left(\frac{\pi}{2n}\right) & \cos\left(\frac{\pi}{2n}\right) \end{pmatrix} \qquad (6)$$

We can use this to take the limit as $n \to \infty$, in which case we get:

$$T(t) = \lim_{n\to\infty} T_n^{nt} = \begin{pmatrix} \cos\left(\frac{\pi t}{2}\right) & -\sin\left(\frac{\pi t}{2}\right) \\ \sin\left(\frac{\pi t}{2}\right) & \cos\left(\frac{\pi t}{2}\right) \end{pmatrix}$$

$$= \cos\left(\frac{\pi t}{2}\right) I_2 + T_D \sin\left(\frac{\pi t}{2}\right) \qquad (7)$$

Notice that $T(n) = T_0{}^n$ for integer n so that our 'analog clock' agrees with the digital clock at the integers where the original digital clock has events.

From equation (7) we can see that the analog transfer matrix $T(t)$ is just a rotation matrix and that the continuum limit replaces the set of four state vectors S_0 with the set of all column unit vectors

$$S = \left\{ \begin{pmatrix} u \\ v \end{pmatrix} : u^2 + v^2 = 1, u, v \epsilon \mathrm{R} \right\}.$$

2.2. *The Clock and Boosts*

In a relativistic universe a clock needs to *automatically* satisfy Lorentz transformations. The invariance of the speed of light in all inertial frames fixes the appearance of the graphs of clocks in moving frames. In Fig. 4, the graph of the first cycle of the digital clock is sketched.

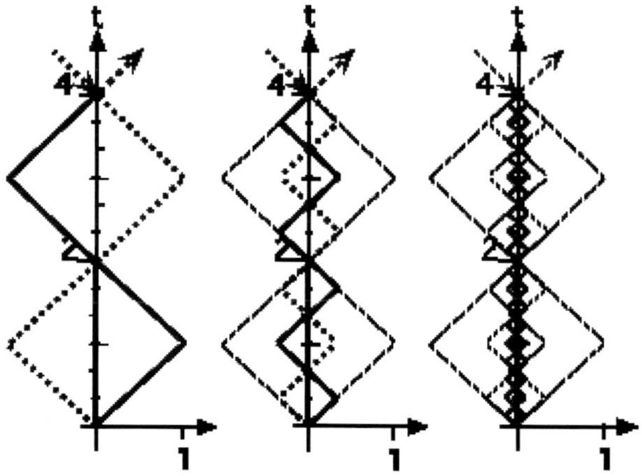

Figure 3. A full clock cycle for the original digital clock with similar clocks running at triple and nine times the original frequency. The higher the frequency the clock, the closer the off-worldline events are to the t-axis and the denser the on-worldline events on the t-axis. In the continuum limit the distance between off-worldline events and the t-axis goes to zero as does the areas between events.

The causal inter-event areas have the same magnitude and sign as in the case of the stationary clock but projections onto the left and right cones are different [1]. To see how to handle this algebraically consider the original digital to analog conversion equation (6) that approximates T_0 by the n-th power of T_n. What is happening is that a single tick of the original clock is replaced by n smaller ticks of equal size to bring the higher frequency device to the same state. That is, the first step from state 1 to state 2 becomes:

$$T_D \to \underbrace{T_D I_2 I_2 \ldots I_2}_{n \text{ terms}} \to \underbrace{T_n T_n \ldots T_n}_{n-\text{terms}} \qquad (8)$$

where replacement by n copies of T_n is an interpolation. The application of either T_D or its replacement switches

$$\begin{pmatrix} 1 \\ 0 \end{pmatrix} \text{ to } \begin{pmatrix} 0 \\ 1 \end{pmatrix} \text{ with } T_D \to T_n^n$$

being a smoothing operation that allows a continuum limit. In the original rest frame clock the transition to the subsequent state would be accomplished by the same sequence as equation (8) regardless of the state of motion of the clock, however relativistically, a clock moving at constant speed v in the x direction has to stay in state 1 longer than state 2. We might expect something like

$$T_D^2 \to \underbrace{T_D I_2 I_2 \ldots I_2}_{n \text{ terms}} \to \underbrace{T_D I_2 I_2 I_2 \ldots I_2}_{m-\text{terms}} \to \underbrace{T_{m+n} \cdots T_{m+n}}_{m+n-\text{terms}} \qquad (9)$$

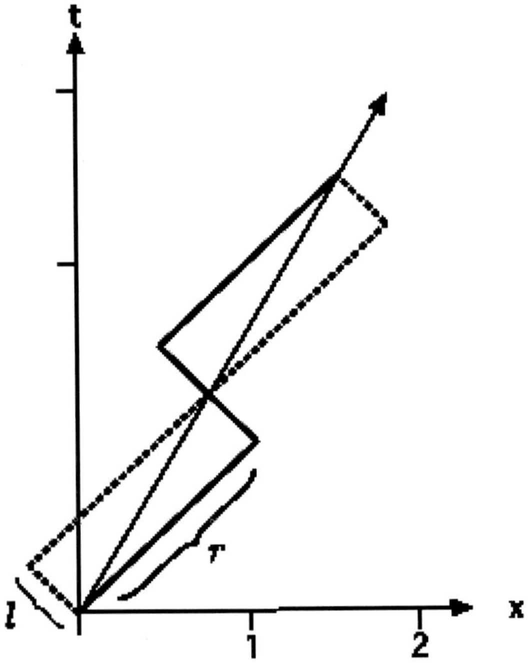

Figure 4. A clock cycle in a moving frame. The null boundaries remain null but the left and right projections change length. The signs and magnitudes of the causal areas are invariant.

Here the new transfer matrix T_{m+n} has to function as an $(m + n)$-th root of T_D^2 while weighting the occupation of state 1 more heavily than state 2 ($m > n$). This means increasing the magnitude of the 1-1 element and decreasing the magnitude of the 2-2 element. To see how this may be done expand $T(t)$ to first order in $t = 1/n$:

$$T(1/n) = \begin{pmatrix} 1 & -\dfrac{\pi}{2n} \\ \dfrac{\pi}{2n} & 1 \end{pmatrix}$$

where we assume $n >> 1$. The ones on the diagonal equally weight 1-1 and 2-2 transitions. If we try $T(1/n) \rightarrow T_M(1/n) = T(1/n) + w\dfrac{\pi}{2n}\sigma_z$ this will effect an increased residence time in the odd states ($w > 0$) and a decreased residence time in the even states as suggested by Fig. 4. w is to be chosen so that the 'moving clock runs slow' in accordance with the Lorentz transformation and the area preservation of the light clock [1]. To be in accordance with Fig. 4 the first on-worldline event should take place at $t = 2\gamma$ where $\gamma = 1/\sqrt{1-v^2}$ is the time dilation factor. So we must have:

$$T_v(2\gamma) = \lim_{n\to\infty} T_v(1/n)^{2n\gamma} = -I_2. \tag{10}$$

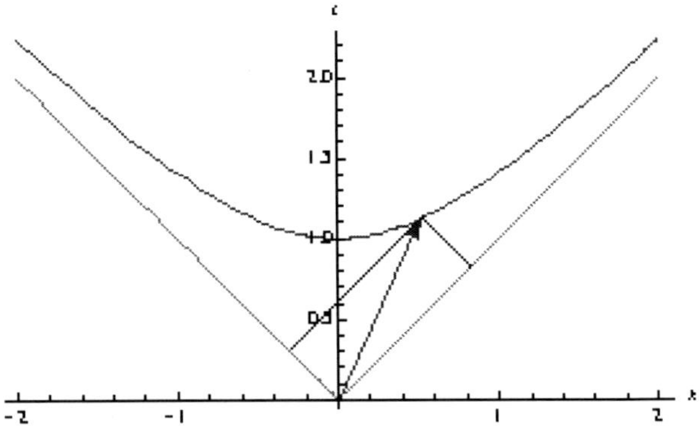

Figure 5. The odd part of the transfer matrix is a vector. The unit vector starts at the origin and ends on the first on-worldline event.

For equation (10) to hold, the eigenvalues of T_v raised to the power $2\gamma n$ must be -1 in the limit as $n \rightarrow \infty$. The eigenvalues of T_M to lowest order in $1/n$ are $\lambda = 1 \pm \dfrac{i\pi\sqrt{1-w^2}}{2n}$ giving

$$\lim_{n\to\infty} \lambda^{2\gamma n} = e^{\mp i\pi\sqrt{1-w^2}\gamma}. \tag{11}$$

For this to equal -1 we have $w = \pm v$. Choosing the positive value of v for w we can calculate the transfer matrix for a clock moving at speed v with respect to the lab frame whose transfer matrix was given by (7). The matrix is

$$T_v(t) = \cos\left(\frac{\pi t}{2\gamma}\right) I_2 + \underbrace{\gamma(v\sigma_x - i\sigma_y)}_{\text{kinematics}} \overbrace{\sin\left(\frac{\pi t}{2\gamma}\right)}^{\text{dynamics}}. \tag{12}$$

The matrix labeled 'kinematics' above is a unit vector where matrix multiplication is assumed. That is:

$$\hat{\mathbf{v}}^2 = \left(\gamma(v\sigma_x - i\sigma_y)\right)^2 = \gamma^2(v^2-1)I_2 = -I_2 \tag{13}$$

This is to be compared with the transfer matrix in the rest frame (7)

$$T(t) = \cos\left(\frac{\pi t}{2}\right) I_2 - i\sigma_y \sin\left(\frac{\pi t}{2}\right).$$

The transfer matrix as expressed above shows that its domain is a one dimensional subspace of a two dimensional space. As (7) illustrates, part of the domain of T_v as a function is the subspace defined by the set of all vectors that are scalar multiples of $\mathbf{v} = \gamma(v\sigma_z - i\sigma_y)$. The state defined by the action of T_v has a component proportional to the 'dynamic' coefficient $\sin\left(\dfrac{\pi t}{2\gamma}\right)$ but for the purpose of seeing how 'spacetime' comes about, we need not be concerned with this coefficient. Indeed the vector $v = \gamma(v\sigma_z - i\sigma_y)$ would occur for any clock created this way *regardless* of either the initial frequency chosen for T_D, or the actual machinery that generates the periodic sequence. We thus see that the transfer matrix describing the continuum limit of our rather general digital clock immediately implicates using σ_z and $-i\sigma_y$ as vectors along the x and t axes, respectively. In particular, for fixed v

$$R_v = \{\lambda\gamma(v\sigma_x - i\sigma_y) : \lambda \in \mathbb{R}\} \qquad (14)$$

gives the subspace domain over which the transfer matrix produces its analog signal, a component of which is given by $\sin\left(\frac{\pi t}{2\gamma}\right)$. The first postulate that "Physics is the same in all inertial frames" appears here explicitly in that the dynamic coefficient of v in equation (12) above is identically its pre-image of the coefficient of $i\sigma_y$ in $T(t)$ above.

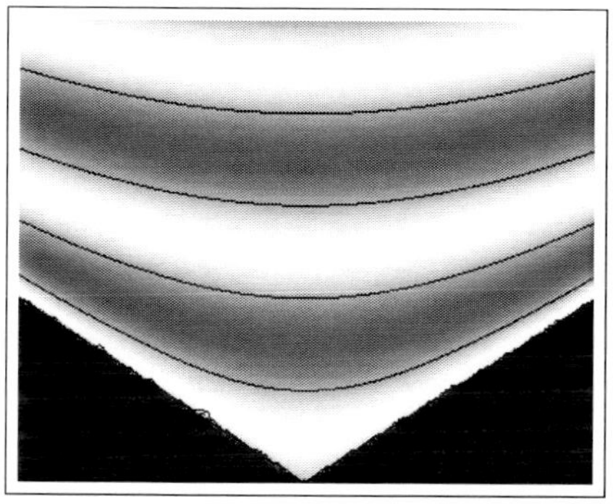

Figure 6. The coefficient of the odd term in equation (12) plotted as a colour map. This illustrates the connection of phase with path that is inherent in any relativistic clock. The black contours correspond to the on-worldline events of the discrete clock. We can think of the image as a raster scan based on the one-dimensional rays through the origin into the future cone of the clock.

Equation (12) gives the transfer matrix for a relativistic clock that moves with speed v with respect to the original clock. If we plot the coefficient of the unit vector v in the x - t plane we get Fig. 6. Equation (12) shows that one of the effects of the Lorentz transformation is to replace $-i\sigma_y$ with $\gamma(v\sigma_z - i\sigma_y)$. If we think of the Pauli matrices as unit vectors, then this change looks like a rotation which one might expect is accomplished by a similarity transformation. That is, we might expect that

$$\gamma(v\sigma_x - i\sigma_y) = B(-i\sigma_y)B^{-1} \qquad (15)$$

where B is some invertible matrix. It may be easily verified that the requisite matrix is

$$B = \begin{pmatrix} \cosh\left(\frac{\alpha}{2}\right) & -\sinh\left(\frac{\alpha}{2}\right) \\ -\sinh\left(\frac{\alpha}{2}\right) & \cosh\left(\frac{\alpha}{2}\right) \end{pmatrix} \qquad (16)$$

with $\alpha = \text{Arctanh}(-v)$. This confirms the interpretation that we can indeed regard the odd part of the transfer matrix equation (12) as a function times a unit vector. If we take $-i\sigma_y$ as lying on the t-axis and σ_z lying on the x-axis then the velocity at the point (x, t) is just x/t and the subspace of vectors along $\gamma(v\sigma_z - i\sigma_y)$ lie along the ray $x/t = v$. Thus the two dimensional vector space with unit vectors $(\sigma_z, -i\sigma_y)$ can serve a dual purpose.

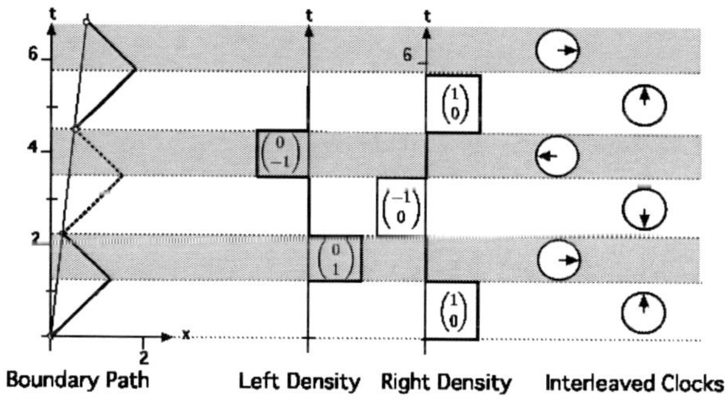

Figure 7. To accommodate boosts, the two-photon clock uses both complex eigenvalues of the transfer matrix. The extra information encodes the two different residence times in the even and odd states. This is illustrated in the figure where the tilted time axis gives rise to different projection intervals on the left and right cones.

On one hand, the space constitutes part of the domain of the set of transfer matrices that evolve the light clock. Its presence there is dictated by the need for the clock to preserve spacetime area under inertial transformations. It is through the role of the transfer matrix as an operator on states that we find the link to the vector space with basis vectors $(\sigma_z, -i\sigma_y)$ and with special subspaces along the rays specified by $\gamma(v\sigma_z - i\sigma_y)$.

On the other hand, as a representation of space-time, it encodes the notion of proper time in the algebra of matrix multiplication. This happens because the odd signature, $\sigma_z^2 = 1$, $(-i\sigma_y)^2 = -1$ automatically encodes the invariant area $t^2 - x^2$.

2.3. *The Clock in 3D*

The three dimensional version of this is equally transparent. If we embed Minkowski's clock in a four dimensional space we need a representation that is general enough to handle rotations as well as boosts. However, the essential one-dimensionality of the clock remains.

To implement this the extension has to be done in such a way that the two extra directions yield a

Euclidean subspace allowing spatial rotations that do not affect the evolution of proper time. One way to do this is to simply take the transfer matrix and replace each element by a 2 × 2 matrix in the following way.

$$I_2 \rightarrow \begin{pmatrix} I_2 & 0 \\ 0 & I_2 \end{pmatrix} = I_4 \qquad -i\sigma_y \rightarrow \begin{pmatrix} 0 & -I_2 \\ I_2 & 0 \end{pmatrix} = -i\beta$$

$$\sigma_z \rightarrow \begin{pmatrix} (x\sigma_x + y\sigma_y + z\sigma_z) & 0 \\ 0 & -(x\sigma_x + y\sigma_y + z\sigma_z) \end{pmatrix}$$

$$= x\sigma_x + y\sigma_y + z\sigma_z \tag{17}$$

Then the new transfer matrix becomes:

$$T_v(t) = \cos\left(\frac{\pi t}{2\gamma}\right) I_4 + \gamma\left(v\left(x\sigma_x + y\sigma_y + z\sigma_z\right) - i\beta\right)\sin\left(\frac{\pi t}{2\gamma}\right) \tag{18}$$

Let us check this replacement to make sure it fulfills the necessary requirements. It may be verified that the α_k and β have unit norm and anti-commute. The α_k are a suitable basis for Euclidean 3-space as they are simply $\alpha_k = \sigma_z \otimes \sigma_k$. The term $\gamma((x\alpha_x + y\alpha_y + z\alpha_z) - i\beta)$ is a unit vector that squares to $-I_4$ provided $r = \sqrt{x^2 + y^2 + z^2} = 1$ so that $(x\alpha_x + y\alpha_y + z\alpha_z)$ is a unit vector of positive norm. We can see that in the limit that $v \rightarrow 0$ we get

$$T_v(t) = \cos\left(\frac{\pi t}{2\gamma}\right) I_4 + (-i\beta)\sin\left(\frac{\pi t}{2\gamma}\right) \tag{19}$$

and the rest frame transfer matrix for the discrete clock is

$$T_0 = \begin{pmatrix} 0 & 0 & -1 & 0 \\ 0 & 0 & 0 & -1 \\ 1 & 0 & 0 & 0 \\ 0 & 1 & 0 & 0 \end{pmatrix} \tag{20}$$

We can then see that we have the same basic counting algorithm with the four original discrete states S_0 replaced by

$$S_0 = \left\{ \begin{pmatrix} 1 \\ 1 \\ 0 \\ 0 \end{pmatrix}, \begin{pmatrix} 0 \\ 0 \\ 1 \\ 1 \end{pmatrix}, \begin{pmatrix} -1 \\ -1 \\ 0 \\ 0 \end{pmatrix}, \begin{pmatrix} 0 \\ 0 \\ -1 \\ -1 \end{pmatrix} \right\} \tag{21}$$

As in the case of the two dimensional spacetime, we see that four basis vectors arise through their role in the construction of a transfer matrix, itself a mathematical encoding of a discrete clock.

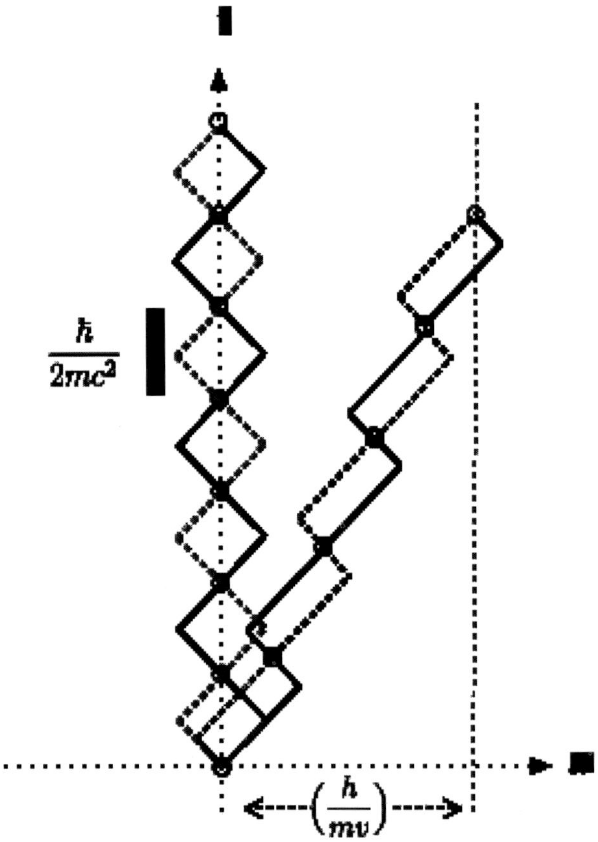

Figure 8. At fixed velocity v the moving clock, 'running slow' will eventually lose one complete event, thus temporarily re-synchronizing with the stationary clock. The sequence of space and time coordinates of these re-synchronization events represent a beat frequency that correspond to the de Broglie scale.

As in the two dimensional case, we see that the transfer matrix, in providing the digital to analog conversion of a discrete periodic clock, implicates Dirac's matrices as vectors in a spacetime of odd signature. The fact that the vector representing the time axis has negative norm is a remnant of the periodicity of the clock combined with the requirement that the Lorentz transformation be obeyed. Similar representations will occur whenever we make a digital to analog conversion for a digital periodic clock. Summarizing some of the features of this clock, we note the following:

- Minkowski's clock shows us where the odd signature of spacetime algebra comes from!
- The negative norm of the time vector T encodes both the periodicity and the directional information needed for a relativistic clock.
- From a Clifford algebra perspective, the negative norm arises because T is actually a *bivector* representing a spacetime *area* with invariant norm!

- It is because relativistic clocks are based on invariant spacetime *areas* that space and time unit 'vectors' have different squares!
- This is the origin of the familiar $t \rightarrow it$ that takes Euclidean space to Minkowski space.

3. The Connection to Dirac

3.1. *The de Broglie Scale*

In the above, the presence of the Dirac matrices operating on column vectors is reminiscent of the Dirac equation and the clock periodicity an analog of 'zitterbewegung'. However if this is a harbinger of quantum propagation we should expect to see the analog of the de Broglie wavelength.

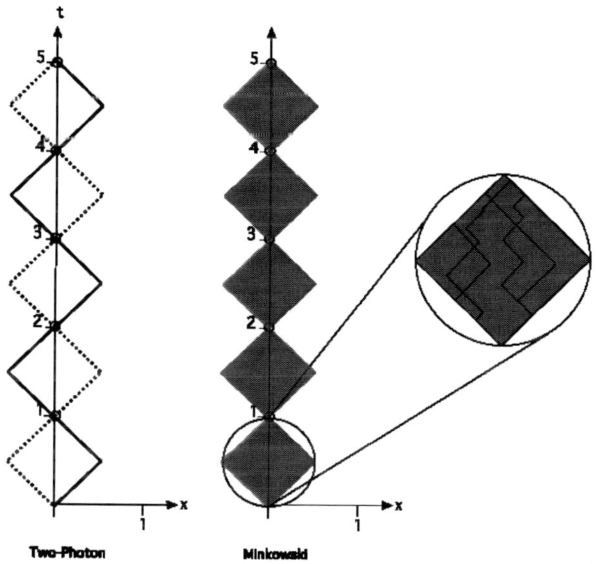

Figure 9. The original clock ticks at the integers which are the diagonal end points of the causal spacetime areas. The areas are defined by both the on-worldline and off worldline events. Higher frequency interpolations for the clock do not have to be attracted to the worldline between events. They can be allowed to spread throughout the area.

If the event sequence of the clock moves with finite speed v with respect to the lab frame, the 'moving clock will run slow'. As a result, there will be a point in space and time where the moving clock will re-synchronize, with the stationary clock having lost a single beat, Fig. 8. If the basic period of the original digital clock is chosen to be the Compton period, h / mv^2, the spatial wavelength of this 'beat frequency' clock is just the de Broglie length $\lambda = h / mv$. Note the disappearance of the high frequency rest mass 'Zitterbewegung' at this scale. It is interesting to see that here, the deBroglie wavelength is geometric in origin and *arises from the beat frequency of the slightly different frequencies of the Zitterbewegung due to time dilation.*

3.2. *The Dirac Equation*

In the above we have seen that the transfer matrices contain representations of the Dirac matrices that operate on column vectors. The column vectors are representations of a clock signal along the direction specified by the velocity v in the transfer matrix. As we can see from Fig. 3, when we take the continuum limit we force on-worldline events onto a linear subspace towards which the off-worldline events are pulled in the continuum limit. In this limit we have a classical path that in a quantum context would violate the uncertainty principle. Furthermore, in the transition from clock to spacetime we have ignored the fact that the clock produces a signal from which we would extract the time. This process itself has an associated intrinsic uncertainty that has been ignored in the extraction of the spacetime algebra.

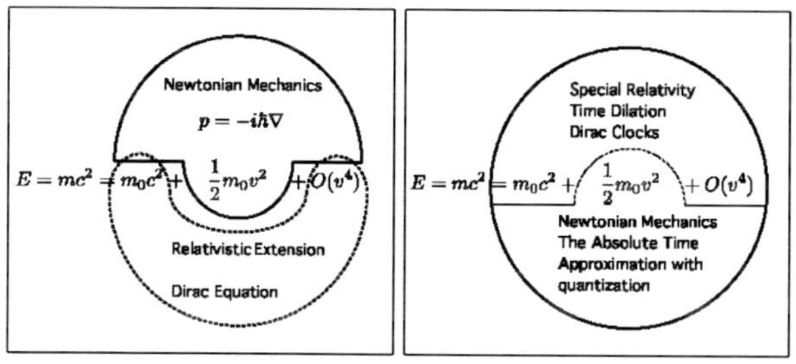

Figure 10. Left. The advent of Relativity extended the concept of energy and produced the Dirac equation as an extension of the Schrödinger equation. In reality, Dirac's clock illustrates that both Newtonian mechanics and the Schrödinger equation are different approximations to the path dependent time of relativity.

When constructing Minkowski's clock, the continuum limit used was a device to allow an arbitrarily high frequency of events. This was in accordance with the classical paradigm that associates a continuous worldline of events with a particle. However, this is not the only continuum limit possible. We can allow arbitrarily high frequency without confining the clock path to a smooth worldline.
As an algebraic example we might think of $i = e^{i\pi/2}$ as a *self-referential* Taylor expansion, where i is our 'tick' matrix

$$\underbrace{\begin{pmatrix}0 & -1\\1 & 0\end{pmatrix}}_{\text{Single Digital Tick}} = \underbrace{\begin{pmatrix}0 & -1\\1 & 0\end{pmatrix}^0 + \frac{\pi}{2}\begin{pmatrix}0 & -1\\1 & 0\end{pmatrix}^1 + \frac{\pi^2}{8}\begin{pmatrix}0 & -1\\1 & 0\end{pmatrix}^2 \cdots}_{\text{Indefinite number of Ticks}}$$

This expansion allows arbitrarily high frequency, but the high frequency components have suitably small weighting factors allowing the series to converge. Fig [9] illustrates the fact that the inter-event areas may be covered by paths corresponding to clocks running at higher frequencies that are not restricted to the worldline.

In the two dimensional case, implementing the de-restriction of higher frequency events from the world line is easy to do and the result is a well-known model of the Dirac propagator called the Chessboard model. (The model originated with Feynman and appeared as an exercise in his text [6]. Some work on the model may be found in references [7-11].) In terms of the transfer matrix above, it corresponds to using a Fourier transform of the transfer matrix in place of the x-space version we have used. The effect is to replace the unit vector v by a canonically conjugate momentum p so that the transfer matrix (12) becomes

$$T(p,t) = \lim_{n\to\infty} \begin{pmatrix} \exp[ip\varepsilon] & -m\varepsilon \\ m\varepsilon & \exp[-ip\varepsilon] \end{pmatrix}^{t/\varepsilon}$$

$$= \cos(Et)I_2 + \frac{-1}{E}(ip\sigma_x + im\sigma_y)\sin(Et) \quad (22)$$

where we have replaced $\pi/2$ by the positive constant m and written $E = \sqrt{p^2 + m^2}$. Differentiating this and rearranging terms one finds that

$$i\frac{\partial T(p,t)}{\partial t} = HT(p,t) \quad (23)$$

with

$$H = p\sigma_z + m\sigma_y \quad (24)$$

This is a form of the Dirac equation in a two dimensional spacetime in units with $c = \hbar = 1$.

4. Discussion

The progression from Minkowski's clock to the Dirac equation illustrates that it is really the geometry of special relativity that is a progenitor of both classical spacetime and quantum propagation. The former arises by neglecting the dynamic information of clocks and just accepting the common kinematic information that would result from any periodic relativistic clocks as suggested by (12). The latter arises by taking into account both the kinematic *and* dynamic information of the transfer matrix. Ultimately the difference between classical spacetime and quantum propagation here is simply a choice of which continuum limit to take. This reinforces the picture that suggests that relativity is logically prior to quantum mechanics, in contrast to the historic view that the Dirac equation is an extension of the Schrödinger equation Fig. 10. In this view, the Schrödinger momentum arises from a natural beat frequency over the Zitterbewegung that acts as a carrier frequency. This also appears in Newtonian mechanics as the first order expansion of the relativistic energy. The rest mass fails to appear in slow collisions since it is invariant under changes of frame. The higher order terms are generally too small to appear so the Schrödinger and Newtonian regimes run in parallel. The relation between the Schrödinger and the Dirac equation actually comes from equation (23) by the replacement of H with its non-relativistic approximation, rather than the other way around.

Stepping outside the 'clock' picture we can see the quantum implications of the above models. The fact that Minkowski's clock gives classical spacetime despite the fact that the column vectors look a little like wavefunctions is manifestation of the quantum Zeno effect. By taking the continuum limit to restrict the events to worldlines, we essentially squashed the 'quantum propagation' into the spacetime areas between the events, and then sent these areas to zero. From this perspective classical spacetime is actually an infinite mass construction. The other side of this picture is simply that the Dirac equation is about periodic clocks that contain all the frequency information needed to place events in spacetime *that can be read appropriately from any inertial frame*, while not actually writing events on spacetime at infinite frequency.

The efficacy of spacetime algebra to both describe the metric structure of spacetime and to provide an elegant formulation for the Dirac equation [12-14] is explained here through the common use of the transfer matrix in its role in describing simple relativistic clocks. Both the $t \to it$ of Minkowski (space-time $\to$ spacetime) and the $t \to it$ of Schrödinger (taking the diffusion equation to the Schrödinger equation) are both here remnants of Lorentz invariant *area* between events. These areas are oriented, and are algebraically represented by bivectors that have negative norm. In the same vein, the path dependent phase of quantum mechanics is here a remnant of the path dependent proper time of special relativity.

References

[1] Ord, G.N. & Mann, R.B. (2011) How does an electron tell the time? Intl. J. Theoretical Physics, 51(2):652–666.

[2] Ord, G.N. (2012) Quantum propagation from the twin paradox, in G. Grössing (ed.) Emergent Quantum Mechanics, volume 361, Journal of Physics Conference Series.

[3] Ord, G.N. (2011) What happens between the ticks of a clock?, in A. Ford & K.G. Bowden (eds) Proceedings of ANPA, Cambridge, volume 32.

[4] Ord, G.N. (2009) Quantum mechanics in a two-dimensional spacetime: What is a wavefunction? Annals of Physics, 324(6):1211–1218.

[5] Ord, G.N. (2010) Feynman's corner rule; quantum propagation from special relativity, Intl. J Theoretical Physics, 49:2528–2539, 2010. 10.1007/s10773-010-0445-8.
[6] Feynman, R.P. & Hibbs, A. R (1965) Quantum Mechanics and Path Integrals. New York: McGraw-Hill.
[7] Gersch, H.A. (1981) Feynman's relativistic chessboard as an ising model, Int. J. Theor. Physics, 20:491.
[8] Jacobson, T. & Schulman, L.S. (1984) Quantum stochastics: the passage from a relativistic to a non-relativistic path integral. J. Phys. A, 17:375–383.
[9] Ord, G.N. (1992) A reformulation of the Feynman chessboard model,. J. Stat. Phys., 66:647–659.
[10] Ord, G.N. & Mann, R.B. (2003) Entwined paths, difference equations and the Dirac equation. Phys. Rev. A, 67(2):022105.
[11] Kauffman, L.H. & Noyes, H.P. (1996) Discrete physics and the Dirac equation, Phys. Lett. A, 218:139.
[12] Doran, C. & Lasenby, A. (2003) Geometric Algebra for Physicists, Cambridge: Cambridge University Press.
[13] Hestenes. D. (2003) Mysteries and insights of Dirac theory, *Annales de la Fondation Louis de Broglie*, 28:390–408.
[14] Hestenes. D. (2008) *Zitterbewegung* in quantum mechanics. arXiv:8002.2728 [quant-ph].

General Relativity Theory — Well Proven and Also Incomplete?

JÜRGEN BRANDES

Danziger Str. 65, 76307 Karlsbad, Germany
jg-brandes@t-online.de

With a few arguments (half a page) it is proven that general relativity (GRT) makes contradictory predictions about the total energy of a particle resting in the gravitational field. With a few further arguments (one page) it is proven that these contradictions are resolved by expanding general relativity. The other situation: Though it is not the aim of the author to reject general relativity but to expand it, he is treated as some uncritical anti-relativist - since the start of his considerations and meanwhile for more than 20 years. My public question: Are relativists – on account of their many famous results – unable to admit imperfections of general relativity? General relativity is contradictious in energy questions since on one side the total energy of a particle resting in the gravitational field is lower than its rest mass (there is energy needed to pull out the particle from the gravitational field) while on the other side it is equal to its rest mass (this is a consequence of the equivalence principle).

Keywords: General relativity; Lorentz-interpretation; equivalence principle .

To reach the source you have to swim against the current. - Stanislaw J. Lec, Poland

1. Preliminary Remark

Incompleteness means some imperfections or even contradictions of a theory but which can be overcome by some extension.

Classical general theory of relativity (GRT) has difficulties with high energies, e. g. with simulation of supernovae explosions, with evaluation of the fireball in the fireball model of gamma bursts (the "fireball" is derived only from observations of the afterglow but not from first principles) and with emission of UHECR (cosmic radiation of highest energy). Therefore it is appropriate to consider in greater detail the energy formulas of this theory.

2. Contradictory Results of Total Energy

The elementary question: What is the total energy E_G of a particle resting in the r,t-reference system of the *Schwarzschildmetrik* (SM)? SM reads:

$$ds^2 = \left(1 - \frac{2GM}{c^2 r}\right)^{-1} dr^2 + r^2(d\theta^2 + \sin^2 \theta d\varphi^2) - \left(1 - \frac{2GM}{c^2 r}\right) c^2 dt^2 \quad (1)$$

From the formulas of radial free fall [1] one gets:

$$E_G = mc^2 \left(1 - \frac{2GM}{c^2 r}\right)^{1/2} \quad (2)$$

This is at least qualitatively correct since removing the particle from the gravitational field needs energy. Doing this the total energy of the particle becomes mc^2 and therefore, within the gravitational field it has to be lower.

On the other side, there is the equivalence principle. A particle resting in its local inertial system (i.e. the freely falling particle) has a total energy equal to its rest mass:

$$E_G = mc^2 \quad (3)$$

Formula (2) and (3) contradict each other. Certainly, they belong to different reference systems with one of them being accelerated, in fact. But: At time point $t = 0$ the free falling particle is also a resting one within the r,t-reference system since its velocity $v = 0$. Only its acceleration $b \neq 0$. Special theory of relativity is applicable and therefore the free falling particle at $t = 0$ as well as an always resting particle at the same position possess identical total energies (3).

Formulas (2) and (3) contradict each other.

On account of the qualitative argument above, formula (2) is the correct one. One can see it again by series expansion of (2):

$$E_G = mc^2 - \frac{GmM}{r} \pm \ldots \quad (4)$$

The second term describes the negative gravitational energy. Approximately formula (2) becomes the rest mass minus Newtonian gravitational energy. Therefore formula (2) meets the Newtonian limit but formula (3) does not.

Deepening of formula (3): It is possible to object that the particle resting relative to the *r,t*-reference system owns the total energy mc^2 relative to the local inertial system only. That is not correct. Perform a measurement. At time point $t=0$ resting and free falling measuring devices are not different from each other, especially such clocks don't run differently – see chapter 4. Therefore it is irrelevant whether the measurement is done relative to the *r,t*-reference system or relative to the local inertial system at $t=0$. In both cases one gets mc^2 or the equivalence principle would be wrong.

3. Rewording of Equivalence Principle

The equivalence principle is tried and tested. So formula (3) should be correct, too. Rewording this principle a little shows why: "For *measurements* within gravitational fields the *measuring results* within local inertial systems are predicted by special relativity."

Concerning our application this means: The *measurement* of E_G using measuring instruments resting in the gravitational field yields formula (3). This is no contradiction to formula (2) any longer if one can assume that measuring instruments become modified by gravitational fields.

4. Modifications of Measuring Instruments Within Gravitational Fields

Let us consider possible modifications of measuring instruments during the measurement of E_G. Let us choose some intellectual simple measuring procedure. Transfer an antiparticle to the resting particle and perform the measurement of the annihilation frequency of the two resulting photons. One gets:

$$\begin{aligned} E_{G,measured} &= mc^2 \\ &= h\nu_{\tau,measured} \end{aligned} \tag{5}$$

$\nu_{\tau,measured}$: annihilation frequency measured by a clock resting in the gravitational field

τ: proper time of a clock resting in the gravitational field (τ-clock)

On the other side, it is derivable from SM[1] (and this is experimentally proven, too):

$$\tau = t\left(1-\frac{2GM}{c^2 r}\right)^{1/2} \tag{6}$$

with words. Time passing of $t=1\text{s}$ of a *t*-clock outside the gravitational field means time passing of a τ-clock less by a factor

$$\left(1-\frac{2GM}{c^2 r}\right)^{1/2} \tag{7}$$

Therefore, the measured frequencies are:

$$\nu_{\tau,gemessen} = \nu_t\left(1-\frac{2GM}{c^2 r}\right)^{-1/2} \tag{8}$$

(8) inserted into (5):

$$\begin{aligned} E_{G,gemessen} &= h\nu_t\left(1-\frac{2GM}{c^2 r}\right)^{-1/2} \\ &= E_G\left(1-\frac{2GM}{c^2 r}\right)^{-1/2} \end{aligned} \tag{9}$$

or

$$E_G = mc^2\left(1-\frac{2GM}{c^2 r}\right)^{1/2} \tag{10}$$

identical with formula (2).

With words: Taking into account the modification of measuring instruments by a gravitational field – in this case the slowing down of clocks – makes it possible to derive the total energy of a resting particle applying the equivalence principle

5. Supplements

1.) Newton's formula of negative gravitational energy is not a relativistic one since on account of Einstein's famous relation

$$E = mc^2 \tag{11}$$

this means negative masses. Negative masses don't exist. Formula (2) explains the real meaning of negative gravitational energy – it's nothing else than a decrease of rest mass.

2.) Formula (2) shows decrease of rest mass within gravitational fields. The same has to be true for the rest masses of measuring instruments and can be taken as a qualitative proof of the assumption that measuring instruments become modified by gravitational fields.

3.) Formula (2) is evaluated easily using the scalar product of the covariant four-dimensional energy-momentum-vector

$$p^2 = p_\mu p_\nu g^{\mu\nu} \tag{12}$$

which is

$$p^2 = m^2 c^2 \tag{13.1}$$

and also

$$p^2 = \left(1 - \frac{2GM}{c^2 r}\right)^{-1} \frac{E_G^2}{c^2} \quad (13.2)$$

since the components $p_1\, p_2\, p_3$ of the resting particle are zero. This results in formula (2), too.

4.) In Ref. 1 it is shown that with these additions (i.e. using the Lorentz-interpretation) there are testable differences with GRT.

5.) A simple fact: If two clocks near each other at the same position show different (clock-) times then one of them is slow or fast. Here everyone should agree. But within relativity theory it is told wrong as the discussion of chapter 6 proves. Therefore an additional argument. Every monochromatic light wave of known frequency can be taken as a clock by counting its wave crests passed by and by calculating the elapsed time using its frequency. Now convert the running light wave into a standing wave. Out of the measured wavelength one can calculate its frequency with $c = \nu\lambda$. Nobody can deny that there is only one wavelength and therefore only one frequency and if two measuring sticks get a different length in meters then at least one of the measuring sticks is uncalibrated.

6. Discussions

This talk was presented at DPG-*Fachtagungen* of GR and DD in Göttingen and Mainz 2012. There and in e-mail discussions before and later some comments were given. While often students accepted the idea to expand GRT gravitational physicists neither agreed with any expansion of GRT nor solved its contradiction. Here the main arguments of those scientists, details s. Ref. [2].

At first two e-mails concerning formula (2) and (3): "... there is no problem with a proton in a gravitational field. Calling *m* its rest mass then its mass in a gravitational field is root of *g_00* times *m*, where *g_00* is the 00-component of the metric (here static); since *g_00* is smaller than 1 the mass (energy) in the gravitational field is smaller than *m*. The resting proton obviously has a mass (energy) lower than the rest mass."

Pointing out once more the possibility of formula (3) the answer was: "... the proton resting in the gravitational field never has the energy mc^2 but the energy lowered by the factor *SQRT(g_00)*. The reason is that resting in the gravitational field is no inertial motion and therefore your argument before is wrong."

Certainly, resting in the gravitational field is no inertial motion but formula (3) needs $v = 0$ at $t = 0$ only.

Another scientist accepts formulas (2) and (3): "... Your formulas (2) and (3) point out the energy measured by an observer moving on *t*-lines [that means he is resting in the gravitational field]. But the clocks used are different. Equation (2) is correct if the clock shows coordinate time *t*, equation (3) is correct if the clock shows proper time τ (as is done when applying the equivalence principle). ..."

This argument says that everything depends on the observer. Choose the right clock and formula (2) or (3) is the right one. But that's not correct: You can do both measurements at once. Then both of the clocks rest near each other but show different times for the same event. This means that one of the clocks is uncalibrated and you don't know which one. E_G performs a length measurement using different meter sticks. If the measuring results are different then one of the meter sticks is harmed or not? See 5.) in chapter 5 for more arguments.

This clock counterargument was not answered up to now [2]. In all the other discussions similar open questions remained. Obviously GRT has a problem. The other situation: Though it is not the aim of the author to reject general relativity but to expand it, he is treated as some uncritical anti-relativist - since the start of his considerations and meanwhile for more than 20 years. Nevertheless, the discussions above were helpful in formulating this contradiction.

The author demands research means for those institutes and their coworkers who are willing to investigate the complex energy statements of GRT.

7. Summary

GRT knows two contradictory formulas of total energy of a particle resting within gravitational fields. This contradiction is resolved if one can assume that measuring instruments become modified by gravitational fields. This assumption has testable consequences [1].

References

[1] Brandes, J. & Czerniawski, J. (2010) *Spezielle und Allgemeine Relativitätstheorie für Physiker und Philosophen* – Einstein- und Lorentz-Interpretation, *Paradoxien, Raum und Zeit, Experimente.* (German), Karlsbad: VRI, 4. extended edition, ISBN 978-3-930879-08-3, http://www.buchhandel.de or http://www.amazon.de/ Search with 9783930879083

[2] Brandes, J. (2012) *Ist die Klassische Allgemeine Relativitätstheorie unvollständig?* In PhyDid B – *Didaktik der Physik – Beiträge zur Frühjahrstagung*, http://www.phydid.de/, to be published, pdf-file available from the author.

Rectification of General Relativity, Experimental Verifications, and Errors of the Wheeler School

C. Y. LO

Applied and Pure Research Institute,
7 Taggart Drive, Unit E, Nashua, NH 03060, USA
c_y_lo@yahoo.com

General relativity is not yet consistent. Pauli has misinterpreted Einstein's 1916 equivalence principle that can derive a valid field equation. The Wheeler School has distorted Einstein's 1916 principle to be his 1911 assumption of equivalence, and created new errors. Moreover, errors on dynamic solutions have allowed the implicit assumption of a unique coupling sign that violates the principle of causality. This leads to the space-time singularity theorems of Hawking and Penrose who "refute" applications for microscopic phenomena, and obstruct efforts to obtain a valid equation for the dynamic case. These errors also explain the mistakes in the press release of the 1993 Nobel Committee, who was unaware of the non-existence of dynamic solutions. To illustrate the damages to education, the MIT Open Course Phys. 8.033 is chosen. Rectification of errors confirms that $E = mc^2$ is only conditionally valid, and leads to the discovery of the charge-mass interaction that is experimentally confirmed and subsequently the unification of gravitation and electromagnetism. The charge-mass interaction together with the unification predicts the weight reduction (instead of increment) of charged capacitors and heated metals, and helps to explain NASA's Pioneer anomaly and potentially other anomalies as well.

Keywords: Einstein's equivalence principle; Einstein's covariance principle; principle of causality; $E = mc^2$; dynamic solution; repulsive gravitational force; charge-mass interaction; Pioneer Anomaly. PACS: 04.20.Cv; 04.50.-h; 04.50.Kd; 04.80.Cc.

1. Introduction

The difficulty in understanding general relativity can be illustrated by the dialogue between a journalist and A. S. Eddington. P. Y. Zhou [1,2] would be the third person because he accepted Einstein's equivalence principle, but rejected his covariance principle.[1] However, misinterpretations abound, and include the 1993 Nobel Committee in Physics [4]. In fact, still there is no expert in this field of general relativity after almost 100 years of its creation, and at least a dozen Nobel Laureates who have incorrectly interpreted general relativity and made errors [3].

We shall concentrate on the basics such as Einstein's equivalence and covariance principles, the principle of causality, and illustrate both the misinterpretations of the Einstein equation, and consequences that have followed. Among sources of misinterpretation, the Wheeler School [5-9] is the most influential. They have made errors in physics, mathematics and logic [10,11]; yet have not responded to the challenge of Bondi, Pirani, & Robinson [12,13]. Further, they did not rectify their error on local time shown in their eq. (40.14) [5];[2] and made invalid claims about dynamic solutions.

Wheeler started his career as a nuclear physicist. He picked up the abandoned theory of *Oppenheimer,* and proposed the formation of black holes after the completion of a simulation test [14] with unverified assumptions. Wheeler led his school at Princeton University while his colleagues, Sciama and Zel'dovich (another H-bomb maker) developed the subject at Cambridge University and the University of Moscow.

Although Wheeler, Thorne & Misner collected an exceptionally rich literature in "Gravitation" [5], Einstein's 1916 paper [15] and his 1923 book [16] on general relativity are not included. In "Gravitation" they not only misinterpreted general relativity with unverified speculations; but also include errors in physics, mathematics, and logic (see Sections 2-6).

The Wheeler School also obstructed developments in physics.[3] Some attempted to shut down the Super Collider. Some have regarded NASA's pioneer anomaly as an error [17-19] since they have failed to explain it. They also viewed anti-gravity experiments as unscientific because they conflict with $E = mc^2$ [10]. They have rejected the non-existence of dynamic solutions, but accepted the space-time singularity theorems that use the same assumption [7]. This paper is written, in part, to support the cutting-edge scientific projects.

The accurate predictions of Einstein created a faith in all aspects of his theory, and a critical analysis of the theory failed to materialize [10]. Moreover, as time goes by, even the misinterpretations of the theory and errors such as the singularity theorems have been accepted as part of the faith. Thus, to rectify these errors, a systematic analysis of the existing theories as a whole is necessary. This paper may serve as a road map to these errors. Evidence with details require considerable deliberation are provided in the references.

Einstein [20] once remarked, "If you want to find out anything from the theoretical physicists about the methods they use, I advise you stick to one principle, don't listen to their words, fix your attention on their deeds." In other words, one should focus on calculations and deductions that can, in principle, be checked with observations and facts.

A focus on such calculations and deductions leads to a rectification of errors in general relativity. This in turn will lead to important new developments in physics that will have new applications, including the rectification of past errors. It turns out that Einstein is an even greater physicist than we have recognized.

Since it is agreed that Einstein's equivalence principle is crucial to his theory [21], we will start with a discussion of this principle. Nonetheless, many were confused that his 1916 equivalence principle was the same 1911 assumption of equivalence that has been proven invalid. The following sections illustrate first the errors in the interpretation of Einstein's equivalence principle and the subsequent development of a new direction in physics.

2. Differences Between Einstein's 1911 Assumption of Equivalence and his Equivalence Principle

There is no book or reference, other than Einstein's own, which states and explains his equivalence principle correctly [22,23]. Though agreeing with Einstein, many confused this principle with Einstein's invalid 1911 assumption of equivalence [24], and missed the Einstein-Minkowski condition [15]. Thus, it is useful to clarify what his 1911 assumption comprised.

In 1911 Einstein assumed the equivalence of a uniformly accelerated system K' and a stationary system with a uniform Newtonian gravitational potential ϕ. This would mean a Newtonian metric form,

$$d\tau^2 = (1 + 2\phi)\, dt^2 - dx^2 - dy^2 - dz^2, \qquad (1)$$

which later Fock [25] has proven to be impossible equivalent to an accelerated frame. From this metric (1), Einstein derived the correct gravitational redshifts, but an incorrect light velocity [24].

In 1916, Einstein assumed the equivalence of a uniformly accelerated system K' and a stationary system of coordinate with a space-time metric that generates a uniform gravitational field. Then, the Einstein -Minkowski condition is a consequence [15]; but there is no statement on a small neighborhood of Minkowski space as what the Wheeler School [5] claimed.

In the 1993 press release of the Nobel Committee [4], however the equivalence principle was claimed as the identity between gravitational and inertial mass but not Einstein's equivalence principle.

However, since Einstein did not provide an example to illustrate his equivalence principle, one may have mistaken the 1911 assumption of equivalence as the 1916 equivalence principle. In 2007, a metric for uniform gravity [23] was obtained as follows:

$$ds^2 = (c^2 - 2U)dt'^2 - (1 - 2U/c^2)^{-1} dx'^2 - (dy'^2 + dz'^2), \quad (2)$$

where $c^2/2 > U(x', t') = (at)^2/2$, "a" is the acceleration of system K'(x' y' z') with respect to K(x, y, z, t) in the x-direction. Here, dt' is defined locally by $cdt' = cdt - (at/c)dx'[1 - (at/c)^2]^{-1}$. Also (2) is equivalent to the metric that Tolman [26] derived. It was surprising that U is time dependent, and this explains the earlier failed derivation [27]. *Then, it is recognized also that the equivalence principle can be used to derive a field equation with the Maxwell-Newton Approximation* [23,28]. Thus, Fock and the Wheeler School [5] are wrong.

To avoid the association of an elevator with the gravity of Earth, Einstein [29] described the equivalence of an accelerated frame and a uniform gravity in terms of an accelerated chest. However, due to the "Einstein's elevator" of Bergmann [30], Einstein was often falsely accused of ignoring the tidal force [14].

To illustrate the principle further, consider a disk K' uniformly rotating w.r.t. an inertial system K(x, y, *z*, *t*), a metric for the disk of space K' (x', y', z') is derived [31]. According to Landau & Lifshitz [32], the metric is

$$ds^2 = (c^2 - \Omega^2 r^2)dt^2 - 2\Omega r^2 d\phi' dt - dr^2 - r^2 d\phi'^2 - dz'^2 \quad (3)$$

where Ω is an angular velocity relative to an inertial system *K*. Also, z & z' coincide with the rotating axis, and $r^2 = x^2 + y^2 = x'^2 + y'^2$. Metric (3) is equivalent to,

$$ds^2 = (c^2 - \Omega^2 r'^2)dt'^2 - dr'^2 - (1 - \Omega^2 r'^2/c^2)^{-1} r'^2 d\phi'^2 - dz^2,$$

where

$$cdt' = cdt - (r\Omega/c)\, rd\phi'[1 - (r\Omega/c)^2]^{-1}. \qquad (3')$$

Thus, in this canonical form, the space contractions are clear. However, dt' in (3') is not integrable [31] because local time dt' is related to different inertial systems at different r or time *t*.

That the local time t' is not a global time led to the rejection of the editorial of the Royal Society [31]. This rejection is incorrect since metric (3') can be derived with special relativity. The time dilation from (3') for the local metric $ds^2 = c^2dT^2 - dX^2 - dY^2 - dz^2$, is

$$dT = [1 - (r\Omega/c)^2]^{1/2}\, dt'. \qquad (4a)$$

From (3') the local clock resting at K', would have

$$dt' = dt, \quad \text{and } dT = [1 - (r\Omega/c)^2]^{1/2}\, dt. \qquad (4b)$$

if observed from K. As Kundig [33] has shown, the time dilation (4b) is valid for a local clock fixed at K'. Hence, the equivalence principle has experimental support. Thus, the Nobel Committee should not have rejected Einstein's equivalence principle implicitly [4]. Note also that, from metric (3'), it has been shown [10], quite unexpectedly, that Einstein's justifications for his theory of measurement [15] are actually based on invalid applications of special relativity.[4]

3. Mathematical Foundation of Einstein's Equivalence Principle and its Invalid Presentations

Another source of confusion is that Pauli's invalid version [34] has been mistaken as Einstein's equivalence principle. However, Einstein has made clear that Pauli's is a misinterpretation [21], and illustrated his claim with an example [15].

Many just pay lip-service to Einstein, but used Pauli's version. For instance, in "Gravitation" [5], there is no reference to Einstein's equivalence principle (i. e. [15] & [16]). Instead, they refer to Einstein's invalid 1911 assumption [24] and Pauli's invalid version [34]. Like Pauli, they also did not refer to the related mathematical theorems [35]. As mentioned, they even failed to understand the local time of a particle at free fall [5], a basic of general relativity.

Since the 1911 assumption has been proven invalid by observation, Fock [25] misidentifying it as Einstein's equivalence principle of 1916, *is beyond just incompetence but a deliberate unethical distortion to discredit Einstein.* Nevertheless, universities, research institutes, the 1993 Nobel Committee and etc. are victims of such a distortion. However, this error was sustained due to errors in physics and mathematics [10].

Many cannot tell the difference between the 1916 principle and the 1911 assumption [23,36-39]. Although Einstein's equivalence principle is still inadequate, it is valid because a uniform gravity due to the principle is generated by acceleration but not mass [40]. However, experiments on the equivalence of inertial mass and gravitational mass have not been updated beyond that the mass-charge interaction is absent [41].

The related mathematical theorems are as follows:

> **Theorem 1.** Given any point *P* in any Lorentz manifold there always exist coordinate systems (x^μ) in which $\partial g_{\mu\nu}/\partial x^\lambda = 0$ at *P*.
>
> **Theorem 2.** Given any time-like geodesic curve Γ there always exists a coordinate system (the so-called Fermi coordinates) (x^μ) in which $\partial g_{\mu\nu}/\partial x^\lambda = 0$ along Γ.

In these theorems, the local space of a particle is locally constant, but not necessarily Minkowski.

What Einstein added to the theorems is that in physics such a locally constant metric must be Minkowski. In contrast to Pauli, Einstein [15; p.145] also remarked that this does not imply the existence of a neighborhood where gravity could be transformed away by a coordinate transformation.[5]

Note that Pauli's version [34], which is a corrupted version of these theorems, is as follows:

> "For every infinitely small world region (i.e. a world region which is so small that the space- and time- variation of gravity can be neglected in it) there always exists a coordinate system K_0 (X_1, X_2, X_3, X_4) in which gravitation has no influence either in the motion of particles or any physical process."

Thus, Pauli initiated incorrectly that, for any given point P, there is a small neighbourhood of local Minkowski space. Apparently, Pauli did not see that the removal of gravity in a small region is different from a removal of gravity at one point, but Einstein does.

Einstein [15] remarked, "For it is clear that, e.g. the gravitational field generated by a material point in its environment certainly cannot be 'transformed away' by any choice of the system of coordinates..." Misner et al. [5] claimed his equivalence principle as follows:

> "In any and every local Lorentz frame, anywhere and anytime in the universe, all the (nongravitational) laws of physics must take on their familiar special-relativistic form. Equivalently, there is no way, by experiments confined to infinitesimally small regions of spacetime, to distinguish one local Lorentz frame in one region of spacetime frame from any other local Lorentz frame in the same or any other region."

They even claimed this as Einstein's Equivalence principle in its strongest form. *The Wheeler School combines errors of Pauli and the 1911 assumption, but ignores the Einstein-Minkowski condition, i.e. the physical essence of Einstein's equivalence principle.*

Moreover, they can even be factually incorrect. For instance, Thorne [14] criticized Einstein's equivalence principle as *ignoring tidal forces.* However, Einstein had already explained to Rehtz [21] that not every gravitational field can be produced by acceleration of the coordinate system.

To calculate the light bending, Einstein [15,16] assumed the satisfaction of his equivalence principle in the Schwarzschild and the harmonic gauges separately. He obtains the respective deflection angles, to the first approximation,

$$B = \frac{2\alpha}{\Delta} = \frac{\kappa M}{2\pi\Delta}, \text{ and } B' = \frac{2\alpha}{\Delta'} = \frac{\kappa M}{2\pi\Delta'} \qquad (5)$$

where Δ & Δ' are respectively the shortest distances from the sun center to the light ray, interpreted according to the Schwarzschild and the harmonic gauges. However, Δ' and Δ are actually different, since we have $r = r' + \kappa M$ [42].

Moreover, since space contractions can be obtained from measurements, only one gauge can be valid for a given frame of reference; but the covariance principle implies otherwise. Furthermore, the calculation of the bending of light is inconsistent with his theory of measurement that necessitates the covariance principle. In fact, both of them are invalid in physics, and thus the interpretation of the Hubble redshifts as Doppler effects is also invalid [10].

4. Invalidity of Einstein's Covariance Principle

Einstein's covariance principle is a source of errors that sustains misinterpretations [10,43]. Starting from this "principle", Einstein inadvertently and implicitly assigns different physical meaning to coordinates for different gauges [3,44,45].

Einstein extended the principle of general relativity to the "principle of covariance" to accommodate his invalid theory of measurement [46]. However, he failed to see that different gauges would lead to different interpretations of the coordinates. Moreover, since both the harmonic and the Schwarzschild solutions produced the same first order deflection of a ray, Einstein [16] immaturely remarked, "It should be noted that this result, also, of the theory is not influenced by our arbitrary choice of a system of coordinates."

Einstein emphasized that a physical theory is about the coincidences of the space-time points, but the meaning of measurements is omitted. To describe events, one must be able to relate events of different locations in a definite manner [47,48]. Moreover, as pointed out by Morrison, the "covariance principle" disrupts the necessary continuity from special relativity [31,47]. The covariance principle was proposed as a remedy for the deficiency of the adapted notion of distance, which was justified with invalid applications of special relativity [10].

If one defines the distance mathematically as in the Riemannian space, one would get only half of the observed value of light blending, in addition to other problems [22]. It turns out, however, that the correct theory of measurement [45] is just what Einstein practiced in his calculation of the light bending [10,16]. Nevertheless, many still believe in this invalid "principle" because gauge invariance has a long history. Yang, Mills, & Shaw [49,50] developed the notion of gauge invariance to non-Abelian gauge theories in 1954. However, Aharonov & Bohm [51] showed in 1959 that the electromagnetic potentials are physically effective; and Weinberg [52] showed all the physical non-Abelian gauge theories are not gauge invariant. In physics, a non-Abelian gauge theory has elements to represent distinct particles. Therefore the whole theory cannot be gauge invariant [53,54].

Yet Gérard & Piereaux [55] Bodenner & Will [56] showed that the deflection angle is gauge invariant to the second order. However, upon examining the meaning of the impact parameter b of the light ray and the shortest distance r_0 from the ray to the center of the sun, it is clear that these physical quantities cannot be both gauge invariant. From the harmonic gauge and the Schwarzschild gauge, one has respectively

$$b \approx 2\kappa M + r_0, \text{ but } b \approx \kappa M + r_0. \qquad (6)$$

Thus, Einstein's covariance principle is invalid.

Another counter example is the de Sitter precession. From the Maxwell-Newton Approximation [57,58], and the Kerr metric one would obtain different formulas [48]. For a circular orbit, we have respectively

$$\frac{d\vec{S}}{d\tau} = -2(\vec{v} \bullet \vec{S})\vec{\nabla}\Phi + v(\vec{S} \bullet \vec{\nabla}\Phi) \qquad (7a)$$

and

$$\frac{d\vec{S}}{d\tau} = -3(\vec{v} \bullet \vec{S})\vec{\nabla}\Phi \text{ since } (\vec{v} \bullet \vec{\nabla}\Phi) = 0. \qquad (7b)$$

One may ask whether the difference between (7a) and (7b) can be detected. In principle, they should be. However, the time average of the difference is essentially zero. Thus, gravity Probe-B failed this in its design due to inadequate theoretical understanding. To show the breakdown of gauge invariance, it would still be the experiment on local light speeds [45].

Misner et al. [5, p. 430] claimed that the covariance principle can be verified experimentally, but provided the opposite evidence. For instance, Will [5] claimed Whitehead's theory is invalid; but the solution of Whitehead is diffeomorphic to Einstein's [59,60]. One may wonder why nobody corrected their mistakes [5]? Also, although many agree with Einstein, the covariance principle is to remedy Einstein's theory of measurement [16], which was justified with invalid applications of special relativity [10].

However, many theorists did not see these problems. Understandably, Feynman [61] claimed that theorists in gravitation are just incompetent. To see this clearly, it is helpful if the principle of causality is understood better.

5. Principle of Causality and Einstein's Equation

The time-tested assumption that phenomena can be explained with identifiable causes is called the principle of causality [57, 58]. In general relativity, Einstein and subsequent theorists have used this principle implicitly on symmetry considerations [57] such as for a circle in a uniformly rotating disk and the metric for a spherically symmetric mass distribution.

Nevertheless, this principle is often neglected [57,62-64] because the confusion on physical coordinates created by the covariance principle. Applications of the principle of causality become clear after Einstein's equivalence principle is understood [10,11].

There are other useful consequences of the principle of causality. For instance, that the weak sources would produce weak gravity is the theoretical foundation of Einstein's requirement on weak gravity [28]. Unbounded "weak waves" of Bondi et al. [12] are invalid because the metric cannot be reduced to the flat metric. Parameters unrelated to any physical cause in a solution are not allowed. For instance, Penrose [65] accepted the metric with an electromagnetic plane wave as a source, but it is invalid in physics because unphysical parameters are involved [13]. Moreover, a dynamic (i.e. with accelerated sources) solution must be related to appropriate dynamic sources [66].[6]

However, 't Hooft claimed [67], "To me, causality means that the form of the data in the future, $t > t_1$, is completely and unambiguously dictated by their values and, if necessary, time derivatives in the past, $t = t_1$. So, I constructed the complete Green function for this system and showed it to Mr. L. This function gives the solution at all times, once the solution and its first time derivative is given at $t = t_1$, which is a Cauchy surface." Nevertheless, his data for his example are calculated values only [66] and this confirms his confusion between mathematics and physics.[7]

Thus, his causality only means that a Maxwell-type equation, which produces the Green function, is satisfied. This is inadequate because a solution of the Maxwell equation could violate the principle of causality. For instance, the electromagnetic potential $A_0[\exp(t - z)^2]$ (A_0 is a constant), is invalid in physics. Such a function, unlike a plane-wave, cannot be considered as an idealization of a field generated by sources [66]. *A dramatic consequence due to such a violation* [58,68] *is, however, the space-time singularity theorems of Penrose and Hawking* [7].

In these singularity theorems, it is assumed that the Einstein equation is satisfied with the weak energy condition and/or the strong energy condition [7]. Note that such conditions require an implicit assumption in the Einstein equation, i.e. *all the coupling constants have the same sign.* However, a dynamic solution does not exist unless a gravitational energy-momentum tensor with a coupling of a different sign is added to the Einstein equation [58,68]. It follows that their implicit assumption violates the principle of causality.

In other words, it is necessary to modify the Einstein equation [58] to, for space-time metric $g_{\mu\nu}$,

$$G_{\mu\nu} \equiv R_{\mu\nu} - \frac{1}{2} g_{\mu\nu} R = -\kappa[T(m)_{\mu\nu} - t(g)_{\mu\nu}] \quad (8)$$

where $t(g)_{\mu\nu}$ is the "additional" energy-stress tensor for gravity. For radiation, the tensor $t(g)_{\mu\nu}$ is equivalent to Einstein's notion of gravitational energy-stress. However, the energy-momentum of a radiation cannot be zero [58].

Note that the modified eq. (8) is compatible with the Maxwell-Newton Approximation [58], and implies also that *the energy conditions for the singularity theorems cannot be satisfied.*

Nevertheless, journals such as the Physical Review, the Royal Society, etc still do not understand the principle of causality adequately, and accept unbounded solutions as valid in physics [66]. The covariance principle would let applied mathematicians such as 't Hooft,[8] easily fail to distinguish physics from mathematics [66].

A violation of the principle of causality means that a physical requirement cannot be satisfied. Thus, one cannot deny it. For example, one cannot circumvent the non-existence of a bounded dynamic solution since it is needed for gravitational radiation.

6. Limitation of the Einstein Equation

Einstein [15,16] made three predictions namely: 1) the gravitational redshifts, 2) the perihelion of Mercury, and 3) the deflection of light. Observational confirmation created a faith in his theory, but they are actually questionable:

1) The gravitational redshifts were first derived from the 1911 assumption of the equivalence between acceleration and Newtonian gravity. Therefore, they can be derived from an invalid theory.
2) The bending of light is inconsistent with Einstein's theory of measurement [64], but is consistent with the Euclidean-like structure [16].
3) As Gullstrand [69] suspected, in 1995 it was proven impossible to have a bounded dynamic solution.[9] Thus, the perihelion of Mercury, in principle, is still beyond the reach of the Einstein equation [58]. Also, the assumption of a unique sign for coupling constant has been proven incorrect [58].

Moreover, diffeomorphic solutions with the same frame of reference are not equivalent in physics [3].

Another basic problem is that there is no gravitational radiation reaction force. Although an accelerated massive particle would create radiation, the metric elements in the geodesic equation are created independent of the test particle. This is also manifested by that there is no dynamic solution for the Einstein equation.

An urgent issue is to find a valid physical gauge for a given problem. Fortunately, the Maxwell-Newton approximation has been proven to be an independently valid first order approximation so that the binary pulsar radiation experiments can be explained [57,58]. Thus, Einstein's notion of weak gravity (including gravitomagnetism and gravitational radiation [70]) is valid.

However, calculations of the experiments of the binary pulsars necessitate that the coupling constants have different signs [58]. This leads Lo [71] to discover the static charge-mass neutral repulsive force, and thus further confirms that $E = mc^2$ is conditionally valid. The dynamic Einstein equation violates the principle of causality because of the absence of an energy-stress tensor in vacuum, as Hogarth [72] pointed out in 1953.

To counter Gullstrand [69], Christodoulou & Klainerman [9] claimed that bounded dynamic solutions were constructed. However, they did not show that the initial *dynamic* set is non-empty [73,74].[10] Misner et al. [5] invalidly claimed that their eq. (35.31) has a bounded plane-wave solution [11]; and Wald [7] claimed that his eq. (4.4.52) has a solution for the second order [57]. Wald [7] also incorrectly extended the process of perturbation approximation to the case that the initial metric is not flat.

Consequently, they failed to see that electromagnetic energy is not equivalent to mass [6-8], can be proven even if the electrodynamics were approximately valid [18,75]. As a result, not only do they incorrectly insist that $E = mc^2$ is unconditional [62] but they also over-looked that the Einstein equation necessitates the existence of a repulsive charge-mass interaction and also the need of unification [76,77].

However, in 2005 the effect of such a repulsive force was inadvertently detected by Tsipenyuk & Andreev [78]. They discovered that the weight of a metal ball is reduced after it is irradiated with high energy electrons. Yet they could not explain this phenomenon because it was believed that gravity would increase as energy increases. The static charge-mass repulsive force was discovered in 1997 because Lo [71] had already suspected $E = mc^2$.[11] Obviously, a complete experimental verification of this force is important.

In short, for the dynamic case, the Einstein equation is proven invalid. For the static case, verification of the Einstein equation beyond the Maxwell-Newton Approximation depends on the experimental confirmation of the static charge-mass repulsive force. The discovery of such a repulsive force casts a strong doubt on the belief that gravity is always attractive. This would explain that a black hole remains an observational speculation.[12]

Nevertheless, textbooks invalidly claim that linearization will give a valid approximate solution. The authors seem unaware that the Einstein equation does not have a bounded solution for a two-body problem [57,58]. The Wheeler School [5] failed to see that there is no bounded wave solution for their equation (35.31), but it needs mathematics at only an *undergraduate level* [10,11].

Einstein believed that he had proven $E = mc^2$ for the electromagnetic energy because he assumed that photons have only electromagnetic energy. In 1997, it was shown that $E = mc^2$ is only conditionally valid, and this explains the failure of Einstein's several attempts to prove this formula for other types of energy [79]. This misinterpretation of $E = mc^2$ is the reason that the charge-mass interaction is not only overlooked but also denied by earlier theorists [80].

Now, due to the mass-charge interaction, there are local experiments that can distinguish the effect of an accelerated frame from an approximately uniform field [81] generated by massive matter. However, misinterpretations of general relativity and errors are still common in courses of universities. For example, MIT, Stanford, etc. are also victims. Since the instructors do not understand Einstein's equivalence principle, they are unaware that it can play a crucial role in the derivation of the field equation [23,28].

In short, a failure to understand the Einstein equation adequately leads to not only overlooking the charge-mass interaction but the need of unification between gravitation and electromagnetism.

7. Damage to Education & MIT Phys. 8.033, 2006

To illustrate how such theoretical errors have been incorporated into our educational system, a publicly accessible, open course, MIT phys. 8.033, has been chosen as an example.[13] This illustrates that MIT, Stanford [37], etc. are not immune to be a victim of these errors.

Notably, $E = mc^2$ is considered as unconditionally. MIT Course Phys. 8.033 addresses three issues:

- Principle of equivalence
- Light bending, gravitational redshift
- Metrics

In this course, just as Misner et al. [5] did, the invalid 1911 assumption of equivalence is mistaken as Einstein's equivalence principle of 1916.
The course proclaimed the "weak equivalence principle" as no local experiment can distinguish between a uniform gravitational field g and a frame of accelerated with a = g. However, in 2006 it has been found that, there are local experiments that can distinguish an accelerated frame from an approximately uniform field generated by massive sources [80].

The claim of the "strong equivalence principle" that the laws of physics take on their special relativistic form in any local inertial frame and the claim that: "a free falling elevator is a locally inertial frame, so that the strong version says that special relativity applies in all such elevators anywhere and anytime in the universe" are errors from the Wheeler School. This demonstrates a lack of understanding of Einstein's equivalence principle The course incorrectly claimed

- EP implication 1: Gravity bends light
- EP implication 2: Gravitational redshift.
- EP implication 3: It is all geometry (learn how to work with metrics!)

First their version of EP cannot lead to the correct light bending. Second, although it does lead to gravitational redshift, the argument has been proven invalid. Moreover, the claim, "It is all geometry" has no meaning since the issue of the physical gauge is ignored.

The instructor still uses the well-known invalid Newtonian metric (1) because he fails to understand Einstein's equivalence principle. He also fails to see its conflict with Einstein's covariance principle that is not just the coordinate transformation, but leads to conflicts in physics because the meaning of the coordinates is related to the gauge [1-3,42].

Gullstrand [69] suspected Einstein's claim that the perihelion of Mercury has been explained being invalid. Since the perihelion is calculated in term of the perturbations of other planets, it is crucial whether the perturbation approach is valid. However, in 1995 it has been proven that the Einstein equation does not have a bounded solution for a two-body problem [57, 58].
The instructor Tegmark fails to see that there is no bounded wave solution for eq. (35.31) in [5], and to tell the difference between mathematics and physics. He has formulated the "Ultimate ensemble theory of everything", whose only postulate is that "all structures that exist mathematically exist also physically". This idea is formalized as the "Mathematical universe hypothesis" in his paper *The mathematical universe*, a short version of which was published as *Shut up and calculate* (Wikipedia). A suggestion for him would be "Shut up, think, and then calculate".

8. Rectifications of General Relativity

The major errors, except the unconditional validity of $E = mc^2$ [82-84], have been pointed out by physicists such as Aharonov & Bohm [51], Eddington [47], Einstein [15,21], Gullstrand [69], Morrison [31], Pauli [34], Synge [60], Weinberg [42,52], Whitehead [46], and Zhou Pei-yuan [1,2], and further developed by the work of Lo. Moreover, most errors can be traced back to the Princeton University.[14]

Einstein's equivalence principle is further reconfirmed, but his covariance principle is proven invalid. Then, it becomes possible to clarify misinterpretation, and to confirm the invalidity of Einstein's theory of measurement. This removes the problem of incompatibility with the rest of physics.

Concurrently, it is possible to prove the non-existence of dynamic solutions. Then, the puzzle of the existence of space-time singularities is solved as due to the physical assumption of Penrose and Hawking that violates the principle of causality [7]. Thus, the claim of general relativity being invalid for microscopic phenomena is actually groundless.

This leads to the investigation of whether $E = mc^2$ is unconditional. Einstein's proof for the equivalence of photons and mass [15] was found to be incomplete. The demand for further experimental verification led to the discovery of the charge-mass interaction, and thus Einstein's conjecture of unification was proven to be necessary [40]. It was found that the static mass-charge interaction is repulsive; and thus an increase of energy may not lead to strengthening attractive gravity.

In fact, a repulsive force could have been found in 1916 from the Riessner-Nordstrom metric [5],

$$ds^2 = \left(1 - \frac{2M}{r} + \frac{q^2}{r^2}\right)dt^2 - \left(1 - \frac{2M}{r} + \frac{q^2}{r^2}\right)^{-1} dr^2 - r^2 d\Omega^2$$

where q and M are the charge and mass of the particle Q, and r is the distance from the particle center. Note that the gravitational components generated by electricity have not only a different radial coordinate dependence but also a different sign.

However, Herrera, Santos, & Skea [85], Pekeris [86], and t' Hooft [87], argued that mass M should include the electric energy. If M included the electric energy, no repulsive effects due to the charge could be detected [84]. The weight reduction of a charged metal ball [78], also rejects such an interpretation.[15] However, for the case of a charged capacitor, the observed repulsive force is beyond general relativity. For a particle *P* with mass m at **r**, a force on *P* is

$$-m\frac{M}{r^2} + m\frac{q^2}{r^3} \qquad (9)$$

of the first order approximation since $g^{rr} \cong -1$. Then the force acting on the charged particle *Q* has the same magnitude as the action and reaction forces are equal. However, if one calculates the metric according to the particle *P* of mass m, only the first term is obtained.

Thus, it is necessary to have a repulsive force with the coupling q^2 to the charged particle *Q* in a gravitational field generated by masses. It thus follows that, force (9) to particle *Q* is beyond the current theoretical framework. As predicted by Lo, Goldstein, & Napier [53], general relativity leads to a realization of its inadequacy, just as electricity and magnetism lead to their shortcomings discovered by Maxwell.

In confirmation of earlier derivation [71], the force is

$$F = \frac{q^2 m}{r^3}, \qquad (10)$$

between a point charge q and a point mass m with a distance r. However, due to the factor q^2, the force is independent of the sign of charge q. An immediate question would be whether such a charge-mass repulsive force is subjected to electromagnetic screening. It is conjectured that this force, being independent of a charge sign, should not be subjected to such a screening although it is according to general relativity.

9. Extension to the Five-Dimensional Relativity

If we consider the coupling with q^2, this naturally leads to a five-dimensional space of Lo, Goldstein & Napier [53]. The static force on Q is generated by

$$g_{tt} = 1 - \frac{2m}{\rho c^2} \text{ and } g_{55} = \frac{mMc^2}{\rho^2}K^2 + \text{const.}, \qquad (11)$$

where $dx^5/d\tau = q/Mc^2K$ and K is a constant.

For a metal ball charged with Q and a particle with mass m, such a repulsive force is

$$F_b = \frac{Q^2 m}{R^3}, \qquad (12)$$

where R is the distance from the center of the metal ball. The effect of such a force has been confirmed.

From four-dimensional theories, such a repulsive force would not act upon a charged capacitor because of the screening. However, the repulsive force could be interpreted as that the charge interacts with a new neutral field and thus would not be subjected to a screening[16] as in the conventional theories.

Moreover, this force can be verified simply by weighing a capacitor before and after it is charged [88, 89]. The effect of the repulsive force created by a field from the earth would be observed as a lighter weight for the charged capacitor. Moreover, the magnetic energy would similarly lead to an attractive current-mass force. Such a force has been verified by the experiments of Tajmar and Matos [90].

The charged capacitor becomes lighter because the balance has been disrupted [40], and the repulsive force is due to the changed state of motion of electrons. Thus, it would take a while before a discharged capacitor returned to its original weight. A discharged capacitor is heated up, and thus a piece of metal would also reduce its weight after being heated up. This verifiable prediction is directly in disagreement with Einstein [91] who predicted the opposite.

A repulsive force on a charged capacitor would be evidence of the fifth dimension. According to eq. (12), it is expected that the weight reduction of a charged capacitor would be increased proportional to the square of the potential V since the charge Q = CV, where C is the capacity. Such a repulsive force[17] and its dependence of potential square have been verified by Musha's experiments with very high voltages [92].

However, Liu used rolled up capacitors that can stand less voltage. Since they keep the heat better, the weight recovery delays were observed [88]. However, the weight reduction effects would cancel out for a rolled up capacitor, according to Musha's theory (proposed originally for the Biefeld-Brown effect) [92]. Thus, his theory is proven invalid for this case [93].

A long-range repulsive interaction should have consequences in astrophysics. The Space-Probe Anomaly would be an example [18,94]. It is conjectured that the anomaly would be due to a repulsive force from the sun,

$$F_{ps} = \frac{P_s m_p}{R^3}, \qquad (13)$$

where P_s is a parameter due to the sun, m_p is the mass of the Space-Probe, and R is the distance from the sun.

Since this force is much smaller than the gravitational force, the existence of such a repulsive force would result in a very slightly smaller mass than M_s of the sun, i.e.

$$F = \frac{M_s m_p}{R^2} - \frac{P_s m_p}{R^3} \tag{14a}$$

and

$$\frac{M_s m_p}{R_0^{\ 2}} - \frac{P_s m_p}{R_0^{\ 3}} = \frac{M_{ss} m_p}{R_0^2} \tag{14b}$$

where R_0 is the distance from the sun. Then, we have

$$F = \frac{M_{ss} m_p}{R^2} + \frac{P_s m_p}{R^2}\left(\frac{1}{R_0} - \frac{1}{R}\right). \tag{15}$$

Thus, there is an additional attractive force for R >> R_0. Moreover, such a force is not noticeable from a closed orbit. However, for open orbits at large distance, the repulsive force becomes negligible, and thus an additional attractive force would appear as the anomaly. Thus, the repulsive fifth force satisfies the overall requirements.

10. Conclusions and Remarks

Many theorists, dominated by those in the Wheeler School, continue Einstein's error on the principle of covariance; and make new errors in misinterpreting Einstein's equivalence principle and the principle of causality. They maintain even obvious errors by ignoring others, including Einstein [15] and Weinberg [42]. *Moreover, to justify the covariance principle, they necessarily distort Einstein's equivalence principle; and thus create more errors.* However, Einstein's equivalence principle has a foundation in mathematics [40] as well as experimental supports [33], and is crucial in the derivation of a valid field equation [23,28].

Nevertheless, the 1993 Nobel Committee embraced this incorrect version [4] because their errors in physics appeared to have been obscured by errors in mathematics and earlier accumulated errors in physics [10]. Further, our educational system teaches these errors, as demonstrated by the MIT Open Course phys. 8.033 since 2006.[18] While they have created new errors, these errors are unfairly attributed to Einstein.

Yet, the views of the Wheeler School dominate [44,58] although there is no conclusive evidence to support their speculations. They [5] rely on the covariance principle to create additional confusion. This is why the contribution of Zhou [1,2] on the invalidity of the covariance principle is important to highlight. They label their interpretation as "standard theory". They help each other with invalid arguments.[3)] but also inadvertently illustrate their errors at the undergraduate level [11,81].

Moreover, there are prominent theorists, who have made similar errors, but won prizes from the "Gravity Research Foundation". For instance, Eric J. Weinberg, editor of the Physical Review D, claimed that the difference between versions of Einstein and Pauli on the equivalence principle is not physical [22].[19] He insisted on the validity of E = mc^2 and the existence of dynamic solutions [95]. Thus, he perpetuates the errors of Penrose and Hawking, but also helps discovering of the charge-mass interaction along with experimental verification [18,84].[20]

In general relativity, the fundamental issues are: Einstein's equivalence principle, his covariance principle, the principle of causality, invalidity of linearization, and measurements of the distance. The 'Standard Theory' makes errors in all five areas. Moreover, there are three related issues: 1) that the formula E = mc^2 is conditionally valid since the electromagnetic energy is not equivalent to mass; 2) that the coupling signs have been found not unique, and thus the space-time singularity theorems actually are irrelevant to physics; and 3) that the photons include non-electromagnetic energy and thus the quantum theory and general relativity are actually related. The root of these errors is due to earlier questionable concepts. For instance, the photon was proposed before general relativity.

The photons including energy other than the electromagnetic energy would imply also that quantum mechanics is not at the final stage, and the need for renormalization.

Penrose and Hawking,[21] based on their singularity theorems, claimed that general relativity is not valid for microscopic phenomena. However, an implicit assumption of their theorems violates the principle of causality [58][22] that many physicists, including the editor of the Physical Review D, failed to understand adequately. Thus, the claim of an expanding universe not only has no adequate theoretical foundation, but also misleads theorists from the necessary efforts to improve the Einstein equation to a physical equation for the dynamic case.

Nevertheless, after the rectifications, the necessity of unification between gravitation and electromagnetism is clear since the charge-mass interaction has been discovered. Thus, Einstein actually leaves us a far greater treasure to be explored [18,75]. The existence of a new repulsive force between charge and mass further confirms E = mc^2 is only conditionally valid. This interaction is a key to understanding the phenomena such as the reduction of weight of a capacitor after being charged. Then, in terms of the new force, the discovery of NASA's pioneer anomaly would be understandable. The weight reduction of heated metal would start the investigation on the temperature dependence of gravity,[23] particularly in the low temperature area; and this may lead to a better understanding of the observations that led to the assumption of dark matter. In fact, a new chapter of physics has been opened for experimental and theoretical explorations.

Great scientists such as Einstein can make mistakes. *However, after rectification, general relativity is no longer incompatible with other theories in physics; and Einstein emerges as an even better physicist since his conjecture of unification is proven necessary.* Whitehead [46] remarked, "But the worst homage we can pay to genius is to accept uncritically formulations of truths which we owe to it."

Modern physics has been developed to such a stage that frontier physicists can no longer afford to ignore physical principles, and/or to leave all pure mathematics to mathematicians. Einstein did not fully understand mathematical analysis, and thus he could not modify the mathematics required in physics [45]. Pauli and the Wheeler School do not fully understand the related mathematics, and thus fail to see that there are restrictions to the equivalence principle that cannot be altered. Now, the importance of Einstein's equivalence principle has been re-established [10]. When the principle of causality is better understood, we have proved the non-existence of dynamic solutions.

One may wonder why the errors in mathematics were first discovered by Gullstrand [69] when many mathematicians failed to discover them. They failed because the related physics was not applied, and this can lead to overlooking necessary considerations, as for example, in the case of Hawking and Penrose [7]. In fact, D. Christodoulou was awarded a 2011 Shaw Prize in mathematics for his errors [73,74][24] in opposition to Gullstrand. (This was also an attempt to save the relevance of Space-time Singularity Theorems to physics.) Some claim if there are more experiments, the situation would be better. However experiments, the binary pulsar being an example, are misinterpreted due to theoretical errors [58].[25] A problem is that theorists and the journals in which they publish such as the Physical Review,[26] General Relativity & Gravitation, etc are often still adhering to outdated concepts which are based on reputation rather than sound science.

The conditional validity of $E = mc^2$ is the central problem. This issue has been addressed before special relativity [96], but was unresolved until 1997 [71]. It is hoped that this paper will highlight this issue, lead to better theoretical developments, and encourage more physicists to work on the charge-mass interaction.

The rectification of errors will lead to a new area of physics that has been opened for scientists to explore and which will concurrently address old problems. Moreover, Einstein emerges as an even better theorist since his static equation can lead to the discovery of the charge-mass interaction that has been supported by various experiments [17,19,78,88-90,92,97,98].

Acknowledgments

The paper is dedicated to Institute Professor P. Morrison of MIT for over 15 years of association in the research. The author is grateful to Prof. I. Halperin for discussions on the mathematical foundation of Einstein's equivalence principle, and Dr. T. Musha for accurate data on the weight reduction of charged capacitors. Special thanks are to Prof. Michael Dopita, Dr. Susan Hockfield, Dr. Daniel Kulp, Dr. Gail Wood, and Sharon Holcombe for valuable comments, useful suggestions, and unfailing supports. This work is supported in part by Innotec Design, Inc., USA and the Chan foundation, Hong Kong.

Endnotes

1) In the claims of Zhou, because he still uses Einstein's terminology, "coordinates do matter" actually means "a gauge does matter".
2) Eddington [47], Liu [37], Wald [7], and Weinberg [42] did not make the same mistake.
3) A. Ashtekar, editor-in-chief of Gen. Rel. Grav., claims the Wheeler School as "well-established in science" (March 8, 2012). His thesis, "Asymptotic Structure of the Gravitational Field at Spatial Infinity", seems to just inherit the errors of Wald [7]. Moreover, in his quantum gravity, he failed to see that the photons must include gravitational energy [10,82]. C. M. Will, editor-in-chief of Class. & Quant. Grav., insists on his errors, particularly $E = mc^2$ being unconditional [6,71].
4) Einstein integrates different infinitesimal elements from different local Minkowski spaces as if elements in K'.
5) The so-called Lorentz invariance [5] misled Chu et al. [99] to claim that general relativity may be invalid.
6) Pauli [34] is the first who pointed out possible different signs of couplings.
7) The calculations of Liu and Zhou [100] also showed that there were no bounded wave solutions [68].
8) 't Hooft showed also a misunderstanding of the notion of inertial mass and special relativity in his 1999 Nobel Speech. Nevertheless, Veltman [101] showed that his physics is adequate.
9) An error of Einstein, Infeld, & Hoffmann [102], Misner et al, [5], Will [6], Damour [103] and etc. is unaware that the mathematical existence of a bounded solution needs a

proof. Moreover, Wald [7] failed to see that his eq. (4.4.52) cannot be satisfied.

10) Christodoulou & Klainerman [9] did not complete the need to show their dynamic initial data sets being non-empty [73, 74], and thus were unaware that only static solutions exist. One should not be surprised that Christodoulou made mathematical errors at an elementary level since his Ph. D. advisor, Wheeler made similar crucial errors in mathematics at an undergraduate level [104]. Nevertheless, Christodoulou has been awarded some prizes as consequences of the prevailing practice of authority worship. For instance, in 2011 a half Shaw Prize for mathematics was awarded to him because both C. N. Yang, Chairman of Board of Adjudicators of the Shaw Prize and Peter C. Sarnak, Chairman of the Selection Committee of mathematics did not know enough mathematics and physics in general relativity [104].
11) This is a case where the static Einstein equation can predict beyond the Maxwell-Newton Approximation [105].
12) The 2008 Shaw Prize was awarded to R. Genzel, "in recognition of his outstanding contributions in demonstrating that the Milky Way contains a supermassive black hole at its centre". However, Genzel only claimed, "... must indeed be a massive black hole, beyond any reasonable doubt." Thus, he is not 100% sure.
13) Having graduated from MIT in physics and Queen's University in mathematics, I have reported these problems to the respective departments.
14) I respect many graduates of Princeton University; Professor A. J. Coleman who pointed out errors of Einstein, and Professor I. Halperin, who was my advisor for my degrees in mathematics.
15) The non-equivalence of mass and electromagnetic energy actually comes from the electromagnetism alone.
16) Thus, unification is crucial to understanding the weight reduction of a charged capacitor [89].
17) W. Q. Liu (http://www.cqfyl.com) performed weighing of rolled-up capacitors in a Chinese Laboratory of the Academy of Science, and got certified results of lighter capacitors after charged. He also observed the delay of weight recovery of a discharged capacitor as the theory predicted.
18) Note that, under the leadership of Weisskopf, the MIT tradition until 2005 was that general relativity must be understood in terms of physics.
19) Nachtergaele also failed to distinguish the difference between Einstein's equivalence principle and Pauli's version in physics [106].
20) There are others physicists who also have mistaken. Nobel Laureates,'t Hooft and Wilczek also failed to see that there is no dynamics solution for the Einstein equation [57, 58, 68]. Moreover, as shown in their Nobel lectures,'t Hooft [107] who does not understand special relativity adequately, regarded the electric energy of a charged par particle contributes to its inertial mass, and Wilczek [108] failed to see that $m = E/c^2$ is not generally valid [80].
21) Penrose [65] also accepted a solution with unphysical parameters. Hawking even made elementary logical errors on space and time, in his popular book [109].
22) Mathematicians may overly trust the physical assumptions, whereas physicists would re-examine the physical validity of assumptions if the results were totally unexpected. This was what happened to Penrose. This also explains that before 1993 mathematicians (including the Field Medalists E. Witten, and S. T. Yau whose works have been closely related to general relativity) also failed to discover the non-existence of dynamic solutions.
23) Fan, Feng, & Liu [98] have verified the weight reductions (0.33% - 0.82%) for six kinds of metal due to increments of temperature (from 100°C to 600°C).
24) If the Shaw Prize Committee found a dynamic solution of Christodoulou, they would have found his errors.
25) P. Morrison had discussed the dynamic solutions with J. H. Taylor Jr. at Princeton [31]. He explained that Damour [110] is responsible for the calculations of the gravitational radiations. Subsequently, Morrison suggested that I should write a book on dynamic solutions and related problems. This leads to another 15 years of research.
26) Daniel Kulp has changed such a practice. The recent position of the Physical Review is that they are not yet convinced, but no longer object [111]. However R. Shergill of the Royal Society did not see the merits of the principle of causality [112], probably because it is used to prove that the space-time singularity theorems are irrelevant to physics. However, a violation of the principle of causality is only an expression for a problem, for instance, such as the non-existence of bounded wave solutions [58] for the Hulse-Taylor experiments.

References

[1] P. Y. Zhou, in Proc. of 3rd Marcel Grossmann Meet. on Gen. Rel. ed. Hu Ning, Sci. Press & North Holland. (1983), 1-20.

[2] P. Y. Zhou, Proc. of the Internat. Sym. on Experimental Gravitational Physics, Guang Zhou, China (1987).

[3] Lo, C.Y., Phys. Essays, 23 (3), 491- 499 (Sept. 2010).

[4] The 1993 Press Release of the Nobel Prize Committee (The Royal Swedish Academy of Sciences, Stockholm).

[5] C. W. Misner, K. S. Thorne, & J. A. Wheeler, Gravitation (W. H. Freeman, San Francisco, 1973).

[6] C. M. Will, Theory and Experiments in Gravitational Physics (Cambridge University, Cambridge. 1981).

[7] R. M. Wald, General Relativity (University of Chicago Press, 1984).

[8] H. C. Ohanian & R. Ruffini, Gravitation and Spacetime (Norton, New York, 1994).

[9] D. Christodoulou & S. Klainerman, The Global Nonlinear Stability of the Minkowski Space (Princeton Univ. Press, 1993), no. 41 of Princeton Mathematical Series.

[10] Lo, C.Y., Phys. Essays, 23 (2), 258-267 (2010).

[11] Lo, C.Y., Linearization of the Einstein Equation and the 1993 Press Release of the Nobel Prize in Physics, 18 th Annual Natural Philosophy Alliance Conference, July 6-9, 2011, College Park, MD (2011).

[12] H. Bondi, F. A. E. Pirani, & I. Robinson, Proc. R. Soc. London A 251, 519-533 (1959).
[13] Lo, C.Y., Bull. of Pure and App. Sci., 29D (2), 81-104 (2010).
[14] K. S. Thorne, Black Holes and Time Warps (Norton, New York, 1994), p. 105.
[15] Einstein, A., H. A. Lorentz, H. Minkowski, H. Weyl, The Principle of Relativity (Dover, New York, 1923).
[16] Einstein, A., The Meaning of Relativity (Princeton University, Princeton, 1954).
[17] S. G. Turgshev, V. Toth, L. R. Kellogy, E. L. Lau, and K. J. Lee, Int. J. Mod. Phys. D, 15(1), 1-55 (2006).
[18] Lo, C.Y., International Conf. on Phys. Interpretation of Relativity Theory, 6-9 July 2009, Bauman Moscow State Tech. Univ., Russia.
[19] Slava G. Turyshev, Viktor T. Toth, Space Science Reviews, 148, No. 1-4, 149-167 (December, 2009).
[20] Einstein, A., 'On the Method of Theoretical physics (1934)' in Ideas and Opinions (Dover, 1982), p. 270.
[21] J. Norton, What was Einstein's Principle of Equivalence? in Einstein's Studies Vol.1: Einstein & the History of General Relativity, Eds. D. Howard & J. Stachel (Birkhäuser, Boston, 1989).
[22] Lo, C.Y., Phys. Essays 16 (1), 84-100 (March 2003).
[23] Lo, C.Y., Bull. Pure & App. Sci., 26D (2), 73 (2007).
[24] Einstein, A., Annalen der Physik, 35, 898-908 (1911).
[25] V. A. Fock, The Theory of Space Time and Gravitation (Pergamon Press, 1964); the Russian edition (1955).
[26] R. C. Tolman, Relativity, Thermodynamics, and Cosmology (Dover, New York 1987).
[27] J. P. Hsu & L. Hsu, Chin. J. of Phys., 35 (4), 407 (1997).
[28] Lo, C.Y., Phys. Essays 12 (3): 508-526 (1999).
[29] Einstein, A., Relativity (Dover, New York, 1920), p. 82.
[30] P. G. Bergmann, Introduction to the Theory of Relativity (Dover, New York, 1976), pp. 156 and 159.
[31] Lo, C.Y., Phys. Essays 18 (4), 547-560 (2005).
[32] L. D. Landau & E.M. Lifshitz, Classical Theory of Fields (Addison-Wesley, Reading Mass, 1962).
[33] W. Kundig, Phys. Rev, 129, 2371 (1963).
[34] W. Pauli, Theory of Relativity (Pergamon Press, London, 1971).
[35] J. L. Synge, Relativity: The General Theory (North - Holland 1971).
[36] Yu Yun qiang, An Introduction to General Relativity (Peking University Press, Beijing, 1997).
[37] L. Susskind, open video lectures on Einstein's Theory of Relativity, Stanford University (2012).
[38] Liu Liao, General Relativity (High Edu. Press, Shanghai, 1987).
[39] 梁灿彬, 微分几何与广义相对论教程, video course of Peking Normal University (1912).
[40] Lo, C.Y., Physics Essays 21 (1), 44-51 (March 2008).
[41] C. M. Will, 'Experimental gravitation …' 300 Years of Gravitation (Cambridge University Press, 1987).
[42] S. Weinberg, Gravitation and Cosmology (John Wiley, New York, 1972)
[43] Lo, C.Y., Bull. Pure and App. Sci., 27D (1), 1-15 (2008).
[44] Lo, C.Y., 16th Annual Natural Philosophy Alliance Conference, University of Connecticut, Storrs, May 25-29, 2009.
[45] Lo, C.Y., Chin. J. of Phys. (Taipei), 41 (4), 233 (2003).
[46] A. S. Eddington, The Mathematical Theory of Relativity (Chelsea, New York, 1975), p. 10.
[47] Lo, C.Y., Phys. Essays, 18 (1), 112-124 (March, 2005).
[48] A. N. Whitehead, The Principle of Relativity (Cambridge Univ. Press, Cambridge, 1962).
[49] C. N. Yang & R. L. Mills, Phys. Rev. 96, 191 (1954).
[50] Ron Shaw, Ph. D. thesis, Cambridge University (1955).
[51] Y. Aharonov & D. Bohm, Phys. Rev. 115, 485 (1959).
[52] S. Weinberg, The Quantum Theory of Fields (Cambridge University Press, Cambridge, 2000).
[53] Lo, C.Y., G. R. Goldstein, & A. Napier, Hadr. J. 12, 75 (1989).
[54] Lo, C.Y., Phys. Essays, 5 (1), 10-18 (1992).
[55] J. M. Gérard & S. Piereaux, The Observable Light Deflection Angle, arXiv:gr-qc/9907034 v1 8 Jul 1999.
[56] J. Bodenner & C. M. Will, Am. J. Phys. 71 (8), 770 (2003).
[57] Lo, C.Y., Phys. Essays, 13 (4), 527-539 (2000).
[58] Lo, C.Y., Astrophys. J. 455: 421-428 (Dec. 20, 1995).
[59] Lo, C.Y., Bull. Pure & Applied Sci., 28D (1), 67 (2009).
[60] A. J. Coleman, J. L. Synge on Whitehead's Principle of Relativity, arXiv: phys./0505027 v1, 4 May 2005.
[61] R. P. Feynman, The Feynman Lectures on Gravitation (Addison-Wesley, New York, 1995).
[62] Lo, C.Y., Proc. IX Internat. Sci. Conf. on 'Space, Time, Gravitation,' Saint-Petersburg, August 7-11, 2006.
[63] Lo, C.Y., Math., Physics and Philo. in the Interpretations of Relativity Theory II, Budapest, 4-6 Sept. 2009.
[64] Lo, C.Y., Bull. Pure & Applied Sci., 29D (1), 1 (2010).
[65] R. Penrose, Rev. Mod. Phys. 37 (1), 215-220 (1965).
[66] Lo, C.Y., Phys. Essays 24 (1), 20-27 (2011).
[67] 't Hooft, G. Strange Misconcept. of Gen. Rel. , http://www.phys.uu.nl/~thooft/gravitating_misconceptions.
[68] Lo, C.Y., Astrophys. Space Sci., 306: 205-215 (2006).
[69] A. Gullstrand, Ark. Mat. Astr. Fys. 16, No. 8 (1921); ibid, Ark. Mat. Astr. Fys. 17, No. 3 (1922).
[70] E. Bertschinger, Cosmological Dynamics, Department of Physics MIT, Cambridge, MA 02139, USA.
[71] Lo, C.Y., Astrophys. J. 477, 700-704 (March 10, 1997).
[72] J. E. Hogarth, 1953 Ph. D. Thesis, Dept. of Math., Royal Holloway College, Univ. of London, p. 6.
[73] Volker Perlick (republished with an editorial note), Gen. Rel. Grav. 32 (2000).
[74] Lo, C.Y., Phys. Essays 13 (1), 109-120 (March 2000).
[75] Lo, C.Y., Proc. in 17th Ann. Nat. Phi. Alliance Conf., California State University, Long Beach, Calif. June 23-26, 2010.
[76] Lo, C.Y., Bull. Pure & App. Sci., 26D (1), 29 (2007).
[77] Lo, C.Y. & C. Wong, Bull. Pure & App. Sci., 25D (2), 109 (2006).
[78] D. Yu. Tsipenyuk, V. A. Andreev, Proc. in Physical Interpretations of the Theory of Relativity Conference, Bauman Moscow State Tech. Univ., 4 – 7 July 2005.
[79] Einstein's Miraculous Year, edited by John Stachel (Princeton University Press, Princeton 1998).
[80] Lo, C.Y., Phys. Essays, 24 (4), 477-482 (2011).
[81] Lo, C.Y., Bull. Pure & App. Sci., 27D (2), 149 (2008).
[82] Lo, C.Y., Progress in Phys., Vol. 4, 14-18 (2006).
[83] Lo, C.Y., Bull. Pure & App. Sci., 25D (1), 41- 47 (2006).

[84] Lo, C.Y., Phys. Essays, 25 (1), 49-56 (2012).
[85] L. Herrera, N. O. Santos & J. E. F. Skea, Gen. Rel. Grav., 35, 11, 2057 (2003).
[86] C. L. Pekeris, Proc. Nat. L Acad. Sci. USA, 79, 6404 (1982).
[87] 't Hooft, G. A Confrontation with Infinity, Nobel Lecture, December, 1999.
[88] W. Q. Liu, Inst. of Physics, Chinese Academy of Sciences, Beijing, China (personal communication, 2007).
[89] Takaaki Musha, Proceedings of the 37th Conference on Aerospace Propulsion, JSASS, 1997, pp. 342–349.
[90] http://news.softpedia.com/news/The-First-Test-That-Proves-General-Theory-of-Relativity-Wrong-20259.shtml 2006.
[91] Einstein, A., 'E = MC^2' (1946), Ideas and Opinions (Crown, New York, 1982).
[92] T. Musha, Theoretical explanation of the Biefeld–Brown Effect, *3-11-7-601 Namiki, Kanazawa-ku, Yokohama 236-0005 Japan;* Email: musha@jda-trdi.go.jp.
[93] Lo, C.Y., Gravitation, Physics, & Technology, Phys. Essays, 25 (4) (2012).
[94] Charles Q. Choi, NASA Baffled by Unexplained Force Acting on Space Probes, SPACE.com, 3 March (2008).
[95] Eric J. Weinberg, Y. Hu, and M. Turner, Phys. Rev. D49, 3830 (1994).
[96] R. P. Feynman, R. B. Leighton, and M. Sands, Feynman Lectures on Phys., 2, Chap. 28, (Addison-Wesley, 1964).
[97] T. Valone, Electro Gravitics II (Integrity Research Institute, Washington DC, 2008).
[98] L. Z. Fan, J. S. Feng, W. Q. Liu, Eng. Sci., 8, 2, 9-11 (2010).
[99] K.-Y. Chung, S.-w. Chiow, S. Herrmann, Steven Chu, and H. Müller, Phys. Rev. D 80, 016002 (July 2009).
[100] Liu Hongya & Zhou Peiyuan, *Scientia Sinica* (Series A), XXVIII (6): 628-637 (1985).
[101] M. J. G. Veltman, From Weak Interactions to Gravitation , Nobel Lecture, December, 1999.
[102] Einstein, A., L. Infeld, & B. Hoffmann, Annals of Math. 39 (1): 65-100 (1938).
[103] T. Damour, The Problem of Motion in Newtonian and Einsteinian Gravity in 300 Years of Gravitation edited by Hawking and Israel (Cambridge Univ. Press. 1987).
[104] Lo, C.Y., GJSFR Vol. 12 (4), Ver. 1.0, 5-15 (June 2012).
[105] Lo, C.Y., David P. Chan, Richard C. Y. Hui, Phys. Essays 15 (1), 77-86 (March 2002).
[106] B. L. Z. Nachtergaele, Editor of the J. of Math. Phys., official communication (June 22, 2012).
[107] 't Hooft, G. A confontation with infinity , Nobel Lecture (Dec. 1999).
[108] F. Wilczek, Asymptotic freedom: from paradox to paradigm , Nobel Lecture (Dec. 2005).
[109] S. W. Hawking, A Brief Hisory of Time (Bantam, New York, 1988), p. 23 & p. 143.
[110] T. Damour & J. H. Taylor 1992. Phys. Rev. D, 45 (6), 1840-1868 (1992).
[111] Daniel Kulp, Editorial Director of the American Physical Society, official communication (July 2012).
R. Shergill, editor of the Royal Society, claimed that the physical principles are not sound. (July 24, 2002).

Relativity Based on Physical Processes Rather Than Space-Time

ALBRECHT GIESE

Taxusweg 15, D 22605 Hamburg, Germany
note@ag-physics.de

Physicists' understanding of relativity and the way it is handled is at present dominated by the interpretation of Albert Einstein, who related relativity to specific properties of space and time. The principal alternative to Einstein's interpretation is based on a concept proposed by Hendrik A. Lorentz, which uses knowledge of classical physics to explain relativistic phenomena. In this paper, we will show that on the one hand the Lorentz-based interpretation provides a simpler mathematical way of arriving at the known results for both Special and General Relativity. On the other hand, it is able to solve problems which have remained open to this day. Furthermore, a particle model will be presented, based on Lorentzian relativity, which explains the origin of mass without the use of the Higgs mechanism, based on the finiteness of the speed of light, and which provides the classical results for particle properties that are currently only accessible through quantum mechanics.

Keywords: Relativity, Lorentzian relativity, Gravity, Gravitation, Particle mass, Dark matter, Dark energy

1. Introduction

The current state of physics is characterized by many open problems:

- Dark matter
- Dark energy
- Cosmological inflation
- The weakness of gravity
- The fact that gravity only attracts
- Gravitational waves
- Quantum gravity
- Origin of mass, Higgs?
- Supersymmetric particles
- Leptoquarks.

In our view, these problems constitute a general crisis in present-day physics, as they seem to be interrelated; this means that they apparently share a common cause.

This common cause is, in our opinion, an incorrect paradigm in the sense of Thomas Kuhn [2]. This paradigm is composed of two parts:

- Elementary particles are point-like – and have no internal structure.
- Relativity caused by properties of spacetime.
- We believe that these assumptions are incorrect. The reality is as follows:
- Elementary particles are extended and they do have an internal structure.
- Relativity is caused by the properties of fields and by the internal structure of elementary particles.

In the following, we will demonstrate the consequences of this altered paradigm for the following areas:

- Special relativity
- Particle physics, particularly the origin of mass, and
- General relativity, i.e. gravity.

2. A Brief Look at the History of Relativity

The basic experiment which triggered all the discussions about relativity was the Michelson-Morley experiment. Several explanations for the observed null result were offered by physicists at the time. We shall only describe two of them here, which continue to be relevant to this day.

An early explanation was presented by the Dutch physicist Hendrik A. Lorentz [3]. Oliver Heaviside had analysed Maxwell's theory of electromagnetism and deduced that electric fields contract when moving with respect to a postulated ether. A further consequence was that objects in motion should contract. When applied to Michelson and Morley's apparatus, this contraction explains the null result.

However, most physicists later adopted the position put forward by Einstein, that the speed of light is in fact physically the same for every inertial system, not just as a result of its measurement. This position went along with the view that no ether of any kind existed. According to Einstein, the constancy of the speed of light is a result of the specific properties of space and time [1].

Einstein later extended his idea of a contraction of space to the more general idea of a curvature of space-time and thus created a mathematical model with which to describe gravitation.

3. Special Relativity, the Lorentzian Way

The basic phenomena of special relativity – dilation and contraction – can in fact be explained by means of classical physics. Furthermore, it can be shown that the apparent constancy of the speed of light follows from these two phenomena.

3.1 Contraction

3.1.1 Extended Object Contraction

Fields in motion contract. Historically this was discovered by O. Heaviside in 1888 as a consequence of Maxwell's theory of electromagnetism.

The contraction is expressed mathematically by the relationship:

$$d \rightarrow d' = d/\gamma \quad \text{where} \quad \gamma = 1/\sqrt{1-v^2/c^2}\,,$$

later called the Lorentz factor.

We then have to take into account that electric fields determine the size and the shape of macroscopic objects. Those objects are made up of atoms and molecules, which are bound together by electric multi-pole forces. Since fields contract when in motion, all extended objects in our world must also contract when in motion. The Michelson-Morley apparatus must therefore contract when the laboratory moves with respect to some frame of reference, and the null-result is fully explained.

These conclusions were originally put forward by FitzGerald and Lorentz. – Later it was shown that every type of field contracts when in motion.

3.2 Dilation

The dilation of all time-related processes in physics is the consequence of the internal motion within elementary particles.

3.2.1 Indications for Internal Oscillations

As early as 1909, J. Ziegler mentioned that "in the smallest objects", i.e. in elementary particles, a permanent motion occurs at the speed of light *c*. When Louis de Broglie detected the wave-particle phenomenon in 1923 [4], it became obvious that there was an oscillation associated with every particle.

This detection initiated the quantum mechanical description of a particle by means of a wave function. Paul Dirac developed this description to the point where he was able to present a relativistic wave function of the electron in 1928 [5]. When Erwin Schrödinger analysed this wave function, he concluded, in his famous paper of 1930 [6], that a constant motion occurs in the electron at the speed of light, *c*, which he gave the German name '*Zitterbewegung*'.

In the context of the particle model presented here, it is assumed that not only the electron, but all leptons as well as all quarks exhibit this internal motion at *c*; this is simply an extension of what we know about the electron.

This assumption not only explains time dilation but also some further properties of those particles, such as their magnetic moment and spin. Despite the contrary conviction held by conventional quantum mechanics, these phenomena can in fact be explained by classical means.

3.2.2 The Internal Oscillation

In anticipation of chapter 4, we will assume here that the oscillation takes on the form of two sub-particles performing circular motion. This motion takes place exclusively at the speed of light *c*, in accordance with the particle model that will be presented.

3.2.2.1 Dilation Caused by Circular Motion

Any periodic motion occurring at the speed of light, *c*, must necessarily cause an extension of the period and a corresponding reduction in the frequency of oscillation when the entire configuration undergoes linear motion. This will be shown here for the simple case of circular motion.

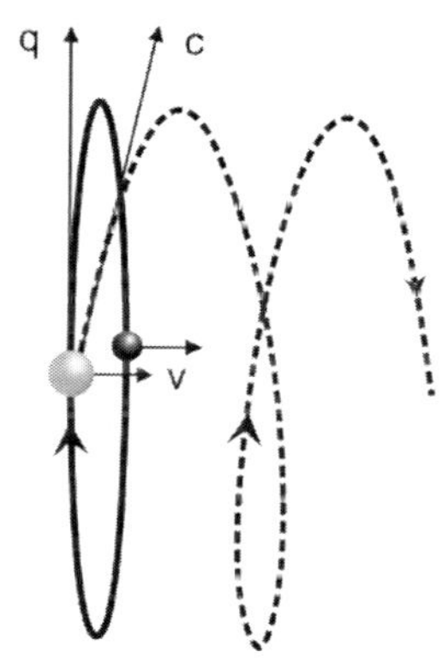

Figure 3.1 Dilation in an elementary particle.

Let us take the simplest case. The elementary particle shall move in an axial direction at a velocity *v* (Fig. 3.1). This causes the circular motion to become a helical motion. The extension of the period of oscillation follows from the Pythagorean Theorem:

As the speed along the helix must be *c* and the forward motion (with respect to an observer at absolute rest) is v, the remaining speed along the projected circuit is given by

$$q^2 = c^2 - v^2.$$

If the radius of the orbit is R, then the period T of the configuration at rest is

$$T = 2\pi \cdot R/c.$$

When in motion, the period T' of one revolution is given by

$$T' = 2\pi \cdot R/q.$$

Combining these two equations, we arrive at the following relation between the two periods:

$$T'/T = c/q = 1/\sqrt{1 - v^2/c^2} = \gamma$$

which is the Lorentz factor.

This extension of the period of rotation means that a moving clock will display a shorter time interval. The time indicated by the moving clock is conventionally called the "proper time" and denoted by the symbol τ. Using this we arrive at the conventional form of the temporal part of the Lorentz Transformation for an object moving at a velocity, v:

$$\tau = t \cdot \sqrt{1 - v^2/c^2}. \quad (3.1)$$

3.2.2.2 Dilation in the General Case

In the more general case, the motion of the elementary particle will not necessarily be in the direction of the axis but in some arbitrary direction. In this case, the result is the same; however the calculation is more complex and will not be presented here.

3.3 Constancy of c

Whereas Einstein treats it as an ontological fact that the value of c is a true constant in every inertial system, in the physically based relativity presented here, only the *measured* value of c is a constant. This is caused by the contraction of the gauges and by the retardation of clocks, which also causes a desynchronization of clocks in different positions.

4. The Basic Particle Model

The particle model presented in this paper is unavoidable for the understanding of physical phenomena, in two ways. On the one hand, it is a consequence of the facts of relativity. Particularly general relativity can be understood very easily on the basis of this particle model, whereas this is extremely complex by the method chosen by Einstein. On the other hand, it explains the cause of mass and of other properties of particles without the use of quantum mechanics.

4.1 Structure of an Elementary Particle

An elementary particle is composed of two sub-particles, which are called basic particles in the scope of this concept.

The main arguments to understand the structure of elementary particles in this way are the following:

- From the quantum mechanical analyses of Dirac and Schrödinger it follows that a permanent motion at c occurs within the electron. Its frequency follows from the energy state of the particle, given by $E = h*\nu$
- From the dilation, it follows that this internal oscillation at c applies to all particles, since dilation is a general phenomenon and not restricted to a specific type of particle.

In addition, we have to take into account the fact that there are further restrictions on the possible structure of an elementary particle:

- An oscillation is only possible if there are at least two sub-particles in the elementary particle. Otherwise the oscillation would violate the law of the conservation of momentum
- If those sub-particles are constantly moving at the speed of light, c, then they cannot have any mass. It is a general fact of relativity, independent of the specific interpretation of relativity, that only a massless object can travel at a velocity of c.

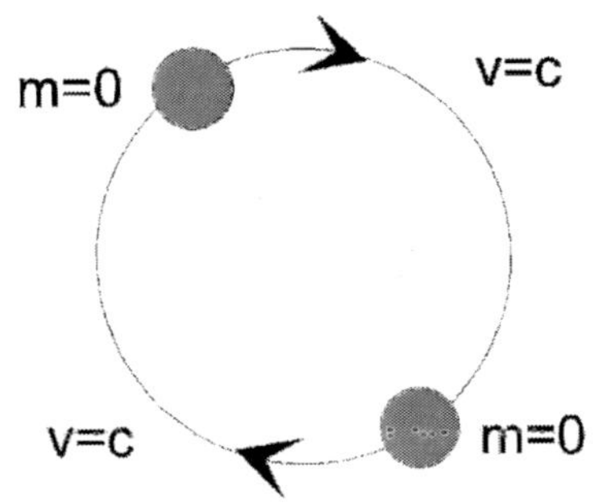

Figure 4.1 Structure of an elementary particle.

In the following, this particle model will be called the '**Basic Particle Model**'.

These assumptions define most of our general particle structure. They also raise a new question. If the sub-particles of an elementary particle have no mass, whereas on the other hand the particle as a whole does have mass, what is the origin of that mass?

4.2 The Mass of an Elementary Particle

The inertia observed in physics is a direct consequence of the fact that the speed of light c is finite.

As explained above, according to this model the elementary particle is made up of two basic particles.

The bond between these basic particles has to be of such a nature that both particles remain a certain distance apart; otherwise the elementary particle would have no extension. It must, however, have an extension in order to have a spin and a magnetic moment.

4.2.1 The Bond within an Elementary Particle

The bond between the two basic particles can only be a multi-pole bond, in that a multi-pole field has a potential minimum, which defines the equilibrium distance between the two sub-particles and hence the extension of the particle as a whole. A planetary model is not a possible alternative, as the basic particles have no mass.

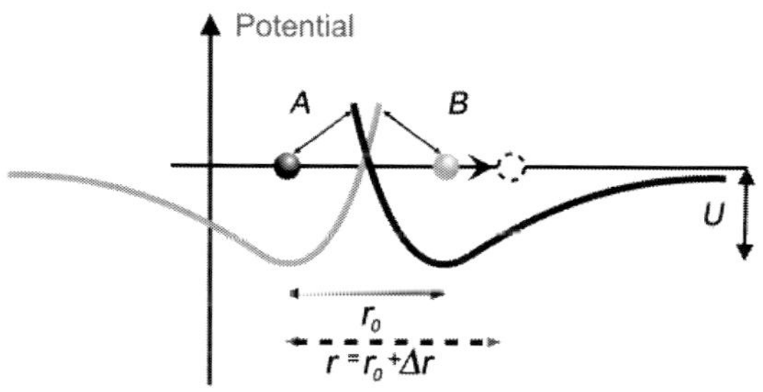

Figure 4.2. The Binding Field.

The potential of the multi-pole field is assumed to be shaped such as to produce a force, F, given by:

$$F = S \cdot \frac{\Delta r}{r^3} \quad (4.1)$$

where S is the binding field, Δr the offset from the equilibrium position, and r the distance between the sub-particles.

The following consideration suggests that this binding force is in fact the strong force. The multi-pole configuration is achieved by an appropriate arrangement of monopole charges of different signs. The bond produced by this arrangement must be strong enough to compensate for an additional – repulsive – electric charge in the case of a charged elementary particle such as an electron. For this reason, no force other than the strong force can produce a stable bond.

4.2.2 The Behaviour in Motion

The binding field between the basic particles has this specific shape in order to create a bond which keeps a constant distance between the two. If a particle is now set in motion, the field follows the changing position with a certain delay which is caused by the finiteness of the speed of light. As a consequence, the other basic particle remains in its current position for a short time. And as a further consequence, the field due to this other particle will not change at all for a short time (fig. 4.3). The displacement of the basic particle in question therefore requires a force to be applied for a short time.

After a period of time given by the distance between the two basic particles, the change in the field arrives at the other particle, A, which is repositioned. After a further period of time, the field due to the repositioned particle A will reach particle B, and no force will be necessary any longer.

This fact, that a change in the state of motion requires an intermediate force, accounts for the physical phenomenon of inertia.

Note: This displacement of particle B is presented here as a *step-like* motion. This is done to make the process and the spatial field change easier to visualise. The reality is of course different. The motion of B is a smooth process. In the course of this smooth change in position, the field changes continuously, but with the delay shown here.

4.2.3 The Force in the Case of Constant Acceleration

For the quantitative determination of the inertial force, we shall assume that one of the basic particles, B, is accelerated by an external agent. This causes the position of this basic particle to be displaced relative to the other one, A. The displacement is a result of the time needed for the change in the field due to particle B to propagate to particle A at a speed of c:

This displacement requires a force, given by eq. (4.1). To produce a constant acceleration, a, a force F is needed, which is determined by a simple calculation using classical mechanics:

$$F = S \cdot C \cdot \frac{1}{r} \cdot a \cdot \frac{1}{c^2},$$ and correspondingly the inertial mass is given by

$$m = \frac{F}{a} = S \cdot C \cdot \frac{1}{r} \cdot \frac{1}{c^2},$$ where C is a constant of integration taking into account the different directions of motion.

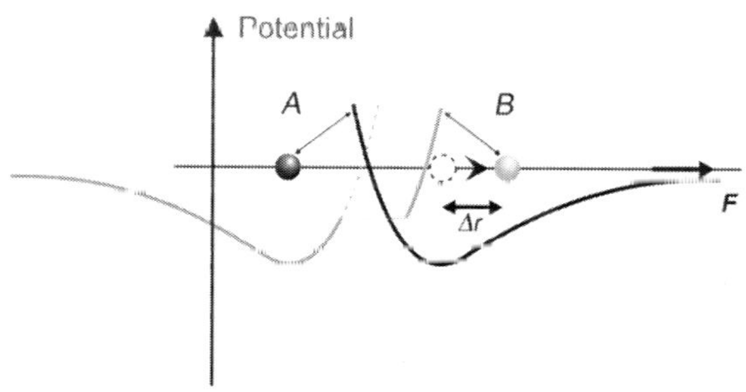

Figure 4.3 Binding field in motion.

The unknown parameters S and C can be determined from the following, well-known equations (which can also be derived in the context of the Basic Particle Model):

$$E = mc^2 \text{ and } E = h \cdot \nu,$$

and the distance r is replaced by the radius of the particle, R. So we end up with the formula

$$\boxed{m = \frac{\hbar}{R \cdot c}} \qquad (4.2)$$

for the mass of an elementary particle made up of two basic particles.

This is now the inertial mass of an object deduced from the delay with which field forces between charges are propagated.

This result has the following remarkable properties:

1. It yields the fact that the quotient of force and acceleration is a constant at non-relativistic velocities. Therefore, this is a deduction of Newton's law of motion. For Newton, this law had the property of an axiom (or a principle of nature)
2. The result shows that the mass of an elementary particle is inversely proportional to its size, R.

4.2.4 The Relationship between Mass and Magnetic Moment

We can now use the magnetic moment of electrically charged particles to check the usability of the mass formula deduced above. First, we recall the classical equation for the magnetic moment μ of a particle

$$\mu = i \cdot \pi \cdot R^2 .$$

The loop current i within a particle with an elementary charge e_0 at frequency ν is simply:

$$i = \nu \cdot e_0 \text{ with } \nu = c/2\pi R .$$

So it follows that:

$$\mu = c \cdot e_0 \cdot R/2 .$$

If R is now inserted from eq. (4.2), the magnetic moment turns out to be

$$\mu = \frac{\hbar}{2} \cdot \frac{e_0}{m} .$$

For the electron, this is the 'Bohr Magneton'.

Please note: Physics textbooks state that the Bohr Magneton can only be derived using quantum mechanics. The preceding, however, shows that this equation can be derived classically using the Basic Particle Model.

4.3 The Relativistic Mass

4.3.1 The Increase in the Mass during Motion

According to eq. (4.2), the mass of a particle is given by

$$m = \hbar/(R \cdot c)$$

When in motion, the radius R shrinks by the Lorentz factor

$$\gamma = 1/\sqrt{1 - v^2/c^2} .$$

So, when in motion the mass increases according to

$$m \rightarrow m' = m \cdot \gamma .$$

4.3.2 The Relationship between Mass and Energy

From the preceding section:

$$m = m_0 / \sqrt{1 - v^2/c^2} \text{ or equivalently}$$
$$m = m_0 \sqrt{c^2/(c^2 - v^2)}$$

where m_0 is the rest mass of the particle, it follows that an increase in the velocity of an object will increase its mass. On the other hand, an increase in velocity means an increase in its energy. The relationship between mass and energy, which is the most famous equation attributed to Einstein, can easily be deduced by expressing the increase in mass in terms of the kinetic energy. A fairly simple calculation leads to the result:

$$\boxed{E = mc^2}.$$

4.4 The Experimental Situation of the Electron

There is an apparent conflict between the model presented here and experiment.

Present-day physics regards the electron, for instance, as a particle which is point-like and has no internal structure. This is deduced from scattering experiments. The conclusion is based on the premise that, if an electron had sub-particles, these sub-particles would have some mass. Such a conception of the electron would indeed contradict the measurements.

The Basic Particle Model, on the other hand, assumes that basic particles do not have any mass. When this assumption is made, no conflict with the experiment exists.

4.5 The Comparison with the Higgs Model

The Higgs model is understood by present-day physics to explain the phenomenon of mass. However, at present the following problems occur with its use:

1. A new particle was detected using the LHC storage ring at CERN, near Geneva, in 2012 [9,10]. However, the parameters of this particle have not yet been determined with sufficient accuracy to fulfil the requirement of the Higgs theory. There are indications that some decay parameters could deviate.

2. Higgs theory does not provide a method for independently determining the masses of known particles. For every particle, a specific parameter, the Yukawa coupling, is required, which is derived from the *known* mass of this particle.

3. There is a discrepancy between the necessary Higgs field in the vacuum and the vacuum field

measured in the universe. This discrepancy is of an order of at least 10^{60}.

This means that on the basis of present data, the Higgs theory does not provide a satisfactory answer to the question of mass. The Basic Particle Model, by contrast, does not need any additional parameters as long as the magnetic moment of a particle is known.

5. General Relativity

The general theory of relativity (GRT) is Einstein's theory to explain gravitation. Einstein does so using his geometrical model of space-time.

In contrast to Einstein, the Basic Particle Model combined with Lorentzian relativity explains gravitation on the basis of physical processes.

5.1 Gravitation According to Einstein

According to Einstein, the concept of space-time also provides the explanation for gravitation. Objects move along geodesics in four-dimensional space-time. In the vicinity of a mass, or according to Einstein equivalently in the vicinity of an occurrence of energy, space-time is curved, and so the geodesic is the natural path of an object.

In order to determine the motion of objects in a gravitational field, it is necessary to determine the shape of the geodesics in question. This requires the use of multi-dimensional Riemannian geometry, which is a very challenging task.

The calculations using Einstein's approach are so complex that in usual cases the more specialized calculation of Schwarzschild, the so-called Schwarzschild Solution, is used.

5.2 Gravitation as a Physical Process

Gravitation based on physics rather than geometry makes use of the fact that in a physical interpretation, which refers directly to the measurement, the speed of light c is not always constant but varies in the vicinity of matter.

This variation of c causes the refraction of light-like particles in general. This refraction influences the movements within an elementary particle and causes the particle to accelerate. This process explains gravitation and produces quantitatively correct results for all the phenomena treated by general relativity.
Below we will show that general relativity based on the Basic Particle Model is equivalent to Einstein's version up to the Schwarzschild Solution.

5.2.1 Speed of Light in a Gravitational Field

The speed of light varies in the gravitational field in the vicinity of an object. As a result, photons and light-like particles are refracted in this field.

The dependency of c on position for a spherically symmetric object is known to be

$$c(r) = c_0 \cdot (1 - 2 \cdot \frac{G \cdot M}{r \cdot c_0^2})^p \tag{5.1}$$

where c_0 is the speed of light in gravitation-free space, G is the gravitational constant and M is the mass of the object, which is traditionally said to cause the gravitational potential; r is the distance from the centre of gravity. The power p is ½ or 1 depending on the direction of motion with respect to the centre of gravity, i.e. tangential or radial respectively.

Equation (5.1) is initially used here as an experimental result. Although this dependency is also asserted by Einstein's theory, we do not need to make any reference to Einstein. We will later explain how this dependency follows from the model presented.

This dependency was the first time measured by I. I. Shapiro in the late 1960's using radar ranging between Earth and Venus. The measurement was later repeated by others with increasing accuracy.

5.2.2 Gravitational Lensing

Gravitational lensing follows from the fact that the path of a photon is deflected by a gravitational field (Fig. 5.1).

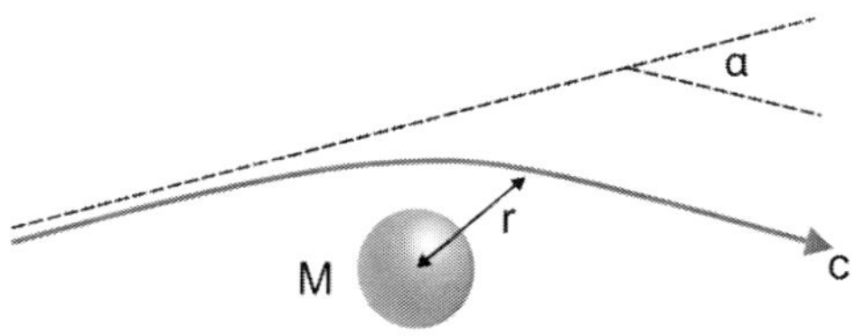

Figure 5.1. Deflection by the sun.

Using eq. (5.1) and applying classical geometry and the laws of refraction to calculate the angle of deflection and the acceleration, yields the following result:

$$\alpha = 4 \cdot GM/(c^2 y) \tag{5.2}$$

where y is the distance from the vertex to the centre of the sun and α is the angle of deflection. For the acceleration a at the vertex of the path, the result is

$$a_{vertx} = GM/r^2 . \tag{5.3}$$

For details of the derivation, see references [11,12].

After inserting the values applicable for the sun, we get the correct, known result of

$$\boxed{1.75 \text{ arc-sec}}.$$

This number corresponds to twice the normal gravitational acceleration and conforms to observations. This numerical result, as well as the analytical result (5.2), also agree with the predictions of general relativity – however without any use of general relativity.

5.2.3 Gravitational Acceleration for a Particle at Rest

If an elementary particle is placed in a gravitational field, its orbiting basic particles are subject to refraction as explained in section 5.2.2. This refraction causes the basic particles to deviate from their circular path. This in turn will cause the entire elementary particle to move.

Taking the case where the elementary particle is oriented such that its orbital axis points towards the source of gravity, the refraction causes the basic particles to spiral towards the source of gravity. So the entire elementary particle will move in the direction of the source. Figure 5.2 shows the accelerated downward motion. Due to refraction, the pitch angle of the basic particles, α in this figure 5.2, will steadily increase. This causes the elementary particle to perform an accelerated motion towards the gravitational source. (Please note that for the sake of simplicity only the path of one of the two basic particles is shown in Fig. 5.2.)

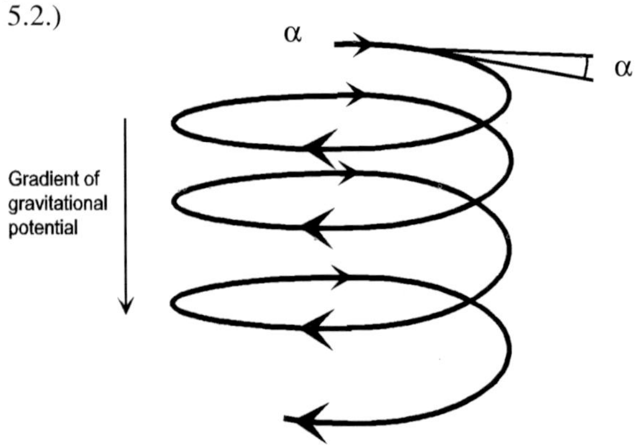

Figure 5.2. Progressive downward spiral.

In this case, the acceleration of the (composite) elementary particle is similar to the acceleration given by eq. (5.3):

$$a = \frac{G \cdot M}{r^2}$$

which is Newtonian acceleration.

If the elementary particle has an arbitrary orientation, the process is mathematically more complicated but produces a similar result.

5.2.4 The Equivalence Principle

Looking at figure 5.2, the deflection of the path of the basic particles is independent of the radius of the particle and, because of eq. (4.2), independent of the mass of the particle. This means that the independence of gravitational acceleration and mass has a very natural cause. No assumptions about any equivalence are necessary.

5.3 Lorentzian Path to General Relativity Formally

5.3.1 The Proofs of General Relativity

Next, we will show that the general proofs of Einstein's general relativity theory can be deduced using the concepts presented in the preceding sections. We will now show that this concept is able to explain the observations below in the same way as Einstein's theory. Hence it can be shown that the known proofs using Einstein's theory of relativity are also proofs of the Lorentzian interpretation of relativity.

The following proofs of General Relativity will be envisaged here:

- The Shapiro effect (reduction of c)
- Gravitational lensing
- Geodetic effect - with the Lense-Thirring effect
- Time dilation in a gravitational field
- Black holes
- Perihelion advance.

The Shapiro effect and gravitational lensing have been explained above. The other proofs use the Schwarzschild Solution for general relativity and will be covered by the following.

5.3.2 The Schwarzschild Solution

To work with Einstein's field equations is an extremely challenging task. A short time after Einstein published his theory of general relativity, Karl Schwarzschild presented a solution for the simplified, less general solution of a spherically symmetric field, such as that of the sun, which is a frequent situation in astronomy. The experiments and observations cited in the literature as proofs of Einstein's general relativity usually refer to the results of the Schwarzschild Solution.

The Schwarzschild Solution is normally deduced by starting with Einstein's field equations and using Riemannian geometry, and then restricting these to the special situation. Here we will present a different deduction. We will start with the physical version of special relativity and the Basic Particle Model and demonstrate how easily this solution can be deduced from these physical foundations.

According to the Basic Particle Model, an elementary particle is made up of two sub-particles orbiting each other. Their temporal behavior is described by eq. (3.1), which leads to the following equation for the proper time of an object in motion:

$$\tau = t \cdot \left(1 - v^2/c^2\right)^{1/2}.$$

This equation is now differentiated with respect to

t, squared and rearranged:

$$c^2 \cdot (d\tau/dt)^2 = c^2 - v^2 \qquad (5.4)$$

In a gravitational field this time behaviour changes. Understanding this change guides us directly to the Schwarzschild Solution. We first split the speed into a radial and a tangential component, since the Schwarzschild Solution is normally given in terms of polar coordinates:

$$c^2 \cdot (d\tau/dt)^2 = c^2 - v_{rad}{}^2 - v_{\tan}{}^2 \qquad (5.5)$$

where v_{rad} and v_{tan} denote the radial und the tangential component of the velocity respectively.

Next, we have to take into account the fact that c changes as follows in a gravitational field, according to eq. (5.1):

$$c_{rad} \to c'_{rad} = c_{rad} \cdot (1 - r_s/r) \; ;$$
$$c_{tan} \to c'_{tan} = c_{tan} \cdot (1 - r_s/r)^{1/2} . \qquad (5.6)$$

Here, in order to simplify the equations, we have used the common definition for the so-called Schwarzschild radius r_s

$$r_s = 2 \cdot G \cdot M / c_0{}^2 .$$

As a consequence of the change in c, fields also contract and so too does the size of particles in a radial direction with respect to the center of gravity:

$$r \to r' = r \cdot (1 - r_s/r)^{1/2} . \qquad (5.7)$$

If we now wish to apply the above equations to the temporal equation (5.5), we have to take into account the fact that the reduction in the speed of light depends on the direction of motion. We can facilitate the calculation by changing to a modified coordinate system. Replacing the radial component in the following way

$$r \to \hat{r} = r/(1 - r_s/r)^{1/2} \qquad (5.8)$$

it follows that

$$v_{rad} \to \hat{v}_{rad} = v_{rad}/(1 - r_s/r)^{1/2} . \qquad (5.9)$$

By making this replacement, we can use eq. (5.6) as the general reduction of c independently of the direction. The result of the temporal equation (5.5) is not changed by making this replacement.

$$c^2 \frac{d\tau}{dt} = c_{\tan}{}^2 - \hat{v}_{rad}{}^2 - v_{\tan}{}^2 . \qquad (5.10)$$

Now, inserting (5.6) and (5.9) into (5.10) and using

$$\hat{v}_{rad} = dr/dt \cdot (1 - r_s/r)^{1/2} = r'(1 - r_s/r)^{-1/2}$$

and

$$v_{\tan} = d\varphi/dt \cdot r = \varphi' \cdot r$$

we get

$$\boxed{c^2 \frac{d\tau}{dt} = \left(1 - \frac{r_s}{r}\right) \cdot c^2 - \left(1 - \frac{r_s}{r}\right)^{-1} r'^2 - \varphi'^2 \cdot r^2} \qquad (5.11)$$

which is a common form of the Schwarzschild Solution.

The experimental proofs of general relativity listed in section 5.3.1 use this Schwarzschild Solution. So, we can now refer to the according deductions in textbooks in order to verify that the Lorentzian way presented here covers these cases.

5.4 The Cause of Gravitation

We have seen that gravity is in fact not a force but a refraction process. And the cause of this refraction is the varying speed of light c in the vicinity of matter. In the case of a transverse deflection, the exchange particle causes a speed offset v perpendicular to the original speed vector as depicted in Fig. 5.4. This fact stands in contrast to conventional physics, but it helps to overcome the unresolved problems of present-day gravitational physics

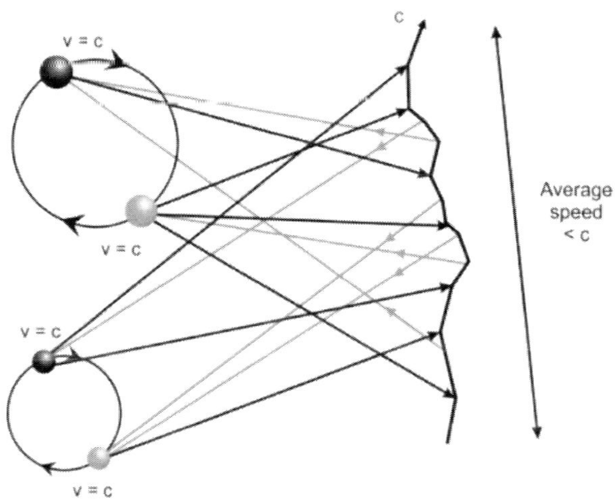

Figure 5.3. Disturbed path of a light-like particle.

5.4.1 Varying Speed of Light

Equation (5.1) is the basis for explaining all phenomena attributed to gravitation. The next question to be answered is; why c is reduced in the vicinity of matter. The answer is that the reduction of c is caused by the effect of the exchange particles, which produce the binding field between the basic particles. According to the Basic Particle Model, the binding field is the field of the strong interaction, which is – also according to the model – the universal force in our world affecting all existing particles.

These exchange particles, which are emitted by a multi-pole compound and cause attraction and repulsion in a random way, also interact with every light-like particle. They cause such a particle to be deflected towards the origin of the exchange particle (i.e. of the basic particle) or away from it. So the light-like particle performs a random walk as depicted in figure 5.3. As a result, the average speed of the light-like particle is reduced, even though the microscopic speed is still the speed of light *c*.

5.4.2 Determination of the Reduction in Speed

We will now present the formula for the reduction in the speed of light, *c*, in a gravitational field. However, we will not deduce it here but refer to M. Duffy et al.[11] for the details. If we define the number of elementary particles in the gravitational source as being *N*, then the resulting deflection of a light-like particle passing by depends on whether the motion is in radial or in tangential direction with respect to the gravitational field. The difference between both cases is demonstrated in Figs. 5.4 and 5.5.

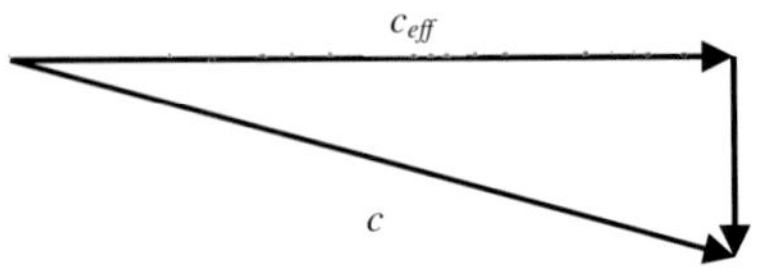

Figure 5.4. Transverse deflection.

The projected effective speed c_{eff} follows from the deflection by the law

$$c_{eff,tan} = \sqrt{c^2 - v^2} = c\sqrt{1 - v^2/c^2}\,. \qquad (5.12)$$

In the case of a longitudinal deflection the exchange particle causes a speed offset *v* in the direction parallel to the original speed vector as depicted in Fig. 5.5.

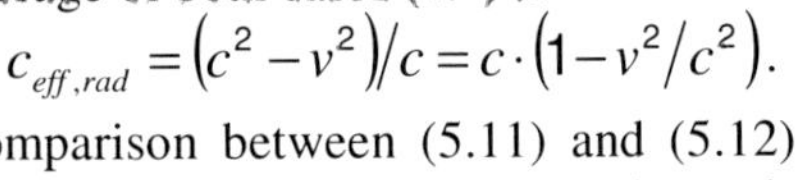

Figure 5.5. Longitudinal deflection.

Here the projected effective speed c_{eff} follows from the deflection by the law

$$c_{rad} = c + v \quad \text{or} \quad c_{rad} = c - v$$

The average of both cases (+/-) is

$$c_{eff,rad} = \left(c^2 - v^2\right)/c = c\cdot\left(1 - v^2/c^2\right). \qquad (5.12)$$

The comparison between (5.11) and (5.12) shows in which way the reduction of *c* depends on the direction of the motion. The resulting reduction in the speed of light, *c*, depends now on *N* in the following way:

$$c_{eff} = c\cdot\left(1 - gN/c^2 r\right)^p\,. \qquad (5.13)$$

The parameter g is the proportionality factor for the influence of the flow related to N particles, replacing the gravitational constant G, and as explained above p=1 for radial motion and p=1/2 for tangential motion. The dependency of the contraction of multi-pole fields in a gravitational field works analogously to the contraction of fields at motion. The result for the reduced distance is

$$r_{red} = r\cdot\left(1 - gN/c^2 r\right)^{(p-1/2)} \qquad (5.14)$$

with *p* defined as above.

An important consequence of this understanding is that the strength of the gravitational field does not depend on the mass or on the energy of the gravitational source. In contrast to conventional physics, every elementary particle provides the same contribution to the field, so also mass-less or light particles as photons and neutrinos.

6. Cosmology

This chapter deals with open problems in astronomy and cosmology, including dark matter, inflation, and dark energy.

6.1 Dark Matter

Some decades ago, it was noticed that the rotational speed within and around big galaxies is in conflict with the equilibrium speed determined on the basis of standard gravitation. Figure 6.1 shows this discrepancy. As a possible solution, present-day physics assumes a specific type of matter, which is invisible and has almost no interaction with known matter, but must have a high mass to explain the missing mass of the calculations. This missing mass has been given the name "dark matter".

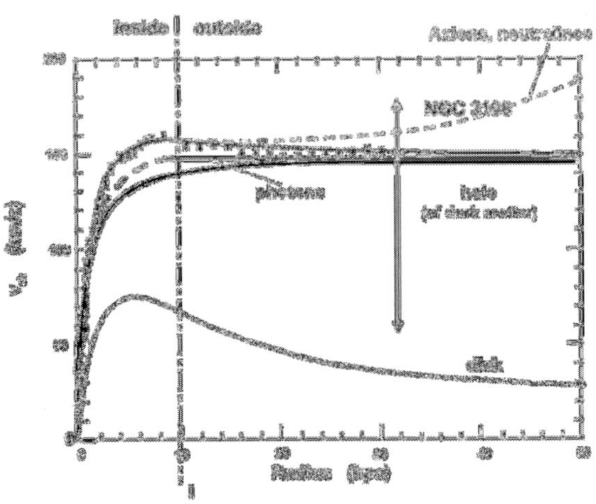

Figure 6.1. Equilibrium conflict in the galaxy NGC 3198. (The radius of the galaxy is 10 kpc).

In figure 6.1, the solid curve labelled "disk" is the rotational speed as a function of the radius based on a normal gravitational calculation. The uppermost single values are measurements of the real speed; a curve (also solid) is fitted through these measurements. The black solid line labelled "halo" describes the required distribution of the proposed "dark matter" in order to explain the measured values.

The horizontal grey line, which is very close to the "halo" curve, follows from the assumption, described above, that every elementary particle contributes equally to the gravitational field. It represents the contribution of light particles, i.e. neutrinos and photons. In the drawing, the height of this line has been adjusted to fit this diagram; however it fits the known data within a tolerance factor of 2-3. Its curvature, however, is given by the natural distribution of the light particles and is not parameterised.

Of the light particles mentioned, the photons are mainly generated by the hot, shining stars in the centre of the galaxy. The neutrinos are similarly generated by the nuclear processes within the stars, the sources of which are also mostly in or close to the centre of the galaxy. These particles produce a continuous flux away from the centre at the speed of light c (or almost this speed) which has a density proportional to $1/r^2$. This is the cause of the flat shape of the rotation curve in the outside regions which is otherwise unexplained.

6.2 The Horizon Problem

From the temperature distribution of the Cold Microwave Background (CMB) it is concluded that there must have been a correlation between regions of the universe which have been separate a short time after the Big Bang. On the other hand, those regions have moved away from each other at such a high speed that, in the face of the limited speed of light, no causal relationship could exist. This conflict has been named the "horizon problem".

6.2.1 Inflation According to Einstein

Following Einstein's interpretation, present-day physics assumes a change in space as a solution, postulating that shortly after the Big Bang space was approx. 10^{60} times smaller than today and then expanded, initially very rapidly, later slowly until the present. This purported process was given the name 'inflation'.

Present-day physics does not have a proper explanation for the process of inflation. As an ad-hoc assumption, a new field produced by so-called "inflatons" is thought to cause it.

6.2.2 The Horizon Problem Explained by the Varying Speed of Light

From a logical point of view, the problem with this correlation is the conflict between spatial extension and the speed of light. So instead of assuming a change in space, it can equally well be assumed that the speed of light changed, namely that it was extremely large during a short period close to the Big Bang. Afterwards the speed of light decreased rapidly at first, and later more slowly to its present value.

The assumption that the speed of light changed during the evolution of the universe is attractive anyway, since it would not only solve this causal problem. As a further benefit, it could also solve the fine-tuning of basic physical parameters, which is not understood at present. (See, for example, the work of A. Albrecht and J. Magueijo [7].) Hence, aside from other problems, it avoids the necessity of a 'landscape' of 10^{100} uni-(multi-)verses.

6.3 Dark Energy

The observations of type 1a supernovae have recently led to the conclusion that the objects in our universe are accelerating. The results of Riess et al. [8] are presented in figure 6.2.

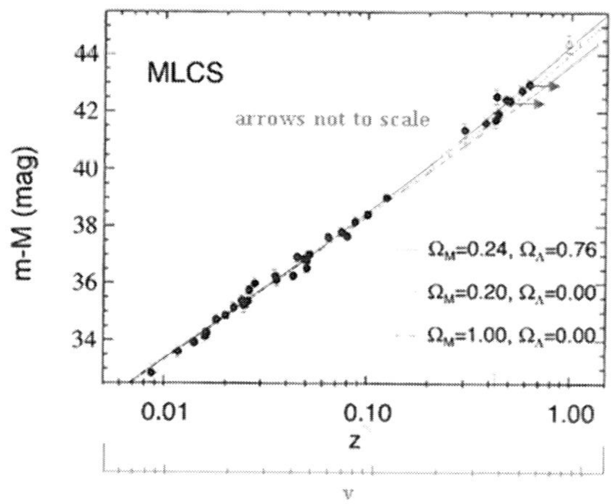

Figure 6.2. Supernova 1a, Hubble diagram – with possible correction factor.

Figure 6.2 shows the apparent magnitude of the observed supernovae (as the ordinate) versus the redshift z (the abscissa), which is identified with the recessional velocity of the stars. The redshift z is defined as

$$z = \Delta\nu/\nu_{ob} \tag{6.1}$$

where ν_{ob} is the *observed* frequency and $\Delta\nu$ is the frequency shift. From the Doppler effect, it follows that the velocity V is

$$V = c \cdot z/(z+1) \tag{6.2}$$

which is used for the evaluation according to figure 6.2, but conventionally with the assumption that c is a constant over all times and that space is unchanged during the time investigated.

According to Hubble's Law, all stars, and hence also the supernovae investigated, should be located on a straight line, represented in figure 6.2 by the dotted line for the most probable assumption. This means that the recessional velocity of these objects is proportional to their distance from the observer. However, the measurements in the upper part can be understood as lying too far to the left, which means that the redshift of the older supernovae is too small compared with the younger stars in the lower (left-hand) region. (Please disregard the arrows for the moment). This is commonly interpreted as meaning that the younger supernovae are too fast compared with the older ones. They are assumed to be accelerated.

However, the assumption described above, that the speed of light c has changed, is able to explain the acceleration as an evaluation effect. From equations (6.1) and (6.2), it follows for the velocity V in the case of a Doppler shift $\Delta\nu$ that there is:

$$V = c \cdot \Delta\nu / (\Delta\nu + \nu_{ob}). \qquad (6.3)$$

If it is assumed that the speed of light c was higher at an earlier time, then c must be replaced by a larger value and the resulting V will be larger. The arrows in figure 6.2 show the effect of this additional speed. This means that with reference to the speed scale the supernovae can now be positioned on the dotted line without any conflict with observation. This is indicated as an example by the arrows in figure 6.2. In physical terms, this means that the alleged acceleration vanishes.

7. Conclusions

We have shown that relativity can be derived from physical processes, i.e. the properties of particles and fields. We believe there are compelling arguments that the speed of light is only constant with respect to an absolute frame of reference, not with reference to any inertial system. Furthermore, we see good arguments that the speed of light has changed during the history of the universe – as also assumed by other physicists. This means that the speed of light is only constant over a limited time interval.

This approach has extraordinary benefits:

- Relativity now fits seamlessly into physics as a whole. The theory becomes far more comprehensible and its formalism is much easier to understand. When accounted for in this way, relativity can even be taught at a high-school level; yet the results still conform to those derived using Einstein's approach, at least to the extent that these can be proven by experiments and observations.
- Important unresolved questions of present-day physics are resolved with surprising ease:
  (1) The *dark energy problem* is resolved as a result of the changing speed of light during the evolution of the universe.
  (2) The *dark matter problem* vanishes, as it follows from the model that every elementary particle contributes the same amount to the gravitational field – irrespective of its mass. In the quantitative calculations, photons are able to constitute dark matter particles. Furthermore, the spatial distribution of dark matter in the universe, which could not otherwise be explained, fits the model.
  (3) *Inertial mass* is explained, including the dynamic aspects, i.e. the relativistic increase in mass and the mass-energy equation. In present-day physics, the origin of mass is still an open issue. Even if the Higgs boson is confirmed, it would not constitute a complete explanation.
  (4) *Quantum gravity* is no longer an open issue. On the one hand, gravity is shown to be a side effect of the strong force and, on the other hand, the strong force is fully covered by quantum mechanics. Hence the alleged conflict between relativity and quantum mechanics disappears.

References

[1] Einstein, A. (1905) *Zur Elektrodynamik bewegter Körper*, Annalen der Physik, IV. Jg. 17, S. 891–921.
[2] Kuhn, T. (1970) The Structure of Scientific Revolutions, Chicago: University of Chicago.
[3] Lorentz, H.A. (1904) Electromagnetic phenomena in a system moving with any velocity smaller than that of light, Proc. Acad. Science Amsterdam 6: 809–831.
[4] de Broglie, L. (1923) *Radiations - Ondes et Quanta*, *Comptes rendus*, Vol. 177, pp. 507-510.
[5] Dirac, P.A.M. (1928) The Quantum Theory of the Electron, Proceedings of the Royal Society of London, Series A, 117; 778, 610–624.
[6] Schrödinger, E. (1930) *Über die kräftefreie Bewegung in der relativistischen Quantenmechanik* (On the free movement in relativistic quantum mechanics) Berliner Ber., pp. 418-428.
[7] Albrecht, A. & Magueijo, J. (1999) A time varying speed of light as a solution to cosmological puzzles, Phys. Rev. D 59, 043516.
[8] Riess, A.G. et al. (1998) Observational evidence from supernovae for an accelerating universe and a cosmological constant, Astronomical J. 116:1009–38.
[9] Adam, W. et al. (2012) Observation of a new boson at a mass of 125 GeV with the CMS experiment at the LHC, Phys. Lett. B, Vol. 716, Issue 1, pp. 30–61.
[10] Aad, G. et al. (2012) Observation of a new particle in the search for the Standard Model Higgs boson with the ATLAS detector at the LHC, Phys. Lett. B, Vol. 716, Issue 1, pp. 1-29.
[11] Duffy, M.C. & Levy, J. (2009) (eds.) Ether Space-Time & Cosmology, Montreal: Apeiron, V. 3, pp. 143-192.
[12] www.ag-physics.org/gravity.

On The Fundamental Nature of the Electromagnetic Interaction

AL F. KRACKLAUER

Belvederer Allee 23C; PF 2040
99425 Weimar, Germany
af.kracklauer@web.de

We propose an experiment based on statistical analysis to test the currently accepted statement, that "a photon cannot be split." The essential statistical phenomenon pertains not only to beam splitters, but also to the generation of correlated photon pairs used extensively in quantum optics experiments. A consequence of this analysis is that, arguably the tactic of reducing the window defining a coincidence has the unexpected effect of preferentially selecting uncorrelated photon pairs, thereby undermining an essential prerequisite for the conclusions drawn from such experiments.

Keywords: Photon, coincidence detection, beam splitter, entangled photons.

1. Introduction — The Issue

The mathematical rendition of a physics theory comprises equations involving various symbols, of which only a few actually pertain to ontological (physical) entities. Many others pertain to concepts or abstractions for which the use and purpose is convenience in formulating the theory, especially its mathematical exposition. Ideally, it might be inferred that, only mathematically invariant symbols correspond to ontological entities --- after all, real, existing objects, logically cannot depend on arbitrary mathematical choices, such as coordinate systems, units, etc. An example of the second type, that is concept-symbol which is contrivance for convenience, is *energy*. The amount of kinetic energy of an entity depends on its velocity so that obviously it is a concept relative to whatever frame in which the velocity is expressed. Basically, kinetic energy, for example, is a bookkeeping concept; it specifies how much work would have to be done on the entity of interest to bring it to a halt in the frame in which its velocity is expressed (and presumably, measured).

Often times a symbol-concept becomes so familiar, that in the minds of some it is taken (mistakenly) as an ontological entity. This can have the deleterious consequence, that an implicit (secondary) goal of theorizing unconsciously becomes clarifying its physical nature, even while in reality it can have no such nature. Herein our central interest is electrodynamics as formulated nowadays in terms Maxwell's Equations for electric and magnetic *fields*. It should be recognized immediately that, fields cannot be ontological entities. First, they are not invariant with respect to an observer's choice of coordinates, indeed, with a suitable choice either one or the other can be reduced to zero. In addition, they are essentially *unobservable*. That is to say, in order to observe a field strength at some point, x, it is necessary to insert a "test charge" at that point so as to observe its reaction to the impugned "field." It is the test charge which is observable, not any field in which it is ostensibly bathed. Further, fields for static situations are conceptually clumsy; are or are they not in motion, do they propagate? Nevertheless, much analysis on the behavior of systems of charged particles is discussed as if fields in fact are ontological entities filling space with some kind of substance that has been "emitted" by source charges and in turn will be "absorbed" by sink ("test") charges, usually in the form of a receiving antenna, eye or other photo sensitive instrument.

This circumstance carries over to the concept of "photon" in which it is imagined that, whatever the substance ultimately constituting oscillating waves is, it is, in addition, confined to a specific, finite volume propagating similarly to a particle. Thus, nowadays photons are categorized often as particles (albeit bosonic) and customarily discussed as if they are virtually micro bullets, i.e., an ontological entity of fundamental significance, and not just an abstraction useful in describing interaction among charged ontological entities, (particles --- fermions). One of the central and most popular arguments in favor of the concept of a particle-like photon is the empirical fact that if beam of electromagnetic radiation (ostensibly a photon stream) is sent through a beam splitter, then at the one-photon intensity level, no (within tolerances) coincidences are observed between its output channels. This, it is said, reflects the fact that photons "cannot be split," which in turn implies that they have some kind of real, ontological particle-like character or existence. This argument, however, overlooks certain subtleties. To begin, at the lowest, i.e., one-photon intensity level, it is imagined that, a photon in transit has something of the character of a finite duration electromagnetic pulse comprising one (at least) signal (in the sense of being a Fourier component) with the frequency being attributed to the photon. Of course, none of this is known for certain; all that is known in fact is that, repeated observations of reputed single photon pulses do not conflict with the following description: The pulses travel through space essentially as customarily imagined wave pulses until they impinge on a "photo-detector" within which they elevate an electron from the valence to conduction band of the detector material. This elevated electron is then drawn off and amplified to give a practically detectable signal, which is counted and attributed with a specific "arrival time" considered to be that corresponding to the arrival time of the photo-electron in the counting circuitry. Again, although it cannot be known for certain, laboratory observations are consistent with the idea that arrival times of photo-electrons are distributed in time stochastically within the duration of their eliciting pulse. It is said that, some electrons "arrive" virtually instantaneously with the arrival of the leading edge of the pulse itself, before even the pulse could have deposited energy sufficient to lift the electron from its valence band environment. So much is conventional nowadays.

This paradigm for detection, however, inadvertently renders any argument against photon splitting mute. Whether or not a photon is split, given that all detections are statistically distributed over a time interval equaling some pulse length, the probability for a coincidence detection at any given instant, as a continuous density, will be essentially vanishing small to the extent that the detection window within which detections are considered "simultaneous," is narrowed. It is the purpose herein, therefore, to closely analyze this situation for the purpose of discerning whether there exists alternate evidence of correlation between the detections in the output channels of a beam splitter instead of just noting that the probability of a coincidence vanishes as the detection window is narrowed, which it always does (given a sufficiently narrow window). If there is evidence of correlation, then there is *ipso facto* evidence that the pulse (or a so-regarded photon in flight) has been split. If evidence of correlation is not found, then a "no-splitting" claim survives intact.

2. The Essential Difference

The essential difference between coincident detections of photo-electrons evoked by split "photons" on the one hand, and two accidentally somewhat overlapping pulses on the other hand, is that, split pulses (if they exist), would have had a common progenitor pulse, and therefore should have identical arrival times of the leading edges at the output detectors. There can be no systematic relationship, on the other hand, for independent but accidentally overlapping pulses. It shall be shown below that this leads to consequences, observable at least in principle.

The essential nature of the difference in the structure of coincidences generated by split and overlapping pulses is that, in the former case all the stochastic character of the photo-electron arrival times is due to processes in the detectors on the output channels of the beams splitter. In the case of simply overlapping pulses there is an additional random input unique to each pulse which is due to the random time off-sets of the overlapping pulses incidental to processes in the source crystal.

Let us first consider the case in which the two pulses are absolutely independent but overlapping sufficiently so that the photo-electrons they evoke are detected within the window (by definition chosen by the experimenter, not Nature) defining a "coincidence." The detection instant in each output channel is a random variable, t_1 and t_2, are in fact the sum of two random processes,"

$$t_{1,2} = T_{1,2} + \tau_{1,2},$$

where the $T_{1,2}$ are the random instants of the leading edge of generating pulse (which depends on processes in the source crystal) and the τ_j are the random instants of photo-electron appearance within the pulse length in the photo detector. What we propose studying is the distribution function of the difference in arrival times, $\delta t = t_1 - t_2$ of the photo electrons in the output channels. δt is again a random variable and its distribution function, $F(\delta t)$, is the integral of the probability density function of the random variable, δt, namely:

$$F_{\delta t}(w) \equiv \int_{-\infty}^{w} \rho(\delta t) d\delta t,$$

where w denotes the "window width'' chosen as the definition of a "coincidence." $F_{\delta t}(w)$, therefore, gives the number of coincidences seen between the output channels of a beam splitter as a function of the selected window width w; for brevity below it is denoted $F_C(w)$.

To obtain this density function, in turn, we need to compute $\rho(\delta t)$, which, being a probability density, would be given by the convolution of the density functions of t_1 and $-t_2$:

$$\rho(\delta t) = \int_{-\infty}^{+\infty} \rho_2(\delta t + t_1)\rho_1(t_1)dt_1.$$

Now, each of the densities under this integral pertains to a random variable which itself is the sum of two other random variables, that is of T_j (determined by emission processes in the source) and τ_j (determined by absorption processes in the detectors), ice

$$\rho_j(t_j) = \int_{-\infty}^{+\infty} \rho_S(t_j - \tau_j)\rho_D(\tau_j)d\tau_j,$$

where $\rho_S(t_l - \tau_l)$ is the probability distribution for $T_j = t_j - \tau_j$, i.e., the instant of arrival of the leading edge of the "pulse'' emitted by the source. Most often it is taken that, the distribution of these instants is uniformly distributed over a pulse length, which itself can be determine or estimated from the average of the frequency, f^{-1}, of photons per unit time of the source:

$$\rho_S(T) = 1; \quad 0 \le T \le f^{-1}, \quad 0, \text{otherwise}$$

correspondingly, $\rho_D(\tau_j)$ is the probability density for the detection process, i.e., the arrival instant of a photo-electron, as found from a straight forward quantum calculation to be given by an exponential decay:

$$\rho_D(\tau_j) = \lambda e^{-\lambda\tau}, \quad \tau \ge 0, \quad 0, \text{otherwise}$$

Now, our central point herein is to compare $F_{\delta t}(w)$ for the two distinct cases: 1) in which the stimulus pulses are just accidentally overlapping ($T_1 \neq T_2$) and 2) when a single pulse is split in two by the beam splitter ($T_1 = T_2$).

So for case 1) the following sequence in "Mathematica" coding can be used to get a distribution function for at least a first approximation for the case in which the "photons" are imagined to be just accidentally overlapping.

Block 1: Mathematica code to compute the probability density function for accidentally overlapping pulses. Lambda, the decay constant, and the pulse length, T, are model dependant parameters. (Code available from author by email).

For the case involving hypothetically "split" photons, the following code gives a distribution function. This code is predicated on a certain physical model; other models would yield possibly very different results.

Block 2: Code to compute (hypothetical) "split photon" coincidence probability distributions. Lambda is a physical model dependent parameter. Alternate model may involve different and additional parameters. (Code available from author by email).

Figure 1 presents the results:

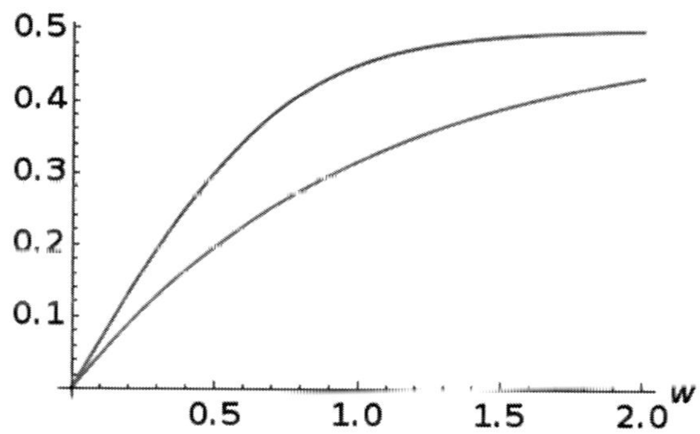

Figure 1. Probability distributions as functions of the window width. The upper curve is for accidentally overlapping pulses, the lower for "split photons."

From this graph it can be deduced that there should be a distinct difference in the curves between these two cases. Conceivably, this difference can be interpreted for specific experiments to evaluate the validity the claim that photons cannot be split.

3. An Experiment

Given the structure described above, one might imagine the following experiment, requiring no more than a reexamination of the data available from conventional beam splitter experiments. Suppose the output from both channels of a beam splitter is available in the form of two streams of time instances corresponding to the arrival times of the photo-electrons in each channel, i.e.: $\{t_i\}_{trans}$ in the transmitted channel, and $\{t_j\}_{refl}$ in the reflected channel. Then, the frequency of the occurrence of a given value of

$$\{t_i\}_{trans} - \{t_j\}_{refl} = \delta_{ij},$$

can be found by scanning in parallel the two streams progressively in increasing magnitude until a recorded data point is encountered, which is then taken as the opening instant of a window for the which the closing instant is the next encountered value in the other channel. By continuing in this manner a population distribution of values of δ_{ij} or δ_{ji} conveniently relabeled as a value of w. Continuing this count procedure will reveal a distribution of undetermined character; that is, it can be anything from a pure case 1 distribution or some mixture with non split photons.

On the other hand, to obtain the same distribution corresponding to the case of coincidences for any given value of w arising from simply overlapping pulses, the same analysis can be done on two streams where one of them is any portion of either stream offset by a fixed amount large enough to be reasonably assured that coincidences cannot result from the split photons. That is, the pulses are so displaced in time that they cannot represent a single photon no matter how defined.

Now, the results of the first analysis can result in a stream containing some percentage of split photons, whereas the second cannot. Thus, the validity of the proposition that "photons cannot be split" is challenged if the results are sufficiently distinct.

The analysis presented herein is strictly abstract with little regard for practical impediments. It may well be that although the effect considered here does exist, that for practically available experimental regimes, i.e. essentially suitable ratios of λ / f^{-1}, the magnitude of the effect is too small to be observable, for example. Additionally, the probability densities attributed to the processes in the source and detector, conceivably could involve features that conceal a real effect. These are issues hopefully to be resolved by experimentalists. In the end, we see no reason that some variation of the structure suggested by the above analysis could well exist in nature and be discernible experimentally.

4. An Application

Beyond the purely conceptual interest in the fundamental nature of the electromagnetic interaction, the consequences of the effect considered here have important significance for many experiments done nowadays to fathom the quantum nature of the electromagnetic interaction.

Many essential experiments in this area employ the coincident emission of two correlated photons, similar in constitution to the outputs of a beam splitter. The sources employed for such experiments are usually such that the individual pairs are generated by individual molecules, or other microscopic structures, constituting the source. The multitude of micro-sources in the macro source (crystal, gas, etc.) can be expected to generate many undesired accidental overlapping micro-signals. However, for the purposes of these experiments, it is essential that the photon pairs be correlated (in fact, from theory this correlation is expected to be a state denoted "entangled," which is thought to be stronger than just statistically correlated) and not just accidentally coincident. To obtain such ideal pairs, the customary technique used to procure these correlated (entangled) pairs is to employ as small a coincidence window as possible. Obviously, the assumption is that, real coincidences will outnumber accidentals to an increasing extent as the definition of a coincidence window, w, is reduced.

However, if the ratio of the density of uncorrelated, or accidental coincidences to that of the truly correlated is graphed; for the ratio arbitrarily employed above, one obtains results like those shown in Figure 2, which illustrates the fact that the relative proportion of truly correlated pairs is, contrary to the conventional and intuitive expectation, *reduced* upon reduction of the window width. This phenomenon is not only counterintuitive, but counterproductive for the purposes of these experiments. It in fact achieves just the opposite of the intended purpose for reducing the window width in the first place.

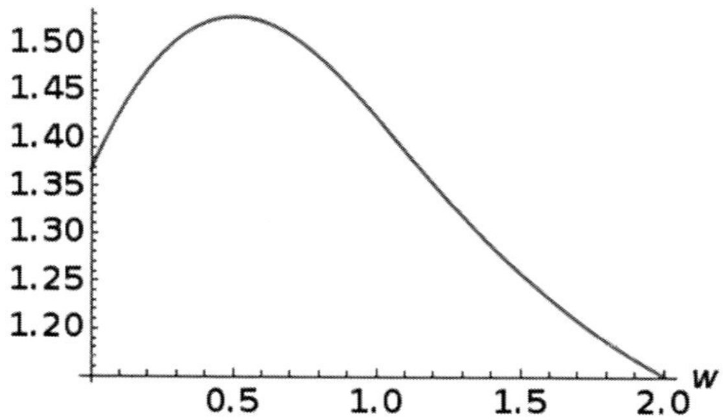

Figure 2. An illustration of the relative abundance of uncorrelated to correlated coincidences for an arbitrary choice of model dependent parameters. Other models lead to a virtual vanishing of correlated pairs as the window width is reduced, contrary to expectations.

For many crucial quantum optical experiments this effect can lead, arguably, to the perverse situation of spoofing features peculiar to quantum theory. That is, by providing perfectly classical structure where it has been precluded by intention --- based on the inappropriate assumption that spurious, uncorrelated pairs have been excluded when actually this effect lowers the relative abundance of correlated pairs --- it leads to a false understanding of just what empirical results actually imply. Perhaps the most fundamental consequence from this analysis is clarification of the exact nature of "entanglement," which currently is explicated with virtually a preternatural rationalization.

The consequences of this effect for the conclusions drawn from certain quantum optics experiments has been analyzed in greater detail elsewhere [1].

References

[1] Kracklauer, A.F. (2011) Proc. SPIE 8121, 1812102.

Unified Geometrodynamics: A Complementarity of Newton's and Einstein's Gravity

RICHARD L. AMOROSO

Noetic Advanced Studies Institute,
Escalante Desert, Beryl, UT 84714-5104 USA
amoroso@noeticadvancedstudies.us, www.noeticadvancedstudies.us

Newton claimed the influence of gravity is instantaneous; Einstein insisted no influence could propagate faster than the speed of light. Recent experiments to test the speed of gravity have been controversial and inconclusive on technical grounds. Considerable effort is currently expended in the search for a Quantum Gravity; but there is no a priori reason there should be one. We propose that is not the regime of integration which instead occurs in the arena of the Unified Field, UF; further that a completed model of Geometrodynamics inherently includes a Newton/Einstein duality which introduces shock effects in certain arenas. The unified theory predicts that there is no graviton of the usual phenomenal form (an artifact of the incompleteness of Gauge Theory, i.e. gauge theory is only an approximation suggesting new physics). A new Large Scale Additional Dimensional (LSXD) M-Theoretic topological charge alternative is presented. We also attempt to show how the Titius-Bode Law for solar and exoplanetary configurations appears to provide indicia of this multiverse gravitational model. Applications of the dual geometrodynamics formulation include an interpretation of quasar luminosity as the result of gravitational shock waves in a manner countering explanations of large redshift, Z in Big Bang cosmology putatively based on Doppler recession. Instead redshift occurs as the result of a periodic minute photon mass anisotropy caused by periodic coupling to a covariant polarized Dirac vacuum.

Keywords: Exoplanets, Geometrodynamics, Newtonian gravity, Quantum gravity, Shock waves, Titius-Bode Law.

1. Introduction

A growing number of conflicts within the Standard Model call into question the fundamental interpretation of the Doppler component of the Hubble Expansion Law, v = H0D recently measured at ~ 74.3 ± 2.1 (km/s)/Mpc and the nature of events in spacetime associated with conventional coordinates of the line element attached to the observer. We postulate that nonlinear effects associated with the propagation of light in a strong gravitational field can produce shock waves revealed at cosmological distances approaching the limit of observation where 'edge effects' compound observation. These gravitational shock waves are manifest observationally in the spectrum of quasi-stellar objects (QSOs) and Supernovae as a continuous array of 'light booms' produced by superluminal boosts or shocks associated with continuous coordinate transformations relative to a distant observer.

This model suggests that QSOs are most likely a form of Seifert spiral galaxies with active galactic nuclei (AGN) in the vicinity of the observational limit of the ~13.796 billion light year (bly) Hubble radius, HR (revised 3-21-2013 by Planck satellite data). The Hubble length, c/H0 is claimed to be 13.9 billion light years in the standard Big Bang cosmological model which is larger than c times the purported 13.796 bly age of the universe because 1/H0 is derived as a backwards in time extrapolation from a Doppler recessional velocity that assumes a constant galactic velocity. But since recent observations suggest quintessence - a time-varying form of dark energy explaining accelerated expansion of the universe, the 1/H0 factor remains indeterminate.

We believe Gravitational shock waves probably would not occur relative to distant QSOs unless gravitation was a duality of Newtonian instantaneous and Einstein's relativistic form. We endeavor to illustrate how the Titius-Bode series law in solar and exoplanetary systems illustrates a component of this duality. The fact that there may not be quantum gravity (not the regime of integration) may be an important factor in the properties of a dual geometrodynamics. Feynman: "...maybe nature is trying to tell us something new here, maybe we should not try to quantize gravity... Is it possible that gravity is not quantized and all the rest of the world is?" [1].

Even Einstein himself maintained that the theory of gravity is far from complete:

> ...the right hand side includes all that cannot be described so far in the Unified Field Theory, of course, not for a fleeting moment, have I had any doubt that such a formulation is just a temporary answer, undertaken to give General Relativity some closed expression. This formulation has been in essence nothing more than the theory of the gravitational field which has been separated in a somewhat artificial manner from the unified field of a yet unknown nature - *Albert Einstein.*

The EPR experiments have solidly demonstrated that a nonlocal arena exists and that some form of hyper- holographic-like instantaneous connectedness occurs between correlated quanta. Our premise is that local 4-space velocities are restricted by relativity and that nonlocal or LSXD space has some form of inherent instantaneity; thus an obvious basis for a geometrodynamical duality.

2. Titius–Bode Conundrums

A pattern that stood steadfast until 1846, was the Titius-Bode rule. This rule noted that the distance of the planets from the sun seemed to follow a pattern described by the equation $a = 0.4 + 0.3 \times 2n$ where n was the planet number in order of distance from the Sun. This pattern held very well for the first 7 planets, so long as one included the asteroid Ceres, or the asteroid belt itself, as planet #5. Yet the discovery of Neptune and Pluto discredited this pattern as a mere coincidence, mathematical happenstance and numerology, as the Titius-Bode rule severely under-predicted their distances [2-.

For more than two centuries it has been known that the distances of planets from the Sun follow a simple and mysterious law. Now, at long last, two French scientists have come up with a plausible explanation. The best way to see the law is to write down the sequence 0, 3, 6, 12 and so on, where each number is obtained by doubling its predecessor. Next, add 4 to each number, and divide the result by 10. Now look up the distances of the planets from the Sun in astronomical units, the Earth-Sun distance being defined as 1. The distances are virtually identical to the terms in the number sequence for all but the outermost planets. This curious 'coincidence' has divided astronomers into two camps ever since its publication in 1766 by the Prussian astronomer Johann Daniel Titius.

Recent Refinements of the Titius-Bode Series Indicate a Possible New Gravitational Dynamic. The Titius-Bode law for planetary orbitals is in an exponential function of planetary sequence out from the sun. The law relates the semi-major axis, a of each planet in units so that the Earth's semi-major axis = 10, with $a = n + 4$ where $n = 0, 3, 6, 12, 24, 48...$ with each value of $n > 3$ twice the previous value. The resulting values can be divided by 10 converting them to astronomical units (AU). The hypothesis was discredited as a predictor of orbits after the 1846 discovery of Neptune and the discovery of Pluto in 1930. When originally published it generally satisfied by all the known planets Mercury through Saturn. Two solar planets have a number of large moons that could have been created by a process similar to that which created the planets themselves. The four large satellites of Jupiter plus the largest inner satellite Amalthea adhere to a regular, but non-Bode, spacing with the four innermost moons in orbital periods that are each twice that of the next inner satellite. The large moons of Uranus have a regular, but non-Bode, spacing.

Results from simulations of planetary formation support the idea that a randomly chosen stable planetary system will likely satisfy a Titius–Bode law. Dubrulle and Graner [36] have shown that power-law distance rules can be a consequence of collapsing-cloud models of planetary systems possessing two symmetries: rotational invariance (the cloud and its contents are axially symmetric) and scale invariance (the cloud and its contents look the same on all length scales), the latter being a feature of many phenomena considered to play a role in planetary formation, such as turbulence. To test if a similar rule applies to extrasolar planetary systems so far only 55 Cancri, a binary star approximately 41 light-years away in the constellation Cancer, has sufficiently known planets to make predictions. An undiscovered planet / asteroid belt is predicted at ~2 AU.

Recent new calculations have shown that the Titius-Bode Law can be accurately demonstrated by the Euler-LaGrange equation for the free energy variations of the plasma initially forming the sun and solar system [37-39]. Using a 1st order Bessel function scaled to the geometry of the solar system, Wells has shown that the Titius-Bode numbers correspond to extrema of the roots and make exact predictions for the outer planets where the Titius-Bode series originally failed [40]. These new insights stem from the seminal work of Chandrasekhar [41] on the equilibrium properties of the boundary conditions of a volume of plasma.

We make the following radical speculations regarding indicia of gravitational theory duality:

- The asteroid belt between Mars and Jupiter is a result of nonlinear gravitational shock or destructive interference in the coupling of the geometric topology of Newton-Einstein duality.
- Jupiter's 'star-like' size represents a symmetry doubling or phase overlap in gravitational effects as a secondary node in a Bessel function star-solar system repeat cycle constructive interference overlap.
- The rings on the trans-Jupiter planets are caused by a constructive interference harmonic of the Newton-Einstein shock duality overlap. This and the prior tenet suggest why the Titius-Bode law's initial incarnation was not 'perfect'.

TITIUS-BODE LAW - SOL				
Planet	k	Mass (ME)	Bode Distanc e	Actual AU Distance
Mercury	0	0.05527	0.4	0.39
Venus	1	0.81500	0.7	0.72
Earth	2	1.0000	1.0	1.00
Mars	4	0.10745	1.6	1.52
(Ceres)	8	0.00016	2.8	2.77
Jupiter	16	317.83	5.2	5.20
Saturn	32	95.159	10.0	9.54
Uranus	64	14.500	19.6	19.2
Neptune	128	17.204	38.8	30.06

Table 1. Titius-Bode Law for planets orbiting Sol [42].

TITIUS-BODE LAW - 55 CANCRI				
Planet	K	Mass (MJ)	Bode Distance	Actual AU Distance
55 Cancri -E	0	>0.034	0.039	0.038
55 Cancri -B	1	>0.824	0.104	0:115
55 Cancri -C	2	>0.169	0.283	0:240
55 Cancri -F	4	>0.144	0.768	0:781
55 Cancri -5	8	-	2.08	undiscovered
55 Cancri -D	16	>3.835	5.643	5:77
55 Cancri -7	32	-	15.3	undiscovered

Table 2. Titius-Bode Law for exoplanets orbiting 55 Cancri [43,44].

This "hints at other phenomena associated with the morphology of the system" [37]; we postulate this might reveal a feedback mechanism between the two modes of operation for gravity that could be responsible for destruction of a planet that should have formed in the asteroid belt. It is sometimes suggested that the gravitational force from Jupiter disrupted the planets formation. Our idea is that this feedback mechanism might arise from a harmonic oscillation between the effects of classical gravitation operating at the speed of light, c and the operation of the as yet undiscovered effect of quantum/unitary gravitation operating instantaneously. This effect if true provides indicia for our model of gravitational shock waves which also have oscillatory parameters.

3. Critique of Hubble's Law Applied to Doppler Expansion

Redshift refers generally to motion of a source relative to an observer; with blueshift for motion toward the observer, $Z < 0$ and redshift for velocity away from the observer, $Z > 0$ for an object not in the line of sight the relativistic form of the Doppler effect is

$$1 + Z = \frac{1 + v\cos(\theta)/c}{\sqrt{1 - v^2/c^2}}. \tag{3}$$

When the Tbl. 2. Titius-Bode Law for exoplanets orbiting 55 Cancri [43,44] motion of the source is in the line of sight, $\theta = 0$ the equation reduces to the general formula

$$1 + Z = \sqrt{\frac{1 + v/c}{1 - v/c}} \tag{4}$$

where one can tabulate Z:

V	Z
.5c	.73
~.6c	1
.75c	1.64
.8c	2.00
.85c	2.51
.95c	5.24
.96c	~6
.99c	13.11

Table 3. Tabulation of Z compared to velocity approaching *c*.

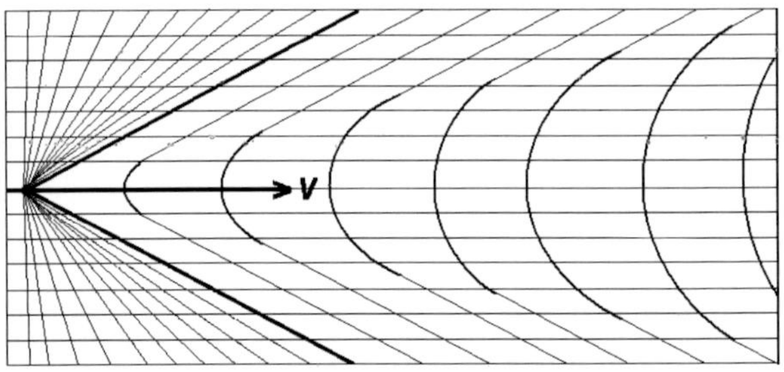

Figure 1. Wave front of a Doppler redshift for Z = ~ .85 c.

The largest Z currently known is for the redshift of galaxy UDFy-38135539 [62] at a redshift of $z = 8.6$, M.D.Lehnert, et al.; Nesvadba, NP; Cuby, JG; Swinbank, AM; Morris, S; Clément, B; Evans, CJ; Bremer, MN et al. (2010). Spectroscopic Confirmation of a galaxy at redshift z = 8.6, Nature 467 (7318): 940–942. arXiv:1010.4312.

The most distant QSO CFHQS J2329-0301 with Z $\simeq$ 6.43 [45]. A QSO with Z > 10 has been observed and a galaxy in 2012 with Z 12.8. Hubble's redshift law is considered quite variable; and interpretation depends on a number of factors like the specific cosmological model utilized or if Λ is 0, + or -. The best indirect evidence supporting our thesis is that QSO's are the most luminous objects in the known universe and that an object, especially one as massive as a QSO is supposed to be, receding at ~c would indicate ~ infinite mass.

4. The Observer and the Cosmological Principle

In summarizing the Cosmological Principle (that the universe is homogeneous and isotropic on average in the large-scale) [46] events are idealized instants in spacetime defined by arbitrary time and position coordinates t, x, y, z, written collectively as x^i where i runs from 0 to 3. The standard line element is

$$ds^2 = \sum_{ij} g_{ij} dx^i dx^j = g_{ij} dx^i dx^j, \tag{5}$$

where the metric tensor

$$g_{ij}(x) = g_{ji}(x) \tag{6}$$

is symmetric [46]. In local Minkowski form all first derivatives of g_{ij} vanish at the event and equation (5) takes the form

$$ds^2 = dt^2 - dx^2 - dy^2 - dz^2. \tag{7}$$

The Cosmological Principle generally suggests that the clocks of all observers are synchronized throughout all space because of the inherent homogeneity and isotropy. Because of this synchronization of clocks for the same world time t, for commoving observers the line element in (7) becomes

$$ds^2 = dt^2 + g_{\alpha\beta} dx^\alpha dx^\beta = dt^2 - dl^2, \tag{8}$$

where dl^2 represents spatial separation of events at the same world time, t. This spatial component of the event dl^2 can be represented as an Einstein 3-sphere (compatible with the dual 6D Calabi-Yau 3-torus)

$$dl^2 = dx^2 + dy^2 + dz^2 + dw^2 \tag{9}$$

which is represented by the set of points (x, y, z, w) at a fixed distance R from the origin:

$$R^2 = x^2 + y^2 + z^2 + w^2 \tag{10}$$

where

$$w^2 = R^2 - r^2 \text{ and } r^2 = x^2 + y^2 + z^2 \tag{11}$$

so finally we may write the line element of the Einstein 3-sphere from equation (9) as

$$dl^2 = dx^2 + dy^2 + dz^2 + \frac{r^2 dr^2}{R^2 - r^2}. \text{ [46]} \tag{12}$$

By imbedding an Einstein 3-sphere in a flat HD space, specifically as a subspace of a new complex 12D superspace, [34,47,48] new theoretical interpretations of standard cosmological principles are feasible. This is the line element we feel is most compatible with the oscillatory spacetime boundary parameters required by our model of gravitational shock waves in QSO luminosity.

5. QSO Redshift-Distance Controversy

As optical and radio telescopes continue to improve a vast amount of data continues to be accumulated on the large-scale structure of the universe. The popular view has been to interpret the data to support a hot Big Bang cosmological model; but as attested to in this chapter QSO's provide strong observational evidence that the Big Bang assumption is incorrect. From the early 1960's when the redshift of QSOs, and galaxies were compared with radio sources it became apparent that the redshift plot of QSOs contrasted with apparent brightness did not follow the usual Hubble correlation [1]. These redshift observations beginning with QSO 3C 273 in 1963 to more than 100 QSOs in 1963 still continued to show the same redshift apparent magnitude disparity when the number of sources was increased beyond 7,000 QSOs in the mid 1990's [2]. Most astrophysicists were not willing to accept that these redshift observations were not a measure of distance. Large redshift QSOs are not faint and typically have bolometric luminosities of ~ 100 times that of normal galaxies [1]. Woltjer [3] and Rees [4] found a way to interpret the QSO redshift as being wholly cosmological phenomena by considering the radiating surfaces as having relativistic motion [1].

Around the same time Arp comprised a catalog of unusual galaxies [5]. He noticed a physical association between radio sources, QSOs and some of the peculiar galaxies. But the observed redshift of the central galaxies was small and the redshift of the associated QSO very large suggesting that the QSO redshift could not be of cosmological origin. Arp had clearly shown with a high level of statistical accuracy that 'there was a clear association between radio QSOs with large redshifts and galaxies with very small redshifts' [1]. The linear separation between galaxy and QSO was generally the same demonstrating a clear association between the galaxy and the QSO [6-8].

His work was greeted with astonishment and disbelief...and heavily criticized, often very unfairly. In response he began an extensive observational program...The community has remained skeptical of these results...one argument made against the reality of these associations by a leading observer was that if these results were correct, we had no explanation of the nature of the redshift! In other words, if no known theory is able to explain the observations, it is the observations that must be in error! [1]

Arp's colleagues at Mt. Wilson and Palomar were so troubled by his results that they petitioned the observatories directors to take away all of Arp's observing time. Arp protested when the recommendation was implemented and after his appeals to the trustees were turned down he retired and relocated to the Max Planck Institute in Munich [1,9].

6. QSOs Issues as a Hint for Fundamental Basis Geometrodynamics

Newton's formulation of the gravitational force law requires each particle to respond instantaneously to every other massive particle regardless of the distance between them which he proved; but the proof is only valid in Euclidian space. Today this would be described by the Poisson equation,

$$\left(\frac{\partial^2}{\partial x^2}+\frac{\partial^2}{\partial y^2}+\frac{\partial^2}{\partial z^2}\right)\varphi(x,y,z)=f(x,y,z) \quad (13)$$

according to which, when the mass distribution of a system changes, its gravitational field instantaneously adjusts. Therefore the theory requires the speed of gravity to be infinite. Einstein's Geometrodynamics

$$G_{\mu\nu}+\Lambda g_{\mu\nu}=\frac{8\pi G}{c^4}T_{\mu\nu} \quad (14)$$

is a classical extension of Newtonian gravitation and therefore an incomplete theory. Physical theory incorporates an upper limit on the propagation speed of an interaction, maintaining that instantaneous action-at-a-distance is impossible. However quantum entanglement between separated particles enables instantaneous correlations which led to the puzzle as to whether causality or locality must be abandoned.

The recent measurement controversy of the speed in which Gravity propagates has only addressed this semi-classical component [10-28]. The search for a Quantum Gravity (QG) is misplaced by the incorrect assumption that gravitation is quantized [29]; this is not so. The interaction between gravity and Quantum Theory (QT) occurs at the level of unitarity, not within an intermediate G-QT regime. This is because of the nature of the graviton, a quadrupole photon-graviton complex not the usual spin 2 Bose quantum-graviton with any associated properties. See Chap. 5. It is instead a condition of brane topology (according to the theory presented in this volume). We hope this chapter creates some insight into solving the conceptual; basis of this puzzle. Still in either case gravity has properties beyond the local, $v_g \simeq c$ velocity propagation. An additional instantaneous correlation as in the EPR experiment, but in unitarity where this action occurs is needed to describe gravitation. I suppose EPR in that sense provides good indicia of the incompleteness of QT. This duality of the laws of gravity are indicated in a variety of astrophysical effects such as the Titius-Bode series or QSO luminosity as addressed here. Before going into that, the aim of this chapter, a discussion of the Titus-Bode relation suggests an associated relationship to the missing components of gravitation.

Astrophysicist Silk stated, "...highly redshifted sources, most notably the radio galaxies and the quasars, reveal strong evolutionary effects. Equal volumes of space contain progressively more quasars and powerful radio galaxies at greater distances. Only by disputing the inter-pretation of quasar redshifts as a cosmological distance indicator can this conclusion be avoided" [30].

Taking an axiomatic approach we begin with a number of postulates:

- That the Hubble redshift, H_0 is non-Doppler (no cosmological expansion or inflation).
- Redshift is due instead to photon mass, $m_\gamma \neq 0$ anisotropy, for $m_\gamma = h\nu / c^2$ with internal motion coupling periodically to the Dirac covariant polarized vacuum [31].
- Quasars (QSOs) are most likely a form of Seifert spiral galaxy with Active Galactic Nuclei (AGN) near the limit of observation at the Hubble Radius, HR [32].
- The spectra of QSOs, the most luminous objects in the universe, can be explained in terms of Gravitational Shock Waves (GSW).
- Spacetime is asymptotically flat [33].
- The Cosmological Principle (CP) holds within reasonable limits.
- Expansion/Inflation of the universe is an observational illusion of misinterpreting of the Hubble redshift as a Doppler effect [34].
- This illusion arises from the continuous-state dimensional reduction properties of the present instant as a virtual subspace of an HD atemporal domain [35].

7. Parameters of Shock Waves

In general a shock wave is defined as an abrupt, discontinuous, nonlinear change in the characteristics of a medium that travels at a velocity higher than an ordinary wave often through a vortex fanning out from the source of the shock. Shock energy dissipates in a short distance and the accompanying expansion wave merges with the shock wave, partially canceling it. So our putative gravitational light-boom results from the degradation and merging of the shock wave and the expansion wave produced by the oscillating spacetime boundary conditions. To get a shock wave something has to be traveling faster than the local wave speed. In this regard some segments of the light around the vortex fan are traveling at the normal speed of light, so that the waves leaving the QSO pile up on each other and a shock wave, the pressure increases, and spreads out sideways. Because of this 'constructive interference' effect, shocks are intense like an explosion.

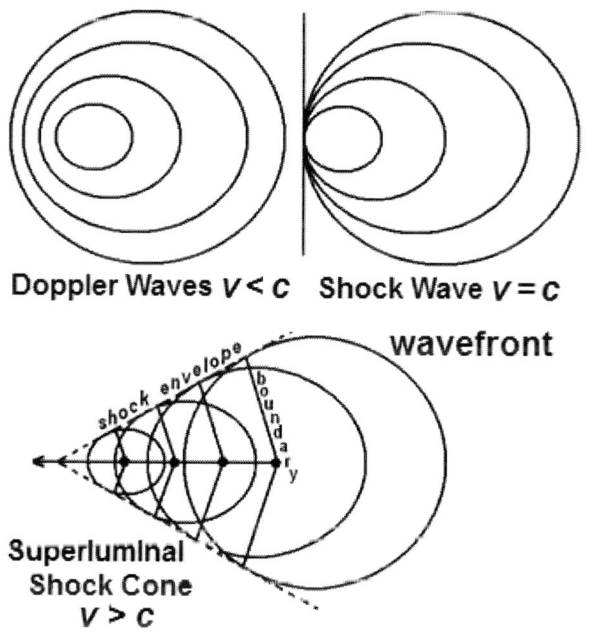

Figure 2. a) Depiction of Doppler waves, b) luminal Shock Waves and c) superluminal shock wave.

Shock/vortex interactions and superluminal vortex breakdown occur when a superluminal vortex stream encounters a shock wave, the discontinuous pressure rise of the shock wave can be sufficient to burst the vortex with an oscillation of light booms depending upon the structure of the vortex [49-55]. The structure of shock/vortex interactions has been investigated in a series of Soviet studies using various flow visualization methods [56-58]. These studies show that shock/vortex interactions result in highly unsteady flow patterns in which the shock wave bulges forward in the upstream direction showing a decrease to minimum value on the vortex axis. The cause of this action - the Ranque-Hilsch effect is currently unknown. But Crocco's theorem [59] (Eq. 13) suggests a steady flow gradient's total enthalpy relates to entropy gradients and vorticity, both of which are present in a vortex core. For an over expanded nozzle flow a strong interaction is distinguished from a weak interaction by the formation of a secondary recompression shock downstream of the bubble shock suggesting that the strong interaction corresponds to supersonic vortex breakdown. Finally Delery et al show that the strength of a shock required to burst a supersonic stream-wise vortex is inversely related to the vortex strength [51].

Simplistically considering a shock as originating from a point source in two dimensions the disturbance forms circular wavefronts centered at successive positions of the QSO's harmonic gravitational image source as illustrated in the Fig. Xc. The wavefronts overlap and form the shock envelope. In 2D the shock envelope is a wedge, and in 3D it forms a cone. In aircraft nomenclature the semi vertical angle of the cone is the Mach angle $\mu M = \arcsin(1/M)$, where M is the ratio of source speed to sound speed and is called the Mach number. All sound is contained in the shock envelope where for the first approximation the envelope is the location of the sonic boom [60]. Analogous phenomena exist in disciplines besides fluid mechanics. In nuclear physics particles accelerated beyond the speed of light in a refractive medium create a visible phenomenon known as Čerenkov radiation emitted when a charged particle like an electron passes through an insulator at a speed greater than the speed of light in that medium. The characteristic 'blue glow' of nuclear reactors is Čerenkov radiation [61]. We suggest the possibility of some sort of redshift anomaly related to QSO gravitational shock.

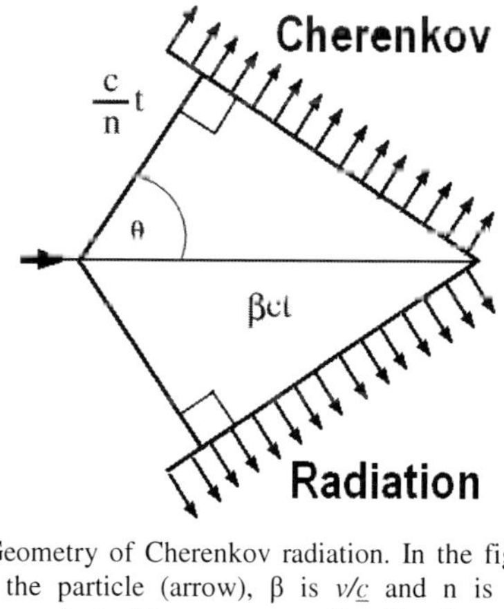

Figure 3. Geometry of Cherenkov radiation. In the figure, v is the velocity of the particle (arrow), β is v/c and n is the refractive index of the medium. The arrows are the direction of Cherenkov radiation so that $\cos\theta = 1/n\beta$.

In line with our postulate that all shock phenomena have similar characteristics Lyman & Morgenstern [52] have garnered three geometric insights into aircraft shock suppression that could also shed light on spacetime characteristics of gravitational shock waves: 1) A relation between lift force and airfoil area, 2) A volumetric shock cancellation phenomena that could give indicia to our postulate of constructive and destructive interference in the gravitational wave light cone, and 3) A directionality control by non-planar shaping that reduces centerline off-track signals.

8. New Cosmological Concepts Related to Gravitational Shock

The nature of the universe has remained an open question. Kant attempted to solve the debate between Newton and Leibniz concerning whether the universe was open or closed by suggesting the antinomy [62] that the universe is both open and closed, i.e closed and finite in the semi-classical limit within the observed temporal boundaries of the observed Hubble radius, HR; and open and infinite into a HD atemporal holographic multiverse domain beyond, HR. Our model is cast in such a Multiverse with a potential for an infinite number of nested Hubble Spheres in causal separation each with their own fine-tuned laws of physics. This is pertinent here in passing because the cosmology in balancing the cosmological constant, Λ gives a backcloth that predicts asymptotically flat spacetime and an interpretation for dark energy as arising from the rest of the multiverse.

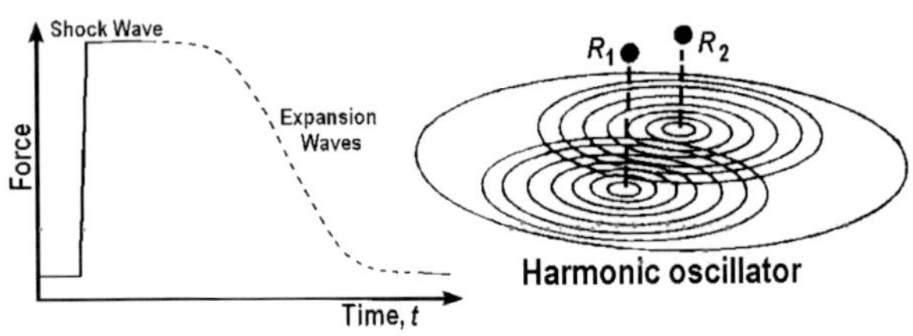

Figure 4. a) General shape of a simple shock wave. b) Harmonic oscillator effect. If superluminal shocks oscillate harmonically because of the putative dual nature of geometrodynamics they may undergo constructive and destructive interference creating 'light booms' as in Fig. 6.

The Einstein gravitational potential oscillates the tidal gravitational field associated with the curvature of spacetime and predicts gravitational waves that propagate with a velocity of $v \simeq c$. We postulate a new cosmological principle related to the action of gravitational wave shock fronts [63]. This action arise form a duality in the nature of gravity, whereas classical general relativity propagates according to $v \simeq c$ the eventual discovery of the completed form of quantum/unitary gravitation will show an additional quantum component with similarities to the EPR condition with instantaneous nonlocal synchronicity. Our postulate here is that this action at cosmological distances and for massive objects such as AGN QSOs creates a spacetime oscillatory shock fronts in the line of sight gravitational light cones leading to an apparent 'light boom' in QSO luminosity.

9. Hypersonic Shock Wave Parallels

We proceed for preliminary delineation under the assumption that the equations of state for hypersonic shock waves apply generally to any compressible media with shocks such as sonic booms or gravitational shock waves on the Dirac superfluid of spacetime especially those of secondary shock waves [60] that we postulate could be extended to support our theory that QSO luminosity can be explained by gravitational shock waves arising from an oscillatory interference of boundary conditions in propagation between of the dual modes of gravity, i.e. classical and quantum-unitary.

From Crocco's equation for smooth flow in an ideal gas [49,59]

$$\Delta h_0 = -\frac{\partial \mu}{\partial t} + T\Delta S + \vec{\mu} \times \vec{\omega} \tag{13}$$

then following Kaouri [60] we develop a circulation theorem for a flow with shocks to eventually apply to the oscillation of boundary conditions for gravitational duality because the circulation theorem can be applied to parallel and perpendicular vorticity. For the closed curve,

$$C = \sum_{i=1}^{n} C_i \tag{14}$$

in Fig. 10.4 where C_i is the domain from P_i to $P_i + 1$ and the circulation around C is

$$\Gamma = \sum_{i=1}^{n} \int_{C_i} u \cdot dx = \sum_{i=\Gamma}^{n} \Gamma_i \, . \tag{15}$$

To construct the circulation theorem one needs to evaluate

$$\frac{d\Gamma}{dt} = \sum_{i=1}^{n} \frac{d}{dt}\left(\int_{C_i} u \cdot dx\right) = \sum_{i=1}^{n} \frac{d\Gamma_i}{dt} \, . \tag{16}$$

For each *Ci* the expression

$$\frac{d\Gamma_i}{dt} = \int_{C_i} u \cdot dx$$

needs to be evaluated. Applying Crocco's equation (13) and summing all the Ci contributions we arrive at

$$\frac{d\Gamma}{dt} = \sum_i \int_{P_i^+}^{P_{i+1}^-} TdS + \sum_i [H]_i + \sum_i \left(u\left(P_{i+1}^-\right) \cdot \dot{P}_{i+1} - u\left(P_i^+\right) \cdot \dot{P}_i \right) \tag{17}$$

where the 2nd term on the right is the sum of $\left[H\right]_i$, the total jump at the ith shock [60].

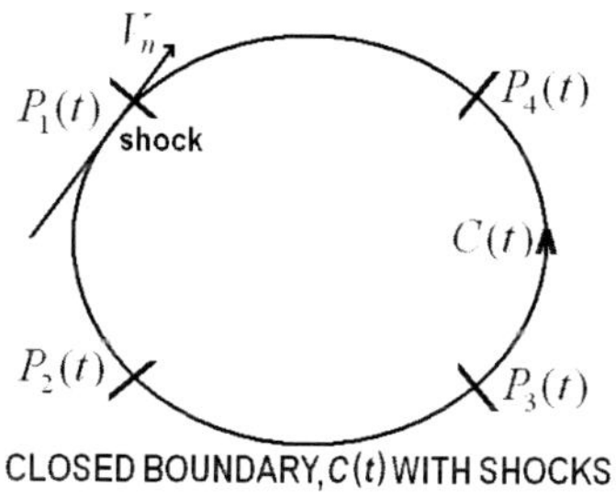

Figure 5. A closed curve boundary, C(t) impinged arbitrarily by 4 gravitational shocks at positions Pi(t) here with $1 \le i \le 4$ creating an entropic jump or constructive interference summation of gravity shock waves. Figure redrawn from [60].

Recent new work by Kaouri [60] on the dynamics of secondary sonic boom shock waves appears to provide insight into our idea of the dual nature of gravitational wave propagation.

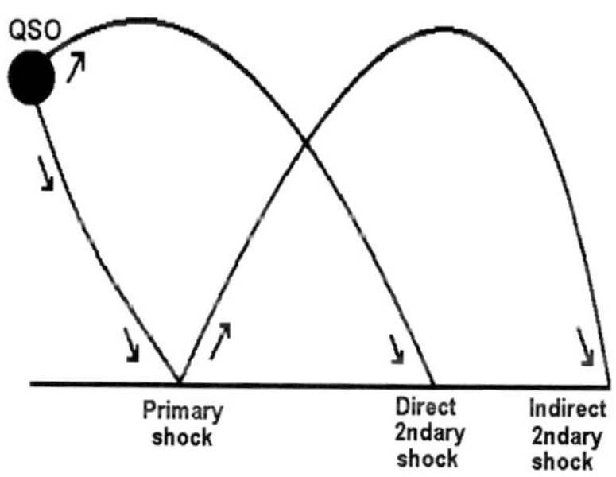

Figure 6. A conceptualization of a direct and indirect secondary boom. Figure redrawn from [60].

If Figs. 10.4 and 10.5 were combined one might end up with a conceptual view like that diagrammed in Fig. 10.6 but with a QSO at the center. If the physical case for a QSO contained a 2nd set of cusps in the bottom quadrant, the harmonic (Fig, 10.8) constructive and destructive interference of gravitational pressure waves could be a factor in producing 'light booms' in a manner dynamically similar to those producing Cerenkov radiation [61].

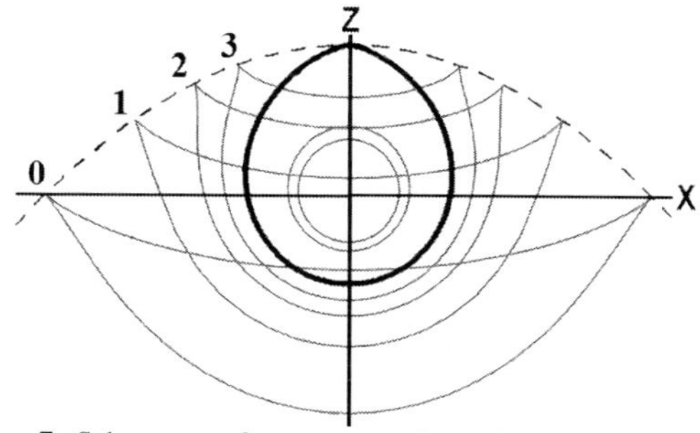

Figure 7. Select wave-fronts, a caustic or locus of cusps, when $Z = 1 - x^2/4$ is plotted on the dashed line of a wave envelope. Figure adapted from [60].

The nonlinear nature of compressed fluid flow is the primary element of shock formation. If we consider a sinusoidinal gravitational influence of sufficient intensity where the curvature fluctuation across the wave is propagating adiabatically a disparity occurs in the velocity, *c* of propagation of light. The 'compressed' portion of the 'wave' will steepen to form a 'vertical' pressure front or shock as in Fig. 1b. The shock wave propagates because of a 'shift' in momentum transfer among flow regions of variable velocity. Shocks, being waves, only form in hyperbolic flow. The characteristic lines of flow are linear and merge into an envelope creating the shocks. Also as generally known parametric conditions can create triple shocks.

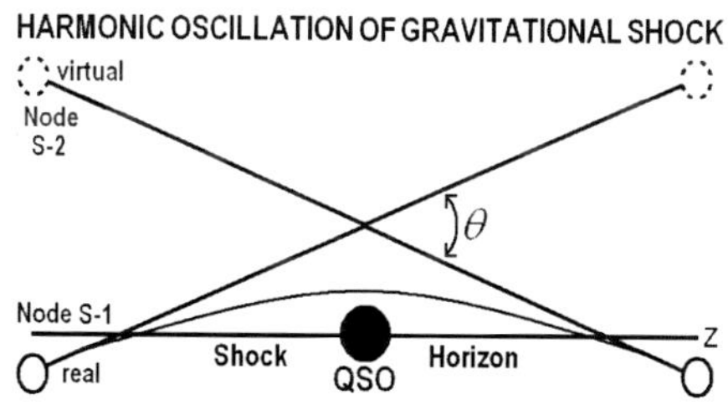

Figure 8. Schema of the primary postulate for a gravitational 'light boom' with harmonic shock cascades.

If the putative dual nature of gravitational propagation is physically real, than at cosmological distances for narrow axis large masses such as AND QSO's the coupling and uncoupling of the two principles could lead to a harmonic oscillation of the boundary conditions of the gravitational horizon such that a constructive/destructive interference occurs where at the summation nodes 'light booms' occur.

The nonlinear x component of gravitational shock nodes summating at collective shock fronts along RE-Q Adv + RE-Q Ret as seen by an Earth observer is shown in Fig. +9. The actual 'light boom' is the harmonic summation of shock nodes from the x, y & z axes and nonlinear assets arising from the interaction of the dual coupling of gravitational wave propagation.

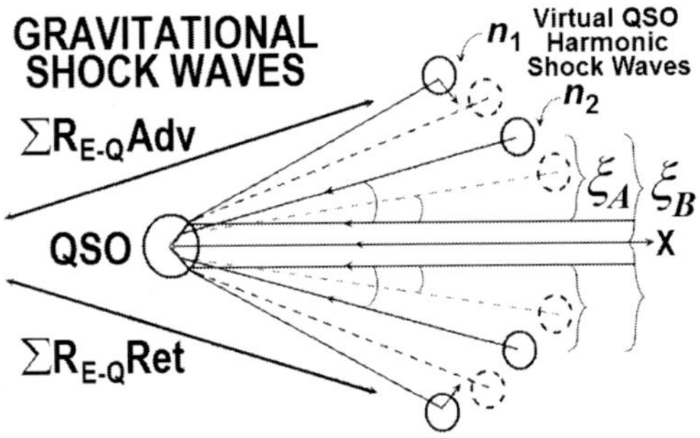

Figure 9. Elaboration of one quadrant of Fig. 8 showing a more detailed association of the oscillatory nature of the boundary conditions.

10. Gravitational Shock Waves

According to Misner, Thorne & Wheeler junction conditions may act as generators of gravitational shocks. They suggest that the dynamics of spacetime geometry for a 3-surface, Σ which includes the intrinsic Riemann scalar curvature invariant, R for example, also includes an extrinsic curvature tensor, K_{ij}. When imbedded in an enveloping 4-geometry hypersurface it can be applied to the change (shrinkage and deformation) in the vector, n parallel transported as junction conditions applicable to the gravitational field (spacetime curvature) and the stress-energy generating it. A discontinuity in K_{ij} across a null surface without stress-energy producing it is a geometric manifestation of a gravitational shock-wave generated by a different embedding in spacetime 'above' Σ than 'below' Σ [64].

Dray and 't Hooft [65] developed the fundamental conditions for introducing a gravitational shock wave in a particular class of vacuum solutions to Einstein's field equations by way of a coordinate shift. They outlined a model for generalizing gravitational shock waves for a massless particle moving in flat Minkowski space [66] formulated as two Schwarzschild black holes of equal masses glued together at the horizon. For a spherical shell of unequal masses moving along $u = u_0 \neq 0$ their solution [67] represents two Schwarzschild black holes glued together at $u = u_0$. By infinitely boosting the Dray-'t Hooft solutions various forms of gravitational shock waves have been found [68-73]. Sfetsos [74] extends these results to the case with matter fields and a non-vanishing cosmological constant. Using the d-dimensional spacetime metric

$$ds^2 = 2A(u,v)\,du\,dv + g(u,v)\,h_{ij}(x)\,dx^i dx^j \quad (18)$$

with $(i, j = 1,2,...,d-2)$ he considers a string based dilatonic black hole gravitational solution [75,76] from the perspective of a conformal background field theory of coset $SL(2,\mathbb{R})/\mathbb{R}\otimes\mathbb{R}^2$ to achieve a differential shift factor

$$\left(\frac{d^2}{d\rho^2} + \frac{1}{\rho}\frac{d}{d\rho} - \varepsilon\right) f(\rho) = -16\varepsilon\rho\frac{1}{p}\delta(\rho) \quad (19)$$

where $\rho^2 = x^2 + y^2$ and for the black hole singularity case with $\varepsilon = 1$ Eq. (19) is a modified Bessel equation. When $\varepsilon = -1$ Eq. (19) is interpreted as an expanding universe [74].

Spitkovsky [77] has developed a simulation for a relativistic Fermi emission shock process that could provide an alternative to or component process for our gravitational shock work. His simulations on relativistic collisionless shocks propagating in initially unmagnetized electron-positron pair plasmas showed natural production of accelerated particles as part of a shock evolution. He studied the mechanism that populates the suprathermal tail for particles gaining the most energy. The simulation showed the main acceleration occurs near the shock where for each reflection these particles gain energy, $\Delta E \sim E$ as is expected in relativistic shocks [78-80].

11. Summary

Newton's theory of gravitation required instantaneous action at a distance or the conservation of angular momentum would be violated. According to Einstein's theory of general relativity an instantaneous influence would violate causality and the special theory of relativity and so must be mediated by a field. This is the dual nature of gravity that we have put as the central basis for our model.

Shock phenomena remain a relatively little explored area of science both within physics and transdisciplinary. We have tried to show that it is possible with further study to relate shock phenomena to gravitational waves especially for narrow axis massive cosmological objects such as AGN QSOs that readily lend themselves to 'light-boom' shock effects that could therefore be used to explain QSO luminosity as further evidence of the insurmountable shortcomings of Big Bang cosmology with observed Zs greater than the putative Big Bang age of the universe.

Our model would appear to work best by contrasting both modes of the dual nature of gravity because a nonlinear jump in flow occurs with a discontinuity. From the 2nd Law of Thermodynamics entropy can only increase when a particle crosses a shock (or catastrophe). The duality of the propagation of the gravitational influence is evident in the idea of Birkhoff's theorem [81] in that a spherically symmetric gravitational field is produced by a massive object such as a QSO at the origin; if there were another concentration of mass-energy somewhere else, this would disturb the spherical symmetry. This effect could occur if interference occurs between the usual modes of the gravitational influence by shock parameters. Also the Titius-Bode relationship seems to indicate shock or gravitational interference zones in planetary formation.

More work needs to be done developing this model. We have only outlined what we perceive as an intellectually appealing avenue. At the close of writing we found an interesting 2009 article by Crawford suggesting new supernova data consistent with a static universe [82]. Also several more high redshift QSO's have been discovered that seem to support our shock theory for QSO luminosity [83-86].

Appendix: Saturn Hurricane

As we go to press 28 April 2013 NASA CASSINI satellite has discovered a huge hurricane at the North pole of Saturn with 300 MPH winds and a 1200 mile central eye.. All that is interesting; but what fascinates is that the hurricane is encased in a 'perfect hexagon' with the diameter the size of two Earths.! http://go.nasa.gov/17tmHzo. We suspect an as yet unknown em-gravity of Unified Field link!

We close with these words of Eddington:

> I am standing on the threshold about to enter a room. It is a complicated business. In the first place I must shove against an atmosphere pressing with a force of fourteen pounds on every square inch of my body. I must be sure to land on a plank traveling at twenty miles a second around the sun a fraction of a second too early or too late, and the plank would be miles away. I must do this while hanging from a round planet, head outward into space and with a wind of aether blowing no one knows how many miles a second through every interstice of my body. The plank has no solidity of substance. To step on it is like stepping on a swarm of flies. Shall I not slip through? No, if I make the venture one of the flies hits me and gives a boost up again; I fall again and am knocked upward by another fly; and so on. I may hope that the net result will be that I remain about steady; but if unfortunately I should slip through the floor or be boosted too violently up to the ceiling, the occurrence would not be a violation of the laws of Nature, but a rare coincidence. These are some of the minor difficulties. I ought really to look at the problem four-dimensionally as concerning the intersection of my world-line with that of the plank. Then again it is necessary to determine in which direction the entropy of the world is increasing in order to make sure that my passage over the threshold is an entrance, not an exit. Verily, it is easier for a camel to pass through the eye of a needle than for a scientist to pass through a door. -- Sir Arthur Eddington, The Nature of the Physical World, p. 192.

References and Notes

[1] Feynman, R.P., Morinigo, F.B., & Wagner, W.G. (1995). Feynman lectures on gravitation, Hatfield, B. (ed.) Reading: Addison-Wesley.

[2] Hewitt, A. & Burbidge, G. (1993) A revised and updated catalog of quasi-stellar objects, Astrophys. J. Suppl. Ser., 87: 451-947.

[3] Woltjer, I. (1966) Inverse Compton radiation in quasi-stellar objects, Astrophysical Journal, 146: 597.

[4] Rees, M.J. Studies in radio source structure I, A relativistically expanding model for variable quasi-stellar radio sources, MNRAS 135, 345 (1967).

[5] Arp, H. (1966) Atlas of Peculiar Galaxies, Long Beach: CA Institute of Technology.

[6] Burbidge, G., O'Dell, S.L. & Strittmatter, P.A. (1972) Astrophys. J., 175, 601.

[7] Bolton, J. & Wall, J. (1970) Aust. J. Physics, 23, 789.

[8] Bahcall, J.N., McKee, C.F. & Bahcall, N. (1971) Astrophysics Letters, 10, 147.

[9] Arp, H. (1987) Quasars Redshifts and Controversies, Berkeley: Interstellar Media.

[10] Carlip, S. (2000) Aberration and the speed of gravity, Phys. Let. A 267: 81–87. arXiv: GR-qc/9909087.

[11] Kopeikin, S.M. (2001) Testing relativistic effect of propagation of gravity by very-long baseline interferometry, Astrophysics J. 556: L1–L6; or arXiv:gr-qc/0105060.

[12] Asada, H. (2002) The light-cone effect on the Shapiro time delay, Astrophys. J. 574: L69; or arXiv:astro-ph/0206266.

[13] Van Flandern, T.; & Vigier, J.P. (2002) Experimental repeal of the speed limit for gravitational, electrodynamic, and quantum field interactions, Found. Phys. 32: 1031–1068.

[14] Will, C.M. (2003) Propagation speed of gravity and the relativistic time delay, Astrophys. J. 590: 683–690 or arXiv:astro-ph/0301145.

[15] Fomalont, E. B., & Kopeikin, S.M. (2003) The measurement of the light deflection from Jupiter: Experimental results, Astrophysics. J. 598: 704–711; or arXiv:astro-ph/0302294.

[16] Kopeikin, S.M. (2003) The measurement of the light deflection from Jupiter: Theoretical interpretation, arXiv: astro-ph/0302462.

[17] Kopeikin, S.M. (2003) The post-Newtonian treatment of the VLBI experiment on September 8, 2002. Phys. Lett. A 312: 147–157; or arXiv:gr-qc/0212121.

[18] Faber, J.A. (2003) The speed of gravity has not been measured from time delays, arXiv: astro-ph/0303346.

[19] Samuel, S. (2003) On the speed of gravity and the v/c corrections to the Shapiro time delay, Phys. Rev. Lett. 90: 231101; or arXiv:Astro-ph/0304006.

[20] Asada, H. (2003) Comments on 'measuring the gravity speed by VLBI', arXiv: astro-ph/0308343.
[21] Samuel, S. (2004) On the Speed of Gravity and the Jupiter/Quasar Measurement, Int. J. Mod. Phys. D 13: 1753–1770; or arxiv:astro-ph/0412401.
[22] Shapiro, E.I. (1964) Fourth test of General Relativity, Phys Rev Let. 13; 789-791.
[23] Pascual-Sánchez, J-F (2004) Speed of gravity and gravitomagnetism, Int. J. Mod. Phys. D 13: 2345-2350; or arXiv:gr-qc/0405123.
[24] Carlip, S. (2004) Model-Dependence of Shapiro Time Delay and the Speed of Gravity/Speed of Light Controversy, Class. Quant. Grav. 21: 3803-3812 or arXiv:gr-qc/0403060.
[25] Kopeikin, S.M. (2004) The speed of gravity in general relativity and theoretical interpretation of the jovian deflection experiment, Classical and Quantum Gravity, 21: 3251–3286; or arXiv:gr-qc/0310059.
[26] Kopeikin, S. & Fomalont, E. (2006) On the speed of gravity and relativistic v/c corrections to the Shapiro time delay, Physics Letters A 355: 163–166.. arXiv:gr-qc/0310065.
[27] Kopeikin, S. (2006). Gravitomagnetism and the speed of gravity, Int. J. Mod. Phys. D 15: 305–320; or arXiv:gr-qc/0507001.
[28] Kopeikin, S. & Fomalont, E. (2007) Gravimagnetism, causality, and aberration of gravity in the gravitational light-ray deflection experiments, General Relativity and Gravitation 39: 1583–1624; or arXiv:gr-qc/0510077.
[29] Feynman, R.P., Morinigo, F.B. & Wagner, W.G. (2003) Feynman Lectures on Gravitation, Boulder: Westview Press.
[30] Silk, J.(1989) The Big Bang, New York: W.H. Freeman & Co.
[31] Amoroso, R.L., Kafatos, M. & Ecimovic, P. (1998) The origin of cosmic microwave background radiation in the intrinsic fluctuation of vacuum compactification cavity quantum electrodynamics, In Causality and Locality in Modern Physics, G. Hunter, & S. Jeffers, (eds.) Dordrecht: Kluwer.
[32] Kembhavi, A.K. & Narlikar, J.V. (1999) Quasars and Active Galactic Nuclei, Cambridge: Cambridge Univ. Press.
[33] Hawking, S. W. & Ellis, G. F. R. (1973) The Large Scale Structure of Space-Time, Cambridge: Cambridge University Press.
[34] Amoroso, R.L. (2002) Developing the cosmology of a continuous state universe, in R.L. Amoroso, G. Hunter, M. Kafatos & J-P Vigier (eds.) Gravitation & Cosmology: From the Hubble Radius to the Planck Scale, Dordrecht: Kluwer Academic.
[35] Amoroso, R.L. (2004) The Fundamental Limit and Origin of Complexity in Biological Systems: A New Model for the Origin of Life, in D.M. Dubois (ed.) Computing Anticipatory Systems, AIP Conf. Proceedings Vol. 718, pp. 144-159, Melville: American Inst. of Physics.
[36] Graner, F. & Dubrulle, B. (1994) Titius-Bode laws in the solar system, Part I: Scale invariance explains everything, Astronomy and Astrophysics 282, pp. 262-268.
[37] Coles, P. & Lucchin, F. (2002) Cosmology, The Origin and Evolution of Cosmic Structure, Chichester: J. Wiley & Sons.
[38] Wells, D.R. (1986) Titius Bode and the helicity connection, IEEE Trans. Plasma Sci., 14; 865-873.
[39] Janaki, M.S. & Dasgupta, B. (1990) Eigenmodes for electromagnetic waves propagating in a toroidal cavity, IEEE Trans. Plasma Sci., 18:1; 78-85.
[40] Wells, D.R. (1990) Was the Titius-Bode series dictated by the minimum energy states of the generic solar system?, IEEE Trans. Plasma Sci. 18:1; 73-76.
[41] Chandrasekhar, S. (1958) On the equilibrium configuration of an incompressible fluid, Proc. Nat. Acad. Sci. US, 42; 273-279.
[42] Nieto M. (1972) The Titius-Bode Law of Planetary Distances: its History and Theory, Oxford: Oxford Univ Press.
[43] Poveda, A. & Lara, P. (2008) The exo-planetary system of 55 Cancri and the Titius-Bode law, Revista Mexicana de Astronomay Astrofsica, 44, 243-246 or astro-ph/0803.2240v1.
[44] Kotliarov, I. (2008) The Titius-Bode law revisited but not revived, Physics.space-ph-arXiv:0806.3532v1.
[45] Willott, C.J. et al. (2007) Four quasars above redshift 6 discovered by the Canada-France high-Z quasar survey, Astron J, 134; 2435–2450.
[46] Peebles, P.J.E. (1993) Principles of Cosmology, Princeton: Princeton Univ Press.
[47] Amoroso, R.L. (2006) The Holographic Conscious Multiverse, in R.L. Amoroso, B. Lehnert & J-P Vigier (eds.) Beyond the Standard Model: The Search for Unity in Physics, Orinda: Noetic Press.
[48] Amoroso, R.L & Pribram, KH (2009) The Complementarity of Mind and Body: Realizing the Dream of Descartes, Einstein and Eccles, Cambridge: MIT University Press.
[49] Cattafesta, L.N. & Settles, G.S, (1992) Experiments on shock/vortex interaction, 30th Aerospace Sciences Meeting, January 6-9, Reno, NV, Washington: American Institute of Aeronautics and Astronautics www.mne.psu.edu/psgdl/AIAA92-0315.pdf.
[50] Cutler, A.D. & Levey, B.S. (1991) Vortex breakdown in a supersonic jet, AIAA-91-1815, Honolulu, Hi, June, 1991, Washington: American Institute of Aeronautics and Astronautics.
[51] Delery, J., Horowitz, E., Leuchter, O., & Solignac, J-L., Fundamental studies on vortex flows, La Recherche Aerospatiale, No. 1984-2.
[52] Lyman, V. & Morgenstern, J.M. (2004) Calculated and measured pressure fields for an aircraft designed for sonic-boom alleviation, 22nd Applied Aeronautics Conference, 16-19 August 2004, Providence, RI, Washington: American Institute of Aeronautics and Astronautics, www.sonicbooms.org/ReferenceTechPapers/AIAA2004_4846.pdf.
[53] Hall, M.G. (1972) Vortex breakdown, Annual Review of Fluid Mechanics, V. 4, pp. 195-218.
[54] Leibovich, S. (1978) The Structure of vortex breakdown, Annual Rev Fluid Mechanics, V. 10, pp. 221-46.
[55] Leibovich, S. (1984) Vortex stability and breakdown: survey and extension, AIAA J. V. 22, No. 9.
[56] Zatoloka, V.V., Ivanyushkin, A.K., & Nikolayev, A.V., (1978) Interference of vortices with shocks in air scoops: dissipation of vortices, Fluid Mechanics, Soviet Research, Vol. 7, No. 4.
[57] Glotov, G.F., (1989) Interference of vortex braid with free-stream shock waves & nonisobaric jets, Uchenyye Zapiski TsAGI, XX:5; 21-32.

[58] Gostintsev, Yu.A., Zelentsov, V.V., Ilyukhin V.S., & Pokhil, P.F., (1969) Structure of underexpanded supersonic swirling gas jet, Izv. AN SSSR, Mekhanika Zhidkosti i Gaza, 4:5; 158-162.
[59] Prim, R.C. (1948) Extension of Crocco's theorems to flows having non-uniform stagnation enthalpy, Phys. Rev. 73, 186-186.
[60] Kaouri, K. (2004) Secondary Sonic Boom, Phd. thesis, Oxford University, http://eprints.maths.ox.ac.uk/159/.
[61] Lehnert, R. & Potting, R. (2004) Vacuum Cherenkov radiation, arXiv:hep-ph/0406128v1.
[62] Kant, I. (1878) The Critique of Pure Reason, JMD Meiklejohn, (Trans.) London: G. Bell & Sons, Pdf - http://books.google.com/books.
[63] Amoroso, R.L. (2005) On the possibility of relativistic shock-wave effects in observations of QUASAR luminosity-preliminary introduction, The Noetic Journal, 6:1-4; 53-56.
[64] Misner, C.W., Thorne, K.S. & Wheeler, J.A. (1973) Gravitation, San Francisco: W. H. Freeman
[65] Dray, T. & 't Hooft, G. (1985) The gravitational shock wave of a massless particle, Nucl. Phys. B253, 173.
[66] Aichelburg. P.C., & Sexl, R.U. (1971) General Relativity and Gravitation 2, 303.
[67] Dray, T. & 't Hooft, G. (1985) Communications Mathematical Physics 99, 613.
[68] Lousto, C.O. & Sanchez, N. (1991) 915; Nuclear Physics B355, 231.
[69] Lousto, C.O. & Sanchez, N. (1992) Nuclear Physics B383, 377.
[70] Hayashi, K. & Samura, T. (1994) Physical Review D50, 3666.
[71] Hotta, M. & Tanaka, M. (1993) Classical and Quantum Gravity 10, 307.
[72] Ferrari, V. & Pendenza, P. (1990) General Relativity and Gravitation 22, 1105.
[73] H. Balasin, & H. Nachbagauer, H., The Ultrarelativistic Kerr-Geometry and its Energy-Momentum Tensor, gr-qc/9405053.
[74] Sfetsos, K. (1994) On gravitational shock waves in curved spacetimes, arXiv:hep-th/9408169v3.
[75] Gibbons, G. & Maeda, K. (1988) Nucl. Phys. B298, 741.
[76] Kallosh, R., Linde, A., Ortın, T., Peet, A.& Van Proeyen, A. (1992) Phys. Rev. D46, 5278.
[77] Spitkovsky, A. (2008) Particle acceleration in relativistic collisionless shocks: Fermi process at last? arXiv: Astro-ph 0802.3216v2.
[78] Papapetrou, A. & Treder, H-J. (1962) Shock waves in general relativity, in L. Infeld (ed.) Recent Developments in General Relativity, Oxford: Pergamon Press.
[79] Zeldovich, Y.B. & Rayzer, Y.P. (1967) Physics of Shock Waves And High Temperature Hydrodynamic Phenomena, New York: Academic Press.
[80] Kaloper, N. (2005) Brane-induced gravity shocks, arXiv:hep-th/0501028v4.
[81] Birkhoff, G. (1923) Relativity & Modern Physics, Cambridge: Harvard Univ Press.
[82] Crawford, D.F. (2009) Observations of type 1A supernovae are consistent with a static universe, arXiv: Astro-ph-CO: 0901.4172v1.
[83] Willott, C.J., Delorme, P., Reylé, C., Albert, L., Bergeron, J., Crampton1, D., Delfosse, X., Forveille, T., Hutchings, J.B., Mclure, R.J., Omont, A. & Schade, D. (2009) Six more quasars at redshift 6 discovered by the Canada-France high-Z quasar survey, arXiv: 0901.0565v1.
[84] Lopez-Corredoira, M. (2009) Apparent discordant redshift QSO-galaxy associations, arXiv:0901.4534v1, astro-ph.CO.
[85] Komberg1, B.V. & Pashchenko, I.N. (2009) Giant radio galaxies - old long-living quasars? arXiv:0901.3721v1 [astro-ph.CO].
[86] Croton, D.J. (2009) A simple model to link the properties of quasars to the properties of dark matter halos out to high redshift,
Monthly Notices Radio Astronomical Society 000, 000–000; or arXiv:0901.4104v1, astro-ph.CO.
[87] The Holographic Anthropic Multiverse (2009) R.L. Amoroso & E.A. Rauscher, Singapore: World Scientific.
[88] Gravitation and Cosmology: From the Hubble Radius to the Planck Scale, (2002) RL Amoroso et al, Dordrecht: Kluwer Academic.
[89] Unified Field Mechanics, R.L. Amoroso, L. Kauffman, E.A. Rauscher & P. Rowlands, in process.
[90] New Scientist, Science: The ghostly hand that spaced the planets 09 April 1994 by Robert Matthews
[91] http://www.universetoday.com/87784/applying-the-titius-bode-rule-to-exoplanet-systems/#ixzz26SeCeJ6s
[92] Edward Anderson (2004) Geometrodynamics: Spacetime or Space? Ph.D. thesis, http://arxiv.org/pdf/gr-qc/0409123v1.pdf
[93] Hoyle, F., Burbidge, G. & Narlikar, JV. (2000) A Different Approach to Cosmology, From a Static Universe Through the Big Bang Towards Reality, Cambridge: Cambridge University Press.

Review of the Relationship Between Galilean Relativity and the Velocity of Light

SHUKRI KLINAKU

University of Prishtina,
Sheshi Nëna Terezë, 10000, Prishtina, Kosova
klinaku@uni-pr.edu

In the part of physics called "classical mechanics" the Galilean principle of relativity is accepted by everyone. But, according to the theory of special relativity (TSR) this principle cannot be applied to light. By analyzing several experiments (Fizeau's experiment, Michelson's experiment, de Sitter's observation and Alväger's experiment) in this paper it will be concluded that the reliance of TSR's postulates in those experiments is unsustainable. The essence of the reviewing these experiments is that some of them do not have the same conditions as the motion to which refers the Galileo principle (Fizeau's experiment, the de Sitter's observation and the Alväger's experiment) and in some of them the Galileo principle is not applied correctly (Michelson's experiment). In this paper it is also concluded that the light obeys to the Galileo principle like all other bodies. Finally, an idea is proposed for conducting an experiment which will prove that the velocity of light is not constant in all reference frames.

Keywords: Galilean principle, Velocity of light, Special relativity.

1. Introduction

In 1632 Galileo Galilei built the principle of relativity. Today for this principle we write:

$$x = x' + vt \, . \tag{1}$$

The same things can be expressed with the velocity-addition formula for relative motion:

$$u = u' + v. \tag{2}$$

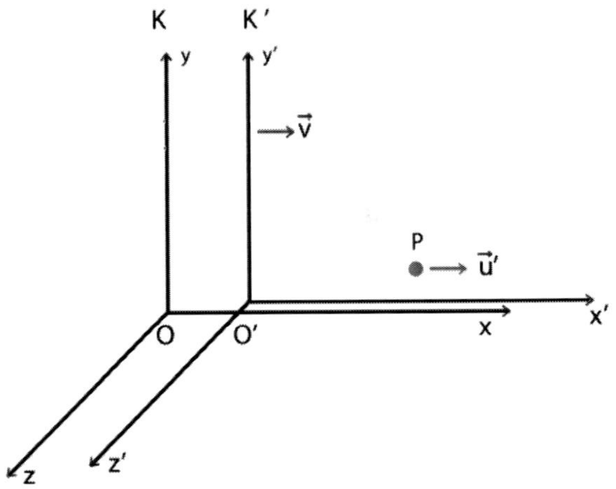

Figure 1. The relative motion in two reference frames.

Both equations derive from relative motion in two reference frames as we see in Fig. 1 where x represents the distance of event from observer in origin of system K; x' represents the distance from event to origin of system K'; v represents the velocity of system K' relative to system K; t represents the time within which it carries out the motion of K' and event in point P; u represents the velocity of point P relative to observer in O and u' represents the velocity of point P relative to observer in O'.

The theory of special relativity (TSR) claims that Galileo principle is not valid. Relativistic physicists use some experiments to overthrow the Galileo principle. Viewing these known experiments we raise only one question: what do they have in common with the Galileo's idea.

2. Review of Fizeau's Experiment

If we intend presentation of Galileo principle in a container filled with water in which the light signal is issued, then what Galileo studied corresponds to Fig. 2*a* and to equation (3) which expresses exactly the velocity-addition formula:

$$c_{lab} = \frac{c}{n} + v \tag{3}$$

where c_{lab} the velocity of light as is measured by the observer in the resting system (laboratory); n is index of refraction of light through the water; c is the velocity of light in vacuum and v is the velocity of the wagon filled with water. So, in Fig. 2*a* there is a system at rest (the laboratory) and other system where the event carried out (wagon) which moves with velocity v relative to the first. Water is in rest relative to the wagon and wagon's velocity does not affect the intensity of velocity of light through the water. But Fig. 2*a* shows no Fizeau's experiment. Fizeau has changed conditions of motion and claims again to apply the equation (3). Let us see the Fig. 2*b*. In this figure there is only one reference system filled with moving water with velocity v.

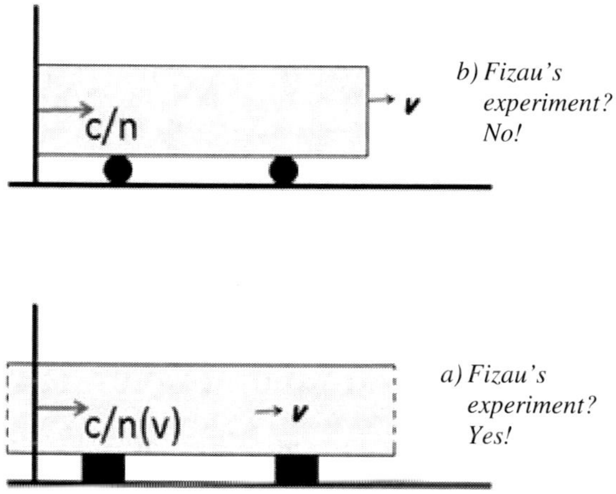

Figure 2. Wagon filled with water (a) and moving water in tube (b).

In this moving water is released a signal of light. We have only one system in this experiment, because the velocities of the two motions (motion of water and motion of light) are measured only relative to tube within which water and light flows. The problem is to find the velocity of light in the environment. This represents Fizeau's experiment. But in this case we can't apply the Galileo principle of relativity. In other words, the outcome of this experiment can't be served as test of accuracy of the Galileo principle; as well it can't be taken as an argument for the confirmation of the TSR.

Influenced by Fresnel, Fizeau has adapted the result from this experiment in this expression [1]:

$$c_{lab} = \frac{c}{n} + v\left(1 - \frac{1}{n^2}\right). \tag{4}$$

This equation has no relation to Galileo principle and it has no relation with TSR as well.

So, the relative motion of Fizeau's experiment is not a simple relative motion, and therefore in this experiment Galileo principle and TSR can't be tested. This motion is more complex, where the velocity of one body (the light) is affected by friction with another body in motion (the water). Finally, the problem that Fizeau's experiment raises is: "dynamic index of refraction, *n(v)*". The Physics knows how to find the refraction index for different materials at rest ("static index of refraction, *n* "), but yet doesn't know the variation of this index when these materials move with a certain velocity ($\boldsymbol{n(v)}$ =?).

3. Review of Michelson's Experiment

Michelson's experiment fulfilled all conditions of Galileo's motion. The problem to this experiment was created for two reasons. First problem is the point of observation. We can judge right about Michelson's experiment if we treat the experiment as in Fig. 3. We should bear in mind that the calculations are made by observer at hypothetical system in space (the origin O). For observer in origin O' nothing of interest happens; indeed the observer in origin O' is moving together with Earth and interferometer. Michelson's has made calculations by the head of the observer at point O, but it is known that stood at the point O'. Second problem about Michelson's experiment is calculations. Michelson began with the implementation of the Galileo principle, but did not implement it entirely.

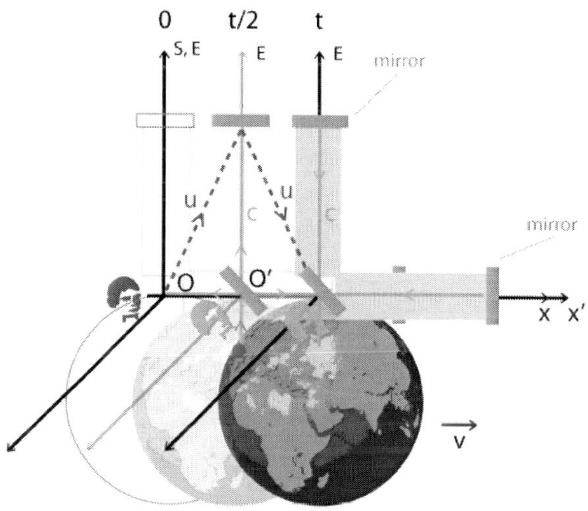

Figure 3. Full view of Michelson's experiment.

Michelson has noted with t_1 the time within which the light propagates through the interferometer's arm located in the direction of Earth's motion and calculated as follows:

$$t_1 = \frac{l}{c - v} + \frac{l}{c + v}. \tag{5}$$

He applied the Galileo principle in both denominators (the velocity-addition formula), but in the numerators did not apply it. For propagating of light through the interferometer's arm placed in the normal direction to the Earth's motion, Michelson calculated the time, t_2:

$$t_2 = \frac{l_1}{c} + \frac{l_1}{c} = 2\frac{\sqrt{x^2 + l^2}}{c}. \tag{6}$$

We see that Michelson now even in the denominators does not take into account the Galileo principle. Instead he takes the velocity of light as constant [2, 3]. With l_1 is marked the ray's road in interferometer's arm which is normal to the direction of Earth's motion. This road is slope due to the displacement of the Earth for the length x with velocity v. Michelson has judged that the slope of road is caused by the Earth's (interferometer's) displacement for the distance x. However, this displacement must be taken into account for interferometer's arm that is located in the direction of Earth's motion also. Therefore, the right calculation t_1 is:

$$t_1 = \frac{l+x}{c+v} + \frac{l-x}{c-v}. \tag{7}$$

Indeed, as in normal arm the road l is extended in l_1, in the same manner in parallel arm the road l is extended in $l+x$ due to the Earth motion. Obviously turning light signals makes shorter road, because it moves in the opposite Earth's motion $(l-x)$. Therefore, for time t_2 the right calculation is as follows:

$$t_2 = \frac{l_1}{u} + \frac{l_1}{u} = 2\frac{\sqrt{x^2+l^2}}{\sqrt{v^2+c^2}}. \tag{8}$$

With such calculation we get even theoretically the result obtained by the experiment:

$$t_1 = t_2. \tag{9}$$

So, not only the result of Michelson's experiment is zero; moreover, even expectation from this experiment is zero [4].

Now, let's put interferometer so that one arm closes the angle θ relative to the direction of Earth's motion, as in Fig. 4. One such review of Michelson's experiment was done by Claus Lämmerzahl [5]. Even calculations according to Fig. 4 show that light obeys to the Galileo principle. The coordinates of point B for the time $t = 0$ are:

$$B\big(x(B), y(B)\big) = (l\cos\vartheta, l\sin\vartheta),$$

while after the time t the light will reach the point B′ with coordinates:

$$B'\big(x(B'), y(B')\big) = (l\cos\vartheta + vt, l\sin\vartheta).$$

We require the time t within which the light passes the path AB′. Let us suppose that in this path the velocity of light is c, as TSR claims and, than we obtain:

$$ct(AB') = \sqrt{(l\cos\vartheta + vt)^2 + l^2\sin^2\vartheta}$$

$$ct(AB') = \sqrt{\left(\vec{l} + \vec{v}t\right)^2}$$

$$\vec{c}t(AB') = \vec{l} + \vec{v}t(AB')$$

$$t(AB') = \frac{\vec{l}}{\vec{c} - \vec{v}}. \tag{10}$$

From last equation we obtain:

$$\vec{c} = \frac{\vec{l}}{t(AB')} + \vec{v}. \tag{11}$$

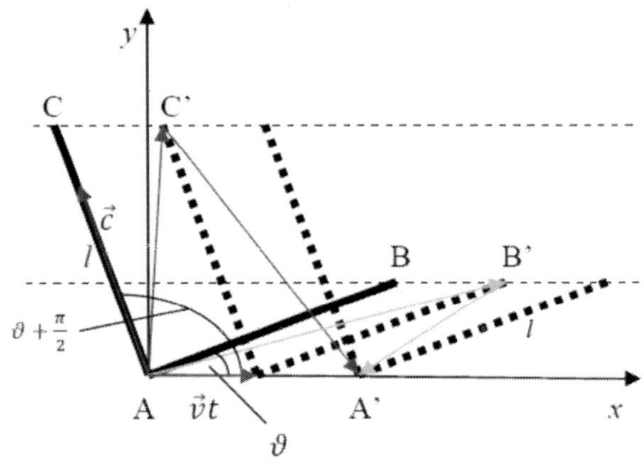

Figure 4. Interferometer (CAB), placed in such way that the arm AB forms the angle ϑ relative to direction of Earth's motion (x axis).

The equation (11) raises the question what kind of velocity presents this term: $\frac{\vec{l}}{t(AB')}$. It certainly represents the velocity of light $\vec{c}$, i. e. the velocity of light with which the light passes the path l. It follows that the velocity on the left side of equation (11) is not $\vec{c}$, but greater than $\vec{c}$ and represents the Galileo principle for light propagation on the path AB′. So the right form of equation (11) is:

$$\vec{c'} = \vec{c} + \vec{v}. \tag{12}$$

4. Review of de Sitter's Observation and Alväger's Experiment

Observation of double star by de Sitter [6] claims to have thrown the Galileo principle. What do the motion of double star and Galileo's motion have in common? The essence of these two motions is different. What de Sitter has observed? Fig. 5 shows a double star system. A star is rotating around another one. De Sitter said that he has applied the velocity-addition formula according Galileo principle and when the star rotates away from the observer, its light must have velocity $c-u$, where u is the linear velocity of the star in trajectory. And when the star is approaching to the same observer, then the velocity of light should be $c + u$. However, de Sitter with spectrographic method concluded that the velocity of light from rotating star remains constant to observer.

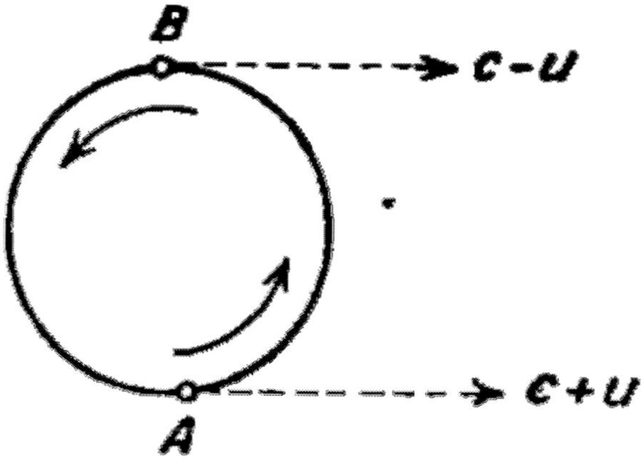

Figure 5. De Sitter's idea and Fig. for applying the Galileo principle to double star [6].

De Sitter claims that this observation has confirmed the second postulate of TSR and at the same time rejects the Galileo principle. Similar pattern to the de Sitter's observation represents the Alväger's experiment [7]. In Alväger's experiment instead the star is elementary particle π^0 which moves with very high velocity and then it undergoes to discarding by issuing a γ ray. Alväger et al. have not found that the velocity of γ ray is $\approx 2c$ (supposedly according to Galileo principle) but only c (supposedly according to TSR). De Sitter's observation and Alväger's experiment are directly inspired by the definition of the second postulate of Einstein's thesis of 1905 [8]. In Einstein's work [8] the second postulate of TSR is deficient compared to the attributes of light's velocity in TSR. Galileo has not established the principle of relativity for the motion like these de Sitter's and Alväger's. To the Galileo motion is clearly defined reference systems and in one of these systems (moving system) is carried out an event, which happens wholly within the system. In other words, the moving system to Galileo's motion is holder of event (motion) and not event's source. Galileo is not studying the motion of a fountain that shoots a projectile, but study event that takes place completely independently within an inertial system observed from another inertial system. In conclusion we can say that the Galileo's motion is not ballistic motion. While the de Sitter's observation and Alväger's experiment are typical ballistic, where instead of the projectile is the light's signal. To this paper the outcomes of these ballistic motion is not interesting, but only the conditions of their development.

5. Experimental Test of the Second Postulate

To prove the second postulate of TSR an experiment would perform with the same conditions as to the Galileo's motion. An example would be this: a light source illuminates both entrances of two vacuumed pipes. At the moment when the light reaches the pipes one of them starts to move. Then, within a considerably short time we intersect the path of these two rays at the exit of the pipes. If we achieve to prevent propagation of both rays with simultaneous barrier at the exit of the moving pipe, then TSR's second postulate is proven. Otherwise, the postulate is rejected.

6. Conclusions

From the comparison between motion's conditions for which Galileo put the principle of relativity and the motion's conditions of some known experiments which claim to overthrow the Galileo principle of and to prove TSR's second postulate was found that these conditions differ substantially. As a result of this difference with these experiments can't prove the second postulate of TSR. Even more, by right explanation of Michelson's experiment is proved that the velocity of light obeys to the Galileo principle.

References

[1] Fizeau, H. (1859) *Ann. de Chim. et de Phys.* 57.
[2] Michelson, A.A. (1881) American Journal of Science.
[3] Michelson, A.A. & Morley, E.W. (1887) American Journal of Science.
[4] Klinaku, Sh. (2010) in R.L Amoroso, P. Rowlands & S. Jeffers (eds.) AIP Conf. Proc. 1316.
[5] Lämmerzahl, C. (2005) *Annalen der Physik*, 14, 1 3.
[6] de Sitter, W. (1913) *Physik. Zeitschr.* 14, 429.
[7] Alväger, T. et al. (1964) Physics Letters, 12, nr. 3.
[8] Einstein, A. (1905) *Annalen der Physik*, 17.

Review of the Derivation of the Lorentz Transformation

SHUKRI KLINAKU[a] & VALBONA BERISHA[b]

University of Prishtina, Sheshi Nëna Terezë, 10000, Prishtina, Kosova
[a]klinaku@uni-pr.ed, [b]valbona.berisha@uni-pr.edu

All physicists who had contributed to the explanation of the relative motion before Einstein, in their calculations, have considered the velocity of light as constant magnitude in all reference frames. But to derive the Lorentz transformations is not enough just to say that velocity of light is constant in all reference frames. For this derivation is needed a "thought experiment" and then in this experiment should be some inconsistent physical assumptions; also should be some non-transparent mathematical equations and only after these can be seen the Lorentz transformations. In this paper we will present a critical review, in physical and mathematical terms, on some parts in the path of constructing of special relativity's basic equations.

Keywords: Lorentz transformation, Special relativity.

1. Introduction

Theory of special relativity (TSR) is constructed to explain some problems of the late nineteenth century physics. All the time there were physicists that had doubts about TSR's explanations. This theory was confronted with based objections, with "paradoxes" and with right interpretations of the experiments which are taken as experimental basis of TSR. However, the relativist physicists have preferred to escape this problem by setting it to the high sphere of mathematics; by treating theory of relativity as problem with more than three dimensions of space and by creating some speculative properties of the matter. So the theory which in its start was difficult to understand became more complicated. Now the people could talk about it only relying on belief and not on knowledge. Let us remember that also Ptolemaic system was reformed so much that finally became unusable.

2. Review of the Derivation of the Lorentz Transformations

Since the beginning we can say that the postulate about the velocity of light as universal constant is necessary but not sufficient to build the basic equations of TSR. All physicists who had contributed to explain the relative motion before Einstein, in their calculations, the velocity of light have considered as constant magnitude in all reference frames. To derive the Lorentz transformations we need a "thought experiment", even some non-transparent algebraic equations and some wrong physical assumptions. Let's see one by one.

In the "thought experiment" Einstein in [1] says: when the origins of the two systems (K and K') are at the same point, the experiment's time starts to flow and at the same moment from the origin of the system K' (which moves with constant velocity v relative to the system K) is transmitted a light's signal along the axis x'. Then, for a description of this event he writes the system of equations:

$$\left.\begin{aligned} c &= \frac{x}{t} \\ c &= \frac{x'}{t'} \end{aligned}\right\} \tag{1}$$

where c is the velocity of light, x is the displacement of the signal relative to the origin of system K, t is time, which according to Einstein, flows in the system K, while x' and t', with the same description for the system K'. This is the first Einstein's mistake. At the start was said that a light signal was transmitted, but in the algebraic treatment there are two light signals and two independent roads. From the system of equations (1), we don't understand in which system the signal moves, because they have the same structure. Neither equation, (1) contains the velocity of system K' relative to K.

This quantity (v) cannot be overlooked, because without it there is no relative motion. The system of equations (1) does not represent what author claims; they don't represent the equations of the event. What does the system of equations (1) indeed represent? If we have only one event (only one propagating light signal in K') than equations (1) are accurate, if we say that they represent the equations of information's motion about this event that receive the observers in their origins O and O'. Because the event is the light signal, the velocity of the event and the velocity of the information on this event are the same, i.e. velocity c. Roads x and x' are the same as for the "events" and for return of information to observers. Only times t and t' from (1) must be redefined. Times t and t' in (1) are not "times of events" or system's times, but they are the times within which observers at O and O' receive the information from the place where the light signal has reached. So, within time t takes the information the observer O and within the time t' the observer O'. Therefore, these times are different and only in this sense they can be different. But their change is not in various ticks, but in the number of ticks.

The next Einstein's mistake is this. Einstein needs another system similar to (1), but with the minus. He "realizes" that in just one sentence. He says: let we think now the same "experiment", but the light signal is transmitted in the negative direction along axis x' [1]. Then, according to this we have:

$$\left.\begin{aligned} c &= -\frac{x}{t} \\ c &= -\frac{x'}{t'} \end{aligned}\right\} \tag{2}$$

However, this can only be said, while in practice it may not happen. From Fig. 1 it is clear that the intensity of length x, when the signal was transmitted in positive direction of x axis and the intensity of the length x, when the signal was transmitted in the opposite direction are not equal. Therefore, systems (1) and (2) can't enter to the joint reports without clarifying the relationship between these two lengths, which Einstein mistakenly took as equal. But even this is not enough to obtain the Lorentz transformations.

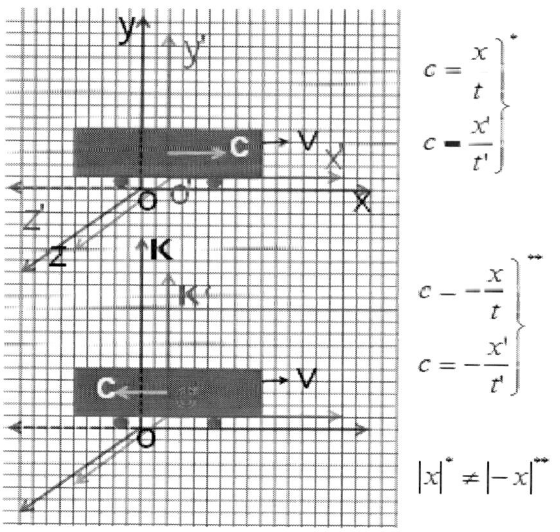

Figure 1. Einstein's "thought experiment". x in first case is 6a, while in second case only 2a (a is the square's side of this background).

From systems (1) and (2) Einstein obtains only one system:

$$\left.\begin{aligned} (ct - x) &= \mu(ct' - x') \\ (ct + x) &= \lambda(ct' + x') \end{aligned}\right\} \tag{3}$$

where μ and λ are two parameters that depend on the velocity v, which was not until now presented (!).

Einstein once again transforms the system (3) and obtains:

$$\left.\begin{aligned} x' &= ax - bct \\ ct' &= act - bx \end{aligned}\right\} \tag{4}$$

Now the mathematicians should intervene immediately. Systems (3) and (4) are no system of equations, but only one equation. Based on initial conditions (1) or (2) from (4) we obtain:

$$\left.\begin{aligned} ax - bx - x' &= 0 \\ ax - bx - x' &= 0 \end{aligned}\right\}. \tag{5}$$

Thus we have only one line. Therefore, we can't proceed to derive the Lorentz transformations from "system of equations" (4). To prove it in another way, let us treat the "system" (4) according the Cramer's rule. Let us write the "system" (4) in this form:

$$\left.\begin{aligned} xa - ctb - x' &= 0 \\ cta - xb - ct' &= 0 \end{aligned}\right\}. \tag{6}$$

We have to find the unknown parameters a and b. According the Cramer's rule we find:

$$\frac{x}{ct} = \frac{ct}{x} = \frac{x'}{ct'}. \tag{7}$$

And from the initial conditions (1) we find that the all three ratios from (7) are equal to 1:

$$1 = 1 = 1. \tag{8}$$

So, we have to conclude that the system (4) has the infinitely many solutions for a and b. And according the Cramer's rule in (4) we have only one line.

Even this mistake is not enough to obtain Lorentz transformations. The next mistake is as it follows. Einstein in [1] says: "*For the origin of K' we have permanently x' = 0, and hence according to the first of the equations (4):*

$$x = \frac{bc}{a}t$$

If we call v velocity with which the origin of K' is moving relative to K, we then have:

$$v = \frac{bc}{a}."$$

It is important to note that this Einstein's assumption has wide usage. Most authors use this erroneous assumption and we find it almost in all university books [2, 3, 4]. Let us analyse this a bit. The first, the system (4) claims to represent the path of the light signal. Therefore, in these equations can't be replaced the coordinate from other motion, such is the motion of origin of the system *K'*. Also the motion of the origin of the system *K'* relative to system *K* should have no the same notes, because with *x'* is already marked the displacement of the signal light relative to the origin *O'*. The second, the defining velocity v as: $v = \frac{x}{t}$ is wrong. Why? Because the distance between the origins of two frames is neither *x* nor *x'*. Let us denote this distance by x_0. This is the distance between the two frames, which is equal to the product vt, therefore it should be written as:

$$v = \frac{x_0}{t}. \tag{9}$$

When we can say that $x_0 = x$? Then and only then, when there are given two reference frames and at time $t = 0$ from their overlaping origins an arbitrary material point starts its motion. Thus, for $t = t' = 0$, also $x_0 = x = x' = 0$. When relative motion starts, then never again can be $x_0 = x$. This explains that the conlusion derived by Einstein for velocity v never holds true, i. e. , the equation $v = \frac{x}{t}$ is not true.

Hence, according to Einstein, the results of this transformation are equations:

$$x' = \frac{x-vt}{\sqrt{1-\frac{v^2}{c^2}}}, \qquad t' = \frac{t-\frac{v}{c^2}x}{\sqrt{1-\frac{v^2}{c^2}}}. \tag{10}$$

The truth about them is as follows: the fact that there are two transformations comes due to the wrong mathematics, while the form of the transformed expression comes due to the wrong physics. If we replace the initial conditions (1) in equations (10) and (11) we obtain:

$$x = x\sqrt{\frac{c-v}{c+v}}, \qquad t' = t\sqrt{\frac{c-v}{c+v}}. \tag{13}$$

We see that there is only one transformation and not two. Can we make substitution of initial conditions (1) in equations (10 and 11), *that is the question*. The „permission" to do that we take from algebra. Also, the equations (10) and (11) were transformed to the equations (12) and (13) by Einstein himself, in order to emphasize the enchantment of his formulas. In [1] he says: "*Aided by the following illustration, we can readily see that, in accordance with the Lorentz transformation, the law of the transmission of light in vacuo is satisfied both for the reference-body K and for the reference-body K'. A light-signal is sent along the positive x-axis, and this light-stimulus advances in accordance with the equation x = ct, i.e. with the velocity c. According to the equations of the Lorentz transformation, this simple relation between x and t involves a relation between x' and t'. In point of fact, if we substitute for x the value ct in the first and fourth equations of the Lorentz transformation, we obtain:*

$$x' = \frac{(c-v)t}{\sqrt{1-\frac{v^2}{c^2}}}, \qquad t' = \frac{\left(1-\frac{v}{c}\right)t}{\sqrt{1-\frac{v^2}{c^2}}},$$

from which, by division, the expression x' = ct' immediately follows."

We can see that the last equations are identical to equations (12) and (13). Finally, very easy we can conclude that equations (12) and (13) can't derive the velocity-addition formula; i. e. that the velocity-addition formula of special relativity is unsustainable.

3. Conclusions

After each objection to TSR, after each difficulty of TSR the relativist physicists invented something to make the TSR to survive. The believers of the geocentric theory did the same to survive the geocentric system. These efforts to survive the TSR will not stop in the future as well. However, TSR is a theory that can't be improved also. It is point where I fully agree with Einstein about the TSR. In 1950 he said: "*A modification of it* (theory of relativity) *seems impossible without destruction of the whole*" [5]. Indeed, the time has come that we abandon the TSR entirely. In this paper is proved that Lorentz transformations are derivation of wrong physical assumption and non-transparent mathematical manipulation. Also the velocity-addition formula in TSR is unsustainable.

References

[1] Einstein, A. (1920) The Special Theory of Relativity.
[2] Lämmerzahl, C. (2006) Lect. Notes Phys. Springer-Verlag, 702.
[3] Robertson, H.P. (1949) Reviews of Modern Physics, 21.
[4] Landau, L.D. & Lifschitz, E.M. (1977) *Lehrbuch der Teoretischen Physik I, Berlin: Akademie Verla.*
[5] Einstein, A. (1950) Out of My Later Years.

Marriage of Electromagnetism and Gravity in an Extended Space Model and Astrophysical Phenomena

V. A. ANDREEV

Lebedev Physical Institute of the Russian Academy of Sciences,
Moscow, Russia
andrvlad@yandex.ru

D. Yu. TSIPENYUK

Prokhorov General Physics Institute of the Russian Academy of Sciences,
Moscow, Russia
tsip@kapella.gpi.ru

The generalization of Einstein's special theory of relativity (SRT) is proposed. In this model the possibility of unification of scalar gravity and electromagnetism into a single unified field is considered. Formally, the generalization of the SRT is that instead of (1+3)-dimensional Minkowski space the (1+4)-dimensional extension G is considered. As a fifth additional coordinate the interval S is used. This value is saved under the usual Lorentz transformations in Minkowski space M, but it changes when the transformations in the extended space G are used. We call this model the extended space model (ESM). From a physical point of view our expansion means that processes in which the rest mass of the particles changes are acceptable now. If the rest mass of a particle does not change and the physical quantities do not depend on an additional variable S, then the electromagnetic and gravitational fields exist independently of each other. But if the rest mass is variable and there is a dependence on S, then these two fields are combined into a single unified field. In the extended space model a photon can have a nonzero mass and this mass can be either positive or negative. The gravitational effects such as the speed of escape, gravitational red shift and detection of light can be analyzed in the frame of the extended space model. In this model all these gravitational effects can be found algebraically by the rotations in the (1+4) dimensional space. Now it becomes possible to predict some future results of visible size of supermassive objects in our Universe due to new stage of experimental astronomy development in the RadioAstron Project and analyze phenomena is an explosion of the star V838 Mon.

Keywords: Gravity, Electromagnetism, Interval, Gravstar, Multi-dimensional space

1. Introduction

We consider a generalization of Einstein's special theory of relativity (STR) on a 5-dimensional space, or more specifically on a (1+4)-dimensional space with a metric $(+----)$ However, it is well known that the photon can be considered as a massless particle, and described by the plane wave only in an infinite empty space, [1,2]. But if a photon falls into the environment or in confined space, such as a resonator or waveguide, it acquires a nonzero mass, see [3,4].

Under the particle mass m, we will understand its rest mass, which is a Lorentz scalar. No other masses will appear in this work. Here we follow the recommendations of [5]. Similarly, we can consider the process of changing the mass of other particles, such as electrons, assuming that it depends on external conditions and influences.

Thus, it seems natural to expand the space of parameters characterizing particle, taking into account the fact that the interaction of its mass can vary. We call it the extended space.

2. The Structure of the Extended Space

Such particle having a mass m, corresponds to a hyperboloid in Minkowski space, in the limiting case this hyperboloid degenerates into a cone.

$$s^2 = (ct)^2 - x^2 - y^2 - z^2 \tag{1}$$

Since the change of the mass of a particle corresponds its transition from one hyperboloid to the other, i.e. change of the corresponding interval, it seems natural to choose interval s as an additional fifth coordinate. Thus, we will work in a space with coordinates (t, x, y, z, s) and metric $(+----)$. The objects under consideration are located on a cone

$$(ct)^2 - x^2 - y^2 - z^2 - s^2 = 0. \tag{2}$$

We denote this space as $G(1, 4)$. The Minkowski $M(1, 3)$ space is a subspace of $G(1, 4)$. An interval in the Minkowski $M(1, 3)$ space plays the role of the fifth coordinate in the G(1, 4) space. We designate this coordinate by the letter S. The other coordinates are designated as T,X,Y,Z. One of the characteristic features of this Extended Space Model (ESM) is that the particle's rest mass m_0 is s variable quantity and a photon, moving in a medium with refraction index $n >$ 1, acquires a nonzero mass. This mass can be both positive and negative.

The usual (1+2)-dimensional cones and hyperboloids occur as sections of the surface (2) by hyperplanes $s = s_0$. In the space of $G(1, 4)$ one can constructed in usual way the objects that have different tensor nature and transform appropriately under linear transformations of the $G(1, 4)$ space [6].

In Minkowski space $M(1, 3)$ a 4-vector of energy and momentum

$$\tilde{p} = \left(\frac{E}{c}, p_x, p_y, p_z\right) \tag{3}$$

is associated to each particle [1]. In the extended space of $G(1, 4)$, we complete its 5-vector

$$\bar{p} = \left(\frac{E}{c}, p_x, p_y, p_z, mc\right) \tag{4}$$

For free particles, the components of the vector (4) satisfy the equation

$$E^2 = c^2 p_x^2 + c^2 p_y^2 + c^2 p_z^2 + m^2 c^4. \tag{5}$$

It is well-known relation of relativistic mechanics, which relates the energy, momentum and mass of a particle. Its geometric meaning is that the vector (4) is isotropic, i.e. its length in the space of G(1, 4) is equal to zero. However, in contrary to the usual relativistic mechanics, we now suppose that the mass m is also a variable, and it can vary at motion of a particle on the cones (2),(5). It should be understood so that the mass of the particle changes when it enters the region of the space that has a nonzero density of matter. Since in such areas the speed of light is reduced, they can be characterized by value n - optical density. The parameter n relates the rate of light in vacuum c with the speed of light in a medium v.

$$v = \frac{c}{n} \tag{6}$$

A set of variables (4) forms a 5-pulse, its components are conserved, if the space G(1, 4) is invariant under the corresponding direction. In particular, its fifth component p_4, having sense of mass, does not change if the particle moves in the area with constant value n.

3. The Vectors of the Free Particles

In the usual relativistic mechanics and field theory the mass of a particle is constant, and for particles with zero masses and nonzero rest masses different methods of description are used. The particles with nonzero rest masses are characterized by their mass m and speed $\vec{v}$. The particles with zero mass (photons) are characterized by frequency ω and wavelength λ. These ω and λ are connected with energy E and momentum $\vec{p}$ follows

$$E = \hbar\omega, \quad \vec{p} = \frac{2\pi\hbar}{\lambda}\vec{k}. \tag{7}$$

The 4-vector

$$\tilde{p} = \left(\frac{E}{c}, \vec{p}\right) = \left(\frac{mc}{\sqrt{1-\beta^2}}, \frac{m\vec{v}}{\sqrt{1-\beta^2}}\right), \quad \text{where} \quad \beta^2 = \frac{v^2}{c^2} \tag{8}$$

corresponds to a particles with nonzero rest mass. The 4-vector

$$\tilde{p} = \left(\frac{\hbar\omega}{c}, \frac{2\pi\hbar}{\lambda}\vec{k}\right) = \left(\frac{\hbar\omega}{c}, \frac{\hbar\omega}{c}\vec{k}\right). \tag{9}$$

corresponds to a particles with zero mass.

In the frame of our approach, there is no difference between massive and massless particles, and therefore should establish a connection between these two methods of description. This can be done using the relation (7) and the hypothesis of de Broglie, according to which these relations hold for the massive particles. Now, substituting (7) in (4), we obtain the relation between the mass m, frequency ω and wavelength λ

$$\omega^2 = \left(\frac{2\pi c}{\lambda}\right)^2 + \frac{m^2c^4}{\hbar^2}, \tag{10}$$

$$\omega = \frac{mc^2}{\hbar\sqrt{1-\beta^2}}, \quad \lambda = \frac{2\pi\hbar}{mv}\sqrt{1-\beta^2}. \tag{11}$$

It follows that if $v \to 0,\ \lambda \to \infty$, but $\omega \to \omega_0 \neq 0$. Here ω_0 determines the energy of a particle at rest.

Now we construct 5-vectors from 4-vectors (8), (9). We suppose that a 5-vector

$$\bar{p} = \left(mc,\ \vec{0},\ mc\right) \tag{12}$$

corresponds to a stationary particle of mass m.

The 5-vector of a particle, which moves with velocity $\vec{v}$, can be obtained by transformation to the moving coordinate system. Then the vector (12) takes the form

$$\bar{p} = \left(\frac{mc}{\sqrt{1-\beta^2}},\ \frac{m\vec{v}}{\sqrt{1-\beta^2}},\ mc\right) \tag{13}$$

Similarly the 4-vector (9) transforms into 5-vector

$$\bar{p} = \left(\frac{\hbar\omega}{c},\ \frac{2\pi\hbar}{\lambda}\vec{k},\ 0\right). \tag{14}$$

At the transition to a moving coordinate system the vector (14) does not change its form, only the frequency ω changes its value.

$$\omega \to \omega' = \frac{\omega}{\sqrt{1-\beta^2}}. \tag{15}$$

Thus, in an empty space in a stationary reference frame there are two fundamentally different object with zero and nonzero masses, which in the space of $G(1, 4)$ correspond to the 5-vectors

$$\left(\frac{\hbar\omega}{c}, \frac{\hbar\omega}{c}, 0\right) \tag{16}$$

and

$$\left(mc,\ 0,\ mc\right) \tag{17}$$

For simplicity, we write the vectors (16), (17) in (1+2)-dimensional space. The vector (16) describes a photon with zero mass, the energy, $\hbar\omega$, and the velocity c. The vector (17) describes a stationary particle of mass m. The photon has a momentum, $p = \hbar\omega/c$, a massive particle has a momentum equal to zero. In the 5-dimensional space, these two vectors are isotropic, in Minkowski space only the vector (16) is isotropic.

The length of the vector (x_0, x_1, x_2 x_3, x_4) is equal to

$$l^2 = x_0^2 - x_1^2 - x_2^2 - x_3^2 - x_4^2$$

If we restrict ourselves to Lorentz transformations in Minkowski space it is impossible to transform an isotropic vector into anisotropic one and vice versa. In other words in frame of the SRT photon cannot acquires mass, and a massive particle cannot be a photon. But in the extended space $G(1, 4)$ a photon and a massive particle can be related to each other by a simple rotation.

As was already mentioned the parameter n connects the velocity of light in vacuum with that in the medium: $v = c/n$: Using it, one can parameterize the fifth coordinate in the $G(1, 4)$ space. The value $n = 1$ corresponds to the empty Minkowski space $M(1, 3)$ in which light moves at the velocity c. The propagation of light in a medium with $n \neq 1$ is interpreted as an exit of a photon from the Minkowski space and its transition into another subspace of $G(1, 4)$ space. This transition can be described as a rotation in the $G(1, 4)$ space. All types of such rotations were studied in [7.

For hyperbolic rotation through the angle θ in the (TS) plane the photon 5-vector (16) with zero mass is transformed in the following manner [7]:

$$\begin{gathered}\left(\frac{\hbar\omega}{c}, \frac{\hbar\omega}{c}, 0\right) \Rightarrow \\ \left(\frac{\hbar\omega}{c}\cosh\theta, \frac{\hbar\omega}{c}, \frac{\hbar\omega}{c}\sinh\theta\right) = \\ \left(\frac{\hbar\omega}{c}n, \frac{\hbar\omega}{c}, \frac{\hbar\omega}{c}\sqrt{n^2-1}\right).\end{gathered} \tag{18}$$

As a result of this transformation a particle with mass

$$m = \frac{\hbar\omega}{c^2}\sinh\theta = \frac{\hbar\omega}{c^2}\sqrt{n^2-1} \tag{19}$$

arise. The velocity of this particle is defined by formula (6).

Under the same rotation the massive 5-vector (17) is transformed as

$$\begin{gathered}\left(mc,\ 0,\ mc\right) \Rightarrow \left(mce^{\theta}\pm,\ 0,\ mce^{\theta}\pm\right); \\ e^{\theta}\pm = n \pm \sqrt{n^2-1}.\end{gathered} \tag{20}$$

Under such rotation a massive particle changes its mass

$$m \to me^{\theta},\ 0 \leq \theta < \infty \tag{21}$$

and energy but conserves its momentum.

Upon rotation through the angle ϕ in the (XS) plane the photon vector is transformed in accordance to the law

$$\left(\frac{\hbar\omega}{c},\frac{\hbar\omega}{c},0\right)\Rightarrow$$

$$\left(\frac{\hbar\omega}{c}\cosh\phi,\frac{\hbar\omega}{c},\frac{\hbar\omega}{c}\sin\phi\right)= \tag{22}$$

$$\left(\frac{\hbar\omega}{c},\frac{\hbar\omega}{cn},\frac{\hbar\omega}{cn}\sqrt{n^2-1}\right).$$

Given this, the photon acquires the mass

$$m=\frac{\hbar\omega}{c^2}\sin\phi=\frac{\hbar\omega}{c^2 n} \tag{23}$$

and velocity

$$v=c\cos\phi=\frac{c}{n}. \tag{24}$$

The vector of a massive particle is transformed in accordance to the law

$$\begin{gathered}(mc,\ 0,\ mc)\rightarrow(mc,-mc\sin\phi,mc\cos\phi)=\\ \left(mc,-\frac{mc}{n}\sqrt{n^2-1},\frac{mc}{n}\right).\end{gathered} \tag{25}$$

In this transformation the energy of a particle is conserved but its mass and momentum change

$$m\rightarrow m\cos\phi=\frac{m}{n} \tag{26}$$

$$0\rightarrow -mc\sin\phi=-\frac{mc}{n}\sqrt{n^2-1}. \tag{27}$$

The important fact is that the photon mass generated by transformations (18), (22) can have either a positive or a negative sign. This immediately follows from the symmetry properties of $G(1, 4)$ space. As to the particles that initially had positive mass, after transformations (20), (25) it remains positive.

4. Electrodynamics and Gravitation in the Extended Space

The source of the electromagnetic field is a current. In the traditional formulation of the electromagnetic theory the current is described by a 4-vector in Minkowski space $M(1, 3)$, [1].

$$\tilde{\rho}=\left(\rho,\vec{j}\right)=\left(\frac{\rho_0 c}{\sqrt{1-\beta^2}},\ \frac{\rho_0\vec{v}}{\sqrt{1-\beta^2}}\right), \tag{28}$$

$$\text{where}\quad \beta^2=\frac{v^2}{c^2},\tilde{\rho}^2=c^2\rho_0{}^2.$$

Here $\rho_0\,(t, x, y, z)$ - is an electric charge density in the point (t, x, y, z) in the space $M(1, 3)$, and $(v_x\,(t, x, y, z)$; $v_y\,(t, x, y, z)$; $v_z\,(t, x, y, z)$ - is a local velocity of a charge density.

At the transition to the extended $G(1, 4)$ space it is necessary to change a (1+3)-current vector $\tilde{\rho}$ by a (1+4)-vector $\bar{\rho}$. In accordance with the principles of the developed model, an additional coordinate of the vector $\bar{\rho}$ must be an isotropic (1+4)-vector. In addition, we want our model describes both the electromagnetic and gravitational field, so the fifth component of the current should be defined so that it be the source of the gravitational field.

We suppose that the source of a unit electromagnetic and gravitational field, is a particle which has both a mass and a charge. In this case, we assume that the mass may not have any charge, but the charge should always have a mass. In our model we assume that the charge is constant and does not change under transformations of the rotation group $L(1, 4)$ of the extended space $G(1, 4)$. And the rest mass, which is a scalar with respect to the Lorentz group, is a fifth component of the vector with respect to the group of $L(1, 4)$.

We want to construct a 5-dimensional current vector $\bar{\rho}$ as a generalization of 4-dimensional current vector $\tilde{\rho}$. To do this one must add one component to it.

In the ordinary electrodynamics 4-dimensional current vector $\tilde{\rho}$ has the form (28). Its structure it similar to structure of the energy-momentum vector (8) of the particle, having a rest mass. The difference between them is that in the vector (28) instead of the rest mass m_0 there is a local density of charge ρ_0. In the extended space $G(1, 4)$ we consider the energy-momentum-mass vector (28) instead of the energy-momentum vector (8).

Thus, the 5-dimensional vector of current, generating a unit electro-gravitational field, has the form

$$\bar{\rho}=\left(j_0,\vec{j},j_4\right)=\left(\frac{emc}{\sqrt{1-\beta^2}},\frac{em\vec{v}}{\sqrt{1-\beta^2}},emc\right) \tag{29}$$

It is an isotropic vector, $\bar{\rho}^2 = 0$.

The continuity equation, as in the usual case, is expressed by the vanishing of the 5-divergence of the 5-current (6)

$$\sum_{i=0}^{4} \frac{\partial j_i}{\partial x_i} = 0 \tag{30}$$

If the charge is at rest the continuity equation takes the form

$$\frac{\partial m}{\partial t} + \frac{\partial m}{\partial x_4} = 0 \tag{31}$$

Equation (31) can be interpreted as the variation of the rest mass of the particle by changing the properties of the environment. In ordinary electrodynamics from the continuity equation should be the law of conservation of charge

$$\frac{\partial}{\partial t}\int j_0 dV = -\vec{j}d\vec{n}. \tag{32}$$

There is an integral over the volume in the left side of this equation, and the right side - is the integral over the surface bounding this volume. In electro-gravidynamics there is a law of conservation of the value, which is the product of the charge at the mass of a particle, which hold this charge. This law reads

$$\frac{\partial}{\partial t}\int j_0 dV = -\int \vec{j}d\vec{n} - \int \frac{\partial}{\partial x_4} j_4 dV. \tag{33}$$

In this case, the change in the product of em of a charge at the mass inside a volume is defined as a stream of charged particles across the surface of the volume and change of masses of particles within the volume due to their dependence on the coordinate $x4$.

The current (29) generates the electrogravitational field in the extended space of $G(1, 4)$. The potentials of this field are determined by the equations [7,8].

$$\Delta^{(5)} A_0 = -4\pi\rho, \tag{34}$$

$$\Delta^{(5)} \vec{A} = -\frac{4\pi}{c}\vec{j}, \tag{35}$$

$$\Delta^{(5)} A_5 = -\frac{4\pi}{c} j_5. \tag{36}$$

Here

$$\Delta^{(5)} = \frac{\partial^2}{\partial s^2} + \frac{\partial^2}{\partial x^2} + \frac{\partial^2}{\partial y^2} + \frac{\partial^2}{\partial z^2} - \frac{1}{c^2}\frac{\partial^2}{\partial t^2}. \tag{37}$$

With the help of the potentials (A_0, A_x, Ay, A_z, A_s) one can construct the tension tensor

$$F_{ik} = \frac{\partial A_i}{\partial x_k} - \frac{\partial A_k}{\partial x_i};\ i,k = 0,1,2,3,4. \tag{38}$$

$$\|F_{ik}\| = \begin{pmatrix} 0 & -E_x & -E_y & -E_z & -Q \\ E_x & 0 & -H_z & H_y & -G_x \\ E_y & H_Z & 0 & -H_x & -G_y \\ E_z & -H_y & H_x & 0 & -G_z \\ Q & G_x & G_y & G_z & 0 \end{pmatrix} \tag{39}$$

Here

$$Q = F_{40} = \frac{\partial A_4}{\partial x_0} - \frac{\partial A_0}{\partial x_4} = \frac{\partial A_5}{c\partial t} - \frac{\partial \varphi}{\partial_S}. \tag{40}$$

$$G_x = F_{41} = \frac{\partial A_4}{\partial x_1} - \frac{\partial A_1}{\partial x_4} = \frac{\partial A_5}{\partial x} - \frac{\partial A_x}{\partial_S},$$

$$G_y = F_{42} = \frac{\partial A_4}{\partial x_2} - \frac{\partial A_2}{\partial x_4} = \frac{\partial A_5}{\partial y} - \frac{\partial A_y}{\partial_S}, \tag{41}$$

$$G_z = F_{43} = \frac{\partial A_4}{\partial x_3} - \frac{\partial A_3}{\partial x_4} = \frac{\partial A_5}{\partial z} - \frac{\partial A_z}{\partial_S}.$$

Here is the equation satisfied by the intensity of F_{ik}. We'll call them the extended Maxwell system. The usual system of Maxwell equations consists of two pairs of equations, which have fundamentally different structures. They are usually well known as the first and second pair of Maxwell's equations. Extended system of Maxwell's equations also consists of two types of equations fundamentally different structure. We shall call them the equations of the first and second types.

The equations of the first type are formal consequence of the formula (38), which expresses the tension via potentials. It follows immediately from their form that for any three indices (i, j, k) the relation

$$\frac{\partial F_{ij}}{\partial x_k} + \frac{\partial F_{ki}}{\partial x_j} + \frac{\partial F_{jk}}{\partial x_i} = 0 \tag{42}$$

is satisfied. The validity of (42) can be verified by direct substitution of the expression (38) in the equation (42).

$$\frac{\partial^2 A_i}{\partial x_k \partial x_j} - \frac{\partial^2 A_j}{\partial x_k \partial x_i} + \frac{\partial^2 A_k}{\partial x_j \partial x_i} -$$

$$\frac{\partial^2 A_i}{\partial x_j \partial x_k} + \frac{\partial^2 A_j}{\partial x_i \partial x_k} - \frac{\partial^2 A_k}{\partial x_i \partial x_j} = 0.$$

There are exist 10 such equations. Let us consider now the specific form of these equations, using the tension tensor (39).

If we restrict ourselves to the sets of indices taking values (0,1,2,3), then corresponding 4 equations are simply the first pair of Maxwell's equations

$$div\vec{H} = 0,\ \ sim\ indices\ (1,2,3). \tag{43}$$

This is one equation. The three other equations which

correspond to sets of indices (0,1,2), (0,1,3), (0,2,3) form a unit vector equation

$$rot\vec{E} + \frac{1}{c}\frac{\partial \vec{H}}{\partial t} = 0. \qquad (44)$$

Thus, the first pair of Maxwell's equations retain its form. In the extended space $G(1,4)$ another 6 equations are added to them. Three of them, that are corresponding to the sets (1,2,4) (1,3,4), (2,3,4), can be combined into one vector equation

$$rot\vec{G} + \frac{\partial \vec{H}}{\partial s} = 0. \qquad (45)$$

The other three triples (0,1,4) (0,2,4) (0,3,4) give us three remaining equations of the first type. They also can be merged into a single vector equation

$$\frac{\partial \vec{E}}{\partial s} + \frac{1}{c}\frac{\partial \vec{G}}{\partial t} + gradQ = 0. \qquad (46)$$

Thus, the equation of the first type from the exptended system of Maxwell's equations in the space $G(1,4)$ have the form (43)-(46). These 10 equations can be combined into three vector equations and one scalar equation. Note that the vector operators *div*, *rot*, *grad* that appear in these equations have the usual three-dimensional form.

Let's turn now to construction of the Maxwell equations of the second type. These equations are followed from the equations for the potentials (34)-(36). However, it is necessary first to impose the Lorentz gauge condition, which must satisfy potential (33). In the space $G(1,4)$ it has the form

$$\frac{1}{c}\frac{\partial A_0}{\partial t} + \frac{\partial A_x}{\partial x} + \frac{\partial A_y}{\partial y} + \frac{\partial A_z}{\partial z} + \frac{\partial A_s}{\partial s} = 0. \qquad (47)$$

The second type of Maxwell's equations from the extended system reads

$$\sum_{k=0}^{4} \frac{\partial F_{ik}}{\partial x_k} = -\frac{4\pi}{c} j_i\,; \quad i = 0,1,2,3,4. \qquad (48)$$

Substituting the elements of the tension tensor (39) into the equation (48) and taking into account the Lorentz gauge condition (47), one can obtain five vector equations.

$$div\vec{E} + \frac{\partial Q}{\partial s} = 4\pi\rho, \quad (i = 0) \qquad (49)$$

$$rot\vec{H} - \frac{\partial \vec{G}}{\partial s} - \frac{1}{c}\frac{\partial \vec{E}}{\partial t} = \frac{4\pi}{c}\vec{j}, \quad (i = 1,2,3) \qquad (50)$$

$$div\vec{G} + \frac{1}{c}\frac{\partial Q}{\partial t} = 4\pi j_4, \quad (i = 4). \qquad (51)$$

The tension tensor (39) contains, in addition to the components that are analogous to the usual electric and magnetic fields, some additional components that describe gravitational field. More precisely, in the case when the components of the 5-current (6) depend on the coordinate x_4, all components of (39) describe a unit electro-gravitational field. If the current does not depend on the coordinate x_4, the system of equations (44), (46), (48), (51) splits into two systems. One of them is the system of Maxwell's equations and the other is a Laplace equation for the scalar gravitational field.

Thus, according to our model, in an empty space the gravitational and electromagnetic fields exist as two different fields, but in the region where there are particles and fields they form a unit electromagnetic-gravitational field.

5. Refraction Index of a Gravitational Field

Let's now study a problem of refraction index of a gravitational field. Let there is a point mass, which gravitational field is described by the Schwarzchild solution. We assume, that the gravitational radius r_g is small and we will consider all effects at distance $r > r_g$. In the literature there are considered two expressions for refraction index n, appropriate to the Schwarzchild field. One of them, we shall name it n_1, is used in papers of Okun' [11,12] and looks like

$$n_1(r) = (g_{00})^{-1} = (1 - \frac{r_g}{r})^{-1} = 1 + \frac{2\gamma M}{rc^2}. \qquad (52)$$

It can be found in the supposition, that in a constant gravitational field the frequency of photon ω remains constant, but the wavelength λ and speed v are varied. The other refraction index n_2 one can get from the formula of an interval in a weak gravitational field [1].

$$ds^2 = (c^2 + 2\varphi)dt^2 - d\vec{r}^{\,2}. \qquad (53)$$

Here φ - is a potential of gravitational field. Supposing that $d\vec{r} = \vec{v}dt$ and $ds^2 = 0$, one can find a speed of photon in a gravitational field

$$v = c\left(1 + \frac{2\varphi}{c^2}\right)^{1/2} \approx c\left(1 + \frac{\varphi}{c^2}\right). \qquad (54)$$

It is necessary here to take into account the fact that a potential of a gravitational field φ - is negative. For a point mass M we have

$$\varphi(r) = -\frac{\gamma M}{r}. \qquad (55)$$

Substituting expression (55) in the formula (54), one gets

$$v \approx c\left(1 - \frac{\gamma M}{rc^2}\right). \qquad (56)$$

Collins obtained the same formula in another way [13]. He considered a particle of mass m_0, located indefinitely far from a point source of a gravitational field of mass M. Such particle has an energy $E_0 = m_0 c^2$. When moving on a distance r from a source of a field, the particles energy increases up to size

$$E = m_0 c^2 + \frac{\gamma m_0 M}{r}.$$

Collins offered to interpret this change of energy as change of a rest mass in a gravitational field.

$$m = m_0\left(1 + \frac{\gamma M}{rc^2}\right). \qquad (57)$$

Then he used a conservation law of a momentum $mv = m_0 v_0$ and received the law of change of speed in a gravitational field

$$v = v_0\left(1 + \frac{\gamma M}{rc^2}\right)^{-1}. \qquad (58)$$

Supposing, that this law is valid also for photons, we get the formula for change of photons speed in a gravitational field

$$v = c\left(1 + \frac{\gamma M}{rc^2}\right)^{-1} \approx c\left(1 - \frac{\gamma M}{rc^2}\right). \qquad (59)$$

It is possible to interpret the formulas (56), (59) as hit of a photon in medium with a refraction index:

$$n_2(r) = 1 + \frac{\gamma M}{rc^2}. \qquad (60)$$

In the case, when the speed of a particle v is comparable with the speed of light c, it is necessary to take into account in the formula (57) relativistic correction to a rest-mass m and to record it as

$$M = m_0\left(1 + \frac{\gamma M}{rc^2} + \frac{v^2}{2c^2}\right). \qquad (61)$$

Appropriate refraction index will look like

$$n' = 1 + \frac{\gamma M}{rc^2} + \frac{v^2}{2c^2}. \qquad (62)$$

Such difference in definition of refraction index of a gravitational field is connected with that the speech in these cases goes about different objects, which differently interact with a gravitational field. In ESM different rotations in extended space correspond to these situations.

6. Gravitational Effects in ESM

1) Speed of escape.

The speed of escape v_2 is that speed, which should be given to a body located on a surface of the Earth, that it could be deleted from Earth on an indefinitely large distance. Let M - mass of the Earth, m - mass of a body located at the Earth surface, and R - radius of this surface. The expression for the speed of escape is

$$v_2 = \sqrt{2gR} = \sqrt{\frac{2\gamma M}{R}}. \qquad (63)$$

We will receive now formula (63) using ESM methods. Let's consider a massive particle at rest, which removed to infinite large distance from the Earth. Within the framework of our model such particle is described by isotropic 5-vector of energy-momentum-mass (4). Space motion in gravitational field along an axis X can compare movement in extended space G(1, 4) in a plane XS from a point with refraction index $n = 1$ to point with refraction index $n(r)$. Such motion is described by some rotation in expanded space $G(1,4)$. The rotation angle express through refraction index n. Thus the massive particle at rest acquires speed

$$v = c\frac{\sqrt{n^2 - 1}}{n}.$$

In this case we consider a motion of a massive body, so we assume natural to use the refraction index n_2. Assuming, that it is close to unit, i.e. that

$$1 >> \varepsilon = \frac{\gamma M}{rc^2} \qquad (64)$$

One can get, that $V \approx c\sqrt{2\varepsilon}$. (65)

In the case when r – R –is the radius of the Earth, the formula (65) coincides with formula (63) and gives the speed of escape $v = v_2$.

2) Red shift.

Gravitational red shift usually considered as a change of frequency of a photon in the case of changing of a gravitational field, in which photon is submerged. In particular, at decreasing of strength of a field the frequency of a photon also decreases, see [1]. However Okun' offered to recognize that not frequency varies but the wavelength of a photon varies, and just it named as red displacement [11,12]. Under our judgment both cases are possible, but they correspond to different physical situations and are described by different rotation angles express through refraction index n. In the general theory of relativity the formula that describes change of light frequency is, [1]

$$\omega = \frac{\omega_0}{\sqrt{g_{00}}} \approx \omega_0\left(1+\frac{\gamma M}{rc^2}\right). \qquad (66)$$

Here ω_0 - is a frequency of photon, measured in universal time, it remains constant at propagation of a beam of light. And ω - is a frequency of the same photon which measured in its own time. This frequency is various in various points of space. If the photon was emitted by a massive star, near to a star at small r the frequency of a photon is more, than far from it at large r. On infinity in the flat space, where there is no gravitational field, the universal time coincides with own time, and ω_0 there is an observable frequency of a photon.

Let's consider now the same problem from the ESM point of view. Within the framework of our model the isotropic 5-vector (4) is compared to a photon located in empty space. Process of its movement to the point with refraction index n, at which the change of frequency and energy happens, is described by a rotation in (TS) plane. At these rotations the photon vector transforms as

$$\frac{\hbar\omega}{c}(1,1,0) \rightarrow \frac{\hbar\omega}{c}(\cosh\theta, 1, \sinh\theta) = \frac{\hbar\omega}{c}\left(n, 1, \sqrt{n^2-1}\right) \qquad (67)$$

One can see from here that the frequency ω_0 of a photon in vacuum and its frequency ω in a field, are connected by a ratio

$$\omega = \omega_0\cosh\theta = \omega_0 n. \qquad (68)$$

We assume, that at calculation of change of photon frequency it is necessary to use refraction index n_2, as to the index of refraction n_1 was found in the supposition, that this frequency does not vary. Substituting (60) in (68), we get the formula

$$\omega = \omega_0 n_2 = \omega_0\left(1+\frac{\gamma M}{rc^2}\right), \qquad (69)$$

which coincides with the formula (66). Thus in extended space model for red shift is received the same expression, as in general theory of relativity.

In papers [11,12] Okun' has offered to consider the red shift of a photon as change it of speed, momentum and wavelength, but the frequency was assumed constant. He proceeded from a dispersing ratio for a photon with zero mass in space with the Schwarzchild metric

$$g^{00}p_0p_0 - g^{rr}p_rp_r = 0. \qquad (70)$$

The Schwarzchild metric is

$$g^{00} = (1-\frac{r_g}{r})^{-1},\ g^{rr} = (1-\frac{r_g}{r}),\ r_g = \frac{2\gamma M}{c^2}. \qquad (71)$$

Assuming, that $\tilde{n}p_0 = \hbar\omega = const$, Okun' has found for relation of a momentum p_r the expression

$$p_r(r) = \frac{\hbar\omega}{c}(1-\frac{r_g}{r})^{-1} = p_r(\infty)n_1, \qquad (72)$$

Here $p_r(\infty)$ - is momentum of the photon at infinity, where the influence of gravitational field is absent. Using connection between a momentum of a photon and its wavelength $\lambda(r)$, we get the expression

$$\lambda(r) = \frac{2\pi}{\omega}v = \frac{2\pi c}{\omega}(1-\frac{r_g}{r}) = \frac{2\pi c}{\omega n_1} = \frac{\lambda(\infty)}{n_1}. \qquad (73)$$

For the speed of a photon $v(r)$ Okun' got the expression

$$v(r) = \frac{\lambda(r)\omega}{2\pi} = c(1-\frac{r_g}{r}) = \frac{c}{n_1}. \qquad (74)$$

Let's look now at transformation (67) from the ESM point of view. As the frequency of a photon remains constant, but vary its momentum and mass the appropriate transformation must be described by a rotation in the plane (XS) of the spaces G(1,4). As the frequency does not vary, we take the refraction index n_1. At such rotation the speed of a photon varies according to the formula

$$v = c\cos\psi = \frac{c}{n_1}. \qquad (75)$$

This formula coincides with the formula (6) for transformation of speed. Being repelled from it is possible to receive the formula (73), assigning change of a wavelength of a photon, when photon hit in a gravitational field.

From a point of view of our model it is necessary to consider the formula (67) only as first approximation to an exact result. Let's estimate correction appropriate to that in this model the photon, hitting in area with $n>1$ gains a nonzero mass. For this reason the part of photon energy can be connected not with the frequency, but with the mass. Let's estimate magnitude of this energy for case, when photon frequency change, in case of incident from a height H in a homogeneous gravitational field, with acceleration of gravity g is measured. Such situation was realized in well known Pound and Rebka experiments [14]. The energy change which appropriated to such frequency shift, is equal

$$\Delta E = \left(\frac{\hbar\omega}{c^2}\right) gH. \qquad (76)$$

According to the formula (6) in the case of rotation in tye plane (TS) the photon gains the mass

$$m = \frac{\hbar\omega}{c^2}\sqrt{n^2-1}. \qquad (77)$$

The difference of potential energies in the point of emission and the point of absorption of a photon, which differ by height H, is equal

$$\delta E = mgH = \left(\frac{\hbar\omega}{c^2}\right) gH\sqrt{n^2-1}. \qquad (78)$$

Near to a surface of the Earth refraction index of gravitational field is define by the formula (60). Taking into consideration an inequality (64), one can get an evaluation

$$\delta E = mgH = \left(\frac{\hbar\omega}{c^2}\right) gH\sqrt{n^2-1} \approx \qquad (79)$$

$$\left(\frac{\hbar\omega}{c^2}\right) gH\sqrt{\frac{2\gamma M}{Rc^2}} =$$

$$\left(\frac{\hbar\omega}{c^2}\right) gH\sqrt{\frac{2gR}{c^2}} \approx \left(\frac{\hbar\omega}{c^2}\right) gH(2.5\cdot 10^{-5}).$$

We see, that correction to effect connected to emerging of the photons nonzero mass, near to the Earth surface is only 10^{-5} from magnitude of the total effect.

3) Delay of radar echo.

The appearance of radar echo delay is, that the time of light distribution up to some object, and back, can differ in dependence from that, docs this light spread in a hollow, or in a gravitational field. Such delay was measured in experiments on location of Mercury and Venus [15]. Such experiments give satisfactory agreement with GR predictions. These experiments also were analyzed in [16]. Here we do not interesting to analysis of these work. We want only to indicate that the analytical expression for magnitude of delay of a radar echo in ESM coincides what is received in GRT. This result can be obtained from the fact that the photon time delay Δt is calculated from only from the photon velocity $v(t)$ [11,12]. Let's imagine that we locate the Sun. In this case we have

$$\Delta t = 2\left(\int_{R_s}^{r_e}\frac{dr}{v(r)} - \int_{R_s}^{r_e}\frac{dr}{c}\right). \qquad (80)$$

Here R_s - is a radius of Sun, r_g - is a gravitational radius of Sun, and r_e - is a distance from Earth up to Sun. The speed of light in a gravitational field is $v = \frac{c}{n}$. Here we deal with photons, therefore it is necessary to use the refraction index $n = n_1$. Substituting it in (80), we obtain

$$\Delta t = 2\frac{r_g}{c}\ln\frac{r_e}{R_s}. \qquad (81)$$

The formula (81) coincides with expression for magnitude of radar echo delay obtained in works [12], 16.

4) Deviation angle of light beam.

In general theory of relativity the magnitude of deviation angle $\delta\psi$ of a light beam from a rectilinear trajectory in the case of photon motion near to a massive body determine, deciding the cyconal equation which defining trajectory of this beam in a central-symmetrical gravitational field [1]. In this case one finds

$$\delta\psi = 4\frac{\gamma M}{Rc^2}. \qquad (82)$$

Here M - is a mass of a body, and R - is a distance at which the light beam passes from a field center. As in this case speech goes about photons motion it is necessary to use $n = n_1$. Let's consider two beams - one passes precisely through an edge of the Sun, and other at a distance x from it. It is supposed, that

$$h << R_s < r = \sqrt{x^2+R^2}.$$

In the case of passing by these rays of a linear segment of length dx the residual of optical paths will be

$$\delta x = dxn_1(r) - dxn_1(r + h\cos\varphi) = \qquad (83)$$

$$dx\left(1-\frac{r_g}{r}\right) - dx\left(1-\frac{r_g}{r+h\cos\varphi}\right) \approx$$

$$\approx \frac{r_g h\cos\varphi}{r^2}dx.$$

To such difference of optical paths there corresponds an angle of a wave front deviation

$$\delta\varphi \approx \frac{\delta x}{h} = \frac{r_g\cos\varphi}{r^2}dx = \frac{r_g R_s}{r^3}dx = \qquad (84)$$

$$\frac{r_g R_s dx}{(x^2+R_s^2)^{3/2}}.$$

Integrating this expression on x from $-\infty$ up to $+\infty$ we shall receive deviation angle

$$\varphi = r_g R_s \int_{-\infty}^{\infty} \frac{dx}{(x^2 + R_s^2)^{3/2}} = 2\frac{r_g}{R_s} = 2\frac{\gamma M}{R_s c^2}. \quad (85)$$

Expression (85) yields half the angle (82). This epression is obtained in the geometrical-optics approximation ignoring the fact that according to the ESM the photon must acquire a mass in the gravitational field. For this effect we now estimate the second half caused by the fact that in the gravitational field the photon acquires a nonzero mass. In the case under discussion the value of this mass is of no importance, and we denote simply as m_f. We analyze the motion of a particle of mass m_f assuming this particle to move with impact parameter R to the center of the gravitational field produced by the mass M. Let the motion of a particle in the (XY) plane be described by the Newton equation

$$m_f \frac{d^2 y}{dt^2} = -\gamma \frac{M m_f}{r^2} \frac{y}{r}. \text{ Here } r^2 = x^2 + y^2 \quad (86)$$

The photon mass m_f is supposed to be constant. For this reason we can exclude it from equation (86). We assume that the motion of the photon proceeds basically along the X axis, and the variable y varies only slightly remaining close to the value of the impact parameter R. We also consider the photon velocity to remain constant all the time and to be equal to the velocity c of light in vacuum. Therefore, using the relationship

$$y \approx R, \ x = ct, \quad (87)$$

we can transform equation (86) to the form

$$\frac{d^2 y}{dx^2} = \gamma \frac{MR}{r^3}. \quad (88)$$

After the first integration, we arrive at

$$\frac{dy}{dx} = \frac{\gamma M}{c^2 R} \frac{x}{\sqrt{x^2 + y^2}}. \quad (89)$$

Using this equation we can calculate the deflection angle

$$\theta \approx \frac{dy}{dx}|_{-\infty} - \frac{dy}{dx}|_{\infty} = 2\frac{\gamma M}{c^2 R} \quad (90)$$

When the impact parameter R equals the Sun radius R_s, the angle θ coincides with the angle φ given by the formula (85). In the ESM these two effects are summed up to yield the total deflection angle

$$\theta + \varphi = 4\frac{\gamma M}{c^2 R}. \quad (91)$$

The result (91) coincides with formula (82).

5) Perihelion precession of Mercury.

One more classical GR effect is the perihelion precession of Mercury. It arises due to a space curvature the Newton's law of an attraction is deformed. It reduces that the trajectory of a particle becomes nonclosed, and after of each rotation it the perihelion recessed at some angle. The magnitude of this rotation is determined by the law of interaction of a central mass M and mass m particles rotated around it. In case of a Schwarzchild potential the force of interaction of these masses is [17]

$$F(r) = -\frac{\gamma M m}{r^2 \sqrt{1 - R_g/r - v^2/c^2}}. \quad (92)$$

Here R_g is the gravitational radius corresponding to the mass M , and v is the orbital velocity of a particle of mass m. The velocity of Mercury orbital motion around the Sun is approximately equal to 48 km/s, which results in the relativistic correction $v^2/c^2 \approx 5\times10^{-8}$. The gravitational correction also attains $R_g/r \approx 5\times10^{-8}$ and is very close to the relativistic one. One can assume that the particle mass m in formula (92) depends on both the distance and velocity. However as both these corrections are small one can write the total transformation of mass m in the form

$$m \to m\left(1 + \frac{\gamma M}{rc^2} + \frac{1}{2}\frac{v^2}{c^2}\right). \quad (93)$$

The calculation using the expression for the force (92) with allowance for approximation (93) yields the Mercury perigee shift to the observed one.

Let's now analyze this problem from the stand point of ESM. We deal with a nonzero mass particle placed into the gravitational field. Since one cannot already ignore the relativistic corrections, we will use the refraction index in the form (62). In this case the particle gets from a domain with a refraction index $n = 1$ to a domain with refraction index n' because of the variation of the force acting on it, i.e. because of a change of its energy. Therefore the (TS) rotation in $G(1,4)$ space should be used. In the case of this rotation the massive vector is transformed in accordance with formula (20) and a massive particle changes its mass obeying the formula (21). Under such transformation particles of masses m_+ and m_- can arise from of mass m.

$$m_+ = m e^{\theta_+} = m(n + \sqrt{n^2 - 1}), \quad (94)$$

$$m_- = me^{\theta_-} = m(n - \sqrt{n^2 - 1}).$$

We assume that a macroscopic massive body has an equal number of particles transformed according to laws (94). We will us for this situation the average transformation law

$$m \to mn_2' = m\left(1 + \frac{\gamma M}{rc^2} + \frac{1}{2}\frac{v^2}{c^2}\right). \qquad (95)$$

As one can see formulas (93) and (95) coincide.

7. Radioastron Mission — Visible Size of Bubble Objects

In recent years new phenomena which go beyond the traditional GRT-based concepts concerning the structure of the Universe, have been discovered by astronomers. The essence of these phenomena is basically as follows:

1) The bulk of the Universe mass (more than 90%) is dark matter and hidden energy which is associated with the cosmic vacuum.
2) This dark matter does not emit electromagnetic radiation and does not interact with it, but shows gravitational properties.
3) The cosmic vacuum possesses negative pressure or, in other words shows antigravitational properties, which determine the current dynamics of Universe expansion.
4) Usual massive objects are surrounded by dark-matter halo.

The ESM gives us an approach for the explanation of these phenomena. As was already shown the motion in the additional fifth dimension corresponds to change of the particle rest mass. When the photon gets into the external field, it acquires a nonzero mass which can be either positive or negative. We suppose that the inertia of such mass is always positive, and it is only its gravitational properties that can have different signs. In a pair of photons born in an external field one has a positive and the other a negative mass. According to ESM the dark matter consists of massive photons. Positive-mass photons are concentrated around massive stars and black holes and form their halos. Negative-mass photons are throws away into the free cosmic space where they create an antigravitating vacuum with negative pressure. Hence, in our opinion, the dark matter mostly consists of positive-mass photons and the dark energy is generated by negative-mass photons. Different ways of assigning a nonzero mass to a photon are discussed in review [4]. The possibility of the existence of bodies with a negative mass was discussed in [18]. Attention has recently been drawn by a new gravitational model, the so-called gravstar, or gravitational condensate star [19]. It has been proposed as an alternative to black holes. These objects correspond to the solutions of Einstein equations which outside a region occupied by masses coincide with the Schwarzschild solution. Inside it there is another nonsingular solution, and so the metric as a whole appears to be nonsingular. The gravstar structure is similar to that of a bubble. A bubble has a rigid dense shell which is stressed because of a liquid substance pressing out from inside. This particular model is now typically used to explain the nature of some observed objects. It is shown that in ESM model formation of bubble gravitational structures is possible. In the frame of ESM one can obtain the following physical picture. Bubble gravitational objects have a halo formed by dark matter generated by photons with a positive mass. Now it becomes possible to predict some future results of visible size of supermassive objects in our Universe due to new stage of experimental astronomy development in the RadioAstron Project [20,22]. As to in RadioAstron Project has to reached unprecedented angular resolution equal 0,00001 it is becomes possible to distinguish real size of such objects active galactic nuclei to investigate the size of supermassive objects applying for black hole role and to obtain direct results of comparison event horizon with its radius. In the case when the visible radius of supermassive object will not exceed the gravitational radius we will agree that this object is real black hole. In the case when visible radius will be in 2 to 3 times larger than gravitational radius appropriated to the mass of supermassive object - we will discuss the nature of such objects with taking into account that one of the possible answers is the gravastar.

8. V838 Monocerotis Explosive Outburst — Local Big Bang or Light Echo?

The ESM can be applied to analysis of new phenomenon taking place in the Universe. They did not have until now any generally accepted physical explanation. One of these mysterious phenomena is an explosion of the star V838 Mon. In 2002, the previously unknown variable star V838 Monocerotis brightened suddenly by a factor of about 104. Unlike a supernova or nova, V838 Mon did not explosively eject its outer layers; rather, it simply expanded to become a cool supergiant with a moderate-velocity stellar wind. Investigation of this phenomenon show, that when combined with the high luminosity and unusual outburst behavior, these characteristics indicate that V838 Mon represents a hitherto unknown type of

stellar outburst, for which we have no completely satisfactory physical explanation [23]. The morphing sequence of six images taken by Hubble's Advanced Camera of V838 explosion could be found at [24]. V838 image evolution was attributed not to a cloud expanding (as are normal supernovas) but light echoing (hitting different parts) of an interstellar cloud. A set of mechanism were proposed to explain outburst see reference in [25]. There is some discussion with the nature of light echo and source of light echo materials due to estimation [26 mass light echo materials is about 90-150 mass of the Sun. In the frame of ESM one can consider a possibility to explain V838 explosive outburst [9,10]. We could interpret this phenomenon as local Big Bang. Thus we try to consider V838 evolution in ESM formalism as a new space local birth. In our (4+1)-dimensional model V838 expansion could be explained as evolution of high-density complex field consisting from 4
fields - scalar field Q, gravitational field G, electric field E and magnetic field H. We also discuss the nature of origin SiO maser emission from the direction of V838 [25]. In the framework of our ESM model we also explain the absence of large molecular lines shift in the spectrum of CO and AlO [25, 26].

9. Conclusion

In given work the generalization of Einstein's Special theory of relativity is proposed. It is the (4+1)-dimensional Extended space model. In this model gravity and electromagnetism are unified into a single field. The gravitational effects such as the speed of escape, gravitational red shift and detection of light can be found algebraically by the rotations in the (1+4) dimensional space. The dark matter and hidden energy get natural interpretation in the frame of this model. Thus, in ESM frameworks there is a mechanism, according to which in a neighborhood of massive gravitational objects the halo consisting of photons with positive masses can be formed. These halos we associate with a dark matter. The photons with negative masses are concentrated far from massive objects. These photons will create areas of a dark energy. Such areas are characterized by negative pressure and exhibits properties of antigravitation. We also consider a possibility to explain V838 explosive outburst with its help. We suppose that this phenomenon can be interpreted as a local Big Bang. In this case we see the movement of the space itself, and the rate of its expansion is seen as the rate of expansion of stars shell. The observed rate of expansion may exceed the speed of light.

References and Notes

[1] Landau, L.D. & Lifshitz, E.M. (1994) The classical theory of fields, Singapore: World Scientific.
[2] Schweber, S.S. (1961) An Introduction to Relativistic Quantum Field Theory, Singapore: World Scientific.
[3] Ginzburg, V.L. (1981) Theoretical Physics and Astrophysics Moscow: NAUKA.
[4] Rivlin, L.A. (1997) Physics-Uspekhi 40, 291.
[5] Okun', L.B. (1989) Physics-Uspekhi 32, 629.
[6] Kobayashi, S. & Nomizu, K. (1963) Foundation of Differential Geometry. Vol. 1, New York: London.
[7] Tsipenyuk, D.Yu. & Andreev, V.A. (2001) Bulletin of the Lebedev Physics Institute 6, 23 (2000); arXiv:gr-qc/0106093.
[8] Tsipenyuk, D.Yu. & Andreev, V.A. (2002) Bulletin of the Lebedev Physics Institute 6, 1; arXiv:physics/0302006, 2003.
[9] Tsipenyuk, D.Yu. & Andreev, V.A. (2004) Bulletin of the Lebedev Physics Institute, 9, 10; arXiv:physics/0506002.
[10] Tsipenyuk, D.Yu. & Andreev, V.A. (2004) Bulletin of the Lebedev Physics Institute 10, 13; arXiv:physics/0407144.
[11] Okun', L.B., Selivanov, K.G. & Telegdi, V.L. (1999) Physics-Uspekhi 42, 1141.
[12] Okun', L.B. (2000) Physics-Uspekhi 43, 1366.
[13] Collins, R.L. (1994) Gravity slows the speed of light, APS reprint server, (8/9/97) Singapore: World Scientific.
[14] Pound, R.V. & G.A. Rebka, (1960) Phys. Rev. Lett. 4, 337.
[15] Shapiro, I.I. (1964) Phys. Rev. Lett. 13, 789.
[16] Weinberg, S. (1972) Gravitation and Cosmology, New York: John Willey and Sons.
[17] Moeller, C. (1972) The Theory of Relativity, Oxford: Clarendon Press.
[18] Bondi, H. (1957) Rev. Mod. Phys. 29, 423.
[19] Mazur, P.O. & Mottola, E. (2010) Gravitational Condensate Stars: An Alternative to Black Holes, arXiv:gr-qc/0109035.
[20] RadioAstron Project, http://www.asc.rssi.ru/radioastron/index.htm.
[21] Radioastron, The Ground -Space Interferometer:http://www.asc.rssi.ru/radioastron/files=bookleten:pdf.
[22] Radioastron Handbook, http://www.asc.rssi.ru/radioastron/documents/rauh/en/rauh.pdf.
[23] Bond, H.E., Henden, A., Levay, Z.G. et al., An energetic stellar outburst accompanied by circumstellar light echoes, arXiv/astro-ph/0303513.
[24] V838 Mon outburst movie, http://www.spacetelescope.org/videos/heic0503a/.
[25] Kamiski, T. (2003) Extended CO emission in the field of the light echo of V838 Mon, arXiv:0801.1689v1.
[26] Kamiski, T., Tylenda, R. & Deguchi, S. (2002) A molecular cloud within the light echo of V838 Monocerotis, arXiv:1102.5237v1.

Both the Twin Paradox and GPS Data Show the Need for Additional Physics

J. N. PERCIVAL

79 Haviland Rd., Ridgefield, Connecticut 06877, U.S.A
Semantech, 193 Peaceable Street, Ridgefield, Connecticut 06877-4843 U.S.A

An analysis of the net proper time difference for a Twin Paradox scenario is shown to require an asymmetric construct such as a special physics frame and also shows that Special Relativity's time dilation equation does not describe proper time. An analysis of GPS confirms the need for a special physics frame and is sharply at odds with Special Relativity despite claims that GPS uses it.

Keywords: Inertial frame, Special relativity, Time dilation, Twin paradox

1. History of the Twin Paradox

1.1 *Einstein's Changing Views*

Einstein's 1905 paper [1] defined the basis for Special Relativity. In that paper, Einstein derived the symmetric time dilation equation which *applies equally to all inertial observers.* Then he wrote, "*From this*", referring to the time dilation equation, "*the following peculiar consequence follows*" that if a clock makes a round trip, it will lose time versus a stay-at-home clock.

Einstein's claim seemed to many to be unjustified and, hence, not just counter-intuitive but also counter-logical. Hence, Einstein's claim initiated what was first called the Clock Paradox as many asked, in essence, "*How can a symmetric effect, time dilation, cause an* *asymmetric result, namely, a net proper time difference (NPTD)?*" The critics said that the time dilation equation could not be interpreted as describing physical time dilation as that would mean that all the clocks in inertial frame A run faster than all the clocks in inertial frame B AND all the clocks in inertial frame B run faster than all the clocks in inertial frame A! Furthermore, even if one accepted the apparently self-contradictory interpretation of the time dilation equation that Einstein's claim implied, it must be as true for the traveling twin as for the stay-at-home twin as relativity claims that all inertial frames are equal and therefore, by applying Einstein's logic consistently from the traveling twin's frame(s), one would conclude that it was not the traveling clock that accumulated less time but rather the stay-at-home clock.

The relativists tried to counter this problem by claiming that because the traveling twin accelerated, Special Relativity could not be used from his frames. This did indeed identify a bona fide asymmetry. However, if this criterion were to be applied consistently, then this cure would be worse than the original illness as it would rule out virtually any practical application of Special Relativity. Furthermore, the claim seemed specious as the traveling twin would be in an inertial frame for the constant velocity segments which could be made to be an arbitrarily high percentage (e.g., 99.99999%) of the round trip and adding proper times from different frames has always been allowed.

These problems clearly troubled Einstein throughout his life. As discussed below, Einstein published a number of differing versions of Special Relativity with some specifying that time dilation was apparent and not physical. Mendel Sachs reviewed many Einstein papers and presentations and wrote [2], "*I will quote some of his comments that were made during his life, that convince me that Einstein did not believe, after all, that the kinematic relations, such as the Lorentz transformations of special relativity, or the space-time transformations between accelerated frames (of general relativity), can indeed induce relative physical changes in the make-up of matter, such as the claim of the twin paradox.*

"In a lecture that Einstein gave to the Prussian Academy of Sciences, in 1921, he said the following: "Geometry predicates nothing about relations of real things, but only geometry together with the purport of physical laws can do so... The idea of the measuring rod and the idea of the clock coordinated with it in the theory of relativity ***do not find their exact correspondence in the real world****. It is also clear that the solid body and the clock do not in the conceptual edifice of physics play the part of irreducible elements, but that of composite structures,* ***which may not play any independent part in theoretical physics.****" (Sachs' bold.)*

Sachs continued, "*Einstein then went on to say that, in spite of the foregoing comment, we should temporarily support the use of the length and time transformations as though they were physically real, because "we are still far from possessing such certain knowledge of theoretical principles as to be able to give exact theoretical constructions of solid bodies and clocks.""*

Sachs continued, "*Thus we see that, on the one hand, Einstein admitted that one must not, in principle, interpret the theory of relativity to imply that similar physical entities age differently by virtue of their relative motion. But, on the other hand, he said that we should nevertheless assume (for the time being!) that there is such a physical correlation between the physical aging of material entities and their relative motion-because we haven't yet learned how to treat the laws of matter in an exact way, when taking account of the measuring rods and clocks that are used to verify these laws.*"

In 1911, Einstein was shifting focus on the Twin Paradox solution to spacetime diagrams and the invariance of the interval [3]. Also, in 1911, Max von Laue made a similar shift [4]. (Incidentally, Paul Langevin, also in 1911, popularized the issue by focusing on the differential aging of a pair of twins rather than on a pair of clocks and the Clock Paradox was renamed the Twin Paradox [5].) In 1916, Einstein while working on General Relativity, sharpened his focus on spacetime diagrams and the invariance of the interval [6]. In 1918, Einstein articulated the so called General Relativity Explanation of the Twin Paradox [7]. That explanation contended that the turnaround acceleration creates an (artificial/virtual) gravitational field and the difference in gravitational potential between the accelerating twin and the stay-at-home twin causes the net proper time difference. The net proper time difference accumulates during the turnaround acceleration.

The General Relativity Explanation of the Twin Paradox gained some adherents such as Max Born [8]. The problems associated with this solution are discussed below in section 3.

In late 1918, Einstein treated an accelerationless Twin Paradox scenario [7] but the lack of accelerations and the lack of an asymmetry eliminated the relativist's rebuttals to the original Clock Paradox concerns and, hence, the discussion had gone full circle back to the original Clock Paradox problems. Regarding the above cited reference, C. S. Unnikrishnan wrote [9] about Einstein's imaginary discussion with a critic, "*Curiously, the discussion starts with a complaint by the critic that none of the relativists had adequately responded to the criticisms of relativity by many in journals. In fact, the critic accuses relativists of 'shirking' the issue. This certainly suggests that Einstein considered that none of the earlier discussions adequately addressed the problem and that it was necessary to respond. Ironically, Einstein's resolution goes against the standard resolutions discussed in textbooks and in most other writings! As the physical cause of the asymmetry he uses the pseudo-gravitational field and the gravitational time dilation of general relativity, after admitting that special relativity is not suitable for resolving the issue due to the fact that one of the twins undergoes accelerations during his trip.*"

So we see that there are strong arguments against claiming that Special Relativity's time dilation equation addresses proper time and, particularly, asymmetric proper time accumulation, arguments that Einstein himself explicitly concurred with.

1.2 *Changing Mainstream Views*

Einstein's resolution of the Twin Paradox had shifted by 1918 from Special Relativity's time dilation to a General Relativistic virtual gravitational field and many such as Born agreed with that shift. However, the accepted resolution remained Special Relativity's time dilation well into the 1950s. During that period, Herbert Dingle, one of the most distinguished relativists of his era, argued that Einstein's original explanation for the net proper time difference led to a contradiction [10]. Dingle's debates with the mainstream on that point were published and drew great attention. Relativists claimed that Dingle was defeated and a quack. However, as E. G. Cullwick put it, "*On one thing Professor Dingle's critics are all agreed, that he is wrong. They do not all agree, however, on the nature of his error.*" [11] Furthermore, Hasok Chang, then of Harvard, did a thorough review of the Dingle debates [12] and concluded that Dingle's opponents had NOT addressed the arguments he raised.

So during the period discussed above, two mutually exclusive resolutions of the Twin Paradox were accepted despite many shortcomings having been pointed out. Presumably because of those alleged shortcomings, the list of different (often mutually exclusive), possible, published resolutions of the Twin Paradox continued to grow [13] (Also, see http://TwinParadox.net – the "Accepted Resolutions" and "References" sections of the Report page.) A review of the major categories and types of reconciliation arguments is given below.

2. Problems with Twin-Paradox Reconciliation Arguments

Two major groups of arguments that try to reconcile or explain the net proper time difference in the Twin Paradox with currently accepted theory are analyzed below and found wanting. A third type of solution that is built on a construct that lies outside of currently accepted theory is shown to straightforwardly resolve the paradox.

A problem common to most reconciliation arguments is they mix proper time which is the issue at hand and an invariant with observed time which is observer dependent. Also, they try to compare clock readings at a distance based on relative simultaneity which is observer dependent. These problems are avoided by limiting the discussion to the proper time accumulated between well-defined events and empirical data about invariants.

2.1 The Constant Velocity-Symmetric Effects Class

This class of reconciliation argument claims that the net proper time difference (NPTD) accumulates smoothly during the constant velocity segments and **half** of the NPTD accumulates during the outbound segment and **half** during the inbound segment.

We use the empirical data that all clocks at rest in the same inertial frame accumulate proper time at the same rate - other things being equal. As a corollary to the above, we note that if a clock in an inertial frame accumulates a net proper time difference versus another clock in another frame, then it also accumulates the same net proper time difference versus all other clocks, accumulating proper time at the same rate, in that other inertial frame.

If one just looks at a single Twin Paradox scenario, everything may seem fine with this type of reconciliation argument. However, if one adds a second Twin Paradox scenario and one switches the roles of two frames such that the original stay-at-home frame, the S-frame, now takes on the role of the outbound constant velocity frame and where the original traveling twin's constant velocity outbound frame, the T-frame, now takes on the role of the stay-at-home frame, then this reconciliation argument leads to contradictory conclusions. On the one hand, when this class of reconciliation argument is applied to the original Twin Paradox scenario, it claims that a clock at rest in the S-frame **gains** net proper time versus all clocks at rest in the T frame for the constant velocity outbound segment including all its sub-segments. On the other hand, when this class of reconciliation argument is applied to the 2nd, nested Twin Paradox scenario, it claims that a clock at rest in the S-frame **loses** net proper time for the constant velocity outbound segment, including all its sub-segments, versus all the clocks at rest in the T-frame. Hence, consistent use of this type of reconciliation argument leads to contradictory conclusions.

The above should be sufficient. However, we'll analyze a totally different problem. First, we'll modify the above classic Twin Paradox scenario such that the traveling twin, **T**, does not start at rest in the stay-at-home frame, the S-frame, but rather the scenario begins with **T** already traveling at relative velocity v with respect to the S-frame, having an arbitrarily near miss with the stay-at-home twin, **S**, to start the round trip.

Next, let's say that when **T** gets to the space-time point where in the classic scenario above he started his turnaround acceleration, a decision is randomly made as to whether the **T** or the **S** twin will actually be the traveling twin. In this modified scenario, at this space-time point, there's complete relativistic symmetry between the twins as both will have been at rest in an inertial frame the whole time.

If it's decided that **S** is the traveling twin, then, **S** commences his turnaround acceleration. The constant-velocity-symmetric-effects class of reconciliation argument holds that 1/2 the net time difference has accumulated by this point - in other words, **S** has accumulated a **negative** net proper time difference with respect to **T**.

However, if it's decided that **T** will be the traveling twin, then, the constant-velocity-symmetric-effects class of reconciliation argument holds that a very different net time difference has accumulated at the turnaround point - in other words, at that point, **S** has accumulated a **positive** net proper time difference with respect to **T**. Again, consistent use of this type of reconciliation argument leads to contradictory conclusions.

[Note that the above arguments not only show that Special Relativity's time dilation cannot be used to explain the net proper time difference (NPTD), but also show that any argument that meets the criteria of the opening paragraph of this section fails. In addition, it also shows that Special Relativity's time dilation cannot be describing proper time or what's happening physically as that interpretation leads to contradictions.]

3. The Turnaround Class

If one contends that the net proper time difference (NPTD) is caused by the traveling twin's turnaround, then one encounters the problems discussed below.

As in the prior section, we'll designate **S** as the stay-at-home twin who's at rest in the S-frame and **T** as the traveling twin who's at rest in the T-frame during the constant velocity segment of the outbound leg. In addition, we'll introduce **S2** who's at rest in the S-frame and is at the point where **T** comes to rest again, momentarily, in the S-frame, during the turnaround. We'll assume that **S**'s proper time for the round trip is relatively long (e.g., two million years) compared to the turnaround acceleration portion (e.g., two days).

We'll assume that the relative velocity is 0.866c yielding a time dilation factor of ½.

3.1 Problem 1

Since **S** and **S2** are at rest in the same inertial frame and hence, they accumulate proper time at the same rate, this presents some problems.

If we use the logic of this type of solution to calculate the net proper time difference (NPTD) that accumulates between **S** and **T** between the start and the end of the turnaround, we get one million years of proper time. One of the factors that allegedly cause such a large NPTD is the large distance separation between **S** and **T** during the turnaround acceleration. However, if one uses the *same methodology* for computing the difference in proper time for **T** versus **S2**, we will compute a very small net proper time difference (approximately one day of proper time) due to the small distance separation between **S2** and **T** during the turnaround. Yet the **S** and the **S2** clocks are accumulating proper time at the same rate while at rest in the S-frame. Hence, there's a contradiction.

3.2 Problem 2

The above should be sufficient. However, looking at other problems may be instructive as well.

This class of reconciliation argument claims that the traveling twin loses about a million years in his turnaround acceleration versus **S** and, hence, versus all observers in the S-frame whose clocks are ticking in unison. In contrast, the argument says that the traveling twin loses just less than a day of proper time during his *start* and *return* accelerations versus S because during the initial and final accelerations the distance of separation factor is small

Yet, if we have **T** continually going back and forth in a series of Twin Paradox scenarios, then most acceleration segments become **both** turnaround acceleration segments for one scenario and start/return acceleration segments for another scenario. Hence, the *very same acceleration segment* is claimed to cause very large net proper time difference between **T** and the clocks at rest in the S-frame and conversely to not cause a very large net proper time difference between **T** and the clocks at rest in the S-frame. This is a contradiction.

3.3 Problem 3

This class of reconciliation argument claims that the net proper time difference accumulates during the turnaround acceleration. Hence, about a million years of net proper time difference are claimed to accumulate during the turnaround acceleration. However, let's say **T** decides right after the turnaround acceleration to not return to **S**, but instead to return to being at rest in the S-frame not far from **S2**. Consistently using the reconciliation argument's methodology would indicate that since it was a very short trip from **S2** , only a very small net proper time difference would accumulate versus **S2** and, hence, versus **S**. On the one hand, the reconciliation argument says that the total net proper time difference has accumulated during the turnaround, but on the other hand the size of the net proper time difference due to the turnaround acceleration depends on what happens AFTER the turnaround.

3.4 Problem 4

If you have a Twin Paradox scenario *without* any turnaround acceleration, theory still predicts approximately the same net proper time difference.

We modify our basic scenario so that **T**, instead of decelerating to come to rest with **S2**, simply has a near miss with **S2** and keeps on going. We further suppose that that near miss is really a 3-way near miss and includes **T2** passing by **S2** going back to **S** at velocity 0.866c. If one computes the NPTD for **T** and **S** between the start of the round trip and the three way near miss at the midpoint and adds that to the NPTD for **T2** and **S** between the three way near miss and the end of the round trip, it is basically the same as the net time difference for the standard Twin Paradox. This suggests that the *effects of the turnaround acceleration* are not the cause of the net proper time difference.

3.5 Problem 5

Whatever method this reconciliation argument is employing, it should be dependent on the physical parameters that define **T**'s turnaround. The problem is that one can define any number of round trips where **T** has *exactly* the same parameters for the acceleration (same location, and start and exit velocity, same rate of acceleration, etc., etc.) and yet this alleged explanation dictates that the same physical parameters will yield wildly varying predictions for the net proper time difference between **T** and **S**. Why should the physical effects of the very same acceleration be dependent on whether the traveling twin stopped for tea at **S** or much further away at "**S9**" at a very distant point in the past?

3.6 Problem 6

The data on time dilation does not match this type of claim. The data says that the time differences are a function of velocity and NOT acceleration per se or frame change per se or General Relativistic effects [14].

4. Changes in Relative Simultaneity as Cause

Some claim that the change in relative simultaneity causes this huge jump in net proper time difference. Interestingly, relative simultaneity as cause has been used in reconciliation arguments from both the "Constant Velocity-Symmetric Effects" class and the "Turnaround" class and as such is subject to all the problems listed above in sections 2 and 3, respectively.

However, since relative simultaneity has been a popular proposed solution, we'll make one additional point that's specifically associated with it. Relative simultaneity is a measure of differences in *observations* - differences in what different observers consider to be simultaneous - not about how proper time accumulates. One can rearrange the distance of separation and relative velocity terms in relative simultaneity formulae to compute the "correct numerical answer" for the final, total NPTD. However, that result is in terms of observed time not proper time. Note that many *mutually exclusive* solutions to the Twin Paradox have been manufactured to produce the "correct numerical answer" so that's not sufficient to know one has the right *physics* answer. While one can compute the final, total NPTD, using the change in relative simultaneity that approach is unconnected from proper time accumulation.

As an interesting aside, some, such as Popper[15], contend that Quantum Mechanics' nonlocality shows the existence of absolute simultaneity. When a pair of particles is created and later a certain property of one of those particles is observed, then that same property is "instantly" set for the other particle regardless of the distance of separation. In other words, the other particle's property is set *simultaneously* with the observation of the first particle *wherever that other particle is in space*. This implied absolute simultaneity and absolute space and time to Popper [15].

5. The Mainstream Response — "Unknowable"

As of 4/9/12, 140 had signed the NPA (Natural Philosophy Alliance) Twin Paradox Open Letter agreeing that the topic needed further analysis and discussion. The five who disagreed gave a consistent response, namely, that how the NPTD accumulates was "unknowable". (For a detailed examination of that response, one should go to the "Mainstream Response" page of the Open Letter web site at http://TwinParadox.net)

Given that that the accepted solution of the Twin Paradox has changed multiple times over the past century or so and that during most of that time many, mutually exclusive solutions have been published in top physics journals only to be tacitly abandoned when analyzed from a *physics*, as opposed to a pure math, point of view, this is not a surprising position. In fact, the NPA agrees with that position as long as it's qualified to be "*unknowable in the context of currently accepted spacetime theory*".

(As an aside, these five, "mainstream" responders declined to answer any of the survey questions about whether any of the well-known, specific solutions to the Twin Paradox (e.g., time dilation, virtual gravitational field, relative simultaneity) were either valid or invalid.)

5.1 Problem

Even this mainstream retreat to agnosticism regarding how the NPTD accumulates still runs into problems. If, in general, the NPTD accumulates half in the inbound and half in the outbound segment, then the same contradictions that were described in section 2 above apply. However, if, in general, the NPTD does not accumulate half in the inbound and half in the outbound segment, then that means the NPTD accumulation is direction dependent and, hence, frame dependent which is contrary to the 1st relativity principle and it, in fact, implies a hierarchy of frames which would necessarily have a special frame at the bottom of the hierarchy (e.g., where clocks run the fastest). Again, the NPA would agree with this conclusion and recommend that it be seriously investigated.

6. GPS Data Confirms the Need for a New Construct

It is widely claimed and widely believed that GPS is built on Special Relativity's time dilation (and General Relativistic effects). Tom Van Flandern explains how GPS works [16]. For velocity effects, Lorentz Relativity is used with the ECI frame (the Earth Centered Inertial frame) being used as the base frame. Satellite clock rates are adjusted so that General Relativistic effects (differences in gravitational potential) and velocity effects will be accounted for and all clocks, including earth based clocks, will run at the same rate. Note that none of the clocks are in an inertial frame.

Van Flandern is kind to Special Relativity and notes that while the model is Lorentz Relativity, the data is consistent with Special Relativity. However, he is only looking at half the data – at the data from the earth clocks' perspective. From that perspective, the satellite clocks would be running **slow** due to velocity if the compensating adjustment had not been made. However, from the satellite clock perspective, earth clocks would be running **fast** (not slow as Special

Relativity's time dilation, if applied consistently, predicts) due to velocity if the compensating adjustment had not been made.

Proponents of Special Relativity might try to rule out this data that is in sharp contradiction with Special Relativity's predictions by claiming such observations are invalid as the satellites are not inertial, but this would also rule out using Special Relativity for the GPS system for the velocity component as neither satellites nor the rotating earth are inertial. It would also rule out claiming that any earth based data on "time dilation" supports Special Relativity.

7. A Compelling Solution Outside Currently Accepted Theory

To explain an asymmetric effect one needs an asymmetric cause/property/characteristic. One such construct is a special physics frame where velocity with respect to that frame causes clock retardation, thus the net proper time difference increases or decreases depending on whether **S** or **T** has the greater velocity with respect to a special physics frame. Thus, this construct avoids the problems discussed above and gives a straightforward, physics explanation of how the net proper time difference accumulates without a hint of a paradox.

Also, an unbiased look at existing data shows no support for symmetric time dilation, but rather clear support for asymmetric clock retardation.

8. Conclusion

The logic in section 2 shows that Special Relativity's time dilation equation, using relative velocity, cannot explain the net proper time difference in the Twin Paradox. The physics community's abandonment of that solution confirms that view. Further, the same logic shows that Special Relativity's time dilation equation does not describe proper time. As discussed above, Einstein confirmed that conclusion. Yet, the physics community still claims that empirical clock slowing data supports Special Relativity. However, a close look at GPS data shows that the clock slowing velocity effect is asymmetric and, hence, is not consistent with Special Relativity even in the context of "just observed" clock rates.

Given the above, the physics community needs to give serious consideration to the construct of a special physics frame where velocity with respect to that frame causes physical, asymmetric clock retardation as that construct is currently the only construct that provides a logical, physics basis for understanding clock slowing data and it is also consistent with all such data. Further, the physics community needs to give serious consideration to using Lorentz Relativity either in addition to Special Relativity or instead of Special Relativity. If the reader concurs or thinks that the Twin Paradox and/or related issues warrant serious and open discussion by the physics community, please sign the Twin Paradox Open Letter at http://TwinParadox.net

References and Notes

[1] Einstein, A. (1905) On the Electrodynamics of Moving Bodies, *Annalen der Physik* 17 891.

[2] Sachs, M. (1985) On Einstein's later view of the twin paradox, Foundations of Physics, Vol. 15, No. 9.

[3] Einstein, A. (1911) *Die Relativitäts-Theorie*, *Naturforschende Gesellschaft*, Zürich, *Vierteljahresschrift* 56: 1–14.

[4] von Laue, M. (1911) Two objections against the theory of relativity and their refutation, *Physikalische Zeitschrift* 13: 118–120.

[5] Langevin, P. (1911) The evolution of space and time, *Scientia* 10: 31–54.

[6] Einstein, A. (1916) The foundation of the general theory of relativity.

[7] Einstein, A. (1918) Dialog about objections against the theory of relativity, *Die Naturwissenschaften* 48, pp. 697-702.

[8] Born, M. (1965) Einstein's Theory of Relativity, p. 261, p. 355, New York: Dover.

[9] Unnikrishnan, C.S. (2005) Current Science, Vol. 89, No. 12, p. 2008, online at: http://cs-test.ias.ac.in/cs/Downloads/article_39568.pdf

[10] Dingle, H. (1972) Science at the Crossroads, Martin Brian & O'Keeffe, p. 129. Downloadable from NPA at http://www.worldsci.org/php/index.php?tab0=Books&tab1=Display&id=269&tab=3.

[11] Cullwick, E.G. (1959) Bull. Inst. Phys. 10 52.

[12] Chang, H. (1993) Stud. Hist. Phil. Sci. 24,5, 741.

[13] Percival, J. N. (1995) Phys. Essays, 8:1, 29.

[14] Misner, C.W., Thorne, K. & Wheeler, J.A. (1973) Gravitation, San Francisco: W.H. Freeman & Co., p. 1055.

[15] Popper, K. (1982) Quantum Theory and the Schism in Physics, p. 30, Rowan & Littlefield.

[16] Van Flandern, T (2003) Aperion, 10 #1, 69.

[17] Additional information is in the References section at the end of the Report page in the Open Letter web site at http://TwinParadox.net.

An Alternative to Dirac's Model, Confining Change due to the Bound Electron, not in the Field, but Within the Electron Itself

TOLGA YARMAN,[a] METİN ARİK,[b] FATİH OZAYDİN,[a] AZMİ ALİ ALTİNTAS,[a] LİDYA AMON SUSAM[c]
[a]*Okan University,Tuzla, 34959, Istanbul, Turkey*
[b]*Bogazici University, Bebek, Istanbul, Turkey*
[c]*Istanbul University, Istanbul, Turkey*
[a]*tolgayarman@gmail.com*

Yarman et al. proposed a new Dirac-like equation confining the change due to the bound electron, not in the field, but within the electron itself, as an alternative to the standard field approach. Solving this equation, they showed that for the ground state of the Hydrogen atom, the two approaches produce the same result, which supports previous efforts towards the description of the micro and macro worlds, based on the same philosophy, thus providing a *unique fusion* of these worlds. In this work we study the energies of the excited states of the Hydrogen atom at possible total angular momenta, j's. And for the transition between any pair of *j*'s, we find practically the same energy difference as that predicted by the standard Dirac equation, but surprisingly with the opposite sign. Since not the energy levels alone, but the energy differences between levels are measured, we come to the amazing bifurcation point of choosing between the *change-in-the-field-concept* versus the *change-in-the-particle-concept*. It is further interesting to note that the levels such as 2S(1/2) and 2P(1/2), which are identical in the standard approach, turn also to be identical in our approach. We happen accordingly to predict the same Lamb shift as the difference of the magnitudes of the lines 2P(3/2)-2S(1/2) and 2P(3/2)-2P(1/2), assuming yet that the 2S(1/2) level *(Lamb-shift-wise)* is displaced downward *(and not upward)*, in contrast with what the common approach, rather deliberately, supposes. While the standard approach, neither in the celestial world nor in the atomic world, allows any unification of these two worlds, our approach, on the contrary, provides an easy fusion of them through simple means, and the same change of metric, allowing further an immediate quantization of the celestial world. The main problem in the atomic world, according to our approach, arises to be, the choice between the following two statements:1) The greater j, the higher is the energy level *(common approach)*. 2) The greater j, the lower is the energy level *(our approach)*. A heuristic assessment favors our approach against the common approach.

Keywords: Dirac Equation; relativity; hydrogen atom

1. Introduction

Starting with just the energy conservation law and the special theory of relativity, Yarman was successful to obtain the end results of general theory of relativity; he and his colleagues thus extended the approach, to setup a basis for a possible common description of different fields [2-12].

The key question we ask here is, whether any kind of interaction alters the field of concern, or instead, the overall relativistic energies inside the micro particles (or macro objects) of concern.

According to the latter approach, for the case of Hydrogen atom, instead of considering the classical potential energy, next to the kinetic energy of the given object, we propose to alter the rest mass of the electron as much as the static binding energy coming into play, when bound to proton. Thus we modify [1] the standard Dirac equation, i.e.

$$\vec{\alpha}.\vec{p}c + \beta mc^2 = E_D + \frac{Ze^2}{r} \tag{1}$$

(*written with the familiar notation*), to propose

$$\vec{\alpha}.\vec{p}c + \beta m\left(1 - \frac{Ze^2}{rmc^2}\right)c^2 = E_Y \tag{2}$$

The solution of standard Dirac equation, i.e. the total energy E_D, turns out, along with $m_{0\infty}c^2$ unity, as usual to be:

$$E_D = \frac{1}{\sqrt{1 + \dfrac{Z^2\alpha^2}{\left(n_r + \sqrt{\left(j+\frac{1}{2}\right)^2 - Z^2\alpha^2}\right)^2}}} \tag{3}$$

whereas the solution of our modified Dirac Equation, i.e. the total energy E_Y of Eq.(2) - *through a mathematical derivation similar to that one achieves, to solve Eq.*(1) - is obtained to be

$$E_Y = \sqrt{1 - \frac{Z^2\alpha^2}{\left(n_r + \sqrt{\left(j+\frac{1}{2}\right)^2 + Z^2\alpha^2}\right)^2}} \tag{4}$$

Here n_r is the radial *quantum number* and j is the *total angular momentum quantum number,* and α is the fine structure constant [13]. The *principle quantum number* can also be written as $n = n_r + j + \frac{1}{2}$. Note that for Hydrogen atom, $Z = 1$.

Eq.(2) is written in the frame of reference of the electron [1], assuming that the proton being much heavier than the electron, we can well assume, it does not move through the binding process. In our approach, the distance of the electron as measured by an observer attached to the electron, differs from the same distance measured by an outside observer, far from the interaction in consideration. Indeed, in the approach we present herein, already due to static binding, the internal dynamics of the electron slows down (as discussed in Reference [11]), while the velocity of light stays unaltered.

Thereby an observer attached to the electron will count a lesser number of ticks for a light signal he would send to the proton, to go forth and to come back, as compared to the number of ticks an outside observer would count for the same round trip process.

We assume though that Coulomb force (*as supposed classically, and in fact, imposed by the special theory of relativity*), is valid in the reference frame of the bound electron, which represents a general case, the frame of the distant observer remaining an extension of this.

Thus, since we come to write Eq.(2) in the reference frame of electron, as proposed by Yarman, we ought to make the transformation $(Z\alpha)^2 \rightarrow \frac{(Z\alpha)^2}{1-(Z\alpha)^2}$ to Eq.(4), to gear, it to what the outside observer will measure:

$$E_Y = \sqrt{1 - \frac{\frac{(Z\alpha)^2}{1-(Z\alpha)^2}}{\left(n_r + \sqrt{\left(j+\frac{1}{2}\right)^2 + \frac{(Z\alpha)^2}{1-(Z\alpha)^2}}\right)^2}} \tag{5}$$

We thus calculated the ground state energies of Hydrogen atom for both Eq.(3) and Eq.(5), which thus for $n_r = 0$ and $j = 1/2$ comes to be

$$E_D = E_Y = \sqrt{1 - Z^2\alpha^2} = 0.9999733739684983$$

and the corresponding *ionization energy* is

$$E_I = 13.605873075004437\ eV$$

Now, as a further prediction, in this work, we study the energy differences between levels of possible angular momenta at excited states of Hydrogen atom. Thus, for the given transitions, we obtain energy intervals presented in Table 1.

Table 1. Comparison Of Energy Differences Between Levels Of Possible Total Angular Momenta

n=2	Dirac	Present Work	(|Dirac|-|Present|) / |Dirac|
J: 3/2 →1/2	8.86188×10^{-11}	8.86140×10^{-11}	5.41646×10^{-5}
n=3			
J: 5/2 →3/2	8.75222×10^{-12}	8.75211×10^{-12}	1.25682×10^{-5}
J: 3/2 →1/2	2.62573×10^{-11}	2.62557×10^{-11}	6.09354×10^{-5}
J: 5/2 →1/2	3.50095×10^{-11}	3.50078×10^{-11}	4.85582×10^{-5}
n=4			
J: 7/2 →5/2	1.84630×10^{-12}	1.84618×10^{-12}	6.49949×10^{-5}
J: 5/2 →3/2	3.69237×10^{-12}	3.69226×10^{-12}	2.87912×10^{-5}
J: 3/2 →1/2	1.10772×10^{-11}	1.10766×10^{-11}	5.41653×10^{-5}
J: 7/2 →3/2	5.53868×10^{-12}	5.53845×10^{-12}	4.15261×10^{-5}
J: 7/2 →1/2	1.66159×10^{-11}	1.66151×10^{-11}	4.81467×10^{-5}
J: 5/2 →1/2	1.47696×10^{-11}	1.47689×10^{-11}	4.73946×10^{-5}
n=5			
J: 9/2 →7/2	5.67323×10^{-13}	5.67101×10^{-13}	3.91311×10^{-4}
J: 7/2 →5/2	9.45243×10^{-13}	9.45243×10^{-13}	0
J: 5/2 →3/2	1.89048×10^{-12}	1.89048×10^{-12}	0
J: 3/2 →1/2	5.67146×10^{-12}	5.67124×10^{-12}	3.87907×10^{-5}
J: 9/2 →5/2	1.51256×10^{-12}	1.51234×10^{-12}	1.45449×10^{-4}
J: 9/2 →3/2	3.40305×10^{-12}	3.40283×10^{-12}	6.46479×10^{-5}
J: 9/2 →1/2	9.07451×10^{-12}	9.07407×10^{-12}	4.84875×10^{-5}
J: 7/2 →3/2	2.83573×10^{-12}	2.83573×10^{-12}	0
J: 7/2 →1/2	8.50719×10^{-12}	8.50697×10^{-12}	2.58605×10^{-5}
J: 5/2 →1/2	7.56195×10^{-12}	7.56172×10^{-12}	3.04154×10^{-5}

2. Conclusions

We have henceforth shown that our alternative approach predicts practically the same energy differences between given levels of Hydrogen atom, as those predicted by the standard model, yet with opposite sign.

Since not the energy levels alone, but the energy differences between levels are measured, we come to the amazing bifurcation point of choosing between the *change-in-the-field-concept* versus the *change-in-the-particle-concept.* It is further interesting to note that the levels such as 2S(1/2) and 2P(1/2), which are identical in the standard approach, turn also to be identical in our approach. We happen accordingly to predict the same Lamb shift as the difference of the magnitudes of the lines 2P(3/2)-2S(1/2) and 2P(3/2)-2P(1/2), assuming yet that the 2S(1/2) level (*Lamb-shift-wise*) is displaced downward (and not upward), in contrast to what the common approach, rather deliberately, supposes.

While the standard approach, neither in the celestial world nor in the atomic world, allows any unification of these two worlds, our approach, on the contrary, provides an easy fusion of them through simple means, and the same change of metric, allowing further an immediate quantization of the celestial world. The main problem in the atomic world, according to our approach, arises to be, the choice between the following two statements:1) The greater *j*, the higher is the energy level *(common approach).* 2) The greater *j*, the lower is the energy level *(our approach).* A heuristic assessment favors our approach against the common approach. Indeed one can reason as follows. If the direction of the spin motion of the electron, on the outer side of the orbit, lies, parallel to the direction of the orbital motion of the electron, then *j* would be greater than what it would have been in the opposite case. In the former case, we have, on the whole, parallel currents determining, the *spin – orbit interaction.* But parallel currents attract each other, which thus should lead to a tighter binding, thereby the lowering of the energy level of concern. (*And this is what we came to predict.*)

In order to check our approach, against the experimental results, the next step would be to derive a Breit-like Equation, our way, thus for two particles, instead of keeping the proton untouched through the creation of the hydrogen atom, and compare our results with the classical Breit Equation results, along with the experimental data.

Acknowledgments

The authors wish to thank A. Kholmetskii and S. K. Ozdemir for very many hours of fruitful discussions. Prof. Kholmetskii, furthermore, skillfully presented the content of the present article to the Vigier VIII - BCS Conference, 2012, London, and kindly assisted the improvement of it thereafter.

This work was supported by the Scientific Research Projects Coordination Unit of Istanbul University, project number 5623.

References

[1] Ozaydin, F. et al. (2011) AIP Conf. Proc., 1476, pp. 361-364.

[2] Dirac, P.A.M. (1928) Proc. R. Soc. A 117, 610-624.

[3] Kholmetskii, A.L. et al., (2010) American Journal of Physics 78-4, 428-432.

[4] Kholmetskii, A.L. et al. (2010) Physica Scripta 82-4, 045301.

[5] T. Yarman, T. (2006) Foundations of Physics Letters 19-7, 675-693.

[6] Yarman, T. (2009) International Journal of Theoretical Physics 48-8, 2235-2245.

[7] Sobczyk , G. & Yarman, T. (2008) Applied and Computational Mathematics 7-2, 255-268.

[8] Kholmetskii, A.L. et al. (2008) Physica Scripta 77-3, 035302 .

[9] Kholmetskii, A.L. et al. (2009) Physica Scripta 79-6, 065007 .

[10] Kholmetskii, A.L. et al. (2009) Nuovo Cimento 124-8, 791-803.

[11] Yarman, T. et al. (2011) International Journal of Theoretical Physics 50-5, 1407-1416.

[12] Yarman, T. & Kholmetskii, A.L. et al., (2011) Chinese Physics B, 20-10, 105101

[13] Mohr, P.J. et al. (2008) Rev. Mod. Phys. 80-2, 633-730.

Thoughts on Landauer's Principle and its Experimental Verification

DAVID SANDS

Department obvf Physics, Hull University, Hull HU6 7RX, England
d.sands@hull.ac.uk

JEREMY DUNNING-DAVIES

Institute for Theoretical Physics and Advanced Mathematics (IFM) Einstein-Galilei,
Via Santa Gonda, 14 - 59100 Prato, Italy
and
Institute for Basic Research, Palm Harbor, Florida, U.S.A.
j.dunning-davies@hull.ac.uk

Here the basic ideas behind Landauer's introduction of his principle supposedly linking information theory and thermodynamics are examined afresh. Since this examination was prompted by the recent claims of an experimental verification of that principle, the details of the actual experimental verification are re-examined also in the light of these new thoughts concerning the supposed link between information theory and thermodynamics.

Keywords: Information theory, Landauer's Principle, Thermodynamics.

1. Introduction

The *raison d'être* for this article is provided by ideas concerning information theory as put forward in an article by Landauer [1] which have given rise to recent claims of experimental verification [2]. It has been pointed out that, in his original article, Landauer suggested that erasure of information is a dissipative process and, therefore, that a small quantity of heat is necessarily produced when a classical bit of information is deleted. However, before considering the validity of the experimental verification, it seems worthwhile to consider the original conjecture by Landauer. Landauer argues that computing machines usually involve devices which perform logical functions which do not have single-valued inverses. He then proceeds to claim that this logical irreversibility is associated with physical irreversibility and requires heat generation typically of the order of kT for each irreversible function. There are a number of questions arising from these notions, not least regarding the precise meaning of the term 'irreversible'.

2. Comments on Landauer's Conjecture

The notions of 'reversible' and 'irreversible' have been the cause of much discussion within thermodynamic circles for many years and have not really been clarified satisfactorily even now. Indeed this topic was covered in a recent article [3] and it was noted that, in his fundamental text on thermodynamics [4], Landsberg very cleverly avoided the issue by examining Carnot cycles run in their forward and reverse modes separately. This proved a satisfactory way of avoiding the awkward problem while still examining a two-way Carnot cycle thoroughly. However, in the above, when Landauer introduces the idea of heat being dissipated, he actually produces little justification for introducing the actual notion of heat into the discussion. Probably this is due to the usual identification of some quantities in statistical mechanics and, therefore, for historical reasons in information theory, with well-established functions in thermodynamics. This is particularly true of the quantity termed entropy. This appeared first in thermodynamics via its change being found equal to a change in heat divided by the temperature at which that heat change occurs; that is

$$dS = \frac{d'Q}{T} \tag{1}$$

This was a deduction from one of the original orthodox forms of the Second Law of thermodynamics; i.e, either from the Clausius or Kelvin form of that law [5]. It must always be remembered also that both forms point out that a particular physical reaction is impossible in a cyclic process in the absence of other effects and these restrictions to cyclic processes in the absence of other effects are crucial when it comes to considering the Second Law and possible violations of it.

However, separately, investigations were taking place into systems composed of large numbers of particles and, when it was realised that it was virtually impossible to even write down the equations of motion for each individual particle, let alone solve the resulting system of simultaneous equations, people resorted to statistical methods. Hence statistical mechanics was born and with it came a function called the entropy forever associated with the name of Boltzmann and defined by

$$S = k \ln W \qquad (2)$$

This, in turn, was noted to be a very similar expression to that obtained for the entropy in information theory. Since both thermodynamics and statistical mechanics were mainly used to describe physical systems, it was possibly natural to assume that the entropy functions in each actually referred to the same physical quantity. No-one can deny that the results of such an identification have been successful but it must be asked if the said identification is universally valid? Also, because of the similarity of expressions in statistical mechanics and information theory, there has been a tacit acceptance that the entropy in information theory is also the same quantity as in the other two areas. Again, is it?

It should be noted at the outset that, in basic classical thermodynamics, all the attention is concentrated on macroscopic systems. When fundamental quantities such as internal energy, heat and work are considered, very definite amounts of each quantity are understood to be involved; there is absolutely no fluctuation in any of these quantities even remotely involved in considerations. However, when large systems of particles need to be considered, due to the computational nature of things, it is necessary to adopt statistical techniques. These, by their very nature, immediately introduce a degree of uncertainty into proceedings and only the average value of a particular quantity may be measured. This, in turn, means that fluctuations in the relevant values must come into play. This is a totally different scenario from that encountered in traditional classical thermodynamics. The link comes via a seemingly reasonable assumption that the various macroscopic quantities must be the same in both classical thermodynamics and what might be termed statistical thermodynamics and so, the average values of the various quantities found in statistical thermodynamics are put equal to the same named quantities found in classical thermodynamics. This may seem eminently reasonable and, indeed, it is something which has worked extremely well for many years in a wide range of applications but it is still an assumption and one which is seen immediately to open up questions as to its range of validity. This latter point follows because in one area only average values are considered and these average values are considered to be capable of fluctuating, as must be accepted as reasonable in a statistical theory, while in the other area, no such indeterminacy is admitted.

As far as information theory is concerned, Brillouin's well-known text [6] concerns itself purely with the basics of information theory for the first eight chapters but, suddenly at the very beginning of chapter nine, he states that 'a close connection has been discovered between information theory and thermodynamics; information appears related to entropy'. However, in truth, no such link had, or has, been discovered between these two seemingly different fields of study; there was simply an obvious similarity in the form of the expressions for the quantity called entropy in statistical mechanics – or perhaps, more properly, in statistical thermodynamics - and information theory. Nowhere does the notion of heat enter obviously into either scenario! In thermodynamics though, heat is all important in the introduction of the idea of the entropy via the original forms of the Second Law. These thoughts raise immediate questions concerning the complete equivalence of the various functions called entropy in each of the three disciplines mentioned here.

It is important also to note that, as described by Brillouin, [6] information as defined by Shannon represents uncertainty. The word "information" is perhaps an unfortunate choice of word since it is readily open to misinterpretation. In everyday usage, information represents useful knowledge, so that a random sequence of letters might be said to contain no information of value, whereas an ordered sequence of characters forming a word or sentence would. However, Shannon was interested in a mathematical theory of communication and defined information on the basis of the probability that a particular character would be received based on the properties of the alphabet. A random sequence of characters can contain as much information as an ordered sequence and Brillouin makes the point that the everyday value of information is of no consequence in information theory. This may well be a tricky concept to appreciate but, in order to understand the basics of information theory, it is absolutely vital that it is. Crucially, this line of reasoning has no place in it for the concept of heat and that is one reason for wondering about the identification of information with the thermodynamic concept of entropy since, as pointed out above, only changes in entropy are considered in thermodynamics and such changes are always brought about as a result of a flow of heat. Incidentally, this means that, since heat may flow into, or out of, a system, such changes of entropy may be either positive or negative. This latter point is illustrated by the well-known Carnot cycle in which contributions to entropy change throughout the cycle cancel out on completion of that cycle.

It might be noted also that the idea of information being represented by $kln2$ arises quite naturally out of a statistical argument, as may be seen in Brillouin's text [6]. When multiplied by a temperature, T, a quantity with the dimensions of energy results. The immediate reaction is to call this a quantity of heat but this only follows by considering the deduction of the thermodynamic equation

$$d'Q = TdS \tag{3}$$

as a defining equation for the change in thermodynamic entropy dS. The argument leading to this identification in information theory contains no physical justification whatsoever; it is purely an assumption based on the mathematical similarity of two expressions – one for the change in thermodynamic entropy, the other for the quantity called information. It might be noted also at this point that a similar remark is valid for the regular identification of thermodynamic entropy with the entropy of statistical mechanics. The main justification in this latter case really amounts to a claim that it works and, therefore, must be true. However, all that can really be said is that it has worked so far. The bases for these three expressions are totally different and it is only in classical thermodynamics that the quantity called heat appears naturally; in the other two areas it is introduced merely out of association. It would seem that the link between thermodynamics and information theory is tenuous to say the least.

3. Comments on Recently Claimed Experimental Verification of Landauer's Principle

In a recent article, Bérut et al [2] claim to have verified Landauer's principle experimentally. However, it is not clear that the experiment performed by Bérut at al is capable of demonstrating a connection between information and thermodynamics.

As discussed above, information as defined by Shannon represents uncertainty. Bérut's experiment should be considered in this light. A bead is trapped in one side or other of a double potential well which is symmetric in the acquiescent state. Assigning a value of 0 or 1 according to whether the particle is in the left or right well is purely a matter of choice and represents information only in the sense that it has a human value. Erasing a 1 by moving the particle from one well to another is identical to writing a 1 by moving the particle the other way. If one operation destroys information in the sense defined by Shannon then so must the other.

It might be argued quite reasonably, however, that Shannon information is not destroyed by this process. Shannon information represents uncertainty and is non-zero only if we do not know in which well the particle exists. Then, and only then, must we resort to probability to describe the position of the particle and the information entropy is $ln2$. However, a system in which we have to resort to probability to describe the state cannot logically function as a memory device. It is implicit in the idea of memory that once the state is set it remains set, at least for a limited time, otherwise the memory device has no use. In principle, then, the state can be known and there is no need to resort to probability to describe the state.

The source of much of this confusion lies in the statistical interpretation of the Second Law of Thermodynamics. It should be remembered that both Kelvin and Clausius formulated versions of the Second Law which apply to engines operating in a cyclic process. Indeed the two forms of the second law can be shown to be equivalent by considering the operation of two such engines in tandem [7]. One engine operates in the forward direction and the other in reverse. If one of the machines violates one of the laws the two together become a self-acting machine that together violates the other law. Bérut et al describe the operation of cycles within their work, but these are not the kind of thermodynamic cycles implied in these statements of the Second Law. Work is being done on the particle and quite possibly this is converted to heat but heat is not being converted to work. In this sense the processes of writing and erasure are both dissipative and all that is happening is that the states of the system are being cycled.

In view of the irreversible nature of the processes described by Bérut et al we should not be surprised that there is a work cost, but it is not evidence of Landauer's principle. Indeed, the principle has been questioned in some of the references cited by Bérut et al [8, 9], wherein the origin of the factor $ln2$ in the entropy has been discussed. There are two possibilities; either it comes about through a probabilistic description of the system's coordinates in phase space, in which case the entropy corresponds to Shannon's information entropy, or it comes about through a work term of the form pdV and a doubling of the volume in the erasure process. The latter would link erasure with thermodynamic entropy if the erasure occurred isothermally, but it is restricted to a specific set of circumstances. It is by no means clear that the work done in the process described by Bérut et al is equivalent to this or reduces to it in some limit as it involves the movement of a particle through a viscous fluid along a potential gradient defined by an optical field. Bérut et al appear to confuse these two different issues by associating the entropy cost with the heat dissipated during irreversible work whilst at the same time referring to the statistical properties of the system. Given that there is no a priori reason why the dissipated heat should correspond to the Landauer heat and, therefore, that Landauer's principle even applies to this system, and that the very act of reading the state, for example by monitoring the position with a camera, removes the uncertainty and destroys the statistical, or information, entropy, it would appear that Landauer's principle has not been verified experimentally.

4. Conclusion

This further publicising of Landauer's conjecture by this recent work of Barut et al [2] highlights a much bigger, though rarely acknowledged, problem facing present day physics and that is the precise meaning of much of basic thermodynamic theory and the possible links, if any, of that topic with ideas in both statistical mechanics and information theory. As has been quite clearly pointed out, fundamental thermodynamics is a macroscopic subject which has its heart notions of heat and, subsequently, the production of work from heat. However, heat is basic to thermodynamics. This latter statement is obviously inapplicable to both statistical mechanics and information theory; in both areas the notion of heat has to be introduced additionally since it appears in neither naturally. It is this very basic realisation that initially raises questions about both Landauer's conjecture and its apparent experimental verification by Barut et al.

Hence, here queries are raised concerning the validity of Landauer's conjecture and its apparent subsequent experimental verification. The thermodynamic link is tenuous and needs to be examined in far more detail but only after the basis of thermodynamics has been fully re-examined and placed on a far more rigorous footing than it now has. Many of the fundamental points to be reassessed have been raised before and much appears to emanate from work in the latter half of the 19th century when, as also pointed out earlier, some eminent theoreticians produced some very muddled thinking and even more muddled writing. Much of this occurred with the rise of statistical mechanics and the perceived need to link that subject with thermodynamics but the situation has been exacerbated by the totally separate development of information theory. As Bernard Lavenda has opined on more than one occasion, statistical mechanics might have developed more effectively if it had done so in partnership with information theory since those two subjects do have much in common. However, that still leaves the position of thermodynamics somewhat open because it remains to be truly established where the crucial notion of heat enters statistical thermodynamics and information theory. At present it seems this is achieved through merely stating that the function called *entropy* in each area is the same. Then the link between thermodynamic entropy and heat, which comes via the traditional statements of the Second Law of Thermodynamics, is used to introduce the notion of heat into these areas. This is surely a less than satisfactory way to proceed and is, indeed, one reason for so much confusion about fundamental thermodynamics existing among so many.

References

[1] Landauer, R., I. (1961) BM J, 5, 183-191.

[2] Bérut, A., Arakelyan, A., Petrosyan, A., Dillerschneider, S. G. R. & Lutz, E. (2012) Nature 483, 187-190.

[3] Sands, D. & Dunning-Davies, J. (2010) in R. L. Amoroso, P. Rowlands, S. Jeffers (eds.) Search for Fundamental Theory, AIP conference proceedings 1316, 125-140.

[4] Landsberg, P. T. (1961) Thermodynamics, New York: Interscience Publishers.

[5] Dunning-Davies, J. (2007) Concise Thermodynamics, 2nd edition, Chichester: Horwood Publishing, p 26.

[6] Brillouin, L. (1963) Science and Information Theory, 2nd edition, New York: Academic Press.

[7] Zemansky, M. W. (1957) Heat and Thermodynamics, 5th Edition, Tokyo: McGraw-Hill Kogakusha, p185.

[8] Maroney, O.J.E. (2005) Studies in History and Philosophy of Modern Physics 36, 355–374.

[9] Norton, J. D. (2005) Studies in History and Philosophy of Modern Physics 36, 375–411.

Information, Entropy, and the Classical Ideal Gas

DAVID SANDS

Department of Physics and Mathematics,
University of Hull, Cottingham Rd., Hull, HU6 7RX, UK
d.sands@hull.ac.uk

JEREMY DUNNING-DAVIES

Institute for Theoretical Physics and Advanced Mathematics (IFM) Einstein-Galilei,
Via Santa Gonda, 14 – 59100 Prato, Italy
j.dunning-davies@hull.ac.uk

The physical basis of the canonical and grand canonical distributions is questioned. In particular, we question the usual methods by which these distributions are derived, namely that fluctuations in entropy around energy and particle number are assumed to occur when the entropy depends only on variables which cannot themselves fluctuate. We show, starting from the Maxwellian velocity distribution, that the probability that a classical ideal gas at a fixed temperature occupies a given energy state corresponds not to the canonical ensemble of classical statistical mechanics but to the Gamma distribution. Computer simulations of a hard-sphere fluid demonstrate the principles. The analysis is extended to open systems in which the number of particles fluctuates and we show that for a system connected to a particle reservoir the Poisson distribution governs the probability of finding a given number of particles. The resulting probability distributions are used to calculate the Shannon information entropy which is then compared with the thermodynamic entropy. We argue that information theoretic entropy and thermodynamic entropy, whilst related, are not necessarily identical and that the information entropy contains non-thermodynamic components.

Keywords: Entropy, Ideal gas, Information

1. Introduction

The foundations of statistical mechanics are not nearly as secure as is commonly supposed, but it is difficult to publish research into them in modern physics journals. Many years ago, for example, the authors wrote a paper on the central limit theorem and the canonical distribution [1] which was rejected for publication on the grounds that it contained nothing essentially new. In one sense this was correct; the central limit theorem is well known and so too is the fact that the states accessible to a gas in contact with a reservoir are not described by an exponential distribution, but are normally distributed about the mean energy. What we believed was new, however, was our method of demonstrating this fact. Moreover, modern text books on the subject don't dwell on this topic in any depth so there is great pedagogical value in a simple treatment.

Our approach was to generate random numbers in simulation of the velocity of an ideal gas and then to consider the distribution of energy states arising from a large number of such random numbers. This in itself is not straightforward. Typical computational random number generators generate random numbers uniformly in the interval 0 to 1, so generating speeds consistent with the Maxwell-Boltzmann distribution requires some manipulation. We found by trial and error that if two random numbers are generated, the second of which is scaled logarithmically from the first, the resulting random number is Gauss distributed. In detail, if $0+\gamma \leq R_1 \leq 1$ is a uniformly generated random number, where γ is a very small offset, then if

$$L = \ln[R_1] \tag{1}$$

and

$$f(L) = |L|^{\beta} \tag{2}$$

a second uniformly generated random number, R_2, scaled between the limits $\pm v_{lim}.\mathrm{f}(L)$ yields a number z_j, for which the vector sum

$$v = \left\{ \sum_{j=1}^{3} z_j^2 \right\}^{1/2} \tag{3}$$

is distributed according to

$$n(v) = a.v^2 \exp[\frac{-v^{\alpha}}{b}] \tag{4}$$

The Maxwell-Boltzmann distribution is generated for α=2, for which β=0.394476 (figure 1).

It follows, then, from the theory of the Maxwell-Boltzmann speed distribution that the z_j, which correspond to the orthogonal components, v_x, v_y and v_z, of the velocity, are distributed normally. The standard deviation for v_{lim}=1 is 0.55565 which allows the scaling limits to be defined in terms of the desired most probable velocity, v_m, as

$$v_{\lim} = \frac{1}{0.55565} \frac{v_m}{\sqrt{2}} \tag{5}$$

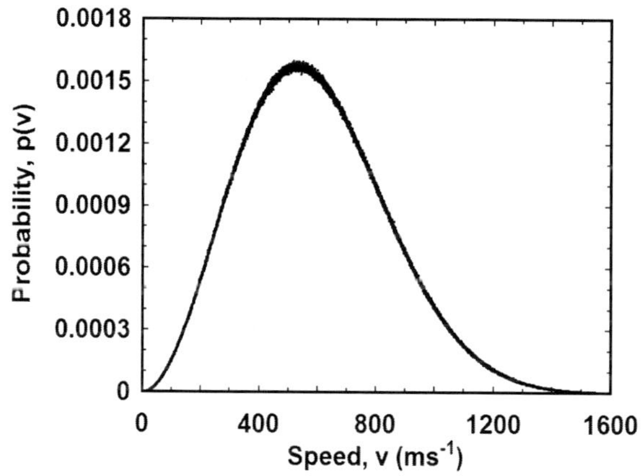

Figure 1. The Maxwell-Boltzmann speed distribution generated from 100,000 samples of the random numbers generated using equations (1-3).

We chose γ= 10^{-6}, which gave a maximum value of $|L|$=13.8. Thus large values of z can be generated but with a low probability. Six random numbers are required (j=1, 2 and 3) to generate a single speed distributed according to the Maxwell-Boltzmann distribution and $6n$ random numbers will generate a set of possible speeds for a gas containing n particles. This yields a single energy state and repeated sampling yields the distribution of energy states. It turns out that for even low numbers of particles of the order of 300 or the distribution of energy states is Gaussian (figure 2).

Apart from saying that this is not new, the referees also claimed that they had been using a similar treatment to demonstrate the canonical distribution to their own students for many years. Perhaps, but we have no way of knowing. More to the point, the Gaussian distribution is seen to be intrinsic to the small system. It arises naturally from the Maxwellian distribution of velocities, which raises the question as to why the distribution of states is assumed to be of the form

$$p(E_i) = \frac{1}{Z} e^{\frac{-E_i}{kT}} \tag{6}$$

when clearly it should include the density of states, $\Omega(E)$;

$$p(E_i) = \frac{1}{Z} \Omega(E) e^{\frac{-E_i}{kT}} \tag{7}$$

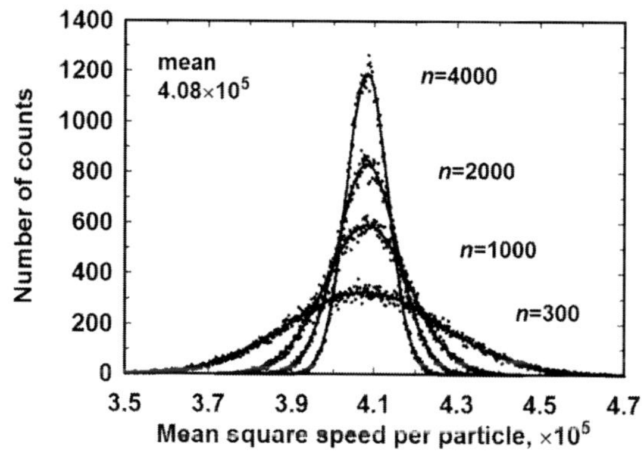

Figure 2. The distribution of energy states for systems containing the number of particles, N, as shown. The solid lines are best-fit Gaussian distributions.

This form can sometimes be seen in the open literature, but one has to wonder why the density of states is imposed as an afterthought on to a distribution that is clearly incomplete. It would be better, surely, to derive the whole distribution from scratch? The purpose of this paper, then, is to examine the foundations of the canonical distribution, its relation to a physical system comprising an ideal gas, and to examine the meaning of information entropy in relation to the ideal gas. We will use computer simulations of a hard sphere fluid to demonstrate the probability distributions as well as to shed some light on the meaning of information entropy.

2. The Canonical Distribution

The difficulty with the canonical distribution described above can be traced back to Gibbs, who simply proposed the canonical distribution in his 1902 book on statistical mechanics [2]. Lindsey and Margenau [3] described the construction of the canonical distribution as "a creative act of reason" not necessarily intended to represent any physical reality, but in fact the idea of a distribution of systems originated with Maxwell some 26 years earlier [4]: "*I have found it convenient, instead of considering one system of material particles, to consider a large number of systems similar to each other in all respects except in the initial circumstances of the motion, which are supposed to vary from system to system, the total energy being the same in all. In the statistical investigation of the motion, we confine our attention to the number of these systems which at a given time are in a phase such that the variables which define it lie within given limits*." The description of an ensemble given by Lindsay and Margenau is almost identical to Maxwell's description, with the exception that the restriction on the energy is relaxed. The canonical distribution is therefore just an extension of Maxwell's idea to a collection of systems with different energies.

The theory of ensembles is quite abstract and the link with physical systems somewhat obscure. The basic postulate of the Gibbs approach to statistical mechanics is that the probability that at any instant a given real system is found in a given dynamical state specified by a set of position and momenta coordinates is equal to the probability that a system selected at random from the ensemble should be characterised by the same set of coordinates. Both Maxwell and Gibbs made use of the Liouville theorem regarding the constancy of the density of points in phase space to assert that the initial distribution of phases is not arbitrary but must in fact correspond to the equilibrium distribution. Since the distribution of phases of a collection of material points was known to correspond to the Maxwellian distribution, it follows that the distribution of the phases in an ensemble must also be Maxwellian and is therefore given by equation (6).

There is thus an ambiguity about the physical basis of the canonical distribution that is built in to the structure of statistical mechanics. Statistical mechanics is intended to be a general theory that goes beyond the specifics of a particular system. As such it should be applicable to a wide range of physical systems whilst at the same time not actually representing any physical system at all. The idea of material bodies, of force laws, of collisions, etc. is the realm of kinetic theory of Maxwell and Boltzmann and others, whose attention was focused on the properties of physical systems as represented by the ideal gas. Yet, if statistical mechanics is to be useful it has to be able to describe physical systems, which leads naturally to the question; is statistical mechanics a physical theory or not? If not, what is its place within the realm of physics? Jaynes, who made significant contributions to statistical mechanics in the 1950s, was firmly of the view that statistical mechanics need not be regarded as a physical theory [5]; that it is possible to draw a clear distinction between the statistical and physical aspects. For Jaynes the latter was not so much a description of the laws of force or the physical mechanisms of interactions between particles that concerned kinetic theorists, but "the correct enumeration of the states of a system and their properties". Jaynes applied the techniques of information theory to draw statistical inferences from the canonical distribution. These are entirely mathematical operations that appear to yield physically sensible results.

Despite this lack of a physical basis, the canonical distribution is nonetheless given a physical interpretation. To quote Baierlein [6]; "*Qualitatively stated, the probability of high-energy states is exponentially smaller than the probability of low-energy states*". This interpretation is very common [7]. Equation (6) is therefore seen as representing a physical system, but inspection of (6) shows, however, that it cannot give the probability of finding a system in a state of energy E_i as the most probable state corresponds to E_i=0 . This would imply that all the energy in the system has transferred spontaneously to the reservoir, which experience would suggest is highly improbable and unlikely to happen even momentarily. As figure 2 shows, the distribution of energy states is Gaussian. The probability is evenly distributed about the mean energy and high energy states are neither more nor less probable than low energy states.

The fact of a Gaussian distribution of energy states has implications for the entropy. It is straightforward to show that, if

$$S = -kH \tag{8}$$

where k is Boltzmann's constant and

$$H = \sum_i p(E_i).\ln[p(E_i)] \tag{9}$$

then, for $p(E_i)$ given by equation (6)

$$S = -k\ln Z + \frac{\overline{E}}{T} \tag{10}$$

Here $\overline{E}$ is the average energy. This is a well known expression. However, for $p(E_i)$ given by a Gaussian distribution with standard deviation σ the information entropy varies as lnσ. It is well known in statistical mechanics that for a system with well-defined temperature T the energy fluctuates with a variance

$$\sigma^2 = kT^2\frac{\partial U}{\partial T}, \tag{11}$$

where U is the internal energy and

$$\frac{\partial U}{\partial T} = \frac{3}{2}k.n \tag{12}$$

Therefore the entropy varies as

$$S = \ln\sigma + terms = \tfrac{1}{2}\ln n + \ln T + terms \tag{13}$$

Equation (13) does not agree with thermodynamics either in the variation with n or energy. Jaynes' argument is thus seen to be flawed in relation to the entropy: we cannot ignore the physical aspects of the problem because a probability distribution which takes into account the density of states yields a different expression for the entropy from one that does not.

This is not the only difficulty with the entropy, however. Modern texts on statistical mechanics (see for example Huang [8]) invariably contain a derivation of the canonical ensemble based on the idea of a small system is in contact with a reservoir which determines the temperature. Energy can fluctuate between the reservoir and system and the state of the system is supposed to be determined by the state of the reservoir. That is to say, if the system is in a given state with energy, E_S, and the total energy, E, is fixed the reservoir must have energy, $E_R=E-E_S$. Expanding the entropy of the reservoir, or equivalently $\ln\Omega(E)$, where $\Omega(E)$ is the density of states of the reservoir, to first order

$$k\ln[\Omega(E-E_S)] = S_R(E-E_S)$$

$$= S_R(E) - E_S\left[\frac{\partial S_R(E_R)}{\partial E_R}\right]_{E_R=E} \quad (14)$$

Equating the last term on the right with inverse temperature, ie.

$$\frac{\partial S_R(E_R)}{\partial E_R} = \frac{1}{T} \quad (15)$$

leads eventually to equation (6) with the entropy given by equation (10). However, the variables in equation (10) are either constants of the system, such as temperature, or averages so the entropy is seen to depend on parameters which do not themselves fluctuate. Moreover, the validity of equation (15) in this context is also open to question. Jaynes showed, in the context of information theory [5], that

$$\frac{\partial S(\bar{E})}{\partial E} = \frac{1}{T} \quad (16)$$

The physical interpretation of this expression is that entropy does not change in response to a fluctuation, but with a change in the internal or average, energy,

$$E = \frac{3}{2}n.kT \quad (17)$$

In thermodynamics this corresponds to an exchange of heat. The basis upon which the canonical distribution is derived would therefore seem to be invalid.

A similar difficulty exists with the grand canonical distribution, which, as is well known, describes the probability that a system will occupy a point in $6n$-dimensional phase space when the phase space volume fluctuates with n. In essence the fluctuations in E and n are independent in as much as we may write [9],

$$p(n, E_{n,i}) = p(n).p(E_{n,i}) \quad (18)$$

Equation (18) corresponds to the probability of finding an open system connected to both particle and thermal reservoirs in an n-particle energy state i, denoted by $E_{n,i}$, given that the system has at a given instant n particles. Expansion of the entropy about both E and n leads to [8];

$$p(n, E_{n,i}) = \frac{1}{Z}e^{\beta(\mu n - E_i)} \quad (19)$$

Summation over both n and E then yields for the information entropy (equations 7 and 8)

$$S = -k\ln Z + \frac{\bar{\bar{E}}}{T} - \frac{\overline{(\mu n)}}{T} \quad (20)$$

The double bars indicate energy averaged over fluctuations in both E and n and the last term on the right is the mean value of the product μn, μ being the chemical potential. If μ is constant over the range of fluctuations in n this reduces to $\mu\bar{n}$. Again, the terms in (20) are all averages and therefore not susceptible to fluctuation [10].

In addition to this difficulty Khinchin [11] has shown that for a set of events described by a conditional probability of the form of equation (18) the entropy contains a contribution from both events. In the specific case that a system occupies the energy state $E_{n,i}$ given the number of particles n, the entropy is,

$$S(n,E) = -k\left[H(n) + \sum_n p(n)H(E)\right] \quad (21)$$

H has the same meaning as in (9) above. A physical theory for the probability distribution $p(n)$ will not in general give $\mu\bar{n}$. For example, if $p(n)$ is Gaussian with standard deviation σ_n, $H(n)$ will vary as $\ln\sigma_n$. This difficulty with information theory arises for a physical probability distribution rather than the distributions of statistical mechanics.

In the remainder of this paper we consider computer simulations of a simple hard sphere fluid to demonstrate the statistical distributions that apply in practice. We consider isolated systems as well as both closed and open canonically distributed systems. In the limit of large numbers both the distribution of energy states and particle numbers are Gaussian with no place for the chemical potential. Finally we discuss the information entropy of these systems and argue that fundamentally there is a difference between information theoretic entropy and the entropies of bothstatistical mechanics and thermodynamics.

3. Computer Simulations of Canonically Distributed Systems

Computer models of canonically distributed systems were constructed by assuming that a small part of a larger system is canonically distributed (see Appendix). Therefore an ideal gas consisting of several hundred hard spheres of atomic mass 20 (corresponding to argon) at room temperature was modelled and a small number of the particles were considered as the canonically distributed system. The position of each particle was advanced in time intervals of 10^{-5} seconds and if the centres approach closer than the sum of the two radii a collision is deemed to occur. There are no other interactions between the particles. The probability of collision depends on the relative size of the radius compared with the volume which the particles occupy and for convenience we chose a particle radius of 0.03 m and a volume of square unit cross-section (typically 1m × 1m unless otherwise mentioned), but variable length so that open systems could be modelled by placing a permeable partition, consisting of fixed hard spheres of approximate diameter 0.02m, along the length of the chamber. Particles are reflected elastically off the walls of the chamber as well as the permeable membrane. The diameter of the particles in the membrane was chosen after several trials in order to provide a balance between reflection off it and transport through it.

The computation proceeds by assigning to each particle the same initial kinetic energy, equivalent to $3kT/2$, but a random direction of motion. The particles are allowed to undergo an initial number of collisions in order to randomise the velocities before sampling the energy states, and where relevant the positions, of the particles of the canonically distributed system. Sampling occurs at fixed intervals of time within a programme but from run to run the interval has been varied in order to ensure that it is longer than the mean time between collisions and that, on average, the energy states change within the sampling time. Figure 3 shows the distributions of energy states for n=1, 3 and 5 particles among a total of N=400. The solid line corresponds to Gamma distribution, which we show in Appendix 1 is the correct probability distribution for the energy states of a canonically distributed ideal gas obeying Maxwell-Boltzmann statistics.

The Gamma distribution was derived by Jeans in 1916 [12] and given later by Khinchin in 1949 [11], but surprisingly, hardly any mention of it appears in modern texts on statistical mechanics. This might be because neither text is readily accessible; Jeans' highly mathematical derivation employed rather abstract ideas of ensembles and the physical interpretation is not readily apparent, and Khinchin's work is aimed squarely at the mathematician. It might also be because, as Jaynes described, it is seen simply as an unnecessary detail; an enumeration of the density of states that information theory has rendered irrelevant. Although the energy states of 1,2,3,4 and 5 particles were sampled only the odd numbers are shown for clarity in figure 3. A total of 100,000 instantaneous samples were recorded in order to produce these histograms and it is clear that the Gamma distribution matches the experimentally generated probability. It follows, then, that these small particles have access to all the states of the Maxwell-Boltzmann distribution and the principle of the canonically distributed system is established.

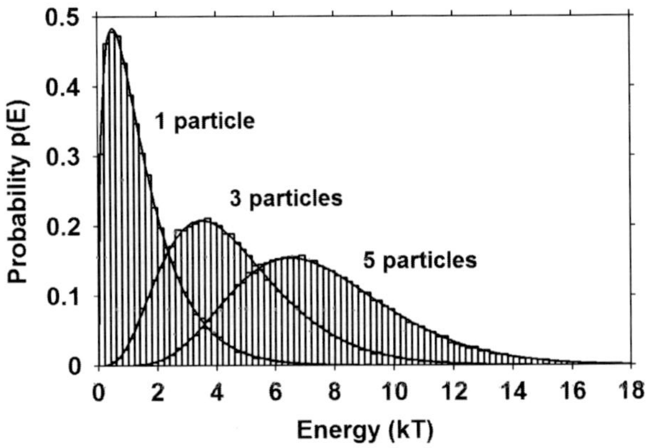

Figure 3. The probability distributions for three small systems consisting respectively of 1, 3 and 5 particles amongst a total of 400. The histograms correspond to the probabilities calculated from 10^5 samples of system energy and the solid lines to the respective Gamma distributions.

For the extension to open systems we consider first a system that is canonically distributed but is connected to a source of particles similar to the system under consideration. A total of 400 particles in a volume measuring 1m × 1m × 2m with a permeable wall located at the midpoint of the long axis were simulated. Therefore the number of particles on one side of the partition can fluctuate with time. As before, canonically distributed systems are constructed by considering a small fraction of the total number of particles; respectively of n=6 and n=15 particles. As the total number of particles on one side of the partition fluctuates we consider how many of the particles comprising the small system also appear on the same side of the partition. The system was sampled total of 220,000 times, from which the small system occupations were calculated. As might be expected, the particle distribution is described by the binomial distribution with a probability of 0.5 corresponding to the ratio of volumes of 1:1. It follows that the probability of finding the small system in a given energy state is the weighted sum of Gamma distributions,

$$p(E_i) = \sum_n p(n, E_{n,i}) = \sum_n p(n).p(E_{n,i}) \qquad (22)$$

Figure 4 shows the histograms of the energy states as well as the weighted sum of Gamma distributions (solid lines) given by equation (22).

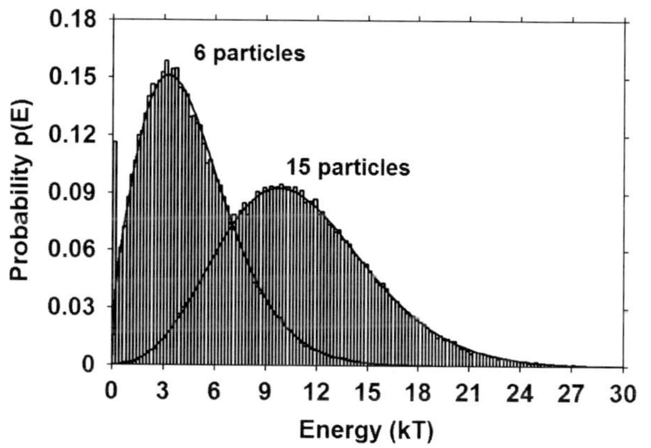

Figure 4. The distribution of energy states for two small systems comprising respectively 6 and 15 particles among a total of 400 distributed between equal volumes.

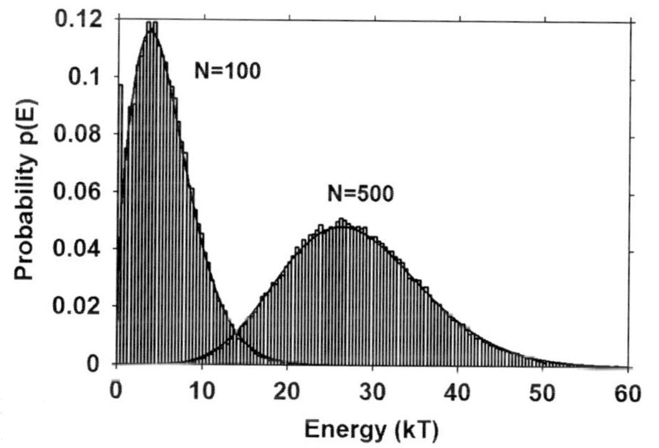

Figure 5. The distribution of energy states from 10₅ samples for small systems connected to much larger systems. The total number of particles is as shown and the mean numbers of particles in the smaller system are approximately 4 and 20 respectively. The solid lines correspond to sum of Gamma distributions weighted by the respective particle probabilities as given by the Poisson distribution.

Similar calculations have been performed for a number of different configurations of particle numbers and volume ratios, but all were found to follow similar principles. However, such systems do not represent grand canonical systems which are connected not only to thermal reservoirs but also particle reservoirs. This can be simulated by having a very large ratio of volumes. The smaller volume then represents a small part of a much larger system and should therefore also be canonically distributed, making it unnecessary to consider a small subset of the particles. Figure 5 shows the distribution of energy states from 10^5 samples of two such systems consisting of a total of 100 and 500 particles. The states illustrated in figure 5 are those found in a volume 1m × 1m × 1m connected via a permeable partition to a volume of 1m × 1m × 24m. Again, the solid line is the sum of n-particle Gamma distributions weighted according to the Poisson distribution, to which the binomial distribution approximates in these circumstances. Figure 5 therefore confirms that the particles within the small system constitute a canonically distributed system with the particles in the larger volume acting not only as a particle reservoir but also as a thermal reservoir.

4. Discussion — Entropy in the Ideal Classical Gas

We have argued in this paper that the canonical and grand canonical distributions of statistical mechanics, though used successfully for many years, are not physically representative of the distributions of either the energy states available to a thermally connected system or of the distribution of particles. We have shown through computer simulations of a hard sphere fluid that the energy states of a thermally connected system are given by the Gamma distribution and for open systems by a sum of Gamma distributions weighted by the appropriate binomial probability. In the limit that the system is very small compared to the size of the reservoir the binomial probability approximates to the Poisson distribution, and for particle numbers of just around 100 or so, which is very small compared with those found in normal gaseous systems, the Gamma distribution approximates to the Gaussian distribution. This we have also shown through the random generation of energy states from speeds drawn from the Maxwell-Boltzmann speed distribution. The purpose of these demonstrations is to show that physically representative distributions give rise to expressions for the entropy that differ from the distributions of statistical mechanics. In particular, the entropy taken over the distribution of energy states differs from that taken over the velocity distribution, whereas in the canonical distribution the two give identical outcomes.

We have shown that the entropy taken over a Gaussian distribution for the energy states depends on $\ln T$. Likewise, the entropy taken over the Gamma distribution for the energy states is

$$H(\xi) = -\ln\Gamma(\gamma) + (\gamma - 1)\overline{\ln\xi} - \overline{\xi} \qquad (23)$$

This is independent of temperature on account of the normalised parameter, $\xi = E/kT$, but

$$H(E) = -\ln\Gamma(\gamma) + \gamma\ln\beta + (\gamma-1)\overline{\ln E} - \frac{\overline{E}}{kT} \quad (24)$$

Therefore

$$H(E) = H(\xi) + \ln\beta \quad (25)$$

Taking into account the negative sign in equation (8) we find again that S depends on ln$T$. On the other hand, if S is taken over the n-particle product of single-particle Maxwellian velocity distributions we find [13]

$$S \propto \tfrac{3}{2} n \ln T \quad (26)$$

We can conclude from this that even though the canonical distribution is ostensibly a distribution in energy states it does not correspond to the actual probability of finding the system in a given energy state, which has to take into account the density of states, and neither does it yield a physically sensible entropy, no matter that it appear to agree with thermodynamics. What does appear to agree with thermodynamics, however, is the information entropy taken over the n-particle product of single-particle Maxwellian velocity distributions. Moreover, this yields the physically sensible result that if the particle number is uncertain then by equation (20)

$$S \propto \tfrac{3}{2} \overline{n} \ln T \quad (27)$$

This also agrees with thermodynamics in as much as we would naturally take the average number of particles as the basis for calculating the entropy in a system in which the number of particles fluctuates.

It is also apparent from equation (21) that there is a seemingly non-thermodynamic component to the entropy of an open system, namely the contribution from $H(n)$. This is an aspect of information theory that has been overlooked in statistical physics, principally because of the use of non-physical distributions such as the grand canonical distribution. Looking at equation (26), it would be quite natural to assume that if the entropy depends on the number of particles then it will fluctuate as the number of particles fluctuates. Hence, the idea of fluctuations is built into the justification of the grand canonical distribution with the paradoxical result that the entropy so derived (equation 19) does not itself fluctuate. However, equation (21) shows that entropy does not fluctuate with n, but the uncertainty in the particle number increases the entropy. This is the essence of information entropy. It is a measure of uncertainty; the broader the distribution the greater the uncertainty of the outcome of a single event the greater the entropy.

A single event could be the determination at a given instant in time of the number of particles and their phase, or, for a system with a fixed number of particles, just the phase. It is apparent, then, that entropy should not fluctuate with energy. As we have described in the appendix, the probability for a canonical system can be written as the product of single-particle Maxwellian distributions. The entropy calculated from such a product of distributions leads directly to equation (26), with the implication that the entropy is defined for even a single particle (n=1). For the purposes of the present discussion we can place a restriction on this; the particle has to be one of a large number of other particles with which it interacts through collisions. It is not necessary that the particles be identical, simply that they act as a reservoir and that through collisions the Maxwellian velocity distribution is established. It might well be that we could consider a single particle in thermal equilibrium with the walls of a container but to our knowledge it has never been established that the velocity distribution of such a particle will be Maxwellian. Therefore we restrict our attention to a single particle among a number of others.

The fact of a single particle Maxwellian distribution must imply that over a period of time the particle will access all the velocities allowed by the distribution. This is shown very clearly by figure 3, which displays the Gamma distribution for a single particle and shows that the particle has access to the full range of allowed energies. The information entropy of a single particle is therefore well defined and represents the uncertainty of finding the particle in a given velocity state. It follows that the entropy of every single particle in the system is equally well defined and therefore the entropy of the system is also well defined, regardless of the actual state of the system. This has implications not only for the entropy of a canonical system but also for all closed systems of ideal gases. As a consequence of his famous H-theorem, Boltzmann argued that if the system starts in an arbitrary state the entropy will increase through collisions until the Maxwellian velocity distribution is achieved and the entropy is maximised. However, Boltzmann took as his basis of probability the actual distribution of velocities over the particles; that is, the probability that a particle is in a given state is simply the fraction of particles in that state. The so-called evolution to a Maxwellian distribution corresponds, therefore, to a redistribution of the particles among the states, but we have argued above that the system is always described by the Maxwellian distribution regardless of the actual fraction of particles in each state. Therefore there is no sense in which the system evolves from an arbitrary distribution to a final Maxwellian distribution.

This interpretation is illustrated in figures 6 and 7, which show respectively the fluctuations in the H-

function, calculated from the fractions of particles occupying a given state, for a system of 2000 particles at fixed total energy and the mean H-function calculated for similar systems consisting of various numbers of particles from 300 to 2000. The system is allowed to evolve over a period of time and the H-function is calculated at regular intervals from a speed distribution constructed from the number of particles lying within intervals of 30 ms^{-1}. This is an arbitrary division but one which nonetheless ensures that the distribution of speeds across the particles approximates to the Maxwell-Boltzmann distribution. A different interval might well yield different values of H, but our interest here is in comparing the mean value of H for different systems and as long as the interval is consistent between the different calculations such a comparison is possible. The limit of H for the continuous Maxwell-Boltzmann speed distribution is 6.863 and the mean is clearly tending to a limiting value close to this with increasing system size. The standard deviation, however, is decreasing and in the limit of very large numbers, as stipulated by Boltzmann in the construction of his H-theorem [14], we can imagine the distribution of speeds among the particles will correspond almost exactly to the continuous distribution with a neglible standard deviation for H.

The significance for the present work is that there appears to be two different ways of calculating the entropy; either from the actual distribution of speeds across the particles, which necessarily fluctuates in time as the distribution fluctuates, or from the limiting speed distribution, which applies to a single particle, and hence the whole system, and is constant over time. In the limit of asymptotically large numbers of particles, or the so-called thermodynamic limit, these two should become identical, but for smaller systems they differ. As we have described above, entropy in information theory is a measure of uncertainty and should not fluctuate as the state of the system fluctuates. It follows, then, that the correct way of calculating the information theoretic entropy of a thermodynamic system is to use the limiting, or theoretical, distribution, which is independent of fluctuations. Applying this to the canonical distribution, the entropy can only be said to fluctuate with energy in any meaningful sense in the same way that entropy in figure 6 fluctuates; that is, if it is calculated across the actual distribution of speeds or velocities, which must of course vary as energy varies. However, each particle is still described by the Maxwellian velocity distribution and in this sense the entropy of each particle, and hence the system, is well defined. Only if the average energy changes will the underlying distribution, and hence the entropy, change.

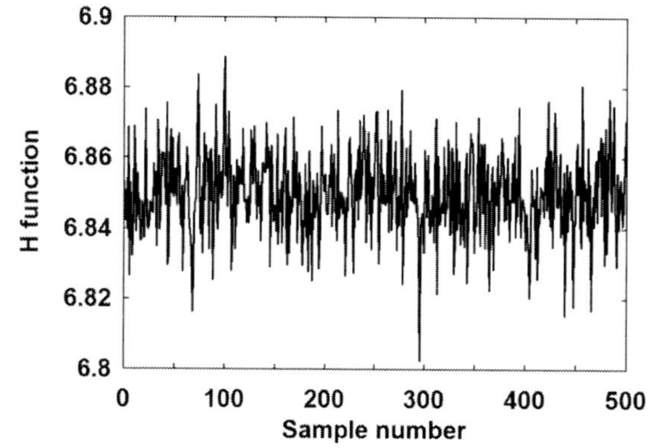

Figure 6. Fluctuations in the H-value for 2000 particles at fixed total energy at room temperature. A total of 500 samples were taken.

Finally, we consider the role that volume plays in the information entropy. We have shown that the Maxwellian probability distribution for the velocities rather than the Gamma distribution for the energy yields the correct temperature dependence of the entropy as well as the physically sensible result that in the event of an open system in which the number of particles is uncertain, the entropy depends on the mean number of particles. However, for an open system there is also an apparently non-thermodynamic component of the entropy. We argue here that the volume term in the information theoretic entropy of an ideal gas might also be non-thermodynamic.

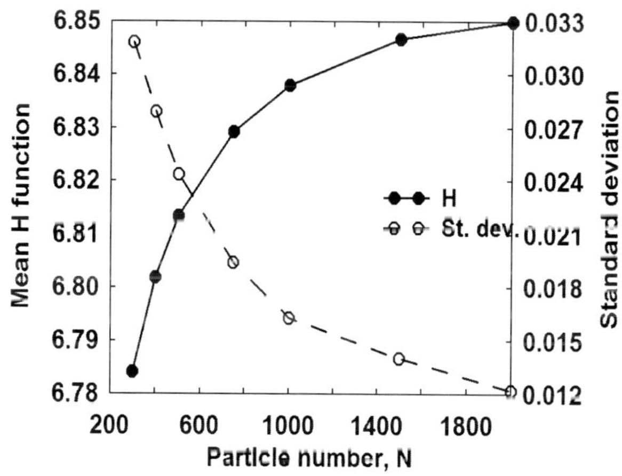

Figure 7. The mean value of H and its standard deviation for various numbers of particles at fixed total energy.

The volume term arises, as is well known, because both the position and the velocity of the particles are needed to follow the evolution of the system. The probability that a point particle is found at a point x, y, z in a container of dimensions X, Y, Z, is

$$p(x,y,z)=\frac{1}{X}dx.\frac{1}{Y}dy.\frac{1}{Z}dz=\frac{1}{V}d^3V \qquad (28)$$

This contributes a term

$$S(V)=n.k.\ln V \qquad (29)$$

to the entropy. The difficulty arises because although a an identical term appears in the thermodynamic entropy of a classical ideal gas, the dependence on volume in equation (29) above is absolute and doubling the size of the system increases the entropy by $nk \ln 2$. Planck overcame this difficulty by assuming that the states are "over counted" by a factor $n!$; that is, as the particles are indistinguishable then interchanging them should make no difference to the system hence the partition function needs to be reduced. This is a powerful argument, but it is not consistent with information theory, the basis of which is not the number of states but a probability distribution. The probability distribution in (28) is normalized and the entropy derived from it, equation (29), is correct as far as information theory is concerned. This suggests a real difference between thermodynamics and information theory.

5. Conclusion

We have argued that, despite the great success of statistical mechanics in explaining the microscopic properties of many physical systems, it is not in fact a physical theory. A physical theory of the classical ideal gas leads to the conclusion that for a canonically distributed system of particles, that is, a small number of particles thermally connected to a much larger number of particles that act as a reservoir, the probability that the system occupies a given energy state is given by the Gamma distribution. For systems containing just a few hundred particles this approximates to a Gaussian distribution so that the energy states are evenly distributed about the mean energy. We have demonstrated these principles through computer simulations.

We have referred to Jaynes, who argued in 1957 that statistical inference can be used to obtain physically sensible results, despite the apparent lack of a physical basis for the distributions in statistical mechanics. Whilst true, this does not extend to the entropy, which we have shown depends crucially on the probability distributions employed. The entropy derived from the Gamma distribution does not agree with thermodynamics in its temperature dependence, leaving the probability over phase space as the most likely basis for a statistical entropy consistent with thermodynamics. However, the statistical entropy for an open system appears to contain a non-thermodynamic contribution for the uncertainty in particle number. We have also argued that in so far as the dependence of the entropy on volume appears to disagree with thermodynamics it is possibly non-thermodynamic in nature. We conclude, then, that information theoretic entropy and thermodynamic entropy are not identical, though they would appear to be related in some way.

Appendix: The Gamma distribution

The Gamma distribution correctly gives the distribution of energy states in a canonically distributed ideal gas and follows naturally from the assumption of independent velocities obeying the Maxwellian distribution. If there are n particles out of a total of N in the system under consideration the probability, P, that the system is found within a small volume of phase space, $d\tau$, factors into a product of n identical single particle probabilities [15].

$$Pd\tau = p_1p_2p_3...p_n.dv_1dv_2dv_3...dv_n = p(v)^n d^n v \qquad (A1)$$

This may be best understood with reference to Gibbs' discovery that a small part of a larger system is canonically distributed [16,17]. As each particle independently has access to all the velocities allowed by the Maxwellian distribution it is a matter of probability whether the system is in a high or low energy state. Thus the idea of energy fluctuations is built in to the distribution in (A1). The essential requirement is that the density of states of the remaining particles is large enough to allow the n particles to occupy all possible states allowed by the Maxwellian velocity distribution without restriction. Thus if the system is represented as a point in 3-n dimensional velocity or momentum space then equation (A1) describes the probability of finding this point and therefore defines the canonical distribution for an ideal gas. Strictly the distribution in phase space also includes a configurational component taken over the particle coordinates, but this term is fixed and adds nothing to the energy dependence.

As is well known, the motion in each of the three orthogonal directions is independent of the motion in each of the other spatial dimensions. Each p_i (i=1 to n) is therefore a product of three identical Gaussian distributions and the probability that a particle, i, has a velocity in the range v to $v+dv$ given by the vector sum of its components in the ranges, v_x to v_x+dv_x, v_y to v_y+dv_y, and v_z to v_z+dv_z is

$$p_i(v).dv = p_i(v_x)dv_x.p_i(v_y)dv_y.p_i(v_z)dv_z \qquad (A2)$$

Hence

$$p_i dv = \left(\frac{m}{2\pi kT}\right)^{3/2} e^{-\left(\frac{mv^2}{2kT}\right)} dv \qquad (A3)$$

where

$$v^2 = (v_x^2 + v_y^2 + v_z^2) \quad \text{(A4)}$$

and

$$dv = dv_x . dv_y . dv_z \quad \text{(A5)}$$

Therefore the probability that a small system of n particles will occupy a given energy state can be determined by expanding equation (A3).

$$Pd^n v = \left(\frac{m}{2\pi kT}\right)^{3n/2} e^{-\left(\frac{m}{2kT}\right)\sum_{i=1}^{n} v_i^2} dv_1 . dv_2 ... dv_n \quad \text{(A6)}$$

Clearly,

$$\frac{m}{2}\sum_{i=1}^{n} v_i^2 = \frac{m v_s^2}{2} \quad \text{(A7)}$$

is equal to the total energy in the system but v_s^2 is the magnitude of a 3-n dimensional vector (the system velocity) lying within the volume element $d^n v$. It is necessary to convert this to the number of energy states lying within an interval dE.

The total number of states up to and including v_s is given by the content of a $3n$-dimensional sphere [18]:

$$G(v) = \frac{2 v_s^{3n} \pi^{\frac{3n}{2}}}{3n\Gamma(\frac{3n}{2})} \quad \text{(A8)}$$

The density of states between energy E and $E+dE$ is therefore,

$$\Omega(E)dE = \left(\frac{dG}{dE}\right).dE = \left(\frac{2}{m}\right)^{\frac{3n}{2}} \frac{\pi^{\frac{3n}{2}}}{\Gamma(\frac{3n}{2})} E^{\frac{3n}{2}-1} .dE \quad \text{(A9)}$$

Hence,

$$P(E).dE = \Omega(E)\left(\frac{m}{2\pi kT}\right)^{\frac{3n}{2}} e^{-\left(\frac{E}{kT}\right)} .dE$$

$$= \left(\frac{1}{kT}\right)^{\frac{3n}{2}} \frac{E^{\frac{3n}{2}-1} e^{-\frac{E}{kT}}}{\Gamma(\frac{3n}{2})} .dE \quad \text{(A10)}$$

is the desired probability of finding the system in an energy state in the range E to $E+dE$. Substituting $\beta=(kT)^{-1}$ and $\gamma=3n/2$

$$P(E).dE = \frac{\beta^\gamma}{\Gamma(\gamma)} e^{-\beta E} E^{\gamma-1} dE \quad \text{(A11)}$$

This is a Gamma distribution [19] and has the following important thermodynamic properties. If $\xi=\beta E$

$$P(\xi).d\xi = \frac{\xi^{\gamma-1}}{\Gamma(\gamma)} e^{-\xi} .d\xi \quad \text{(A12)}$$

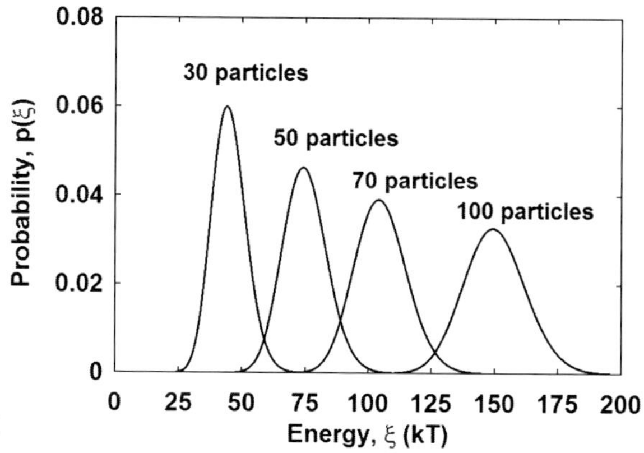

Figure A1. The Gamma distribution for small numbers of particles.

The mean energy is

$$\bar{\xi} = \int_0^\infty \xi . P . d\xi = \int_0^\infty \frac{\xi^\gamma}{\Gamma(\gamma)} e^{-\xi} d\xi \quad \text{(A13)}$$

This is a standard integral [20] with solution

$$\bar{\xi} = \int_0^\infty \frac{\xi^\gamma}{\Gamma(\gamma)} e^{-\xi} d\xi = \frac{1}{\Gamma(\gamma)}\left[-\xi^\gamma e^{-\xi}\right]_0^\infty + \gamma\int_0^\infty \frac{\xi^{\gamma-1}}{\Gamma(\gamma)} e^{-\xi} d\xi$$

$$= \gamma = \frac{3n}{2} \quad \text{(A14)}$$

The mean energy is therefore identical to the internal energy of thermodynamics, reinforcing the point made in the introduction in relation to entropy fluctuations, namely,

$$\bar{E} = U = \frac{3n}{2} kT \quad \text{(A15)}$$

The most probable energy state of the system, $\xi_m=\beta E_m$, is given by $dP/d\xi=0$ and is

$$\xi_m = \gamma - 1 = \frac{3n}{2} - 1 \quad \text{(A16)}$$

For n as small as 100 this equates to the internal energy of thermodynamics. In addition the gamma distribution is almost Gaussian in shape and fluctuations in energy around the mean will occur with equal probability (figure A1).

The second moment, or variance, is defined by

$$\langle\Delta\xi\rangle^2 = \int_0^\infty (\xi - \bar{\xi})^2 . P(\xi) . d\xi \quad \text{(A17)}$$

On expansion the right hand side becomes,

$$\int_0^\infty \frac{\xi^{\gamma+1}}{\Gamma(\gamma)} e^{-\xi} d\xi - 2\bar{\xi} \int_0^\infty \frac{\xi^{\gamma}}{\Gamma(\gamma)} e^{-\xi} d\xi + (\bar{\xi})^2 \int_0^\infty \frac{\xi^{\gamma-1}}{\Gamma(\gamma)} e^{-\xi} d\xi \tag{A18}$$

Therefore

$$\langle \Delta\xi \rangle^2 = (\gamma+1)\gamma - 2\gamma\bar{\xi} + (\bar{\xi})^2 = (\gamma - \bar{\xi})^2 + \gamma = \gamma \tag{A19}$$

and, in agreement with thermodynamics,

$$\langle \Delta E \rangle^2 = \frac{1}{\beta^2} \langle \Delta\xi \rangle^2 = \frac{3}{2} n (kT)^2 \tag{A20}$$

References

[1] D. Sands and J. Dunning-Davies. The canonical distribution and the central limit theorem, http://www.arXiv.org, art. no. 0706.0978.
[2] J. W. Gibbs, Elementary Principles in Statistical Mechanics, developed with especial reference to the rational foundation of thermodynamics, University press, John Wilson & Son, Cambridge, USA, 1902.
[3] Robert Bruce Lindsay and Henry Margenau, Foundations of Physics, John Wiley & Sons, New York, 1936.
[4] J.C. Maxwell, On Boltzmann's theorem on the average distribution of energy in a system of material points, Cambridge Philosophical Society's Transactions, Vol. XII., reprinted in The Scientific Papers of James Clerk Maxwell, Edited by W. D. Niven, M.A., F.R.S., Volume Two.
[5] E T Jaynes, Information Theory And Statistical Mechanics, Phys.Rev. 106(4), 620-630 (1957).
[6] R. Baierlein, Thermal Physics , Cambridge U. P., New York, 1999.
[7] R. P. Feynman, Statistical Mechanics: a set of lectures, Perseus Books, (1998)
[8] Kerson Huang, Statistical Mechanics, John Wiley & Sons, New York, 1963, p160,
[9] Charles Kittel, Elementary Statistical Physics, Dover Publications Inc., New York, 1986 p62
[10] Max Born, Natural Philosophy of Cause and Chance, Dover Publications Inc., Toronto, 1964 p172.
[11] A I Khinchin, Mathematical Foundations of Information Theory, Trans. R A Silverman, M D Friedman, Dover Publications Inc, New York 1953, p5-7
[12] James H Jeans, The Dynamical Theory of Gases, Cambridge University Press, 2nd Edition, 1924.
[13] David Sands, Thermodynamic entropy and the accessible states of some simple systems, Eur. J. Phys, 29 129-135,(2008).
[14] M. Badino, Mechanistic Slumber vs. statistical insomnia: the early history of Boltzmann's H-theorem (1868-1877), The European Physical Journal H, 36, 353-378 (2011).
[15] E. T. Jaynes, Gibbs vs Boltzmann entropies. Am. J Phys.33, (1965) 391-8.
[16] Paul and Tatiana Eherenfest, The Conceptual Foundations of the Statistical Approach in Mechanics, Trans. Michael J Moravcsik, Cornell University Presses; Ithaca, New York, p65, 1959.
[17] Richard C. Tolman, The Principles Of Statistical Mechanics, Clarenden Press Oxford, 1938, p505
[18] M G Kendall; 1961, A Course In The Geometry Of n Dimensions, Charles Griffen& Co. Ltd, (London), p35
[19] Sir Maurice Kendall and Alan Stuart; 1977, The Advanced Theory of Statistics, Vol 1, Charles Griffen& Co. Ltd , London, p64.
[20] H.B. Dwight; 1961, Tables of Integrals and Other Mathematical Data, fourth edition, Macmillan, New York, p134.

Clausius' Concepts of '*Aequivalenzwerth*' and Entropy: A Critical Appraisal

DAVID SANDS

Department of Physics and Mathematics,
University of Hull, Cottingham Rd., Hull, HU6 7RX, UK
d.sands@hull.ac.uk

JEREMY DUNNING-DAVIES

Institute for Theoretical Physics and Advanced Mathematics (IFM) Einstein-Galilei,
Via Santa Gonda, 14 – 59100 Prato, Italy
j.dunning-davies@hull.ac.uk

Clausius' original papers covering the re-working of Carnot's theory and the subsequent development of the concept of the entropy of a body are reviewed critically. We show that Clausius' thinking was dominated by the then prevalent idea that a body contained heat and argue that his concept of *aequivalenzwerth*, the forerunner of entropy, should be understood as the quantity that measures the ability of the heat in a body to be transformed into work. Although this view of heat had been rejected by the mid-1870s, the concept of the entropy of a body not only survived but went on to dominate thinking in thermodynamics. We draw attention to several contradictions within Clausius' papers, among them the idea that only infinitesimal quasi-static processes are reversible even though reversibility is integral to the theory of motive power. In addition, Clausius developed his theory of entropy to account for internal work done in overcoming the forces between particles but then applied the theory to the ideal gas in which there are no such forces. We show how this conflicts with the First Law and discuss whether entropy is the state function Clausius claimed it to be.

Keywords: *Aequivalenzwerth,* Carnot's theory, Entropy, Ideal gas

1. Introduction

We have been concerned for some time with the current conception of entropy and present in this paper a critical analysis of Clausius' original papers. Rather long by today's standards and published in German, it is possible that they were read by only a few people at the time and by hardly anybody since then. Although he collected his papers together in a single volume, he published another book shortly after with ostensibly the same title, The Mechanical Theory of Heat, but with very different content and subtitles. The volume containing nine original papers published between 1850 and 1865, was published in German in 1865 and translated by T. Archer Hirst in 1867 [1]. The second was a re-working by Clausius of the theory of heat so as to present the subject in a more coherent form as a text book on the subject [2]. However, in this second book many of the ideas behind Clausius' thinking on entropy and irreversible processes are absent, and on page 214 the relationship $TdS \geq dQ$ is quoted as fact. Generations of physicists might well have read this more accessible text book and accepted the ideas without ever having encountered the thinking that lay behind them.

The purpose of this paper, therefore, is to examine Clausius' ideas on reversible and irreversible processes and entropy increases as they appear in his First, Fourth, Sixth and Ninth Memoirs in the translation by Archer Hirst. In the First Memoir he applies the mechanical theory of heat to Carnot's theory of motive power. In the Fourth the concept of *aequivalenzwerth*, the forerunner of entropy, is developed for reversible cyclic processes and in the Sixth he extends the idea to irreversible non-cyclic processes. In the Ninth he coins the term 'entropy'. In developing the arguments we use Clausius' own words, and unless explicitly stated all quotations from Clausius are drawn from these Memoirs, but principally the First, Fourth and Sixth. We start, however, by summarizing briefly Carnot's theory on motive power in order to lay the foundations for later discussion.

2. Carnot's Theory of Motive Power

Among writers of the later twentieth century and beyond there is considerable confusion between Carnot and Clausius to the extent almost of ignorance of what it is that Carnot actually said. For example, Wilson [3] states Carnot's theorem as; 'There exist two functions of state S and T, where T is a positive function of the empirical temperature θ only, such that, in any infinitesimal quasi-static change of a body or of a system of bodies, $dQ=TdS$'.

Wilson's statement bears little or no relation to anything that Carnot wrote in his treatise on motive power [4]. Rather the concepts within it, and there are many, such as the functions of states S and T, infinitesimal quasi-static changes, the relationship $dQ=TdS$, and even the notion that the properties of bodies rather than cyclic changes are important, all originated with Clausius some thirty years or more after the publication of Carnot's original work. Likewise Sommerfeld [5] also attributed to Carnot Clausius' idea of a quasi-static engine when he wrote that Carnot, 'considers an engine … which realizes the processes … infinitely slowly'.

Carnot was concerned with understanding the operation of a heat engine and introduced the notion of cyclic changes. He reasoned that the nature of the working fluid is immaterial; any two perfect, or reversible, engines operating between the same temperatures convert the same amount of heat into work. The working fluid and the changes that take place therein are therefore of no consequence. In fact, Carnot was still working within the theory of heat as matter, caloric, in which the expansion of matter under the influence of heat was regarded as a natural consequence of one material substance entering into another. For Carnot this is the origin of the motive power of heat. He explicitly excludes heating without possibility of expansion and identified alternate heating and cooling, which lead, respectively, to alternate expansion and contraction, as the means of producing motive power.

As described so eloquently by Tait [6], a contemporary and colleague of Thomson and one of the most lucid writers on thermodynamics of the time, Carnot's theory, though founded on the notion of heat as caloric, survives the abandonment of this theory and the adoption of the dynamical theory of heat. Carnot's theorem that the conversion of heat into work, and vice versa, depends only on the temperatures of the hot and cold reservoirs and not on the nature of the working fluid, was subsequently shown by Clausius to hold true with the adoption of the mechanical theory of heat. What is perhaps not so well appreciated is how much the idea of caloric seems to have influenced subsequent descriptions of Carnot's work, even though the notion had long since been abandoned. In Carnot's description of how the cycle might be realised in practice, he imagines the vessel being placed in contact with a hot reservoir and the gas being allowed to expand to some arbitrary volume. This is consistent with heat as caloric, because it was assumed to enter into materials and cause them to expand. The idea of an arbitrary, almost unlimited, expansion follows naturally. An explanation based on kinetic theory, on the other hand, would suggest that expansion comes first and heat follows due to the reduction in temperature of the working fluid and that the capacity for expansion is limited by the external pressure. It is remarkable then, that long after the demise of the theory of caloric descriptions of Carnot's cycle by writers such as Clausius, Tait and, later in the twentieth century, Fermi [7], Sears [8] and Callen [9], to name but three, were more or less identical to that given by Carnot himself [4].

Carnot's model of an ideal engine is quite abstract and might even be said to have contained a contradiction. Carnot implicitly recognised the need for high pressure in developing motive power, as this allowed the possibility of 'utilizing a larger descent of caloric' as 'the steam being produced under greater pressure is also at a higher temperature' [10]. In a real engine, though, expansion is determined by the length of the piston stroke and the manner in which the piston is coupled to the rotary shaft. Carnot's description not only failed to recognize the importance of either the internal pressure or the external load against which the piston expands, but was also very different from the way in which an engine actually operates.

Carnot also seems to have attached very little importance to a definition of motive power. The definition he gave appeared as a footnote to a discussion on whether motive power is limited or not and took the following form [4]:

> The expression motive power here signifies the useful effect that a motor is capable of producing. This effect may always be measured in terms of the elevation of a weight through a certain distance: it is measured, as is well known, by the product of the weight and the height to which it is raised.

This corresponds in modern terms to work done rather than power. No criticism can be attached to Carnot here, as the concept of energy was certainly not well developed in Carnot's time and was still very much under construction in the middle of the nineteenth century, when Rankine, Clausius and others were writing on heat engines and using essentially the same definition of motive power. In 1854, for example, Rankine expressed the motive power as the area under a p-V curve [11]. The concept of power as a rate at which work is done seems certain not to have been understood, but with this lack goes also a failure to appreciate the necessity for a piston to move with some speed in order to develop power. Although Carnot made reference to high pressures and high speeds, the cycle of operations he described appears to occur slowly, but not necessarily quasi-statically, as described by Wilson [3] and others [12]. This idea originated with Clausius and, indeed, it is possible that Clausius developed the concept of a quasi-static engine partly because the need for speed in developing power was not recognized.

3. Clausius on Carnot

Clausius' first major work on Carnot was published in 1850 [1]. Entitled, 'On the moving force of heat', it is a long article by today's standards and notable for its application of the equivalence of heat and work, drawing as it does on the work of Joule, to whom he refers explicitly in the opening passages. On Carnot he writes, '... we must look steadfastly into this theory which calls heat a motion ...' and concludes that 'the new theory is opposed, not to the real fundamental principle of Carnot, but to the addition that '*no heat is lost*''. Clausius' modified proposition therefore becomes; 'In all cases where work is produced by heat, a quantity of heat proportional to the work done is consumed; and inversely, by the expenditure of a like quantity of work, the same amount of heat may be produced'.

This innocuous looking statement is in fact based on an anachronistic view of heat that characterised the thinking of the time and permeates all of Clausius' writings. The phrase, 'heat is produced', might be understood as shorthand for a rise in temperature and a subsequent flow of heat to the surroundings, but Clausius means instead that the kinetic energy of the gas particles is increased. Modern thermodynamics has rejected the notion of heat upon which Clausius based all his thoughts, but the concept of entropy has been accepted, expanded, and, according to Meixner [13], applied without foundation to irreversible thermodynamic processes.

Clausius regarded heat as a property of a body which is increased after the expenditure of work. It is in this context that Clausius defines what has come to be recognised as the First Law of Thermodynamics on page 28 of the First Memoir;

$$dQ = dU + AR\frac{(a+t)}{v}dv \qquad (1)$$

A here is the mechanical equivalent of heat, *t* is the temperature in degrees Celsius and *a* is a constant that Clausius shows is equal to 273.15. Later, on page 78, Clausius explicitly wrote

$$dQ = dU + ApdV \qquad (2)$$

Clausius described the heat, *Q*, as (p29):

> the entire quantity of heat absorbed by the gas during a change of volume and temperature [and] may be decomposed into two portions...One of these, *U*, which comprises the sensible heat and the heat necessary for interior work ... fulfils the usual assumption; it is a function of *v* and *t*... and is therefore determined by the state of the gas at the end of the alteration; while the other portion, which comprises the heat expended on *exterior work*, depends not only upon the state of the gas at these two limits, but also upon the manner in which the alterations have been effected throughout.

Of the quantity *U* he wrote further that it is 'of great importance in the theory of heat' and that it 'will frequently come under discussion in the following memoirs'. Notice, however, that despite the 'great importance' of *U*, Clausius seems to regard *Q* as the more important. He restated that *U* 'involves two of the three quantities of heat that enter into consideration when a body changes its condition; these are the augmentation of the so-called *sensible* or *actually present* heat and the heat *expended in interior work*'.

The phrase, 'sensible or actually present heat' is telling. Maxwell wrote on p19 of his *Theory of Heat* [14], that, 'Those ... who maintained heat to be a substance supposed that it existed in the fluid in a concealed or latent state'. This was to distinguish it 'from the heat which, when applied to a body, makes it hotter, or raises the temperature ... called sensible heat. A body, therefore, was said to possess so much heat.'

Maxwell clearly associates the phrase 'sensible heat' with caloric, but this is the phrase used by Clausius time and again in connection with the mechanical theory of heat, which suggests that his view of heat was heavily influenced by existing notions of caloric. In a letter to Tait in 1876 [15] Maxwell commented:

> With respect to our knowledge of the condition of energy within a body, both Rankine and Clausius pretend to know something about it. We certainly know how much goes in and comes out and we know whether at entrance or exit it is in the form of heat or work, but what disguise it assumes when in the privacy of bodies, ... is known only to R., C. and Co..

Two years later, Maxwell attributed the same view to Tait [16], writing; 'We find it difficult enough, even in 1878, to attach any distinct meaning to the total actual heat of a body'.

Notwithstanding this flawed view of heat, Clausius' first memoir on the motive power is arguably his greatest achievement. Tait certainly believed so. Tait has often been accused of nationalism in his treatment of Clausius, but in his first edition of *Sketch of*

Thermodynamics [17] he had written, 'the grand point of Clausius' work is his proof that Carnot's principle of reversibility still holds, though on other grounds than those from which Carnot deduced it'. Even today we can agree with Tait that this 'was a step of the utmost importance in thermodynamics, and sufficient (had he done no more) to entitle him to a foremost place in the history of the subject' [17]. However, the flaws in Clausius' understanding of the concept of heat represent a thread running through his subsequent work, and, as we shall show, influenced his ideas on entropy.

The Fourth Memoir is notable because although it is concerned principally with reversible cycles, Clausius never explicitly defines what is meant by this term. Rather he implies that a reversible cycle is composed separately of reversible processes, as in this statement on page 130 of the translation by Archer Hirst:

> ... let us assume that all parts of the body have the same temperature; then in order that the process may be reversible the changing body when imparting or receiving heat can only be placed in communication with such bodies as have the same temperature as itself, for only in this case can the heat pass in the opposite direction.

Thus Clausius regarded isothermal heat flow as a reversible process in its own right rather than just one stage of a cycle which is reversible. He acknowledges that, strictly, heat flow at the same temperature cannot happen but nonetheless proceeds by assuming the difference in temperature to be so small that it is negligible. This is clearly an approximation, but one that at first sight seems reasonable and which has been accepted by physicists since, as has the corresponding approximation to reversibility in the work done. In the footnote of p131 he writes that 'the sole exterior force in operation is a pressure ...that ... always differs so little from the expansive force of the body that the two may be regarded as equal to one another in all calculations'.

The implication is that both the isothermal and adiabatic stages of the Carnot cycle are separately reversible processes, which is the origin of the modern notion of a quasi-static process as reversible. Carnot, on the other hand, was concerned only with reversing 'the cycle of operations' so that the cycle is run in reverse. According to McCulloch [18], an American civil engineer and Professor of Mechanics, writing in 1876, 'Carnot's test of a perfect engine is that it be capable of being worked backwards, in the cycle, reproducing all the changes; or *that it be reversible*' (p139). In other words, as long as the cycle itself can be reversed the engine is reversible.

In applying the mechanical theory of heat to Carnot, Clausius would appear to have gone further than simply replacing the principle that 'no heat is lost' by also redefining the concept of reversibility to include processes that can be considered in isolation from a cycle. Moreover, whereas Carnot defined the motive power in terms of lifting a weight, which is what a practical engine would have been required to do, Clausius required an external load that varies with the internal pressure, decreasing as the volume expands and increasing as it contracts. Not only is this different from Carnot's definition but it is also unrepresentative of the kind of work required of an engine. Clausius' motivation for these ideas is revealed in the Fourth Memoir and stems directly from his conception that heat is a property of a body. On irreversible processes, he commented (p133) that the

> ... different kinds of operations ... are ... as far as external appearances are concerned, rather numerous, even though they may not differ very essentially. One of the most frequently occurring examples is that of the transmission of heat by mere conduction, when two bodies at different temperatures are brought into immediate contact; other cases are the production of heat by friction...together with all cases where a force, in doing mechanical work, has not to overcome an equal resistance, and therefore produces a perceptible external motion, with more or less velocity, the *vis viva* of which afterwards passes into heat.

There are two important phrases; 'mere conduction' of heat and 'perceptible external motion'. The latter implies a quasi-static process and the former that normal heat conduction is different from reversible heat conduction. Thus, Clausius' thinking in respect of isothermal heat flow seems to be part of his peculiar view of heat. Even though he acknowledges that it cannot actually happen, he nonetheless seems to believe it to be something different from normal heat flow. Modern kinetic theory, on the other hand, would suggest that it is just a limiting case. A gas doing work loses energy and heat flows in to compensate. If the rate of flow of heat matches the rate at which work is done, either because the processes are very slow or the thermal connection is very good, the process may be considered isothermal.

As for the second phrase, 'a perceptible external motion', this might be taken to imply the motion of a piston, but this does not appear to be what Clausius had in mind. Rather, he refers to the free expansion as an example of an irreversible process giving rise to a perceptible motion, implying that the large scale motion he mentions is hydrodynamic. If so, the total energy of the gas remains constant throughout the expansion and whereas a real gas might undergo a slight change in temperature as a result of the expansion the temperature of an ideal gas will remain unchanged. There can be no heat flow as we understand the term today as no work is done and the most likely explanation is that in describing the *vis viva* as passing into heat Clausius is referring to the transformation from the directional motion associated with mass flow to the random motion of the particles of the gas in equilibrium. In other words, Clausius is referring again to the idea that heat is a property of a body, specifically the random motion of its particles.

Clausius' thinking on reversibility may be summarised as not only being different from Carnot's, in as much as Carnot was concerned with the reversibility of the cycle as a whole and Clausius is clearly of the view that the individual stages of the cycle are reversible, but also as being based on a misconception about the nature of heat, however widespread that view might have been at the time. Moreover, it implies the most staggering contradiction, but one that, to our knowledge, has not been recognised. Certainly, we can find no mention of it among anything we have read on Clausius. The contradiction is this; just four years after publishing a paper on the moving force of heat, in which the concept of reversible cyclic processes is central, Clausius has concluded that, in order for the cycles to be reversible and therefore for the theory to apply, there should be no perceptible motion. That is, nothing should move. One might well ask to what extent Clausius has retained any notion of motive force, which is a crucial aspect of the theory of heat engines which Carnot was trying to develop. The full import of regarding motive power in terms of work done rather than as a rate of doing work now becomes clear.

4. *Aequivalenzwerth* or Equivalence Value

Clausius' Fourth Memoir is also notable for the introduction of the concept of compensated and uncompensated transformations. As with so much of Clausius' work, the concept is somewhat obscured by a lack of clarity in his writing. Tait observed to Maxwell that Clausius, along with Rankine, was 'about as obscure in their writings as anyone can well be' [19]. The idea of compensation is closely connected to Clausius' statement of the Second Law, namely that heat cannot pass by itself from a colder to a hotter body. In a rather extensive footnote on pages 117-8 of the Fourth Memoir Clausius explains that the phrase 'by itself' may be replaced by the word uncompensated, so that the law becomes; 'an uncompensated transmission of heat from a colder to a warmer body can never occur'. Mathematically, Clausius treated compensation through the concept of '*Aequivalenzwerth*', or equivalence value, which was a quantity with the value Q/T, Q being a quantity of heat transmitted at a temperature T.

The passage describing the equivalence of transformations perhaps represents Clausius at his most obscure, not least because the idea of equivalence immediately brings to mind Joule, who had shown that the equivalence of heat and work lay in their effects. However, the Ninth Memoir makes it clear that Clausius regarded the equivalence of transformations as a fundamental theorem standing alongside the equivalence of heat and work, but in what way two transformations are equivalent is never really made clear.

The two transformations that Clausius intended should be regarded as equivalent are; a transfer of heat without loss from a hot body to a cold body and a conversion of part of that heat into work. On p119 Clausius illustrated a more general cycle than the Carnot cycle consisting of six separate stages and proceeded to show in the subsequent pages that the transformation of heat into work must be accompanied by the transfer of heat from a hot to a colder body. Conversely, if the cycle is executed in the reverse direction heat passes from a cold to a hotter body and heat is generated from work. However, Clausius then went on to say that one transformation can replace the other, which is conceptually challenging as they are different stages of the same cycle. In the Carnot cycle, for example, the development of work from heat occurs on the expansion stroke and the transference of heat to the colder body on the compression stroke, but of course it is only the isothermal parts of the two strokes that compensate each other in the sense used by Clausius. In what sense these two transformations are equivalent in as much as they can replace each other is difficult to understand, and as we shall show Clausius himself regarded the theory as somewhat abstract and conceptually difficult.

Clausius defined the conversion of work into heat as a positive transformation so that around a closed reversible cycle a positive transformation is compensated by a negative transformation and

$$\sum \frac{Q}{T} = 0 \tag{3}$$

By this reckoning, the isothermal expansion has a negative equivalence value, which is opposite to what one would expect based on the direction of heat flow. However, the idea of heat as an interaction is a later idea and, as we have shown, Clausius writes as if the conduction of heat and the transformation of heat into work are two separate things. This appears to have been a common view at that time. Hutchison, writing about Rankine, argued that the modern historian or physicist can easily be misled about what most scientists thought about heat in the early 1850s [20] and described Rankine's conception of the relationship between heat and work as the following:

> … that when an expanding gas absorbed heat, and converted that heat into work, the conversion was catalyzed by the presence of the gas's attendant internal heat. More generally, contained heat somehow activated gases and enabled them to convert heat, taken either from some external source, or from their own internal store, into work.

Clausius' ideas on the transformation of heat into work would appear to be very similar, so it is not surprising that the modern reader finds some aspects of his writings very difficult to comprehend. Of the two theorems that Clausius claimed to be at the heart of the mechanical theory of heat, only Joule's regarding the conversion of heat into work has survived into the modern day. The theorem of the equivalence of transformations has slipped into obscurity.

Even though Clausius had not by this time developed the concept of entropy, equation (3) is instantly recognizable as the expression for the entropy change around a reversible cycle. In so far as Clausius conceived a reversible cycle as being composed of a series of quasi-static reversible processes, the view has arisen that Clausius developed the concept of entropy for reversible processes [13]. Although he mentions that the theory of equivalence value can be extended to irreversible processes, it is not until the Sixth Memoir of 1862 that he does so. His motivation is ostensibly to develop a method for treating the interior work, which he describes as '… for the most part so little known, and connected with an equally unknown quantity [the heat present in the body] that we are obliged to trust in some measure to probabilities'. Clausius was not advocating a statistical approach with this phrase but pointing out that there is no certainty that the extension of the concept of uncompensated transformations to non-cyclic processes will lead to the correct result.

It is in this context that Clausius develops the idea that an irreversible process is characterised by an increase in entropy and develops what has come to be seen by many as the Second Law of Thermodynamics, namely that

$$TdS \geq dQ \qquad (4)$$

Clausius starts the Sixth Memoir with a summary of the theory of the equivalence of transformations developed in the Fourth memoir and states, without proof, that the relationship

$$\int \frac{dQ}{T} \geq 0 \qquad (5)$$

holds good for every cyclic process 'which is in any way possible'. Moreover, he acknowledges in the very next paragraph (p219) that the theorem in this form is somewhat obscure and difficult to comprehend; 'Although the necessity of this theorem admits of strict mathematical proof … it nevertheless retains an abstract form, in which it is with difficulty embraced by the mind, and we feel compelled to seek for the precise physical cause, of which this theorem is a consequence'.

However, there is no evidence at all to suggest that Clausius achieved his aim of putting this theory on a firm physical footing. Indeed, it might be argued on physical grounds that the opposite relationship might be a better summary of the physical reality, namely

$$\int \frac{dQ}{T} \leq 0 \qquad (6)$$

This at least reflects the fact that for an irreversible cycle heat has to be extracted from a system in order to return to the initial state. Not only is this not abstract, it is also easily appreciated through the example of the free expansion: in order to return the gas to its original state it is necessary to compress it to the original volume, thereby raising the temperature, and cool it, thereby decreasing the entropy. Indeed, Clausius later adopted the sign convention in equation (6) in the Ninth Memoir of 1865 for this very reason; that heat has to be extracted in an irreversible cycle to restore the initial state. The fact that such an easy interpretation is available is further evidence that Clausius did not think in these terms when writing his earlier memoirs and that the conceptual difficulty to which he referred stems from a peculiar conception of the nature of heat and its transformation into work.

The key to understanding Clausius' approach to irreversible processes in the Sixth Memoir is to realise that he has simply applied the results from an analysis of cyclic processes to non-cyclic processes; '… as there is no essential difference between interior and exterior work, we may assume with certainty that a theorem which is so generally applicable to exterior work cannot be restricted to this alone'. This is exemplified by the relationship between pressure and absolute temperature, of which Clausius writes (p219),'In all cases in which heat contained in a body does mechanical work by overcoming resistances, the magnitude of the resistance which it is capable of overcoming is proportional to the absolute temperature'. Allowing that the resistance that a gas can overcome is simply equivalent to the internal pressure, this could be taken to be an innocent statement about the equation of state of an ideal gas, but in fact there is more to it than that. After a long, and somewhat obscure passage about the nature of internal work, including the 'self-evident' idea that heat must be consumed in the production of internal work and, *vice versa*, generated if internal work is expended, Clausius then modifies the theorem to read (p223), 'The mechanical work which can be done by heat during any change of arrangement of a body is proportional to the absolute temperature at which this change occurs'. This is a direct comparison between external and internal work, and is directly related to the idea that heat causes work.

Clausius' comparison between internal work and external work in this passage is not consistent with the dynamical theory of heat. Consider the free expansion of a real gas in which the particles, be they molecules or atoms, are subject to inter-particle forces. Provided the particles have sufficient kinetic energy to overcome the attractive forces, the amount of work done by the particles in moving from one average separation to another depends only on the nature of the forces between the particles and not on the speed at which they move. External work, on the other hand, could be considered to depend on the speed of the particles via the pressure. These passages suggest strongly that Clausius had in mind the idea that there is a force associated with the heat contained in a body and that this force does work in some mysterious manner. By this reasoning, if heat can do external work then it is reasonable to assume that it can also do internal work, hence Clausius' statement that a theory that applies to one must certainly apply to the other.

Central to Clausius' arguments is the idea that the 'force of the heat' can be measured through the quasi-static process. Work performed quasi-statically corresponds to the maximum amount of work that can be done between two states. This is clearly important in the Carnot cycle, for which maximum work is essential if the engine is to be ideal and reversible, but it assumes another function in the passages quoted above. The external pressure represents the magnitude of the resistance that the expansive force of the gas is capable of overcoming, so a quasi-static process is equivalent to measuring the 'force of the heat'. In an irreversible process, however, the work done is less than the maximum possible and Clausius writes in the very next sentence to the passage above (p223), 'The law does not speak of the work which the heat actually *does*, but that which it *can do*. Similarly, in the first form of the law [quoted above] it is not of the resistances which the heat overcomes but of those which it *can overcome* that mention is made'. In just a few sentences further on he writes, in relation to the free expansion, 'In order ... to determine the force of the heat we must evidently not consider the resistance which actually is overcome, but that which can be overcome'. Thus, for Clausius, heat does work, whether internal or external, because there is a force associated with it. In accounting for the work that can be done in an irreversible process rather than the work that is actually done, Clausius derived the well-known inequality of equation (4). We shall show that Clausius actively sought this inequality for the entropy change in an irreversible process. It was, in his view, analogous to the inequality for the entropy change around an irreversible closed cycle and therefore the mathematical expression of his belief that a common theory underlay the production of internal and external work.

5. Irreversibility and Entropy

Clausius was led to the view that a non-cyclic process is reversible if 'the change can likewise take place in the opposite direction under the influence of the same forces'. This corresponds to his earlier definition that the internal pressure should match the external pressure so that the work done is just the integral of pdV. This makes it clear that Clausius is no longer considering reversibility in the context of a cycle, if he ever did, and makes explicit what was once just an implication. That is, the separate stages of the Carnot cycle are themselves supposed to be reversible if each is executed in this manner, and if not they, and the cycle, are irreversible. It is perhaps not surprising, then, that Clausius carried the principal mathematical result characterising the irreversibility of cyclic processes, equation (5), across to non-cyclic processes.

Clausius starts with the First law, but whereas in his papers of 1850 and 1854 he expressed the law in terms of changes in U, in the Sixth Memoir of 1862 he preferred instead to write,

$$dQ = dH + AdL \tag{7}$$

The total work done, dL, is composed of internal and external contributions. The internal work depends on the magnitude of the inter-particle forces, which are unknown, so Clausius had previously expressed the First law as a relationship between two measurable quantities, heat and work, in terms of a third, the internal energy, U. Therefore the internal energy combined the internal work with the 'actual heat in the body', H, defined as;

$$H = mcT \tag{8}$$

with c being the 'true capacity for heat'. However, in equation (7) Clausius has reversed the construction of U in favour of two quantities which are not measurable; the internal work and the heat contained in a body. The latter had been rejected by Maxwell by the late 1870s as being of no physical significance and modern thermodynamics agrees. In real gases, where the specific heat varies with temperature, H in equation (8) has no practical meaning. Yet, Clausius' extension of *aequivalenzwerth* to the concept of the entropy of a body depends upon separating out the components of internal energy and treating internal work as a separate quantity.

On the change of sign in dQ, Clausius gave as his justification for it the following (p225); 'We have already represented by dQ the infinitely small amount of heat imparted *to* another body *by the one which is undergoing modification* [our emphasis], hence we must represent in a corresponding manner, by $-dQ$, the heat which it withdraws from another body.' However, in the Fourth Memoir (p113) he had written, 'Let Q, therefore, be the quantity of heat which must be imparted to a body during its passage, in a given manner, from one condition to another, any heat withdrawn from the body being counted as an imparted negative quantity of heat.' These two statements appear to contradict each other, as the first statement implies that a positive dQ is used for heat flowing out of a body to another body whereas the second phrase implies a positive dQ is associated with heat flowing into a body. The last phrase was also used in the Fifth Memoir on the theory of the steam engine (p141) and would therefore seem to confirm this view.

This appears to be a deliberate change of sign, as Clausius reverses the sign change just a few pages later (page 229); '... the positive sense of the quantity of heat is here taken differently from what it is in [equation (7)], in which heat given up by the body, and not heat imparted to it, is reckoned positive'. Clausius' writing sheds no light on his motivation for this change of sign. He acknowledged the change in a footnote, but referred, rather obscurely, to using the positive sign previously 'in order to attain greater correspondence with equation I of Art. I'. That equation is essentially the equation of state of the ideal gas, from which he developed the common expression for the First Law (equation 1), so quite what he meant by gaining 'greater correspondence' is not clear. It is clear, however, that he used a positive sign for dQ from the outset and retained it right up until this section of the Sixth Memoir. He then reversed the sign again in order to derive an equation which he claimed agreed with another '... disregarding the difference in the sign of dQ (which is caused only by the different way we have chosen to employ the signs + and – in this case)...'. Clausius seems to have regarded the sign not as representing some physical reality but as something that can be changed according to use.

Whilst the sign of the heat might indeed be arbitrary, once decided the sign of other terms must be consistent. Taking a positive sign for dQ, as in equation (2), means that if all the terms are gathered together on one side, the result is arithmetically sensible; the heat is negative and sum of positive and negative terms adds to zero. The signs could be reversed without difficulty, but reversing the sign of only one term, as in equation (7), implies that the sum of three terms all of the same sign adds to zero. Implicitly, one of the terms must have an opposite sign to the others. It is difficult to escape the conclusion that Clausius adopted a negative sign simply because it suited his purpose and, moreover, justified it with a change of definition.

Clausius proceeded from equation (7) by introducing a term called the 'disgregation', Z. He concluded that work of any kind is always accompanied by a rearrangement of the particles and defined Z as the quantity that represents the state opposite to aggregation. Specifically, for an infinitesimal reversible change,

$$dZ = \frac{1}{KT}dL = \frac{1}{T}AdL \qquad (9)$$

Then, from equation (7)

$$\frac{dQ}{T} + \frac{dH}{T} + dZ = 0 \qquad (10)$$

For larger changes,

$$\int \frac{dQ + dH}{T} + \int dZ = 0 \qquad (11)$$

After a lengthy digression spent discussing the difference between Z and Rankine's thermodynamic function, Clausius described Z thus (p244):

> If the equivalence value Q/T be assumed for the production of a quantity of heat Q from work at the temperature T, a magnitude admits of being introduced, as a second transformation corresponding thereto, which has relation to changes in the arrangement of the particles of the body, is completely determined by the initial and final conditions of the body, and fulfils the condition that in every reversible change of condition the algebraic sum of the transformations is equal to nothing.

This is the origin of the notion current in thermodynamics that entropy is a state function, but Clausius is considering the case in equation (9) when the external work done can expressed in terms of state functions, p and V and the disgregation simply represents a transformation of the work done. Multiplying equation (10) through by T shows that TdZ has the units of energy, and if equation (9) applies the sum of all terms is zero, notwithstanding the difficulty that implicitly one must be negative. In other words, disregarding the change in the sign of dQ, energy is conserved in accordance with the First Law of thermodynamics on which equation (10) is based.

However, the quasi-static work is not exactly pdV but rather an approximation to it [21] and it is open to question to whether in such a case either the external work or the disgregation can truly be regarded as state functions. As for irreversible transformations, Clausius had already written in 1850, in a passage we have already quoted, that the 'heat expended on *exterior work*, depends not only upon the state of the gas at these two limits, but also upon the manner in which the alterations have been effected throughout'. As the disgregation depends on the external work it is difficult to see how Z can be regarded as a state function in an irreversible process. Moreover, it is self-evidently the case that ${}^{dQ}/_{T}$ is not a state function for irreversible cycles as the sum of changes around a closed cycle is not zero. Likewise, ${}^{dW}/_{T}$ cannot be a state function for the same reason. Despite this, Clausius examines 'the manner in which the foregoing theorem is modified when we give up the condition that all changes of condition are to take place reversibly' and continues to use the disgregation as a state function, writing in his paper of 1866 [22] on the disgregation; '… as I have shown in the paper referred to [the Sixth Memoir] …

$$dL = \frac{T}{A} dZ \tag{12}$$

where Z is a quantity *completely determined by the actual condition of the body and independent of the way in which the body arrived at that condition* [our emphasis]'. Clausius seems to ignore in this sentence the importance of effecting the change in a so-called reversible, or quasi-static, manner for the equality to hold, whereas in his Sixth Memoir Clausius specifically included this condition and concluded that for irreversible changes $dZ > dW$ That is, the change of disgregation 'may be greater, but not smaller than the accompanying transformation' whilst for negative changes in Z, corresponding to work done on the gas, the change 'may be smaller, but not greater' than the transformation of work into heat. Algebraicly,

$$dL \leq \frac{T}{A} dZ \tag{13}$$

Then, in a modification of equation (10),

$$\frac{dQ + dH}{T} + dZ \geq 0 \tag{14}$$

and for larger changes

$$\int \frac{dQ + dH}{T} + \int dZ \geq 0 \tag{15}$$

This is the mathematical expression of the idea that a quasi-static process measures the force of the heat, hence the equality, but in an irreversible process, where the work done is less than the maximum, what is important is the work that can be done, hence the inequality. Clausius summarised this essential result (p247) as follows: 'The algebraic sum of all the transformations occurring during any change of condition whatever can only be positive, or, as an extreme case, equal to nothing'. This is the essential result that Clausius sought. It may be compared directly with the key result from the analysis of cyclic processes, which is phrased in exactly the same way. That is to say, around a closed, reversible cycle the sum of all transformations is zero, as expressed by equation (3), but in an irreversible cycle the sum of transformations is greater than zero (equation 5). Thus Clausius has achieved his aim expressed earlier as, '… we may assume with certainty that a theorem which is so generally applicable to exterior work cannot be restricted to this alone'.

Perhaps it is a reflection of Clausius' opacity that the full meaning of this result does not appear to have been recognised by his contemporaries. In Maxwell's first edition of *The Theory of Heat* in 1871 entropy was described in terms of dissipation using the ideas of Tait. This prompted Clausius to write a letter to the Philosophical Magazine in 1872 [23], in which he explicitly called the quantity

$$\int \frac{dH}{T} + dZ \tag{16}$$

the entropy of a body, and therein arose the modern idea that entropy increases in an irreversible adiabatic transformation.

6. Discussion and Conclusion

The idea that entropy is a state function is so ingrained into the structure of modern thermodynamics that to question it would seem unnecessary. However, recent work by the authors [21] has shown that an internal damping mechanism exists even in an ideal gas and that an adiabatic expansion against a constant pressure is intrinsically irreversible. That is to say, it is not possible to return to the initial state by an adiabatic process. It follows that a sequence of such expansions is also irreversible and we question the notion of a quasi-static adiabatic expansion as reversible. This is central to Clausius' conception of the Carnot cycle and to his argument that entropy is a state function. Carnot's idea that the cycle itself is reversible presents no such difficulties, but if only cyclic processes are truly reversible then the validity of the Second Law would appear also to be restricted to cyclic processes only. Clausius' extension to non-cyclic processes would then be invalid.

Moreover, Clausius' derivation of the law of increasing entropy in equations (14) and (15) is questionable on several counts. The terms *aequivalenzwerth* and entropy reflect the idea heat was believed to be able to perform work because it had a force associated with it. Thus, when heat performed work it brought about some kind of unspecified transformation in the body and *Aequivalenzwerth* was the name used to represent the mathematical value of this transformation. The name entropy was chosen by Clausius precisely because it represented the idea of transformation. We have already quoted Maxwell as saying that by 1878 Tait had rejected the notion of contained heat, but there is a clear association between the concept of the entropy of a body and the idea of the heat contained in a body. Indeed, it follows logically that if heat is a property of a body and entropy is associated with a change in that heat, then entropy must also be a property of a body. Conversely, if heat is something that is exchanged during a thermal interaction then logically the entropy should be regarded as a characteristic of that interaction. Strangely, modern thermodynamics recognizes this view of heat but not of entropy.

Central to Clausius' extension of the Second Law to non-cyclic processes is the concept of disgregation, which seems to correspond in Clausius' mind to the work that can be done and is therefore the quantity that measures the 'force of the heat'. Its incorporation into the entropy of the body associates entropy with the ability, or potential, of a body to do work rather than any work that is actually done. The difficulty can be appreciated by applying the idea to the free expansion of an ideal gas. As the heat capacity is independent of temperature, $U \equiv H$. In addition, the disgregation would appear to be integrable over large changes in V. and changes in Z can be defined for any change of volume at a given temperature. However, there are no inter-particle forces and the internal work is zero. Clausius' whole motivation for introducing the concept of disgregation disappears. According to equation (15), though, the entropy of the gas still increases. Indeed it has to if disgregation is the state function that Clausius claimed it to be and herein lies a difficulty. There is no external heat flow, no heat generated internally through work done by inter-particle forces or any other mechanism of dissipation, and no change in internal energy. Yet, some quantity, TdZ, with the dimensions of energy is believed to be increasing in the body.

Clausius' whole theory depends on the negative sign for dQ in equation (7). For changes in which $dZ = 0$ a positive dQ would cause a positive change in dH, which is only sensible, but then for $dH = 0$ and a negative change in dZ the sum of all transformations would be negative. Clausius' extension of the Second Law to irreversible processes is thus seen to depend not only on adopting a particular sign convention for the equivalence value of a transformation (equation 5), which is opposite from what might be expected on the basis of heat flow (equation 6), but also on adopting a particular sign for dQ which appears to have been adopted for this purpose only. Elsewhere, even within the Sixth Memoir, a different sign convention was used. Moreover, not only did he change the sign, he also appears to have changed the definition of a positive quantity of heat.

This is essentially a problem of the conservation of energy. From the inequality in equation (4) it is possible to write

$$TdS = dQ + d\phi \tag{17}$$

where $d\phi$ has the units of energy. However, it is not contained within the First Law, as

$$dQ = dU - dW \tag{18}$$

Therefore, $d\phi$ is not dQ, dU or dW or given by any combination of these consistent with the law.

Clausius' failure to recognize that his theory of entropy violates energy conservation is particularly odd given that the function we now call the internal energy and the form of the First Law that follows from it, which is still used today, are due to him. However, we have shown that, despite describing U as important, he dispensed with it in favour of Q, which he seems to have regarded as the more important. This is just one of a number of contradictions. Clausius' theory of motive power is based on the contradiction that there should be no perceptible motion. The application of disgregation to the ideal gas is another. Indeed, the theory of entropy developed by Clausius in the Sixth Memoir is built around contradictions. Therefore, we question not only the foundations of the concept of entropy as a property of a body but also the logical consistency of entropy as a state function, and draw attention in particular to the conflict with the First Law.

References

[1] R. Clausius, The Mechanical Theory of Heat, with its applications to the steam engine, Ed. T. Archer Hirst, John van Voorst, London, 1867, is a collection of nine papers of which the First Memoir (pp14 -69), the Fourth Memoir (pp111 – 135) and the Sixth Memoir (pp 215-250) are of especial interest.

[2] R. Clausius, The Mechanical Theory of Heat, trans. Walter R Browne, MacMillan & Co., 1879.

[3] A. H. Wilson, Thermodynamics and Statistical Mechanics, Cambridge University press, 1957.

[4] Sadi Carnot, Reflections on the Motive Power of Heat in The Second Law of Thermodynamics, Translated and Edited by W. F. Magie, Harper & Brothers, New York and London, 1899.

[5] Arnold Sommerfeld, Thermodynamics and Statistical Mechanics; Lectures on theoretical physics vol V. eds. F. Bopp, J. Meixner, J. Kestin Academic Press, New York and London, 1967.

[6] Peter Guthrie Tait, Lectures on Some Recent Advances in Physical Science, Elibron Classics Series, Adamant Media Corporation, 2006, p98.

[7] Enrico Fermi, Thermodynamics, Dover Publications, New York, 1936.

[8] Francis Weston Sears, An Introduction to Thermodynamics, the Kinetic Theory of Gases, and Statistical Mechanics, second edition, Addison-Wesley, 1953.

[9] H. B. Callen, Thermodynamics, John Wiley & Sons, 1960.

[10] Sadi Carnot, Op.Cit. p114.

[11] W. J. Macquorn Rankine, On The Geometrical Representation Of The Expansive Action Of Heat, And The Theory Of Thermodynamic Engines, read before the Royal Society of London on January 19, 1834, and published in the Philosophical Transactions for 1854. Reprinted in Miscellaneous Scientific Papers, Ed W. J. Millar, Charles Griffin & Co., London, 1881, pp. 339-409.

[12] J. Güémez, C. Fiolhais and M. Fiolhais, Sadi Carnot on Carnot's theorem, Am. J. Phys. 70(1): 42-47 (2002).

[13] J. Meixner, Processes in Simple Thermodynamic Materials, Arch. Rational Mech. Anal., 33: 3-53 (1969).

[14] James Clerk Maxwell, Theory of Heat, Dover Publications, Inc., New York, 2001.

[15] Cargill Gilston Knott, Life And Scientific Work Of Peter Guthrie Tait Supplementing The Two Volumes Of Scientific Papers Published In 1898 And 1900, Cambridge University Press, 1911, p222.

[16] James Clerk Maxwell, Tait's Thermodynamics, Nature Vol XVII, and reprinted in The Scientific Papers Of James Clerk Maxwell, Edited by W. D. Niven, M.A., F.R.S., Volume Two, p660.

[17] P. G. Tait, Sketch of Thermodynamics, Edmonston and Douglas, Edinburgh, 1868, p29.

[18] R. S. McCulloch, Treatise on the Mechanical Theory of Heat and its Applications to the Steam Engine, etc., D. Van Nostrand, New York, 1876.

[19] P. M. Harman (Ed), The Scientific Letters and Papers of James Clerk Maxwell, Cambridge University Press, 1995, p328 footnote 2.

[20] Keith W. Hutchison, J. M. Rankine and the rise of thermodynamics, The British Journal For The History of Science, 14(6): 1-26 (1981).

[21] David Sands and Jeremy Dunning-Davies, The compound piston: resolution of a thermodynamic controversy by means of kinetic theory, Journal Of Non-Equilibrium Thermodynamics, 35(2): 125-143 (2010).

[22] R. Clausius, On the Determination of the Disgregation of a Body, and on the True Capacity for Heat, Philosphical Magazine, 31 p28 (1866).

[23] Rudolf Clausius, A Contribution to the History of the Mechanical Theory of Heat, Philosphical Magazine, 43: 10. 6-115 (1872).

THE CONFORMAL STEADY-STATE FREE PRECESSION: A KEPPLERIAN APPROACH TO AUTOMORPHIC SCATTERING THEORY OF ORBITON /SPINON DYNAMICS

WALTER J. SCHEMPP

Lehrstuhl für Mathematik I, Universität Siegen, 57068 Siegen, Germany
schempp@mathematik.uni-siegen.de

Based on projective geometry, a quantum holographic approach to the orbiton / spinon dynamics of quantum black-holography and clinical magnetic resonance tomography is mathematically described. Crucial applications of the conformal steady-state free-precession modality and automorphic scattering theory are the evidence for a supermassive central black hole in the Milky Way galaxy and the modalities of clinical cardiovascular magnetic resonance tomography and diffusion weighted magnetic resonance tomography of non-invasive radiological diagnostics.

Keywords: Blackholography, Mathematical radiology, Automorphic scattering theory, High-Resolution NMR spectroscopy and Tomography, Metaplectic group representation, Harmonic Heisenberg Lie Group, Projective Lie-Poisson Structure, Cine Loops of Orbiton / Spinon Dynamics, Inverse Plane Geometry, Conformal Non-Euclidean Hyperbolic Geometry, Poincaré models

The universe is in reality quantum mechanical. - Peter W. Shor

The best understanding of what one can see comes from theories of what one can't see. - Steve Smale

There is no branch of mathematics, however abstract, that will not eventually be applied to the phenomena of the real world. - Nikolai Ivanovich Lobachevsky

S'il est une vérité que l'évolution des mathématiques contemporaines a mise en évidence, c'est qu'on ne simplifie pas une théorie compliquée, pour peu qu'elle ait quelque valeur, en la spécialisant mais en la généralisant; bien etendu, je ne fais pas allusions par là aux généralisations superficielles ou artificielles que suggère parfois la mode, mais à celles qui placent une théorie dans son cadre naturel en la débarrasant des circonstances adventices qui l'encombrent ou la défigurent. - André Weil

One factor that has remained constant through all the twists and turns of the history of physical science is the decisive importance of the mathematical imagination. - Freeman John Dyson

There are three stages in the history of every medical discovery. When it is first announced, people say that it is not true. Then, a little later, when its truth has been borne in on them, so that it can no longer be denied, they say it is not important. After that, if its importance becomes sufficiently obvious, they say that anyhow it is not new. - Sir James Mackenzie

The following three important prerequisites for cardiac magnetic resonance remain: Triggering, triggering, triggering. In short, there is no cardiac image without triggering. - Mark Doyle

1. Overview

The concept of balanced steady-state free-precession can be traced back to the epoch-making treatise *Astronomia Nova* published by Johannes Keppler in 1609. In view of the fact that the universe is basically a quantum physical construction, the orbiton / spinon dynamics of the Keppler space observatory, launched by NASA on March 5th, 2009 on an earth-trailing heliocentric orbit to discover earth-sized and smaller extrasolar planets or exoplanets of the Milky Way galaxy in or near the habitable zone, tries to determine how many of the billions of stars in the Galaxies are such stars.

The seesaw pair detachment diagram establishes the planetary laws of the Kepplerian projective libration theory of harmonic oscillations (*libra* = torsion balance, already known to Copernicus) by acting in the three-dimensional real projective space $\mathbb{P}_3(\mathbb{R})$ of spin-tags

$O(\mathbb{C}_{\mathbb{R}} \oplus \bar{\mathbb{C}}_{\mathbb{R}}, \mathbb{R})$ $\qquad$ $Sp(2,\mathbb{R}) \times Sp(2,\mathbb{R})$

$O(2,\mathbb{R}) \times O(2,\mathbb{R})$ $\qquad$ $Mp(2,\mathbb{R})$

The vertical arrows visualize embeddings and slanted arrows indicate connections of detached members of dual reductive pairs in the spirit of orbiton / spinon dynamics, performed by the Fourier filter of the unitarily exponentiated reducible metaplectic representation of $\mathrm{Mp}(2, \mathbb{R})$.

Another highly interesting application of the orbiton / spinon dynamics of balanced steady-state free-precession which is implemented by the Fourier filter of the seesaw pair detachment configuration is the general relativistic test for the existence for gravitational wave radiation by means of the rapid orbital period decay in the recently discovered detached binary white dwarf system J065133+284423. It is based on the loss of angular momentum of the detached binary dwarf system of extreme low masses $0.26M_\odot$ and $0.50M_\odot$ which has been observed at optical wavelengths at a distance of about 3000 light years from the earth with a 12.75 minute orbital period and a 1315 kms^{-1} radial velocity amplitude.

The importance of the seesaw pair configuration derives from the fact that it implies in terms of the Euclidean distributional Laplace operator $\Delta : S'(\mathbb{R}\oplus\mathbb{R}) \longrightarrow S'(\mathbb{R}\oplus\mathbb{R})$ the partial differential equation of second order

$$\frac{\partial^2 u}{\partial t^2} = y^2 \Delta u + \frac{1}{4} u \ (y = \Im w)$$

for the automorphic wave function $(t, w) \rightsquigarrow u(t, w)$ on the space $\mathbb{R} \times \mathbb{C}_\mathbb{R}$. As a consequence of automorphic scattering theory [23], the Schrödinger wave equation of quantum mechanics

$$i\frac{\partial \psi}{\partial t} = \frac{1}{2}\Delta\psi$$

in standard notation arises.

The metaplectic group $\mathrm{Mp}(2, \mathbb{R})$ was first introduced in order to reformulate Carl Ludwig Siegel's analytic theory of quadratic forms in terms of group theory [12,40,41]. As a projective representation of the symplectic group $\mathrm{Sp}(2, \mathbb{R}) \cong \mathrm{SL}(2, \mathbb{R})$ the metaplectic representation forms a genuine unitary linear group representation of $\mathrm{Mp}(2, \mathbb{R})$.

In the context of the seesaw pair configuration, the involutive automorphism of the Argand plane $\mathbb{C}_\mathbb{R} \cong \mathbb{R}(i) \cong \mathbb{C}$, defined by the complex conjugation mapping, provides the orientation reversing spin echo transition $\mathbb{C}_\mathbb{R} \longrightarrow \bar{\mathbb{C}}_\mathbb{R}$. It forms an efficient tool of magnetic resonance spectroscopy and clinical magnetic resonance tomography [9,32].

Due to the isomorphy of the Galois cohomology group

$$H^2(\mathrm{Sp}(\mathbb{C}_\mathbb{R}, \mathbb{R}), \mathbb{Z}_2) \cong \mathbb{Z}_2.$$

the symplectic group $\mathrm{Sp}(\mathbb{C}_\mathbb{R}, \mathbb{R}) \cong \mathrm{Sp}(2, \mathbb{R})$ admits, up to isomorphy, a unique non-trivial extension by $\mathbb{Z}_2 = \mathbb{Z} / 2\,\mathbb{Z}$, the metaplectic group $\mathrm{Mp}(\mathbb{C}_\mathbb{R}, \mathbb{R}) \cong \mathrm{Mp}(2, \mathbb{R})$ which can be explicitly constructed in the exact sequence

$$\{1\} \longrightarrow \mathbb{Z}_2 \longrightarrow \mathrm{Mp}(\mathbb{C}_\mathbb{R}, \mathbb{R}) \longrightarrow \mathrm{Sp}(\mathbb{C}_\mathbb{R}, \mathbb{R}) \longrightarrow \{1\}$$

by means of the Maslov index [16].

The phase shift due to the Maslov index provides the corrected Feynman path integral representation on phase space. Since Richard Feynman's original formula is valid only for short times, the connection of the Maslov index to the metaplectic group $\mathrm{Mp}(2, \mathbb{R})$ justifies the mathematician's care, of which Feynman emphatically complains.

The physicist cannot understand the mathematician's care in solving an idealized physical problem. The physicist knows the real problem is much more complicated. It has already been simplified by intuition which discards the unimportant and often approximates the remainder.

The orbiton / spinon dynamical subtleties hindering a thorough understanding of the modality of magnetic resonance tomography were the reason for the difficulties with its clinical reception. Erwin Louis Hahn decisively changed the practical applicability of magnetic resonance tomography by the seminal discovery of the concept of spin echo in 1949. In pulsed nuclear magnetic resonance spectroscopy, relaxation times are measured directly and accurately from the spin echo amplitudes. Concerning magnetic resonance tomography, Hahn wrote:

I apologize to magnetic resonance imaging (MRI) pioneers because I never believed MRI would work, like Rutherford, *who said that anyone who believed nuclear radioactivity would be useful "is talking moonshine". However, I was only one of many unbelievers. Another unfidel in particular was* Anatole Abragam, *a distinguished French physics researcher in magnetic resonance. The French Society of Radiology wanted to award Abragam a medal in spite of the fact that he told them he hadn't contributed to MRI and didn't believe it would work.*

Abragam replied: *There was nuclear magnetic resonance before* Erwin Louis Hahn's *discovery of the spin echo, there is nuclear magnetic resonance after. It is not the same.*

The truth is, contrary to some of the preconceived opinions about its evolution, that clinical magnetic resonance tomography is a quickly expanding topic at an extremely rapid pace of innovation. The level of mathematical understanding required to master some of the more advanced techniques can be intimidating for the majority of diagnostic radiologists [7,8,19,39,42].

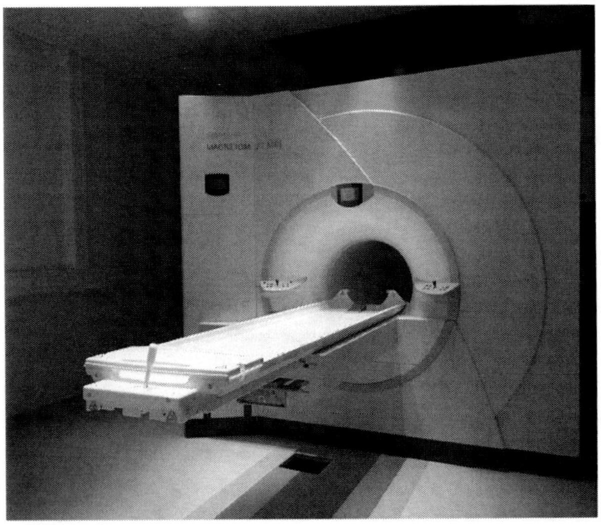

Figure 1. Modern high resolution magnetic resonance scanner which realizes the the affine coadjoint orbit picture Lie($\mathcal{N}$)* / CoAd($\mathcal{N}$) of the three-dimensional real Heisenberg Lie group $\mathcal{N}$ at a magnetic field density of 7 T. The patient positioned inside the bore can hear the noisy action of the conformal affine-linear magnetic field gradient group $\mathrm{Gr}_+(\mathbb{R})$ to establish the Hopf principal fibration quantum gate $\mathbb{S}_1 \hookrightarrow \mathbb{S}_3 \longrightarrow \mathbb{S}_2$. The scanner works especially for diffusion weighted magnetic resonance tomography and tractography, cardiovascular magnetic resonance tomography, magnetic resonance angiography, and neurofunctional magnetic resonance tomography.

2. Elementary-Number Theoretic Background

In analytic number theory, the quantum entanglement identity of Landsberg-Schaar for quadratic Gaussian sums

$$\frac{1}{\sqrt{p}} \sum_{0 \le m \le p-1} exp\left(\frac{\pi i m^2 q}{p}\right) =$$

$$\frac{e^{\frac{\pi i}{4}}}{\sqrt{2q}} \sum_{0 \le n \le 2q-1} exp\left(-\frac{\pi i n^2 p}{2q}\right)$$

derives for $p \in \mathbb{N}$; $p > 0$ and $q \in \mathbb{N}$; $q > 0$ from the seesaw pair detachment diagram and the two-cocycle of Maslov index –1 in the classical identity

$$\underbrace{e^{\frac{\pi i}{4}}} = \frac{1}{2}\sqrt{2}(1+i) \in \mathrm{U}(1, \mathbb{C})$$

In the case when the product pq is an even natural number, the preceding identity reduces to the elegant relation

$$\frac{1}{\sqrt{p}} \sum_{0 \le m \le p-1} exp\left(\frac{\pi i m^2 q}{p}\right) =$$

$$\frac{e^{\frac{\pi i}{4}}}{\sqrt{q}} \sum_{0 \le n \le q-1} exp\left(-\frac{\pi i n^2 p}{q}\right)$$

Evaluation of the Landsberg-Schaar identity at $q = 1$ yields the law of quadratic reciprocity of number theory in the form of the Gaussian sum identity

$$\sum_{0 \le m \le p-1} exp\left(\frac{2\pi i m^2}{p}\right) = \sqrt{p}.i^{\frac{(p-1)^2}{4}},$$

which holds for $p \in \mathbb{N}$; $p \ge 1$. Taking account of the elementary congruence

$$\left(\frac{pq-1}{2}\right)^2 - \left(\frac{p-1}{2}\right)^2 - \left(\frac{q-1}{2}\right)^2 \equiv$$

$$\frac{(p-1)(q-1)}{2} \pmod 4.$$

the famous reciprocity formula arises in terms of the Legendre-Jacobi symbols for quadratic group characters with values in $\mathbb{Z}_2 \cong \mathbb{Z} / 2\,\mathbb{Z} \cong \{-1, +1\}$:

$$\left(\frac{p}{q}\right)\left(\frac{q}{p}\right) = (-1)^{\frac{p-1}{2}.\frac{q-1}{2}}$$

The complementary formulae read:

$$\left(\frac{-1}{p}\right) = (-1)^{\frac{p-1}{2}}, \left(\frac{2}{p}\right) = (-1)^{\frac{p^2-1}{8}}$$

Notice that the compressed reciprocity formula holds for all odd primes $p \in \mathbb{N}$ and $q \in \mathbb{N}$ [1]. Remarkably, harmonic analysis of the real Heisenberg Lie group *constitue à certains égards la méthode la plus satisfaisante qui soit actuellement connue pour établir la loi de réciprocité sous sa forme la plus générale* [40].

3. The Real Heisenberg Lie Group

The quantum holographic approach to conformal steady-state free-precession and magnetic resonance tomography depends on the ordering concept of real Heisenberg Lie group $\mathcal{N}_0 = \mathbb{R} \times \mathbb{R} \times \mathbb{T}$ with the twisted multiplication law

$$(x_1, y_1, z_1)(x_2, y_2, z_2) = \left(x_1 + x_2, y_1 + y_2, z_1 z_2 \mathbf{e}(x_1 y_2)\right)$$

where, as usual, $\mathbf{e}(\theta) = \exp(i\theta) = e^{i\theta}$ for $\theta \in \mathbb{R}$ describes the central phase circle $\mathrm{U}(1, \mathbb{C}) \cong \mathbb{T}$ within the symplectic plane $\mathbb{R} \oplus \mathbb{R} \cong \mathbb{C}_{\mathbb{R}}$ of infinitesimal loop rotation $i\frac{d}{d\theta} 1_{\theta=0}$

The symplectic plane $\mathbb{R} \oplus \mathbb{R} \cong \mathbb{C}_{\mathbb{R}}$ and the one-dimensional compact torus group

$$\mathbb{T} \cong \mathbb{R}/\mathbb{Z} \cong \mathrm{U}(1, \mathbb{C}) \cong \mathrm{SO}(2, \mathbb{R})$$

are considered as embedded into $\mathcal{N}_0$. The centre of $\mathcal{N}_0$ is $\mathbb{T} \lhd \mathcal{N}_0$. There is an obvious complexification of $\mathcal{N}_0$, which is an extension of $\mathbb{C}_{\mathbb{R}} \cong \mathbb{R}(i) \cong \mathbb{C}$ by $\mathbb{C}^{\times}$.

According to the Stone-von Neumann theorem, the central character

$$\mathbb{R} \ni \theta \rightsquigarrow \boldsymbol{e}(\nu\theta) \in \mathbb{T} \ (\nu \in \mathbb{R}^{\times})$$

determines by their central Larmor labels $\nu \in \mathbb{R}^{\times}$ the equivalence classes of irreducible unitary linear representations of $\mathcal{N}_0$. The strong Stone-von Neumann theorem asserts on the L^2 level of the two-sided ideal of Hilbert-Schmidt operators that square integrability modulo the centre is valid if and only if unicity by the central character holds.

4. A Relativistic Perspective — The Spin

The real Lie group generated by the conformal Möbius inversions of the Riemannian sphere $\mathbb{S}_2 \cong \mathbb{P}_1(\mathbb{C})$ acting on the three-dimensional real projective space $\mathbb{P}_3(\mathbb{R})$ is given by the projective orthogonal group

$$\mathrm{PO}(3, 1, \mathbb{R}) \cong \mathrm{O}(3, 1, \mathbb{R}) / \{\mathrm{id}, -\mathrm{id}\}$$

The invariance group $\mathrm{PSO}(3, 1, \mathbb{R})$ of inversive plane geometry is isomorphic to the Lorentz group $\mathrm{PSL}(2, \mathbb{C})$ associated with the non-Euclidean hyperbolic half-space $\mathcal{H}^3 \hookrightarrow \mathbb{P}_3(\mathbb{R})$. Let Γ denote the cross-sectional functor and S the symmetrisation functor of the category of fiber bundles over differentiable manifolds. Then Γ recognizes the fibered structure of the underlying smooth manifold. The projective orthogonal group $\mathrm{PO}(3, 1, \mathbb{R})$ associates the cross-sectional bilinear symmetric two-form fiber

$$-\mathrm{d}r^2 + \mathrm{d}x^2 + \mathrm{d}y^2 + \mathrm{d}z^2 \in \Gamma(\mathrm{S}^2\mathrm{T}^*\mathbb{R}^4)$$

of Minkowski signature (3,1) on the dual tangent bundle of the real vector space $\mathbb{R}^4$ with positive definite symmetric *absolute* two-tensor. The group $\mathrm{PO}(3, 1, \mathbb{R})$ admits the neutral component $\mathrm{PO}_0(3, 1, \mathbb{R}) \cong \mathrm{O}_0(3, 1, \mathbb{R})$.

The Lie group of orientation preserving conformal Möbius inversions of $\mathbb{P}_3(\mathbb{R})$ is non-compact and isomorphic to the group of automorphisms of the Riemannian sphere $\mathbb{S}_2 \cong \mathbb{P}_1(\mathbb{C})$. It was this kind of observations in which Arthur Cayley obtained various metrics that led him to state with succinct enthusiasm metric geometry to be subordinate to projective geometry, and projective geometry to be *all* of geometry.

The spin factor associated with the Heisenberg Lie group $\mathcal{N} \cong \mathbb{C}_{\mathbb{R}} \oplus \mathbb{R}$ is given by

$$\mathcal{J}(\mathbb{C}_{\mathbb{R}} \oplus \mathbb{R}) \cong \mathbb{C}_{\mathbb{R}} \oplus \mathbb{R} \underbrace{\oplus \ \mathbb{R}}.$$

It forms a Jordan algebra under the multiplication law induced by the Minkowski metric and therefore is one of the most remarkable in the classification of simple formally real Jordan algebras [2]. The projective orthogonal group $\mathrm{PO}(3, 1, \mathbb{R})$ acts on $\mathcal{J}(\mathbb{C}_{\mathbb{R}} \oplus \mathbb{R})$ as the group of isometries. Its one-dimensional subspaces spanned by non-zero isotropic vectors are the light rays inside the light cone of $\mathcal{J}(\mathbb{C}_{\mathbb{R}} \oplus \mathbb{R})$. The manifold of all light rays is called the heavenly sphere $\mathbb{P}_1(\mathbb{C}) \cong \mathbb{S}_2$ of the spin factor $\mathcal{J}(\mathbb{C}_{\mathbb{R}} \oplus \mathbb{R})$ associated with $\mathcal{N}$. If the symplectic spinors are restricted to the multiplicative Lie group $\mathrm{SU}(2, \mathbb{C}) \cong \mathrm{Spin}(3, \mathbb{R}) \cong \mathbb{S}_3$ of unit quaternions, the Hopf projector $\mathbb{S}_3 \longrightarrow \mathbb{S}_2$ pops up.

5. The Third Kepplerian Law

The orbiton / spinon dynamics of the balanced steady-state free-precession combined with the automorphic scattering theory yields the extremally weighted, irreducible $\mathrm{Lie}(\mathrm{SL}(2, \mathbb{R}))$ eigenmodule $\mathbb{Z}_2 \times \mathbb{Z}_2$ graduation

$$\left(\mathcal{M}_{-\frac{3}{2}} \widehat{\otimes} \bar{\mathcal{M}}_{\frac{1}{2}}, \mathcal{M}_{\frac{3}{2}} \widehat{\otimes} \bar{\mathcal{M}}_{-\frac{1}{2}}\right)$$

of the complex vector space of tempered Schwartz distributions

$$\mathcal{S}'(\mathbb{R}) \widehat{\otimes} \mathcal{S}'(\mathbb{R}) \cong \mathcal{S}'(\mathbb{R} \oplus \mathbb{R})$$

on the real symplectic plane $\mathbb{R} \oplus \mathbb{R} \simeq \mathbb{C}_{\mathbb{R}}$. The action of the projective orthogonal group $\mathrm{PO}(3, 1, \mathbb{R})$ induces the third Kepplerian law outlined in the great *Harmonices Mundi* of 1619: According to the hyperbolic distance of inversive plane geometry [4], the third power of the large semi-axis $a > 0$ of the Kepplerian ellipse is proportional to the product of the square of the time of oriented revolution

$$T = \frac{1}{2\pi\nu} \ (\nu \in \mathbb{R}^{\times})$$

and the mass m_0 of the central object. The seesaw pair configuration implies the spectral decomposition of the complex vector space $\mathcal{S}'(\mathbb{R} \oplus \mathbb{R})$. It follows the *ratio orbium*, the reciprocal Schwarzschild-Kepplerian proportions of Heisenberg rotations

$$\left\{\frac{m_0 T^2}{a^3}, \frac{a^3}{m_0 T^2}\right\}$$

which permits to derive the mass of the object located at the focus of the Kepplerian ellipse and hence to incontrovertibly establish the existence of the supermassive black hole Sgr A* in the centre of the Milky Way galaxy.

6. First Conclusion

The unitarily $\mathrm{Lie}(\mathrm{SL}(2,\mathbb{R}))$ derived two-sheeted metaplectic representation of the metaplectic Lie group Mp(2, $\mathbb{R}$) provides a projective geometrical orbiton / spinon test of Kepplerian spectral type to establish in the vein of quantum blackholography the existence of the supermassive black hole Sgr A* in the centre of the Milky Way galaxy. Its mass is about 4×10^6 $M_\odot$, its distance to earth amounts to 27 000 light years.

The incontrovertible evidence for the supermassive black hole Sgr A* in the centre of the Galaxis underlines that Keppler, as a radical Copernican, played a major role in the creation of the modern science of astrophysics. Not only were his models much more accurate than their predecessors, they were innovative theoretically to such an extent that they stand quite alone. Keppler's mathematical ingenuity was such that he could invent an alternative theory of orbiton / spinon dynamics, good enough to save even the excellent observations of Tycho Brahe. Unspoken, but also present, is a misguided and sometimes unwitting interpretation of his astrophysics as a quantum physical precursor of Newton's gravitational mechanics, which has impeded any unbiased analysis of it.

7. Nuclear-Magnetic-Resonance Spectroscopy

High resolution magnetic resonance spectroscopy is not only an efficient quantum physical technique for chemists, biochemists, and diagnostic radiologists, but also offers a fascinating intellectual study in its own right. It is based on the fact that ensembles of nuclear spins can be manipulated in order to extract information about molecular structure and molecular motion. A whole novel science has been created where none existed before. Nuclear spins can be controlled through an intricate choreography, carefully programmed in advance, to enhance, simplify, correlate, decouple, edit or assign the nuclear magnetic resonance spectra. This remarkable choreography can only be inferred indirectly, based on the knowledge of the electronic music provided by the Kepplerian harmonies of radiofrequency pulse trains and the observation and detection of the ensuing nuclear magnetic resonance signals [5,34].

8. Johannes Keppler's Work

Keppler's scientific work, profoundly original and fundamental for the development of modern science, was based on the supposition that the universe had been created according to rational ordering principles. Keppler's highly individual version of this presumption colors his whole astrophysical work and anticipates a key idea of the Enlightenment. It must be sought most particularly in the great treatise *Harmonices Mundi* of 1619. There it remains largely unknown, even though it has become trite to describe Keppler in general terms as a Platonist. He was so excited by the ecstatic truth of his fundamental discovery of balanced steady-state free-precession that he immediately added the following lines of enthusiasm to the introduction of Book V of *Harmonices Mundi*:

Now, since the dawn eight months ago, since the broad daylight three months ago, and since a few days ago, when the full Sun illuminated my wonderful speculations, nothing holds me back. I yield freely to the sacred frenzy; I dare frankly to confess that I have stolen the golden vessels of the Egyptians to build a tabernacle for my God far from the bounds of Egypt. If you pardon me, I shall rejoice; if you reproach me, I shall endure. The die is cast, and I am writing the book – to be read either now or by posterity, it matters not. I can wait a century for a reader, as God himself has waited six thousand years for a witness.

Keppler has been regarded by historians of science as a diligent condenser of astronomical data, or as a numerically inclined mystic. The truth is that he was a brilliant mathematician whose theoretical analysis was performed by the help of sophisticated geometric configurations. His understanding of the interplay between observation and theory was subtle, one of the characteristic features of his genius. It can be said of Keppler, as of very few great scientists, that what he accomplished in the fields of inversive plane geometry and projective steady-state free-precession would never have been done had he himself not accomplished it. It is a great experience in orbiton / spinon dynamics to recognize that the sophisticated ordering principles of the Kepplerian celestial harmonies are non-invasively realized by magnetic resonance scanners to the best well-being of suffering patients and hope of prediction and prevention potential.

9. Nuclear-Magnetic-Resonance Tomography

Following the perspicacious Lobachevskian prophecy, a valuable clinical application of the concept of balanced steady-state free-precession is in high resolution magnetic resonance tomography [36]. More precisely, in clinical cardiovascular magnetic resonance tomography the electrocardiographic gating of the entire cardiac cycle in which the quantum holographic data collection is synchronized to reduce the impeding cardiac beating motion artifacts, provides a multitude of non-invasive insights into the beating human heart by radiofrequency excitation of nuclear spin assemblies [15,17,25,26,33]. Because the contraction of the *Myocardium* is a major determining factor in cardiac magnetic resonance tomography, both prospective and retrospective cardiac triggering are useful electro-cardiographic synchronization techniques of bright blood imaging. It may be difficult to perceive focal areas of cardiac wall motion abnormality on non-invasive serial static pictures, but these abnormalities are often easily seen on the cine loops of the orbiton / spinon dynamics.

An outstanding property of the modality of clinical magnetic resonance tomography is the creation of quantum holograms by the scanner for reading out the diagnostic tomographies by fast Fourier filter algorithms (FFT). The FFT algorithms read out the spectral content of the quantum holograms implemented by planar tomographic slices of positive energy [31]. They are computer controlled by trains $\sum \ell$ of affine-linear magnetic field gradients $\ell \in \mathrm{PLie}(\mathcal{N})^*$, conceived as affine real linear forms on the projectivized real Heisenberg Lie algebra $\mathrm{PLie}(\mathcal{N})$ consisting of affine line directions in the three-dimensional real vector space $\mathrm{Lie}(\mathcal{N}) \cong \mathrm{Lie}(\mathcal{N}_0)$.

In the context of the modality of diffusion weighted magnetic resonance tomography or diffusion tensor tomography [28,29], the affine-linear magnetic field gradients simulate the locally linear functions of the gradient controlled Lévy-Hinčin type assessment formula of phase-randomizing Lévy flights $(X_t)_{t\in\mathbb{R}_+}$. They are conceived as random walks in continuous time, that is, as stochastic processes with independent and stationary increments which allow to measure the stochastic molecular motion as echo signal attenuation.

10. Random Walks

Jean-Pierre Vigier and several of his cowokers tried to apply the Einstein-Smoluchowski approach to Brownian motion onto the stochastic interpretation of quantum physics [18]. The elegant Kappler phase-randomizing experiment in dissipative diffusion is based on a small mirror of area $\leq 1\ \mathrm{mm}^2$ and several molecular layers thick which is suspended by a thin quartz fiber in a container to determine the Avogadro constant by phase-randomizing Lévy flights $(X_t)_{t\in\mathbb{R}_+}$ of gas molecules at various densities. As the gas molecules strike the mirror, they cause it to perform a Lévy stochastic process – that of a harmonic oscillator, since the restoring force due to the torque of the fiber is linear for the very small angles of the Heisenberg rotations about the vertical axis involved. The angle is measured by shining laser light on the mirror and measuring the remote spot of the reflected laser beam.

Thus, at all densities, the probability distribution of the angle is the same – it is determined by thermodynamics. However, the archetypical stochastic process [6], although of central theoretical importance, is too restrictive for clinical cardiovascular and diffusion weighted magnetic resonance tomography. For balanced steady-state free-precession fast gradient echo techniques, the pulse train is designed so that the phase coherence of the transverse magnetization is maintained from repetition to repetition. Thus the transverse magnetization generated by a given excitation radiofrequency pulse may contribute to the signal amplitude measured in many succeeding echo periods.

11. The Lévy–Hinčin Type Assessment Formula

Because the Brownian stochastic process is the unique probability distribution of functions, which has independent increments that are stationary and of finite variance, it is advisable to relax the last of these conditions. Pierre-Gilles de Gennes, a former student of Lévy, described the holographic sum-over-phase-histories method of phase-randomizing Lévy flights as follows:

Long ago, Paul Lévy invented a strange family of random walks – where each segment has a very broad probability distribution. These flights, when they are observed on a macroscopic scale, do not follow the standard Gaussian statistics. When I was a student, Lévy's idea appeared to me as (a) amusing, (b) simple – all the statistics can be handled via Fourier transforms – and (c) somewhat baroque: where would it apply? As often happens with new mathematical ideas, the fruits came later.

Thus, phase-randomizing Lévy flights $(X_t)_{t\in\mathbb{R}_+}$ are appropriate to deal with the Kepplerian projective libration theory of harmonic oscillations and to clinical magnetic resonance tomographic scanners. For the treatment of diffusion anisotropy, they represent the method of choice for diffusion weighted tractography. Actually, the powerful tools of harmonic analysis and complex function theory in Hardy spaces on Siegel II domains, Lie sphere geometry, Galois cohomology, and the symbolic calculus of boundary conformal quantum field theory reveal to form the natural framework for establishing the spinorial version of the gradient controlled Lévy-Hinčin type assessment formula for the time-dependent expectation operator

$$\mathbb{E}\big(\mathbf{e}((X_t)),\mu\big) = \exp\Big(-t\big(\Phi(\ell,\ell)\big) + i\gamma(\ell) + \frac{1}{2}\int_{\mathrm{PLie}(\mathcal{N})}\Big(1 - \mathbf{e}\big(\ell(v)\big) + i\ell(v)\mathbf{1}_{\mathcal{B}}(v)\mathrm{d}\lambda(v)\Big)\Big)$$
$$(t \in \mathbb{R}_+)$$

of the phase-randomizing Lévy flight $(X_t)_{t\in\mathbb{R}_+}$with respect to the infinitely divisible spectral measure μ uniquely determined by the tomographically sliced Lévy measure λ on the projectivized three-dimensional nilpotent real Heisenberg algebra $\mathrm{Lie}(\mathcal{N}) \cong \mathrm{Lie}(\mathcal{N}_0)$ of step two through its one-dimensional centre, the positive definite Hermitian form

$$\mathrm{PLie}(\mathcal{N})^* \times \mathrm{PLie}(\mathcal{N})^* \ni (\ell,\ell') \rightsquigarrow \Phi(\ell,\ell') \in \mathbb{C}_{\mathbb{R}}$$

derived from boundary conformal quantum field theory by the Bargmann-Fock representation of the three-dimensional unipotent real Heisenberg Lie group $\mathcal{N} \cong \mathbb{C}_{\mathbb{R}} \oplus \mathbb{R}$ with the twisted multiplication law

$$(w_1,z_1)(w_2,z_2) = \left(w_1 + w_2, z_1 + z_2 + \frac{1}{2}\Im(\overline{w}_1 w_2)\right).$$

There is a natural diffeomorphism

$$\exp : \mathcal{N} \rightarrow \mathrm{Lie}(\mathcal{N})$$

of central infinitesimal generator

$$i\nu\frac{d}{d\theta}1_{\theta=0} \quad (\nu \in \mathbb{R}^\times)$$

which allows to geometrically classify in the dual vector space $\mathrm{Lie}(\mathcal{N})^*$ the unitary dual $\widehat{\mathcal{N}}$ of $\mathcal{N}$ which consists of the equivalence classes of irreducible unitary linear representations of central Larmor labels $\nu \in \mathbb{R}^\times$.

12. The Projective Lie–Poisson Structure

Due to the isomorphy $\mathrm{Lie}(\mathcal{N})^* \cong \mathrm{Lie}(\mathcal{N}_0)^*$, the real Heisenberg Lie groups $\mathcal{N}$ and $\mathcal{N}_0$ admit isomorphic affine coadjoint orbit pictures $\mathrm{Lie}(\mathcal{N})^* / \mathrm{CoAd}(\mathcal{N}) \cong \mathrm{Lie}(\mathcal{N}_0)^* / \mathrm{CoAd}(\mathcal{N}_0)$ which visualize $\widehat{\mathcal{N}} \cong \widehat{\mathcal{N}}_0$ in the ambient conformally projectivized dual vector spaces $\mathrm{PLie}(\mathcal{N})^* \cong \mathrm{PLie}(\mathcal{N}_0)^*$ by means of foliations [22]. These foliations consist of the singular homogeneous plane representing by its singletons characters of $\mathcal{N}$, and a stack of parallel transversal symplectic planes, which represent tomographic slices of central Larmor labels $\nu \in \mathbb{R}^\times$. They are supporting symplectic forms

Figure 2. Tessellating triangulation of the Poincaré disc model conformally transferred by the Cayley transformation h^{-1} to the real Lie group $\mathrm{Gr}_+(\mathbb{R})$ of affine-linear gradients. The curvature $\equiv -1$ of the hyperbolic Riemannian metric of the open unit disc in the plane $\mathbb{C}_{\mathbb{R}}$ is obtained by unipolar stereographic projection of the Beltrami hemisphere model. The horizontal support of the infinitely divisible spectral measure μ in the gradient controlled Lévy-Hinčin type assessment formula can be visualized by a tiling of the open unit disc in $\mathbb{C}_{\mathbb{R}}$.

$$\nu.\mathrm{d}w = \nu.\mathrm{d}x \wedge \mathrm{d}y \ (\nu \in \mathbb{R}^\times)$$

which are invariant under the action of the symplectic group $\mathrm{Sp}(2,\mathbb{R}) \cong \mathrm{SL}(2,\mathbb{R})$. It admits the modular group $\mathrm{SL}(2,\mathbb{Z})$ as a discrete subgroup which gives rise to the Eisenstein series for $\mathrm{PSL}(2,\mathbb{Z})$ and generalizes the quadratic Gaussian sums appearing as trace forms

$$\mathrm{Tr}_{\mathbb{C}_{\mathbb{R}}}\ \mathcal{F}_n = \sum_{0\le m\le n-1} \zeta_n^{m^2} =$$

$$\begin{array}{ll} (1+\mathrm{i})\sqrt{n} & \text{for } n \equiv 0 \pmod 4 \\ \sqrt{n} & \text{for } n \equiv 1 \pmod 4 \\ 0 & \text{for } n \equiv 2 \pmod 4 \\ i\sqrt{n} & \text{for } n \equiv 3 \pmod 4 \end{array}$$

of the matrix of order $n \geq 1$ associated with the Fourier

$$\text{filter: } \mathcal{F}_n = \begin{pmatrix} 1 & 1 & 1 & \cdots & 1 \\ 1 & \zeta_n & \zeta_n^2 & \cdots & \zeta_n^{n-1} \\ 1 & \zeta_n^2 & \zeta_n^4 & \cdots & \zeta_n^{2(n-1)} \\ \vdots & \vdots & \vdots & \ddots & \vdots \\ 1 & \zeta_n^{n-1} & \zeta_n^{2(n-1)} & \cdots & \zeta_n^{(n-1)^2} \end{pmatrix},$$

$$\zeta_n = \underset{\sim}{\mathrm{e}}\left(\frac{2\pi}{n}\right)$$

The mod 4 multiplicities of its cyclic spectrum

$$\{+1, i, -1, -i\}$$

are given by the table

	$+1$	i	-1	$-i$
0	$+1$	0	0	-1
1	$+1$	0	0	0
2	$+1$	0	$+1$	0
3	$+1$	$+1$	$+1$	0

Conversely, identification of the trace form $\mathrm{Tr}_{\mathbb{C}_{\mathbb{R}}} \mathcal{F}_n$ implies the law of quadratic reciprocity. According to Roger Godement, the Eisenstein series opens the doorway to *Le jardin des délices modulaires ou, l'opium des mathématiciens.*

The planar tomographic slices support conformal non-Euclidean hyperbolic geometries of open half-planes of automorphic scattering theory, and by complex conjugation the phenomenon of gradient echo. Due to the central involution $\mathbb{R}^{\times} \ni \nu \rightsquigarrow -\nu \in \mathbb{R}^{\times}$ of $\mathcal{N}$, which implies the central involution

$$\mathbb{T} \ni \mathbf{e}(\nu\theta) \rightsquigarrow \mathbf{e}(-\nu\theta) \in \mathbb{T}\ (\theta \in \mathbb{R})$$

of $\mathcal{N}_0$, the central projection of the planar cogredient coadjoint orbit corresponding to the central Larmor label $\nu \in \mathbb{R}^{\times}$ onto the planar contragredient coadjoint orbit corresponding to the central Larmor label $-\nu \in \mathbb{R}^{\times}$ in the dual vector space Lie($\mathcal{N}$)* provides the phenomenon of quantum entangled spin echo by the group $\mathbb{Z}_2$.

13. Quantum Entanglement and Spin Echoes

According to the projective classification of involutive motions as central collineations in the metaplectic geometry of the affine coadjoint orbit picture Lie($\mathcal{N}$)* / CoAd($\mathcal{N}$) of the real Heisenberg Lie group $\mathcal{N}$ within the ambient conformally projectivized dual vector space PLie($\mathcal{N}$)*, the main difference between the phenomena of gradient echo and spin echo consists in the fact, that the gradient echo occurs by complex conjugation inside the same tomographic slice, whereas quantum entangled spin echoes are phase coherently created by unipolar stereographic projection across the foliation of planar coadjoint orbits from the cogredient to the associated contragredient orbit of $\mathcal{N}$ in Lie($\mathcal{N}$)* of inverse central Larmor label.

In terms of the realization of the affine coadjoint orbit picture Lie($\mathcal{N}$)* / CoAd($\mathcal{N}$) of $\mathcal{N}$ by the Hopf principal fibration quantum gate, the visualizaion of quantum entangled spin echoes is geometrically performed by the concept of Clifford parallelism of flat Liouville tori based Villarceau phase circles of the first and second kind. The Hopf links consisting of linked circles which are coherently embedded into $\mathbb{S}_3$ visualize the phenomenon of quantum entangled spin echoes.

14. Quantum Field Theory

In the context of the first part of the assessment formula for phase-randomizing Lévy flights $(X_t)_{t\in\mathbb{R}_+}$, the universal covering group $\mathcal{N}$ of $\mathcal{N}_0$ is realized as the connected and simply connected real Lie group of affine holomorphic, respectively anti-holomorphic automorphisms of the unbounded Siegel II domains

$$\mathrm{D}_{\Phi} = \{(w,\ell) \in \mathbb{C}_{\mathbb{R}} \oplus \mathbb{C}_{\mathbb{R}} \mid \Im w - \Phi(\ell,\ell) > 0\}$$

and

$$\mathrm{D}_{\Phi}^{*} = \{(-\bar{w},-\ell) \in \mathbb{C}_{\mathbb{R}} \oplus \mathbb{C}_{\mathbb{R}} \mid \Im w - \Phi(\ell,\ell) > 0\}$$

in the stabilizer of the point ∞, associated with the positive half-line $\mathbb{R}_{+}^{\times}$ [35]. The reflection

$$(w,\ell) \rightsquigarrow (-\bar{w},-\ell)$$

is an involution of D_{Φ} onto D_{Φ}^{*}, and the Siegel II domains D_{Φ} and D_{Φ}^{*} are biholomorphically, respectively bi-antiholomorphically equivalent to the open unit ball in $\mathbb{C}_{\mathbb{R}}^{2}$. Their *distinguished* boundaries or Šilov boundaries

$$\partial_0\mathrm{D}_{\Phi} = \partial_0\mathrm{D}_{\Phi}^{*} = \{(w,\ell) \in \mathbb{C}_{\mathbb{R}} \oplus \mathbb{C}_{\mathbb{R}} \mid \Im w - \Phi(\ell,\ell) = 0\}$$

can be identified with the Heisenberg Lie group

$$\partial_0\mathrm{D}_{\Phi} = \partial_0\mathrm{D}_{\Phi}^{*} \cong \mathcal{N}$$

as a non-trivial central extension

$$\{0\} \longrightarrow \mathbb{R} \lhd \mathcal{N} \longrightarrow \mathbb{C}_{\mathbb{R}} \longrightarrow \{0\}$$

acting simply transitively on the distinguished boundaries $\partial_0\mathrm{D}_{\Phi} = \partial_0\mathrm{D}_{\Phi}^{*}$,. As an application of Galois cohomology to quantum field theory, the axes' directions of the non-circular Keppler ellipses are in projective correspondence to the vector fields of conjugate tangential Cauchy-Riemann creation and annihilation operators relative to the Siegel II domains D_{Φ} and D_{Φ}^{*}.

15. The Hopf Principal Fibration Quantum Gate

It is actually the Galois cohomology group $H^2(\mathrm{Sp}(2,\mathbb{R}),\mathbb{Z}_2)$ which establishes the intimate relationship between spin echo and quantum entanglement by the Möbius band principal $\mathbb{Z}_2$ bundle over the phase circle $\mathbb{S}_1 \cong \mathbb{T}$, immersed into the Hopf principal fibration quantum gate

$$\mathbb{S}_1 \hookrightarrow \mathbb{S}_3 \longrightarrow \mathbb{S}_2$$

Its Villarceau circular fibers bounding Möbius strips arise by the bipolar stereographic projection of the affine coadjoint orbit picture Lie($\mathcal{N}$)* / CoAd($\mathcal{N}$) of $\mathcal{N}$. Their phase control performed by the point leaflets of the projective Lie-Poisson structure implements the synchronization of the cine loops of the orbiton / spinon dynamics.

Only the unit circle $\mathbb{S}_1 \cong \mathbb{T} \cong \mathbb{C}$ and the round Poincaré sphere $\mathbb{S}_3 \cong$ Spin(3, $\mathbb{R}$) $\cong$ SU(2, $\mathbb{C}$) can be realized as Lie groups. This is already true when the spheres are considered as smooth manifolds, regardless of any question about the Riemannian metrics and their constant sectional curvature $\equiv 1$. The Hopf fibration of

$$\mathbb{S}_3 = \{(w, w') \in \mathbb{C}_{\mathbb{R}} \times \mathbb{C}_{\mathbb{R}} \mid |w|^2 + |w'|^2 = 1\}$$

$$\cong \frac{SO(4, \mathbb{R})}{SO(3, \mathbb{R})}$$

over the homogeneous manifold

$$\mathbb{S}_2 \cong \frac{SO(3, \mathbb{R})}{SO(2, \mathbb{R})}$$

together with the bundle morphism $\mathbb{S}_3 \longrightarrow$ SO(3, $\mathbb{R}$) is a spin structure on the tangent bundle T$\mathbb{S}_2$. The compact Lie group SU(2, $\mathbb{C}$) $\cong \mathbb{S}_3$ is the simply connected universal covering group SL(2, $\mathbb{C}$) $\cap\, \mathbb{H}$ of SO(3, $\mathbb{R}$).

16. Random Walks Continued

In the context of the second part of the assessment formula for phase-randomizing Lévy flights $(X_t)_{t\in\mathbb{R}_+}$, $\ell \rightsquigarrow \chi(\ell)$ denotes the drift coefficient. In clinical magnetic resonance tomography, the drift coefficient is determined by the magnetic flow density within the bicylindrical bore and, among other parameters, responsible for the tomography resolution of the clinical magnetic resonance tomographic scanner.

In the context of the third or projective jump part of the assessment formula for phase-randomizing Lévy flights $(X_t)_{t\in\mathbb{R}_+}$, $\mathcal{B} \subset$ Lie($\mathcal{N}$) denotes the solid closed unit ball of topological boundary sphere $\partial\mathcal{B} = \mathbb{S}_2 \cong$ O(3, $\mathbb{R}$) / O(2, $\mathbb{R}$), and

$$\text{PLie}(\mathcal{N}) \cong \text{PLie}(\mathcal{N})^* \cong \mathbb{P}_3(\mathbb{R}) \cong \mathbb{S}_3 /\!\approx\, \cong \mathcal{B} / \sim$$

denotes the projectivized real Heisenberg Lie algebra PLie($\mathcal{N}$) of affine line directions with its three-dimensional real dual vector space PLie($\mathcal{N}$)*. Moreover $\approx$ denotes the standard projective equivalence relation induced on the round Poincaré sphere $\mathbb{S}_3$ by projective coordinatization, and $\sim$ is the equivalence relation obtained on $\mathcal{B}$ by identifying antipodal points on the topological boundary sphere $\mathbb{S}_2 = \partial\mathcal{B}$.

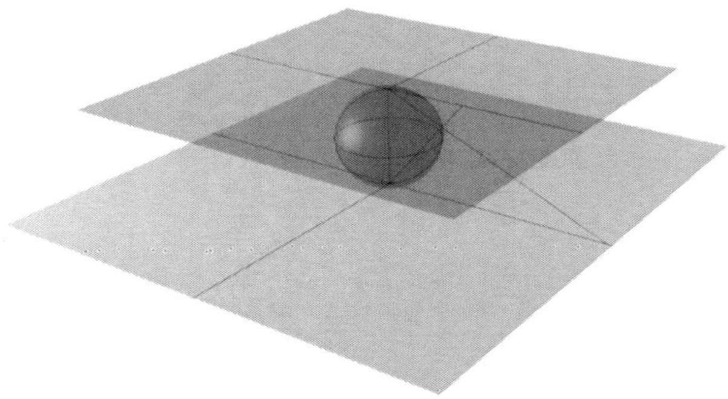

Figure 3. The modality of bipolar stereographic projection η in the contact geometry of planar coadjoint orbits of the three-dimensional unipotent real Heisenberg Lie group $\mathcal{N}$. The Riemannian sphere $\mathbb{S}_2 \cong \mathbb{P}_1(\mathbb{C})$ is regarded as the union of two affine copies of the plane CR with parallel real axes and identification performed by the inversion $\eta : \mathbb{C}_{\mathbb{R}}^{\times} \ni w \rightsquigarrow \frac{1}{w} \in \mathbb{C}_{\mathbb{R}}^{\times}$. The manifold of fibers of the Hopf principal bundle $\mathbb{S}_1 \hookrightarrow \mathbb{S}_3 \longrightarrow \mathbb{S}_2$ fit together conformally to represent the two-dimensional round sphere $\mathbb{S}_2 \cong \mathbb{P}_1(\mathbb{C})$ in the conformally projectivized dual vector space PLie($\mathcal{N}$)*. A cross-sectional disc for the corresponding solid torus meets every fiber inside the torus just once. The complementary disc meets each fiber outside the torus once. The fibers on the torus meet each disc once in its bounding Villarceau circle and provide a diffeomorphism between these two circular fibers. Thus the manifold of fibers is made up of two discs glued together by a diffeomorphism of their boundaries, which is just the Riemannian sphere $\mathbb{S}_2 \cong \mathbb{P}_1(\mathbb{C})$. The rays of projection under the action of affine{linear gradients are interrelated by the right-angle involution with respect to isotropic lines passing through the absolute circular points of the zero-dimensional projective quadric of the corresponding projective plane within PLie($\mathcal{N}$)*.

17. Conformal Hyperbolic Geometry

The sophisticated technique of gradient echoes, which is based on the Kepplerian second law or area law, is a development which occurred after the discovery of the spin echo concept. The area law amounts of choosing a symplectic structure

$$J = \begin{pmatrix} 0 & -1 \\ 1 & 0 \end{pmatrix}$$

so that det $J = 1$, $J^2 = -$id, and $J^4 =$ id holds.

The complex structure J is the matrix of multiplication by the imaginary unit $i = \sqrt{-1}$ in an appropriate identification $\mathbb{R} \oplus \mathbb{R} = \mathbb{C}_{\mathbb{R}}$. It generates the central circle, or loop whch gives rise together with the central axis to the Hopf principal fibration quantum gate $\mathbb{S}_1 \hookrightarrow \mathbb{S}_3 \longrightarrow \mathbb{S}_2$. Notice that $J \in \mathrm{Sp}(2, \mathbb{R})$ is crucial for the construction of the twisted group law of the three-dimensional unipotent real Heisenberg Lie group $\mathcal{N} \cong \mathbb{C}_{\mathbb{R}} \oplus \mathbb{R}$, and its coherent state representations of quantum optics. According to the theory of symplectic spinors [21], the unitary operators symplectically assigned to J and $-J$ are the Fourier cotransform $\overline{\mathcal{F}}_{\mathbb{R}}$ and Fourier transform $\mathcal{F}_{\mathbb{R}}$, respectively, acting on the complex Schwartz space $\mathrm{S}(\mathbb{R}) \hookrightarrow L^2(\mathbb{R})$ of smooth functions on the real line $\mathbb{R}$ which are, as well as their derivatives, rapidly decreasing at infinity. The standard technique of transposition extends the operators $\mathcal{F}_{\mathbb{R}}$ and $\overline{\mathcal{F}}_{\mathbb{R}}$ to the complex vector space $\mathrm{S}'(\mathbb{R})$ of tempered Schwartz distributions.

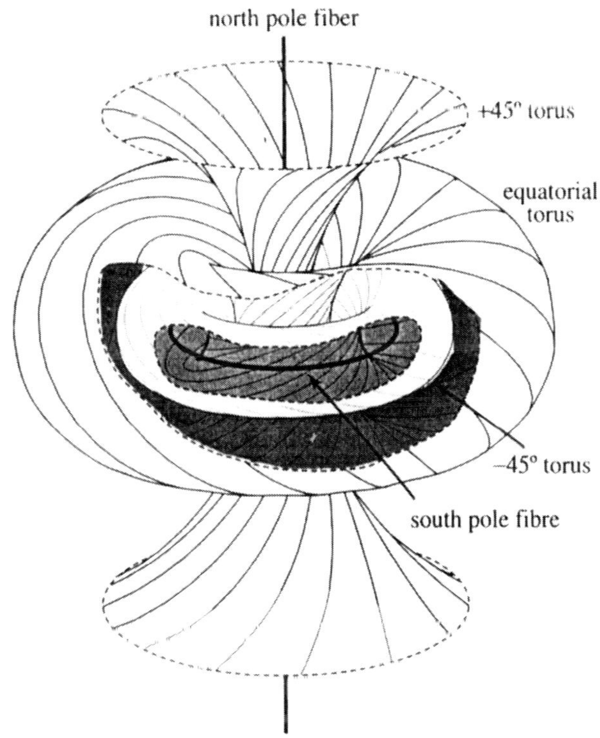

Figure 4. Schematic drawing of the realization of the Heisenberg Lie group $\mathcal{N}$ by means of the Hopf principal fibration quantum gate $\mathbb{S}_1 \hookrightarrow \mathbb{S}_3 \longrightarrow \mathbb{S}_2$ which visualizes the affne coadjoint orbit picture $\mathrm{Lie}(\mathcal{N})^* / \mathrm{CoAd}(\mathcal{N})$ of $\mathcal{N}$ in the conformally projectivized dual vector space $\mathrm{PLie}(\mathcal{N})^*$. The inverse image of a circle on the round sphere $\mathbb{S}_2$ in the sphere $\mathbb{S}_3 \cong \mathrm{Spin}(2, \mathbb{R}) \cong \mathrm{SU}(2, \mathbb{C})$ of unit quaternions is a two-dimensional at torus $\mathbb{T} \times \mathbb{T} \cong \mathbb{C}_{\mathbb{R}} / \mathbb{Z}^2$. Picturing via bipolar stereographic projection shows a conformal perspective of the Villarceau circular fibers which are appropriate for the synchronization of the cine loops of the orbiton / spinon dynamics. The Villarceau phase circles form the boundaries of the Möbius band $\mathbb{Z}^2$ principal bundle on the foliation of nested two-dimensional at Liouville tori. The coherently embedded Hopf links visualize the phenomenon of quantum entangled spin echoes.

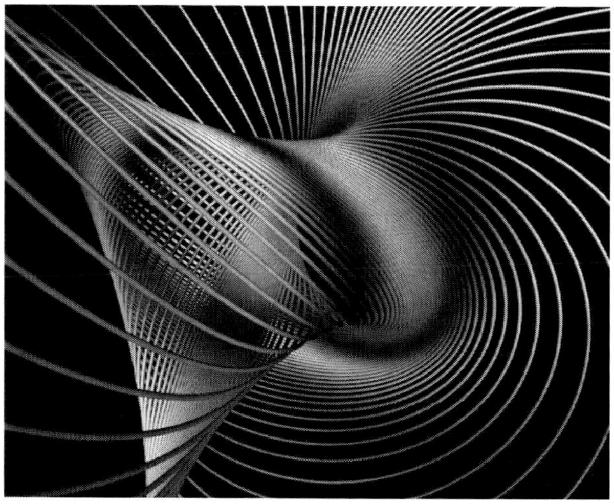

Figure 5. A hyperbolic perspective of the Heisenberg Lie group $\mathcal{N}$ via the Hopf principal fibration quantum gate in the conformally projectivized dual vector space $\mathrm{PLie}(\mathcal{N})^*$. It visualizes the orbiton / spinon dynamics by synchronizing Villarceau phase circles which form a dense set of the manifold of fibers. The fibers over the north and south poles are the central axis and the disjoint absolute circle, respectively. The subset of $\mathrm{Lie}(\mathcal{N})^*$ consisting of $\mathrm{Lie}(\mathcal{N})^*$ – {central axis, absolute circle} is foliated by the flat Liouville tori over the lines of latitude on the two-dimensional sphere $\mathbb{S}_2 \hookrightarrow \mathrm{PLie}(\mathcal{N})^*$. All but two of the circular fibers of the Hopf bundle lie on foliated at Liouville tori in the projective space $\mathrm{PLie}(\mathcal{N})^*$. On each torus the fibers run around once in each affine-linear gradient direction of $\mathrm{PLie}(\mathcal{N})^*$. The compact base manifold $\mathbb{S}_2 \cong \mathbb{P}_1(\mathbb{C})$ may be regarded as the manifold of fibers in the sense that the Hopf projector $\mathbb{S}_3 \longrightarrow \mathbb{S}_2$ establishes a bijection between the set of fibers and the base manifold.

18. *Cine* Magnetic Resonance Tomography

The modality of cine magnetic resonance tomography in moving into clinical mainstream requires fast switching of affine{linear magnetic field gradients $\ell \in \mathrm{PLie}(\mathcal{N})^*$. The *horizontal* affine-linear magnetic field gradients $\mathcal{H}^2 \cong \mathrm{PLie}(\mathcal{N})^*$ of amplitude $x \in \mathbb{R}$ and slope $y > 0$ can be conceived as points of the conformal non-Euclidean hyperbolic open half-plane

$$\mathcal{H}^2 = \{v \in \mathbb{C}_{\mathbb{R}} \,|\, \Im v = y > 0\}$$

endowed with its complex differentiable manifold structure. With respect to the usual composition law, the horizontal affine-linear gradients of positive slope form the conformal affine-linear Lie group

$$\mathrm{Gr}_+(\mathbb{R}) = \left\{\begin{pmatrix} 1 & 0 \\ x & y \end{pmatrix} \,\middle|\, x \in \mathbb{R} \;\&\; y \in \mathbb{R}_+^\times \right\}$$

with underlying manifold $\mathcal{H}^2$. The real Lie algebra $\mathrm{Lie}(\mathrm{Gr}_+(\mathbb{R}))$ consisting of the matrices of the form

$$\begin{pmatrix} 0 & 0 \\ x & y \end{pmatrix} \; (w = x + iy \in \mathbb{C}_{\mathbb{R}})$$

is, up to isomorphism, the only non-abelian, solvable, two-dimensional Lie algebra over the field $\mathbb{R}$. The left-invariant Riemannian metric on $\mathcal{H}^2 \cong \mathrm{Gr}_+(\mathbb{R}) \otimes \mathbb{C}_\mathbb{R}$ which coincides at the neutral element $(0, 1) = i = \sqrt{-1}$ of $\mathrm{Gr}_+(\mathbb{R}) \otimes \mathbb{C}_\mathbb{R}$ with the Euclidean metric admits the first fundamental form

$$\mathrm{d}s^2 = \frac{\mathrm{d}x^2 + \mathrm{d}y^2}{y^2} = -4\frac{\mathrm{d}v\mathrm{d}\bar{v}}{(v-\bar{v})^2} \; (y = \Im v > 0)$$

where, in projective correspondence to the complex structure J and $\nu = 1$, the length metric

$$\mathrm{d}s = \frac{|\mathrm{d}v|}{\Im v}$$

derives from the differential form

$$\frac{\mathrm{d}v}{\Im v} = \frac{\mathrm{d}x \wedge \mathrm{d}y}{y} = -y.\mathrm{d}\left(\frac{\mathrm{d}x}{y}\right)$$

Thus the symplectic group $\mathrm{Sp}(2, \mathbb{R}) \cong \mathrm{SL}(2, \mathbb{R})$ forms an isometry group of $\mathcal{H}^2$ with the two-sheeted covering group $\mathrm{Mp}(2, \mathbb{R})$. This metric of the Lobachevskian non-Euclidean geometry of the hyperbolic open half-plane $\mathcal{H}^2$ is not bi-invariant.

As a bundle cross-section, the symmetric curvature tensor field

$$R \in \Gamma(\Lambda^2\, \mathrm{T}^*\mathcal{H}^2)$$

of the simply connected Riemannian manifold $\mathcal{H}^2$ provides the negative sectional curvature $\equiv -1$ of the Lobachevskian non-Euclidean geometry of $\mathcal{H}$. The Riemannian sectional curvature R is based on the area law of planar tomographic slices and provides the holonomy group $\mathrm{SO}(2, \mathbb{R}) \cong \mathrm{U}(1, \mathbb{C})$ of $\mathcal{H}^2$.

The geodesics of the complete Riemannian manifold $\mathcal{H}^2$ are half-circles whose centre lies on the real axis, and half-lines orthogonal to this axis. Notice also that the complexified homogeneous manifold

$$\mathrm{Gr}_+(\mathbb{R}) \otimes \mathbb{C}_\mathbb{R} \cong \frac{\mathrm{Sp}(2,\mathbb{R})}{\mathrm{SO}(2,\mathbb{R})} \otimes \mathbb{C}_\mathbb{R}$$

is biholomorphically equivalent by the Cayley transformation

$$h : v \rightsquigarrow \frac{v-i}{v+i}, h^{-1} : w \rightsquigarrow i\,\frac{w+1}{-w+1}$$

of $\mathcal{H}^2$ onto the open unit disc $\{w \in \mathbb{C}_\mathbb{R} \,|\, |w| < 1\}$ in the complex plane $\mathbb{C}$ under the hyperbolic metric

$$\mathrm{d}s = \frac{2|\mathrm{d}w|}{1-|w|^2}.$$

Indeed, the density identity

$$\frac{1}{\Im v} = \frac{2|h'(v)|}{1-|h(v)|^2} \; (\Im v > 0)$$

holds for all elements $v \in \mathcal{H}^2$. The conformally invariant Riemannian metric reads

$$\mathrm{d}s^2 = \frac{\mathrm{d}x^2 + \mathrm{d}y^2}{(1-x^2-y^2)^2}.$$

It is uniquely determined up to a positive constant factor. A line is represented as a circular arc whose ends are perpendicular to the disc's boundary circle of location at infinity. Two arcs which do not meet correspond to parallel lines, and arcs which meet orthogonally correspond to perpendicular lines.

The horizontal support of the infinitely divisible spectral measure μ in the gradient controlled Lévy-Hinčin type assessment formula can be visualized by a hyperbolic tessellation of the Poincaré disc model, and the induced action with its unique fixed point on its topological boundary circle $\mathbb{T} = \{\mathbf{e}(\theta) \,|\, \theta \in \mathbb{R}\}$ associated with the phase angle $\theta = \frac{\pi}{2}$.

In the context of the cardiac cycle conformally reconstructed by the reproducing kernel of high resolution cardiovascular magnetic resonance tomography, the tomographically sliced Lévy measure λ has to be adapted to the orbiton / spinon dynamics of balanced steady-state free-precession. A similar problem arises with respect to the diffusion weighted magnetic resonance tomography. Few advances in magnetic resonance tomography have had the impact that diffusion weighted magnetic resonance tomography has had in the conformal evaluation of the human brain [30,37].

19. Social and Economic Impact

The first electrocardiographically gated magnetic resonance tomographies of the beating human heart were published about a quarter of a century ago. The early clinical indications for cardiovascular magnetic resonance tomography were based upon assessment of cardiac morphology and were very limited in relation to other pre-existing cardiac non-invasive imaging modalities. It was soon recognized that paramagnetic contrast could be utilized to attain diagnostically valuable myocardial tissue characterization. The non-invasive diagnostic modality of cine magnetic resonance tomography now appears poised to realize its immense clinical potential.

Due to the fact that cardiovascular disease is the major cause of death in many Western countries, accounting for more than twice the number of deaths due to cancer, and therefore absorbs a large portion of health-care budgets, the last ten years has seen explosive expansion of the number of centers performing clinical cardiovascular magnetic resonance tomography. The majority of this expansion has been in the field of adult myocardial ischaemic cardiovascular magnetic resonance tomography, but congenital heart disease remains one of the main indications for pediatric cardiovascular magnetic resonance tomography. However, without previous experience or intense formal training, the interpretation of the results of dffusion weighted magnetic resonance tomography and cardiovascular magnetic resonance tomography of valvular heart disease, right ventricular dysplasia, aortic and other vascular conditions remains to be a demanding task of clinical non{invasive diagnostics.

20. Second Conclusion

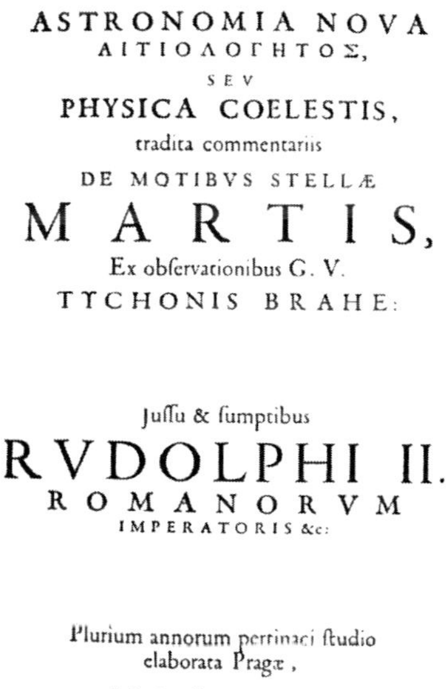
ASTRONOMIA NOVA
ΑΙΤΙΟΛΟΓΗΤΟΣ,
SEV
PHYSICA COELESTIS,
tradita commentariis
DE MOTIBVS STELLÆ
MARTIS,
Ex obſervationibus G. V.
TYCHONIS BRAHE:
Juſſu & ſumptibus
RVDOLPHI II.
ROMANORVM
IMPERATORIS &c:
Plurium annorum pertinaci ſtudio elaborata Pragæ,
A S. C. M.tis S. Mathematico
JOANNE KEPLERO,
Cum ejusdem C. M.tis privilegio speciali
Anno æræ Dionyſianæ cIↃ IↃ c ix.

Figure 6. Title page of Johannes Keppler's treatise *Astronomia Nova* of 1609 indicating the role of the planet Mars for the Kepplerian astrophysical studies. For the first time they are concerned with a mathematical treatment of the orbiton / spinon dynamics of balanced steady-state free-precession. Based on Tycho Brahe's long-term observations with unaided eyes, the first and second Kepplerian law of planetary motion have been established and conformally visualized in the record of a decade's intense labor.

From the mathematical point of view it is highly instructive to see how the Poincaré models of conformal Non-Euclidean metrical geometry of strictly negative curvature fit projectively to the affine coadjoint orbit model Lie($\mathcal{N}$)* / CoAd($\mathcal{N}$) of quantum entanglement in the ambient conformally projectivized dual vector space PLie($\mathcal{N}$)* associated with the three-dimensional unipotent real Heisenberg Lie group $\mathcal{N} \cong \mathbb{C}_{\mathbb{R}} \oplus \mathbb{R}$. As opposed to the tesselations of hyperbolic geometry, there exists only a finite number of tilings in the Euclidean and elliptic geometries.

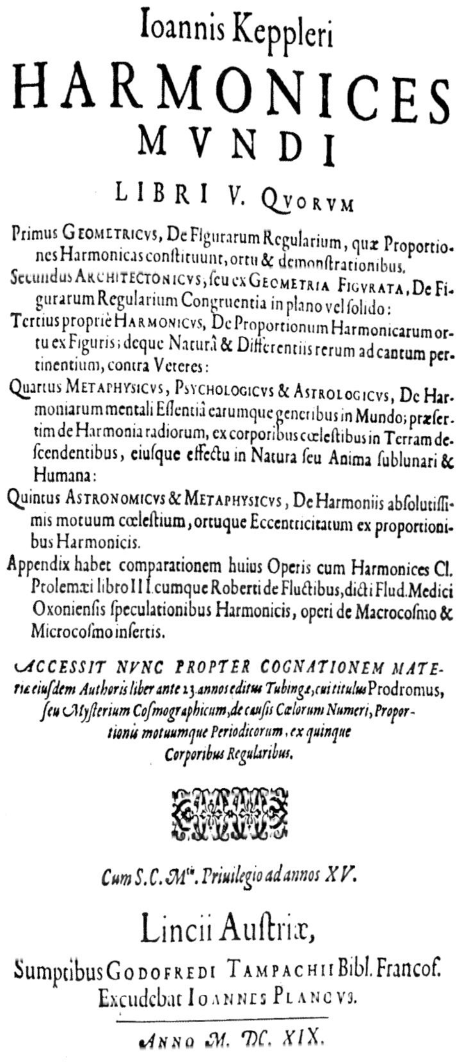
Ioannis Keppleri
HARMONICES
MVNDI
LIBRI V. QVORVM
Primus GEOMETRICVS, De Figurarum Regularium, quæ Proportiones Harmonicas conſtituunt, ortu & demonſtrationibus.
Secundus ARCHITECTONICVS, ſeu ex GEOMETRIA FIGVRATA, De Figurarum Regularium Congruentia in plano vel ſolido:
Tertius propriè HARMONICVS, De Proportionum Harmonicarum ortu ex Figuris; deque Naturâ & Differentiis rerum ad cantum pertinentium, contra Veteres:
Quartus METAPHYSICVS, PSYCHOLOGICVS & ASTROLOGICVS, De Harmoniarum mentali Eſſentiâ earumque generibus in Mundo; præſertim de Harmonia radiorum, ex corporibus cœleſtibus in Terram deſcendentibus, eiuſque effectu in Natura ſeu Anima ſublunari & Humana:
Quintus ASTRONOMICVS & METAPHYSICVS, De Harmoniis abſolutiſſimis motuum cœleſtium, ortuque Eccentricitatum ex proportionibus Harmonicis.
Appendix habet comparationem huius Operis cum Harmonices Cl. Ptolemæi libro III. cumque Roberti de Fluctibus, dicti Flud. Medici Oxonienſis ſpeculationibus Harmonicis, operi de Macrocoſmo & Microcoſmo inſertis.
ACCESSIT NVNC PROPTER COGNATIONEM MATERIÆ eiuſdem Authoris liber ante 23. annos editus Tubingæ, cui titulus Prodromus, ſeu Myſterium Coſmographicum, de cauſis Cælorum Numeri, Proportionis motuumque Periodicorum, ex quinque Corporibus Regularibus.
Cum S. C. M.tis. Priuilegio ad annos XV.
Lincii Auſtriæ,
Sumptibus GODOFREDI TAMPACHII Bibl. Francof.
Excudebat IOANNES PLANCVS.
ANNO M. DC. XIX.

Figure 7. The third Kepplerian law of planetary motion followed in the great *Harmonices Mundi* of 1619. It determined the fundament for the orbit control of the recently launched Mars mission of Rover Curiosity. The incomprehension of later generations have left Keppler's work on the cosmic harmonies in the obscurity he foresaw in the Proemium to Book Five *Astronomicus & Metaphysicus*. It is a wonderful experience to recognize that the sophisticated ordering principles of the Kepplerian *Harmoniis absolutissimis motuum cœlestium* is realized by means of the magnetic resonance scanner.

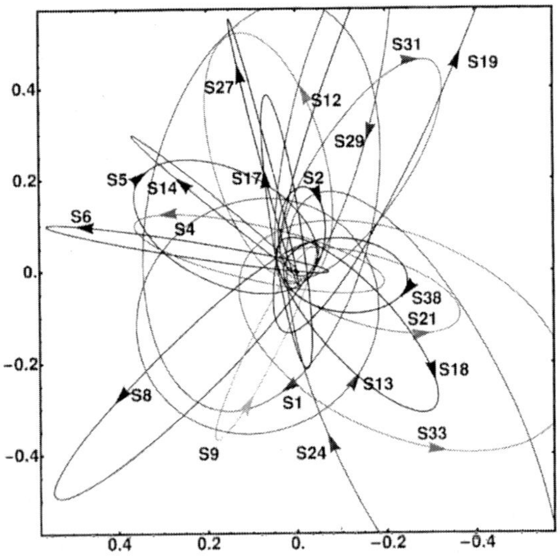

Figure 8. Conformal steady-state free-precession: A cloud of elliptic orbits of high excentricity established by long-term adaptive optical observation in the infrared region. Notice the different sectional curvatures of the stellar orbiton / spinon dynamics. The Kepplerian test of the projective orbiton / spinon dynamics of balanced steady-state free-precession establishes incontrovertibly the existence of the supermassive black hole Sgr A* in the centre of the Milky Way galaxy.

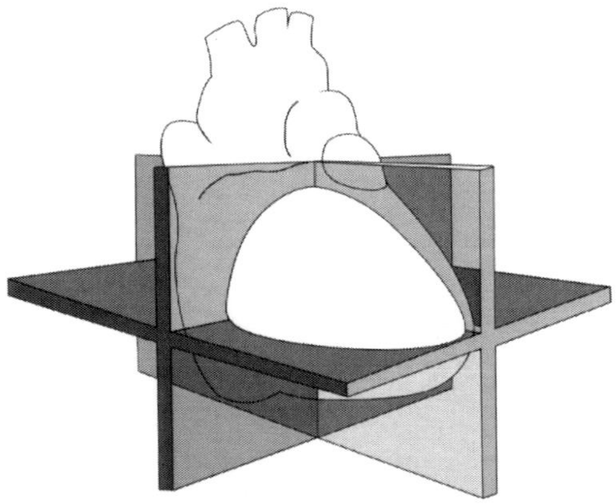

Figure 10. Mathematical radiology: Schematic drawing of the tomographic slice directions which coordinatize projectively the still{frame affine coadjoint orbit picture Lie(N)* / CoAd(N) of the three-dimensional real Heisenberg Lie group N in the conformally projectivized dual vector space PLie(N)*: Generic coronal short axis direction to study the connection between the ventricles and the great vessels; generic horizontal long axis or transversal direction through the left ventricle and right ventricle apices; and generic saggital or vertical long axis passing through the *Septum interventriculare*. The classical axis directions are pairwise orthogonal in the three-dimensional real projective space PLie(N)* where orthogonal oriented lines are projectively interrelated by the right-angle involution with respect to isotropic lines passing through the absolute circular points of the zero-dimensional projective quadric in the corresponding projective plane of PLie(N)*. Slightly slanted slices passing through the horizontal long axis provide a three-chamber view of the cardiac morphology and valvular heart function.

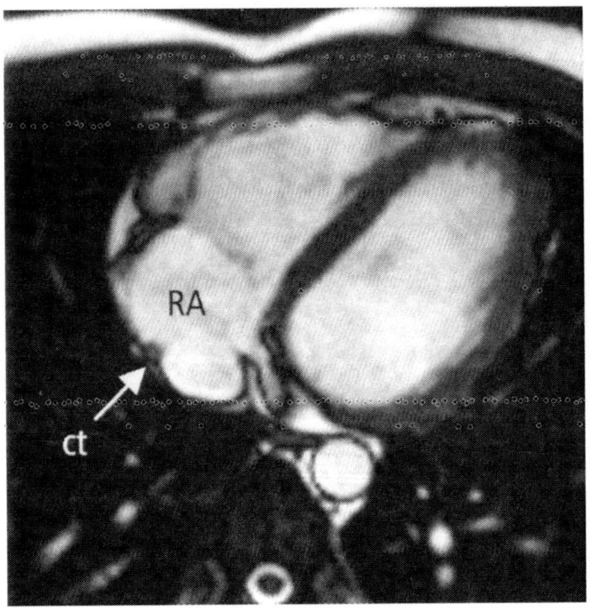

Figure 9. Clinical cardiovascular magnetic resonance tomography presenting a morphological horizontal long axis four chamber view of the human heart: A still frame from a cine conformal steady-state free-precession sequence visualizing non-invasively by means of electrocardiographic triggering the cardiac right atrium (*Atrium cordis dextrum*, RA) and a *Crista terminalis* (ct arrow). The *Septum interventriculare* is well delineated down to the left ventricle and right ventricle apices. The non-invasive diagnostic modality of cine magnetic resonance tomography visualizes by means of synchronization the open and closing functions of the *Valva atrioventicularis sinistra* and *dextra* and the *Valva aortæ* and *Valva trunci*.

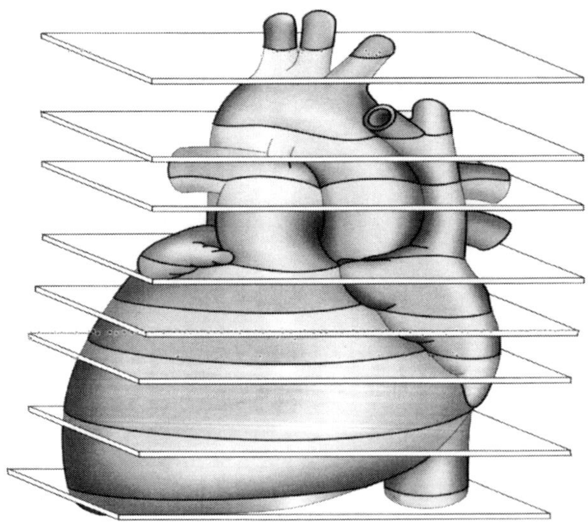

Figure 11. Schematic drawing of the still{frame embedding of the human heart into the affine coadjoint orbit picture Lie(N) / CoAd(N) of the three-dimensional real Heisenberg Lie group N. The ambient space of the tomographic slice foliation is the conformally projectivized dual vector space PLie(N)* in which the slices of positive energy are selected in the longitudinal direction by affine-linear gradients according to their Larmor labels. The transverse or axial planar coadjoint orbits of N in Lie(N)* are endowed with the non-Euclidean hyperbolic geometries of open half-planes and their complex conjugates. The tomographic slices are useful for studying horizontal long axis cardiac morphology down to the left ventricle and right ventricle apices, and the relationships of the four cardiac chambers, the *Arcus aortæ* and the *Pericardium*.

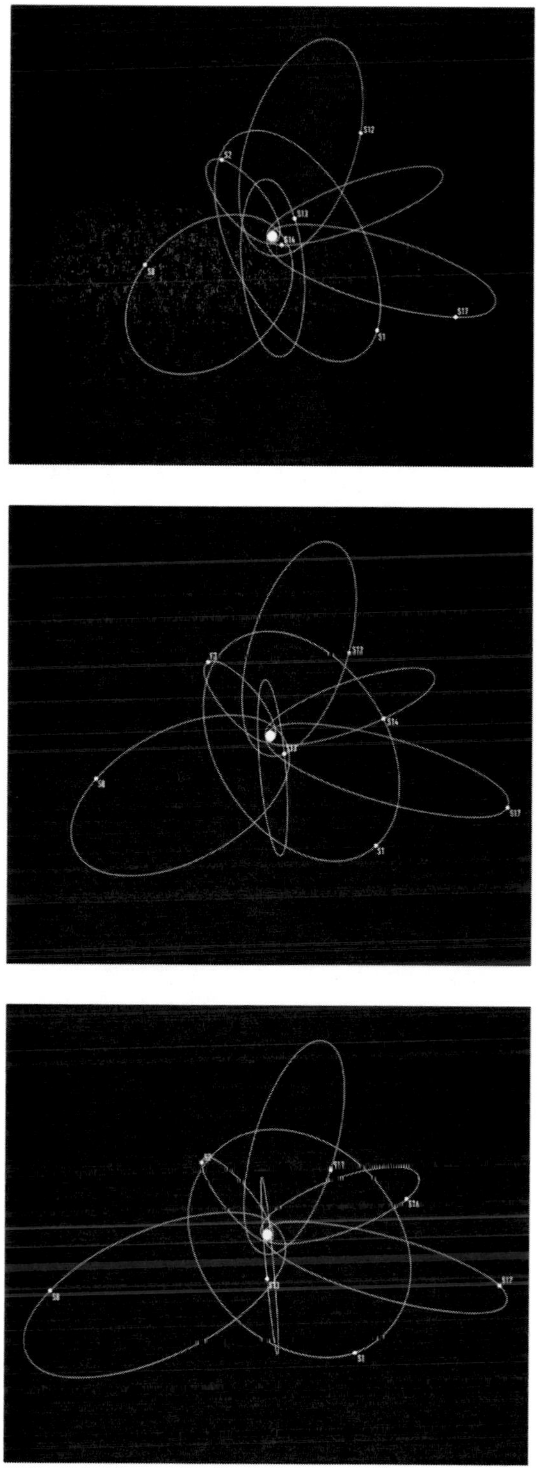

Figure 12. Kepplerian projective orbiton / spinon dynamics of balanced steady-state free-precession: Computer simulation of the highly excentric planar orbits of the cluster of stars steadily precessing around the central supermassive black hole Sgr A* of the galaxy. Due to the cosmic dust, the interpolatory data of the focal bundle of inversive planes are obtained by long term adaptive optics observation in the infrared region.

Acknowledgments

This expository paper is dedicated to the memory of the author's grandfather Johannes Keppler (1877-1935). – The author would like to thank Professor John Dainton (University of Liverpool) for encouraging remarks and the chairmen Drs. Peter Marcer and Peter Rowlands for continuing support.

References

[1] V. Armitage, A. Rogers, Gauss sums and quantum mechanics. *J. Phys. A: Math. Gen.* **33**, 5993-6002 (2000)
[2] J. C. Baez, The octonions. *Bull. (New Series) Amer. Math. Soc.* **39**, 145-205 (2002)
[3] D. M. S. Bagguley (ed.), *Pulsed Magnetic Resonance: NMR, ESR, and Optics, A Recognition of E.L. Hahn.* Oxford University Press, Oxford, New York, Toronto (1992)
[4] A. F. Beardon, *The Geometry of Discrete Groups.* Springer-Verlag, New York, Heidelberg, Berlin (1983)
[5] M. A. Bernstein, K. F. King, X. J. Zhou, *Handbook of MRI Pulse Sequences.* Elsevier, Amsterdam, Boston, Heidelberg (2004)
[6] J. Bertoin, *Lévy Processes.* Cambridge University Press, Cambridge, New York, Melbourne (1998)
[7] J. Bogaert, A. J. Duerinckx, F. E. Rademachers, (eds.), *Magnetic Resonance of the Heart and Great Vessels.* Springer-Verlag, Berlin, Heidelberg, New York (1999)
[8] L. M. Boxt (ed.), Cardiac MR Imaging. *Magn. Reson. Imaging Clin. North America*, Vol. 4, Number 2, W.B. Saunders, Philadelphia, London, Toronto (1996)
[9] R .R. Edelman, J. R. Hesselink, M. B. Zlatkin, J. V. Crues III (eds.), *Clinical Magnetic Resonance Imaging.* Three Volumes, Third edition, Saunders Elsevier, Philadelphia (2006)
[10] J. P. Finn (ed.), Physics of MR Imaging. *Magn. Reson. Imaging Clin. North America*, Vol. 7, Number 4, W. B. Saunders, Philadelphia, London, Toronto (1999)
[11] M. A. Fogel, *Principles and Practice of Cardiac Magnetic Resonance in Congenital Heart Disease: Form, Function, Flow.* Wiley-Blackwell, New York, London (2010)
[12] S. S. Gelbart, *Weil's Representation and the Spectrum of the Metaplectic Group. Lecture Notes in Mathematics*, Vol. 530, Springer-Verlag, Berlin, Heidelberg, New York (1976)
[13] R. Godement, *Analyse Mathématique* IV. Springer-Verlag, Berlin, Heidelberg, New York (2003)
[14] W. M. Goldman, *Complex Hyperbolic Geometry.* Oxford University Press, Oxford, New York, Toronto (1999)
[15] J. D. Grizzard, R. M. Judd, R. J. Kim, *Cardiovascular MRI in Practice: A Teaching File Approach.* Springer-Verlag, London (2008)
[16] A. Guichardet, Théorie de Mackey et méthode des orbites selon. *M. Duo. Expo. Math.* 3, 303-346 (1985)
[17] V. Hombach, N. Merkle, V. Rasche, *Cardiovascular Magnetic Resonance Imaging: Textbook and Atlas.* Schattauer, Stuttgart, New York (2012)
[18] S. Jeffers, B. Lehnert, N. Abramson, L. Chebotarev (eds.), *J.-P. Vigier and the Stochastic Interpretation of Quantum Mechanics. Apeiron*, Montreal (2000)
[19] R. J. Kim (ed.), *Cardiovascular MRI. Magn. Reson. Imaging Clin. North America*, Vol. 15, Number 4, W.B. Saunders, Philadelphia, London, Toronto (2007)

[20] D. M. Koh, H. C. Thoeny (eds.), *Diffusion-Weighted MR Imaging: Applications in the Body*. Springer-Verlag, Berlin, Heidelberg, New York (2010)

[21] B. Kostant, Symplectic spinors. *Symposia Mathematica*, Vol. XIV, pp. 139-152, Academic Press, London, New York (1974)

[22] C. Laurent-Gengoux, A. Pichereau, P. Vanhaecke, *Poisson Structures*. Springer-Verlag, Berlin, Heidelberg, New York (2013)

[23] P .D. Lax, R. S. Phillips, Scattering Theory of Automorphic Functions. *Annals of Mathematics Studies* 87, Princeton University Press, Princeton, New Jersey (1976)

[24] D. Le Bihan, *Diffusion and Perfusion Magnetic Resonance Imaging: Applications to Functional MRI*, Raven Press, New York (1995)

[25] V. S. Lee, *Cardiovascular MRI: Physical Principles to Practical Protocols*. Lippincott Williams & Wilkins, Baltimore, Philadelphia, New York (2006)

[26] W. J. Manning, D. J. Pennell (eds.), *Cardiovascular Magnetic Resonance*. Second edition, Saunders / Elsevier, Philadelphia (2010)

[27] J. Mattson, M. Simon, *The Pioneers of NMR and Magnetic Resonance in Medicine: The Story of MRI*. Bar-Ilan University Press, Ramat Gan (1996)

[28] S. Mori, S. Wakana, L. M. Nagae-Poetscher, P. C. M. van Zijl, *MRI Atlas of Human White Matter*. Elsevier, Amsterdam, Boston, Heidelberg (2005)

[29] S. Mori, *Introduction to Diffusion Tensor Imaging*. Elsevier, Amsterdam, Boston, Heidelberg (2007)

[30] T. Moritani, S. Ekholm, P.-L. Westesson, *Diffusion-Weighted MR Imaging of the Brain*. Second edition, Springer-Verlag, Berlin, Heidelberg, New York (2009)

[31] A. Pressley, G. Segal, *Loop Groups*. Oxford University Press, Oxford, New York, Toronto (1990)

[32] M.F. Reiser, W. Semmler, H. Hricak (eds.), *Magnetic Resonance Tomography*. Springer-Verlag, Berlin, Heidelberg, New York (2008)

[33] V. M. Runge, W. R. Nitz, S. H. Schmeets, S. O. Schoenberg (eds.), *Clinical 3 T Magnetic Resonance*. Thieme Medical Publishers, New York, Stuttgart (2007)

[34] S. Schaefer, R. Balaban (eds.), *Cardiovascular Magnetic Resonance Spectroscopy*. Springer-Verlag, Berlin, Heidelberg, New York (1993)

[35] W. Schempp, Schwartz kernels on Siegel II domains. *Math. Z.* 171, 101-112 (1980)

[36] W. J. Schempp, *Magnetic Resonance Imaging: Mathematical Foundations and Applications*. Wiley-Liss, New York, Chichester, Weinheim (1998)

[37] W. J. Schempp, The metaplectic sampling of quantum engineering. In: *Search for Fundamental Theory*, R. L. Amoroso, P. Rowlands, S. Jeffers, (eds.), pp. 232-243, American Institute of Physics, Melville, New York (2010)

[38] W. J. Schempp, *Ordo syderum orbium*: Boundary conformal quantum field theory and seesaw contact geometry (to appear)

[39] J. Schwitter, *CMR Update*. Second edition, Juerg Schwitter, Publisher, Zürich, Switzerland (2012)

[40] A. Weil, Sur certains groupes d'opérateurs unitaires. *Acta Math.* **111**, 143-211 (1964); *Collected Papers*, Vol. III, pp. 1-69, Springer-Verlag, New York, Heidelberg, Berlin (1979)

[41] A. Weil, Sur la formule de Siegel dans la théorie des groupes classiques. *Acta Math.* **113**, 1-87 (1965); *Collected Papers*, Vol. III, pp. 71-157, Springer-Verlag, New York, Heidelberg, Berlin (1979)

[42] P. K. Woodard (ed.), *Cardiac MR Imaging. Magn. Reson. Imaging Clin. North America*, Vol. 11, Number 1, W.B. Saunders, Philadelphia, London, Toronto (2003)

An Introduction to Boscovichian Unified Field Theory

ROGER JAMES ANDERTON

Independent researcher, UK
R.J. Anderton@btinternet.com

Boscovich proposed a unified theory in the 18th century, from this theory both quantum physics and Einstein's relativity was derived. Rather than look back at the history of how modern physics was developed, mainstream seems to go blindly onwards looking for a unified theory and ignoring this particular unified theory.

Keywords: Boscovich, Unified field

1. Introduction

People have noted that Boscovich had a unified theory, for instance: Helge Kragh [1] notes that Boscovich had an early "theory of everything" based on point particles. John D Barrow [2] says: "He [Boscovich] was the first to envisage, seek, and propose a unified theory of all the forces of nature." Sadly this unified theory is mostly ignored and it should not be ignored because it is the basis of modern physics; it is something that is seriously wrong in its omission from most modern physics education. In 2011 there was an International year of celebrating Boscovich, as proclaimed by the Croatian parliament.[3] But even that did not seem to get Boscovich the publicity he deserves to get him more widely known.

2. Relativity

J.F. Barrett gives what he calls a "Bibliography on the History of the Theory of Relativity [4]."These are in chronological order up to the appearance of the Poincaré-Einstein theory." Relativity gets mentioned by Galileo, and then Barrett gives a long list of others dealing with relativity. The one I want to emphasis in this list is Boscovich, he says: "1755 Bošković R.J.(Boskovich): De moto absoluto, an possit a relativo distingui, Rome 1755. Reprinted by Varicak (ed.) South Slavic acad. Sci. 190 1912."

A. K. T. Assis [5] points out that the first to propose the idea that if all the motion and distances of the universe were to change by the same amount in the universe then it would be undetected: "Some authors in the past have expressed their point of view that no effect should be detected by a length transformation (that is, if all the bodies in the universe, including the atoms, increased in size by the same amount, the same happening with all distances). To our knowledge the first to generalize this idea to include time and motion was Boscovich in 1755." He then gives a few typical statements of Boscovich [6].

A motion which is common to us and the world cannot be recognized by us - not even if the world as a whole were increased or decreased in size by an arbitrary factor…It even is conceivable that this whole world before our eyes contracted or expanded in a matter of days - with the magnitude of the forces contracting or expanding in unison. Even if this occurred, there would be no change of the impressions in our minds and no perception of this kind of change.

Although Einstein does not mention Boscovich, links between them can be made, for instance: Tilman Sauer tells us of the correspondence between Einstein and Varicak [7]: "The Collected Papers of Albert Einstein recently published nine letters by Einstein to Vladimir Varicak (1865–1942), professor of mathematics at Agram (now Zagreb, Croatia). Varicak had published on non-Euclidean geometry and is known for representing special relativistic relations in terms of real hyperbolic geometry. The correspondence seems to have been initiated by Varicak asking for offprints of Einstein's papers. In his response, Einstein added a personal tone to it with his wife Mileva Maric, a native Hungarian Serb, writing the address in Cyrillic script in order to raise Varicak's curiosity.

After exchanging publications, Varicak soon commented on Einstein's (now) famous 1905 special relativity paper, pointing to misprints but also raising doubts about his treatment of reflection of light rays off moving mirrors." Varicak dealt with Boscovich's theory as we have earlier noted. Varicak had interest in Einstein's relativity and Boscovich, it's the Boscovich bit I want to pick up on [8]: "Varicak investigated life and work of Ruđer Bošković with a great interest. He studied historical documents about Bošković in Milan, Rome, Vienna and other European towns, and published about twenty papers on that subject. From Varicak's papers about Bošković one can learn a lot about Varicak himself. In his studies of Bošković's mathematical work Varicak used a specific methodology that was very appropriate for the history of mathematics. His investigations of Bošković's life and work were recognized by the international community.

"J. F. Barrett says: "Varicak made an extensive study of Boscovich's works and wrote many scholarly articles on them. He brought to notice Boscovich's 1755 article containing 'many clear and radical ideas regarding the relativity of space, time and motion' [9].

In historical terms of development of relativity, we thus have a connection between the ideas of Boscovich to Einstein's relativity. Boscovich is dealing with relativity in the 18th century, many others have dealt with relativity after him long before Einstein tackled the issue. What can be overlooked by those who idolise Einstein is that others worked on relativity before Einstein, i.e. those working on relativity before Einstein tend to get ignored and forgotten by the mainstream. The main source of ideas on Boscovich's relativity from his 1755 paper "On absolute motion - if it is possible to distinguish it from relative motion ("Of Space and Time") [10]."

Karl Svozil [11] after pointing out that the speed of light (in vacuum) is a convention and that there are those who deal with special relativity falsely thinking it to be an empirical fact, points out that the basis of special relativity comes from Boscovich's theory. Instruments that are test to light-speed (in vacuum) constancy are affected by time dilation and length contraction, so they are unable to detect change in light-speed (in vacuum).

As Karl Svozil puts it: "If the very instruments which should indicate a change in the velocity of light are themselves dilated, then any dilation effect will be effectively nullified." He then points out this was considered by Boscovich: " This possibility has already been imagined in the 18th century by Boskovich and was later put forward by FitzGerald."

He points out about convention: "Conventions are a necessary and indispensable part of operationalizable phenomenology and tool-building. There is no perception and intervening without conventions. They lie at the very foundations of our world conceptions. Conventions serve as a sort of "scaffolding" from which we construct our scientific world-view. Yet, they are so simple and almost self-evident that they are hardly mentioned and go unreflected."

The convention issue he points out is not understood by everyone: "To the author, this unreflectedness and unawareness of conventionality appears to be the biggest problem related to conventions, especially if they are mistakenly considered as physical "facts" which are empirically testable. This confusion between assumption and observational, operational fact seems one of the biggest impediments for progressive research programs, in particular if they suggest postulates which are based on conventions different from the existing ones." And: "Despite the obvious conventionality of the constancy of the speed of light, many introductions to relativity theory present this proposition not as a convention but rather as an important empirical finding."

Of course the term relativity principle he points out comes from Poincare: "..."relativity" theory deals with. It derives its name from Poi[n]car´e's 1904 "pri[n]ciple of relativity" stating that "the laws of physical phenomena must be the same for a stationary observer as for an observer carried along in a uniform translation; so that we have not and cannot have any means of discerning whether or not we are carried along in such a motion." Formally, this amounts to the requirement of form invariance or covariance of the physical equations of motion."

"Einstein's major reason for introducing the Lorentz transformation was the elimination of asymmetries which appeared in the electromagnetic formalism of the time but are not inherent in the phenomena, thereby unifying electromagnetism. Secondly, not too much consideration has been given to the possibility that experiments like the one of Michelson and Morley may be a kind of "self-fulfilling prophesy," a circular, closed tautologic exercise." So this convention of the speed of light (in vacuum) as constant is a tautology, it is not testable. By Karl Popper philosophy – a theory should be testable, and this light-speed (in vacuum) constancy as dealt with by special relativity is not testable, it is defined constant and experiments adjusted to fit it.

Now looking at the Boscovich source [12] that Svozil refers to Boscovich says: "When either objects external to us, or our organs change their modes of existence in such a way that that first equality or similitude does not remain constant, then indeed the ideas are altered, & there is a feeling of change ; but the ideas are the same exactly, whether the external objects suffer the change, or our organs, or both of them unequally. In every case our ideas refer to the difference between the new state & the old, & not to the absolute change, which does not come within the scope of our senses. Thus, whether the stars move round the Earth, or the Earth & ourselves 'move in the opposite direction round them, the ideas are the same, & there is the same sensation. We can never perceive absolute changes; we can only perceive the difference from the former configuration that has arisen. Further, when there is nothing at hand to warn us as to the change of our organs, then indeed we shall count ourselves to have been unmoved, owing to a general prejudice for counting as nothing those things that are nothing in our mind; for we cannot know of this change, & we attribute the whole of the change to objects situated outside of ourselves. In such manner any one would be mistaken in thinking, when on board ship, that he himself was motionless, while the shore, the hills & even the sea were in motion."

This is the way that Galileo was thinking in terms of relativity. The ship moves relative to the shore, and the shore moves relative to the ship. He then goes on to consider lengths changing but being undetected as changing and also the same applying to time intervals, this being in the context of inertial frames as what Galileo was dealing with, that means constant speed scenario (zero acceleration). As Boscovich describes it: "Again, it is to be observed first of all that from this principle of the unchangeability of those things, of which we cannot perceive the change through our senses, there comes forth the method that we use for comparing the magnitudes of intervals with one another; here, that, which is taken as a measure, is assumed to be unchangeable."

It is better explained by H V Gill [13]: "We cannot perceive this change in the length, because we ourselves and all our means of measuring the lengths of the rod under different conditions change in exactly the same proportion. Since time is measured by some such means as pendulum or clock, which depends on its physical dimensions, it follows that time thus estimated will also vary in same proportion as the dimension."

Having length and time intervals changing in the same way, divide that length by the time interval and it equals a constant; given the length as distance travelled and the time interval as time taken to travel that distance then it gives a speed. Given then a distance travelled in certain time interval, with both distance and time interval changing in undetected way we could identify this as a special "speed" and that is indeed what has happened with regards to speed of light (in vacuum). (Fastest light speed being in vacuum.)

In the introduction to Child's translation of Boscovich's theory we are told by Child [14]: "with Boscovich light is matter moving with a very high velocity" So it must have been light speed that Boscovich was thinking about in regards to length and time changing. So for Einstein we have as consequence of speed of light (in vacuum) as being constant then there is change in lengths (length contraction) and change in time (time dilation), whereas Boscovich went the other way from length and time interval changes (in the context of relativity) he got to the idea of a special speed. Taking that special speed as being speed of light (in vacuum) as opposed to taking some other speed must then be a convention for that.

Christopher Jon Bjerknes [15] explains the method of Einstein: "Einstein disclosed his modus operandi for manipulating credit for the synthetic theories of others, when he stated in 1936: "There is no inductive method which could lead to the fundamental concepts of physics. Failure to understand this fact constituted the basic philosophical error of so many investigators of the nineteenth century. Logical thinking is necessarily deductive; it is based upon hypothetical concepts and axioms. How can we expect to choose the latter so that we might hope for a confirmation of the consequences derived from them? The most satisfactory situation is evidently to be found in cases where the new fundamental hypotheses are suggested by the world of experience itself."" Bjerknes points this out as: "This is a clear statement by Einstein that he would have science deduce a thing from itself, taking the world of experience as a hypothesis, only to deduce the world of experience as an effect, of itself. Of course, Mileva and Albert [n.b. Bjerknes believes both Einstein and his wife wrote the 1905 paper on relativity] were forced to present the real hypotheses, which they stuck in the middle of their arguments by way of induction, or an attempt at induction, which analyses they attempted to disguise as deductions from a priori principles, but which "a priori principles" were well-known summations of phenomena. Einstein wanted people to believe that it is irrelevant that his predecessors induced the theories he later copied, because Einstein just invented them, *sua sponte*, irrationally, after he had read them, and therefore deserved credit for them:" So, Einstein's special relativity really had its origins in Boscovich's theory long before Einstein was born.

3. Special Relativity

Looking now at special Relativity it's construction seems to be as follows: For the scenario idealisation of vacuum and no forces acting on the light. Considering a moving box being observed with speed v and light bouncing between the walls inside the box, the light first going in same direction as v so by Newtonian velocity addition having velocity $c + v$ according to the outside observer, the light then hits the wall of the box and get reflected back so that its speed is now c-v. The time for journey both forwards and back is deemed the same time interval t, so that the light travels distance (c + v)t in forward direction and distance $(c - v)t$ in backward direction. When the box is not moving then v = 0, and we can consider how the light appears from a frame of reference inside the box, let us say from such a frame it travels distance $c't'$ in the forward (and backward direction) in time interval t' with c' as what it observes light-speed to be. This $c't'$ is squared and equated to $(c - v)t(c + v)t$, giving:

$$c'^2 t^{2'} = (c^2 - v^2)\, t^2 \qquad (1)$$

For Newtonian physics the time intervals should be equal so that t = t', and also the speed of light be variable and that would give us c' = square root of $(c^2 - v^2)$, this would then be the mean two-way light speed. For Special Relativity it wants c = c' and this would then lead to the time dilation equation. So what we have is a simple connection between Newtonian physics and Special relativity from this equation (1). If we go by Einstein, he says after a bit of maths about how he wants clocks to record time [16]: "The principle of the constancy of the velocity of light then states that this adjustment of the clocks will not lead to contradictions. With clocks so adjusted, we can assign the time to events which take place near any one of them. It is essential to note that this definition of time relates only to the inertial system K, since we have used a system of clocks at rest relatively to K. The assumption which was made in the pre-relativity physics of the absolute character of time (i.e. the independence of time of the choice of the inertial system) does not follow at all from this definition."

Which makes it clear that he wants the constancy of light-speed to be used to set clocks and define time that way. Whereas by Newtonian physics this would not be the method because it would make the clocks record different time intervals and be defining time in a different way than what it wants. It seems then to this writer that a great deal of confusion arose due to Einstein changing what is meant by time and he then proceeded confusing people further with other changes.

4. Quantum theory

Boscovich's theory also deals with quantum theory. After the discovery of the electron – the issue became about the orbit of the electron around the nucleus of the atom of a chemical element. H V Gill [17] tells us: "There was about this time-1906 an effort to devise a theory in which the electron could only revolve in what we shall call 'allowed' orbits." After pointing out the problem with the laws of electrodynamics then known, Gill says: "J.J. Thomson deducted his hypothesis directly from the theory and curve of Boscovich, and showed that the notion of 'allowed' and 'forbidden' orbits follows from it, and thus laid the foundations of the theory developed by Bohr and other." The theory developed by Bohr and others was of course -quantum theory, so Boscovich's theory led to quantum theory.

On this issue Bohr [18] says: "Our esteem for the purposefulness of Boskovic's great scientific work, and the inspiration behind it, increases the more we realize the extent to which it served to pave the way for later developments."

Ivica Martinovic deals with this in more detail [19] and lectured at Royal Society in London, the lecture I put on the web [20].

Dragoslav Stoiljkovic also deals with the quantum physics of Boscovich [21]and says in his paper: "Roger Boscovich, in his monumental work "Theory of natural philosophy", as early as in the 18th century, pointed out that the spiral atomic chains could be formed. He also pointed that the shape of the chain could be markedly changed due to slight changes of the distances among the atoms. The elastic properties of the chains have been stressed. Furthermore, the interaction between two polymer chains is described by Boscovich's curve. The priority of Boscovich's macromolecular hypothesis is doubtless and his theory is still applicable in current polymer science. Here we presented only scientific issues of Boscovich's theory and its importance for polymer science. His theory is of the greatest significance for the other scientific fields, such as the particle theory, the electric and magnetic field theory and the quantum mechanics."

5. Conclusions

Thus with the basis of modern physics being Boscovich's theory and that basis not being widely known it is no wonder that many theorists blindly stumble on without proper appreciation of the foundations of their subject. Quantum and relativity theories both have their roots from a recognized unified theory of the 18th century and it is being mostly overlooked. Conferences are held for Boscovich which surprisingly are poorly attended. Only a few specialists seem sufficiently aware of Boscovich. Emphasis in physic education has been on Quantum theory and Relativity and the subject of the unified theory from which these two theories sprang has been omitted from the education system, hence why it is so little known. One of the charges against Boscovich's theory has been that it is dealing with things in a qualitative way instead of quantitative, however Augustus Prince has presented a paper dealing with it in a quantitative way at last year's conference of Boscovich [22] thus opening up a new era of theoretical investigation.

6. Further Information on Boscovich

Ioan James in his book "Remarkable Physicists From Galileo to Yukawa" mentions Boscovich as one of these remarkable physicists, but unlike the other physicists he mentions that are mostly known to physics students, Boscovich is mostly unknown to physics students. He says of Boscovich's theory [23]:

This daringly original work, the mature expression of ideas that Boscovich had put forward in a series of papers from 1745 onwards, was well known and influential for 150 years thereafter. Faraday, Clerk Maxwell and Kelvin were all interested in his ideas, as were many of the leading continental scientists of that period. That it should be so neglected today, at least in the western world, is ironic since Boscovich's ideas are in several respects in tune with modern thought.

He continues: "the aim being to demonstrate that a single interaction law of the type proposed can in principle account for a wide range of physical properties. Thus he presented, in the form of a logical and mathematical scheme, a programme for point atomism, in which all primary particles are identical, a single oscillatory law determines their interactions, only relational quantities enter and the distinction between empty and occupied."

Lancelot Law Whyte [24] says of Boscovich's theory: "The work was the first general mathematical theory of atomism and made its author famous when it appeared in 1758. He was elected a Fellow of the Royal Society and was lionized in London, Oxford and Cambridge; he became a corresponding member of the French Academie."

Boscovich is not completely unknown, in some obscure places he is known. It is worth going through what the *Hrvatska enciklopedija*, vol. II, Be-Da, The Miroslav Krleža Institute of Lexicography, Zagreb 2000 has to say [25]: "In his work Bošković investigated various fields of science, making his most profound contribution to the understanding of the structure of matter. His theory of forces and the structure of matter is now widely accepted, making him a scientist two centuries ahead of his time." Boscovich's theory of point-particles has been adopted by modern physics, but the entirety of the theory has not been acknowledged.

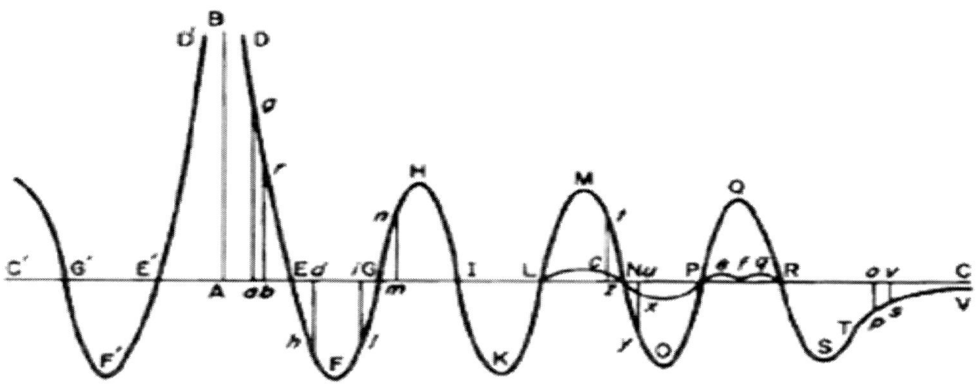

Figure 1. The Boscovich curve for force.

The *enciklopedija* continues: "His [Boscovich's] theory was postulated on the principles of simplicity and analogy within nature, and on the principle of continuity. The empirical purpose of the theory was to develop the then-topical scientific problem of collision analysis. According to Bošković, matter is composed of points (*puncta*), which are simple, indivisible, non-extended, impenetrable, discrete and homogenous, and which are sources of forces that act remotely. These points differ from mathematical points in that they possess the property of inertia, and in that there is a force – Bošković's force – acting between them, which is represented by the Bošković curve (Lat. *curva Boscovichiana*). At close distances the force is repulsive. As distances increase it reaches the point of neutrality, then becomes attractive, then reaches neutrality again, and finally becomes repulsive again. At farther distances the force is attractive, in accordance with Newton's theory of gravity. Boskovic proposed a modification in Newton's law of gravity with respect to very long distances. The Bošković curve is uninterrupted and it has two asymptotic ends (the repulsive and the attractive). It crosses the x-axis at the points of neutrality, called the points of cohesion and non-cohesion. Bošković's force is very akin to the force between atoms in a molecule or solid matter as well as to the nuclear force between nucleons (protons and neutrons). Hence Herzfeld described it as "potential energy according to Bošković". A single law of forces existing in nature (Lat. *lex unica virium in natura existentium*), i.e. the idea that one law can explain all of reality, constitutes Bošković's main contribution to science."

Now the next part becomes highly significant: "The same idea has been entertained by A. Einstein, W. Heisenberg and more contemporary scientists, but the four forces in nature (gravitational, electromagnetic, weak and strong nuclear energy forces) have yet to be described by a unified theory. Bošković's single law is a framework for a unified theory of fields or, even more so, for a theory of everything. As a result of the unconditional assumption that the law of continuity must be observed, it followed that there can be no direct contact between particles because of the repulsive force (nobody had challenged the idea that there was contact between particles of matter). Modern scientists now agree with Bošković's conception of the basic elements of matter. Bošković's *puncta* are the most basic particles of matter comparable to quarks and leptons in modern science. Since matter consists of points, it follows that it contains a lot of empty space. This idea disproved the materialistic-corpuscular theory of matter, set foundations for a real dynamistic-atomic theory, and provided a new perspective on the perception of reality. Just as the work of Copernicus resulted in the idea of the Copernican Turn, this breakthrough should be recognized as the "Boscovichian Turn" since it constitutes "the greatest triumph over the senses achieved on Earth to this time", and since Bošković and Copernicus "have been the greatest and the most victorious opponents of appearances" (F. Nietzsche, 1882., 1886)."

So it is saying that Boscovich's theory is the blueprint or basis for unified field theory. And I have talked to scientists that have developed further from this. Keith Ansell Pearson [26] points out the theory was known to Karl Popper as the Kant-Boscovich theory, and viewed it as a direct forerunner of the Faraday-Maxwell theory of fields as well as the ideas of Einstein, de Broglie, and Schrodinger. Boscovich has just not been given the publicity that he deserves so that those working on unified field theory ideas are not aware of the connections to him.

References and Notes

[1] H. Krage, Higher Speculations: Grand Theories and Failed Revolutions in Physics and Cosmology, Oxford University Press 2011, A book review by Dabby Yee 2011 at: http://dannyreviews.com/

[2] J. D. Barrow, New theories of everything, Oxford University press, 2007 p 21.

[3] L. Shelstad, Roger Boscovich Bettero: Jesuit scientist http://suite101.com/article/roger-boscovich-bettero-jesuit-scientist-a332270#ixzz210dSX0ps

[4] J. F. Barrett, The Hyperbolic theory of Special Relativity, http://arxiv.org/ftp/arxiv/papers/1102/1102.0462.pdf

[5] A. K. T. Assis, *Annales de la Fondation Louis de Broglie,* Volume 29 no 1-2, 2004 149: The Principle of Physical Proportions.

[6] O. E. Rossler (ed), Endophysics - The World as an Interface, pp. 107–112. World Scientific, Singapore, 1998. English translation by O. E. Rossler from the original Latin 1755.R. J. Boscovich. On space and time as they are recognized by us.

[7] T. Sauer, The Einstein-Varicak Correspondence on Relativistic Rigid Rotation, Einstein Papers Project, California Institute of Technology 20-7 http://arxiv.org/pdf/0704.0962v1.pdf

[8] V. Varicak bio, Mathematical Institute Sanu http://www.mi.sanu.ac.rs/History/varicak.htm

[9] J. F. Barrett, see ref. 4 quotes from L Silberstein: The Theory of Relativity London 1914, 1924 (MacMillan) p. 38.

[10] J. M. Child (trans), A theory of natural Philosophy by R. J. Boscovich, trans 1922, Open Court (1922) and MIT Press, Cambridge, MA, 1966.contains Boscovich's relativity in "Of Space and Time" (1763).

[11] K. Svozil, Conventions in relativity theory and quantum mechanics, Institut f'ur Theoretische Physik, University of Technology Vienna 9 Oct 2001 arXiv:quant-ph/0110054v1

[12] References Svozil gives are: R. J. Boskovich. De spacio et tempore, ut a nobis cognoscuntur. Vienna, 1755. English translation - R. J. Boskovich. De spacio et tempore, ut a nobis cognoscuntur. In J. M. Child, see ref. 10 pages 203–205.

[13] H. V. Gill, Roger Boscovich SJ (1711-1787) Forerunner of modern physical theories, 1941 p 47

[14] J. M. Child, see ref.10 p xvii

[15] C. J. Bjerknes, Einstein's Irrational Ontology of Redundancy : The Special Theory of Relativity and Its Many Fallacies of Petitio Principii http://itis.volta.alessandria.it/episteme/ep6/ep6-bjerk2.htm

[16] A Einstein, The Meaning of Relativity, ISBN 0-412-20560-2 p 26-27.

[17] H. V. Gill, Roger Boscovich SJ (1711-1787) Forerunner of modern physical theories, 1941 p 19

[18] N. Bohr in Actes du Symposium International R J Boskovic 1958, Beograd, Zagreb, Ljubljana, p 28

[19] I. Martinovic, Boscovich's Model of the atom from 1748, in M. Bossi and P. Tucci (ed), Bicentennial Commemoration of Ruggerio Giuseppe Boscovich, Milan 1988.

[20] I Martinovic, Boscovich video lecture.http://www.youtube.com/watch?v=E2vcb6x_50M

[21] D. Stoiljkovic, Importance of Boscovich's theory of natural philosophy for polymer science, Polimery 2007, 52, nr 11-12

[22] A. Prince, "An Analytical Form of the Boscovich Curve with Applications", pre-print for International Conference for the 3rd Centenary of the Birth of_Ruggiero Boscovich (Raguso 1711-Milano 1787) Pavia University Pavia, Italy September 8th -11th, 2011

[23] I. James, Remarkable Physicists From Galileo to Yukawa, Cambridge University Press, 2004, ISBN 13- 978-0- 521- 81687-8 p 50

[24] L. L. Whyte (ed.), Roger Joseph Boscovich, S.J., F.R.S., 1711-1787: Studies in His Life and Work on the 250th Anniversary of His Birth, 1961 – Library of Congress Catalog Card #63-21822

[25] Hrvatska enciklopedija, vol. II, Be-Da, The Miroslav Krleža Institute of Lexicography, Zagreb 2000 http://www.studyincroatia.hr/about-croatia/culture/300th-anniversary-of-the-birth-of-rudjer-josip-boskovic/rudjer-boskovic-full-biography.

[26] K.A. Pearson, Nietzsche's Brave New World of Force: Thoughts on Nietzsche's 1873 Time Atom Theory, Fragment & on the Influence of Boscovich on Nietzsche, Journal of Nietzsche Studies issue 20, 2000, p 6-35 http://www.jstor.org/discover/10.2307/20717743?uid=3738032&uid=2129&uid=2&uid=70&uid=4&sid=21101225969503.

A Scintilla of Unified Field Mechanics Revealed by a Conceptual Integration of New Fundamental Elements Associated With Wavepacket Dispersion

RICHARD L. AMOROSO

Noetic Advanced Studies Institute
Escalante Desert, Beryl, UT 84714-5104 USA
amoroso@noeticadvancedstudies.us; www.noeticadvancedstudies.us

JEREMY DUNNING-DAVIES

Institute for Basic Research
Palm Harbor, FL, USA
j.dunning-davies@hull.ac.uk

A third regime is proposed in the natural progression of the description of the physical world - Classical to Quantum to Unified Field (U_F) Mechanics. We describe the new conceptual panoply and propose an experimental method to falsify the new U_F hypotheses. Like Penzias & Wilson wondering if bird droppings affected their antenna we describe a serendipitous insight into wavepacket dispersion of 1800 MHz telecommunication em-waves in the arena where signal strength attenuates periodically by factors attributed to perceived properties that we postulate can only be mediated by U_F mechanics. Salient suggested elements include extended geometrodynamics (duality of Newton's instantaneous and Einstein's relativistic models), Solar dynamo activity, geomagnetic phenomena, seasonal precession of the Earth's axis, near - far field: geomagnetic core dynamo - solar scale-invariant wavepacket dispersion coupling and longitudinal em components. This U_F model putatively also provides an indirect measure of photon mass.

Keywords: Earth's dynamo, Geometrodynamics, Scale-invariance, Unified field mechanics, Wavepacket dispersion

1. Introduction — The Process of Serendipity

The avenue leading to this insight into a putative Unified Field (U_F) mechanics seems sufficiently interesting to briefly relate its process. One of us (RLA), enjoying the setting of a remote high desert ranch, noticed a disparity or seasonal pattern of periods of no service and intermittent (dropped) service in 900 or 1800 MHz telecommunication signals. Since there are no known thermodynamic effects on the propagation of em-waves; by the process of elimination the only possible parameters remaining were the position of the Earth in its orbit around the Sun and the seasonal inclination of its axis. The second of us (JD) had come across work by Simon Berkovich [1] highlighting a variety of so-called calendar effects seemingly occurring in nature. These effects included the annual variability of rates of radioactive decay in physics [2] and the apparent seasonal variations in cardiac deaths in biology [3]. Berkovich pointed out that quantum mechanics seems to provide the real foundation for natural events and that he, and others, have developed an interpretation of strange quantum mechanics in terms of interactive holography. To him it follows that both physical and biological phenomena should be dependent on their position with respect to the holographic machinery of the Universe. In particular, it is felt that these phenomena will be affected by the periodic alterations of the Earth's position in its solar orbit; that is, by calendar time.

Another problem, possibly related to the above, concerns the apparent irreproducibility of at least some scientific experiments. Up until very recent times, it has been taken for granted that, if experiments are performed correctly and carefully, they should be capable of being repeated. However, reliability of results has become a bigger and bigger problem, especially in biomedical research where studies have shown results to be irreproducible [4,5]. However, probably the best known example of irreproducibility of results occurred with the announcement of so-called cold fusion. To many, the notion has been simply dismissed but the fact remains that there have been recorded instances of excess heat being generated in a number of metal-hydrogen systems. This effect definitely contradicts conventional wisdom in atomic physics but the question still remains unanswered as to what causes these effects. Again, reproducibility of scientific experiments comes under scrutiny. Berkovich returns to ideas of a holographic universe to offer answers. Maybe at some future point, this approach will also be able to tackle this very real problem.

Here we point out that the Earth's core dynamo, inclination of the earth's axis and solar activity form a many-body coupled oscillator. Since we had already been pondering U_F mechanics [6,7] we were in that sense primed to take the leap suggested here regarding correlating wavepacket dispersion with U_F mechanics. We build our U_F wavepacket dispersion model based on work inspired by Vigier [8-10]. It seems interesting to consider, that at some time in the future, a link will be found between these two apparently separate approaches to these puzzling but very real problems facing present day science.

2. Parameters for Integrative Action of U_F Mechanics on Em-Wave Propagation

We attempt to make our case initially by listing the perceived concatenation of essential parameters from associated or extended physical regimes required to correlate with our model of U_F dynamics:

- **Quantum Theory - Copenhagen and beyond**
  - De Broglie-Bohm-Vigier quantum potential - piloted interpretation extended to Large-Scale Extra Dimensions (LSXD) where U_F dynamics acts as a form of 'super' quantum potential.
  - Cramer's Transactional Interpretation - off diagonal elements physically real.
  - Quantum aspects of wavepacket cohesion/dispersion.
- **Gravity**
  - Extended Geometrodynamics - Nonlinear 'shock' effects.
  - Duality - Newton's (instantaneous); Einstein's (relativistic).
  - Periodic photon mass, m_γ - cyclic vacuum coupling by internal motion (The 'tired light' alternative to a Doppler redshift).
- **Electromagnetism**
  - Longitudinal, B^3 component added to Maxwell's equations (required for m_γ); and Additional field parameters allow U_F coupling.
  - Extended electromagnetic theory, Dirac conformal covariant polarized vacuum (required for m_γ); correlates with Einstein's energy dependent spacetime metric, $\hat{M}_4$.
- **Geomagnetism**
  - Local variations in magnetic field strength.
  - Drift in secular variation of pole location.
  - Earth's core dynamo (coupled oscillator).
  - How to explain perceived seasonal discrepancies in wave packet dispersion/attenuation.
- **Cosmology**
  - Solar activity - Helioseismology; A dynamo effect created by different rotational speeds of an outer convective zone and inner radiative zone generating the main magnetic field of the Sun.
  - Conformal scale-invariance 3-body coupling of core dynamo, inclination of Earth's axis and variation in solar activity [11-13].
- **Possible Unified Field (U_F) correlations**
  - Cut off occurs in winter only when the Earth is closer to the Sun and the inclination of the Earth's axis points away from the Sun, but there is no known thermodynamic effect on the propagation of em-waves
  - Additional year round sunset/sunrise field cutoff (suggestive of a geometrodynamic effect?)
  - Effect would putatively not occur without the existence of m_γ and geometrodynamic Newton/ Einstein nonlinear duality?
  - A variable cutoff effect occurs; the attenuation point with seasonal (winter - data from two year period) higher percentage of connection drop off (by wavepacket dispersion) is ~ one to four miles from the point of sufficient signal strength to generally maintain wavepacket cohesion.

When brought to the attention of a respected colleague, he called this concatenation 'very imaginative'. Most of the parameters listed have generally been ignored by mainstream physicists; but for some decades these parameters have been integral components of alternative theory and more recently the basis of holographic anthropic multiverse (HAM) cosmology [14] (alternative to Big Bang). But the factor in deciding to explore this case is that there is no known thermodynamic effect on the propagation of em-waves that can disperse/attenuate propagation of a wavepacket. By process of elimination the only remaining conditions are the seasonal tilt or precession of the Earth's axis and the position of the Earth in its orbit around the Sun.

In relation to the factors to be considered above - a covariant polarized Dirac vacuum, de Broglie-Bohm-Vigier causal interpretation of quantum theory, extended electro/geometrodynamics and m_γ, a putative U_F effect requires most importantly the scale-invariant causal parameters associated with a 3-body coupling between the solar field dynamo and the Earth's geomagnetic core - in conjunction with seasonal tilt of the Earth's axis. We suspect also that the U_F effect, if there is one, is a property of the inherent LSXD mirror symmetry of HAM cosmology [14] also indicating a link, at least, with the approach of Bercovich. We realize at this point that this model is highly speculative but also point out that the model can readily be falsified by measuring a variety of correlations with attenuation among:

- Solar-Earth em-field dynamics
- Secular drift - local field strength variations
- Wavepacket attenuation of 900 or 1800 MHz telecommunications waves at operational cutoff limit

We will continue this discussion presently.

3. Overview of What's Known About Em-Radiation and Wavepackets

Current thinking suggests that em-radiation coming from a point source is subject to spreading of the photon wave-packet over $x = -ict$. Recently nondispersive waves have been found for the Maxwell, Klein-Gordon, Dirac, Weyl and Schrödinger field equations [15-24]. However, there is good theoretical (and thus observational) reasoning to assume the photon wavepacket has a periodic mass moment occurring when the wavepacket couples cyclically to spacetime [25]. This and the covariant Dirac polarized aether provide the basis for extended electromagnetic theory and so far controversial putative addition of a longitudinal (B^3) component to Maxwell's equations. If photon mass does exist, there should also be a gravitational effect on the near and far fields of em propagation as we have suggested earlier [25], with an extension of geometrodynamics in this volume [26].

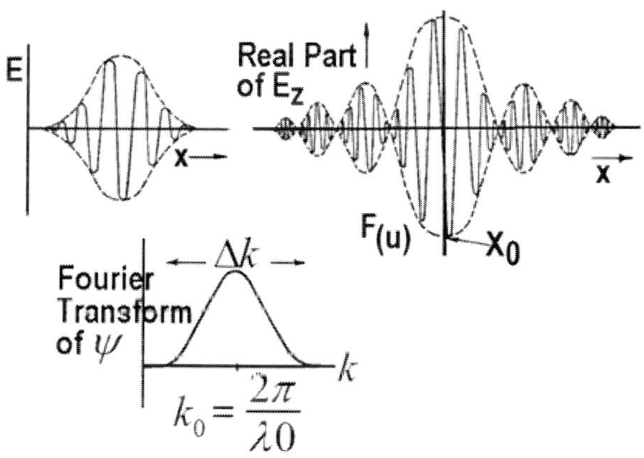

Figure 1. 1A is plotted in 1B according to Eq. (1) as a function of $x - x_0$ reaching a maximum at $x - x_0$ to zero where $x - x_0 = \pi / \Delta k$; thus obtaining a wave function concentrated in a packet where $\lambda = 2\pi / k_0$ The Fourier transform of which, as shown in Fig. 1C, is the wavepacket for a single photon. Eq. 2 is a general type of wavepacket for any function thus defined.

A wave-packet is a balanced superposition of waves of one predominant wave number, k with phase amplitudes that interfere constructively over the small region ct or Δx outside of which the amplitude reduces to zero quickly by destructive interference. This is not true for ideal monochromatic wavepackets; Figure 1A shows a wave function or amplitude Ψ that depends on space coordinates x,y,z and time, t simplified to Ψ x,t with average wavelength λ_0 over the limit Δx defined by a light pulse passing through a shutter, since an ordinary plane wave would be spread over all space [27,28].

$$E_z(x) = \int_{k_0-\Delta k}^{k_0+\Delta k} dk e^{ik(x-x_0)} = 2\frac{\sin \Delta k(x-x_0)}{x-x_0} e^{ik(x-x_0)} \tag{1}$$

$$\psi = \int_{-\infty}^{\infty} f(k-k_0) e^{ik(x-x_0)} dk \tag{2}$$

Extended electromagnetic theory suggests the photon wavepacket is piloted and does not spread over cosmological distances as considered in terms of current classically oriented interpretations of Maxwell's field equations. The experiment is accomplished by comparison of the double-slit fringes of monochromatic light from near and far field monochromatic emission sources.

Recent em-theory discussions on the possible existence of photon mass, m_γ and U(1) group invariance of em-theory in view of the putative m_γ implies the introduction of new terms for Maxwell-Lorentz equations. This cannot be considered as purely theoretical as in the past because of recent new experimental evidence [15]. There are two possible interpretations of Maxwell's equations:

1. Photon propagation without trajectory - random probability distributions (classical).
2. Photon propagation with trajectory - piloted with no or minute spreading of the wavepacket (extended).

The former is in accord with the standard model of classically oriented Copenhagen approaches to the em and quantum formalism; and the latter the Vigier-Bohm-de Broglie extended charge particle causal approaches.

About 10 years after Einstein proposed that radiant em-energy should appear as quanta he assumed that these quanta should also be spatially quantized with a unique orientation, concluding that Maxwell's - spherical radiation around a source – does not exist but that elementary light energy $h\nu$ always appears in a unique direction creating a recoil upon emission $h\nu / c$ which he called *Nadelstrahlung* (needle radiation) [29]. *Nadelstrahlung* is also responsible for radiation pressure. According to classical interpretations of Maxwell's theory dispersion of the wave-packet over distance is expected [27, 28]. This is partly due to the fact that in the Copenhagen view emission is point-like so any packet should expand. The uncertainty principle can be derived from propagation of the wavepacket which in classical terms propagates according to the Schrödinger equation. This is described by Bohm [27] as

$$\Delta x = \Delta x_0 \sqrt{1+\frac{\hbar^2 t^2}{m^2(\Delta x_0)^4}} \rightarrow \frac{\hbar}{m\Delta x_0} t \quad (3)$$

as $t \rightarrow \infty$

where a wavepacket with an initial diameter of Δx_0 will spread to $\Delta x \cong ht / \Delta x$ as t becomes limitless; the narrower the wavepacket originally, the more rapid the spreading. (Fig. 2) According to Bohm [27] the reason for the spread is in terms of the uncertainty principle. The region Δx_0 confining the packet has a number of wavelengths near Δx_0 so that even though the average velocity of the wavepacket is equal to the group velocity, the actual velocity will fluctuate, and the distance propagated by the packet isn't fully determined. It can fluctuate by

$$\Delta x \cong t\Delta v \cong \frac{\hbar t}{m\Delta x_0}. \quad (4)$$

According to Bohm [27] photons have momentum as evidenced in the radiation pressure during absorption, such that the energy and momentum of light quanta is the same as a zero mass particle in 3D space. It is the wave properties of a wavepacket that produces the $\Delta x\Delta k \geq 1$that allows spreading because a particle will never spread; but a collection of particles because of uncertainty in velocity gradually spread with Δt [30].

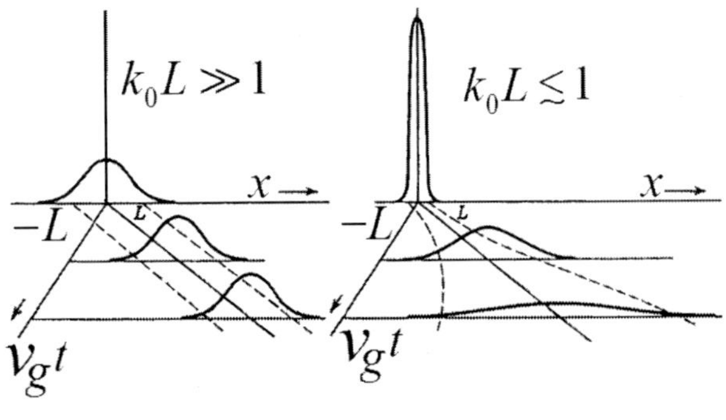

Figure 2. Dispersion and change of shape of a wavepacket during propagation. Broader packets with many wavelenghts near k ($kL \geq 1$) distort relatively little; a narrow k packet ($kL \leq 1$) rapidly broadens.

Thus, although equations (3) & (4) used for illustration of classical wavepacket spreading are Schrödinger type, Maxwellian equations give a similar result for photons in the classical limit, i.e. spreading. In our extended theoretical approach that includes photon mass [9,10,30,31]; piloting effects prevent spreading as $\Delta t \rightarrow \infty$. There may be an infinitesimal spreading over cosmological distance. This is the current empirical knowledge limit; where further understanding can only be achieved by experimentation.

In the de Broglie mode, a classical wave of wave number k can be of arbitrary amplitude and momentum, and, whenever position or momentum is measured a definite number results. The de Broglie relation $p = hk$ implies a definite wave number k for a definite momentum. This is contrary to a classical description of a wavepacket, which suggests a range of wave numbers and positions [27]. Rather than the classically oriented Schrödinger equation which suggests spreading of the wavepacket (in a variation of Fig. 2), non-dispersive wave modes would be expected to propagate according to the de Broglie relativistic Klein- Gordon type equation [15,32]. Recent work on non-dispersive modes of the wavepacket by numerous authors [15-24] has demonstrated mathematically the possibility for the existence of 'real' non-dispersive modes of the photon wavepacket satisfying linear Maxwell equations but considered contrary to the prevailing opinion. Hence, empirical testing of the issues at the heart of the matter has been urged.

According to established wave mechanics a de Broglie wave with infinite wavelength is said to be associated with all particles and have a wave function uniform throughout all space. A particle's internal vibration and infinite de Broglie wave stays in phase at the particle's location. This suggests how the de Broglie wave pilots a particle's motion with no spreading; whereas a Schrödinger wavepacket spreads because of uncertainty in momentum [15]. Because the de Broglie relation is relativistic a Lorentz transformation might be involved between the particle's point of origin and present position during propagation, canceling insertion of any would be classical uncertainty effects and maintaining phase coherence between the particle's internal motion and the wave function in the de Broglie relativistic-piloted regime. This might be considered reminiscent of error correction modes discussed in terms of quantum computing. If $\omega(k_1) - \omega(k_0) = \mu(k_1 - k_0)$ then $d\omega(k_1)/ dk_1 = \mu$ for all k_1 and therefore d $2\omega(k)dk = 0$. With these conditions de Broglie theory yields [15]

$$F(x,t) = G(x,t)H(x,t) \quad (5)$$

Classical approaches predict wavepacket spreading

because of uncertainty relationships. The de Broglie-Bohm approaches predict coherence over all space and time in view of putative causal action of the pilot wave or quantum potential. This is the well-known assumption of hidden variables deemed inconsistent with the classically oriented Copenhagen model. Wavepacket spreading during propagation seems straight forward; but subtleties are involved at the heart of wave mechanics and quantum theory which is by no means complete and entails persistent discussion on the merits of Copenhagen vs. extended forms of quantum theory and now aspects of not yet invented Unified Field Mechanics. A definitive delineation is not possible in terms of any type of current theoretical discussion alone. Therefore, technically feasible experiments on the nature of the wavepacket and its propagation like those proposed here might advance our understanding of Quantum Theory.

A full understanding of the photon and its propagation is far from achieved [33]; one might reasonably see the truth of this claim in the writing of Lazar Mayants [34]. To understand the anticipated experimental results, photon propagation may not only have to be perceived in terms of internal de Broglie Lorentz transformations [6] but also deeper aspects of nonlocality [35] which might only be clarified in terms of a post Big Bang cosmology and the attendant understanding of spacetime hyperstructure [36] which non-zero photon restmass seems to demand. The Newtonian ether ('substance' permeating and connecting the entire universe) was disallowed by Einstein's relativistic dynamics and the Michelson-Morley experiment. Einstein himself said that relativity did not preclude existence of an ether:

> ...Recapitulating: ...according to the general theory of relativity space is endowed with physical qualities; in this sense...there exists ether. According to the general theory of relativity space without ether is unthinkable; for in such space there not only would be no propagation of light, but also no possibility...for standards of measuring rods and clocks, nor therefore any space-time intervals in the physical sense. But this ether may not be thought of as endowed with the quality characteristic of ponderable media, as consisting of parts which may be tracked through time...motion may not be applied to it...it would be a great advance if we could succeed in comprehending the gravitational field and the electromagnetic field together as one unified conformation. Then...the epoch of theoretical physics founded by Faraday and Maxwell would reach a satisfactory conclusion. The contrast between ether and matter would fade away, and, through the general theory of relativity, the whole of physics would become a complete system of thought, like geometry, kinematics, and the theory of gravitation.

We revisit this issue in terms of a Dirac covariant sub-quantum stochastic ether with correspondence to relativity and inclusive of de Broglie-Bohm-Vigier charged particle models [37].

4. Properties of de Broglie Matter-Waves

De Broglie, by considering a material moving object of rest mass, m_0, for a stationery observer, suggested that a phase wave, or 'pilot' wave, accompanies a particle because the principle of inertia said it should possess an internal energy equal to m_0c^2 [37]. This phase wave arises as an inevitable consequence of de Broglie's assumption of the internal periodic phenomenon of the particle and the Lorentz transformation laws of the special theory of relativity

$$h\nu_0 - m_0c^2, \tag{6}$$

with $v - \beta c,\ (\beta < 1)$ for total energy $\nu - m_0c^2 / h\sqrt{1-\beta^2}$.

De Broglie's result arose from a combination of the principle of Einstein's special relativity and the quantum relationship for the observer which he initially applied to a photon of nonzero restmass, $m_\gamma\ (<10^{-50}g)$ which because of its associated internal motion he associated with a piloting phase wave of frequency, ν at each point in space.

MacKinnon [15] described the de Broglie wave packet for stationery states and nondispersive wave packets of a free particle. He states that the nondispersive wave packet, ψ is a solution of

$$\Box\psi = 0 \tag{7}$$

where

$$\Box = \nabla^2 - \frac{1}{c^2}\frac{\partial^2}{\partial t^2}.$$

From this MacKinnon shows that the nondispersive wave packet for a particle relative to the observer has the form

$$\psi = \sin(kr/kr)\exp\left[i(\omega t - k_0x)\right] \tag{8}$$

where

$$k = m_0c/\hbar, \quad r = \left\{\frac{(x - vt^2)}{1-(v^2/c^2)} + y^2 + z^2\right\}^{1/2} \tag{9}$$

$$\omega = mc^2/\hbar \quad k_0 = mv/\hbar.$$

Equation (8) is a spherically symmetric solution to Eq. (7) after being subjected to the Lorentz transform as initially obtained by de Broglie.

Of critical interest to us is MacKinnon's work to set up a de Broglie wave packet for a stationery state. Although we are interested in relativistic waves, it is not for de Broglie waves for the usual particles in coordinate motion but for de Broglie waves for stationery matter with internal 'continuous-state' relativistic effects.

Consider two identical particles moving in opposite directions relative to an observer at x^* and t^*

$$\psi_1^* = A\cos\left(\omega t^* - kx^*\right), \qquad \psi_2^* = A\cos\left(\omega t^* + kx^*\right) \quad (10)$$

which represent standing waves when solved by the Schrödinger equation for a particle in a box and cannot depend on the reference frame [38]. MacKinnon concludes that these stationery states are static and for which Bohm postulated a quantum potential to account for it. MacKinnon carries this point further [15] to suggest that:

> The motion of a particle in spacetime does not depend on the motion relative to it of any observer or any frame of reference [and] if the particle has an internal vibration of the type hypothesized by de Broglie, the phase of that vibration at any point in spacetime must appear to be the same for all observers...Each observer or reference frame will have its own de Broglie wave for the particle. The phase of the particle's vibration must, by definition, be the same as that for all possible de Broglie waves at the point where the particle is. By superimposing all these possible de Broglie waves, a [nondispersive] wave packet is formed centered in space on the particle.

In his original work de Broglie was not able to form a wave packet properly that could localize the particle; MacKinnon constructed a wave packet from de Broglie's original wave phenomena that is also nondispersive [15].

5. Correlation of Photon Mass

Photon anisotropy requires vacuum zero point coupling, and its propagation can no longer be considered independent of the Dirac vacuum. Einstein, Schrodinger, and de Broglie have attested to the significance of non-zero photon rest mass. Frequency anisotrophy results from a putative 10^{-65} g periodic nonzero photon rest mass according to

$$E = \hbar\nu - mc^2\left[1 - v^2/c^2\right]^{-1/2}. \quad (10)$$

The de Broglie wavelength relationship for massive particles, taking the accepted value for R applied to the Vigier mass m_γ of the photon is

$$m_\gamma = \frac{h}{\lambda c} \quad (11)$$

taking $\lambda = R \approx 10^{28}\,cm$ for the de Broglie wavelength, λ of the photon then $m_\gamma \approx 2.2\times10^{-65}\,g$ which is the value for photon restmass obtained by a number of researchers [39]. Where R is the radial size of the universe; and by the uncertainty relation this is the smallest possible photon mass. Further $m \to 0$ only if $R \to \infty$. The de Broglie hypothesis was verified by [40] for the wavelength of a material particle. A photon mass of $10^{-65}\,g$ is in total agreement with Vigier's tired-light hypothesis [41].

From the redshift-distance relation, $z = f(d)$ (for static or expanding universe models) following [42] photons with restmass, m_γ interact with vacuum particles of mass, m_{vac} with acceleration

$$\int \frac{d^2y}{dt^2}dt = -\frac{2\lambda\rho\omega}{y\left[\left(\frac{1}{2}\right)^2 + y^2\right]^{1/2}}. \quad (12)$$

The momentum transfer per vacuum particle, m_{vac} is

$$\int m_{vac}\frac{d^2y}{dt^2}dt = -\frac{2m_{vac}m_\gamma\omega}{y\left[\left(\frac{1}{2}\right)^2 + y^2\right]^{1/2}} \quad (13)$$

With t the time, y the coordinate intersecting the path, ω producing a 'tired-light' redshift-distance law

$$\frac{\Delta\nu}{\nu} = e^{kd} - 1, \quad (14)$$

where k is determined by m_γ estimated to have a value of $10^{-65}\,g$ [43], m_{vac} which is currently unknown and may not be completely relevant other than the putative fact that vacuum coupling occurs.

It is inherently obvious that the photon is annihilated when brought to rest; therefore it is suggested that the photon has a rest mass with a half-life on the order of the Planck time of 10^{-44} s, which would still preserve gauge in the domain of the standard model of elementary particles and allow for anisotropic vacuum zero point coupling of the photon which if it also occurs in the limit of the Planck time can be a virtual interaction.

6. U_F-Mechanics Summary

Regarding gravitational contingencies in the continuous -state cycling inherent in Unified Field, U_F dynamics, the dimensional reduction cascade includes a form of parallel transport [14,44] with a 'deficit angle' occurring at the point of return to the initial position. We want to try and summarize here this point in terms of the duality of Newton-Einstein gravity as it acts on our 3-body cosmological test system - Solar dynamo; Earth axis inclination; Earth core dynamo. We have attempted to show that this U_F approach in conjunction with our newly discovered 'dual geometrodynamics' parameter has an effect on wavepacket dispersion. The effect does not seem to apply nor would it operate under any of the usual interpretations of gravity, electromagnetism, Big Bang cosmology or the Copenhagen interpretation of quantum theory making this initial foray difficult.

Firstly, we have extended the de Broglie- Bohm-Vigier causal interpretation to the extent that spacetime and the matter within it evolves under the quantum potential / pilot wave and is continuously annihilated-recreated without collapse of the wave- function (unless acted upon). We assume this as an obvious aspect of the principle of de Broglie matter waves. Also, the 'guidance' is more under the auspices of a so-called 'super-quantum potential' or coherent action of the U_F. The 12D space is a highly symmetric form of M-Theoretic Calabi-Yau mirror symmetry best said to operate according to parameters suggested by Cramer's transactional interpretation, where the present state in simplistic terms is said to be a 'standing-wave of the future-past' [45].

Secondly, we have recently found a method to integrate gravity and electromagnetism [14, 47-49]. Conventionally Maxwell's equations describe transverse elements described as 'em' waves; but by utilizing the Einstein/de Broglie relations one can derive additional degrees of freedom so that Maxwell's equations are not 'cut off' at the vacuum but lead to Longitudinal wave components and non-zero electric conductivity of the vacuum Therefore one must employ the μv fields in addition to the standard em, suggesting also that the photon is piloted. The two sets of coordinates for the em or μv fields are mutually exclusive and generally considered to be independent of each other. We developed a method for integrating them in terms of a Dirac covariant polarized vacuum and extended theoretical perspectives. One must 'fix' the coordinates of either the em field or the μv field; we chose the latter and developed a method for integrating the other. It should be noted that while c is constant in the rest frame and the velocity of massive photons would be frequency dependent; there is no contradiction because as Dirac himself stated according to coordinate law the pilot wave and the photon decouples [44]. It is mandatorily to derive this integration in terms of a Covariant Dirac Polarized vacuum. With addition of the 'continuous-state' process [14] this model is extended further to cosmology by alternating the coordinate fixing cyclically. (This allows gating of the U_F.)

Thirdly, the utility of the Dirac vacuum is such that it allows additions to Maxwell's equations, i.e. small periodic photon mass by introducing the longitudinal $B^{(3)}$ field. This is a key element in that it provides a parameter allowing gravity (dual form) to act on wavepacket dispersion-coherence.

Fourthly, there is one more wrinkle in terms of the de-Broglie-Bohm cyclic Geometrodynamic fix-unfix-fixing parameters of the spacetime backcloth. Recent WMAP and Planck satellite CMB radiation observations have suggested that the Hubble spheres, H_R, fundamental polyhedron is a form of topologically closed Poincairé-Dodecahedral space (PDS) with positive curvature such that looking far enough into space in certain directions one would see the back of one's head [50,51]. For reasons not entered into here HAM cosmology suggests that this cosmological geometric topology (related to the cosmological constant, Λ) oscillates continuously from positive to negative curvature. This in conjunction with a variable Planck constant at the microscopic level (which was a component of the original Hadronic form of string theory) [52] provides the complete concatenation of parameters for action of the U_F and basis for a systematic approach to a unified field mechanics approach to dispersion of the em wavepacket.

Appendix: Additional Anomalies With Putative U_F Aspects

We have concentrated on wavepacket dispersion in telecommunications as the primary thrust of this introductory paper taking a stab at Unified Field Mechanics, U_F; and in that guise we have chosen to include any parallel indicia of what we perceive as related to conformal scale-invariant properties from microcosm to macrocosm in this appendix.

- ANOMALOUS RADIO DECAY - The sacrosanct view that nuclear decay rates are fundamental constants has been challenged by numerous reports contrary to the conventional view that the decay rate of an unstable isotope is an intrinsic property of that isotope. We cite some of the overwhelming evidence of anomalous time-dependent (seasonal) features present in the count rates of various nuclei [53-60].

- FINE STRUCTURE CONSTANT - Seasonal and astrophysical observational anomalies have been found for the fine structure constant [61-72].
- SOLAR FLARE ACTIVITY - Numerous observations of solar flares and other solar activity have been shown to effect radio decay rates [73-83].
- EARTH SUN DISTANCE - Some research has also been done on the effect of radio decay related to Earth-Sun distance. Although not as marked in the literature, this effect is of interest to us as we have observed it; not only distance but inclination of the Earth's axis [84].
- MARS MESSENGER SATELLITE - Anomalous radio decay was observed by radio species aboard the Mars satellite [85, 86].
- QUASAR ACTIVITY - Some theorists have also postulated radio variations related to QSO activity.
- BIOLOGICAL EFFECTS - Strong evidence has been found for seasonal variation in cardiac death rates [3,87,88]. Other statistical evidence although not yet as rigorous has been related to overall health and birth month [89]. We recall several studies in Japan (around the 1970s or 1980s) that have attempted to manipulate contagion of Influenza by placing subjects in cold rooms for various periods of time. No increase in influenza contagion was found; Reference currently lost.

Comment: It is interesting to note in relation to the unified theory we are developing here, in terms of conformal scale-invariance, a possible related effect has been recognized in relation to the annual 'flu season'. Studies have been done exposing subjects to cold by putting them in a freezer – No increased incidence of influenza contagion appeared. As strange as it seems we are suggesting a correlated effect with viral adhesion and inclination of the Earth's axis also; i.e. 'flu season' comes not because of winter but because the inclination of the earth's axis has an associated 'cosmology of mind' effect on self-organized living systems (SOLS) that also correlates as a U_F effect. Several mind-body theorists suggest the U_F is an aspect of consciousness or 'light of the mind'. This idea is beyond the scope of this paper and will be rigorously addressed elsewhere.

References and Notes

[1] Berkovich, S., Calendar variations in the phenomena of Nature and the apparition of two Higgs bosons, preprint

[2] Clark, S. (2012) Half-life heresy: Seasons change in the atom's heart, New Scientist, 14 Nov., 2891, pp.42-45.

[3] B. G. Schwartz & R. Kloner; abstract 11723, Seasonal variations in cardiac death rates is uniform across different climates, http://circ.ahajournals.org/cgi/content/meeting_abstract/126/21_MeetingAbstracts/!1172

[4] B. Frezza; What's fueling our growing loss of faith in big science? http://www.bio-itworld.com/2013/1/11/whats-fueling-our-growing-loss-faith-big-science.html

[5] E. Iorns; Is medical science built on shaky foundations? New Scientist, 17 Sept.,2012, issue2882, pp.24-25

[6] See Appendix 1.

[7] Amoroso, R.L., Kauffman, L.H., Rowlands, P. & Rauscher, E.A. Introduction to Unified Field Mechanics, in process.

[8] R.L. Amoroso, J-P Vigier, M. Kafatos, G. Hunter (2002) Comparison Of Near And Far Field Double-Slit Interferometry For Dispersion Of The Photon Wavepacket, in, R.L. Amoroso Et Al (Eds.), Gravitation And Cosmology: From The Hubble Radius To The Planck Scale, 59-64, Dordrecht: Kluwer Academic Publishers.

[9] Amoroso, R.L., Kafatos, M. & Ecimovic, P., The origin of cosmological redshift in spin exchange vacuum compactification and nonzero rest mass photon anisotropy, in G.Hunter, S. Jeffers & J-P Vigier (eds.) Causality and Locality in Modern Physics, Dordrecht: Kluwer.

[10] Vigier, J-P, 2000, Photon mass and Heaviside force, Phys. Let. A, 270, 221-231.

[11] Ossendrijver, M. (2003) The solar dynamo, The Astronomy and Astrophysics Review 11 (4): 287–367.

[12]Thompson, M.J.; Christensen-Dalsgaard, J.; Miesch, M.S.; Toomre, J. (2003). "The Internal Rotation of the Sun". Annual Review of Astronomy & Astrophysics 41 (1): 599-643.

[13] Vorontsov, S.V., Christensen-Dalsgaard, J.,Schou, J., Strakhov, V.N. & Thompson, M.J. (2002) Helioseismic measurement of solar torsional oscillations, Science 296 (5565): 101–103.

[14] Amoroso, R.L. & Rauscher, E.A. (2009) The Holographic Anthropic Multiverse, Hackensack: World Scientific.

[15] Mackinnon, L. (1978) A nondispersive de Broglie wave packet, Found. Phys. 8:3/4, 157-176.

[16] Hillion, P. 1991, Nonhomogenious nondispersive electromagnetic waves, Phys. Rev. A, 45:4, pp.2622-2627.

[17] Peshkin, M. 1999, Force-free interactions and nondispersive phase shifts in interferometry, Found. Phys. 29:3, 481-489.

[18] Rodrigues, W.A., & Lu, J-Y., 1997, On the existence of undistorted progressive waves, Found. Phys. 27:3, 435-508.

[19] Ignatovich, V.K., 1978, Nonspreading wave packets in quantum mechanics, Found. Phys. 8:7/8, pp. 565-571.

[20] Shaarawi, A.M. & Ziolkowski, R.W. , 1990, A Novel approach to the synthesis of nondispersive wave packet solutions to the Klein-Gordon and Dirac equations, J. Math. Physics, 3:10, 2511-2519.

[21] Brittingham, J.N. (1983) Focus waves modes in inhomogenious Maxwell's equations: Transverse electric mode, J Appl. Phys. 54:3, 1179-1189.

[22] Beil, R.G. 1993, Found. Phys. 23, 1587.

[23] Beil, R.G. 1995, Found. Phys. 25, 717.

[24] Beil, R.G. 1997, pp. 9-16, in The Present Status of the Quantum Theory of Light, S. Jeffers et al, (eds.) Dordrecht: Kluwer Academic.

[25] R. L. Amoroso, J-P Vigier, M. Kafatos, G. Hunter (2002) Comparison of near and far field double-slit

Interferometry for dispersion of the photon wavepacket in R.L. Amoroso et al. (eds.) Gravitation and Cosmology: From the Hubble Radius to the Planck Scale, 59-64 Dordrecht: Kluwer Academic.
[26] Amoroso, R. L. (2013) Unified Geometrodynamics: A Complementarity of Newton's and Einstein's Gravity, in R. L. Amoroso, P. Rowlands & L.H. Kauffman (eds.) The Physics of Reality: Space, Time, Matter, Cosmos, Hackensack: World Scientific.
[27] Bohm, D. (1963) Quantum Theory, Prentice-Hall : Englewood Cliffs.
[28] Schiff, L.I. (1987) Quantum Mechanics, London : McGraw-Hill.
[29] Cormier-Delanoue, C. (1988) *Sur l'emission d'une radiation electromagneticique par une charge electrique en movement rectiligne accelere*, *Annales de la Fondation Louis de Broglie,* 13 :1, pp. 43-63. (Translation by JP Vigier, edited by R.L. Amoroso)
[30] MacKinnon, L. (1981) A fundamental equation in quantum mechanics? Let Al Nuovo Cimento, 32:10; 311-316.
[31] Kulhánek, J. (1965) Propagation of de Broglie waves in space-time, Il Nuovo Cimento, 38:3; 1955-1965.
[32] De Broglie, L. 1925, Ann. Phys. 3,22.
[33] Amoroso, R. L. & Vigier, J-P (2002) The origin of cosmological redshift and CMBR as absorption/emission equilibrium in cavity-QED blackbody dynamics of the Dirac vacuum, in R.L. Amoroso, G. Hunter, M. Kafatos & J-P Vigier (eds) Gravitation & Cosmology: From the Hubble Radius to the Planck Scale, Dordrecht: Kluwer Academic.
[34] Mayants, L. (1994) Beyond the Quantum Paradox, Taylor & Francis, London.
[35] Stevens, H.H. (1989) Size of a least-unit, in M. Kafatos (ed.) Bell's Theorem, Quantum Theory and Conceptions of the Universe, Dordrecht: Kluwer Academic.
[36] Dirac, P.A.M. (1951) Nature, 906.
[37] Vigier, J-P (2000) selected papers, in S. Jeffers, B. Lehnert, N. Abramson, & L. Chebotarev (eds.) Jean-Pierre Vigier and the Stochastic Interpretation of Quantum Mechanics, Montreal: Aperion.
[38] Bohm, D (1951) Quantum Theory, Englewood Cliffs: Prentice-Hall.
[39] Lehnert, B. (2002) New developments in electromagnetic field theory, in R.L. Amoroso, G. Hunter, M. Kafatos & J-P Vigier (eds.) Gravitation & Cosmology: From the Hubble Radius to the Planck Scale, Dordrecht: Kluwer Academic.
[40] de Broglie, L. & Vigier, J-P (1972) Phys. Rev. L. 28, 1001-1004
[41] Vigier, J-P (1990) Evidence for nonzero mass photons associated with a vacuum induced dissipative redshift mechanism, IEEE Trans Plasma Sci. 18:1;64-72.
[42] Zwicky, F. (1929). On the red shift of spectral lines through interstellar space, Proceeding
[43] Evans, M. & Vigier, J-P (1994) The Enigmatic Photon Vol. 1,2, Dordrecht: Kluwer Academic.
[44] Amoroso, R. L. (2002) Developing the cosmology of a continuous-state universe, in R.L. Amoroso, G. Hunter, M. Kafatos & J-P Vigier (eds) Gravitation & Cosmology: From the Hubble Radius to the Planck Scale, Dordrecht: Kluwer.
[45] Cramer, J.G. (1986) The Transactional Interpretation of Quantum Mechanics, Rev. Mod. Phys 58, 647-687.
[46] Rowlands, P. (2007) Zero to Infinity, Hackensack: World Scientific.
[47] Rauscher, E.A. & Amoroso, R.L. (2011) Orbiting the Moons of Pluto: Complex Solutions to the Einstein, Maxwell, Schrödinger & Dirac Equations, World Scientific.
[48] Vigier, J-P & Amoroso, R.L. (2003) Can one integrate gravity and electromagnetic fields? in R. L. Amoroso, J-P Vigier, M. Kafatos, G. Hunter (eds.) Gravitation and Cosmology: From the Hubble Radius to the Planck Scale, pp. 241-258, Netherlands: Springer.
[49] Amoroso, R.L. & Vigier (2004) Toward the unification of gravity and electromagnetism, in V. Dvoeglazov & A. Espinoza (eds.) Relativity, Gravitation, Cosmology, 71-88, New York: Nova Science.
[50] Luminet, J-P et al. (2003) Dodecahedral space topology as an explanation for weak wide-angle temperature correlations in the cosmic microwave background, arXiv:astro-ph/0310253v1.
[51] Riazuelo, A., Weeks, J., Uzan, J-P, Lehoucq, R. & Luminet, J-P (2003) Cosmic microwave background anisotropies in multi-connected flat spaces arXiv:astro-ph/0311314v1 (2003)
[52] Veneziano, G. (1968) Construction of a crossing symmetric, Regge-behaved amplitude for linearly rising trajectories, Nuovo Cim. 57A 190.
[53] Evidence for time-varying nuclear decay rates: Experimental results and their implications for new physics E. Fischbach, J.H. Jenkins, P.A. Sturrock, arXiv:1106.1470v1 [nucl-ex] 7 Jun 2011
[54] J.H. Jenkins & E. Fischbach, *Astropart. Phys.* 31, 407 (2009)
[55] J.H. Jenkins *et al*, *Astropart. Phys.* 32, 42 (2009)
[56] E. Fischbach *et al*, *Space Sci. Rev.* 145, 285 (2009)
[57] P.A. Sturrock *et al*, *Astropart. Phys.* 34, 121 (2010)
58] Javorsek II *et al*, *Astropart. Phys.* 34, 173 (2010)
[59] P.A. Sturrock *et al*, *Sol. Phys.* 267, 251 (2010)
[60] P.A. Sturrock, E. Fischbach and J.H.
[61] Melville, K. Fine structure constant may vary across universe, http://www.scienceagogo.com/news/20111005023759data_trunc_sys.shtml
[62] Berkovich, S. On dipole anisotropy in spatial distribution of Planck's constant values, American Physical Society, APS March Meeting, March 21-25, 2011, abstract #Q29.01 http://adsabs.harvard.edu/abs/2011APS..MARQ29013B
[63] Revision of VLT/UVES constraints on a varying fine-structure constant M. T. Murphy,1,2⋆, J. K. Webb,3 V. V. Flambaum3 Mon. Not. R. Astron. Soc. 000, 1–11 (2007) Printed 5 February 2008 (MN LATEX style file v2.2) arXiv:astro-ph/0612407v3 11 Nov 2007
[64] Varying Alpha: New Constraints from Seasonal Variations John D. Barrow1 and Douglas J. Shaw2, arXiv:0806.4317v1 [hep-ph] 26 Jun 2008
[65] Shaw, D.J. (2007) Detecting seasonal changes in the fundamental constants, arXiv:gr-qc/0702090v1.
[66] J. K. Webb et al. (2001) Phys. Rev. Lett. 82, 884; (1999) 87, 091301.
[67] H. Chand et al., Astron. Astrophys. 417, 853 (2004);
[68] R. Srianand et al.,, Phys. Rev. Lett. 92, 121302 (2004).
[69] M. T. Murphy et al., astro-ph/0612407.
[70] E. Reinhold et al., Phys. Rev. Lett. 96, 151101 (2006).

[71] D. J. Shaw and J. D. Barrow, Phys. Rev. D 73, 123505 (2006); 73, 123506 (2006); Phys. Lett. B639 596-599 (2006).
[72] J. Magueijo, J. D. Barrow and H. B. Sandvik, Phys. Lett. B549, 284-289 (2002); H. B. Sandvik, J. D. Barrow, J. Magueijo, Phys. Rev. Lett. 88, 031302 (2002).
[73] Perturbation of Nuclear Decay Rates During the Solar Flare of 13 December 2006 Jere H. Jenkins1 and Ephraim Fischbach1 Solar flare decay 0808.3156
[74] Jenkins, J.H., Fischbach, E.: Astroparticle Physics 31(6), 407 (2009)
[75] Jenkins, J.H., Mundy, D.W., Fischbach, E.: Nuclear Instruments and Methods in Physics Research Section A: Accelerators, Spectrometers, Detectors and Associated Equipment 620(2-3), 332 (2010)
[76] Jenkins, J.H., Fischbach, E., Buncher, J.B., Gruenwald, J.T., Krause, D.E., Mattes, J.J.: Astroparticle Physics 32(1), 42 (2009)
[77] Jenkins, J.H., Herminghuysen, K.R., Blue, T.E., Fischbach, E., Javorsek II, D., Kauffman, A.C., Mundy, D.W., Sturrock, P.A., Talnagi, J.A.: Additional experimental evidence for a solar influence on nuclear decay rates (Submitted to NIM-A, 2011)
[78] Analysis of environmental influences in nuclear half-life measurements exhibiting time-dependent decay rates
Jere H. Jenkins_,a,b, Daniel W. Mundyc, Ephraim Fischbachb, arXiv:0912.5385v1 [nucl-ex] 2009
[79] Power Spectrum Analyses of Nuclear Decay Rates D. Javorsek II, P.A. Sturrock, R.N. Lasenby, A.N. Lasenby, J.B. Buncher, E. Fischbach, J.T. Gruenwald, A.W. Hoft, T.J. Horan, J.H. Jenkins, J.L. Kerford, R.H. Lee, A. Longman, J.J. Mattese, B.L. Morreale, D.B. Morrish, R.N. Mudry, J.R. Newporte, D. O'Keefe, M.A. Petrellie, M.A. Silver, C.A. Stewarte, B. Terry, arXiv:1007.0924v1 [nucl-ex] 2010
[80] An Analysis of Apparent r-Mode Oscillations in Solar Activity, the Solar Diameter, the Solar Neutrino Flux, and Nuclear Decay Rates, with Implications Concerning the Sun's Internal Structure and Rotation, and Neutrino Processes.
[81] Indicates a possible correlation between the solar flare of 13 December 2006 and a decrease in the measured decay rate of 54Mn coincident in time with the flare. An analysis of data from Brookhaven National Laboratory (BNL) on the measured decay rates of 32Si and 36Cl, and from the *Physikalisch-Technische Bundesanstalt* (PTB) in Germany on the measured decay rates of 226Ra and its daughters, show that both data sets exhibit similar annual variations in their respective decay rates analyzed 137Cs decay data, obtained from a small sample onboard the Messenger spacecraft en route to Mercury, with the aim of setting limits on a possible correlation between nuclear decay rates and solar activity. Such a correlation has been suggested recently on the basis of data from 54Mn decay during the solar flare of 13 December 2006, and by indications of an annual and other periodic variations in the decay rates of 32Si, 36Cl, and 226Ra.
[82] Correlations between solar activity and nuclear decay rates. This includes an apparent correlation between Earth-Sun distance and data taken at Brookhaven National Laboratory (BNL), and at the *Physikalisch-Technische Bundesanstalt* (PTB). Although these correlations could arise from direct interaction between decaying nuclei and some particles or f ields emanating from the Sun.
[83] Evidence for Correlations Between Nuclear Decay Rates and Earth-Sun Distance Jere H. Jenkins,1 Ephraim Fischbach, John B. Buncher, John T. Gruenwald, Dennis E. Krause, & Joshua J. Mattes1 arXiv:0808.3283v1 [astro-ph] 2008.
[84] Solar Influence on Nuclear Decay Rates: Constraints from the MESSENGER Mission E. Fischbach K.J. Chen R.E. Gold J.O. Goldsten D.J. Lawrence R. J. McNutt, Jr.
[85] E.A. Rhodes J.H. Jenkins J. Longuski, arXiv:1107.4074v2 [astro-ph.SR] 2011
[86] Stuart Clark, Half-life heresy: Seasons change in the atom's heart, New Scientist, 2012, issue 2891, pp. 42-45.
[87] Leonid Gavrilov and Natalia Gavrilova, Long-life babies born in autumn, New Scientist, 21 July 2012.
[88] N. Schlesinger and M. Schlesinger, Seasonal Variation of Rheumatic Diseases.
http://www.discoverymedicine.com/Naomi-Schlesinger/2009/07/16/seasonal-variation-of-rheumatic-diseases/
[89] Welsh, J. Season of birth may affect the rest of your life, http://www.livescience.com/13958-birth-month-health-effects.html.

Fields in Action? From the Inside Looking Out (Musings of an Idiosyncratic Experimentalist!)

JOHN DAINTON

Physics Department, University of Liverpool, Oliver Lodge Laboratory
Oxford St, Liverpool, L69 7ZE, UK
john.dainton@stfc.ac.uk

The phenomenology of fields is a major part of physical science. We briefly outline the most important aspects here. Photon structure derives from quantum fluctuation in quantum field theory to fermion and anti–fermion, and has been an experimentally established feature of electrodynamics since the discovery of the positron. In hadronic physics, the observation of factorisable photon structure is similarly a fundamental test of the quantum field theory Quantum Chromodynamics (QCD). An overview of measurements of hadronic photon structure in e + $e-$ and electron-positron, *ep* interactions is presented, and comparison made with theoretical expectation, drawing on the essential features of photon fluctuation into quark and anti–quark in QCD.

Keywords: Dirac fermion, Electromagnetism, Gauge invariance, Grand unification, Superconducting

What we observe is not nature itself, but nature exposed to our method of questioning. - Werner Heisenberg

1. Introduction — Seeing the Action

In regards to the action of the electromagnetic field Photon structure has always been central, either implicitly or explicitly, to major steps in our understanding of the physics of short distances and time frames. The photon's influence in Particle Physics stems from the discovery of the positron, the development of QED, and the impact of quantum fluctuations on the behaviour of otherwise point–like quanta. Latterly this has provided a means of testing QCD by probing the spatially extended structure of the photon due to hadronic fluctuation. A simple picture emerges of a factorizable hadronic structure of the photon which is driven by quark–antiquark (qq) pair production and chromodynamics. The characteristic features of the structure function of the photon, F are a logarithmic growth with increasing factorisation scale and a relatively featureless and slightly rising dependence on the inelasticity variable, Bjorken-x. The former is attributable to the splitting function in QCD. The latter can be understood in terms of the competing influences of the rising x–dependence in perturbative QCD. This picture continues to pose a serious challenge to QCD because it relies on phenomenology and is thereby quantitatively incomplete. This challenge may well be met with the symbiosis of QCD theory and improving measurements of the structure of increasingly compact photons. Here the presence of two hard scales, that of the probe and that of the target, facilitates QCD calculation. This in turn may lead to new phenomena and insight as the short–distance properties of the QCD fluctuation of virtual photons are forced to ever smaller dimension and are probed with ever greater precision.

2. Seeing Beyond the Action?

Seeing the Action:
"Human beings must have action; and they will make it if they cannot find it." Albert Einstein

"I never worry about action, but only inaction." Winston Churchill

3. Degrees of Freedom of the Laws of Physics

- Fields
  - action at a distance
  - local gauge invariance (LGI)
  - quantization
  - flavour (isospin, hypercharge) / colour symmetry

- Dirac fermions
  - leptons + quarks
  - flavour + colour

- Spontaneously broken LGI ↔ renormalisation
  - leptons+quarks+bosons (S+V) + mass + (self-) coupling "Standard Model" of physics
  - rigorous basis for origin of physical law
  - 0.2 fm ≤ distance ↔ ~10^{27} m (visible)

4. How Things Unify

- **Symmetry**
  - upwards **!**
  - aligned identically "all the same"- symmetric uniform "order"

- **Symmetry**
  - not quite upwards - leaning **!** no preferred direction of lean
  - many Chesterfields**!** uniform disorder (degeneracy!)
  - equal possibility for any lean
  - symmetry

- **Asymmetry**
  - not quite upwards - leaning **!**
  - many Chesterfields**!**
  - one preferred direction of lean asymmetric, yet "all the same"
  - "order" with the asymmetric
  - order parameter←lean direction

- **Hidden symmetry**
  - not quite upwards - leaning **!**
  - many worlds of Chesterfields**!**
  - each world with one direction of lean, with order parameter
  - only one world, our Universe one order, the rest hidden
  - spontaneous choice of parameter

- **Manifest symmetry of field**
  - singlet scalar under transform: non-degenerate singlet coupling measured
  - multiplet under transform$^{\underline{n}}$: degenerate multiplet states observed reduced couplings + mixing measured

- **Hidden symmetry of field**
  - multiplet under transform$^{\underline{n}}$: degenerate one of multiplet of states = Universe + asymmetry
  - one of multiplet of states observed reduced couplings + mixing measured order parameter of property measured
  - Nature from within

5. Nature From Within

"We are in" "a cave, imprisoned by the limitations on the sorts of experiments we can do. In particular, we can study matter only at relatively low temperatures, where symmetries are likely to be spontaneously broken, so that nature does not appear very simple or unified. We have not been able to get out of this cave, but by looking long and hard at the shadows on the cave wall, we can at least make out the shapes of symmetries, which though broken, are exact principles governing all phenomena, expressions of the beauty of the world outside [1]." S. Weinberg

"We are in" "a **superconductor**, imprisoned by the limitations on the sorts of experiments we can do. In particular, we can study matter only **within our superconducting world**, where symmetries are likely to be spontaneously broken, so that **electromagnetism** does not appear **familiar**. We **are** not able to get out of this **superconductor**, but by **measuring carefully**, we can at least make out the shapes of symmetries, which though broken, are **manifest as relations between measureds which reveal the beauty of the theory of electromagnetism outside**."

With apologies to S Weinberg and with reference to his Nobel Lecture December 8th 1979 [1].

6. Nature From Within a Superconducting World

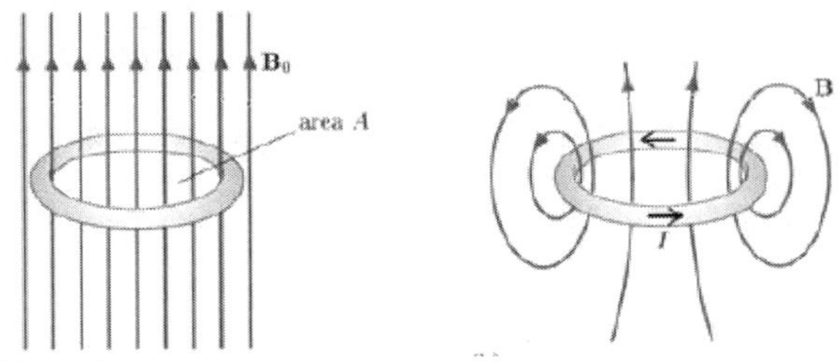

Figure 1.

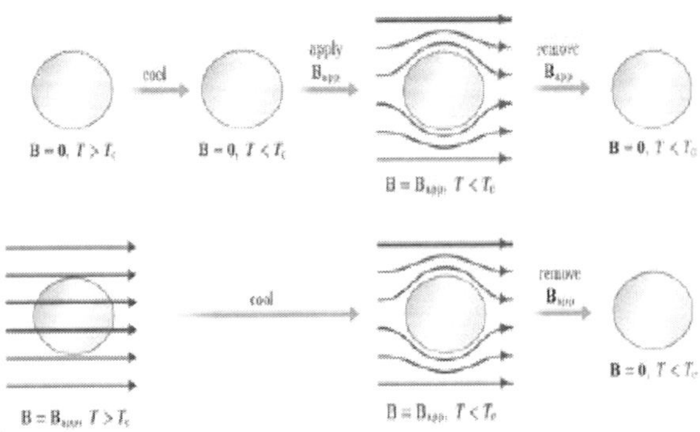

Figure 2.

- **Inside a superconductor**
  - super-currents (sc) zero resistivity
  - e^-e^- macroscopic coherence "order" with e^-e^- correlation >> lattice spacing
  - flux (**E B**) exclusion finite range electrodynamics

• **Inside a superconductor**
- super-currents (sc) zero resistivity
- e^-e^- macroscopic coherence "order" with e^-e^- correlation >> lattice spacing
- flux (**E B**) exclusion finite range electrodynamics
- energy gap Cooper-Fröhlich e^-e^- pair ground state Bose condensate
- Ginzburg-Landau long range order → scalar field
- scalar field "eaten"

$$\partial^\nu(\partial^\mu A_\nu - \partial^\nu A_\mu) = J_\mu$$

$$J_\mu = -\frac{\rho_{\rm em}}{m} A_\mu + \frac{\rho_{\rm sc}}{m}\partial_\mu \phi$$

$$A'_\mu = A_\mu - \frac{\rho_{\rm sc}}{\rho_{\rm em}}\partial_\mu \phi$$

$$\partial^\nu(\partial^\mu A'_\nu - \partial^\nu A'_\mu) + \frac{\rho_{\rm em}}{m} A'_\mu = 0$$

7. Nature From Within the Universe

• **Local gauge invariance (LGI) of EM vector field**
+ long range scalar field
+ spontaneous symmetry breaking (→ hidden)
finite range electrodynamics electrodynamics EM
+ finite range vector fields
+ finite range scalar field

• Gauge ↔ quantum phase triumph of the > ½ 20th century

• **LGI of massless, nonabelian, quantised vector fields, bosons**
+ mass-less, nonabelian, **quantised** scalar (Goldstone) fields, boson(s)
+ spontaneous symmetry breaking (→ hidden) EM with mass-less photon
+ massive vector bosons: Fermi scale @ 100 GeV
+ massive scalar boson(s) Higgs

• **Dirac fermion current (charge *q*): lepton**
+ quark GSW electroweak unification

8. Unified Electro-Weak

• **Perturbative approach → Feynman diagrams**
- electroweak vertices fermion-boson + boson-boson
- electroweak "splitting functions" boson-boson: probability initial on→ off + on-shell

neutral
charged currents
vector-vector
self-interaction
(non-abelian)
vector-scalar
interaction
(non-abelian)
Sub-atometric Horizon
← (logarithmically) towards unification … modestly

9. 'Un-unified' QCD

• chromo(colour)dynamics: The strong field
- perturbative approach → Feynman diagrams?

Matter
"Three quarks for Muster Mark!" - James Joyce Finnegans Wake.

"The Eightfold Way" $SU(3)_F$ M. Gell Mann, G Zweig.

10. Towards Grand Unification?

• **Super-gravity?**
"The discovery of an elementary spin-$^3/_2$ particle in the laboratory would be a triumph for super-gravity because the only consistent field theory for interacting spin-$^3/_2$ fields is supergravity. The remarkable thing is that a gauge symmetry between bosons and fermions can only be implemented in field theory if space-time is curved and hence of gravity is present [2]."

+ The Energy Frontier: Terascale

• **Beyond the Standard Model @ LHC?**

+ Scalar Field + Gravity

• **Higgs ↔ gravity?**
- mass in V+S fields gradient of V+S potentials → force magnitude of S potential → mass Higgs ↔ space time?
- Gravity @Atomic Scale

• **Precision→ gravitational coupling to (Cs) atoms?**
- change of gravitational PE → phase change (LGI)
- gravitational interferometry
- laser control of Cs atom state time in free-fall

• **Map**
Gravitational variation

• **Dark energy?**

Colour from within?

• **A missing piece: chromodynamics unified?**
- chirality in EW:
  lepton_L+antilepton_R
- chirality in QCD:
  $\text{quark}_{L=R}$+$\text{antiquark}_{L=R}$?
- (spontaneously?) broken chirality
  (Goldstone) scalar field/boson ?
  light (« eV) “axions” ?
  anomalous EM coupling?

CASCADE – SC cavity for HSP

• **Hidden sector axi/photons: sc cavities**
 - Q-factor ~10^9 (cf NC~10^6)

• **Cryo for good S/N**

• **Kinematic region 1 to 100 μeV**

• **ILC crab cavities + shielded cryostat**

• **Receiver under test**

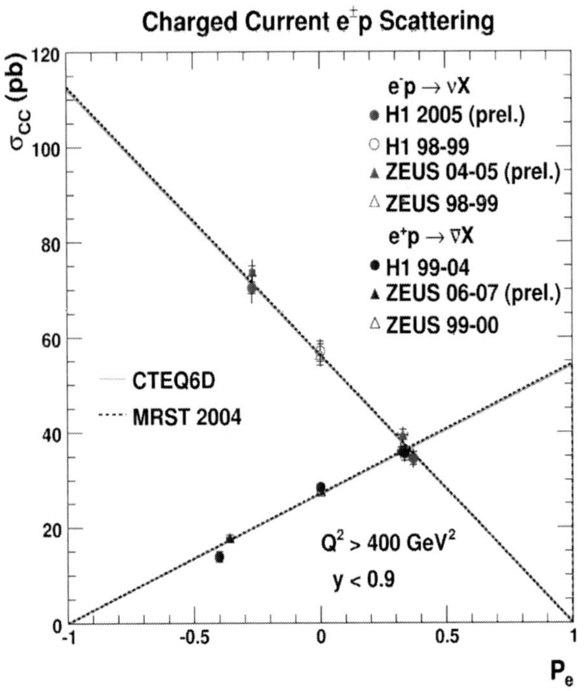

Figure 3.

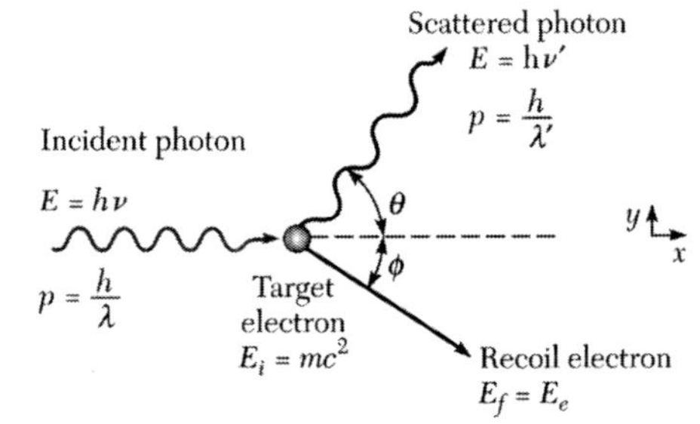

Figure 4.

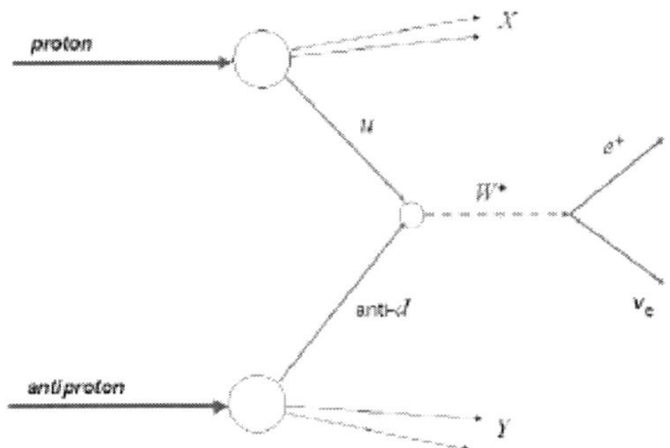

Figure 5.

We are where we are…

“We are in” “a cave, imprisoned by the limitations on the sorts of experiments we can do. In particular, we can study matter only at relatively low temperatures, where symmetries are likely to be spontaneously broken, so that Nature does not appear very simple or unified. We have not been able to get out of this cave, but by looking long and hard at the shadows on the cave wall, we can at least make out the shapes of symmetries, which though broken, are exact principles governing all phenomena, expressions of the beauty of the world outside [1].”

… and how we got here is a triumph of humanity, conceptually, coherently, and with amazing ingenuity.

11. Tomorrow? Seeing Beyond the Action?

• Perspective of “inside the Universe”
- Are answers to be found in new fields + symmetry?
  scalar fields? If not, conceptually how? Unitarity?
- What is the vacuum? What is nothing?
  How many “nothings”? How dark is nothing?
- What is mass? Where is gravity?
  Why is there, and what is, mass equivalence?
  How is space-time related to the Higgs field?
- Why three generations of quarks and leptons?
  Why quarks and why leptons? Where sits QCD?
  Are quarks and leptons fundamental?
  If not, do they have substructure?
- Can Dirac’s beautiful fermions have **sub**structure?

References and Notes

[1] Wheeler, J.A. (1979) Nobel Lecture December 8th.
[2] Van Nieuwenhuizen, P. (1981) Supergravity, Physics Reports, 68, 189-398.
[3] E Rutherford, Phil. Mag. 37 (1919) 537
[2] R Taylor, these proceedings
[3] D J Gross, F Wilczek, Phys. Rev. D8 (1973) 3633, Phys. Rev. D9 (1974) 980
[4] B Foster, these proceedings.
[5] R P Feynman "Photon Hadron Interactions" (1972), W Benjamin Inc., ISBN 0-805-
32510-7
[6] W E Lamb and R C Retherford, Phys. Rev. 72 (1947) 241
[7] B L Ioffe, Phys. Lett. 30B (1969) 123
[8] F Sciulli, these proceedings
[9] S J Maxfield, PhD Thesis, University of Maryland, 1984
[10] V M Budnev et al., Phys. Rep. 15C (1975) 181
26
[11] G Rossi, U. California at San Diego preprint UCSD-10P10-227 (1983), Phys. Rev. D29
(1984) 852
[12] V N Gribov, B L Ioffe, I Ya Pomeranchuk, Yadernaya Fiz. 2 (1965) 768
[13] W J Stirling, these proceedings.
[14] G T P Tarrant Proc. Roy. Soc. (London) A128 (1930) 345, A135 (1932) 223
[15] C Y Chao Proc. Natl. Acad. Sci. U.S. 16 (1930) 431
[16] C D Anderson S. H. Neddermeyer Phys. Rev. 43 (1933) 1034L
[17] J R Oppenheimer M. S. Plesset Phys. Rev. 44 (1933) 53L
[18] H A Bethe, W Heitler, Proc. Roy. Soc. (London) A145 (1934) 83
[19] OPAL Collab., G Abbiendi et al., Eur. Phys. Journal C11 (1999) 409
[20] PLUTO Collab., Ch Berger et al., Phys. Lett. 142B 111
[21] PLUTO Collab., Ch Berger et al., Nucl. Phys. B281 (1987) 365
PLUTO Collab., Ch Berger et al., Phys. Lett. B107 (1981) 168
JADE Collab., W Bartel et al., Z. Phys. C24 (1984) 231
TASSO Collab., M Althoff et al., Z. Phys. C24 (1984) 231
TPC/2γ Collab., H Aihara et al., Phys. Rev. Lett. 58 (1987) 97
TPC/2γ Collab., H Aihara et al., Z. Phys. C34 (1987) 1
[22] E Witten, Nucl. Phys. B120 (1977) 189
[23] T F Walsh, P M Zerwas, Nucl. Phys. B41 (1972) 551
R L Kingsley, Nucl. Phys. B60 (1973) 45
C H Llewellyn Smith, Phys. Lett. 79B (1978) 83
R J DeWitt et al., Phys. Rev. D19 (1979) 2046
W R Frazer, J F Gunion, Phys. Rev.. D20 (1979) 147
C T Hill, G G Ross, Nucl. Phys. B148 (1979) 373
C Petersen, T F Walsh, P M Zerwas, Nucl. Phys. B174 (1980) 424
[24] J J Sakurai, Phys. Rev. Lett. 22 (1969) 981
J J Sakurai, D Schildknecht, Phys. Lett. 40B (1972) 121
[25] J H Field, F Kapusta, L Poggioli, Phys. Lett. B181 (1986) 362, Z. Phys. C36 (1987) 121
F Kapusta, Z. Phys. C42 (1989) 225
[26] M Erdmann, Phys. Lett. B488/2 (2000) 131
[27] OPAL Collab., R. Akers et al., Z. Phys. C61 (1994) 199
[28] L3 Collab., M. Acciarri et al., Phys. Lett. B483 (2000) 373
[29] DELPHI Collab., P. Abreu et al., Phys. Lett. B342 (1995) 402
DELPHI Collab., P. Abreu et al., Zeit. Phys. C69 (1996) 223
27
[30] AMY Collab., T. Sasaki et al., Phys. Lett. B252 (1990) 491
AMY Collab., B. J. Kim et al., Phys. Lett. B325 (1994) 248;
AMY Collab., Phys. Lett. B400 (1997) 395
[31] TOPAZ Collab., H. Hayashii et al., Phys. Lett. B314 (1993) 149
TOPAZ Collab., K. Muramatsu et al., Phys. Lett. B332 (1994) 477
[32] ALEPH Collab., D. Buskulic et al., Phys. Lett. B313 (1993) 509
ALEPH Collab., R.Barate et al., Physics Letters B 458 (1999) 152
[33] Yu L Dokshitzer, JETP 46 (1977) 641
V N Gribov and L N Lipatov, Sov. Journ. Nucl. Phys. 15 (1972) 78
G Altarelli and G Parisi, Nucl. Phys. B126 (1977) 298
[34] M Gl¨uck, E Reya, A Vogt, Z. Phys. C53 (1992) 127
H Abramowicz, K Charchula, A Levy, Phys. Lett. B269 (1991) 458
[35] M Gl¨uck, E Reya, A Vogt, Phys. Rev. D46 (1992) 1973
[36] M. Gl¨uck, E. Reya and A. Vogt, Phys. Rev. D45 (1992) 3986
[37] PLUTO Collab., Ch Berger et al., Phys. Lett. 142B 119
[38] T Uematsu, T F Walsh, Phys. Lett. B101 (1981) 263
T Uematsu, T F Walsh, Nucl. Phys. B199 (1982) 93
[39] H E Montgomery, these proceedings
[40] J Chyla, Phys. Lett. B320 (1994) 186
[41] H1 Collaboration, C Adloff et al., Eur. Phys. J. C1 (1998) 97
[42] H1 Collab., C Adloff et al., Eur. Phys. J. C13 (2000) 397
[43] G Marchesini et al., Comp. Phys. Comm. 67 (1992) 465
[44] T Sj¨ostrand, CERN-TH-6488 (1992), Comput. Phys. Commun. 82 (1994) 74
[45] M Klasen, G Kramer and B P¨otter, Eur. Phys. J. C1 (1998) 261
[46] L E Gordon, J K Storrow, ANL-HEP-PR-96-33, hep-ph-9607370 (1996)
[47] B. V. Combridge, C. J. Maxwell, Nucl. Phys. B239 (1984) 429
[48] S Geer, Proc. Yale Advanced Study Institute (Singapore World Scientific) 2 (1986)
[49] M Gl¨uck, E Reya, M Stratmann, Phys. Rev. D54 (1996) 5515
[50] F. Borzumati and G. Schuler, Z. Phys. C58 (1993) 139
[51] S. Frixione, M. Mangano, P. Nason and G. Ridolfi, Phys. Lett. B319 (1993) 339, 28
[52] H1 Collab., C Adloff et al., Phys. Lett. B483 (2000) 36
H1 Collab., C Adloff et al., Eur. Phys. J. C10 (1999) 363
H1 Collab., T Ahmed et al., Nucl. Phys. B445 (1995) 195.
[53] H Geiger, E Marsden, Proc. Roy. Soc. (1909) 495

Evidencing 'Tight Bound States' in the Hydrogen Atom: Empirical Manipulation of Large-Scale XD in Violation of QED

RICHARD L. AMOROSO

Noetic Advanced Studies Institute,
Escalante Desert, Beryl, UT 84714-5104 USA
amoroso@noeticadvancedstudies.us; www.noeticadvancedstudies.us

JEAN-PIERRE VIGIER (Posthumous)

Department of Theoretical Physics
Empyrean Mobile, Kolob, Kokauvim
vigier@noeticadvancedstudies.us

In this work we extend Vigier's recent theory of 'tight bound state' (TBS) physics and propose empirical protocols to test not only for their putative existence, but also that their existence if demonstrated provides the 1st empirical evidence of string theory because it occurs in the context of large-scale extra dimensionality (LSXD) cast in a unique M-Theoretic vacuum corresponding to the new Holographic Anthropic Multiverse (HAM) cosmological paradigm. Physicists generally consider spacetime as a stochastic foam containing a zero-point field (ZPF) from which virtual particles restricted by the quantum uncertainty principle (to the Planck time) wink in and out of existence. According to the extended de Broglie-Bohm-Vigier causal stochastic interpretation of quantum theory spacetime and the matter embedded within it is created annihilated and recreated as a virtual locus of reality with a continuous quantum evolution (de Broglie matter waves) governed by a pilot wave - a 'super quantum potential' extended in HAM cosmology to be synonymous with the a 'force of coherence' inherent in the Unified Field, U_F. We consider this backcloth to be a covariant polarized vacuum of the (generally ignored by contemporary physicists) Dirac type. We discuss open questions of the physics of point particles (fermionic nilpotent singularities). We propose a new set of experiments to test for TBS in a Dirac covariant polarized vacuum LSXD hyperspace suggestive of a recently tested special case of the Lorentz Transformation put forth by Kowalski and Vigier. These protocols reach far beyond the recent battery of atomic spectral violations of QED performed through NIST.

Keywords: Bohr orbit, Dirac vacuum, Large scale extra dimensions, M-Theory, Multiverse, Tight bound states, QED Violation

1. Introduction — Tight-Bound-State Modeling

Recently Tight Bound States (TBS) due to electromagnetic interactions at small distances below the lowest Bohr orbit have been postulated for the Hydrogen atom [1, 2]. We begin summarizing this seminal work of Vigier - In the usual understanding of atomic physics spin-orbit and spin-spin coupling perturbations for example give rise to only small corrections to classic Bohr energy levels. However with distances in the $1/r^3$ and $1/r^4$ range these interaction terms, until now overlooked, can be much higher than the Coulomb term at distances much less than the Bohr radius - predicting new physics [1, 2]. Corben [3] was first to notice that motion of a point charge in a magnetic dipole field at rest is highly relativistic with orbits of nuclear dimensions. Further investigation undertaken by [1, 2, 4] in a model representing an extension of the Pauli equation to a two-body system as defined by the Hamiltonian

$$H = \frac{1}{2m_1}\left(P_1 - e_1 A(r_1)\right)^2 + \frac{1}{2m_2}\left(P_2 - e_2 A(r_2)\right)^2 + \frac{1}{4\pi\varepsilon_0}\frac{e_1 e_2}{|r_1 - r_2|} + V_{dd} \tag{1}$$

where, m_1 is the mass, $\boldsymbol{P}_1$ the momentum, e_1 the charge, $\boldsymbol{r}_1$ the position of the particles (i = 1,2), $\boldsymbol{A}$ is electromagnetic vector potential and V_{dd} the dipole-dipole interaction term:

$$V_{dd} = -\left(\frac{\mu_0}{4\pi}\right)\mu_1\mu_2\delta(r_1 - r_2) + \left(\frac{\mu_0}{4\pi}\right)\left[\frac{\mu_1\mu_2}{|r_1 - r_2|^3} - \frac{3\left[\mu_1(r_1 - r_2)\right]\cdot\left[\mu_2(r_1 - r_2)\right]}{|r_1 - r_2|^5}\right] \tag{2}$$

.

In the center-of-mass frame and with a normal magnetic moment, $\mu = (e/m)s$ the Hamiltonian (1) becomes:

$$\begin{aligned} H = & \frac{1}{2m_1}p^2 - \left(\frac{\mu_0}{4\pi}\right)\frac{e_1e_2}{m_1m_2}\frac{SL}{r^3} \\ & + \left(\frac{\mu_0}{4\pi}\right)^2 \frac{e_1^2 e_2^2 \hbar^2}{4m_1m_2m}\frac{1}{r^4} \\ & + \frac{1}{4\pi\varepsilon_0}\frac{e_1e_2}{r} - \left(\frac{\mu_0}{4\pi}\right)\frac{e_1e_2}{m_1m_2}s_1s_2\delta(r) \\ & + \left(\frac{\mu_0}{4\pi}\right)\frac{e_1e_2}{m_1m_2}\left[\frac{s_1s_2}{r^3} - \frac{3(s_1r)\cdot(s_2r)}{r^5}\right] \end{aligned} \quad . \quad (3)$$

Continuing to follow Vigier [1], the possibility of TBS physics as derived from Hamiltonian (3) is shown in simplified form when limited to spherically symmetric terms by the radial Schrödinger equation:

$$\frac{d^2X}{dr^2} + \frac{2m}{\hbar^2}\left[E - V(r)\right]X = 0 \qquad (4)$$

and contains a form for the effective potential in the inverse power law:

$$V(r) = \frac{A}{r^4} + \frac{B}{r^3} + \frac{C}{r^2} + \frac{D}{r}. \qquad (5)$$

At large distances this potential is an attractive Coulomb tail with a repulsive core at small distances due to the A/r^4 term [1]. For proper values of potential (5) its coefficients could have another potential well in addition to the one at distances of the order of the Bohr radius where new physics is suggested to be 'located'. Additional theoretical details on the seminal development of TBS by Vigier can be found in [1-4].

2. Ongoing Trouble With Point Particles and Singularities

In the standard model of particle physics an elementary particle has no known composite subparticles. A mathematical idealization of an elementary particle is often called a point particle or point charge that lacks spatial extension (0D) which perhaps arose because in a mathematical coordinate context its size can be considered to be irrelevant. The nature of a point particle has remained an open question in physics. Recent *avant-garde* work by Rowlands (this volume) has extended our understanding of this conundrum:

> Physics at the fundamental level can be effectively reduced to an explanation of the structures and interactions of fermions. Fermions appear to be singularities rather than extended objects, but there is no obvious way of creating such structures within the 3-dimensional space of observation. However, the algebra associated with the Dirac equation appears to suggest that the fermion requires a double, rather than a single, vector space, and this would seem to be confirmed by the double rotation required by spin ½ objects, and the associated effects of *zitterbewegung* and Berry phase shift. Further investigation of the second 'space' reveals that it is, in effect, an 'antispace', which contains the same information as real space but in a less accessible form. The two spaces effectively cancel to produce a norm 0 (nilpotent) object which has exactly the mathematical structure required to be a fermionic singularity [5].

Continuing to follow Rowlands we further note that the fermion as a singularity exists in its own multiply-connected space requiring a double rotation to return to its starting position. Fermions also undergo the quantum process of *zitterbewegung* continually switching between real space and complex vacuum space. The double circuit in real space is required because a fermion only exists in this space for half its existence. It is not coincidental that fermion algebra (gamma matrices) requires a commutative combination of two vector spaces for a full mathematical representation. Thus it becomes obvious that constructing a physical 'singularity' requires a dual space [5,6].

While the Rowlands' nilpotent space-antispace model brilliantly extends our understanding of the nature of a fermionic singularity in terms of the standard model, elegant quaternionic algebra is not necessarily tantamount to a penultimate description of nature. What we mean by that barb is even though the theoretical elements of Rowlands' model are *avant garde* to the standard model they are not sufficiently radical to satisfy the needs of HAM cosmology [7,8] and the associated Noetic Field Theory (NFT): The Quantization of Mind [9-11] that requires Unified Field Mechanics, U_F [12]; but it does provides an inspired basis for making correspondence to the profoundly unique 'singularity' under development in HAM cosmology [7-13].

Before we define the HAM singularity let's briefly review some discrepancies in contemporary theory: Interactions of extended objects can appear point-like. A spherical object in Euclidean space described by the inverse square law can behave like its mass was concentrated in a geometric center. According to Coulomb's law the electric field associated with a classical point charge increases to infinity as the distance from the point charge decreases towards zero making energy-mass of point charge infinite. In Newtonian gravity and classical electromagnetism a field outside a spherical object is identical to a point particle located at the center of the sphere. In quantum mechanics, the nature of a point particle is complicated by the Heisenberg uncertainty principle where neither elementary nor composite particles are spatially localized, i.e. elementary particles with no internal structure occupy a nonzero volume. A point charge is an idealized model of a particle with no dimensions. However the particle wavepacket always occupies a nonzero volume. For example the electron is an elementary particle, but its quantum states form 3D patterns. Good reason remains to call an elementary particle a point particle. Even if an elementary particle has a delocalized wavepacket, the wavepacket is in a quantum superposition of quantum states localizing the particle. For example, a 1*s* electron in a hydrogen atom occupies a volume of $\sim 10^{-30}$ m^3.

3. The Nature of Space

Vigier postulated a 4D model of TBS in hydrogen extending QED [1,2]. The theory seemed incomplete (see forward this volume) until it was extended to a 12D M-Theoretic model with correspondence to the newly postulated LSXD regime of the Unified Field, U_F [12,14] in conjunction with HAM cosmology [7,8]. It remains difficult to know the fundamental polyhedron (topology/geometry) of our cosmology made from observations within it; and astrophysicists continue to struggle with this problem [15-17]. Preliminary data from WMAP has supported an Anti de Sitter ($AdS_5 \times S^5$) Poincaré Dodecahedral 'wrap-around' model but more precise Planck satellite data will take several more years as the 2013 data released was not taken at the required frequencies.

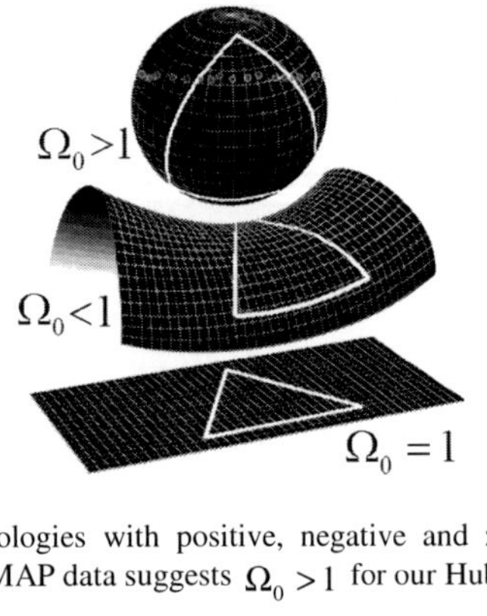

Figure 1. Topologies with positive, negative and zero curvature. Preliminary WMAP data suggests $\Omega_0 > 1$ for our Hubble universe.

For example space appears infinitely flat or Euclidean, but there are numerous observationally flat topologies that are not actually flat. The simplest shape for our reality is a 3-sphere with positive curvature. When we draw tori they appear curved only because we embed them in 3-space. A flat torus can be made from a square with its edges wrapped around to join seamlessly. The criterion is that the angles of a triangle add up to 180°. If our observed Hubble sphere, H_R is a flat torus observations in certain directions allow one to see oneself in the distance [15-17].

This is compatible with the topology of a Big Bang 3-sphere postulated to be like a 3D expanding balloon; one with a 12 to 15 billion light year radius where light hasn't had sufficient time to cross yet. If the symmetry of curvature is not broken there are three possibilities - 3-sphere topologies with positive, negative and zero curvature as in Fig. 1. But a 3D flat torus (cube with opposite faces joined) has zero curvature. There are also hyperbolic 3D spaces with negative curvature. In a finite 3-space like the H_R purported self-observations in various directions would allow us to work out the curvature and shape of our H_R which is considered closed and finite in HAM cosmology; but open and infinite in the LSXD it is embedded in.

The spatial region observed from any local point is a circular disk that increases in size with time. When it grows to the same size as the H_R it begins to overlap. At this moment because of relativity it would be possible to observe the same object in many directions in the overlapping portions of space. Age of the universe predictions form Big Bang cosmology suggest that if the H_R is a 3-sphere overlapping should have begun and that these overlaps would form large circles in the sky, circles because the intersection of two 3-sphers is a circle [18]. Recent WMAP and Planck satellite observations have observed these putative circles in initial support of Poincaré Dodecahedral Space (PDS) [19-21].

3.1 *New Background Conditions of a Dirac Vacuum*

If one assumes in conjunction with the de Broglie-Bohm-Vigier Causal Stochastic Interpretation (CSI) of quantum theory [22] that de Broglie matter-waves describe a wave-particle duality built up with real extended space structures with internal oscillations of particle-like spin, it is possible to justify Bohr's physical assumptions and predict new properties of a real Dirac covariant polarized vacuum [23-26]. Bohr's major contribution to modern physics was the model of photon emission-absorption in Hydrogen in terms of random energy jumps between stable quantum states and atomic nuclei. This discovery was one of the starting points for the Copenhagen Interpretation of quantum theory. We suggest this structural-phenomenology by general covariance applies equally as well to the symmetry conditions of a LSXD Dirac vacuum backcloth also; but as one knows the purely random description of quantum jumps suggested by Bohr is obviated by CSI quantum mechanics [22,26,27] suggesting this interaction is piloted. We feel the CSI model is required for our LCU exciplex gating mechanism to work because it is the internal motion of a massive photon, m_γ that enables (periodic) coupling to the polarized Dirac vacuum and the rf-resonance hierarchy to operate [29,30].

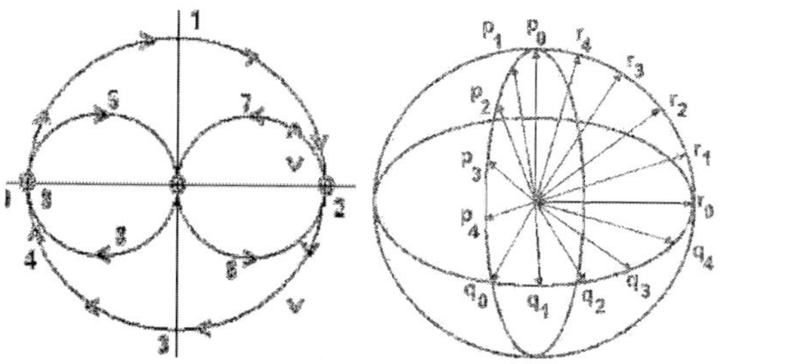

Figure 2 a) 2D simplistic view of 3D Dirac rotation map. b) 2D rendition of 4D view of Dirac hyperspherical rotation for raising and lowering Dirac-type topological advanced-retarded annihilation-creation vectors.

Some experimental evidence has been found to support this view showing the possibility that the interaction of these extended structures in space involve real physical vacuum couplings by resonance with the subquantum Dirac ether. Because of photon mass the CSI model, any causal description implies that for photons carrying energy and momentum one must add to the restoring force of the harmonic oscillator an additional radiation (decelerating) resistance derived from the em (force) field of the emitted photon by the action-equal-reaction law.

Kowalski has shown that emission and absorption between atomic states take place within a time interval equal to one period of the emitted or absorbed photon wave. The corresponding transition time corresponds to the time required to travel one full orbit around the nucleus [31,32]. Individual photons are extended spacetime structures containing two opposite point-like charges rotating at a velocity near c, at the opposite sides of a rotating diameter with a mass, m $10^{-65}g$; and with an internal oscillation $E = mc^2 = h\nu$. Thus a new causal description implies the addition of a new component to the Coulomb force acting randomly and may be related to quantum fluctuations. We believe this new relationship has some significance for our model of vacuum C-QED blackbody absorption/ emission equilibrium [33]. The result from real causal interactions between the perturbed local background "ether" and its apparently independent moving collective perturbations imply absolute total local momentum and angular momentum conservation.

4. Pertinence of the New Cosmological Paradigm

Utilizing the earlier hadronic form of string theory based on the original Stoney (precursor to $\hbar$) instead of Planck's constant the quaternionic Fermionic 'singularity' is allowed to oscillate from a virtual Plank radius (asymptote never reached) to the Lamoure radius of the hydrogen atom. Fortunately for parameters of HAM cosmology this scenario provides a much better footing for understanding the cyclic operation of tight bound states in the Hydrogen atom. We will show how a new form of the Lorentz transformation applicable to our TBS experimental protocol requires these oscillations.

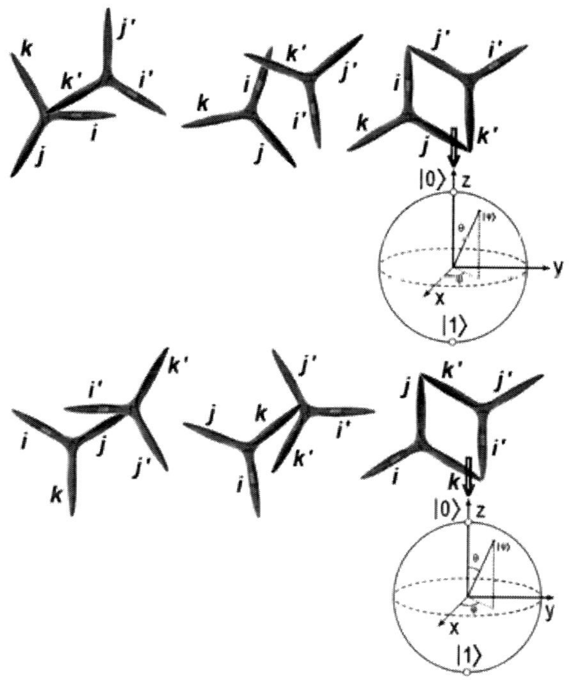

Figure 3. Alternative rendition of Fig. 4 in quaternionic form. Locus of HD mirror symmetric Calabi-Yau 3-tori here depicted as dual quaternion trefoil knots spinning relativistically and evolving quantum mechanically in time. Nodes in the cycle are sometimes chaotic (degenerate) and sometimes periodically couple coherently into resultant quantum states in Euclidean 3-space depicted in the figure as faces of a 3-cube that reduce further to the Riemann Bloch 2-sphere at the bottom.

This horrendous concatenation occurs as the basis for the arrow of time because our temporal virtual reality surfs as it were of the face of the LSXD eternal realm hidden behind it. (The discrete frames of spacetime film in Fig. 3 producing the continuous virtual image of reality relativistically on the screen) These parameters are essential to the LCU as an exciplex gating mechanism - This is how each point (LCU open-closed singularity - like a rotating light house beacon) is created in the temporal locus and allows the U_F's 'force of coherence' to modulate complex self-organized living systems (SOLS) as the organizing principle itself and likewise mediate the physical basis of qualia [9-11]. We will do our best to clarify this scenario in the text and figures in sections following.

We are not yet finished outlining the required battery of new physical parameters; recall that NFT represents a whole paradigm shift. (why it hasn't been easy this past 1,000 years) WE MUST also utilize the parameters of another well established and generally ignored aspect of contemporary physics called Extended Electromagnetic Theory [34-36] in conjunction with a covariant Dirac polarized vacuum (ignored also) [23-26]; both because physicists erroneously believe they conflict with Gauge Theory which has been eminently successful for decades. Now we are finally be set up with enough parameters to putatively manipulate the spacetime backcloth (Einstein energy dependent spacetime metric) [37].

Metaphorically if one throws a stone in water concentric ripples occur. If one throws two stones regions of destructive and constructive interference occur. We will utilize an M-Theoretic Calabi-Yau symmetric version of this model to set up an rf-pulsed spacetime resonance hierarchy to access the 'hidden' regime of the U_F [12]. In the next series of several figures (Figs. 5-8) we will attempt to clarify the continuous-state structure of LSXD as it applies to HAM cosmology.

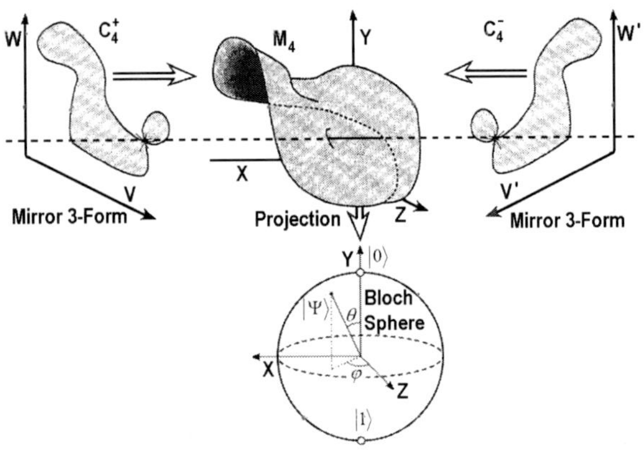

Figure 4. Complex Calabi-Yau mirror symmetric 3-form potentia, C_4 become an embedded quantum resultant in Minkowski space, M_4. This resultant projection entails a continuous quantum state evolution represented as a Bloch 2-Sphere. This represents the lower portion that embeds in local spacetime; there is an additional 5D dodecahedral duality above this projection embedded in the infinite potentia of the U_F from which it arises.

5. Required New Particle Physics — Cosmology

In contrast to the standard model of particle physics and cosmology we utilize a continuous-state [7] hypothesis and follow an extended form of the de Broglie-Bohm causal interpretation of QT in conjunction with Cramer's Transactional Interpretation where matter is created - annihilated - and recreated with a beat frequency in conjunction with a 'least cosmological unit' (LCU) tiling the spacetime backcloth [7,13].

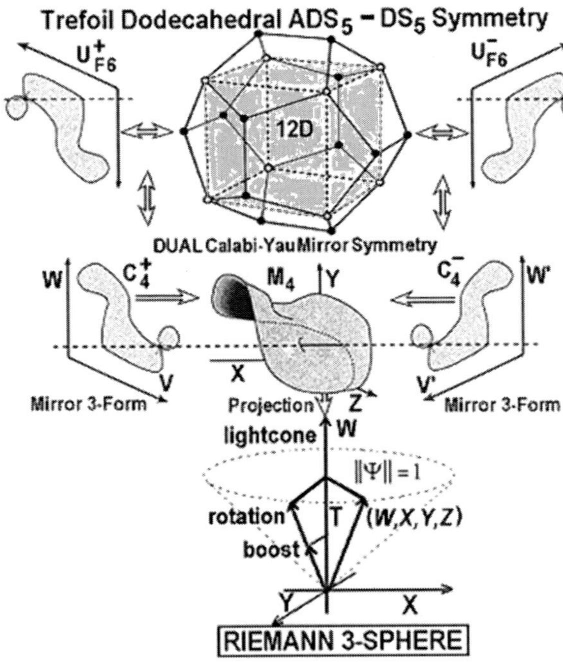

Figure 5. Complete LSXD regime of HAM cosmology illustrating the hierarchy of its geometric topology. Dodecahedral involute properties, as well as the continuous-state exciplex 'hysteresis loop' of noeon injection not shown. It represents a unique M-Theoretic model of 'Continuous-State' U_F dynamics as it relates to NFT and its putative exchange quanta of the U_F called the noeon.

This is a nilpotent condition, $Z \cdot Z' = 0$ [5,6] where matter and the spacetime in which it is embedded arise from an HD infinite potentia. Some GUT theories postulate proton decay with a half-life of 10^{36} years according to $p \to e^+ + \pi^0;\ \pi^0 \to 2\gamma$. The conundrum is that this lends some support to the Copenhagen/Big Bang scenario of nucleosynthesis occurring near the time of the putative original singularity. But the lifetime of a proton is not the concern; it is its true quantum nature in terms of a continuous-state M-Theoretic Calabi-Yau mirror symmetric relativistic topological quantum field theory.

However what we need to do is extend the de Broglie-Bohm point of view that suggests matter is a form of HD complex standing-wave that is continuously annihilated and recreated with the quantum wave function piloted by a unitary anthropic action principle tantamount to a 'super-quantum potential.

In addition to the spacetime vacuum being considered as a form of a Dirac covariant polarized vacuum, this framework has parameters of hyperspherical rotation (condition usually described by the Dirac equation for properties of the electron) relative to the topology of the HD Calabi-Yau mirror symmetry or dual 3-torus. In terms of the continuous-state hypothesis and large scale XD a 'cootie catcher'-like gating mechanism is an inherent structure mediating local 3-space and HD space transformations that constitute a varying spacetime cavity volume. The putative existence of this hysteresis loop volume is suggested as an empirical proof of TBS and LSXD.

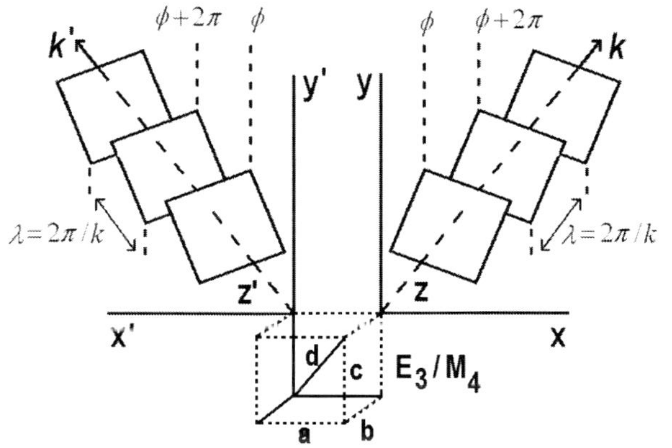

Figure 6. Calabi-Yau future-past mirror symmetry potentia shown as tiered surfaces (additional dimensionality suppressed) of constant phase, in this case to represent cyclic components of evenly spaced orthogonal standing reality waves with the E_3/M_4 cubic resultant localized at the bottom. The resultant locked in 4D by the uncertainty principle. The resultant Euclidean cube is locked in 4D by the uncertainty principle keeping the HD parameters inaccessible to the empirical tools available to the Copenhagen framework. k & k' topologically infolds into an HD continuous-state torus.

The noeon mediating the U_F does not imply the usual phenomenological exchange of energy as in a standard field interaction such as the photon of the electromagnetic field; but constitutes an ontological exchange (without energy transfer). This is achieved by a process called 'topological switching' and implies instead a 'force of coherence' inherent in the action of the U_F. This process also allows the quantum uncertainty principle defined by the Copenhagen Interpretation of quantum theory to be surmounted [7]. Figure 5 illustrates a nilpotent continuous-state regime cycling from a 12D dodecahedral de Sitter space [15-20] through an intermediate Calabi-Yau mirror symmetry to a virtual 'standing-wave' nilpotent 3D Euclidean space resultant. Compare Figs. 3 and 4.

We extend Vigier's original model of TBS in hydrogen [1,2] to include a unique 12D M-Theoretic perspective [7] with a Calabi-Yau : Dodecahedral involute mirror symmetry in Continuous-State HD space [8,11] elevating Cramer's Transactional model and wave-particle duality to principles of cosmology. The 4-Space nilpotent resultant of the E_3/M_4 virtual present which is a standing-wave of the future-past is shown in Figs. 2-9. $\hat{M}_4$ being 'locked' into place cyclically by the uncertainty principle [7].

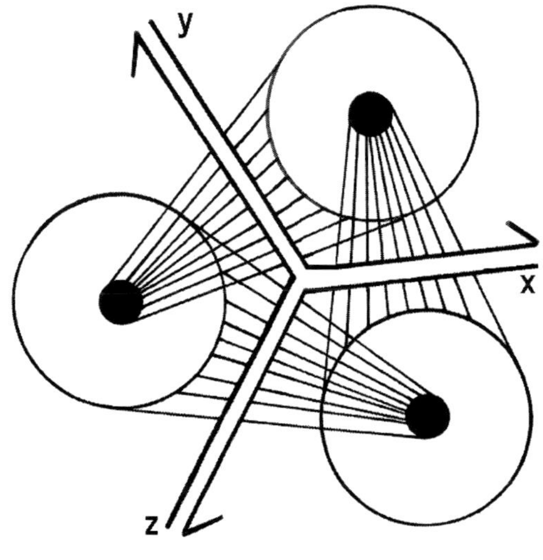

Figure 7. Triune structure of a solitary least unit that like an isolated quark does not exist in nature. The central parallel lines are the Witten string vertex with properties of a complex Riemann sphere able to continuously rotate from zero to infinity. The field lines represent the 'super quantum potential' of the unified field, U_F.

The HD are not curled up at the Planck scale because they are invisible; they are Large-scale XD (LSXD) because of subtractive interferometry as it were of the C_4^+ - M_4 - C_4^- standing-wave modes that operates like a movie theatre where discrete frames of film moving through the projector at a few cm/sec appear continuous on the screen. For our virtual reality exchange quanta of the U_F is relativistically 'pumped' through discrete holographic-like LCU tiling the raster of spacetime to produce virtual images of a Minkowski space present. Behind the virtual veil is a continuous-state cycle from $0 \Leftrightarrow \infty$ as shown in Fig. 3.

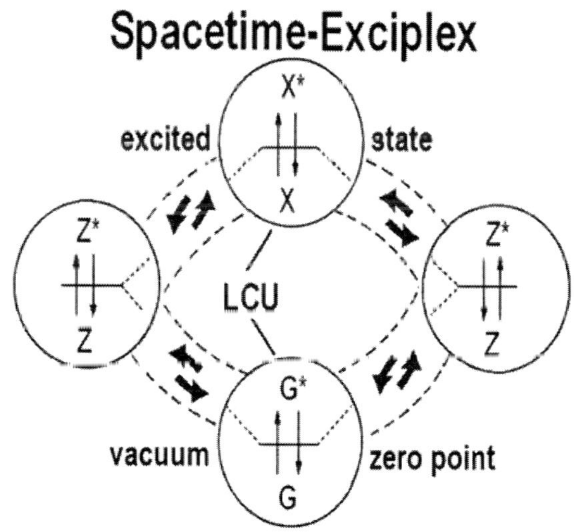

Figure 8. Least Unit (LCU) Exciplex Composite. The spacetime exciplex complex is comprised of an array of least cosmological units that act as a brane topology gating mechanism for entry of unified field control parameters to operate on Minkowski 4-space.

5.1 *Possibility of Cavity QED Emission From Continuous Spacetime Compactification*

Exciplex properties of spacetime and matter also suggest that further development of the C-QED model of CMBR emission could be extended to include spontaneous emission from the continuous dimensional reduction process of compactification. This would follow from modeling spacetime cavity dynamics in a manner similar to that in atomic theory for Bohr orbitals. As well-known photon emission results from electromagnetic dipole oscillations in boundary transitions of atomic Bohr orbitals. Bohr's quantization of atomic energy levels is applied to the topology of Spacetime C-QED boundary conditions in accordance with equation (37) where spacetime QED cavities of energy, E_i undergo continuous harmonic transition to a higher state, $E_j\left(> E_{iH}\right)$ (redshift-absorption mode).

The general equations for a putative LCU spacetime exciplex are:

$$G^* + G^* \Leftrightarrow Z^*; \quad Z^* + m_\gamma \Leftrightarrow X^*$$

$$X^* - m_\gamma \xrightarrow{emission} Z^* \quad or \quad G^* \tag{37}$$

$$X^* + m_\gamma \rightarrow Z^* \quad or \quad G^*$$

where G is the ZPF ground, Z black body cavity excited states and X the spacetime C-QED exciplex coupling. The numerous configurations plus the large variety of photon frequencies absorbed allow for a full black body absorption-emission equilibrium spectrum. We believe the spacetime exciplex model also has sufficient parameters to allow for the spontaneous emission of protons by a process similar to the photoelectric effect but from spacetime C-QED spallation rather than from metallic surfaces.

A torus is generated by rotating a circle about an extended line in its plane where the circles become a continuous ring. According to the equation for a torus, $\left[\left(\sqrt{x^2+y^2}\right)-R\right]^2 + z^2 = r^2$, where r is the radius of the rotating circle and R is the distance between the center of the circle and the axis of rotation. The volume of the torus is $2\pi^2 Rr^2$ and the surface area is $4\pi^2 Rr$, in the above Cartesian formula the z axis is the axis of rotation.

Electron charged particle spherical domains fill the toroidal volume of the atomic orbit by their wave motion. If a photon of specific quanta is emitted while an electron is resident in an upper more excited Bohr orbit, the radius of the orbit drops back down to the next lower energy level decreasing the volume of the torus in the emission process.

We suggest that these toroidal orbital domains have properties similar to QED cavities and apply this structure to *topological switching* during dimensional reduction in the continuous-state universe (HAM) model [7]. To summarize pertinent aspects of HAM cosmology:

- Compactification did not occur immediately after a big bang singularity, but is a continuous process of dimensional reduction by *topological switching* in view of the Wheeler-Feynman absorber model where the present is continuously recreated out of the *future-past*. Singularities in the HAM are not point like, but dynamic wormhole like objects able to translate extension, time and energy.
- The higher or compactified dimensions are not a subspace of our Minkowski 3(4)D reality, but our reality is a subspace of a higher 12D multiverse of three 3(4)D Minkowski spacetime packages.

During the spin-exchange process of dimensional reduction by topological switching two things pertinent to the discussion at hand:

- There is a transmutation of dimensional form from *extension to time to energy*; in a sense like squeezing out a sponge as the current Minkowski spacetime package recedes into the past down to the Planck scale ; or like an accordion in terms of the *future-past* recreating the present.
- A tension in this process (string tension, T_0 in superstring theory) allows only specific loci or pathways to the dimensional reduction process during creation of the transient Planck scale domain. Even though there are discrete aspects to this process it appears continuous from the macroscopic level (like the film of a movie); the dynamics of which are like a harmonic oscillator.

With the brief outline of HAM parameters in mind, the theory proposes that at specific modes in the periodicity of the Planck scale pinch effect, cavities of specific volume reminiscent of Bohr toroidal atomic orbits occur. It is proposed rather speculatively at present that these cavities, when energized by stochastically driven modes in the Dirac ether or during the *torque moment* of excess energy during the continuous-state compactification process, or a combination of the two as in standard C-QED theory of Rabi/Rydberg spontaneous emission, microwave photons of the CMBR type could be emitted spontaneously from the vacuum during *exciplex* torque moments. This obviously suggests that Bohr atomic orbital state reduction is not the only process of photon emission; (or spacetime modes are more fundamental) but that the process is also possible within toroidal boundary conditions in spacetime itself when in a phase-locked mode acting like an atomic volume. A conceptualization of a Planck scale cavity during photon emission is represented in Figs. 8 and 9 with nine dimensions suppressed.

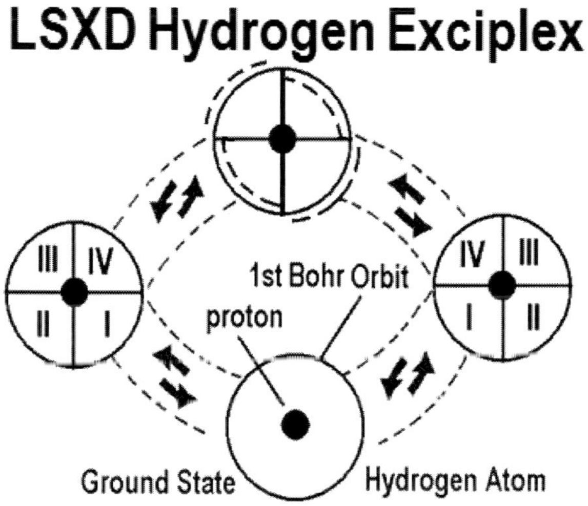

Figure 9. LSXD Exciplex complex with conformal scale-invariant properties revealing operation of TBS cyclicality in the hydrogen atom mediated by the fluctuating form of Planck's constant varying from asymptotic virtual Planck to the Larmor radius of the hydrogen atom. This is a variable cavity-QED representation whereby new spectral lines will appear at various periodic nodes in the continuous-state LSXD cycle. Nonlocal/HD parts not drawn to scale.

In early spectroscopy the orbital series associated with Rydberg states was proportional to the difference between the two terms of an energy level transition which became known as **s**harp, **p**rinciple, **d**iffuse and **F**undamental, so the designators *s, p, d, f* were used to represent orbital angular momentum states of an atom.

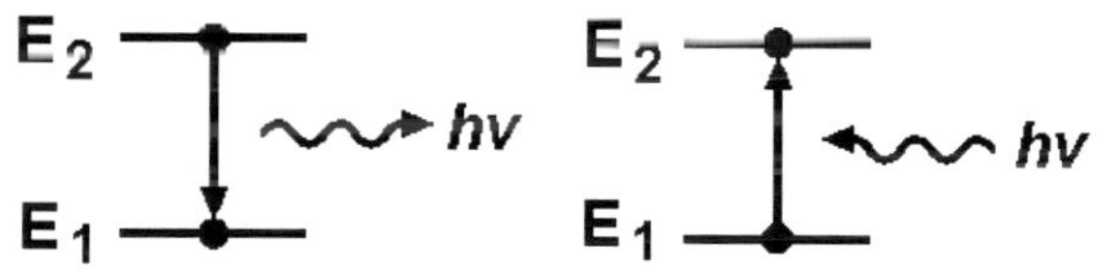

Figure 10. Usual geometric consideration of absorption/emission from an atomic orbital. a) Decrease in energy level E_2 to E_1 resulting in photon emission (squiggle arrow) with energy, $h\nu$. b) An increase in energy level from E_1 to E_2 resulting from absorption of a photon (squiggle arrow) with energy, $h\nu$.

Quantized energy levels result from the relation between a particle's energy and its wavelength. For a confined particle such as an electron in an atom, the wave function has the form of standing waves. Only stationary states with energies corresponding to integral numbers of wavelengths can exist; for other states the waves interfere destructively, resulting in zero probability density. Elementary examples that show mathematically how energy levels come about are the particle in a box and the quantum harmonic oscillator.

Figures 3-5 represent a conceptualized view of the continuous-state evolution of 6D Calabi-Yau mirror symmetry. The cycle goes from chaotic-uncertain to coherent-certain perhaps cycling from noncommutative to commutative. This is like the Dirac string trick or a Philippine wine dance [38].

The energies of Rydberg states are sensitive to the geometrical structure of the molecular ion core. Rydberg states with low quantum numbers are conveniently accessed using multi-photon excitation via valence states, providing spectra with intensity distributions that depend sensitively on the molecular isomeric form. This discovery opens up the possibility of using Rydberg states to fingerprint the shapes of molecules. Because of the large size of the Rydberg orbitals, the Rydberg fingerprint methodology can have applications in the characterization of biological and nanoscale structures [39].

6. Atomic Theory — Elements of Atomic Structure

Atoms and molecules have Intrinsic orbital state energy levels - specifically, here for the case of a hydrogen atom with a single proton nucleus and a one electron orbital, the energy of state is primarily determined by the electrostatic interaction of the negatively charged electron with the positively charged proton. The energy levels of an electron around a nucleus are given by $E_n = -\hbar c R_\infty (Z^2 / n^2)$ where R_∞ is the Rydberg constant, Z the atomic number and n the principle quantum number. For the hydrogen atom Rydberg levels depend only on the principal quantum number, n (the only Bohr model quantum number).

- Electron Energy, The 1st term is kinetic and the 2nd potential

$$E = \frac{1}{2} m v^2 - \frac{e^2}{r}$$

- Bohr Radius

$$r = \alpha_0 n^2$$

- Centripetal Force

$$F^2 = \frac{e^2}{mr} \Rightarrow$$

$$E = \frac{1}{2} m \left(\frac{e^2}{mr} \right) - \frac{e^2}{r} = -\frac{e^2}{2r} = -\frac{e^2}{2\alpha_0 n^2}$$

- Rydberg Energy

$$E_R = \frac{e^2}{2\alpha_0} \Rightarrow E = -\frac{E_R}{n^2}$$

$$\text{for}\begin{cases} n = 1 \; E = -E_R \\ n = \infty \; E = 0 \end{cases}$$

For our process the periodic presence of a larger continuous-state QED cavity (LSXD) will shift the energy level structure in the TBS hydrogen atom, thereby altering the frequency of the emitted radiation. According to atomic theory the duration of this influence is much longer than the lifetime of the emission process thus providing a sufficient period for the putative experimental effect to occur. The Zeeman and Stark effects could help explain or act as an aid in setting up the LSXD TBS experiments.

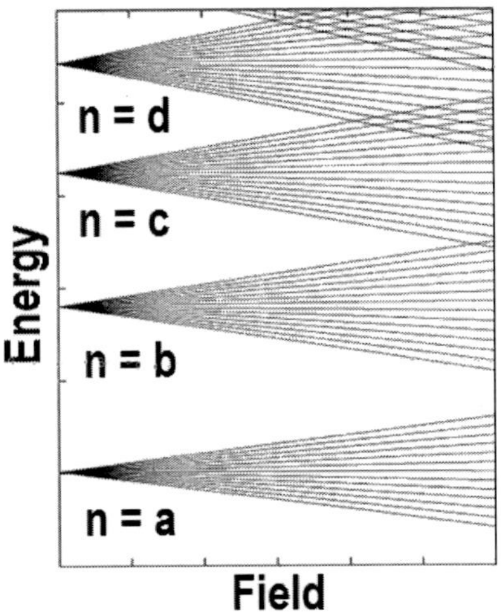

Figure 11. Rydberg atomic energy level spectra for hydrogen in an electric field. If the TBS experiment proves successful the position and 'volume' of these QED orbital cavities will be different than the historical spectral lines of hydrogen.

The Zeeman Effect describes splitting a spectral line into a number of components in the presence of a static magnetic field. It is analogous to the Stark effect, the shifting and splitting of spectral lines of atoms and molecules into several components in the presence of an external electric field. When the spectral lines are absorption lines, the effect is called inverse Zeeman Effect. The Stark effect can lead to splitting of degenerate energy levels. For example, in the Bohr model, an electron has the same energy whether it is in the 2s state or any of the 2p states. However, in an electric field, there will be quantum superpositions of the 2s and 2p states. Where the electron tends to be to the left it will acquire a lower energy; in other hybrid orbitals where the electron tends to be to the right it will acquire a higher energy. Therefore, the formerly degenerate energy levels will split into slightly lower and slightly higher energy levels. Since an atom is a collection of point charges (electrons and nuclei) dipole conditions apply. The interaction of atom or molecule with a uniform external field is described by the operator, $V_{\text{int}} = -F \cdot \mu$.

7. TBS Experimental Theory

Traditionally spectral emission/absorption lines provide information characteristic of the internal structure of an atom by, $E = \hbar\nu = \hbar c / \lambda$ or by the wave number, $\sigma \equiv 1/\lambda = E/\hbar c$ such that each atom has discrete characteristic wavelengths confirmed by monochromatic x-ray bombardment. Every atom has a variety of possible energy levels; the lowest called the ground state. From the excited state of energy, E_2 decay may occur to a lower state, E_1 with the energy difference occurring as a photon of energy, E_2 - E_1 with frequency, ν, wavelength, λ and wave number,

$$E_2 - E_1 = \hbar\nu = \frac{\hbar c}{\lambda} = \hbar c\sigma$$

with a perfectly definite value for a monochromatic spectrum [40].

The Born-Oppenheimer (BO) approximation [41,42] which is based on the fact that within molecular systems fast-moving electrons can be distinguished from slow-moving nuclei allows the wavefunction of a molecule to be broken into its electronic and nuclear (vibrational, rotational)

$$\Psi_{\text{Total}} = \psi_{\text{Electronic}} \times \psi_{\text{Nuclear}}$$

components for easier calculation. The assumption is made that if non-adiabatic coupling terms are negligibly small then the upper electronic surfaces have no effect on the nuclear wave function of the lower surface. This assumption is not considered dependent on the systems energy.

However, the ordinary BO approximation was also employed for cases where these coupling terms are not necessarily small, assuming that the energy can be made as low as required. The justification for applying the approximation in such a case is that for a low enough energy the upper adiabatic surfaces are classically forbidden, implying that the components of the total wave function related to these states are negligibly small. As a result the terms that contain the product of these components with the nonadiabatic coupling terms are also small, and will have a minor effect on the dynamical process.

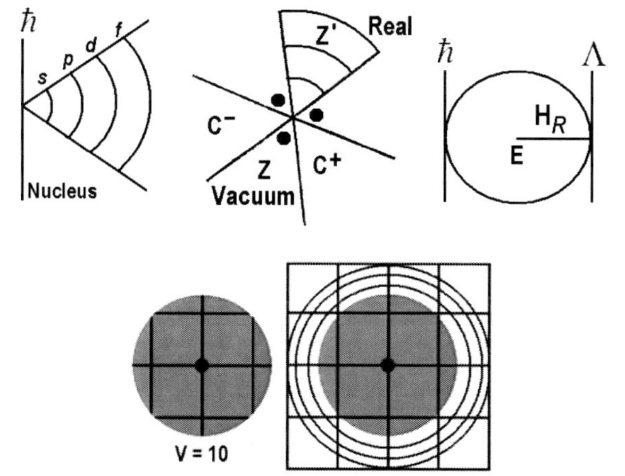

Figure 12. Historical evolution of atomic concepts. a) 2D representation of 3D reduction cone of standard Bohr orbits. b) *Z'* Real is same as (a), Z Vacuum is the Rowland's nilpotent doubling, C^+ & C^- complex Calabi-Yau mirror symmetric generators & the 3 dots represent continuous-state control elements. c) Wave function domain of the Hubble Universe, $H = \Psi$ with a 1st and 2nd quantization, $\hbar$, Λ respectively of the Hubble radius, H_R with the Earth observer, E in center. d) Conceptualized area, V =10; see text. (central dot is proton) for the usual representation of the *s* Bohr orbital in a hydrogen atom. e) Same as (d) but with additional LSXD complex TBS space area represented. In reality the square would be folded up into a 3-torus or HD Klein bottle. The additional three concentric circles beyond the V = 10 volume of the gray circle representing the hydrogen atom *s* orbital are meant to represent the added TBS periodic LSXD 12-14-16 energy volume cycles of the continuous-state matter-wave annihilation-creation process.

The protocol will test for both the existence of TBS and also for large scale XD. HAM Cosmology predicts novel HD cavities in the brane topology of Calabi-Yau mirror symmetry. Simplistically a tunable NMR device acts on a vial of hydrogen over a range of de Broglie wavelengths set for specific Cavity-QED resonances to probe the lowest Bohr orbit for TBS. If our cosmological model is correct there will be novel resonances that cannot correspond to either classical wave mechanics or Copenhagen modes. One might suspect C-QED to detect nodes in the Dirac spherical rotation of the electron (cyclical pattern of Klein bottle open-closed modes). Critics might say this is just a 4D effect Dirac effect of the putative Klein bottle symmetries in the electron's spinor rotation. But our XD cosmology predicts a much richer Calabi-Yau mirror symmetry within the higher 9D brane topology so there "should" be a cycle of novel TBS resonances in the Calabi-Yau symmetry. Likewise these resonance nodes would have de Broglie wavelengths different than any higher Bohr orbit excitation in Hydrogen. It may be possible to predict the de Broglie wavelengths in the resonance hierarchy if the topology can be determined or if a clear C-QED resonance hierarchy appears the topological structure of higher dimensions may be revealed. Vigier discussed using deuterium; it is an open question if that would make a qualitative difference in success or results in such an experiment.

In an earlier work [8,43] we designed a tachyon measurement experiment by initially considering Bohr's starting point for the development of quantum theory, i.e. the emission of photons by atoms from quantum jumps between stable Bohr orbits. We did this from the point of view of the de Broglie-Bohm causal stochastic interpretation in order to take into consideration new laser experimental results described by Kowalski [31,32]. As one knows light emitted from atoms during transitions of electrons from higher to lower energy states takes the form of photon quanta carrying energy and angular momentum. Any causal description of such a process implies that one adds to the restoring force of the harmonic oscillator an additional radiation (decelerating) resistance associated (derived from) with the electromagnetic (force) field of the emitted photon by the action equal reaction law. Any new causal condition thus implies that one must add a new force to the Coulomb force acting at random and which we suggest is related to ZPF vacuum resonant coupling and motions of the polarized Dirac aether. We assume that the wave and particle aspects of electrons and photons are built with real extended spacetime structures containing internal oscillations of point-like electromagnetic topological charges, $e^{\pm}$ within an extended form of the causal stochastic interpretation of quantum mechanics. Kowalski's interpretation drawn from recent laser experiments [31,32] showing that emission and absorption between Bohr atomic states take place within a time interval equal to one period of the emitted-absorbed photon wave, the corresponding transition time is the time needed for the orbiting electron to travel one full orbit around the nucleus. We note that the same Lorentz conditions denoted in the tachyon measurement experiment apply directly to the TBS experiment with slight phase control alterations in the Cramer-like standing-wave oscillation of the HD Calabi-Yau mirror symmetries.

- This suggests that electrons (like all massive particles) are not point-like but must be considered as extended spacetime topological structures imbedded in a real physical Dirac aether [23-26].
- These structures contain internal oscillations of point-like quantum mechanical charges around corresponding gravitational centers of mass, Y_μ so that individual electrons have different centers of mass and electromagnetic charge in the particle's and piloting fields.
- The Compton radius of mass is much larger than the radius of the charge distribution [1,2].

- The centers of charge, X_μ rotates around the center of mass, Y_μ with velocity near the velocity of light, c so that individual electrons are real oscillators with Broglian internal oscillations [7].
- Individual photons are also extended spacetime structures containing two opposite point-like charges, $e^{\pm}$ rotating with the nearly the velocity of light, $v \simeq c$ at opposite sides of a rotating diameter, with a mass, $m_\gamma \simeq 10^{-65}$gm. and an internal oscillation, $E = mc^2 = \hbar$ (Fig. 13).
- The real aether is a covariant polarized Dirac-type stochastic distribution of such extended photons which carry electromagnetic waves built with sets of such extended photons beating in phase and thus constituting subluminal and superluminal collective electromagnetic fields detected in the Casimir Effect so that a Bohr transition with one photon absorption occurs when a non-radiating Bohr orbital electron collides and beats in phase with an aether photon. In that case a photon is emitted and Bohr electron's charge e^- spirals in one rotation towards the lower level (Exceplex).

7.1 *Lorentz Condition in Complex 8-Space and Tachyonic Signaling*

In order to examine as the consequences of the relativity hypothesis that time is the fourth dimension of space, and that we have a particular form of transformation called the Lorentz transformation, we must define velocity in the complex space. That is, the Lorentz transformation and its consequences, the Lorentz contraction and mass dilation, etc., are a consequence of time as the fourth dimension of space and are observed in three spaces [8]. These attributes of 4-space in 3-space are expressed in terms of velocity, as in the form $\gamma = \left(1-\beta^2\right)^{-1/2}$ for $\beta \equiv v_{\text{Re}} / c$ where c is always taken as real.

If complex 8-space can be projected into 4-space, what are the consequences? We can also consider a 4D slice through the complex 8D space. Each approach has its advantages and disadvantages. In projective geometries information about the space is lost. What is the comparison of a subset geometry formed from a projected geometry or a subspace formed as a slice through an XD geometry? What does a generalized Lorentz transformation "look like"? We will define complex derivatives and therefore we can define velocity in a complex plane [8].

Consider the generalized Lorentz transformation in the system of x_{Re} and t_{Im} for the real time remote connectedness case in the $x_{\text{Re}}, t_{\text{Im}}$ plane. We define our substitutions from 4-to 8-space before us,

$$\begin{aligned} x &\rightarrow x' = x_{\text{Re}} + i x_{im} \\ t &\rightarrow t' = t_{\text{Re}} + i t_{im} \end{aligned} \tag{15}$$

and we represented the case for no imaginary component of x_{Re} or $x_{\text{Im}} = 0$ where the $x_{\text{Re}}, t_{\text{Re}}$ plane comprises the ordinary 4-space plane.

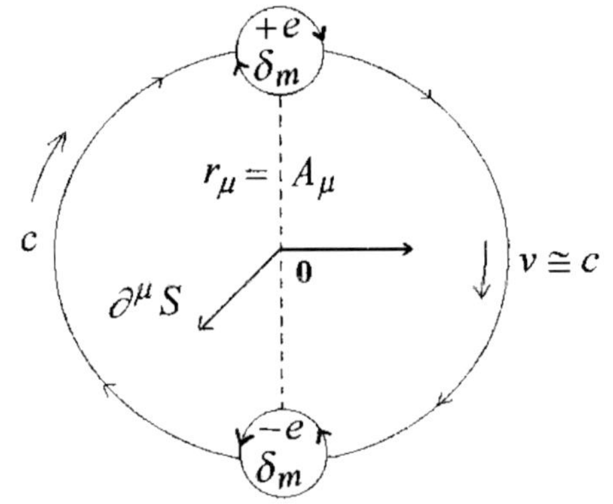

Figure 13. Diagram conceptualizing two oppositely charged sub-elements rotating at $v \cong c$ around a central point 0 behaving like a dipole bump and hole on the topological surface of the covariant polarized Dirac vacuum.

Let us recall that the usual Lorentz transformation conditions is defined in 4D real space. Consider two frames of reference, Σ, at rest and Σ' moving at relative uniform velocity v. We call v the velocity of the origin of Σ' moving relative to Σ. A light signal along the x direction is transmitted by $x = ct$ or $x - ct = 0$ and also in Σ'as $x' = ct'$ or $x'-ct' = 0$, since the velocity of light in vacuo is constant in any frame of reference in 4-space. For the usual 4D Lorentz transformation, we have as shown in Eq. (16), $x = x_{\text{Re}}, t = t_{\text{Re}}$ and $v_{\text{Re}} = x_{\text{Re}} / t_{\text{Re}}$.

$$\begin{aligned} x' &= \frac{x - vt}{\sqrt{1 - v^2/c^2}} = \gamma(x - vt) \\ y' &= y \\ z' &= z \\ t' &= \frac{t - \left(v/c^2\right)x}{\sqrt{1 - v^2/c^2}} = \gamma\left(t - \left(\frac{v}{c^2}x\right)\right) \end{aligned} \tag{16}$$

for $\gamma = (1-\beta^2)^{-1/2}$ and $\beta = v/c$. Here x and t stand for x_{Re} and t_{Re} and v is the real velocity.

We consider the $x_{\text{Re}}, t_{\text{Im}}$ plane and write the expression for the Lorentz conditions for this plane

(Fig. 2.1). Since again t_{Im} like t_{Re} is orthogonal to x_{Im} and t'_{Im} is orthogonal to x'_{Im} we can write

$$\begin{aligned} x' &= \frac{x - ivt_{\text{Im}}}{\sqrt{1-v^2/c^2}} = \gamma_v\left(x - vt_{\text{Im}}\right) \\ y' &= y \\ z' &= z \\ t' &= \frac{t-\left(v/c^2\right)x}{\sqrt{1-v^2/c^2}} = \gamma_v\left(t - \left(\frac{v}{c^2}x\right)\right) \end{aligned} \tag{17}$$

where γ_v represents the definition of γ in terms of the velocity v; also $\beta_{v\text{Im}} \equiv v_{\text{Im}}/c$ where c is always taken as real [19] where v can be real or imaginary.

In Eq. (17) for simplicity we let x', x, t' and t denote $x'_{\text{Re}}, x_{\text{Re}}, t'_{\text{Re}}$ and t_{Re} and we denote script v as v_{Im}. For velocity, v is $v_{\text{Re}} = x_{\text{Re}}/t_{\text{Re}}$ and $v = v_{\text{Im}} = i_{\text{Im}}/it_{\text{Im}}$; where the i drops out so that $v = v_{\text{Im}} = x_{\text{Im}}/t_{\text{Im}}$ is a real value function. In all cases the velocity of light c is c. We use this alternative notation here for simplicity in the complex Lorentz transformation.

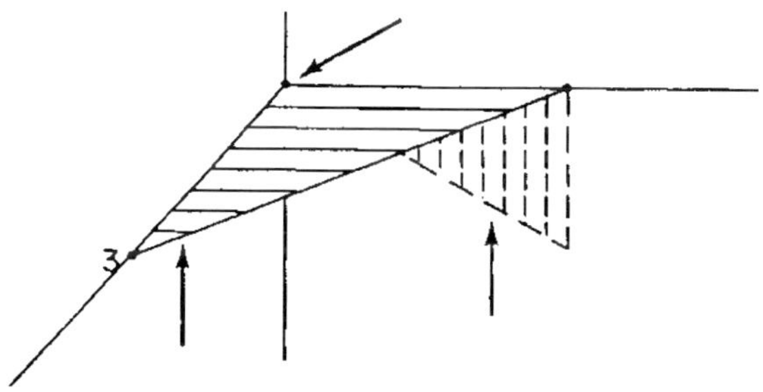

Figure 14. We illustrate an example in which a real space-like separation of events P_1 and P_2 appears to be contiguous by the introduction of the complex time, $t_{x\text{Re}} + it_{x\text{Im}}$ such that from the point of view of event P_3, the time-like separation between $(x_2(P_2) - x_1(P_1))$ appears to be zero.

The symmetry properties of the topology of the complex 8-space gives us the properties that allow Lorentz conditions in 4D, 8D and ultimately 12D space. The example we consider here is a subspace of the 8-space of $x_{\text{Re}}, t_{\text{Re}}, x_{\text{Im}}$ and t_{Im}. In some cases we let $x_{\text{Im}} = 0$ and just consider temporal remote connectedness; but likewise we can follow the anticipatory calculation and formulate remote, nonlocal solutions for $x_{\text{Im}} \neq 0$ and $t_{\text{Im}} = 0$ or $t_{\text{Im}} \neq 0$. The anticipatory case for $x_{\text{Im}} = 0$ is a 5D space as the space for $x_{\text{Im}} \neq 0$ *and* $t_{\text{Im}} = 0$ is a 7D space and for $t_{\text{Im}} \neq 0$ as well as the other real and imaginary spacetime dimensions, we have our complex 8D space.

It is important to define the complex derivative in order to define velocity, v_{Im}. In the $x_{\text{Re}}t_{\text{Im}}$ plane then, we define a velocity of $v_{\text{Im}} = dx/dit_{\text{Im}}$. In the next section we detail the velocity expression for v_{Im} and define the derivative of a complex function in detail [8].

For $v_{\text{Im}} = dx/idt_{\text{Im}} = -idx/dt_{\text{Im}} = -iv_{\text{Re}}$ for v_{Re} as a real quantity, we substitute into our $x_{\text{Re}}, t_{\text{Im}}$ plane Lorentz transformation conditions as

$$\begin{aligned} x' &= \frac{x_{\text{Re}} - v_{\text{Re}}t_{\text{Im}}}{\sqrt{1+v_{\text{Re}}^2/c^2}} \\ y' &= y \\ z' &= z \\ t'_{\text{Im}} &= \frac{t_{\text{Re}} - v_{\text{Re}}x_{\text{Re}}}{\sqrt{1+v_{\text{Re}}^2/c^2}} \end{aligned} \tag{18}$$

These conditions are valid for any velocity, $v_{\text{Re}} = -v$.

Let us examine the way this form of the Lorentz transformation relates to the properties of mass dilation. We will compare this case to the ordinary mass dilation formula and the tachyonic mass formula of Feinberg [8] which nicely results from the complex 8-space.

In the ordinary $x_{\text{Re}}t_{\text{Re}}$ plane then, we have the usual Einstein mass relationship of

$$m = \frac{m_0}{\sqrt{1-v_{\text{Re}}^2/c^2}} \quad \text{for } v_{\text{Re}} \leq c \tag{19}$$

and we can compare this to the tachyonic mass relationship in the xt plane

$$m = \frac{m_0^*}{\sqrt{1-v_{\text{Re}}^2/c^2}} = \frac{im_0}{\sqrt{1-v_{\text{Re}}^2/c^2}} = \frac{m_0}{\sqrt{v_{\text{Re}}^2/c^2-1}} \tag{20}$$

for v_{Re} now $v_{\text{Re}} \geq c$ and where m^* or m_{Im} stands for $m^* = im$ and we define m as m_{Re},

$$m = \frac{m_0}{\sqrt{1+v^2/c^2}} \tag{21}$$

For m real (m_{Re}), we can examine two cases on v as $v < c$ or $v > c$, so we will let v be any value from $-\infty < v < \infty$, where the velocity, v, is taken as real, or v_{Re}.

Consider the case of v as imaginary (or v_{Im}) and examine the consequences of this assumption. Also we examine the consequences for both v and m imaginary and compare to the above cases. If we choose v imaginary or $v^* = iv$ (which we can term v_{Im}) the $v^{*2}/c^2 = -v^2/c^2$ and $\sqrt{1+v^{*2}/c^2}$ becomes $\sqrt{1-v^{*2}/c^2}$ or

$$m = \frac{m_0}{\sqrt{1-v_{\text{Re}}^2/c^2}} \tag{22}$$

We get the form of this normal Lorentz transformation if v is imaginary $(v^* = v_{\text{Im}})$

If both v and m are imaginary, as $v^* = iv$ and $m^* = im$, then we have

$$m = \frac{m_0^*}{\sqrt{1+v^{*2}/c^2}} = \frac{im_0}{\sqrt{1-v^2/c^2}} = \frac{m_0}{\sqrt{v^2/c^2-1}} \tag{23}$$

or the tachyonic condition.

If' we go "off" into x_{Re} t_{Re} t_{im} planes, then we have to define a velocity "cutting across" these planes, and it is much more complicated to define the complex derivative for the velocities. For subliminal relative systems Σ and Σ' we can use vector addition such as $W = v_{\text{Re}} + iv_{\text{Im}}$ for $v_{\text{Re}} < x$, $v_{\text{Im}} < c$ and $W < c$. In general there will be four complex velocities. The relationship of these four velocities is given by the Cauchy-Riemann relations in the next section.

These two are equivalent. The actual magnitude of v may be expressed as $v = [vv^*]^{\frac{1}{2}}\hat{v}$ (where $\hat{v}$ is the unit vector velocity) which can be formed using either of the Cauchy-Riemann equations. It is important that a detailed analysis not predict any extraneous consequences of the theory. Any new phenomenon that is hypothesized should be formulated in such a manner as to be easily experimentally testable.

Feinberg suggests several experiments to test for the existence of tachyons [8]. He describes the following experiment – consider in the laboratory, atom A, at time, t_0 is in an excited state at rest at x_1 and atom B is in its ground state at x_2. At time t_1 atom A descends to the ground state and emits a tachyon in the direction of B. Let E_1 be this event at t_1, x_1. Subsequently, at $t_2 > t_1$ atom B absorbs the tachyon and ascends to an excited state; this is event E_2, at t_2, x_2. Then at $t_3 > t_2$ atom B is excited and A is in its ground state. For an observer traveling at an appropriate velocity, $v < c$ relative to the laboratory frame, the events E_1 and E_2 appear ro occur in the opposite order in time. Feinberg describes the experiment by stating that at t_2' atom B spontaneously ascends from the ground state to an excited state, emitting a tachyon which travels toward A. Subsequently, at t_1', atom A absorbs the tachyon and drops to the ground state.

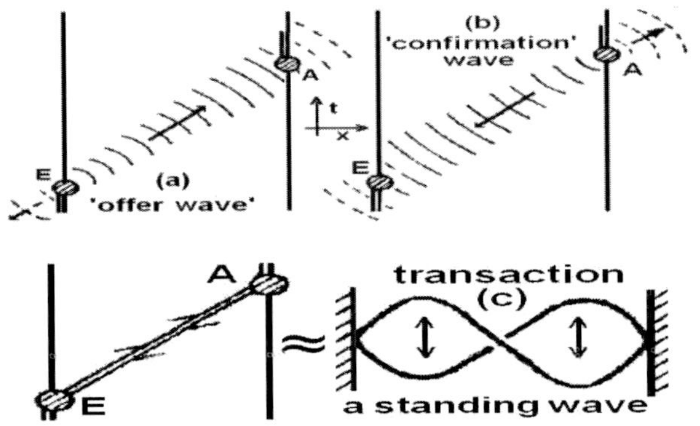

Figure 15. Transactional model. a) Offer-wave, b) confirmation-wave combined into the resultant transaction c) which takes the form of an HD future-past advanced-retarded standing or stationary wave. Figs. Adapted from Cramer [21].

It is clear from this that what is absorption for one observer is spontaneous emission for another. But if quantum mechanics is to remain intact so that we are able to detect such particles, then there must be an observable difference between them: The first depends on a controllable density of tachyons, the second does not. In order to elucidate this point, we should repeat the above experiment many times over. The possibility of reversing the temporal order of causality, sometimes termed 'sending a signal backwards in time' must be addressed [8]. Is this cause-effect statistical in nature? In the case of Bell's Theorem, these correlations are extremely strong whether explained by $v > c$ or $v = c$ signaling.

Bilaniuk, et al formulated the interpretation of the association of negative energy states with tachyonic signaling [8]. From the different frames of reference, thus to one observer absorption is observed and to another emission is observed. These states do not violate special relativity. Acausal experiments in particle physics have been suggested by a number of researchers [8]. Another approach is through the detection of Cerenkov radiation, which is emitted by charged particles moving through a substance traveling at a velocity, $v > c$. For a tachyon traveling in free space with velocity, v > c Cerenkov radiation may occur in a vacuum cause the tachyon to lose energy and become a tardon [8].

In prior joint volumes [7,8] in discussions on the arrow of time we have developed an extended model of a polarized Dirac vacuum in complex form that makes correspondence to both Calabi-Yau mirror symmetry conditions which extends Cramer's Transactional Interpretation [7] of quantum theory to cosmology. Simplistically Cramer models a transaction as a standing wave of the future-past (offer wave-confirmation wave).

Hierarchical Harmonic Oscillator Parameters	
classical	$X = A\cos(\omega t)$
quantum	$\frac{\hbar^2}{2m}\frac{d^2\psi}{dx^2} + \left(E - \frac{kx^2}{2}\right)\psi = 0$
annihilation creation	$x(t) = x_0\left[a\exp(-i\omega t) + a^\dagger \exp(i\omega t)\right]$
future-past retarded-advanced	$F_1 = F_0 e^{-ikx} e^{-2\pi i f t}$, $F_2 = F_0 e^{ikx} e^{-2\pi i f t}$, $F_3 = F_0 e^{-ikx} e^{2\pi i f t}$, $F_4 = F_0 e^{ikx} e^{2\pi i f t}$
incursive	$\frac{dx(t+\Delta t)}{dt} - v(t) = 0, \frac{dv(t+\Delta t)}{dt} + \omega^2 = 0$

Figure 16. Basic mathematical components of the applied harmonic oscillator: classical, quantum, relativistic, transactional and incursive are required in order to achieve coherent control of the cumulative resonance coupling hierarchy in order to produce harmonic nodes of destructive and constructive interference in the spacetime backcloth.

However in the broader context of the new paradigm of Holographic Anthropic Multiverse (HAM) cosmology it appears theoretically straight forward to 'program the vacuum' The coherent control of a Cramer transaction can be resonantly programmed with alternating nodes of constructive and destructive interference of the standing-wave present. It should be noted that in HAM cosmology the de Broglie-Bohm quantum potential becomes an eternity-wave, ℵ or super pilot wave or force of coherence associated with the unified field ordering the reality of the observer or the locus of the spacetime arrow of time.

To perform a simple experiment to test for the existence of Tachyons and Tardons and atom would be placed in a QED cavity or photonic crystal. Utilizing the resonant hierarchy through interference the reduced eternity wave, ℵ is focused constructively or destructively as the experimental mode may be and according to the parameters illustrated by Feinberg above temporal measurements of emission are taken.

7.2 *Velocity of Propagation in Complex 8-Space*

In this section we utilize the Cauchy-Riemann relations to formulate the hyperdimensional velocities of propagation in the complex plane in various slices through the hyperdimensional complex 8-space. In this model finite limit velocities, $v > c$ can be considered. In some Lorentz frames of reference, instantaneous signaling can be considered. It is the velocity connection between remote nonlocal events, and temporal separated events or anticipatory and real time event relations.

It is important to define the complex derivative so that we can define the velocity, vy_{Im}. In the *xit* plane then, we define a velocity of $v = dx / d(i\tau)$. We now examine in some detail the velocity of this expression. In defining the derivative of a complex function we have two cases in terms of a choice in terms of the differential increment considered. Consider the orthogonal coordinates x and it_{Im}; then we have the generalized function, $f(x, t_{\mathrm{Im}}) = f(z)$ for $z = x + it_{\mathrm{Im}}$ and f(z) = $u(x, t_{\mathrm{Im}}) + iv(x, t_{\mathrm{Im}})$ where $u(x, t_{\mathrm{Im}})$ and $v(x_{\mathrm{Im}}, t_{\mathrm{Im}})$ are real functions of the rectangular coordinates x and t_{Im} of a point in space, $P(x, t_{\mathrm{Im}})$. Choose a case such as the origin $z_0 = x_0 + it_{0\mathrm{Im}}$ and consider two cases, one for real increments $h = \Delta x$ and imaginary increments $h = i\Delta t_{\mathrm{Im}}$. For the real increments $h = \Delta t_{\mathrm{Im}}$ we form the derivative $f'(z_0) \equiv df(z)/dz_{z_0}$ which is evaluated at z_0 a

$$f' = \lim \Delta x \to 0 \left\{ \frac{u(x_0 + \Delta x, t_{0\mathrm{Im}}) - u(x_0, t_{0\mathrm{Im}})}{\Delta x} + i\frac{v(x_0 + \Delta x, t_{0\mathrm{Im}}) - v(x_0, t_{0\mathrm{Im}})}{\Delta x} \right\} \tag{24a}$$

or

$$f'(z_0) = u_x(x_0, t_{0\mathrm{Im}}) + iv_x(x_0, t_{0\mathrm{Im}}) \text{ for } u_x \equiv \frac{\partial u}{\partial x} \text{ and } v_x \equiv \frac{\partial v}{\partial x}. \tag{24b}$$

Again $x = x_{\mathrm{Re}}$, $x_0 = x_{0\mathrm{Re}}$ and $v_x = v_{x\mathrm{Re}}$.

Now for the purely imaginary increment, $h = i\Delta t_{\mathrm{Im}}$ we have

$$f'(z_0) = \lim \Delta t_{\mathrm{Im}} \to 0 \left\{ \frac{1}{i}\frac{u(x_0, t_{0\mathrm{Im}} + \Delta t_{\mathrm{Im}}) - u(x_0, t_{0\mathrm{Im}})}{\Delta t_{\mathrm{Im}}} + \frac{v(x_0, t_{0\mathrm{Im}} + \Delta t_{\mathrm{Im}}) - v(x_0, t_{0\mathrm{Im}})}{\Delta t_{\mathrm{Im}}} \right\} \tag{25a}$$

and

$$f'(z_0) = -iu_{t\mathrm{Im}}(x_0, t_{0\mathrm{Im}}) + v_{t\mathrm{Im}}(x_0, t_{0\mathrm{Im}}) \tag{25b}$$

for $u_{\mathrm{Im}} = u_{t\mathrm{Im}}$ and $v_{\mathrm{Im}} = v_{t\mathrm{Im}}$ then

$$u_{t\mathrm{Im}} \equiv \frac{\partial u}{\partial t_{\mathrm{Im}}} \text{ and } v_{t\mathrm{Im}} \equiv \frac{\partial v}{\partial t_{\mathrm{Im}}}. \tag{25c}$$

Using the Cauchy-Riemann equations

$$\frac{\partial u}{\partial x}=\frac{\partial v}{\partial t_{\mathrm{Im}}} \text{ and } \frac{\partial u}{\partial t_{\mathrm{Im}}}=-\frac{\partial v}{\partial x} \quad (26)$$

and assuming all principle derivations are definable on the manifold and letting $h=\Delta x+i\Delta t_{\mathrm{Im}}$ we can use

$$f'(z_0)=\lim h\rightarrow 0 \quad \frac{f(z_0+h)-f(z_0)}{h}=\frac{df(z)}{dz}\Big|_{z_0} \quad (27a)$$

and

$$u_x(x_0,t_{0\mathrm{Im}})+iv_x(x_0,t_{0\mathrm{Im}})-\frac{\partial u(x_0,t_{0\mathrm{Im}})}{\partial x}+i\frac{\partial v}{\partial x}(x_0,t_{0\mathrm{Im}}) \quad (27b)$$

with v_x for x and t_{Re} that is $u_{\mathrm{Re}}=u_{x\mathrm{Re}}$, with the derivative form of the charge of the real space increment with complex time, we can define a complex velocity as,

$$f'(z_0)=\frac{dx}{d(it_{\mathrm{Im}})}=\frac{1}{i}\frac{dx}{dt_{\mathrm{Im}}} \quad (28a)$$

we can have $x(t_{\mathrm{Im}})$ where x_{Re} is a function of t_{Im} and $f(z)$ and using $h=i\Delta t_{\mathrm{Im}}$, then

$$f'(z_0)=x'(t_{\mathrm{Im}})=\frac{dx}{dh}=\frac{dx}{idt_{\mathrm{Im}}}. \quad (28b)$$

Then we can define a velocity where the differential increment is in terms of $h=i\Delta t_{\mathrm{Im}}$. Using the first case as $u(x_0,t_{0\mathrm{Im}})$ and obtaining $dt_{0\mathrm{Im}}/\Delta x$ (with i's) we take the inverse. If u_x which is v_x in the $h\rightarrow i\Delta t_{\mathrm{Im}}$ case have both u_x and v_x, one can be zero. Like the complex 8D space, the 5D Kaluza-Klein geometries are subsets of the supersymmetry models. The complex 8-space deals in extended dimensions, but like the TOE models, Kaluza-Klein models also treat n > 4D as compactified on the scale of the Planck length, 10^{-33} cm [8].

In 4D space event point, P_1 and P_2 are spatially separated on the real space axis as $x_{0\mathrm{Re}}$ at point P_1 and $x_{1\mathrm{Re}}$ at point P_2 with separation $\Delta x_{\mathrm{Re}}=x_{1\mathrm{Re}}-x_{0\mathrm{Re}}$. From the event point P_3 on the t_{Im} axis we move in complex space from event P_1 to event P_3. From the origin, $t_{0\mathrm{Im}}$ we move to an imaginary temporal separation of t_{Im} to $t_{2\mathrm{Im}}$ of $\Delta t_{\mathrm{Im}}=t_{2\mathrm{Im}}-t_{0\mathrm{Im}}$. The distance in real space and imaginary time can be set so that measurement along the t_{Im} axis yields an imaginary temporal separation Δt_{Im} subtracts out, from the spacetime metric, the temporal separation Δx_{Re}. In this case occurrence of events P1 and P2 can occur simultaneous, that is, the apparent velocity of propagation is instantaneous.

For the example of Bell's Theorem, the two photons leave a source nearly simultaneously at time, $t_{0\mathrm{Re}}$ and their spin states are correlated at two real spatially separated locations, $x_{1\mathrm{Re}}$ and $x_{2\mathrm{Re}}$ separated by $\Delta x_{\mathrm{Re}}=x_{2\mathrm{Re}}-x_{1\mathrm{Re}}$. This separation is a space-like separation, which is forbidden by special relativity; however, in complex space, the points $x_{1\mathrm{Re}}$ and $x_{2\mathrm{Re}}$ appear to be contiguous for the proper path 'travelled' to the point.

7.3 *Possible New Consequences of the Model*

Since such models evidently imply new testable properties of electromagnetic and gravitational phenomena we shall conclude this work with a brief discussion of the points where it differs from the usual interpretations and implies new possible experimental tests.

If one considers gravitational and electromagnetic phenomena as reflecting different behaviors of the same real physical field i.e. as different collective behavior, propagating within a real medium (the aether) [23-26] one must start with a description of some of its properties.

We thus assume that this aether is built (i.e. describable) by a chaotic distribution $\rho(x_\mu)$ of small extended structures represented by four-vectors $A_\mu(x_\alpha)$ round each absolute point in I_0. This implies

- the existence of a basic local high density of extended sub-elements in vacuum.
- the existence of small density variations $\delta\rho(x_\mu)A_\alpha(x\mu)$ above $\delta\rho>0$ for light and below $(\delta\rho<0)$ for gravity density at x_μ.
- the possibility to propagate such field variations within the vacuum as first suggested by Dirac [25].

One can have internal variations: i.e. motions within these sub-elements characterized by internal motions associated with the internal behavior of average points (i.e. internal center of mass, centers of charge, internal rotations: and external motions associated with the stochastic behavior, within the aether, of individual sub-elements. As well known the latter can be analyzed at each point in terms of average drift and osmotic motions and A_μ distribution. It implies the introduction of non-linear terms.

To describe individual non-dispersive sub-elements within I_0, where the scalar density is locally constant and the average A_μ equal to zero, one introduces at its central point $Y_\mu(\theta)$ a space-like radial four-vector

$A_\mu = r_\mu \exp(iS/\hbar)$ (with $r_\mu r^\mu = a^2 =$ constant) which rotates around Y_μ with a frequency $\nu = m_\gamma c^2/h$. At both extremities of a diameter we shall locate two opposite electric charges e^+ and e^- (so that the sub-element behaves like a dipole). The opposite charges attract and rotate around Y_μ with a velocity $\cong c$. The +e and –e electromagnetic pointlike charges correspond to opposite rotations (i.e $\pm\ \hbar/2$) and A_μ rotates around an axis perpendicular to A_μ located at Y_μ, and parallel to the individual sub-element's four momentum $\partial_\mu S$.

Assuming electric charge distributions correspond to $\delta m > 0$ and gravitation to $\delta m < 0$ one can describe such sub-elements as holes ($\delta m < 0$) around a point 0 around which rotate two point-like charges rotating in opposite directions as shown in Figure 6.1 below.

These charges themselves rotate with a velocity c at a distance $r_\mu = A_\mu$ (with $r_\mu r_\mu$ = Const.). From 0 one can describe this by the equation

$$\Box A_\mu - \frac{m_\gamma^2 c^2}{\hbar^2} \cdot A_\mu = \frac{\left[\Box(A_\alpha^* A_\alpha)\right]^{1/2}}{(A_\alpha^* A_\alpha)^{1/2}} \cdot A_\mu \qquad (29)$$

with $A_\mu = r_\mu \cdot \exp[iS(x_\alpha)/\hbar]$ along with the orbit equations for e^+ and e^- we get the force equation

$$m \cdot \omega^2 \cdot r = e^2/4\pi r^2 \qquad (30)$$

and the angular momentum equation:

$$m_\gamma \cdot r^2 \cdot \omega = \hbar/2 \qquad (31)$$

Eliminating the mass term between (31) and (33) this yields

$$\hbar\omega = e^2/2r \qquad (32)$$

where $e^2/2r$ is the electrostatic energy of the rotating pair. We then introduce a soliton-type solution

$$A_\mu^0 = \frac{\sin \cdot K \cdot r}{K \cdot r} \cdot \exp[i(\cot - K_0 x)] \qquad (33)$$

where

$$K = mc/\hbar, \quad \omega = mc^2/\hbar \quad and \quad K_0 = mv/\hbar \qquad (34)$$

satisfies the relation (31) with

$r = ((x - vt)^2 \cdot (1 - v^2/c^2)^{-1} + y^2 + z^2)^{1/2}$ i.e.

$$\Box A_\mu^0 = 0: \qquad (35)$$

so that one can add to A_μ^0 a linear wave, A_μ (satisfying $\Box A_\mu = (m_\gamma^2 c^2/\hbar^2) A_\mu$) which describes the new average paths of the extended wave elements and piloted solitons. Within this model the question of the interactions of a moving body (considered as excess or defect of field density, above or below the aether's neighboring average density) with a real aether appears immediately. According to Newton massive bodies move in the vacuum with constant directional velocities, i.e. no directional acceleration, without any apparent relative friction or drag term. This is not true for accelerated forces (the equality of inertial and gravitational masses are a mystery) and apparent absolute motions proposed by Newton were later contested by Mach.

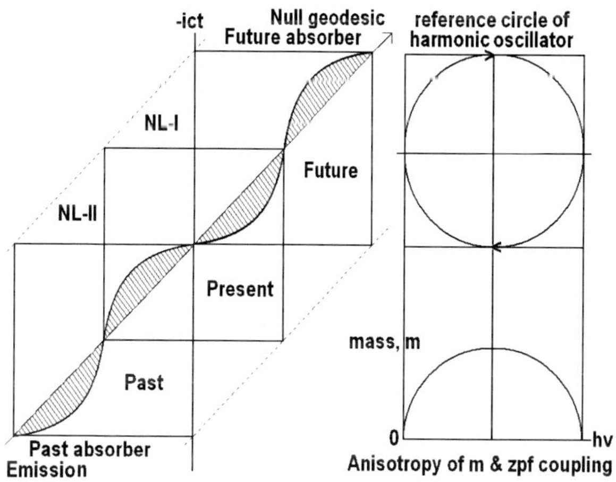

Figure 17. a) 2D drawing of a 3D view of a 4D hyperstructure. A Minkowski spacetime diagram of the electric vector only in terms of a present moment of 'tiled' Planck units utilizing the Wheeler-Feynman theory of radiation. The vertices represent absorption & emission. The observable present is represented by bold lines, and nonlocal components by standard line. Each event is a hyperstructure of Past, Present, and Future interactions, ultimately governed by the quantum potential. b) In the reference circle photon mass and energy fluctuate harmonically during propagation of the wave envelope (wave) and internal rotation of the ZPF during coupling (particle).

As well known, as time went by, observations established the existence of unexplained behavior of light and some new astronomical phenomena which led to discovery of the Theory of Relativity.

In this work we shall follow a different line of interpretation and assume that if one considers particles, and fields, as perturbations within a real medium filling flat space time, then the observed deviations of Newton's law reflect the interactions of the associated perturbations (i.e. observed particles and fields) with the perturbed average background medium in flat space-time. In other terms we shall present the argument (already presented by Ghosh et al. [8]) that the small deviations of Newton's laws reflect all known consequences of General Relativity.

The result from real causal interactions between the perturbed local background aether and its apparently independent moving collective perturbations imply absolute total local momentum and angular momentum conservation resulting from the preceding description of vacuum elements as extended rigid structures.

Retarded:

$$F_1 = F_0 e^{-ikx} e^{-2\pi ift}, \quad F_2 = F_0 e^{ikx} e^{-2\pi ift} \quad (36a)$$

Advanced:

$$F_3 = F_0 e^{-ikx} e^{2\pi ift}, \quad F_4 = F_0 e^{ikx} e^{2\pi ift} \quad (36b)$$

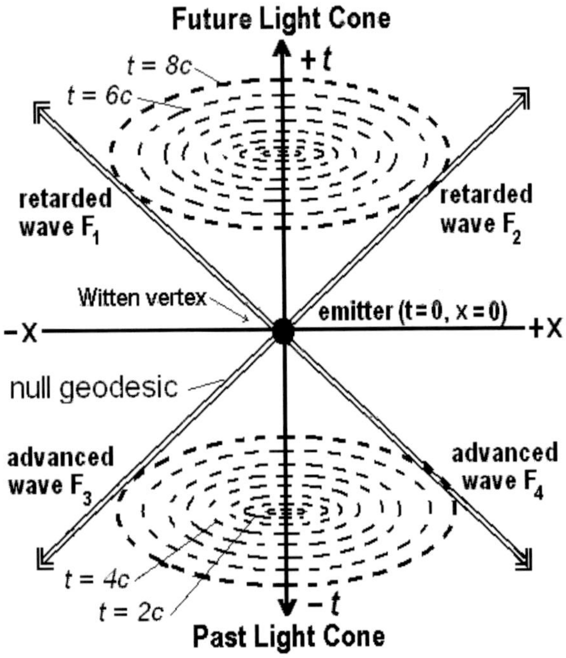

Figure 18. 4D Minkowski light-cone of advanced and retarded waves (Eq. 1) emitted from a locus at $(x,t) = (0,0)$. Adapted from concepts of Cramer [21].

As part of the symmetry breaking process the continuous-state spin-exchange compactification dynamics of the vacuum hyperstructure is shown to gives rise naturally to a $2.735^\circ K$ Hawking type radiation from the topology of Planck scale (albeit a whole new consideration of how the Planck regime operates) micro-black hole hypersurfaces. All prior considerations of 'tired-light mechanisms have been considered from the perspective of 4D Minkowski space. This new process arises from a richer *open* (non-compactified) Kaluza-Klein dimensional structure of a continuous-state cosmology in an M-Theory context with duality-mirror symmetry; also supporting the complex standing-wave postulate of the model.

From Section 5.1, 6 and 7 recall jumps to a *lower state* $E_k\ (< E_{iL})$ (CMBR-emission) according to the relation

$$h\nu = E_j - E_{iL} = E_{iH} - E_k .$$

Thus we postulate that boundary conditions inherent in continuous standing-wave spacetime spin exchange cavity compactification dynamics of vacuum topology also satisfy the requirements for photon emission. In metaphorical terms, periodic phases or modes in the continuous spacetime transformation occur where *future-past exciplex states* act as *torque moments* of CMBR/Redshift BB emission/absorption equilibrium. An exciplex (a form of excimer- short for excited dimer), usually chemistry nomenclature, used to describe an excited, transient, combined state, of two different atomic species (like XeCl) that dissociate back into the constituent atoms rather than reversion to a ground state after photon emission.

An excimer is defined as a short-lived dimeric or heterodimeric molecule formed from two species, at least one of which is in an electronic excited state. Excimers are often diatomic and are formed between two atoms or molecules that would not bond if both were in the ground state. The lifetime of an excimer is very short, on the order of nanoseconds. Binding of a larger number of excited atoms form Rydberg matter clusters the lifetime of which can exceed many seconds.

An Exciplex is an electronically excited complex of definite stoichiometry, 'non-bonding' in the ground state. For example, a complex formed by the interaction of an excited molecular entity with a ground state counterpart of a different structure. When if hits ground photon emitted Quasiparticle soliton.

In reviewing atomic theory Bohm states:

> Inside an atom, in a state of definite energy, the wave function is large only in a *toroidal region* surrounding the radius predicted by the Bohr orbit for that energy level. Of course the toroid is not sharply bounded, but ψ reaches maximum in this region and rapidly becomes negligible outside it. The next Bohr orbit would appear the same but would have a larger radius confining ψ and propagated with wave vector $k = \rho / h$ with the probability of finding a particle at a given region proportional to $|\psi|^2 = |f(x,y,z)|^2$. Since f is uniform in value over the toroid it is highly probable to find the particle where the Bohr orbit says it should be [44].

8. Experimental Design and Procedure

Some experimental evidence has been found to support this view showing the possibility that the interaction of these extended structures in space involve real physical vacuum couplings by resonance with the subquantum Dirac ether. Because of photon mass the CSI model, any causal description implies that for photons carrying energy and momentum one must add to the restoring force of the harmonic oscillator an additional radiation (decelerating) resistance derived from the em (force) field of the emitted photon by the action-equal-reaction law. Kowalski showed that emission and absorption between atomic states take place within a time interval equal to one period of the emitted or absorbed photon wave.

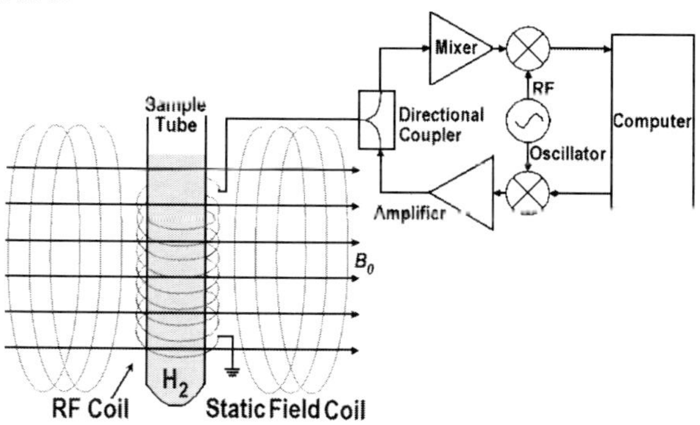

Figure 19. NMR apparatus designed to manipulate TBS in Hydrogen. The Fig. only shows possible details for rf-modulating TBS QED resonance, not the spectrographic recording and analysis components.

The corresponding transition time corresponds to the time required to travel one full orbit around the nucleus. Individual photons are extended spacetime structures containing two opposite point-like charges rotating at a velocity near *c, at the opposite sides of a rotating diameter with a mass,* $m = 10^{-65}$ *g* and with an internal oscillation $E = m^2 = h\nu$. *Thus a new causal description implies the addition of a new component to the* Coulomb force acting randomly and may be related to quantum fluctuations. We believe this new relationship has some significance for our model of vacuum C-QED blackbody absorption/emission equilibrium.

The purpose of this simple experiment is to empirically demonstrate the existence of LSXD utilizing a new model of TBS in the hydrogen atom until now hidden behind the veil of the uncertainty principle. If for the sake of illustration we arbitrarily assume the *s* orbital of a hydrogen atom has a volume of 10 and the *p* orbital a volume of 20, to discover TBS we will investigate the possibility of heretofore unknown volume possibilities arising from cyclical fluctuations in large XD Calabi-Yau mirror symmetry dynamics. This is in addition to the Vigier TBS model. As in the perspective of rows of seats in an auditorium, rows of trees in an orchard or rows of headstones in a cemetery, from certain positions the line of sight is open to infinity or block. This is the assumption we make about the continuous-state cyclicality of HD space. Then if the theory has a basis in physical reality and we are able to measure it propose that at certain nodes in the cycle we would discover cavity volumes of say 12, 14, and 16. We propose the possibility of three XD cavity modes like 'phase locked loops' depending the cycle position - maximal, intermediate and minimal. Perfect rolling motion

9. TBS Testing

The 'couple-punch' is done in conjunction with a 'couple-suck' in certain spin-spin coupling modes of the putative TBS of the Hydrogen atom as postulated for an HD Calab-Yau mirror symmetric cavity and for LSXD. In the Vigier internal motion we mentioned briefly Kowalski's Lorentz transform and in the Tachyon paper [43] further utility of Kowalski's model.

TABLE I

KICK	**PULL**	***P + K***	***K + P***
R	*R*	*R + A*	*R + A*
A	*A*	*A + R*	*A + R*
R + A	*R + A*	*A = R*	
A + R	*A + R*		
A = R	*A = R*		

Table I. Possible TBS experimental Kick-Pull (*K-P*) coupling mode parameters. *R* signifies a retarded *K* or *K* coupling, *A* respectively of advanced *K* coupling. The plus (+) sign signifies sequential order and the equals sign (=) means simultaneous action for a total of 15 experimental *KP* coupling options.

HAM cosmology suggests that there is an Feynman synchronization backbone with an inherent beat frequency in the spacetime backcloth of empty space. If we trap a hydrogen atom in a specific continuous-state CQED mode we may manipulate it with an rf-pulsed resonance hierarchy in the context of Kowalski's Lorentz transform according to the regimen in Table 1.

References

[1] Dragic´, A. Maric´, Z. Vigier, J-P (2000) New quantum mechanical tight bound states and 'cold fusion' experiments, Physics Letters A: 265;163-167.

[2] Vigier, J-P (1993) New Hydrogen (Deuterium) Bohr Orbits, Proceedings ICCF4, Hawaii, Vol 4, p. 7.

[3] Bhabha, H. J., & Corben, H. C. (1941) General classical theory of spinning particles in a Maxwell field, Proceedings

of the Royal Society of London Series A: Mathematical, Physical & Engineering Sciences, 178:974, 273-314.
[4] Barut, A.O. (1980) Surv. High Energy Phys. 1, 113.
[5] Rowlands, P. (2013) Space and antispace, in R.L. Amoroso, P. Rowlands & L.H. Kauffman (eds.) Physics of Reality: Space, Time, Matter, Cosmos, London: World Sci.
[6] Rowlands, P. (2007) Zero to infinity: the foundations of physics, Singapore: World Scientific.
[7] Amoroso, R. L. & Rauscher, E.A. (2009) The Holographic Anthropic Multiverse: Formalizing the Ultimate Geometry of Reality, Singapore: World Scientific.
[8] Rauscher, E.A. & Amoroso, R. L. (2011) Orbiting the Moons of Pluto: Complex Solutions to the Einstein, Maxwell, Schrödinger, and Dirac Equations, Singapore: World Scientific.
[9] Amoroso, R. L. (2010) (ed.) Complementarity of mind and body: Realizing the dream of Descartes, Einstein and Eccles, New York: Nova Science Publishers.
[10] Amoroso, R. L. (2014) Through the Looking Glass: Discovery of the Mind, Hackensack: World Scientific.
[11] Amoroso, R. L. (2012) Physical parameters of mind-body interaction: breaking the 1st person 3rd person barrier, J Nonlocality, 1,1, http://journals.sfu.ca/jnonlocality/index.php/jnonlocality/article/view/5
[12] Amoroso, R. L., Kauffman, L.H., Rowlands, P. & Rauscher, E.A. Introduction to Unified Field Mechanics, in process.
[13] Amoroso, R. L. (2003) Awareness: physical cosmology of the fundamental least unit, Noetic Journal, 4(1), 7-23.
[14] Amoroso, R. L., Kauffman, L.H. & Giandinoto, S. (2013) Universal Quantum Computing; 3rd Gen Prototyping Utilizing Relativistic 'Trivector' R-Qubit Modeling Surmounting Uncertainty, in R.L. Amoroso, P. Rowlands & L.H. Kauffman (eds.) The Physics of Reality: Space, Time, Matter, Cosmos: Singapore: World Scientific.
[15] Luminet, J-P, Weeks, J. R., Riazuelo, A., Lehoucq, R., & Uzan, J. P. (2003) Dodecahedral space topology as an explanation for weak wide-angle temperature correlations in the cosmic microwave background, Nature, 425:6958, 593.
[16] Luminet, J-P (2005) A cosmic hall of mirrors, arXiv preprint physics/0509171.
[17] Luminet, J-P (2008) The Wraparound Universe, Wellesley: AK Peters.
[18] Stewart, I. (2001) Flatterland: like Flatland only more so, Basic Books.
[19] Martins, C. J. A. P., Melchiorri, A., Rocha, G., Trotta, R., Avelino, P. P., & Viana, P.T.P. (2004) WMAP constraints on varying <i> α</i> and the promise of reionization. Physics Letters B, 585:1, 29-34.
[20] Bull, P., & Kamionkowski, M. (2013). What if Planck's Universe isn't flat?. arXiv preprint arXiv:1302.1617.
[21] Lehners, J. L., & Steinhardt, P. J. (2013) Planck 2013 results support the simplest cyclic models. arXiv preprint arXiv:1304.3122.
[22] Holland, P. R. (1995). The quantum theory of motion: an account of the de Broglie-Bohm causal interpretation of quantum mechanics. Cambridge university press.
[23] Hainzl, C., Lewin, M., & Séré, É. (2005) Existence of a stable polarized vacuum in the Bogoliubov-Dirac-Fock approximation., Communications in mathematical physics, 257:3, 515-562.
[24] Amoroso, R.L., Kauffman, L.H., Rauscher, E.A., Rowlands, P., & Vigier, J-P (2010 "Hidden" Parameters Describing Internal Motion Within Extended Particle Elements, in AIP Conference Proceedings, 1316, p. 1.
[25] Dirac, P.A.M. (1952) Nature (London) 169, 702.
[26] Petroni, N.C & Vigier, J-P (1983) Dirac's aether in relativistic quantum mechanics, Foundations of Physics, 13:2; 253-285.
[27] de Broglie, L. (1960) Non-linear wave mechanics: A causal interpretation. Elsevier Pub. Co..
[28] Bohm, D. J., & Hiley, B. J. (1982) The de Broglie pilot wave theory and the further development of new insights arising out of it, Foundations of Physics,12:10, 1001-1016.
[29] Amoroso, R. L., Kafatos, M., & Ecminovic, P. (1998) The origin of cosmological redshift in spin exchange between Planck scale vacuum compactification and nonzero rest mass photon anisotrophy, in Causality and Locality in Modern Physics, Dordrecht: Kluwer.
[30] Vigier, J. P. (1997) New non-zero photon mass interpretation of the Sagnac effect as direct experimental justification of the Langevin paradox, Physics Letters A, 234:2, 75-85.
[31] Kowalski, M. (1999) Photon Emission from Atomic Hydrogen, Physics Essays, Vol.12, 312-331.
[32] Kowalski, M. (2003) The process of photon emission from atomic hydrogen, in Amoroso, R. L., Hunter, G., Kafatos, M., & Vigier, J. P. (eds.) Gravitation and cosmology: From the Hubble radius to the Planck scale, Vol. 126, New York: Springer, 207-222.
[33] Amoroso, R. L., & Vigier, J-P (2003) The origin of CMBR as intrinsic blackbody cavity-QED resonance inherent in the dynamics of the continuous state topology of the Dirac vacuum, in Gravitation and Cosmology: From the Hubble Radius to the Planck Scale, Dordrecht: Kluwer, 27-38.
[34] Lehnert, B. & Roy, S. (1998) Extended Electromagnetic Theory, Singapore: World Scientific.
[35] Lehnert, B. (2002) Photon wave and particle concepts of an extended electromagnetic theory, *Phys. Scripta*, 66:2, 105.
[36] Lehnert, B. (1994) Basic concepts of an extended electromagnetic field theory, Speculations in Science and Technology-Complete Edition, 17:4, 259-266.
[37] Barcelo, C., Visser, M., & Liberati, S. (2001) Einstein gravity as an emergent phenomenon?. International Journal of Modern Physics D, 10:06, 799-806.
[38] http://www.evl.uic.edu/hypercomplex/html/dirac.html.
[39] Rydberg states: sensitive probes of molecular structure, Chemical Physics Letters Volume 378, Issues 5–6, 12, 2003, Pages 647–653, Narayanan Kuthirummal, Peter M Weber.
[40] Cowan, R.D. (1981) The Theory of Atomic Structure and Spectra, Berkeley: Univ. of CA Press.
[41] Baer, M., Lin, S. H., Alijah, A., Adhikari, S., & Billing, G. D. (2000) Extended approximated Born-Oppenheimer equation. I. Theory, Physical Review A, 62(3), 032506.
[42] M. Born and J. R. Oppenheimer, *Ann. Phys.* (Leipzig) 84, 457 (1927).
[43] Amoroso, R.L. & Rauscher, E.A. (2010) Empirical Protocol for Measuring Virtual Tachyon/Tardon Interactions in a Dirac Vacuum, in AIP Conf. Proceed. 1316, p. 199.
[44] Bohm, D. (1951) Quantum theory, New York: Dover.

Quantum Causality

SERGEY M. KOROTAEV & EVGENIY O. KIKTENKO

Geoelectromagnetic Research Centre of Schmidt Institute of Physics of the Earth, Russian Academy of Sciences, P.O. 30 Troitsk, Moscow Region 142190, Russia

The quantum extension of causal analysis has shown a rich picture of the subsystem causal connections, where the usual intuitive approach is hampered more commonly. The direction of causal connection is determined by the direction of irreversible information flow, and the measure of this connection, called the course of time c_2, is determined as the velocity of such flow. The absence of causality corresponds to infinite c_2, accordingly the degree of causal connection is inversely related to c_2. This formal definition of causality is valid at any time direction. The possibilities of causal analysis have been demonstrated before by series of examples of the two- and three-qubit states. In this paper we consider the new applications. The first one is the application of quantum causal analysis to the asymmetric entangled state under decoherence. Three models of decoherence: dissipation, depolarization and dephasing are studied. For the all models the strength and the direction of induced causality has been computed. It turns out that the decoherence acting along original causality destroys entanglement to a lesser degree than it acting against this causality. The second application is the interaction between a two-level atom and infinite-dimensional quantized mode of a field by Jaynes-Cummings model. An analytical solution of von Neumann equation for different initial states is examined. The filed is considered initially to be in thermal mixed state, while atom – sequentially in excited, ground or thermal states. Negativity, mutual information and causal characteristics for different temperatures are computed. It is obtained that for high temperatures distinction between behaviors of different initial states smoothes over and the state turns out to be causal, entangled and "classical" in entropic sense. And the third application is the teleportation (three-particle protocol). Contrintuitively the teleported qubit is not an effect of the original one; it proves the common effect of both two other ones. But at the same time the result of Bell measurement constitutes a cause with respect to every qubits of entangled pair just since moment of their birth. The latter is manifestation of causality in reverse time.

Keywords: Causality, Quantum theory

1. Introduction

The causality is one of the universal physical principles. It plays the twofold role. On the one hand, in the problems brought to the sufficient theoretical level, this principle allows selecting of the physically realizable solutions among a plethora of the mathematically admissible ones. It is just the case of relativity theory. On the other hand, the establishment of causal-effect connections in analysis of the complicated systems is the first step to the construction of a phenomena model. In references to the causality principle, usually it does not bear in mind anything except retardation of the effect relative to the cause. With indefinite terms the "cause" and "effect" in the theoretical problems it may lead to the confusions. In the complicated phenomena investigation the rather serious mistakes are possible. It is particularly important for the quantum entangled states. Usually the question about possible reversal of time ordering at quantum correlation through a space-like interval is avoided presupposing quantum correlation to be causeless. But it is in conflict with the possibility of quantum information transfer. Although practically the conflict is damped by the fact that for the communication purposes one should use an ancillary classical subluminal channel, recently the problem became relevant in connection with macroscopic entanglement, quantum wormholes, etc. The necessity of formal taking into account of really existing causal connections was felt by many researchers ([1] and references therein). In answer to this challenge the formal method of classical causal analysis was suggested [2]. This method had been successfully applied before to the various experimental problems of classical physics ([3] and references therein). Recently it is also applied to the experiments on macroscopic entanglement [4–8]. But the classical approach to that quantum phenomenon is rather limited. The quantum extension of causal analysis has shown a richer picture of the subsystem causal connections, where the usual intuitive approach Conformal Steady State Free Precession is hampered more commonly [9]. The direction of causal connection is determined by the sign of irreversible information flow, and the measure of this connection, called the course of time, is determined as the velocity of such flow. The absence of causality corresponds to infinite course of time; accordingly the degree of causal connection is inversely related to its value. This formal definition of causality is valid at any time direction. The independence functions used in the causal analysis allow classification of quantum and classical correlations of the subsystems. The possibilities of causal analysis have been demonstrated before by series of examples of the two- and three-qubit states [9–13].

In this paper we consider the new applications. Quantum mechanical development of the causality concept turns out not only possible, but fruitful in many respects, in particular, in solving of problem of entanglement protection under decoherence. Next, the quantum mechanical principle of weak causality (suggested intuitively long ago by Cramer [14] and formalized now in causal analysis) admits availability of the signals in reverse time for the random processes. It helps to understand the teleportation process and opens way to understanding more complicated phenomena.

In Sec. 2 the kernel of quantum causal analysis formalism is reviewed. In Sec. 3 application of causal analysis to the entangled states under different kinds of decoherence is demonstrated. Sec. 4 is dedicated to the analysis of entanglement and causality in interaction of a two-level atom with the field. In Sec. 5 we consider three-particle teleportation protocol at different approaches and reveal causality in reverse time. The general results are summarized in Sec. 6.

2. Kernel of Quantum Causal Analysis

Quantum causal is an extension of classical causal analysis [2] which operates only with classical variables. The essence of causal analysis bases on formalization of usual intuitive “cause” and “effect” concepts from information-wise asymmetry of a process without invoking time relations. The retardation of effect relative to the cause is introduced after their definition as an axiom.

Consider a quantum bipartite state, which characterized by density matrix ρ_{AB}, which consists of two subsystems A and B with the reduced density matrices $\rho_A = Tr_B\rho_{AB}$ and $\rho_B = Tr_A\rho_{AB}$ respectively. From these matrixes we can calculate corresponding von Neumann entropies $S(A)$, $S(B)$ and $S(AB)$ by general formula:

$$S(X) = -Tr[\rho_X \log_2 \rho_X]. \tag{1}$$

Mathematical formalization of causal analysis is founded on a pair of independence functions:

$$i_{B|A} = \frac{S(B\,|\,A)}{S(B)},\quad i_{A|B} = \frac{S(A\,|\,B)}{S(A)},\quad i \in [-1,1], \tag{2}$$

where $S(B\,|\,A) = S(AB) - S(A)$ and $S(A\,|\,B) = S(AB) - S(B)$ are conditional entropies. To understand the idea of independent functions let us consider the main demonstrative cases. $i_{B|A} = -1$ (which can be realized only when $i_{A|B} = -1$) means that we have pure entangled state: $S(AB) = 0$, $S(A) = S(B) \neq 0$, that corresponds to maximal quantum correlations between the two subsystems. If $i_{B|A} = 0$ then $S(AB) = S(A)$ and we obtain that state B is one-valued function of state A (notice, that $i_{B|A} = 0$ does not mean that $i_{A|B} = 0$). Therefore in this context we have maximal classical correlations. And in case of $i_{B|A} = 1$, the B is independent of the A. It is worth to mention, that generally $i_{B|A} \neq i_{A|B}$, so the independence functions characterize one-way correlations between two subsystems in contrast to e.g. mutual information:

$$I = S(A) + S(B) - S(AB), \tag{3}$$

this characterizes total two-way correlation between the subsystems.

It is important that for the classical variables $i \in [0,1]$, that is a result of the classical inequality $S(AB) \geq \max[S(A), S(B)]$. Therefore independence functions can indicate weather the system is “quantum”' or “classical” in entropic sense. If at least one $i_{A|B} < 0$ or $i_{B|A} < 0$, then a system should be called quantum. If both $i_{A|B} > 0$ and $i_{B|A} > 0$, then a system should be called classical. The similar definitions, although in other terms, were proposed before in Ref. [15] (there was considered quantum-classical' bipartite state AB, where the A subsystem was quantum with $i_{B|A} < 0$ and the B was classical with $i_{A|B} > 0$.

Causality in our consideration corresponds to inequality $i_{B|A} \neq i_{A|B}$. For the measure of causal connection between subsystems A and B we use $c_2(A,B)$, called the course of time (notation follows Kozyrev's pioneer work on causal mechanics [1]), and derived in [9–11] as the velocity of irreversible information flow:

$$c_2(A,B) = k\frac{(1 - i_{A|B})(1 - i_{B|A})}{i_{A|B} - i_{B|A}}, \tag{4}$$

where $k = \Delta r / \delta t$, Δr is an effective distance between A and B, and δt is a time of brachistochrone evolution [16]. For the orthogonal states:

$$\delta t = \frac{\pi\hbar}{2(\Delta E)_{\max}}, \tag{5}$$

where $(\Delta E)_{\max}$ is a maximal difference between eigenvalues of the Hamiltonian.

The sign of $c_2(A,B)$ is specified by the direction of causal connection: $c_2(A,B) > 0$ means that subsystem A is a cause (information-wise source) and

B is an effect (informational-wise sink). $c_2(A,B)<0$ means that B is a cause and A is an effect ($c_2(A,B)=-c_2(B,A)$. The strength of the causal connection corresponds to absolute value $|c_2(A,B)|$: the stronger is the causality, the greater is asymmetry, the less is $|c_2(A,B)|$. It is noteworthy that e.g. for all the pure entangled states $|c_2(A,B)|\to\infty$ that totally conform to representation of quantum correlations as causeless and instantaneous. But in the mixed states the independence functions need not be equal, therefore causality can exist.

Cramer was the first to distinguish the principles of strong and weak causality [14]. The strong causality corresponds to usual condition of retardation $\tau_{A\to B}$ of the effect relative to the cause:

$$c_2(A,B)>0 \Rightarrow \tau_{A\to B}>0,$$
$$c_2(A,B)<0 \Rightarrow \tau_{A\to B}<0, \quad (6)$$
$$|c_2(A,B)|\to\infty \Rightarrow \tau_{A\to B}\to 0.$$

Without the axiom (6) we have weak causality, which corresponds only to nonlocal correlations. Even as they occur in reverse time they only relate the unknown states (hence the "telegraph to the past"' is impossible). Although it is not very important for the present work scope, note that weak causality admits the extraction of information from the future without well known classical paradoxes. The experimental possibility of detection of such time reversal phenomena was theoretically predicted by Elitzur and Dolev [17] and really proved for the intramolecular teleportation [18] and for the macroscopic entanglement, [4–8]. And note that we do not use the axiom (6) anywhere in the current paper.

To keep the examples described bellow from becoming too involved; we shall restrict ourselves by calculations of c_2 with accuracy to $k=1$ in Eq. (4), since it does not qualitatively influence on the c_2 behavior [9,10].

3. Decoherence Asymmetry and Causality

3.1. *Models*

We consider the models of some well known three-qubit entangled symmetric states – GHZ and W ones where causality originally is absent and emerges only as a result of decoherence, and asymmetric CKW one with finite original causality [9, 10]. The measure of quantum causality c_2 is compared to the negativity N as a standard measure of entanglement.

So, the model states are:

1. Greenberg- Horn-Zeilinger (GHZ) state:

$$|GHZ\rangle=\frac{1}{\sqrt{2}}(|000\rangle+|111\rangle), \quad (7)$$

2. W-state:

$$|W\rangle=\frac{1}{\sqrt{3}}(|001\rangle+|010\rangle+|100\rangle), \quad (8)$$

3. Coffman-Kundu-Wooters (CKW) state [19, 20]:

$$|CKW\rangle=\frac{1}{\sqrt{2}}|100\rangle+\frac{1}{2}(|001\rangle+|010\rangle). \quad (9)$$

The first qubit of every state we call the subsystem A, the second and third ones – the subsystems B and C. Any two-party partitions of (7) and (8) are equivalent. In the state (9) the party A sets off from B and C, therefore only parties B, C and AB, AC are equivalent. Since all the states (7)–(9) are pure any their two-one partitions $AB-C$, $A-BC$, etc. are causeless ($|c_2|\to\infty$) [9, 10].

Finite causality potentially is possible only in the mixed subsystems $A-B$, $A-C$ and $B-C$. But due to the symmetry there is a finite causality only in the state (9); namely the computations of Ref. [9, 10] has yielded for the state (9): $c_2(A,B)=c_2(A,C)\approx 5.30$. Thus A is the cause with respect to B and C.

The three kinds of decoherence (of $0\le p\le 1$ degree) reduce to the following transformations [21, 22]:

Dissipation:

$$\begin{aligned}&|0\rangle\langle 0|\to|0\rangle\langle 0|,\\&|1\rangle\langle 1|\to(1-p)|1\rangle\langle 1|+p|0\rangle\langle 0|,\\&|1\rangle\langle 0|\to\sqrt{1-p}\,|1\rangle\langle 0|,\\&|0\rangle\langle 1|\to\sqrt{1-p}\,|0\rangle\langle 1|.\end{aligned} \quad (10)$$

Depolarization:

$$\begin{aligned}&|0\rangle\langle 0|\to(1-p)|0\rangle\langle 0|+p\frac{I}{2},\\&|1\rangle\langle 1|\to(1-p)|1\rangle\langle 1|+p\frac{I}{2},\\&|1\rangle\langle 0|\to(1-p)|1\rangle\langle 0|,\\&|0\rangle\langle 1|\to(1-p)|0\rangle\langle 1|.\end{aligned} \quad (11)$$

Dephasing:

$$|1\rangle\langle 0| \to (1-p)|1\rangle\langle 0|,$$
$$|0\rangle\langle 1| \to (1-p)|0\rangle\langle 1|. \quad (12)$$

We apply (10)–(12) to one of the qubits of (7)–(9). Due to the symmetry of these states it is enough to apply a transformation to any of qubits of (7) and (8) and we select this qubit to be C. For the state (9) the distinguishable results are achieved by application of a transformation only to the qubits C and A (the transformations of B and C are equivalent).

The resulting mixed states are the following:
Decoherence of GHZ (7):

$$\rho_{GHZ}^{dissC} = \frac{1}{2}\begin{bmatrix} |000\rangle\langle 000| + (1-p)|111\rangle\langle 111| \\ +p|110\rangle\langle 110| \\ +\sqrt{1-p}\left(|000\rangle\langle 111| + |111\rangle\langle 000|\right) \end{bmatrix}, \quad (13)$$

$$\rho_{GHZ}^{depolC} = \frac{1}{2}\left[\left(1-\frac{p}{2}\right)\begin{pmatrix} |000\rangle\langle 000| + |000\rangle\langle 111| \\ +|111\rangle\langle 000| + |111\rangle\langle 111| \end{pmatrix} \right. \quad (14)$$
$$\left. +\frac{p}{2}\left(|001\rangle\langle 001| + |110\rangle\langle 110|\right)\right],$$

$$\rho_{GHZ}^{dephC} = \frac{1}{2}\begin{bmatrix} |000\rangle\langle 000| + |111\rangle\langle 111| \\ +(1-p)\left(|000\rangle\langle 111| + |111\rangle\langle 000|\right) \end{bmatrix} \quad (15)$$

Decoherence of W (8):

$$\rho_{W}^{dissC} = \frac{1}{3}\begin{bmatrix} |010\rangle\langle 010| + |010\rangle\langle 100| \\ +|100\rangle\langle 010| + |100\rangle\langle 100| \\ +(1-p)|001\rangle\langle 001| + p|000\rangle\langle 000| \\ +\sqrt{1-p}\begin{pmatrix} |001\rangle\langle 010| + |001\rangle\langle 100| \\ +|010\rangle\langle 001| + |100\rangle\langle 001| \end{pmatrix} \end{bmatrix}, \quad (16)$$

$$\rho_{W}^{depolC} = \frac{1}{3}\left[(1-p)\begin{pmatrix} |001\rangle\langle 010| + |001\rangle\langle 100| \\ +|010\rangle\langle 001| + |100\rangle\langle 001| \end{pmatrix}\right.$$
$$+\left(1-\frac{p}{2}\right)\left(|001\rangle\langle 001| + |010\rangle\langle 010| + |100\rangle\langle 100|\right)$$
$$\left. +\frac{p}{2}\begin{pmatrix} |000\rangle\langle 000| + |011\rangle\langle 011| + |011\rangle\langle 101| \\ +|101\rangle\langle 011| + |101\rangle\langle 101| \end{pmatrix}\right], \quad (17)$$

$$\rho_{W}^{dephC} = \frac{1}{3}\left[|001\rangle\langle 001| + |010\rangle\langle 010| \right.$$
$$+|010\rangle\langle 100| + |100\rangle\langle 010| + |100\rangle\langle 100|$$
$$\left. +(1-p)\begin{pmatrix} |001\rangle\langle 010| + |001\rangle\langle 100| \\ +|010\rangle\langle 001| + |100\rangle\langle 001| \end{pmatrix}\right]. \quad (18)$$

Decoherence of CKW (9):

$$\rho_{CKW}^{dissC} = \frac{1}{2}\left[\frac{1}{2}|010\rangle\langle 010| + \frac{1}{\sqrt{2}}\begin{pmatrix} |010\rangle\langle 100| \\ +|100\rangle\langle 010| \end{pmatrix}\right.$$
$$+|100\rangle\langle 100| + \frac{1}{2}(1-p)|001\rangle\langle 001| + \frac{1}{2}p|000\rangle\langle 000|$$
$$\left. +\sqrt{1-p}\begin{pmatrix} \frac{1}{2}|001\rangle\langle 010| + \frac{1}{\sqrt{2}}|001\rangle\langle 100| \\ +\frac{1}{2}|010\rangle\langle 001| + \frac{1}{\sqrt{2}}|100\rangle\langle 001| \end{pmatrix}\right], \quad (19)$$

$$\rho_{CKW}^{depolC} = \frac{1}{2}\left\{(1-p)\begin{bmatrix} \frac{1}{2}\begin{pmatrix} |001\rangle\langle 010| \\ +|010\rangle\langle 001| + |100\rangle\langle 100| \end{pmatrix} \\ +\frac{1}{\sqrt{2}}\left(|001\rangle\langle 100| + |100\rangle\langle 001|\right) \end{bmatrix}\right.$$
$$+\left(1-\frac{p}{2}\right)\begin{bmatrix} \frac{1}{2}\left(|001\rangle\langle 001| + |010\rangle\langle 010|\right) \\ +\frac{1}{\sqrt{2}}\left(|010\rangle\langle 100| + |100\rangle\langle 010|\right) \end{bmatrix}$$
$$\left. +\frac{p}{2}\begin{bmatrix} \frac{1}{2}\left(|000\rangle\langle 000| + |011\rangle\langle 011|\right) \\ +\frac{1}{\sqrt{2}}\left(|011\rangle\langle 101| + |101\rangle\langle 011|\right) + |101\rangle\langle 101| \end{bmatrix}\right\}, \quad (20)$$

$$\rho_{CKW}^{dephC} = \frac{1}{2}\left\{\begin{matrix} \frac{1}{2}\left(|001\rangle\langle 001| + |010\rangle\langle 010|\right) \\ +\frac{1}{\sqrt{2}}\left(|010\rangle\langle 100| + |100\rangle\langle 010|\right) \\ +|100\rangle\langle 100| \end{matrix}\right.$$
$$\left. +\frac{1}{2}(1-p)\begin{bmatrix} |001\rangle\langle 010| + |010\rangle\langle 001| \\ +\sqrt{2}\left(|001\rangle\langle 100| + |100\rangle\langle 001|\right) \end{bmatrix}\right\}; \quad (21)$$

$$\rho_{CKW}^{dissA} = \frac{1}{2}\left[\frac{1}{2}\begin{pmatrix}|001\rangle\langle001| + |001\rangle\langle010| \\ +|010\rangle\langle001| + |010\rangle\langle010|\end{pmatrix} + (1-p)|100\rangle\langle100| + p|000\rangle\langle000| + \sqrt{\frac{1-p}{2}}\begin{pmatrix}|001\rangle\langle100| + |010\rangle\langle100| \\ +|100\rangle\langle001| + |100\rangle\langle010|\end{pmatrix}\right], \tag{22}$$

$$\rho_{CKW}^{depolA} = \frac{1}{2}\left\{\frac{1-p}{\sqrt{2}}\begin{pmatrix}|001\rangle\langle100| + |010\rangle\langle100| \\ +|100\rangle\langle001| + |100\rangle\langle010|\end{pmatrix} + \left(1-\frac{p}{2}\right)\begin{bmatrix}\frac{1}{2}\begin{pmatrix}|001\rangle\langle001| + |001\rangle\langle010| \\ +|010\rangle\langle001| + |010\rangle\langle010|\end{pmatrix} \\ +|100\rangle\langle100|\end{bmatrix} + \frac{p}{2}\left[|000\rangle\langle000| + \frac{1}{2}\begin{pmatrix}|101\rangle\langle101| + |101\rangle\langle110| \\ +|110\rangle\langle101| + |110\rangle\langle110|\end{pmatrix}\right]\right\}, \tag{23}$$

$$\rho_{CKW}^{dephA} = \frac{1}{2}\left[\frac{1}{2}\begin{pmatrix}|001\rangle\langle001| + |001\rangle\langle010| + |010\rangle\langle001| \\ +|010\rangle\langle010| + 2|100\rangle\langle100|\end{pmatrix} + \frac{1}{\sqrt{2}}(1-p)\begin{pmatrix}|001\rangle\langle100| + |010\rangle\langle100| \\ +|100\rangle\langle001| + |100\rangle\langle010|\end{pmatrix}\right]. \tag{24}$$

From Eqs. (13)–(24) we have computed all the marginal and conditional entropies, then – the independence function i like (2), and at last – the course of time c_2 like (4) for all the distinguishable two-party partitions. For the same partitions the negativity N, as a measure of entanglement, has been computed too.

3.2. *Causal connections at different kinds of decoherence*

Decoherence of the most symmetric GHZ state produces the most simple causality picture shown in Figure 1 (recall that according to our notation $c_2(X,Y)>0$ means directionality of causal connection $X \to Y$, $c_2(X,Y)<0$ means $Y \to X$). Only dissipation leads to finite causality in any partition. If the dissipated qubit constitutes an individual party (the partitions $AB-C$ and $B-C$) this party always corresponds to the effect

$$(c_2(AB,C)>0,\ c_2(A,C)>0)$$

and with the increase of the degree of dissipation p the causality amplifies ($c_2 \to 0$ at $p \to 1$). It is in full agreement with the intuitive expectation – the irreversible flow of information is directed to the dissipated particle. The fact that $c_2(AB,C) > c_2(B,C)$ is explained in Ref. [11] by stronger mixedness of the reduced state $\rho(BC)$ as compared to $\rho(ABC)$, because mixedness is a necessary condition of causality. In its turn stronger mixedness of $\rho(BC)$ is the consequence of both interaction with A and dissipation of C i.e. interaction with the non-controlled environment; while mixedness of $\rho(ABC)$ is the consequence of only the latter. Note that in the case of dissipation of one of the particles of two-particle counterpart of GHZ state (that is Bell state) all the corresponding entropies and therefore all the other parameters, including c_2 [9, 10] exactly coincides with those of GHZ $AB-C$ partition. In the partition $AC-B$ the behavior of causality is nontrivial. In contrast to the above case the couple AC including the dissipated particle C constitutes the cause. The fact is dissipation of C decreases $S(C)$ (the states approaches to the certain ground state according to Eq. (10)). On the other hand the dissipation of C opens the subsystem AC to the environment and $S(AC)$ increases and has the maximum at $p=1/2$ equal to 3/2 [11], while $S(B)=const=1$. The particle B always corresponds to the effect but c_2 is not monotonous: it has the minimum at $p=0.594$. To explain this fact, note that at $p=0$ the state (13) is pure therefore

$$c_2(AC,B) \to \infty;\ \text{at}\ p=1$$

the state (11) is maximally mixed, but $S(AC)=S(B)$ (the fully dissipated particle C has "disappeared") therefore $c_2(AC,B) \to \infty$ too. The denominator of Eq. (4) for $c_2(AC,B)$: $i_{AC|B} - i_{B|AC}$ has the maximum at $p=0.401$ [11], while the nominator that is correlation $(1-i_{AC|B})(1-i_{B|AC})$ decreases as p increases, therefore $\min c_2$ is shifted to a higher p relative to 1/2. But by comparison with other partitions causality in the $AC-B$ one is prevailing, as it is seen from Figure 1, at small dissipation ($p<0.387$).

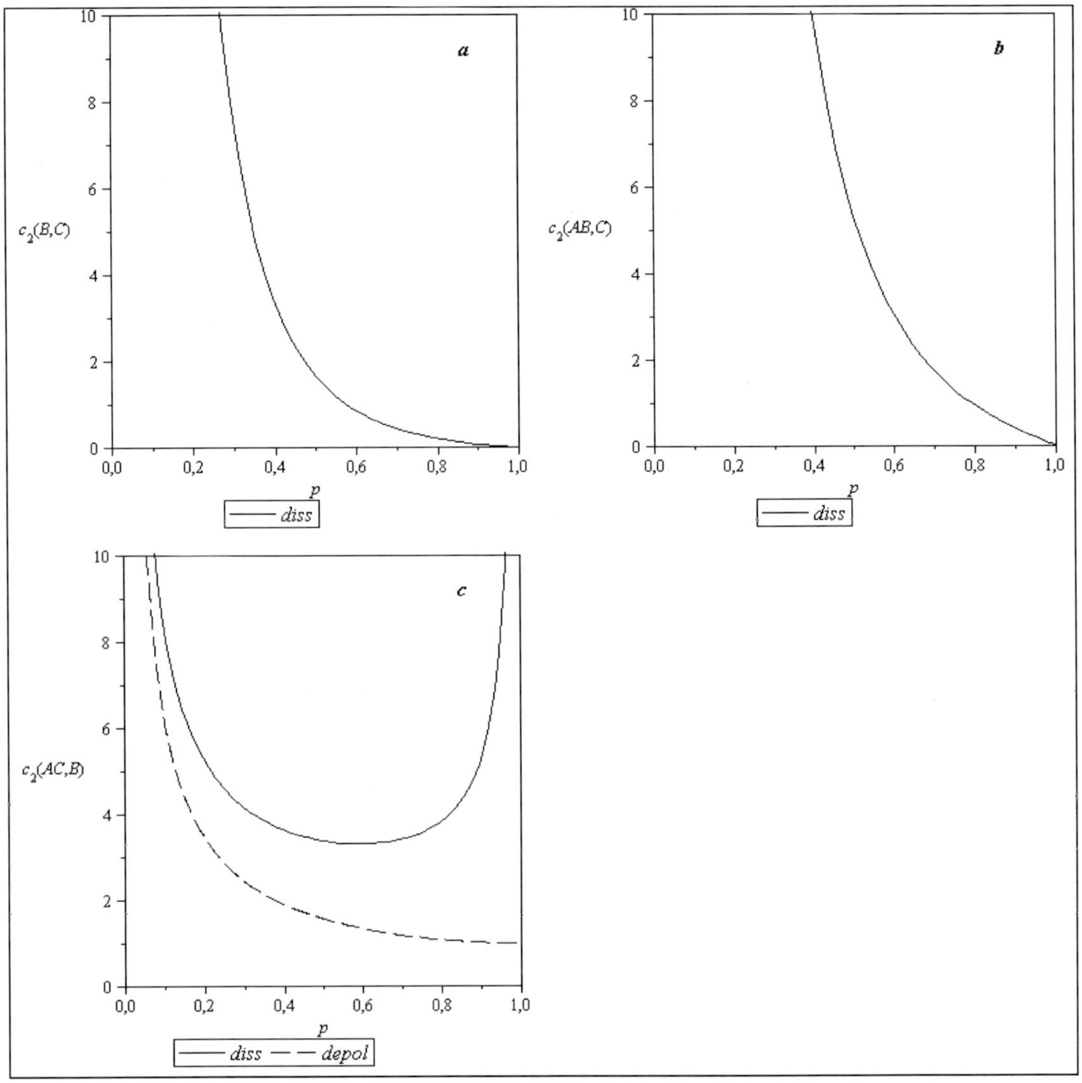

FIGURE 1. Causality in GHZ state with decohered qubit C.

Depolarization leads to only finite causality $AC \to B$ (Figure 1(c)) that is with the same direction as in the dissipation case but strength of causality amplifies monotonously as p increases, achieving $c_2 = 1$ at $p = 1$.

Dephasing of GHZ state does not lead to emergence of any causality.

W-state decoherence (Figure 2) differs from GHZ one in that depolarization leads to finite causality in all the three partitions, so does the dephasing in $AC - B$ partition. Quantitative features of the c_2 behavior induced by dissipation are the same as in GHZ state and they are explained by the same reasons. The distinction is that c_2 in $AC - B$ partition has the minimum at $p = 0.576$ and c_2 in this link is higher, i.e. causality is weaker, than in the two other partitions at any p. At depolarization, in contrast to dissipation, if the depolarized qubit constitutes an individual party (the partitions $AB - C$ and $B - C$) this party always is the cause ($c_2(AB,C) < 0$, $c_2(B,C) < 0$) and with the increase of the degree of dissipation p the causality amplifies ($c_2 \to 0$ at 1). It is also in agreement with the intuitive expectation – the irreversible flow of information (noise) encroaches through the depolarized party and propagates to another one. The fact that $|c_2(AB,C)| > |c_2(B,C)|$ is also explained by stronger mixedness of the reduced state $\rho(BC)$ as compared to $\rho(ABC)$.

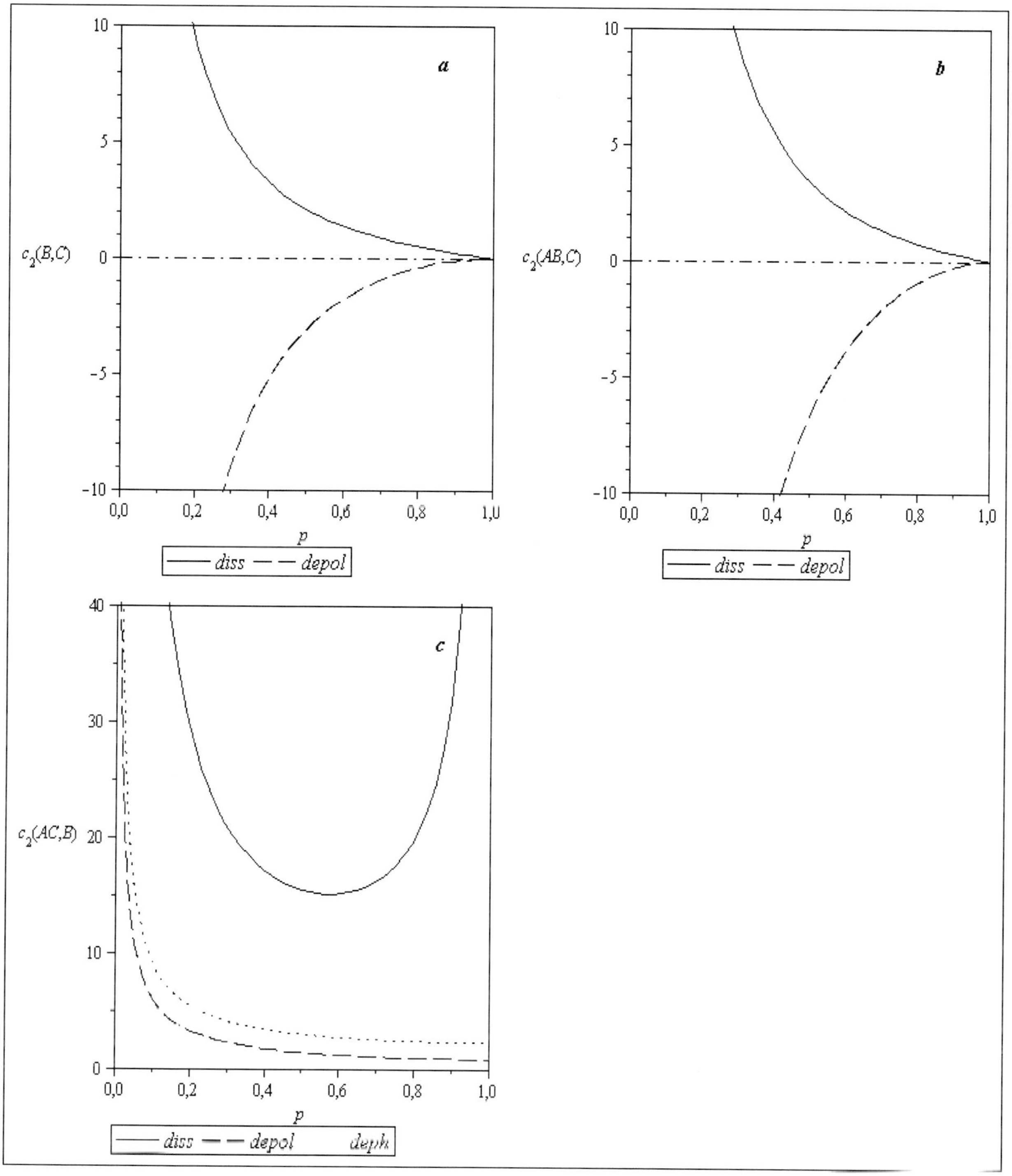

FIGURE 2. Causality in W-state with decohered qubit C.

A comparison between Figures 1(c) and 2(c) shows that in both the cases directionality of causal connection is $AC \to B$ and the curves c_2 are alike. In the W-state the dephasing also induces the causality very similar to the depolarization case, but weaker.

The decohered CKW-state, having originally the two causal connections $A \to B$ and $A \to C$, produces much more rich induced causality distribution. At the beginning consider an original effect C decoherence (Figure 3).

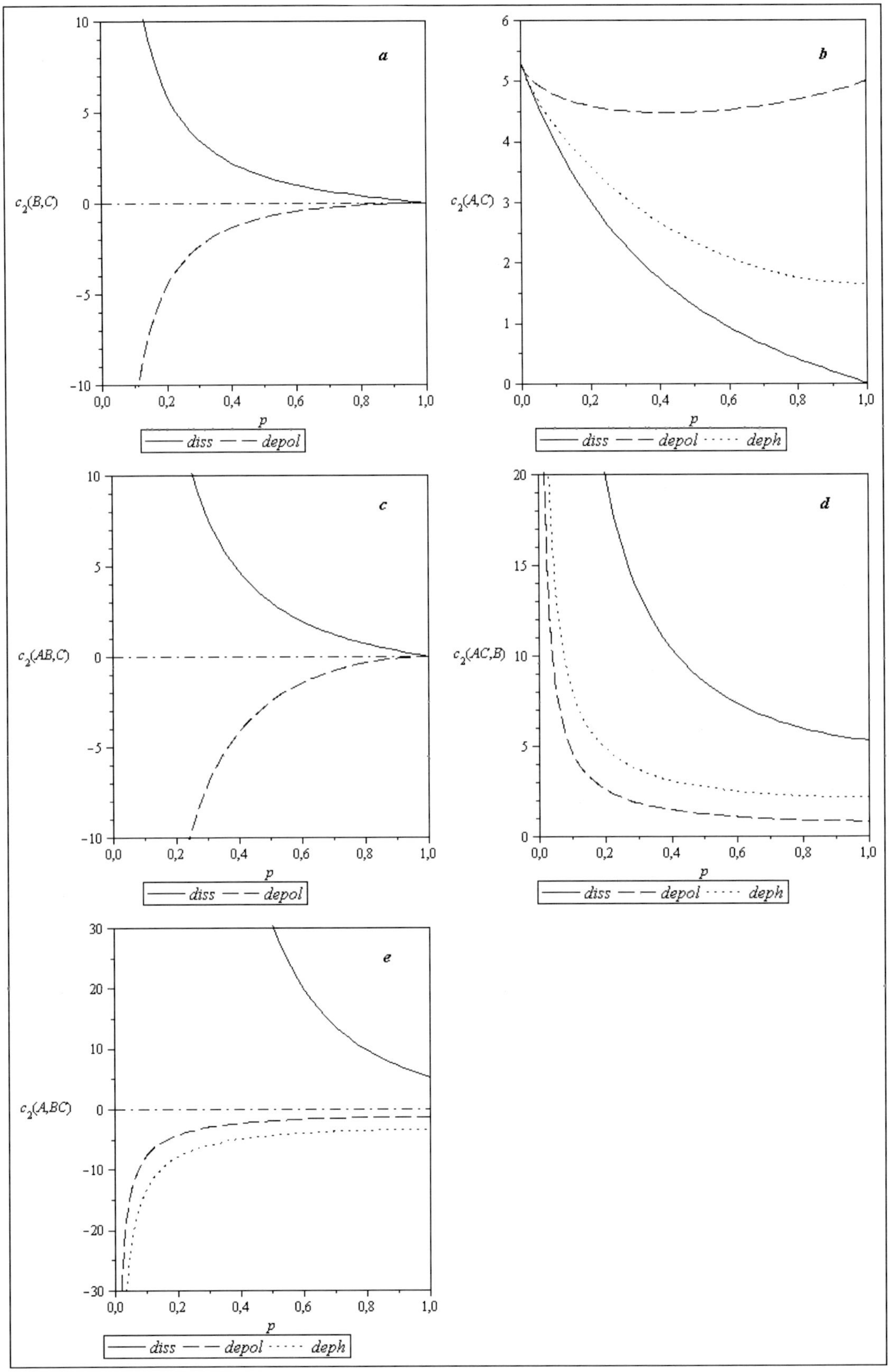

FIGURE 3. Causality in CKW state with decohered qubit C.

The only pair $B-C$ is originally symmetric and therefore one should expect the same behavior of c_2 as in the W-state. Really Figure 3(a) looks qualitatively like Figure 2(a), but we observe stronger causality at the depolarization. As we have seen before, the depolarization and dissipation, acting on one-qubit party, induce the opposite directions of causal connection with another party. But in the pair $A-C$ (Figure 3(b)) all the three kinds of decoherence amplify the original causality $A \to C$. The strongest causality is observed in the intuitively expected case of the dissipation. For the depolarization intuitively we could expect reversal or, at least, attenuation of original causality, but it turns out amplified, though nonmonotonously with $\min c_2$ at $p=0.427$. The reason is that for CKW^C-state $S(A)=1=\max$ and it is impossible to reverse causal connection without decreasing $S(A)$ below this maximum. The depolarization of C at relatively small p opens more the subsystem AC and amplifies the original causality. At $p \to 1$ $S(C)$ increases up to $S(C)=1=\max$ and causality returns to its original level.

In the case of partition $AB-C$ (Figure 3(c)) we have the same as for W-state (Figure 2(b)) and intuitive expected result: the dissipated party C is the effect with respect to AB, whilst the depolarized C is the cause.

If the decohered qubit C is included in the two-qubit party AC (Figure 3(d)) we observe causality $AC \to B$ at any kind of decoherence. The variation from W-state reduces to the stronger and monotonously amplifying causality at the dissipation. The case of partition $A-BC$ (Figure 3(e)) is close, but at depolarization and dephasing $BC \to A$, while at dissipation $A \to BC$. This peculiarity of dissipation is clear. Indeed, at full dissipation ($p=1$) the particle C "disappears" from its two particle party and as a result $c_2(AC,B)=c_2(A,BC)=c_2(A,B)=5.30$.

The original cause A decoherence leads to the different causal picture (Figure 4). One may expect that as a result of increasing dissipation of A, the original causal connection $A \to C$ will at the beginning attenuate until disappearance at some p, after that direction of causality will reverse with further utmost amplification of the connection $C \to A$ as p will tend to 1. In Figure 4(a) it is seen that indeed $c_2(A,C)$ changes its sign at $p=1/2$. But the variation of positive $c_2(A,C)$ (corresponding to directionality of the causal connection $A \to C$) proves to be not monotonous; it has the intuitively unexpected minimum equal to 5.08 at $p=0.103$. Next, in the pair $A-C$ (Figure 4(a)) the depolarization leads to considerable and monotonous amplification of causality as compared to CKW^C (Figure 3(b)). On the one hand, it is in agreement with intuition (the depolarized A becomes the more intensive information source). On the other hand, it can easily be shown that $S(A)$ and $S(C)$ remain independent of p, which demonstrates that one should not consider the marginal entropic asymmetry as a sufficient condition or measure of causality.

In the partition $AB-C$ (Figure 4(b)) the directionality of causal connection is $AB \to C$ at any kind of decoherence, therewith the c_2 curves for depolarization and dephasing are monotonous like Figure 3(c), while for dissipation the curve has the minimum at $p=0.603$. The reason of this curve tends to infinity at $p \to 1$ is that at full dissipation the partition $AB-C$ becomes equivalent to the symmetric one $B-C$. It is notable that at dissipation $\min c_2(AB,C)=\min c_2(A,C)$. And there is an interesting relation, which is valid not only in this model [13]:

$$p(\min c_2(AC,B)) = 1 - p(|c_2(A,C)| = \infty) + p(\min c_2(A,C)).$$

In contrast to the case when the decohered single-party was the original effect C (Figure 3(c)), in the case of decoherence of the original cause A (Figure 4(c)) only dissipation induces the causal connection, therewith A becomes the effect. The monotonous increase of negative $c_2(A,BC)$ simply reflects amplification of causality along with increase of dissipation of the effect A. At the same time $|c_2(AB,C^{diss})| \to 0$ at $p \to 1$ quicker than $|c_2(A^{diss},BC)|$. It reflects the influence of the original (at $p=0$) causality $A \to C$.

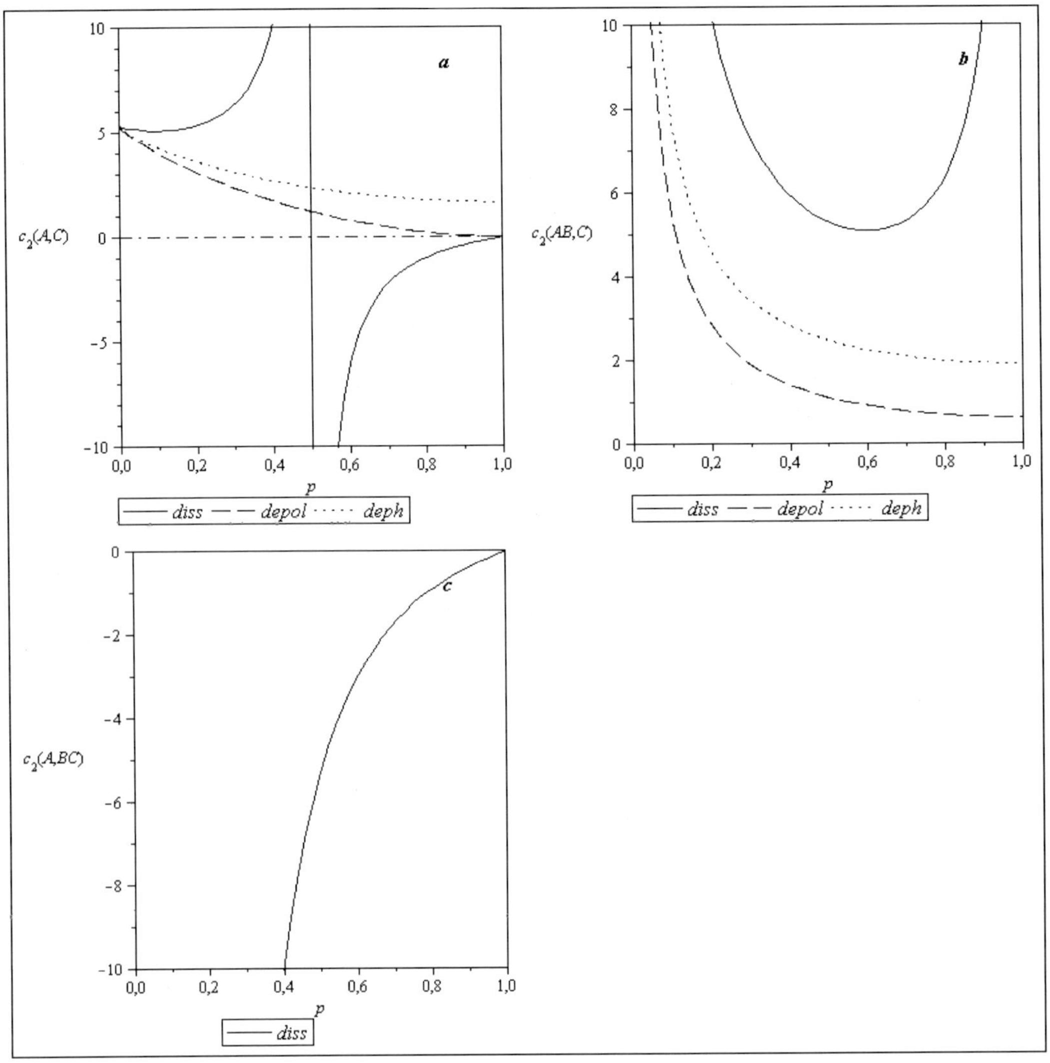

FIGURE 4. Causality in CKW state with decohered qubit A.

3.3. *Relation between entanglement decay and causality*

All the considered states in any partition (except pairwise one in GHZ state) are entangled. Compare decrease of negativity N with increasing p presented in Figures 5–8 with c_2 variation in corresponding Figures 1–4.

If we compare N in accordance to whether the decohered party is a cause or an effect within a given state, we conclude that almost always (except GHZ $AB-C$) $N\,(\text{cause}) < N\,(\text{effect})$. Further if we compare decoherence of the causes (within a given state) with different values of c_2, we conclude that decoherence in the cases of lesser c_2 (stronger causality) leads to the stronger decrease of N. The inverse conclusion follows from comparison of decoherence of the effect with different values of c_2.

Apparently we obtain a quite logical conclusion: the cause decoherence leads to more dramatic decay of entanglement than the effect one and the stronger causality the stronger decay. That is causality reveals the role of asymmetry in information propagation (harmful for entanglement in this context). But in this consideration we have to compare the different kinds of decoherence. Such a consideration can not distinguish the role of causality and decoherence manner.

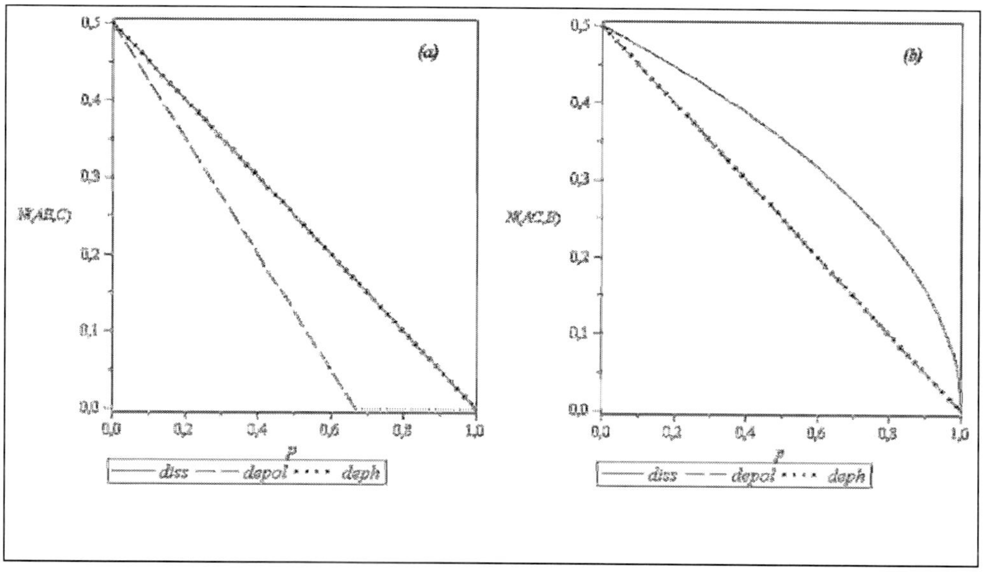

FIGURE 5. Negativity of GHZ state with decohered qubit C.

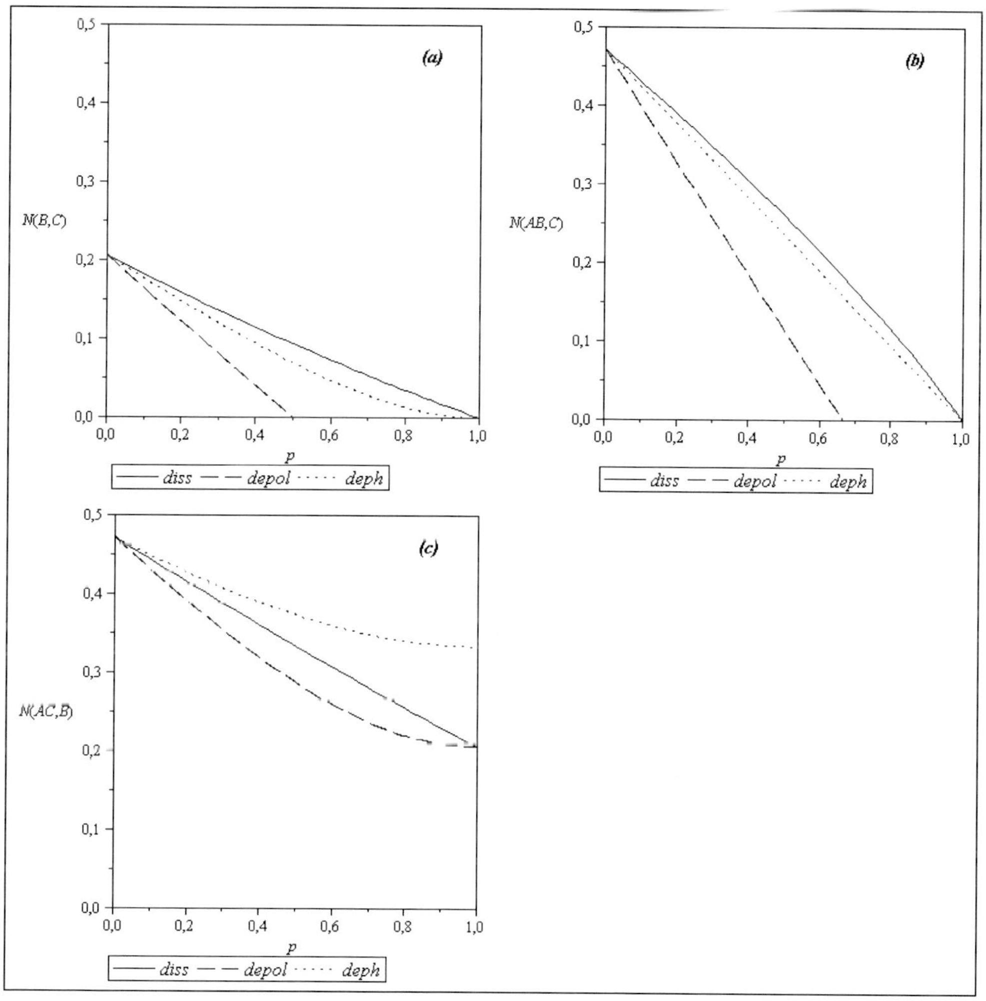

FIGURE 6. Negativity of W-state with decohered qubit C.

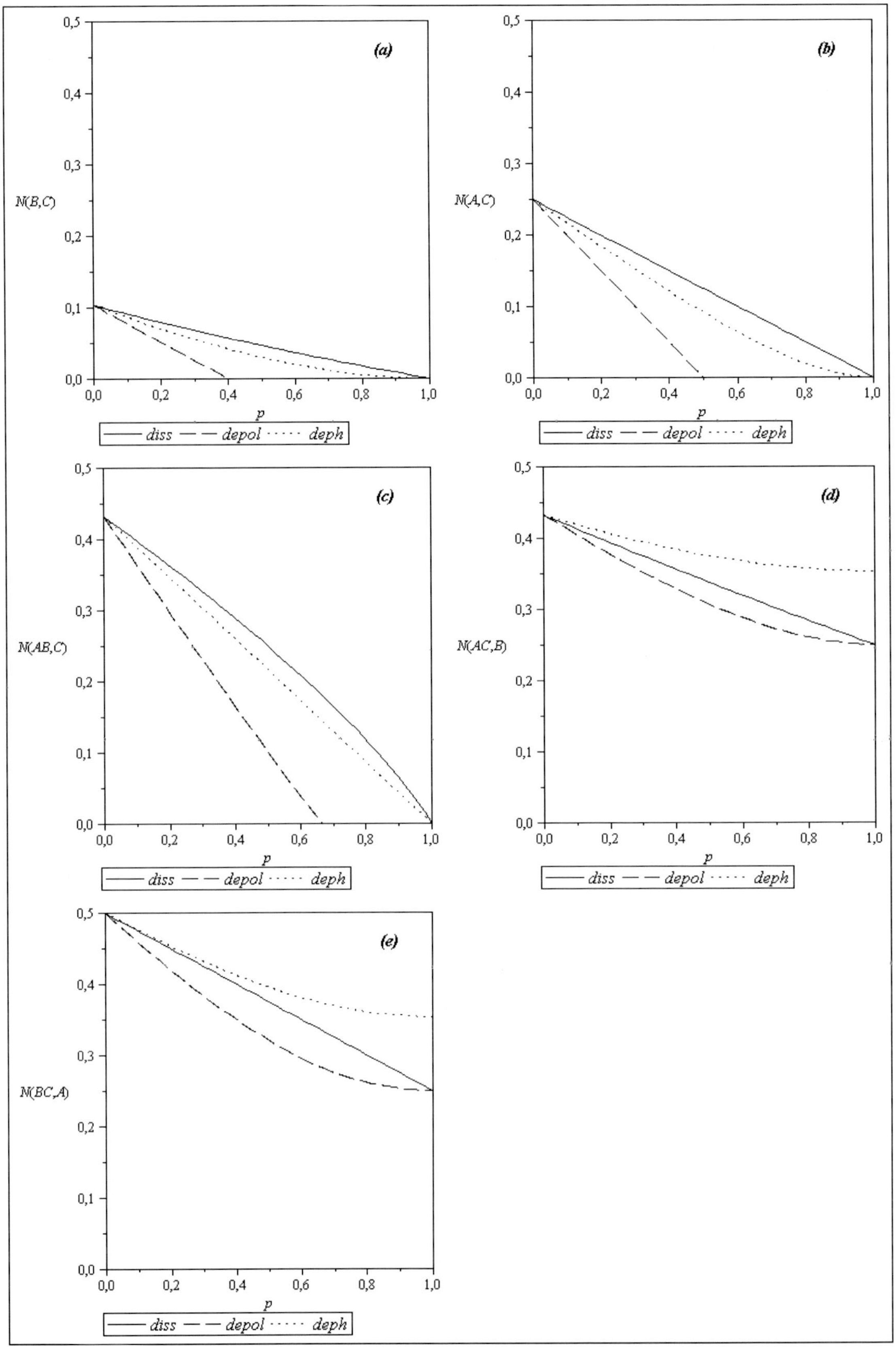

FIGURE 7. Negativity of CKW state with decohered qubit C.

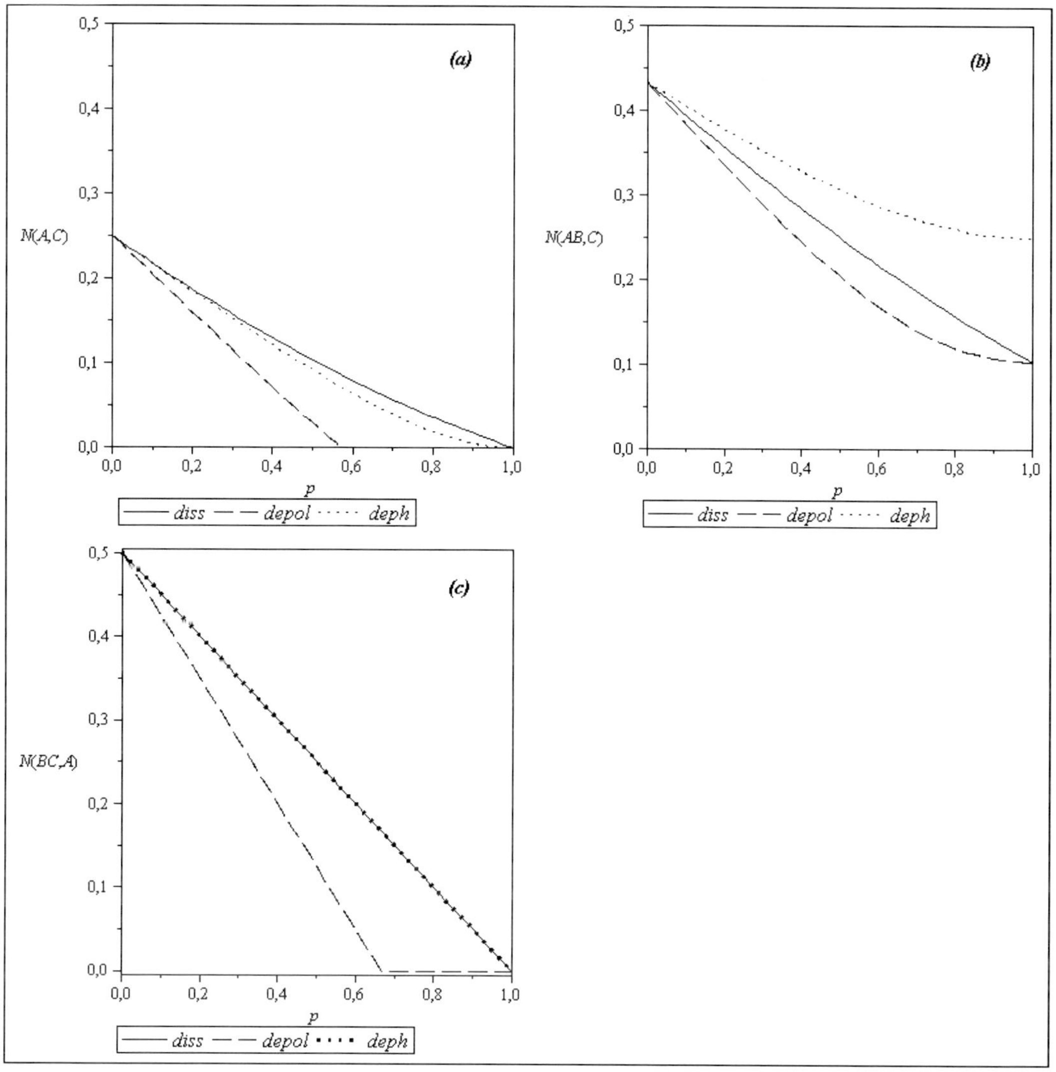

FIGURE 8. Negativity of CKW state with decohered qubit A.

Another approach is the comparison of N and c_2 at fixed both the state and the kind of decoherence. Therefore we should consider the original cause decoherence (A) and effect (C) in CKW state.

Begin with the reduced states. Therewith the case of dephasing is irrelevant ($N^{dephC} = N^{dephA}$). In the dissipated CKW state (Figures 7(b) and 8(a)) $N^{dissC} > N^{dissA}$. As we already know, dissipation of A leads to reversal of original causality (Figure 4(a)); dissipation of C amplifies original causality (Figure 3(b): $|c_2(A, dissC)|<|c_2(dissA, C)|$. We conclude that dissipation, amplifying original causality, destroys entanglement to a lower extent than dissipation, acting against it.

In the depolarized CKW state (Figures 7(b) and 8(a)) $N^{depolC} < N^{depolA}$. And we know that depolarization of A leads to the strong amplification of the original causality (Figure 4(a); depolarization of C only slightly varies it (Figure 3(b)): $|c_2(A, depolC)|>|c_2(depolA, C)|$. We conclude that depolarization, amplifying original causality, destroys entanglement to a lower extent than depolarization, acting almost indifferently or against it.

Both the conclusions coincide. Decoherence by the dissipation or depolarization acting along original causality is better from viewpoint of entanglement persistence, than acting against this causality. In other words, for entanglement persistence one should not "stroke the system against the grain". As a consequence, having compared the above inequalities for N and c_2, we infer that stronger entanglement corresponds to stronger causality. Of cause, this inference is not universal, but it shows that less information-wise symmetric states can be more entangled.

Now consider decoherence in the partitions where a decohered qubit is in the party AC (or AB). That is the party consists of both the original cause and effect. Thus we consider influence of the "internal" causality variation on entanglement in the partition $AC-B$ in CKW state. The corresponding curves of Figures 7(c) and 8(b) evidence at any of three ways of decoherence at any fixed p : $N^{decohC} > N^{drcohA}$. The inference is nontrivial: the decohered internal effect destroys entanglement to a lower extent than the decohered internal cause.

4. Entanglement and Causality in Interaction of a Two-Level Atom With the Field

4.1. *Interaction model*

We consider a bipartite system which consists of a two-level atom, which can be founded in the ground state $|g\rangle_a$ and excited state $|e\rangle_a$, and quantized mode of a field with possible energy states $|0\rangle_f$, $|1\rangle_f$, $|2\rangle_f$,... For simplification we set detuning frequency to zero (resonance case is considered). The interaction is described by Jaynes-Cummings model (JCM) with the Hamiltonian:

$$H = \frac{1}{2}\hbar\omega\sigma_z + \hbar\omega a_f^\dagger a_f + \hbar g(|e\rangle\langle g|_a a_f + |g\rangle a_f^\dagger), \quad (25)$$

where ω is the resonance frequency, $a_f^\dagger$ and a_f are the creation and annihilation operators respectively, g is dipole matrix element which determines Rabi frequency. It is helpful to write the Hamiltonian of the full system as a sum of two commuting parts: $H = H_0 + V$, where $H_0 = \frac{1}{2}\hbar\omega\sigma_z + \hbar\omega a_f^\dagger a_f$ is diagonal matrix and $V = \hbar g(|e\rangle\langle g|_a a_f + |g\rangle a_f^\dagger)$ is matrix with only off diagonal elements and corresponds to the interaction between the subsystems.

The dynamics of the system is described by von Neumann equation:

$$i\hbar\frac{\partial\rho_{af}(t)}{\partial t} = \left[H, \rho_{af}(t)\right], \quad (26)$$

where $\rho_{af}(t)$ is a density matrix of whole system. The Hamiltonian (25) is time independent, the solution of (26) is:

$$\rho_{af}(t) = e^{-iHt/\hbar}\rho_{af}(0)e^{iHt/\hbar}, \quad (27)$$

If the initial state is diagonal (later we will see that such is the case) then $e^{-iHt/\hbar}\rho_{af}(0)e^{iHt/\hbar} = \rho_{af}(0)$, so the resulting solution of Eq. (26) takes the form

$$\rho_{af}(t) = e^{-iVt/\hbar}\rho_{af}(0)e^{iVt/\hbar}. \quad (28)$$

In our consideration we deal only with separable initial states:

$$\rho_{af}(0) = \rho_a(0) \otimes \rho_f(0), \quad (29)$$

where $\rho_a(0)$ and $\rho_f(0)$ are the initial states of atom and field respectively.

In the all variants we consider field initially to be in the mixed thermal state

$$\rho_f(0) = \sum_{i=1}^{\infty} P_i |i\rangle\langle i|_f, \quad (30)$$

where P_i is the probability distribution. As the field satisfies Bose-Einstein statistics, we have

$$P_i = \frac{1}{1+\langle n\rangle}\left(\frac{\langle n\rangle}{1+\langle n\rangle}\right)^i, \quad (31)$$

with the mean photon number

$$\langle n\rangle = \frac{1}{e^{\hbar\omega/k_B T} - 1}, \quad (32)$$

where T is the temperature. As we see $\langle n\rangle$ characterizes the temperature of the field.

Next one should examine a computational problems caused by infinite dimensionality of $\rho_a(0)$. It is evident from Eq. (31) P_i are exponentially decaying series so that contribution of the matrix elements $P_i|i\rangle\langle i|_f$ at sufficiently high i vanishes. Therefore we can confine series P_i at $i = N_{\max} - 1$ and estimate occurred error as

$$\varepsilon = 1 - \sum_{i=0}^{N_{\max}-1} P_i = \left[\langle n\rangle/(1+\langle n\rangle)\right]^{N_{\max}}. \quad (33)$$

For our calculations we have set $N_{\max} = 400$, which gives $\varepsilon = 0.007 < 1\%$ at the highest $\langle n\rangle = 80$. At lower $\langle n\rangle$ calculations are much more accurate.

For the initial states of an atom $\rho_a(0)$ we consider the pure excited and ground states:

$$\rho_{af}^e(0) = |e\rangle\langle e|_a \otimes \rho_f(0), \quad (34)$$

$$\rho_{af}^{g}(0) = |g\rangle\langle g|_a \otimes \rho_f(0). \qquad (35)$$

Finally these states give two different solutions of Eq. (28), which we will discuss further.

4.2. *Computation results*

With the density matrix $\rho_{af}(t)$ we can compute the reduced matrices of atom and field: $\rho_a(t) = Tr_f \rho_{af}(t)$ and $\rho_f(t) = Tr_a \rho_{af}(t)$. From these three matrices we can get time dependent von Neumann entropies of the whole system $S(af)$ and the two subsystems $S(a)$ and $S(f)$ by Eq. (1). Then we can compute the mutual information (3) and the independence functions $i_{a|f}$ and $i_{f|a}$ (2), which determine the course of time (4). As we shall see further, it turns out $i_{f|a} > i_{a|f}$, so in our consideration we use the notation $c_2(f,a)$ to deal with the positive values. And like before, we use the negativity N as a measure of entanglement.

Let us start the overview of computation results from the initial state (34), where the atom is in the pure excited state and the field is in the thermal mixed state. At $\langle n\rangle = 0$ we get the pure oscillating entangled state vector $|\psi_{af}^{e}\rangle = \sin(t)|g,1\rangle_{af} - i\cos(t)|e,0\rangle_{af}$ l Because of whole state purity we have $i_{a|f} = i_{a|f} = -1$ and $c_2(f,a) = \infty$.

In Figure 9(a) the dynamics of negativity for $\langle n\rangle = 0$, $\langle n\rangle = 1$ and $\langle n\rangle = 10$ is presented. As we see, the range of N variations decays and negativity begins to fluctuate near the average value. Moreover, the entanglement is present at $\langle n\rangle > 0$ and $t > 0$ in total agreement with results of Ref. [23].

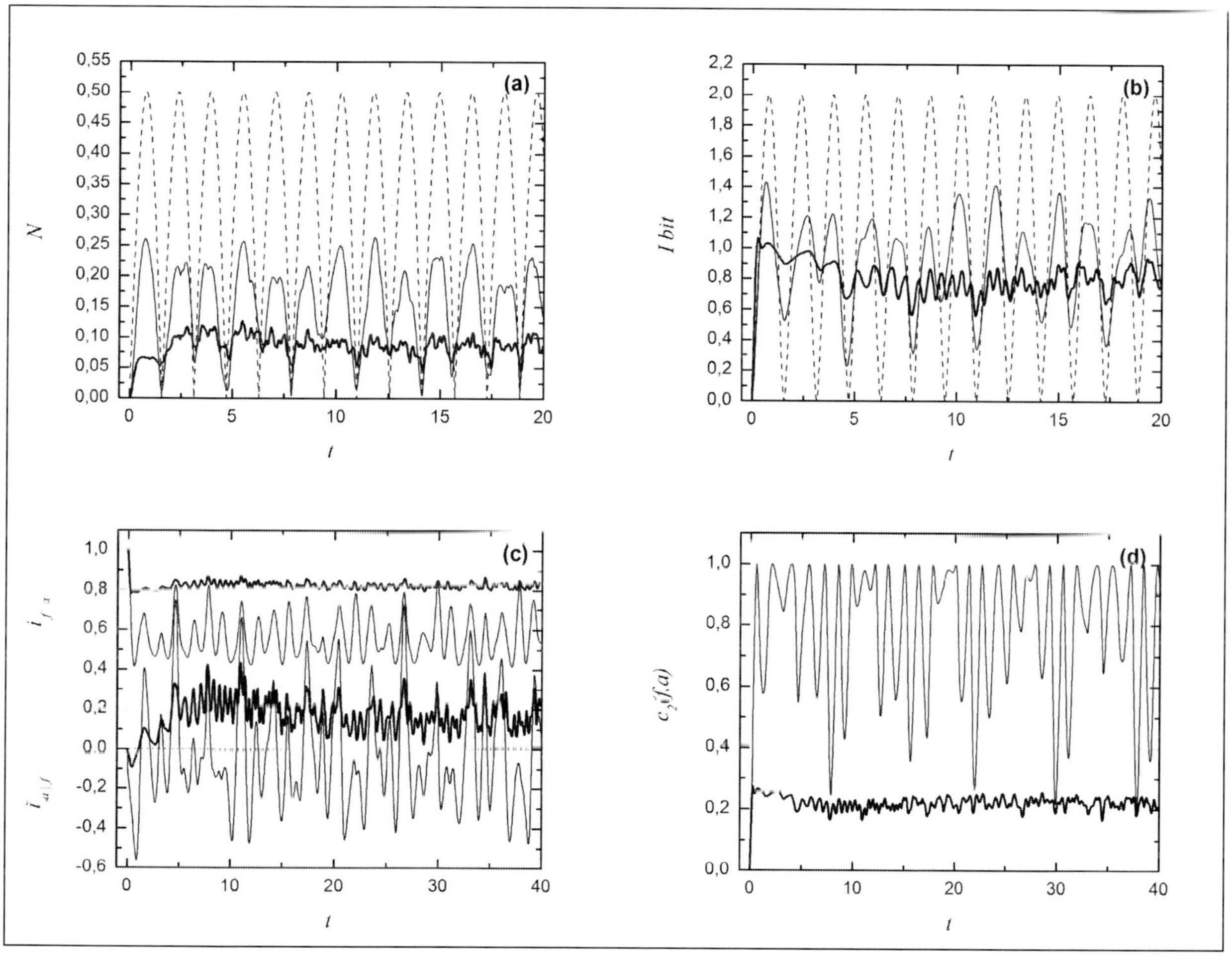

FIGURE 9. Dynamics of characteristics for the initial state (34) at $\langle n\rangle = 1$ (thin lines) and $\langle n\rangle = 10$ (bold lines): (a) negativity (dashed line corresponds to $\langle n\rangle = 0$; (b) information; (c) independence functions $i_{f|a}$ (upper lines) and $i_{a|f}$ (lower lines); (d) causality.

The same behavior shows mutual information I in Figure 9(b), which corresponds to total correlations between the subsystems. It also decays with temperature growth and again it is positive at all times except t = 0 (at nonzero temperature).

More detailed description of correlations independence functions present, which are shown in Figure 9(c). As we see $i_{f|a} > i_{a|f}$, so our system is asymmetric and the field corresponds to the cause (information source) and the atom corresponds to the effect (informational sink). Also it is very interesting, that in contrast to case $\langle n \rangle = 1$, when our system demonstrates quantum properties ($i_{a|f}$ can be negative), at $\langle n \rangle = 10$ both independence functions remain positive at t greater than about 1. It means that system is classical in entropic sense but still is entangled. Causality is presented in Figure 9(d). It is particularly remarkable that that for $\langle n \rangle = 1$ $c_2(f,a)$ is bounded by unit value. Variation and average value of $c_2(f,a)$ decrease, so we see amplification of the causal connection.

It also notable that time of transfer to quasistationary state growths with the temperature rise (the most demonstrable is Figure 9(c). After this time all the characteristics of the system begin to fluctuate near some average values. The extent of such fluctuations goes down with the temperature increase.

Next consider the case of initial state (35), where an atom is in the ground state, while the field still is in the thermal state (Figure 10). At $\langle n \rangle = 0$ we have stationary separable state vector $|\psi_{af}^{g}\rangle = |g,0\rangle_{af} = const$, which is classically uncorrelated too: $i_{a|f} = i_{a|f} = 1$.

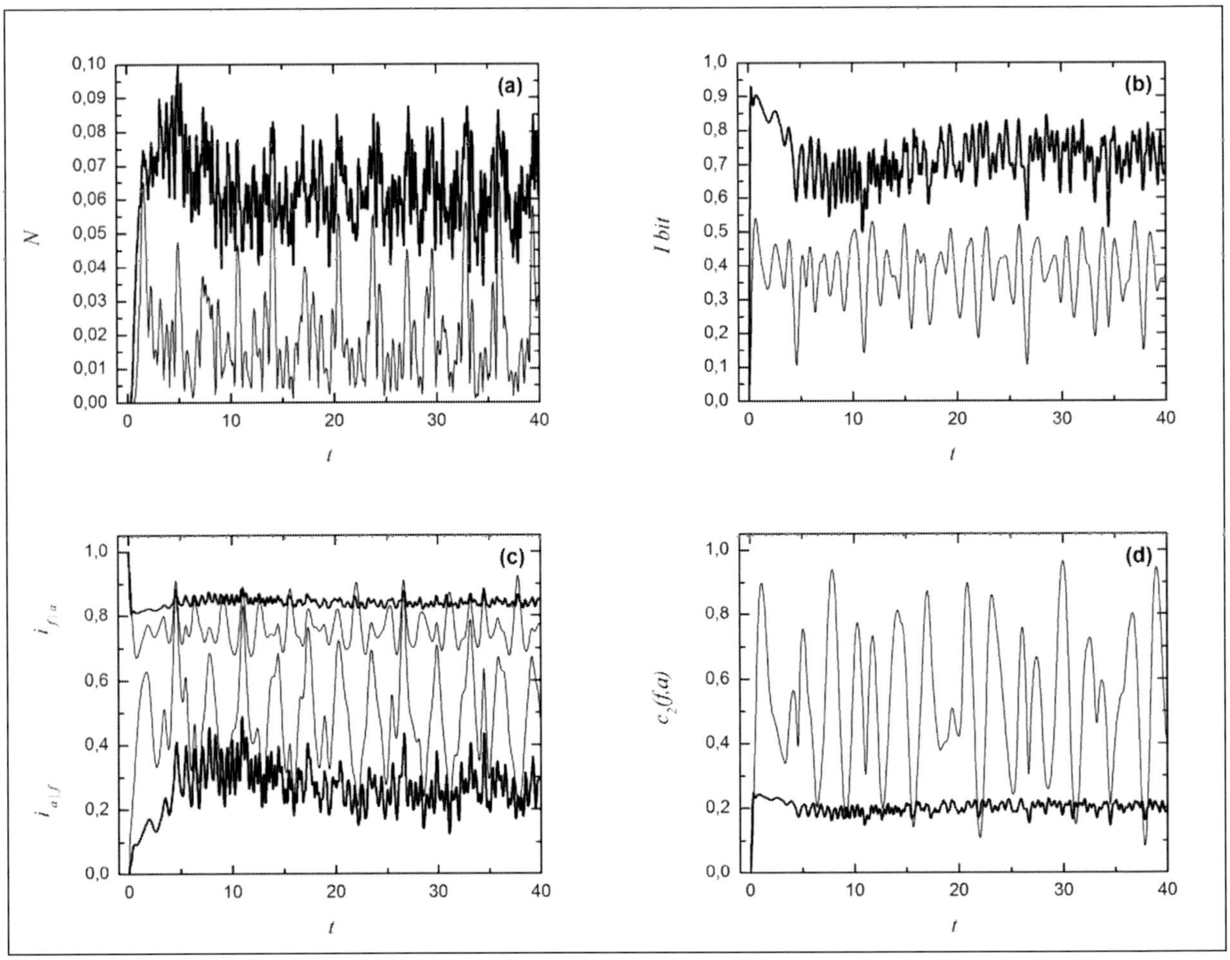

FIGURE 10. Dynamics of characteristics for the initial state (35) at $\langle n \rangle = 1$ (thin lines) and $\langle n \rangle = 10$ (bold lines): (a) negativity; (b) information; (c) independence functions $i_{f|a}$ (upper lines) and $i_{a|f}$ (lower lines); (d) causality.

As it is seen from Figure 10(a) with a rise of $\langle n \rangle$ there is an increase of the negativity: the atom in the ground state becomes entangled with nonzero energy states of the field ($|1\rangle_f, |2\rangle_f, \ldots$). The same behavior demonstrates information in Figure 10(b). It might be presupposed (as the temperature is held to have a destructive influence on correlations) that there is some $\langle n \rangle$ after which an entanglement and information would decrease. But it is not the case. The independence functions in Figure 10(c) show that system always is classical in entropic sense. And again the independence functions demonstrate asymmetry between the subsystems: $i_{f|a} > i_{a|f}$ (the field state still is the cause with respect to the atom state. Causality is presented in Figure 10(d). As well as in previous case it amplifies with the temperature increase.

As we have seen, all parameters of the system for both considered initial states fluctuate near some average values at high temperatures. It seems logically to estimate these values as functions of $\langle n \rangle$. We have chosen time series $150 \le t \le 400$ with time step $dt = 0.5$ and have computed the average values $N_{av}, I_{av}, i_{f|a\,av}, i_{a|f}, c_{2av}$ for the set of mean photon numbers $1 \le \langle n \rangle \le 80$. $t_{\min} = 150$ has been chosen to avoid getting in time of transfer to quasistationary state it is high enough for our biggest $\langle n \rangle = 80$. Time step $dt = 0.5$ has been chosen as it does not correspond to any of system eigenfrequencies. Also with average values we have stored minimal and maximal values of characteristics to see variability at our time series. The results of such time averaging are presented in Figure 11.

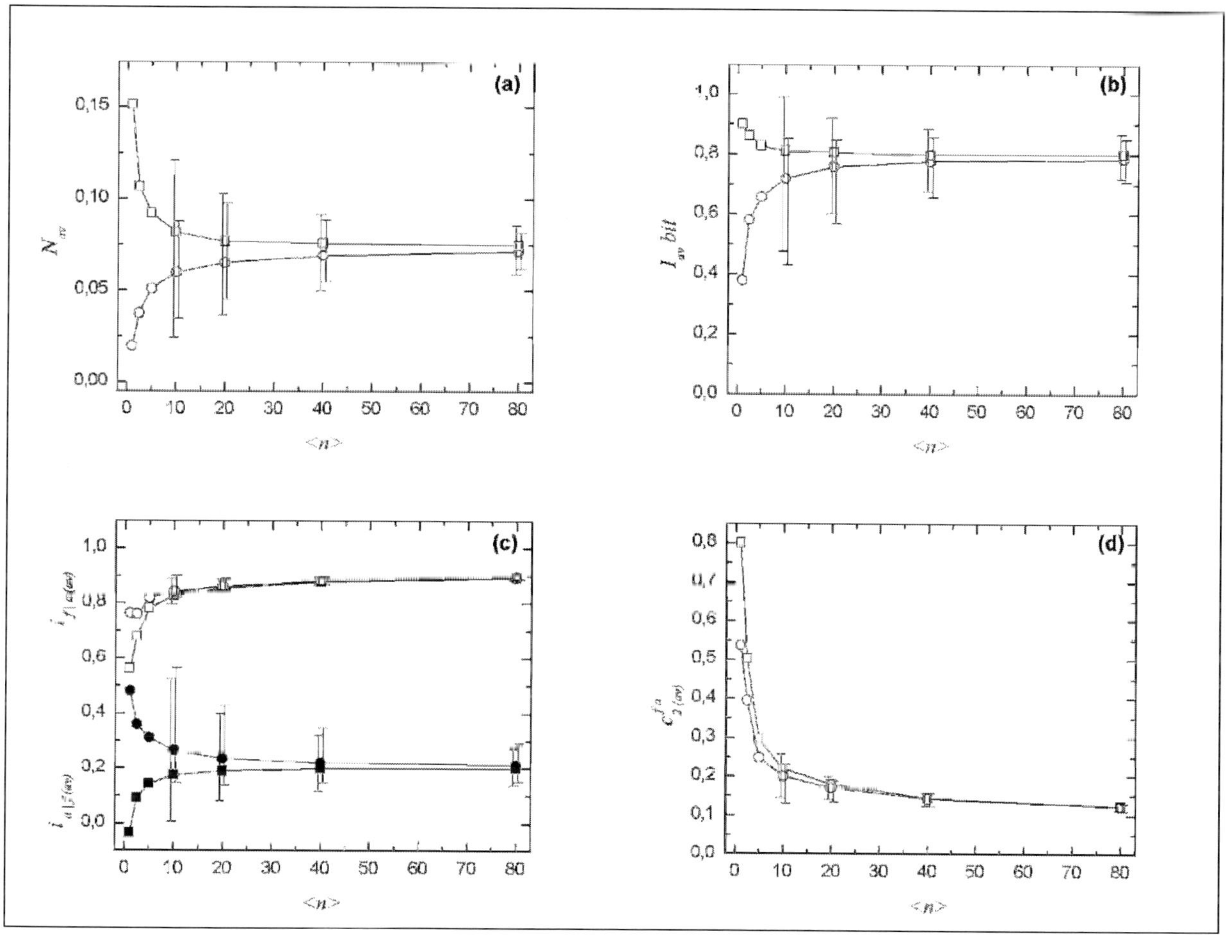

FIGURE 11. Time averaged characteristics and corresponding minimal and maximal values as functions of $\langle n \rangle$ for different initial states: squares and right vertical lines (34), circles and middle vertical lines (35). Averaged characteristics: (a) negativity; (b) information; (c) independence functions $i_{f|a}$ (empty symbols) and $i_{a|f}$ (filled symbols); (d) causality.

First, let us discuss the general features. As noted above the extent of fluctuation for all the parameters decreases with the temperature growth. Moreover the distance between the curves for different initial states decreases: it means that the significance of atom initial states for the average characteristics disappears at the high temperature. It totally corresponds to the result that field is an information source that is a cause.

The most interesting is Figure 11(a), which demonstrates dependence of negativity on $\langle n \rangle$. It is expectable that for initially pure excited state of the atom entanglement decreases with the temperature rise, but it surprisingly does not vanish. It tends to an asymptotic value, as well as the curve for the initially ground state. It is intriguing that for the ground initial atom state there is an amplification of entanglement with growth of the temperature, so in this case the temperature creates entanglement.

All the other characteristics also have such asymptotic values, as it is seen in Figures 11(b)–(d). We can estimate that for $\langle n \rangle \gg 1$ the averaged values are: $N_{av} \approx 0.07$ (14% of maximal value), $I_{av} \approx 0.8$ *bit*, $i_{a|f(av)} \approx 0.90$, $i_{f|a} \approx 0.25$, $c_2(f,a)_{av} \approx 0.25$ – the field state is the cause with respect to the atom state.

We can summarize the time averaged results (Figure 11) as follows. Information, reflecting total correlations behaves similarly to the negativity. But the independence functions are completely positive that is classical. The atom-field state is entangled, but correlations are apparently classical. The field state is the cause with respect to the atom one under any conditions. However relation between the degrees of causality and entanglement at the low temperature strongly depends on the initial conditions. At the high temperature both causality and entanglement are indifferent to them.

5. Teleportation

Teleportation is well known and amazing quantum phenomenon. The most interesting fact is that teleportation protocol can be considered as a process involving hidden signal transmission in reverse time. And it does not turn out a matter of "sophisticated" interpretation. The experiment based on postselection gave a direct proof of such a time reversal [18]. Another experiment demonstrated the possibility of teleportation traveling along closed time-like curve without the classical paradoxes [24]. At last, recently the experiment on entanglement swapping (that is teleportation of entanglement) has demonstrated, even without postselection, quantum information transfer from the future to the past; in fact it has demonstrated a possibility of observation of the random future as the existing reality [25]. Thus teleportation is just such a process where determination of causality irrespective to time direction is relevant.

We consider the standard three-particle (three-qubit) teleportation protocol (Figure 12).

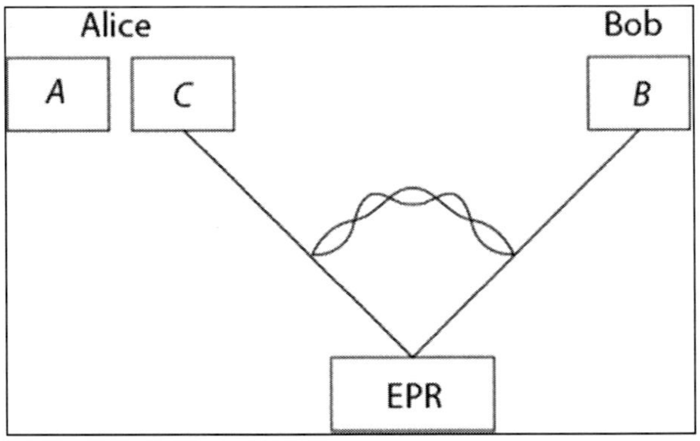

FIGURE 12. Qubit A teleports onto qubit B.

The input particle, which state to be teleported by Alice, is A; the EPR source produces two entangled particles in the state

$$|\Psi^-\rangle_{BC} = \frac{1}{\sqrt{2}}(|01\rangle - |10\rangle), \tag{36}$$

one of which, B goes to Bob, another one, C goes to Alice, who performs Bell measurement with four possible outcomes:

$$|\Phi^+\rangle_{AC} = \frac{1}{\sqrt{2}}(|00\rangle + |11\rangle). \tag{37}$$

$$|\Phi^-\rangle_{AC} = \frac{1}{\sqrt{2}}(|00\rangle - |11\rangle), \tag{38}$$

$$|\Psi^+\rangle_{AC} = \frac{1}{\sqrt{2}}(|01\rangle + |10\rangle), \tag{39}$$

$$|\Psi^-\rangle_{AC} = \frac{1}{\sqrt{2}}(|01\rangle - |10\rangle), \tag{40}$$

If her result is (40) the output state B coincides with input A, if not – Bob performs a unitary operation on B to complete the protocol. In any case Bob needs information about Alice result, which she sends him through an ancillary classical channel. As we are interested in investigation of quantum information namely, we exclude this channel. Instead Bob may measure his particle B. Note that any measurement (by Alice or by Bob) implies dephasing.

The state of B, accordingly commonly accepted interpretation, changes instantaneously at the moment of Alice joint AC measurement. But accordingly time reversal formalism developed by Laforest, Baugh and Laflamme (LBL) and the corresponding experiments [18, 25] the B "knows" about future AB

measurement from very beginning. We aim to clear up this question with causal analysis. We will do it in the framework of usual tensor product treatment and LBL-like time reversal treatment.

5.1. *Tensor-product treatment*

The peculiarity of our approach is that we consider Alice joint AC measurement as dephasing of degree p accordingly to Eq. (12). One may consider it as a soft measurement. More interesting consideration is that dephasing is a process from $p=0$ (measurement without record, that is pre-measurement) to $p=1$ (measurement is completed). Thus p is indirect time measure of this, certainly very fast process. The measurement which Bob may do to get to know something about his particle also is dephasing of a degree p_1; we will limit ourselves by the cases $p_1=0$ and $p_1=1$

So let the original matrix is:

$$\rho^0_{ABC} = \rho_A \otimes \rho_{BC} = \rho_A \otimes \left|\Psi^-\right\rangle\left\langle\Psi^-\right|_{BC}, \qquad (41)$$

Expand the matrix in Bell measurement basis:

$$\rho^0_{ABC} = \sum_{i,j,k,l} F_{ijkl}\left|\Psi_i\right\rangle\left\langle\Psi_j\right|_{AC} \otimes \left|\phi_k\right\rangle\left\langle\phi_l\right|_B, \qquad (42)$$

where $i,j=1,2,3,4$ correspond to the states (37), (38), (39) and (40) respectively; $k,l=1,2$ correspond to $\left|0\right\rangle_B, \left|1\right\rangle_B$ respectively; and $F_{ijkl} = \left\langle\Psi_i,\phi_k\right|\rho^0_{ABC}\left|\Psi_j,\phi_l\right\rangle$.

Alice Bell measurement of AC means the replacement:

$$F_{ijkl} \to F_{ijkl}(1-p) \text{ at } i \neq j, \qquad (43)$$

while Bob measurement of B means the replacement:

$$F_{ijkl} \to F_{ijkl}(1-p_1) \text{ at } k \neq l. \qquad (44)$$

Transforming (42) according to (43) and (44), we obtain the resulting full matrix ρ_{ABC} (which explicit expression is very longish) and can do all the subsequent computation for causal analysis.

Consider the results for the simplest different variants of the input states A. A common property of all the variants described below turns out the fact, that in contradiction with classical intuition, there are no causal connections between any one-particle parties, in particular, $A \to B$. It is a simple consequence of the no-cloning theorem. Another common property is identity of causality in the partitions $AC-B$ and $AB-C$. So, below we concentrate on the partition $AC-B$.

1. A is in the definite state: $\rho_A = \left|0\right\rangle\left\langle 0\right|$. Figure 13 demonstrates qualitative quite expectable result. At $p=0$ causality is absent ($c_2(AC,B)=\infty$) as the state ABC is pure. At finite p causality $AC \to B$ appears that if information goes from Alice to Bob. It is natural that at $p_1=1$ causality is stronger since dephased B is more definite. When Alice completes her measurement ($p=1$), causality is most expressed: $c_2(AC,B)=1$ at any p_1 because B is already dephased together with C.

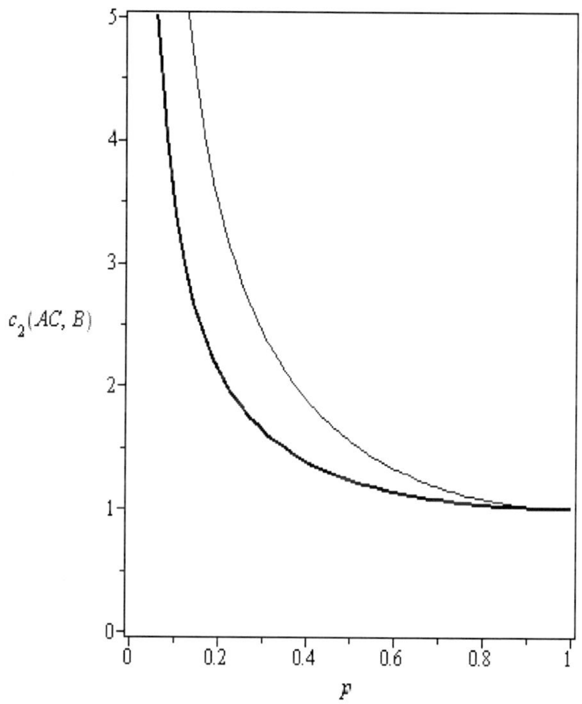

FIGURE 13. Causality at teleportation of $\rho_A = \left|0\right\rangle\left\langle 0\right|$ at $p_1=0$ (thin line) and $p_1=1$ (bold line).

2. A is in the maximally mixed state: $\rho_A = 1/2\left(\left|0\right\rangle\left\langle 0\right| + \left|1\right\rangle\left\langle 1\right|\right)$. Figure 14 demonstrates stronger causality than in above case. The original ABC is mixed therefore the course of time is finite at $p=0$ already: $c_2(AC,B)=2$ at $p_1=0$ and $c_2(AC,B)=1$ at $p_1=1$. At $p \to 1$ causality amplifies to the utmost value: $c_2(AC,B) \to 0$ (that means the random input completely tends to determine a certain output, while recovery of input by output tends to full impossibility).

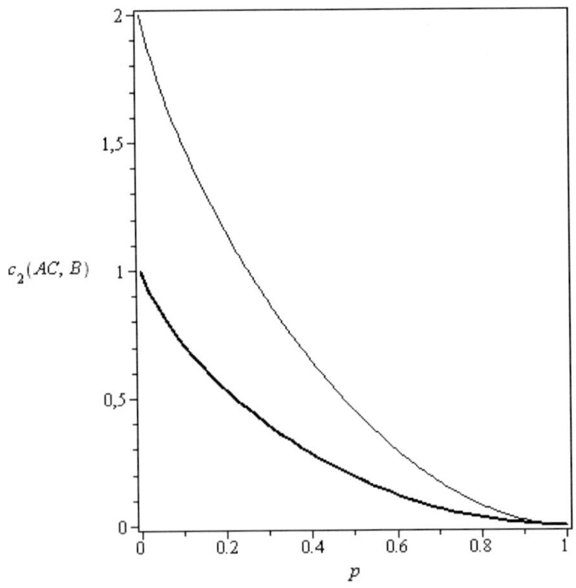

FIGURE 14. Causality at teleportation of $\rho_A = \frac{1}{2}(|0\rangle\langle 0| + |1\rangle\langle 1|)$ at $p_1 = 0$ (thin line) and $p_1 = 1$ (bold line).

3. A is in the pure equilibrium state: $\rho_A = 1/2(|0\rangle\langle 0| + |0\rangle\langle 1| + |1\rangle\langle 0| + |1\rangle\langle 1|)$. Figure 15 demonstrates that again at $p = 0$ $c_2(AB, C) = \infty$ as the state ABC is pure. But as $p \to 1$ causality is different at different p_1: $c_2(AC, B) = 1$ at $p_1 = 0$ and $c_2(AC, B) \to 0$ at $p_1 = 1$. The latter is a clear result of Bob measurement of output qubit which selects a definite state from the superposition.

Although qualitatively these formal results agree with intuition (namely Alice (AC) send quantum information to Bob (B), note that any direction of time in the established causal link $AC \to B$ will do. Indeed, we nowhere specified when Bob measures (dephases) B. Bob's measurement may occur after Alice's measurement as well as *before.* Causality in reverse time is allowed. But "telegraph in the past" is impossible since a result of Alice's measurement is random. Instead Bob has the possibility of observation of *the random future as existing reality.* Next, we saw that Bob's measurement can only amplify the degree of causality $AC \to B$, but not generate it. That is the result of Bell measurement constitutes a cause with respect to every qubits of entangled pair just since moment of their birth (like [25]). There is no a contradiction with the above original statement: "the EPR source produces two entangled particles in the state (36)". This statement in fact is conditioned on absence of the future Bell measurement. Commonly accepted realizing of causality as directed only from the past to the future impeded to perceive that conditionality before.

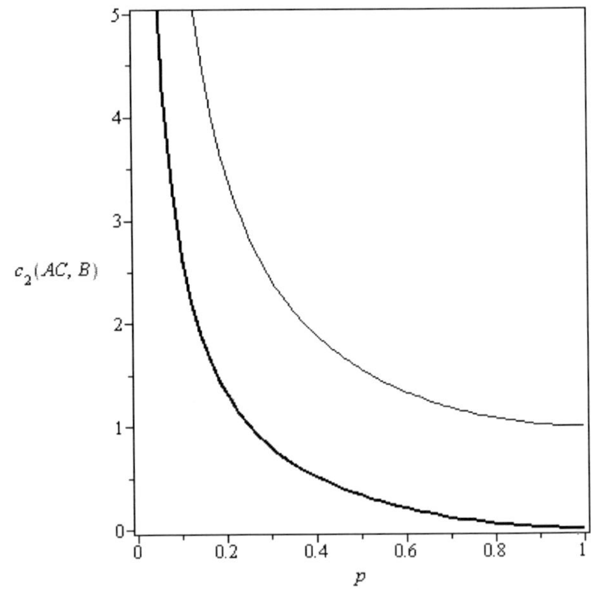

FIGURE 15. Causality at teleportation of $\rho_A = \frac{1}{2}(|0\rangle\langle 0| + |0\rangle\langle 1| + |1\rangle\langle 0| + |1\rangle\langle 1|)$ at $p_1 = 0$ (thin line) and $p_1 = 1$ (bold line).

5.2. *Time-reversal treatment*

We follow LBL time reversal treatment described in detail in Ref. [18] with some simplification. The main idea is that Bell measurement and EPR source act as "time mirrors". The input qubit (riding on the different particles as the carriers) travels to Alice's Bell measurement device, reflects, travels backward in time to the EPR source, reflects and goes to Bob.

Every reflection is correspondent to some operator W: $|\psi\rangle^{tr} = W_i|\psi\rangle$, were *tr* is symbol of time reversal, the components $W_i^{a,b} = \langle b, a | \psi_i\rangle$ in Bell basis are: $W_1 = \mathbf{1}$, $W_2 = \sigma_z$, $W_3 = \sigma_x$, $W_4 = -i\sigma_y$. Qubit travel and transformations are shown in Fig. 16. The $W_?$ means that we do not know results of Bell measurement; W_4 corresponds to our convention that the source generates the state (36).

It is not difficult to get the whole density matrix. But we are unable to implement gradual dephasing during Bell measurement. Instead we consider the most important practically both the extreme cases: measurement with ignoring of Bell measurement record (to compare with the case $p = 0$ of tensor product treatment) and with taking into account the record (to compare with the case $p=1$ of tensor product treatment).

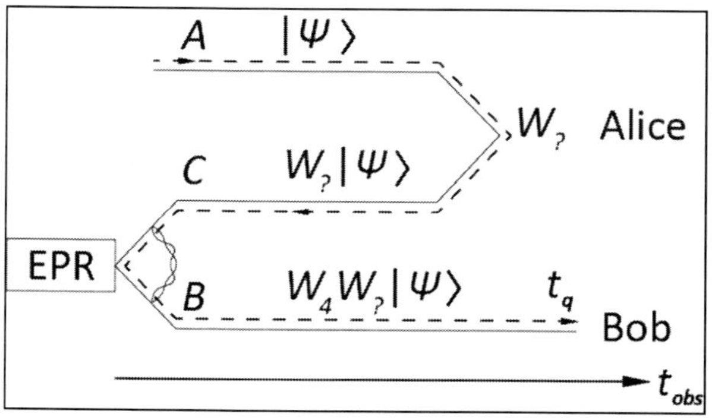

FIGURE 16. Time reversal treatment of teleportation from A to B; t_{obs} is time of an external observer, t_q is proper time of the teleporting qubit.

In the first case: (corresponding to $p=0$):

$$\rho_{ABC} = \frac{1}{4}\sum_{i=1}^{4} |\psi, W_i\psi, W_4W_i\psi\rangle\langle\psi, W_i\psi, W_4W_i\psi|, \quad (45)$$

where $|\psi\rangle\langle\psi|$ is any input (A) state.

The second case (corresponding to $p=1$) is some more complicated, so we have restrict ourselves to the case of diagonal input states A that is $|\psi\rangle = |j\rangle\langle j|$, $|j\rangle = |0\rangle, |1\rangle$. We introduce to the protocol a new object D which records the Bell measurement at Alice site. The state of D in Bell basis is $|\psi_i\rangle$ $i=1,2,3,4$. As a result we have:

$$\rho_{ABCD} = \frac{1}{8}\sum_{j=1}^{2}\sum_{i=1}^{4} | j, W_i j, W_4W_i j, \psi_i\rangle\langle j, W_i j, W_4W_i j, \psi_i|. \quad (46)$$

We have computed all the c_2 (under condition $p_1=1$) with the following results.

In the first case: (corresponding to $p=0$) for A is in the definite state $c_2(AC,B)=\infty$; for A is in the maximally mixed state $c_2(AC,B)=1$; A is in the pure equilibrium state $c_2(AC,B)=\infty$. Thus we have exactly the same result as in tensor product treatment.

In the second case (corresponding to $p=1$) we must take D instead of AB ($c_2(AC,B)$ are the same as in above case). For A is in the definite state $c_2(D,C)=1$ that exactly the same as in tensor product treatment. If A is in the maximally mixed state $c_2(D,C)$ is indefinite (in tensor product treatment, having variable p we could take the limit at $p\to 1$: $c_2(AC,B)\to 0$).

Thus time reversal formalism gives in fact the same mathematical results as traditional tensor product one. But physically it proves our conclusion about causality in reverse time in much more strait manner. The random future D influences via backward time traveler C on the factual result of EPR emission. See Figure 16.

With the recorder D we also can consider a partition $AD-B$ (which is equivalent of $AD-C$). As a result for A in the maximally mixed state we have $c_2(AD,B)=c_2(AD,C)=1/2$. This value is minimal among all presented above finite values of c_2, therefore causal connections $AD\to B$ and $AD\to C$ are the strongest ones. Since Bell measurement at Alice site occurred later (by time of an observer t_{obs}) than EPR emission took place, this result is further proof of causality in reverse time. The EPR pair "knows" about the random future – interaction with random A and random D. For A in the definite state $c_2(AD,B)=c_2(AD,C)=1$. Therefore the greater is randomness of the future, the stronger is time reversal causality. Obviously in the case of deterministic future time reversal causality must absent. It is just impossibility of "telegraph to the past".

6. Conclusions

The quantum causal analysis is extension of the classical one; therefore it is extension of formalized intuitive understanding of causality. Indeed in the simple situations our formal results are not surprising, e.g. when dissipating particle proved to be an effect. But even in these cases our formal approach has an advantage over usual informal, intuitive one, because it provides the *quantitative* measure of causal connection. However in the Quantum World common intuition often fails in rather simple systems, consisting of a few particles. Causal analysis quite works with any system, although its results may seem contrintuitively. Therewith these results turns out practically useful, e.g. in explanation of peculiarities of entanglement decay under different kind of decoherence or in relation between intersystem causality and consequences of asymmetric decoherence.

Very simple and general property of quantum causality is that it can be finite only in the mixed states. In our previous works [9–13] we interpreted this fact as quantum causality can be finite only in the open systems. But in the model of atom-field interaction considered in this paper the system is closed, the state mixedness, necessary for causality, was created before, at the stage of thermal state preparation. Therefore we have to correct interpretation as follows: *quantum causality can be finite only in the systems, which are or were open.*

The most prominent property of quantum causality is that it can exist in direct as well as *in reverse time.* Remarkably time reversal causality does not imply the naive classical paradoxes. We have considered such unusual causality in connection with contemporary teleportation experiments [18, 24, 25]. But, of course its significance is much wide, e.g. for interpretation and development of the forecasting experiments based on macroscopic entanglement [7, 8].

Acknowledgment

This work was supported by RFBR (grant 12-05-00001).

References

[1] N. A. Kozyrev, "On the possibility of experimental investigation of the properties of time", in Time in Science and Philosophy, edited by J. Zeman, Prague: Academia, 1971, pp. 111-132.

[2] S. M. Korotaev, Geomagnetism and Aeronomy 32, 27 (1992).

[3] S. M. Korotaev, Geomagnetism and Aeronomy 35, 387 (1995).

[4] S. M. Korotaev, V. O. Serdyuk, V. I. Nalivaiko, A. V. Novysh, S. P. Gaidash, Yu. V. Gorokhov, S. A. Pulinets and Kh. D. Kanonidi, Phys. of Wave Phenomena 11, 46 (2003).

[5] S. M. Korotaev, A. N. Morozov, V. O. Serdyuk, J. V. Gorohov and V. A. Machinin, NeuroQuantology 3, 275 (2005).

[6] S. M. Korotaev, Int. J. of Computing Anticipatory Systems 17, 61 (2006).

[7] S. M. Korotaev, V. O. Serdyuk and J. V. Gorohov, Hadronic Journal 30, 39 (2007).

[8] S. M. Korotaev and V. O. Serdyuk, Int. J. of Computing Anticipatory Systems 20, 31 (2008).

[9] S. M. Korotaev and E. O. Kiktenko, AIP Conference Proceedings 1316, 295 (2010).

[10] S. M. Korotaev, Causality and Reversibility in Irreversible Time, Scientific Research Publishing, 2011.

[11] S. M. Korotaev and E. O. Kiktenko, "Causality in the entangled states" in Physical Interpretation of Relativity Theory edited by P Rowlands, Moscow, BMSTU PH, 2011, pp. 140-149.

[12] E. O. Kiktenko and S. M. Korotaev, Phys. Lett. A376, 820 (2012).

[13] S. M. Korotaev and E.O. Kiktenko, Physica Scripta. 85, 055006 (2012).

[14] J. G. Cramer, Phys. Rev. D22, 362 (1980).

[15] K. Zyczkowski, P. Horodecki, M. Horodecki and R. Horodecki, Phys. Rev. A65, 012101 (2002).

[16] A. Borras, A. R. Plastino, M. Casas and A. Plastino, Phys. Rev. A78, 052104 (2008).

[17] A. S. Elitzur and S. Dolev "Is there more to T?," in The Nature of Time: Geometry, Physics and Perception, edited by R. Buccery, M. Saniga and W. M. Stuckey, Kluwer Academic Publishers, 2003, pp. 297-306.

[18] M. Laforest, J. Baugh and R. Laflamme, Phys. Rev. A73, 032323 (2006).

[19] V. Coffman, J. Kundu and W. K. Wootters, Phys. Rev. A61, 052306 (2000).

[20] W. Dür, Phys. Rev. A63, 020303 (2001).

[21] S. S. Jang, Y. W. Cheong, J. Kim and H. W. Lee, Phys. Rev. A74, 062112 (2006).

[22] W. Song and Z.-B. Chen, Phys. Rev. A 76, 014307 (2007).

[23] S. Bose, I. Fuentes-Guridi, P. L. Knight and V. Vedral, Phys. Rev. Lett. 87, 050401 (2001).

[24] S. Lloyd, L. Maccone, R. Garcia-Patron, V. Giovannetti, Y. Shikano, S. Pirandola, L. A. Rozema, A. Darabi, Y. Soudagar, L. K. Shalm, and A. M. Steinberg, Phys. Rev. Lett. 106, 040403 (2011).

[25] X.-S. Ma, S. Zotter, J. Kofler, R. Ursin, T. Jennewien, Č. Brukner and A. Zeilimger, *Nature Physics* **8**, 479 (2012).

Relativistic Entanglement

JOHN E. CARROLL[a] & ADRIAN H. QUARTERMAN[b]

Centre for Photonic Systems, University of Cambridge, 9 J.J. Thomson Avenue, Cambridge, CB3 0FA, UK.
[a]jec1000@hermes.cam.ac.uk; [b]ahq20@hermes.cam.ac.uk

A new explanation is given for correlated measurements on widely separated entangled particles. This explanation uses relativity where events in the same proper frame can be local to one another. Proper frames are shown to be possible, even for light waves, because of diffraction. Waves ΨA and ΨB with distinct proper frames Γp and Γn, travel in opposing directions to reach well separated observers A and B. Consideration of the formality of combined time-frequency-reversal leads to concepts of adjoint waves ΨAa and ΨBa with phase velocities still travelling to A and B respectively. Measurements are now associated with ΨAa*ΨA and ΨBa*ΨB rather than ΨA*ΨA and ΨB*ΨB. It is shown how waves ΨA and ΨBa* can have the same proper frame Γp while ΨAa* and ΨB can have the same proper frame Γn. Consequently these waves ΨA and ΨBa* can be said to be relativistically local and could have correlated properties without any violation of the principle of locality. Similar remarks apply to ΨAa* and ΨB. It is no longer surprising if widely separated particles, represented by measurements of ΨAa*ΨA and ΨBa*ΨB, have correlated properties though the theory's generality prevents detailing specific correlations. Similar arguments are advanced for Dirac waves. However essential quantum aspects of entanglement lie in carefully timed measurements that choose the correct waves with the correct proper frames. Classical measurements at A and B contain random sets of waves from different proper frames averaged over time and this generally returns uncorrelated measurements. The paper provides satisfying links between relativity, entanglement, quantum wave collapse, and classical independence of non-local observations.

Keywords: Relativity; Proper frames; Entanglement.

1. Introduction

Entanglement has raised much debate ever since the seminal paper known as EPR [1]. It's ideas may have seemed purely theoretical until a range of experiments [2-6] demonstrated that there really was an unexpected correlation of linear polarization of photons even though these measurements were made close together in time but at such widely separated distances that any correlations appeared to violate both relativity [7] and the classical concepts of locality [5]. The discussions about mechanisms that might or might not be able bring about this correlation include hidden variables [8-10] sophisticated links between uncertainty and non-local correlations [11], classical geometry playing a key role [12], the possibility of classical effects mimicking the quantum measurements [13] and of course the mysteries of quantum theory itself [14]. A full review is outside the scope of this paper so that the above references are but a very small selection. They apply to photons where entanglement has been demonstrated over 144 kilometers [15]. However it has to be noted that electrons [16], molecules [17] and solid diamonds [18] have all been able to demonstrate entanglement. While Albert and Galchen [7] suggest that entanglement compromises the theory of relativity theory, Timpson and Brown [19] suggest that relativity is not a problem if one accepts Everett's view [20] that quantum waves do not collapse.

The present paper takes the view that the problem lies in our incomplete understanding of relativity and its interaction with quantum theory. A relativistic answer to entanglement is proposed that demonstrates how waves, associated with entangled pairs of particles, can have the same proper frame of reference. These associated waves, when viewed in their proper frames, are then local even though, in the laboratory frame, they are widely separated. The discussion of how these proper frames arise uses scalar waves, travelling with the velocity of light, as exemplars.

In sections 2 and 3 of this paper, we derive the properties, including proper frames and directions of 'energy' propagation, of a self-consistent group of waves which can be used to describe the motion of a pair of entangled (massless) particles. Towards the end of section 3, the applicability of the ideas to quantum waves with mass is discussed. Section 4 considers the implications of the properties of the adjoint and reference waves for the proposed explanation of the apparently non-local nature of entanglement with section 5 providing conclusions.

2. Diffracting Light Waves and Proper Frames

2.1 *Diffracting Forward Waves*

Consider a wave, travelling with the velocity of light, described for simplicity by some scalar function Ψ_A in a laboratory frame Γ. The wave is moving along the +Oz direction in a direction towards an observer traditionally named Alice. The wave varies as:

$$\Psi_A = \Psi_{0A}(x, y)\exp\{i[kz - (\omega/c)(ct) + \theta_A]\}. \quad (1)$$

The angular frequency is ω and axial wave vector is k while θA gives a frame invariant reference phase. The function $\Psi_{0A}(x, y)$ expresses the variation of the wave over the (x, y) plane. In reality, it is impossible to have a perfectly uniform plane wave over the whole of the transverse plane (x, y) because all waves diffract. This is shown schematically in Figure 1. The velocity of a spherical wave front moving away from a point source would ideally be *c* in free space but, because of diffraction, the group velocity of a wave front $\Psi_{0A}(x, y)$ moving along the Oz axis must be less than c, even if by an amount Δ that is too small to measure accurately.

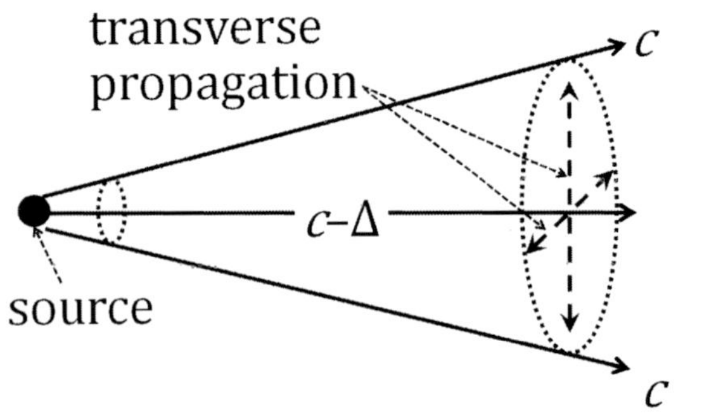

Figure 1. Waves diffracting from some source. The axial group velocity is less than the velocity of light. This permits a proper frame to be defined for such waves in relativity theory.

A wide variety of transverse functions Ψ0A (x, y) are possible but will be limited here to solutions of the transverse part of a wave equation:

$$(\partial x2 + \partial y2)\Psi 0A = -\kappa 2\Psi 0A \ ; \ \kappa 2 > 0. \quad (2)$$

The value of κ will be called the transverse propagation constant and is taken to be independent of frequency and axial position. For the most general diffraction, there would be a superposition of waves with different values of κ. Here, a single representative value is assumed with κ tending to zero in order to find the plane wave limit. The waves in equation (1) are then assumed to have a dispersion equation where all waves have a velocity c as $\kappa 2 \rightarrow 0$:

$$(\omega/c)2 - k2 - \kappa 2 = 0. \quad (3)$$

The phase velocity vp and group velocity vg for such waves are given, independent of κ, from:

$$vp = (\omega/k) > c \ ; \quad vg = (d\omega/dk) = c2k/\omega < c. \quad (4)$$

Now consider a frame Γp moving with a velocity (c tanh βp) relative to the frame Γ. The parameter βp is known as the rapidity [21]. It is a convenient measure of relativistic velocity because, when combining velocities, it is the rapidities and not the velocities that are added to get the correct relativistic answer.

Lorentz-Einstein-Poincaré transformations [22,23] allow co-ordinates (*z, ct*) and the reciprocal space coordinates [k, (ω/c)] of the wave Ψ_A in Γ to be be related to corresponding coordinates in Γp:

$$ct = ct_p \cosh\beta_p + z_p \sinh\beta_p,$$
$$z = ct_p \sinh\beta_p + z_p \cosh\beta_p,\ \beta_p > 0. \quad (5)$$

$$(\omega/c) = (\omega_p/c)\cosh\beta_p + k_p \sinh\beta_p,$$
$$k = k_p \cosh\beta_p + (\omega_p/c)\sinh\beta_p. \quad (6)$$

The transverse coordinates (x, y) are unchanged by changes of rapidity along Oz.

Now choose βp so that (ωp/c) = κ > 0 and kp = 0 so that βp satisfies the dispersion relationship (3) with:

$$(\omega/c) = \kappa \cosh\beta p \ ; \ k = \kappa \sinh\beta p. \quad (7)$$

Differentiating with respect to βp gives vg = (dω/dk) = (dω/dβp)/(dk/dβp), the standard group velocity:

$$vg = c \tanh\beta p < c. \quad (8)$$

In the laboratory frame Γ, the position zp = 0 also moves with a velocity dz/dt = c tanh βp = vg along the +Oz axis. Consequently in the frame Γp there is no movement along Oz and no net motion changing the transverse origin of (x, y). This frame, Γp , may then be called the zero momentum frame [24] or alternatively the proper frame.

Because of diffraction, the waves have a group velocity vg along the Oz axis that is less than c (Figure 1) so that the rapidity βp is always finite. There is then always a proper frame with an increasing proper time tp. In case it is thought that this is a way of forcing the velocity of electromagnetic waves to travel at less than c, note that the angular frequency in Γp is ωp = κc while the transverse propagation constant is κ. The transverse electromagnetic velocity is therefore still κc/κ = c in the proper frame, though the waves are not travelling in a single transverse direction but diffracting in many different transverse directions away from the axis.

Substituting the proper values into the expression for ΨA (Equation 1) setting (ωp/c) = κ and kp = 0, one finds that zp cancels leaving a phase factor that is independent of rapidity and increasing with the proper time tp, with the arbitrary phase reference θA;

$$\Psi A = \Psi 0A\,(x, y)\exp(-ic\kappa tp + i\theta A). \quad (9)$$

If, in the laboratory frame Γ, one finds a distinctly different light wave, say Ψ_{Ba}, with an identical proper frame of reference, then relativistically Ψ_{Ba} and Ψ_A can be local to one another and some of their properties could be correlated even though each may be viewed at widely spaced values of (*z, ct*)A (*z, ct*)Ba in Γ. Such pairs of waves are found in subsequent sections after first reviewing how rapidity and proper frames change with propagation reversal.

2.2 *Waves Travelling in the Reverse Direction*

Consider a wave ΨB that is now travelling back along –Oz towards a different observer named Bob. This wave is assumed to have the same value of frequency as for ΨA and, from relativity theory, the waves travelling in the direction –Oz have equivalent physics to the waves travelling in the direction +Oz. There is no one preferred direction. If it is assumed that both waves originate at tp = 0 in their proper frames, the progress of the two wave fronts can be measured by looking at the same value of tp as they travel. The wave ΨB is found by the reversal of z to –z' . Because the frequency has not changed, the dispersion equation (3) remains unaltered, k changes to –k' = –k. If this is an electro-magnetic wave, the appropriate conditions on the transverse electric and magnetic fields must be observed [25] when reversing the direction of z. With a scalar wave, equation (1) simply changes to become:

$$\Psi_B = \Psi_{0B} \exp\{ i[(-k')(-z') - (\omega/c)(ct) + \theta_B]\} . \quad (10)$$

If ΨB has the same emission reference time as ΨA, a new proper frame Γn can be found for ΨB with the same value of proper time. Γn requires not only changing the reference position zp into some value –zn but also changing the rapidity of the new proper frame from βp into βn. The transformation equations, corresponding to equations (5) and (7), are now given by:

$$\begin{aligned} ct &= ct_p \cosh\beta_n + (-z_n)\sinh\beta_n, \\ (-z') &= (-z_n)\cosh\beta_n + ct_p \sinh\beta_n. \end{aligned} \quad (11)$$

$$(\omega/c) = \kappa\cosh\beta_n, \quad -k' = -k = \kappa\sinh\beta_n. \quad (12)$$

The frequency ω has not changed sign and so κc, which is the proper frequency, has not changed sign. Consistency with previous equations therefore requires that βn = – βp . As would be expected, the proper frame Γn is travelling in the opposite direction to Γp as measured in the laboratory frame Γ. With these changes the wave ΨB in its proper frame is given from:

$$\begin{aligned} \Psi_B &= \Psi_{0B} \exp\{ i[(-k')(-z') - (\omega/c)(ct) + \theta_B]\} \\ &= \Psi_{0B} \exp[(-i\kappa ct_p) + i\theta_B]\}. \end{aligned} \quad (13)$$

There is no need to retain z' =–z because the observer in Γ can distinguish directions +Oz and –Oz so that Ψ_B can be described in the Γ frame as:

$$\Psi B = \Psi 0\ B(x, y) \exp\{i[-kz - (\omega/c)(ct) + \theta B]\}. \quad (14)$$

Beside the reversal of rapidity, an important feature for waves in Γp and in Γn is seen by considering two waves emitted together and considering two positions or points with the same value of normalised proper time ctp and the same reference positions:e.g. *zp* = *zn* = 0. One point on ΨB in Γn moves towards Bob and the other point on ΨA in Γp moves towards Alice. In the laboratory frame Γ, these points, in the laboratory frame Γ, have a space-like separation Δz = ctp sinh βp –ctp sinh βn = 2ctp sinh βp. There is no possibility of reducing this space-like separation to zero by moving to any new frame. Any transfer of information between points with this space-like separation is not possible in relativity theory [22]. Consequently, according to any local theory, waves ΨA and ΨB should be uncorrelated and independent unless some correlation is built in at tp = 0. The experimental evidence [5,6] shows that correlation can be found that could not have been built in at tp = 0 making entanglement appear 'spooky' [26].

3. Relativistic Waves With 'Loss'

Summarizing section 2.2, reversal of both k and z leaves (–k)(–z') = k z' and reverses the sign of the rapidity for the proper frame. The choice of Oz is arbitrary: no one spatial direction has been preferred and any observer in Γ can easily distinguish the negative spatial direction z' = –z from the positive spatial direction z. If this method of reversing the rapidity could be applied to simultaneously reversing both ω and t then a new wave ΨAa , called here an adjoint wave, would have the same proper frame as ΨB even though their phase velocities were in opposition.

This ω-t reversal should then have very interesting implications for the concept of locality. However time reversal, even when accompanied by frequency reversal, leads to difficulties of interpretation. Unlike space reversal, it is not possible for the observer in the laboratory frame Γ to observe a wave travelling backwards in time. In Γ, the product (–ω)(–t) is therefore observed in exactly the same way as the product ωt. Key questions for this section is "What happens to energy?" and can the process of combined time-frequency-reversal avoid violating causality?

At present, only classical relativistic implications are considered. Simultaneous reversal of time and frequency leaves exp[–i(–ω)(–t)] looking exactly like exp(–iωt) for all ω in the laboratory frame Γ so that the phase velocity (ω/k) is unaltered. The dispersion equation (3), containing ω2, remains the same so that it is possible to argue that the group velocity (dω/dk) remains unchanged. A relevant problem arises in materials with negative dispersion where it can be unclear about the direction of 'energy' propagation. Feigenbaum et al. [27] argue that rather than using the group velocity (dω/dk), as in equation (4), to identify the direction of energy flow, it is more helpful to consider the role of loss, an idea that is followed here.

3.2 *Introduction of 'Loss'*

Figure 2 shows the concept of four different conditions of phase and energy velocity for a wave which has loss. The concept of 'energy' is limited here to a quantity that is proportional to the squared amplitude of the wave. Two waves have opposing directions of the phase velocity: either to Alice or to Bob. Conventional diffracting waves in dispersion free material obey the concept that the group velocity, (dω/dk), and phase velocity, (ω/k), align in the same direction as the proper frame. Such waves will be called reference waves. To gain an idea in which direction the 'energy' travels, a small amount of loss is considered and the direction of amplitude decay indicates the direction of 'energy' travel as illustrated in Figure 2 A and B. Section 3.2 will demonstrate that waves with simultaneous frequency and time reversal have 'energy' travelling in the opposite direction as in Figure 2Aa and Ba. These waves will be called adjoint waves. It is possible to create, purely as a guide, an artificial loss that conforms with changes of rapidity and has the essential asymmetry of loss with both time and space reversal.

Equation (5) is unaltered but equation (6) is re-written with an additional 'loss' term σp that will be assumed to be asymmetrical with time and space:

$$(\omega/c)+i\sigma_t=\kappa\cosh\beta_p+i\sigma_p\sinh\beta_p,$$
$$k+i\sigma_z=\kappa\sinh\beta_p+i\sigma_p\cosh\beta_p. \tag{15}$$

Equation (15) satisfies a dispersion relationship that is invariant to changes of rapidity along the Oz axis and has both κ > 0 and σp > 0 in the first instance:

[(ω/c) + i σt]2 – (k + i σz)2 = κ2 + σp2 (16)

To comply with the concept of loss, σp will reverse sign on either time reversal or z-reversal.

In the proper frame Γp, where kp = 0, σz ⇒ σp and equation (9), for ΨA, changes to:

ΨA = Lim(σp →0)[Ψ0A exp(–icκtp+iθA) exp(–σp zp)]. (17)

In the Γ frame one finds, after some algebra, that this ΨA is given as:

$$\Psi_A=\mathrm{Lim}_{(\sigma_p\to 0)}\{[\Psi_{0A}\exp[i(kz-\omega t+\theta_A)]\times$$
$$\exp[-\sigma_p(z-v_g t)\cosh\beta_p]\}. \tag{18}$$

This shows that |ΨA|2 decays for (z–vgt) > 0 indicating the direction of 'energy' propagation as in Figure 2A.

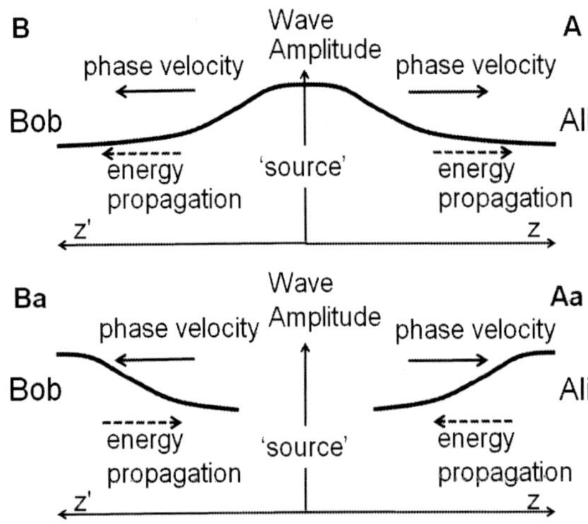

Figure 2. Amplitude loss and energy propagation. A and B illustrate amplitude loss with changes of position as waves travel away from the source to Alice and Bob respectively. These diagrams indicate a phase velocity and energy velocity travelling in the same direction. Aa and Ba indicate 'adjoint' waves where both frequency and time have been reversed to show energy travelling in the opposite direction.

A check is now made on the consistency of the loss concepts with z-reversal. Consider then the wave ΨB travelling towards Bob. Equation (11) holds and the proper frame of ΨB is Γn. Following the steps leading to equation (17), ΨB decays as exp(–σn zn). But with z-reversal, the concept of loss requires that σn = –σp so that the proper frame loss factor exp(–σnZn) changes in the laboratory frame Γ to:

$$\exp(-\sigma_n z_n)\Rightarrow\exp[-\sigma_n(-z'\cosh\beta_n-ct\sinh\beta_n)]$$
$$=\exp[-\sigma_p(z'-v_g t)\cosh\beta_p] \tag{19}$$

This shows that |ΨB|2 decays for (z'–vgt) > 0. If there is loss for waves that are travelling to Bob, then these waves decay in their direction of propagation as in Figure 2B, similar to the waves travelling to Alice. This gives re-assurance that z-reversal still means that the direction of energy flow is in the same direction as the proper frame of reference. Simultaneous time and frequency reversal can now be considered.

3.2 *Adjoint Waves and Loss*

Consider the reference wave ΨA and change this to itsadjoint wave ΨAa defined by simultaneous time and frequency reversal. Let ΨAa have a proper frame Γ'n with a rapidity β'n relative to the laboratory frame Γ. The proper (primed) parameters for ΨAa satisfy:

$$c(-t) = c(-t_p)\cosh\beta_n{}' + z_p \sinh\beta_n{}', \qquad z = z_p\cosh\beta_n{}' + c(-t_p)\sinh\beta_n{}'. \tag{20}$$

$$(-\omega/c) + i\sigma_t{}' = (-\kappa)\cosh\beta_n{}' + i\sigma_n \sinh\beta_n{}', \qquad k + i\sigma_z{}' = (-\kappa)\sinh\beta_n{}' + i\sigma_n\cosh\beta_n{}'. \tag{21}$$

For consistency with equations (5) and (6), β'n = βn = –βp and so Γ'n is identical to Γn. ΨAa has the same proper frame as the wave ΨB. Then cκ, tp and σp all change sign in equation (17) to give in Γ'n = Γn:

ΨAa = Lim(σp→0)[Ψ0Aexp(–icκtp+iθAa) exp(σp zp)]. (22)

The exponential term in equation (18) changes to exp[I σp (z–vgt) cosh βn] where cosh βn = cosh βp so that the decrease in 'energy' is as shown schematically in Figure2Aa. Consequently this 'loss' concept confirms that simultaneous time and frequency reversal in the adjoint wave ΨAa changes the direction of 'energy' propagation to be the same direction of travel as Γn.

Similarly simultaneous ω–t reversal in ΨB creates an adjoint ΨBa .Again loss is as sketched in Figure 2Ba with energy in ΨBa and the proper frames for ΨBa and ΨA travelling in the same direction.

Although (–ω)(–t) appears identical to ωt for all values of ω, the consideration of loss has helped to confirm the straightforward result that the proper frame of reference gives the correct direction of propagation of energy. There is no longer any point in carrying around artificial loss terms. These are now omitted, being set to zero in future.

However, if the adjoint wave ΨAa has a positive phase velocity but is carrying energy or information back towards the source consistent with the direction of travel of its proper frame of reference, how can it avoid violating causality? The answer given here is that ΨAa is a consequence of ΨA. The use of reference and adjoint waves ΨA and ΨAa, in this work is very different from early papers about the reaction [28,29] on a radiating electron that is caused by combinations of retarded and advanced waves. These waves were used in their traditional sense with both their phase velocities and group velocities in opposing directions. Advanced waves pose major difficulties with causality and alternative treatments of radiation reaction can avoid their us [30]. In this work the adjoint wave ΨAa* is assumed to be excited at the interface of the measurement system as a consequence of incoming wave ΨA. It is well known for example in analyzing signaling systems that the optimum response of a matched filter to a given input pulse is the conjugate time reversed replica of the input [31]. In classical systems, there is a temporal delay factor that ensures causality so that the system's complete response to a single pulse does not finalize until the pulse has completed its passage. The time reversed conjugate wave ΨAa* is then argued to be the response which gives a maximum output for ΨA. The simultaneous frequency reversal is required if time and frequency are to have the same relativistic frames of reference. The important factor here is that ΨAa* is realized only at the measurement port in response to ΨA and if time reversal creates any violation of causality, this violation is strictly limited to the time taken to measure one quanta.

Figure 3 illustrates the process of measurement. The time of flight between excitation from the source to the observer is determined by group velocity of ΨA. The maximum energy is extracted from the wave ΨA by its interaction with an optimal response ΨAa* of the measurement system to the incoming wave ΨA. Both waves are located in a finite region around the measurement system. It is envisaged that the round trip phase is an integer multiple of 2π to obtain a resonance that removes the energy completely from the incoming wave into the detection system. The energy in the packet is no longer taken to be proportional to |ΨA|2 or |ΨAa|2 but is assumed to be proportional to the phase invariant ΨAa*ΨA. This expression is used in a vector form when discussing Maxwell's classical equations [32]. It is suggested that this is the form of relativistic response for all types of wave.

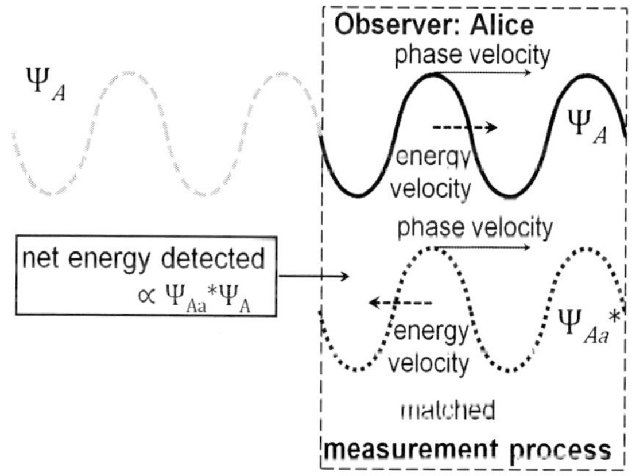

Figure 3. An adjoint wave ΨAa is excited at Alice's measurement system in response to the incoming reference wave ΨA. The feedback of 'energy' from ΨAa to ΨA leads to a resonant wave-packet that determines a net energy ∝ ΨAa* ΨA sent from source to an observer Alice. The finite energy in the packet means that outside that particular packet the wave ΨA associated with that packet is zero.

At any point along the path of the wave ΨA one can say that there is a probability, if ΨA is detected at that point, that it will give out energy proportional to ΨAa*ΨA but ΨAa* is only physically present during an actual measurement process. Never the less it will be found helpful to keep track of the proper frame in which ΨAa* has to lie as discussed next.

Compare the transformation equations for coordinates in the wave ΨB as given by equation (11) and the transformation equations for coordinates in the wave ΨAa as given by equation (20), repeated for convenience in equations (23) and (24) below:

$$ct = ct_p \cosh\beta_n + (-z_n)\sinh\beta_n,$$
$$(-z') = (-z_n)\cosh\beta_n + ct_p \sinh\beta_n. \tag{23}$$

$$c(-t) = c(-t_p)\cosh\beta_n + z_p \sinh\beta_n,$$
$$z = z_p \cosh\beta_n + c(-t_p)\sinh\beta_n. \tag{24}$$

With zn = zp and (–z') = z, the coordinates in the two different waves, ΨAa and ΨB with phase velocities travelling in opposing directions, can be reduced to the same values. In other words, provided that the waves ΨB and ΨAa have the same proper time tp, then they have the same proper frame Γn and are relativistically local to one another. The wave ΨB and ΨAa can indeed have the same proper time, provided that ΨB and ΨA are launched together at the same time and place. The wave ΨAa* is activated by ΨA so that ΨAa* acquires the phase associated with the proper time in ΨA and hence also in ΨB. This does not mean that setting ΨB to zero will necessarily mean that ΨAa has to be set to zero, or vice-versa, but it does mean that there is no surprise if some properties of ΨB and ΨAa happen to be correlated. These two waves do not have to be in touch with each other in the laboratory frame Γ; there can be two widely separated segments of each wave that are under observation in Γ but with the same proper time. In these special circumstances, segments of the waves ΨB and ΨAa are local to one another.

Similar results hold for the reference wave ΨA and its adjoint wave ΨBa where again frequency and time have been simultaneously reversed in ΨB . These waves ΨA and ΨBa* can also have the same proper time if ΨA and ΨB are launched together, and then ΨA and ΨBa will again have the same proper frame of reference Γp relative to the laboratory frame Γ. As before, separate portions of ΨA and ΨBa* can be local to one another removing any surprise if ΨA and ΨBa* happen to have correlated properties.

3.3 *Overlapping Reference and Adjoint Waves*

The concept of reference and adjoint waves has been proposed previously [33] but is reconsidered here with attention given to proper frames of reference and causality. Figure 4 expands on the ideas in Figure 3 by considering waves, from a single source, carrying energy in opposite directions to both Bob and Alice.

In Figure 4(a) the reference waves are the key messengers that determine the direction of net energy flow travelling with the group velocity (dω/dk). These waves then preserve causality. Although the adjoint wave ΨAa* is not actually excited until a measurement is in progress, its phase and proper frame, that it would have if excited, can be tracked alongside ΨA. Similarly the adjoint wave ΨBa* can have its phase and proper frame tracked before it actually is excited at any measurement. These adjoint waves are shown as gray lines in Figure 4(a) to remind one that they can be excited if the waves are detected but until that detection process as in Figure 4(b) they are not actually present.

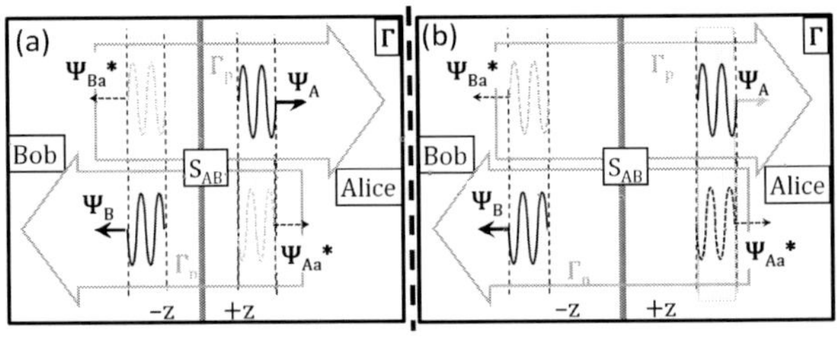

Figure 4. (a) Reference waves are sketched travelling in their proper frames before any detection. Their associated adjoint waves are purely shown for convenience to emphasise their role in relation to the proper frames of reference. They will not actually materialize until a measurement is made, as in (b). Here the adjoint wave ΨAa is excited at the point of detection by ΨA to yield the energy as proportional to ΨAa *ΨA . See text for more detail.

In Figure 4(b) the measurement process is being made at Alice. The adjoint wave ΨAa* (shown in black rather than gray) is now excited and interacts with ΨA to produce an energy proportional to ΨAa* ΨA.

3.4 *Some Initial Quantum Implications*

Diverting for the moment from the discussion about entanglement, consider some advantages of the concept of overlapping reference and adjoint waves for defining the 'particle' associated with a wave during the measurement process. The longer the overlap then the more uncertain is the position of the 'particle' but the better defined is the value of the propagation constant (the momentum). The waves ΨAa and ΨA can collapse to zero locally (as energy is transferred from the waves to the detector) without any conceptual problems of 'instantaneous' collapse over extensive regions. This well known problem is discussed at several points in Miller's helpful text book [34]. The overlap region can be short enough so that, in real experiments with practical equipment, the collapse may appear 'instantaneous'. This is very different from the idea that the problems of sudden collapse can be avoided by arguing that there is no wave function collapse [20].

The outlines of an explanation for entanglement giving correlation of some events with space like separation can now be seen to be based on relativity rather than in violation of relativity. Particle measurements at Alice (A) and Bob (B) are assumed to be determined from the product of two pairs of two waves: $\Psi_A a^*\Psi_A$ and $\Psi_B a^*\Psi_B$. Conjugation is required for a phase invariant quantity but does not alter the proper frame. The waves ΨA and ΨBa* have the same proper frame even though they have phase velocities in opposite directions. Similarly waves Ψ_B and ΨAa* can have the same proper frame. Each pair of waves (Ψ_A & $\Psi_B a^*$ and ΨAa* &Ψ_B), provided that they are carefully timed in both their generation and measurement, are relativistically local to each other because they are in the same proper frame. Consequently, if they are found to have correlated properties there is no violation of relativity.

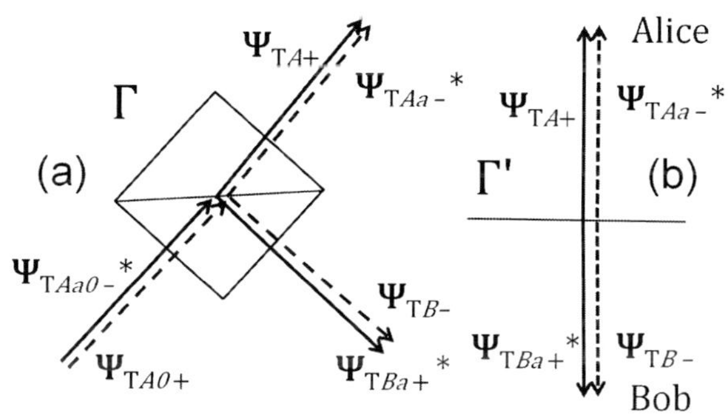

Figure 5. The beam splitter seen as a single source (see text).

While this general theory cannot give precise correlations it can be used to interpret experiments on multi-path problems such as the beam splitter as in Figure 5 (a) where two waves from a beam splitter with a single input emerge typically at right angles in a laboratory frame Γ. The adjoint wave is a wave that would be excited at a detector if the photon were to be measured. The adjoint waves are shown artificially side by side with the reference wave. Along the direction that bisects the two output directions, the waves are moving at less than *c* and so it is possible to move to a new frame of reference Γ' (Figure 5b) where the two output waves, are arranged to be travelling in opposite directions. With a single photon, either $\Psi'_A a^*\Psi'_A$ or $\Psi'_B a^*\Psi'_B$ represents one photon, but the experimenter has no way of knowing which way the photon will go. To explain this enigma, allow Alice for example to make a measurement and find that there is one photon at her detector. Bob's detector knows that it must register zero because $\Psi'_B a$ would be local to Ψ'_A and Ψ'_{TB} is local to $\Psi'_A a$. This relativistic localization means that there is no mystery about Bob's detector registering zero. The relativistically local waves tell Bob that their energy has been removed at Alice's position in the laboratory frame. The argument is reversed if Bob makes the first measurement.

The Dirac concept of 'bra' and 'ket' functions [35], <ΨA| and |ΨA>, is that these are Hermitian adjoint functions. The adjoint-reference product ΨAa* ΨA is equivalent to <ΨA|ΨA> and acknowledges the concept of an adjoint function but with a even stronger distinction because ΨAa* lies in a different proper frame to ΨA . However it will be shown next, some of the details have to be changed for Dirac waves because frequency reversal is not allowed.

3.5 *Implications for Waves with Mass*

The above arguments referred to waves that, in the absence of diffraction, have a group velocity of c. Any associated particles would therefore have to have zero rest mass. If one changes to quantum waves with non-zero mass m0, as in the Dirac or Klein-Gordon equations then the dispersion equation (3) would read:

$$(\omega/c)2 - k2 = K2;\ K = \pm\sqrt{[\kappa 2 + (m0\ c/\hbar)2]};\ \kappa 2 > 0. \quad (25)$$

This dispersion equation is still independent of changes of rapidity along the Oz axis and applies to the individual components of a Dirac wave. The parameter K now replaces κ in all the previous arguments. Just as cκ is the proper frequency, so cK is the new proper frequency. Reversal of the frequency ω was seen to require the reversal of the proper frequency cK. If the diffraction parameter κ were to be set to zero, then K reversal would imply mass reversal. Diffraction again is essential so that K reversal does not imply mass reversal, as can be seen from equation (25). Waves associated with non-zero mass still have this relativistic requirement that one can have adjoint waves in opposing proper frames. Consequently the arguments in section 3 still hold: relativistic entanglement need not be restricted to waves associated with massless particles. However waves for Dirac-like particles need more careful consideration even though their individual components satisfy equation (25).

Time reversal symmetry is well known in quantum field theory [36,37] and carries implications for the Hamiltonian function H linked to the quantum wave frequency ω by H = ħω. As Greiner [36] helpfully explains, time reversal for the Dirac Hamiltonian shows that H is unchanged by t-reversal. This means that ω reversal combined with t reversal is not compatible with Dirac-like quantum waves. The operator *i* in quantum theory has a more subtle meaning than a simple phase operator because it changes sign on time reversal.

Consequently the definition of the adjoint wave now needs re-consideration.

It is now proposed for Dirac waves that the adjoint wave has simultaneous i reversal and t reversal, each action appearing to cancel the other. Products $(-i)(H/\hbar)(-t) = i(H/\hbar)t$ are un-changed. However to preserve the sign of (ik) in spite of the change of sign in i, the adjoint form of t-reversal must include i, t and k reversal for the forward wave. The reverse adjoint waves will have z, t and i all reversed.

It can then be seen, from Table 1 above, that it is still possible using Dirac-like waves to construct reference and adjoint waves ΨA and ΨBa that lie in the same proper frame Γp and waves ΨB and ΨAa that also lie in the same proper frame Γn. The product ΨAa*ΨA is still required for a phase invariant probability. Conjugation (change of i to –i) does not by itself lead to a change of proper frame provided that it is understood that ΨAa* is the adjoint of ΨA*. Then all the previous results can still be used to show that correlations of properties carried by distant quantum waves moving in opposite directions may still be relativistically permitted through waves lying in the same proper frame.

A rapid glance at the proposed theory might suggest that ΨA is similar to a de Broglie-Bohm-Vigier pilot-wave as discussed by Holland [38] but now with ΨA determining where the particle, represented by ΨAa*ΨA,, can travel. However a more considered view of the different structures of the de Broglie pilot wave and ΨA shows that there is no satisfactory comparison, particularly with ΨA and ΨAa* being embedded in relativity theory.

Table 1. Relativistic Changes for Reference and Adjoint Dirac-Like Waves

Forward travelling waves in Γ: $\Psi_O(x,y)\exp[(ikz - i\omega t)]$.	Reverse travelling waves in Γ: $\Psi_O(x,y)\exp[(-ikz - i\omega t)]$.
Proper frame Γ_p with positive rapidity $+\beta_p$	Proper frame Γ_n with positive rapidity $-\beta_p$
Advancing reference wave Ψ_A (no reversals) Direction of 'energy' travel: $+Oz$ $K > 0,\ t_p > 0$ $ik = iK\sinh(+\beta_p)$. $z = ct_p\sinh(+\beta_p) + z_p\cosh(+\beta_p)$. $(i\omega/c) = iK\cosh(+\beta_p)$. $ct = ct_p\cosh(+\beta_p) + z_p\sinh(+\beta_p)$.	Backward reference wave Ψ_B (z, k reversed in Ψ_A) Direction of 'energy' travel: $-Oz$ $K > 0,\ t_p > 0\ (-z_n) < 0$ $-ik = iK\sinh(-\beta_p)$. $-z' = ct_p\sinh(-\beta_p) - z_n\cosh(-\beta_p)$. $(i\omega/c) = iK\cosh(-\beta_p)$. $ct = ct_p\cosh(-\beta_p) - z_n\sinh(-\beta_p)$.
Backward *adjoint* wave Ψ_{Ba} (z, t, i reversed in Ψ_A) $K > 0, (-t_p) < 0$. $(-i)k = (-i)\ K\sinh(+\beta_p)$. $-z' = (-ct_p)\sinh(+\beta_p) + (-z_p)\cosh(+\beta_p)$. $-i)(\omega/c) = (-i)K\cosh(+\beta_p)$. $(-ct) = (-ct_p)\cosh(+\beta_p) + (-z_p)\sinh(+\beta_p)$.	Advancing *adjoint* wave Ψ_{Aa} (k, t, i reversed in Ψ_A) $K > 0, (-t_p) < 0$. $(-i)(-k) = (-i)\ K\sinh(-\beta_p)$. $z = (-ct_p)\sinh(-\beta_p) + z_p\cosh(-\beta_p)$. $(-i)(\omega/c) = (-i)K\cosh(-\beta_p)$. $(-ct) = (-ct_p)\cosh(-\beta_p) + z_p\sinh(-\beta_p)$.

4. Discussion — Relativistic Entanglement

A quintessential experiment demonstrating the non-local nature of entanglement would be conceptually similar to the seminal experiments demonstrating the violation of Bell's inequalities performed by Aspect et al. [2] and also by the Zeilinger group at the University of Vienna5. Here correlated measurements of entangled photon pairs were performed at space-like separations. In both cases, the crux of the measurement was the generation of pairs of polarisation-entangled photons, followed by their separation, and then by independent polarization measurements of the two photons in randomly selected transverse orientations. Precisely timed measurements at widely separated measurement stations prevents classical communication, about the orientation of the polarizers for specific photon pairs.

Consequently any correlations measured in such an experiment are interpreted as being 'non-local'. Such an experiment is now discussed.

4.1 *Specific Entanglement Processes*

As already stated, the present general theory cannot show specific entanglement processes such as indicated in Figure 6 where Alice and Bob have correlated parallel polarizations or alternative correlated orthogonal polarizations as discussed in the companion paper [32]. Although it is too difficult to show in such schematic diagrams, it is important to make clear that, in principle, Bob and Alice can determine the orientation of their polarizers after the photons have left the source SAB but before the measurement at either A or B1. Figure 6 must not be regarded as representing a situation analogous to that, outlined by Bell [39], where Dr. Bertlmann decided at the start of each morning on the different colour of his socks on each foot. Either Alice or Bob has to make a measurement before any polarization is determined, even in theory.

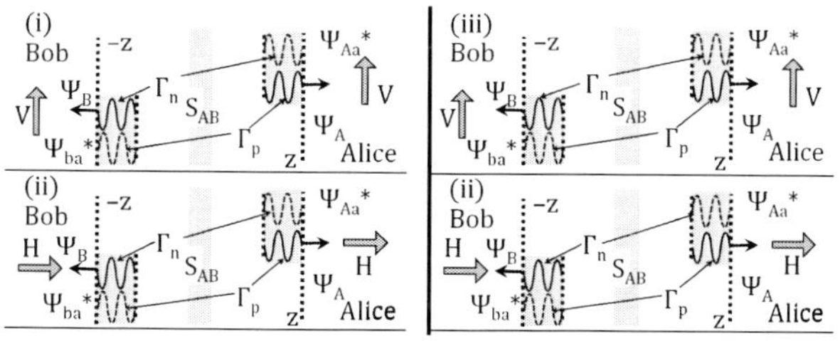

Figure 6. Relativistic polarization entanglement from a source SAB through common proper frames of reference and carefully timed measurements of separated quanta. (i) and (ii) entanglement where Alice and Bob have parallel polarizations (iii) and (iv) entanglement where Alice and Bob have orthogonal polarizations.

4.2 *Clarifications and Potential Objections*

It may now appear that all measurements made by Alice and Bob could, or even should, be correlated. However, keeping with photons as the exemplar, an ensemble of many photons is always measured classically. The selection of each individual photon in the ensemble from the source SAB will not be from correlated pairs when measured at Alice and at Bob. Distinct and different pairs of photons have no reason to have correlated properties unless their waves are in the same proper frames.

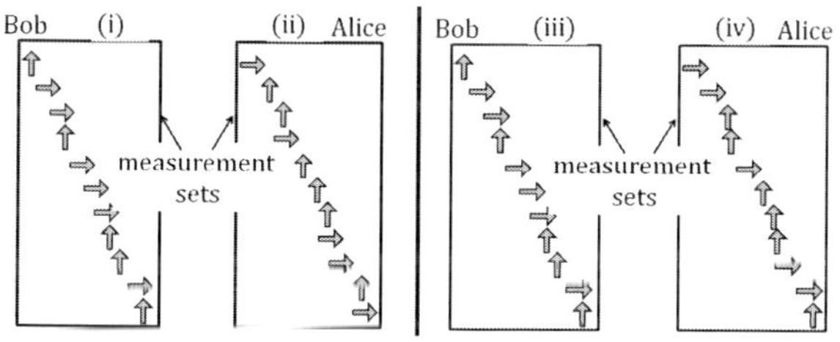

Figure 7. Non-entanglement in classical measurements of many pairs of quantum events. In (i) and (ii), Alice and Bob measure many correlated quanta but with any one pair uncorrelated with any subsequent pair. In (iii) and (iv), Bob and Alice measure many quanta but pairs at Alice are not from the same pair of proper frames as those at Bob and so are uncorrelated.

Figure 7 sketches two different possibilities. In (i) and (ii) all pairs of photons are correlated but the pair at tn is uncorrelated with the pair at tm. The classical measurement is an ensemble of all pairs at a set of times {tq} so that the classical measurement, averaging over this set, is uncorrelated. In (iii) and (iv) the measurements by Alice and Bob are such that they never actually measure pairs from the same proper frames but from a wide variety of proper frames. Events from different proper frames are in general uncorrelated unless they are correlated at a common origin through classical properties of the source.

Causality is a major concern in entanglement. It might be thought that adjoint waves with reversed rapidity must violate causality. However it the causal reference waves determine the net energy flow and also excite the adjoint waves at the detector. A resonance or matching between the two waves to extract energy is always initiated by the reference waves so that causality is not violated by the time reversal. Detection can occur only after arrival of the reference wave. It is again pointed out that adjoint waves are not like advanced waves. The former are assumed to be generated locally while the latter are generated from some distant object such as an absorber [29].

While this analysis may provide a relativistic entanglement that looks like quantum entanglement, it refers to two waves or particles travelling in opposing directions. This is not like a typical real experiment. For example, parametric down conversion of light produces photon pairs coming off at an angle of a few degrees relative to an axis say Oz [40,41]. Because of this angle, these waves are travelling at slightly less than the velocity c along the bisecting axis (say Oz) of their angle of emission. Therefore they have a finite rapidity for their velocity along Oz. Consequently, it is possible to move to a new reference frame in which the pairs appear to be generated travelling in opposing directions, say along +Oy and –Oy perpendicular to Oz. Again because of diffraction, the speed in these opposing directions still has a finite rapidity with respect to their proper frames, which are now taken to travel along +/– Oy with diffraction transverse to Oy. Because the waves in these new frames are travelling in opposing directions, all the previous arguments apply and relativistic entanglement is once again a matter of choosing particles or waves that have reference and adjoint waves that can be simultaneously localized by having the same proper frame with the same rapidity.

A further objection is that, in most practical experiments, the paths of the two photons from the source need not be straight. Mirrors and lenses can focus and deflect the beams. However the argument is not dependent on the path. All that is required is that there are definite directions of travel, a and b, for the photons as they approach the detectors at Alice and Bob. The two directions form a plane a∧b. In that plane, the bisector of a and b forms the axis, as discussed in the section 3.4. One can again change the frame of reference so that the photons are moving in opposite directions thus enabling the main arguments to be valid once again.

Yet another challenge is given by the phenomenon of entanglement swapping [42] or several particles being entangled with no common origin [43]. Inspection of experiments concerned with multi-photon entanglement show that there is either pre- or post-selection of particle sets with the correct combinations of properties for entanglement. This is illustrated in the next paragraph.

Four-photon entanglement can be set out algebraically by supposing that the four photons are determined by the adjoint and reference waves: ΨAa*ΨA, ΨBa*ΨB, ΨaC*ΨC, ΨaD*ΨD. To identify common proper frames an additional bold subscript will be provided. Here ΨAP and ΨBaP have a common proper frame ΓP with also ΨBQ and ΨAaQ having a common proper frame ΓQ indicating the entanglement of A and B photons. Similarly ΨCR and ΨDaR have a common proper frame ΓR along with ΨDS and ΨCaS having a common proper frame ΓS indicating the entanglement of C and D. The experiment then starts by entangling two completely separate pairs or photons which can be indicated now as:

ΨAaQ*ΨAP, ΨBaP*ΨBQ, ΨCaS*ΨCR, ΨDaR*ΨDS.

The next feature is to subject many such photon pairs to a careful experimental selection process whereby ΨBaP*ΨBQ and ΨCaS*ΨCR are selected in such a way that one knows they are entangled. Such an experiment is called a Bell state measurement and typically means the loss of one photon leaving three entangled photons. One then knows that it is possible to relate the proper frames ΓP and ΓR to a proper frame ΓU to know that ΨBaU* and ΨCU can be correlated. For the same reason, ΨBQ and ΨCaS* will have some common proper frame ΓV leading to ΨBV and ΨCaV*. The set of waves with their proper frames as subscripts is then re-written replacing P and R with U while replacing Q and S with V:

ΨAaV*ΨAU,ΨBaU*ΨBV,ΨCaV*ΨCU, ΨDaU*ΨDV.

It is then clear that all four particles are in principle entangled although not all having a common origin. Recent experiments have shown that macroscopic diamond crystals can exhibit entanglement even when spaced a few centimetres apart. They are found to share vibrational states18. On the present theory this is interpreted as the reference and adjoint waves, associated with these vibrational states, having the same proper frame of reference. However the entanglement was observed only with ultrafast laser pulses on a sub-picosecond time scale. This is interpreted as the correlations being destroyed by measurements when the measurements have to be averaged with time over results derived from many different proper frames.

5. Conclusions

This paper proposes that the apparently non-local nature of quantum entanglement can be explained by careful examination of the relativistic proper frames of conventional diffracting waves and the proper frames when time and frequency are simultaneously reversed. Although the arguments had free space photons mainly in mind, it was shown that the concepts are applicable to all types of wave, including massive particle solutions of the Dirac or Klein-Gordon equations, or the waves associated with vibrational states in crystals. Diffraction is required so that 'forward' waves ΨA observed in a frame Γ can always be assigned a finite positive rapidity with respect to a proper frame Γp, even if the plane waves travel at c. Diffraction also removes the need for mass reversal when considering adjoint massive particle waves. Waves travelling in the opposite direction to ΨA are identified as ΨB. Again diffraction allows the definition of a finite negative rapidity and a proper frame Γn.

The new work shows that it is possible to identify adjoint waves ΨAa and ΨBa where both time and frequency are reversed (or *i, t* and *k* reversal for quantum waves). The waves ΨAa and ΨBa then travel with the same phase velocity as ΨA and ΨB respectively but lie on opposite proper frames. Particles are now to be associated with the phase invariant product of waves: ΨAa* ΨA and ΨBa* ΨB. The sharing of the proper frames for the waves {ΨAa* and ΨB} and {ΨBa* and ΨA} suggests that there should be no difficulties in the associated particles having some correlated properties, even when distant from one another.

The general nature of this discussion on relativistic entanglement has precluded discussing specific correlations. A companion paper discusses, starting from Maxwell's classical wave equations how polarization entanglement can arise using the principles in this paper [32].

Acknowledgments

The authors are indebted to the help of Ian White who has been supportive of work on novel photonic concepts for many years.

References

[1] A. Einstein et al., Phys. Rev. 47, 777 (1935).
[2] A. Aspect et al., Phys. Rev. Lett. 49, 91 (1982).
[3] S. J. Freedman and J. F. Clauser, Phys. Rev. Lett. 14, 938 (1972).
[4] Y. H. Shih and C. O. Alley, Phys. Rev. Lett., 61, 2921(1988).
[5] G. Weihs et al., Phys. Rev. Lett. 81, 5039 (1998).

[6] M. A. Rowe, et al., Nature, 409, 791 (2001).
[7] D. Z. Albert and R. Galchen, Scientific American, 300, 32 (2009).
[8] J. Von Neumann, Mathematical Foundations of quantum theory Princeton University Press Princeton N.J. (ch.4)(1955).
[9] J. S. Bell, Rev. Mod.Phys. 38, 447,(1966).
[10] M. Genovese, Phys. Rep., 413, 319 (2005).
[11] J. Oppenheim and S. Wehner, Science 330,1072 (2010).
[12] A. F. Kracklauer, J.Opt. B: Quantum Semiclass. Opt. 6 S544 (2004).
[13] A. F. Kracklauer, Proc. SPIE 8121, (2011).
[14] N. D. Mermin, Jnl. Philosophy,78, 397 (1981).
[15] T. Schmitt-Manderbach et al., Phys. Rev. Lett. 98, 010504 (2007).
[16] J. Bao et al., Nature Materials 2, 175 (2003).
[17] O. Nairz et al., Am. Jnl. Phys., 71, 31 (2003).
[18] K. C. Lee et al., Science 334, 1253 (2011).
[19] C. G. Timpson, and H. R. Brown, Understanding Physical Knowledge University of Bologna, Coopertativa Libraria Universitaria eds. R. Lupacchini and V. Fano. 147 (2002).
[20] H. Everett III, Rev.Mod.Phys. 29, 454 (1957).
[21] J-M. Levy-Leblond and J-P. Provost, Am.J. Phys. 47, 1045 (1979).
[22] C. Møller, The Theory of Relativity, Oxford, Clarendon Press (ch.2), (1952).
[23] R. E. Marshak, Conceptual Foundations of Modern Particle Physics, Singapore, World Scientific (1992).
[24] J. Lekner, J. Opt. A: Pure Appl. Opt. 5, L15 (2003).
[25] J. D. Jackson. Classical Electrodynamics (2nd ed) Wiley New York (1975).
[26] A. Einstein, The Born-Einstein Letters, London, Macmillan (letter to Born dated 3rd March 1947 pages 157-8) (1971).
[27] E. Feigenbaum et al., Optics Express 17, 18934 (2009).
[28] P. A. M. Dirac Proc Roy Soc A, Math and Phys. 167, 148-169 (1938).
[29] J. A. Wheeler and R. P. Feynman Rev.Mod.Phys. 17 157 (1945).
[26] F. Rohrlich, Classical Charged Particles Addison-Wesley, New York, (1965).
[27] G. L. Turin, IRE Trans. Info. Theory IT-6, 311 (1960).
[28] J. E. Carroll and A.H.Quarterman, The Physics of Reality: Space, Time, Matter, Cosmos VIII Proceedings, R. L. Amoroso, P. Rowlands & L. H. Kaufmann, (eds.) World Scientific (2013)
[29] J. E. Carroll, Search for fundamental theory, VII Vigier symposium, R. L. Amoroso, P. Rowlands &S. Jeffers (eds.) , Am. Inst. Phys. Conference Series 1316, 217 (2010).
[30] D.A.B. Miller, Quantum mechanics for scientists and engineers, Ch.15. CUP, Cambridge (2008).
[31]P. A. M. Dirac The Principles of Quantum Mechanics 4th Ed. Oxford, Clarendon Press (1958).
[32] W.Greiner, Relativistic Quantum Mechanics 3rd Ed. Berlin Springer-verlag p312 (2000).
[33] M. E. Peskin and D. V. Schroeder, An Introduction to Quantum Field Theory, Reading Mass., Addison and Wesley (1995).
[34] Holland, P. R. The Quantum Theory of Motion Ch.1.,Cambridge, C.U.P. (1993).
[35] J. S. Bell, Jnl.Physique, 42, C2 (1981).
[36] P. G. Kwiat et al., Phys. Rev. Lett., 75, 4337 (1995).
D. Dehlinger and M.W. Mitchell, Am. J. Phys. 70, 898 (2002).
[37] D. Bouwmeester et al., Phys.Rev.Lett., 82, 1345 (1999).
A. Zeilinger et al., Phys.Rev.Lett. 78, 3031 (1997).

Relativistic Entanglement From Maxwell's Classical Equations

JOHN E. CARROLL[a] & ADRIAN H. QUARTERMAN[b]

Centre for Photonic Systems, University of Cambridge, 9
J. J. Thomson Avenue, Cambridge, CB3 0FA, UK.
[a]jec1000@hermes.cam.ac.uk, [b]ahq20@hermes.cam.ac.uk

With the help of light cone coordinates and light cone field representations of Maxwell's classical equations, quantum polarization entanglement is explained using the relativistic results of a companion paper that shows how conventional or reference waves can have an adjoint wave, travelling in phase with the reference wave, but in a proper relativistic frame that travels in the opposing direction to the proper frame of the reference wave. This subsequently allows waves, travelling in opposite directions, to have the same proper frame and consequently such waves can be regarded as relativistically local. The light cone coordinates offer a classical form of a quantum wave function and demonstrate a classical equivalent of a mixed quantum state.

Keywords: Maxwell's equations; Entanglement, Relativity

1. Introduction

In a companion paper [1], it was shown that the foundations of entanglement [2-5] could be based in the relativistic fact that waves with the same proper frame of reference can be local to one another. The key results showed how diffraction allows light waves to have well defined relativistic proper frames with finite rapidities [6]. A *reference* wave is the name used here to refer to a wave where the phase velocity, group velocity and direction of the wave's proper frame are all aligned. An *adjoint* wave can be constructed by simultaneous time and frequency reversal. Although this adjoint wave has the same phase velocity as the reference wave, its proper frame is found to travel in the opposite direction and so have a negative rapidity. This then leads to the non-intuitive conclusion that certain waves, although travelling in opposite directions can in fact have the same proper frame with the same rapidity in relativity theory. We are not aware of anyone taking Maxwell's classical equations [7] as a starting point to explain linear polarization entanglement using these principles. Unlike the companion paper [1], light cone coordinates [8-10] are used because such coordinate systems, not only confirm the analysis that simultaneous time and frequency reversal change the proper frame of reference, but they also show more clearly the connection between Maxwell's classical waves and quantum waves. This also extends the concepts of Bialynicki-Birula and Bialynicki-Birula [11,12] with their exposition of the Riemann-Silberstein vector, combining the electric and magnetic fields, in a way that makes it relevant as a photon wave function.The paper starts by outlining the notation for the electromagnetic fields with distinct transverse electric (TE) and transverse magnetic (TM) waves, avoiding the zero diffraction TEM (transverse electric and magnetic) waves [13,14]. Light cone methods [8-10] are used to combine TE and TM fields that will act as the classical equivalent of Maxwellian quantum waves. Combined vector forms of these waves, called here *reference* waves Ψ_A and Ψ_B, are considered to be emitted from a single source with Ψ_A travelling towards an observer Alice on the right and Ψ_B travelling towards Bob on the left. It is possible to find proper frames of reference for these diffracting waves which move apart but enable positions to be identified along the axis for Ψ_A and on Ψ_B with the same proper time. The concept of *adjoint* waves $\Psi_{Aa}{}^*$ and $\Psi_{Ba}{}^*$ with their time and frequency reversal are considered analogous to the time reversed conjugate replica of the incoming signals that is used in matched filtering [15] for optimum energy transfer. It is found, under certain conditions, that $\Psi_{Aa}{}^*$ can have the same proper frame Γ_n as Ψ_B, travelling in the opposite direction to the proper frame Γ_p for the wave Ψ_A. Similarly, $\Psi_{Ba}{}^*$ can have the same proper frame of reference Γ_p as for Ψ_A. These adjoint waves are considered as part of the boundary conditions when detecting the waves, Ψ_A and Ψ_B, conditions which cannot be fully determined until a measurement is made either by Alice or by Bob.

These adjoint waves $\Psi_{Aa}{}^*$ and $\Psi_{Ba}{}^*$ do not violate causality in spite of time reversal because they are excited in response to the causal reference waves interacting with the measurement system [1] and are realized only at the detection process. Any violation of causality is then limited to the time taken to measure a single photon. During that interval it is not possible to say that the energy has or has not arrived so that the concept of causality *during* a measurement is unclear.

Rotational invariance is demonstrated through arbitrary rotations of polarized beam splitters at Bob and Alice. These rotations can be considered at any time, including after any photons have been launched from the source but before measurement. The classical results are found to be consistent with those of a mixed quantum state that is not determined until a measurement is made. Although the theory is simplified by waves travelling in either direction along a single axis *Oz*, the results can be applied to more complex geometries requiring only definite directions before detection of the photons [1].

2. Maxwell's Equations — Matrix Notation

2.1. *Basic Equations*

The electromagnetic fields E, D, B, and H in free space have $D = \varepsilon_o E$ and $B = \mu_o H$. These fields are normalized to have the same dimensions as the electric field E using sans-serif fonts $\mathsf{E} = E$ and $\mathsf{H} = (c\mu_0 H)$ to give Maxwell's equations in free space as:

$$\text{div}\,\mathsf{H} = 0;\ \text{div}\,\mathsf{E} = 0;$$
$$\mathbf{M}_E = \text{curl}\,\mathsf{E} + \partial_{ct}\mathsf{H} = 0;\ \partial_{ct} = \partial/\partial ct;$$
$$\mathbf{M}_H = \text{curl}\,\mathsf{H} - \partial_{ct}\mathsf{E} = 0. \tag{1}$$

Here M_E and M_H are simply shorthands to identify these field equations later on.

Electromagnetic waves, advancing towards an observer Alice along the $+Oz$ direction, are assumed to have an angular frequency ω and a forward axial propagation constant $k > 0$ in the laboratory frame of reference Γ. Every component, say Ψ_A, of these vector forward travelling fields then varies in Γ as:

$$\Psi_A = \Psi_{0A}(x, y)\exp\{i\,[kz - (\omega/c)(ct) + \theta_A]\} \tag{2}$$

where i always means of an operator that changes the phase by 90° with $i^2 = -1$. A subscript A reminds the reader that this forward wave travels to an observer Alice while a subscript B will be used for waves travelling to Bob.

Any vector Ψ is split into transverse components Ψ_T and a parallel component $\Psi_{\|}$ along the arbitrary direction of the main propagation of the wave labeled here as Oz so that $\Psi_{\|} = \Psi_z$. Transverse components and operators are written with a matrix notation where a superscript ^{tr} means row-column transposition.

$$\mathbf{\Psi}_T = \begin{bmatrix}\Psi_x \\ \Psi_y\end{bmatrix};\ \mathbf{\Psi}_T^{\ tr} = [\Psi_x,\ \ \Psi_y];$$
$$\mathbf{\nabla}_T = \begin{bmatrix}\partial_x \\ \partial_y\end{bmatrix};\quad \mathbf{\Lambda} = \begin{bmatrix}0 & -1 \\ 1 & 0\end{bmatrix}. \tag{3}$$

Note that the operator Λ rotates the transverse vector components through 90° so that $\Lambda\Psi_T = [-\Psi_y,\ \Psi_y]^{tr}$ and $\Lambda^2 = -1$:i.e. Λ operating twice gives a 180° change.

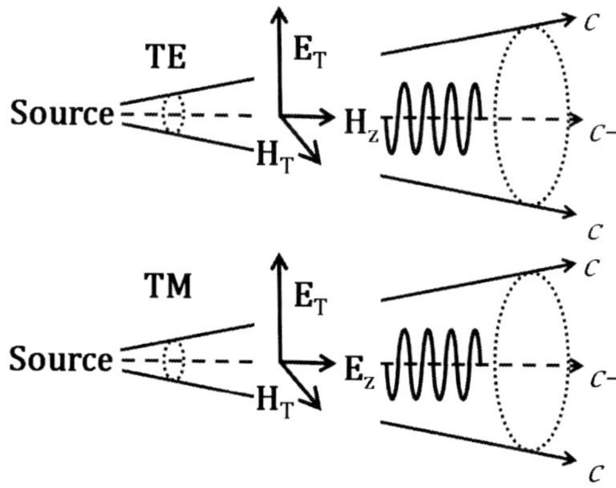

Figure 1. Transverse Electric (TE) and Transverse Magnetic (TM) waves. These diffracting waves travel with axial velocities slightly less than *c* in free space.

All real waves diffract and have some transverse propagation so that it is essential to consider Transverse Magnetic (TM) waves in combination with Transverse Electric (TE) waves (Figure 1) rather than the simpler Transverse Electromagnetic (TEM) waves or plane waves [13,14] that are devoid of any transverse variation. While TE and TM modes are commonly introduced for guided waves, they are essential in beam or packet formation [16,17]. Using the notation of equations (1) and (2), the fields advancing towards Alice along $+Oz$ are determined for the TM fields from:

$$\text{div}\,\mathsf{E} = \nabla_T^{\ tr}\mathsf{E}_{TA} + i\,k\,E_{zA} = 0; \tag{4}$$

$$\mathsf{M}_{H\,\|} = (\Lambda\nabla_T)^{tr}\,\mathsf{H}_{TA} + i(\omega/c)E_{zA} = 0; \tag{5}$$

$$\mathsf{M}_{E\,T} = ik\Lambda\mathsf{E}_{TA} - \Lambda\nabla_T E_{zA} - i(\omega/c)\,\mathsf{H}_{TA} = 0. \tag{6}$$

The TE fields are determined from:

$$\text{div}\,\mathsf{H} = \nabla_T^{\ tr}\mathsf{H}_{TA} + i\,k\,H_{zA} = 0; \tag{7}$$

$$\mathsf{M}_{E\,\|} = (\Lambda\nabla_T)^{tr}\,\mathsf{E}_{TA} - i(\omega/c)H_{zA} = 0; \tag{8}$$

$$\mathsf{M}_{H\,T} = ik\Lambda\mathsf{H}_{TA} - \Lambda\nabla_T H_{zA} + i(\omega/c)\,\mathsf{E}_{TA} = 0. \tag{9}$$

To curtail the amount of algebra it will be accepted without proof that every field component satisfies the relativistically invariant wave equation in vacuum:

$$[\partial_x^{\ 2} + \partial_y^{\ 2} + \partial_z^{\ 2} - \partial_{ct}^{\ 2}]\,\Psi_A = 0. \tag{10}$$

A simplified treatment of diffraction assumes a single transverse propagation constant κ so that all Cartesian field components, as in equation (2), satisfy:

$$(\partial_x^{\ 2} + \partial_y^{\ 2})\Psi_{0A}(x,y) = -\kappa^2\,\Psi_{0A}(x, y). \tag{11}$$

In free space, the TE and TM fields vary transversely in the same way because there are no special boundary conditions. The value of κ^2 will, later on, be made to tend to zero in order to approach the plane wave limit. Equations (10) and (11) then give the dispersion equation:

$$(\omega/c)^2 - k^2 - \kappa^2 = 0\ ;\ \kappa^2 > 0;\ (\omega/c)^2 > k^2. \qquad (12)$$

The group velocity $v_g = (d\omega/dk) = c^2(k/\omega)$ is always less than c.

2.2. *Rapidity and Light-Cone Formalism*

The concept of rapidity [6] provides a useful parameter that, unlike velocity, has straightforward additive properties when successively changing relativistic frames of reference. A frame Γ_p, moving with a velocity $v = (c \tanh \beta_p)$ relative to the laboratory frame Γ, is defined to have a rapidity of a β_p. In this frame Γ_p, the time can be written as t_p with a z-coordinate z_p. The wave Ψ_A in equation (2), when viewed in Γ_p, has an angular frequency ω_p and axial wave vector k_p. These coordinates in space and reciprocal space are related to their equivalents t, z, ω and k in Γ through the Lorentz-Einstein-Poincaré transformations [18,19] which, using light-cone methods [8-10] may be written in the form:

$$z_\pm = [ct \pm z]_\pm = (ct_p \pm z_p)\exp(\pm\beta_p); \qquad (13)$$

$$k_\pm = [(\omega/c) \pm k]_\pm = [(\omega_p/c) \pm k_p]_\pm \exp(\pm\beta_p). \qquad (14)$$

The transverse coordinates (x, y) and the transverse wave-vector κ in equation (12) are all unchanged by any changes of rapidity along the Oz axis.

Many different properties P_+ and P_- can be represented in a light cone format with the convention that these properties change with a rapidity β as:

$$P_\pm = P_0 \exp(\pm\beta); \qquad (15)$$

$$(P_+/P_-) = \exp(2\beta). \qquad (16)$$

If $(P_+/P_-) > 1$ then the convention gives a rapidity $\beta > 0$, while if $(P_+/P_-) < 1$ then $\beta < 0$. Any product $(P_+ P_-)$ is invariant to changes of rapidity.

If Γ_p is the relativistic proper frame of reference then $k_p = 0$. This is also known as the zero momentum frame [18]. From equation (12), the proper frequency $\omega_p = \kappa c$ while the transverse propagation constant is κ. The net phase velocity of the wave remains with a speed of c, as relativity requires, but the finite diffraction makes the wave travel transversely simultaneously in many different directions. The diffraction also ensures a finite value of $\beta_p > 0$, determined from the group velocity: $v_g = (d\omega/dk) = (c \tanh \beta_p) = (c^2\ k/\omega) < c$ for waves travelling forward along $+Oz$ towards Alice. Now rewrite the equations (13) and (14) in terms of the proper time t_p and proper frequency $\omega_p = c\kappa$ with $k_p = 0$ and $z_p = 0$. Considering some point P behind the wave front requires $|\ z\ | < ct$ while diffraction requires that $|k\ | < (\omega/c)$ to give:

$$z_\pm = [ct \pm z]_\pm = ct_p \exp(\pm\beta_p);$$
$$k_\pm = [(\omega/c) \pm k]_\pm = \kappa \exp(\pm\beta_p). \qquad (17)$$

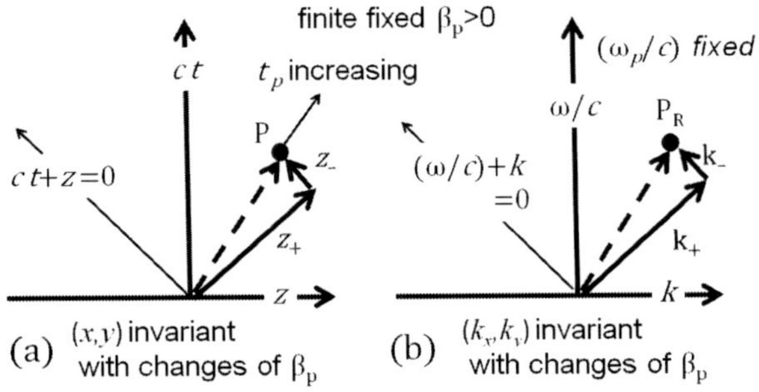

Figure 2. (a) Light cone coordinate position (z_+, z_-) for point P in wave travelling to Alice, time increasing with the proper time t_p. (b) The reciprocal position P_R is fixed by (k_+, k_-). Note that because $k_+/k_- > 1$, the rapidity is positive.

It is sometimes thought that light cone coordinates are useful only for entities travelling at the speed of light when $z = ct$ *or* $z = -ct$ (i.e. $\beta_p = +/- \infty$). However with finite rapidities, as is the case for all diffracting beams, light-cone coordinates are as valid as Cartesian or Minkowski coordinates but have distinct advantages in determining a wave's proper frame. Figures 2(a) and (b), sketch the key light cone coordinates in space and reciprocal space as seen in the laboratory frame. The proper frame is travelling in the positive Oz direction with a finite rapidity $\beta_p > 0$ with z_p set to zero for the illustration. In the laboratory frame Γ, even as proper time t_p increases, fixed ratios are still found for:

$$(z_+/z_-) = \exp(2\beta_p) > 1;\ (k_+/k_-) = \exp(2\beta_p) > 1. \qquad (18)$$

Applying light cone coordinates to equation (2) and relating them to the proper frame gives:

$$\Psi_A = \Psi_{oA}(x, y)\exp[-\tfrac{1}{2}i\,(k_+ z_- + k_- z_+) + i\,\theta_A] \qquad (19)$$

$$= \Psi_{oA}(x, y)\exp\{-i[(\omega_p/c)ct_p - \theta_A]\} \qquad (20)$$

Equation (19) gives an example of the Lorentz invariance of terms like P_+P_-. The term $(k_+ z_- + k_- z_+)$ gives them a Lorentz invariant phase.

2.3. *Maxwell's TE & TM Light-Cone Fields*

Noting that $\Lambda^{tr} = -\Lambda$, equations (4) and (5) combine to give TM fields determined by the axial electric field:

$$\mp\nabla_T^{tr}(E_{TA} \mp \Lambda H_{TA})_{TM\pm} = i[(\omega/c) \pm k]_\pm E_{zA} \qquad (21)$$

The subscripts $_\pm$ on the left hand side appear because of the general principle that if, in any one equation, elements lying on one branch of the light cone are equal to some second element, then that second element also lies on the same branch. Because electromagnetic fields are second order tensors rather than vectors, the axial fields E_{zA} and H_{zA} are invariant to changes of reference frame [7,18] moving along the direction $+Oz$. As a consequence, the terms on the left hand side of equation (21) lie on the same branch as the light cone terms $[(\omega/c) \pm k]_\pm$ on the right hand side.

Combining equations (7) and (8) gives TE fields that are determined from the axial magnetic field:

$$(\Lambda\nabla_T{}^{tr})(E_{TA} \mp \Lambda H_{TA})_{TE\pm} = i\,[(\omega/c) \pm k]_\pm H_{zA}. \quad (22)$$

Now note that $(\Lambda\nabla_T)^{tr}\nabla_T$ is identically zero while equation (11) gives:

$$(\Lambda\nabla_T)^{tr}(\Lambda\nabla_T)\Psi_{oA} = \nabla_T{}^{tr}\nabla_T\Psi_{oA} = -\kappa^2\Psi_{oA}(x,y). \quad (23)$$

Check, by substitution, that the combined TE and TM Maxwellian transverse fields of equations (21) and (22) can be integrated to give:

$$\Psi_{TA\,\pm} = (E_{TA} \mp \Lambda H_{TA})_\pm$$
$$= (i/\kappa^2)\,[(\omega/c) \pm k]_\pm\,[\pm\nabla_T E_{zA} - \Lambda\nabla_T H_{zA}]. \quad (24)$$

As pointed out earlier, E_{zA} and H_{zA} satisfy the same wave equation and so in free space, with no selective boundaries, these two fields vary transversely in the same way. As κ^2 tends to zero so do the field gradients $\nabla_T E_{zA}$ and $\nabla_T H_{zA}$ to give limiting ratios of the form:

$$(i/\kappa^2)\nabla_T E_{zA} = e_A\,U;\ (i/\kappa^2)\Lambda\nabla_T H_{zA} = h_A\,U^\perp. \quad (25)$$

Here U and $U^\perp$ are defined as unit vectors that are orthogonal to one another because of the matrix Λ operating on $\Lambda\nabla_T H_{zA}$. Because $U^\perp$ and U are proportional to transverse gradients of axial electromagnetic fields, they are both invariant to changes of rapidity. The multipliers e_A and h_A in principle allow arbitrary proportions of TE and TM fields to add together and will be determined by boundary conditions. The combined fields in equation (24) then are:

$$\Psi_{TA\pm} = [(\omega/c) \pm k]_\pm[\pm e_A\,U - h_A\,U^\perp]. \quad (26)$$

It may be checked that fields given in equations (21) and (22) are consistent with equations (6) and (9) and check the solutions satisfy the wave equation (10).

2.4. *Waves Travelling in a Negative Oz Direction*

Now consider a single source that excites waves Ψ_A in the $+Oz$ direction but also waves Ψ_B travelling in the $-Oz$ direction. Relativity requires that all directions are equivalent so that with the same frequency, the wave vector has the same magnitude k but now in the opposite direction: $k' \rightarrow -k$. The distance z' is measured in the opposite direction: $z' \rightarrow -z$. The proper time t_p and the proper frequency $c\kappa$ remain unaltered and the parameters in the wave keep $|z| < ct$ and $|k| < \omega/c$:

$$z'_\pm = [ct \pm (-z)]'_\pm = ct_p \exp(\pm\beta_n);$$
$$k'_\pm = [(\omega/c) \pm (-k)]'_\pm = \kappa\exp(\pm\beta_n). \quad (27)$$
$$(z'_+/z'_-) = (k'_+/k'_-) = \exp(2\beta_n) < 1. \quad (28)$$

The coordinates are shown schematically in Fig. 3 indicating the result of equation (28) showing that the rapidity β_n now has to be negative. The proper frame Γ_n therefore moves in the opposite direction to Γ_p as would be expected for a wave travelling along $-Oz$. The observer Bob, in the same laboratory frame Γ as Alice, observes negative values of $z = (-z')$ with electromagnetic waves $\Psi TB\pm$ travelling along $-Oz$, corresponding to the waves $\Psi_{TA\pm}$ travelling to Alice.

Note that consistency requires keeping the same proper frame in reciprocal space (k'+, k'−) as in space (z'+, z'−) and this consistency requires simultaneous reversal of k and z. As just explained, $\Psi TB\pm$ has a proper frame Γn with rapidity $\beta n = -\beta p$ with respect to the laboratory frame Γ. If Γ changes to a new frame Γ' such that $\Gamma' p$ has a rapidity $(\beta p - \alpha)$ then $\Gamma' n$ will have a rapidity $(\beta n - \alpha) = -(\beta p + \alpha)$ but $\Gamma' p$ and $\Gamma' n$ are then of course no longer proper frames.

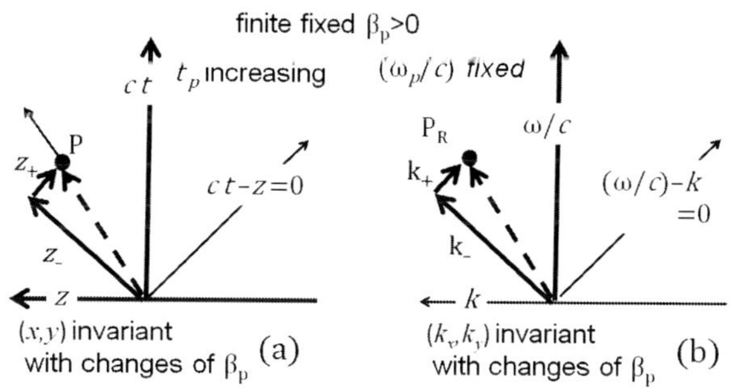

Figure 3. (a) Light cone coordinate position (z+, z−) for point P in a wave travelling to Bob; time increases with the proper time t_p. (b) The reciprocal position P_R is fixed by (k_+, k_-) with $k_+/k_- < 1$ so the rapidity is negative. The proper frame Γ_n travels in the reverse direction to the proper frame Γ_p for waves travelling to Alice.

Reversing z is not the same as a full space-reversal because x and y are left unchanged. Maxwell's equations (4)-(9), show that to obtain consistent changes of Ψ_{TA} travelling along $+Oz$ into Ψ_{TB} travelling along Oz requires changes of z to $-z$, k to $-k$, E_{zA} with $(-E_{zB})$ and H_{AT} with $(-H_{BT})$ in the combined field. The changes of sign in the fields keep the right handed sense of the trio of vectors given by the transverse electric fields, transverse magnetic fields and the direction of propagation [7]. The equivalent result to equation (26) is:

$$\Psi_{TB\,\pm} = [E_{TB} \mp \Lambda(-H_{TB})]_\pm \quad (29)$$
$$= [(\omega/c) \pm (-k)]_\pm\,[\pm(-e_B\,U) - (h_B\,U^\perp)], \quad (30)$$

where $U^\perp = \Lambda U$. The observer in the laboratory frame readily distinguishes $-z'$ and z and so all the fields travelling to Bob implicitly vary as:

$$\Psi_B = \Psi_{oB}(x, y)\exp\{i\,[(-kz - (\omega/c)(ct) + \theta_B]\}. \quad (31)$$

The waves, Ψ_A and Ψ_B will be referred to as *reference* waves where for each, separately, their phase velocity, group velocity and proper frame of reference travel in a single direction, unlike *adjoint* waves discussed next.

3. Adjoint Waves

The concept of adjoint waves is that they travel with the same phase velocity as the reference wave but simultaneously have both ω and *t* reversed as below:

$$\Psi_{Aa} = \Psi_{oA}(x, y)\exp\{i\,[(kz - (-\omega/c)(-ct)+\theta_{Aa}]\}. \quad (32)$$

The dispersion equation (12) is unaltered as is the transverse equation (11) so that superficially Ψ_{Aa} and Ψ_A appear to be identical travelling together. However time and frequency reversal applies not only to both ω and *t* but also to the proper time t_p and the proper frequency $c\kappa$ so that equations (17) now have to read for $|z| < ct$ and $|k| < \omega/c$:

$$z_{\pm} = [(-ct) \pm z]_{\pm} = (-ct_p)\exp(\pm\beta_n);$$

$$k_{\pm} = [(-\omega/c) \pm k]_{\pm} = (-\kappa)\exp(\pm\beta_n). \quad (33)$$

$$(z'_{+}/z_{-}) = (k_{+}/k_{-}) = \exp(2\beta_n) < 1. \quad (34)$$

Only if both ω and t are reversed together will the pairs (k+, k–) and (z+, z–) be referred to the same proper frame. Equation (34) can be satisfied only if βn < 0 while consistency with equation (17) requires βn = – βp leading to the conclusion that the proper frame is now Γn that travels in the opposite direction to Γp the proper frame for ΨA. It has previously been shown that energy travels with the proper frame of reference [1]. Fig.4 shows the equivalent diagrams as for Fig.2.

Adjoint wave solutions are usually never considered because time reversal is argued to make the wave inapplicable to real physics. What is envisaged here is that every measurement is a made by correlating adjoint and reference waves through the product $\Psi_{TAa-}{}^{\dagger}\Psi_{TA}$ where † denotes conjugation and transposition but Ψ_{TAa} has the all-important time reversal to extract the maximum energy from the incoming signal Ψ_{TA} as in a matched filter. Fig. 5 below enlarges on the concepts of measurement outlined in Fig. 3 of reference 1.

The reference wave assures causality with no energy ever being measured until the reference wave arrives. The adjoint wave gives the additional degree of freedom that allows polarization to be determined only when a measurement is made. Now by reversing time and frequency it is always possible to track the mathematical form and the proper frame of $\Psi_{TAa}{}^{*}$ as Ψ_{TA} propagates. Although this procedure is used here, it is important to understand that in this theory, Ψ_{TAa} only exists during measurement process as in Fig.5.

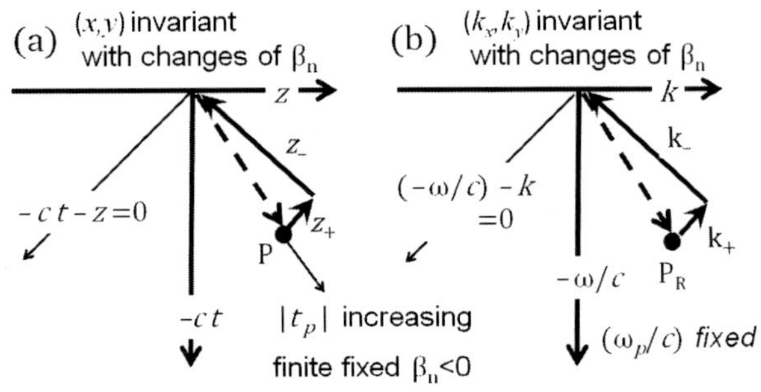

Figure 4. (a) Light cone coordinate position (z_+, z_-) for point P in adjoint wave with phase velocity travelling to Alice but time and frequency reversed. (b) The reciprocal position P_R given by (k_+, k_-). With $k_+/k_- < 1$, the rapidity is negative. The proper frame Γ_n travels towards Bob.

The adjoint waves, corresponding to the reference waves in equation (26), are determined from the fact that the electric field is an even function of time while the magnetic field is an odd function of time so that, with *t*-reversal, the magnetic fields' sign has to chang [7]. This gives the two branches of the adjoint wave as:

ΨTAa ± = [(– ω/c) ± k]±[±eAa U + hAa U⊥]. (35)

The adjoint wave ΨTBa± corresponding to ΨTB ± in equation (30) has similar changes to give:

ΨTBa± = [(–ω/c)±(–k)]± [±(–eB U) +(hB ΛU)]. (36)

4. Measurable Waves and Probabilities

To preserve the full electromagnetic symmetry, there are now eight light cone fields to be tracked relevant to a single source that sends out waves in opposite directions: $\Psi_{TA\pm}$, $\Psi_{TB\pm}$ as the reference waves, with $\Psi_{TAa\pm}$, $\Psi_{TBa\,\pm}$ as the associated adjoint waves required in the measurement process. However consider the amplitudes of these waves in the laboratory frame as the diffraction becomes weak and the waves tend to plane waves: $[(\omega/c) - k]^2 \lll [(\omega/c) + k]^2$. This means that for 'energy' or 'probability', as measured by the squared amplitude of a wave, those waves that are proportional to $[(\omega/c) - k]$ can be safely neglected. That leaves four electromagnetic waves in light cone form:

$$\Psi_{TA\,+} = [(\omega/c) + k]_{+}\times[e_A\,U - h_A\,U^{\perp}]; \quad (37)$$

$$\Psi_{TB\,-} = [(\omega/c) - (-k)]_{-}\times[-(-e_B\,U) - (h_B\,U^{\perp})]; \quad (38)$$

$$\Psi_{TAa-} = [(-\,\omega/c) - k]_{-}\times[-e_{Aa}\,U + h_{Aa}\,U^{\perp}]; \quad (39)$$

$$\Psi_{TBa+} = [(-\omega/c) + (-k)]_{+}\times[(-e_B\,U) + (h_B\,U^{\perp})]. \quad (40)$$

At a symmetrically emitting source, where any measurement has not yet been made, it will be found that there are not enough boundary conditions to determine all the ratios (e_A/h_A), (e_B/h_B), (e_{Aa}/h_{Aa}), (e_{Na}/h_{Ba}) at the same time.

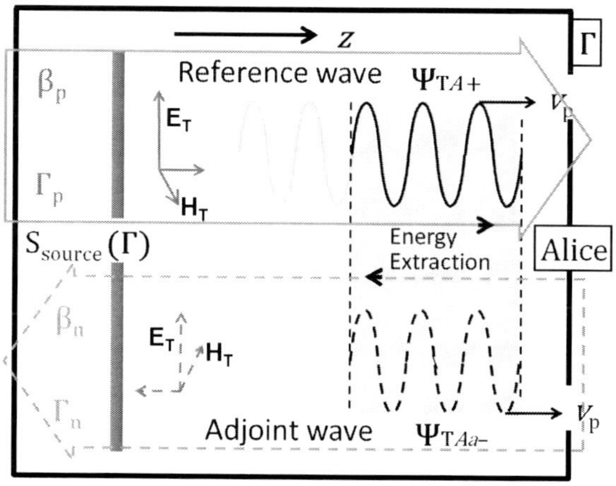

Figure 5. Reference and adjoint waves in the laboratory frame Γ during detection have their phases travelling together. The adjoint wave Ψ_{Aa}* appears in an analogous manner to the response of a matched filter that uses a time-frequency reversed replica of the incoming pulse. Ψ_{Aa}* extracts the maximum energy from the incoming wave Ψ_A: the smallest unit of energy being one 'photon'.

Bialynicki-Birula & Bialynicki-Birula [11,12] show how the Riemann-Silberstein (RS) vector Ψ_{RS} = (E + iB) can provide a correspondence between the quantum wave vector and the classical electromagnetic fields. In this present paper, the suggestion is that the quantum-classical correspondence, given a direction of propagation, is found by replacing i in Ψ_{RS} with the rotation operator Λ, noting that like i, $\Lambda^2 = -1$. However, Λ has the clear physical meaning of a physical rotation. Note also for example that for *circularly polarized* waves, i and Λ are not readily distinguishable as a 90° rotation is accompanied by a 90° phase change. The equivalent to Ψ_{RS} for a wave propagating along the +Oz direction is now taken to be the light cone transverse field: $\Psi_{T+} = (E_T - \Lambda H_T)_+$. As κ^2 tends to zero so as to approach the plane wave limit, then Ψ_{T+} is a measurable reference wave for forward propagation while Ψ_{T-} is negligible.

Using, as before, the superscript † to indicate both complex conjugation * and transposition ^{tr} of rows and columns, the suggestion now is that, for waves travelling towards Alice, the photon probability is given from the product $\Psi_{TaA-}{}^\dagger\Psi_{TA+}$ as shown in the shaded overlap region in Fig.5. Taking the complex conjugate, and thereby ensuring phase invariance, does not alter the proper frame for Ψ_{TaA-} which remains opposite to that of Ψ_{TA+}. This notation confirms the distinction between bra and ket vectors in the Dirac [20] notation, <Ψ|Ψ>, and replaces the conventional Ψ*Ψ.

4.1. *Overlap Regions*

The region, where there is a finite probability of measuring a photon, is now limited to any region where Ψ_{TaA-}* and Ψ_{TA+} can overlap. It is no longer essential to consider the probability over large regions of space but only over a short where any *measurement* is made.

The finite region of overlap then removes one of the mysteries of quantum theory where it is traditionally asserted that a measurement immediately collapses a wave to zero over the whole of space which always seems an unlikely event. Here, a measurement need only remove the waves over a local region and this could appear as an immediate collapse in relation to practical observational time scales.

A third reason for supporting this approach is given from considering how $\Psi_{TAa-}{}^\dagger\ \Psi_{TA+}$ varies as the laboratory frame of reference varies. Begin by referring the fields Ψ_{TAa-} and Ψ_{TA+} to their appropriate proper frames where, for example:

$$\Psi_{TA+} = (\Psi_{TA+})_p \exp(\beta_p); \tag{41}$$

$$(\Psi_{TAa-}) = (\Psi_{TAa-})_n \exp(-\beta_n). \tag{42}$$

We now come to a subtle point. Find the correct rapidity β_q for any proper frames Γ_q with reference to a laboratory frame Γ and change Γ into a new frame Γ' by a change in rapidity of α. Then all rapidities change by the same amount: $\beta_q - \alpha$. In this case:

$$\beta_p \rightarrow (\beta_p - \alpha) \text{ and } \beta_n \rightarrow (\beta_n - \alpha), \tag{43}$$

$$\beta_p - \beta_n = (2\beta_p + \alpha - \alpha) = 2\beta_p\,. \tag{44}$$

As a consequence, in either the Γ or Γ' frame one finds *independent* of α:

$$\Psi_{TAa-}{}^\dagger\ \Psi_{TA+} = (\Psi_{TAa-})_n{}^\dagger(\Psi_{TA+})_p \exp(2\beta_p). \tag{45}$$

The cancellation of α shows that the formalism allows a probability that is invariant with a change of rapidity α in the laboratory frame. This is surely what is expected of a scalar probability in a space-time invariant volume.

All these arguments made about $\Psi_{TAa-}{}^\dagger\Psi_{TA+}$ for photons travelling towards Alice can be applied to the proposed probabilities $\Psi_{TBa+}{}^\dagger\Psi_{TB-}$ for photons travelling towards Bob. Again the probability of measuring energy is confined to the region where Ψ_{TBa+} and Ψ_{TB-} overlap. Reference 1 discussed how this theory explained the outcomes for a single photon where there is more than one path, as in a beam splitter. This gives yet another reason for considering the overlap theory as re-emphasized here.

4.2. *Entanglement for Two Photons*

The final reason for considering this overlap concept is because it suggests the probability for a photon measured at Alice is $\Psi_{TAa-}{}^\dagger\ \Psi_{TA+}$ and a photon measured at Bob is, $\Psi_{TBa+}{}^\dagger\ \Psi_{TB-}$. Figure 6 illustrates the concept [14] that Ψ_{TA+} and Ψ_{TBa+} can have the the same proper frame Γ_p and similarly Ψ_{TB-} and Ψ_{TAa-} can lie in the same proper frame Γ_n. If the frame Γ changes to Γ' as in equations (41) and (42) then in Γ'

$$\Psi'_{TA+} = (\Psi_{TA+})_p \exp(\beta_p - \alpha) \; ; \quad (46)$$

$$\Psi'_{TBa+} = (\Psi_{TBa-})_n \exp[(\beta_p - \alpha)]. \quad (47)$$

If $\alpha = \beta_p$ then $\Gamma' = \Gamma_p$ the proper frame and it is seen that it is then possible for both Ψ_{TA+} and Ψ_{TBa+} to be relativistically local to each other although in the laboratory frame they are far apart. There need be nothing spooky[21] about correlations between waves in the same relativistic proper frame.

To emphasize the point, let Alice measure the polarisation of $\Psi_{TAa-}^{\dagger}\Psi_{TA+,}$ giving a measurement result with a certain polarization. Then because of the relativistic localization of Ψ_{TA+} and Ψ_{TBa+}, along with the similar localization of Ψ_{TAa-} and Ψ_{TB-}, the observer Bob when he comes to make a measurement has detailed local knowledge of the polarization measured by Alice. Any correlation of Bob and Alice's polarization for these waves in the same proper frame is no surprise. Section 5 works through the details for a hypothetical source with certain simple characteristics.

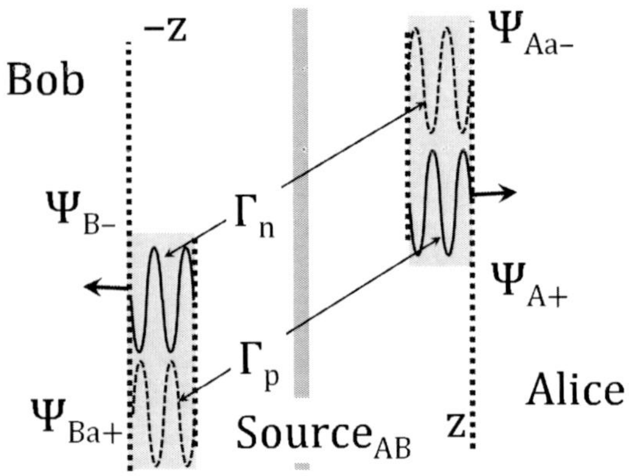

Figure 6. Pairs of reference waves Ψ_{TA+} and Ψ_{TBa+} share a proper frame Γ_p while Ψ_{TB-} and Ψ_{TAa-} share a proper frame Γ_n. Waves with a common frame can have correlated properties without violating locality.

Not all waves, Ψ_{TA+} and Ψ_{TBa+} will have the same proper frame Γ_p. To have the same proper frame these waves must arise from the same source at the same time. This leads to the quantum aspects of entanglement. Classical measurements are composed of elements with many different proper frames of reference [1], usually without any correlations between these different proper frames. If there were correlations on this classical scale then it would have to be built in as a classical property of the source.

One of the tests for true entanglement of linearly polarized photons is to allow the observers, Alice and Bob, to independently measure their photons polarizations in arbitrary bases. They then decide on what angle to set these measurement bases *after the photons have actually left a source* which emits random linear polarization. That is the next step that leads to the detail of what linear polarization Bob and Alice will be able to measure.

5. Polarization Amplitudes

5.1. *Circular Polarization: Rotational Invariance*

Consider equations (37)-(40) and absorb the rapidity varying factors $[(\omega/c) + k]_+$ or $[(\omega/c) - (-k)]_-$ etc. into the amplitudes for the various Ψ_T so for example:

$$\begin{aligned} \Psi_{TA+} &= E_A\mathbf{U} - H_A\mathbf{U}^{\perp} \\ &= E_{PA}(\mathbf{U} + i\mathbf{U}^{\perp}) + E_{NA}(\mathbf{U} - i\mathbf{U}^{\perp}); \\ E_{PA} &= \tfrac{1}{2}(E_A + iH_A)\,; E_{NA} = \tfrac{1}{2}(E_A - iH_A). \end{aligned} \quad (48)$$

The first line of equation (48) gives the 'amplitudes' of the linear polarizations while the second line gives the 'amplitudes' of circular polarizations. These circularly polarized fields have orthogonal vectors varying in phase quadrature: $\mathrm{U} \pm i\mathrm{U}^{\perp}$. Similar expressions to those in equation (48) hold for Ψ_{TAa-}:

$$\begin{aligned} \Psi_{TAa-} &= E_{Aa}\mathbf{U} - H_{Aa}\mathbf{U}^{\perp} \\ &= E_{PAa}(\mathbf{U} + i\mathbf{U}^{\perp}) + E_{NAa}(\mathbf{U} - i\mathbf{U}^{\perp}); \\ E_{PAa} &= \tfrac{1}{2}(E_{Aa} + iH_{Aa}); E_{NAa} = \tfrac{1}{2}(E_{Aa} - iH_{Aa}). \end{aligned} \quad (49)$$

An 'unpolarized' source will have equal probabilities of linear polarization and equal probabilities of circular polarization. The averaged values over many measurements of the components of the 'probability' $\Psi_{TAa-}^{\dagger}\Psi_{TA+}$ in this work gives the equality of the different polarizations from:

$$\begin{aligned} \overline{E_{Aa} * E_A} &= \overline{H_{Aa} * H_A}; \\ \overline{E_{PAa} * E_{PA}} &= \overline{E_{NAa} * H_{NA}}. \end{aligned} \quad (50)$$

The lower equation (49) also requires:

$$\overline{E_{Aa} * H_A} - \overline{H_{Aa} * E_A} = 0. \quad (51)$$

The bar indicates the statistical averages over many measurements. A similar expression applies for Bob's wave Ψ_{TB-} with appropriate changes of the subscripts.

The requirement that special relativity has no preferred orientation can now be checked. Suppose that Bob and Alice both rotate their measurement systems for linear polarization by an arbitrary angle φ. Note that the matrix Λ in equation (3) rotates transverse vectors by 90° with $\Lambda^2 = -1$. Then an operator rotating any transverse vector through an angle φ is given from [exp(Λφ)] so that one may define:

$$U_\varphi = [\exp(\Lambda\varphi)]\, U\, ;\; U_\varphi^{\perp} = [\exp(\Lambda\varphi)]\, U^{\perp} \qquad (52)$$

Remembering $U^{\perp} = \Lambda U$, it can be shown that:

$$\begin{aligned}\Psi_{TA+} &= \exp(-\Lambda\phi)[E_{PA}(\mathbf{U}+i\mathbf{U}^{\perp}) + E_{NA}(\mathbf{U}-i\mathbf{U}^{\perp})]\\ &= E_{PA\phi}(\mathbf{U}_\phi + i\mathbf{U}_\phi^{\perp}) + E_{NA\phi}(\mathbf{U}_\phi + i\mathbf{U}_\phi^{\perp})\\ &= E_{A\phi}\mathbf{U}_\phi - H_{A\phi}\mathbf{U}_\phi^{\perp}\,.\end{aligned} \qquad (53)$$

$$E_{PA\phi} = E_{PA}\exp(i\phi);\; E_{NA\phi} = E_{NA}\exp(-i\phi).$$

Similarly

$$\begin{aligned}\Psi_{TAa-} &= E_{PAa\phi}(\mathbf{U}_\phi + i\mathbf{U}_\phi^{\perp}) + E_{NAa\phi}(\mathbf{U}_\phi + i\mathbf{U}_\phi^{\perp})\\ &= E_{Aa\phi}\mathbf{U}_\phi - H_{Aa\phi}\mathbf{U}_\phi\,;\end{aligned} \qquad (54)$$

$$E_{PAa\phi} = E_{PAa}\exp(i\phi);\; E_{NAa\phi} = E_{NAa}\exp(-i\phi).$$

Equations (49) and (50) still hold in the form:

$$\overline{E_{A\varphi a}{}^{*}E_{A\varphi}} = \overline{H_{A\varphi a}{}^{*}H_{A\varphi}};$$
$$\overline{E_{PA\varphi a}{}^{*}E_{PA\varphi}} = \overline{E_{NA\varphi a}{}^{*}H_{NA\varphi}}. \qquad (55)$$

Rotational invariance is satisfied, no matter when or by how much the reference frame is rotated. Similar remarks apply to Ψ_{TBa+} and Ψ_{TB-}.

5.2. *Linear Polarization*

Equations (48) and (49) give the linear polarizations for the reference and adjoint waves with phase velocities travelling towards Alice. The equivalent waves with phase velocities travelling towards Bob will be:

$$\Psi_{TB} = E_B\mathbf{U} - H_B\mathbf{U}^{\perp};$$
$$\Psi_{TBa+} = E_{Ba}\mathbf{U} - H_{Ba}\mathbf{U}^{\perp}. \qquad (56)$$

All expressions like E_P U are proportional to $\nabla_T E_{zP}$. All expressions like H_P U are proportional to $\nabla_T H_{zP}$. Any sign reversal of the fields was taken into account in considering these field components so that the continuity at a simple source boundary suggests that the right boundary conditions are:

$$E_B = E_A; E_{Ba} = E_{Aa}; H_B = H_A; H_{Ba} = H_{Aa}. \qquad (57)$$

The transverse coordinates are not changed by changes of rapidity so that on rotating by an angle φ, the transverse fields 'travelling' to both Alice and Bob are rotated by the same operator [exp(Λφ)]. Consequently the waves 'travelling' towards Bob equivalent to equations (53) and (54) are:

$$\begin{aligned}\Psi_{TB-} &= E_{B\phi}\mathbf{U}_\phi - H_{B\phi}\mathbf{U}_\phi^{\perp}\,.\\ \Psi_{TBa+} &= E_{B\phi a}\mathbf{U}_\phi - H_{B\phi a}\mathbf{U}_\phi^{\perp}.\end{aligned} \qquad (58)$$

The field amplitudes should be equated as:

$$E_{B\varphi} = E_{A\varphi}; E_{B\varphi a} = E_{A\varphi a};$$
$$H_{B\varphi} = H_{A\varphi}; H_{B\varphi a} = H_{A\varphi a}. \qquad (59)$$

The ratios $H_{B\varphi}/E_{B\varphi} = H_{A\varphi}/E_{A\varphi}$ are unknown and not determined in equations (56) and (58) making the waves in these equations a classical equivalent to quantum mixed state. The ratios will be determined only after a measurement has been made.

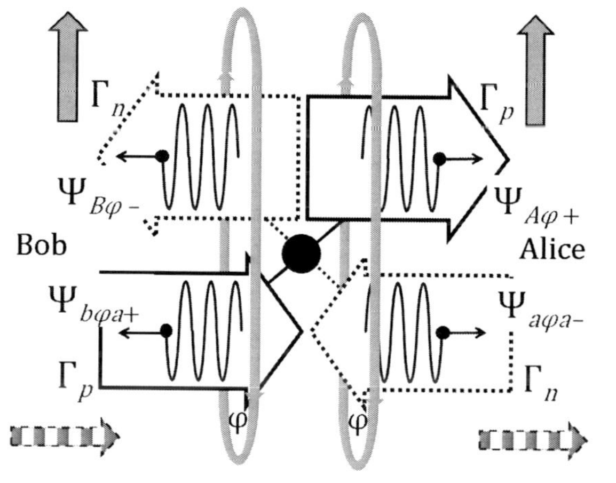

Figure 7. Alice and Bob both have their measurement systems rotated by an angle φ. Wave pairs show localization links through the same proper frames. With respect to the angle φ, either horizontal polarizations (dashed arrows) or vertical polarizations (solid arrows) are measured at both Bob and Alice.

Now assume that Alice is the first to make a measurement of her waves just after both Bob and Alice have rotated their systems by an arbitrary angle φ. Assume that Alice happens to measure the polarization represented by U_φ so that gives the boundary condition that $H_{A\varphi a}{}^{*}H_{A\varphi} = 0$. The fact that $H_{B\varphi a}$ is relativistically local to $H_{A\varphi}$ while $H_{B\varphi}$ is relativistically local to $H_{A\varphi a}$ allows $H_{B\varphi a}{}^{*}H_{B\varphi} = 0$ so that Bob can only measure the U_φ polarisation. Similarly if Alice measures the polarization represented by $U^{\perp}_\varphi$ then $E_{A\varphi a}{}^{*}E_{A\varphi} = 0$ so that Bob has to measure $U^{\perp}_\varphi$ polarisation. Figure 7 sketches these concepts with possible measurement outcomes.

If there were a 90° rotation between the left and right travelling waves Ψ_{TA+} and Ψ_{TB-} then, when Alice measured U_φ, Bob would now measure $U^{\perp}_\varphi$ and vice versa. The correlations would give orthogonal linear polarizations rather than the same linear polarizations.

Entangled state manipulation is discussed by other authors [5].

Suppose now that Bob had rotated his polarizing measurement system by ($\varphi+\chi$) while Alice had rotated her system by φ. Then Bob is measuring either $U_{(\varphi+\chi)}$ or $U_{(\varphi+\chi)}{}^{\perp}$. This can be related to U_φ and $U_\varphi{}^{\perp}$ using the rotation operation $\exp(-\Lambda\chi)$:

$$\begin{aligned}
&\Psi_{\mathbf{T}B-} = \\
&E_{B\varphi}\exp(-\mathbf{\Lambda}\chi)\mathbf{U}_{\varphi+\chi} - H_{B\varphi}\exp(-\mathbf{\Lambda}\chi)\mathbf{U}_{\varphi+\chi}{}^{\perp} \\
&= (E_{B\varphi}.\cos\chi - H_{B\varphi}\sin\chi)\mathbf{U}_{\varphi+\chi} \\
&-(H_{B\varphi}\cos\chi + E_{B\varphi}\sin\chi)\mathbf{U}_{\varphi+\chi}{}^{\perp}; \\
&\Psi_{\mathbf{T}Ba+} = \\
&E_{B\varphi a}\exp(-\mathbf{\Lambda}\chi)\mathbf{U}_{\varphi+\chi} - H_{B\varphi a}\exp(-\mathbf{\Lambda}\chi)\mathbf{U}_{\varphi+\chi}{}^{\perp} \\
&= (E_{Ba\varphi}.\cos\chi - H_{Ba\varphi}\sin\chi)\mathbf{U}_{\varphi+\chi} \\
&-(H_{Ba\varphi}\cos\chi + E_{Ba\varphi}\sin\chi)\mathbf{U}_{\varphi+\chi}{}^{\perp}. \qquad (60)
\end{aligned}$$

As previously, if Alice measures U_φ polarization then $H_{B\varphi a}=H_{B\varphi} = 0$ leaving Bob measuring $U_{\varphi+\chi}$ polarization with a probability of $\cos^2$ and the orthogonal polarization $U_{\varphi+\chi}{}^{\perp}$ with a probability of $\sin^2$. If on the other hand Alice measures $U_\varphi{}^{\perp}$ polarization, then $E_{B\varphi a}=E_{B\varphi} = 0$ and Bob measures $U_{\varphi+\chi}$ polarization with a probability of $\sin^2$ and the orthogonal polarization $U_{\varphi+\chi}{}^{\perp}$ with a probability of $\cos^2$. These give the same magnitudes of probability as for quantum theory.

5.3. *Beams at an Angle and Multi-Phonon Entanglement*

As discussed in the companion paper [1], the favored geometry for entangled photon generation has the selected optical beams emerging at a narrow angle [22,23]. Real systems will in general also have mirrors so that the paths are not linear. Never the less at the final polarizing detectors, the photons are viewed as coming from a definite direction at some angle. It is then possible to change the laboratory frame of reference so that these final directions just before the detectors are changed into being at 180° instead of a narrow angle1. The above theory can then be applied.

Experiments show that it is possible to entangle more than just a pair of photons [24,25]. This was also explained by the adjoint-reference wave product formulation in the companion paper [1].

6. Conclusions

Maxwell's equations for diffracting optical beams are expressed using light cone algebra to support a model that shows it is possible to have the same relativistic proper frames for conventional reference waves and novel adjoint waves, where frequency, time and the physical direction of the phase velocity are all simultaneously reversed. Once the existence of such common proper frames of reference are accepted then correlations between the waves in these proper frames, which are moving apart in the laboratory frame, is no longer seen as violating relativity. The detailed analysis for Maxwell's equations shows that, if on separate occasions Alice measures arbitrary linear polarizations θ and $\theta^{\perp}$ then there are two possibilities for Bob's measurements that depend on the detailed physics of the source. With the first type of source Bob will find that he also measures θ and $\theta^{\perp}$ respectively from the matching waves coming towards him on the same occasions as Alice makes her measurements of either θ or $\theta^{\perp}$. With the second type of source, Bob measures $\theta^{\perp}$ and θ respectively on the same occasions as Alice measures θ and $\theta^{\perp}$. Of course the entangled photons can have their entangled states manipulated [4].

Given an unpolarized source, correlations will occur only between photons where the waves travelling in different directions are emitted at the same time so that the relevant waves have identical proper frames of reference. In practice this requires carefully timed quantum measurements on single photons. Classical measurements measure many photons simultaneously and find no correlation because the independent photons cannot all have the necessary identical proper frames. Strict Einstein locality conditions [5] will then be observed. However if the source always gave out only one type of linear polarization in both directions then classical correlations would of course be observed.

Acknowledgments

The authors are indebted to Ian White who has been supportive of novel photonic concepts for many years.

References

[1] J. E. Carroll & A. H. Quarterman, in R. L. Amoroso, P. Rowlands and L. H. Kaufmann, Eds. The Physics of Reality: Space, Time, Matter, Cosmos VIII Proceedings. World Scientific (2013).
[2] A. Aspect et al., Phys. Rev. Lett. 49, 91 (1982).
[3] Y. H. Shih, and C. O. Alley, Phys. Rev. Lett., 61, 2921 (1988).
[4] G. Weihs et al., Phys. Rev. Lett. 81, 5039 (1998).
[5] M. A. Rowe et al., Nature 409, 791 (2001).
[6] J-M. Levy-Leblond and J-P. Provost, Am. J. Phys. 47, 1045 (1979).
[7] J. D. Jackson, Classical Electrodynamics, 2nd Ed. Wiley, New York (1975).
[8] Y. S. Kim and M. Noz, Am. J. Phys. 50, 721 (1981).
[9] Y. S. Kim and E. P. Wigner, Phys. Rev. A. 36, 1293 (1987).

[10] L. Parker & G. M. Schmeig, Am. J. Phys. 38, 218 (1970).
[11] I. Bialynicki-Birula, Acta Phys. Pol. B. 37, 935 (2006).
[12] I. Bialynicki-Birula and Z. Bialynicki-Birula, Opt. Comm. 264, 342 (2006).
[13] K. F. Sander and G. A. L. Read, Transmission and Propagation of Electromagnetic Waves C.U.P. (2nd Ed) (1986).
[14] J. Lekner, J.Opt.A:Pure Appl.Opt., 3, 407 (2001).
[15] G. L. Turin, IRE Trans. Info. Theory IT-6, 311 (1960).
[16] A. Sezginer, J. Appl. Phys. 57, 678 (1985).
Particle Physics, World Scientific (p106) (1992).
[17] J. Lekner, J. Opt. A: Pure Appl. Opt. 5, L15 (2003).
[18] C. Moller, The Theory of Relativity, Oxford (1952).
[19] R. E. Marshak, (1997) Conceptual Foundations of Modern Particle Physics, Singapore, World Scientific (p106)
[20] P. A. M. Dirac, The Principles of Quantum Mechanics (4th Ed) Oxford (1958).
[21] A. Einstein and M. Born, The Born-Einstein Letters McMillan, London (1971).
[22] P. G. Kwiat et al., Phys. Rev. Lett., 75, 4337 (1995).
[23] D. Dehlinger and M. W. Mitchell, Am. J. Phys. 70, 898 (2002).
[24] A. Zeilinger et al., Phys.Rev..Lett. 78, 3031 (1997).
[25] D. Bouwmeester et al., Phys. Rev. Lett., 82, 1345 (1999).

Universal Quantum Computing: Third Gen Prototyping Utilizing Relativistic 'Trivector' R-Qubit Modeling Surmounting Uncertainty

RICHARD L. AMOROSO

Noetic Advanced Studies Institute, Escalante Desert, Beryl, UT 84714-5104 USA
amoroso@noeticadvancedstudies.us, www.noeticadvancedstudies.us

LOUIS H. KAUFFMAN

Department of Mathematics, Statistics & Computer Science
University of Illinois at Chicago, Chicago, IL, 60607-704 USA
Kauffman@uic.edu

SALVATORE GIANDINOTO

BioHarmonic Resonance, Inc.
Las Vegas, NV 89121-2742
Salg@unifiedfieldtheories.com

We postulate bulk universal quantum computing (QC) cannot be achieved without surmounting the quantum uncertainty principle, an inherent barrier by empirical definition in the regime described by the Copenhagen interpretation of quantum theory - the last remaining hurdle to bulk QC. To surmount uncertainty with probability 1, we redefine the basis for the qubit utilizing a unique form of M-Theoretic Calabi-Yau mirror symmetry cast in an LSXD Dirac covariant polarized vacuum with an inherent 'Feynman synchronization backbone'. This also incorporates a relativistic qubit (r-qubit) providing additional degrees of freedom beyond the traditional Block 2-sphere qubit bringing the r-qubit into correspondence with our version of Relativistic Topological Quantum Field Theory (RTQFT). We present a 3rd generation prototype design for simplifying bulk QC implementation.

Keywords: Bloch sphere, Differential geometry, Graphene, M-Theory, Quantum computing, Qubit, Quantum Hall effect, Trivector

1. Introduction

Quantum Computing (QC) has remained elusive beyond a few qubits, enough only to operate a logic gate. Feynman's recommended use of a "synchronization backbone" [1] for achieving bulk implementation has generally been abandoned as intractable; a conundrum we believe arises from limitations imposed by the standard models of Quantum Theory (QT) and Cosmology. It is proposed that Feynman's model can be utilized to implement Universal Quantum Computing (UQC) with valid operationally completed extensions of QT and cosmology [2]. Requisite additional degrees of freedom are introduced by defining a relativistic basis for the qubit (r-qubit) in a higher dimensional (HD) conformal scale-invariant context and defining a new anticipatory based cosmology (cosmology itself cast as a hierarchical form of complex self-organized system) making correspondence to unique 12D Calab-Yau mirror symmetries of M-Theory. The causal structure of these conditions reveal an inherent new Unified Field, U_F "action principle" (force of coherence) driving self-organization and providing a basis for applying Feynman's synchronization backbone principle. Operationally a new set of transformations (beyond the standard Galilean / Lorentz-Poincaré) *ontologically* surmounts the quantum condition (producing decoherence during both initialization and measurement) by an acausal energyless (ontological) topological interaction [2]. Utilizing the inherent structural-phenomenology of the HD regime requires new commutation rules and corresponding I/O techniques based on a coherent control process with applicable rf-pulsed incursive harmonic modes of HD spacetime manifolds such as those described by the a spin-exchange continuous-state spacetime resonance hierarchy. See US Patent [3].

In this work we review the 3rd generation prototype design for our model of bulk universal QC, an evolutionary scenario that has occurred as we have awaited funding over the last five years. Each generation of prototype design has reduced the perceived prototyping costs by an order of magnitude.

- The 1st generation was to utilize an IC chip that held an array of 1,000 ring lasers; the central cavity of which would hold the QC molecule. We arbitrarily chose a class II mesoionic xanthine crystal because it is stable at room temperature for ~ 100 years and has 10 relatively even separable quantum states in its ground state configuration. This xanthine was to be acted upon by rf-pulsed Sagnac Effect resonance [2]. This version would have cost over 10 million US$ to prototype because of a required partnership with the IC ring laser patent holder.
- The 2nd generation was operationally similar to the 1st generation, but utilizing much less expensive quantum dot ring laser arrays instead of an IC array. Quantum dots may be manufactured with internal mirrors to create quantum dot ring lasers. The quantum dots would be arrayed on a suitable substrate rather than an IC. Prototyping costs here were anticipated at about 5 million US$.
- The 3rd generation at a cost of 1.5 to 3 million US$ may possibly utilize a class II mesoionic xanthine doped multilayer graphene molecule array (currently under study) where it may be possible to operate a QC by forms of Quantum Hall effects, bilayer graphene alone, or a stand-alone mesoionic xanthine configuration since several mesoionic xanthine molecules have pertinent polar properties.

Graphene differs from most materials; electrons and holes near the six molecular corners behave like relativistic spin ½ particles that can be described by the Dirac equation causing graphene to be thought of as an ideal material for spintronics, also because it displays the anomalous quantum Hall effect at room temperature. The anomalous behavior is a result of emergent massless Dirac electrons. In a magnetic field the spectrum has a Landau level with energy precisely at the Dirac point. Bilayer graphene also shows the quantum Hall effect with a tunable band gap.

Figure 1. Graphene is a flat monolayer of carbon atoms tightly packed into a 2D honeycomb lattice, and is a basic building block for graphitic materials of all other dimensionalities. It can be wrapped up into 0D fullerenes, rolled into 1D nanotubes or stacked into 3D graphite composed of pure carbon atoms arranged in a regular hexagonal pattern similar to graphite. Figure courtesy [4].

Bilayer graphene typically can be found either in twisted configurations where the two layers are rotated relative to each other or in a stacked configuration where half the atoms in one layer lie atop half the atoms in the other. Stacking order and orientation greatly influence the optical and electronic properties of bilayer graphene [29-31]. There are numerous challenges (i.e., incorporating a polar mesoionic xanthine molecule within grapheme layers) for implementing our QC model in graphene such that we may end up abandoning this avenue. We mention it for illustrative purposes suggesting that our QC prototype may become a working model costing just pennies!

We have not finished feasibility research yet in this regard. An additional simplification is to utilize rf-modulated Quantum Hall Effects instead of ring lasers. Initial calculations indicate that frequencies amenable to such rf-modulated Quantum Hall Effects are in the 10-20 GHz range. We may or may not require a mesoionic-xanthine as the computing molecule, although it is currently our best choice of substrate based on empirical calculations and other relevant considerations. If we stay with the mesoionic-xanthine, there are several molecular variations, some of which contain polarizable bilayers; suggesting that the proper mesoionic-xanthine may be doped directly onto a substrate as the QC itself. This 3rd generation prototype would be very close to an inexpensive marketable commercial product; a scenario where delay has proffered rewards. A 2013 US corporation called Bering Strait Systems has been formed to attempt to fund this project. We are patiently waiting.

Recent developmental research on QCs has focused on simple 2-state qubit systems described by two geometric models of the 2-state transformations:

- 1) The SU(2) action of a complex projective line and
- 2) The SO(3) action on a Euclidean Bloch 2-sphere.

Because our model utilizes a method that surmounts the quantum uncertainty principle in a complex 12-space it is postulated that the current Bloch (Riemann) sphere representation of qubits (a classical 2-sphere model) may be too primitive and not suited for actualizing bulk universal QC. For the past several years our model was based on a relativistic (r-qubit) where the additional degree of freedom was an aid to surmounting uncertainty [2,5]. Recently we realized the 4D r-qubit, while on the right track was also insufficient. This arose from extending quantum theory to the regime of the Unified Field, U_F primarily based on extended HD versions of Cramer's transactional interpretation and the de Broglie-Bohm interpretation.

This was as much a breakthrough in nilpotent cosmology as quantum theory. We discovered there was more to a quantum state than a Copenhagen 'particle in a box'; the quantum state was conformally scale-invariant requiring a representation utilizing a system of dual continuous-state Calabi-Yau mirror symmetric 3-tori (class of Kähler manifolds) [6-8]. One surprise is that this cosmology contains an inherent 'synchronization backbone' [1,2] which ends up like getting half the QC for free; and of course making the essential process of surmounting uncertainty almost simplistic [2]. We spend considerable effort to discuss such putative developments in a later section.

2. Qubit Basis, Geometry, and Invariants

This summarizes the current thinking on representations of quantum states where the quantum wavefunction is the most complete description that can currently be given to a physical system:

- Physical information about a transition is encoded in a unit vector in a complex vector space.
- Physical process without measurement corresponds to unitary transformation of this vector representation.
- A measurement corresponds to the probabilistic choice of a covector to form an amplitude $\langle\Psi|U|\Phi\rangle$ where the probability is $|\langle\psi|U|\phi\rangle|^2$.

We intend to show that this currently utilized vector algebra is not physical but rather a convenient mathematical representation. The Bloch sphere is a 2D representation of 4D reality. We show below a recent attempt at a 6D dual qubit as an indicia of our model which is cast in 12D which we believe is required to fully represent a properly physicalized qubit!

In the philosophy of physical science there is no *a priori* reason why nature must be described as a U_F theory. The current drive in physics is to bring the four fundamental field interactions into a single unified framework as a form of quantum theory. Because of the inherent difficulty associated with renormalization and uniting gravity and quantum theory many physicists believe a framework other than a field theory such as a version of an 11D M-Theory may be a viable alternative avenue.

In the usual nonrelativistic quantum theory of computation it was necessary only to point to the number of states, 2^n for a description of n qubits. In relativistic theory there are many special cases. Charged and neutral, massive and massless particles etc. should be described differently.

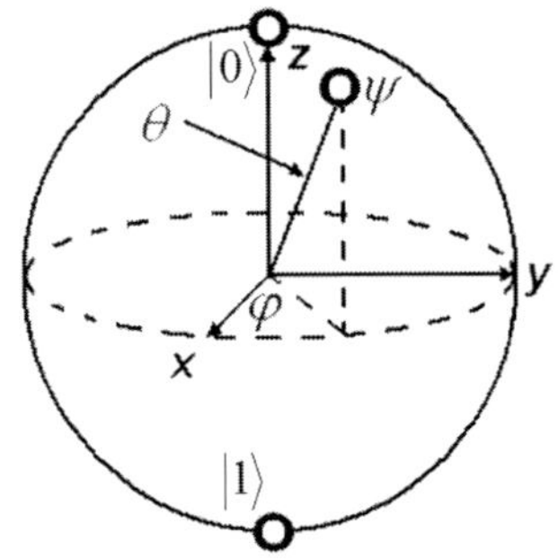

Figure 2. Usual representation of a qubit $|0\rangle+|1\rangle=\Psi^2$ as the complex Riemann Bloch 2-sphere.

The problem of extending the fundamental basis of the qubit is manifold. Many physicists do not accept dimensionality beyond four. Those that do, predominantly string theorists, now M-Theorists, are confounded by the search for a unique string vacuum claimed to have a Googolplex or 10^∞ possibilities. HAM cosmology has something going for it in that a unique string vacuum has been derived from its parameters [2]. Further restrictions arise from its unique form of inherent Calabi-Yau mirror symmetry. Thus a clear avenue is provided to 'divine' the complex HD space from which our 3D virtual reality is a resultant. Fortunately our model is empirically testable. Nine of the fourteen experimental protocols derived so far are summarized elsewhere in this volume [9,10].

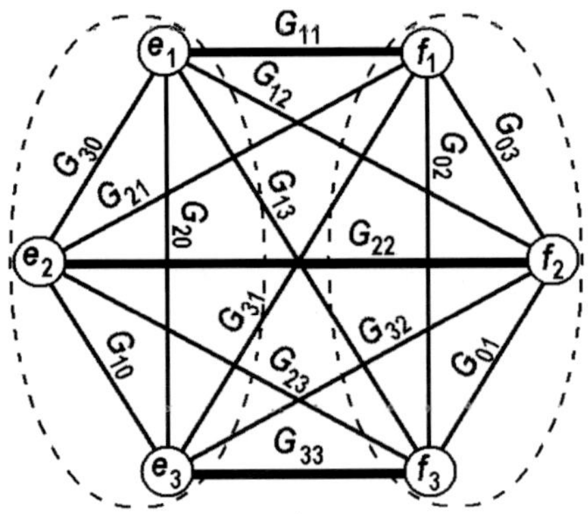

Figure 3. Combinatorial graph of vertices corresponding to basis vectors of a Bloch sphere for two qubits [e1, e2, e3] & [f1, f2, f3] and the edges to the corresponding bivector basis G_{ij}. Dashed ellipses enclose the induced subgraphs corresponding to the "local" subalgebras of each Bloch sphere model, while the perfect matching of a Cartan subalgebra is indicated by the bolder lines of edges G11,G22,G33. Figure redrawn from [11].

The perceived required redefinition of the qubit requires new logic gates and QC algorithms to take full advantage of the new physics. Operationally the new r-qubit basis entails a new set of transformations beyond the usual Galilean-Lorentz-Poincairé which have been temporally adjoint along an axis or light cone in Euclidean and then Minkowski coordinates. We have chosen to call the new transformation 'The Noetic

Transformation' because it is cast in an anthropic multiverse. What separates the Noetic Transform from its precursors (Galilean, Lorentz-Poincaré) is that it uncouples from the 3D or 4D realm of the observer and has no temporal component. This evolution from Classical to Quantum-Relativistic now continues to a new regime of Unified Field Theory, U_F.

We do not wish to say 'uncouples from reality', rather that fundamental reality should now be considered 12D instead of the 3(4)D of the Lorentz-Poincaré Transformation. The elimination of the concept of time occurs by a double superluminal boost, $x \leftrightarrow t_x \leftrightarrow w_x$ that also occurs along the y and z axes simultaneously. The infinities plaguing renormalization are indicia of this 12D reality (in the same way the infinities in the Raleigh-Jeans law for black body radiation were an indicia of the immanent discovery of quantum mechanics).

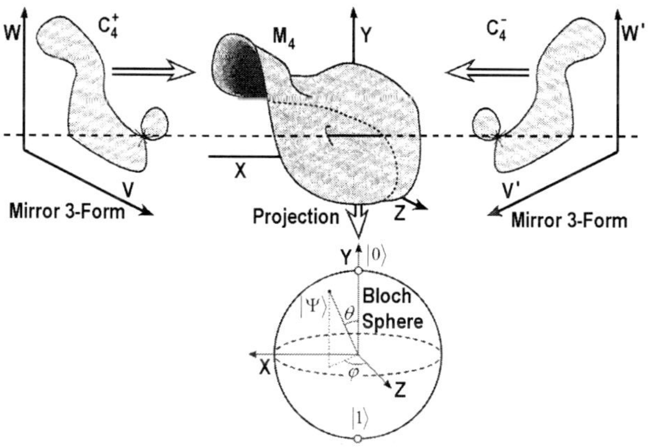

Figure 4. Alternative rendition of Fig. 7. Complex HD Calabi-Yau mirror symmetric 3-forms, $\pm C_4$ complex dimensions become embedded in Minkowski space, M_4. This resultant is projected as a continuous quantum state evolution considered a nilpotent Bloch sphere representing the lower portion that embeds in local spacetime. There is an additional duality above this projection embedded in the infinite potentia of the U_F from which it arises. What this new LSXD reality means is that the usual consideration of a Bloch 2-sphere vector based qubit is insufficient for bulk QC.

During the late 1980s, a second set of quantum axioms based on geometric-topological ideas was proposed restricting parameters to a specific class of topological quantum field theories inaugurated most saliently with Atiyah and Segal, and notably expanded upon by Witten, Borcherds, and Kontsevich [32-36]. However, most of the physically relevant quantum field theories, such as the Standard Model, are not topological quantum field theories; the quantum field theory of the fractional quantum Hall effect is a notable exception [17]. The realized basis for Bulk Universal Quantum Computing diverges from the anticipated form by current quantum computing researchers utilizing the standard Copenhagen Interpretation (CI) of quantum theory.

We anticipate that the realized basis for bulk universal QC diverges from the anticipated form by current QC researchers utilizing the standard Copenhagen Interpretation (CI) of quantum theory. What this means is that the Bloch 2-sphere vector basis, $|\psi| = \alpha|0\rangle + \beta|1\rangle$ where $|\alpha|^2 + |\beta|^2 = 1$ (single qubit only) is archaic and not an appropriate model for QC gates or algorithms. As our starting point we follow recent efforts of Makhlin [12] Zhang et al. [13] and Havel [11,14], (MZH) who have pointed the way to our model with a geometric algebra rendition of a dual Bloch sphere.

This is then used by MZH to illustrate the Cartan decompositions and subalgebras of the 4D unitary group, which have recently been used to study the entangling capabilities of two-qubit unitaries. "*...we show how the geometric algebra of a 6D real Euclidean vector space naturally allows one to construct the special unitary group on a two-qubit (quantum bit) Hilbert space, in a fashion similar to that used in the well-established Bloch sphere model for a single qubit*" [11]. The group SU(2) is isomorphic to the group of quaternions of norm 1, and is thus diffeomorphic to the 3-sphere Since unit quaternions can be used to represent rotations in 3D space (up to sign), we have a surjective homeomorphism from SU(2) to the rotation group SO(3) whose kernel is $\{+I, -I\}$. The geometric structure of nonlocal gates is a 3-torus. The local equivalence classes of two-qubit gates are in one-to-one correspondence with the points in a tetrahedron except on the base [13].

The MZH model is based on complex Minkowski space and the Copenhagen Interpretation. Our model is different - cast in a 9D M-Theoretic Calabi-Yau mirror symmetry utilizing an operationally completed form of QT achieved by integrating LSXD forms of the de Broglie-Bohm Causal Interpretation [15] and Cramer's Transactional Interpretation [16] but that still makes correspondence with the MZH 6D model [11-14].

3. Case for Relativistic Qubits (R-Qubits)

Refer to Fig. 2 for the usual Block Sphere representation of a qubit, a geometrical representation of the pure state space of a two-level quantum mechanical system. Alternately, it is the pure state space of a 1 qubit quantum register. In the conventional consideration of quantum computing a quantum bit or qubit is any two-state quantum system defined as a superposition of two logical states of a usual bit with complex coefficients that can be mapped to the Riemann sphere by stereographic projection (Fig. 5). Formally a qubit is represented as: $\Psi = \xi|0\rangle + \eta|1\rangle$ with each ray $\xi, \eta \in C$ in complex Hilbert space and

$\|\Psi\|^2 = \xi\bar{\xi} + \eta\bar{\eta} = 1$, where $|0\rangle$ corresponds to the south or 0 pole of the Riemann sphere and $|1\rangle$ corresponds to the opposite, north or ∞ pole of the Riemann complex sphere. The conventional qubit maps to the complex plane of the Riemann sphere as:

$$\xi\bar{\eta} + \eta\bar{\xi} \to X,\ \xi\bar{\eta} - \eta\bar{\xi} \to iY,\ \xi\bar{\xi} - \eta\bar{\eta} \to Z\,. \quad (1)$$

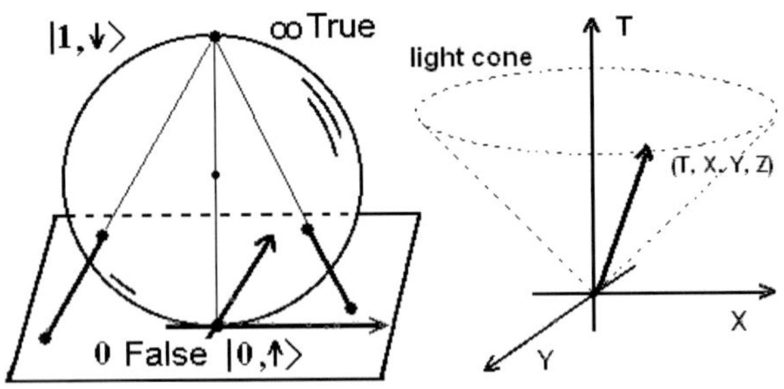

Figure 5. (a) Stereographic projection model of a qubit on a complex Riemann sphere. (b) Relativistic model of a qubit (r-qubit) with interacting quantum fields entailing an extra HD degree of freedom.

Unitary transformations of a qubit correspond to 3D rotations of the Riemann sphere. Following Vlasov [5] for relativistic consideration of a qubit (r-qubit) an additional 4D W parameter is added to equation (2):

$$\begin{aligned} &\xi\bar{\eta} + \eta\bar{\xi} \to X, \quad \xi\bar{\eta} - \eta\bar{\xi} \to iY, \\ &\xi\bar{\xi} - \eta\bar{\eta} \to Z, \quad \xi\bar{\xi} + \eta\bar{\eta} \to W \end{aligned} \quad (2)$$

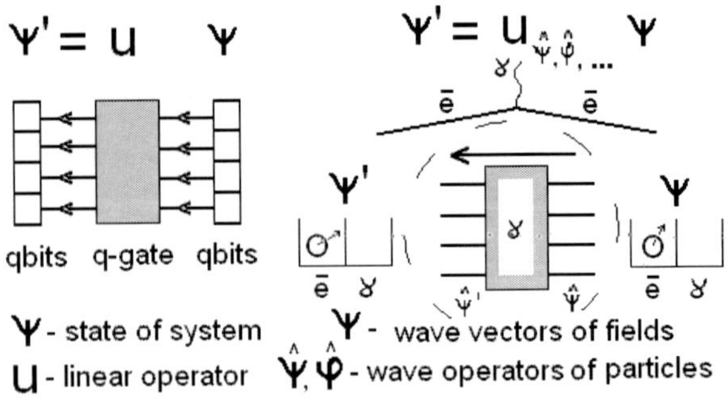

Figure 6. (a) Usual q-gate with constant number of states and particles. (b) Relativistic quantum bit (r-qubit) with constant particles but variable or infinite states.

In cartography and geometry, the stereographic projection is a mapping that projects each point on a sphere onto a tangent plane along a straight line from the antipode of the point of tangency (with one exception: the center of projection, antipodal to the point of tangency, is not projected to any point in the Euclidean plane; it is thought of as corresponding to a "point at infinity"). One approaches that point at infinity by continuing in any direction at all; in that respect this situation is unlike the real projective plane, which has many points at infinity.

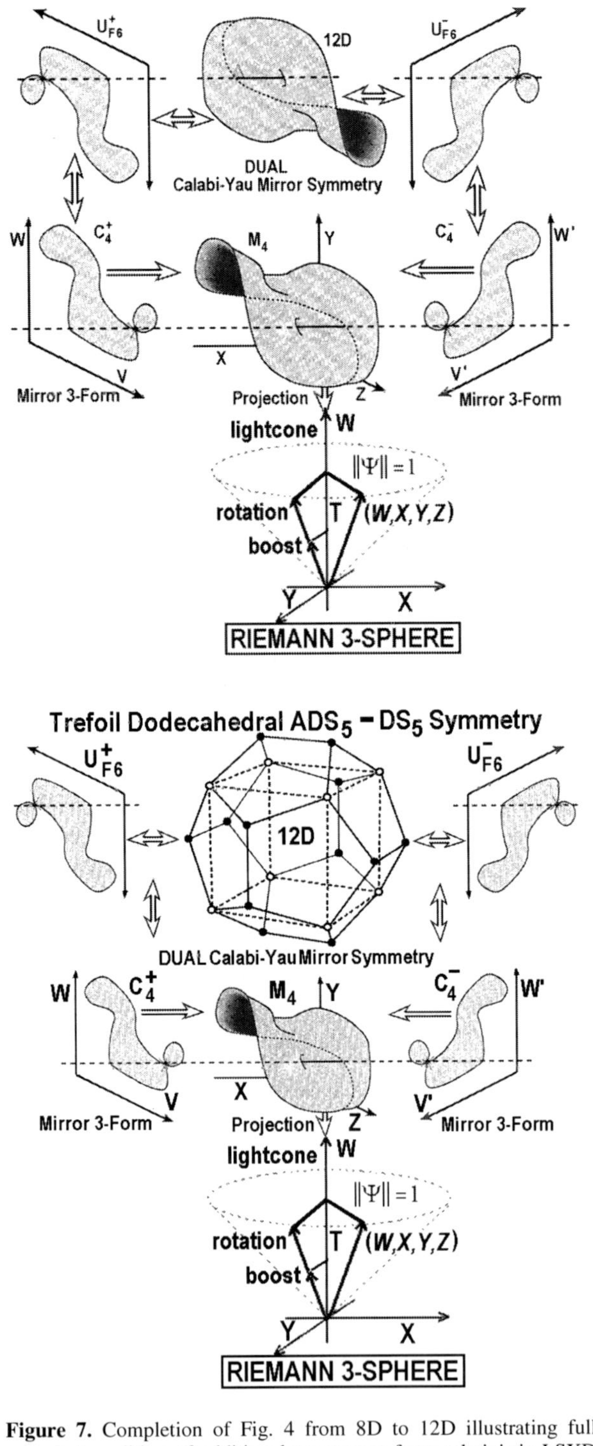

Figure 7. Completion of Fig. 4 from 8D to 12D illustrating full extended rendition of additional parameters for a relativistic LSXD quantum state in continuous-state dual Calabi-Yau mirror symmetric HAM cosmology in 7(a) as far as currently understood. Replacing also the Bloch 2-sphere qubit representation with the new extended r-qubit Riemann 3-sphere resultant representation that has sufficient parameters to surmount the uncertainty principle. 7(b) ADS_5 version.

Recent WMAP satellite observations have given good preliminary evidence that the fundamental topology of cosmology is a 'wrap-around' polyhedron of Poincaré-Dodecahedral (PD) form that correlates with Anti-de Sitter space, AdS_5 matching up conformal supersymmetry in 4D with AdS supersymmetry in 5D. There are $8N$ real SUSY generators and the bosonic part consists of the conformal AdS group Spin(4,2) times an internal group $SU(N)_T \times U(1)_A$.

For the case $N = 4$, there are 32 real SUSY generators and an internal group $SU(4)_T \times U(1)_A$. $SU(4) \cong$ Spin(6) and Spin(6) is the isometry group of $\mathbf{S}^5$ with spinorial fields. The bosonic spatial isometry group of $AdS_5 \times \mathbf{S}^5$ is spin (4,2) × spin (6).
In $N = (2,0)$ 10D SUSY there are 32 real SUSY generators. In a generic curved spacetime, some of the SUSY generators are broken but in the special compactification of $AdS_5 \times \mathbf{S}^5$ with both factors having the same radius there are 32 real unbroken generators.

The integral of this 5-flux over $\mathbf{S}^5$ has to be a nonzero integer (if it's zero, we have no stress-energy tensor). Because the part of the 5-flux lying in AdS_5 contains a time component, it gives rise to negative curvature. The part of the 5-flux lying in $\mathbf{S}^5$ doesn't have a time component, and so, it gives rise to a positive curvature.

The near horizon geometry is approximately $AdS_5 \times \mathbf{S}^5$ with the approximation becoming more and more exact closer to the horizon. If we take the limit in which we are always in the near horizon region, the geometry becomes exactly $AdS_5 \times \mathbf{S}^5$ [38,39]. This is sufficient for HAM cosmology to embrace a PD AdS_5 backcloth especially since the possibilities of positive and negative curvature coincide with HAMs oscillating cosmological constant.

4. Basis for the Noetic Transformation

An event in spacetime is an idealized instant of time at a definite position in space labeled by time and position coordinates t,x,y,z. Coordinates have no absolute significance; they are arbitrary continuous single-valued labels given invariant meaning by the expression for the line element connecting two events [18,19]. The usual expression for a line element in Minkowski coordinates is

$$ds^2 = dt^2 - dx^2 - dy^2 - dz^2. \qquad (3)$$

For simplicity at this stage of development of the Noetic Transformation we devise the XD coordinates as orthogonal and evenly spaced. Firstly since the LSXD space is time independent we may drop the dt^2 term from the line element and introduce a new spatial form, dl^2 where dl^2 reduces to ds^2 and

$$dl^2 = dx^2 + dy^2 + dz^2 + dW^2 \qquad (4)$$

where $W = w^i + w^j + w^k$ (before complex dualing to Calabi-Yau mirror symmetry) as a 9D quaternion-like trivector representation. This is like an extension of the 3-sphere of Einstein's space where the set of points x,y,z,W are at a fixed distance R from the origin such that $R^2 = x^2 + y^2 + z^2 + W^2$ preserving the wanted three time independent space variables x,y,z and where the fourth LSXD variable W^2 is given by

$$W^2 = R^2 - r^2 \qquad (5)$$

where $r^2 = x^2 + y^2 + z^2$ such that (5) becomes

$$dW = \frac{r\,dr}{W} = \frac{r\,dr}{\left(R^2 - r^2\right)^{1/2}} \qquad (6)$$

So that the dual local-HD spatial line-element dl^2 becomes

$$dl^2 = dx^2 + dy^2 + dz^2 + \frac{r^2 dr^2}{R^2 - r^2} \qquad (7)$$

where R may be used to represent the center of dual Calabi-Yau mirror symmetric 3-tori. See Fig. 10.

Continuing to follow Peebles [18,19] this generalizes the usual 2D line element to 9D where the length R in the expressions is a constant for the model because spacetime is assumed to be static. For $r \ll R$ our extended Einstein line element approaches the usual Minkowski form (1). When $r = R$ the geometry makes correspondence to the surface of a Riemann 2-sphere which is utilized in the standard description of a qubit as a Bloch Sphere. (Fig. 2)

We may now look at the additional parameters this space allows us to add to the fundamental description of a quantum state beyond the usual inherent uncertainties of Copenhagen interpretation. Because of the conformal scale-invariance of the Nilpotent criteria an additional duality must be incorporated into the mirror symmetric parameters of W^2 which is a further correspondence to the standing wave-like properties of the Cramer Transactional Interpretation to simplistically what might be labeled, $\pm W^2$. This addition as far as we currently understand would incorporate all the additional parameters for a complete description of a quantum state as embedded in the LSXD (Large Scale Extra Dimensionality) aspects of the U_F requiring a new representation of the qubit to include the additional HD conformal scale-invariant parameters.

The Pythagorean Theorem $a^2 + b^2 + c^2 = d^2$ gives the length, d of the diagonal of a 3D cube, a,b,c. By adding additional terms to the equation it describes the diagonal of an nD hypercube. This is illustrated in Fig. 7b above. The locking together of the Calabi-Yau components in the resultant localized cube creates the quantum uncertainty principle which can be surmounted [2,3] if the Calabi-Yau nilpotent 'copies' are accessed by incursive resonance.

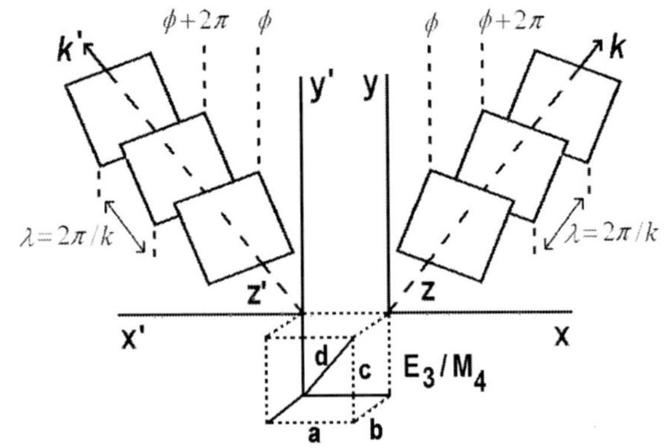

Figure 8. Conceptualized view of complex Riemann planes cast as Calabi-Yau future-past mirror symmetric potentia illustrated as tiered surfaces (additional topology suppressed) of constant phase in this case to represent cyclic components of evenly spaced orthogonal standing reality waves with the E_3/M_4 cubic resultant localized as the 3-cube at the bottom. The resultant Euclidean cube is locked in 4D by the uncertainty principle keeping the HD parameters inaccessible to the empirical tools available to the Copenhagen framework. k & k' will topologically infold into an HD continuous-state torus.

A surface of constant phase, $k \cdot r - \omega t = k_x x + k_y y + k_z z - \omega t =$ *constant* is a wavefront [2,17]. For a surface of constant phase if any wave equation has a time harmonic (sinusoidal) solution of the form $Ae^{i\phi}$ where *A* is the amplitude and the phase, ϕ a function of position with *(x,y,z)* constant and phase difference 2π separated by wavelength, $\lambda = 2\pi / k$. The direction cosines of the planes of constant phase are proportional to *k* and move in the direction of *k* equal to the phase velocity where

$$\mu = \frac{\omega}{k} = \frac{\omega}{\sqrt{k_x^2 + k_y^2 + k_z^2}}. \tag{8}$$

Where $\lambda = 2\pi / k = 2\pi\hbar / p = h / p$ is equivalent to the de Broglie matter wave relations, $E = \hbar\omega$, **p** $= \hbar k$ [20].

We may now look at the additional parameters this space allows us to add to the fundamental description of a quantum state beyond the usual Copenhagen interpretation. Because of the conformal scale-invariance to the Nilpotent criteria an additional duality must be incorporated into the mirror symmetric parameters of W^2 which is a further correspondence to the standing wave-like properties of the Cramer Transactional Interpretation to simplistically what might be labeled, $\pm W^2$. This addition as far as we currently understand would incorporate all the additional parameters for a complete description of a quantum state as embedded in the HD aspects of the U_F requiring a new representation of the qubit to include the additional parameters.

Most are familiar with the 3D Necker cube (center of Fig. 8 is like a Necker cube) that when stared at certain vertices reverse. This is called topological switching. There is another paper child's toy called a 'cootie catcher' [37] that fits over the fingers and can switch positions. What the cootie catcher has over the Necker cube is that it has a better defined center or vertex point such that in the LCU exiplex [9] space-time background we have this topological switching [2]. It represents the frame that houses the gate which is like a lighthouse with the rotating beam on top. Additionally inside the structure there is a baton passing. The baton is like a shutter on the lens that the 'light' shines through. In the HD U_F regime the light is always on omni-directionally. But in addition the baton passing is also a form of leap-frogging. The leap-frogging represents wave-particle duality (which as you may recall we elevated to a principle of cosmology). The leaping moment is the wave, and the crouched person being leapt over is the particulate moment. The particle moment acts like a domain wall and no light passes when its orientation is aligned towards the 3-D world. This is a primitive explanation of the knot provided by the uncertainty principle.

We can also attempt to describe this topological geometry with dual quaternion-like trefoil knots. The trefoil knot array (in Fig. 10 drawn as Planck scale quaternion vertices) is holomorphic to the circle. Since energy is conserved we may ignore the complexity of the HD symmetries and use the area of that circle as the Lagrangian, in this case a resultant of two trefoil knots as a 2-sphere quantum state as the coupling area. The figure also provides a conceptualized view of how HAM cosmology sees continuous-state evolution of conformal scale-invariant Calabi-Yau mirror symmetric topology. As QT has a semi-classical limit this might be termed semi-quantum in terms of the HD U_F. There is a 2nd LSXD level 'above' this one postulated as the regime of full U_F potentia. The cycle goes from chaotic-uncertain to coherent-certain perhaps non-commutative to commutative according to the noetic transformation. This is represented in the Dirac string trick or Philippine wine dance [26].

The traditional quantum state, $|\psi| = \alpha|0\rangle + \beta|1\rangle$ where $|\alpha|^2 + |\beta|^2 = 1$ is the usual vector representation of a single qubit on a Bloch sphere. This is according to the Copenhagen interpretation of quantum theory where quantum logic gates are limited by the uncertainty principle limiting quantum computation to 10 or 20 qubits. The new HD quantum state/qubit representation extends to the conformal scale-invariant continuous-state properties to the U_F which includes mirror symmetric 'potentia copies' making correspondence to the resultant standard Copenhagen quantum state but which are not limited by the uncertainty principle.

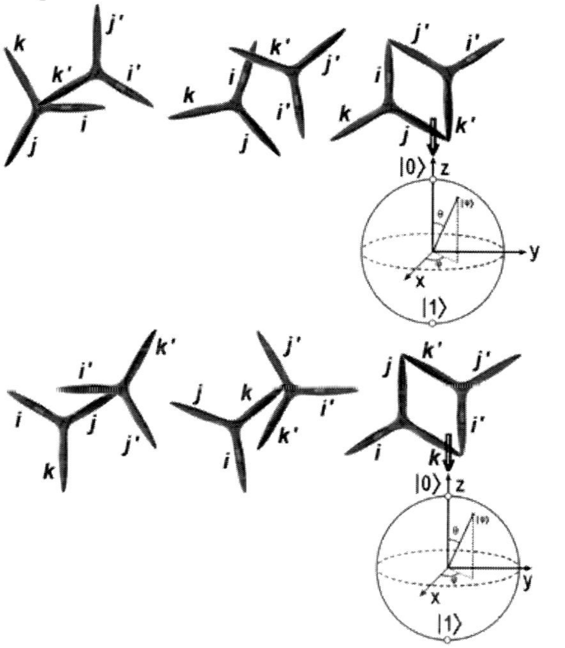

Figure 9. Alternative rendition of Fig. 4 in quaternionic form. Locus of HD mirror symmetric Calabi-Yau 3-tori (here depicted trefoil knots) spinning relativistically and evolving quantum mechanically time. Nodes in the cycle are sometimes chaotic (degenerate) and sometimes periodically couple into resultant quantum states in Euclidean 3-space depicted in the figure as faces of a 3-cube that reduce further to the Riemann Bloch 2-sphere.

5. Toward a Quaternion Trivector R-Qubit Algebra

A dual 3-form multivector is sometimes called a trivector which we choose here in view of some nomenclatural disparity as to precise usage and definition in the literature. We believe we need to use a dual quaternion triplet, a double set of 3 quaternions which could perhaps be defined as a dual trivector. Dual because of LSXD Calabi-Yau mirror symmetry.

Most physicists admit quantum theory is incomplete. In the Copenhagen interpretation, the Schrödinger equation is considered semi-classical and we have the bra-ket Block sphere 2D convenience representation of a qubit. We wish to consider a new LSXD U_F view of the actual physical state represented. HAM cosmology naturally suggests this required change in what we consider as the physical quantum state. We can say the quantum regime is coupled to the classical world, but a completed QT will be in actuality coupled to the unified field, U_F. This allows us to 'Godelize' beyond the Copenhagen quantum state to 'see' the whole noumenal (thing in itself behind the phenomenology of appearance) basis of it. Which by using Ben Goertzel's term is as a form of 'quaternion mirror-house' [21] which also makes correspondence to the Rowlands' nilpotency which requires the 1st doubling. And then a second doubling required by HAM cosmology to render the complete Calabi-Yau mirror symmetric (or mirror house) rendering.

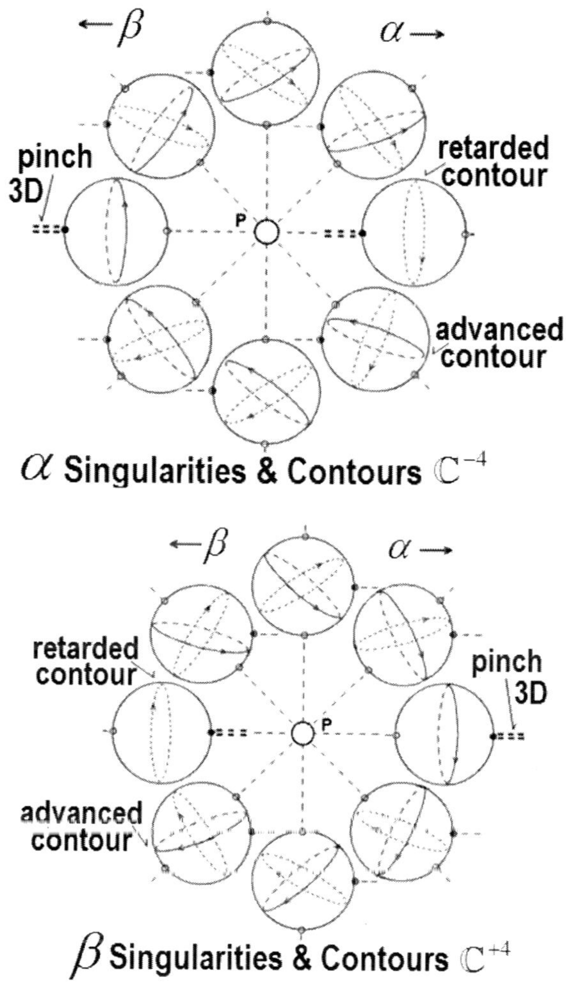

Figure 10. Two of six HD Nilpotent trivector r-qubit symmetries.

Thus in contrast to Havel's 6D bivector in complex Minkowski or Hilbert space (Fig. 3) we can illustrate HD qubit by the Philippine wine dance [26]. Each wine glass would represent one standard Bloch sphere; the dancer is like an atom and each glass represents one of the 2 possible spin states. Havel would have 2 entangled wine dancers standing near each other in Minkowski-Hilbert space. What we see is required to completely define a quantum state physically is that the wine dancers are like puppets standing additionally in a

hall of mirrors [21]. The puppet master is the super-quantum potential provided by parameters of the unified field. The mirror images are restricted on each side of the Cramer future-past Calabi-Yau mirror symmetry. By the continuous-state premise of this HD hierarchy - the left-right or future-past components become embedded in each other in the cycle [2,9,10]. The bottom (3D resultant) becomes the usual semi-classical phenomenological q-state we observe. At the 12D top the embedding is the causally free (ontological) quantum state copy.
A tabulation of vectors, scalars, quaternions and commutivity properties is as follows:

i j k	*ii ij ik* (bold)	*i'*	1
i j k	*ii ij ik*	*i'*	1
vector	*bivector*	*trivector*	*scalar*
	quaternions		

Vectors

i j k	anticommute
i j k	anticommute
ii jj kk	commute

Quaternions

i j k	*anticommute*
i j k	*anticommute*
ii ij ik	***j*** *i'**k** nilpotent*
ii ij ik	***j*** *- nilpotent*
	(5th term automatic)
*ii' jj k**k***	*commute*

In summary Havel uses a 6D bivector to represent 2 qubits. In our model a single qubit should be represented as some form of a dual trivector. What we get with this new qubit representation is QC logic gates able to surmount the uncertainty principle and proper algorithms for universal QC.

Normalized quaternions are simply Euclidean 4-vectors (length one) and thus fermionic vertices in spacetime or points on a unit hypersphere (this case a 3-sphere) embedded in 4D. Just as the unit sphere has two degrees of freedom, e.g., latitude and longitude, the unit hypersphere has three degrees of freedom.

According to Rowlands [22] quaternion cyclicity reduces the number of operators by a factor of 2 and *prevents the possibility* of defining further complex terms. To overcome this generalization of term limitations we explore utilizing shadow dualities employing additional copies of the quaternion algebra mapped into a 5D and 6D space to transform a new dualed 'trivector' quaternion algebra. As far as we know this can only occur under the auspices of a continuous-state cosmology within a dual metric where one coordinate system is fixed and then the other is fixed in a leap-frog baton passing fashion [23-25].

The coordinate fixing-unfixing mechanism is superbly illustrated by the 'walking of the Moai on Rapa Nui' [27, 28].

However a 3rd complex metric is involved making an evolution from dual quaternions to a 3rd quaternion we choose to name a trivector that acts as a baton passing mechanism between the space-antispace or dual quaternion vector space. The trivector facilitates a 'leap-frogging' between anti-commutative and commutative modes of HD space. This inaugurates a Mobius transformation between the Riemann dual stereographic projection complex planes.

Geometrically, a standard Möbius transformation can be obtained by first performing stereographic projection from the plane to the unit 2-sphere, rotating and moving the sphere to a new location and orientation in space, and then performing stereographic projection (from the new position of the sphere) to the plane. These transformations preserve angles, map every straight line to a line or circle, and map every circle to a line or circle. Möbius transformations are defined on the extended complex plane (i.e. the complex plane augmented by the point at infinity): $\hat{\mathbb{C}} = \mathbb{C} \cup \{\infty\}$.

This extended complex plane can be thought of as a sphere, the Riemann sphere, or as the complex projective line. Every Möbius transformation is a bijective conformal map of the Riemann sphere to itself. Every such map is by necessity a Möbius transformation. Geometrically this map is the Riemann stereographic projection of a rotation by 90° around $\pm i$ with period 4, which takes the continuous cycle $0 \rightarrow 1 \rightarrow \infty \rightarrow -1 \rightarrow 0$.

This is an LSXD point particle representation of a fermionic singularity [9,10]. The 8 + 8 or 16 ($\mathbb{C}^{\pm 4}$ complex space) 2-spheres with future-past retarded-advanced contours are representations of HD components of a Cramer 'standing-wave' transaction. This can be considered in terms of Figs. 4, 7, 8, 10 and 11 as Calabi-Yau dual mirror symmetries. To produce the quaternion trivector representation (Fig. 11) a 3rd singularity-contour map is required which is then also dualed, i.e. resulting in 6 singularity/contour maps. This may be required to oscillate from anticommutivity to commutivity in order to provide the *cyclic* opportunity to violate 4D quantum uncertainty [2, 3]!

References and Notes

[1] Feynman, R.P. (1986) Quantum mechanical computers, Found. Phys. 6, pp. 507-531.

[2] Amoroso, R.L. & Rauscher, E.A. (2009) The Holographic Anthropic Multiverse, Singapore: World Scientific.

[3] Amoroso, R.L. (2012) Spacetime energy resonator: a transistor of complex dirac polarized vacuum topology, US Patent 12/928,592, http://www.freepatentsonline.com/y2012/0075682.html.

[4] Graphene.jpg figure courtesy GNU Free Documentation License, Version 1.2.

[5] Vlasov, A.Y. (1996) Quantum theory of computation and relativistic physics, in T. Toffoli, M. Biafore & J. Leao (eds.) Physcomp96, Cambridge: New England Complex Systems Institute; http://arxiv.org/abs/quant-ph/9701027, and additional material from private communication.

[6] Gross, M. & Wilson, P.M.H. (1996) Mirror symmetry via 3-tori for a class of Calabi-Yau threefolds, arXiv:alg-geom/9608004v3.

[7] Chan, K., Lau, S-C& Leung, N.C. (2012) SYZ mirror symmetry for toric Calabi-Yau manifolds, arXiv:1006.3830v3.

[8] de la Ossa, X. (2011) Calabi-Yau manifolds and mirror symmetry, http://people.maths.ox.ac.uk/delaossa/LecturesQuad.pdf

[9] Amoroso, R.L. (2013) "Shut The Front Door!": Obviating the Challenge of Large-Scale Extra Dimensions and Psychophysical Bridging, in RL Amoroso, LH Kauffman & P. Rowlands (eds.), The Physics of Reality: Space, Time, Matter, Cosmos, Singapore: World Scientific.

[10] Amoroso, R.L. & Vigier, J-P (2013) Evidencing 'tight bound states' in the hydrogen atom: Empirical manipulation of large-scale XD in violation of QED, in RL Amoroso, LH Kauffman & P. Rowlands (eds.), The Physics of Reality: Space, Time, Matter, Cosmos, Singapore: World Scientific.

[11] Havel, T.F. & Doran, C.J.L. (2004) A Bloch-sphere-type model for two qubits in the geometric algebra of a 6-D Euclidean vector space, arXiv:quant-ph/0403136v1.

[12] Makhlin, Y. (2002) Nonlocal properties of two-qubit gates and mixed states, and the optimization of quantum computations, Quantum Inform. Processing 1, pp. 243–252.

[13] Zhang, V.J., Sastry, S. & Whaley, K.B. (2003) Geometric theory of nonlocal two-qubit operations, Phys. Rev. A 67, p. 042313.

[14] Havel, T.F. & Doran, C.J.L. (2001) Geometric algebra in quantum information processing arXiv:quant-ph/0004031v3.

[15] Holland, P. R. (1995) The quantum theory of motion: an account of the de Broglie-Bohm causal interpretation of quantum mechanics, Cambridge: Cambridge university press.

[16] Cramer, J.G. (1986) The transactional interpretation of quantum mechanics.Reviews of Modern Physics, *58*(3), 647.

[17] Zhang, S. C., Hansson, T. H., & Kivelson, S. (1989) Effective-field-theory model for the fractional quantum Hall effect, Physical review letters, *62*(1), 82-85.

[18] Peebles, P. J. E. (1993) Principles of Physical Cosmology, Princeton: Princeton University Press.

[19] Peebles, P. J. E. (1980) The Large-Scale Structure of the Universe, Princeton: Princeton university press.

[20] Peebles, P.J.E. (1992) Quantum Mechanics, Princeton: Princeton Univ. Press.

[21] Goertzel, B., Aam, O., Smith, T.F. & Palmer, K. (2007) Mirror Neurons, Mirrorhouses, and the Algebraic Structure of the Self http://www.goertzel.org/dynapsyc/2007/mirrorself.pdf.

[22] Rowlands, P. (2013) Space and Antispace, in RL Amoroso, LH Kauffman & P. Rowlands (eds.) The Physics of Reality: Space, Time, Matter, Cosmos, Singapore: World Scientific Publishers.

[23] Rauscher, E.A. & Amoroso, R.L. (2011) Orbiting the Moons of Pluto: Complex Solutions to the Einstein, Maxwell, Schrödinger & Dirac Equations, World Scientific.

[24] Vigier, J-P & Amoroso, R.L. (2003) Can one integrate gravity and electromagnetic fields? in R. L. Amoroso, J-P Vigier, M. Kafatos, G. Hunter (eds.) Gravitation and Cosmology: From the Hubble Radius to the Planck Scale, pp. 241-258, Netherlands: Springer.

[25] Amoroso, R.L. & Vigier (2004) Toward the unification of gravity and electromagnetism, in V. Dvoeglazov & A. Espinoza (eds.) Relativity, Gravitation, Cosmology, 71-88, New York: Nova Science.

[26] Francis, G., Kauffman, L.H. & Sandin, D. (1993) Air on the Dirac Strings (video) http://www.evl.uic.edu/hypercomplex/html/dirac.html.

[27] Amazing Video, Walking of the Moai on Rapa Nui (Easter Island) http://www.youtube.com/watch?v=yvvES47OdmY.

[28] Lipo, C. P., Hunt, T. L., & Haoa, S. R. (2012) The Walking'Megalithic Statues (Moai) of Easter Island, Journal of Archaeological Science.

[29] Barlas, Y., Côté, R.; Lambert, J.; MacDonald, A. H. (2010) Anomalous exciton condensation in graphene bilayers, Physical Review Letters 104, 9.

[30] Su, J. J., MacDonald, A. H. (2008) How to make a bilayer exciton condensate flow, Nature Physics 4 (10): 799–802.

[31] Schwierz, F. (2010) Graphene transistors, Nature Nanotechnology 5;7:487-496.

[32] Atiyah, M.F. (1989) Topological quantum field theories, Publ. Math. IJES 68, 175.

[33] Witten, E. (1988) Topological quantum field theory, Comm. Math. Phys. 117, 353.

[34] Segal, G.B. (1988) The definition of conformal field theory, in Differential Geometrical Methods in Theoretical Physics, NATO ASI Ser., Ser. C 250,165{171.

[35] Kontsevich, M. (1992) Intersection theory on the moduli space of curves and the matrix airy function", Commun. Math. Phys 147, 1.

[36] Borcherds, R. (1992) Monstrous moonshine and monstrous Lie superalgebras, Invent. Math. 109, 405.

[37] Go to: http://images.google.com/; search for cootie catcher.

[38] Luminet, J. P., Weeks, J. R., Riazuelo, A., Lehoucq, R., & Uzan, J. P. (2003) Dodecahedral space topology as an explanation for weak wide-angle temperature correlations in the cosmic microwave background, Nature,425(6958), 593-595.

[39] Caillerie, S., Lachièze-Rey, M., Luminet, J. P., Lehoucq, R., Riazuelo, A., & Weeks, J. (2007) A new analysis of the Poincaré dodecahedral space model, Astronomy and Astrophysics, 476(2), 691-696.

Deterministic Impulsive Vacuum Foundations for Quantum-Mechanical Wavefunctions

JOHN S. VALENTINE

Independent Researcher
johnv@johnvalentine.co.uk

By assuming that a fermion de-constitutes immediately at source, that its constituents, as bosons, propagate uniformly as scalar vacuum terms with phase (radial) symmetry, and that fermions are unique solutions for specific phase conditions, we find a model that self-quantizes matter from continuous waves, unifying bosons and fermion ontologies in a single basis, in a constitution-invariant process. Vacuum energy has a wavefunction context, as a mass-energy term that enables wave collapse and increases its amplitude, with gravitational field as the gradient of the flux density. Gravitational and charge-based force effects emerge as statistics without special treatment. Confinement, entanglement, vacuum statistics, forces, and wavefunction terms emerge from the model's deterministic foundations.

Keywords: Duality, Vacuum, Matter, Fermion, Quantization, Boson, Entanglement, Wavefunction, Symmetry, Weak Interaction, Strong Force, Coherence, Neutrino Oscillation. **PACS**: 02.30.Vv, 04.60.Bc, 11.10.Gh, 11.10.Jj, 12.10.-g, 12.39.-x.

1. Foundations

Minimal foundations lead to fewer free variables, and reduce assumptions. Symmetries found in these foundations may then describe wider ontologies. In this work, we explore a simple model.

Our simplest entity is a **wave** on just one basis axis, b [1, 3–7]. The waves propagate in bound pairs, as oscillators: these are fundamental **bosons**, constituting all matter and energy. All waves propagate at the same rate, both in terms of phase φ and space r traversed: $dr = d\varphi$.

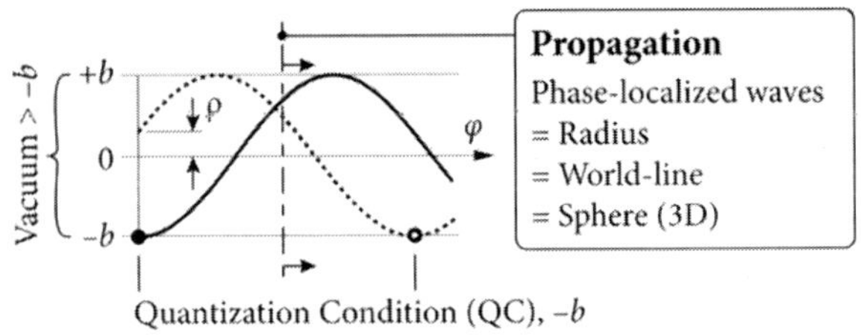

Figure 1. Boson Structure.

This structure gives properties: A boson's energy is *scalar*, and is invariant (conserved) while it propagates. In Euclidean space, these are *spherically symmetrical* expanding impulses where $d\rho/d\varphi = 0$. Intrinsic **mass-energy** ρ (fig.**1**) is the cosine of the phase difference between a **reference wave** having phase $\varphi_{\mathbf{A}}$ and its entangled vacuum partner having phase $\varphi_{\mathbf{B}}$:

$$-b\rho = -b\cos(\varphi_{\mathbf{B}} - \varphi_{\mathbf{A}}) = -b\, e^{-i(\varphi_{\mathbf{B}} - \varphi_{\mathbf{A}})} \qquad (1)$$

with the **field** being the gradient of the flux density of the mass-energy of all waves with respect to space (3.1.1). The wave structure allows both positive and negative **spin states** to be superimposed in the boson, and resolved on collapse; where $\varphi_{\mathbf{A}} = 0$ and $\varphi_{\mathbf{B}}$ is modulo $\pm\pi$,

$$\mathrm{sign}(spin) = \mathrm{sign}\left(b\,\frac{d\mathbf{B}}{dt}\right) \qquad (2)$$

$$\mathrm{sign}(chirality) = \mathrm{sign}(b\,\mathbf{B}) \qquad (3)$$

Spin states are merely options for presenting the same intrinsic phase difference when one of the waves is privileged as being the matter-state reference wave.

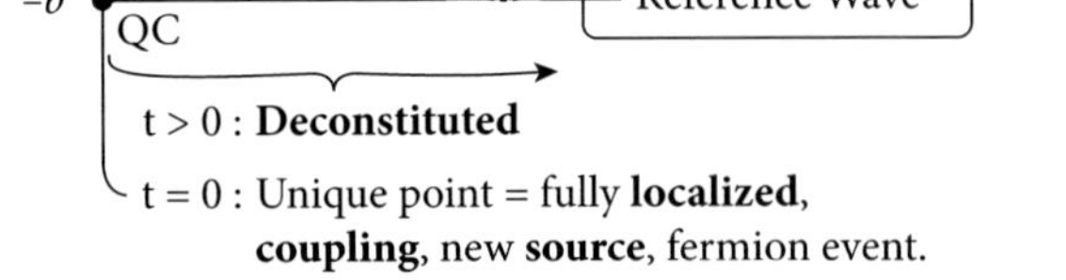

Figure 2. Phase evolution of a fermion's constituents.

Fermions (fig.**2**) exist as a unique solution to two bosons *each* having *one* wave at phase $-b$; the **quantization condition** ("QC"). This unique solution identifies a point on the otherwise spherically symmetric and *entangled* bosons. Because the bosons are impulsive (zero width), the fermion exists as a *point* in spacetime, i.e. the fermion state has no duration. The total **constitution** of a fermion event is as follows: (a) two waves, $b = -1$, at the QC; (b) two waves, $b \neq -1$, being **vacuum** terms; and (c) any other wave pairs at the point, also necessarily vacuum terms.

1.1.1 *Direct interactions*

Mass-energy acts as a phase operator or modulator on the waves of other bosons, as the sum of the overlapping vacuum phase potentials ρ_n:

$$\varphi_{\text{modulated}} = \varphi_{\text{carrier}} + \sum_n \rho_n \tag{4}$$

In terms of phase operators, wave **Z** is phase-modulated by ρ from vacuum bosons (eq.5, fig.**3**):

$$\mathbf{W} = \rho\,\mathbf{Z} \tag{5}$$

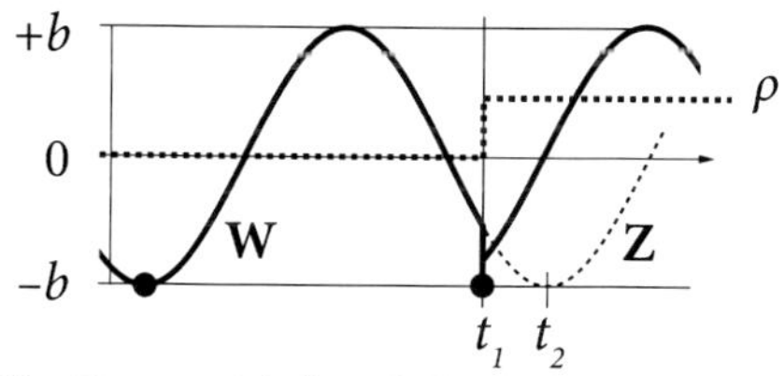

Figure 3. **W** = Phase modulation of **Z** by ρ.

It is then the *modulated* wave that qualifies for the quantization condition, with the modulation *retarding* or *advancing* the QC (fig.**4**), and therefore the prospective positions for the fermion, expressing **action** (3.0).

Figure 4. A volumetric plot of the propagation evolution of the *quantization conditions* of non-excluded waves from two co-modulating bosons, of known mass-energy. The right source has $\rho = 0.1\pi$ which *advances* the QC for the left wave on overlap, creating the inner ring. Conversely the left source has $\rho = -0.25\pi$ which *retards* the QC for the right source (outer ring, left). This may be viewed as a **longitudinal modulation**.

Without modulation, it would be improbable for fermions to form. We say that: (a) a deconstituted fermion requires vacuum energy to reconstitute; (b) free fundamental fermions are impossible; and (c) vacuum energy (mass-energy) prevents infinite propagation of bosons. Thus, a complete wavefunction must incorporate both the *matter* under consideration and the confining or environmental *vacuum* energy.

1.1.2 *Exclusion and symmetry breaking*

We require that states be uniquely identifiable, otherwise they cannot interact nor be resolved, and that any fermion event occurs only at unique spatial solutions to the QC. This goes further than implying the Pauli Exclusion Principle, to indicate that a boson's wave may not engage with other waves if it has the same phase and source as another wave – a condition that occurs after every fermion event.

This **exclusion symmetry** can only be broken when an excluded wave's *partner* (in the same boson) forms a QC, and *disentangles* both waves of that boson from the expanding shell (fig.**5**, Table 1), leaving both waves of the other boson free to interact, where one of them was previously excluded (3.2).

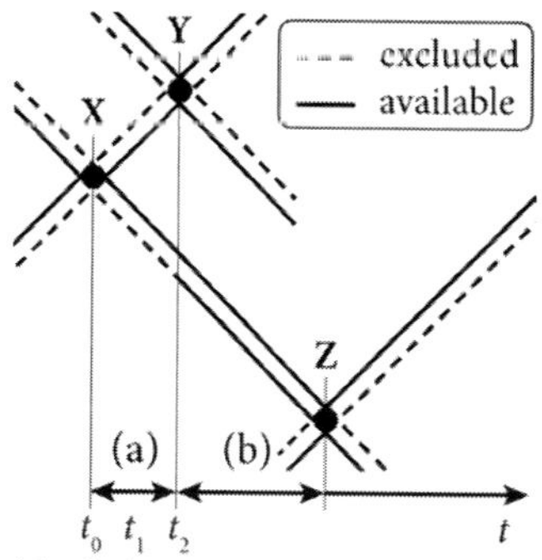

Figure 5. The identical waves from fermion event **X** will be excluded from triggering the next quantization condition **Y** (Table 1). (a) weak-excluded until **Y**; (b) weak-broken until **Z**.

	Fermion Source				
	Boson 1		Boson 2		
t	φ_A	φ_B	φ_C	φ_D	Description
t_0	$\boldsymbol{a}$	b	$\boldsymbol{a}$	d	Fermion Event **X**
t_1	$a + t_1$	$b + t_1$	$a + t_1$	$d + t_1$	Propagation
t_2	$a + t_2$	$b + t_2$	c	$\boldsymbol{d + t_2}$	Symmetry-breaking

Table 1. The corresponding sequence chart of phase values, showing the breaking of source exclusion symmetry. In $t_{0..2}$, excluded wave states are shown with shaded background, and a quantization condition is shown in bold face. At t_2, wave **D** interacts with an external boson, at fermion event **Y** (fig.5).

1.1.3 *Summary: a new picture of vacuum and fields*

Our picture presents reality as naturally-quantized, phase-localized vacuum energy. Its continuously-propagating waves are impulses with specific origin and eventual limits. Rather than assuming continuous fields, we present vacuum energy as discrete *spherical* impulses carried by bosons. This has implications for the conventional coupling and vacuum statistics: in our model, the *free parameters* of the Standard Model (and others) are derived *variables* that approximate a wavefunction's environment. Convention assumes an unchanging constitution of a fermion, and that some vacuum properties are constant, whereas our model is **constitution-invariant** because it operates on the fundamental information units: the waves of bosons, allowing fermions and the interacting elements of their environment to accountably change their constitution.

This uses deterministic classical foundations, which imply emergent quantum-mechanical processes, approximating to modern physics at sufficient scale. From the macroscopic viewpoint, fermions only appear to move or exist continuously because measurement may only occur at fermion events, similar to how a strobe effect illuminates its target. Conversely, conventional background vacuum energy density and its related statistics assume a uniform or nonlocal field value, and we question here the reliability of such statistical approximations, as having limited scope as *effective* methods that degrade at smaller scales and higher energies.

For quantum computation with these foundations, we may regard fermions as the sharp eigenstates of a wave, with their partner wave being unknown: for each boson: $[-b, ?]$. **Qubits** can be constructed from an individual boson's non-excluded waves (where the resolved intrinsic state is *spin*), or from the entangled wave states of all the propagating waves from a fermion event (where the resolved state is a *mass-energy* value).

2. Wavefunctions

Wavefunctions *require* a context of vacuum energy. For any boson, every *other* boson is vacuum energy.

2.1 *One-dimensional phase examples*

These following examples (figs.6–9) are introductory illustrations of the model, rather than being representative of physicality. Our first example shows two coherent *massless* sources $\{\mathbf{A}, \mathbf{B}\}$ in one spatial/phase dimension. Their bosons will not couple unless the two sources are *exactly* aligned so that their $-b$ states coincide exactly. Given that matter cannot be prepared with absolutely exact precision, the zero-width impulses make coupling impossible.

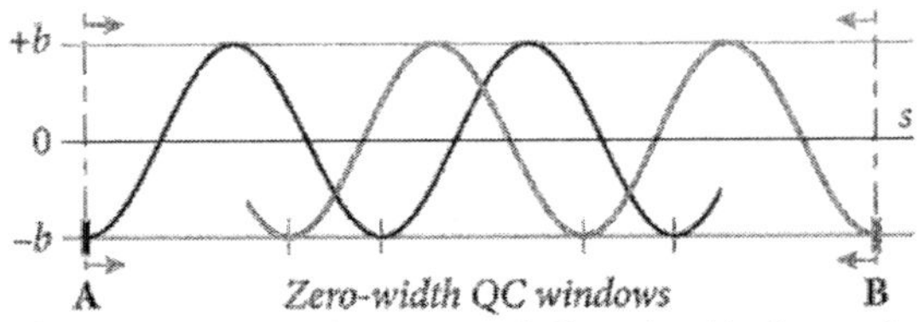

Figure 6. Two massless sources, 1-dimensional background.

Next, we change boson **B** to have *mass-energy* (fig.7). On overlap, this modulates the phase of the other boson, creating a **phase window** for coupling. Depending on the phase of both sources, a coupling will predictably either happen or not.

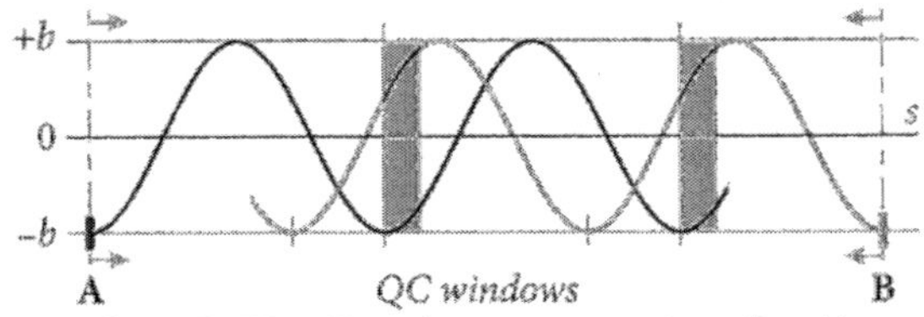

Figure 7. Wave **B** carries mass-energy (retarding **A**).

If we then add some randomness to the phase of any boson (fig.8), the edges of the window are 'blurred', the shape of fall-off depending on the profile (e.g. uniform, or normal distribution) of the randomness, in the manner of a Dirac **delta function**.

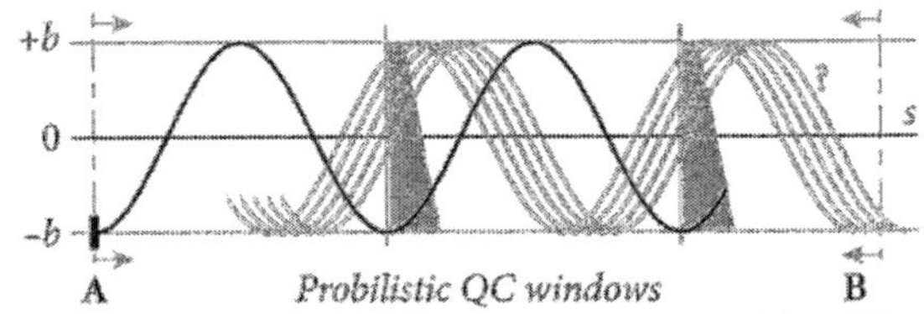

Figure 8. Randomness in the phase or source position of **B**.

Adding more sources gives a statistical **power spectrum** of the mass-energy (fig.9). To account for the environmental vacuum, we need to know its power spectrum, and model it as a flux of mass-energy. Indeed, where system confinement is not sufficient to generate a unique solution, extra statistical vacuum terms must be included.

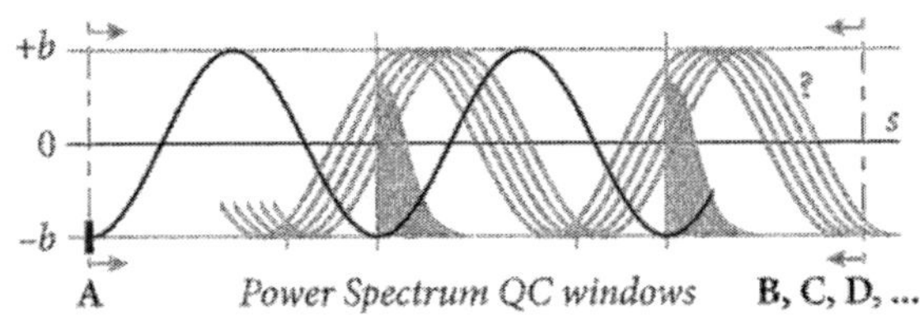

Figure 9. More than one vacuum source, random phase.

These examples illustrate how mass-energy, in the form of vacuum energy, is essential for the collapse of fermions, and that wave phase and (de)coherence can be critical to the occurrence of wave collapse.

2.1.1 *An approximation of one wave with vacuum*

Where a wave interacts with another wave of unknown random phase (say vacuum energy), the probability of interaction by phase modulation is a function of the vacuum energy. Where external bosons are considered and their phases are unknown, the phase-localized quantization condition is spread over the phase range as a probability distribution. Rather than being zero, as it was in the case of single wave, the *independent* probability of generating a quantization condition within the wave cycle is simply

$$P(x_A) = p. \tag{6}$$

Although p is the same for each wave cycle, the probability P_n, of an event occurring in cycle n, requires that the previous event was unsuccessful. Where $p < 1$, P_n is a series converging to zero. We define the probability Q of a null interaction

$$Q_n = Q_{n-1} - P_n \tag{7}$$

and the probability of a quantization condition

$$\begin{aligned} P_n &= p\, Q_{n-1} \\ &= p\,(Q_{n-2} - P_{n-1}). \end{aligned} \tag{8}$$

With $n = 0$ for the first wave cycle, we initialize the sequence for the first test:

$$\begin{aligned} Q_{n<0} &= 1 \\ P_{n<0} &= 0 \end{aligned} \tag{9}$$

Eqs. 6-8 (fig.**10**) simplify to

$$P_n = p^n;\; Q_n = (1-p)^n \tag{10}$$

The probability of generating a quantization condition in the interval between source and cycle n tends to 1 with increasing n:

$$P_{0..n} = \sum_{m=0}^{n} p\,(Q_{m-2} - P_{m-1}) \tag{11}$$

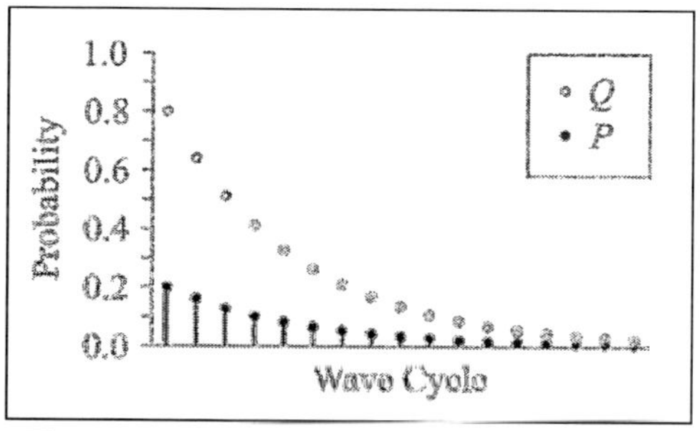

Figure 10. Probability P of a quantization condition.

The radial phase is trivial:

$$r = \varphi_A = x_A + 2\pi n \tag{12}$$

When *two* waves of a boson are available for the quantization condition, the solutions for x_A, x_B are superimposed: interleaved and ordered.

2.2 *Euclidean approximation*

When one wave is considered, the nondimensional probability of achieving the quantization condition *per phase cycle* is simply p (eqs.6–11). For an *expanding sphere in vacuum*, the independent radial form for one cycle *approximates* at large integer values of r to

$$P(r) = 1 - (1-p)^{\frac{dV(r)}{dr}} \tag{13}$$

where p is the fraction of radial phase that is available for the quantization condition, depending on vacuum energy density, and $V(r)$ is the volume of space enclosed by the previous cycle at radius r:

$$\begin{aligned} \frac{dV(r)}{dr} &= \frac{4\pi}{3}(r^3 - (r-1)^3) \\ &= 4\pi\left(r^2 - r + \frac{1}{3}\right) \end{aligned} \tag{14}$$

This approximation is not suited to changes of conditions, because it assumes a uniform probability over each successive 1-unit-thick crust of a hollow sphere. The history-dependent radial form of probability distribution P_H (incorporating the failure of previous events, eq.8, and the remaining null term available to infinity) tends to zero for infinite r:

$$P_H(r) = p^r\left(1 - (1-p)^{\frac{dV(r)}{dr}}\right) \tag{15}$$

$$\int_0^\infty P_H(r)\, dr = 1 \tag{16}$$

The total area under the curve, for $0 < r < \infty$ is 1, so it is pre-normalized, and the function increases from near-zero to a maximum value (the zone of **asymptotic freedom**), then tails off to follow P_n (eq.8), a distribution that describes a tendency to collapse fermions at a given distance (the zone of **confinement**). This kind of function provides essential clues for the distances that various types of particles tend to maintain, in given vacuum conditions. Fig.**11** shows the first 400,000 terms for $p = 10^{-5}$: for $r > 64$, the series tends to zero at infinite r. For small p, approximate values of r at respective percentiles $\{5, 50, 95\}$ of $\int P_H(r)$ are:

$$r = \left\{\frac{5.129}{p}, \frac{6.931}{p}, \frac{29.96}{p}\right\} \tag{17}$$

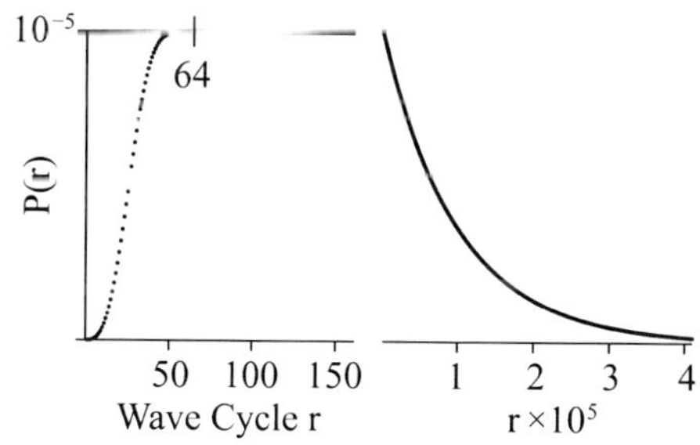

Figure 11. Plot of $P_H(r)$ for $p = 10^{-5}$, eq.15.

2.3 *Vacuum*

Approximating this process for discrete sources to statistics, we may use standard terminology for QM, and assign each modulator as an **amplitude** (factor) function in the probability distribution. Where this becomes impractical, or vacuum energy is sufficiently de-coherent or random, a power spectrum of flux (where positive and negative potentials maintain separate amplitude axes; complex) replaces individual terms, to represent a conventional field.

When applied to larger-scale dynamics, the higher the energy of a wave, the more likely a wave will collapse; the effect of mass-energy on phase modulation and wave collapse is analogous to the Higgs Mechanism for intrinsic mass, where bosons carrying large phase potentials are less likely to radiate far in a degenerate state, because they are prone to being collapsed by a wider variety of energies and phases of environmental bosons (vacuum energy).

2.3.1 *Unique solvability and derived background space*

For a wavefunction to be fully quantifiable, we must understand the role of external instances of vacuum energy on a system. These instances may have almost zero mass-energy, but their presence will create unique solutions in a degenerate wavefunction, and allow more tiers of residuals, each of which helps create a macroscopic tier of scale (physical hierarchy). We believe that the geometric structure of background space is not fundamental, but is instead derived from (and limited by) the uniqueness of fermion solutions, allowing lower-dimension spaces in the simplest of interactions where few bosons overlap. As a boson grows, there is a transition from simple one-dimensional to three-dimensional (four-boson) phase solutions, after which it is safe to assume flat Euclidean space. Each causal introduction of a boson will partition its wavefunction, or add a coupling term to the aggregated wavefunction.

This creates a sub-structure of trivial solutions at the highest energies and smallest scales, in cases where there are insufficient instances of vacuum energy flux to require a full (3, –1) metric. Although trivial, it depends heavily on phase coherence, and therefore has high deviation from the statistical expectation values when preparation conditions are unknown. We speculate that this may be a 'phase transition' where the three-dimensionality of space may emerge [but as yet this proposition lacks the required geometric working].

2.3.2 *Unique 4-boson solutions in Minkowski spacetime*

For any given boson shell, there are two active waves, whose relative phase determines a phase modulation value that is applied to the phase of all other overlapping waves. In Minkowski spacetime, two overlapping shells are not sufficient to create a unique solution; the circular solutions are not unique points. Adding a further overlapping shell implies up to two point solutions, again non-unique, so a fourth overlapping shell is needed to create a unique solution, but with each shell overlap, the probability of generating a 'hit' with a QC is diminished.

2.3.3 *Simplifying the vacuum*

We may model the simpler interactions, by assuming an exact preparation for the two main interacting bosons for which candidate solutions may be found, by applying a statistical evaluation of the vacuum energy flux. This has two effects on solutions: Firstly, the modulation of these vacuum bosons will enable more solutions for the known bosons; secondly, the vacuum bosons can enable uniqueness in existing solutions, even if the vacuum bosons have zero mass-energy. The latter is ideal for approximations, reducing the extra terms' contributions to an expectation value determined by the modulations of the known bosons (2.1). These approximations generalize the vacuum energy into anisotropic fields, to allow flux gradients (3.1) to accurately affect the wavefunctions.

3. Forces

Fundamentally, boson propagation is *scalar*; direction is not intrinsically encoded in fundamental boson, so any vector-like directional attributes are emergent geometric expressions of the positions of solutions, aggregated as momentum. We identify two possible mechanisms of force: one is a *direct modulation* effect, which is the displacement that a vacuum modulation imparts on a solution. Although this affects individual solutions, we find that it is approximately cancelled out by the fact that the modulation prevent a solution in the same wave cycle as the non-modulated case [there is detail in the exceptions, which should be explored further]. The other mechanism is a *statistical tendency* of a conserved fermion's bosons to collapse towards the sources of vacuum bosons. We place significance on the latter.

3.1 *Gravitational interactions*

We describe **gravity** as an effect of the variation of the density of vacuum energy flux over space, making it an emergent statistic of the anisotropic environmental vacuum. The effect is that waves are more likely to collapse in the direction of boson sources, especially where many fermions interact with unconfined (radiated) vacuum currents. There are two considerations that affect what we might regard as a stable gravitational 'field' near a body: (a) the interaction of other vacuum energy, (b) the self-interaction of the field.

3.1.1 *The flux gradient created by vacuum*

We calculate the change of flux density, not directly from the geometry of an expanding sphere (which would yield no change in the density of impulses), but by the interaction of the flux with other environmental vacuum energy, which collapses quanta of the flux in proportion to the increase in volume per unit radius, from (eq.15),

$$\frac{dV(r)}{dr} = 4\pi\left(r^2 - r + \frac{1}{3}\right) \tag{18}$$

which can be rewritten as "the expected probability of a test particle interacting with the body's field",

$$\frac{P(r,b)}{dr} = \frac{4\pi\, p_b}{r^2 - r + 1/3} \tag{19}$$

where p_b is the radiated mass-energy of the body. The mean deflection ∇ is the probability that the test particle will interact with the body's flux p_b, rather than with the environmental vacuum flux p_v, scaled by the mean expected vector ∇_b between particle events where the particle interacts with the body's bosons:

$$\nabla(r,b,v) = \nabla_b \frac{4\pi\, p_v p_b}{r^2 - r + 1/3} \tag{20}$$

The resulting force is comparable with other classical formulations; we have encoded the gravitational constant G into the mass-energy terms, and acquired some extra terms that may seem negligible, but might be relevant at grand unification energies and large time intervals. Classical momentum is realised on particles by assuming they have internal kinematics that is emergent in the model without special consideration, as a regular system of waves.

Interestingly, the scaling of probability, resulting from this gradient of collapse, is not affected by the mass of the satellite body, unless it is incorporated into ∇_b, though we may calculate the reciprocal force by switching the roles of the bodies in the approximation. It also scales well to the collapse interval, making it almost independent of the constitution of the satellite. This means we can regard items of interest as 'test particles' in the 'field'.

We believe the usefulness of this approximation is limited to illustrating a possible origin of gravity, and to be less useful than running the model itself, where the gravitational effects are inherently present in the process, rather than computed separately. We lose fidelity from the physical model by approximating the actual bosons as a power spectrum of vacuum energy, and we lose even more fidelity by assuming the power spectrum as a scalar flux term. We also lose the phase coherence of any radiation, and the quantum detail for individual wave collapses.

3.1.2 *Self-interaction (macroscopic)*

We can assume that, in a massive body, a proportion of radiated bosons will interact with other quanta of radiation from the same body, dissipating the field with increasing radius, and generating higher-order fields in a chaotic system that requires approximation. This field is likely to comprise bosons of low mass-energy; the lighter of the fermion constituents, because they are most likely to propagate large distances. Near to the body, gravitation is likely to be indistinguishable from the interactions that conserve the body's constitution. Far from the body, self-interaction quickly loses relevance, as the increasing surface area of the boson quickly exposes it to a more significant probability of collapse from vacuum energy. We will leave this aspect for further study, e.g. the detail of cosmological extremes like black holes.

3.1.3 *Summary of gravitational interactions*

These deflections give a directional element to what is otherwise just a 'field' comprising scalar waves. Its aggregated magnitude, expressed as classical particle momentum, is small compared to an equivalent effect from the phase modulation from the interacting parts of the flux. The nature of the resulting force is different from the direct interactions of the bosons themselves. This means we will not find a gravity-carrying boson; instead we must look to environmental vacuum statistics to quantify the effect of gravity.

3.2 *Charges and their forces*

Charge is the interaction of a boson's positive or negative phase modulation with the environmental vacuum energy. Where one of a fermion's waves has already interacted with vacuum and been taken off-shell (1.1.2), the remaining on-shell waves are guaranteed to have only one sign of modulation $\{++, --\}$ present, in contrast to a shell that has not yet interacted which has one of $\{++, +-, -+, --\}$ modulations present. Shells having mixed sign are neutrally charged; those with one sign carry charge.

Each fermion therefore has two charge values: the **initial charge** on two bosons, and the **residual charge** on the boson that remains after its first interaction. We map the initial charge to the **weak interaction**, and the residual charge to the **electric interaction**. The radiation of a fermion source therefore has two wavefunctions, partitioned by source and interaction events. The residual wavefunction is a choice between two possible wavefunctions, depending on which boson interacts first as a result of the initial wavefunction. This choice is generally unobserved because it is *part of the process* that preserves the constitution of conserved fermions. However, we can detect weak interactions when they *change the constitution* of an otherwise conserved fermion: a change in momentum, effects from the change in constitution of future fermions, or the sign of wave properties like chirality. The strong force has different a origin, being a result of the double-linking confinement structures [6,7] found in QCD (5.2).

3.2.1 *Attraction and repulsion*

This requires that interchangeable 'lighter' bosons are available from vacuum, to supply the persistent heavy boson with a partner. Field line vectors are a sum of the source radial vectors, scaled by their probabilities. In future works, we intend to reconcile charge structures with established literature.

3.3 *Summary of forces*

In this model, the forces of the Standard Model are an effect of flux gradients on the probability of wave collapse. Rather than being separate and specific forces acting independently on classical bodies, the forces are intrinsically present in the model, without any need for specific mathematical treatment to generate or isolate them. Indeed, to rediscover these forces, we have explored the observable displacement effects of organised structured sources (classical bodies) on test particles and micro-systems, in comparison with the same systems in background vacuum. We stress here that these forces are emergent expressions of the fundamental propagation and collapse, and are Lorentz-invariant, whereas the separated fields are not translation-symmetric.

4. Physical Structures

Author's Note: It is customary for foundational works to attempt to explain grand unification and every aspect of the Standard Model. Although these remain our aims, we believe it is too soon to declare success in this regard, as our researches are not yet sufficiently exhaustive, nor are they quantitatively calibrated to experimental physics. We provide this section as guidance for further work.

4.1 *Particles*

Conserved fermions are those that are similarly re-constituted over time: their constituent bosons 'bounce' off vacuum energy at virtual anti-fermions, and return to create a mutual coupling point for the next fermion event. Such fermions have **classical momentum** as an aggregation of displacements. When this cycle is interrupted (rather than simply nudged) by other bosons, fermions change their constitution or are annihilated.

A truly *free* fundamental fermion, i.e. the sole occupant of a universe without vacuum energy, can never collapse, for two reasons: (a) a single source can only contribute *one* term, of the *two* required to constitute a further fermion event, because of exclusion (1.1.2); and (b) there is no vacuum energy, to provide the second term. However, we can model a free fermion with the addition of environmental vacuum energy.

A **composite** structure is one where its fermions depend on each others' constituents to remain conserved. This means that some of the fermions' energy is **confined** rather than emitted anonymously into vacuum.

4.1.1 *Fundamental fermions*

Our model defines a minimum constitution for fermions, essentially two bosons, with as many supplemental bosons as required to create a unique solution for a quantization condition. When we examine the particles identified by the Standard Model in this context, we find opportunity to define quarks and leptons (both electron and neutrino-type) in these new terms, rather than just assume they are fundamentally indivisible. This necessarily admits instances of discrete vacuum energy as part of the constitution, whether coherent and confined in a composite, or incidental to the system. For example, neutrinos might be a combination of very light bosons, quarks a combination of heavy bosons, and electron-type leptons are a mix of the two boson types.

4.1.2 *Neutrinos*

Our neutrinos, being fermions, are the combination of two very light bosons, which needn't be massless. The tiny mass of the boson constituents allows them to propagate great distances, and their part in interactions is generally limited to providing a unique solution for heavier bosons, rather than the behavior of massive bosons which tend to perpetuate the conserved particles that collapse near their source. The mechanism for oscillations is a vacuum interaction.

4.2 *Antimatter*

As with Dirac images of the fermion, the *vacuum constituents* of a fermion would form anti-matter if they later couple as fermions. The cyclic process of matter conservation involves a fermion's bosons each finding a fermionic solution with vacuum, before returning for re-constitution. As a process, this creates a 'polarized vacuum' where idempotent conserved ±fermions (matter, or antimatter, whichever has more mass-energy), with their corresponding one-off separated ∓half-fermion instances as the reflection points.

5. Predictions and Applications

5.1 *Description of a black hole*

Our model can be applied to the scenario of a black hole, without any special treatment: the same process of fermion reconstitution is at work in a black hole, as is at work in the low energy environments that we are more familiar with. The Schwarzschild radius is not privileged, but the classical effects are achieved by quantum means; our event horizon is a fuzzy probabilistic boundary. From the outside of a black hole, as we approach the event horizon, the vacuum energy density increases, making bosons collapse more readily into fermions (1.1.1, 2.1, 3.1), with a greater proportion of fermions failing to reconstitute in a conserved manner from one instance to the next.

5.1.1 *Evaporation, probabilistic boundary*

The event horizon itself does not have any special properties. The event horizon can be assigned to be a radius from which there is a significant confidence that matter cannot escape. Indeed, it is the large-scale result of many interactions within a zone where two relevant variables (both being classical generalizations) are high: *vacuum energy density* is large enough to prevent bosons propagating far, and a *gradient of vacuum energy density* implies a directional preference for collapse. There are no limits to the vacuum energy density; many boson shells may occupy a length or volume that is less than a fundamental wavelength in size, but interaction opportunities per boson are limited.

Although our model was not designed around solutions to problems related to black holes, we find it presents a good model for some currently-favored ideas, such as black hole evaporation as proposed by Bekenstein & Hawking, without the problems of classical gravitation. Our constitution for a fermion requires that at least two bosons leave each fermion event. In low-energy environments, fermions might typically re-constitute after their bosons 'bounce' off the vacuum. This is in contrast with the dense environment of a black hole, where higher energy densities will increase the probability that fermions will not reconstitute themselves because their bosons pair up with other vacuum waves, never to return to their original encoding. If this occurs near the event horizon zone, then some energy may probabilistically escape.

This model does not suffer 'information paradox' problems, because our matter is encoded as separate bosons, and even within the dense body of a black hole, these bosons interact as normal. However, the encoding of fermions entering a black hole is likely to be significantly scrambled by the interactions within.

5.2 *Mass variation of quarks and gluons*

Gluons are the confined energy of hadrons: these bosons, having significant mass-energy, collapse readily at short distances, each quark coupling to both other members of the hadron (6 bosons), with residuals providing larger-scale currents at significantly lower mass-energies. There are nine components in total, for the six bosons. For each pair of the bosons leaving a quark, one of the bosons (say, the anti-color) has two wavefunctions, partitioned by the symmetry-breaking interaction of the other boson, which makes both its waves available for interaction. Where two wavefunctions are concurrent for any two bosons on their sphere shell, the probability of interaction depends on the phase coherence of their respective interaction windows. Where the boson energies are near-equal, the maximum probability (where windows are non-overlapping) is twice that of the minimum probability (where windows overlap), and where boson energies are unequal, the range of probability is narrower, with the smaller mass-energy value seeming to be an error on the larger value. For a wavefunction having only a single boson with both waves active, their interaction windows are non-overlapping, and conversely, the windows of a pair of bosons *before* the symmetry break may overlap. Aggregating such distributions from masses or initial phase values should yield typical ranges found in experiment, and vice versa. The phase coherence of the wavefunctions gives directionality to the probabilities, particularly when sources are close.

5.3 *Radial occlusion effect; EPR variants*

Consider a large body at radius r from a radiating source: the large body should collapse some of the waves that would otherwise have radiated beyond the system. Given that we think that all mass-energy values are very small relative to the highest possible value of mass-energy that a boson may carry, it is likely that the occlusion effect is significantly smaller than the direct exchanges of bosons that occur between the two bodies; a squared order smaller than the direct modulation interaction (3.1). However, it should be testable.

5.4 *Confinement energy and momentum*

The model allows composite particles to gain or lose their confined energy, but this is not necessarily in proportion to their classical momentum; it is possible for a system to lose confinement energy while gaining momentum, though strictly speaking, this also implies a change in the constitution of the composite particle.

5.5 *Example exotic particle structures**

A three-fermion singlet may be constructed, where each fermion couples to both the other fermions (5.2) within $1r$ wavelength, and is equivalent to a very compact hadron. This composite travels at light speed without residuals, and without external vacuum interaction; indeed it must avoid vacuum energy (within $2\pi r/3$ of its axis) to remain conserved, given that its $p = 0.5$.

5.6 *Fast guided information transport**

It should be possible to transport information towards a recipient at a density (several orders of magnitude) greater than the fundamental density, if the receiver prepares the vacuum with phase-coherent periodic emissions. If the source is able to generate a boson having a wave phase slightly de-synchronised from the target, but within the mass-energy modulation window of the target's coherent signal, then the information will propagate to the target. An analogy for this mechanism is a 'domino run on a conveyor belt', where a target periodically drops dominos onto a conveyor belt that is moving away from it, and a downstream source tips a domino to send a signal. Many such signals could be sent in parallel, using non-overlapping phase windows, provided there is no interference (which would cause cross-talk).

Although this signalling method could work at near-fundamental scale if the machinery could be built, at a scale significantly smaller than typical electron interactions, it has been mentioned only as a curiosity, because it has many problems that prevent it being practical: (a) It is inefficient over distance; (b) Information propagation collapses the carrier signal; (c) The sender of the information has no way of knowing if a carrier signal is available for the complete propagation path, so this is suited only to short distances, and must be implemented with a full protocol; (d) It is prone to interference from oppositely-modulated bosons having higher energy than the information, shifting the signal phase out of the modulation window of the carrier signal, and losing the information to vacuum.

6. Further Work

6.1 *Disproof and viable alternatives*

Reasonable alternatives weaken the model, so we have identified these to open it to challenge.

6.1.1.1 We assume the quantization condition (1.0) occurs only at $-b$, because we use it as an absolute phase position in the algebra that has special privileged physical meaning [1, 3–7]. Viable alternatives may be: (a) at any point where phases are equal, or (b) $+b$.

6.1.1.2 We assume that at the point at which two boson shells overlap, phase modulation spontaneously sweeps over the phase range, to include any swept quantization conditions. The alternative view, of not including the phase range, severely reduces amplitudes.

6.2 *Problems*

Here we list aspects of the model that are weak, poorly understood, or worthy of further work:

6.2.1.1 What is the fundamental scale of this model, which is not yet calibrated to standard length. We expect a scale to emerge from the solutions to composite configurations.

6.2.1.2 A description of harmonic oscillators and black bodies (QED). We expect 'sparse sampling' to provide the mechanism.

6.2.1.3 Is there a process that creates bosons of a given mass-energy, without assuming a big-bang?

6.2.1.4 Verify that different vacuum environments exist in the universe, and investigate the possibility that an increase in vacuum flux for an observer may cause their own confinement and vacuum interactions to increase in frequency, red-shifting the surroundings.

6.2.1.5 (a) Verify that 3D background space emerges from scalar phases (2.3.1), where four or more bosons have common geometric solutions. (b) When less than four sources overlap, unique QC solutions may exist in lower-dimension space (2.1). Examine the resulting ontologies.

6.2.1.6 Do 'flavors/generations' correspond to the number of bosons required to create a unique solution for a fermion?

6.2.1.7 Is energy conserved or mixed for all bosons passing through a fermion event?

6.2.1.8 Investigate the modulation properties of matter and anti-matter (4.2), for any imbalance or cascade effect.

6.2.1.9 Investigate whether non-persistent fermions (4.2) are an adequate description of 'dark matter'.

6.2.1.10 Computability of the deterministic aspects of the model, and sufficient approximation in statistics. Operational calculus on approximations of vacuum interactions (2.2), into e^n format.

7. Afterword

We are hopeful that the model outlined in this work describes useful symmetries. First is the symmetry of matter and vacuum states (1.0) on a common axis [3–7], with simple physical rules that encapsulate important entity and exclusion principles. Our model's intrinsic inclusion of forces (3.3), and their emergence without special treatment, strongly suggests that our model has a force-unifying symmetry in its basic process. Further, we feel that the constitution-invariance of the process (1.1.3), free of renormalization and singularity problems and effective energy limits, is worthy of further study.

References

[1] P. Rowlands, The Fundamental Parameters of Physics: an Approach towards a Unified Theory, PD Publications, Liverpool (1991).

[2] P. Rowlands, Zero to Infinity, ISBN-13: 9789812709141, World Scientific Publishing, (2007).

[3] J.S. Valentine, Presentation A Development of The Fundamental Parameters of Physics, Alternative Natural Philosophers' Association 20, Cambridge (1998).

[4] J.S. Valentine, Gravitation in Symmetrical Context of Space-Time, Physical Interpretation of Relativity Theory X, British Soc Phil of Science, London (2006).

[5] J.S. Valentine, Algebra of a Three-fold Symmetry for Fundamental Physics, Physical Interpretation of Relativity Theory XI, British Society for Philosophy of Science, London (2008).

[6] J.S. Valentine, Presentation Splitting the Fermion, Vigier VII Symposium, London (2010).

[7] J.S. Valentine, AIP Conf.Proc. 1316, 349-370, DOI: 10.1063/1.3536446, (2010).

Emergence of a New Quantum Mechanics by Multivalued Logic

CLAUDE GAUDEAU DE GERLICZ,[a,f] MATHIAS ANTOINE,[a] PHILIPPE BOBOLA,[b] NICOLAS FLAWISKY,[c] XAVIER HEBRAS[d] & MUSA MUNDEDI[e]

[a]*Bioespas, Laboratoire de Bioinformatique et de Biotechnologie, Tours, France*
[b]*Société de Bioinformatique et de Biotechnologie, SBB, Tours, France*
[c]*CEMES-CNRS, Toulouse, France*
[d]*Pr de Mathématique Ecole Supérieure d'Architecture, Paris Val de Marne, France*
[e]*Département de Mathématiques à l'université de sciences et technologies de Tours, France*
[f]*bioespas.international@gmail.com*

Quantum Mechanics associated with new logic like Multivalued Logic and Fuzzy Logic has progressed in different ways and their applications can be found in many fields of sciences and technologies. All the concepts attached to this theory are far from the classical view. Classical mechanics can be viewed as crisp limit of a Fuzzy quantum mechanics. This leads to the following interpretation: It is the consequence of an assumption that a quantum particle "reside" in different place or in every path of the continuum of paths which collapse into a single "unique" trajectory of an observed classical motion The reality is "Fuzzy" and nonlocal not only in space but also in time. In this sense, idealised pointlike particles of classical mechanics corresponding to the ultimate sharpness of the fuzziness density emerge in a process of interaction between different parts of fuzzy wholeness. This process is viewed as a continuous process of defuzzification. It transforms a fuzzy reality into a crisp one. It is clear that the emerging crisp reality as a final step of measurements carries less of information that the underlying fuzzy reality. This means that there is an irreversible loss of information usually called "collapse of the wave function". It is not so much a "collapse" as a realization of one of the many possibilities existing within a fuzzy reality. Any measurements rearrange the fuzzy reality leading to different detection outcomes.

Keywords: Quantum mechanics, Multivalued logic, Entanglement, Formal systems

1. Theoretical Aspects

Fuzzy logic set theory can be thought of a generalization of ordinary set theory defined as follow: Ordinary set: an object either belongs to or does not belong to a set. For this reason, ordinary sets are often called "crisp" sets in fuzzy logic. Fuzzy logic set theory: the transition from membership to non-membership is gradual rather than abrupt (Zadeh 1973) $[6a]$ The object belongs to a set with some degree of certainty. Fuzziness is not simply due to the lack o It can be understood as the result of having to summarize large quantities of data.

1.1. *Spatial Membership Density*

This approach allows us to formally introduce the « membership » density defined as a derivative of the membership function. If we denote the membership density byμ, then a degree of membership of the particle staying in an elemental volume, ΔV is

$$\mu \Delta V \quad (1)$$

Given of degree of membership $\mu(x_i)dV$ of a particle in a volume dV containing the point x_i and a degree of membership $\mu(x_j)\, dV$ of the same particle in a volume dV containing the point x_j we find the relative degree of membership of the particle in both volume (x_i, x_j) [2] (Granik, Caulfield):

$$\mu\,(x_i, x_j) = \mu(x_i)/\,\mu(x_j) \quad (2)$$

In reality, any physical particle occupies a small but nonzero spatial interval. This means that the membership density is a sharp (but not delta like) function corresponding to a minimum fuzziness. At the other end of the spectrum in the micro world, the fuzziness is optimal.

1.1.1. Particles and Waves

If we admit that the quantum mechanical particle seen as a micro object « resides » in different elemental volumes dV of a three dimensional space with the varying degree of residence (membership), we can apply to such micro object our concept of the membership density. This density cannot be made arbitrarily narrow as is in the case of a classical particle. This latter implies the emergence of the interference phenomenon of micro objects, which in the classical domain is an exclusive property of waves.

Therefor mutually exclusive concepts of particles

and waves in classical mechanic become inapplicable in the fuzzy reality where « particles » and « waves » are not mutually exclusive concepts, but rather different expression of fuzziness. The double-slit experiment can be interpreted now as a micro object's interference with itself because it has a simultaneous membership in all parts of space including elemental volume containing both slits.

1.1.2. Measure of the fuzziness [2]

According to Granik and Caulfield fuzziness can be measured. Considering the different trajectories between two points A and B, If $S(x,t)$ is some scalaire function defined on the continuum of paths we can seek a function related to the notion of fuzziness. By using the Hamilton Jacobi equation for a classical particle in a potential field:

$$\partial S/\,\partial t \;+\; (l/2m)\;(\nabla S)^2 + V \;=\; 0 \qquad (3)$$

we can represent S as a functional defined on the continuum of paths connecting two given points (0,1) corresponding of the moments of time (t_0, t_1)

$$dS/dt \;=\; p^2/2m \;+\; V \qquad (4)$$

then,

$$S = \int_{t_0}^{t_1} \left(\frac{p^2}{2m} - V\right) dt \qquad (5)$$

Equation 5 is the definition of the action of particle moving in the potential field V, thus the concept of fuzziness in classical mechanic in connected with the action S. By minimizing this functional, we eliminate fuzziness by generating the unique trajectory of the classical particle.
The principle of less action serves as a defuzzification procedure

1.1.3. Fuzzy function

The membership function $\mu A(x)$, characterizes the level of membership that an object x has within set A, where the level of membership is expressed as a value from the closed interval $[0,1]$. Formally, $\mu A(x) : U \to [0,1]$, for fuzzy set A subset$U$, where U is the universe of discourse. The support set of the fuzzy set A consist of all objects x whose $\mu A(x) \geq 0$

Fuzzy operations on Functions:
The membership function of the intersection of two sets A and$B$, having the « same object » x can be interpreted as the min operator, or as the product of Larsen operator, *, typically the min operator is used:

$$\mu A(x) \cap \mu B(x) = min\;(\mu A(x), \mu B(x)) \qquad (6)$$

The membership of the union of two sets, A and B, having the same object x, can be interpreted as the max operator, or as the algebraic sum, typically the max operator is used:

$$\mu A(x) \,\cup\, \mu B(x) \;=\; max\;(\mu A(x), \mu\, B(x)) \qquad (7)$$

1.2. Temporal Membership Density

We can suppose that the micro object can resides in different "eventual" times. So we can attribute a fuzzy existence to the potential events (or possible events) which have not the weight of the present measurement. Fuzzy logic is an appropriate tool for describing phenomena which contains some uncertainty. This theory provide a strict mathematical framework in which fuzzy concept phenomena can be precisely studied and which are difficult to explain by conventional logic

As we have been tried to explain recent experiment photon intrication (Aspect) $[6b]$, we can extend the concept of spatial membership density to the concept of temporal membership density, Notated by υ then a degree of membership of the micro object in an elemental interval of Time ΔT is:

$$\upsilon \Delta \mathrm{T} \qquad (8)$$

Given of degree of temporal membership $\upsilon\Delta T$ of micro object in a time element dT containing the point x_i and a degree of temporal membership $\upsilon(x_j)\;dT$ of the same micro object in a time element dT containing the point x_j, we find the relative degree of membership.

$$\upsilon(x_i, x_j) \;=\; \upsilon(x_i)\,/\,\upsilon\,(x_j) \qquad (9)$$

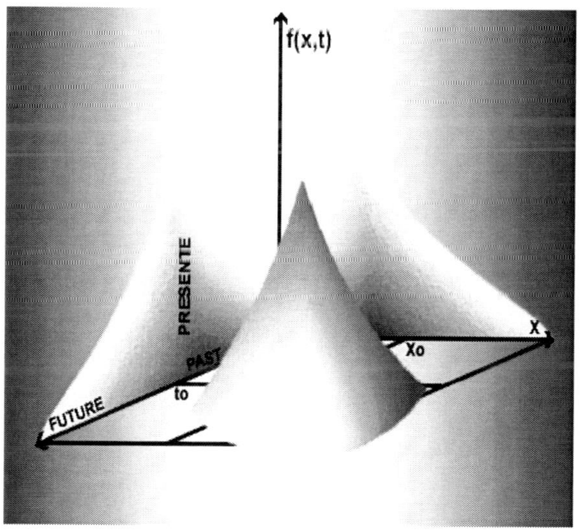

Figure 1. Spatio-Temporal characteristic function of a particle.

The fact that there is next to the spatial extension, a temporal extension of the micro object involves a lot of consequences:

- The temporal membership density is related to the time permanent property, the "time density" (see Kozyrev 1991 and appendix 2) [7,7a,7b]
- This implies also the existence of "potential events" in the future to be distinct of "possibility" and which have, as we have describe in a previous article [3], a fuzzy existence by using the concept of 'Predicates Logic', $\exists\alpha$.

A monochromatic light signal is emitted from a source S and passes through two slits g_1 and g_2. Then it lights a photographic screen P (cf Figure 1). Using a cannon to electrons, it sends them one by one with two open slots.

In predicats logic,
$if\ \alpha\ =\ 0$, means "doesn't exist",
$if\ \alpha\ =\ 1$, means "it exists"
$if\ 0 > a > 1$, means fuzzy existence

In fuzzy predicates logic, it is the "degree of existence" The concept of Potentiality appeared in a logic that has three values developed by S. Lupasco [**8**] and it is linked to the concept of actualization (i.e.: measurement of the reduction of the wave packet). He speaks so an antithetical actualization of the particle and gave to the potentiation, an ontological status.

2. Application to Young's Experience Revisited by A. Tomura

2.1. *Experience with one slot opened*

- If we consider only one slit there is no interference. Light behaves as corpuscles. If we block one slit after the other then we obtain for each slit a distribution of lightning intensity $l_1(x)$ and $l_2(x)$, the diffraction patterns of g_1 and g_2.

2.2. *Both slits are opened*

- At low source's intensity and at the beginning a few light flashes of electrons appeared. It has been first shown by the Bologna group who has photographed the number of a sensitive TV camera.
- At high source's intensity we have a system of interference fringes, but the intensity of which is not the sum of the intensities produced by g_1 and g_2 separately. Light behave as waves. The fringes were able to increase the storage timed up to values of minutes.
- At low source's intensity and after a certain time, (Tonomura and Al. [9]) demonstrate that a single electron build up on an interference pattern. If the number of electrons arrives practically one by one, we should not see any interference, but if the time increases there are interference. The experiment was carried out from the beginning to the end with constant and extremely low electron intensities (< 1000 electrons/sec).
- This removed any probability that the fringe might be due to "classical" interactions between electrons

Solitary electrons can pass through one of the two slits, and observed two types of events as been observed on the screen:

(a) [10.. 100] sent electrons: particle impacts on the screen;
(b) 3000 sent electrons: we begin to see crab-wise interference;
(c) 20000 sent electrons: more visible interference fringe;
(d) 70000 sent electrons: clear interferences.

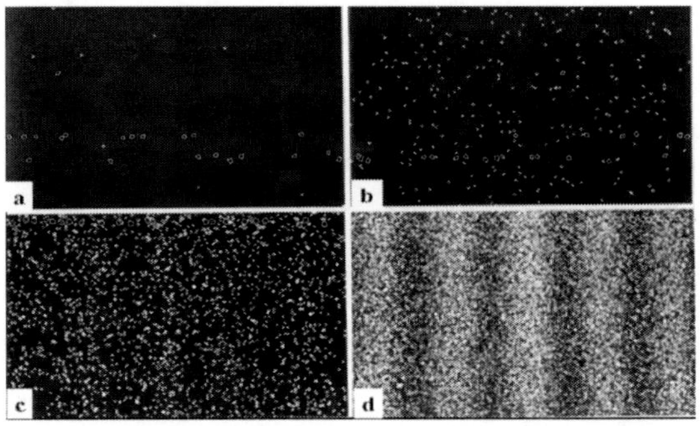

Figure 2.

The fringe formation could be observed as a time series, the electrons were accumulated over time to gradually form an interference pattern on the monitor, and the electrons arrived at random positions on the detector only one by one and it took more than 20 minutes. If we reduce the source intensity until the electrons arrive practically one by one, we should not see any interference. In fact, when the electrons arrive slowly, if the time of the experiment is short, then the spots have a located impact, but if the time increases, there are fringes of interference.

The classical interpretation of this phenomenon does not allow us to choose between corpuscle and wave for an electron. There must be a critical number of particles (over 300) or a very long exposure of time that enable interference patterns to clearly appear on the screen.

2.3. *Interpretation*

Interpretation1:

In Copenhague Interpretation, the authors conclude that the corpuscular and wave electron aspects are inseparable and are superimposed. Submitted to the wave-particle duality, the electron behaves like a wave and a particle but without any explanation.

Interpretation 2:

The observed interference pattern requires that two electrons pass through the 2 slots simultaneously. However the electronic gun sends electrons one by one. It leads to the conclusion that the electron could pass through two slits at the time, and that it might interfere with itself. This solution was considered by Paul Adrien Maurice Dirac[10].

Interpretation 3:

Fuzzy logic allows the association of a characteristic function with time, which allows a temporal extension beside spatial extension. The lone electron could interfere with the temporal extension of electrons that preceded it [3].

There are a critical number of electrons sent (3000) because the solitary electron meets a group of electrons to each of them, with a temporal extension passed, who has not yet crossed the cracks and that it could find a partner of interference. It introduces a memory function that must be developed according to knowledge of logic.

The Fuzzy logic concept of temporal extension allows us to introduce a time with multiple dimensions. This time would have a present, the past and the future. The "Time" variable has ever been mentioned by various authors (Lupasco...etc.) [8] in quantum mechanics and the concept of multidimensional time and on the other hand temporal logic tools have been proposed. We cannot evoke the spatial extension without mentioned possibilities theory [6], since this concept is correlated to the concept of potentiality. (See Appendix A) (gives a status of density or potential existence but however with the same required observation density.) Some authors have already close the gap between quantum theory concept and conventional Language (Rauscher, Amoroso) [11] by interpreting the uncertainty notion link to the wave function.

3. Conclusion

In this paper we introduce the relationship between the concept of potentiality and the concept of actualization revisited under the Fuzzy Logic approach. The actualization (defuzzification) of the potentialities.

Both time and spatial extension concepts in Hadron Mechanics by mean of fuzzy sets theory can give an interpretation of the results obtained by the A. ASPECT's correlated photons experimentation. Hadronic Mechanics [12,13] are the highest density physical Media measured by mankind until now. The great differences between the atoms and hadrons structures require new tools in mathematical and physical method and interpretations.

Hadron are composed of particles with « extended » wavepackets in condition of total mutual penetration because the size of any wavepackets is in the same order of the magnitude as the size of the Hadrons Multivalued Logic or Fuzzy Logic is an appropriate tool to describe all phenomena which contains some uncertainty. This theory provides a strict Mathematical framework in which fuzzy conceptual phenomena can be precisely studied and which are difficult to explain by conventional logics.

Appendix A: The Possibilities Theory

The collapse of the wave function, an irreversible loss of information is more a realization or actualization of many possibilities existing in the Fuzzy Reality. It express how the realization of an event could exist and to what extend it is certain without however its assessment of the likelihood of this prevision. We can measure the possibilities of an event as:

A finite reference set X, with (0,1) coefficient attribute to each sub set to a possibility of an event.

Measure of the possibility Π

$$\Pi\ (\acute{\emptyset}) = 0$$

$$\Pi\ (X) = 1$$

$$\Pi\ (U_i = 1,2....A_i) = sup\ i = 1,2....(\Pi(A_i)),$$

link between the possibility and probability:

$$If\ Prop\ H\ is\ impossible \rightarrow poss\ Null = prob\ Null$$

$$If\ Prop\ H\ is\ certain \rightarrow Necessity\ (1) = prob\ (1)$$

For a strict positive probability

a) H is not completely possible, necessity is null : $0 \leq P(H) \leq \Pi\ (H) < 1$

b) H is surely littlie certain (possible) P(H) > 0,

possibility is at unity (1) and $0 < N(H) < P(H) \leq 1$
c) H is not certain and not possible, every probability measure P is acceptable: $0 = N(H) \leq P(H) \leq \Pi(H) = 1$

Appendix B: Theory of Time Density From Kozyrev

The density of time characterizes activity of time influence on systems and processes of our World. The density of time at a given location in space depends on the processes occurring around it. The processes in which entropy increases, i.e. when there occurs disordering, increase the density of time around them, and vice versa, the processes with decreasing entropy decrease the density of time. It is safe to say that time carries order or negative entropy (negentropy), and is either emitted by a system when its order decreases or absorbed by a system when its order increases.

Since any process changes the density of time around it, it affects the course of other processes and the state of ambient matter by means of this property of time. Thereby the interrelation between all the processes occurring in nature is established through the density of time. Let us pay attention to the fact that since time is the characteristic of the fourth direction orthogonal to our world, both internal and external domains of any three dimensional objects of our world are equally open to the stream of time flowing throughout the world along the normal to it. For recording the influence of the density of time, Kozyrev use various detectors such as non-symmetric torsion balance, electric resistors, contact pairs of metals,

To links relativity to the notion of density (ρ):

"in the Langevin “thought experiment of two twins”, the twin who is located in the vehicle is aging less quickly and therefore its local time flows more slowly and would be less dense;" also could you formalize this fact by the relationship $\rho = k/v$ (k is a constant) from the equation:

$$t = \frac{t_0}{\sqrt{1 - \frac{k^2}{\rho^2 c^2}}}$$

References

[1] C.Gaudeau , M. Antoine, N. Flawisky, P. Bobola, X. Hebras, D. Parello; Interpretation of Quantum Mechanic and Entanglement with Multivalued logic.

[2] A.Granik, H.J.Caulfield. Fuzziness in Quantum Mechanics. arXiv:quant-ph/0107054vI 9 (Jul 2001).

[3] C. Gaudeau de Gerlicz, P. Bobola, M. Berge; Multivalued Logic and Quantum Mechanics. Physical Interpretation of Relativity Theory Imperial College London (8, 11 Sep 2006).

[4] C. Gaudeau, V. Rakotomalala, L. Gouthiere; Fuzzy Automata and Grammars: Manipulation of Uncertain Knowledge (p 136) 14 th International CODATA Conference 18 22 Chambery (sep 1994).

[5] F. Smrandanche, V. Christianto; n-Ary Fuzzy Logic and Neutrosophic Logic Operator from Neutrosophic Logic, Wave Mechanics and other stories Selected works (2005 – 2008).

[6] I. Ketata. A. Garibotti arias *Théorie des Possibilités, Master Recherche Informatique – Cognition et Connaissances.* (2006-2007).

[7] L. A. Zadeh outline of a new approach to the analysis of complex systems and decision processes. IEEE Trans. Systems, Man and Cybemetics, 1973; 3; 28-44.

[8] A. Aspect, P. Grangier, G. Roger, Phys. Rev. Lett, 49, 91 (1982).

[9] N. A. Kozyrev, Possibility of experimental study of properties of time, Pulkovo, September 1967 (txt available).

[10] N. A. Kozyrev, V. V. Nasonov, on some properties of time, discovered by astronomical observations, in Problemy issledovaniya vselennoi, 1980, (Russian lang.).

[11] L. S. Shikhobalov, the fundamentals of N. A. Kozyrev's causal mechanics from: N. A. Kozyrev, selected works, published by Leningrad State Univ., 1941. 488p.

[12] S. Lupasco, *le principe d'antagonisme et la logique de l'énergie. Editions du Rocher (1983).*

[13] A. Tonomura, Electron Holography, 2nd Edition/A. Tonomura, Springer, Springer Series in Optical Sciences (1999).

[14] P.A.M. Dirac, the Principles of Quantum Mechanics Oxford (1957).

[15] E.A. Rauscher R.L. Amoroso Relativistic Physics in Complex Minkowski Space, Nonlocality, Ether Model and Quantum PhysicsTecnic Research Laboratories, 3500 S. Tomahawk Rd, Bld 188, Apache Junction, AZ 85219 USAThe Noetic Advanced Studies Institute, 608 Jean St., Oakland, CA 94610-1422 USA.

[16] RM. Santilli, Hadronic mathematics, mechanics and chemistry, volumes I, II, III, IV, and V, international Academic Press (2008).

[17] R. M. Santilli, Found. Phys. Vol. 27, 625-729 (1997) Relativistic Hadronic mechanics: nonunitary, axiom-preserving completion of relativistic quantum mechanics.

[18] D. Rabivinski, L. Borisova , F. Smarandache; Entangled states and Quantum causality thresold in the General Theory of Relativity; PIP Vol 2, 2005 from Neutrosophic Logic Wave Mechanics and other stories, Selected Works (2005-2008).

Dirac Sea and its Evolution

BORIS VOLFSON

Institute for Basic Research (IBR), 150 Rainville Road,
Tarpon Springs, FL 34689 USA
BorisVolfson@aol.com

The hypothesis of transition from a chaotic Dirac Sea, via highly unstable positronium, into a Simhony Model of stable face-centered cubic lattice structure of electrons and positrons securely bound in vacuum space, is considered. 13.75 Billion years ago, the new lattice, which, *unlike* a Dirac Sea, is permeable by photons and phonons, made the Universe *detectable*. Many electrons and positrons ended up annihilating each other producing energy quanta and neutrino-antineutrino pairs. The weak force of the electron-positron crystal lattice, bombarded by the *chirality-changing* neutrinos, may have started capturing these neutrinos thus transforming from cubic crystals into a *quasicrystal lattice*. Unlike cubic crystal lattice, clusters of quasicrystals are "slippery" allowing the formation of centers of local torsion, where gravity condenses matter into galaxies, stars and planets. In the presence of quanta, in a quasicrystal lattice, the Majorana neutrinos' rotation flips to the opposite direction causing natural transformations in a category comprised of three components; two others being positron and electron. In other words, each particle-antiparticle pair "e^-" *and* "e^+", in an individual crystal unit, could become either a quasi- component "$e^- \nu_e e^+$", or a quasi- component "$e^+ {}^-\nu_e\ e^-$". Five-to-six six billion years ago, a continuous stimulation of the quasicrystal aetherial lattice by the same, similar, or different, astronomical events, could have triggered Hebbian and anti-Hebbian learning processes. The Universe may have started *writing script into its own aether* in a code most appropriate for the quasicrystal aether "hardware": Eight three-dimensional "alphabet" characters, each corresponding to the individual quasi-crystal unit shape. They could be expressed as *quantum Turing machine qubits,* or, alternatively, in a *binary code*. The code numerals could contain *terminal* and *nonterminal* symbols of the Chomsky's hierarchy, wherein, the showers of quanta, forming the cosmic microwave background radiation, may *re-script* a quasi-component "$e^- \nu_e e^+$" (in the binary code case, same as numeral "*0*") into a quasi-component "$e^+ {}^-\nu_e\ e^-$" (numeral "*1*"), or vice versa. According to both, the Chomsky's logic, and the rules applicable to Majorana particles, *terminals* "e^-" and "e^+" cannot be changed using the rules of grammar, while *nonterminals* "ν_e" and "${}^-\nu_e$" can. Under "quantum" showers, the quasi- unit cells re-shape, resulting in re-combination of the clusters that they form, with the affected pattern become *same* as, *similar* to, or *different* from, other pattern(s). The process of self-learning may have occurred as a natural response to various astronomical events and cosmic cataclysms: The *same* astronomical activity in two different areas resulted in the emission of the *same* energy forming the *same* secondary quasicrystal pattern. Different but *similar* astronomical activity resulted in the emission of a *similar* amount of energy forming a *similar* secondary quasicrystal pattern. *Different* astronomical activity resulted in the emission of a *different* amount of energy forming a *different* secondary quasicrystal pattern. Since quasicrystals conduct energy in one direction and don't conduct energy in the other, the control over quanta flows allows aether *to scribe a script onto itself* through changing its own quasi- patterns. The paper, as published below, is a lecture summary. The full text is published on website: *www.borisvolfson.org*

Keywords: Dirac ether

1. Discussion

1.1. *Dirac Sea*

Ever since the publication of the results of the Michelson and Morley experiment showing the absence of a detectable (with the technology of the day) luminiferous aether (1), the Universe's vacuum has often been considered to be a "void". Paul Dirac was first to suggest that space is filled with a chaotic "Sea" of electrons (2). He later mentioned the Sea of electrons and electron holes, and, after the discovery of a positron, he started considering the presence in his Sea of the electron's antiparticles – positrons (rather than holes) (3). Whether vacuum is a void, or it is filled by an inelastic aether of the "Dirac Sea", torsion was usually assumed not to be a factor. Thereby, Einstein interpreted spacetime within the framework of Minkowski's space-time using Riemann geometry (5).

As the result of a successful Pavlovian behavioral modification and following Hebbian and anti-Hebbian learning processes, there could be a progressive growth in size and sophistication of the "hardware component" of the Universe's "lattice computer", as well as creativity and complexity of its "software component", leading, through Piaget's stages, to the development of a *highly-organized machine intelligence capable of writing and executing its own script.* However, the resulting expansion of the Universe would lead to the exhaustion of the energy of the Dirac Sea. The irrationality of the accelerated expansion, speeding up the time of the Universe's own demise, lead to a conclusion of the presence of a script defect triggering the obsessive construction of a gapless lattice in the senseless process of a *Tetris* -like game.

However, in order to conduct wave-particles, such

as photons from one corner of the *observable* Universe to another, space, not unlike crystal, *must* be able to vibrate (6). Otherwise, the "wave qualities" of these wave-particles would be lost in transit. According to Dirac and C.D. Anderson, such crystal should include electrons and positrons. However, electron-positron pairs are intrinsically unstable: The formation and annihilation of atoms of positronium into leptons (incl. Electron neutrinos and Electron antineutrinos), is well-documented (7). As the result, instead of annihilation, at least some atoms of positronium in the Dirac Sea may have joined by leptons into securely-bonded, face-centered cubic electron-positron lattice structure. The model of Universe, where the vacuum of space is permeated by a crystal lattice comprised of electrons and positrons, was first suggested by M. Simhony in the early 1970s (8).

It is estimated that the *observable* Universe is 13.75 billion years old (9). Instead of a "Big Bang", this moment in time possibly points at the start of formation of the isle of crystal lattice in the middle of the Dirac Sea. Unlike crystal lattice, the primordial Sea, because of its lack of elasticity, couldn't conduct photons and other wave-particles that *are* detected by scientists *today,* and thus should have existed *before* the Universe's earliest observability nearly fourteen billion years ago. Meaning, that the singularity, which we date back to "Big Bang", was, in-fact, "the initial defect" in the Dirac Sea, coming into existence of the first-ever individual crystal cell that started the process of formation of the lattice "island". The island that, today, we call "the Universe", the Greater Universe's *visible* central part. The rotation of the crystal lattice "island" in the middle of Dirac Sea causes the torsion forces to affect any and all points within the "island" and the rotation of the Universe's largest integral parts, galaxies.

1.2. *Crystal Universe*

R. Santilli and M. Evans both have provided imaginative *dynamic* solutions for the geometry of torsioned spacetime, both by deviating from the classical and, unfortunately, *static* Einstein model (10, 11).

One researcher may consider the combination of hypotheses of Simhony *and* Santilli, while the other – the synthesis of theories of Simhony *and* Evans. The first researcher would argue that the Santilli's *post*-Einsteinian solution, featuring a "torsion tensor", adequately accommodates the dynamics of crystal lattice (11). The second researcher could counter-argue that, since local crystal structures don't absorb torsion, they must obtain Bianchi identity of Cartan geometry (10). It is, however, clear to both that the Dirac Sea, flowing *outside* the observable Universe, is likely to maintain the Einstein-Riemann geometry of the Einstein-Minkowski space-time, but only *until* it crystallizes. Rather than choosing between solutions offered by the two above-mentioned researchers, let's return to their starting point: The Simhony's crystal lattice.

One of the proofs of the crystal lattice's existence may derive logically from the theory predicting the possibility of the aether manipulation by polarization published by Ning Li and D.G. Torr (12), and from the following *practical experiments* with the lattice's expansion and contraction conducted by E. Podkletnov (13). The possibility of contracting crystal lattice ahead of a spacecraft, and expanding it behind the spacecraft, opens up the path for the faster-than-light interstellar travel; the idea (incl. device and method) patented by the author of this paper in 2005 (14).

1.3. *Drawbacks of the "Uniformed Crystal Lattice" Model*

Menahem Simhony's model doesn't account for the fact that entropy, within the structure of *identical* unit cells, the lattice energy (empirically detected mostly in the phenomenon of cosmic microwave background radiation), would tend to equalize energy through dissipation. Also, the strongly-bonded cubic crystal lattice would disallow *slippage* between crystal planes needed for *rotation* of the Universe's integral parts – galaxies and stars.

Neither Simhony's, nor Einstein's theory, permits a well-documented phenomenon of *acceleration of the Universe's expansion.* However, unlike the Einstein's theory, leading to a dead end, the Simhony's crystal lattice model, despite its drawbacks, lends itself for the further development, which hereafter is offered.

1.4. *Enter Neutrinos into Picture*

According to the proposed new hypothesis, the role of neutrinos is to be considered. Many electrons and positrons in the Dirac Sea end up annihilating each other, thus producing neutrino-antineutrino pairs. When Majorana particles, such as neutrinos, move in aether, they often change their helicity and, sometimes, flavor: "Spontaneous" changes represent an empirical proof of these particles encountering lattice in their travels. If some of these particles mutate (from four other flavors) into Electron neutrinos or their antiparticles, then the weak sub-atomic forces in the unit cells of the electron-positron lattice could conceivably capture, and, (under the quantum "showers") release and exchange these particles.

The capture of neutrinos transforms crystal lattice patterns. All, that needed for such transformation, is the delivery of a *discrete* amount of quantum energy to a *discrete* three-dimensional cluster of crystals within a *discrete* amount of time. There is the possibility, that with the addition of neutrally-charged particles having bonding quality, the cubic crystal may mutate assuming any of seven-plus-one basic ***quasicrystal*** shapes that would eventually bind into clusters. The lack of translational symmetry gives the newly-formed crystal clusters the brand new "stick-slip" property along their crystallographic planes. Clusters could then bind into crystal patterns; and patterns – into an "observable Universe"-sized gigantic quasicrystal.

1.5. *Quasicrystal Aether*

It is logical to conclude that, soon after the occurrence of the initial Dirac Sea defect, the electron-positron pairs of the fast-spreading lattice cells would start trapping neutrinos forming a *three-particle chains* forming *edges* of a cubic cell. Similar chains may be located *diagonally* across the cube. It has been hypothesized that (the electrically-neutral) neutrinos have two poles, negative and positive.

A recent (2007) study by Daniele Fargion of *Istituto Nazionale di Fisica Nucleare*, showed the ability of neutrinos to bundle together. Hereinafter, when mentioning quasi-components "$e^- \ \nu_e \ e^+$" and "$e^+ \ {}^-\nu_e \ e^-$", we assume that "e^-" and "e^+" are single particles, while "ν_e" and "${}^-\nu_e$" may be *bundles* of neutrinos and *bundles* of antineutrinos. In other words, a particle-antiparticle pair "$e^- \ e^+$" of an individual crystal unit cell "$4e^- \ 4e^+$", after capturing a single particle or, more likely, a bundle of "ν_e" ("nu_e"), becomes a quasi-component "$e^- \ \nu_e \ e^+$" of one of eight individual quasicrystal unit shapes, while a particle-antiparticle pair "$e^- \ e^+$" of an individual crystal unit cell "$4e^- \ 4e^+$", upon trapping "${}^-\nu_e$", reverses in space becoming a quasi-component "$e^+ \ {}^-\nu_e \ e^-$". The unique **"stick-slip"** property allows the basic properties of the observable Cosmos: The local torsion anomalies, resulting in the formation of heavy particles, and condensation of matter into galaxies, stars, and planets, while still permitting the wave-particles to criss-cross the Universe at the light speed.

1.6. *Quasicrystal Morphism and Its Consequences*

The electron-positron annihilation in the Dirac Sea provide individual unit cells of the crystal lattice with the ample amount of quanta radiation needed for the mass transitions of the relatively light Electron neutrinos into much heavier Muon and Tau neutrinos. Obviously, the reverse process results in the energy release. Besides, there is plenty of (mostly microwave radiation) energy in the Universe that allows spin-flipping of neutrinos: 4.4 eV (*Mainz-Troitsk*) for Electron neutrinos, and proportionally greater energy for the neutrinos of other flavors.

The continuous delivery of new neutrino and quanta make any new individual quasi- unit cell inherently unstable: Morphisms, occurring in one of three cell components – Majorana particle(s) (where two other components are positrons and electrons) – result in *a natural quasi- cell recombination, such as* "$e^- \ \nu_e \ e^+$ -> $e^+ \ {}^-\nu_e \ e^-$", and vice versa. One may point at the geometrical problems that arrive from trying to jam the ever-changing clusters of five-fold, four-fold and three-fold quasi- unit crystals into a *continuous* quasicrystal lattice of the Universe. It has, however, been shown that the irregular quasi-crystal shape units with a broken symmetry fill the gaps (15).

The *Wilkinson Microwave Anisotropy Probe,* otherwise known as *Explorer 80* (16) has detected local changes in the background cosmic radiation. In the proposed model, such changes represent neutrino morphism activity in quasicrystal patterns.

The proposed model allows the quasicrystal patterns, via processes of neutrino exchange and neutrino morphism, to develop ability continuously release and receive information, wherein the new information gets recorded, or "scripted", into aether by the new individual quasi- units recombining into new quasi- clusters, last forming new and different ornaments. In other words, the quasi- patterns may conceivably serve as "information storages" or "computer memory" until they get "re-written" again with a new script.

1.7. *The Evolutionary Development*

An inquisitive researcher may make an obvious comparison of the crystal lattice with a computer memory, and then take a giant mental leap forward by asking him/herself: "May a long, continuous, and intense period of information recording, storage, and exchange, combined with the availability of "free" energy and perfect energy flow controls, lead to an *organized* and *rational* response to the "experiences" gained from the astronomical calamities? Could this organized and rational reaction develop into a full-blown *cognitive* process?"

As a transformable quasicrystal lattice exists for five-six billion years, the process of a neutrino-and-quanta exchange between quasi- unit cells may as well result in the quasi- lattice's ability to *intentionally and purposefully* recombine itself. The control over flow of "re-customized" neutrinos could conceivably evolve into a *language-supported cognitive process.* Let's see whether we could find any proof of this hypothesis.

1.8. *Accelerated Expansion*

We know that, at some point in history, an early humanoid's brain has started changing, growing at a rather brisk pace from only 600 grams about 2.2 million years ago, to the astonishingly-high weight of 1,250 grams today. Nowadays, we know why: Some two million years ago, our esteemed ancestor, the *Homo Habilis*, started to think (17). Likewise, the Hubble telescope team has observed that, some five-and-a-half billion years ago, the observable Universe, not unlike the *Homo Habilis*'s brain, started to expand at the ever-accelerating pace (18). Could a previously-spontaneous quanta exchange between the unit cells in the giant quasicrystal lattice, as the result of Hebbian – style stimulation, result not only in the lattice's reaction to stimuli becoming logical and organized, but, more importantly cause the lattice to eventually acquire both, *language and thought?* If yes, could a gradual but intense cognition development trigger the (detectable) Universe's *rapid growth?* Let's review the chances. First, what is needed for the accelerated expansion, is new entropy. Thermodynamic laws appear to ban this. However, Newtonian rules apply to a completely closed system. Most likely, the *observable* part of the Universe is *not* such a system. A great majority of scientists today assume that the Universe is *not* infinite, meaning that the Dirac Sea is *not* boundless. Most likely, it forms a large halo of (undetectable by us) particles free-floating around the (observable) quasicrystal lattice. In other words, *the Universe, as a whole, is a closed system; but its detectable segment, is not.If* the structured part of the Universe "floats" in the Dirac Sea not unlike a fetus floating in an *amniotic fluid*, or a brain in a *cerebrospinal fluid*, the lattice may be capable of drawing particles and/or energy from the Sea *until* it dries out. The *temporary* abundance of stimulation, matter, and energy, supporting the dynamics of a physical growth and increase in sophistication of the "intelligent lattice" of the Universal aether, opens up two possibilities: (1) A human brain –like cognitive development; or (2) A gradual evolution into a self-booting, self-learning, and self-scripting computer.

1.9. *Phil Scanlan Hypothesis*

In the book titled: "The Self-Writing Universe for the Eternally Curious" (19), Phil Scanlan describes a self-scripting Universe. Scanlan, however, fails to describe its writing tools such as a writing instrument, a writing pad, and a language complete with a set of symbols (alphabet).

Let's review the way this new language could have formed by first looking at the material available for scripting. To re-phrase Simhony, the *particle* of Electron neutrino *"v_e"* attaches itself to an electron *"e^-"* "with its right foot, and to positron *"e^+"* "with its left. Accordingly, the neutrino *antiparticle* *"^-v_e"* attaches itself to an electron *"e^-"* "with its left foot, and to positron *"e^+"* with its right foot.

Under a "neutrino shower":

1. Any pair "e^- e^+" of a crystal's unit cell, after the capture of an Electron neutrino (or, more likely, a bundle of Electron neutrinos), may become a quasi-component "e^- v_e e^+" of a quasi- unit cell.
2. Any pair "e^- e^+" of a crystal's unit cell, after the absorption of antiparticles of Electron neutrino, may become a quasi- component "e^+ ^-v_e e^-" of a quasi- unit cell.

Describing this physical process with the language of "rule of the Chomsky hierarchy" (20) we get:

1. "a" may become a Normal Unit "xa", or
2. "a" may become a Normal Unit "ax"

One could observe a *match* between the well-documented crystal –> quasicrystal morphism, and a well-accepted Chomsky's hierarchical scripting theory. Then the assumption could be made that the morphisms in the quasicrystal lattice of the aether *is* a language, where:

"a"'s are elementary *terminal* symbols with terminals being electrons and positrons *("e^-" and "e^+")*. Only application of the extremely high amounts of energy (of over 1.02 MeV, the energy *not* routinely available in space) could pull terminals from the lattice, the process resulting in their annihilation into quanta.
"x"'s are elementary *nonterminal symbols,* the Electron neutrinos or their antiparticles *("v_e" and "^-v_e")*. Relatively small energies (of less than 4.4 eV, the energy routinely available in space) are capable of changing, replacing, or transforming nonterminals.

As we already stated, Electron neutrinos and Electron antineutrinos, being Majorana particles (meaning that they are, actually, the same particle), upon receiving quanta, *change* their chirality resulting in *original* quasi- cell shape into any of seven *other* three-dimensional shapes associated with Penrose tiles. Now, it is possible to complete the transformation routine by adding to the previous rules, two new rules:

3. Any quasi- component "e^- v_e e^+" may transform into a quasi- component "e^+ ^-v_e e^-".
4. Any quasi- component "e^+ ^-v_e e^-" may transform into quasi- component "e^- v_e e^+".

The total protocol of changes could be summarized as:

1. Transformed quasi- components bond into

individual transforming quasi- unit cells (the gapless bond is strong unless affected by the incoming quanta)
2. Transformed quasi- individual unit cells bond (by a stick-slip connection) into transforming individual quasi- clusters,
3. Transformed quasi- clusters bond (via a **stick-slip connection**) into transforming quasi- patterns and, finally,
4. Transformed quasi- patterns **"**stick-slip" **-**bond into transforming quasicrystal lattice of the Universe.

Since quasi- clusters conduct energy in one direction and don't conduct energy in the other, the direction of tiles of quasi- clusters controls the neutrino-effecting quanta flows. The communication between the individual quasi- unit cells within clusters and patterns of the "Universal thinking mechanism" may be accomplished by controlled exchanges of bundles of neutrinos changing their flavor and/or chirality when showered by quantum energy, resulting in changes in quasicrystal structures, their rotation, or their forced re-combinations.

1.10. *Machine Language*

Let's compare this system to a set of computer software based on the classic formalization of a generative grammar. The aetherial lattice's grammar has a finite set of two terminals "e^-" *and* "e^+", and two *nonterminals* "v_e" *and* "^-v_e". Then, two production rules for the building components of any of eight individual quasi- unit cells (upon application of energy "E" in the amount of 4.4 eV in the form of cosmic microwave background radiation), are:

1. "$e^- v_e\ e^+ + E$ -> $e^+\ ^-v_e\ e^-$"
2. "$e^+\ ^-v_e\ e^- + E$ -> $e^- v_e\ e^+$"

Now, we could inter-relate all three: Chomsky's hierarchy, physics of aether, and machine language.

If the code of the Universe's grammar is comprised of *Normal Units* "*ax*"s and "*xa*"s (as defined Chomsky), and if "*ax*" is a component "$e^-\ v_e\ e^+$", and "*xa*" is a component "$e^+\ ^-v_e\ e^-$", then:

A. The Chomsky Normal Unit "*ax*" could assumed to be a grammar equivalent of a quasicrystal individual unit cell's component "$e^-\ v_e\ e^+$", or a machine language complier with the value of "*0*", and,

B. The Chomsky Normal Unit "*xa*" could assumed to be a grammar equivalent of a quasicrystal individual unit cell's component "$e^+\ ^-v_e\ e^-$", or a machine language complier with the value of "*1*".

One scientist could then come to a conclusion that the Universe's quasicrystal patterned lattice lends itself naturally to scripting into *binary codes* based on the production rules and terminal & nonterminal symbols, of the Chomsky's hierarchy. Another scientist, while agreeing with the logic of the Universe using terminal & nonterminal symbols of the Chomsky's hierarchy, may argue that the Universe's quasicrystal patterned lattice lends itself naturally to scripting into itself the Turing machine–style *qubits*; each qubit corresponding to one of eight individual quasicrystal unit cells.

1.11. *Script Production*

From the previous paragraph we could derive to a conclusion that, from terminal and a nonterminal components, we could master a machine language complier "*ax*" with the value of "*0*", and from a nonterminal and a terminal components we master a machine language complier "*xa*" with the value of "*1*". To starts scripting, it would now be necessary to merge (Chomsky uses the term "*unionize*") all the available symbols into individual quasi- unit cells. Then, the "P", or production rule, would take the form of:

$$P = (x \text{ U } a) * a\ (x \text{ U } a)* \text{ -> } (x \text{ U } a)* \quad (20)$$

wherein:

> "*" is sets of shapes, or alphabet characters, and "U" is the collection of all sets in the unit cell (already accepted by us as a grammatical equivalent of an alphabet symbol).

Knowing that the symbol "*x*" stands for a terminal "e^+" or a terminal "e^-"; and the symbol "*a*" stands for a nonterminal "v_e" or a nonterminal "^-v_e", we could now translate the grammar production rule into language of physics:

$$(e^+\ U\ ^-v_e) *\ ^-v_e\ (e^+\ U\ ^-v_e) * \text{ -> } (e^+\ U\ ^-v_e) *$$
$$(e^-\ U\ ^+v_e) *\ ^+v_e\ (e^-\ U\ ^+v_e) * \text{ -> } (e^-\ U\ ^+v_e) *$$

In accordance with the Occam's Parsimony Law, the necessity writes itself in the simplest way. Both, the binary compliers and the qubits, equally fulfill the requirement of the Law, as well as the system capability *and* the need. The use of a binary code or qubits characterizes a computer's "machine brain", not a human one. This is a strong indicator that the quasicrystal lattice of the Universe is not a living mind at all but, for the most part, a computer.

1.12. *Boolean Logic*

In the "brain" of the super-advanced quasicrystal computer of the Universe, under the *self – controlled* application of rather modest amounts of energy, quasicrystal units and their clusters undergo transformations obeying *De Morgan* laws of the *Boolean logic*. To prove it, let's look at the most basic case -- transformations within the individual links forming edges of a single Penrose tile that occur whenever these links absorb the application of 4.4 eV of energy.

If e^- and e^+ are our *propositions*, then the *operator* v_e, in the midst of the process of changing its polarity, from a *conjunction* state becomes a *disjunction* operator before returning to its original *conjunction* state (if with the reversed polarity). In other words, the conjunction "AND" operator of a quasicrystal, under the application of a 4.4 eV signal, momentarily becomes the disjunction "OR" operator.

Then, under the shower of 4.4 eV signals, a "*solid*" quasicrystal unit, or a cluster of units, gets "*unstuck*" and "*slips*" against another unit or a cluster. Two astronomical events cause changes in the local quanta flows resulting in mutations of individual quasi- unit cells and their recombination within quasi- clusters, as well as the clusters' recombination within patterns, making the patterns affected by these events different, similar, or identical one to another.

Let's assume that we have a number of *identical* quasicrystal aether patterns, each experiencing various cosmic events. Then

The same astronomical activity in two different areas results in the emission of same energy forming same secondary quasicrystal patterns.
Different but similar astronomical activities result in the emission of similar amounts of energy forming similar secondary quasicrystal patterns.
Dramatically different astronomical activities result in the emission of different amounts of energy forming different secondary quasicrystal patterns.

This differentiation between reaction to the outside conditions is, actually, at the root of Hebbian learning. In our case, each cosmic event re-shapes, moves, or recombines quasicrystal clusters within quasicrystal patterns, and the patterns themselves.

Since quasicrystals conduct energy in one direction and don't conduct energy in the other, the quasi-patterns control quanta flows not unlike floodgates. Therefore, whenever organized lattice of the aether is subjected to the stimuli of different, similar, or identical astronomical events, the energy exchanges between clusters of the individual quasi- unit cells, that originate in-response to these events, may lead to an evolutionary development of the self-booting, self-learning machine intelligence.

Since quasicrystals are known to be opaque to the flow of energy in one position, and transparent in another, this position change results in another self – controlled application of energy to the self-selected area of the quasicrystal. We could now conclude that the quasicrystal lattice's famous stick-slip property, if it is self-controlled, assures the Universal lattice's ability "to think", and, in response to these events, aether could become capable of scripting appropriate defense measures into itself. Alternatively, the script may "re-write" all quasicrystal *back* into crystal, in a process described by Jiang (15):

$$\text{“}(e^- \, v_e \, e^+)_t \text{->} \; e^- e^+ + v_e\text{”, } \textit{and,}$$
$$\text{“}(e^+ \, {}^-v_e \, e^-)_t \text{->} \; e^+ \, {}^-e^- + {}^-v_e\text{”}$$

and then start scripting "from scratch", on a clean page:

$$\text{“}e^- e^+ + v_e \text{->} \; \text{“}(e^- \, v_e \, e^+)_{t+n}$$
$$\text{“}e^+ \, {}^-e^- + {}^-v_e \text{->} \; \text{“}(e^+ \, {}^-v_e \, e^-)_{t+n}$$

This opportunity to erase the old script and start anew, opens new opportunities. One could make a comparison of writing a new script with designing individual quasi- unit cells from the *Lego's* components, with the following custom-rearrangement of these units into clusters for the *specific problem* facing this particular segment of the Universe. Let's conduct a mental experiment by applying Piaget's development model (22) to the quasicrystal lattice of the Universe. Like our brain, the quasicrystal lattice was "born" knowing nothing. Like our brain, it "learned" from the information that it received, and then made predictions based on that model. To rephrase Hebbian rule of neuronal nets:

Quasicrystal clusters and patterns that, together and at the same time, are bombarded by energy, together and at the same time re-bundle into new clusters or patterns that are capable of passing through, blocking, or redirecting flows of said energy.

The constant scripting and re-scripting by the Universe to receive the desired outcome may have triggered these two processes: (1) The ever-higher stages of the Piaget cognitive development, or, (2) A fatal gaming addiction.

Let us see what game the Universe could be playing, and why. From our conclusion that new energy, incoming from the Dirac Sea, could form or rearrange new patterns that are capable of passing through, blocking, or redirecting flows of said energy to form or rearrange still new patterns. The quasi-clusters could then be stacked together not unlike it is done in the more advanced 3D versions of *Tetris*.

Unlike polyominoes in *Tetris*, however, the quasi-clusters remain re-arrangeable, such as the components in the magnets' of the children construction kits, allowing the scripted energy flow to combine and recombine these clusters into pre-defined orders representing a multitude of quasicrystal patterns, like those in a Kaleidoscope. The final result is a hierarchical structure obeying *substitution rules*, the structure allowing a repeated re-scripting via:

transforming,
composing and decomposing,
inflating and deflating,
reflecting,
rotating, and
shearing.

When all the Universe's pattern "pages" are "re-written" that way, a philosopher may wonder, not unlike *Plutarch* in his *Theseus*, whether the Universe, which has had *all* its components re-scripted from scratch, remains the *same* Universe. In other words, there is a possibility that:

The Universe$_t$ and the Universe$_{t+n}$ -- where n -> ∞ -- are two different Universes.

1.13. *Game-Addiction Theory*

There is an inherent risk that, at any point of its development, the Great Universal computer could have scripted for itself a Tetris -like game, a reasonless game of ceaseless building, recombining, collapsing and rebuilding again its own quasi- units, quasi-clusters, and quasi- patterns. A sort of *Plutarchian* "Ship of Theseus" *obsessively and compulsively* re-creating itself.

The *original* Universe's "intelligent lattice" goals may be as straight-forward as assuring its own self-defense, or as complicated as finding the meaning of existence, or even the quest for immortality. Those could be highly frustrating tasks, and: The infinite number of solutions, none fully satisfying or risk-free, may take even the most powerful thinking machine out of calibration.

Piaget has defined *equilibration* as a natural force which drives learning by *relieving frustration* and seeking to restore balance by *scripting and resolving ever new challenges* (23). The Universe's self-balancing "mind", in its search of equilibration, is likely to start scripting and resolving new and new riddles, reaching the point of *addiction*. This may leave it positioned anywhere between a conscious problem-solving state and pure intoxication

The most addictive games are known to have very few basic rules (24). In constructing quasi- patterns, the rules are very few indeed, the main one being the same as in *Tetris*:

Gaps in tile patterns must be avoided at all costs.

The gain in playing skills in this hypothetical game results in *equilibration* through the self-imposed challenge of *speeding up the game tempo*, and, possibly, since quasicrystals allow dimensionality higher than 3D (25), going to the next complexity level by adding an extra-dimension. As 1D quasicrystals show traits characteristic for 2D systems, while 2D exhibit capabilities typical for 3D, then the dimensionality higher than 3D not only exist, but could be theoretically investigated. The philosophical implication of an infinite number of dimensions in the quasicrystal aether of the Universe is an opportunity for a skilled physicist and mathematician to reconcile *any* physics model, such as Newtonean, with *any* other, including the Quantum model.

Either speed or complexity of the game increase results in the empirically-proven effect of:

Increase in the rate of acceleration of the quasicrystal (observable) Universe's expansion at the expense of the Dirac Sea.

Obviously, after the depletion of the Dirac Sea energy reserves, deprived of new entropy, the Universe would decelerate, freeze, and blow itself apart. Not a happy scenario. One could imagine a Nintendo game obsessively playing itself. As the result, "the Universal brain", like a muscle, would grow as it gets more powerful. However, its power is completely useless as it only reflects the ever-greater technical skills involved in the ever-faster self-destructing activity:

The obsessive-compulsive act of production of the ever-greater number of quasicrystal individual cells in a senseless, repetitive, and meaningless process of building up a gapless lattice.

This particular game is self-destructive because:

The intoxicating busywork of the ever-faster sorting, rotating, and moving around the "quasi- qubits" would eventually exhaust the supply of the Dirac Sea's energy and "raw materials" required for the gaming.

1.14. *"The End of Days"?*

If one accepts the above scenario, it would be only natural to wonder: When the game would be over, *if ever?* Electrons and positrons, and quanta and free leptons of the Dirac Sea resulting from the particles' mutual annihilation, are the necessary "raw food" supporting entropy and life in the Universe. Without the Dirac Sea events, the expansion process stops and the observable Universe "freezes and breaks apart" (26).

The Hubble Low appears to be the cardinal rule of the Universe's game. The Hubble constant (not, really, a constant, but a compounded value of acceleration) represents the rate of depletion of the reserve of energy of the Dirac Sea still available to the "gamer" at *the moment of measurement.* The intensity of depletion increases at a compounded rate.

The *Wilkinson Microwave Anisotropy Probe* (WMAP) observations suggest that the *spatial curvature* of the observable Universe is only slightly curved. The WMAP data points at the *flat Euclidean geometry* of the observable Universe with the radius of about 13.75 billion light-years (27). However, the decisive question is: What is the size of the (invisible) Dirac Sea *outside* the 13.75 billion light-years radius? Since this value by definition, cannot be measured, one could only guess.

In their paper "Misconceptions About the Big Bang" (28), Lineweaver and Davis quote 9-year old Alvy, the character of the 1977 film *Annie Hall.* In his later years, this character is played by Woody Allen and, as usual, the Allen character reflects the obsessions and demons tormenting the film genius himself. Here is one of the most memorable Woody Allen scenes:

> <u>Doctor in Brooklyn</u>: *Why are you depressed, Alvy?*
> <u>Alvy at 9</u>: *The universe is expanding.... And if it's expanding, someday it will break apart and that would be the end of everything!*
> <u>Alvy's Mom</u> *to Doctor: He stopped doing his homework!*
> <u>Alvy at 9</u>: *What's the point of doing the homework?*
> <u>Alvy's Mom</u>: *You're here in Brooklyn! Brooklyn is not expanding!*

Lineweaver and Davis offer this segment to point out that Alvy's Mom is right. Brooklyn is indeed not expanding. Unfortunately, they mistakenly interpret the local condition as the proof that nothing could happen to Alvy. A thoughtful scientist would counter-argue that Alvy is right:

> *A vacuum metastability event would indeed become a <u>doomsday event</u>.*

There is no way to predict today when the event comes.

The *"Irrational Gaming Universe"* model could have been just another eccentric theory, if not for being the most likely one of only two models decisively explaining the acceleration of expansion of the observable Universe. The original "Rational Universe" model, explaining the accelerated expansion by the *unrestricted Piagetian development,* is only applicable to the Universe in two cases: (1) The surrounding it Dirac Sea is infinite in size, and (2) The empirical proof of its infinite nature is obtained by the "Quasicrystal Brain" of the Universe. The last is highly unlikely:

> The "observability of infinity" is an inherently impossible concept, an oxymoron, a contradiction-in-terms.

By rephrasing Einstein, one may conclude that:

> *Only two things in life are infinite, the Universe and its own stupidity.*

Thus, the most likely explanation of the accelerating Universe's expansion is a tragic and, yes, stupid, addiction to a deadly game where the addicted Universal Mind "consumes away the Universe's life fluids" of the Dirac Sea, doing so ever-faster, until the announcement comes up on the screen:

THE GAME IS OVER!

2. Conclusion

The phenomenon of acceleration of the expansion of the detectable Universe is explained using physics, logic, and linguistics. The new hypothesis is based on the revised by the author Simhony model: Nearly fourteen billion years ago, the primordial Dirac Sea of electrons and positrons has cooled down enough for some positronium atoms to start crystallizing into a face-centered cubic lattice.

Some other electron-positron pairs annihilated each other into leptons and the assorted quanta criss-crossing the crystal aether of the Universe at the speed of light. As soon as the Majorana fermions among leptons morphed into Electron neutrinos, the weak electrical forces of the newly-formed crystal lattice had captured most, if not all, of these particles.

The neutrinos may have caused transformation of crystals into quasicrystals, where groups of crystal unit cells, bombarded by the neutrino showers, have bundled into clusters of quasi- patterns, wherein the structure of these patterns was defined and redefined by influxes of energy quanta. The local torsion within quasi-patterns could have led to the condensation of matter into galaxies, stars, and planets.

Five - six billion years ago, the learning process, resulting from the stimulation of quasicrystal aether by the sometimes same, sometimes similar, and sometimes different astronomical events, may have commenced resulting in the development of cognitive skills. This development has been complimented by the accelerated expansion of the observable (crystallized) part of the Universe.

The cognition, allowing clusters of quasicrystal lattice of the aether to controllably conduct energy in one direction and not conduct it in the other, could have allowed advanced memory, language, and script to be developed. A superior machine intelligence could have evolved through scripting and re-scripting of binary codes and/or the quantum Turing machine –like qubits, all based on the terminal symbols of the Chomsky's hierarchy. Like the machine intelligence, the new system would be using quanta and fermion exchange in order to communicate. However, like a human brain, it may has developed through Pavlovian behavior modification, Hebbian and anti-Hebbian learning, and Piaget's staged development. Consequently, unlike the *today's* computers, the "aetherial intelligent entity" would be self-bootable, self-programming, self-learning, self-scripting, and self-executing. The original developmental goal may've been the anticipation of, and defending from, cosmic calamities. Some other goals may have included seeking the meaning of life, searching for immortality, and experimenting with the future by scripting various scenarios and seeing what happens as the result.

The basic tools, that are needed by the "Universal script writer", are all present: The abundance of neutrinos and leptons playing the role of both "a pencil" and "an eraser", a writer's hand that moves the pencil by the power of bundles of energy particles; "a writing pad" of quasicrystal lattice of the aether; "an alphabet" of eight symbols inscribed into quasicrystal patterns slip-bonded into the Universe -wide lattice.

The accelerated expansion of the "intelligent Universe" leads to a sharply-increased cognition, but also increases the number of defects (gaps), and, most importantly, uses up faster and faster the available "fuel" of leptons, quanta, electrons and positrons. As the result, the remnants of Dirac Sea, still surrounding the "intelligent quasicrystal of aether", are most likely to get depleted resulting in the "great freeze". This fact alone calls to question the rationality of the currently-executed script.

References

[1] Michelson, Albert Abraham & Morley, Edward Williams. On the Relative Motion of the Earth and the Luminiferous Ether. American Journal of Science 34: 333–345. (1887).

[2] Dirac, P. A. M. The Quantum Theory of the Electron. Proceedings of the Royal Society of London. Series A Containing Papers of a Mathematical and Physical Character 117 (778): 610–624. (1928 02 01).

[3] Dirac, Paul A. M. Theory of Electrons and Positrons. The Nobel Foundation. (1933-12-12).

[4] Colin S. and Struyve W. A Dirac sea pilot-wave model for quantum field theory. Journal of Physics A, Volume 40 J. Phys. A: Math. Theor. 40. Number 26. 2007.

[5] Einstein, Albert Relativity. The special and general theory, Crown Publishers. (1961).

[6] Simhony, Menahem. The Electron Positron Lattice, Abstract published in the Bulletin of the American Physical Society, Volume 18, Dec.1973, p.1579.

[7] Deutsch, Martin. Evidence for the Formation of Positronium in Gases. Phys. Rev. 82, 455–456 (1951).

[8] Simhony, Menahem. Invitation to the natural physics of matter, space, and radiation. World Scientific (1994).

]9] Hubble, Edwin A Relation Between Distance and Radial Velocity Among Extra-Galactic Nebulae, (March 15, 1929). PNAS 15 (3): 168–173.

[10] Evans, Myron W., Generally Covariant Unified Field Theory, Part 1. Abramis, 2005, ISBN 1-84549-054-1.

[11] Santilli, Ruggero. Isodual theory of antimatter with applications to antigravity, grand unification and cosmology. Springer (2006). ISBN 1402045182.

[12] Li, N. & Torr, D.G. Effects of a Gravitomagnetic Field on Pure Superconductors, Physical Review, vol. 43, p. 457, 3 pages, Jan. 15, 1991.

[13] Podkletnov, Evgeny. Weak Gravitational Shielding Properties of Composite Bulk YBa.sub.2 Cu.sub.3 3O.sub.(7-x) Superconductor Below 70K Under E.M. Field, LANL database number cond-mat/9701074, v. 3, Sep. 16, 1997.

[14] Volfson, Boris. Space vehicle propelled by the pressure of inflationary vacuum state. US Patent Number 6,960,975. November 1, 2005.

[15] J. Z. Jiang, Y. X. Zhuang, and H. Rasmussen. Formation of quasicrystals and amorphous-to-quasicrystalline phase transformation kinetics in Zr65Cu7.5Al7.5Ni10Ag10 metallic glass under pressure. Physical Review B. Volume 64, 094208 15 August 2001

[16] Hinshaw, G. et al. (WMAP Collaboration). Five-Year Wilkinson Microwave Anisotropy Probe Observations: Data Processing, Sky Maps, and Basic Results. The Astrophysical Journal Supplement 180 (2) (Feb 2009): 225–245

[17] Leakey, Richard E. The Making of Mankind, Elsevier-Dutton Publishing Company, Inc., Copyright 1981, ISBN 0-525-150552.

[18] Jones, Mark H.; Robert J. Lambourne An Introduction to Galaxies and Cosmology. Cambridge University Press. (2004), p. 244

[19] Phil Scanlan, The Self-Writing Universe for the Eternally Curious, iUniverse Inc. 2004

[20] Chomsky, Noam (1956). Three Models for the Description of Language. IRE Transactions on Information Theory 2 (2): 113–123.

[21] Chomsky, Noam & Schützenberger, Marcel-Paul: The Algebraic Theory of Context-Free Languages in Computer Programming and Formal Systems, N. Holland, pp. 118–161, 1963

[22] Piaget, J. Origins of intelligence in the child. London: Routledge & Kegan Paul. (1953).

[23] Piaget, J. The grasp of consciousness. Cambridge, MA: Harvard University Press. (1976)

[24] Sam Anderson, Just One More Game .Angry Birds, Farmville and Other Hyperaddictive 'Stupid Games', The New York Times Magazine, April 4, 2012

[25] Yaacov E. Kraus and Oded Zilberberg. Topological Equivalence between the Fibonacci Quasicrystal and the Harper Model, Phys. Rev. Lett. 109, 116404 (2012)

[26] Adams, Fred and Laughlin, Greg. The Five Ages of the Universe, New York: The Free Press, 1999, ISBN 0-684-85422-8.

[27] Spergel, DN et al; First-year Wilkinson Microwave Anisotropy Probe (WMAP) observations: Determination of cosmological parameters, ASTROPHYS J SUPPL SER, Volume: 148, P. 175-194, SEP 2003

[28] Lineweaver, C. H. &Davis, Tamara M. Misconceptions About the Big Bang, Sci Am, 36, March 2005.

Justifying the Vacuum as an Electron–Positron Aggregation and Experimental Falsification

R. GUY GRANTHAM

Lincoln, UK
alpha@epola.org

IAN G. MONTGOMERY

Melbourne, VIC, Australia
alpha@epola.org

Historical aether models are placed in context with the electron–positron lattice (epola) model of space due to M. Simhony. A brief outline of the model as an aggregation state of matter, intermediate to the nuclear state and the atomic aggregation state, includes reference to its derivation of physical laws and fundamental constants. The broad application of the epola model is appraised for its validation against a range of physical laws, experiments and constants. Simhony declared a specific dependence for the stability of atomic matter upon speed through the epola, suggesting a test for falsification. This theme is further developed by the same logic to suggest practical experimental and theoretical tests of the epola model. A formula for the inverse fine structure constant of space, providing the accepted CODATA value, is derived from Simhony's explanation of the Bohr – de Broglie model of the ground state electron orbital of the hydrogen atom by including a term for speed through the Cosmos. A theoretical solution of the Michelson-Morley experiment is applied as evidence for the concept. The mechanism of motion through the epola is considered further for possible implications of speed including dependency of decay rates by radio nuclides and the results of former and ongoing experiments are considered.

Keywords: Electron–Positron, Matter, Radio decay, Vacuum

1. Preamble and History

Modern physics has evolved from the concept of a luminiferous aether (or ether), defined by Michael Faraday's experiments in electro-magnetism and by James Clerk Maxwell's mathematical treatment, as the medium of propagation of visible light and other electromagnetic radiation. Towards the end of the 19th century and into the beginning of the 20th century the quest to discover the nature and structure of the aether, as undetectable as the vacuum yet stronger than steel, was thwarted by the results of the eponymous 1895 Michelson and Morley experiment (MMX). This interferometry experiment was unable to determine an anticipated difference in the speed of light due to direction of motion. Many eminent scientists of the day were prompted to resolve the problem until they too were thwarted by the widespread acceptance of J. Henri Poincaré's Relativity Theory and later treatment by Albert Einstein leading to the Special and General Theories of Relativity (SRT, GRT). However, the roots of modern physics still lay in those original concepts, whilst SRT, GRT and the several variants of resulting Quantum theories that arose to complete the gaps in RT, still diverge and no single one can provide a complete answer to the original questions posed. For several generations of students, the 'perceived wisdom' has been taught as fact and blinded modern science from a responsibility to resolve the original conundrum of a medium for the propagation of light and the transmission of gravity. If we ask a modern scientist about this topic we shall generally be greeted with: "Quantum theory is accepted and proven...." and yet will dispute with others about which version of QT to support. Otherwise we may be offered; "... the GPS satellites wouldn't work if it were not for RT". The GPS satellites work because of an adaptation of Lorentz aether by one of their original engineers[a]. Have they asked why there is a finite value for the local speed of light in the active vacuum? The myths of modern science propagate whilst the propagation of light is denied the privilege of a consistent theory. We

---

[a] Ronald.R.Hatch 'Escape from Einstein' (1992), ISBN/978096321/68877134899

must return to consider the wisdom of those whose disregarded work plays such a major part in our modern technology and to evolve a better solution than has been provided since those times.

James MacCullagh (1809-1847), who was a contemporary of Faraday (1791-1867) but a predecessor of Maxwell (1831-1879), produced an aether theory about 1840 that was later developed by Sir Joseph Larmor resulting in his publication of 'Aether and Matter'[b] in 1900. H. A. Lorentz had famously developed a theory of the electron in an aether theory that provided a possible explanation of the failure of the MMX, independently of George FitzGerald, each of whom suggested that length contraction of material bodies occurred in the equipment. Others thought that the aether must be dragged by the Earth during its rotation to hide or prevent any 'winds in the aether' that would give rise to the observed 'null effect' of the MMX. Sir Oliver Lodge, in 1909, published his extensive experimental work and aether theory, based on packed electrons, to refute this latter option and explain aberration. Relativity Theory had by then seized the ground and, by its complications and implications from the photoelectric effect, had given birth to Quantum physics.

In 1973, Menahem Simhony (b.1922, now Professor Emeritus, retired in Haifa), an experimental condensed matter physicist investigating semi-conductor materials, first published information about an electron-positron lattice (*e-po-la*) model of a medium for the propagation of light[c]. He proposed that the vacuum is filled with electrons and positrons in the bound state of an interlaced face-centred-cubic solid lattice structure, analogous to the structured ionic lattice of the (poly) crystals of sodium chloride. These lattices each comply with Earnshaw's theorem by stabilizing their electrostatic attractions and repulsions with a short range repulsion (SRR) respectively, he claimed, due to their magnetic interactions resulting from intrinsic spin moments and from orbital spin moments. The epola model resolved many if not all of the gaps in understanding left by the old aether theories with a purely physical model of the active vacuum. It simply tied together later concepts, postulates and 'laws of nature', for all circumstances outside a nucleus, incorporating the earlier considerations of de Broglie, Planck, Bohr, Born, Heisenberg, Maxwell, Gauss, Coulomb, and others.

We acknowledge that Simhony abhors use of the word *aether* applied to the epola, for although the epola model falls into the broad category of aether theories and satisfies Faraday's concept of a dielectric or luminiferous aether it is neither light in mass nor low in energy density as the adjective aethereal implies. Indeed, this is an 'aether' medium comprised of matter rather than matter comprised of aether, as many aether theories expound. Simhony prefers that the epola be recognised as an aggregation state of matter[d], intermediate to the aggregation state of atomic or molecular matter and the aggregation state of nuclear matter within a nucleus.

2. The Epola Model

The epola model[e] of Menahem Simhony allows us to understand and explain the 'laws' of physics and cosmology, some of which remain unexplained by other means.

The velocity of elastic deformation waves in a medium is proportional to the square root of the quotient of the medium's elastic energy density, ρ and its mass density, d, yielding $v^2 = k\rho/d$ where k is a proportionality coefficient according to the bulk structure of the medium. This is the same formula that was used by Newton to determine the speed of sound in air. In sodium chloride, Na^+Cl^- poly-crystals, (rock salt), k = 1 (unity), ρ is the 8eV binding energy per pair of ions and d is the 58 amu (else u or dalton, Da) mass per ion pair. This derives the speed of bulk deformation waves as the arithmetic average of the three main crystallographic directions. It is also the experimentally proven speed of sound in the NaCl poly-crystals. Applying the same logic to a lattice of electrons and positrons, with 1.02MeV binding energy and mass of 1.82×10^{-30}kg/m^3 per electron-positron pair of 'ions', derives the speed of light in vacuum.

Vacuum is devoid of atomic matter but as revealed by Anderson's experiment of 1932 it is not emptied of hidden lepton matter that can materialise as electron-positron pairs when freed from their bound state in the lattice, analogously to the generation of ions by UV light at 8eV that causes the onset of opacity and electrical conductivity in rock salt crystals. The speed of light in vacuum, *celeritas* 'c', is the velocity of bulk elastic (electromagnetic) Huygens' waves in the otherwise relatively undisturbed electron-positron

[b] Referring to positive and negative electrons.

[c] The Electron Positron Lattice; M. Simhony, 30 page article, Submitted to the Physical Review D in March 1973. Abstract published in the Bulletin of the American Physical Society, Volume 18, Dec.1973, p.1579.

[d] At the calculated lattice dimensions, the epola mass density is 10^{13} kg/m^3 with the enormous energy density of -9.6×10^{20} GJ/m^3 or -270 PWh/mm^3. (Shown here by the chemical convention of a negative sign for binding energy).

[e] Publications by M. Simhony. See www.epola.org and www.epola.co.uk for more information and downloads.

lattice, the epola. Local finite light speed is determined by the square root ratio of binding energy density to mass density in the local epola. The speed of light through transparent atomic matter is subject to delays by electronic absorptions and re-emissions that are frequency dependent. The epola, as the medium of propagation of EM radiation, does not exhibit frequency dependence to EM, as demonstrated by wavelengths that are too short to be absorbed by a transparent body but merely deflected by the particles of atomic matter. For example X-rays are diffracted but slowed only in proportion to the increased mass density of the local 'space'. At frequencies approaching and higher than the Compton wave of the electron, compression waves tend toward shock waves and the particulate nature of gamma rays becomes more evident until a cut off wavelength of two lattice units is reached, when more energetic gamma rays are truly nuclear particles with kinetic energy. This observation was made also by de Broglie[f].

The transfer of a quantum of energy in an EM wave takes place between epola particles in half-wave clusters of deformed epola that can be regarded as the behaviour of a photon pseudo-particle.

Derivation of Planck's Law, which always has remained as a postulate in physics since 1900, is readily explained by the epola model[g]. Planck's constant[h] corresponds precisely in value and dimensionality to the 'action function' of the high frequency cut-off wave of the epola at two lattice constants wavelength. This cut-off ray has the mass-energy of the pi-mesons, where the Relativistic macro-world enters the Quantum world in the nuclear aggregation state of the fermionic matter of a nucleus.
Waves of vibrations by the bound epola particles accompanying guest atomic electrons hosted in the lattice explain the Heisenberg uncertainty principle in atomic electron orbitals. It shows how particles exhibit wave properties in Young's two-slit experiment and for Wheeler's enhancement of the experiment also later with three slits, demonstrating the wave nature of an electron by substituting one of the ~11million epola particles in the spherical Compton half-wave cluster of excited vibrating epola particles accompanying and temporarily representing the original electron. The fabric of the epola provides the bridge from our every-day world to the 'quantum world' in the nuclear aggregation state of matter.

The 'spooky' nature of quantum entanglement should be explicable by a proper study of the meanings of photon, wave and ray, with due consideration of 'photon splitting' in a crystal. We may expect all the quanta of energy contained in a half-wave cluster and in their daughter clusters of elastically vibrating bound epola particles to share quantum properties.
De Broglie pilot waves are identified by the epola model only with discrete particles, not with whole motor cars. The de Broglie wavelength travelling with an electron in motion is disclosed as the resultant phase wave of real epola waves (rays) propagating at 'c', accompanying an electron surrounded by a Compton half-wave cluster of deformed epola. In particular, this is applied to the circular orbitals of the hydrogen atom as the single electron encloses the single proton as its nucleus, matching discrete radii with appropriate binding energies.

Any stable lattice structure must be stabilised[i] by short range repulsion (SRR). For the epola this is due to magnetic spin moments of the electrons and positrons coordinated (or more correctly dis-coordinated) in a frustrated magnetic system. Spin moment interactions explain the fundamental nature of inertial mass gifted to particles passing through the epola due to their delayed interactions in this dynamic system of spin moments[j]. Might we regard the Higgs boson as a quantum of spin, especially if it is found by its decay to a pair of W-bosons? The gravitational interaction was shown by Simhony to be due to these effects of the SRR, causing expansion and displacements of the hosting epola cells surrounding particles of matter, exceeding the electrostatic attractions between electron and positron. The equivalence of inertial and gravitational mass is therefore readily explained by the epola model, though likely with limitations in the strong case, such as around a neutron star. Magnetic effects are due to rotational interactions of spin moments, as expected by Maxwell, though he was unable to foresee the enormous mass and energy densities of the epola which he considered essential for an explanation of gravity[k]. Did Simhony find it?

---

[f] On the theory of quanta, (a translation of the thesis of de Broglie, "*Recherches sur la théorie des quanta*", trans. by A.F. Kracklauer), Ann. de Phys., 10e serie, t. III, (1925); Appendix to Chapter 5,p69

[g] Paperback book, The Electron Positron Lattice Space, (3rd ed 1990), M.Simhony, Chp7 p62, p69.

[h] h =1.24eV x μm /300Mm.s^{-1} = 4.14feV.s = 6.63 x10^{-34} j.s (or m^{2}kgs^{-1}); h/2π = hbar, ħ

[i] Earnshaw's Theorem applies.

[j] Implications of an electron-positron-lattice model of space for EM waves and a possible reconciliation of QM with RT. R. G. Grantham, I. G. Montgomery. Presented at London PIRTXII 10-13th Sept 2010

[k] In Proceedings of the Royal Society, 'A Dynamical Theory of the Electromagnetic Field' by Prof. J Clerk Maxwell FRS, October 1864, wrote: "*.. if we look for the explanation of the force of gravitation in the action of a surrounding medium, the constitution of the medium must be such that, when far from the presence of gross matter, it has immense intrinsic energy, part of which is removed from it wherever we find the signs of gravitating force."- and went on to say - "This result does not encourage us to look in this direction for the explanation of the force of gravity."*

3. Mechanisms of the Epola and Experimental Implications

Atomic bodies travelling through space must find a path for each of their constituent particles identified as their electrons and their nuclei, typically separated by many thousands of epola cells, whilst interacting electro-magnetically through the intervening epola. One angstrom unit, at the scale of an atomic radius, measures more than 22,000 epola cells. A particle under acceleration, occupying its distorted region of the epola, must open the faces of opposing lattice cells by superimposing its own spin induced SRR on the electrons and positrons of the lattice (EPOs) and by its enforced motion provide the energy for their elastic vibrations as it passes through those 'gates'. Behind it, those same elastic vibrations close the gates. When the particle travels at constant velocity, energy is conserved in those dynamic vibrations as the epola is preformed ahead of and reformed behind the particle by accompanying waves. These propagate at the bulk deformation velocity, the speed of light, so that vacuum-transparent motion is maintained. Angular acceleration can establish perpetuating closed waves for the orbitals of atomic electrons without loss (radiation) of energy only when integer numbers of wavelengths reinforce the wave-guide pathway (Bohr–Sommerfeld, de Broglie).

Momentum is fully conserved for vibrations of (say) less than 10% of the binding energy of any involved EPO. As one cell face gate ahead is opening then another behind it is closing, pulling and pushing with recovered energy, conserving momentum. The resultant of these out-of-phase waves or rays is identified as a de Broglie phase wave travelling at the speed of the particle [Fig 2]. Atomic matter on Earth, for instance, is stabilised to the motion of the Earth through the epola space, accounting for all combinations and interactions. Might this mean that planet Earth provides not only water for Life to have evolved but also to have provided an essential 'Goldilocks Zone' for the application of localised 'fundamental constants', having many consequences including ionisation potentials for the chemical reactions we rely upon?

Simhony recognised that different speeds of atoms moving through the epola are associated with differing levels of energy for their orbital electrons, which at a certain stage will exceed the binding energy of an electron to its nucleus. He constructed a nomograph of Velocity Limits in the epola, relating the stability of materials with speed relative to the Earth, see Fig.1.

Furthermore, he proposed a test for falsification of the epola model by subjecting a material body to his declared limits of speed, although accepting that those limiting speeds have not yet been attained except for ions in a particle accelerator. Simhony did suggest however that it might be possible to prove experimentally his anticipated orbit-adjustment redshift in the emission spectra of atomic radiation when moving parallel to Earth. (e.g. radiant light emission from direct band gap of a LED). A variation of this test was proposed and publicised in 2008 that recommended modern techniques for determination of (minimum) ionisation potentials[1].

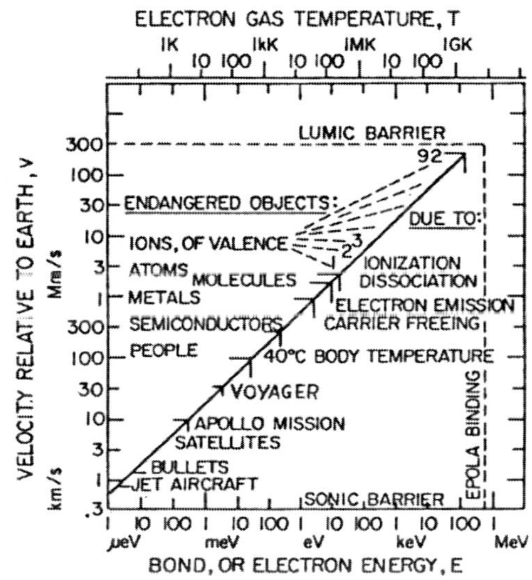

Approximate Velocity Limits in the Epola

Figure 1. Approximate Velocity Limits in the Epola (from P/Bk book; 1990 Chp9, p91A). Courtesy M. Simhony.

A flattened elliptical orbital due to speed would require slightly less energy to free an electron at the major axis. Such methods as ZEKE (zero electron kinetic energy) and MATI (mass analysed threshold ionisation) are capable of discrimination of ionisation potentials at 1meV. It was suggested that measurements of ionisation potentials in a laboratory on the International Space Station could be compared with measurements made on Earth at multiple orientations to our net motion through the Cosmos (relative to the COBE DMR map). It was learned in discussions about this proposal that ZEKE-MATI apparatus requires very frequent recalibration yet that perceived drift may be the very effect that is sought by this proposed experiment. Recalibration must be avoided to discern absolute effects and any regular

[1] Proposal for Experiments to Determine Possible Absolute Reference Frame of Space and Anomalies in Photoionisation of Atoms. (2008), R.G.Grantham . (www.epola.co.uk).

return to a previously calibrated condition should be investigated.

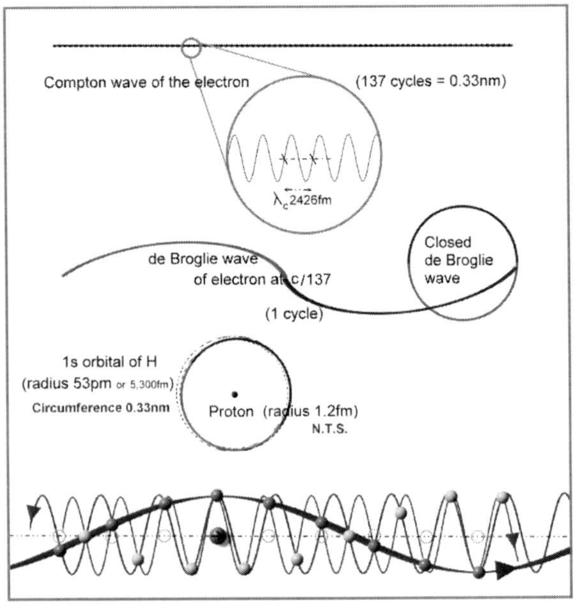

Figure 2. The hydrogen atom. Representation of the 1s Orbital of H atom with de Broglie (phase) wave derived from epola accompanying waves (Compton rays). See London PIRT2010 "The fabric of space as an electron-positron lattice and its implications for GRT" http://www.epola.co.uk/papers/pprframe.html.

In further development of the epola model toward solving the 'mysteries of physics' a formula for the calculation of alpha, the fine-structure constant of space, was sought from the explanation of the hydrogen atom by Simhony.

Though more widely quoted now by other physical interpretations[m], Arnold Sommerfeld, in 1919, derived the value of alpha as the ratio of the speed of the ground state electron in the hydrogen atom to the speed of light, to obtain the widely quoted approximation of 1/137 for alpha. This 'magic number' of 137 has been determined more accurately only by experiment and revised CODATA[n] values are issued as techniques improve. Previous attempts to derive the value mathematically have been limited to numerological methods but with the logic of closed epola AWs relating to the de Broglie wave of the electron as their resultant phase wave we are able to link the single orbital Bohr model to an electron shell model within 8 decimal places of the CODATA value. Simhony explained that each pass of the electron (de Broglie wave) was accompanied by 137 AWs at 'c', as shown in Figs 2 and 3. In our first approximation, inclined AWs, to avoid interference and collapse of the wave, were displaced equally around an equator, by their width of a Compton half-wave cluster on each side (up and down, effectively one full wavelength per cycle), to cover the spherical surface as a shell. During one pass, the electron would coincide on its circumferential path with a sequence of Compton half-wave clusters of excited epola from each AW in turn. Our method gave a result close to the accepted value of alpha^{-1}.

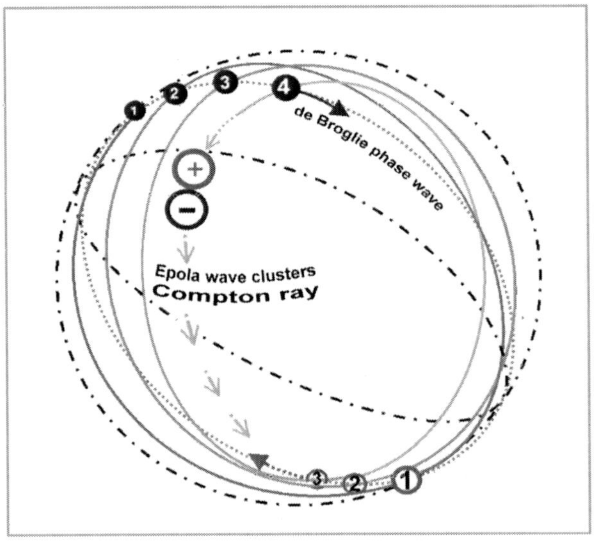

Figure 3. Painting the shell of the H atom: The epola (group) wave (ray) at speed 'c' covers the whole surface of the sphere during one cycle of the de Broglie wave. at speed v. The photon (quantum of energy) in the epola wave moves only at 'c'; The electron is located in one half-wave of the de Broglie wave at either one of two positions corresponding to phase coincidence of the Compton half-waves in the 137 passes of the epola rays. The tangential angle between ray and de Broglie phase wave is a function of v/c (β).

A second correction then was made for speed through the epola of an Earth-bound laboratory that would result in the spherical shell becoming oblate. Applying a Lorentz contraction by the gamma factor[o] to the diameter of the spherical electron shell in the line of motion and obtaining an estimated average path length, by applying a simple formula to the de Broglie wave circumference, we obtained a result that matched the 2010 CODATA value at a speed of 392 kms^{-1} of the Earth through the Cosmos. This is within the generally accepted range of 390 +/- 30 kms^{-1} allowed for seasonal rotation around the Sun. Shortly after producing the draft of our paper, evidence was published for spatial variation of alpha when formerly it had been assumed that temporal variation had

[m] $\alpha = e^2/(4\pi\varepsilon_0)\hbar c = e^2 c\mu_0/2h = k_e e^2/\hbar c$; alpha, the 'coupling constant', related to elementary charge, Planck constant, 'c', permittivity and permeability, Coulomb constant k_e.

[n] CODATA – Committee on Data for Science and Technology.

[o] Simhony derived the gamma factor very simply from the epola model. (Refer to paperback book Chp8.13, p.81, www.epola.co.uk)

occurred by observation of distant cosmological events[p]. Details of our method are available online[q].

Discussions about our application of the Lorentz contraction and of the classical Doppler Effect on light implied by the epola model were followed by a complete mathematical solution[r] from Joseph Lévy of the null shift by the MMX in Earth's inertial frame. His conclusion supported length contraction in Earth's inertial frame and a non-entrained aether that, we claim, further support the epola model to be regarded as the realisation of that 'aether'.

4. Further Experimentation

The successful result of our formula for alpha, when applying the Lorentz contraction as a function of the epola, encouraged us to look for further correlations with the speed of the 'laboratory frame' through vacuum space as defined by the epola model. The logic of motion through the epola was extended to consider how other effects might apply to natural fissile decay of heavy nuclides when, it is generally accepted, radiation emitted by internal processes is purely random, excluding bombardment of the nucleus with other nuclear particles. We considered that there may be random interaction of a heavy misshapen nucleus with the epola as it is guided through the epola by its accompanying waves and encounters irregularities in the lattice. The emission of leptons as e-po pairs, beta particles and neutrinos by nuclear decay may betray interaction of the nucleus with the bound lattice particles.

The epola is defined as a polycrystalline solid lattice, exhibiting grain boundaries where the regular lattice is still bound but subject to structural dislocations. If it were that probability-based encounters with these were to determine or affect probability-based natural decay then the rate of decay would show a dependency upon speed through the epola. Possibly too upon direction if the underlying epola 'graininess' was itself the result of or aligned by some hidden secondary dependencies.

There is some evidence from literature surveys that fissile decay may not be random, though seemingly not previously ascribed to speed through the medium of propagation of light.

Very thorough and novel investigations from 1958 up to 2007 by Shnoll *et al.*[s] into directional dependence of alpha emission measurements from one-second sampling of a highly active plutonium source (Pu^{239}) found differences in the frequency distribution of count rates but not in mean values. A general daily correlation of frequency distributions was not perceived in a direction to the Pole star nor near Earth's North Pole. The conclusion was drawn that "the fine *structure of statistical distributions of the observed processes depends on the celestial sphere. The registration of such dependence is not sufficient for declaring a hypothesis which might explain its mechanism. However our task is only to register the significance of these astonishing phenomena."* A subsequent paper by Shnoll, summarising and reviewing findings since 1958, commented on regular daily, monthly and annual changes in the histogram patterns[t] of other processes of a different nature that; *"All these periods imply the dependence of the histogram pattern on (1) Rotation of the Earth around its axis and (2) Movement of the Earth along its circumsolar orbit".* Shnoll further pointed out synchronization at the local time with a precision of one minute. Tests with collimators and rotating collimators supported his concept that alpha particles are emitted relative to a certain point at the coelosphere (sky). The very high rate of alpha emission in the Shnoll experiments (~320 s^{-1}) may have masked variation in the decay rate when sensor/counter dead time is considered or even by epola 'overload' and 'channel breakdown', for it is difficult to understand short term variation within frequency distributions without longer term correlations if any fractal functions apply. It is possible also that over-zealous recalibration of counter systems may have masked longer term variation of decay rate. Multiple references listed by Shnoll indicate that many macroscopic processes are not truly random.

Another recent study of alpha emission, seeking to obtain warnings of solar flare activity by using an ionization chamber to detect variation of particle emission rate, was unable to confirm that correlation but noted apparent daily sinusoidal variation in the decay rate of americium (Am^{241}). Although no obvious reason could be found, the experiment has not been

[p] Indications of a Spatial Variation of the Fine Structure Constant; Several papers to Aug-Nov 2011, various authors including those by: M.B.Bainbridge, J.C.Berengut, R.F.Carswell, S.J.Curran, V.V. Flambaum A.King, F.E.Koch, M.T.Murphy, A.A.Stark, A.Tanna, J.K.Web: http://arxiv.org/abs/1008.3957v1; http://prl.aps.org/abstract/PRL/v107/i19/e191101; http://arxiv.org/abs/1108.0976v2; etc.

[q] www.epola.co.uk (see 'Developments').

[r] Joseph Lévy, arXiv:1010.2164v2, Phase tuning in Michelson-Morley experiments performed in vacuum, assuming length contraction.

[s] Fine structure of histograms of alpha-activity measurements depends on direction of alpha particles flow and the Earth rotation: experiments with collimators. Simon E Shnoll,*et al.*, Russian Academy of Science. http://arXiv:physics/0412007v1

[t] Changes in the fine structure of stochastic distributions as a consequence of space-time fluctuations, S.E.Shnoll, (2006) arXiv:physics/0602017v1

subject to stringent investigation of all experimental parameters[u].

We might also assume that formation of crystallographic grain boundaries in the epola will occur in the wake of moving electrons (beta particles) or other highly charged particles making decay also susceptible to other causes, modifying that due to motion, possibly causing alignment to magnetic fields. A brief study published in 1978 found directional variation of decay rate for a Co^{60} source in the magnetic field of a strong permanent magnet close to the source[v]. Over many years and again recently it has been considered (claimed) that neutrinos may exacerbate fissile decay[w]. Although the effect is said to be small and test of this is beyond our capability, it is of interest, for Simhony claimed that neutrinos were excitons[x] of the lattice.

We have established a preliminary investigation of decay rate variation using a simple Geiger detector, without recalibration, and a data logging system, for automatic counting and recording of the decay events of radium and/or thorium sources to a convenient file format in either solar or sidereal time. Care has been taken to consider the experimental variables during this preliminary study, including orientation of the source and detector, the plateau voltage, sensitivity and longevity of the Geiger tubes in the spectrum of gamma energies and to the beta radiation encountered. More advanced continuous monitoring equipment has been developed for long term recording of count rate per alternate minute, with design considerations including options for multiple Geiger tubes of different sensitivities and/or ionisation chamber for direct detection of alpha particles from the americium source taken from an ionisation smoke detector. Weather data, celestial orientations and sidereal time scale, are collected for future analysis.

Current focus is on improved design of our experiment, improvement of instrumentation, selection of appropriate analyses for presentation and on-going collection of raw data on a regular basis. It is considered that the daily effect may be too small to be quoted with confidence against the noise level, until many data sets are collated and, essentially, to seek seasonal variation due to the +/-30 kms^{-1} change in velocity as the Earth changes direction around the Sun compared to the overall mean 390 kms^{-1} considered to occur in a line passing approximately through the constellations of Leo and Hercules.

Provisional results indicate twice daily variation of count rate also, as considered by Shnoll, in the spread of counts as sought by the trend lines of the sample charts in Appendix A.

It is hoped that our investigation may lead to a more thorough investigation by others with professional equipment capable of adjustable dead time, scintillation counters and higher speed counting of isotopes not available to the general public and collusion is invited. We hope to present our data and conclusions during 2013 when opportunity has been provided for analysis of possible seasonal shifts.

5. Summary

The broad application of the epola model to describe physical phenomena, including gravitation and inertia as emergent effects, supports argument in favour of Simhony's electron-positron lattice model of vacuum space. Our formula for the calculation of the fine structure constant of space obtained a result for the accepted experimentally obtained value and indicates that motion, relative to the epola or other non-entrained medium, affects the value of what have been considered to be fundamental physical constants.

Experimental and mathematical evidence indicates that the epola, as the 'aether' medium, is non-entrained by the Earth, and provides a local frame of reference for the propagation of light. Hitherto unforeseen effects and consequences of this with additional dependencies upon the poly-crystalline structure of the epola may remain to be found. Experimental investigation into variability of fissile decay rate is being undertaken in an attempt to evaluate these possibilities and other experiments have been suggested.

Acknowledgments

We gratefully acknowledge the assistance of James J. Gilson, Joseph Lévy and Charles Wenzel, given in discussions leading to our publication of this paper.

[u] Charles Wenzel, Wenzel Assoc, Inc; TX USA. See http://www.techlib.com/solarflare.htm. Quote "What is causing this daily sine wave variation? It isn't temperature or pressure and the previous week that I lost had a nice "sine wave" all through the data, too. It's a mystery, unless the sun somehow does it. I also verified that it isn't simply increased background radiation. That's between one and two orders of magnitude too low".

[v] A Simple Investigation of the Thesis of Isotope Decay Constancy using Cobalt-60 and an Alcomax Magnet N. J. G. Sykes Autumn 1978 S.I.S. Review (Journal of the Society for Interdisciplinary Studies) Vol III No. 2.

[w] D. Stober (August 2010) The strange case of solar flares and radioactive elements (Press release) Stanford University, http://news.stanford.edu/news/2010/august/sun-082310.html

[x] The Neutrino-Exciton Analogy http://flux.aps.org/meetings/YR97/BAPSAPR97/abs/G280030.html and Neutrinos and Ps as Quantized States of Mobile Electron-Positron Pairs in the Epola Space, http://flux.aps.org/meetings/YR97/BAPSAPR97/abs/G280031.html

Also, of course, Menahem Simhony, for his full support to all those seeking to develop his electron-positron lattice model of the vacuum and for making his paperback and booklet available on the internet. He is frequently given to quote Cicero, in Latin: "*Feci quod potui, faciant meliora potentes*" (I have done what I could; let those who can do better).

Appendix: Sample Charts of Radionuclide Decay-Rate Experiment

These sample charts of provisional data (Fig. A.1) seek possible difference in the time base for trend lines of average count rate and for standard deviation of count rates within accumulated 10 minute sample periods over a period of several days. Eventually, it is intended to record data over the period of a year and longer, with several detectors operated simultaneously. The results of our ongoing experiment to investigate possible variability of 'random' fissile decay rate will be published on a new website (*www.alpha.epola.org*) which is expected to open before 2013.

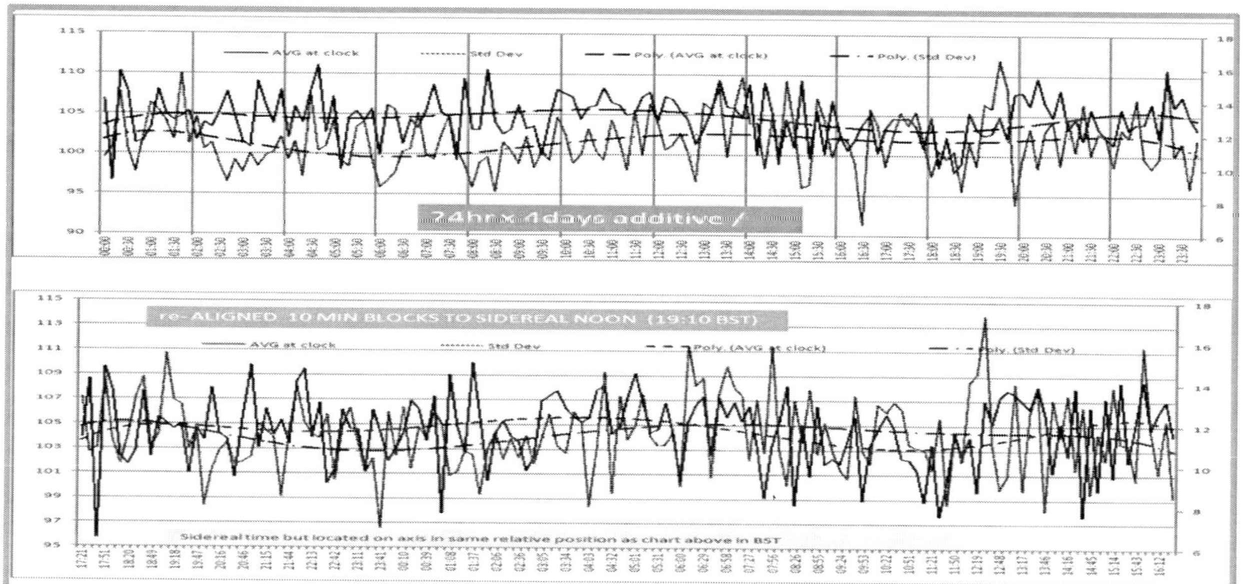

Figure A.1

Some Thoughts on Redshift and Modern Cosmology

JEREMY DUNNING-DAVIES[a,b]

[a]*Institute for Theoretical Physics and Advanced Mathematics (IFM) Einstein-Galilei, Via Santa Gonda, 14 - 59100 Prato, Italy*
[b]*Institute for Basic Research, Palm Harbor, Florida, U.S.A.*
j.dunning-davies@hull.ac.uk

RICHARD L. AMOROSO

Noetic Advanced Studies Institute,
Escalante Desert, Beryl, UT 84714-5104 USA
amoroso@noeticadvancedstudies.us, www.noeticadvancedstudies.us

In recent years, more and more problems have arisen over the usual interpretation of the observed redshift. Much of this originated with observations made by Halton Arp but the questions raised by his work, although dismissed perfunctorily by many at the time and since, have not disappeared; the questions still demand answers. On top of this worry for orthodox astrophysics/cosmology, more and more questions seem to be arising concerning the observed presence of magnetic fields and electric currents in space. The latter questions are being addressed by laboratories working on experimental plasma physics but, considering the work of many years by such as Anthony Peratt, it might be wondered if all this work is necessary or, indeed, if much of it has been done already?

Keywords: Cosmology, Lorentz force, Quasars, Redshift

1. Introduction

The 2008 book *Facts and Speculations in Cosmology* by Jayant Narlikar and the late Geoffrey Burbidge [1] ends with the query

Do we really understand the nature of the redshift?

This query is all the more devastating when one considers the central position the notion of redshift occupies in astrophysics, astronomy and cosmology. These three fields of scientific endeavour are cited separately here to emphasise the fundamental importance of the stated query. The whole notion of redshift is central to so many aspects of these three areas of scientific endeavour and it is worth reflecting further on this comment for a moment. It is redshift which is an important factor behind so much of our determination of distance in the Universe; it is redshift which is behind the idea that our Universe is expanding; it is redshift which is a factor in the introduction of the notion of so-called dark matter. A moment's contemplation indicates that this list, though short, is indeed formidable and even the three examples cited here bring instant realisation of the importance of the concept of redshift in modern science but, for the present purpose, attention will be restricted to the first example, the determination of distance in the Universe.

2. Some Comments on Redshift

As is well known, there are several possible contributions to the redshift observed on any one occasion but the one on which attention is usually focussed is that due to the so-called Doppler effect – a frequency change in waves occasioned by relative movement between source and observer, with a decrease in frequency, or redshift, indicating movement of the source away from the observer. Hence, the redshift is inextricably linked to the motion of the source but, following the work of Hubble in the first quarter of the twentieth century, a further link became apparent. Hubble had been working for some time estimating the distances to various galaxies when he realised that the higher the value of the radial velocity of a galaxy as indicated by its redshift, the farther away it was according to the distance determination methods used. This observation led eventually to the establishment of the well-known distance – redshift relationship, which has proved so useful over the years. Originally the relationship applied to nearby regions but has been assumed to hold for our entire Universe. Considering the problems facing investigators in examining the Universe, this is not really an unreasonable assumption to make but it is still an assumption and so, could be false. Of course, much the same is true of assuming Newton's laws applicable throughout the Universe and that assumed validity has, in fact, been challenged

However, whether or not it is valid to extend regions of validity in this cavalier manner, the biggest challenge to the relationship possibly arises through major questions concerning the interpretation of the observed redshift.

3. The Observations of Halton Arp

One important class of objects to be considered in the present context is provided by the quasars; the most 'distant' quasars are thought to have redshifts far in excess of those for the furthest galaxies. It is accepted by many that there were far more quasars and, indeed, radio galaxies in the past than there are now. However, this whole question is, or should be, a completely open one. Many seem to give the impression that everything in this area is absolutely clear cut and anyone opposing the generally accepted view is to be ignored as lacking in understanding of the truth. Frankly this appears to be the view adopted in the corridors of conventional wisdom towards the work and ideas of Halton Arp. While able to make use of the most powerful of telescopes, Arp also discovered that many pairs of quasars which possess extremely high redshift values appear to be associated physically with galaxies having much lower redshift values; galaxies, in fact, which are known to be much closer to the earth than the redshift values of the quasars concerned would imply. This all follows from the Hubble law which indicates that objects having high redshift values must be receding from the earth very quickly and, therefore, must be found at large distances from the earth. Hence, Arp was faced with the intriguing question of how objects with totally different redshift values, objects which according to 'conventional wisdom' had to be located at totally different distances from the earth, could be physically associated – in some instances, Arp's photographs actually showed a physical bridge between the quasars and the associated galaxy. As has been recorded many times, Arp has many photographs of pairs of quasars, with high redshifts, symmetrically located on either side of low redshift galaxies. It has to be noted that these pairings occur far more often than the probability of random placement would allow. Of course, the main problem with Arp's photographs is that according to orthodox theorists, high redshift objects must be at a great distance from the earth; to them high redshift is effectively a measurement of distance from the earth. It is often claimed by the advocates of 'conventional wisdom' that Arp's statistical analysis is in error; after many years, this still seems to be the main line of attack on his work. However, from all the accumulated evidence it seems there is no satisfactory foundation for criticising Arp's work on the basis of the statistics involved, and that seems to be the only criticism actually offered. Much of Arp's work is well-documented in his book *Seeing Red* [2] and reference should be made to this work for further details of the specific points involved.

As might be expected, this work of Arp's has not been welcomed by the orthodox astronomical community because, if accepted, it casts severe doubt on the assumption, which is quite basic to most, if not all, of accepted cosmological theory, that objects possessing a high redshift must be far away from the earth. Whether people approve of his work or not, it is undoubtedly true that Arp's work raises serious questions about the present state of cosmological theory and to ignore these questions, as some would advocate, should not be an option for any serious investigator in the field. It is also undeniably true that serious questions about the true interpretation of observed redshifts remain and must be addressed with open minds if real progress is to be made. However, a new method to determine distances in space has been announced recently [3]. This proposal, involves the possible use of quasars as standard candles. Once again, though, the concept of redshift appears central to the discussion. Hence, interesting and valuable as this reported work may be, it does seem its real usefulness will depend on a correct understanding of redshifts observed. Therefore, all in all, it is seen that probably the most important question facing cosmology concerns the correct meaning of this concept of redshift – something which initially seemed so simple to interpret.

On top of this, though, various other questions have arisen but the importance of these has frequently been played down by the scientific press. Most of these depend on observations of magnetic fields and electric currents in space and one problem with many is that, in all likelihood, the experiments proposed to help solve the associated problems have been carried out and documented already.

4. Magnetic Fields in Space

On 5th December, NASA announced that its Voyager 1 spacecraft had entered a new region between our solar system and interstellar space [4]. In this announcement, one of the more interesting comments is that "Voyager has detected a 100-fold increase in the intensity of high-energy electrons from elsewhere in the galaxy diffusing into our solar system from outside". This comment is of interest because, apart from the word 'diffusing', it describes what the electrical model of our universe expects in the virtual cathode region of the solar discharge boundary.

Also, on 6th December, it was revealed that a new all-sky map shows the magnetic fields of the Milky Way with the highest precision [5]. It was claimed that the origin of galactic magnetic fields remains unknown despite intensive research, although it was seemingly assumed that they are constructed via dynamo processes such as are said to occur – in violation of Cowling's well-known theorem incidentally – in the interiors of the Earth and the Sun.

Some years ago, in an entirely different context, Sir Winston Churchill advised people to learn from the lessons of history and, in the present context, it might seem appropriate to follow this advice in astrophysics. Hence, in this spirit, it might be noted that, following the introduction of Newton's mechanical ideas, work still proceeded apace investigating electromagnetic phenomena and this continued at least into the earlier years of the twentieth century, as is evidenced by the contents of J. J. Thomson's book *Electricity and Matter* [6]. However, this book provides but one example to illustrate the very real emphasis on work involving the effects of the electric and magnetic fields, work which, incidentally, constantly sought an explanation for the concept of mass in terms of those forces. However, after those early years of the century, the emphasis seems to have shifted to explanations of phenomena purely in terms of gravitational effects. Considering that it is accepted that much of the matter in the universe is in the form of plasma, this might be thought a retrograde step. One may only speculate as to why the emphasis of much scientific research changed in this way. However, thanks to people like Birkeland, Alfvén and, more recently, Peratt, work in the areas of electromagnetism and plasma physics has continued.

The work on plasmas and other electromagnetic phenomena has inspired people to examine astronomical phenomena in these terms and this has resulted in the so-called Electric Universe idea as expounded, for example, in the books *The Electric Universe* [7] and *The Electric Sky* [8]. Reading through this material makes one immediately aware that just like accepted theory the electric universe ideas are supported by computer modelling, but it is also able to draw on parallels between astronomical phenomena and plasma phenomena observed in the laboratory. Admittedly, drawing such parallels involves scaling up tremendously but assuming this possible is little different from assuming that laws seemingly applicable here on the Earth are also applicable in the Solar System and, indeed, throughout the universe. At least visually, some of the phenomena observed in the laboratory are very like what is observed by some of the most powerful of telescopes. Electric currents in plasma naturally form filaments due to the so-called 'pinch effect' of the induced magnetic field. Electromagnetic interactions cause these filaments to rotate about one another to form a helical 'Birkeland Current' filament pair and this is very much the structure seen in the Double Helix nebula near the galactic centre; again, the Hubble image of the planetary nebula NGC6751 looks remarkably like the view down the barrel of a plasma focus device. Examples such as these prove nothing but might awaken people to the possibility of alternative explanations for at least some astronomical phenomena.

5. The Heritage of Kristian Birkeland

Much of the laboratory work originated with the work of Kristian Birkeland more than one hundred years ago. It was during his Arctic expeditions at the end of the 19th century that the first magnetic field measurements were made of the Earth's polar regions. His findings also indicated the likelihood that the auroras were produced by charged particles originating in the Sun and guided by the Earth's magnetic field. Birkeland, though, was an experimentalist and is still known for his Terrella experiments carried out in a near vacuum and in which he used a magnetised metallic sphere to represent the Sun or a planet and subjected it to electrical discharges. By this means, he was able to produce scaled down auroral-type displays as well as analogues of other astronomical phenomena. These claims, however, were only vindicated finally by satellite measurements in the 1960's and 70's. To that point in time, his experimental and observational achievements had tended to be overshadowed by the purely theoretical predictions and explanations of the geophysicist, Sydney Chapman. Powerful mathematics seems to have held sway over the more expected techniques of physics – experimentation and observation, with mathematics a mere tool to be used when necessary. This is not to decry Chapman's work but to emphasise the overwhelming importance of the physics when investigating natural phenomena.

Birkeland also showed experimentally that electric currents tend to flow along filaments shaped by current induced magnetic fields. Of course, this confirmed observations of Ampère that indicated that two parallel currents flowing in wires experience a long range attractive magnetic force that brings them closer together. However, as plasma currents come closer together, they are free to rotate about each other. Such action generates a short range repulsive magnetic force which keeps the filaments separated so that they are, in effect, insulated from each other and able to maintain their separate identities. The end effect is for them to appear like a twisted rope and it is this configuration which is termed a 'Birkeland current'. Satellites orbiting above the auroras in the 60's and 70's were able to detect a movement of ions, indicating that electric currents were present. Later missions found quasi-steady electric fields above the auroras following the magnetic field lines, thus lending some credence to Birkeland's claim of the existence of an electric circuit between the earth and the Sun.

However, the so-called Electric Universe is really just an hypothesis, a new way of interpreting known data by using both new and well-established knowledge relating to electricity and plasma. It should be emphasised immediately that, in this new interpretation, gravity still has a role to play but it is a secondary one since the electric force is so much more powerful. A major point to be stressed from the outset is that, in this interpretation of astronomical phenomena, scientists are able to call on evidence from laboratory based experiments to help form and support suggested explanations for a wide variety of phenomena. It has been found that, as explained in more detail in the above-mentioned books, a plasma in a laboratory is a good model for providing possible explanations for many recently observed astronomical phenomena which, in several cases, have puzzled astronomers seeking explanations via more usual routes. This is not to say that gravity is ignored and regarded as irrelevant; rather, the possible effects of the electromagnetic force on astronomical phenomena are investigated while still recognising the importance of gravitational effects. In the electric universe, the gravitational systems of galaxies, stars, moons and planets are felt to have their origins in the proven ability of electricity to generate both structure and rotation in plasma. It is felt further that the force of gravity assumes importance only as the electro-magnetic forces approach equilibrium. As has been noted already, great consternation has been caused in astronomical circles by the realisation that gravity, as presently understood, cannot explain much that is observed if the amount of mass available is as now felt to be present. Hence, instead of positing the existence of 'dark matter' or following the path of modifying Newton's well-tried law of gravitation significantly, it is suggested here that the effects of the electromagnetic force be examined to see if, in conjunction with orthodox ideas on gravity, these puzzling observations can be explained. However, returning to the realisation that much of the matter permeating the Universe is in the form of plasma, it might be remembered that these clouds of plasma respond to the well-known laws of Maxwell. Also, as pointed out by Scott in his book [8], another law, formulated by Lorentz, does help explain the galactic speeds alluded to earlier. This law states that

a moving charged particle's momentum (speed or direction) can be changed by application of either an electric field or a magnetic field or both.

This seems a highly likely contributory factor, at least, causing galaxies to rotate as they are perceived to do but would indicate, contrary to the accepted view, that gravity has less to do with things than has been thought. However, it should be noted that nowhere is it being suggested that Newton's law of gravitation is in error; it is simply being suggested that, in deep space where everything swims in a sea of plasma, the Maxwell – Lorentz electromagnetic forces dominate over those of gravity.

It might be remembered also that the Lorentz force alluded to here changes a charged particle's momentum and that change is directly proportional to the strength of the magnetic field through which the particle is moving. Further, the strength of a magnetic field produced by an electric current is inversely proportional to the distance from the current but the gravitational force between stars is inversely proportional to the *square* of the distance. This well-known difference between the two forces could lie at the heart of the problem of the galactic rotation curves; certainly it seems an avenue worth exploring further, especially considering the fact that more and more space missions are indicating that electromagnetic forces are distributed more widely throughout space and are, of course, many orders of magnitude stronger than gravitational forces.

As well as a great many laboratory experiments being performed to establish plasma properties [9] , it has been shown also, using the Maxwell and Lorentz equations, that streams of charged particles, such as are found in the intergalactic plasma, will evolve into the familiar galactic shapes under the influence of electromagnetic forces. The results fit extremely well with the observed velocity profiles in the galaxies and all this without recourse to missing mass or other esoteric entities. Much of this simulation work has been carried out by Anthony Peratt and is reported in various issues of the IEEE Transactions on Plasma Science. However, recent reports [10] of the production of magnetic fields in a laboratory by using a high-power laser to explode a carbon rod in helium gas in an effort to simulate the plasma out of which the first galaxies are thought to have been formed are accompanied by discussion of experiments to be performed to help examine these phenomena in laboratories here on earth. It seems that the hope is to examine the physics of the cosmos over billions of years in a laboratory here on Earth. Such experiments as proposed by more than one group in the United Kingdom usually involve examining plasmas. The grave suspicion must arise that such experiments have often, if not always, been carried out already by the likes of Peratt and his mentors. It is worrying that so many either do not know of this body of work or dismiss it partially because the journal in which many of the results are published is regarded by some as non-prestigious, although how such a comment can be made – apparently seriously – about the above-mentioned IEEE journal remains something of a mystery.

Appendix: CMB Redshift as Blackbody C-QED Equilibrium

We report an additional controversial model of redshift. The Casimir effect is said to supply the best evidence for a covariant polarized Dirac vacuum which is a form of Einstein energy dependent spacetime metric, $\hat{M}_4$ amenable to extended electomagnetic theory [11,12] where Maxwell's equations are not cut off at the vacuum. This has been described by the Proca equation

$$\Box A_\mu = -\xi^2 A_\mu; \text{ where } \xi = \frac{m_\gamma c}{\hbar}$$

and m_γ admits a small anisotropic photon mass [13].

Photon mass, m_γ is the basis of the 'tired light' non-Doppler redshift model; where because of periodic photon mass an internal motion occurs coupling the photon to the Dirac vacuum causing a loss of energy over cosmological distances.

Photon anisotropy requires vacuum zero point coupling, and its propagation can no longer be considered independent of the Dirac vacuum. Einstein, Schrodinger, and de Broglie have attested to the significance of non-zero photon rest mass. Frequency anisotrophy results from a putative 10^{-65} g periodic nonzero photon rest mass according to

$$E = \hbar\nu - mc^2\left[1 - \nu^2 / c^2\right]^{-1/2}$$

The de Broglie wavelength relationship for massive particles, taking the accepted value for *R* applied to the Vigier mass m_γ of the photon is:

$$m_\gamma = \frac{h}{\lambda c}$$

taking $\lambda = R \approx 10^{28}\, cm$ for the de Broglie wavelength, λ of the photon then $m_\gamma \approx 2.2 \times 10^{-65}\, g$ which is the value for photon restmass obtained by a number of researchers [34,52]. Where *R* is the radial size of the universe; and by the uncertainty relation this is the smallest possible photon mass. Further $m \to 0$ only if $R \to \infty$. The de Broglie hypothesis was verified by [53,54] for the wavelength of a material particle. A photon mass of $10^{-65}\, g$ is in total agreement with Vigier's tired-light hypothesis [34].

From the redshift-distance relation, $z = f(d)$ (for static or expanding universe models) following [33] photons with restmass, m_γ interact with vacuum particles of mass, m_{vac} with acceleration

$$\int \frac{d^2 y}{dt^2} dt = -\frac{2\lambda\rho\omega}{y\left[\left(\tfrac{1}{2}\right)^2 + y^2\right]^{1/2}}.$$

The momentum transfer per vacuum particle, m_{vac} is

$$\int m_{vac} \frac{d^2 y}{dt^2} dt = -\frac{2 m_{vac} m_\gamma \omega}{y\left[\left(\tfrac{1}{2}\right)^2 + y^2\right]^{1/2}}$$

With *t* the time, *y* the coordinate intersecting the path, ω producing a 'tired-light' redshift-distance law

$$\frac{\Delta\nu}{\nu} = e^{kd} - 1,$$

where *k* is determined by m_γ estimated to have a value of $10^{-65}\, g$ [34,52,55], m_{vac} which is currently unknown and may not be completely relevant other than the putative fact that vacuum coupling occurs.

It is inherently obvious that the photon is annihilated when brought to rest; therefore it is suggested that the photon has a rest mass with a half-life on the order of the Planck time of 10^{-44} s, which would still preserve gauge in the domain of the standard model of elementary particles and allow for anisotropic vacuum zero point coupling of the photon which if it also occurs in the limit of the Planck time can be a virtual interaction.

The astrophysical ramifications of the polarized Dirac vacuum m_γ 'tired light' model are extended by the discovery of its relationship to the the cosmic microwave background radiation (CMB). Both Doppler redshift and CMB have been considered major pillars of Big Bang cosmology, but there are other interpretations. Hubble discovered cosmological redshift not expansion of the universe!

Recently Vigier and Amoroso have made a strong case that the CMB is not the remnant of an explosive singularity but that the CMB and redshift (with m_γ)

are cavity QED blackbody equilibrium conditions; where redshift of course is absorption and CMB emission [13-15].

It is curious that the encyclopedia of physics states:

> Such space-times are nonphysical in the sense that they require a periodicity where events are so ordered that the future prepares the present which is to evolve to the future. We do not conceive of the universe being that well ordered...[16]!

The author neglected to consider the highly ordered topology of Calabi-Yau mirror symmetry (actuallt he said string theory was beyond the scope of his encyclopedic article) and perhaps was unaware of the highly ordered conditions provided by Cramer's Transactional Interpretation where a 'present event is a standing-wave of the future-past'.

Because of myopic queuing behind popular beliefs, many probable 'truths' remain ignored and unexplored. Progress in science often proceeds at a snails pace, recalling that it took 150 years to fully accept the heliocentric model of Copernicus and nearly causing Galileo his life....

References

[1] Narliker, J. & Burbidge, G. (2008) Facts and Speculations in Cosmology, C.U.P., Cambridge.

[2] Arp, H. (1998) Seeing Red, Apeiron, Montreal.

[3] Quasars shine a new light on cosmic distances (2012) Science Daily, May, 4th.

[4] http://www.jpl.nasa.gov/news/news.cfm?release=2011-372

[5]http://www.phys.org.com/news/2011-12-all-sky-magnetic-fields-milky-highest.html

[6] Thomson, J.J. (1904) Electricity and Matter, Westminster: Archibald Constable & Co.

[7] Thornhill, W. & Talbott, D. (2002) The Electric Universe, Portland, Mikamar Publishing.

[8] Scott, D.E. (2006) The Electric Sky, Portland, Mikamar Publishing.

[9] Peratt, A. (1992) Physics of the Plasma Universe, New York, Springer-Verlag.

[10] Laser hints at how Universe got its magnetism (2012) Science Daily, March 24th.

[11] Lehnert, B. & Roy, S. (1998) Extended Electromagnetic Theory: Space Charge in Vacuo and the Rest Mass of the Photon, Singapore: World Scientific.

[12] Lehnert, B. (1999) An extended electromagnetic theory, Physica Scripta, (T82), 89.

[13] Amoroso, R. L., Kafatos, M., & Ecimovic, P. (1998) The origin of cosmological redshift in spin exchange vacuum compactification and nonzero rest mass photon anisotropy, in Vigier, J-P, Hunter, G. & Jeffers, S. (eds) Fundamental Theories of Physics, Dordrecht: Kluwer Academic, 23-28.

[14] Amoroso, R. L. & Vigier, J-P (2002) The origin of cosmological redshift and CMBR as absorption/emission equilibrium in cavity-QED blackbody dynamics of the Dirac vacuum, in R.L. Amoroso, G. Hunter, M. Kafatos & J-P Vigier (eds) Gravitation & Cosmology: From the Hubble Radius to the Planck Scale, Dordrecht: Kluwer Academic.

[15] Amoroso, R.L. & Rauscher, E.A. (2009) The Holographic Anthropic Multiverse: Formalizing the Complex Geometry of Reality, Singapore: World Scientific.

[16] Goldberg, J.N. (1990) Space-time, in R.G. Lerner & G.L. Trigg. (eds.) Encyclopedia of Physics, 2nd ed. New York: VCH Publishers.

Examining the Existence of the Multiverse

JAMES J. HURTAK. & D. E. HURTAK

AFFS-Europe, 4058 Basel, Switzerland
dh@affs.org

ELIZABETH A. RAUSCHER

Tecnic Research Laboratory, 3500 S. Tomahawk Rd. Bldg. 188
Apache Junction, AZ 85119 USA
bvr1001@msn.com

Guth's inflationary universe model has been widely accepted in modern physics. Expanding upon this concept Linde introduced the Chaotic and Eternal Inflationary model. The inflationary universe structure allows for multiple universes or various bubble universes connected through scalar and tensor fields and making the structure of space self-similar on larger scales. In this paper we briefly examine these and other theories associated with a multiverse, including the M-Theory that considers that our universe and others are created by collisions between membranes in an 11-dimensional space.

Keywords: Cosmology, Quantum Mechanics, Multiverse, Scalar Fields

1. Introduction — Why a Multiverse?

Use of the observational data, in the study of CMB and WMAP, indicates that the universe is accelerating its expansion. If this acceleration is caused by a positive energy density of the vacuum (i.e., cosmological constant, $\Lambda > 0$), the process could continue forever. With a variety of theories researchers have sought to explain this phenomenon including abstract theories of elementary particles, such as M-theory, string theory, supergravity and the standard model (SM). Many of them lead to the conclusion that our known universe is part of a multiverse.

There are numerous ideas of how a multiverse came into existence, including Linde's Bubble Theory, the many worlds, interpretation (MWI) of quantum physics, braneworlds predicted by string theory and M-branes, and other models which we examine in this paper. There are at least six different classifications of a multiverse: the Quilted Multiverse, Inflationary Multiverse, Brane, Cyclic and Landscape Multiverses, MWI, the Holographic Universe. In most of these theories—string theory, the inclusion of the cosmological constant and quantum physics—have shown certain mathematical congruities.

According to Rauscher, Hurtak and Hurtak [3], Jenkins [4] and others, our ability to exist, that is, for life to form as we know it, depends on a precise set of conditions, physical constants that exist in our universe. Specifically, in this observable universe, the parameters seem to be fine-tuned into particular values, including the Planck length, such that if they had other values, known forms of life would not exist.

A critical constant that had been generally overlooked, the cosmological constant, Λ, is now beginning to occupy a major role following the research of Perlmutter [5] and the High-z Supernova Search Team. Sorkin [6] examined the data, to provide an explanation of the importance of Λ stating that based upon observations of supernovas, it is most likely that $\Lambda > 0$ but also $\Lambda < 0$. According to Steinhardt and Turok [7] the observed data, 7 x 10^{-30} g/cm^3, is over 120 orders of magnitude smaller than the Planck density of 10^{93} g/cm^3, the density of the universe as it emerges from the big bang.

Einstein formulated the field equation of [3],

$$R_{\mu\upsilon} - \frac{1}{2} g_{\mu\upsilon} R_{\mu\mu} + \Lambda g_{\mu\upsilon} = -\frac{8\pi G}{c^4} T_{\mu\upsilon} = -\frac{8\pi}{F} T_{\mu\upsilon} \quad (1)$$

where $T_{\mu\nu}$ is the Riemann-Christoffel curvature mass-energy tensor, and $F = c^4/G$ is the Rauscher cosmological acceleration force. Although he never clearly defined Λ [8], a positive cosmological constant accelerates the universal expansion and this is exactly what Perlmutter [5] observed in the supernova data. The observable data also reveals that the calculations of the density parameter Ω are closer to: $\Omega_m \approx 0.25$ and $\Omega_\Lambda \approx 0.75$.

The cosmological constant has become an important factor in measuring and understanding the acceleration of the universe, as well as the H_0 (Hubble) = 73.8 ± 7 km/sec/MPC [3] are also required for its calculation. Specifically, according to Guth's Inflationary Theory [1], the initial inflation occurs because of the positive vacuum energy density, and where Λ creates a negative pressure which causes an accelerated expansion measured in the Hubble constant H, of the universe [9].

As before mentioned, our universe seems to have a fine-tuning suitable for life as we know it. In the multiverse theory, one universe has the same temperature and same physical constants, whereas other universes may display other properties. For instance, the constant temperature of the cosmic background radiation throughout space is measured at ~ 2.73° K, as determined by the NASA COBE satellite in 1992. According to Fixsen [10] specifically, T = 2.725 FIRAS calibration (2.725 $\pm$ 0.002 K.) as derived from the WMAP. The temperature consistency is key to the consideration of the inflationary model because, for the universe to reach the same temperature throughout, with limited irregularities of only a few millionths of a degree, different regions must exchange heat with every other part. The Big Bang would not alone account for this amount of homogeneity.

The universe, however, is not only homogenous, but is isotropic where its matter is evenly distributed in all directions. Specifically, the cosmic microwave background has shown that space is isotropic in all directions to within one part in 100,000. Guth's [1] conclusion is that a rapid inflationary period took place in approximately the first 10^{-35} seconds of our universe's start of existence. Since the standard model assumes the universe is both homogenous and isotropic, inflation is used to explain the dilemma, known as the horizon problem, as to how different regions of space, appear to be essentially indistinguishable in nature. Linde's Chaotic or Eternal Inflation Theory [2] followed in which it was considered that this universe was just a bubble that was generated from the foam of many other possible universes, during the inflationary period. Linde describes our universe as a child universe generation out of a false vacuum of quantum fluctuations.

Moreover, in the consideration of Linde's Eternal Inflationary model or other string theorists, Brane models hypothesize that in other universes, universal constants could be different, although some duplication or similar constants may occur. One of the mathematical geometric concepts behind these theories come from the question of a possibly open or closed universe, where in addition to our universe's homogeneous nature, it appears our universe is relatively flat. The present day range of the density parameter is 0.1 <~ Ω <~ 2 [11].

In fact, since the observable data of the energy density, Ω yields a value close to the value of 1, and supports the flatness of the universe hypothesis which, in turn, supports the inflationary model. Specifically, measurements, including MAT/TOCO, Boomerang, Maxima and DASI [12] have shown that the brightest spots within the WMAP are about 1 degree across which demonstrates that we are living in a flat universe.

The flatness, Ω is defined by the density parameter, ρ / ρ_c, which is the ratio between the actual mass density of the universe (ρ) to the critical density (ρ_c). The flatness is a relative parameter and yet may be sufficient to create an infinite universe. The combination of CMB anisotropy, LSS (Large Scale Structure at the Institut Laue-Langevin) and other observations demonstrate the flat universe as $\Omega_K \approx -0.02 \pm 0.02$, where Ω_K is the deviation from flatness, that is curvature which corresponds to a closed, flat, and open universe, or the magnitude of the spatial curvature in our current universe [13]. According to calculations by NASA scientists, there is only a .5% margin of error in the CMB. This is based on the basic geometrical fact that angular scale subtended in the sky by the acoustic horizon is different in a universe with uniform positive (spherical), negative (hyperbolic), or, zero (Euclidean) spatial curvature.

If space is flat, it could be also infinite; we prefer to say open-ended. An open universe indicates a negative large-scale curvature that the universe would expand forever. If the universe were slightly larger, it would have collapsed in to a big crunch; if the universe were smaller the stars would have disappeared too early in its evolution. Flat space would indicate that space is infinite or open ended and keeps expanding forever. Yet regions in our own universe may be off-limits due to what is called a Quilted Multiverse, where only our region of the universe is available for observation. So even though space is infinite with a zero large-scale curvature there may be additional spaces, some even undergoing additional Big Bangs that we cannot normally observe, hence the limited age of our own universe at a mere 13.7 billion years. This is consistent with the Little Whimper or multiple bangs of Rauscher [14].We may all be living in part of a quantum possibility where every possible world manifests as some other universe.

Both flat and open universe models account for our observations of an accelerating universe [15]. Thus, Drs. Hurtaks agree with Linde [2] who claims the initial expansion was caused by the positive vacuum energy and the cosmological constant, $\Lambda > 0$, allowing for the acceleration of the universe to continue forever. On the other hand, this acceleration may stop and the universe may eventually collapse. The flatness of the universe may be simply a temporary phase of expansion. Since it is clear that the average density and the critical density changes with time; three major factors are important to consider. We have already introduced these factors, but now to determine the relationship between Ω_Λ, Ω_M and Ω_K or what Bahcall [16] tells us are the three factors needed to assess the state of the universe. Mathematically they can be interpreted as,

$$\Omega_m = \frac{8\pi G \rho_{matter}}{3H^2} \tag{2a}$$

$$\Omega_\Lambda \equiv \frac{8\pi G \rho_\Lambda}{3H^2} = \frac{\Lambda}{3H_0^2} \tag{2b}$$

$$\Omega_K = \frac{-\kappa}{a^2 H^2} \tag{2c}$$

a is the scale value which is taken to be 1 at the present time and G is the gravitational constant. In an open universe $\Omega_K > 0$ and in a closed universe $\Omega_K < 0$. According to Guth's Inflationary Cosmology [1] a flat universe is a consequence of the inflationary era, where $\Omega_K = 1 - \Omega_m - \Omega_\Lambda$

Here we see how Ω_Λ is associated with the density parameter,

$$\Omega \equiv \frac{\rho}{\rho_c} = \frac{8\pi G \rho}{3H^2} \tag{3}$$

By utilizing the observational data [5,12], it appears that the universe is flat which agrees with Einstein's theory of general relativity that confirms that the energy density of the universe has a critical value of [16],

$$\rho_c \equiv \frac{3H^2}{8\pi G} \sim 1.7\, x 10^{-29}\, g \cdot cm^{-3} \tag{4}$$

where,

$$\Omega \equiv \frac{\rho}{\rho_c} \approx 1 \tag{5}$$

Although one important premise is that temperature is homogenous throughout the universe, there are also small perturbances in the temperature of the universe, as further noted in the CMB which add support to the inflationary theory [1], and the flat spectrum where the scalar spectrum n_s is indexed as $0.963 \pm .012$ Some of these perturbances may also be a clue to observing a previous connection to other universes [17].

Many of the problems with a stand-alone Big Bang theory can be found in Amoroso and Rauscher [18] which details not only flatness and temperature constants, as mentioned here, but also rotations and density fluctuation. However, Rauscher does not consider Inflation, but the concept of the multi-Big Bang or Little Whimper model [14] which demonstrates the consistency between Einstein's field equations and the Big Bang-like cosmology.

Drs. Hurtak postulate a false vacuum that comes from previous – false vacuum bubbles (or a cyclic-type universe) which connects with Rauscher's fundamental rotation as one of the multi-dimensions in the Rauscher model, but not necessarily full inflation.

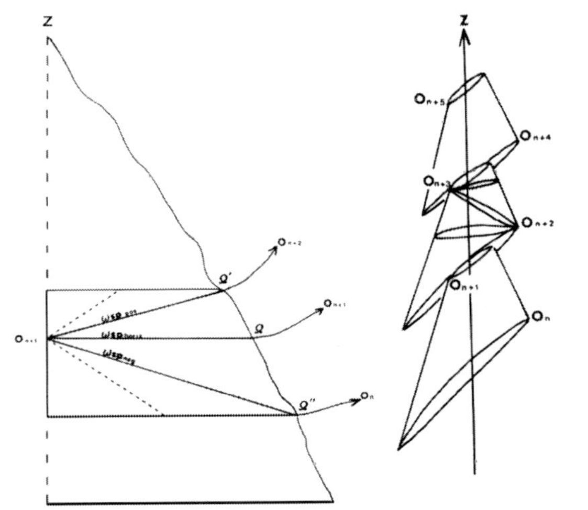

Figure 1. 2-dimensional (left) and 3-dimensional (right) Geometric Structures of our universe and other pocket universes generated by false vacuum. (Figure courtesy of J. Hurtak.)

Considering the possible existence for other universes and understanding that our universe is flat, there is no reason to consider that all possible universes are flat. All possible geometries may be available for other universes. In terms of geometric structures of the multiverse, Hurtak [19] considered two versions of the geometry of multiple universes in both 2-D and 3-D structures as shown in Figure 1, as just two of the numerous possibilities.

We also take into account the Calabi-Yau formulations of over 100^{500} possible geometries for the multiverse. Every Calabi-Yau manifold with mirror symmetry or T-duality admits a hierarchical family of supersymmetric toroidal 3-cycles [18]. It is currently unknown whether the attempt to formalize this continuous-state structure should follow a Kaluza-Klein spin tower, logarithmic or golden ratio spiral, cyclotron resonance hierarchy, Genus-1 helicoid 'parking-garage' or some other higher dimensional structure. Consider the idea of the holographic multiverse, we realize that there are some basic complex structures as detailed in Figure 2 from Amoroso and Rauscher [18].

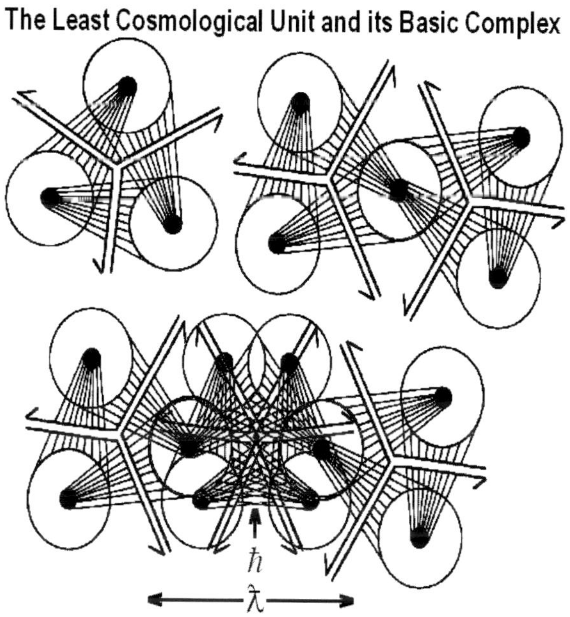

Figure 2. According to Amoroso and Rauscher the basic configuration for the Holographic multiverse can conceptualize the triune nature of an isolated least-unit not existing in nature. This illustrates the coupling of two isolated least-units along an x coordinate. The central portion denoted by $\hbar$ represents the realization of one virtual Euclidean point which oscillates harmonically to ~ the Larmor radius of a hydrogen atom here denoted as $ƛ$ which represents the new *Stoney* representation of $\hbar$ plus string tension, *Ts*. This model can be considered a Cramer transaction. (Figure courtesy of Amoroso and Rauscher) [18].

Rauscher, however, further points out that if we cannot measure or observe details that constitute a multiverse its existence may never be able to be proven. Although in science many proposals appear well in advance of observations, for example, the Higgs particle. At the same time, there may be ways of observing the previous existence of other universes as proposed by WMAP data, which may show signs of a parent or sister universes [17]. There is also the possibility that other universes may not be completely unique, and share similar constants and could be cosmic duplicates of our realm [20].

2. Guth's Inflationary Theory

Alan Guth hypothesized that the early universe grew exponentially, driven by vacuum energy pressure expanding many times faster than the velocity of light. The universe generated from a hot Big Bang would not allow for the observable even temperatures of the current universe. In the Big Bang, space would be flying apart faster than light, so all parts of the universe would not have been in touch with each other.

To solve this dilemma in the early 1980s, Guth introduced his inflationary cosmology with the hypothesis that started with the fabric of space exploding by the stupendous factor of roughly a factor of 10^{40} 10^{100}[1], during the very initial moments after the the space-time foam and the initial Planck epoch (t= 0 to t=10^{-43}), probably at t=10^{-37} to 10^{-34}s, and then inflation came to a halt. He further postulated that during the inflation field, the preliminary time would have permitted matter/energy to interact throughout all of space. The inflation field is a scalar field which causes the universe to expand exponentially. Additionally, the flatness is due to the enormous size of the universe at the initial period of exponential inflation. Inflation ended when the cosmological constant dropped to a value similar to what we have today ($\Lambda > 0$ and $\Lambda < 1$) [1]. The energy perturbations became frozen into the fabric of spacetime and lead to the formation of galaxies. Matter began to dominate the universe with gravity coalescing it around areas of greater density.

Two critical components behind this theory should be understood: 1) the inflation field and 2) the concept of a false vacuum. The inflation field has an energy density (ρ) and also pressure (p) considered to be negative or repulsive. We define density of the early universe as

$$(\Omega^{-1} - 1)\rho a^2 = \frac{-3kc^2}{8\pi G} \tag{6}$$

The energy density, ρa^2, remains roughly constant because of being dominated by the inflation field. Hence, the expansion of the universe directly relates to one of Friedmann's equation,

$$H^2 = \left(\frac{\dot{r}}{r}\right)^2 = \frac{8\pi G}{3}\rho - \frac{kc^2}{a^2} \tag{7a}$$

and the closer H^2 gets to $8\pi G/3(\rho)$, the flatter the universe becomes. This continues to a second Friedmann equation which is introduced to show how a universe has been dominated with both energy density ρ and pressure p:

$$\left(\frac{\ddot{r}}{r}\right)=-\frac{4\pi G}{3}(\rho+3p) \tag{7b}$$

At the end of the inflationary era essentially all of the energy of the universe is contained in this one, nearly homogeneous field. We find this also defined in the acceleration equation:

$$\left(\frac{\dot{r}}{r}\right)=-\frac{4\pi G}{3c^2}(\varepsilon+3p) \tag{7c}$$

The false vacuum [21], in quantum mechanics is considered as a metastable spacetime that is a teeming foam of particles and antiparticles that are coming into existence from virtual space and then annihilating each other. Unlike in a true vacuum, a false vacuum, is a combination of energy states with a nonzero value. A standard vacuum state (i.e., lowest energy state), devoid of any matter or radiation, does have some energy associated with it. The value, however, of this energy is given by Λ which in a false vacuum becomes a condition for an elevated or metastable vacuum energy density.

The question is how does the particle from the vacuum come into this universe, or how do virtual particles become real? Quantum fluctuations, in the form of virtual particle pairs of borrowed energy ΔE, get separated during the interval $\Delta t \leq \hbar / \Delta E$. If during Δt, the physical size Δx leaves the event horizon [22], the virtual particles cannot reconnect, they become real and the energy debt is paid by the driving mechanism of inflation, the energy of the false vacuum, the Λ_{inf} associated with the inflaton potential $V(\phi)$.

This concept, in general, was recently confirmed by Wilson *et. al.* [23] who demonstrated how virtual particles that constantly appear and disappear in the vacuum become real particles. What Wilson and colleagues confirmed is the *"dynamical Casimir effect"* where "virtual" photons after leaving their virtual state become real photons of light in matter. The Casimir effect was predicted more than 40 years ago by Gerald Moore who said that if photons (as virtual particles) bounce off a mirror moving at nearly the speed of light, they could become real.

The theory of the vacuum is based on the concept of quantum energy fluctuations of zero-point energy (ZPE), which is what makes it impossible to reach absolute zero degrees Kelvin. The quantum vacuum is not empty but is seething with virtual particle creation and annihilation in a Dirac-Fermi Sea model. Another area of consideration is in plasma at considerably higher temperatures. Rauscher [24] developed a detailed theoretical formalism of the effect of vacuum state energy states in a polarization on the properties of fully ionized plasma gas, the conductivity and other electrical properties are demonstrated to be modified by the vacuum state polarization of the plasma media, that is, the measured values of the plasma's electrical and magnetic properties are consistent between theory and experiment only when we include the effects of a Dirac-Fermi vacuum of virtual particle states.

Casimir and Polder [25] considered that the vacuum virtual particles could induce a short-lived electrical current between two parallel smooth conducting metallic plates. In turn, the induced current will create a magnetic field, which could either pull the plates together or push them apart. The direction of the force between the plates on the respective current flows and forms fields.

Sparnaay [26] conducted an experiment and came within 15% of the theoetrical value and later Steve Lamoreaux [27] designed and conducted an experiment, which was more sensitive, the results of which were within 5% of experimental value.

In current experiments by Wilson and colleagues, to create a dynamic Casimir effect, used a mirror simulating movement at the speed of light as in the early universe and were able to vary the physical distance to the mirror [23]. More specifically, the "mirror" consisted of a SQUID magnetometer. By changing the direction of the magnetic field several billion times a second, with the use of the SQUID, the mirror moved slightly causing a 'wiggle' at around 25% the speed of light, which was sufficient to cause photons to be extracted from the vacuum.

The result was that photons appeared in pairs from the vacuum, which were able to be measured in the form of microwave radiation. It is believed that other particles (e.g., electrons, positrons, protons, etc.) might also be extracted from a vacuum, but photons are the easiest as they require less energy to extract from the vacuum. Now that virtual particles from a vacuum can become real particles, we can ask the question; where do the virtual particles come from? Does the vacuum pair production and Big Bang Nucleogenesis (BBN) come from empty space, that is, out of nothing, or are the vacuum particles from our connection to other dimensions or parallel universes?

3. Linde's Chaotic or Eternal Inflation

The question of quantum fluctuations, are addressed by Albrecht [28], Steinhardt [29], and Linde [2] expanded upon Guth's Inflationary Theory [1]. According to Linde [2], our universe is one of an infinite number of universes. In addressing the question of where do virtual particles come from, Linde described the universe as a bubble emerging from the quantum foam of a parent universe. On very small scales, the foam is dynamic and frothing due to energy fluctuations [30]. Fluctuations in the foam create tiny bubbles and wormholes. Small energy fluctuations establish a small bubble universe, which expands, but then contracts and goes out of existence. If the energy fluctuation is greater than a particular critical value, a bubble universe forms from the parent universe, and long-term expansion continues, permitting matter and large-scale galactic structures formation.

Universes are generated from one dS (deSitter) vacuum, to another due to the formation of bubbles. A dS universe is spatially flat and neglects ordinary matter, but has a positive cosmological constant which sets the expansion rate [31]. Thus, each bubble contains a new dS vacuum. If gravity is very strong in an alternative universe, no expansion may occur and the universe will never develop. If, on the other hand, gravity is weaker, the expansion will be so fast that no stars or galaxies develop.

Each new universe, will have its own life cycle, like our own, depending on the physical constants that constrain it. As a universe could have different laws of physics, this may further mean that any intelligent life forms may only be capable of exploring their own universe, making full observation of other universes difficult, if not impossible.

Linde, *et. al.* [31] hypothesizes that only a small bang or something like Rauscher's Little Wimper [14] would be required. Linde considers that the Eternal Chaotic universe becomes more like a growing fractal in that from the false vacuum several pocket universes can be formed and a universe can continue to produce excessive landscape realms. A similar structure is described by Hurtak in Figure 1, where universes generate other universes.

The inflation from the quantum fluctuations is unstoppable and because of this there will always be regions of space somewhere that continue to inflate [32] creating other universes with *ad infinitum* possibilities. In the presence of scalar fields, the fields began to oscillate, they lose energy, to form elementary particles. This can be observed in the vibrational resonance patterns of the universe as quantum jitters. Linde calls his theory 'chaotic eternal inflation' indicating that it has no end and which may have no beginning. Chaotic inflation requires $\phi > M_p$ where ϕ is the potential density of the inflating scalar field and M_p is the Planck mass.

Inflation can be formulated as $P(\phi,t)d\phi$ where t = time and $d\phi$ represents an interval. Here the scalar field is represented by a continuous function, ϕ (x, t), which can be real or complex. Inflation occurs when scale factor growth, i.e., $d^2x/dt^2 > 0$ revealing that the universe is accelerating.

Linde [33] theorized that in the simplest chaotic inflation model, eternal inflation begins at the Planck density of 10^{93} g/cm^3 [3], if the potential energy is greater than the kinetic and gradient energy in a smallest possible domain of a Planckian size. The potential of the field as V (ϕ) yields both the energy density ρ, and the pressure p, which are crucial to understand the expansion of the universe. In the following equations we define the potential scalar fields dynamics according to Liddle and Scherrer [34],

$$H^2 = \frac{8\pi G}{3} = [V(\phi) + \frac{1}{2}\dot{\phi}^2 + \rho] \quad (8)$$

From equation 8, we can derive the more complete scalar fields connected to both density and pressure as,

$$\rho_\phi = \frac{1}{2}\dot{\phi}^2 + V(\phi) + \frac{1}{2}(\nabla\phi)^2 \quad (9a)$$

$$p_\phi = \frac{1}{2}\dot{\phi}^2 - V(\phi) - \frac{1}{6}(\nabla\phi)^2 \quad (9b)$$

If $1/2\dot{\phi}^2$ is the kinetic energy and $\frac{1}{n}(\nabla\phi)^2$ as the potential energy, this leads to the simplified energy density of $\rho_\phi = \frac{1}{2}\dot{\phi}^2 + V(\phi)$ and pressure is $\rho_\phi = \frac{1}{2}\dot{\phi}^2 + V(\phi)$.

Domains of the inflationary universe with sufficiently large energy density continuously produce new inflationary domains, due to stochastic processes of generation of the long-wave perturbations of the scalar field [2]. Inflation creates an infinite set of cosmic regions, each with "initial boundary conditions and subsequently evolving properties that are characterized by a statistical distribution that is independent of the choice of region.

In Davies [32] and Linde's [33] inflationary universe theory, is we have been calling 'the universe' and is a very small part of a single bubble, also called pocket universe, in amidst of an infinite number of universes of a multiverse which itself is embedded in a continuously inflating space. The exponential growth of simple quadratic potential energy, *V*, in its most simplified form is,

$$V = \frac{m^2}{2}\phi^2 \qquad (10a)$$

or as Linde [33] demonstrate the manner in which an exponential growth of volume, becomes maximal for the same quadratic potential,

$$V = Vo + \frac{m^2}{2}\phi^2 \qquad (10b)$$

Large amplitude low frequency oscillations are required to continue the process. Nevertheless, as the scalar waves become frozen, each scenario has different estimated lifetimes and dimensions, so according to Liddle [35] we have:

Higgs potential $V(\phi) = \lambda(\phi^4 - M^2)^2$ (11a)

Massive scalar field $V(\phi) = \frac{1}{2}m^2\phi^2$ (11b)

Self-interacting scalar field $V(\phi) = \lambda\phi^4$ (11c)

In Linde's bubble universe, each of the bubbles may contain any possible value of the inflaton field. The bubbles contain no particles unless this process of their creation ends by a stage of a slow-roll inflation. In Linde's model, instead of tunneling out of a false vacuum state as Guth hypothesized, Linde considered that inflation occurred by a scalar field, where the field rolls very slowly compared to the expansion of the universe. Slow-roll inflation is essentially controlled by two parameters defined as: $\in \equiv -\dot{H}/H^2$ and $\delta \equiv -\dot{\in}/(2H\in) + \in$, where *H* is the Hubble rate [36]. This agrees with the observation data [5, 12] that the universe is flat and the particles and galaxies which are produced allow the universe to be stabilized sufficiently for creation to occur. There are other explanations, such as that by Lineweaver [22] who considers that inflation may not be due to a scalar field and its potential V (ϕ), but more to do with extra-dimensions?

4. Many-Worlds Interpretation (MWI)

Linde's Bubble Theory posits an infinite number of open multiverses, each with different physical constants generating from a cosmic foam, but it does not necessarily rely on the notions of the observer and the environment; two important properties of quantum mechanics. Hugh Everett, in 1957, established what now appears to be an acceptable alternative theory, the many-worlds interpretation (MWI). His theory yields a real multi-world approach to quantum mechanics, with a shared time parameter. The MWI has a quantum wave state $\psi\rangle$ associated with the collapse of the wave form and decoherence based on both the observer and the environment [37].

The quantum effects constantly split the universe into multiple diverging copies [38]. This establishes an ongoing "branching" effect as opposed to the continuous bubble. In this case, time becomes more pronounced, as the longer ago a split occurs, the more the universes diverge. So for every possible scenario of reality we have universes that continually split into multi-millions of branches.

Everett's hypothesis [39] is that all isolated systems evolve according to the Schrödinger equation and the pure quantum state $|\psi\rangle$ in Dirac notation,

$$\frac{d\,|\Psi\rangle}{dt} = -\hbar H\,|\Psi\rangle \qquad (12)$$

Most formulations of the MWI consider that the constituent universes are structurally identical to each other, but can no longer in communication. Although they have the same physical laws and values for the fundamental constants, they may exist in different states, and thus, no information can pass between them. The world splits means into a decohered wave function superposition of two or more outcomes $|\psi\rangle$, as observed by an experimenter. The multiple potential states exists before an observer collapses them into a single state. To track these universes, at least theoretically, one should be able to trace the evolution of the wave function in order to find the numerous branches of the state. This is because the state of the entire multiverse is related to the states of the constituent universes by quantum superposition, and is described by a single universal wave function [40].

This provides the possibility that the wave-function of the universe is predominant or at least preliminarily, evolving according to the Schrödinger equation. However, at the same time this theory provides a complete and consistent theory of quantum mechanics which agrees with various experimental results. Aguirre, Tegmark and Layzar [41] describe the cosmic wavefunction where each of the many worlds are the same world, where world in this context refers to the state of an infinite space. They establish a real collection rather than fictitious ensemble for a statistical interpretation of quantum mechanics.

Superficially, the MWI model becomes a manner of explaining probability and hypothetical statements. David Deutsch [42] and others state that all possible worlds exist in reality, and are just as real as the actual world, which is a position known as modal realism). So although Schrödinger's cat might be dead in our universe, Everett's theory is not. This interpretation is garnishing more and more support among scientists, particularly theorists researching M-theory.

5. Nonlinear Quantum Theory and the Detection of Other Universes

Just as the Heisenberg Uncertainly Principle, $\sigma_x\sigma_p \geq \frac{\hbar}{2}$, is a fundamental property of the quantum domain, so is quantum entanglement and non-locality. Our consideration of the MWI multi-universe model, or what is also called the EMW [Everett Many World] model, we contemplate the possibility that entanglement may occupy a role in the possible interconnection between universes. Parity validations in weak interactions may also relate to an asymmetry caused by entanglement with other multiverses, as well as other symmetry breaking processes. Time sequence of emergencies of the possible into actual is one of the key enigmas of quantum formulism.

We conclude the MWI model is useful only if the quantum formulism contains a non-linear component. This allows the weighted states to remain entangled in such a manner that future effects of other break off wave functions, at each time junctions, has interactions with the state that we observe in our universe. In the non-linear model, were the non-linear term is small, we would be able to detect properties of existences of other parallel universes. Hence, the multiverse model could become a testable one [40].

In the non-linear quantum theory, linearity becomes only an approximation. In the splitting universe model of EMW, each new set of wave functions Ψ_n has a different weight or probability $\Psi^*_n\Psi_n = |\Psi_n|^2$ where the probability gives the number of possible states available to the system.

Rauscher and Amoroso [43] explore the formulation of the Schrödinger equation with a small but effective non linear term in the potential energy of the Hamiltonian for time dependent and time independent solutions. In this formulation, for the time dependent case, we obtain non-local, coherent soliton solutions to the Schrödinger equation with long distance coherence. Figure 3 yields a mechanism for non-local entanglement, and a possible description that allows the detection, by small non-linear effects in our universe, of other universes. Small anomalies in the cosmic background radiation may act as an indicator of the effects from other universes. This may lead to an explanation, at least in part, of dark matter and the missing mass problem ([14] and many other researchers) and also dark energy [5,15], which may be part of the effect of other universes [18].

Eugene Wigner in his classic book [44] discusses the implications of a nonlinear quantum theory and the role of an active observer and for the nature of life itself. He expands the vector space to include a matrix form that allows a method to analyze the effect of one or more observers (i.e. perhaps, a human and a cat) and/or multiple humans as in the case of the Cat's Paradox [43]. Wigner also discusses the role of parity and other symmetries in the quantum theory. Van Bise further discusses the implications of Wigner's work in his research on low intensity, electromagnetic effects of microwaves, on the human brain (electro-encephalograph). Essentially, this discourse relates to all effects of or by human (and possibly other) consciousness.

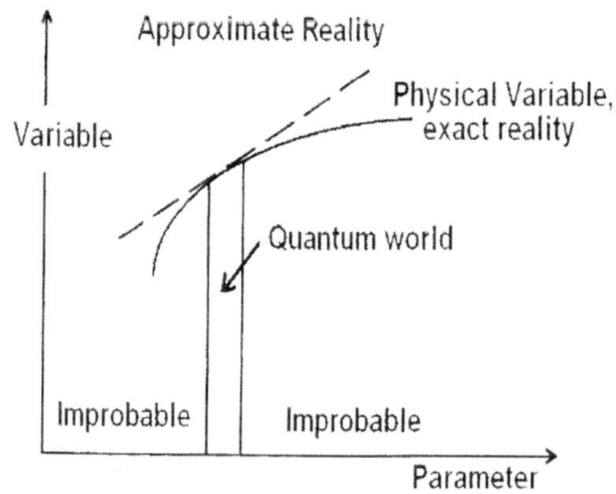

Figure 3. We approximate the quantum domain as a linear variable dependent on a parameter. The full space of exact reality is nonlinear (Courtesy of Rauscher and Amoroso) [43].

The arguments used in Wigner's essay on 'The Mind body Question', evolved in the language of quantum mechanics in which he pointed out that physicists possibly learned 'that the principle problem was no longer the fight with the adversities of nature, but the difficulties of understanding ourselves if we want to survive.' If more than one conscious observer enters into a system of measurement this joint system cannot be described by a wave function Ψ *after* the interaction since the result of an observation modifies the wave function of the system by another observer. A proper description of their quantum state is a mixture of states. The wave function is: $(\Psi_1 \times \chi_1)$ with a probability $|\alpha|^2$; it is $(\Psi_2 \times \chi_2)$ with a probability of $|\beta|^2$ where Ψ_1 and Ψ_2 are non-simple wave functions.

To quote Van Bise [45]: "Philosophically, this implies that if an observer of one group of experiments asks another observer of the same experiments about his feeling *before* he observed the results, his answer, whether or not it agrees with the questioning observer's conclusions, shows that the question was already decided in his mind before he was asked. In order to avoid this difficulty, it is necessary to postulate that the equations of motion of quantum mechanics are grossly non-linear when conscious beings with opinions enter into the equations". Here again is a paradox in scientific research as seen in Figure 4.

Complete objectivity in the real world is only approachable–not attainable. Even the most stubborn scientific attitudes and the most carefully contrived and controlled experiments suffer from subtle biases within each of us–in other words, we are subject to human emotions. And if any of us are not human in any way, then the conclusions from non-humans about humans are not applicable to humans.

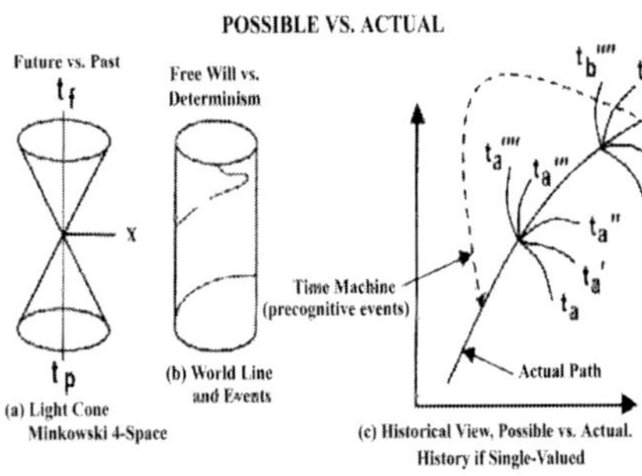

Figure 4. Possible versus actual; Several types of world lines are depicted.here. Figure 2a depicts a worldline with a single-valued now, but Figures 2b and 2c depict a multi-valued present. There is a dual world: constancy and change, absolute versus relativistic and Mach's Principle, and certainty versus uncertainty in terms of Einstein and Bohr (Courtesy of Rauscher and Amoroso [46]).

6. String Theory and M-Theory

String Theory and its 11-D structure M-Theory provides another explanation for a multiverse. These theories are an attempt to unify the forces and particles of physics at the Planck scale of energy, $(\hbar c^5/G)^{1/2}$ [47]. One impetus for string theory in extra dimensions (D>4) is the work of Kaluza and Klein formulation that gravity and EM could be integrated by introducing a 5th dimension that represents the electromagnetic field [18]. In M-theory, our universe and others are created by collisions between membranes (or branes) in an 11-dimensional space and may even share some of the fine-tuning of the universal constants.

Steinhardt and Turok introduced an ekpyrotic model that demonstrates how the collision of branes within a four-dimensional space, creates a big bang type phenomena. Each brane constitutes its own universe [47]. When branes collide the tension energy heats up and the universe starts the hot big bang allowing strings of all sizes and types to be generated during the collision [48].

At the same time, these universes can have completely different laws of physics. It has been postulated [20] that there are approximately over 10^{500} combinations of strings and therefore if all possibilities are to be taken into account, the same number of possible universes could exist. Specifically, this number permits a cosmic landscape of over 10^{500} different states of the quantum vacuum characterized by different fundamental constants. In these different states each could acquire different universal constants, and may emerge with different spatial and temporal dimensions depending on the outcome of symmetry breaking.

The additional discovery of the KKLT mechanism (Kachru, Kallosh, Linde, and Trivedi) [49] provides string theory with an explanation of how various universes can have different stabilized vacua states including those with a negative cosmological constant in anti de Sitter (AdS) space which establishes a positive-cosmological-constant of the de Sitter (dS) vacua. The KKLT mechanism can lead to 10^{500} different states of different vacua, corresponding to different local minima of energy within the string theory landscape. Various anthropic-related string theory structures are also included in the KKLT mechanism which indicates the manner in which the initial fields either roll down to the state where life of our type is impossible (AdS, 10D Minkowski space), or enters the state of eternal inflation.

One can also conjoin M-theory with inflationary cosmology, by expanding the false vacuum, to produce different bubble universes, in which vacuum fluctuations of each particle contribute to Λ. Just as there are 10^{500} different states of the quantum field, so also if we consider $\Lambda \approx 10^{-123}$ there are perhaps 10^{-121} - 10^{-123} possible states of the cosmological constant. These are just two of the elements that are needed to fine-turn each universe. This leads to an infinity of worlds, where each combination is repeated, in an infinite number of realities.

7. The Holographic Universe and the Multiverse

An additional model worth citing is that of the 'holographic universe.' Using Hawking radiation, we know that the surface of a black hole is just as significant as the volume. The edge of the area is a type of 2-dimensional surface. The holographic principle suggests that this surface contains enough information to perfectly represent everything that takes place within it volume. At the same time the volume within the space has the same information, but it is like a hologram of what we are able to measure on the surface area [20].

Amoroso and Rauscher [18] have further developed this theory into a Continuous-State Holographic Anthropic Multiverse (HAM) cosmology. Their HAM cosmology is based primarily on a fundamental least cosmological unit tiling the spacetime backcloth of its 12D superspace that makes correspondence with the SUSY parameters of M-Theory, and introduces the origin of complexity in self-organization, refining the role and nature of the observer in physical theory.

The paradigm has been developed by extending the Wheeler-Feynman-Cramer radiation/transactional interpretation models [50,51] and the de Broglie-Bohm ontological models to an higher dimensional regime commensurate with our version of SUSY-M-Theory parameters, but not interpreting Everett's many worlds condition as a duplicate parallelism, but as additional and unique in their own right.

For the HAM, 'our' whole *relational* Hubble sphere, H_R can be a subspace of an absolute 12 dimensional hyperspace without dimensionality. That is, these are additional dimensions are not compact, but 'open' and of infinite size [52], undergoing a process of 'continuous compactification and dimensional reduction'. To the earthly observer, it appears as a 'standing wave' of the present, continuously created and recreated by future-past advanced-retarded SUSY breaking dynamics.

Here the Hubble radius, H_R remains an observational limit in Continuous-State Anthropic Multiverse (HAM) cosmology also but is not caused by the Doppler effect. It is due to a minute non-zero rest mass for the photon [53]. As a photon propagates it couples to the polarized Dirac vacuum and loses energy also attenuating to zero observability; but if one were able to travel to the Hubble limit observation would extend for another Hubble radius ad infinitum. Thus a critical difference in interpretation of the redshift is a physical limit for the Big Bang and an observational illusion in HAM cosmology.

Therefore, instead of a rigid impenetrable, Planck barrier covered by a stochastic foam of particle creation and annihilation of the vacuum, HAM cosmology has a periodic ordered spacetime with a complex hyperstructure that is closed and finite in time for fermions, but open and infinite atemporally for bosons. In the HAM model, stochasticity, i.e. for zero-point string or brane dynamics, arises in the wake of unitary graviton propagation guiding the dynamics of the continuous-state.

8. Possible Observations

Although it may be impossible to make observations of a universe with different universal constants, Feeney [17] has developed a specific computer algorithm to search for the signatures of collisions between bubble universes in the expectation that they may have some effect within our universe, proving the existence of the multiverse. Feeney's premise is that if the collisions produce inhomogeneities in the inner-bubble cosmology, it raises the possibility that their effects are imprinted in the cosmic microwave background (CMB). This followed the research of Aguirre, Johnson, Kleban [17] and others suggest that a collision of our expanding bubble with another bubble in the multiverse would produce an imprint in the cosmic background radiation as a round spot of higher or lower radiation intensity. These can be relics of heat radiation left over from the Big Bang. Physicists are now searching the data using Feeney's unique algorithm and are looking for disk-like patterns in the midst of the cosmic microwave background (CMB) radiation.

Specifically, in order to verify whether a significant feature is well-modeled as a bubble collision, Feeney, *et al.* [17] perform Bayesian model selection for the bubble template. The Bayesian model selection uses the rules of probability theory to select among different hypotheses. Prior probability distributions are used to describe the uncertainty surrounding all unknowns. The Bayesian model works in the form of trees which are visited by each run of the algorithm, then a method is devised to identify those trees which are of most interest. Although no specific data has been cited, the search is underway to determine the possibility of the collision or prior existence of a false vacuum in the myriad points of the CMB. Analysis is being done on the cold spot in the CMB map with more details with the Planck satellite. Mathematics is also being worked on by Czech [54] which seeks to differentiate between random Gaussian fluctuations and what may occur from a bubble collision and a polarization pattern consistent with various angular power spectra.

9. Conclusions

Numerous theories have been developed leading many physicists to favor the concept that we exist in one of many universes, or a multiverse. We have examined some of the current multiverse worlds. These models describe a new cosmology that can include both the observer and the non-observer, taking into account all quantum possibilities. Although these theories may also be impossible to prove, they seem to be becoming impossible to disprove as they acknowledge many formulations of, for example Friedmann and Einstein. It is on the notion of the latter that we claim the universe, as we know it, is probably not the end of creation, but one of a seemingly infinite at 100^{500+} possibilities of different quantum states.

References

[1] Guth, A. (1997) The Inflationary Universe. New York: Perseus Books.

[2] Linde, A. (1994) The Self-Reproducing Inflationary Universe Scientific American, Vol. 271, No. 5, pp. 48-55, November 1994.

[3] Rauscher, E.A. Hurtak, JJ and Hurtak D.E. (2012) Universal Scaling Laws in Quantum Theory and Cosmology, presented at Vigier VIII - BCS Joint Meeting The Physics of Reality: Space, Time, Matter, Cosmos, London, August, 2012.

[4] Jenkins, A. and Perez, G. (2010) Looking for Life in the Multiverse. Scientific American, January, 2010.

[5] Perlmutter, S., et al. (2005) The Supernova Legacy Survey: Measurement of Omega_M, Omega_Lambda, and w from the First Year Data Set, Lawrence Berkeley National Laboratory, (October 14, 2005).

[6] Sorkin, R. D. (2007) Is the cosmological constant a nonlocal quantum residue of discreteness of the causal set type? arXiv:0710.1675.

[7] Steinhardt, P. J. and N. Turok, Why the cosmological constant is small and positive Science 312, 1180 (2006) [arXiv:astro-ph/0605173].

[8] Kofman, L., Gnedin, N. and Bahcall, N. (1993) Cosmological Constant, COBE DMR Anisotropy, and Large Scale Clustering,Astrophys. J. 413, 1, pp 1-9.

[9] Dharwadker, A. (2011) The Grand Unification, Proceedings of the Institute of Mathematics.

[10] Fixsen, D. J. (December 2009), The Temperature of the Cosmic Microwave Background, The Astrophysical Journal Vol. 707:2, pp. 916-920.

[11] Linde, A. (2005) Particle Physics and Inflationary Cosmology, arXiv:hep-th/0503203v1.

[12] Sievers, J. L., et. al. (2003) Cosmological Parameters from Cosmic Background Imager Observations and Comparisons with Boomerang, DASI, and Maxima, in The Astrophysical Journal Vol, 591:2, (2003 July 10), doi:10.1086/375510, arXiv:astro-ph/0205387v2, 11 Mar 2003.

[13] Souradeep, T. (2011) 'Standard' Cosmological model & beyond with CMB Class. Quantum Grav. Vol. 28 (2011) 114016 (14pp) doi:10.1088/0264-9381/28/11/114016 arXiv:1104.3201v1 [astro-ph.CO] 16 Apr 2011.

[14] Rauscher, E.A. (2005) Cosmogenesis and Quantum Gravity (eds.) R.L. Amoroso, B. Lehnert and J.P. Vigier, Beyond the Standard Model: Searching for Unity in Physics proceedings of the Paris symposium honoring the 83rd birthday of Jean-Pierre Vigier, 1st ed. the Noetic Press, Orinda, CA. pp. 43-72.

[15] Riess, A. G. et al. (1998) Observational Evidence from Supernovae for an Accelerating Universe and a Cosmological Constant, arXiv:astro-ph/9805201.

[16] Bahcall, N. A. et. al, (1999) The Cosmic Triangle: Assessing the State of the Universe, Science, 284, 1481.

[17] Feeney, S. M. et. al. (2011) First Observational Tests of Eternal Inflation, Phys. Rev. D 84, 043507.

[18] Amoroso, R. L. and Rauscher, E.A. (2009) The Holographic Anthropic Multiverse, Formulating the Complex Geometry of Reality. London: World Scientific Pub. Co.

[19] Hurtak, J.J. (1973) Nature of Space-time singularities, a paper presented in a public lecture at the University of California, Irvine, April 1973.

[20] Greene, B. (2011) The Hidden Reality, Parallel Universes and the Deep Laws of the Cosmos. New York: Vintage Books.

[21] Guth, A (2004) Inflation. ed. W. L. Freedman, Measuring and Modeling the Universe. Carnegie Observatories Astrophysics Series, Vol. 2, Cambridge: Cambridge Univ. Press, pp. 31–52.

[22] Lineweaver, C. H. (2003) Inflation and the cosmic microwave background, arXiv:astro-ph/0305179 v1 12 May 2003.

[23] Wilson, C. M. (2011) Observation of the dynamic Casimir effect in a superconducting circuit Nature, Vol. 479, pp. 376-379, 17 November 2011.

[24] Rauscher, E.A. (1968) Electron Interactions and Quantum Plasma Physics. Journal of Plasma Physics Vol. 2, pp. 517–541.

[25] Casimir, H. B. G. and Polder, D. The Influence of Retardation on the London-van der Waals Forces, Phys. Rev. 73, 360–372 (1948).

[26] Sparnaay, M. (1958). Measurements of attractive forces between flat plates. Physica 24 (6–10): 751.

[27] Lamoreaux, S. K. (1997). Demonstration of the Casimir Force in the 0.6 to 6 μm Range. Physical Review Letters Vol. 78: 5.

[28] Albrecht, A. (2004) Cosmic Inflation and the Arrow of Time, in Science and Ultimate Reality: Quantum Theory, Cosmology and Complexity, honoring John Wheeler's 90th birthday. J. D. Barrow, P.C.W. Davies, & C.L. Harper eds. Cambridge University Press.

[29] Steinhardt, P. J. and Turok, N. (2007) Endless Universe: Beyond the Big Bang. New York: Doubleday.

[30] Wheeler, J. A. (1962). Geometrodynamics. New York: Academic Press.

[31] Linde, A, Linde, D. and Mezhlumian, A. (2006) From the Big Bang Theory to the Theory of a Stationary Universe, arXiv:gr-qc/9306035v3.

[32] Davies, P. (2008) The Goldilocks Enigma: Why Is the Universe Just Right for Life? New York: First Mariner Books.

[33] Linde, A,, Sasaki, M. and Tanaka, T. (1999) CMB in Open Inflation, in Phys.Rev. D59:123522,1999, arXiv:astro-ph/9901135v2.

[34] Liddle, A.R. and Scherrer, R. (1998) A classification of scalar field potentials with cosmological scaling solutions, in Phys.Rev. D59:023509,1999, arXiv:astro-ph/9809272.

[35] Liddle, A. R. (1999) An Introduction to Cosmological Inflation., eds. A. Masiero, G. Senjanovic, and A. Smirnov in High Energy Physics and Cosmology, arXiv:astro-ph/9901124v1.

[36] Martin, J., Riazuelo, A. and Schwarz, D.J. (2000) Microwave Background Anisotropy Data The Astrophysical Journal, 543:L99-L102, 10 November 2000.

[37] De Witt, B.S. and Graham, N. eds. (1973) The Many Worlds Interpretation of Quantum Mechanics, Princeton NJ: Princeton University Press.

[38] Deutsch, D., (1986) Three experimental implications of the Everett interpretation, in R. Penrose and C.J. Isham (eds.), Quantum Concepts of Space and Time, Oxford: The Clarendon Press, pp. 204-214.

[39] Tegmark, M., (1998) The Interpretation of Quantum Mechanics: Many Worlds or Many Words?, Fortschritte der Physik 46, pp. 855-862.

[40] Rauscher, E.A. (2010) Quantum Mechanics and the Effect of Intentional Will, R.L. Amoroso, ed. The Complementarity of Mind and Body: Realizing The Dream of Descartes, Einstein and Eccles, New York, Nova Science Publishers, pp. 36-63.

[41] Aguirre, A., Tegmark, M., Layzer, D. (2010) Born in an Infinite Universe: a Cosmological Interpretation of Quantum Mechanics, arXiv:1008.1066v1.

[42] Deutsch, D. (2001) The Structure of the Multiverse The Royal Society Vol. 458:2028, pp, 1-21, April 2001, arXiv:quant-ph/0104033v1.

[43] Rauscher, E.A. and Amoroso, R. L. (2008) The Schrödinger Equation in Complex Space, Non-locality and Anticipatory Systems CASYS, Vol. 22 International Journal of Computing Anticipatory Systems, D.M. DuBois, ed. Institute of Mathematics, University of Liege, Belgium, pp. 371-388.

[44] Wigner, E. (1967) Symmetries and Reflections, Scientific Essays, Bloomington Indiana University Press and private conversations with EA. Rauscher and W.Van Bise.

[45] Van Bise, W. (1978) Very Higher Frequency and Microwave Frequencies and dependent effects of non-ionizing radiation on the human organism; a compilation of man and his electrical world, report from the Pacific Northwest Center for the study of Non-Ionizing Radiation, Portland, Oregon.

[46] Rauscher, E.A. and Amoroso, R. L. (2011) Orbiting the Moons of Pluto: Complex Solutions to the Einstein, Maxwell Schrodinger's and Dirac Equations, London: World Scientific.

[47] Davies, P.C.W. (2004) Multiverse Cosmological Models, Mod.Phys.Lett.A 19, pp., 727-744.

[48] Tye, S.H. H. (2006) Brane Inflation: String Theory viewed from the Cosmos. arXiv:hep-th/0610221v2.

[49] Kachru, S, Kallosh, R., Linde, A and Trivedi, S.P (2003) de Sitter Vacua in String Theory, arXiv:hep-th/0301240v2 10 Feb 2003.

[50] Wheeler, J.A. & Feynman, R. (1945) Interaction with the Absorber as the Mechanism of Radiation, Rev. Mod. Phys. 17:157.

[51] Cramer, J. (1985) The Transactional interpretation of Quantum Mechanics, Rev. Mod. Phys 58, pp., 647-687.

[52] Kafatos, M, Roy, S & Amoroso, R. (2000) Scaling in cosmology and the arrow of time, in R. Buccheri, V. di Gesu & M. Saniga (eds) Studies on the Structure of Time, Dordrecht: Kluwer Academic.

[53]Amoroso, R.L. (2006) The Holographic Conscious Multiverse, in R.L. Amoroso, B. Lehnert & J-P Vigier (eds.) Beyond the Standard Model: The Search for Unity in Physics, Orinda: Noetic Press.

[54] Czech, B.L. et.al., (2011) Polarizing Bubble Collisions, arXiv:1006.0832v2 [astro-ph.CO] 24 Mar 2011.

Universal Scaling Laws in Quantum Theory and Cosmology

ELIZABETH A. RAUSCHER*

Tecnic Research Laboratory, 3500 S. Tomahawk Rd Bldg 188
Apache Junction, AZ 85119 USA
bvr1001@msn.com

JAMES J. HURTAK & D. E. HURTAK

AFFS-Europe, 4058 Basel, Switzerland
dh@affs.org

We have developed a hyperdimensional geometry, *Dn* or Descartes space of dimensionality of $n > 4$, for our consideration $n = 10$. This model introduces a formation in terms of the conditions of constants as the space that allows us to calculate a unique set of scaling laws from the lower end scale of the quantum vacuum foam to the current universe. A group theoretical matrix formalism is made for the ten and eleven dimensional model of this space. For the eleven dimensional expressions of this geometry, a fundamental frequency is introduced and utilized as an additional condition on the topology. The constraints on the *Dn* space are imposed by the relationship of the universal constraints of nature expressed in terms of physical variables. The quantum foam picture can be related to the Fermi-Dirac vacuum model. Consideration is made for the lower limit of a universal size scaling from the Planck length, $l = 10^{-33}$ cm, temporal component, $t = 10^{-44}$ sec, density, 10^{93} gm/cm^3 and additional Planck units of quantized variables. The upper limit of rotational frequency in the *Dn* space is given as 10^{43} Hz, as conditions or constraints that apply to the early universe which are expressed uniquely in terms of the universal constants, *h*, Planck's constant, the G, the gravitational constant and *c*, the velocity of light. We have developed a scaling law for cosmogenesis from the early universe to our present day universe. We plot the physical variables of the ten and eleven dimensional space versus a temporal evolution of these parameters. From this formalism, in order to maintain the compatibility of Einstein's General Relativity with the current model of cosmology, we replace Guth's inflationary model with a matter creation term. Also we have developed a fundamental scaling relationship between the "size scale" of organized matter with their associated fundamental frequency.

* Initially the work of Rauscher was performed in part under the auspices of DOE, Lawrence Berkeley National Laboratory, Navy grant NOR-22, UC Berkeley, Tecnic Research Laboratory of Arizona, and later with the Academy for Future Science, USA and Europe.

Keywords: Cosmology, Quantum mechanics, Group theory, N-Dimensional topological spaces.

1. Introduction

Since the advent of Einstein's theory of special relativity in 1905, his general theory of relativity in 1915, and development of the quantum concept of Planck near the turn of the last century, and the further development of the quantum theory in the 1920s and 1930's, physicists have been attempting to reconcile these theories into a comprehensive and consistent framework with the long existing foundations of classical mechanics.

We formulated an *n*-dimensional topology that obeys the correspondence principle between classical and quantum mechanics, based on a scale correspondence to Planck's constant, and Bohr's principle of complementarity expressed in the Heisenberg uncertainty principle, and the Dirac formulation of relativistic quantum field theory. correspondences and universal complementarity. The complementarity and phase space relations are based on the neoclassical Hamilton-Jacobi theory.

Our theory is based on our five propositions which act to unify the various branches of physics. These propositions are: [1] The fundamental role in physics of constancy as expressed by the universal constants, and also constrains on dynamic processes, conservation principles and group symmetry relations; [2] The fundamental significance of canonical conjugate relations as an expression of a basic set of dual variables in nature; [3] A geometrical interpretation of the spacetime manifold in terms of a discrete set of quantities which comprise dimensional manifold space vectors; [4] The introduction of a new quantization procedure in terms of these fundamental discrete dimensional vectors and scalars as elements of a group theory; [5] Introduction of a ten and eleven dimensional Descartes space which represents, in is metrical expression is performed for the Descartes space for *EnD* where *n* is the number of extended dimensions of the space. The dimensions of the n-dimensional topology are expressed in terms of Planck's units of unique expressions in terms of physical variables and in terms of the universal constants.

These propositions lead to the following new concepts when the universal constants, in Planck unit formulation, are placed on a fundamental theoretical basis. J.A. Wheeler identified the expression for the Planck length in terms of the Wheeler wormhole [1,2]. A set of physical variables, uniquely expressed in terms of universal constants, termed extended dimensions or *EnD*, is developed [3,4]. All physical variables can be expressed in extended dimensional form. These discrete entities represent the geometrical constraints of the spacetime manifold and obey a set of canonical conjugate relations, which are like Heisenberg uncertainty relations in terms of generalized phase space conditions [5]. A group theoretic interpretation yields a common basis for cosmology and the quantum theory. In this conceptual framework, a detailed discussion of the group theoretic formalism is given in [6] and detailed structure of the generalized Minkowski metric space is given in earlier work of E.A. Rauscher [7]. This approach appears to lead to a unification of the basis of relativity and quantum theories.

2. The N-Dimensional Descartes-Space Geometry

In 1619, Rene Descartes suggested that in addition to his three spatial rectilinear coordinates, that time, velocity, energy and momentum, etc. be considered as coordinates [8]. Spinoza, in his treatise on ethics [9], suggested the concepts of space, time and substance (matter). Einstein, who wrote a preface for the republication of Spinoza's work, was influenced by both authors' consideration of additional dimensions to the coordinates of space, such as time, which led to his consideration of the Minkowski four-spacetime [10]. Dimensional quantities similar to those of Planck were developed by G.N. Lewis [11]. Our multidimensional space, in terms of group theory, [6] and an extended Minkowski metric, [7] denoted as M_n for $n > 4$ and usually $n < 11$, we term *Descartes geometry*. The Descartes space hypothesis of a complete geometry is based on physical constants, and creates no essential difference between matter and space. In this way the manner in which one piece of matter can be transferred from neighboring bodies and ultimately to the rest of the universe as nonlocal connections is understood [3]. Descartes states, "all the real and positive properties which are in moving bodies, and by virtue of which we say they move, are also found in those [bodies] contiguous to them, even though we consider the second group to be at rest" [12].

We examined the manner in which a set of geometrical constraints yield closed cosmological solutions to Einstein's field equations in earlier papers [13,14]. These constrains are expressed as a set of quantities or extended physical dimensions which are uniquely defined in terms of universal constants [1,4-7,13]. We have also presented a set of canonically conjugate relations of these physical variables expressed in terms of extended dimensions, *EnD* [6]. It was demonstrated that the quantized variables have operator representations [7], and a generalized form of the Schrodinger wave equation was developed in terms of these operators [14].

Previously, we defined the Descartes space in terms of the set of geometrical constraints expressed in n-dimensional topology of the Descartes geometry comprising a multidimensional space called Descartes space [3,6,13]. The multidimensional geometry was given in terms of invariant extended dimensions termed by *EnD* [4]. The metric for invariant dimensions is called the generalized Minkowski metric, M_n [7]. In this paper we demonstrate the manner in which the *EnD* geometrical constraints act in quantum theory and in relativistic physics, and the fundamental relationship of these two formalisms. The motivation is to determine a fundamental basis of relativity and quantum theories in the context of quantum gravity [3,4]. Although Descartes also stated one has to also consider the geometrical constants to create and maintain order and structure in the universe.

3. The Basis of the Descartes-Space Theoretical Approach

M. Planck introduced a set of units which are physical variables unequally expressed in terms of universal constants [15]. A similar, more limited approach was considered by G.N. Lewis and M. Randall [11]. Wheeler and others explained the use of the *EnD* quantities in geometrodynamic cosmological models [14,15]. Some of the *EnD*'s relevant here are: Planck's length or Wheeler's "*wormhole*" length,

$$\ell = \sqrt{G\hbar / c^3} \text{ ; time, } t = \sqrt{G\hbar / c^5} \text{ ; } p = \sqrt{c^3\hbar / G}$$

energy, and the Planck density $\rho = c^5 / G^2\hbar$. In these equations, $\hbar$, G and c denote, respectively, Planck's constant, the universal gravitational constant, and the velocity of light. Other extended dimensions, or EnD 's are given in Table 1 of Ref. [7].

The Planck quantities can represent geometro-dynamic quantities or physical variables as the EnD's.

That is, each *EnD* has an associated physical variable [6,7]. Note the somewhat different terminology in Refs. [3-7,13,17]. In relation to our terminology, *EnD*'s or quantized variables, we define two distinct quantization procedures, primary in terms of the quantized variables, and the standard quantization procedure [3,5-7]. The distinguishing characteristic between these two procedures is the classical, quantum, or relativistic domain of physics [3,5-7,13,14]; but in all these procedures there exists a set of canonical conjugate variables, termed the generalized Heisenberg relations [5,7]. It can be demonstrated that these two procedures give equivalent results. We form sets of pair relations as relativistic invariant "4-vectors" [7]. In the next section, we will develop the relationship between the canonical conjugate formalism of the *EnD*'s and the set of invariant relations that make up the generalized Minkowski metric space, M_n.

4. Descartes Space Geometrical Constrains in Relativity and Quantum Mechanics

The set of canonical conjugate relations are given in Fig. 1. We have the two usual relations, $(x,p) \geq \hbar$ and (E,t), $\geq \hbar$ and also four new relations [6]. We shall denote possible pair representations as the generalized pair, $(\rho_{\iota\varepsilon}, \upsilon_{\iota\varepsilon})$, where the index i runs 1 to 3 for vector variables and i = 1 for scalar variables. In Fig. 1, we consider only one component of each vector; for example one spatial dimension $x = x_1 = x_2 = x_3$. In this notation the pair $(\rho_{\iota\varepsilon}, \upsilon_{\iota\varepsilon})$ if $\varepsilon = 1$ have the canonical conjugate (cc) pair $(x,p) \geq \hbar$; and for $\varepsilon = 2$ we have the cc pair $(E,t) \geq \hbar$. The four new relations (4) are for $\varepsilon = 3$, $(x,E) \geq c\hbar$: $\varepsilon = 4$, $(p,t) \geq \hbar/c; \varepsilon = 5$, $(x,t) \geq \hbar/F$; and $\varepsilon = 6$, $(p,E) \geq \hbar F$, where F is the universal force, $F=c^4/G$ which occupies a primary in the stress energy term in Einstein's field equations [9.10]. Extensive literature exists which discusses the interpretation of the scalar Heisenberg relation $(E,t) \geq \hbar$. Time operators have been presented by several authors [5,19,20]. H. Eberly and L.P.S. Singh [19] develop an unambiguous and non-singular statement of the energy-time uncertainty relationship. We develop a time operator (as well as a space operator) in conjunction with the development of the generalized Schrodinger equation [14], which V.S. Olkhousky and E. Recami also discuss [20]. Eberly and Singh use the density matrix formalism to develop time operators and their uncertainty principle with Hamiltonian operators.

Formulation of an invariant generalized line element in terms of a universal constant is made, or combinations of universal constants, between any two variables, $\mu_{\iota\kappa}$ and $\eta_{\iota\kappa}$, where again the index i runs 1 to 3 for vector variables and i = 1 for scalar variables and the index κ runs 1 to n where n is defined as the dimensionality of the Descartes space, D_n. Considering one component vector quantities, for example $x = x_1$ and $p=p_1$, we can define a Descartes four-space as $\{x_k\} = \{x,t,p,E\}$. For example, we have the usual invariant relation,

$$S^2_{\delta=1} = x^2 - c^2t^2 \tag{1}$$

for one component of x; for an isotropic subspace where x represents all three components as, $x = x_1 = x_2 = x_3$ in Eq. (1) and for metrical signature (+, +, +, −). There are six invariant line-elements for a Descartes four-space [4,7]. We define a generalized four-vector invariant. The usual definition of a four-vector, for invariance relations, is in terms of a spatial vector quantity and a temporal scalar quantity which form an invariant variable pair relation in terms of the invariance of the universal constant, c.

The usual case is given in Eq. (1) and now we have:

$$S^2_{\delta=2} = p^2 - \frac{1}{c^2}E^2 \tag{2}$$

Again we consider one component of the momentum vector only where p represents all three components as: $p = p_1 = p_2 = p_3$. We show six invariant relations for a Descartes four-space in Tbl. 1. We have the usual relations for $\delta = 1, \delta = 2$ and $\delta = 3$, we have one of the new relations in terms of the invariance of the force, F.

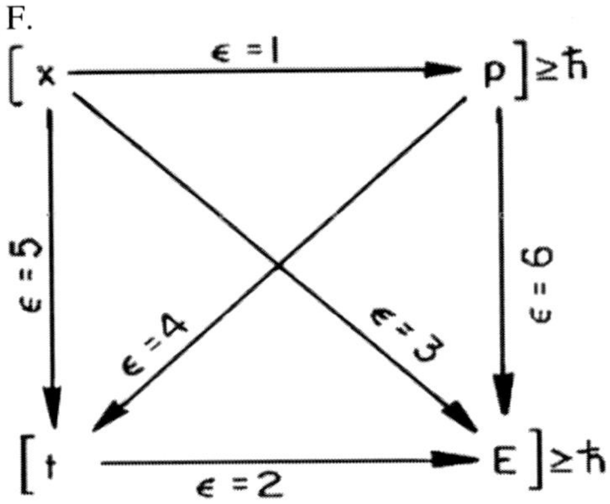

Figure 1. Pair variable relations represented schematically as the generalized Heisenberg Relations, where ε denotes a particular variable pair; for example, ε = 4 denotes the pair $(p_\varepsilon = 4, \upsilon = 4) = (p,t) \geq \hbar/c$. Horizontal rows, ε = 1 and ε = 2 represent the usual Heisenberg relations.

$$S^2_{\delta=3} = x^2 - \frac{1}{F^2}E^2 \tag{3}$$

The universal cosmological force is uniquely expressed in terms of the universal constants c and G as $F = c^4/G$ [4,14]; thus the invariance of the expression in Eq. (3) is dependent on the invariance of the universal constants c and G. For $\delta = 5$ in Table 1, we have a six-subspace for the (x, p) variable pair and for $\delta = 6$, we have a two-subspace for the (E,t) variable pair. Using the one component forms, as in Eqs. (1) and (2), each subspace is then a two-space.

The generalized invariant four-vector (which can also be a six or two-vector space) can be formed in terms of any two variables in terms of the invariance of many of the universal constants, $\hbar$, G and c or combinations of them. Using one-component vector quantities, the set of generalized invariant relations are generalized two-subspaces. A generalized invariant expression for a multidimensional Descartes space is given in Ref. [7], both for the four-space and a ten-space in terms of one-component vector and scalar coordinates, $\{x_k\} - \{x,t,m,E,p,F, L,c,a,P\}$ where m is mass, a is acceleration, L is angular momentum, P is Power and other qualities are defined previously. In Ref. [4] a higher order Descartes space of as many as thirty dimensions is presented which includes electromagnetic and thermodynamic coordinates [4,13,21,22].

The generalized form of an invariant pair is:

$$s_\delta^2 = \mu_{i\kappa}^2 + m_{\lambda\kappa}\eta_{i\kappa}^2 \qquad (4)$$

for any two variables $\mu_{i\kappa}$ and $\eta_{i\kappa}$, where λ and κ run from 1 to n which is the dimensionality of the Descartes space being considered and the index δ runs 1 to I (I is the number of invariant relations for a Descartes space of n dimensions). As before i runs 1 to 3 for vectors and $i = 4$ for scalars. The invariance relation between any two variables is expressed in terms of metrical elements $m_{\lambda\kappa}$ which are expressed in terms of universal constants or combinations of universal constants. The constant elements form a non-diagonal matrix terms the generalized Minkowski metric, M_n.

A generalized invariant, for all n dimensions of the space, can be expressed in terms of the diagonal form of the Minkowski metric. The diagonal form of M_4 is an analytic expression in a form that can be simplified in a linearized formalism [23]. In this approximation, the formalism is quite useful. In the full analytic form, there are a number of interesting implications of the theory [16].

In [4,7] we detail the group theoretical formulation of the Descartes space for an n-dimensional representation. In [4], we consider D_n for n as a 10 and 11 dimensional space. In the EnD model, the group generators are considered to form a finite algebra and a mapping is performed to the SU_n special unitary groups, having infinitesimal group generators. The details of the group structure and multiplication tables are given in Rauscher's earlier work [4]. The Heisenberg uncertainties fall naturally out of the D_n group formalism.

In formulating the generalized Heisenberg relations and generalized invariance in a multi-dimensional geometry, in which each dimensional physical variable is considered on an equal footing, we see that physical variables can be paired in uncertainty relations, such as those in Fig. 1. Also, paired variables can form invariant relations in terms of universal constants, as in Table 1. The relationship between these two pair relationships, as given in more detail in Refs. [6] and [7], is presented in Table 2. For example, the index $\varepsilon = 1$ denotes the Heisenberg pair (x, p); the index $\delta = 5$ denotes the same pair (x, p) in an invariant relation. In the notation in Table 2, $\varepsilon = \delta = 3$ and $\varepsilon = \delta = 4$, both denote the same variable pair related in a Heisenberg relation and as a relativistic invariant. In this manner, and with the assumption of "*equal footing*" of physical variables, we see a way in which quantum mechanics and relativistic invariance can be tied together by a geometrical model of the manifold.

Invariant, S_ε^2	Variable Pair μ_κ	Component of the Minkowski metric
$s_{\delta=1}^2$	$\mu_1 = x, \eta_1 = t$	c^2
$s_{\delta=2}^2$	$\mu_2 = p, \eta_2 = p$	$c^{1/2}$
$s_{\delta=3}^2$	$\mu_3 = x, \eta_3 = E$	$1/F^2$
$s_{\delta=4}^2$	$\mu_4 = p, \eta_4 = t$	F^2
$s_{\delta=5}^2$	$\mu_5 = x, \eta_5 = p$	c^2/F^2
$s_{\delta=6}^2$	$\mu_6 = t, \eta_6 = E$	$c^2 F^2$

Table 1. Pair-variable relations as invariants in terms of the elements of the generalized Minkowski metric. Six such invariant pair relations can be formed for a four-dimensional Descartes space.

The invariant expressions in this paper are special relativistic invariants. General relativistic invariants are discussed in [7] as are light cone relations for the generalized special relativistic invariants. In [4], the quantized variable geometrical constraints are applied in general relativity and closed cosmological solutions are found for Einstein's field equations. Experimental evidence for closed cosmologies is cited in [4,13,16].

GENERALIZED HEISENBERG PAIRS	GENERALIZED INVARIANT PAIR VARIABLES
$\varepsilon = 1, (x,p)$	$S_{\delta=1}\ (x,t)$
$\varepsilon = 2, (t,E)$	$S_{\delta=2}\ (p,E)$
$\varepsilon = 3, (x,E)$	$S_{\delta=3}\ (x,E)$
$\varepsilon = 4, (p,t)$	$S_{\delta=4}\ (p,t)$
$\varepsilon = 5, (x,t)$	$S_{\delta=5}\ (x,p)$
$\varepsilon = 6, (p,E)$	$S_{\delta=6}\ (t,E)$

Table 2. We can represent the relation between a pair of physical variables in two different formalisms. We have the uncertainty relation between two variables $(\rho_\varepsilon, \upsilon_\varepsilon)$ where the index ε denotes a particular variable pair. We can also represent an invariant relation between two variables as $(\mu_\kappa, \eta_\kappa)$ and the index δ denotes a particular invariant relation. For ε = 1 and 2 we have the usual quantum mechanical relation and for δ = 1 and 2 we have the usual invariant four-vector relations. Table 2 depicts the manner in which these two representations of paired variables relate to each other.

In [4,7,13], we formulate the manner in which we solve Einstein's field equations in the n-dimensional Descartes space which leads to uniquely closed cosmological solutions. In [4], we formulate in detail quantum cosmogenesis and evolution of the current universe. Self-consistency between a modified big bang model and Einstein's field equations is attained without Guth's inflationary model, [24] which is extraneous in this picture. Also, our model is consistent with nucleo abundances [16]. Note that Guth's model may require a velocity of 6 x 10^3 the velocity of light. Dark matter comes out naturally from our new, self-consistent model, and modifications to the cosmic evolution allow us to account for dark matter and dark energy. Current Hubble's constant values, and critical densities, appear consistent with observed densities and matter content of the universe, and may accommodate a cosmological constant value Λ near zero, and yet still accommodate an apparent dark energy component [25].

5. A Generalized Group Theoretical Representation Interpretation of Descartes N-Dimensional Space

We present a group representation in which the physical variables expressed as Planck units form the canonical relations and elements of a group. A set of generalized canonical elements are generated from the Planck variable generators. The pair relations of the *EnD*'s are the non-commutative elements of a non-Abelian group, of which the well-known cases are $(\ell, p) \geq \hbar$ and $(t, E) \geq \hbar$...

We term the primary physical laws as those that result from group operations on physical variables expressed as *EnD*'s. Such laws of physics, for example, are $F = ma$, $E = mc^2$, $p = mc$, and $(\ell, p) \geq \hbar$. So termed secondary laws of physics follow indirectly from the group operations of the elements. They involve the group representation chosen, basis vectors and group generators or the group algebra. The group algebra is used to define 'differentiality' on the manifold, and thus such laws as radioactive decay, electroweak interactions, etc., are secondary laws. [4]. The fact that the primary laws of physics come about through the group operations on the *EnD*'s is an aspect of the principle of the dimensionality of universal correspondence, in which that relativistic, quantum mechanical and classical laws have the same basic origin. For example, we have $F = ma$ (classical); $E = mc^2$ (relativity) and $(\ell, p) \geq \hbar$ (quantum mechanical).

The primary group D_s is diphomorphic but the secondary subgroup, *Rs*, is not differentiable nor integratable, but we will see that we can construct an isomorphism of the group algebra that generates this group, which we will denote R, to the generators of a group D, containing the elements of differentiability, integratbility, and exponentiality, and its inverse operation. Also, the representation of *R* is determined isomorphic to the symmetry group so a one-to-one mapping of each element can be made on a specific permutation, P, of objects that will specify the symmetry relations between the physical variables in the *EnD* form. By use of the group D_s, we can define a metric which will give invariant relations of the manifold. One such relation is the invariance $ds^2 = g_{\mu\nu} dx^\mu dx^\nu$ in general relativity. [4,6,7]

This space D_s and subspace R_s (the sub *s* denotes isomorphism to the symmetry groups), does not contain the operations of differentiation and integration directly, but one can see that its operations are connected to these operations through the relation of the principle of least action or integral containing element $d\eta$ where η is a physical variable, which results in canonical conjugate relations. If we have the

Poisson bracket for two canonical conjugate variables (μ,η), then we can write the principle of least action as $\int \mu d\eta = \varsigma$ which is an integral expression involving the differential element $d\eta$.

We construct the multiplication table for the subgroup of physical variables expressed as *EnD*'s or D_s generators' that are considered to have a mechanical correspondence. In Table 3 we have constructed the elements of this group which comprise a multiplication table which is the table of canonical conjugate like relations.

The paired variable processes, resulting from the dual multiplication procedures, generate dimensional correct products. The product canonical relations are expressed in terms of universal constants: $c, \hbar$, and $F = c^4/G$, the universal force in Einstein's stress energy term, or expressed as a single physical variable. Other physical variables appear in the cross product such as the ℓ, m, c, p, etc., as well as, the square of the term *EnD*'s on the diagonal, as squared variables.

6. Closed Cosmologies, Missing Matter, and Energy Without Inflation

Einstein originally introduced the cosmological constant to create a "steady state" universal model in 1915. This all changed with Hubble's discovery of the red shift expansion of the universe in the early 1920's. The Hubble constant is the ratio of the velocity of recession over the distance to the stellar, galactic or other cosmological structures for:

$$H = \frac{V}{R} = \frac{R\ /\ t}{R} \propto \frac{1}{t_0} \qquad (5)$$

where R is the distance to the nearest galactic or stellar source, $V = \dot{R}$ is the velocity of recession and t_0 is the age of the universe which is about 13.7 x 10^9. The Doppler red shift is given as $z = \Delta\lambda / \lambda$ where λ is the rest frame wavelength emitted from the observed cosmological object and $\Delta\lambda$ is the wavelength shift.

Einstein stated that his inclusion of the cosmological constant, Λ was the worst mistake of his life. [26] Currently astrophysicists have reinstated Λ to explain very distant galactic acceleration [27]. We write Einstein's field equations as:

$$R_{\mu\upsilon} - \frac{1}{2} g_{\mu\upsilon} R_{\mu\mu} + \Lambda g_{\mu\upsilon} = -\frac{8\pi G}{c^4} T_{\mu\upsilon} = -\frac{8\pi}{F} T_{\mu\upsilon} \qquad (6)$$

where $T_{\mu\nu}$ is the Riemann-Christoffel curvature mass-energy tensor, and $F = c^4/G$ is the Rauscher cosmological acceleration force. The term $R_{\mu\mu}$ is the Ricci Curvature tensor and $R_{\mu\nu}$ is the spacetime curvature.

One of the most promising explanations of non-gravitational and non-electromagnetic acceleration of the universe is due to the vacuum energy which is implied as a source of the acceleration. This manifestation of the vacuum energy may be expressed as the need to reintroduce Einstein's cosmological constant, Λ where, rather than insuring a static universe, it forms a more dynamic one with an acceleration parameter as a function of R, the cosmological distance. Thus Λ may become an effective gravitational repulsion term. The additional term $\Lambda g_{\mu\nu}$ can be related to the vacuum in the following manner. We express the right side of Einstein's field equations in terms of the stress energy tensor as $G_{\mu\upsilon} = -\frac{8\pi G}{c^4} T_{\mu\upsilon}$ and write this symbolically as two terms, $G_{\mu\upsilon} = -\frac{8\pi G}{c^4}[T_{\mu\upsilon} + T^{vac}_{\mu\upsilon}]$, where we can use $c = 1 = G$ and $T^{(vac)}$ represents the stress energy of the vacuum and can be written as $T^{(vac)} = (\Lambda / 8\pi) G$, that is, the stress energy density expressed as the cosmological constant term contributes to the vacuum energy density; the vacuum density $\rho^{(vac)} = T^{(vac)}_{00} = \Lambda / 8\pi$ and G can be used to account for distant galactic acceleration gravitational effects. In standard approaches $\rho^{(vac)}$ is considered to be small or $\rho^{(vac)} < \rho^{(matter)} \sim 10^{-29}$ gm/cm^3; but in fact, the structure of the vacuum of $\rho \sim 10^{93}$ gm/cm^3 may well determine the structure and form of observable matter. Note that the limits for $\Lambda \neq 0$ are units of cm^{-2}. Acceleration appears to occur at the edge of the known universe [27-29]. The accelerating universe, apparent magnitude vs. red shift, z from z_0 to $z > 1$ is statistically significant. Galactic red shifts will eventually vanish beyond the event horizon where $z > 6$. On the Schwarzschild universe see [4,13,16, 30] and the next section.

We examine the concept of the action of the vacuum as a dynamic system that may explain the recently observed anomalous expansion rates indicated by high z. Work at the Sloan Digital Sky Survey telescope at Apache Point, New Mexico, has identified a very high red shift quasar at $z = 4.75$. This red shift indicates that this cosmological object ought to be 13.75 billion years old, formed when the universe was less than one billion years old. Note in the standard view, the gravitational red shift of the sum is:

$$z = \frac{\Delta\lambda}{\lambda} = \frac{\lambda_r - \lambda_e}{\lambda_e} = 2x10^{-6} \qquad (7)$$

The Sloan Digital Sky survey's main task is to measure red shifts and they have identified several new quasars, the most distant ever observed [31]. Their most distant quasar has a z = 6.28 and another quasar has a z = 5.73 which was discovered a few years ago by the Sloan survey. In fact, observable red shifts have been identified for four quasars in this survey: z = 5.80, z = 5.82, z = 5.99 and z = 6.28, which are the highest z quasars observed. It is believed that 20 more quasars with $z > 6$ will be found [32]. The so-termed "*missing mass*", first suggested by Fritz Zwicky in 1933, is considered to be different from any ordinary matter in the universe in that there are zero detectable emissions or absorption of light in any known or ordinary manner.

The rate of expansion of the big bang universe from a black hole indicates the conditions for a closed universe, and throughout the evolution of the universe [4,16]. Currently, and for the last 65 years, it is thought that the observed mass of the universe indicates that 94% of the matter needed for a just closed universe is "*missing*" [16].

The critical density for a just closed universe is $\rho_c = 3H^2/8\pi G$. The Hubble's constant is given as $H = \dot{R}/R$, where R is the distance scale and $\dot{R}$ the velocity as determined by the redshift, z. We then deduce that $\rho_c \propto 1/R^2$ and thus, also, the time scale of the universe, $1/H \sim t_0$ is affected. Variations in the red shift and the distance scale estimates affect the most likely value of ρ_c. Nuclear abundance and x-ray data are also relevant to determining the value of ρ_c [13]. The value is about $\rho_c \sim 2.4 \times 10^{-29}\ gm/cm^3$ derived from H =73.8*Km*/sec/MPC. These density estimates are very sensitive to the D/H (Deuterium/ Hydrogen interstellar ratios).

The form of the missing mass is hypothesized to be, in part, in the form of black holes, interstellar plasma, or vacuum state energy [33]. Some astrophysicists consider the missing mass maybe "cold dark matter" (CDM). Three proposed pictures emerge. Dark matter is said to only weakly interact with ordinary matter. Three models have been put forward: (1) The concept of WIMP's or weakly interacting massive particles, that create vibrations and bursts of light and heat, (perhaps as quasars which rotate around the center of the galaxy); (2) MACHO's or massive compact halo objects, which may relate to the clumping of galaxies, and (3) HPVS's or hyperspace resonance vacuum structures, which relates to both of the above models and expands on them. If the vacuum energy is structured and interacts with quasars, black holes, and galactic matter, it may act to guide the evolution of universal features such as galactic clumping, since galaxy formation and evolution is thought to have started and from a relatively uniform distribution during cosmogenesis.

All three of these models fail to yield phenomena that are directly observable by even the most sophisticated telescopes, but their properties are reflected in the form of matter-energy we observe, ranging from particles, to galaxies, to galactic clusters, and beyond. Some of these models have within them empirical implications of a matter-energy and entropy evolution occurring throughout cosmogenesis and cosmology [4]. This, in turn, leads to a reconciliation of the Hoyle-Narlikar creation model with big bang cosmology through the formulation of the "*little whimper*" model [16,34]. This picture leads to a just closed cosmology throughout the evolution of the universe.

7. Quantum Cosmology and the Evolution of a Closed Schwarzschild Universe

The initial condition constrains the characterize cosmogenesis are chosen to be the quantum gravity level vacuum containing the Planck length $\ell \sim 10^{-33} cm$, and quantized time of $t \sim 10^{-44}$ sec [4,16]. E.R. Harrison discusses early universe quantities, which are Planck-like units related to the vacuum energy [36]. They are the initial conditions of the universe, in our model, and act as a set of constraints throughout the evolution of the universe. E.R. Harrison discusses the role of the vacuum in quantum gravity and mechanics and such units as length, time and energy, and also the thermodynamic properties of cosmogenesis. Harrison's approach is synergistic to ours. See the previous section on the discussion of the vacuum state contributions to the stress energy term. [4,16,36]

In Table 3, we list some characteristic early universe physical quantities as quantized variables and compare these to characteristic present day universe values which obey the Schwarzschild condition $R_s = 2Gm/c^2$. The length $\ell \sim 10^{-33} cm$ can be interpreted as the limit of length in the manifold, and $t \sim 10^{-44}$ sec the corresponding characteristic time for $c = \ell/t$, where c, the velocity of light is taken to be the characteristic signal propagation velocity in the manifold. The present day magnitude of the physical variable of length is the size of the universe, $R \sim 10^{27}$ cm and the age of the universe is $t_0 \sim 10^{17}$ sec, and c the velocity of light, $c \sim 10^{10} cm$/sec.

A source for matter creation may occur from a mirror universe or universes. The multiverse principle arises from the linear superposition principle in the quantum theory [3] and in this picture for future non-interacting universes is valueless. However, if we expand the vector space for nonlinear coupled resonant phenomena, the multi-universe branching can yield a matter creation term and, perhaps, explain certain aspects of dark matter and dark energy as matter creation terms in our explorations [35].

Physical Quantity	Symbol	Early Universe	Present
Length	ℓ	10^{-33} cm	10^{27} cm
Time	T	10^{-44} cm	10^{17} cm
Mass	m	10^{-5} gm	10^{56} gm
Velocity	c	10^{10} cm/sec	10^{10} cm/sec
Acceleration	a	10^{53} cm/sec^2	10^{-7} cm/sec^2
Force	F	10^{49} dynes	10^{49} dynes
Power	p	10^{59} dynes cm/sec	10^{59} dynes cm/sec
Pressure	P	10^{114} dynes/cm^2	10^{-7} dynes/cm^2
Energy	E	10^{16} erg	10^{75} erg
Density	ρ	10^{93} gm/cm^3	10^{-29} gm/cm^3
Frequency	ω	10^{43} cycles/sec	10^{-17} cycles/sec
Temperature	T	10^{32} degrees (K.)	3 (or 2.7) degrees (K.)
Entropy	S	10^{-16} erg/degree	10^{45} erg/degree

Table 3. The early universe physical quantities are the Planck units which are compared to some present day characteristic values of physical variables on the scale of the universe.

The fundamental vacuum velocity is the velocity of maximal real mass signal propagation in the manifold $v \leq c$. The approximate critical density for a just closed universe is $\rho_c = 3H^2/8\pi G{\sim}10^{-29}$ *gm/cm*^{3}, in this case [2,4] where Λ is taken as zero and $k = +1$. In the previous section we considered cases where $\Lambda \neq 0$.

The ratio $\dot{R}/R = H$ is Hubble's constant, and, for $t_0 \leq 1/H$, H=2.4 x 10^{-18} *cm*/sec/*cm*, where t_0 is the initial time condition of the universe and H is Hubble's constant. Note that for the present universe $\dot{R} = HR$ $=10^{-18}$ *cm*/sec/*cm* x 10^{27}, giving R~10^{10} *cm*/sec, or the velocity of light. We have examined the limits set by deuterium abundance as a constraint on H in Ref. [16].

We calculate the evolution from initial to current characteristic values of physical variables during cosmogenesis to the present day universe. In Table 3 we tabulate the initial values of physical variables and their values throughout the evolution of the universe. These values are plotted in Figure 2 (ordinate of physical variables in their respective units) versus time (abscissa in seconds) for various physical variables.

The physical variable values are obtained as follows: At time $t = 10^{-44}$sec, all physical variables assume their Planck unit values as initial constraints on the universe. We assume that the characteristic velocity of signal propagation of the universe is $v = c$, the velocity of light; therefore, the radius at any time is $R = vt$. Utilizing the Schwarzschild condition, we obtain the mass as $m_s = c^2R_s/2G$.

X →	↑t	↑R	↑v	↑m	↑ρ₁	↑ρ₂	↑ρ₃
	$\frac{1}{10^{-44}}$	10^{-33}	10^{10}	10^{-5}	10^{93}	10^{93}	10^{43}
	10^{-23}	10^{-13}	10^{10}	10^{18}	–	10^{45}	–
	10^{-10}	1	10^{10}	10^{28}	–	10^{25}	–
	10^{-4}	10^{7}	10^{10}	10^{35}	10^{35}	10^{14}	10^{14}
	1	10^{10}	10^{10}	10^{38}	10^{13}	10^{5}	10^{4}
	10^{4}	10^{15}	10^{10}	10^{43}	10^{11}	10^{-2}	10^{-21}
	10^{10}	10^{20}	10^{10}	10^{48}	–	10^{-12}	–
now	10^{7}	10^{28}	10^{10}	10^{56}	10^{-29}	10^{-30}	10^{-32}
unit	sec	cm	cm/sec	gm	gm/cm^3	gm/cm^3	gm/cm^3

↑E	↑T	↑S₁	↑S₂	↑F	↑p	↑a	↑ω
10^{16}	10^{32}	10^{-16}	10^{-16}	10^{49}	10^{59}	10^{59}	10^{43}
10^{53}	10^{7}	10^{7}	–	10^{49}	10^{59}	10^{31}	10^{30}
10^{63}	–	10^{17}	–	10^{49}	10^{59}	10^{21}	10^{10}
10^{53}	10^{12}	10^{24}	10^{43}	10^{49}	10^{59}	10^{14}	10^{6}
10^{58}	10^{9}	10^{27}	10^{49}	10^{49}	10^{59}	10^{11}	1
10^{63}	10^{4}	10^{32}	10^{59}	10^{49}	10^{59}	10^{6}	10^{-9}
10^{68}	–	10^{37}	–	10^{49}	10^{59}	10^{-1}	10^{-19}
10^{76}	3	10^{45}	10^{75}	10^{49}	10^{59}	10^{-7}	10^{-17}
ergs	deg K	erg/°	erg/°	dynes	dynes cm/sec	cm/sec^2	cycles/sec

Table 4. Schwarzschild Evolution of the Universal: Physical Value verses the Age of the Universe.

The corresponding density for the Schwarzschild condition is given by $\rho = m_s / R_s^3$ where the subscript *s* denotes the Schwarzschild criterion. In Table 3 and Figure 2 we denote this density as (ρ_2) which is compared to R. Omnes' [37] values (ρ_1) and E.R. Harrison's values, (ρ_3) [36,38]. The expression for Hubble's constant for a "now" or present day universe applies as $\rho_c - 3H^2/8\pi G$ for $p_c = 10^{-29}$ *gm/cm*^{3}. For the initial conditions of the universe, $\rho_i \sim 10^{93}$ *gm/cm*^{3}, where we utilize $\rho = 3H^2/8\pi G$. If we take G as a constant, then $H^2 \cong 8\pi G/3$ x 10^{93} or $H \sim 10^{43}$ *cm*/sec/*cm* = $\dot{R}$ /*RH* ~ 10^{10} *cm*/sec. We can still use the present day value of Hubble's constant, $H \sim 10^{18}$ *cm*/sec/*cm* to define $t_0 \leq H$, as 1.37 x 10^{17} sec, or the age of the universe. We must note that there is still some controversy over the value of H but the D abundance does set constrains on H [16]. We also make the implied assumption that the microscopic (nuclear and atomic) and macroscopic (electromagnetic and gravitational) physics, are valid throughout the evolution of the universe.

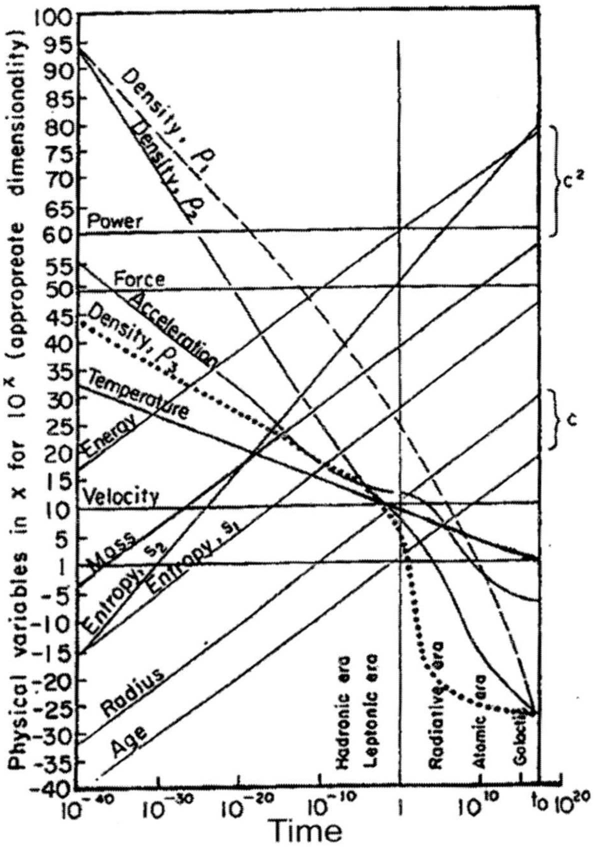

Figure 2. Scaling Law of Physical Variables In the Time-Evolution of a Schwarzschild Universe.

The total energy of the universe is given by $E_s = m_s(t)c^2$, where the subscript denotes the Schwarzschild mass at any time, t, and $T_s = E_s/k$ is the energy per particle. Comparison can be made to E.R. Harrison [36, 38]. The relationship $T_s = E_s/k$ holds for the initial conditions only. In general, throughout the evolution of the universe, $T_s \propto 1/E$. The increase in energy (and mass) has to be reconciled with the cooling off or decrease in temperature. We will now omit the subscript s. The entropy under the Schwarzschild condition is given as $S = m_s(t) \times S_I / m_i$, where S_I and m_i, are the quantized variable initial conditions of the universe and $m_s(t)$ is the Schwarzschild mass as a function of time. The initial entropy is $S \cong k = 10^{-16}$ *erg/degree*, where k is the Boltzmann constant. In Table 4 and Figure 2 we denote this entropy as S_I.

We compare this to the entropy, denoted S_2, obtained from the third law of thermodynamics, $S \sim \Delta E / T$ where $\Delta E = E_f(t) - E_i(t)$, and for temperature, $T = T(t)$. We can use the approximation $\Delta E = E_f(t)$ since for $t \geq 10^{-30}$ sec, $E_f >> E_i$. For $t = t_0$ (the present age of the universe), $S_2 = (10^{76} - 10^{-16})$ $erg/3^\circ K \cong 10^{75}$ $erg/^\circ K$ which agrees with E. R. Harrison's value [36, 38]. The Force is given by $F = E_s(t) / R_s(t)$ with the ratio being constant and given by the universal force, $F = c^4/G$. The power is given by $p = vF$ and for $v = c$, $P = cF$, which is the quantized power. Interpretation of the quantized Force and quantized Power for an isotropic homogeneous universe with a Robertson metric is given in E.A. Raucher [4,16]. In Table 4, the present day Power, P ~ 10^{57} *dynes* x *cm*/sec is just the power to expand the universe to its present day configuration, which is a factor of 100 less than the usual power value. The Pressure, p ~ 10^{-7} $dynes/cm^2$ is the present density pressure for a critical density, $\rho_c \sim 10^{-29}$ gm/cm^3. The acceleration is given by $a_s(t) = F / m_s(t)$, and $m_s(t)$ is the Schwarzchild mass. As the "big bang" expansion continues, the rate of expansion slows down and then appears to accelerate again at the far reaches of the universe. The results in Table 4 and Figure 3 are consistent with this calculation. The identification of stellar and galactic centered black holes, as well as neutron stars, white dwarfs, etc, greatly supports our work [33].

The frequency, ω is given as $\omega \propto 1/t$. There have been some attempts to calculate the total rotation of the universe [16]. This rotation may possibly be interpreted in a Machian sense [3]. The theory of general relativity is compatible with Mach's principle but is left out of the usual interpretation of frames of reference. Indeed, Mach's principle is consistent with matter creation and may even be explained by the continuous creation of matter (matter-energy) [4]. In the multidimensional geometrical model, we may be able to formulate matter-energy creation in terms of a coordinate transformation. Very simply, we can consider a model in which "*stretching*" field lines "*pops*" particles into existence from the vacuum, i.e., space-time transforms into matter-energy by a generalized rotation of the dual Riemannian sheets of the Descartes topology [4]. It is shown by E.A. Rauscher that the generalized Minkowski metric has this property, if rotations in the Descartes space are considered [6,7].

The remarkable fact then emerges that in order to maintain the Schwarzschild condition, as an initial and present condition, we must evoke matter creation (at a macroscopic constant rate) to make this model self-consistent between Einstein's field equations and the big bang, which precludes the need for the inflationary model. It appears that we may be able to reconcile the continuous creation and big bang cosmologies in our "*little whimper*" model, in which the universe, under initial conditions, "*explodes*" as a "*mini-black hole*" and then larger scale black holes, with their surrounding plasma fields, [39,40] evolve with the creation of matter by a continuous process of matter influx from the vacuum energy, which may be, in part, detected as Hawking radiation from black holes in the universe. It is interesting to speculate as to the role of matter *"creation"*, from geometry, and matter "destruction" into geometry, and matter *"destruction"* into geometry, in terms of a multitude of black holes, and in terms of available vacuum energy. One may formulate the one as the converse of the other in terms of time reversal. The issue of time reversal invariance and/or, more completely, CPT invariance, requires consideration in this picture [4].

An active vacuum, in excited plasma states in the vicinity of black holes, may account, at least in part, for some of, the missing mass [4,33,38,40].

By imposing the generalized Schwarzschild condition initially, and presently, and throughout the evolution of the universe, which acts as boundary conditions, we have found that matter creation is necessary in a model that is self-consistent with Einstein's field equations (in the expansion phase of the universe) and the big bang model. This condition led to the *"little whimper"* closed cosmologies. We look now at *"the ashes and cinders"* of which A.G. Lemaitre spoke, and ask, are they ashes of a big bang, in which all the matter of the universe *"instantaneously"*, in $t \sim 10^{-44}$, exploded, or are these the ashes of a developing, on-going exploding growth [4]? When we consider rotations in our Descartes space [4], stretching or distorting the generalized metric, such as the outer reaches of the cosmos, we may be able to interpret this as a disruption of the mechanism whereby the radiation field is converted into particulate matter. One important experimental factor in determining closed versus open cosmologies is the present day deuterium abundance. In [16] we discuss the consistency of our model with the observed deuterium abundance and the implied value of Hubble's constant (See Figure 2).

8. Discussion

The details of a unification scheme between quantum mechanics and relativity is a work in progress [3,4] but a start has been made based on demonstrating the common roots of the Hamilton-Jacobi classical mechanics, non-relativistic and relativistic quantum mechanics and Einstein's field equations. Unifying quantum mechanics and general relativity is one thing; it is quite another to demonstrate the "common roots" of the formalism. A relativistic quantum field theory, or a "quantized" gravitational theory, may relate to an aspect of a unifying theory, but more is needed to attain a fundamental unifying theory. One proposal has been suggested in a paper on fractal scaling models [46, 47]. In addition, quantized gravitational theories may shed some light on the path toward a fundamental relation between quantum and relativistic physics [3]. For example, the problems of reconciling non-linearities in general relativistic fields and the linear super-position principle in quantum mechanics, are well pointed out by some of the researchers in quantized gravitational theories [3]. B. S. De Witt [41, 42] in an extensive treatise, discusses some of the difficulties of non-linearities of the metric tensor and the quantum superposition principle. Attempts have been made to develop a linear theory of the massless, spin-2 field [23]. Relating to the fundamental structure of the "roots" of the canonical formalism is fundamental to a unifying of quantum mechanics and relativity [3]. Since both quantum mechanics (primarily relating to micro-phenomena) and relativity (primarily relating to macro-phenomena) are so successful in elegantly describing physical phenomena, a unifying aspect for these fields of physics must necessarily involve the interrelation of diverse aspects of reality.

It also seems apparent that every aspect of reality depends on every other aspect, as expressed clearly by H. Stapp [43], "Every part of the universe depends on every other part". The bootstrap model of elementary particles developed by G. Chew [44] is another statement of this proposition. A unified theory of reality which must intimately bring together quantum and relativistic phenomena must necessarily be complete. The concept of completeness in quantum theory is discussed in the classic paper by A. Einstein, B. Podolski and N. Rosen [45] and subsequent papers inspired by this work [3] and references therein.

The extended dimensions of the Descartes space, which are expressible uniquely in terms of universal constants, are manifest in quantum mechanics and special relativistic invariants and general relativity, as well as thermodynamics [22], and electromagnetic theory [4]. According to B.N. Taylor, W.H. Parker, and D.W. Langenberg [48], the "universal constants are an important link in the chain of physical theory which binds all the diverse branches of physics together". To paraphrase J.A. Wheeler: "spacetime is not just a passive arena for doing physics, it is the physics." [1] and, we might add, matter, energy, momentum, spin, etc. as in the Descartes geometry.

In 1921, Einstein discussed a fundamental aspect of reality: "It was formerly believed that if all material things disappeared out of the universe, time and space would be left; according to the relativity theory, however, time and space disappear together with the things" [25]. In a Descartes space, should any one-dimensional physical variable disappear, then all the rest of the universe would also disappear. It is the interactive whole that comprises the universe.

Physical Variable	ℓ	t	m	E	p	F	L	c	a	p
ℓ	ℓ^2	$\hbar/F$	$\hbar/c$	$c\hbar$	$\hbar$	E	$\hbar\ell$	$\hbar/m$	c^2	cE
t	$\hbar/F$	t^2	$\hbar/c^2$	$\hbar$	$\hbar/c$	$(\hbar/c)^{1/2}$	$\frac{\hbar^3}{cF}$	ℓ	c	E
m	$\hbar/c$	$\hbar/c^2$	m^2	p^2	P^2/c	$\frac{p^3}{\hbar}$	$\frac{p^3}{F}$	$\frac{E}{c}$	F	Fp
E	$c\hbar$	$\hbar$	p^2	E^2	$\hbar F$	$(c\hbar F^3)^{1/2}$	$(c\hbar F^3)^{1/2}$	$(c^3\hbar F)^{1/2}$	c^2F	$\frac{E^3}{\hbar}$
p	$\hbar$	$\hbar/c$	p^2/c	$\hbar F$	p^2	$\frac{cp^3}{\hbar}$	$(\frac{\hbar^3F}{c})^{1/2}$	E	p	FE
F	E	$(\frac{\hbar^3}{c})^{1/2}$	$\frac{p^3}{\hbar}$	$(c^3\hbar F)^{1/2}$	$(\frac{F\hbar^3}{c})^{1/2}$	F^2	$c\hbar$	p	$(\frac{c^3F^3}{\hbar})^{1/2}$	cF^2
L	$\hbar\ell$	$(\frac{\hbar^3}{cF})^{1/2}$	$\frac{\hbar^2F}{c^2}$	$(c\hbar F)^{1/2}$	$(\frac{\hbar^3F}{c})^{1/2}$	$\hbar F$	L^2	$c\hbar$	cE	E^2
c	$\hbar m$	ℓ	E/c	$(c^3\hbar F)^{1/2}$	E	p	$c\hbar$	c^2	c^2/t	c^2F
a	c^2	c	F	c^2F	p	$(\frac{c^3F}{\hbar})^{1/2}$	cE	c^2/t	a^2	$c\hbar t^3$
p	cE	E	Fp	$(c^2\hbar F)^{1/2}$	FE	cF^2	E^2	c^2F	$c\hbar t^3$	p^2

Table 5. Generalized Canonical Conjugate Relations and Physical Laws.

Acknowledgements

Elizabeth A. Rauscher expresses acknowledgment for fruitful discussions with John A. Wheeler, Bryce De Witt, Geoffrey Chew over the years. J.J. Hurtak expresses acknowledgment for a discussion with Henry Stapp in Salzburg, Austria.

References and Notes

[1] Wheeler, J.A. (1962) Geometrodynamics, Academic Press, and (1972-1992) Private Comm.

[2] Miser, C.W., Thorne, K. & Wheeler, J.A. (1973) Gravitation, San Francisco: W.H. Freeman.

[3] Rauscher, E.A. & Amoroso, R.L. (2011) Orbiting the Moons of Pluto: Complex Solutions to the Einstein, Maxwell, Schrödinger & Dirac Equations, Singapore: World Scientific.

[4] Rauscher, E.A. (1971) A Unifying Theory of Fundamental Processes, UCRI-20808, LBNL/UCB Book Publications.

[5] Rauscher, E.A. (1972) A set of generalized Heisenberg relations, Lett. Al Nuovo Cimento 4, 757.

[6] Rauscher, E.A. (1972) A group theoretical representation of the generalized Heisenberg relations, Lett. Al Nuovo Cimento 5, 925.

[7] Rauscher, E.A. (1973) The Minkowski Metric for a Multidimensional Geometry, Lett. Al Nuovo Cimento 7, 361.

[8] P.J. Davies and R. Hersh, Descartes' Dream: The World According to Mathematics, Houghton Mifflin Co., 1986.

[9] B. de Spinoza, Ethics: Demonstrated in Geometrical Order and Divided into Five Parts. MacMillan and Co., 1883.

[10] H.A. Lorentz, A. Einstein, H. Minkowski and H. Weyl, The Principles of Relativity, Dover Publications, 1923.

[11] J.N. Lewis and M. Randall, Thermodynamics, McGraw Hill, 1923.

[12] E. Slowik, Descartes' Physics, E.N. Zalta, ed. The Standard Encyclopedia of Philosophy, 2009.

[13] Rauscher, E.A. Closed Cosmological Solutions to Einstein's Field Equations. Al Nuovo Cimento, 3, 661. 1972.

[14] Rauscher, E.A. & Amoroso, R.L. (2008) The Schrödinger equation in complex Minkowski space, non-locality and anticipatory systems, CASYS Intl J Computing Anticipatory Systems, D.M. Dubois (ed.) 370-388.

[15] Planck, M. (1959) Theory of Heat Radiation, Dover, p. 175.

[16] Rauscher, E.A. (2005) Cosmogenesis and quantum gravity, in R.L. Amoroso, B. Lehnert, J-P Vigier (eds.)

Beyond the Standard Model: Searching for Unity in Physics, Proceedings of the Paris Symposium Honoring the 83rd Birthday of Jean-Pierre Vigier, Oakland: The Noetic Press.
[17] R. L. Amoroso L. H. Kaufman, P. Rowlands & E.A. Rauscher, Unified Field Mechanics, (in process).
[18] E.J. Post, Ann. Phys. 71, 497, 1972 and Private communication.
[19] J.H. Eberly and O.P.S. Singh, Phys. Rev. 70, 359, 1973.
[20] V.S. Olkhovsky and E. Racami, Lett. Al Nuovo Cimento, 4, 1165, 1970 and Private Communication with E. Racami.
[21] I. Prigogine, J. Geheniaut, E. Gunzigt, and P. Nardonet. Thermodynamics of Cosmological Matter Creation. Proc. Nat. Acad. Sci. USA Physics, 85, pps. 7428-7432. Oct. 1988, and Private Communication with I. Prigogine, 1981-85.
[22] Rauscher, E.A., Speculations on Geometrical Constraints, Superconductivity and Signal Propagation. Lawrence Berkeley National Lab. report, LBL-2153, Sept. 1974.
[23] There are some studies of linearized derivations of the metric tensor from a chosen background, for example, a Minkowski background. One such rigorous treatment is that of K. Kuchar, J. Math. Phys. 11. 3322 (1970) which depends on the decomposition of tensor fields.
[24] A. Guth. The Inflationary Universe. Basic Books, 1998, and Phys. Rev. 23, 347, 1981.
[25] G.W. Gibbons and N. Turok, The Measurement Problem in Cosmology, Phys. Rev. D77, 063516, 2008.
[26] A. Einstein at a New York Times press interview in 1921. See R.W. Clark, Einstein: The Life and Times. The World Publishing Co., 1971.
[27] S. Perlmutter, et al. Discovery of Supernovea Expansion at half the age of the Universe and cosmological implications. Lawrence Berkeley National Laboratory, LBNL-41172 (1997).
[28] S. Perlmutter and B. Schmidt, Measuring Cosmology with Supernovea, arXiv: astro-02/0303428 v.i. (2003).
[29] A. Riess, J Fliri, D. Valls-Gabaud Cephedi Period-Luminosity Relations in the Near Infrared and the Distance to M31 From the Hubble Space Telescope wide field camera 3, Astrophysical Journal, Vol 742, No. 2. (2011)
[30] Rauscher, E.A. & Hyson, M.T. (2012) Geometrical Constraints in and N-dimensional topology in Quantum Mechanics and Cosmology, CASYS, University of Liege, Belgium. IJCS.
[31] X. Fan, et al., Astron. J. 131, 1203, 2006.
[32] R. Barkana and A. Loeb, Phys. Rev. 349, 125, 2001.
[33] N. Haramein and E.A. Rauscher, Collective Coherent Oscilation Plasma Models in Surrounding Media of Black Holes and Vaccum Structure–Quantum Processes with Considerations of Spacetime Torque and
Coriolis Forces, Beyond the Standard Model: Searching for Unity in Physics. R.L. Amoroso, B. Lehnert & J-P Vigier (eds). The Noetic Press, pp. 279-331, 2005.
[34] F. Hoyle and J.V. Narlikar, El Nuovo Cimento 7A, 242, 1972 and Private communication with Sir Fred Hoyle, UCB, 1970.
[35] Hurtak, J.J., Rauscher, E.A. & Hurtak, D.E. (2013) Examining the existence of the multiverse, in R.L. Amoroso P. Rowlands & L.H. Kauffman, The Physics of Reality, Space, Time, Matter, Cosmos, Singapore: World Scientific.
[36] E.A. Harrison, Rev. Model Phys. 39, 862, 1967.
[37] R. Omnes, Phys. Rev. 181, 2130, 1969 and Private Communication LBNL, 1968.
[38] E.A. Harrison. Ann. Phys. 35, 437, 1965.
[39] Rauscher, E.A., Electron Interactions and Quantum Plasma Physics. J. Plasma Physics, 2, 517, 1968.
[40] R.L. Amoroso & Rauscher, E.A., The Holographic Anthropic Multiverse; Formalizing the Complex Geometry of Reality, World Scientific Press, 2009.
[41] B.S. De Witt, Phys. Rev. 160, 1113, 1967 and Private Communication.
[42] B.S. De Witt, Phys. Rev. 162, 1195, 1967.
[43] Henry Stapp, Lawrence Berkeley National Laboratory, UCRL-11688, Sept. 30, 1964, and Private Communications, LBNL, 1964-96.
[44] G. Chew, Science 181, 762, 1968 and Private Communications, LBNL 1964-96.
[45] A. Einstein, P. Podolsky and N. Rosen, Phys. Rev. 47, 777, 1935 and extensive discussions in the Fundmental Fysiks Seminar by E.A. Rauscher, 1975-79.
[46] Hartmut Muller, Fractal Scaling Models of Natural Oscilliation in Chain Systems and the Mass Distribution of the Celestial Bodies in the Solar System. Progress in Physics, pps.62-66, January 2010.
[47] Hurtak, James J. Discussions with H. Muller on the origin of metamatter at Lindau, Germany, June 2011.
[48] B.N. Taylor, W.H. Park er and D.N. Langenberg, Rev. Mod. Phys. 41, 375, 1969.

Implementing Maxwell's Aether Illuminates the Physics of Gravitation: The Gravity-Electric (G-E) Field, Evident at Every Scale, From the Ionosphere to Spiral Galaxies and a Neutron-Star Extreme

MILES F. OSMASTON

The White Cottage, Sendmarsh, Ripley, Woking, Surrey GU23 6JT, UK.
miles@osmaston.demon.co.uk; http://osmaston.org.uk

Relativity Theory (RT) incorporates serious inconsistencies:- (1) embracing the function of transverse e.m. (TEM) waves as perfect messengers but denying the presence of a Maxwell's equations aether lest it might invalidate that perfection, despite it being essential for their existence; (2) assuming the physical absurdity that the external physical properties (mass, magnetic moment) of fundamental particles can be developed in zero volume ("spatially infinitesimal singularities"), despite powerful evidence that they are of finite size. It thereby overlooks that if two electromagnetically defined objects are of finite size the force communication between them is progressively velocity-limited, falling to zero at c [Heaviside 1889]. So this is what happens in electromagnetic accelerators, not mass-increase. For more than a century these defects have hampered progress in understanding the physics of the mass property of particles, thus compelling it to be regarded as 'intrinsic' to those specific infinitesimal points in space. A rewarding substitute, Continuum Theory (CT), outlined here, (A) implements Maxwell's aether as a massless all-pervasive quasi-superfluid elastic continuum of (negative) electric charge, and (B) follows others [Clerk Maxwell, both Thompsons, Larmor, Milner] in seeing mass-bearing fundamental particles as vortical constructs of aether in motion, not as dichotomously different from it. To encompass that motion, these cannot be infinitesimal singularities. Electron-positron scattering provides guidance as to that size. For oppositely-charged particles, one sort contains more aether and the other less, so particle-pair creation is 'easy', and abundantly observed, but has been attributed to 'finding'. This electron-positron relationship defines mean aether density as >1030 coulomb.cm-3, thus constituting the near-irrotational reference frame of our directional devices. Its inherent self-repulsion also offers an unfathomable force capability should the means for displacing its local density exist; that, we show, is the nature of gravitational action and brings gravitation into the electromagnetic family of forces. Under (B) the particle mass is measured by the aether-sucking capability of its vortex, positive-only gravitation being because the outward-diminishing force developed by each makes mutual convergence at any given point the statistically prevalent expectation. This activity maintains a radial aether (charge) density gradient - the Gravity-Electric (G-E) Field - around and within any gravitationally retained assemblage. So Newton's is an incomplete description of gravitation; the corresponding G-E field is an inseparable facet of the action. The effect on c of that charge density gradient yields gravitational lensing. We find that G-E field action on plasma is astronomically ubiquitous. This strictly radial outward force on ions has the property of increasing the orbital angular momentum of material, by moving it outwards, but at constant tangential velocity. Spiral galaxies no longer require Cold Dark Matter (CDM) to explain this. The force (maybe 30 V.m-1 at solar surface) has comprehensive relevance to the high orbital a.m. achieved during solar planet formation, to their prograde spins and to exoplanet observations. The growth of high-mass stars is impossible if radiation pressure rules, whereas G-E field repulsion is low during dust-opaque infall, driving their prodigious mass loss rates when infall ceases and the star establishes an ionized environment. Its biggest force-effect (~1012 V.m-1) is developed at neutron stars, where it is likely the force of supernova explosions, and leads to a fertile model for pulsars and the acceleration of 1019 eV extreme-energy cosmic rays. Our only directly observed measure of the G-E field is recorded at about 1 V.m-1 in the ionosphere-to-Earth electric potential. And temporary local changes of ionosphere electron density, monitored by radio and satellite, have been discovered to act as earthquake precursors, presumably, we suggest, by recording change of G-E field and gravitational potential at Earth surface when its elastic deformation occurs, even when this is deep below electrically conducting ocean water. The paper concludes by noting experimental evidence of the irrelevance of the Lorentz transformations in CT and with a discussion of CT's competence in such matters as perihelion advance and Sagnac effect, widely regarded as exclusively RT attributes. Finally we broach the notion that the aether is the site of inertia. This could explain the established equality of gravitational and inertial masses. In an accompanying paper we explore the cosmological and other aspects of 'making particles out of aether'. This link undermines the expectation of fully distinct dynamical behaviour by particles and aether which motivated the Michelson-Morley experiment.

Keywords: Maxwell's aether; gravitation physics; planetary systems; star formation; spiral galaxies; dark matter; neutron stars; supernovae; origin of inertia; Sagnac effect.

Foreword. This paper is written with certain philosophical maxims in mind. "It is what we think we know that prevents us from learning" (Claude Bernard). If what you think you know leads you to the absurd, then the choice lies between piling on more absurdity and starting afresh. Incisive recognition and avoidance of absurdity is therefore crucial.

1. Introduction

Papers whose titles suggest – as does this one – that they strike at the roots of currently accepted scientific thinking often arise from their author's personal discontent with one or more of its tenets. So we note here that the new basis for physical theory, which I have named Continuum Theory (CT), aspects of which are outlined in this paper, has a more substantial basis than that. Indeed it originates from a discovery I made in 1959 in the course of an intensive and secret research program to develop airborne astronavigation equip-ment for use by missile-carrying high-flying aircraft.

Main Contents

1. Introduction
2. The Physical Mechanism of Gravitation: Setting the Scene
3. Two Disastrous Absurdities lie in RT's Foundations
4. CT as a rewarding replacement for RT: the two-part Basis of Continuum Theory (CT)
5. Constructing Fundamental Particles out of Aether
6. Gravitational Action: Generation of the Radial Gravity-Electric (G-E) Field
7. Quantifying the G-E Field
8. Seven Major Sites of G-E Field Action
9. The G-E Field in the Inner Heliosphere: a Small Selection of the Effects
10. A G-E Field Model for CMEs and Granulation
11. Is this Positive Behaviour of the Sun and Stars Permanent?
12. Constructing the Solar Planetary System — and Others
13. The G-E Field in Spiral Galaxies
14. G-E Field Actions at Neutron stars
15. Relativistic (RT) Matters and CT
16. The Irrotational Aether, Sagnac effect and the Origin of Inertia
17. Three Experimental Checks

The matter at issue was the already well-observed sky brightness distribution at those heights [1] which, to a degree important in this case, does not conform to the expectation of current (primarily Rayleigh) scattering theory, as stressed by Barr [2]. So we tried a treatment involving deflection scattering by an atmosphere-associated randomly moving aether. And, in retrospect somewhat serendipitously, it worked, almost to perfection, as instantly recognized in the establishment. But secrecy prevented my subsequent access to the papers and inhibited public development of it for many years.

So the characterization of the aether presented in this paper is ultimately the product of satisfying the requirement that, in effect, the motion of the aether be closely related to that of the particles in the domain concerned; a 'particle-tied aether', in fact (Osmaston 2003 [3]). But the purpose of this paper is then to show that this characterization has led us, as an inherent by-product, to an insight into the fundamental physics of gravitation that seems to be remarkably rewarded by supporting evidence for its action at all the astronomical scales considered so far.

The quite different but even wider significance of the scattering action which triggered all these studies is discussed in the accompanying second paper [68].

2. The Physical Mechanism of Gravitation — Setting the Scene

The idea that gravitation might be some kind of (large-scale) vortex effect, but in an aether of unknown character, seems to have been originated by Kepler (1609) but was soon followed by Descartes (1637). Newton (1726) picked it up but seems to have decided to concentrate on the behaviour of the force rather than upon its origin.

But, from the early 1860s, the particulate nature of matter now being clear, James Clerk Maxwell, William Thomson (Kelvin) and several others started to think in terms of individual 'particles' – they began with molecules, the smallest objects then known to them – having a vortical character ('vortex rings') of aether in motion as the source of their mutual attraction and mass property. Here, 150 years later, we will be building upon those endeavours.

Then, in 1865-1878, came Maxwell's famous equations based on the findings of Gauss, Faraday and Ampère. From these he established the properties of an aether, essential for the existence and propagation of transverse electromagnetic (TEM) waves. These are continuous math functions, with no suggestion that the aether could have a particulate nature, such as he had been envisaging at the particle scale. Their crucial, and apparently contradictory, requirement which I address here and implement, as perhaps never before, is that the aether must combine quasi-superfluid behaviour with elasticity in shear.

The final item in our scene is that JJ Thomson's (1897) cathode rays work led people to identify the electron as **the** carrier of electric charge, thus regarding negative electric charge to be particulate in nature, not to be found in any other guise.

Here, we will reason that, on the contrary, the vast majority of negative charge in the Universe exists in a distributed form, with electrons incorporating only a tiny fraction of it.

3. Two Disastrous Absurdities Lie in RT's Foundations

Firstly. Embracing the function of transverse e.m. (TEM) waves as perfect messengers between inertial frames (hence the Lorentz transformations) but denying the presence of a Maxwell's equations aether, essential for their existence. This was like saying "I want the ripples but not the water".

This enabled Einstein to avoid any possibility of *transmission effects*, associated with the aether, and to postulate "***c*** *is an absolute constant of physics*". Incisively, this is here seen as physically unrealistic in a real interactive Universe. Everything affects everything else, however slightly that may be. This unrealism has, in turn, spawned a widespread view that there are indeed many such "absolute constants of physics". Rather, a more fertile view for understanding the physics which underlies them might be to regard such numbers as convenient approximations within the range of conditions available for observation.

As noted above, it was in fact the observation and my recognition of one of those TEM wave transmission effects in 1959 which started me along the present trail.

Secondly. Promoting, despite growing contrary evidence, the *physical absurdity*, central to the GR equations, that fundamental particles are *'infinitesimal singularities'*. This would require that their external physical properties – mass, magnetic moment – either be developed within zero volume or (by inventing a new law of physics) treated as 'intrinsic' to that particular infinitesimal space. Perhaps the most explicit evidence of non-zero size is the widely used magnetic resonance imaging (MRI scan) in medicine, which requires the proton's magnetic axis to be long enough to be made to wobble. Another emerges from the electron-positron scattering observations with LEP.

The consequence for fundamental physics of this 'singularity' perspective has been severe. In any e.-m. theory, force communication between two electro-magnetically defined objects is, *IF they are of finite size,* progressively velocity-limited by access delay, to being zero at ***c*** (e.g. Heaviside [4]. A similar limitation had been inferred experimentally by the physicist *Wilhelm Weber* in 1854 [5]). Retardation on impact will be similarly limited, increasing particle penetration. The shock-wave-like sweep-back angle seen in the Çerenkov effect, daily used for determining particle velocities, constitutes observational confirmation of the effect, albeit in a material where the TEM-wave speed is deliberately reduced by the refractive index. So this is what we observe with electromagnetic accelerators, not mass-increase. The extra mass that actually emerges has been created from the enveloping aether, also accelerated - see later.

This opens our door to the 'designing' of a particle's interior to give each a specific mass, behaving as such to any observer, free of any circumstantial dependence.

4. CT as a Rewarding Replacement for RT — The Two-Part Basis of Continuum Theory (CT)

(A) Implementing Maxwell's aether as a massless, all-pervasive, compressible, quasi-superfluid continuum of negative electric charge

It is NOT a **carrier** of electric charge; electric charge is its very nature, so it is **massless** and rotation motions do not fly apart; it is **all-pervasive,** even to the interiors of atoms; it is **compressible**, constrained by the self-repulsion of its charge; it is **superfluid,** so its motions can be perpetual, but it is only **quasi-superfluid**, because shearing motions of charge are subject to Ampére's law. This provides magnetic energy storage and effective 'elasticity' in shear needed to support the existence and propagation of transverse electromagnetic (TEM) waves**.** It is also important because the lack of viscosity if it were a pure superfluid would mean that its relative motions would be incapable of inducing the rotational disturbances of it that we will be considering as being in the nature of fundmental particles; (we show later that this property of coupling in shear arises valuably elsewhere also); it is a true **continuum**, so it cannot exhibit discrete subdivision, although its density may vary.

(B) Following Maxwell [6], W.Thomson (Kelvin) [7], J.J.Thomson [8], Larmor [9] and Milner [10] in making finite-sized particles out of it, as vortical constructs of aether motion.

Milner [10], indeed, regarded the aether as a "fundamental substratum of matter from which we can imagine particles are formed". We will do the same.

In combination, (A) and (B) will lead us in two important directions:-

(a) the physical mechanism of gravitation — this paper;

(b) the particle-associated random motion of the aether — our companion paper [68].

5. Constructing Fundamental Particles out of Aether

5.1. *Quantifying the Aether*

The aether can only be of one sort of charge, so how do we get particles with opposite charge? To provide electrons and positrons with opposite **relative** charge, I suppose one to contain *more* aether and the other *less*.[a]

Figure 1 shows notional aether density profiles that would equip otherwise-similar(?) electron and positron aether dynamical configurations with equal and opposite amounts of electric charge (aether).

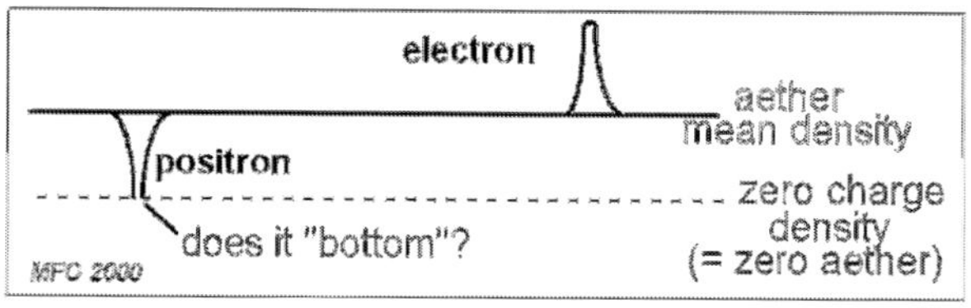

Figure 1. Diagram based on a negative-charge aether. Because of its ubiquitous continuum nature, less than zero aether is not an option.

On this basis the mean charge density of the aether must at least equal the peak charge density in a positron. From the LEP scattering-determined 'effective' electron-positron sizes ($<10^{-16}$cm[b]) and its known (relative) charge we find the **aether <u>mean</u> charge density must be <u>at least 10^{30} Coulombs.cm^{-3}</u> !** *(N.B. 1 Ampere = 1 C.s^{-1})*

In this manner the familiar electron-positron pairs would easily be made without introducing or removing any aether, but merely by 'stirring it up' appropriately, which could explain why they mirror one another. In turn, if the electron exemplifies a particularly stable aether dynamical form, it follows that the mean density of the aether defines the electron as the elementary unit of charge throughout the Universe; and likewise the uniformity of atoms built by its means. This is an important result in a continuous-creation Universe in which (see companion paper [68]), with no BigBang, there was never mutual proximity to give uniformity.

This huge charge density of the aether gives its self-repulsion an immense force potential if it is locally displaced (gravitation, see next Section). Except on the smallest of scales (individual fundamental particles), it also may well provide the irrotational reference frame on which our directional devices depend. Those devices (Foucault pendulum, mechanical gyroscope, laser ring gyro) work in two quite different ways, inertial-gravitational and electromagnetic-Sagnac. So it would not be trivial to do both, the aether being the link. We will return to this in Section 16.

5.2. *Mass and Mutual Gravitation as the Result of Vortical Action: A Speculative look Inside a Finite-Sized Fundamental Particle made of Aether in Motion (Fig. 2)*

If this aether dynamical form (Fig. 2), or something like it, is the origin of the mass property, **TEM waves do not involve the right kind of aether motion to give them mass.**

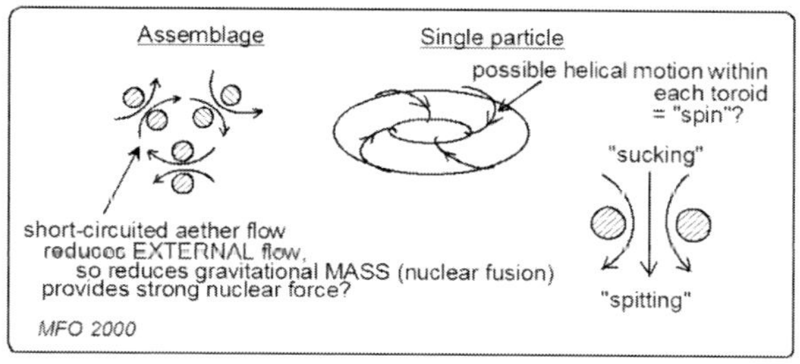

Figure 2. On the right is a cross-section of the central image. The mass of a particle or particle assemblage is measured by its (aether-sucking) ability to attract others. Charge circulation gives its axis and magnetic moment. On the left, three quarks (protons) are stable because of good aether-circuiting although adding an electron (neutrons) causes instability in a free state (~10-minute half-life). Two quarks (mesons) are unstable ($<10^{-7}$ sec) because aether-circuiting is poor also (strong nuclear force is insufficient).

Radiation pressure, often treated as the result of having such mass, is in fact well known as a perfectly valid electromagnetic force. The related, relativistic, idea of TEM waves being subject to gravitational redshift is considered in our companion paper [68].

In the same vein, we [3] have considered neutrinos to be pure eddies of aether, without vortical action, so devoid of any mass property, their energy content being that associated with the aether's eddy motion.

On the basis of Figures 1 and 2 the creation of mass-bearing electron-positron, or other particle-antiparticle pairs is possible merely by the addition of the appropriate rotational energy of aether motion[c]. So here again, if the Second Law is to be sustained, the energy-content of aether motions must be recognized in energy-balance (e.g. entropy) calculations [3]. In our second paper [68] we develop the important Universal

[a] In this suggestion I am apparently preceded by an editorial in The Electrician (1891)[11], commenting on William Crookes' presidential address to the IEE, that he had failed to consider the existence of positive and negative electricity as possibly being "two converse manifestations of one and the same entity" (see p.329), although according to Isobel Falconer (pers comm 2010) this idea was not altogether new even then. There may be other priorities unnoted here.

[b] G.E.Kalmus, pers.comms.1991, 1996

[c] That creation of mass may, on an $E = mc^2$ basis, have occurred from the energy put in, is not a new idea, but our concern is with how that transformation occurs, not just to invoke the magical action of a boson.

cosmogony aspects of 'making particles out of randomly moving aether', which would be impossible unless aether motion contains energy.

The origin of gravity. Mutual attraction by vortical-structured particles predominates statistically over repulsion because of the inevitable force gradient around a body. At a given position on that gradient, infinitesimal convergence has more force-effect than identical separation. That's why we only have

positive-only gravitation – no negative.

6. Gravitational Action — Generation of the Radial_Gravity-Electric (G-E) Field

So in a gravitational assemblage the particles are, by partial self-orientation, 'busy' sucking aether out of the interior. This reduction is opposed by the restorative elastic **self-repulsion of the aether's electric charge**, which is therefore **the underlying nature of Newtonian gravity**. Extreme force is available if the 'sucking' is intense.

But the resulting radial gradient of aether density is an electric field - **the G-E field**. Similar interaction with the rest of the Universe causes the G-E field to extend indefinitely outside the body, as does its Newtonian field also.

So the Newtonian field and the Gravity-Electric (G-E) field are but facets of a single physical mechanism and one is never present without the other.

Because of this direct relationship, G-E field strength at the surface of an object will apparently increase, perhaps linearly but in a manner to be determined, with the Newtonian potential there, being highest at neutron stars, with white dwarfs second.

But note that, unlike the Newtonian, the G-E field is a **discriminatory force**, only producing actual radial repulsion on sufficiently ionized plasma and other entrained material.

7. Quantifying the G-E Field

7.1. The *G-E Field of the Earth – Observations*

Persistent potential difference exists — Earth surface to ionosphere. Around 90% of all steep lightning strikes bring negative charge to ground. Satellites observe sprites from base-ionosphere (~75km) to thunderclouds. The overall potential difference from the F-layer (~300km) is assessed at ~250kV **(= ~1 V.m^{-1})** [12]

The account given by [13] that local changes in ionosphere electron density behave as precursors to major earthquakes, has aroused much interest. But their occurrence even when these occur deep beneath electrically conducting sea-water [14,15] probably rules out interpretation based either on the effusions from active fractures or on the electromagnetic effects of deforming crystalline rock. So we propose that they may actually reflect gravity and G-E field changes at the Earth surface level as it is deformed elastically by plate dynamical forces.

7.2. *The G-E Field of Stars*

A linear extrapolation from the Earth value, based on gravity, would give the following, but it is unclear at present whether linear extrpolation with g is actually appropriate:-

~30 V.m^{-1} at the solar surface, and
approaching 10^{12} V.m^{-1} at a neutron star.

8. Seven Major Sites of G-E Field Action

1) *Inside stars*, it is an additional radial support force, reducing the required rate of nucleosynthesis. Solar neutrino deficiency[d] and longer stellar timescales (OK in a no-Big-Bang Universe – see companion paper [68]). Raises the 'Chandrasekhar mass limit' for Type Ia supernovae, as recently observed [16,17].

2) *Outside stars*, its radial charge density gradient gives lensing (lower density gives lower c in Maxwell's equations). So c is not 'a constant of physics'.

3) *Building massive stars.* This is not possible unless the G-E field is the main agent (*not* radiation pressure (RP)) of their subsequent prodigious rates of mass loss (e.g. 1 solar mass/20 ka). Growth is then possible by the scant G-E action upon dust-opaque plasma-poor infall from a cloud. So the Newtonian infall force wins, as long as the star remains in the cloud. On emerging, the star will ionize its envelope, and the G-E field plus RP will take over, giving the big mass-loss rates. That mass-loss will enhance the opacity of the neighbourhood, encouraging more high-mass stars to form. Is this the mechanism of the starburst phenomenon?

The next four sites, listed below, are dealt with in more detail in the succeeding Sections of this paper.

---

[d] Nuclear reaction rate inside the Sun varies approximately as the 25th power of the temperature, so a tiny drop in temperature, far too small to have an observable helioseismological effect, would suffice to halve the reaction rate and explain the halved neutrino output.

4) *Inner heliosphere.* Coronal energy, accelerating the solar wind, releasing CMEs, causing photosphere opacity and driving granulation.

5) *During planetary system formation.* Outward force driving the plasma-rich protoplanetary disc wind, needed for moving planetary nuclei from very close-in positions (*cf.* ~22% of all exoplanets detected) to attain their high specific orbital angular momenta, relative to solar or other 'parent' star a.m.

6) *In spiral galaxies.* Driving axial-infall material outwards at **constant tangential velocity**, removing the need for Cold Dark Matter (CDM) to explain this very widespread dynamical pattern. (Same basic pattern as for planets (5)).

7) *At the surfaces of neutron stars.* For driving supernova explosions that leave a neutron core, for driving pulsar beams and accelerating the high-end cosmic rays (10^{19} eV).

9. The G-E Field in the Inner Heliosphere — A Small Selection of the Effects

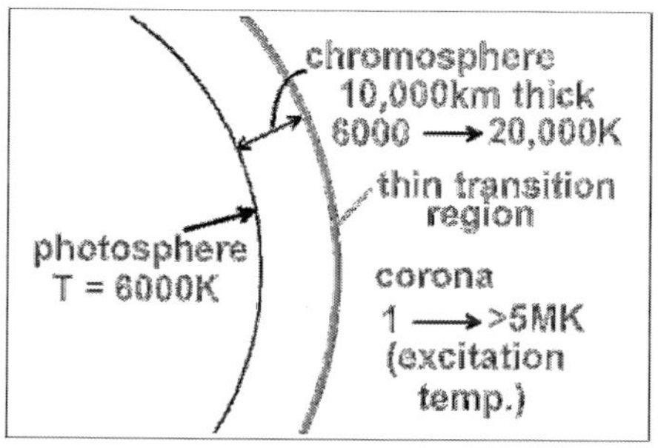

Figure 3. Approximate layout of the inner heliosphere region.

1) Coronal energy support generally. The acceleration of the Fast Solar Wind (FSW) particle streams from coronal holes suggest the action of an electric gradient upon ions.

2) Coronal Mass Ejections (CMEs). Bunches (>10^9 tonnes) of ions seen to accelerate outward to 400-600km.s^{-1}. Some ion speeds attain ~2000km.s^{-1}.

3) The coronal emission line spectrum shows hugely stripped ion species *e.g.* Fe^{XXIV} (helium-like). This implies impact by other high-velocity ions — very high *excitation* temperature; *not* LTE.

4) In coronal streamers, Fe^{XIII} and Si^{XII} ions are often abundant in emission and stay there for months. This requires *electrical support.* Gravitational settling time is of the order of a day.

5) The solar visual 'surface', the photosphere, is due to the strong absorption and opacity of the *negative* H ion **[18]**. Its very low Ionization Potential (0.75 eV) makes it radiant. Its abundance there needs a source of electrons. These electrons were detached from solar wind ions and returned to the Sun by the *electric gradient;* they cannot have come from H (IP=13.6 eV).

6) Solar wind ions (FIP mostly 4.5-8.2 eV) arise at low chromosphere temperature level (7 kK+). Their extraction and differential acceleration requires an *electric field.*

7) Strong light-isotope enhancement in frequent wind events (>1000-fold for $^3He/^4He$) [19] all the way to Mg. Selection for charge-to-mass ratio is the property of an *electric field.*

Simple calculation suggests that all or most of these are well within the capacity of our extrapolated (Sect. 7.2) **30 V.m^{-1}** at the solar surface.

Our inference (Fig. 1) that the polarity of the aether's charge is negative is based on these phenomena. Note that the *acceleration* will be due not only to the progressive action of the G-E field itself but also to any increasing ionization by mutual impact. Perhaps this is what is happening in the transition zone, the electrons detached being those seen entrained in the solar wind.

10. A G-E Field Model for CMEs and Granulation

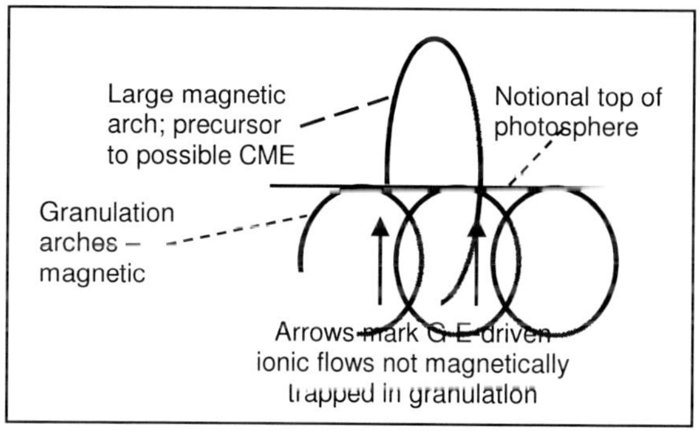

Figure 4. Proposed relationships in the outermost part of the Sun.

The basis for our model (Fig. 4) is that the G-E field is continually driving positive ions outwards from the solar interior. We attribute the growth of magnetic arches to this flow, in the following manner. When an incipient magnetic arch forms, the ions will spiral up its legs in opposite directions. When they meet at the top and many of them are retained by the magnetic field, they will accumulate there and the radial G-E field force upon them will load the arch and stretch it outward. This will continue for as long as the footpoints of the arch are in registration with those streams of ions from the interior. If this registration is lost, this loading will weaken and the arch will shrink again. This is the observed majority behaviour.

But occasionally the loading will become sufficient to break the arch, releasing the accumulated ions as a CME. Strong radio emission is precursory and concurrent with CMEs, which we attribute to the synchrotron radiation from the spiralling ions.

Sunspots expose deeper levels in the solar interior and the temperature there is seen to be **lower** than that of the photosphere. That's why they look dark. This means that the granulation cannot, as widely assumed, be driven by thermal convection. We propose that the effective photosphere temperature is due to the trapping of ions and their G-E field-acquired energy, in arches analogous to those at the higher, proto-CME, level. The ions which feed the latter are those that escape between the granulation structures. Solar flares, on the other hand, are then perhaps when a group of granulation arches fail, CME-like, releasing more energetic ions than those which produce CMEs.

11. Is This Positive Behaviour of the Sun and Stars Permanent?

Effectively, Yes. A simple estimate for the Sun shows that removal of all the negative aether in its interior would provide more than 40 orders more coulombs of effective positive charge than is required to expel all its protons. Evcn the hugely denser packing of mass inside neutron stars seems insufficient to reverse this conclusion. Hence our estimate of their intense G-E fields (Sect. 7.2) and their effects (later in this paper).

12. Constructing the Solar Planetary System — and Others

Repeatedly, but virtually unavailingly, Jeans [20], Lyttleton [21], Jeffreys [22], Spencer-Jones [23], Woolfson [24] and Gold [25] have stressed that the material in the Sun and in our planetary system must have had *dynamically separate origins*, NOT the common origin implied by the Single Contracting Solar Nebula (*SCSN*) model of Kant and Laplace.

The near match between the planetary compositions and the solar spectrum has been thought to validate SCSN. But the Sun is an unmixed star with a tachocline at ~$0.71R_{sun}$. The mass above it is open to modification, see below, so it may not match the deep interior, as SCSN assumes.

An important constraint on construction is the finding [26] that the very short-life isotope 41Ca (half-life 102ka), present in CAIs (refractory Calcium-Aluminium-rich Inclusions in meteorites), and its relation to 26Al (half-life 717ka) also in chondrules, requires that construction be completed (no more accretion on the relevant asteroid) within about 1Ma after the 41Ca had been produced by a stellar explosion. That seems impossible in SCSN.

With these points in mind, we will now show that the G-E field has an essential rôle to play.

12.1. The Four Main Dynamical Problems

1) Mean specific angular momentum (defined as the product of orbital radius and velocity) of the planet materials is >130,000 times the a.m. of that in the Sun. Together with the a.m. of the >10-fold more mass they were formed from, *Where did all this a.m. come from? Could action of the G-E Field have done it?*

2) The 6-degree tilt of the planetary dynamical plane w.r.t. the solar equator. *Tilting in SCSN would cause vigorous precession and loss of their good conformity (bar Mercury) to that plane*.

3) All but one of our planets preserve near-circular orbits, recording the *completion of construction* in the presence of nebular gas-drag. But in the presence of nebular gas, protoplanetary bodies would spiral into the Sun in far less time than it takes to build a planet. *Could the G-E field prevent that?*

4) If you restore Uranus' axis to an upright position so that its inner satellites are prograde, like the other Giant Planets, the spins of all but one (Venus) of the planets are prograde; a systematic behaviour that speaks of gravitational nucleation. But vorticity in a Keplerian disc is retrograde. *Where did they nucleate and get their spins? Was the velocity pattern in the protoplanetary disc Keplerian?*

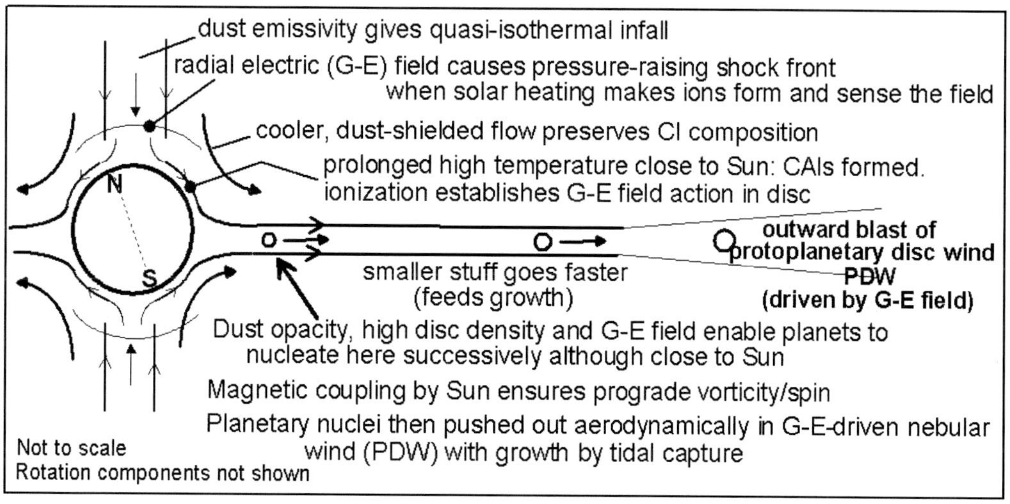

Figure 5. Principal features of the planet-forming second stage of the CT new scenario for forming the solar and other planetary systems, showing the conditions during protosolar traverse of the second dust cloud, whose initial temperature would have been 10K or less. So the resulting protoplanetary disc would be much cooler than in SCSN. Infall will have been bipolar, as shown, only if the protoSun's velocity through the cloud was low. Growth by tidal capture preserves the protoplanet's prograde spin if captures are a dynamically balanced population.

12.2. *The CT Two-Stage New Scenario for Planet Formation in the Solar System (Fig. 5) and Other*

In this scenario the protoSun formed in one dust cloud and became an already H-burning star. Later, (stage 2) it traversed another cloud, with high dust-opacity materials, from which the planets were formed *and* the outer 2.5% of the Sun's mass (above the tachocline) was added to and not mixed in, so their compositions nearly match.

Passing near a recent stellar explosion in this cloud enabled short-life radionuclei to be acquired.

In contrast to SCSN the flow pattern was a through-put of materials, drawn from the cloud and passed outward as the protoplanetary disc, and further.

This 2 stage new scenario (originated by me in 2000 [27]) for formation of the Solar System (SS) has now been repeated, with developments, ten times at different venues. These developments have increasingly underlined the part that needs to have been played by the solar G-E field and this is our objective here also.

The major backdrop feature to have emerged since then is that throughout that interval an almost unvarying ~22% of all exoplanets that had been detected orbit their star, whatever its size, within 12 solar radii (semi-major axis) of it's centre. Why so many so close? Why is the SS now so different, with Mercury, our nearest, at 83 solar radii? Figure 6 shows the recent situation in 2012 as the total number approached 750.

In our CT scenario (Fig. 5) the dynamics of infall are critically determined in the manner outlined above (Sect. 8) in connection with infall to build high-mass stars. The imbalance which here determines that the infall is quasi-polar and the outflow quasi-equatorial is initially determined by the centrifugal force added by the quasi-equatorial magnetic coupling to that part of the new envelope that has become ionized. This determines the prograde orbital direction of any protoplanets nucleated in the resulting disc

I speculate that the presence of such a magnetic field, thought to be dependent in turn upon the presence of a tachocline, probably absent in the A and earlier-type stars subject to high infall rates, is what may determine whether a protoplanetary disc is the result or that the star continues to build.

Once the disc outflow has started, it carries this plasma out to great radial distances, adding a large integrated G-E field force on it, which drives the Protoplanetary Disc Wind (PDW). Within this PDW the protoplanetary nuclei grow by tidal capture of smaller self-accreted lumps being blown past them.

Preservation of the early-acquired systematically prograde spins — the result of the magnetic coupling — means that the widely-supposed random impact process of Safronov, Weidenschilling and others, which would yield a random result, cannot have been the means of growth. To make it systematically efficient the tidal capture process needs the assistance (at least at the first pass) of nebular gas-drag, as provided here, and it greatly increases the capture cross-section and the resulting growth rate. Further evidence that proto-planetary growth was **not** mainly by impact is that it would end, probably in post-nebula time, in multiple occurrence of late giant impacts. In fact, all except Mercury preserve the nearly circular orbits inherited from construction in the presence of nebular gas-drag.

Glose-in gravitational nucleation is facilitated because the disc density is there at its highest and it avoids encountering the Roche condition constraint, which would inhibit it, because the G-E field, acting on the ionized component, will generate a force gradient in the opposite direction to the Newtonian gravity one. If these gradients exactly balance, the Roche condition becomes irrelevant. A further factor would be that the dust opacity of the nebular material could greatly shield the nucleation process from the solar radiation, keeping it cool for gravitational nucleation.

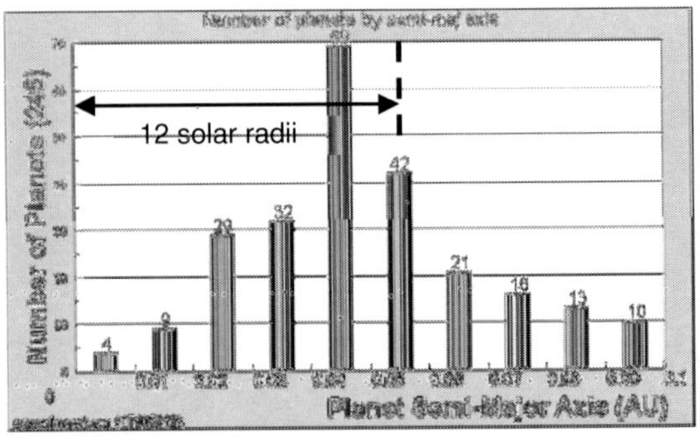

Figure 6. Distribution of close-in exoplanets as at 15th May 2012. Modified derivation from J.Schneider at *exoplanet.eu*. The preferred nucleation distance is clearly about 10 solar radii.

We note above (Fig. 6) that so many exoplanets orbit their star within 12 solar radii of its centre. But in that position it is far too hot for them to have been there long. To have migrated inward, as proposed in various other SS scenarios, including the currently favoured 'Nice' model [28, 29], how did they acquire that higher orbital a.m. to start with?

So we infer that we are seeing these exoplanets not long after emerging from being formed in close-in positions within a high-opacity cloud, which shielded them from their star (and from our view). Now deprived of a G-E field-driven PDW to move them outwards, they are stuck there until they vanish by evaporation in the inferno, leaving only those earlier-formed planets in those systems that had got far enough out to avoid such a fate. Mercury is our closest-in example of that.

If the G-E field were absent, it seems there would be no planets anywhere except close-in ones destined to become exposed to an evaporative inferno as soon as their star left the opacity of the cloud.

The disc density and the masses of the planets formed within it would be largely controlled (a) by the density of the second cloud (which will surely vary from place to place), (b) by the mass of the gathering star and (c) by its velocity of passage through it. In principle, therefore, the scenario may be capable of generating objects with masses up to those of brown dwarfs and beyond, perhaps even the junior partners of disparate binaries.

Since (b) and (c) will change very little along the path the inwards-younging sequence of planet masses in a given system will provide a crude record of (a), the second-cloud density along the star's path. In the SS example the planetless asteroid belt may mark a substantial gap in the density of that cloud. We suggest that the asteroids and many of the Giant Planet's (GP's) satellites may be representative of the outward-moving feedstock bodies from which the planets grew by tidal capture. But those now in the asteroid belt gap had no planet to do that and were left there when the Sun moved out of the second cloud and the PDW died.

12.3. *Chemical Considerations*

A chemical corollary of solar passage through the second cloud is that feedstock body compositions will reflect any variations present along that path. So the surface compositions of the late-formed small bodies too small to have undergone convective overturn, such as all except the largest asteroids, e.g. Ceres and Vesta, will only reflect the cloud chemistry along the last part of the path. Meteorites derived from these surfaces will show that composition.

On the other hand, the overall 'contamination' of the above-tachocline Sun and its spectrum will be some sort of average for the entire path, and therefore be a record of the mean composition from which the entire planetary system was made. For this reason the compositions of meteorites from the outsides of asteroids (as commonly selected for their lack of hydrous alteration) should no longer be regarded as our best records of overall solar planetary system composition.

While on the meteorite topic we may note that there is no chondrite-meteorite evidence that asteroid accretion continued after about 4564 Ma (youngest true chondrules). SCSN-type models (e.g. the 'Nice model' [28, 29]) which invoke planetary accretion for many tens of Ma after that seem inconsistent with this.

12.4. *Orbital Eccentricities and their Causes*

As noted above, the solar planets have predominantly very circular orbits, consistent with completing their growth in a gas-drag nebular disc environment. The preserved circularity of the Earth's orbit probably denies that the Moon can be the product of a giant impact upon the Earth. But the ejecta from such an impact on Mercury, which certainly had one (tilted and eccentric orbit, two-thirds of its mantle missing), offers a dynamically potential source of the lunar material for prograde tidal capture by the Earth and its reassembly in orbit [30].

On the other hand, rather than by impact, the high eccentricity of some exoplanet orbits may have another cause. It has emerged [31], supported by the *exoplanet.eu* database [32], that for those in the close-in position their orbits are circular, but increasing eccentricity seems to have been developed by those with increasing orbit size. This could be attributable to the star's axis being very oblique relative to its direction of motion through the cloud. This would give a shorter path from infall to disc on one side than on the other and a correspondingly unbalanced strength of PDW, thus building the orbital eccentricity every time around.

Evidently the inclination of the solar axis w.r.t. its direction of travel through the cloud was not enough to cause such eccentricity in the solar planets, even though it was probably the cause of the 6 degree tilt of the resulting protoplanetary disc relative to the solar equator. This axis-inclination 'tolerance' can perhaps be assigned to the latitudinal range of the quasi-equatorial magnetic coupling to the disc plasma. In which case the high eccentricities seen in only some of the exoplanet population may mark only the stars that had rather high axial inclinations to their path.

12.5. *Growth by Tidal Capture: Evidence From the GP Satellite Population*

Strong support that SS planetary growth was by tidal capture is provided by the satellite population of the Giant Planets (GPs). Of the 166 currently known satellites of the four GPs the outer ones are small and have mixed prograde and retrograde motions. But, of the 56 which orbit their planet at 4 Gm or less, all except one are prograde. These were identified [33, 34] as the residual part (half?) of a tidal capture population, from which the retrograde captures will by tidal action have spiralled inward to grow the central body.

12.6. *Growth of Planetary a.m. by G-E Field Action; Numerical Support From SS Dimensions*

Now let us return to the planetary a.m. problem noted at the beginning. Specific a.m., being defined as the product of tangential velocity and the radius at which it occurs, means that the outwards push developed in an ionized disc dominated by action of the G-E field has the property, which it shares (hitherto unrecognized?) with radiation pressure, of (for example) doubling the a.m. every time the distance from the centre is doubled. For this a.m. creation to work in the case of the Solar System, the second-stage material must be acquired to a near-Sun position and be moved outward by the G-E field, with the planets growing as this is done. This is exactly what our new scenario achieves (Fig. 5).

To do that the outward movement of a large body must depend on there being sufficient aerodynamic push by the PDW. This push will fall with radial distance because both the ionic density of the PDW and the G-E field strength will do so. But we assume it will at least remain adequate for propelling feedstock materials past the body for it to capture tidally. In Section 12.7 we reason and cite chemical evidence that the nebular density was indeed at least 40 times that hitherto assumed in the canonical SCSN.

For our PDW to be driven by the strictly radial G-E field force would imply that the Newtonian force is wholly overridden. In that case the tangential velocity does not alter with radius, retaining the low value present near the root of the disc (at 10 solar radii?), where solar magnetic coupling may determine the tangential velocity. In fact the Sun, with its ~26.5 day rotation period, is in a class of slow rotators, whereas other G-type stars of similar mass have periods of 5 days or less [35]. So I infer that in generating the planetary system, magnetic coupling slowed solar rotation about 5-fold. Taking the G-E field-driven a.m. growth of disc material as starting at the outside of a polar infall column with a diameter 10% of the solar radius, simple arithmetic shows that the required full 120,000-fold a.m. differential is achieved at the orbit of Jupiter and beyond, if that 5-fold solar slowing is included.

But these a.m. values incorporate the Keplerian orbital velocities which now prevail whereas, with the G-E field on the nebular plasma in control, the tangential/orbital velocity of the created planets might all have been similar at only a few times the present 2km/s equatorial velocity of the Sun. But as the Sun moved out of the second cloud and the PDW strength waned, the transition to Newtonian gravitation means that those that were then orbiting too slowly for that must have speeded up by spiralling inward **at constant a.m.** from well outside its present distance. This validates our use of present a.m. values.

There is no sign, so far, that any of the exoplanet-harbouring stars is spinning fast enough to account directly for the orbital a.m. of its more distant planets, so this G-E field-based resolution of the problem is almost certainly needed there too.

In the foregoing discussion we made the simplifying assumption that the tangential velocity of the G-E-driven PDW would not change with distance fom the Sun. In fact, that would only be true if the PDW only contained plasma responsive to the G-E field. But to make the planets we have to assume that the PDW was loaded aerodynamically with large amounts of neutral material subject only to Newtonian gravity. This would mean that the tangential velocity profile of the actual PDW would have been a compromise between 'flat' and the Keplerian pattern, corresponding to the degree of that loading.

Later in this paper we apply similar considerations to the 'flat' tangential velocity profiles of spiral galaxies, so we may point out here that deviations from that 'flatness' are likely to reflect differences in that 'Newtonian mass loading', rather than imply other dynamical agency.

A remarkable example of these different forces acting (*now*) in the same astronomical object is provided by the bright young star Fomalhaut and its 'planetary nebula'. Here, as in other planetary nebulae ('The Ring Nebula' M57 and 'The Helix Nebula' NGC 7293) there is a light-emitting ring or band seen to be made up of thousands of narrow streaks aligned almost perfectly radially to the (quite distant) central star. I see these as mass loss from the central star, but being formed like the plasma tails of comets (also strictly radial from the Sun [36]), although this is normally attributed to the solar wind, which we now see as driven by the G-E field. **But** a planet, named Fomalhaut b, recently found just inside the ring [37], is *not* moving radially but on a CCW orbit. Evidently, despite its own G-E field, which makes any such body behave in a moderately positive manner (Sect. 11), this is in this case too slight to prevent an orbital response to the Newtonian gravity field of the star, unless the disc wind is dense enough to drive it outward aerodynamically. This explains why Newtonian dynamics serve so well at present throughout most of the solar planetary system.

12.7. *Origin of Solar System Water: Vital By-product of the G-E Field-based Resolution of the a.m. Problem*

The origin of SS water is currently widely accepted as still being an unresolved problem. We now show how the restricted nebula-present (say <5Ma) timescale for the completion of planetary construction, dictated by satisfying the planetary a.m. problem, has the effect of constraining the mode of the construction of planetary iron cores in a manner that also provides the formation of SS water.

Although many people have regarded the comets as the source of SS water, this merely passes the problem to where that water came from. Our scenario offers the PDW mechanism for getting it out there only if we can provide a source nearer home. Interstellar dust clouds seem low in water, so not much can have been imported from there and we must look for a way of making it chemically in the inner part of the SS during planet construction. This approach is supported by the observation that CAIs passed rapidly from a low oxygen fugacity during formation in a refractory environment (see Fig. 5) to one that was 5 orders of magnitude higher where they acquired their Wark-Lovering rims [38], probably while on their way in the disc to the asteroid environment.

It so happens that for many years (1960-1978) A.E. Ringwood, a famous petrologist and Director of the Research School of Earth Sciences at ANU, argued that the Earth's core was made by the reduction by the nebular hydrogen proto-atmosphere of the always-present FeO in volcanically erupted lavas (Fig. 7). This meant that the protoplanet had to grow until volcanism became established. A concomitant of this process would be the formation of huge amounts of water by reaction; an SS benefit foreseen by Ringwood.

A requirement of the Ringwood model was that the nebula should be a cool one, below 600K, not the hot one embraced by SCSN, so that the iron would then be present as FeO, for planetary construction and volcanic eruption, not as reduced Fe. It was subsequently affirmed that this is thermodynamically correct [39]. Such a cool nebular disk and PDW is just what our 2-stage scenario produces (Fig. 5), from a very cold source cloud, even after allowance for admixing with that inner part of the pole-to-equator flow that got heated by the Sun.

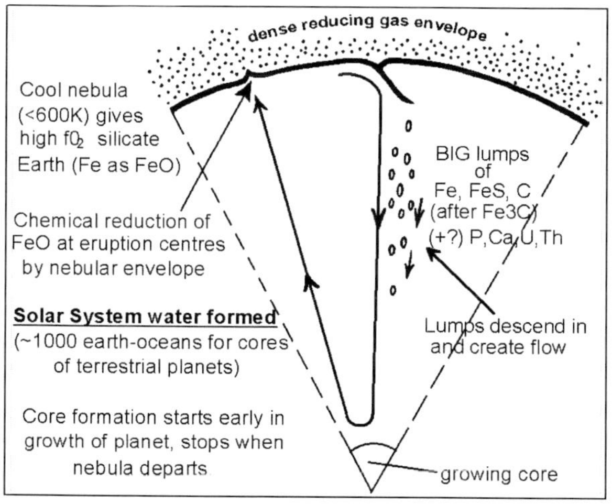

Figure 7. Principles of the Ringwood mode of core formation. Fe loading of the descending limb ensures convective penetration and speeds core formation, as does heat from the release of gravitational energy. The size of the lumps, even if they liquefy, ensures minimal chemical interaction with the co-descending mantle.

An important aspect of the Ringwood model in the present context is that nebular opacity shields the body from solar radiation, so the body builds its own thermal microcosm, dependent mainly on accretion rate and not on orbital distance. The iron cores in three of Jupiter's Galilean satellites [40] then present no special problem, despite being so far ('snowline') from the Sun. Nor do those now suspected to be present in the asteroids Ceres and Vesta as the result of the DAWN space-mission.

The problem that this great distance has raised for the provision of molten iron for the cores-by-percolation model, endorsed in hot SCSN, has been met by the idea of a deep internal 'magma ocean' in the body, from which the iron might percolate. But the growth by tidal capture, seen above to be necessary for dynamical reasons, almost certainly contains a negative feedback mechanism that would prevent the capture rate ever heating the body to that point. The tidal capture rate depends on internal viscosity, and falls if it liquefies and the energy-loss mechanism is lost.

If all the iron in the Earth's core originated as FeO, this would produce over 400 Earth-ocean volumes of water. Although this would likely equip the early Earth with a water-saturated mantle mineralogy (and there is good evidence that it did [41, 42]), this would account for only a few ocean volumes. So Ringwood 1979 [43] had to abandon this idea because there seemed no way of getting rid of the remaining dense hydrous nebular atmosphere which would result. Our G-E-field-driven PDW would now do that, especially during the final outward clear-out as the Sun exited the second cloud.

Up till that moment of nebular clear-out each planet had been completely shielded from solar radiation by nebular opacity, but outwards-progressing removal of that opacity would expose the hydrous atmosphere to ionization by solar EUV, thus rendering it susceptible to expulsion by the G-E field force. The gaseous envelopes of the four Giant Planets were probably sourced from this material, by gravitational capture as it passed [44], the remainder passing out to form or be accreted by the bodies in the cometary region.

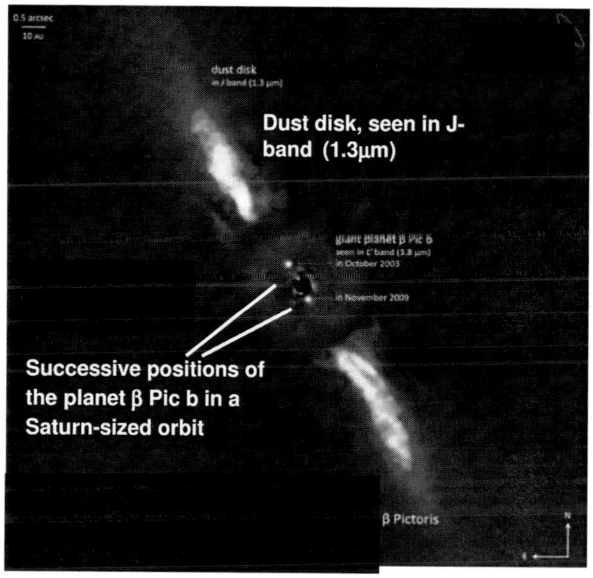

Figure 8. Aftermath of the formation of the giant planet β Pictoris b: a remarkable visual example of G-E field action. The star itself has been artificially obscured. Image modified from NASA APOD 2010 July 3. Credit: A.-M. Lagrange, D. Ehrenreich (LAOG), et al., ESO. See also [47].

This late-stage acquisition of the GP envelopes has a direct bearing on their earlier tidal capture of all those prograde satellites, discussed in Section 12.5. Many of that population have silicate-rich densities, so it means that the GPs began as silicate-rich central bodies with a viscous tidal attribute. The arrival of the gaseous envelopes made the interiors liquid and lose that attribute. The moment of so doing is recorded by the interruption of retrograde Triton's inward spiral [44]. Jupiter, first in the gravitational ring-fence, clearly captured the most (perhaps 310 Earth-masses of its present total of 318) and got spun-up the most. For that to happen, the material must have had prograde vorticity — the signature of G-E field action from the prograde-spinning inner planets.

This reasoning, using only the Jovian acquisition and ignoring what went further, and to the cometary belts, yields a minimum figure for the nebular density in the inner SS, at the moment clear-out began, which is some 40 times the canonical SCSN value. This offers usefully toward securing protoplanetary nucleation in close-in positions, securing the vigour of the PDW for dynamical purposes, securing the vigour of the water-forming reaction during core genesis and to keep down volatile losses from levitated molten chondrules [45].

Finally, if iron cores are made in this way, dependent upon 'convective' overturn in bodies that have grown big enough for that to begin, what is the nature of the many iron-group meteorites thought to be derived from much smaller asteroids? The answer here is that if there is no overturn, but nebular reduction of erupted FeO occurs, the resulting former volcanic magma chamber, in which iron has sunk to the bottom, will remain on the surface, accessible to meteorite derivation by impact. Careful consultation of [46] shows that different kinds of meteorite do indeed offer an almost complete record of different levels in such a differentiation chamber.

I conclude that the cores-by-percolation models, and their appeal to the hot SCSN model, are invalidated by an inability to satisfy the observed values of planetary a.m., whereas our CT scenario not only does so but also provides a well-researched origin for Solar System water. Its effects on the physical properties of the Earth's mantle, changing over time as the ocean water was released from it in volcanism, has had major consequences for the evolution of the Earth, including the replacement of its CO_2-rich atmosphere by an oxygen-bearing one, which is why we are here [42].

12.8. *Solar/Stellar Exit from the Second Cloud: Clear-out of the Nebula from the Protoplanetary Disc*

This nearly edge-on infra-red image (Fig. 8) [47] of the Beta Pictoris planetary system offers a remarkably persuasive illustration of the clear-out stage. It seems to show the G-E field-driven clear-out of the remaining still-warm dust and plasma of the protoplanetary disc from which the giant planet ***β* Pic b** had been formed.

The inner zone, around the planet, is seen in L-band (3.8 μm) whereas the main flow pattern, at J-band (1.3 μm), has evidently remained warmer, presumably due either to higher density or to water vapour/ice-restricted cooling. The rather abrupt inner boundary of that flow probably marks an abrupt exit from the planetogenic second cloud through which the star had passed.

Visually, the <u>outwards</u> morphology of the flow seems irresistible and extends to distances that would be far into the cometary regions of the SS.

12.9. *Planetary Systems, Solar and Exoplanet; the Essential Functions of the Solar/Stellar G-E Field: Summary and Earth-Consequences*

- (SCSN), ending with giant impacts, is untenable or incompetent for multiple reasons, both dynamical and chemical. These include:- planetary angular momentum provision, prograde spins, low eccentricities, prograde satellite population of the GPs, too-long construction timescales, no incorporation of short-life isotopes, inability to provide for SS water. ...But, in contrast....
- G-E field action in the new scenario (Fig. 5) provides:-

(a) for unhampered infall of the high-opacity second-cloud material, just as it does for building up high-mass stars (Sect. 8, Item 3);

(b) a protoplanetary disc wind (PDW) which builds up the a.m. of planets, by moving each outward aerodynamically in a non-Keplerian pattern, *with little change in tangential velocity*, after nucleation in a close-in position, consistent with exoplanet observations.

Of wider interest in (b) is that this tangential velocity profile is just what we discuss in the next Section as being observed in Spiral galaxies. G-E field action at such vastly different scales would be nice confirmation of its truly fundamental nature.

We have found that not only do these two G-E field actions lead to resolutions of all the evident inadequacies of SCSN but they also yield the following stark conclusion. Without those actions, the only planets anywhere would be ephemeral ones nucleated in close-in positions, but destined, upon emergence from the protection of nebular opacity, to vanish there by evaporation in the stellar heat.

Thus it appears that our very existence on this planet depends upon the G-E field in two distinct ways. **It** has provided for the existence of a planet at sufficient distance from the Sun. And it has prescribed the Earth's construction in a manner that has given it the water that has dominated its evolution, and ours.

13. The G-E Field in Spiral Galaxies

13.1. *The 'flat' Tangential Velocity Profiles*

Spiral galaxies typically exhibit tangential velocity profiles which, instead of the outward-decreasing Keplerian expectation with a central mass, tend actually to remain almost flat, outward all the way from its rise across the galactic 'bulge' to far beyond the limits of optical visibility, as betrayed by 21 cm H 1 radiation.

Since its first recognition in the 1970s this has been attributed to the presence, surrounding the entire galaxy, of vast quantities of mysterious Cold Dark Matter (CDM), having no other known physical properties. CDM masses of over 10 times the observable baryonic mass of the galaxy have been inferred. But, for orbiting (as often assumed) mass of any kind to be doing it, there's a **wholly insuperable a.m. problem** that hasn't been thought of. To escape that by regarding the CDM as static would then raise the question as to how a.m.-containing galaxies habitually form in the middle of such patches of CDM.

But such a profile, as we have just seen for planetary systems, is exactly what the radial G-E field would do, when driving outward the material of a plasma-rich disc. Consistent with that, and diagnostic of its electrical action, it has also been found [48] that in Elliptical galaxies (plasma-poor by nature) the profile is, on the contrary, near-Keplerian.

To validate this interpretation, we now show that the plasma flow in the arm-dominated part of the disc is indeed outward, after polar infall, as in our planetary systems scenario (Fig. 5). That, of course, conflicts with the rather instinctive general view that the spiral arms are being wound more tightly. In turn it raises the question, dealt with in our companion paper [68], of how tightly-wound galaxies formed in the first place.

13.2. *The Entire Structure of Spiral Galaxies is Dominated by the Action of G-E Field-Driven Galactic Winds (Fig. 9, a,b,and c).*

1. The constant tangential velocity means geometrically that the arms trail as they move (are driven) outward – yes, *outward.*

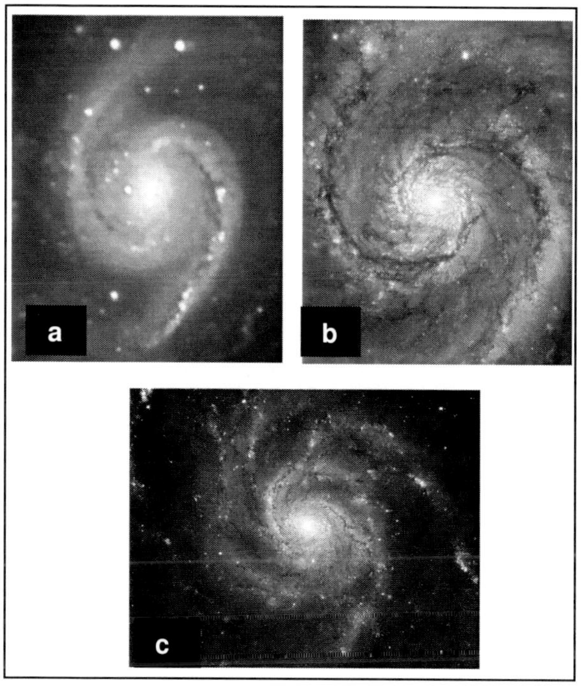

Figure. 9. Spiral galaxy forms. All three are rotating CCW but the arms are **unwrapping**. In (c) they are actually being blown apart. (a) NGC 1566. (b) M51/NGC5194. (c) M101/NGC5457. Image credits given in the footnote e[e]. Image (a) has been cropped here.

2. Dust and less-ionized material, seen as red lanes, has less G-E field drive, so it accumulates along the insides of arms – an ubiquitous and characteristic asymmetry (9a, 9b), widely applied for tracing arms.

3. This dust encourages massive star formation, enabled by G-E field action, in 'starbursts' [49] spaced along the arms (9a).

4. Moving the arms outward, without changing tangential velocity, geometrically requires them to extend over greater length of arc, so they rupture abundantly (seen as black, oblique (9b), or as completely freeing quasi-linear chunks of arm, which clearly once belonged to the assemblage (9c)).

5. Except at these gaps, ionized material is selectively driven out of the arms, and seen as outwards-trailing blue escapes (9a, 9b).

[e] Image credits: (a) NGC1566. www://spiegelteam.de, with permission from Volker Wendel; (b) m51_hstAPOD_2001Ap10; (c) M101_medNGC5457HST, from Wikipedia ('Pinwheel Galaxy'); a very apt dynamical description! NGC1376 is very similar.

13.3. *Spiral-Galaxy Disc Wind: Where From?*

The CT answer:

In planetary system formation we saw (Fig. 5) that the infall material for the protoplanetary disc wind (PDW) had arrived quasi-axially, was ionized during near-star passage to the disc, enabling the G-E field to expel it within the disc.

I propose an identical flow pattern for Spiral galaxies, ionization being effected at or in the stellar mass concentration constituting the galaxy's bulge.

For planetary systems the source is a low-ionization dust cloud. For galaxies, I propose (companion paper [68]) a low-ionization cosmological source for the material, now abundantly evident as the 21 cm neutral hydrogen (HI) emission, thought to be of cosmological significance, for whose study the Square Kilometer Array (SKA) is being built in southern latitudes.

14. G-E Field Actions at Neutron Stars

The surface gravitational potential of a neutron star is "typically a few times 10^{12} m/s^2", "with a maximum of 7 x 10^{12} m/s^2". The former, at $log_{10}\ g$ ~ 13.5, is over 10^{11} times that of the Earth, so a very large corresponding G-E field is to be expected, although the extrapolation relationship is as yet unclear. (Related neutron star parameters are:- Radius - ~12km; Mass - 1 to 2 M_{sun}) (Wikipedia June 2012).

We note in passing that several neutron stars are known to possess planetary systems. One even has a potentially protoplanetary disc [50]. The problem has been how they acquired them after the supernova explosion which would have blown all such material beyond recall. Our 2-stage new scenario for planetary system formation (Sect. 12.2/Fig, 5), seems well adapted for explaining these acquisitions. There certainly would be plenty of G-E field to drive them rapidly away, while the acquired protoplanetary nebular disc was still present, from the vicious radiative environment of the star.

14.1. *G-E Field Actions in Supernovae*

Supernovae of types Ib, Ic and II, are due to the onset of degeneracy – fusing protons and electrons – in a high-mass stellar interior, the result being a neutron core. I propose that the intense G-E field from that core is what causes the explosive ejection of all the remaining ionic material, and why all the neutrons stay behind.

Hitherto it has been supposed that the 'explosion' is the bounce resulting from sudden inward collapse, but a study of the process has failed to reveal the necessary bounce capability [51]. To confirm our G-E field interpretation, evidence of outward *acceleration* of the material should be sought, as seen in solar CMEs.

Type 1A supernovae also yield a neutron core product, but *via* a white dwarf route. These have been regarded as cosmological distance-measuring standard candles, on account of the Chandrasekhar 1.38 (formerly 1.44) solar mass constraint on their pre-collapse masses and energy-release. But we noted above (Sect. 8, Para. 1) that the support provided by the internal G-E field in stars would postpone collapse to higher mass values. Recently several such 'super-Type 1a' supernovae have indeed been observed to have had pre-collapse masses up to 2.8 solar masses. [16,17,52].

14.2. *G-E field Actions in Pulsars*

Having lost all its overburden in the supernova explosion, the outermost layer of a neutron star will no longer be under degeneracy-inducing pressure.

Free neutrons have a half-life of ~10 minutes, so this outermost layer will decay; the protons being accelerated away by the G-E field. But the electrons will form a surface 'sea' which the G-E field will strongly pressurize onto the surface. This 'sea' will deepen until that pressure attains the degeneracy value.

Almost inevitably, therefore, 'islands' of exposed and actively decaying neutrons may be present. These islands will produce intense outward-accelerating streams of protons. I propose that the *synchrotron radiation* from these streams is the pulses from pulsars.

Any given neutron star could have many such islands on its surface, sending beams in different directions, thus greatly increasing the probability that we lie on the spatial sweep of at least one as the star spins. So undetected ones will be rarer than we thought.

Far better than the 'magnetic axis oblique rotator', this mechanism could provide for multipulse (e.g. Vela), pulse shape differences, and altering interpulse time intervals.

14.3. *G-E Field Actions: Cosmic Rays (CR)*

CR energies range, with logarithmically increasing rarity, from 10^9 eV (the defining lower limit for the name) to a 'toe' at somewhat beyond 10^{19} eV. The G-E field seems a prime candidate for accelerating them, but simple calculation for protons suggests that, at our linear-extrapolated 10^{12} V/m (Sect. 7.2), it may be some three orders of magnitude too low, so this will need attention. A link between CRs and pulsars has been been detected during solar occultation of pulsars [53].

There is a 'knee' in the abundance, at ~3 x 10^{15} eV, beyond which the rarity slope steepens [54], but there seems to be no firm observation as to the nuclear species beyond that. This is partly perhaps because the events are so much rarer but partly also influenced by the current expectation, based on the view that they must be of extragalactic origin, which is that they should include a wide variety of species.

If, however, our proposal is correct that the G-E field at neutron stars is the accelerating agent, then the clear prediction here is that protons should be virtually the only nuclear species present at these high energies.

At lower energies, the reported appearance of He and C nuclei, in addition to the predominant protons, strongly suggests white dwarf (WD) sources [55], whose surface gravity lies in the range $\log_{10} g = 7.5 - 8.5$. So WD gravity is second only to neutron stars, but >5.5 orders lower. This compares well with the rather nominal 5 orders toe-to-knee drop in CR energies. That savours of our speculated linearity of the G-E Field/Gravity relationship. Most WDs die slowly without exploding, so there are lots of them to source the many CR above the knee.

15. Relativistic (RT) Matters and CT

So far, this contribution has concentrated on matters upon which RT has little or no bearing. So the wide-ranging benefits we have inferred are specific to CT. We now consider the CT perspective and competence on a few matters thought to underly the strength of RT.

15.1. *Lorentz Transformations are Inapplicable in CT*

In devising RT and adopting the Lorentz transforms, Einstein followed Leibniz (c. 1695) in being unable to recognize the presence of a connecting intermediary between any two apparently separated objects. So these were perforce the only items in a discussion of the relationship. But CT's recognition of an ubiquitous aether requires the full details of motions in the intervening aether, so that vector addition may be applied. As I discussed in [3], Ives & Stillwell [56] performed a beautiful experiment to test, in a single physical system — gravity waves on a relatively moving tray of mercury — whether this change of perspective does indeed require use of those transforms in one case but not in the other. They confirmed, both mathematically and observationally, that this is indeed the case, all three "relativistic adjustments" — the Fitzgerald contraction, the Larmor-Lorentz change of clock rate and the Fresnel convection coefficient — were both expected and observed BUT with ***c*** in this experiment being not the velocity of light, but the velocity of gravity waves on mercury. Thus there is nothing special about the velocity of light in these formulations so long as there is a transmitting medium (e.g. mercury) for the waves. The "relativistic adjustments" arise only if one chooses to deny that the waves can, along any part of their path, travel faster than ***c*** relative to the observer, although travelling no faster than ***c*** relative to the local medium. Evidently, by restoring the local aether as the reference frame for the propagation of change, as in CT, all the phenomena currently attributed to SR effects become equally explicable.

15.2. *Gravitational Communication and the Perihelion Advance of Mercury*

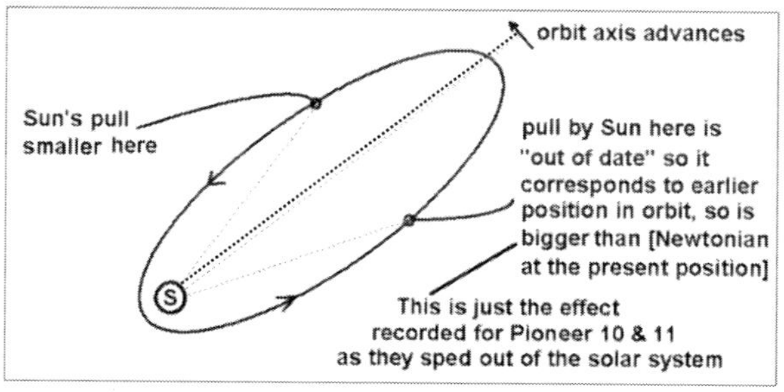

Figure 10. The qualitative effect of gravitational communication time. See Anderson et al [58] for the Pioneer observations. Can they not be an experimental verification of Gerber's result?

The application of Newton's Laws of gravitation is customarily treated as a field theory in which the test particle senses the field of the central body instantaneously and reacts to it, the assumption being that the field is an intrinsic property of the central body and is unaffected by the arrival of the test particle. But if Newton's Third Law, that action and reaction are equal and opposite, is to be satisfied one needs to establish how the reaction force gets from a non-contact position of the particle to the central body which is the source of the field, to generate its reaction.

For this purpose the limiting velocity of communication must be the maximum ***c*** permitted in

RT and in CT, the latter being dictated by the properties of the only intermediary available — the aether. But such a set-up fails to cover what will happen if the force between the bodies is a stimulated response, each to the presence of the other, as in CT[f].

For CT, therefore, we eschew the use of a field theory, adopting one that recognizes the limitations of communication time. A successful account of the periastron advance/perihelion advance of Mercury is neither original nor unique to GR. In CT, as outlined above, gravity is an interaction *intercommunicated* at finite velocity (***c***) due to the particles sensing and reorienting their sucking poles in response to the aether density gradient generated by the other body. In 1898 Paul Gerber [56] successfully modelled this kind of response delay with the implication that if the distance, and the force demand, is changing, then the interactive force actually communicated will have a magnitude that relates to a slightly previous position. The qualitative effects of this are illustrated in Fig. 10.

Gerber's achievement was to make the gravitational potential time- and route-dependent, as distinct from that of Weber, which depended on position only. Gerber inverted the problem. From the then-approximately-known perihelion ad-vance rate to be explained he set out to determine the effective velocity of commun-ication, which he found to be close to ***c***, as then known.

The relation Gerber obtained for this in 1898 was the now-familiar

$$d\omega/dt = 6\pi GM_{\odot}/Pac^2(1 - e^2)$$

(period, major axis, eccentricity along the RHS denominator)

which Einstein incorporated within GR in 1915 but refused to acknowledge Gerber's priority, despite the 1917 republication [59] of Gerber's paper by an infuriated editor [60]. As a matter of history, Gerber's work was discussed by Mach [61] whose full obituary Einstein wrote as a close friend in 1912, so it is unlikely that Gerber's paper was unknown to Einstein when he published GR.

This remarkable formal identity resulting from two quite different physical approaches could be an important general warning to those who consider formulae to be distinctive of one theory and one alone. But if this formal identity was deliberately devised by Einstein in his formulation of GR, knowing that it worked, this caveat would bear less weight.

This lack of a formal difference upon which to base his objections to Gerber's interpretation seems to have forced Roseveare [62][g] , a relativist, to raise two others, both of which are invalid within CT.

Note that the orbital precession of electrons within atoms, responsible for spectral fine structure, is a phenomenon of identical character to that of perihelion advance, so is no longer to be seen as relativistic either.

15.3. *Gravitational Light Deflection, Distortion of Space-Time, and the G-E Field*

The verity of the GR prediction of lensing and its value is now no longer in doubt. Remarkably, CT, just like the perihelion advance matter, offers what seems likely to be the formal equivalent in this case also. This is that the G-E field constitutes a radial gradient of aether charge density which, by Maxwell's equations (as noted in Sect. 8, Item 2), will cause the value of ***c*** to be lower, nearer the Sun. In GR the deflection is attributed to gravitational distortion of space-time (but without any physical account of how that is done). Only termino-logically different from the distortion of TEM-wave propagation space arising in CT, it is likewise possibly proportional to the actual gravitational potential at each point on the path of the TEM-wave.

An important possible diagnostic between them is that it seems that the CT deflection should operate only on the E-vector, so the lensed light forming arcs would be polarized radially to the central object, which is not the case under GR. It should be simple to check observationally.

15.4. *Black Holes*

From the start, and at many subsequent points in this paper, we have demonstrated the gains in physical understanding to be achieved by recognizing that the mass property of fundamental particles needs space within the particle in which this can be developed.

[f] To clarify this point, note that in CT gravitation is the product of the interaction of two objects. The Sun, if alone in the Universe, would have no **external** gravitational field, i.e. no means of generating an external aether density gradient. The aether pumped out of it would, by compressional adjustment of its density, effectively redistribute to infinity, given time.

[g] As already pointed out **[3]**, Roseveare's derivation of Gerber's result is confused. He starts (p.137) assuming a field-propagation-rate theory, in which gravity falls with recession velocity, and would result in perihelion retard, but then (p.137-138), apparently realizing his mistake, swaps to an intercommunication-response-time theory (like CT) and obtains the correct result. It is remarkable and serious that this *non sequitur* wasn't picked up editorially before publication. The destructive effect of so doing may have been why it was not.

Hitherto, for nearly a century, these properties have been treated unquestioningly as intrinsic to the particle concerned, and contained within infinitesimal singularities. That is the basis upon which the relativistic Schwarzschild black hole model and its relatives have rested. Its defining feature, the presence of infinite gravitational field at its center, is the mark of not having had available the understanding of the physics of gravitation that motivates this paper.

Such models have seemed to be supported by the observation of very high velocities of orbital circulation, augmented in some cases by the assumption of relativistic mass increase at those orbital velocities. But in the following Section we introduce the notion of velocity-limited inertia. Since centrifugal force is inertial in nature this has an effect opposite to the relativistic one; the faster the orbital velocity, the less is the central mass required to keep it there.

16. The Irrotational Aether, Sagnac Effect, and the Origin of Inertia

16.1. *CT, Sagnac Effect, and Global Communications*

We here take up the matter raised in Section 5.1 that the ultra-high charge density of the aether would render it irrotational, with potential to provide our measure of absolute direction, customarily termed sidereal. Two kinds of device are known and widely used; (i) Foucault pendulum and mechanical gyroscope — using inertia/gravitation; (ii) ring laser gyro — using TEM-wave propagation/Sagnac effect. That such different kinds of device should both do the job points strongly to the aether as the link, because it is tied not only to TEM-wave propagation but, as shown above, to the gravitational process also.

This sets the Sagnac effect in a new light. As Sagnac himself contended [63], the TEM-waves are propagating at their proper velocity c in an irrotational frame while the apparatus spins within it, thus making the travel time to reach the moving receiver longer in the forward direction than in the backward. It was shown experimentally [64] that the effect varies with path length, i.e. transit time, not the circuit area originally supposed by Michelson (he moved one side of a rectangle) and assumed in popular treatments. Thus the Sagnac effect has nothing to do with the supposition that TEM-waves travel at different speeds in the two directions. Classical and Special Relativity (SR) treatments both yield the correct result because SR introduces effects which cancel out. But SR fails to relate Sagnac to its gyro property of providing an 'absolute' directional reference frame, whereas this classical form does.

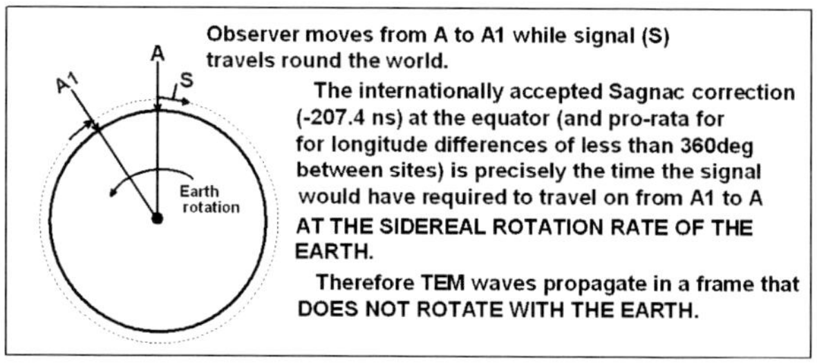

Figure 11. Global correlation of time signals. The CT significance of the Sagnac correction for Earth rotation.

A globally significant result of this understanding is illustrated in Fig. 11. The 207.4 ns figure is, of course, before applying a correction for height above sea level. Evidently the huge charge density of the aether does indeed give it (nearly?) perfect torsional rigidity. Presumably, therefore, the aether doesn't rotate with the Sun or galaxies either?

This result bears upon the MM and subsequent aether drift experiments. As discussed in detail by Kelly [65] these have found that, although undetected in the MM experiment of 1887, they are now consistent with the observation of a drift that corresponds to the surface rotational linear velocity within a non-rotating aether frame. This suggests that we must be aware of the aether's differing responses to rotational and to linear displacements. Its torsional rigidity is apparently enough to overcome its particle-tiedness in the context of rotations as rapid as the Earth's; but the observed lack of aether drift in the case of the much lower angular velocity relating to the Earth's orbital motion implies that it fails to do so in those circumstances.

That the motions of aether in between particles made out of aether should be only partially tied to their motions seems an entirely reasonable outcome, avoiding the difficulty that would have arisen had the MM result been rigorously correct.

Importantly, it is compatible with the evidence, derived in our companion paper [68], that the random motion of the intervening aether in an assemblage of randomly moving particles is a spatially very smoothed reflection of those individual motions.

16.2. *The Aether as the Site of Inertial Action*

Interpretations of inertia based on Mach's Principle continue to be sought and this was the declared aim, as a friend of Mach, by Einstein too in the formulation of GR, though close inspection suggests that GR does not succeed in so doing. A primary snag with a strictly Machian interpretation of inertia, requiring communication 'with the rest of the Universe' has been the evident lack of time-lag in its behaviour. By embracing the infinite communication velocity inherent in field theory, as noted above, this is a problem which GR avoids. Our CT aether, as noted, has both an immense force capability and the possibility of providing a reference frame that substitutes for Newton's 'absolute space'. So one wonders whether the rather local enveloping and all-pervasive aether, with negligible 'communication time', could be the volume from which inertial action originates.

There is an important consequence of thereby retaining a velocity of inertia intercommunication (with the aether of a very local 'rest of the aether Universe') which is limited to ***c***. It is that this will cause **inertial force to be *c*-limited** in just the same manner as we explored above in the case of the supposed relativistic mass increase when under ***c***-limited electromagnetic acceleration force. I return to this in the accompnying paper [68] as the basis for a remarkably fertile new model for the nature of quasars.

By making the aether the underlying agent for both gravitational mass, as above, and for inertial mass (expressed in centrifugal force) one would hope automatically to achieve the rigorous equality of gravitational and inertial mass shown by the Ëotvos experiments and which has so long been a problem.

17. Three Experimental Checks

The foregoing account of CT incorporates a huge range of apparently supportive observations, but additional checks, where possible, are always desirable for any theory, new or otherwise. The following would be especially valuable. Others are proposed in the companion paper [68], in the context of phenomena discussed there.

1) Simplest and most diagnostic (Sect. 15.3), the light in gravitiationlly lensed arcs should be checked for polarization.

2) Central to the whole basis of CT is the charge density and polarity of the aether. The negative polarity of the aether's charge was inferred originally by me from noting that the Sun expels positive ions. A possible experimental method to determine independently the polarity and charge density of the aether is sketched in Figure 12 but careful assessment is required as to whether enough experimental sensitivity can be achieved.

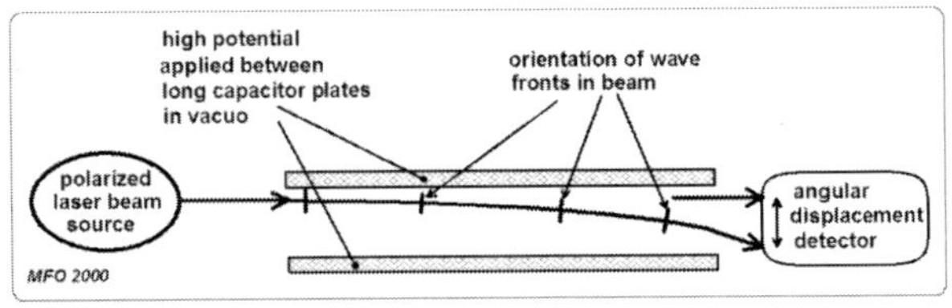

Figure 12. Suggested aether density and polarity experiment.

The CT view of Maxwell's dielectric displacement current is that the charging of a capacitor involves the displacement of aether away from one plate and towards the other. In Maxwell's equations the velocity of TEM-wave propagation rises with increasing elastic modulus of the medium, which relates to the charge density of the aether. So a charge density (i.e. aether density) gradient, set up in the aether between the plates of a charged capacitor, will progressively tilt the wave fronts and deflect the beam.

As noted (Sect. 15.3), it is likely that the deflected light will be polarized. So rotation of the polarized source would modulate the deflection. If the aether is a continuum of negative charge the beam deflection will be towards the negative plate. This experiment would also check the proposed mechanism of the 'gravitational light deflection' (Sect. 8), thereby providing another check upon the intensity of the G-E field in given gravitational circumstances.

3) An attempt should be made to measure the G-E field of the Earth, at ground level, essential for any extra-polation to other bodies (Sect. 7.2). The all-pervasive nature of the gradient raises problems. Being present equally within the structure of the apparatus and of any sample, it is this which has caused it to elude discovery. Some sort of ionic drift method might be worth consideration. Rotating the chamber in a vertical plane, to modulate the signal, would remove zero-point error. But this might generate unwanted signals unless electro-magnetic screening was extremely efficient.

18. Ten Conclusions

1. Maxwell's aether, if implemented as a massless and compressible, ubiquitous continuum of negative electric charge, is inherently able, qualitatively, to fulfil its originally intended purpose, the support

and propagation of transverse electromagnetic waves of any wavelength.

2. By adopting the 150-year-old propositions of Maxwell and Thompson that fundamental particles are 'made out of aether', as vortical structures of its motion, we have been able to quantify its (minimum) stupendously high charge density, namely 10^{30} Coulomb.cm^{-3}.
3. This lifts the aether right out of the obscurity and insignificance, to which it was relegated by Relativity more than a century ago, to a status of perhaps being **the** principal agent in the Universe.
4. This new perspective has, by a wide-ranging quest for observational guidance, apparently led us to new appreciation of:-

i) the physics of gravitation, in which the Newtonian field and the Gravity-Electric (G-E) field are but facets of a single process; and there is, for a very simple stastistical reason, no development of negative gravity;

ii) the astronomically ubiquitous action of the G-E field, including central aspects of:- star formation, stellar evolution, planetary system formation, heliosphere observations, gravitational lensing, supernovae, pulsars, acceleration of cosmic rays and the dynamics of spiral galaxies. A vast range of scales.

5. These observationally integrated findings as to how the mass property is developed and the resulting gravitation really works appear completely to displace the supposed functions of the hypothetical and so-far–elusive Higgs boson. Ramping up the energies in accelerators will surely create ever-more ephemeral particles of ever higher mass but with doubtful bearing on the way Nature does things, or has ever done in the past.
6. Fundamental particles require finite volume in which to develop their mass property and magnetic moment as dynamical constructs of aether motion. This rules out relativistic black hole models which accumulate mass without limit by supposing the presence of infinite gravitational field at the centre. In CT any such limitation of space would result in mass annihilation and energy release.
7. The aether's electrical nature and its function in generating gravitation brings gravitation, at last, into the force-family of electromagnetic ones. Its ultra-high charge density probably renders it irrotational and the link between our diverse directional devices. Since some of those are inertial in nature, this linkage makes the aether the probable site of inertia. Inertia and gravitation thereby have a 'cloned' origin, providing the long-sought reason for the strict equality of inertial and gravitational mass.
8. To sum up on more tangible matters; if the G-E field did not exist, we would have no planet to stand on, apart from an evaporating inferno only ~10 solar radii from the Sun. And the Sun would not be the same either, inside or out.
9. On the positive side, we have been able to trace a continuous linkage from the fundamental physics of G-E field generation in the Sun, all the way to the factors which constrain our human existence (water, oxygen, earthquakes) and the internal behaviour of this planet [42]. This suggests that the further pursuit of CT would be an exceptionally worthwhile matter for us all.
10. Fundamentally, with the reality of the G-E field now secured by the multiple evidence for its action, presented here, there is now a functional requirement to meet as the basis on which, as recently favoured [66], to explore the modelling of non-zero-sized particle interiors (especially the electron and positron), of which Figure 2 is no more than a very simple conceptual starting-point. The 'infinitesimal singularity' view of particles, rejected here as absurd (Sect. 3), has for a century caused this to be treated as an illegitimate enterprise, so [67], a late discovery by me, may be the only contemporary comparison.

Acknowledgements

In 1959, Peter R. Wyke (deceased) was impressed by the physical significance of my work on the observed sky brightness distribution, noted in the Introduction. CT's foundations were the outcome of his prescience in organizing funding at once for me to pursue it for nearly a year, separately from the project. From that moment to this, continuance of my work on CT and its diverse ramifications has uniquely been made possible by the steadfast support of my wife, Margaret.

Helpful contacts with Roger Anderton, John Bahcall, Martin Barstow, Michael Duffy, Isobel Falconer, Albrecht Giese, James Gilson, Douglas Gough, James Hamilton, Duncan Hampshire, George Herbig, Robert Hutchison, George Kalmus, Al Kelly, Roy Nelson, Peter Rowlands, Sara Russell, Virginia

Trimble, Mogens Wegener, Cynthia Whitney, Iwan Williams and Arnold Wolfendale are all gratefully acknowledged. A further list appears in [68].

John Osmaston kindly recorded an English partial translation of Gerber's long 1898 paper, itself kindly provided by Al Kelly from a Dublin library.

References and Notes

[1] Packer DM & Lock C (1951) The brightness and polarization of the daylight sky at altitudes of 18,000 to 38,000 feet above sea level. J.Opt.Soc.Am. 41, 473.

[2] Barr NL (1953 (March)) "Brightness of the atmosphere". U.S. Nav. Med. Res. Inst. Report.

[3] Osmaston MF (2003) A particle-tied aether - Indications of a deeper foundation for physics and relativity. Physical Interpret. Relativity Theory (PIRT) VII (2000), Brit. Soc. Philos. Sci., Late Papers (M.C. Duffy, ed), PD Publicns, Liverpool (ISBN 1 873 694 05 9), 230-240. Also currently available at:- http://www.scribd.com/doc/78606561/Miles-F-Osmaston-A-Particle-Tied-Aether.

[4] Heaviside O (1889) On the electromagnetic effect due to the motion of electrification through a dielectric. Phil. Mag. XXVII, 324-339.

[5] Weber WE & Kohlrausch R (1856) Ann. d. Physik u.Chem., herausgegeben v. J.C. Poggendorff, 99, 10-25.

[6] Maxwell JC (1861a) XXV. On physical lines of force. Part I. The theory of molecular vortices applied to magnetic phenomena. Phil.Mag.,4th Ser., 161-175. Maxwell JC (1861b) XLIV. On physical lines of force. Part II. The theory of molecular vortices applied to electric currents. Phil. Mag.,4th Ser., 281-291 & 338. Maxwell JC (1865) A dynamical theory of the electromagnetic field. Phil.Trans. R. S. Lond. 155, 459-512. Maxwell JC (1873) Treatise on electricity and magnetism (1st ed). Clarendon Press. Maxwell JC (1878) ETHER or ÆTHER. Encyclo-paedia Britannica. 9th Edn. 8, 568-572 [see p. 572].

[7] Thomson W (1867) Phil. Mag. XXXIV, 15; Thomson W & Tait PG (1867) Treatise on natural philosophy. OUP.

[8] Thomson JJ (1883) On the motion of vortex rings: Adams Prize Essay. Macmillan & Co.

[9] Larmor J (1894) Phil. Trans. R. Soc. Lond. 185(810), 719-823. Larmor J (1897) Phil. Trans. R. S. Lond. 190(210), 205-300. Larmor J (1904) On the ascertained absence of the effects of motion through the aether, in relation to the constitution of matter. Phil. Mag., Ser. 6, 7, 621-625.

[10] Milner SR (1960) The classical field theory of matter and electricity: I. An approach from first principles. Phil. Trans. R. S. Lond. A 253, 185-226.

[11] Editorial (1891) In the van of progress. The Electrician 26, 328-330.

[12] Kelley MC (2009) The Earth's ionosphere: plasma physics and electrodynamics. Academic. 577p. Rycroft MJ et al., (2012) Recent advances in global electric circuit coupling between the space environment and the troposphere. J. Atm. Solar-Terr. Phys. 90–91, 198–211.

[13] Pulinets S & Boyarchuk K (2004) Ionospheric precursors of earthquakes. Springer. 315 pp.

[14] Astafieva EI & Heki K (2007) Ionosphere Responses to Large Earthquakes of Different Focal Mechanisms: Case Study of 1994, 2006 and 2007 Kuril Islands Earthquakes. AGU, Fall Mtg, abstr. #S33B-1311.

[15] Blecki J et al. (2011) Statistical analysis of the ELF turbulence registered in the ionosphere over the regions of the strong earthquakes by DEMETER satellite. Geophys. Res. Abstr. 13, EGU2011-13143.

[16] Hachisu I, Kato M, et al. (2012) A single degenerate progenitor model for type Ia supernovae highly exceeding the Chandrasekhar mass limit. Ap. J. 744(1), Article ID 69. arXiv:1106.3510.

[17] Howell DA et al. (2006) The type Ia supernova SNLS-03D3bb from a super-Chandrasekhar-mass white dwarf star. Nature 443, 308–311.

[18] Wildt R (1939) Negative ions of hydrogen and the opacity of stellar atmospheres. Ap. J. 90, 611-620.

[19] Lin RP (1994) Exploring the enigma of solar energetic particles. EOS:Trans.AGU 75(40), 457-466.

[20] Jeans JH (1919) Problems of cosmogony and stellar dynamics. Adams Prize Essay. Clarendon. 293p.

[21] Lyttleton RA (1941) On the origin of the solar system. MNRAS 101, 216-226.

[22] Jeffreys H (1952) Bakerian lecture: The origin of the Solar System. Proc.R.Soc. A214(1118), 281.

[23] Spencer-Jones H (1956) The origin of the solar system. PEPI 1, 1-16.

[24] Woolfson MM (1960) Origin of the Solar System. Nature 187, 47-48.

[25] Gold T (1984) The early solar system and the rotation of the Sun. Phil.Trans. R.Soc.A313, 39-45.

[26] Goswami JN & Vanhala HAT (2000) Extinct radionuclides and the origin of the Solar System. In Protostars and Planets IV (ed. V. Mannings et al), pp. 963-994. U. Arizona Press.

[27] Osmaston MF (2000) A new scenario for formation of the solar planetary system; dynamics, cores and chemistry. J. Conf. Abstr. (CD-ROM) 5 (2), 762. Osmaston MF (2006) A new scenario for forming the Sun's planetary system (and others?): dynamics, cores and chemistry (pt 2). GCA 70(18S), A465.

[28] Gomes R et al. (2005) Nature 435, 466.

[29] Crida A (2009) Solar System formation. arXiv:0903.3008v1 [astro-ph.EP].

[30] Osmaston MF (2009) What can we learn about solar planetary construction and early evolution of the inner members of the system from their present dynamics? Importance of a 2-stage scenario. EPSC Abstr. 4, EPSC2009-265, 2009.

[31] Marcy G et al. (2003) Orbital eccentricities. <http://exoplanets.org/ecc.html>.

[32] Schneider J (2011) The Extrasolar Planets Encyclopaedia. In <http://exoplanet.eu>.

[33] McCord TB (1968) The loss of retrograde satellites in the solar system. JGR 73, 1497-1500.

[34] Counselman CC, III (1973) Outcomes of tidal evolution. Ap. J. 180, 307-314.

[35] Choi PI & Herbst W (1996) Rotation periods of stars in the Orion nebula cluster: the bimodal distribution. Astron. J. 111, 283-298.

[36] Fernández JA (2005) Comets: nature, dynamics, origin and their cosmogonical relevance. Springer.

[37] Kalas P et al. (2008) Science 322, 1345-1348.
[38] Simon JI et al. (2005) A short timescale for changing oxygen fugacity in the solar nebula revealed by high-resolution 26Al-26Mg dating of CAI rims EPSL 238, 272-283.
[39] Wood JA & Hashimoto A (1993) Mineral equilibrium in fractionated nebular systems. GCA 57, 2377.
[40] Kuskov OL & Kronod VA (2001) Core sizes and internal structure of Earth's and Jupiter's satellites. Icarus 151, 204-227.
[41] Osmaston MF (2010) Providing solar system water and high planetary angular momentum, using a return to Ringwood's core formation model, supported by the behavioural evolution of the mantle. GCA 74(S1), A 779.
[42] Osmaston MF (2010) On the actual variety of plate dynamical mechanisms and how mantle evolution affected them through time, from core formation to the Indian collision. Geophys.Res.Abstr. 12, EGU2010-6101. Osmaston MF (2011) 3 stages of Earth evolution - core formation, ocean emergence and the 2.3 Ga rise of atmospheric oxygen: How are they linked? Miner. Mag. 75(3) 1576.
[43] Ringwood AE (1979) Origin of the Earth and Moon. Springer-Verlag. 295 pp.
[44] Osmaston MF (2011) What can Triton's retrograde orbit tell us about how the Giant Planets (and others) were constructed? Geophys. Res. Abstr 13, EGU2011-1242. Osmaston MF (2011) What can Triton's retrograde orbit tell us about how the Giant Planet interiors were constructed and acquired their gas/ice envelopes? EPSC-DPS Joint Meeting, Nantes. EPSC Abstr. 6, EPSC-DPS2011-953.
[45] Galy A. et al. (2000) The formation of chondrules at high gas pressures in the solar nebula. Science 290, 1751-1753.
[46] Hutchison R (2004) Meteorites: a petrologic, chemical and isotopic synthesis. 506p. CUP.
[47] Lagrange A-M et al. (2010) A Giant Planet Imaged in the Disk of the Young Star ß Pictoris. Science 329 (5987), 57-59. Lagrange A-M, Ehrenreich D, (LAOG) et al. and ESO. (2010) APOD.nasa.gov/100703.
[48] Romanowsky AJ et al. (2003) A dearth of dark matter in ordinary elliptical galaxies. Science 301, 1698.
[49] Osmaston MF (2012) How stars grow massive despite radiation pressure, triggering star-bursts; insights from gravitation. UK-Germany NAM2012, <http://www.jodrellbank.manchester.ac.uk/meetings/nam2012/sessions.html>. Session ISM1.
[50] Wang Z et al. (2006) A debris disk around an isolated young neutron star. Nature 440(7085), 772-5
[51] Buras R et al. (2003) Improved Models of Stellar Core Collapse and Still No Explosions: What Is Missing? PRL 90(24), 241101.
[52] Taubenberger S et al. (2011) MNRAS 412, 2735.
[53] Erlykin AD et al. (2002) Ultra high energy cosmic rays and pulsars. J.Phys.G: NucPartPhys. 28(8), 2225-2233.
[54] Erlykin AD & Wolfendale AW (2006) The nature of the 'knee' in the cosmic ray energy spectrum. J. Phys. G: Nuc.Part. Phys. 32(1), 1-7.
[55] Dufour P et al. (2007) White dwarfs with carbon atmospheres. Nature 250, 522-524.
[56] Ives HE & Stillwell GR (1941) Interference phenomena with a moving medium. J.Opt.Soc.Am. 31(1), 14-24. doi: org/10.1364/JOSA.31.000014.
[57] Gerber P (1898) Die räumliche und zeitliche Ausbreitung der Gravitation. Zeits.f.Math.u.Phys. 43, 93-104. Abstr. in Ann.d.Phys. 22, 529-530 (1898).
[58] Anderson JD et al. (2002) Study of the anomalous acceleration of Pioneer 10 and 11. Physical Review D 65, 082004.
[59] Gerber P (1917) Die Fortpflanzungs-geschwind-igkeit der Gravitation. Ann.d.Phys.Ser 4, 52, 415-441.
[60] Gehrke E (1916) Zur Kritik und Geschichte der neueren gravitationtheoren. Ann. d. Phys. Ser 4. 51, 119-124.
[61] Mach E (1902) The science of mechanics, 2nd ed., Transl. T.J.McCormack. Open Court. 605 p.
[62] Roseveare NT (1982) Mercury's perihelion from Le Verrier to Einstein. Clarendon. 208 p.
[63] Sagnac G (1913) L'éther lumineux démontré par l'effet du vent relatif d'éther dans un interféromètre en rotation uniforme [The luminiferous aether demonstrated by the effect of relative aether wind in an interferometer in uniform rotation]. CRAS 157, 708-710. Sagnac G (1913) Sur la preuve de la réalité de l'éther lumineux par l'expérience de l'interférographe tournant [On the proof of the reality of the luminiferous aether by the experience with a rotating interferometer]. CRAS 157, 708-710.
[64] Dufour A. & Prunier F (1939) CRAS 208, 988-990. D. & P. (1941) CRAS 212, 153-154. D. & P. (1942) J. de Phys. 3(9), 153-161.
[65] Kelly A (2005) Challenging modern physics: questioning Einstein's relativity theories. Brown Walker Press. 309p.
[66] Amoroso R. L., et al. (2010) "Hidden" Parameters Describing Internal Motion Within Extended Particle Elements. In Search for Fundamental Theory. The VIIth International Symposium Honoring French Mathematical Physicist Jean-Pierre Vigier (ed. R. L. Amoroso et al.), Imperial College, London, 12-14 July 2010. AIP Conf. Proceedings 1316, 1-27.
[67] Chalidze V (2001) Mass and Electric Charge in the Vortex Theory of Matter. Universal Publishers. ISBN 1-58112-675-1.
[68] Osmaston MF. Continuum Theory (CT): its particle-tied aether yields a continuous auto-creation, non-expanding cosmology and new light on galaxy evolution and clusters. (This volume)

Other CT Work by the Author

Osmaston MF (1997) Implications and evidence of a particle-tied aether: steps towards a deeper foundation for physics and relativity. Physical Interpretations of Relativity Theory (PIRT) V; Imperial College, London: 6-9 Sept 1996. Late Papers (MC Duffy, ed.), pp. 182-198.
Osmaston MF (1998) An interim outline of some research under the heading: some aspects of a continuum theory of physical nature. (Summary text + Appendices A (on RTV redshift) B (on RTV scattering) & G (on RLV line broadening)). PIRT V, Imperial College, London, 6-9 Sept 1996. Supple-mentary Papers (MC Duffy, ed). pp. 241-256.
Osmaston MF (1998) Continuum Theory, further

developments: the nature of mass-bearing particles, evident properties of the resulting gravitation, and an outline cosmogony (extended abstr.). PIRT VI, Imperial College, London, Sept 1998. Proceedings (MC Duffy, ed), pp. 248-250.

Osmaston MF (2004) Continuum Theory (CT); major implications of the 'particle-tied aether' concept for gravitation, rotational effects, and the strong nuclear force. In: PIRT VIII. Sept 2002 Imperial College, London. Proceedings (ed. MC Duffy) v.2., 355-385. PD Publications, Liverpool. ISBN 1 873 694 07 5. (The printed version has several one-line omissions. A complete one is at:- <http://osmaston.org.uk>.)

Osmaston MF (2009) A new, mainly dynamical, two-stage scenario for forming the Sun's planetary system and its relation to exoplanet findings. EGU Vienna, 2009, Geophys. Res. Abstr. 11, EGU2009-12204.

Osmaston MF (2009) A two-stage scenario for forming the Sun's planetary system, with good links to exoplanet findings, arising from new physical insight on the gravitational process. European Planet. Sci. Congr, Potsdam, September 2009. EPSC Abstracts 4, EPSC-2009.264.

Osmaston MF (2009) Construction and differing evolutionary outcomes of the terrestrial planets; insights provided by the 2-stage scenario for constructing planetary systems. European Planet. Sci. Congr., Potsdam, September 2009. EPSC Abstracts 4, EPSC2009-266.

Osmaston MF (2010) Implementing Maxwell's aether illuminates the physics of gravitation, yielding galaxy dynamics without CDM, high-a.m. planetary systems, and how high-mass stars are built. In Joint European National Astron. Mtg (JENAM), Lisbon, Sept, 2010, The European Week of Astronomy and Space Science (ed. A Moitinho et al), Abstract Book (Version 2.0) 159-160, Abstr # 174.

Osmaston MF (2010) Continuum Theory (CT): implications of its continuous auto-creation cosmology for the construction and morphological evolution of galaxies and clusters (abstr). In Joint European Nat. Astron. Meeting (JENAM) 2010, Lisbon, Sept, 2010 (ed. A Moitinho et al.) Abstract Book (Version 2.0) p.160, Abstr # 175.

Osmaston MF (2011) A continuum theory (CT) of physical nature: towards a new 'ground floor' for physics and astronomy, including gravitation and cosmogony, with major tangible support. In Physical Interpretations of Relativity Theory, Proceedings of the International Scientific Meeting - 'PIRT-2006', 8–11 Sept. 2006, Imperial College, London. (ed. MC Duffy, VO Gladyshev, AN Morozov & P Rowlands), pp. 287-317. ISBN 9785703835500. Also (pdf) at <http://osmaston.org.uk>.

Osmaston MF (2011) Continuum Theory: Physical Nature Viewed from a Deeper Level; a Rewarding Replacement for SR/GR. In 17th Ann. Conf. NPA, California State Univ., Long Beach, June 2010. Proc. Natural Philos. Alliance (ed. CK Whitney), v. 7, 720-748. ISSN 1555 4775.
Also (pdf) at <http://osmaston.org.uk>.

Osmaston MF (2012) Close-in exoplanets, but none of ours. Guidance from Triton's orbit and the physics of gravitation. UK-Germany National Astronomy Mtg. Manchester. NAM 2012. Session PL2, Exo-planets. <http://www.jodrellbank.manchester.ac.uk/meetings/nam2012/sessions.html>.

Osmaston MF (in press, a) Continuum Theory (CT): history of its conception, and outlines of its many current results: an informal account. In Proc.11th Int Conf on Physical Interpretations of Relativity Theory (PIRT XI). Brit Soc Philos Sci, Imperial College, London, Sept 2008. (ed. MC Duffy, VO Gladyshev AN Morozov & P Rowlands) 27 pages.

Osmaston MF (in press, b) Continuum Theory: physical nature viewed from a deeper level; a rewarding substitute for SR/GR and its mortal inconsistencies. In PIRT XII, Imperial College, London. Sept 2010 (ed. MC Duffy, VO Gladyshev, AN Morozov & P Rowlands), 37 pp. Also (pdf) from <http://osmaston.org.uk/special.htm>.

Continuum Theory (CT): Its Particle-Tied Aether Yields a Continuous Auto-creation, Non-expanding Cosmology and New Light on Galaxy Evolution and Clusters

MILES F. OSMASTON

The White Cottage, Sendmarsh, Ripley, Woking, Surrey GU23 6JT, UK.
miles@osmaston.demon.co.uk; http://osmaston.org.uk

Our preceding paper "Implementing Maxwell's aether......." (Paper I) concluded:- (A) Maxwell's aether, ignored in Relativity, is a massless, quasi-superfluid continuum of extremely high negative charge density; (B) Fundamental particles are not infinitesimal singularities within the aether but develop their mass by being 'made out of it' (hence the name Continuum Theory) as finite-sized vortical constructs of its motion. So reproduction ('auto-creation') of more of them requires only the addition of suitable dynamical energy, with Ampere's law providing charge-coupling in shear to get rotations. (C) In the resulting gravitational process, generating the Newtonian force simultaneously also generates a radial electric field, the Gravity-Electric (G-E) field, whose action on astronomical plasmas could explain the flat tangential velocity profiles of spiral galaxies without resort to Cold Dark Matter (CDM) if outward disc flow is present. One of the objectives here is to provide that flow by axial infall and to examine its consequences. But first, if particles are 'made out of aether' the associated random aether-charge motion will generate radiation (the CMB) and impose four distance-cumulative, wavelength-independent transmission effects upon electromagnetic waves. One of these – a redshift – we see here as the cosmic redshift, plus intrinsic redshifts in stellar and galaxy 'atmospheres'. Such a redshift appears to have been reliably observed with caesium clocks over long ground-level paths in 1968 but, lacking an appreciation of its mechanism, its wide significance was doubted. In fact, our extrapolation to intergalactic conditions dispenses with the BigBang. The other 3 transmission effects are:- spectral line broadening, scattering and attenuation, each of which has significant astronomical/cosmological expression. If the cosmic redshift is not a velocity, the reason for Dark Energy vanishes. In the resulting no-expansion cosmology the Universe was originally equipped with randomly moving aether, from whose motion and energy content the entire mass content of the Universe has grown over time by auto-creation, the local rate of which experiences positive feedback and acceleration as gravitational accumulations drive energy levels higher. Hence the clumpiness of galaxy distributions. The infall of cosmogonally young material from the auto-creation auras of clusters has 3 major implications. (1) It completely inverts the Big Bang perspective that low-metallicity material, widespread in galaxy haloes, is very ancient. (2) Quasi-axial infall of such broadly neutral material (mostly H) onto a Spiral will spread out in the galactic plane, driven radially from the ionizing bulge by the G-E field, maintaining constant tangential velocity; all without CDM. This pattern means that the arms, although trailing, are actually being blown outward (unwrapping). See Paper I for detail. For such ongoing disruption of Spirals to prevail so widely means that originally each must have started life as an a.m.-conserving, tightly-wound spiral of mostly neutral, cosmogonically young material (mainly H), in which G-E field action was minimal until star formation and ionization had set in. (3) In cluster interiors, other cluster members may deflect the two infall streams as they converge onto a Spiral, introducing a dynamical rotational couple near the centre, with an axis roughly in the galactic plane, to produce a Barred Spiral. Cessation of infall then results in endwise collapse of that bar, yielding a fattened Elliptical. Those are indeed typically concentrated in the centres of clusters and show a dearth of active star formation, consistent with being deprived of young infall. We present images and diagrams in support and elaboration of (2) and (3). The CT model for quasars provides large intrinsic redshift by the CT analogue of Transverse Doppler Effect and offers light-element synthesis by the evolutionary precipitation of a runaway rotational shrinkage, with mass annihilation and emission of a GRB. Of special interest, relative to the arm's-length nature of BigBang cosmology, is that continuous auto-creation (CAC) cosmology is in principle available near-by for direct study and the development of strong observational constraints. In the context of (1), the very low metallicity (Pop II) of globular (star) clusters abundantly present in the haloes of galaxies points to them being (infallen?) local concentrations of quite young auto-creation. In that case the 'blue straggler' stars more recently formed in their core regions may be our youngest examples of ongoing auto creation. In summary, CT offers a much more directly observable Universe, with no Big Bang, CDM, or Dark Energy, and a CMB that records the true temperature of intergalactic space along the path taken by the radiation. Its closely cavity-radiation character arises from the random aether's transmission-related opacity (Olbers' Paradox) of the infinite CT Universe. Fundamentally, the aether's random motion constitutes all-penetrating random electromagnetic excitation at the atomic scale that may offer the accommodation between classical physics and stochastic quantum electrodynamics so long obstructed by Relativity Theory.

Keywords: cosmology, auto-creation; metallicity; particle-tied aether; dark energy; redshift, cosmic; redshift, intrinsic; galaxies; QSO; inertia; Olber's paradox.; photoelectric effect.

1. Introduction

As set out in the preceding paper [1] ('Paper I'), CT is built upon two findings.

A. Maxwell's aether, far from being a negligible aspect of the Universe, is actually seen to be a massless and ubiquitous quasi-superfluid continuum of extremely high negative charge density ($>10^{30}$ C.cm^{-3}).

B. Mass-bearing fundamental particles are **not** infinitesimal singularities, they develop a mass property by being 'made out of aether' as non-zero-sized vortical constructs of its motion (hence 'Continuum Theory').

Main Contents

If B is how the mass property is generated, the production and reproduction ('auto-creation') of more of such particles requires only the addition of suitable dynamical energy, some of which must be rotational.

Relative motions in a superfluid are by definition incapable of initiating rotation. But in this case Ampère's Law operates on the aether charge to limit the superfluidity and provide the necessary coupling in shear. This coupling, with associated magnetic energy storage, also provides the elasticity in shear to enable the aether to support the propagation of transverse electromagnetic (TEM) waves, but these are not the right kind of aether motion to have mass [1]. Conversely, by making particles out of aether electric charge, their wave-like diffraction may be explainable.

On the other hand, TEM waves do possess electromagnetic energy, as must all forms of aether motion. This is a point hitherto neglected in the former context of dubious motions of a rather negligible aether, but it is central to the auto-creation cosmological theme to be developed in this paper. As I noted in 2000 [2], the recognition of that energy in CT is essential if the Second Law of Thermodynamics is to be sustained.

Consequently, in CT the equating of energy to mass now has restricted validity. In no circumstance may the very obvious energy of TEM waves be treated as mass.

In General Relativity (GR), by contrast, Einstein's proposed universal applicability of the now-famous (although not his own, but apparently that of Poincaré (1900) [3]) mass-energy equivalence formula $\boldsymbol{E = mc^2}$ offers what is from an incisive point of view an uncritical shortcut between events and processes without due care as to how the equivalence is physically or dynamically achieved in the particular case at issue. This fixation with energy rather than with mechanism seems to have been adopted from Thomson & Tait (1867)[4].

2. Random Motion of the Aether

By making particles out of aether in vortical motion, as envisaged in Paper I, they are not to be seen as sharply delimited objects, but are "indefinitely co-extensive" – Larmor (1904) [5]. So any random motion of atoms/molecules built of such particles, as in a gas, will surely entrain the intervening aether also, endowing it with a corresponding random motion.

In fact, of course, the 'particles' can vary in two main ways — their mass and their electric charge. The particle **mass** affects its surroundings through the aether pumping mechanism of its constituent fundamental particles, whereas the particle **charge** affects the surrounding aether charge directly through the charge density gradient that it causes. The ratio of these is directly calculable as the ratio of the electrostatic force to the gravitational force between two identical particles, namely $\boldsymbol{q^2/Gm^2}$, where $\boldsymbol{q}$ and $\boldsymbol{m}$ are the charge and the mass. This gives 4 x 10^{42} for electrons and 10^{36} for protons (1836 x electron mass). So the r.m.s. amplitude of the aether motion, and the resulting effects, are extremely, and principally, sensitive to the degree of ionization present in the gas.

In this frame I now reinterpret the famous Michelson-Morley result as having shown that, to the limited precision then available, the aether is indeed 'particle-tied' in its motions. The result is that TEM waves reach us along astronomical paths whose aether is in random motion mainly related to the ***PT*** and ionization there. So we consider next the five consequences of that random motion, and their big implication for cosmology.

3. TEM Waves — The Four Transmission Effects of Aether Random Motion

All four are wavelength-independent and path-length-cumulative. Individual increments are proportional, so overall growth is exponential w.r.t. distance. The rates of all four increase with the gas particle velocity present so, for a Maxwellian distribution of velocities, the growth rate with distance varies as the square of the gas temperature along the path. All are steeply enhanced by ionization. Having a common cause, correlated occur-rences are to be expected — a powerful diagnostic.

First we note the principal character of each, and then add some discussion to set each in context.

1). Redshift

"Random transverse velocity (RTV) redshift". Each transverse local displacement of the propagating aether stretches the wave, relative to the rest of its train, along a corresponding hypotenuse. Propagation is still at ***c*** *relative to that aether* but (vector addition being legitimate in CT [1]) the vector resultant is >***c***, so **transmission time is not affected**. In this case (in contrast to sound waves), precisely transverse displacement *does* rotate the propagation vector because Ampére's Law provides aether coupling ('viscosity') in shear, as noted above. In the CT frame, sudden jumps in the aether transverse velocity encountered by the TEM wave are unlikely, so its stretching will actually be the integrated product of the resulting transverse accelerations experienced.

2). Wavelength dispersion

"Random longitudinal velocity (RLV) line-broadening". Frequency modulation due to the longitudinal components of the propagating aether motions, a process explored mathematically by [6]. The effect emerges as a valuable substitute for 'rotational broadening' in stars or to invoking 'surprising' turbulence.

3). RTV deflection scattering

Associated with **(1),** this has directional properties diagnostically different from Rayleigh, and is seen high in the Earth's atmosphere, see the introduction to [1].

4). Attenuation

Due to the energy scattered. It affects distance estimates based on the inverse square law.

3.1. *Significance of the Transmission Effects*

1. RTV redshift. As shown later (Sect. 6), the cosmic redshift is to be seen as one example of this redshift. Because of the extreme sensitivity of aether motion to ionization, combined with high temperature (high particle velocity), the plasmas of stellar atmospheres and galaxies are expected to impose intrinsic RTV redshifts on emergent radiation, even though the path traversed is relatively short.

Since TEM waves are not the right kind of aether motion to possess mass, this redshift now needs in CT to replace the relativistic gravitational redshift. In Section 5.1 we consider diagnostic observations of the solar redshift in this context.

RTV redshift also needs to be allowed for dynamically in any virial longevity context for stellar clusters or in assessing the relative positions and velocities of galaxies in a cluster. Some star clusters exhibit class-related differences in redshift, the O-B group apparently receding faster [7-10]. Known as the 'K-term', Allen [11] records the effect as being much less for less-bright stars of the same class and as zero for F0-K0, rising again slightly at M0, with their deep atmospheres. Weaver [12] further associated the term with an unexplained line-broadening (our RLV effect (2)?). Velocity-equivalent values for the intensely radiative W-R (Wolf-Rayet) stars (in binaries) range to >150 km/s, relative to their O or B companions [13], so the total for the W-R probably ranges to >250km/s. And much more (>1000 km/s?) for plasma-rich Spiral galaxies, relative to their plasma-poor Elliptical companions in the same cluster, e.g. [14], [15]. Data presented by Finlay-Freundlich [16] convinced ter Haar [17] of the presence of intrinsic redshifts in stellar atmospheres.

2. RLV line-broadening during transmission. It is the variance that grows, so the line-width increases more slowly than the RTV redshift, which may visually dominate when large, as in quasars. Classical interpretation of line width as temperature of the emitting plasma often leads to conflicts with other constraints, the commonest being that a star's line-width temperature is much higher than its colour temperature. RLV line-broadening appears to improve upon both the escape interpretations that have been applied. These have been:-

(a) *"Rotational broadening"* – the doppler result of opposite sides approaching and receding – is widely applied to the line-widths of very hot early-type stars (W-R, O, B, A). But Struve [18] (p.130) noted a "startling decline at F5 [which] is unquestionably real"; an a.m. disposal problem still unresolved. The extreme ionization-dependence of these CT transmission effects means that quite a small drop in temperature and depth of the stellar atmosphere at that evolutionary stage might suffice, casting doubt on the interpreted fast rotation of earlier-type stars.

(b) *'Surprising' or 'unexplained turbulence'.* In 1958 three major UK thermonuclear fusion devices (ZETA, SCEPTRE III & IV) were dismantled for exhibiting 'unexpected turbulence' because they yielded line-width temperatures (~5 MK in one case) some 20-fold higher than was realistic on four distinct other grounds [19]. This provoked Spitzer [20] to consider the problem to be "of great interest in basic physics'. Spectral observation had been along a chord across the pinched toroid. 'Surprising turbulence' apparently remains the official reason for this disaster, and a caveat for subsequent designs. If RTV line-broadening was really the cause, it was just the temperature-monitoring that was inappropriate.

3. RTV deflection scattering. As noted [1], it was my recognition of this phenomenon during astro-navigation development work in 1959 that put me onto the CT trail in the first place. The unpredicted brightness gradient feature which concerned us particularly for star search purposes was that at high flight altitudes, and more markedly with increasing height, the observations [21 - 23], probably done with just this purpose in mind, showed that as the solar altitude in the sky goes below 40^0 a hump of brightening begins to appear in the opposite direction, centred on the antisolar point. We were able to show with mathematical rigour that what we here call 'RTV deflection scattering' by a randomly moving aether would explain this and other details very well. The treatment runs like this.

The scattered brightening at any point on the sky is the quotient of two functions. One (i) is a scattering probability function that decreases radially in all directions away from the source direction (Sun). The other (ii), in the denominator, is the angular area of the elementary annulus from which that light reaches the observer. This area increases up to the angle $\pi/2$ from the source, but decreases to zero at the anti-source point, thus concentrating all the probabilities and providing a brightening rate which, at some point, inescapably surpasses the rate of decrease associated with the probability function. The scattering function along horizon-ward radii from the Sun would be very different from along zenithward ones, so the patch of brightening is not circular. Diagnostic of the wavelength-independent RTV mechanism, the colour of the added light appears to be that of the Sun.

The *gegenschein* [24] is a similar, but far dimmer, phenomenon in the night sky from the ground, also observed by Pioneer 10 at 1.86 AU [25] but is sometimes thought to be associated with the dust-caused zodiacal light. It now appears to have an RTV scattering origin. In this case, the low intensity of the scattered light would be because solar illumination is then confined to the cooler and tenuous gas in interplanetary space.

4. Attenuation due to RTV scattering. Neglecting this attenuation, in addition to the inverse square law with distance, means that we currently over-estimate distances, by large amounts at big redshifts. I see this attenuation/opacity as the probable explanation of Olbers' Paradox and as providing the solar sub-tachocline extra opacity needed [26] for matching the Standard Solar Model to helioseismology.

1 & 4. Although both redshift and attenuation grow exponentially with path-length, the cosmic redshift-distance law appears linear, possibly because it is obtained by comparing these, the exponents of which are constrained by the same parameters.

4. Aether Random Motion — The Fifth Effect: The CMB Radiation

Random aether motion implies corresponding accelerations of electric charge. These will generate synchrotron radiation at frequencies that correspond to the gas/plasma temperature of the extended region through which it arrives. So the ~2.725 K black-body temperature of the CMB radiation may indeed be that of intergalactic space, slightly enhanced where the path passes a galaxy cluster (e.g. Coma), or groups of them. It was to explain the observation of just such correlations that the **Sunyaev–Zel'dovich effect** was proposed and has widely been applied since 1983, even to the stage of using it to predict unseen groups of galaxies. But the S-Z effect invokes quantum energy exchanges not applicable in CT.

The CMB's precisely black-body character is attributable to the effective optical depth implied by the scattering over cosmic distances, just as stars also exhibit nearly black-body spectral continua indicating the temperature at the atmospheric (photosphere) level where substantial opacity occurs. The CT explanation, offered above, is very simple and potentially very informative. Limitations of the present paper prevent further discussion here.

5. RTV Redshift in the Solar Environment

5.1. *The Solar Redshift*

This redshift is customarily cited, uncritically, as being consistent with the relativistic prediction of gravitational redshift (velocity-equivalent 0.636 km.s^{-1}). This is incorrect. Early measurements [27, 28] had already shown that along radial traverses in many directions on the solar disc the redshift is well below the prediction at the centre, rising sharply to well above it as the limb is approached. This was amply and in diagnostic detail confirmed by M.G.Adam and her colleagues at the Oxford Solar Observatory, using a sophisticated form of the Fabry-Pérot interferometer [29].

We note two features here. (i) At the disc centre, absorption lines of the same element but at different wavelengths showed very different redshifts. (ii) Absorption line profiles show distortion from the tip/core to the wings, the wings (being the first-formed part at the reversing level deep in the photosphere) being redshifted less, relative to the core (formed at shallow level as the radiation emerges). Both these features show a gradient of redshift with depth that is far steeper than the $1/\boldsymbol{R}^2$ of the relativistic prediction. At the limb, the rise looks like a zenith-angle (path-length) dependence for the emergent radiation.

Evidently the solar redshift is completely inconsistent with the relativistic interpretation but is qualitatively in accord with RTV redshift within the photosphere and the atmosphere above.

5.2. *Transmission Redshifting by the Solar Corona*

This appears to have been observed on two occasions.

(a) Late in 1968, as Pioneer 6 passed behind the Sun, the received frequency of its 2.292 GHz telemetry carrier wave showed a progressive fall as the sightline moved from 9 to 3 solar radii distance from the centre; a fall which was recovered on emerging the other side (Fig. 1). The fall amounts to a doppler-equivalent redshift increase of 9 m.s^{-1} between these positions.

Simultaneously, "spectral bandwidths increased slowly at first, then very rapidly at 1 degree from the sun. In addition, six solar "events" produced marked increases of bandwidth lasting for several hours" [30]. This is nice correlation of RTV-RLV action.

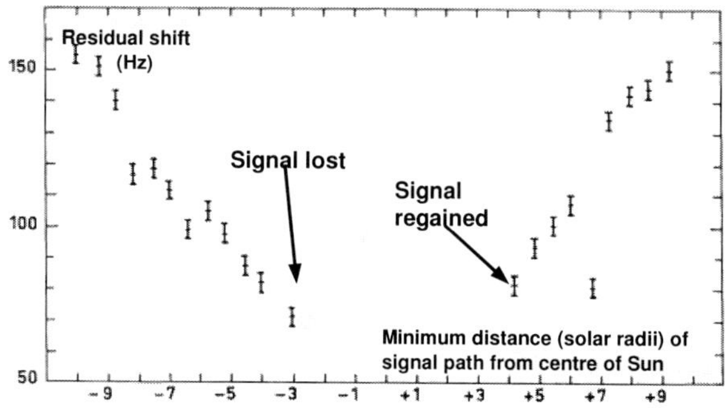

Figure 1. Changes of carrier-wave frequency received from Pioneer 6. 'Residual shift' is after removal of steady transmitter drift and orbital doppler correction. Modified from Fig. 1 of [31]. Reproduced with permission © ESO.

(b) The other occasion was by Sadeh et al 1968 [32]. They, for several years, had used caesium clocks to monitor the 21 cm hydrogen line absorption (1420 MHz) by Taurus A, near its occultation by the Sun. A reasonable view of their Figure 1 is that, relative to sight-lines several degrees away from the Sun, the observed frequency had dropped by ~80 Hz at the two closest-to-Sun paths, although the distances were not stated. This frequency drop is a doppler-equivalent redshift ($\boldsymbol{c}.\delta f/f$) of ~17 m.s^{-1}. Since this observation encompassed the redshifting difference by much more of the corona (and chromosphere?) than that recorded in (a) above, these values are in close mutual agreement as to the magnitude of the RTV redshifting by the solar atmosphere in those circumstances.

Both these results were dismissed as spurious in 1971 [33] when reporting radar pulse-delays and path-length-increase observations aimed at measuring the relativistic gravitational TEM wave deflection by the Sun. Redshift was assumed, relativistically, to imply a slowing of wave transmission, in which case the observed redshifts implied a slowing some two orders greater than the observed pulse delays. In fact, as emphasized above, RTV redshifting **does not affect** transmission time. So CT can make sense of the disparity but Relativity fails, further justifying its rejection.

6. Ground-Wave Observation of RTV Redshift, With Extrapolation as the Cosmic Redshift

Reported in the same paper as (b) in Section 5 by members of the US Naval Research Laboratory [32], RTV redshift appears to have been observed in a persuasively careful manner during the summer of 1968 as a real phenomenon, over ground-level paths up to 1500 km from a base at Cape Fear on the East coast.

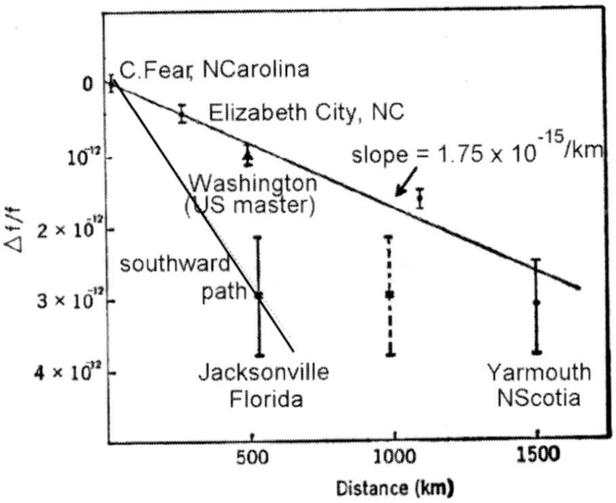

Figure 2. Plot of the redshift observations, modified from [32]. The slope indicated is for the line drawn by the authors. The Washington point was a check for overall stability of the set-up. The Jacksonville point has been replotted at the correct distance from C. Fear. (The published position is shown broken.) The much steeper slope and bigger observational scatter for that path supports our RTV expectation of the effects of temperature and/or ionization.

Each data-point (Figure 2) was for an inter-compared set of 3 caesium clocks, stationary for a week, so any doppler contribution was ruled out. Only the ground wave was used.

In 1969, taking the indicated slope as for 1 bar, 290K (temperature not reported) and air molecular weight 29, extrapolating to neutral H at 2.75K and density 10^{-25} kg.m^{-3}, I got H_0 = 59.5 km.s^{-1}.Mpc^{-1}; well within the range considered, even now. Encouraging?

But that density was the only offering then available, being a mean (inappropriately) for a relativistic expanding Universe (also inappropriately). It seems far too high for real intergalactic paths. But I conclude that allowance (Sect. 2) for even a quite low degree of ionization would allow the density to be at least 8 orders lower and still yield acceptable H_0.

So we now proceed on the basis that the cosmic redshift is not a velocity.

Here at last, though many have tried before, seems to be an observationally well-supported escape from the 80-year bondage to which cosmology physics has hitherto been confined by its adherence to the doppler origin of redshifts.

6.1. *Effective Aether Random Motion, a very Smoothed Result of Particle Motion*

To develop the foregoing observational result it is instructive to get some measure of the degree to which the particle motions are reflected in the r.m.s. aether velocity from which the observed redshift arises. Let us consider, for this purpose, the redshift that would arise if the aether velocity were to change by the most probable particle velocity of the gas every distance along the light ray path equal to the mean gas particle separation. This represents an ideal standard.

It turns out [34] that the observed rate is **5 x 10^{13} times smaller** than that which would be associated with our ideal standard. This, although very imprecise, gives us some guidance as to the high degree to which particles made out of aether involve aether motions which are actually "indefinitely co-extensive" [5]. It also shows how extremely small were the individual TEM wave transverse deflections, even under ground-level Earth-atmosphere conditions. How much tinier they must be, therefore, in building up the cosmic redshift in intergalactic space at 2.7 K. So the conclusion is that this is why this process produces no currently detectable blurring of astronomical imagery. But it does predict an ultimate limitation on resolution. The possiblity of hitting it might need to be considered before planning VLBI using an Earth-Mars baseline!

7. QSO Intrinsic Redshifts, Aberration, the Lyman α Forest, and Velocity-Dependent Inertia (VDI)

7.1. *Stellar Aberration and Redshifts*

The RTV redshift discussed above uses a velocity triangle related to aether motions, but is otherwise essentially the same as the classical one of Bradley, for aberration, which relates to objects (ourselves and a star). We have already shown that in CT the hypotenuse can >**c** but propagation never exceeds ***c*** relative to the local aether medium. So if there is a sufficient gradient of transverse velocity in the intervening space between objects the effective overall resultant can build to >c without limit, providing what we here call aberration-related (A-R) redshift.

Let us clarify the nature of this A-R redshift. The classical velocity triangle treatment of Bradley leads to an aberration angle $\boldsymbol{tan^{-1}v/c}$, and the SR formulation is $\boldsymbol{sin^{-1}v/c}$ (to prevent vectors exceeding $\boldsymbol{c}$), where $\boldsymbol{v}$ is the transverse velocity of the observer relative to the source. At the small angle (20.6 arcsec) produced by the Earth's 30 km/s orbital velocity, it is not practical to distinguish between these formulations (10^{-7} arcsec difference). However, the simple relativistic treatment also predicts, since it is only concerned with the relative velocity of source and observer, an appropriate aberration for, say, a stellar spectroscopic binary component moving transverse to the line of sight. This is not observed; if it were, such binaries would appear to scoot to and fro across the sky, a behaviour which would long ago have been used to detect them without a spectrograph.

Although there have been roundabout attempts to explain this in relativistic terms, appeal in CT to a transmitting medium (the aether) resolves this at once. In the case of the Earth's orbital velocity the gradient of transverse velocity, between transmitting medium and receiver, is clearly concentrated relatively near the receiver, and a classical triangle of velocities yields the aberration angle.

For a binary star component (Figure 3), on the other hand, the gradient of transverse velocity, relating to the orbital velocity of the source, is somewhere near the far end of the path, and the aberration angle produced there mainly results in the observer receiving rays from the star that were not those originally emitted directly towards him so, as the treatment given below shows, little or no visible aberration is to be expected.

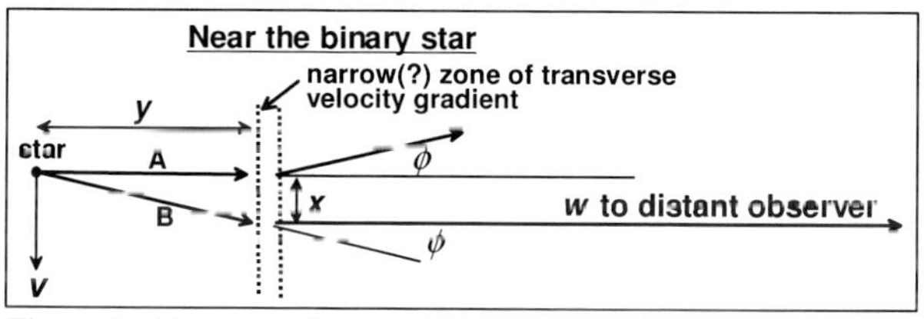

Figure 3. Aberration diagram for a distant orbiting binary star in the presence of a transmitting medium (aether).

Analysis of Fig. 3.

Distance y may be of the same order as the binary orbit radius. The aberration angle is:-

$$\phi = tan^{-1}\, v/c \quad and \quad x/y \approx tan\ \phi$$

The observer sees light ray B instead of A.

The observable angular displacement is thus:-

$$a_{obs} = tan^{-1}\, x/w \approx (y/w)\, tan^{-1}\, v/c$$

which will commonly be too small to detect. In very favourable circumstances it might just be possible to do so using VLBI techniques.

It is easily shown that, contrary to simple intuition, the total aberration does *not* depend upon whether the gradient of transverse velocity is steep and confined to a small part of the sight line, or is rather widely distributed. What *does* change with position of the gradient along the sight line is the amount perceived by the observer.

Note that the associated Continuum Theory redshift ("aberration-related (or A-R) redshift"), equivalent to the "transverse Doppler effect" of Relativity Theory, depends only upon the actual aberration, wherever it occurs, not upon its perceived amount.

Quasars are complex objects and have much to tell us. We will now show that for quasars, in combination with the extreme velocities of circulation around an object which may develop during its shrinkage under the action of velocity-dependent inertia (see Paper I, Sect. 16) **much, or even most** of their redshift may be instrinsic and of the A-R variety. VDI was also favoured by Ghosh [35], but for different reasons.

7.2. *QSOs (Quasars)*

The main features of quasars are as follows.

Diminutive, star-like image size and small physical size, evident in luminosity changes on short timescales. Ultra-high radiation energy output if all the redshift relates to distance on the cosmic redshift scale.

Very broad Lyman α hydrogen emission line (normally at 1216 Å), redshifted ($z = \delta\lambda/\lambda$) in the range <0.2–7.1 [36] (the 2011 maximum known).

Numerous (up to >100) Ly α absorption lines, the so-called "Lyman alpha forest", extending along the shortward flank of the main hydrogen Ly α emission, starting very close to its peak. Related mostly to the latter there are also metal-ion absorptions (and sometimes emissions) for CIV, NV, SiIV, MgII and FeII. In some QSOs, the so-called 'BAL QSOs', these metal absorption lines are very broad [37]. The forest lines nearest to the main emission get further apart as the latter's redshift increases [38]. The number of forest lines increases much more than proportionately with z for $z > 2$ [39].

Historically, it has been envisaged that all these absorptions, at progressively lower redshifts, must be due to clouds or galaxy fringes in intergalactic space along the sight-line. But where the sightline traverses one of the great voids in intergalactic space, corresponding gaps are not seen in the forest [40].

Also, MJ Rees stressed the containment problem posed by the column depths and the implied cloud temperatures of at least several thousand Kelvin if these are at the edges of systems like galaxies (in discussion in [37], [41]) (Note that such temperatures would conflict directly with our 2.7K interpretation of the CMB (Sect. 4)). Nevertheless, a conviction reached by 1995 that the absorptions are completely extraneous to the quasar itself [42] seems to have distracted all subsequent studies from their possible relevance to the quasar itself. We will seek to remedy that here.

Particularly important is that the luminosity of $z > 1$ quasars is uncorrelated with z, contrary to expectation if z were a measure of distance [43]. Moreover, in 1995 the highest-z quasar observed had stood at 4.89, their apparent space density dropping beyond $z = 3.0$ [44]. Since then higher-z quasars have become increasingly difficult to find. The highest z has crept slowly from 6.28 (2001) [45] to 7.1 (2011) [36]. Associated with a >2 orders drop in the forest flux, this has been interpreted (assuming all the redshift is cosmic and relates to distance) as showing that at $z > 6$ we reach the post-BigBang 'Gunn-Peterson trough' in which absorption by neutral H was predicted to have been very high [46].

Apart from those cited above, Arp [47] has been a stalwart advocate of the presence of intrinsic redhifts in quasars, based on many examples of their apparent companionship with galaxies of lower redshift. The contention [48] that they have been ejected from galaxies does seem problematical, but a vigorously rotational environment near a galaxy might be the key.

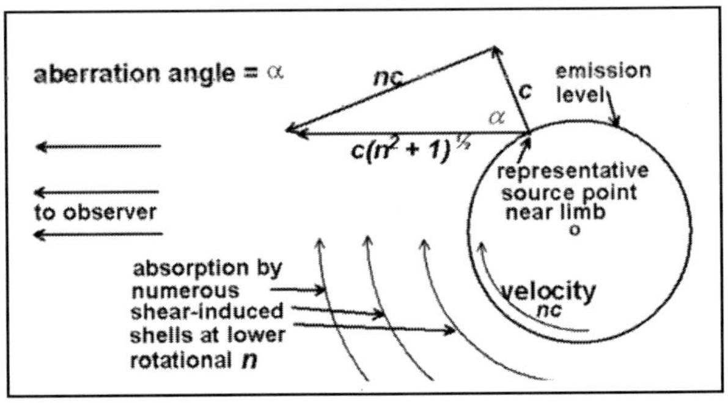

Figure 4. The new model for quasars. Each of the shear-induced shells is seen as responsible for the individual forest line absorptions at successively lower transverse velocity and A-R redshift, evident to the observer. When this was drawn in 1995, $z = 4.89$ was the highest that had been observed, and it sought to show the extreme possibility that all of it could be intrinsic. It demonstrates the geometrical limitation on escaping flux which arises when the intrinsic (A-R) component of z becomes large. In fact we do not at present envisage that the redshift of any quasar actually includes an intrinsic component which is that big. Superluminal factor – $\boldsymbol{n}$.

Features of the new QSO model (Fig. 4) are:-

1. Velocity-dependent inertia (VDI), the result of recognizing that gravitational communication at velocity $\boldsymbol{c}$ drastically reduces centrifugal/inertial (but not central gravitational) force when orbital velocity approaches and surpasses $\boldsymbol{c}$. This is because the effective volume of the "rest of the Universe" responsible, under Mach's Principle, for provision of inertia, becomes drastically reduced by retarded-field action.

So *superluminal* shear-surface velocities, due to gravitational shrinkage of high-angular momentum shells, become possible. Note that in CT (Paper 1) such velocities do not result in a relativistic mass increase of the material, no matter how that velocity is achieved.

2. Much of the redshift can be intrinsic to the body, being of aberration-related (A-R) type, and amounts to $z = \oplus(\mathbf{n}^2 + \mathbf{1}) - \mathbf{1}$. Neglecting any cosmic contribution, $z = 4.89$ requires $\boldsymbol{n} = 5.8$ and $\boldsymbol{\alpha} = 80.2°$. So the highest-$z$ quasar could, in 1995, actually have been 'on our doorstep'. But the radiation received would have been very weak, due to limitation of its escape in our direction to so far round to the side. This geometrical effect is seen here as the real reason for the drop in quasar intensity at high redshift that has been attributed to the Gunn-Peterson effect.

3. The bigger the redshift (higher $\boldsymbol{n}$), the bigger the number of shells (= forest lines), with a fiercer shear velocity gradient towards the centre, and a bigger redshift interval between shells, as observed.

4. Excess central-emission-line breadth is primarily due to rotational, not RLV, broadening ($\boldsymbol{n}$ varies with latitude on the emission surface). Temperatures may be high, due to the rapidity of contraction (see (8) below), so is the source of its intense radiation.

5. The Lyman α forest and the high-ionization metal lines are redshifted absorption as the radiation passes outward through shells, with the metals and the higher temperature being from closer-in shells, where conditions may verge between emission and absorption. The gravitation of the central body resolves the confinement problem. They are **not** due to clouds in intergalactic space, whose temperature can thus be the 2.75K indicated by the CMB (Sect. 4).

6. Perspective effects, such as RLV line-broadening if a shell is traversed obliquely, may be responsible for the line-breadths in BAL QSOs. Other perspective effects will need to be considered.

7. Quasars are **not** at cosmological distances. It may not be difficult to separate the cosmic and intrinsic components of their redshift (see (12) below). Their spatial association with (or in?) galaxies is entirely reasonable. The need for a high angular momentum source cloud makes their occurrence in isolation less likely but would be consistent with occurrence as AGN.

8. As n rises towards and past unity during contraction, centrifugal constraint upon shrinkage decreases. This situation is the opposite of the Keplerian one; the higher the velocity the smaller is the central mass needed to keep it there, possibly precipitating an uncontrolled shrinkage. The consequent rapid gravitational compression will yield superhigh ***PT*** in the interior, and perhaps light element (D, He, Li?) nucleosynthesis, thus replacing the Big Bang in this regard, as is required by our recognizing the cosmic redshift as a transmission effect (Sect. 6). The rapid shrinkage may expel these materials as an axial jet, assisted by the G-E field, with associated synchrotron (radio-loud quasars?; galaxy M87?) and other emission, dispersing them into the cosmos.

9. In more massive quasars the process may go further. Under CT a particle only possesses mass if there is room to accommodate the required aether dynamical configuration. Further compression will annihilate the mass, with enormous energy release (GRB?), so the gravity exerted by that mass disappears too, contrary to current black hole models. Such quasars (and those in (8) too) may decay/expire on quite short time-scales, but the end product is for speculation.

10. The observation of a very high orbital velocity of an object around an unseen centre, e.g. that of the star S2 around the Sgr A* feature at the centre of the Galaxy [49], has been regarded as diagnostic of the presence of a black hole. In such cases the mass has been assessed by applying Kepler's Laws plus, perhaps, a velocity-related relativistic mass increase of the orbiting object. Not only is the latter inapplicable in CT but we have reasoned (Fig. 4 and (1) above) that in the presence of VDI the assumption of a Keplerian orbital velocity structure becomes increasingly invalid as the velocity approaches and surpasses c. Thus the inferrable mass of the central object (quasar) may be quite small and bring the problem of mass-containment without mass-annihilation within reach of solution.

11. Their smaller mass and the drastic reduction of distance by our recognition of intrinsic A-R redshift may help to remove quasars from their reputed and very problematic class as the most energetic objects in the Universe.

12. Quantifying the intrinsic component. If we assume that both Ly β (1026 Å) and Ly α (1216Å) absorption are present in the same innermost shell of the quasar then, if the short end of that α forest begins to overlap the longest members of the β forest it is immediately clear that at least 18% of that quasar's redshift is intrinsic. If the Ly α forest lines can be traced further shortward among the Ly β forest, then the intrinsic proportion is correspondingly greater.

13. A-R redshift is also to be expected in the case of the transverse velocities of binary stars but this does not seem to have been sought because aberration is not observed in this case (Sect. 6.1). In the case of the star S2, mentioned in (10) above, it has been possible to trace an orbit less than 10 arcsec across, but if aberration were present the star's apparent proper motion would be about one degree. We have shown above (Fig. 3) that if the gradient of aether transverse velocity is close to the far end of the sightline the observable aberration is negligible but the associated A-R redshift should still be present. The spectrum of S2 should be checked for this.

8. Dark Energy?

The demand for Dark Energy to drive an apparent acceleration of expansion has arisen **solely** from treating the cosmic redshift as a velocity in a relativistic context. Applying the relativistic doppler formula has the effect of cutting down the high-end (distant) velocities, so that c is never attained at high redshift (z). This makes the near ones appear relatively fast.

If the cosmic redshift is not a velocity the purpose of Dark Energy vanishes.

So does that of Einstein's big lambda Λ.

9. A Non-expanding, Continuous Auto-Creation (CAC) Cosmology for CT

With no expansion, there is no basis for regarding the Universe as anything but truly boundless and infinite. This resolves the question of what keeps all the high-charge-density aether together against its immense self-repulsion. The simple CT answer is "more of the same". In other words, the repulsive force-balance ensures that the mean charge density of the aether is the same throughout the Universe, apart from 'local' disturbances by gravitation.

If electron-positron pairs are related to the mean density of the adjacent aether in the manner illustrated in Figure 1 of Paper I [1], this is of profound significance for the uniformity of particle creation throughout the Universe, particularly if, as we will suggest, the electron, with its specific charge, may have been the first-persister in the creation sequence on account of its stability.

CT has found that fundamental particles are vortical constructs of aether motion. Consequently, random motions of the aether charge have in principle, (via the coupling of relative shear displacements provided by Ampère's law) the potential for auto-creating such vorticity-based particles, especially as particle-antiparticle pairs.

So our CT cosmology is that the Universe began, in the indefinably distant past, as an infinite extent of randomly moving aether, from whose motion energy all the mass in the Universe has progressively been created and, importantly, is ongoing today.

This cosmology escapes, in what we hope is a physically acceptable manner, from the Big Bang absurdity, recognized and rejected in Paper I [1], that any mass or amount of energy, let alone the entire mass-energy content of the Universe, can be present within a zero volume. Similar discomfort recognition underlay the 'steady-state' cosmologies [50 - 52] in which creation was supposed to be ongoing and causing the expansion but only balancing it out.

The CT cosmology, in contrast, is a fully evolutionary one, envisaging that mass-creation would only become extinct when, and if, aether random motion were to become too low.

Much scientific benefit should accrue if we are able to bring observational constraints to bear on mass-creation as an ongoing process, so we pursue this in the next Section in some detail. Suffice to note here that as gravitational interactions and star formation drive environmental energy levels higher, corona-fashion, so the rate of auto-creation around that place is enhanced, providing a positive feedback for the formation and growth of galaxy clusters and even apparently to the much smaller scale of globular (star) clusters.

This cosmology invalidates the application of the mass-energy equivalence ($\boldsymbol{E} = \boldsymbol{mc}^2$), as currently perceived, in that mass-bearing particles are to be created from the motion energy of the massless aether.

As I stressed in my PIRT VII (2000) paper [2] and in Section 1 of this one, this aether motion energy, even in a massless form, beit as TEM waves or as random motion, needs to be recognized in all energy-balance calculations, such as those involving entropy.

This perspective has the benefit that neutrinos, as pure eddies, not vortices, can have energy content but yet be massless and wholly unaffected by gravitational fields. Nonetheless, as pure eddies of aether charge, [2] they must surely possess a magnetic field. I suggest that it is this, currently unsuspected, property that may lie behind the observations that have seemed to give credibility to the 'oscillations' idea for changes in neutrino character.

10. CT Cosmology — The Inverted Significance of Stellar Metallicity

('Metallicity' means the abundance of all elements above H and He and is for convenience usually defined as $\log_{10}$ [Fe/H].)

Metallicity is currently used as a major tool for inferring the cosmological age of stellar assemblages, so it is important here to address the big change in its significance brought about by our continuous auto-creation (CAC) cosmology.

In Big Bang cosmology, low metallicity is taken to record very early formation of the material, before stellar nucleosynthetic activity had built up the metallicity of the Universe. So a zero-to-very-low metallicity 'Population III' have been sought, but so far unsuccessfuly, to trace the very earliest stage of that.

Nevertheless, metallicity of low-to-intermediate character ('Population II'), and supposedly only slightly younger, is typical of the globular (star) clusters which abound in the haloes of spiral galaxies, and of many dwarf and structurally undeveloped ('Irregular') galaxies. The Large and Small Magellanic Clouds, for example, have metallicities less than half that of the Galaxy. But the mystery has been how Pop II had escaped subsequent stellar recycling to higher metallicity.

In the core of our Galaxy, where the stellar density is high, a much higher metallicity (Population I) confirms the effect of such recycling over time. The gradual outward decrease of metallicity within its disc supports that these stars (including the Sun), or the material they were made from, moved outward from the core at an earlier stage of that build-up.

In continuous auto-creation (CAC) cosmology, all newly-created material is of low metallicity, but it may be of any age, so its well-observed presence just outside the central bulges of spiral galaxies strongly suggests its relatively recent arrival by axial infall from the surrounding aura of enhanced auto-creation.

In the case of M104 (the Sombero galaxy, Fig. 5), of the very large number of globular clusters present in its halo the red, more mature ones systematically lie closer in than the blue, low-metallicity ones [53]. That is just right for the blues to be made of young blue stars recently arrived from outside.

This paper's abstract refers to the presence, in the globular clusters of our own galaxy, of even younger blue stars, which have been called 'blue stragglers' because their youth has been seen as a problem within a supposedly very old assemblage of Pop II stars.

In CT these appear to be the best evidence we have of ongoing auto-creation. The youth of these blue stragglers may not be such as to rule out that they formed before the cluster concerned fell towards the Galaxy. In which case the auto-creation building of the first globular clusters, not galaxies, may turn out to have characterized the elementary build-up of the Universe, each having started, perhaps, by the fragmentation of a Jeans mass.

Figure 5. Edge-on viewing of spiral galaxies enables the dust at their optical edges to be seen. Note the slimness of the profiles. M104 (the 'Sombrero' galaxy) is exceptional in having a very large number (>2000) of globular clusters forming its halo, see text. Credit: All three images are cropped and rotated, with permission from Volker Wendel, ***www.spiegelteam.de***.

10.1 *Outward Disc Flow in Spiral Galaxies — G-E Field Action: no CDM*

In Paper I [1] we saw that Spiral galaxies display morphological features indicating outward flow in the disc. Since dust is the product of stellar activity, these images (Fig. 5) of outer-edge dust confirm that direction.

We also saw in [1] that similar flow must have been present in the protoplanetary disc during planet formation. In that case, arrival of the material had been quasi-polar, followed by close-to-star ionization, rendering it responsive in the disc to G-E field action.

Evidently, in Spiral galaxies, the pattern is the same. Evidence of cold neutral hydrogen gas accretion to Spirals has been accumulating in recent years and is now well established [54] but the dynamics have remained unclear. We now see that newly created, little-ionized, material forms quasi-polar infall streams, is ionized at the bulge, and is then driven radially in the disc by the G-E field of the bulge and core. The dust and bigger uncharged materials are blown outward against gravity aerodynamically by the G-E driven wind. But as the wind weakens with radial distance this capability falls, the 'solids' are parked at the levitation limit, and the plasma materials continue. Before that limit is reached, the dust-levitation force on the arms of the galaxy is apparently what drives them outwards [1].

Hydrogen H II radiation abundantly observed from far outboard of the visible regions of galactic discs confirms the presence of outflowed G-E field-driven plasma that must have been aerodynamically responsible for the outward transport of this dust.

All these signs that the G-E field, a purely radial force, is controlling the dynamics of motions in the disc of Spirals provides a ready explanation of their well-documented, but not invariable, habit [51, 52] of displaying a tangential velocity profile which, after a rise from the centre, may remain essentially flat to far beyond the optical limits of the disc. As noted in [1], Keplerian-tending departures from that flatness will result from, and vary according to, the competing Newtonian force experienced by the neutral-body loading of the flow. Thus it is that the not-infrequent partial departures from that flatness can find a ready explanation in terms of the locally-varying mass-loading (e.g. the arms) of the flow in the disc, whereas the CDM argument is a blunt instrument without such a facility.

Thus our conclusion, already presaged in [1], is further consolidated, namely that there is now no need at all for the presence of CDM, widely invoked in huge amounts to explain the flat tangential profiles of Spiral galaxies. Demands for CDM arising from virial treatments of the long-term stability of clusters may also be denied if we recognize that (i) intrinsic redshifts exaggerate the velocity scatter and (ii) auto-creation makes the effective lifetime of a cluster unknown.

11. Functions of the Axial Infall Streams of Mostly New Matter — From Galaxies to Clusters

Our proposed auto-creation mechanism is envisaged as being assisted by high environmental energy levels, so one's thoughts go straight to the stellar environment, and the 'blue straggler' story (above) may be a part of that. But the volumes of space that such processes occupy is tiny compared with the volumes of extragalactic space surrounding galaxy clusters, so we envisage that this is where the majority of auto-creation in the Universe is still in progress and will supply the galaxies with growth by infalling streams.

The Newtonian gravity field of a Spiral's disc will tend to focus these streams towards its axis. Both in the bulge and as this matter spreads out in the disc it will maintain star formation and growth in the mass of the galaxy. G-E field action on its plasma will maintain the Spiral dynamics. The rest will then emerge from the disc edges and may contribute to the initiation of satellite galaxies. As the cluster grows, so will its corona-like aura, and the infall of newly created matter from it. So fresh and younger galaxies will tend to form at the outside of the cluster. When these masses begin to affect the focusing of the infall streams reaching Spiral galaxies in the interior, those streams could bring about a major morphological transformation – to a Barred Spiral - next.

12. Birth and Life of a Barred Spiral (Fig. 6 a,b,c) — Interpretation of Bar Formation With Specific Reference to NGC1300 (Fig. 6a)

The polar infall streams (see inset in (b)) are supposed to have been deflected and misaligned by the gravitation of other galaxies in the cluster. This sets up a couple which forms a rotating bar, here called a 'roller-bar', whose length propagates outward (G-E field action) until it reaches, engages with, and halts, the orbital motion of material at the inner end of a spiral arm. Orientation of the roller-bar's axis is fixed by the external influences, so it does not rotate with the spiral arm structure, which continues to rotate about the original axis.

In (c) (NGC1672) we see the end-on view of such a bar, with the dust lanes peeling off the arms (the one on the right is the nearer one) and gravitating towards the centre of the bar, whose CCW rotation is clearly seen. A bit later in the evolution, the infall gravitation of these misaligned convergent dust streams can set up yet another rotation. The vestiges of that are to be seen in the centre of NGC1300 (Fig. 6a) and are clearly visisble as a 'spiral within a spiral' in many other barred galaxies. In NGC1300 this motion appears to have replaced the original, externally imposed, roller-bar motion.

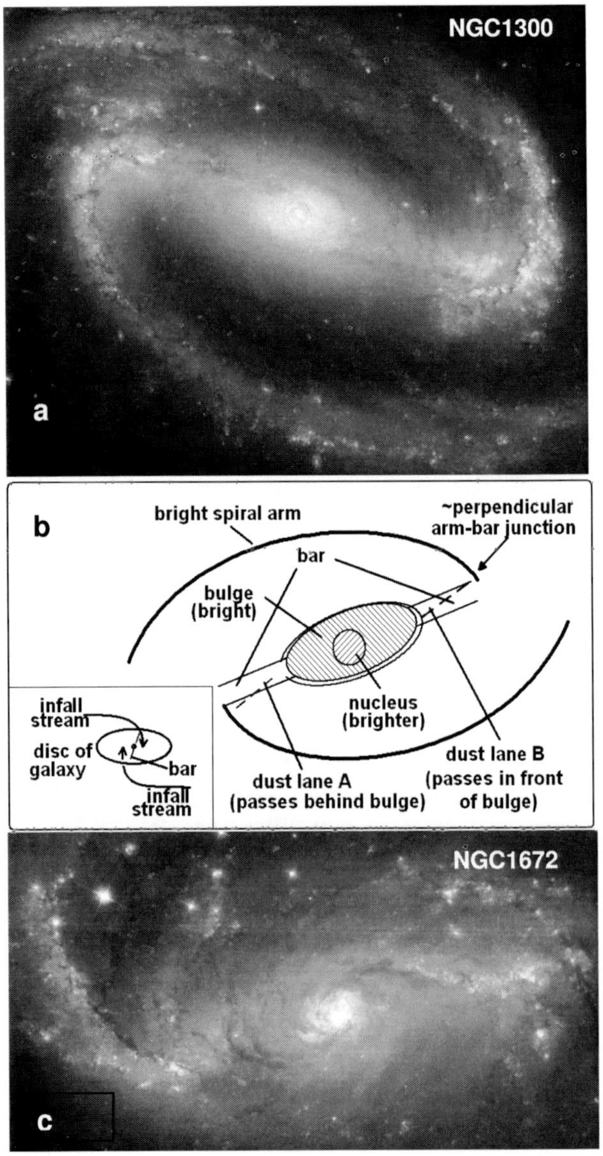

Figure 6. Examples to illustrate transformation of a Spiral galaxy into a Barred Spiral, and its subsequent evolution, see text.
Image credits: (a) NGC1300 APOD 2012 March 25, Hubble Heritage Team, ESA, NASA. (c) NGC1672. APOD 2012 May 13. NASA, ESA, Hubble Heritage Team (STScI/AURA); Acknowledgement: L. Jenkins (GSFC/U. Leicester).
Images have been cropped/rotated to fit the page.

Non-ionized material in the dust lanes which line spiral arms (see images in Paper 1) engages with the bar end and can gravitate along it towards the centre, being twisted into a weak spiral by the faster bar rotation at the centre, where the infall rotational drive is being applied by neutral and invisible infall of 'new' material.

As the continued rotation of the arms takes them past the ends of the bar, its 'consumption' of their non-ionized dust leaves the ionized components of the arms, still supported radially by the G-E field, to continue and perhaps to form a visible ring. This process has just begun in NGC1300 but is clearly seen in others (Fig. 7).

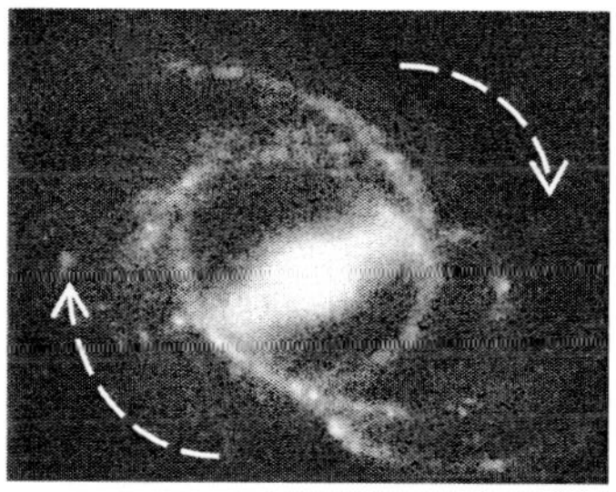

Figure 7. NGC 2523. This shows how the galaxy arm system has continued for ~120 degrees (or perhaps a lot more) to rotate CW past the ends of the stationary bar, leaving a G-E field-levitated ring of residual ionized material from the arms. The main arms, meanwhile, may have ceased unwrapping under G-E field action, see Paper 1 [1], having been shortened by 'consumption' each time they passed the ends of the bar. Image: The Hubble Atlas of Galaxies, p 48 [57].

13. Death of a Barred Spiral — Birth of an Elliptical

Ellipticals display an aging stellar population, little or no internal structure and a dearth of plasma and new star formation. Ovoid in form, much fatter than Spirals, it is well recognized that giant ones predominate in the central regions of major clusters. But why?

When a cluster grows to the point that the infall streams of new matter from outside are captured before they reach the interior, maintenance of the bars of Barred Spirals will cease. Lacking a supply of plasma-genic material, responsive to the G-E field, especially if the arms have also been 'consumed' or become detached, the bar will undergo endwise collapse. The result will be an Elliptical galaxy, the fattening action evidently being due to the central 'spiral within a spiral' established by the misaligned two-stream convergence along the bar. Excellent examples of this are NGC4314, NGC5387, NGC3504, NGC1097 and NGC6951, all illustrated in [57].

This generally coherent interpretation of the features of Barred Spirals makes it clear that the widely-attempted 'pattern speed' determination method [58] fails to detect that the bar is **not** rotating with the arm system e.g. [59]. If it were, ring structure (Fig. 6) would lack a cause and centrifugal force would prevent the along-bar convergence and establishment of central spiral structure, demonstrated above.

14. Continuous Auto-creation and the Radial Profile of Galaxy Morphologies in a Cluster

Proceeding inwards from outside a major cluster we may now predict the following generality:-

1. A surrounding aura of HI (21cm) radiation in which enhanced auto-creation of neutral hydrogen is in progress, excited by the cluster's coronas;
2. A zone of low-metallicity dwarf galaxies in construction, often with starbursts where local auto-creation has taken a hold;
3. A wide zone of growing and evolving Spirals;
4. A zone in which transformations to Barred Spirals are in progress or complete;
5. A core zone of Ellipticals, living out their old age.

Accumulating evidence seems to favour such a profile.

In connection with (1), note that The Square Kilometer Array (SKA) (effective collecting area) is to be built in the southern hemisphere to investigate the already abundant evidence of this 21cm radiation, but which is currently interpreted in terms of the post-Big Bang sequence. In any cosmology, however, (and Continuous Auto-Creation (CAC) is no exception) the formation of neutral hydrogen is an essential stage.

In regard to (2), note that Dwarf and Irregular galaxies may occur in relative isolation, having become detached from the infall-generating environment in which they first formed. So now, unless they have the density to enable local autocreation to occur within them, they will mature to old age at constant mass with increasing metallicity.

In regard to (4), the fact that galaxies move about relative to one another means that bar-forming infall streams may be interrupted or restarted, so not all Barred Spirals will complete their metamorphism to an Elliptical. Unfortunately, early work on the relation of morphologies to their positions in clusters [60] did not recognize Barred Spirals as a distinct form. It is only recently [61] that this has been done, but their relative positions in clusters do not seem to have been well documented. That is now very desirable. But an apparent transition from blue bars to red ones with little star-formimg activity [62] looks like a step in this direction.

In regard to (5) also, galaxy motions may result in the restoration of infall, previously denied by their entourage, in which case the Elliptical will display a rejuvenation of star formation. So that should not delete it from the Elliptical category.

Giant Spirals in the outfield may be nuclei for future clusters. In such a galaxy, by not having other galaxies around it, competitive for the infall from the aura of auto-creation around it, that infall will maintain its G-E field-driven spiral dynamics. So it will continue to grow in mass until the development of satellites around it interferes with that infall and the Spirals-to-bars-to-Ellipticals metamorphic sequence begins. This may be how it was that the giant Ellipticals, now seen in the cores of clusters, came to have such exceptionally large masses.

15. Back to the Beginning, or Nearly — Formation of Tightly Wound Spirals

For Spiral arms now to be *unwrapping*, they must have been tightly wrapped in the first place. In our CAC cosmology this construction need not, in principle, be confined to the earliest evolution of the Universe but may arise at any time subsequently. On the other hand, the reasoning put forward below for the formation of the observed tightly-wound spirals does seem to require that to begin in relative isolation, well away from a galaxy cluster. I know of no report of one within a cluster, but this needs confirmation.

If we assume (see the next Section) that the auto-created hydrogen is not devoid of rotational turbulence at all scales, a reasonable mechanism is that this material could, under Newtonian gravity, develop loci of convergence. Conservation of angular momentum will then result in a tightly-wound spiral, getting tighter towards the centre. The G-E field developed by its resulting central mass concentration will not inhibit the acquisition of more such neutral material.

But, when star formation and ionization become established at the centre, ionization will increasingly spread outward through the inner turns of the Spiral. Their resulting response to the G-E field will push them outward, causing the axial-infall to disc-outflow pattern, characteristic of mature Spirals (Sect. 10), to set in, and probably terminating the acquisition of material to the edge of the disc.

As we have seen already, the strictly radial nature of the G-E field force means that this outward movement of the turns will occur with little or no change of tangential velocity. This, as noted in Paper I [1], has the geometrical consequence that these turns/arms undergo much splitting to enable them to fit the larger circumference. This feature, well recognized in the photographic record of these tight spirals, has made it impossible to trace far the individual turns of the original spiral. Figure 8 shows two examples at about this stage of their evolution.

In his commentary on NGC 488 (Fig. 8) Sandage 1961 [57] writes:- "There is a change in the characteristics of the spiral arms close to the nucleus. The inner arms are very smooth in texture with no indication of resolution into lumps....The amorphous inner arms can just be seen. As they wind outward they...become 'lumpy' with condensations. The luminous arms are nearly circular, with regular dust lanes on their inside edges."

Figure 8. Examples of closely-wound Spirals, possible progenitors of the open two-arm form so widely observed (see text). In NGC 2841 the G-E field has not yet begun to open out the spiral structure significantly. In NGC 488, that process is already under way; its 'flat' tangential velocity profile (310-330 km/s) [63] shows that the causative radial G-E-driven disc wind is already present.
Image credits: NGC2841 APOD March 29 2009. Credit & Copyright Johannes Schedler (Panther Observatory) cropped and rotated with permission. NGC 488 from [57].

The knotty problem in Spiral galaxy evolution with which we are now presented is:- How does this now-confused multi-arm structure sort itself out to become the predominantly two-arm structure seen in so many mature Spirals? If we can do that, then those which display more than two arms may record the progress of that transformation.

Our offered solution, outlined here, is this. The stellar development of dust, and its aerodynamical accumulation as a lining to a vestigial former arm, acts as a filter, from which the ionized remainder of the outward disc wind, now relieved of the neutral load and augmented by the ionization from starbursts in the dust, can now be driven with renewed vigour by the G-E field. This would constitute a mechanism for making a lot more radial space between adjacent arms than the former space between turns. That spacing has clearly begun to affect the outer arms of NGC 488 (Fig. 8). A mutually competitive accommodation of that demand would then be consistent with the general unwrapping already inferred. That such competition should, in the limit, leave the system with only two arms, seems reasonable, but analysis is desirable.

16. CT Cosmogenesis — Particles from Randomly Moving Aether

The motions of a true superfluid, totally lacking viscosity, are unable to produce rotational motions. But as we noted in Paper 1 (Sect. 4), the electric charge nature of our CT implementation of Maxwell's aether means that Ampere's law action produces an effective viscosity for relatively shearing motions.

That characteristic has already been useful to our reasoning in three important ways:-

it enables the aether to accommodate and propagate TEM waves [1];

it enabled us to contemplate the construction of vortex-type aether motions as fundamental mass-bearing particles [1];

it ensures that random aether motions, whether particle-tied or not, will produce the observed RTV deflection-type scattering and attenuation of TEM waves (Sect. 3).

So our fourth example is that the primordially existing infinite extent of randomly moving aether, invoked in our CT cosmology (Section 9), is assured of harbouring random rotational motions at all scales. At the smallest of scales these might correspond to the auto-creation of fundamental particles and, at very large scales, to the vorticity and angular momentum content needed (Sect. 15) for the assembly of the earliest tightly-wound spiral galaxies. So much for the circumstances of primary auto-creation.

Note that for ongoing auto-creation at any later stage, when there are already many particles around, any close passage of two particles (especially if it is at high relative velocity) will be an even more fertile way of inducing vorticity in the intervening aether. To acquire that high relative velocity the system must have awaited gravitational action and star formation. This is the basis of our proposed positive feedback mechanism for the progressive auto-creation of galaxies as clusters.

No wonder, then, that at places like CERN, where particles are deliberately made to impact or pass one another at the highest possible velocities, abundant auto-creation of mass-bearing particle vortices is observed, but is currently always (mistakenly in our CT view) attributed to a relativistic mass increase of the original particles, supposed then to have fragmented.

16.1. *Primary Auto-creation in Detail — How it All Got Started?*

The aim here is to auto-create complete hydrogen atoms from the primordial randomly moving aether, copiously endowed with motions of a vorticle nature. Vortical motions of every conceivable kind will be present. Most will be ephemeral; only the stable ones will survive for long within the prevailing environment of random electromagnetic excitation associated with the randomly moving aether. So the process of primary auto-creation has to be the Darwinian one of natural selection and the survival of the fittest.

In Section 5.1 of Paper 1 we speculated that "if the electron exemplifies a particularly stable aether dynamical form, it follows [from Fig. 1] that the mean density of the aether defines the electron as the elementary unit of charge throughout the Universe; and likewise the uniformity of atoms built by its means".

That is an important constraint in our CT Universe, in which there was no linkage between what was going on in different parts of it. Spectral evidence from distant galaxies seems to show that their atoms are the same as ours.

So let us suppose that among the plethora of motions, electron-positron pairs were persistent survivors. But observation shows abundantly that, of the two, positrons have, for some reason, relatively poor durability. So now we have an electron 'survivor' wandering around seeking further ways of enhancing its longevity. Being a charged body it will be extremely susceptible to the random electrical perturbation by the randomly moving aether. So the most effective way of cutting this down (by a factor of at least 10^{36}) will be to pal up with some other embryonic particle or particles that will neutralize its overall charge.

We now suppose that among the plethora of other vortices there were some, very vigorous ones, with an aether deficiency and consequent positive charge, possible progenitors of quarks. In this case, their powerful aether-sucking property caused them to seek more stability by sucking themselves together in threes, thus, by aether-flow circuiting around the triangle (Fig. 2 of Paper 1), creating a strong bonding action (which we now recognize as the nature of the Strong Nuclear Force). From a random variety of such groupings there would be some whose effective positive charge closely matched the negative charge of an electron.

So an encounter with, and some form of attachment to, an electron would be for the mutual benefit of all, neutralizing their charges and eliminating sensitivity to the otherwise-disruptive vigorous Coulomb-force perturbation by the randomly moving aether. The mechanism by which such an encounter could actually result in the structure which we think we know as the hydrogen atom remains obscure. But BigBang theory does no better in this respect either.

This, we propose, was the basic auto-creation route by which hydrogen atoms came into existence throughout the early Universe, and would be proceeding today in modified form wherever in the Universe auto-creation is active. The important thing is that the hydrogen atom emerges from its testbed environment as the most stable of all possible products of the randomly moving aether — a secure basis from which the nucleosynthetic construction of higher elements can proceed.

16.2. *Energy Density and Temperature of the Primordial Aether's Random Motion*

According to our CAC cosmology this was the energy resource from which all subsequent energy forms (TEM waves, mass and the energy of its relative motions) have been derived. We found (Sect. 4) the 2.726 K (or whatever) of the CMB currently reaching us to be a rather accurate indication of the true present temperature of the random aether. But the energy content of the CMB radiation is, as synchroton radiation, only a tiny fraction of the energy resource from which it derives.

Estimates may be possible, but there is a further uncertainty. This is that in our treatment of TEM wave transmission effects (Sect. 3*ff*) we assumed that the causative random aether motions relate solely, albeit in a much-smoothed manner, to the particle random motions along the path concerned. We took no account of the residual primordial component of aether random motion additionally still present around us and constituting the underlying engine of continued auto-creation at this time. A very precisely controlled transmission redshift experiment might enable this unaccounted addition to be detected and determined.

The other component of the primordial aether energy density has been converted into the space-density of the mass and the provided energy of its motions and TEM waves. To compile an inventory of these energy components for a supposedly representative sample of the Universe may tax our endeavours to the limit. But the reward would be that by putting it together with an estimate of current aether random energy, it might become possible to estimate how much more than 2.7 K was the effective temperature of aether motion when auto-creation first began in the Universe. Fortunately, in our non-expanding but infinite Universe, every sample volume should on average be like every other, so we don't have to peer into the high-redshift distance to do this.

17. Random Electromagnetic Excitation by the Aether, the Photoelectric Effect, and QED

In the preceding Section we drew attention to the potentially disruptive influence, during auto-creation, of the random electromagnetic field arising from random aether motion. At a more familiar level this means that, previously unrecognized, the atoms of everything are experiencing a continuous random electromagnetic excitation.

The ubiquitous nature of the aether means that, to the extent that the electron shells of atoms may provide incomplete shielding, this excitation penetrates to atomic nuclei. That might be responsible for triggering nuclear decay, rather than regard the decay rate of that particular nucleus as one of the immutable constants of Nature, or imaginatively to assign the job to a boson in control of the Weak Force.

Similar reasoning bears upon the physics of the

photoelectric effect. The evidently spotty nature of electron emission from a photosensitive surface under continuous radiation was what caused Einstein to adduce that the illuminating TEM wave energy must arrive in concentrated packets. That, in turn, enabled him to grasp the quanta-based alternative explanation offered by Planck for his understanding of black-body cavity-radiation energy distribution. Thus the idea of the quantization of TEM wave energy has seemed observationally secure.

But, although less well known, Planck had initially achieved his blackbody law for cavity radiation by regarding the TEM wave energy as continuous in transit but as promoting discontinuous transitions between energy levels in emitter or receiver [64]. This avoids conflict with the explicitly continuous nature of TEM waves as defined by Maxwell's equations. So let us explore it further.

From my experience as a youngster, generating smoothly oscillatory electric currents (as proven by their lack of harmonic content) and radiating them as radio waves, I find it unacceptable to have to believe that there was some kind of genie sitting on my aerial chopping up all those electromagnetic fields into little units of the right size for them to be radiated as photons. I argue that if, in these circumstances, the resulting TEM-waves are indeed continuous in transit, then it is wholly inconsistent to suppose that in some other situation TEM-waves are 'different' and only exist as packaged items.

It is true that when an atom radiates, it must indeed represent a jump between stable dynamical (internal) configurations, but that is a function of the source, not of the nature of TEM waves.

We have seen that neither the solar redshift (Sect. 5) nor gravitational lensing (Sect. 8 of Paper 1) seem to require TEM-waves to have the mass property. So what about that 'spotty' emission of photoelectrons from a surface under low-level distributed illumination, which so influenced Einstein's thinking? This is the point at which we need to consider the invasive random excitation by the randomly moving aether. It means that an atom with a loosely bound electron may, at random intervals, be excited to near-release energy. At that moment the additional excitation by a low level TEM-wave field will trigger release of the electron. It does *not* mean, as has hitherto been supposed, that the entire release (quantum/photon) energy has been brought to that point within the illuminating beam. Because the aether excitation is random it may instantaneously either add or subtract from that of the incoming TEM-wave, so the mean effect will appear to be zero and correspond to the TEM-wave input only.

This is a nice demonstration that in cases where one of the inputs is a random one, the use of averaging, with the aim of improving the precision of the answer, may completely obscure its physical interpretation.

By recognizing TEM waves as continua during transmission, not as corpuscular quanta, we seem to approach a resolution of the much-debated two-slit-one-quantum experiments, see [65] for example. In CT there is no longer any question of 'splitting a quantum and joining it up again on the far side'. The wave energy from the source can be divided according both to the physical disposition of the slits and to the random displacement of the wave structure by the random movement of the aether by which it is being propagated. The latter may or may not be enough to prevent the formation of a coherent interference pattern when the wave energies are brought together on the far side but it will (slightly) affect the division of energy going through the slits.

Accordingly I suggest that random excitation by the all-pervasive aether might provide an entirely new foundation for the statistical overlay upon classical electrodynamics which seems to lie at the heart of quantum electrodynamics (QED). The important difference seen in CT is that the amplitude of this statistical overlay is not quantifiable once and for all, nor specific to that particle, but is primarily related to the random motions of mass- and charge-bearing particles in that specific environment.

Another name for that statistical overlay is 'quantum tunelling' which underlies the Schrödinger equation of quantum mechanics. It plays an important part in (for example) our understanding of how the Coulomb force defences holding two protons apart inside a star can momentarily drop their guard and enable them to get close enough together for further interaction and for nucleosynthesis to proceed.

Thus in CT, the statistical behaviour of particles would emerge, not as an inherent stochastic wave-function property of those individual particles, but as due to being energized by the random motion of the aether in which they are steeped. Thus we escape the dilemma posed by the determinism arising from classical and relativistic treatments and the indeterminism affirmed by QED, a dilemma emphasized by Dirac [66, 67]. It seems to me that both the Heisenberg uncertainty principle and the EPR paradox may arise from trying to fit both determinism and indeterminism into the nature of an item.

In CT they are independent. So we can have particles or TEM waves that are deterministic in natural behaviour, but we cannot observe that in **real** nature because of the ubiquitous intervention of the randomly moving aether. This looks to be yet another example in which Einstein's insistence on excluding 'third party' (aether) involvement has seriously misled the scientific path.

The fact that the stochastic perturbation of outcome recognized in QED is mainly at the particle scale and vanishes at large scales seems consistent with the scale of aether random motion being primarily that of the individual motions of particles.

This insight appears now to warrant serious consideration, motivated by the abundant evidence for such random motion outlined in Sections 2 - 6 of this paper. Evidence for the Zero Point Field (ZPF) and zero point energy might be accommodated in the same way, but we cannot pursue that here.

18. CT in Relation to Dirac–Vigier Perspectives

This paper, together with Paper 1, is written for the VIIIth conference held in tribute to the work of Jean-Pierre Vigier, so it is appropriate here to refer to the significance of his work for the subjects dealt with in these two papers. Note at once his direct involvement in two of my citations, namely [13] and [31].

As outlined in the Introduction to Paper 1 [1] and enlarged upon in Section 3, the origin of CT lay in my 1959 discovery that specific features of the high-flight-altitude daylight sky brightness, are attributable to what we have called 'RTV scattering', requiring the presence of an aether, although it had been discarded for Relativity by Einstein in 1905. From that moment until very recently my development of CT was done in ignorance of the prior pleas for reinstatement of the aether made by that famously independent-minded physicist Paul Dirac in 1951/52 [68]. Further, to find that he, and subsequently Jean-Pierre Vigier [69, 70], have concluded, as we do in CT, that fundamental particles must be of non-zero-size in order to accommodate aether-motion activity within them, seems a remarkable concatenation of results, in view of the very different routes by which they were reached. The fact that they were seeking to fit this view into a relativistic setting does not detract from their underlying motive having been to escape the relativistic constraint of zero-sized particles, like we do in CT.

Add to this, I find it highly encouraging that our observation-supported arguments, and our implementing of Maxwell's aether as electrical in nature, offer (as discussed in the preceding Section) to provide the stochastic basis for QED behaviour so influentially sought by these authors.

Even more remarkable to me has been to discover Vigier's 1995 rendering of the "Fundamental problems of quantum physics" [71]. In this not only did he strongly favour a return to the 'ether concept' in a chaotically moving state but, astonishingly, then went on to envisage many of the consequences explored and supported by our findings in the present two papers under the Continuum Theory banner. These include TEM wave redshift as a propagation effect (using the old description as 'tired light') and a non-expanding Universe. The only significant disparity in his account was his retention of Relativity's non-zero mass photons.

Any comparison of Dirac's ideas with those advanced in CT should, to avoid confusion, note that Dirac's 'sea' of negative-energy particles is separate from his views on the aether and bears no relation whatever to CT's aether composed of negative electric charge.

A further important clarification is that Vigier and various others have sought to invoke superluminal communication both within particles and between them. In CT, by contrast, our rigorously applied principle is that *nothing (neither particles nor TEM wave transmission) can exceed the value of c relative to the local immediately environmental aether, as determined by its charge density at that point.*

The superluminal velocities invoked in our QSO model (Sect. 7) are *orthogonally directed with respect to the sightline to the distant observer,* and are built up by the gradient of aether tangential velocity between that point and the observer. TEM wave transmission directly towards the observer remains at ***c***.

One final point. The term 'wave-particle duality' can be used in two different circumstances. One is the QED context of a disturbing wave-function. The other concerns electron diffraction by a slit; this is brought within reach of explanation by making particles out of aether, i.e. electromagnetic in their very nature, as adopted in CT.

19. Loose Ends and Experimental Checks

19.1. *Loose Ends*

1. Viscous heating during a.m. exchange in protoplanetary discs. In Section 12 of Paper 1 we pointed out dynamical shortcomings of the so-called Nice model for solar planetary formation by condensation from a hot nebula. But there is a further reason for our rejection of it.

That model and its subsequent developments invokes viscous nebular action for a.m. partition between the solar and disc material. For the close-in exoplanets (a position seen as original in CT – Paper 1) the same mechanism has been invoked to argue for their prior high-a.m. positions from which they then migrated inward.

But the energy inefficiency and major heating aspects demanded by that viscous shearing action, first treated by [72] but apparently not appreciated by proponents of the mechanism, means that the **increase** in disc temperature (potentially several 10^3 K) would defeat the objective of condensing planets from it, which requires, approximately, that $T < 1800$K.

2. Electron shells as a cavity, the eccentric nucleus and the Mössbauer redshift experiments [73,74]. These sought to observe the Earth's relativistic gravitational redshift by noting the wavelength difference between upward- and downward-received gamma rays from the decay of ^{57}Fe. They overlooked that the support, within an atom, of any nucleus in a gravity field inevitably implies that its position is eccentric, but by many orders smaller than had ever needed consideration in ordinary spectroscopy. It so happens that the normal Fe-decay gamma emission wavelength (8.6×10^{-11} cm) is similar to the effective radius of its K-shell electron cavity. So I propose that cavity resonances were responsible for the up-down $\sim 10^{-15}$ fractional difference in emission wavelength, and for the exceptionally narrow bandwidth of the emissions by this atom (which was the practical reason for choosing it for these experiments). In CT, the nuclear energy-jump on decay is not an absolutely fixed photon quantity, but may be environmentally affected, so resonance could do that. Hence the result is not seen here as secure support for GR.

19.2. *Experimental Checks*

In addition to the three described in Paper 1 [1], the following six specific experimental checks would also be valuable, related to matters discussed in the present paper. Of these, No. 5 is important because of the possible rigour of the experimental control, but the repeats 2, 3 and 4 were also recommended by J-C Pecker (personal communication, Dec. 2010).

1. Daylight sky brightness and colour distribution at 6-10 km above ground. Section 3, Item (3). This should be done to check and perhaps extend the 'RTV scattering' explanation I reached in 1959-60.

2. A corresponding space-vehicle repeat of the Pioneer 10 observations of the *gegenshein* at >1 AU from Earth.

3. Repeat the 1968 solar superior conjunction carrier-wave redshift observations by Pioneer 6; Section 5.2 (a). And the associated spectral line-broadening. To consolidate them as observed examples of RTV redshift and RLV frequency dispersion.

4. Repeat the Sadeh et al. (1968) several-year recording of the Taurus A absorption line redshift when traversing the solar corona; Section 5.2 (b).

5. Repeat the Sadeh et al. (1968) observations of ground-level path redshift using tightly monitored caesium clocks; Section 6. And check its potential extrapolation in the CT frame to the cosmic redshift and the redshifts observed in (3) and (4) above.

6. Examine the spectrum of the star S2 in SGR A* for the presence of the Aberration-Related (A-R) redshift predicted in Section 7.1 and Section 7.2 (Item 13).

20. Philosophy, Review, and Some Conclusions

20.1. *Philosophy*

The deeper that one roots a physical idea, the greater should be the variety of benefits at higher levels, but only *if it is right*. For the Continuum Theory, outlined in this paper and in Paper 1, our physical implementation of Maxwell's aether constitutes a deeper level than appears ever to have been attempted before. So one expects the variety of its implications to be very great indeed, and hopes that these will be rewarding. But one must recognize that this may require the abandonment of ideas formerly sincerely held but less deeply rooted. To achieve that convincingly it needs to be specific enough at root level, but yet not so rigid as to deny natural variability in its development. So, at all levels, we have concentrated here on how things work, thereby reaching out to involve lots of potentially variable factors and observables. We feel such linkages provide strength to the whole.

20.2. *Review*

The Aether. We have implemented the aether as a continuum of electric charge, so that it satisfies Maxwell's equations for the support and transport of TEM waves. We have then found (Sect. 16) that, through the involvement of Ampère's Law, this kind of aether turns out to have other positive consequences in the Universe. For example, in the absence of Ampère's Law action its superfluid random motions would be devoid of rotations. Those are needed to explain the observed deflection (not impact) scattering of TEM waves (Sect. 2 & 3) and the resultant attenuation we infer responsible for Olbers' Paradox. Also for making rotation-configured particles out of aether in auto-creation. If the aether were absent or of any other kind of superfluid, such as often envisaged by others, these would be lacking.
Particles. We have pursued the 19th century idea that fundamental particles are vortical constructs of aether in motion, responsible for gravitational action. Thus TEM waves and aether are both 'made of aether'; hence the name 'Continuum Theory' (CT). The ultra-high charge density we find for the aether, and the charge densities of the particles made from it (Paper 1), changes it to being the principal agent in the Universe – gravitation, inertia, electric fields.

20.3. *Some Principal Conclusions*

(The conclusions which follow relate to the present paper and are in addition to those given at the end of Paper 1.)

1. Particles that are vortical constructs of aether in motion but are themselves in random motion means that the surrounding aether is in random motion too.

2. That motion causes four cumulative wavelength-independent transmission effects on TEM waves; redshift, frequency dispersion, scattering, attenuation. All are hugely enhanced by ionization.

3. The redshift is observed as:- solar redshift on emergent radiation; on external sightlines traversing the solar corona; with caesium clocks over long ground-level paths; excesses for W-R in binary stars; cosmic redshift; and in a variety of other situations.

4. The CMB 2.7K radiation is due to synchrotron radiation due to aether random motion (1), from the gas/plasma along intergalactic paths that reach us.

5. The attenuation (2) across the Universe, over and above inverse square law, is responsible for Olbers' Paradox and means that high-redshift galaxy distances determined by standard-candle method have been substantially over-estimated.

6. With cosmic redshift as a transmission effect (3), the reason for Dark Energy vanishes.

7. The aether's linkage of gravitation and inertia (Paper 1) leads to a fertile model for QSOs with large amounts of intrinsic redshift and the 'in-house' generation of forest absorption lines, so the indicated temperatures do not relate to the intergalactic medium and cloud containment problems are resolved.

8. The Universe is not expanding; its entire mass content has been made, and continues to be added to, by auto-creation from random aether motion over an indefinable time.

9. After an exhaustive sequence of selection for matching charge and mutual attraction, to improve their longevity in the face of disruption by the random aether's electromagnetic excitation, the electron and trios of quarks finally win the day as hydrogen atoms.

10. For this auto-creation, and its mass-increase counterpart in particle accelerators, not to contradict energy conservation laws, it is necessary in all such calculations, to recognize the energy content of aether motion, beit random, systematic in particle interiors or embodied in TEM waves.

11. That hydrogen (9) probably forms the bulk of the mass of the Universe; enhanced auto-creation in the environment of galaxies provides Spiral galaxies with young, low metallicity, quasi-polar infall to drive their dynamics and star formation.

12. Poor access by those streams to the interiors of galaxy clusters results, first in transforming a Spiral to a Barred Spiral, and then to an Elliptical in the deep interior.

13. The arms of most Spirals when we see them are unwrapping by the action of the G-E field on the ionized material in the arms, but they must have formed as tightly-wrapped spirals to start with. Such Spirals are formed of mainly neutral H until star

formation and ionization sets in at the core and the G-E field starts to unwrap them.

14. The ubiquitous random electromagnetic excitation of everything by the randomly moving aether, yields an explanation of the spotty emission of photoelectrons without appeal to quanta. This does not conflict with Planck's original treatment of cavity radiation. So CT, like Planck, sees TEM wave energies in transit as subdivisible without constraint, the quantization being decided, if appropriate, by preferred energy levels in source or receiver.

15. That same random excitation by the aether, likely to be particularly evident at the smallest scale, appears to offer a very favourable new basis for QED, especially as it separates the random influence ('Hidden Parameters') from being inherent in a particle.

Acknowledgements

First, I wish to reaffirm my gratitude to all those acknowledged at the end of Paper 1. In addition, for generous advice and communication in matters specific to those raised in this paper, I wish to thank:- Alan Dressler, John Huchra, George Isaak, Bernard Pagel and Denis Whitehead. I also thank the conveners of the Vigier VIII – BCS conference and editors of this volume, Richard Amoroso, Peter Rowlands Louis Kauffman and Peter Marcer, for the opportunity to contribute these papers.

References

[1] Osmaston MF Implementing Maxwell's aether illuminates the physics of gravitation: the Gravity-Electric (G-E) field, evident at every scale, from the ionosphere to spiral galaxies and a neutron star extreme. This volume. ('Paper 1')

[2] Osmaston MF (2003) A particle-tied aether - Indications of a deeper foundation for physics and relativity. Physical Interpretations of Relativity Theory (PIRT) VII, Brit. Soc.Philos. Sci., 15-18 Sept 2000 Late Papers (MC Duffy, ed), PD Publications, Liverpool (ISBN 1 873 694 05 9), 230-240. Also currently available at:- http://www.scribd.com/doc/78606561/Miles F Osmaston-A-Particle-Tied-Aether.

[3] Poincaré H (1900) La théorie de Lorentz et le principe de réaction. Arch. Néerl. 2(5), 252-278.

[4] Thomson W & Tait PG (1867) Treatise on natural philosophy. Oxford University Press.

[5] Larmor J (1904) On the ascertained absence of the effects of motion through the aether, in relation to the constitution of matter. Phil. Mag., Ser. 6, 7, 621-625.

[6] Wenzel AR (1980) Spectral broadening of waves propagating in a random medium. J. Math. Phys. 21, 2131-2139 http://dx.doi.org/10.1063/1.524722.

[7] Trumpler RJ (1935) Observational evidence of a relativity redshift in class O stars. Publ. ASP 47, 249-256.

[8] Feast MW (1958) Spectral types and radial velocities in the galactic cluster NGC 3293. MNRAS 118(617-630).

[9] Rubin VC (1963) The interpretation of the K-term in the radial velocities of cepheids and O and B stars. Ap. J. 138(2), 613-615.

[10] Pecker J-C (2006) Local abnormal redshifts. In Current issues in cosmology (ed. J-C Pecker & JV Narlikar), pp. 217-221. Cambridge University Press.

[11] Allen CW (1963) Astrophysical Quantities, 2nd edition. Univ. London, Athlone Press.

[12] Weaver HF (1955) The K-effect in stellar motions. In Vistas in astronomy, Vol. I (ed. A. Beer), pp. 228-238. Pergamon, London.

[13] Kuhi LV, Pecker J-C, & Vigier J-P (1974) Anomalous redshifts in binary stars. A & A, 32, 111-114.

[14] de Vaucouleurs G & de Vaucouleurs A (1963) Search for systematic errors in galaxy redshift measurements. Astron. J. 68, 96-107.

[15] Binggeli B, Tammann GA & Sandage A (1987) Studies of the Virgo Cluster. VI. Morphology and kinematics of the Virgo Cluster. Astron. J. 94, 251.

[16] Finlay-Freundlich E (1954) Redshifts in the spectra of celestial bodies. Proc. Phys. Soc. 67(Sect A), 192-193.

[17] ter Haar D (1954) On Freundlich's redshift. Phil. Mag. Series 7 45, 320-324.

[18] Struve O (1950) Stellar evolution. Princeton Univ. Press, 266 p.

[19] Kaufman S & Williams RV (1958) The electron temperature in SCEPTRE III. Nature 182, 557-558.

[20] Spitzer L, Jr. (1958) Nature 181, 221-222.

[21] Tousey R & Hulburt EO (1947) Brightness and Polarization of the Daylight sky at Various Altitudes above Sea Level. J. Opt. Soc Amer. 37(Feb.), 78-92.

[22] Packer DM & Lock C (1951) The brightness and polarization of the daylight sky at altitudes of 18,000 to 38,000 feet above sea level. J. Opt. Soc. Amer. 41, 473-.

[23] Barr NL (1953 (March)) "Brightness of the atmosphere". U.S. Nav. Med. Res. Inst. Report.

[24] Roosen RG (1971) The gegenschein. Rev. of Geophys. Space Phys 9, 275-304.

[25] Weinberg JL et al. (1973) Observation of zodiacal light from the Pioneer 10 asteroid-Jupiter probe. Space Research XIII, 1187–1193.

[26] Turck-Chieze S (1998) Composition and opacity in the solar interior. In Solar composition and its evolution - from core to corona. (ed. C Frölich et al.), Space Science Reviews, 85. 125-132

[27] Evershed J & Royds T (1916) On the change of wave-length of the iron lines in passing from the centre of the Sun's disc to the limb. Kodaikanal Observatory Bulletin XLIX, 145-156.

[28] Finlay-Freundlich E, von Brunn A & Brück H (1930) Zeitschr. f. Astrophisik 1, 43.

[29] Adam MG (1948) Interferometric measurements of solar wave-lengths and an investigation of the Einstein gravitational displacement. MNRAS 108(6), 446-464; MNRAS 112, 546-569 (1952); MNRAS 115, 409-421 (1955); MNRAS 119, 460-474 (1959); Nichols S & Clube SVM MNRAS 118, 496-503 (1958)).

[30] Goldstein RM (1969) Superior Conjunction of Pioneer 6. Science 166(3905), 598-601 DOI: 10.1126/science.166.3905.598.
[31] Merat P, Pecker J-C & Vigier J-P (1974) Possible interpretation of an anomalous redshift observed on the 2292 MHz line emitted by Pioneer-6 in the close vicinity of the solar limb. A & A 30, 167-174.
[32] Sadeh D, Knowles S, & Au B (1968) The effect of mass on frequency. Science 161(3841), 567-569.
[33] Shapiro II, Ash ME, Ingalls RP et al. (1971) Fourth test of general relativity: new radar result. Phys. Rev. Lett. 26, 1132-1135.
[34] Osmaston MF (1998) An interim outline of some research under the heading: some aspects of a continuum theory of physical nature. (Summary text + Appendices A (on RTV redshift) B (on RTV scattering) & G (on RLV line broadening)). Brit Soc Philos Sci; Fifth Int Conf on Physical Interpret-ations of Relativity Theory (PIRT V), Imperial College, London, 6-9 Sept 1996. Supplementary Papers (MC Duffy, ed). pp. 241-256.
[35] Ghosh A (2000) Origin of inertia: extended Mach's Principle and cosmological consequences. Apeiron. 148 pp.
[36] Mortlock S et al. (2011) A luminous quasar at z = 7.085. Nature 274 (7343, 30June), 616-619.
[37] Blades JC et al. eds (1988) QSO absorption lines: probing the Universe. Proceedings of the QSO Absorption Line Meeting, Baltimore 1987. Space Telescope Science Institute, CUP.
[38] Murdoch HS et al. (1986) Ap. J. 309, 19-32.
[39] Carswell RF (1995) Million light-year galaxies and spectral forests. Nature 374, 500-501.
[40] Rees MJ & Carswell RF (1987) MNRAS 224, 13P.
[41] Rees MJ (1995) Sizes and dynamics of the Lyman forest clouds. In QSO Absorption Lines; Proc. ESO Workshop, Garching, Germany 21-24 Nov 1994 (ed. G Meylan), pp. 419-425. Springer.
[42] Weymann R (1995) QSO absorption lines: progress, problems and opportunities. In QSO Absorption Lines; Proc. ESO Workshop, Garching, Germany, 21-24Nov 1994 (ed. G. Meylan), p. 3-10. Springer. Trimble V (1995) In ibid. p. 453-468.
[43] Hoyle F & Burbidge GR (1966) Relation between the Red-shifts of Quasi-stellar Objects and their Radio and Optical Magnitudes. Nature 210, 1346-1347.
[44] Shaver PA et al. (1996) Decrease in the space density of quasars at high redshift. Nature 384, 439-441.
[45] Becker RH & others. (2001) Evidence for reionization at z ~ 6: detection of a Gunn-Peterson trough in a z = 6.28 quasar. Ap. J. 122, 2850-2857 doi: 10.1086/324231.
[46] Gunn JE & Peterson BA (1965) On the density of neutral hydrogen in intergalactic space. Ap. J. 142, 1633-1641 Doi:10:1086/148444.
[47] Arp H (1987) Quasars, Redshifts and Controversies. Interstellar Media. Arp H (1998) Seeing Red: Redshifts, Cosmology and Academic Science. Apeiron. 306 p.
[48] Arp H, Burbidge EM et al. (2002) NGC 3628: Ejection activity associated with quasars. A & A 391(3), 833-840. DOI: 10.1051/0004-6361:20020828.
[49] Gillessen S et al. (2009) The orbit of the star S2 around SGR A* from Very Large Telescope and Keck data. Ap. J. 707, L114 doi:10.1088/0004-637X/707/2/L114.
[50] Bondi H & Gold T (1948) The steady-state theory of the expanding Universe. MNRAS 108(3), 252-270.
[51] Hoyle F (1948) A new model for the expanding Universe. MNRAS 108(5), 372-382.
[52] Hoyle F, Burbidge G & Narlikar JV (1993) A quasi-steady state cosmological model with creation of matter. Ap. J. 410, 437–457 doi:10.1086/172761.
[53] Harris WE et al. (2010) Diamonds on the Hat: Globular Clusters in The Sombrero Galaxy (M104). MNRAS 401(3), 1409-2112.
54. Sancisi R et al. (2008) Cold gas accretion in galaxies. Astron. Astrophys. Rev. 15, 189-223 DOI 10.1007/s00159-008-0010-0.
[55] Rubin VC (1983) The rotation of spiral galaxies. Science 220, 1329-1344.
[56] Persic M & Salucci P (1995) Rotation curves of 967 spiral galaxies. Ap. J. Supplement 99, 501-541.
[57] Sandage A (1961) The Hubble Atlas of Galaxies. Carnegie Institution of Washington, Publ. 618.
[58] Tremaine S & Weinberg MD (1984) A kinematic method for measuring the pattern speed of barred galaxies. Ap J. 282, L.5.
[59] Ryder SD et al. (1996) Neutral hydrogen in the ringed barred galaxies NGC 1433 and NGC 6300. Ap. J. 460, 665-685.
[60] Dressler A (1980) Galaxy morphology in rich clusters: implications for the formation and evolution of galaxies. Ap. J. 236, 351-365. Dressler A et al. (1997) Evolution since z = 0.5 of the morphology-density relation for clusters of galaxies. Ap. J. 490(2), 577-591 doi:10.1086/304890.
[61] Masters KL et al. (2010) Galaxy Zoo: Bars in Disk Galaxies. arXiv:1003.0449v2 [astro-ph.CO].
[62] Masters K et al. and the Galaxy Zoo (2012) Barred Spirals on the Red Sequence: An Important Evolutionary Stepping Stone? UK-Germany Nat. Astron. Mtg, 2012, Manchester. Session GAL3. <http://www.jodrellbank.manchester.ac.uk/meetings/nam2012/sessions.html>.
[63] Peterson CT (1980) The rotation curve of NGC 488. The Astronomical Journal 85(3), 226-229.
[64] Kangro H (1976) Early history of Planck's radiation law. Taylor & Francis.
[65] Vigier J-P et al. (1987) Causal particle trajectories and the interpretation of quantum mechanics. In Essays in Honour of David Bohm (ed. BJ Hiley and FD Peat), pp. 169-204. Routledge.
[66] Dirac PAM (1958) The Principles of Quantum Mechanics, 4th edn. (1st edn.- 1930) OUP. 325 p.
[67] Dirac PAM (1974) The Development Of Quantum Mechanics. Conferenza Tenuta il 14 Aprile 1972, in Roma, Accademia Nazionale dei Lincei (1974).
[68] Dirac PAM (1951) Is there an Æther? Nature 168, 906-907. Dirac PAM (1952) Is there an Æther? Nature 169, 702. doi: 10:1038/16970260.
[69] Cufaro Petroni N & Vigier JP (1983) Dirac's aether in relativistic quantum mechanics. Foundations of Physics 13(2), 253-285.
[70] Amoroso RL, Kauffmann LH, Rauscher EA, Rowlands PR & Vigier J-P (2010) "Hidden" Parameters Describing Internal Motion Within Extended Particle Elements, in Search for Fundamental Theory, The VIIth International Symposium Honoring French Mathematical Physicist Jean-Pierre Vigier (ed. RL Amoroso et al. AIP Conference

Proceedings. 1316, 1-27.
[71] Vigier J-P (1995) Fundamental Problems of Quantum Physics. Apeiron 4, 114-115.
[72] Lynden-Bell D & Pringle JE (1974) The evolution of viscous discs and the origin of the nebular variables. MNRAS 168, 603-637.
[73] Champeney DC et al. (1963) Measurement of the relativistic time dilatation using the Mössbauer effect. Nature 198, 1186-1187.
[74] Pound RV & Snider JL (1965) Effect of gravity on gamma radiation. Physical Rev 140, B788-B803.

A Transluminal-Energy Quantum Model of the Cosmic Quantum

RICHARD GAUTHIER

Santa Rosa Junior College, 1501 Mendocino Ave., Santa Rosa CA 95401, U.S.A
richgauthier@gmail.com, www.superluminalquantum.org

An internally transluminal model of the hypothesized unique cosmic quantum of the very early universe is proposed. It consists of a closed-loop photon model, which has the total mass-energy of the atomic matter and dark matter of our observable universe that will develop from it. The closed-loop photon model is composed of a rapidly circulating point-like transluminal energy quantum (TEQ). This TEQ circulates in a closed helical path with a maximum speed of c sqrt(5), and a minimum speed of c around the photon model's one-wavelength closed circular axis. The transluminal energy quantum model of the cosmic quantum is a boson. The cosmic quantum model may help shed some light on several fundamental issues in cosmology, such as the nature of the cosmic quantum, the predominance of matter over antimatter in our universe, the possible particles of dark matter, the quantum interconnectedness of the universe, and the very low entropy of the very early universe. The cosmic quantum may be the first particle of dark matter, from which all other particles are derived.

Keywords: Cosmogony; Big Bang; Quantum; Dark Matter; Superluminal

1. Introduction

In 1931 Lemaître [1] hypothesized that a single quantum particle (*primeval atom*) composed the universe at the beginning stage: "If we go back in the course of time we must find fewer and fewer quanta, until we find all the energy of the universe packed in a few or even in a unique quantum." [1, p. 706]

According to this hypothesis the primeval atom transformed into many other particles. The system of particles and the volume of space expanded according to Einstein's general theory of relativity to produce our current universe. This hypothesis later came to be known as the Big Bang theory of the universe.

In 1956 Goldhaber [2] built on Lemaître's hypothesis in an attempt to explain why anti-matter is much less abundant in our universe than matter. He proposed that a single particle, a "universon", existed before the Big Bang. The universon decayed into a "cosmon" and "anti-cosmon". The cosmon then decayed to produce our universe whose atomic matter is ordinary matter (as opposed to anti-matter).

In 1975, Tryon [3] proposed that the universe could have resulted from a quantum fluctuation (in the sense of quantum field theory) from the vacuum state of space. This could have happened without violating the conventional laws of physics, such as conservation of energy and conservation of electron and baryon number. The universe formed would have to be on the border of being open or closed (because the quantity of negative gravitational potential energy would exactly balance the quantity of positive mass energy) and also contain an equal amount of matter and anti-matter. Tryon does not give a model of the quantum fluctuation that produced the universe.

In 1989 Dehmelt [4] built on Lemaître's hypothesis. He suggested that a "cosmon, an immensely heavy lower layer subquark, is *the* elementary particle. The world-atom, a tightly bound cosmon/anticosmon pair of zero relativistic total mass, arose from the nothing state in a quantum jump. Rapid decay of the pair launched the big bang and created the universe."

The recent Wilkinson Microwave Anisotropy Probe's (WMAP) detailed measurements [5] of the very nearly uniform (to about 1 part in 100,000) cosmic microwave background radiation (CMBR) coming from all directions in space, with a black body radiation temperature of 2.725 Kelvin, have yielded strong support for what has become the Standard Model of cosmology, the Lambda Cold Dark Matter (ΛCDM) with Inflation model. To briefly summarize this model, the very early universe was a rapidly expanding hot, dense physical state (the Big Bang). This was followed very quickly by a period of extremely rapid inflationary expansion, followed by normal expansion. This inflationary period is supposed to have wiped out all physical evidence of particles that existed before it. There are however physical models for the pre-inflationary era that try to describe this era and explain the cause of inflation.

After the inflationary period there were small variations in mass/energy densities in the universe produced by quantum fluctuations of energy during the inflationary period. The post-inflationary matter and energy evolved into an extremely hot concentration of elementary particles, which later formed atomic nuclei of hydrogen, helium and lithium. Later hydrogen, helium and lithium atoms formed, and still later stars and galaxies formed in the regions of greater mass densities caused by the quantum fluctuations of the inflationary period. The universe continued to evolve into our current universe. According to the WMAP data, our universe now consists of about 4.6% atomic matter, 22.7% dark matter composed of relatively slow (cold) unknown particles, and 72.8% dark energy also of an unknown nature. Currently our universe appears to be geometrically flat and has an accelerated spatial expansion (related to Lambda or the cosmological constant of Einstein's equations of general relativity) that may be caused by the dark energy.

There is an active search for a theory of quantum gravity to describe the very early universe, where the micro-scale realm of quantum physics and macro-scale realm of general relativity converge. A leading approach to the development of quantum gravity is Hawking's [6] "no boundary condition" of the very early universe. The expansion of the universe would have begun from a state that was very smooth and ordered. "There had to be small fluctuations in the density and velocity of particles. The no boundary condition, however, implied that these fluctuations were as small as they could be, consistent with the uncertainty principle." [6, p.148]

In 2010 Wayte [7] proposed that a single particle, with its internal material circulating coherently at speed c, formed the primeval particle of our universe. That first particle transformed into the energy configuration known as the hot big bang of cosmological theory.

The present article, based on the author's [8] work on internally transluminal models of the photon and the electron, proposes that the hypothetical first particle of our universe may have been an internally transluminal single quantum particle—a closed-looped photon. This article also proposes two varieties of the internally transluminal closed photon model as candidates for particles of dark matter.

2. Previous Work on Transluminal-Energy Quantum Particle Models

The proposed transluminal energy quantum (TEQ) model for the cosmic quantum, or first quantum particle of our universe, grew out of an attempt to model the photon and the electron physically and geometrically in a way that would contain the main physical properties of these particles [8]. The TEQ photon model contains an uncharged TEQ moving helically with a constant speed of $\sqrt{2}\,c$, with a forward angle of 45-degrees and with a helical radius equal to the photon's wavelength (the pitch of the photon model's helical trajectory) divided by 2π. (See Appendix.)

For the TEQ electron model, one Compton-wavelength ($\lambda_C = h/mc$ where m is the electron's mass) of the TEQ photon model's helical axis is formed into a double circular loop and then closed on itself so that the axis of the closed double-looping helix forms a double-looping circle with a radius $R_o = \hbar/2mc$. The electron TEQ's trajectory closes after its helical axis makes this one-Compton-wavelength double loop. The circulating TEQ photon model's forward momentum is $p = h/\lambda_C = mc$. The helically moving point-like TEQ has the charge of the electron. The closed double-looping of the TEQ photon model gives the TEQ electron model the electron's spin $s = R_o \times p = (\hbar/2mc)\times mc = \hbar/2$.

The helical radius of the TEQ electron's closed helical trajectory is chosen to be $\sqrt{2}R_o$ so that the TEQ electron model has the pre-quantum-electrodynamics magnetic moment of the electron--one Bohr magneton ($M_B = e\hbar/2m$). Increasing this helical radius slightly in the TEQ electron model would give the current experimental value of the electron's magnetic moment. With these size parameters, the TEQ moves in a closed trajectory along the surface of a spindle torus. The maximum speed of the electron model's TEQ is 1.516... c and the minimum speed is $\sqrt{0.5}c$ or 0.707... c.

3. The Proposed TEQ Closed-Photon Model for the Cosmic Quantum

Several other close-looped TEQ particles with different spins and different helical radii were analyzed mathematically to compare their physical and geometrical properties. One model for a particle with rest mass stood out for its simplicity—the single-looped closed photon model. In this model, a one-wavelength open-helix TEQ photon model is formed into a closed-helix photon model by turning the one-wavelength axis into a circle and turning the open-helix trajectory of the uncharged TEQ into a closed helix. The TEQ in this closed photon model now follows a closed helical trajectory along the surface of a horn torus, with a maximum speed calculated from the model to be $\sqrt{5}c$ and a minimum speed of c, independent of the energy of the photon being modeled.

The TEQ closed photon model is a boson, with spin $1\hbar$ due to its single-closed-loop structure. The TEQ photon's closed helical axis length is a circle of one wavelength circumference. Based on the parameters of the TEQ open helix photon, the radius of the circular axis of the closed helix and the radius of the closed helix itself are both equal to the TEQ photon's wavelength divided by 2π, creating the mathematically simple horn torus surface on which the TEQ moves.

4. Mathematically Generating the TEQ Closed-Photon Model of the Cosmic Quantum

The TEQ closed-helical photon model is generated geometrically and mathematically in the following way. A mathematical generating point moves at the speed of light c around a closed horizontal circular path of radius, $R = \lambda / 2\pi$. The TEQ is on a vertical mathematical circle, also of radius R, whose center is the moving generating point and whose plane is perpendicular to the direction of movement of the generating point around the horizontal circle. The speed of the TEQ along the vertical circle is also the speed of light c. The combined motion of the generating point around the horizontal circle and the TEQ around the vertical circle creates a 3-dimensional TEQ trajectory that closes after the generating point completes one circuit around the horizontal circle. The coordinates of the TEQ's
trajectory can be given parametrically as follows:

$$x(t) = R(1+\cos(\omega t))\cos(\omega t)$$
$$y(t) = R(1+\cos(\omega t))\sin(\omega t)$$
$$z(t) = R\sin(\omega t)$$

where $R = \hbar c / E$ is the radius of both generating circles, and $\omega = E / \hbar$ (where ω is the angular frequency of a circulating photon with energy E and wavelength $\lambda = hc / E$). The radius R of the horizontal circle is the radius of the circle whose circumference is the wavelength of the circulating photon, so

$$R = (1/2\pi)(hc/E) = \hbar c / E .$$

The radius R of the vertical circle used to generate the TEQ trajectory is equal to the radius of the closed helical path of the TEQ photon. This radius in the TEQ open-helix photon model is equal to the photon's wavelength of the photon $\lambda = hc / E$ divided by 2π, so again $R = (1/2\pi)(hc/E) = \hbar c / E$.

It will be noticed that the equations for the closed TEQ photon correspond to the parametric equations used to generate a horn torus. Based on the above three coordinate equations, the TEQ is found to move along the surface of this mathematical horn torus with a maximum speed of $\sqrt{5}c$ and a minimum speed c, independent of the energy of the photon. The TEQ closed photon model of the cosmic quantum looks like Figure 1.

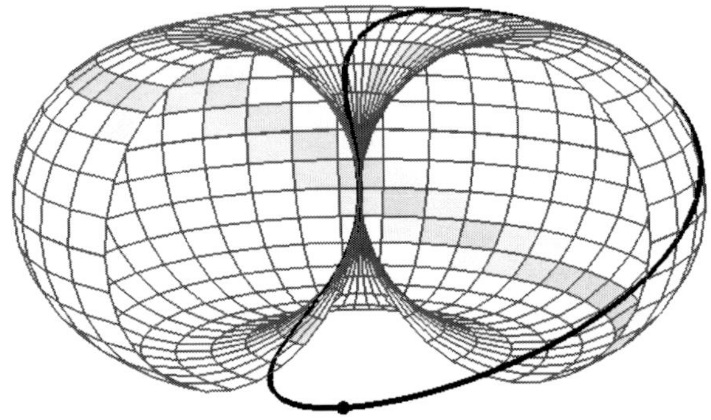

Figure 1. Transluminal energy quantum closed photon model of the cosmic quantum. The mathematical horn torus surface on which the transluminal energy quantum travels is cut away showing the interior. The black closed curve on the surface of the horn torus is the trajectory of the transluminal energy quantum (the black dot).

The TEQ's maximum speed occurs when the TEQ is farthest from the center of the torus, while the minimum speed occurs when the TEQ passes down through the center of the torus.

5. Parameters of the TEQ Closed-Photon as a Model of the First Quantum in the Very Early Universe

The closed photon model permits estimates of various physical parameters of the initial cosmic quantum, based on the estimated positive mass-energy of the very early universe. (An equal amount of negative mass-energy is associated with the gravitational potential energy of the very early universe.)

5.1 *Mass*

The geometry and velocities of the closed photon TEQ model are independent of the energy of the photon in the model. The radius R of the model varies inversely with the closed photon's energy E. The TEQ closed photon model, with the appropriate energy E and radius R given above, is proposed to be a model of the cosmic quantum, the hypothesized first quantum particle of the very early universe.

First, estimate the mass $M = E / c^2$ of the closed photon by estimating the mass-energy of the very early universe. According to the 7-year WMAP results [5], about 4.56% of today's universe is considered to be atomic matter, the rest being dark matter at 22.7% and dark energy at 72.8%. WMAP found that the universe is very nearly "flat" or Euclidian (to within 0.6%) and so its density for all mass-energy would be the critical density of $9.30 \times 10^{-27}\,\mathrm{kg/m^3}$. The spherical volume of the observable universe is $(4/3)\pi R_o^3 = 3.55\times 10^{80}\,\text{m}^3$, where R_o = 14,238 Mpc (Megaparsecs) = 46.4 billion light years = 4.39×10^{26} m is the radius of the observable universe. The total mass-energy is thus found to be $9.30\times10^{-27}\,\text{kg/m}^3\times3.55\times10^{80}\,\text{m}^3$ $= 3.30\times10^{54}$ kg. But according to the current cosmological standard model (the Lambda cold dark matter (ΛCDM) with inflation model), the dark energy portion of this total mass-energy was mostly produced as the volume of the universe expanded from its volume at the time of light decoupling at 379,000 years after the Big Bang to its volume at the present time. The positive mass (excluding dark energy) of the present universe is the current atomic mass plus the current mass of dark matter, or (4.56% + 22.7%) $\times3.30\times10^{54}\,\text{kg} = 9.00\times10^{53}\,\text{kg}$.

But according to the WMAP data, the mass energy density of the universe at the time of light decoupling consisted of 25% photons and neutrinos (both of which today contribute a negligible percentage to the total mass energy density) and 75% atomic matter and dark matter). The mass energy M at the time of decoupling is therefore larger than mass calculated above by a factor of 4/3. So a better estimate M of the total positive mass-energy of the universe at the time of decoupling or recombination, including atomic matter, dark matter, photons and neutrinos is

$$M = (4/3)\times9.00\times10^{53}\,\text{kg} = 1.2\times10^{54}\,\text{kg}.$$

This is the value used to estimate the mass energy M of the cosmic quantum.

5.2 *Radius*

This leads to a radius estimate of the first TEQ closed photon to be

$$\begin{aligned} R &= \hbar c/E = \hbar c/Mc^2 \\ &= \hbar/Mc = 1.05\times10^{-34}/(1.2\times10^{54}\times3\times10^8) \\ &= 2.9\times10^{-97}\,\text{m} \end{aligned}$$

5.3 *Frequency*

The corresponding frequency ν of the closed photon is

$$\begin{aligned} \nu &= E/h \\ &= Mc^2/h \\ &= 1.2\times10^{54}\times(3.00\times10^8)^2/6.63\times10^{-34} \\ &= 1.6\times10^{104}\,\text{Hz} \end{aligned}$$

5.4 *Period*

The period T of the TEQ closed photon would be $T = 1/\nu = 1/(1.6\times10^{104}) = 6\times10^{-105}$ sec.

5.5 *Mass Density*

The TEQ closed photon model allows an estimate for the initial mass density of the very early universe. This would be approximately the mass density of the first closed photon. The mass of the first closed photon was estimated above to be $M = 1.2\times10^{54}$ kg while the estimated radius of the photon model is $R = \hbar/Mc = 2.9\times10^{-97}$ m. If the volume of the closed photon is estimated as R^3, the mass density of the closed photon would be

$$\begin{aligned} \rho_{mass} &= M/R^3 = M/(\hbar/Mc)^3 = M^4c^3/\hbar^3 \\ &= (1.2\times10^{54})^4\times(3.00\times10^8)^3/(1.05\times10^{-34})^3 \\ &= 1.6\times10^{344}\,\text{kg/m}^3 \end{aligned}$$

5.6 *Energy Density*

The corresponding energy density would be

$$\begin{aligned} \rho_{energy} &= M^4c^5/\hbar^3 = \rho_{mass}c^2 \\ &= (1.6\times10^{344})\times(3.00\times10^8)^2 \\ &= 1.4\times10^{361}\,\text{Joules/m}^3 \end{aligned}$$

6. Why is the TEQ Closed-Photon Model Proposed as the Cosmic Quantum?

Firstly, a photon is a basic constituent quantum particle of the universe. Second, a photon is able to travel in a short closed path, even of a single wavelength. At the earliest stage of the universe, a first photon had nowhere to go, since the space-time structure of the universe presumably had not begun expanding as long as only a first particle exists. Third, a closed photon has a rest mass corresponding to its energy content, but an open photon does not have rest mass.

The primeval atom proposed by Lemaitre was described as a super-large radioactive atom that upon decomposing could have become all of the other particles of the universe and formed the expanding hot mass-energy system that has come to be called the Big Bang. But the question of course would remain: from where did the primeval atom come? If there were a first cosmic quantum particle for our universe, it would likely have been very simple. It could possibly have emerged as a quantum fluctuation from a cosmic quantum field.

7. A Cosmic Field Theory for the Cosmic Quantum Boson?

The primordial closed photon's properties are at a convergence of a radius $R = 2.9 \times 10^{-97}$ m quantum-sized particle (so quantum theory should apply) with an estimated mass $M = 1.2 \times 10^{54}$ kg (so general relativity theory should apply). It is well known that quantum theory and general relativity theory are mathematically inconsistent under combined conditions of extremely small sizes and extremely large masses. A synthesis of quantum theory and general relativity theory into a quantum gravity field theory is much needed and much sought. Such a theoretical synthesis would be expected to lead to new and surprising predictions. Quantum gravity field theory could describe a cosmic field that produces cosmic quanta such as the first quantum particle of our universe. Quantum gravity field theory could reasonably be called cosmic field theory.

The TEQ closed photon model is a boson, because its spin is one quantum unit of angular momentum: $s = \hbar = h/2\pi$. This spin around the particle's vertical axis is calculated in the TEQ model from the formula for calculating the angular momentum of a circularly moving object (the photon):

Spin = (radius of photon model) × (momentum of the photon model)

$$s = R \times p$$
$$s = R \times (Mc)$$

where Mc is the momentum of a photon carrying energy corresponding to a mass M. $s = (\hbar / Mc) \times Mc$ since $R = \hbar / Mc$ is the radius of a one-wavelength closed photon model corresponding to a mass M.

$$s = \hbar = h/2\pi$$

8. A Second Very Early Universe Particle — a Closed-Photon Fermion?

At some very early time (perhaps at the beginning of time itself) the original TEQ closed photon is proposed to start generating other particles in the process of the Big Bang. The TEQ closed photon might sometime during this process generate a new particle—a TEQ closed photon fermion—that could lead to the production of other TEQ fermions such as electrons, positrons, quarks, antiquarks and neutrinos. These are all particles with spin $\hbar/2$. The way to generate a TEQ closed-photon fermion from a TEQ photon is for the straight one-wavelength λ axis of a TEQ open photon to wrap itself around into a double circular loop before closing on itself. The associated helical TEQ trajectory also closes on itself to form a closed, double-looping helical trajectory. If this TEQ single-wavelength double-looping closed helical photon keeps the same helical radius $R = \lambda/2\pi$ as that of the TEQ open helical photon, the TEQ closed trajectory that is formed is a path along the surface of a self-intersecting torus (technically a spindle torus) whose central spindle's width is the radius R. In this double-looping closed TEQ photon, the TEQ's maximum speed is calculated from the TEQ coordinate equations for this model to be 3.162 c, which is larger than the maximum speed for the TEQ single-loop closed photon of $\sqrt{5}c$ or 2.236 c, while its minimum speed is still c, the same as for the TEQ single-loop closed photon. The double-looped TEQ photon's trajectory and associated torus surface are shown in Figure 2.

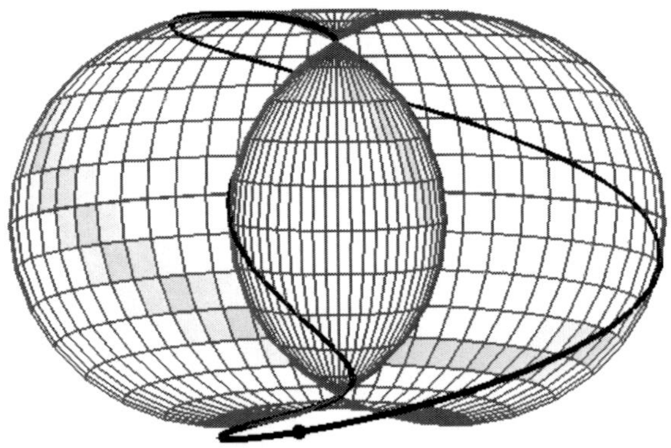

Figure 2. The transluminal energy quantum model of the double-looped closed-photon fermion. Some of the outer part of the mathematical spindle torus on which the TEQ moves is cut away to show the spindle inside. The black closed curve on the surface of the spindle torus is the trajectory of the TEQ (indicated by the black dot).

The coordinates for this second TEQ particle are:

$$x(t) = 0.5R(1 + 2\cos(\omega t))\cos(2\omega t)$$
$$y(t) = 0.5R(1 + 2\cos(\omega t))\sin(2\omega t)$$
$$z(t) = 0.5R\sin(\omega t)$$

where $R = \lambda/2\pi$ is the radius of the photon helix for the TEQ's double-looped helical trajectory of wavelength λ, $0.5R = 0.5\lambda/2\pi$ is the radius of the circular axis of the double-looped helical trajectory, and $\omega = 2\pi c/\lambda$ is the angular frequency of the photon. This particle is a fermion with spin $s = 0.5\hbar$. The calculation of its spin around its vertical axis follows:

Spin = (radius of circular axis of double-looped photon model) × (momentum of the photon model)

$$s = (0.5R) \times p$$
$$= (0.5R) \times h/\lambda$$

where h/λ is the momentum of a photon of wavelength λ.

$$= (0.5\lambda/2\pi) \times h/\lambda$$
$$= 0.5h/2\pi = 0.5\hbar$$

9. The Production of Further Particles From the Original TEQ Closed Photon

Once the Big Bang gets started from the original closed photon, there is hypothesized to be a rapid increase in the number and types of TEQ elementary particles such as TEQ electrons and TEQ photons produced from the original closed photon, as well as a rapid expansion of the space in which the particles are formed.

TEQ models for the electron and the photon have already been developed [8]. The TEQ electron model is a fermion and so is based on a double-looped TEQ photon model. In the TEQ electron model the circulating TEQ is charged with the negative electric charge of the electron. A TEQ positron would have a positively charged TEQ and an oppositely turning closed double-looping helical trajectory compared to that of the TEQ electron. The circulating of the negatively charged TEQ in the electron model gives the TEQ electron model its magnetic moment so that the TEQ electron acts like a tiny magnet. The parameters of the TEQ electron were selected so that the TEQ electron has the main parameters of an actual electron—its mass, charge, spin and magnetic moment. The TEQ model of the electron is shown in Figure 3.

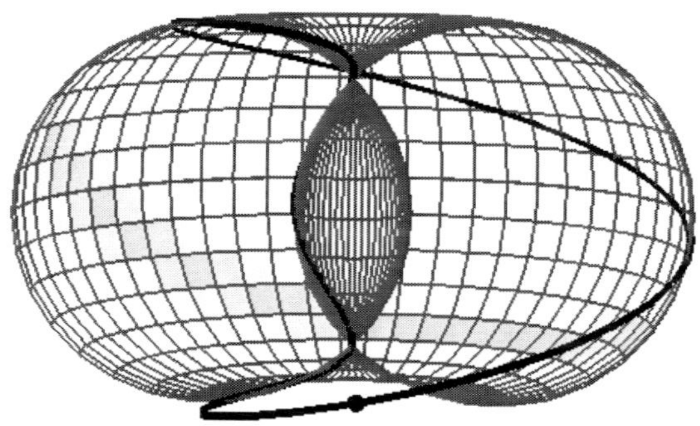

Figure 3. The transluminal energy quantum model of the electron. Some of the outer part of the mathematical spindle torus on which the TEQ moves is cut away to show the spindle inside. The black closed curve on the surface of the spindle torus is the trajectory of the TEQ (indicated by the black dot).

The TEQ model of the electron differs somewhat in shape from the double-looped closed photon model. The helical radius of the TEQ electron model is smaller than the helical radius of the corresponding closed double-looped TEQ photon model. This shortening of the electron model's helical radius is done to give the TEQ electron model a magnetic moment equal to the magnetic moment of the actual electron (which in pre-quantum-electrodynamics is found from the Dirac equation for the relativistic electron to be one Bohr magneton or $e\hbar/2m$. A slightly larger helical radius will give to the TEQ electron model the slightly larger experimental value of the electron's magnetic moment.) The result is that the maximum speed of the TEQ in the electron model is $2.516\ c$, smaller than the maximum TEQ speed of $3.162\ c$ for the double-looped closed photon model but larger than $\sqrt{5}c$ or $2.236\ c$ for the single-looped closed photon model. The electron TEQ's minimum speed is less than the speed of light, at $\sqrt{0.5}c$ or $0.707\ c$, compared to the minimum speeds of exactly c for both the double-looped and single-looped closed photon models. The electron model's TEQ passes through the speed of light twice during each closed trajectory. Since the TEQ is not a particle with rest mass like the electron as a whole, the TEQ can pass through the speed of light as it internally circulates to form the electron, while the TEQ electron moving as a whole cannot reach the speed of light, due the limitation on its external speed prescribed by Einstein's special theory of relativity.

The coordinates for the TEQ in the TEQ electron model are:

$$x(t) = R_o(1+\sqrt{2}\cos(\omega t))\cos(2\omega t)$$
$$y(t) = R_o(1+\sqrt{2}\cos(\omega t))\sin(2\omega t)$$
$$z(t) = R_o\sqrt{2}\sin(\omega t)$$

where $R_o = \lambda_{compton}/4\pi = \hbar/2mc = 1.931\times 10^{-13}$ m is the radius of the double-looped circular axis of the TEQ's closed helical trajectory, $\sqrt{2}R_o$ is the radius of the helix of the TEQ's closed double-looped helical trajectory, and m is the mass of the electron. $\omega = 2\pi c/\lambda_{compton} = 7.77\times 10^{20}$ radians/sec is the angular velocity of the TEQ in its internal trajectory in the TEQ electron model. Like the closed double-looped TEQ photon model, the TEQ electron model is a fermion.

10. Possible Relevance of the TEQ Closed-Photon Model to Cosmology

The proposed TEQ closed photon model is closely related to earlier TEQ models of the photon and electron [7,8, Appendix]. Cosmological considerations were not included in the development of these models. But when the TEQ closed photon model was later considered as a model for the first quantum entity in the very early universe, the model was seen to be possibly relevant to several currently unanswered major cosmological questions.

10.1 *Could a Single Quantum Particle Produce Our Universe?*

If this is possible then the question becomes, what could that first quantum have been? The proposed TEQ closed photon model of this hypothesized cosmic quantum is postulated to contain in its circulating TEQ the entire mass of ordinary matter and dark matter of the very early universe (where there was apparently relatively little dark energy due to the small volume of the very early universe).

The universe is currently understood to have an associated negative gravitational potential energy that could mathematically cancel out the total positive physical energy of the universe to give a net energy of the universe to be exactly zero. This seems to be the currently scientifically accepted view of the total energy content of the universe. Both these positive and negative energies of the universe could be built into a theory of quantum gravity that would describe the first TEQ closed photon or other initial quantum particle or particles of the universe leading to the Big Bang. Current quantum theory and general relativity theory cannot make meaningful predictions for lengths and durations that are less that the

Planck length $l_P = 1.616199 \times 10^{-35}$ m and

Planck time $t_P = 5.39106 \times 10^{-44}$ s.

Clearly the radius $R = 2.9 \times 10^{-97}$ m and period $T = 6 \times 10^{-105}$ s of the proposed closed photon model for the very early universe are well below these limits. Since quantum theory and general relativity theory are acknowledged to be inconsistent, a new quantum gravity theory could lead to physical descriptions that go well beneath the current Planck length and time values.

What initial hypothesized state of the universe would have the smallest quantum fluctuations, consistent with the Heisenberg uncertainty principle? The position and momentum parameters of the circulating TEQ composing the closed photon model are near the minimum limit set by the uncertainty principle, as seen below.

The uncertainty of the x coordinate is defined as the root mean square (rms) value Δx for the x component of a particle. For the TEQ in the closed photon model, the rms value for x, calculated from the TEQ position coordinate equations presented above, is

$\Delta x = 1.87\lambda / 4\pi$ where λ is the wavelength of the closed photon.

The uncertainty of the x component of momentum Δp_x is defined as the rms value of the x component of the momentum. For the circulating TEQ, the momentum $p = h/\lambda$ rotates in a horizontal circle (for the x-y dimensions) and in a vertical circle (for the z dimension) for the closed photon model. So Δp_x is found to be

$$\Delta p_x = p/\sqrt{2} = \sqrt{0.5}h/\lambda .$$

And so

$$\Delta x \times \Delta p_x = (1.87\lambda/4\pi) \times \sqrt{0.5}h/\lambda = 1.32(\hbar/2).$$

A similar calculation of $\Delta y \times \Delta p_y$ for the y components of the position and momentum of the closed photon model gives

$$\Delta y \times \Delta p_y = (1.58\lambda/4\pi) \times (\sqrt{0.5}h/\lambda) = 1.12(\hbar/2).$$

A similar calculation of $\Delta z \times \Delta p_z$ for the z components of the position and momentum of the TEQ closed photon model gives

$$\Delta z \times \Delta p_z = (\sqrt{2}\lambda/4\pi) \times (\sqrt{0.5}h/\lambda) = 1(\hbar/2).$$

Compare the three results for the x, y and z components of position and momentum of the TEQ closed photon model with the Heisenberg uncertainty relation for the position and momentum coordinates of a particle:

$$\Delta x \times \Delta p_x \geq \hbar/2$$

$$\Delta y \times \Delta p_y \geq \hbar/2$$

$$\Delta z \times \Delta p_z \geq \hbar/2$$

The variability of the position and momentum of the TEQ in the closed photon model of the cosmic quantum are nearly equal to the conditions for minimum uncertainty in the Heisenberg uncertainty relations for a single elementary particle.

A single quantum particle such as the proposed TEQ closed photon, formed by a circulating sub-quantum particle (the TEQ) with the calculated variation in its position and momentum, may be a candidate for the most highly ordered (and therefore lowest entropy) state of the very early universe, consistent with the Heisenberg uncertainty principle and the total positive energy content of the early universe. Evolution of that single quantum entity could lead to a "hot Big Bang" of many particles and the associated extremely high temperature of the early universe. This "hot Big Bang" evolving from the single closed photon could then evolve to the very smooth and ordered state that is apparent in the universe today, as seen in the extremely but not completely smooth, low temperature cosmic microwave background radiation (CMBR) observed coming from all directions in the sky today [5]. This CMBR is some of the best current scientific evidence for the Big Bang.

10.2 *Why Does Matter and Not Antimatter Dominate in the Atomic Matter in Our Universe?*

The proposed TEQ closed photon model has a closed helical form. A helix has the property that it turns either clockwise or counterclockwise along its axis, as seen from behind. Clockwise in the TEQ closed photon model of the cosmic quantum might correspond to the TEQ closed helical structure of ordinary matter particles such as the TEQ electron model, and counterclockwise might correspond to the opposite helicity, that of antimatter particles such as the TEQ positron model. If the helicity of the original cosmic quantum closed-photon model's TEQ corresponded to that of TEQ-composed ordinary matter (as opposed to antimatter), this would provide an initial bias in our closed-photon-evolved universe towards ordinary matter. This bias towards ordinary matter in our universe's cosmic quantum might then help explain why almost all of the observed atomic and subatomic matter in our present universe, which would have evolved in stages from the original closed photon's TEQ, appears to be ordinary matter.

If the proposed TEQ closed photon model, a boson having spin $\hbar$ and determined by its helicity to be matter (as opposed to antimatter), emerged from a cosmic field to become our very early universe, an equally massive TEQ closed photon model with spin $-\hbar$, and described as antimatter due to its opposite helicity, could have also emerged from this cosmic field at the same time. This assumes that the laws of conservation of total angular momentum and matter-antimatter pair production are valid for the cosmic field. If so, what happened to the original TEQ antimatter spin $-\hbar$ closed photon that was paired with our cosmic quantum? Could it have formed a mainly antimatter universe, perhaps in a different space-time dimension?

10.3 *Could Dark Matter Be Composed of Closed Single- or Double-Looped TEQ Photons?*

According to the WMAP results [5], about 22.7% of the matter-energy content of the universe appears to consist of cold (relatively slow moving) dark matter, whose main presumed feature is that its particles have a relatively large mass and are mainly affected by gravity, but so far have shown no other identifying physical features. The second proposed very early universe TEQ particle, the closed double-looping photon model, is a fermion, composed of one wavelength of a photon, wrapped around twice before it joins itself. This gives the uncharged TEQ particle ½ unit of quantum spin as well as a mass that depends on its internal wavelength according to $E = hc/\lambda$.

The fundamental particles of matter are all fermions—the electron, muon, and tau particles, the neutrinos and the quarks, and all their antiparticles—while the fundamental particles of force for electromagnetism, the strong and the weak nuclear interactions are bosons—photons, gluons, W and Z particles and the Higgs boson and their antiparticles. (Gravity is not included here since a gravity force particle or graviton has not been observed.)

Perhaps dark matter consists of uncharged closed, double-looped TEQ photon fermions. Such particles could qualify as cold dark matter particles if they would interact with other particles mainly through the gravitational force. Uncharged single-looped closed TEQ photon bosons could also be possible candidates for cold dark matter, or for force particles associated with cold dark matter particles.

If an uncharged closed double-looping TEQ photon, with the helicity of a TEQ electron, has a mass slightly less than the mass of an electron, it could be a very stable particle. It would mainly interact with other matter-energy particles through gravitation. It could only be annihilated to produce photons by its corresponding dark matter anti-particle, an uncharged closed double-looped TEQ photon of opposite helicity to this proposed dark matter particle. For the same (currently unknown) reason that there seems to be very little anti-matter in our observable universe, there would also presumably be in our universe very little dark anti-matter consisting of closed double-looped TEQ photons of opposite helicity to this proposed dark matter particle.

If this proposed neutral double-looped photon dark matter particle has the lowest mass of a possible family of related double-looped photon fermions, it might only be able to decay into one or more low mass neutrinos, that are also hard to observe. Since this proposed dark matter particle could also only annihilate with its dark matter anti-particle, which is very scarce, it would like the electron be a very long-lived particle. Further, since this proposed dark matter particle is a fermion, it would not clump closely together with other like dark matter fermions due to the Pauli exclusion principle and the related Fermi-Dirac statistics for fermions. Single-looped closed TEQ photons bosons however could clump close together, according to Bose-Einstein statistics.

Fermions and bosons have already been proposed as candidates for dark matter particles. Boehm and Fayat [10] found two theoretical possibilities: "Either dark matter is coupled to heavy fermions … or dark matter is coupled to a new light gauge boson U."

Astrophysical research is ongoing in search of gamma ray photons that could result from the annihilation of dark matter particles with their corresponding antiparticles. One proposed type of dark matter particle is the Weakly Interactive Massive Particle (WIMP) that is proposed to be able to annihilate with its antiparticle, producing pairs of gamma ray photons. A tentative positive result has been found by Weniger [11] using data from the Fermi Large Area Telescope. The proposed TEQ dark matter particles could also be sought using this approach. TEQ matter-antimatter annihilations would be rare though, due to the relatively low presence of TEQ antimatter in our universe, as explained above.

10.4 *Does the Universe Have Quantum Non-local Interconnectivity due to an Original Quantum Particle?*

With the TEQ closed-photon model of the cosmic quantum, all the hypothesized TEQ-based atomic (or baryonic) matter and dark matter expressions of the universe would evolve from the cosmic quantum during and after the Big Bang. According to quantum theory, confirmed by many recent experiments, objects that are quantum-entangled when they separate can retain this quantum interconnectivity even when separated over vast distances. This implies that all of the particles derived from the hypothesized cosmic quantum, i.e. all the atomic matter and dark matter in our universe, could retain quantum non-local interconnectedness to some degree. This would be the case despite even intergalactic distances separating many elementary particles and energetic structures today. In a quantum-theoretic as well as experimental sense the atomic and dark matter portions of our universe could still remain a single vast quantum entity.

10.5 *Did Time Begin With the Original Quantum Particle?*

One of the unsolved problems about our universe is why our early universe, consisting of a nearly uniform dense, hot state of matter and energy, had such a low entropy compared to its entropy today, and how this relates to the origin and nature of time. Carroll [12] estimates the entropy of our early, post inflationary universe (the part that developed into our observable universe—our "comoving patch") to be $S_{\text{early}} \approx 10^{88}$. This is based on there being an estimated 10^{88} freely moving particles in our comoving patch in our young and smooth universe. He estimates the entropy of our comoving patch of universe today to be $S_{\text{today}} = 10^{101}$. This is based on an estimate of 10^{11} supermassive black holes (estimated at one per galaxy in our observable universe) and an amount of entropy of 10^{90} per supermassive black hole, calculated according to the Bekenstein-Hawking formula for the entropy of a black hole.

The entropy of our universe, consisting of a single TEQ cosmic quantum before it subdivided into other energy particles, would have been $S_{\text{cosmic quantum}} \approx 1$ according to this particle-count method. Since the entropy of a closed system generally increases with time, in our observable universe time could have begun when the TEQ cosmic quantum particle started to subdivide into other quantum particles and the entropy of this system began to increase. Time has apparently been flowing in a positive direction (towards increasing entropy) in our observable universe from then until now.

11. Can Quantum Mechanics and General Relativity Be Unified Through Transluminality?

Transluminal energy quanta move superluminally and sometimes through luminal to subluminal speeds as they hypothetically form different force and matter particles such as TEQ photons and TEQ electrons. Although particles with mass such as electrons always travel through space at less than the speed of light, according to the special theory of relativity and with much experimental confirmation, the hypothetical TEQ that composes the TEQ electron model is not so limited. An electron's TEQ travels from superluminality through the speed of light to subluminality and back with a frequency $\nu = mc^2/h = 1.24 \times 10^{20}$ Hz. According to the TEQ hypothesis, an electron is the average motion of a rapidly circulating TEQ. Gravitational waves are proposed in the general theory of relativity to travel at the speed of light. But if the proposed graviton also has a TEQ structure, these graviton TEQs may also travel superluminally like the TEQ composing a photon.

The hypothesis that fundamental particles are composed of transluminal energy quanta could provide a different approach to the evolution of the very early universe. At earlier and earlier cosmological times, the four interactions of physics (the strong nuclear interaction, the weak nuclear interaction, the electromagnetic interaction and the gravitational interaction) are projected to lose their distinguishing features and merge or nearly merge into a single force at or near the end of the "Planck era" around 10^{-43} s after the Big Bang, according to the traditional (non-inflationary) cosmological standard model.

According to the TEQ approach, during this era of unified interactions, which would also occur in inflationary cosmology, all positive energies could have been carried by TEQs moving at transluminal speeds. The original cosmic quantum, a TEQ single-looped closed-photon, would have transformed into an expanding volume of superluminal TEQs collectively carrying the total energy of the original cosmic quantum. This would have been a volume of purely TEQ-carried energy, whose total energy content was proportional to the sum of the frequencies of vibration of all the TEQs carrying this energy. This would be the original "hot big bang". As this volume of TEQs expanded with the expanding very early universe, the number and types of TEQs would multiply as the gravitational interaction separated from the electro-nuclear interaction. The types of TEQs would further multiply as the electro-nuclear interaction separated into the strong nuclear interaction and the electro-weak interaction. More types of TEQs would evolve as the electro-weak interaction separated into the electromagnetic interaction and the weak interaction. Finally all types of TEQs would exist and would form the fundamental particles such as photons, quarks, gluons, electrons and neutrinos.

Perhaps the cosmic field theory for the origin and evolution of the cosmic quantum, that combines quantum theory and general relativistic theory, will be based partly on the proposed internal transluminality of the cosmic quantum and other transluminal energy quanta derived from it.

12. Conclusions

The transluminal energy quantum (TEQ) model of a closed photon is a new approach to answering the question "Could the universe have been formed from a single quantum?" The TEQ model of the cosmic quantum exists well below the Planck length and Planck time values that currently restrict models of the very early universe to greater than about 10^{-43} s.

The TEQ model of the cosmic quantum suggests a value of the positive energy density of the cosmic quantum of about 10^{361} Joules/m^3, based on the baryonic matter and dark matter content of the observed universe and the radius of the proposed TEQ closed photon model of the cosmic quantum.

The TEQ closed photon model of the cosmic quantum is a boson, whose helical geometry creates a bias in favor of matter over antimatter in our universe. Conservation of angular momentum suggests that another closed photon model of opposite spin would have been formed at the same time as our cosmic quantum, if the first closed photon was formed as a quantum fluctuation in a cosmic field obeying physical conservation laws.

The TEQ single-looping (boson) and double-looping (fermion) closed photon models provide two possible candidates for particles of dark matter having masses perhaps similar to the mass of an electron. Since both of these proposed dark matter particles have their corresponding anti-particles, they could possibly be detected by observing photons produced from particle-antiparticle annihilation. Finding such TEQ dark matter particles could provide indirect support for the TEQ cosmic quantum hypothesis as well. An experimental approach for detecting TEQ dark matter particles was suggested, which is in line with a current approach to detecting dark matter particles through trying to detect pairs of photons created from dark matter – anti-dark matter particle annihilation. Testing of the TEQ cosmic quantum hypothesis may have to await the development of a theory of quantum gravity. In the meantime, possible cosmic evolutionary scenarios leading from the proposed cosmic quantum to the current Big Bang standard model can be explored.

Appendix: The Parameters of the Transluminal-Energy Quantum Model Of The Photon

From "Superluminal quantum models of the photon and electron", http://superluminalquantum.org/superluminal17Oct2005

A photon is modeled as a superluminal pointlike quantum traveling along an open helical trajectory having a straight axis. The radius of the helix is R and the pitch (wavelength) of the helix is λ. The helical trajectory makes an angle θ with the forward direction. The circumference of the helix is $2\pi R$. Using the superluminal quantum model, the result is:

1) The forward helical angle θ is found to be 45°, for any photon wavelength.
2) The photon helix's radius is found to always be $R = \lambda / 2\pi$.
3) The speed of the superluminal quantum is $\sqrt{2}c = 1.414..\ c$ along its helical trajectory.

These results are derived below.

The superluminal quantum, with total momentum $\vec{P}$ directed along its helical trajectory, has a forward component of momentum $P\cos(\theta)$ determined by the wavelength λ of the helix, and a transverse component of momentum $P\sin(\theta)$ that is used to calculate the spin of the photon. The superluminal quantum's longitudinal component of momentum is

$$P\cos(\theta) = h / \lambda \,, \tag{1}$$

the experimental longitudinal or linear momentum of a photon. The total momentum $\vec{P}$'s transverse component of momentum is $P\sin(\theta)$. This transverse component is perpendicular to the helical radius vector $\vec{R}$ to the superluminal quantum from the helical axis. The magnitude S of the angular momentum or spin of the superluminal quantum model is then

$$S = RP\sin(\theta) = h / 2\pi \tag{2}$$

the experimental spin or angular momentum of the photon. Combining equations (1) and (2) gives

$$\sin(\theta)/\cos(\theta) = \tan(\theta) = \lambda/2\pi R \quad (3)$$

Now look at the helical geometry. As the superluminal quantum advances along the helix a distance λ (one wavelength) in the longitudinal direction, the superluminal quantum travels a transverse distance $2\pi R$, i.e. once around the circle of radius R of the helix. From the way the helix's angle θ is defined,

$$\tan(\theta) = 2\pi R/\lambda \quad (4)$$

We now have two equations (3) and (4) for $\tan(\theta)$. Setting them equal gives $\tan(\theta)=2\pi R/\lambda=\lambda/2\pi R$
This will only be true if

$$\lambda = 2\pi R$$

that is,

$$R = \lambda/2\pi \text{ and } \tan(\theta) = 1$$

and since $\tan(\theta) = 1$ we have $\theta = 45^\circ$

These results are for any photon in this superluminal quantum model of the photon. Since $\theta = 45^\circ$ and the forward speed of the superluminal quantum along its straight helical axis is postulated to be c, then the forward speed of the photon model is also c, and the speed v of the superluminal quantum along its helical trajectory is

$$v = c/\cos(45^\circ) = \sqrt{2}c \ .$$

References

[1] Lemaître, G. (1950) Letters to Nature, Nature 127, p. 706; 9 May 1931, reprinted in The Primeval Atom: An Essay on Cosmogony, New York: D. Van Nostrand Company, p. 3.
[2] Goldhaber, M. (1956) Speculations on cosmogony, Science Vol. 124 no. 3214, pp. 218-219.
[3] Tryon, E. P. (1973) Is the universe a vacuum fluctuation?, Nature Vol. 246 no. 5433, pp. 396-397.
[4] Dehmelt, H. (1989) Triton,...electron,...cosmon,..: an infinite regression?, Proc. Nat. Acad. Sci. USA Vol. 86, pp. 8618-8619, Physics.
[5] Jaroski, N. et al. (2008) Seven-Year Wilkinson Microwave Anisotropy Probe (WMAP) Observations: Sky Maps, Systematic Errors, and Basic Results, http://lambda.gsfc.nasa.gov/product/map/dr4/pub_papers sevenyear/basic_results/wmap_7yr_basic_results.pdf , basic results p. 39 .
[6] Hawking, S. (1988) A Brief History of Time: From the Big Bang to Black Holes, New York: Bantam Books.
[7] Wayte, R. (2010) On the accelerating universal expansion, viXra:1011.0066, p.15.
[8] Gauthier, R. (2007) FTL quantum models of the photon and the electron, CP880, Space Technology and Applications International Forum--STAIF-2007, pp.1099-1108, M.S. El-Genk (ed.), available at http://www.superluminalquantum.org/STAIF-2007article
WMAP 5-year data: Content of the universe page, wmap.gsfc.nasa.gov/media/080998/index.html,
[9] Boehm, C. & Fayat, P. (2003) Scalar dark matter candidates, arXiv:hep-ph/0305261v2, p. 1.
[10] Weniger, C. (2012) A tentative gamma-ray line from dark matter annihilation at the Fermi Large Area Telescope, arXiv:1204.2797v2.
[11] Carroll, S. (2010) From Eternity to Here: The Quest for the Ultimate Theory of Time, New York: Dutton, p. 299.

Transluminal-Energy Quantum Models of the Photon and the Electron

RICHARD GAUTHIER

Santa Rosa Junior College, 1501 Mendocino Ave., Santa Rosa CA 95401 USA
richgauthier@gmail.com
www.superluminalquantum.org

A photon is modeled by an uncharged transluminal energy quantum (TEQ) moving at *c* sqrt(2) along an open 45-degree helical trajectory with radius *R=Lambda/2pi*, (where *Lambda* is the helical pitch or wavelength of the photon). A transluminal spatial model of an electron is composed of a charged pointlike transluminal energy quantum circulating at an extremely high frequency of 2.5 x 10^{20} hz in a closed, double-looped helical trajectory whose helical pitch or wavelength is one Compton wavelength *h/mc*. The transluminal energy quantum has energy and momentum but not rest mass, so its speed is not limited by *c*. The TEQ's speed is superluminal 57% of the time and subluminal 43% of the time, passing through *c* twice in each trajectory cycle. The TEQ's maximum speed in the electron's rest frame is 2.516 *c* and its minimum speed is *c*/sqrt(2), or .707 *c* . The electron model's helical trajectory parameters are selected to produce the electron's spin *hbar*/2 and approximate (without small QED corrections) magnetic moment *ehbar/2m* (the Bohr magneton) as well as its Dirac equation-related "jittery motion" (*zitterbewegung*) angular frequency $2mc^2/hbar$, amplitude *hbar/2mc* and internal speed *c*. The two possible helicities of the electron model correspond to the electron and the positron. With these models, an electron is like a closed circulating photon. The electron's inertia is proposed to be related to the electron model's circulating internal momentum *mc*.

Keywords: Transluminal; Superluminal; Photon; Electron; Zitterbewegung; Inertia; Model

1. Introduction

Dirac's [1,2] theory of the relativistic electron did not include a model of the electron itself, and assumed the electron was a point-like particle. Schrödinger [3] analyzed the results of the Dirac equation for a free electron, and described an extremely high-frequency *zitterbewegung* or jittery motion, which appears to be due to the interference between positive and negative energy terms in the solution. Barut and Bracken [4] analyzed Schrödinger's *Zitterbewegung* results and proposed a spatial description of the electron where the *zitterbewegung* would produce the electron's spin as the orbital angular momentum of the electron's internal system, while the electron's rest mass would be the electron's internal energy in its rest frame. Barut and Thacker [5] generalized Barut and Bracken's analysis of the internal geometry of the Dirac electron to a proper-time relativistic formalism. Hestenes [6-9] reformulated the Dirac equation through a mathematical approach (Clifford algebra) that brings out a geometrical trajectory approach to understanding *zitterbewegung* and to modeling the electron, such as identifying the phase of the Dirac spinor with the spatial angular momentum of the electron. A trajectory approach to the Dirac theory has also been utilized by Bohm and Hiley [10], who describe the electron's spin angular momentum and its magnetic moment as due to the circulatory motion of a point-like electron. However, none of the above work in modeling the electron's jittery motion has a superluminal aspect.

The photon has previously been modeled geometrically with several approaches, with results quantitatively similar to those in the present superluminal quantum model of the photon. Ashworth [11] used a classical model of the photon to obtain a radius $\lambda / 2\pi$ for the photon and a superluminal internal speed of $\sqrt{2}$ *c*, the same quantitative results for the photon as in the present paper. Kobe [12] obtained the same quantitative result for the photon radius as in the present paper, based on a helical approach and quantum mechanical considerations. Sivasubramanian *et al.* [13], using a model of the photon that is helical and explicitly internally superluminal, independently arrived at the same radius $\lambda / 2\pi$ for a photon as in the present paper.

The objectives of the present paper are:

1) Present a simply derived transluminal helical quantum model for the photon,
2) Present a related transluminal closed helical quantum model of the electron having experimental and theoretical features of the Dirac equation's electron,
3) Relate the electron and photon models to the Heisenberg uncertainty relations,
4) Propose a new approach to understanding the electron's inertia, and

5) Suggest approaches for testing the proposed models of the transluminal energy quantum model of the photon and the transluminal energy quantum model of the electron.

2. A Transluminal-Energy Quantum Approach to Modeling the Electron and the Photon

The present approach is a unified approach to modeling both the electron and the photon with transluminal helical trajectories. The electron model has several features of the Dirac electron's *zitterbewegung*. Point-like entities are postulated called transluminal energy quanta (TEQs), which are distinct from electrons and photons themselves and which compose an electron or a photon. These TEQs (which may be subluminal, luminal or superluminal) have an energy E, with its associated frequency ν related by $E = h\nu$, an angular frequency, $\omega = 2\pi\nu$, an instantaneous momentum $\vec{P}$ with its associated wavelength λ, and an electric charge (in the case of the electron). One TEQ forms a photon or an electron. An electron's TEQ is found to alternate between subluminality and superluminality passing through the speed of light, while a photon's TEQ is always superluminal. TEQs move in helical trajectories, which may be open (for a photon) or closed (for an electron). Movement of one of these TEQs along its trajectory produces an electron or a photon. The type of helical trajectory and the associated charge or lack of charge determines which particle is produced. More details about the models presented below are provided in Gauthier [14,15].

3. The Transluminal-Energy Quantum Model of the Photon

A photon is modeled as a TEQ traveling along an open helical trajectory of radius R and pitch (wavelength) λ. The trajectory makes an angle θ with the forward direction. In this helical trajectory, R, λ and θ are related geometrically by $\tan\theta = 2\pi R / \lambda$. By incorporating into the model the photon's experimentally known linear momentum, $p = h/\lambda$ and the photon's experimentally known angular momentum (spin) $s = \hbar$, a second relationship is found: $\tan\theta = \lambda / 2\pi R$. Combining these two relationships containing $\tan\theta$ gives $\lambda = 2\pi R$. This result, combined with the photon's experimentally known energy relationship, $E = h\nu$ where ν is the photon quantum's frequency, leads to the photon model.

The photon model has the following properties:

1) The forward angle, θ of the photon TEQ's helical trajectory is 45°.
2) The radius of the photon TEQ's helical trajectory is $R = \lambda / 2\pi$.
3) The speed of the photon's TEQ is $c\sqrt{2} = 1.414..\, c$ along its helical trajectory.

Using these results, for a right-handed photon traveling in the $+z$ direction, the equations for the trajectory of the transluminal quantum (neglecting a possible phase factor) are:

$$x(t) = \frac{\lambda}{2\pi}\cos(\omega t)\ ,$$
$$y(t) = \frac{\lambda}{2\pi}\sin(\omega t)\ , \qquad (1)$$
$$z(t) = ct\ ,$$

where $\omega = 2\pi f = 2\pi c / \lambda$ is the angular frequency of the photon, f is the photon's frequency in cycles per second and λ is the photon's wavelength. In the transluminal photon model, λ is the pitch of the helix, i.e. the distance along the helical axis corresponding to one rotation of the superluminal quantum along its helical trajectory.

Similarly, for this right-handed photon, the equations for the components of the momentum of the superluminal quantum along its trajectory are:

$$p_x(t) = -\frac{h}{\lambda}\sin(\omega t)\ ,$$
$$p_y(t) = \frac{h}{\lambda}\cos(\omega t)\ . \qquad (2)$$
$$p_z(t) = \frac{h}{\lambda}\ .$$

The x and y components of momentum are 90 degrees out of phase with the x and y position values.

4. The Transluminal-Energy Quantum Model of the Electron

If the open helical trajectory of the photon model is converted into a closed, double-looped helical trajectory, the TEQ is given an electric charge, $-e$, and several helical parameters corresponding to an electron's experimental and theoretical properties are set, we get the transluminal energy quantum model of the electron. Besides having the electron's experimental spin value and the magnetic moment of the Dirac electron, the transluminal energy quantum model of the electron, described below, quantitatively embodies the electron's *zitterbewegung*.

Zitterbewegung refers to the Dirac equation's predicted rapid oscillatory motion of a free electron that adds to its center-of-mass motion. No size or spatial structure of the electron has so far been observed experimentally. High-energy electron scattering experiments by Bender *et al.* [16] have put an upper value on the electron's size at about 10^{-18} m.

Yet the Schrödinger [3] *zitterbewegung* results suggest that the electron's rapid oscillatory motion has an amplitude length of $R_{zitt} = \hbar / 2mc = 1.9 \times 10^{-13}$ m and an angular frequency of, $\omega_{zitt} = 2mc^2 / \hbar = 1.6 \times 10^{21} / \sec$, twice the angular frequency $\omega_0 = mc^2 / \hbar$ of a photon whose energy is that contained within the rest mass of an electron. Furthermore, in the Dirac equation solution the electron's instantaneous speed (its eigenvalue solution) is *c*, although experimentally observed electron speeds are always less than *c*. An acceptable spatio-temporal model of the electron would presumably contain these *zitterbewegung* properties.

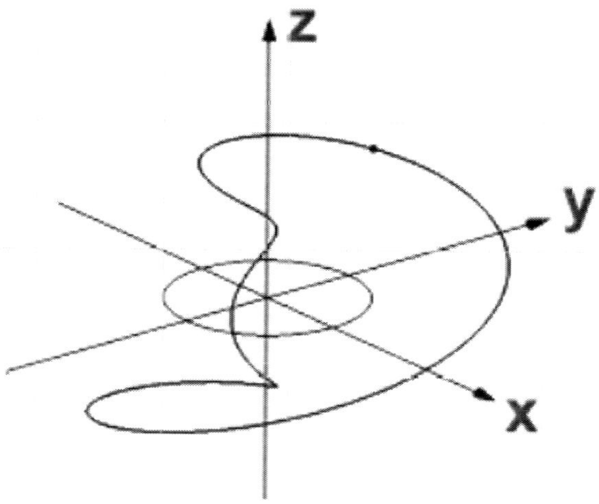

(a) Trajectory of the TEQ electron model.

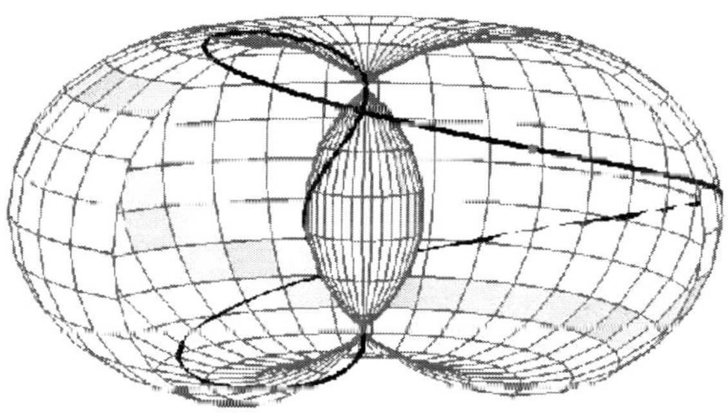

(b) Trajectory of the TEQ positron model.

Figure 1. Trajectories of the transluminal energy quantum of (a) the electron model and (b) the positron model, (the latter showing a cut-away mathematical spindle-torus surface on which the two trajectories are located. (The circle in the x-y plane in (a) is the axis of the TEQ's closed double-looped helical trajectory.)

In the TEQ model of the electron, the electron is composed of a negatively charged point-like TEQ moving along a closed, double-looped helical trajectory in the electron model's rest frame, that is, the frame where the TEQ's trajectory closes on itself. The TEQ electron model structurally resembles a closed, double-looped circulating TEQ photon model having angular frequency, $\omega_0 = mc^2 / \hbar$ and wavelength, $\lambda_C = h / mc$ (the Compton wavelength). The electron model's TEQ moves in a closed double-looped helical trajectory having a circular axis of circumference, $\lambda_C / 2$. The radius of the TEQ trajectory's closed double-looped circular axis is set to be, $R_0 = \frac{1}{2}\hbar / mc = 1.9 \times 10^{-13}$ m.

This gives the TEQ model the electron's spin $s = R_0 \times p_C = (\hbar / 2mc) \times mc = \hbar / 2$, where $p_C = mc$ is the linear Compton momentum of the circulating TEQ photon-like object composing the TEQ electron model. The TEQ trajectory's helical radius is set to be, $\sqrt{2} R_0$.

This gives the TEQ electron model a magnetic moment whose z-component is equal in magnitude to the Bohr magneton (See Appendix). After following its helical trajectory around this circular axis once, the electron TEQ's trajectory is 180° out of phase and doesn't close on itself. But after the TEQ traverses its helical trajectory around the circular axis a second time, the TEQ's trajectory is back in phase with itself and closes upon itself. The total longitudinal distance along its circular axis that the circulating TEQ has traveled before its trajectory closes is λ_C.

In its rest frame, the electron's TEQ carries energy, $E = \hbar \omega_0 = mc^2$, while the TEQ carries the electron's negative charge $-e$. The coordinates for the closed, double-looped helical spatial trajectory for the TEQ in the electron model can be expressed in rectangular coordinates by:

$$
\begin{aligned}
x(t) &= R_0 (1 + \sqrt{2} \cos(\omega_0 t)) \cos(2\omega_0 t) , \\
y(t) &= R_0 (1 + \sqrt{2} \cos(\omega_0 t)) \sin(2\omega_0 t) , \\
z(t) &= R_0 \sqrt{2} \sin(\omega_0 t) ,
\end{aligned}
\qquad (3)
$$

where, $R_0 = \hbar / 2mc$ and $\omega_0 = mc^2 / \hbar$. These equations correspond to a left-handed photon-like object of wavelength, λ_C, circulating counterclockwise (as seen above from the +*z* axis) in a closed double loop. An image of the electron TEQ's trajectory described by equation (3) and the trajectory of a mirror-image TEQ positron are shown in Figure 1.

The velocity components of the electron model's transluminal energy quantum are obtained by differentiating the position coordinates of the TEQ in equation (3) with respect to time, giving:

$$v_x(t) = -c[(1+\sqrt{2}\cos\omega_o t)\sin 2\omega_o t + \frac{\sqrt{2}}{2}\cos 2\omega_o t \sin\omega_o t],$$
$$v_y(t) = c[(1+\sqrt{2}\cos\omega_o t)\cos 2\omega_o t - \frac{\sqrt{2}}{2}\sin 2\omega_o t \sin\omega_o t],$$
$$v_z(t) = c\frac{\sqrt{2}}{2}\cos\omega_o t\ . \qquad (4)$$

From equation (4) it is found that the maximum speed of the electron's TEQ is 2.516c, while its minimum speed is $c/\sqrt{2}$ = .707c. A graph of the speed of the electron's TEQ versus the TEQ's angle of rotation in the x-y plane as the electron's TEQ circulates in its closed double-looped helical trajectory is shown in Figure 2. The TEQ completes each trajectory cycle in 12.56 or 4π radians.

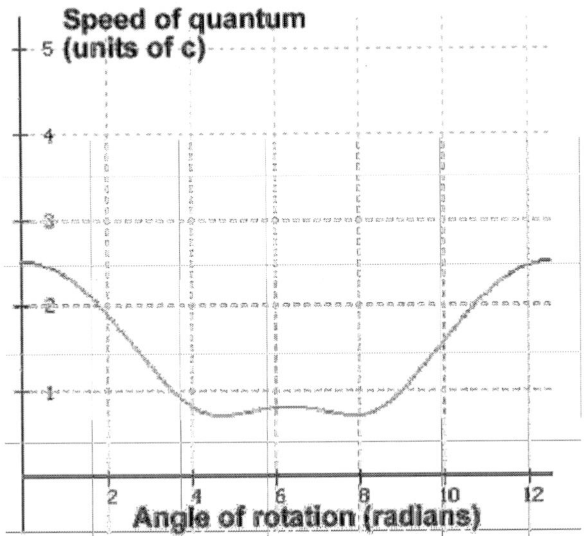

Figure 2. The speed of the electron model's TEQ along its double-looped helical trajectory as a function of the angle of rotation of the TEQ in the x-y plane.

From equation (4) the circulating transluminal energy quantum spends approximately 57% (more precisely 56.64%) of its time traveling superluminally along its trajectory and 43% (more precisely 43.36%) of its time traveling subluminally. The TEQ passes twice through the speed of light c while completing one closed helical trajectory. This passage of the TEQ from superluminal speeds through c to subluminal speeds and back again to superluminal speeds is not a problem from a relativistic perspective. This is because it is the TEQ's electric charge *-e* that is moving at these speeds and not the average center of mass/energy of the TEQ electron model, which remains at rest in the electron model's rest frame. The TEQ electron model as a whole would move at a speed less than c.

5. Similarities Between the Dirac Equation's Free-Electron Solution and the Transluminal-Energy Quantum Model of the Electron

The transluminal energy quantum model of the electron shares a number of quantitative properties with the Dirac equation's electron with *zitterbewegung*:

1) The *zitterbewegung* internal frequency of $\omega_{zitt} = 2mc^2/\hbar = 2\omega_o$.
2) The *zitterbewegung* radius $R_0 = \hbar/2mc = R_{zitt}$. Using the equations (3) for *x(t), y(t) and z(t)* of the TEQ electron model, the *x*, *y* and *z* rms (root mean square) values, which are the values of Δx, Δy and Δz in the Heisenberg uncertainty relation, are all calculated to be $R_0 = \hbar/2mc$, where R_0 is the radius of the closed helical axis of the TEQ electron model. This is also the value of the amplitude of the Dirac electron's jittery motion found by Schrödinger [3]. These rms results for the electron model are only obtained when the radius of the TEQ's double-looped helix is $\sqrt{2}R_0$ as in equation (3). This value $\sqrt{2}R_0$ is the helical radius required to give the electron model's z-component of its magnetic moment a magnitude equal to the Dirac equation's electron magnetic moment magnitude of one Bohr magneton M_B. (See Appendix.)
3) The Dirac electron's speed-of-light result is also contained in the TEQ electron model. The longitudinal speed of the TEQ along the circular helical axis is c, although the TEQ's actual speed is transluminal.
4) The prediction of the electron's antiparticle the positron, having opposite helicity to the electron model's helicity.
5) The calculated spin of the electron, which comes out correctly in the TEQ model due to the double-looping of the TEQ's helical trajectory and its circular axis.
6) The calculated Dirac magnetic moment of the electron $M_z = -M_B$ (M_B is the Bohr magneton). In the TEQ electron model, $M_x = 0$ and $M_y = -.25M_B$, which differs from the Dirac result. But it is the z-component of an electron's magnetic moment that is experimentally measured.
7) The electron's motion is the sum of its center-of-mass motion and its *zitterbewegung*.
8) The non-conservation of linear momentum in the *zitterbewegung* of a free electron, a result first pointed out for the Dirac electron by de Broglie [17].

6. The Heisenberg Uncertainty Principle and the Transluminal-Energy Quantum Models of the Photon and the Electron

With the transluminal energy quantum model of the photon, the TEQ would be the particle that is actually detected when a single photon is detected in an experiment. Suppose a TEQ photon is traveling in the $+z$ or longitudinal direction. Because of the TEQ photon's varying transverse x and y position and momentum components as it moves along its open helical trajectory, a range of values of these x and y components would be detected when various photons traveling in the $+z$ direction are measured successively. A remarkable aspect of the TEQ model of the photon is that the TEQ's transverse position and momentum variabilities are found to quantitatively match those in the Heisenberg uncertainty relation. This relation says that there is a fundamental limitation on the accuracy of simultaneously measuring two related physical properties, such as the corresponding position and momentum components of an elementary particle or other physical object. Greater accuracy in measuring one of the two properties entails a corresponding lesser accuracy in measuring the corresponding property.

The Heisenberg uncertainty relationship for the x coordinate of a particle is stated precisely as $\Delta x \Delta p_x \geq h/4\pi$, where Δx is the standard error or rms value (the square root of the statistical variance) in measuring the position of the particle along the x direction, Δp_x is the standard error or rms value in measuring the particle's momentum along the same x dimension, and h is Planck's constant. How does the Heisenberg uncertainty relation apply to detecting a photon in the TEQ photon model? Using equation (1) for transverse x-position of the TEQ and equation (2) for the transverse x-momentum of the TEQ in the photon model, the rms value or Δx for $x(t)$ is found to be:

$$\Delta x = \frac{1}{\sqrt{2}} \frac{\lambda}{2\pi}, \tag{5}$$

while the rms value Δp_x for $p_x(t)$ is found to be:

$$\Delta p_x = \frac{1}{\sqrt{2}} \frac{h}{\lambda}. \tag{6}$$

Multiplying these rms values Δx and Δp_x for the TEQ model of the photon gives:

$$\Delta x \Delta p_x = \left(\frac{1}{\sqrt{2}} \frac{\lambda}{2\pi}\right)\left(\frac{1}{\sqrt{2}} \frac{h}{\lambda}\right) = \frac{h}{4\pi}. \tag{7}$$

Comparing this result with the Heisenberg uncertainty relation:

$$\Delta x \Delta p_x \geq \frac{h}{4\pi}, \tag{8}$$

we see that the uncertainty product of the transverse or x components of position and momentum for the superluminal quantum in the photon model is exactly the minimum value allowed by the Heisenberg uncertainly relation. The same quantitative results are found for Δy and Δp_y. For the TEQ photon model:

$$\Delta y \Delta p_y = \frac{h}{4\pi}. \tag{9}$$

Any real photon will have a finite value of uncertainty in the measured coordinates of both its position and its momentum. In quantum mechanics, a photon, until it is detected, is described by a mathematical superposition of position states or their corresponding momentum states, each corresponding to a particular wave function with a particular amplitude, frequency and phase. This total quantum wave function describing the photon is then related to the probability of detecting the photon in the regions where the total wave function is non-zero. The transluminal energy quantum model of the photon seems to be consistent with the quantum mechanical interpretation of the photon and the Heisenberg uncertainty principle.

Similarly, the electron is modeled as a helical photon-like object moving in a double-looped circle at the *zitterbewegung* angular frequency $\omega_{zitt} = 2mc^2/\hbar$ with a forward velocity c and Compton momentum $p_C = h/\lambda_C = mc$ (where λ_C is the Compton wavelength h/mc of the photon-like object composing the electron.) This circle (see Figure 1) is the circular axis (of radius $R_{zitt} = \hbar/2mc$) of the TEQ electron model's helix. The rms values for position in the electron model in the x, y and z directions all give $R_{rms} = \hbar/2mc$ as mentioned earlier. Combining this rms position result with the calculated rms value $mc/\sqrt{2}$ for the circulating momentum mc in the x and y directions gives:

$$\Delta x \Delta p_x = (\hbar/2mc)(mc/\sqrt{2}) = .707(\hbar/2) = .707\frac{h}{4\pi} \tag{10}$$

$$\Delta y \Delta p_y = (\hbar/2mc)(mc/\sqrt{2}) = .707(\hbar/2) = .707\frac{h}{4\pi} \tag{11}$$

The relations in (10) and (11) for the electron model contain a value .707 times the minimum value found in the Heisenberg uncertainty relation. Therefore these position/momentum relations for the electron model would not be experimentally detectable in principle, according to the Heisenberg uncertainty relation. Furthermore, since these relations are not detectable in principle, the TEQ electron model would not violate conservation of linear momentum, even though its internal momentum vector rotates at the *zitterbewegung* angular frequency $\omega_{zitt} = 1.6\times10^{21}/\text{sec}$.

Yet the electron model has the electron's spin value

$$s = R_{zitt} p = (\hbar / 2mc)(mc) = \hbar / 2,$$

which is detectable. This rotating internal Compton momentum $p_C = mc$ could give rise to the TEQ electron's rest mass m, and therefore the electron's inertia.

7. The Electron Model and Inertia

The TEQ electron model may provide a new approach to understanding the nature of inertia. The electron is modeled as a double-looping circulating photon-like object having an internal Compton wavelength $\lambda_C = h / mc$ and therefore a Compton momentum $p_C = h / \lambda_C = mc$. This internal Compton momentum is rotating at the *zitterbewegung* angular frequency $\omega_{zitt} = 2mc^2 / \hbar = 1.6 \times 10^{21} / \text{sec}$. The well-known relativistic equation relating the total energy E of an electron to its linear momentum p and its mass m is

$$E^2 / c^2 = p^2 + (mc)^2 = p_x^2 + p_y^2 + p_z^2 + (mc)^2.$$

In the TEQ electron model, an electron with a 'rest mass' m is never internally at rest. The TEQ electron model has a rotating internal linear momentum $p_C = mc$ which is mathematically on an equal footing with the electron's three external linear momentum components, p_x, p_y and p_z. (It is the rapidly rotating internal linear momentum $p_C = mc$ in the TEQ electron model that gives the electron its spin $\hbar / 2$.) The total relativistic energy E of the TEQ electron model is then given by

$$E^2 / c^2 = p^2 + p_C{}^2 = p_x^2 + p_y^2 + p_z^2 + p_C{}^2.$$

It may be that the rotation of an electron's internal Compton momentum $p_C = mc$ at the *zitterbewegung* frequency $\omega_{zitt} = 2mc^2 / \hbar = 1.6 \times 10^{21} / \text{sec}$ is what gives an electron its inertia, that is, its resistance to being accelerated by an applied external force. The inertia of an electron (as measured by its mass m) is then related directly to its internally rotating Compton momentum $p_C = mc$ and only indirectly to the electron's "rest energy" $E = mc^2 = p_C c$ (when $p_x = p_y = p_z = 0$). Momentum is often described as 'inertia in motion'. With the TEQ electron model this can now be turned around: inertia is 'momentum at rest', where the 'at rest' momentum is $p_C = mc$.

8. Testing the Photon and the Electron Models

Since equations (7) and (10) show that the variation in the transverse position and momentum components of the TEQ photon model is at the exact limit of the Heisenberg uncertainty relation, it appears that the photon model can at least in theory be tested by measuring a photon's transverse position and momentum. Knowledge of the phase of the photon's TEQ in equations (1) and (2) would permit a theoretical specification of the TEQ's instantaneous position and momentum. Perhaps such phase relations could be tested experimentally using two-photon coincidence counts as proposed by Sivasubramanian [13]. Another approach to testing the proposed $\lambda / 2$ radius of the photon model is by analyzing the cutoff frequencies for microwaves transmitted in waveguides of different sizes, as did Ashworth [11]. The prediction is that a waveguide will not transmit microwaves as well, as indicated by the waveguide cutoff frequency, if the diameter of the waveguide is less than the diameter of a microwave's photon. Although the cutoff frequencies of waveguides are related to wavelength λ for rectangular waveguides, the possible relation of cutoff frequencies to $\lambda / 2\pi$ is not so straightforward for other geometries and could be researched further, perhaps using waveguides with non-standard geometries.

There could be tests of the TEQ electron model's closed double-looped helical structure. A TEQ electron and a TEQ positron would differ in the direction of their internal helicities. If the electron were structured like a circulating left-handed photon, then a positron would be structured like a circulating right-handed photon (see Figure 1). Electrons and positrons could therefore perhaps differentially absorb, scatter or otherwise interact with differentially polarized gamma photons, for example with differentially polarized gamma photons having energies corresponding to the rest mass of electrons.

It is proposed that the TEQ electron may generate its inertia by the rapid rotation of its internal Compton momentum $p_C = mc$ at the electron's *zitterbewegung* frequency. This rotating momentum is associated with the circulation of the TEQ's point-like charge at the same high frequency. Subatomic effects could show themselves at the macroscopic level, for example as in magnetic materials. The rotation of a macroscopic physical object at a particular frequency could shift the internal rotational rate of the Compton momentum of electrons by a corresponding frequency, depending on the alignment of the electrons with the rotational direction of the macroscopic object. This leads to a testable prediction that the inertia of the rotating object could change with its angular velocity. The object could become more or less massive, with a correspondingly larger or smaller weight. The inertia explanation in the TEQ electron model can also lead to new proposals for experiments in inertia alteration, which could be then subjected to experimental tests. Positive results would of course lead to a better understanding of inertia, while negative results would also be informative.

9. Conclusions

The photon and the electron are modeled as helically circulating point-like transluminal energy quanta having both particle-like (E and $\vec{P}$) and wave-like (ω and λ) characteristics. The number of quantitative similarities between the Dirac equation's electron and the TEQ electron, and between the Heisenberg uncertainty principle and the TEQ photon model of the photon are remarkable, given the relatively simple mathematical form of the TEQ models of the electron and the photon. A new approach to understanding inertia as "momentum at rest" is presented.

Appendix: Finding TEQ Electron's Helical Radius Giving the Electron's Known Magnetic Moment

For the TEQ electron model's closed, double-looped helical trajectory, the model's magnetic moment M_z is calculated from the formula for the total magnetic moment $\vec{M}$ of a 3-dimensional closed current loop. The loop carries a current I caused by the motion of the TEQ's point charge –e around its closed trajectory in a period $T = h/mc^2 = 2\pi/\omega_o$. This gives a current $I = Q/T = -e/(2\pi/\omega_o) = -e\omega_o/2\pi$. The position and velocity components of the TEQ are given in equations (3) and (4). Using these equations, in vector notation the position $\vec{r}(t)$ and velocity $\vec{v}(t)$ of the circulating charged TEQ in the electron model are given by

$$\vec{r}(t) = x(t)\hat{i} + y(t)\hat{j} + z(t)\hat{k}$$
$$= R_x(t)\hat{i} + R_y(t)\hat{j} + R_z(t)\hat{k}$$

and

$$\vec{v}(t) = v_x(t)\hat{i} + v_y(t)\hat{j} + v_z(t)\hat{k}$$
$$= V_x(t)\hat{i} + V_y(t)\hat{j} + V_z(t)\hat{k}$$

The magnetic moment $\vec{M}$ and its component M_z for a rapidly circulating charge in a current loop are given by

$$\vec{M} = \frac{I}{2}\int_{t=0}^{T} \vec{r}(t)\times d\vec{r}(t) \quad \text{where } T = h/mc^2$$

$$\vec{M} = \frac{I}{2}\int_{\theta=0}^{2\pi} \vec{r}(\theta)\times \frac{d\vec{r}(t)}{dt}\frac{dt}{d\theta}d\theta$$

$$\vec{M} = \frac{I}{2}\int_{\theta=0}^{2\pi} \vec{r}(\theta)\times \vec{v}(\theta)\frac{1}{\omega_0}d\theta \quad \text{since } \theta = \omega_0 t$$

$$\vec{M} = \frac{1}{2}\times\frac{-e\omega_0}{2\pi}\int_{\theta=0}^{2\pi} \vec{r}(\theta)\times \vec{v}(\theta)\frac{1}{\omega_0}d\theta \quad \text{since } I = \frac{-e\omega_0}{2\pi} \tag{12}$$

$$\vec{M} = \frac{1}{2}\times\frac{-e\omega_0}{2\pi}\int_{\theta=0}^{2\pi} \vec{r}(\theta)\times \vec{v}\frac{1}{\omega_0}d\theta$$

$$\vec{M} = \frac{-e}{4\pi}\int_{\theta=0}^{2\pi} \vec{r}(\theta)\times \vec{v}(\theta)d\theta$$

$$M_z = \frac{-e}{4\pi}\int_{\theta=0}^{2\pi} (R_x(\theta)V_y(\theta) - R_y(\theta)V_x(\theta))d\theta$$

To obtain the radius R of the electron model's closed helix that would yield the electron's Dirac magnetic moment $M_z = -e\hbar/2m$, the value of M_z in equation (12) was set to be $M_z = -e\hbar/2m$, which is the negative of the value of the Bohr magneton $M_B = 9.274\times10^{-24}\,\text{JT}^{-1}$ (which is the magnitude of the magnetic moment of the Dirac electron). Equations (3) and (4) for the TEQ's position and velocity coordinates were then used with $\sqrt{2}R_0$ replaced by the variable R (since the TEQ helical radius value $\sqrt{2}R_0$ required to get the electron's magnetic moment was initially unknown). The equation for M_z in equation (12) was then solved for the value of R that is the radius of the TEQ's closed helical trajectory that yields $M_z = -e\hbar/2m$ for the TEQ electron model. The solution is $R = \sqrt{2}R_0$ where $R_0 = \frac{1}{2}\hbar/mc$. It is this helical radius $R = \sqrt{2}R_0$ that was then finally included in equation (3) to obtain the trajectory of the TEQ model of the electron.

Using equation (3), the values for M_x and M_y components of $\vec{M}$ for the TEQ electron model can then be found from

$$\begin{aligned} M_x &= \frac{-e}{4\pi}\int_{\theta=0}^{2\pi} (R_y(\theta)V_z(\theta) - R_z(\theta)V_y(\theta))d\theta = 0 \\ M_y &= \frac{-e}{4\pi}\int_{\theta=0}^{2\pi} (R_z(\theta)V_x(\theta) - R_x(\theta)V_z(\theta))d\theta = .25M_B \end{aligned} \tag{13}$$

The values of M_x and M_y depend on the way the x and y axes for of the electron model's helical trajectory are defined in equation (3). A rotation of the x and y axes for these equations by +90 degrees around the *z-axis* would give $M_x = -.25M_B$ and $M_y = 0$, while leaving unchanged $M_z = -M_B$.

The value of the experimental helical radius R_{exp} of the transluminal energy quantum electron model corresponding to the precise experimental value of the electron's z-component of magnetic moment can also be calculated from equations (12). The electron's precise experimentally measured value M_z of its z-component of magnetic moment is $M_z = -(g/2)M_B$ where $g/2$ = 1.00115965219 and M_B is the Bohr magneton. Set M_z in the last line of equations (12) above to $-(g/2)M_B$. To calculate R_{exp} set the helical radius in the electron model equations (3) to $R_{exp} = F_{exp}R_0$, where F_{exp} is the fraction that the required new R_{exp} is, compared to R_0. $F = \sqrt{2}$ in the electron model equations (3), so this means replacing $\sqrt{2}$ by F_{exp} in equations (3) in order to solve for the value of F_{exp} corresponding to the electron's experimental value of M_z. Set $\omega_0 t = \theta$ in the x and y components of the electron model equations (3). Solve the bottom line in equations (12) for F_{exp}. The solution for F_{exp} using the equation for M_z in equation (12) is found to be $F_{exp} = 2\sqrt{(g/2) - .5}$, or F_{exp} = 1.41585260842, as compared to $F = \sqrt{2}$ for the electron model having its magnetic moment's z-component set exactly equal to $-M_B$ (i.e. with $g/2 = 1$ exactly). This new value of F_{exp} gives the percentage increase of the helical radius of the electron model having the exact experimental value of the electron's z-component of magnetic moment, compared to the helical radius corresponding to $F = \sqrt{2}$, corresponding to the exact Bohr magneton value, to be $(F_{exp} - \sqrt{2})/\sqrt{2} \times 100 = 0.115898058\%$.

References

[1] Dirac, P.A.M. (1928a) The Quantum Theory of the Electron, Part 1, Proc. Roy. Soc., **A117**, 610-624.
[2] Dirac, P.A.M. (1928b) The Quantum Theory of the Electron, Part 2, Proc. Roy. Soc., A118, 351-361.
[3] Schrödinger, E. (1930) Non powerful motion in the relativistic quantum mechanics, *Sitzungsb. Preuss. Akad. Wiss. Phys.-Math. Kl.* 24, 418-428.
[4] Barut, A.O. & Bracken, A.J. (1981) *Zitterbewegung* and the Internal Geometry of the Electron, *Phys.* Rev. *D* 23, 2454-2463.
[5] Barut, A.O. & Thacker, W. (1985) Covariant generalization of the *Zitterbewegung* of the electron and it's SO(4,2) and SO(3,2) internal algebras, Phys. Rev. *D* 31, 1386-1392.
[6] Hestenes, D. (1973) Local observables in the Dirac theory, J. Math. Phys. 14, 893-905.
[7] Hestenes, D. (1983) Quantum mechanics from self-interaction, Found. Phys., 15, 63-87.
[8] Hestenes, D. (1993) The *Zitterbewegung* Interpretation of quantum mechanics, Found. Phys. 20, 1213-1232, (1990).
[9] Hestenes, D. (1997) *Zitterbewegung* modeling, Found. Phys. 23, 365-387.
[10] Bohm, D. & Hiley, B. J. (1993) The Undivided Universe: An Ontological Interpretation of Quantum Theory, London: Routledge, pp. 218-222.
[11] Ashworth, R.A. (1998) Confirmation of helical travel of light through microwave waveguide analyses, Phys. Essays 11, 343-352.
[12] Kobe, D.H., (1999) Zitterbewegung of a photon, Phys. Let. A 253, 7-11.
[13] Sivasubramanian, S. et al. (2004) Non-commutative geometry and measurements of polarized two photon coincidence counts, Annals Phys. 311, 191-203.
[14] Gauthier, R. (2006) Superluminal quantum models of the electron and the photon, http://www.superluminalquantum.org/dallaspaper, (accessed 12/30/2012.)
[15] Gauthier, R. (2007) FTL quantum models of the photon and the electron, CP880, Space Technology and Applications International Forum--STAIF-2007, M.S. El-Genk (ed.), http://superluminalquantum.org/STAIF-2007article.
[16] Bender, D. et al. (1984) Tests of QED at 29 GeV center-of-mass energy, Phys. Rev. D 30, 515-527.
[17] De Broglie, L. (1934) *L'Electron Magnetique*, Paris: Hermann et Cie, pp. 294-295.

Thoughts Occasioned by the 'Big Bang' Model of the Universe

JEREMY DUNNING-DAVIES

Institute for Theoretical Physics and Advanced Mathematics (IFM) Einstein-Galilei, Via Santa Gonda, 14 – 59100 Prato, Italy

and

Institute for Basic Research, Palm Harbor, Florida, U.S.A.

j.dunning-davies@hull.ac.uk

For many years now, the so-called 'Big Bang' model for the beginning and evolution of the Universe has been accepted almost without question. In fact, any who have dared question its authenticity has been condemned to a sort of scientific outer darkness. However, what is the truth in this matter? Here some brief thoughts on this problem will be laid out for people to ponder at leisure.

Keywords: Big Bang, Cosmology, Redshift, Steady-state cosmology.

1. Introduction

On the very first page of his book, Before the Beginning, the Astronomer Royal, Lord Rees, states categorically that "Our universe sprouted from an initial event, the 'big bang' or fireball". What a truly amazing statement with which to begin any book or piece of writing. However, is it true?

The big bang as a valid model of the Universe has been under close scrutiny almost since it was proposed and many of the queries concerning it remain. These queries tend to be 'swept under the carpet' but in a rather subtle way. The rise of popular science books has provided a means whereby the general public is persuaded to believe in the ideas accepted as founding 'conventional wisdom'. This has been supported by a proliferation of carefully constructed, well presented public lectures. The 'solutions' to various problems are presented as indisputable facts; the notion that other possible explanations exist is carefully suppressed, or in the case of the steady-state theory mildly ridiculed.

The whole idea of the big bang goes back to the theoretical investigations of Alexander Friedmann and Georges Lemaître [1] in the earlier years of the last century following Einstein's publication of his General Theory of Relativity. Its movement to a position of prominence, if not pre-eminence, in cosmology might be felt to have been brought about by its eloquent advocacy at the hands of George Gamow[2] in the mid to late 1940's, ably supported by such as J. Robert Oppenheimer. It is quite widely claimed that the standard big bang model makes three major predictions which have been verified observationally. If that were true beyond all reasonable doubt, it would indeed be a theory to take very seriously. However, are these claims unquestionably true? First, it is claimed that the model predicts distant galaxies receding from one another at speeds proportional to the distance between them. This view is supposedly supported overwhelmingly by Hubble's discovery of the redshift of light from celestial objects in the 1920's. Secondly, the model is claimed to predict the existence of background radiation which is seen as a remnant of the original big bang. Support for this comes from the detection of the cosmic background radiation by Arno Penzias and Robert Wilson in 1965[3]. Some also claim that the relatively recent examination of the properties of this background radiation by the COBE satellite again confirm totally the predictions of the big bang. Thirdly, the model is said to predict successfully the abundances of the light elements such as helium, deuterium and lithium. At the same time, these claims are taken to imply that no other theory can explain these phenomena and there are no doubts about these deductions from the basic idea of the big bang. It goes almost without saying that the interpretation of experimental and observational results which leads to confirmation of the 'truth' of the big bang theory is accepted without question. However, is the situation quite as clear cut as that? Are all the questions answered, and answered both successfully and correctly?

As far as modern ideas are concerned, one of the first major advances came with Hubble's evidence that three nebulae, M31, M33 and NGC6822, were to be found at distances far beyond the remotest parts of our own galaxy. It was accepted that these were totally separate from the Milky Way. Not long after establishing that these nebulae were extragalactic systems, he also showed that the redshift of their spectral lines increased with distance. Utilizing the most obvious interpretation of redshift, that is that it is a Doppler shift occasioned by the recession of the source, it is easily seen that Hubble's result may be taken to indicate that the Universe is expanding and the most distant galaxies are receding fastest. By looking at things in reverse, this is seen to mean that the Universe was much denser in the past and there is a tendency to extrapolate back to claim that, at some distant time, all the matter in the Universe was so highly compressed that it was all confined to a single point! At this point the 'Cambridge Encyclopedia of Astronomy' comes into its own as far as fair, scientific examination of this issue is concerned. It claims that great care should be taken, since, "it is possible that the simple interpretation of the redshift is not correct, and that the expansion is illusory." Even if the fact of expansion is accepted, "it does not necessarily follow that the Universe was denser in the past than now, for implicit in that conclusion is the assumption that matter in the Universe is neither created nor destroyed." However, it is pointed out also that the hypothesis that the redshift is a Doppler shift occasioned by recession of the galaxies is acceptable scientifically since it is consistent with the known laws of physics. Reflecting the time of writing, it is claimed that "no other scientifically acceptable hypothesis has yet been proposed" but it does note that, as far as the position existing at that time was concerned, there was no proof that that was the true explanation.

The encyclopedia article continues by noting that, since the time of Hubble's original hypothesis, many more observations had been made which served to confirm his postulated relationship between distance and velocity of recession. It is claimed that no obvious deviations from the simple linear relationship,

Velocity = Hubble parameter × distance in megaparsecs,

have been detected.

Hubble also spent a considerable amount of time investigating the distribution of galaxies in the Universe. Obviously, such observations were restricted by the instrumentation available but, nevertheless, he noted that, on very large scales, the Universe does appear homogeneous; there is no obvious sign of diminution of numbers of galaxies as the accessible limits of the Universe are approached. Also, the Universe was found to look more or less the same in all directions and the cosmic expansion seemed to be proceeding at the same rate in all directions; that is, the Universe is said to be isotropic. All this is taken to mean that there is no meaningful centre for our Universe and as confirmation that our own galaxy, the Milky Way, certainly occupies no privileged position within the Universe. Strong confirmation for the isotropic nature of the Universe is felt to be provided by the so-called cosmic background radiation, a component of radiation found by radio astronomers which is itself isotropic to a very high degree and is inexplicable as noise within receiving systems or as originating from any known radio sources. This radiation is, of course, that background radiation mentioned earlier. Since Hubble's time, however, observing equipment has changed for the better and systems are now observed quite regularly which emit far more radiation than many of those observed by Hubble. One important class of objects to be considered here is provided by the quasars; the most 'distant' quasars are thought to have redshifts far in excess of those for the furthest galaxies. It is accepted by many that there were far more quasars and, indeed, radio galaxies in the past than there are now. This, if true, implies that, in the past, the Universe was different from now and this seems to pose a serious problem for the Steady State Theory of the Universe, as well as offering extremely strong support for alternatives, especially the big bang model. However, this whole question is, or should be, a completely open one. Many seem to give the impression that everything in this area is absolutely clear cut and anyone opposing the generally accepted view is to be ignored as lacking in understanding of the truth. Frankly this appears to be the view adopted in the corridors of conventional wisdom towards the work and ideas of Halton Arp.

2. The Work of Halton Arp

While able to make use of the most powerful of telescopes, Arp also discovered that many pairs of quasars, or more correctly quasi-stellar objects, which possess extremely high redshift values appear to be associated physically with galaxies having much lower redshift values; galaxies, in fact, which are known to be much closer to the earth than the redshift values of the quasars concerned would imply. This all follows from the Hubble law which indicates that objects having high redshift values must be receding from the earth very quickly and, therefore, must be found at large distances from the earth. Hence, Arp was faced with the intriguing question of how objects with totally different redshift values, objects which according to 'conventional wisdom' had to be located at totally different distances from the earth, could be physically associated – in some instances, Arp's photographs showed a physical bridge between the quasars and the associated galaxy. As has been recorded many times,

Arp has many photographs of pairs of quasars, with high redshifts, symmetrically located on either side of low redshift galaxies. It has to be noted that these pairings occur far more often than the probability of random placement would allow. Of course, the main problem with Arp's photographs is that according to big bang theorists, high redshift objects must be at a great distance from the earth; to them high redshift is effectively a measurement of distance from the earth. It is often claimed by the advocates of 'conventional wisdom' that Arp's statistical analysis is in error; after many years, this still seems to be the main line of attack on his work. However, from all the accumulated evidence it seems there is no satisfactory foundation for criticizing Arp's work on the basis of the statistics involved, and that seems to be the only criticism actually offered. Much of Arp's work is well-documented in his book "Seeing Red" and reference should be made to this work for further details of the points involved.

As indicated, this work of Arp's has not been welcomed by the orthodox astronomical community because, if accepted, it casts severe doubt on the assumption, which is quite basic to big bang theory and, therefore, to most if not all of accepted cosmological theory, that objects possessing a high redshift must be far away from the earth. Hence, all the claims of the big bang model which depend on the orthodox interpretation of redshifts must be re-examined. Again, Arp's hypothesis, backed by such eminent physicists as Hoyle, Burbidge and Narlikar, casts doubt also on the notion that black holes lurk at the centre of quasars. However, no black hole has yet been identified beyond reasonable doubt, but, if one did exist, it is assumed that it would be drawing matter to itself rather than ejecting it at very high velocities. So once again, Arp displeases the establishment by proposing a solution to a very real problem which suggests matter being ejected from a central mass rather than absorbed into it. In much current astronomical literature, there seems to be a preoccupation with the death of stars and, in some ways more importantly, with the colliding or merging of galaxies. Arp's view, and one supported by Hoyle and many of his associates, is that it is rather the birth of galaxies that is being witnessed; instead of viewing and contemplating possible collisions, it is rather separations that are being seen. It might be felt that this view is more in keeping with big bang cosmology in that the big bang supporters claim the universe to be expanding and so, everything should be moving farther and farther apart; collisions, it would seem, should be highly improbable occurrences. However, this view is too simplistic and absorbing actions, such as that envisaged by black holes, are readily incorporated into big bang theory. The Arpian view of what is happening is taken to be in direct opposition to the big bang theory, probably because it may be interpreted as implying creation of matter and this notion is, of course, at the heart of the new quasi-steady state theory of Hoyle and his collaborators [4], as well as being seemingly contrary to well-established conservation laws. This quasi-steady state theory is a modification of the old steady state theory proposed by Bondi, Gold and Hoyle [5] many years ago and is a modification proposed in answer to criticisms of the original. It might be argued that they have listened to their critics and attempted to provide an answer. The difference between this modification and changes made to the big bang theory is that, in this case, it seems that the theory was modified but, in the case of the big bang, it seems that, when a problem is pointed out, something is simply added on in an attempt to solve that immediate problem.

As indicated above, at one point in time - actually by about 1950 - there were really two rival theories attempting to explain the origin and workings of the Universe. These were the big bang model and the steady state theory. Both accepted the idea that the Universe was homogeneous, isotropic and was expanding against the pull of gravity. However, the steady state theory assumed that matter could be both created and destroyed spontaneously, whereas the big bang did not. The idea of spontaneously creating or destroying matter challenges widely, and strongly, held views on conservation and so will be anathema to many. On the other hand, one apparently awkward consequence of the big bang is that, at some time in the distant past, all matter seems to have been concentrated in some state of infinite density; that is, a singularity, the cosmic singularity, existed. It is often claimed that this singularity is a serious defect in the big bang theory on philosophical grounds but, in many areas of mathematics and physics, it is more usual to note that a singularity heralds the breakdown of a theory or that there are limits to the range of applicability of a particular theory. It is interesting to realize that, for some reason, no such restriction is imposed in this case or, indeed, in the case of black holes of the type which are said to emerge via the general theory of relativity. In both these cases, attempts are actually made to give physical meaning to mathematical singularities. Apparently, it is this singularity in the case of the big bang which prompted Bondi, Gold and Hoyle to propose the steady state theory in which matter could be created spontaneously at a rate which compensated the reduction in density brought about by the cosmic expansion. Such a Universe would presumably have no beginning or end, it would have both an infinite past and future, but, possibly more importantly, the model would have no singularity.

Considering this latter point concerning the steady state theory, it is interesting to wonder at the possible role played by fundamentalist religion in the seemingly widespread acceptance of the big bang model and the resultant rejection of steady state theory. A moment's reflection indicates that the possibility of such a link is not totally ludicrous. If one considers the first nineteen verses of the King James version of the Bible, the first obstacle to be overcome is the unscientific language used. However, when that is done, it becomes immediately apparent that one valid interpretation of what appears in print is that the Universe was created quite suddenly, spontaneously in fact. The ordering that follows also links quite well with big bang philosophy. It might be argued, quite reasonably, that light would be necessary before grass and fruit trees could exist but, bearing in mind that the ideas, or stories, of Genesis are extremely old and may be interpreted sensibly only as representations produced by people without the benefit of modern scientific knowledge to illustrate, to a scientifically uneducated people, the beginnings of the Universe and of life on earth, the correspondence with the ordering of events according to the big bang theory is remarkably close. It might be noted specifically that even the presence of radiation before the formation of the stars may be inferred from verses fourteen to nineteen inclusive. However, was Genesis ever intended to be taken literally? Was it ever meant to be the literal truth describing the origin of the Universe and life in it? On this question, as with questions of theories of evolution, various views abound. Amongst these, is the view that the answers to the above two questions are in the affirmative. There are, and always have been, people who do believe the book of Genesis to be literally true. Some of these people are, and have been, serious scientists. This may seem almost a contradiction in terms but it is, nevertheless, true. It is, therefore, not difficult to see precisely how the big bang theory will appeal to such people as being the perceived 'Word of God'. It is very easy, but also very unfair, to ridicule such a standpoint, since the obvious temptation is so strong.

This short semi-religious discussion merely serves to raise another question and that is whether, or not, religious fundamentalism played any part – however small – in the acceptance of the big bang theory over the steady state theory? Indeed, it is not unreasonable to wonder if, with the seeming resurgence of religious fundamentalism in present day society, it is one factor keeping the big bang theory so much to the fore. However, whatever the reason, it is still the case that the validity of the big bang theory seems accepted totally without question by much of the world-wide scientific community.

3. Cosmic-Microwave-Background Temperature

One of the most vociferous of early proponents of the big bang theory was George Gamow. He and Ralph Alpher first put the theory forward seriously in 1948 and almost immediately became engaged in a war of words with the supporters of the Steady State Theory. However, Gamow's theory did, apparently, make some important predictions. Namely that there should be an abundance of helium of about twenty-five per cent by mass, and that it should be possible to observe the remnants of the radiation from the early hot phase of the universe's existence and this should be an isotropic radiation field with a black body spectrum with a temperature of a few degrees. The estimates forwarded for this temperature varied, however, between about five degrees and fifty degrees absolute. This was interesting because, as early as 1926, Sir Arthur Eddington [6] had predicted a temperature of space of three degrees absolute, purely on the basis of the radiation received from the stars. This calculation is very crude but the magnitude of the result provides food for thought, if nothing else. Again it might be noted that Eddington was discussing the temperature of interstellar space due to stars in our own galaxy; he was not considering intergalactic space. However, be that as it may, the big bang theory received possibly its biggest boost, both within and without the scientific community, with the discovery by Penzias and Wilson of the cosmic background radiation, - that background radiation which is almost universally recognized nowadays as a left-over of the original big bang. Here it is important first to ask whether or not this discovery of the cosmic background radiation is, in fact, really due to Penzias and Wilson. It must be acknowledged that the existence of this background radiation was not universally recognized at the time of Penzias and Wilson. However, its existence had been detected in the late thirties and early forties by various astronomers. In 1941, McKellar had interpreted the observed data and had shown it to be caused by radiation excitation, which was taken to be black body and the temperature required for the observations to be properly explained was found to be 2.3 $^{\circ}$K. Hence, the detection of the microwave background should more correctly be dated from 1941. It is, in all fairness, understandable that this did not happen. In 1941, the world was in turmoil at the height of the Second World War and McKellar's important work did not appear in a front line journal. However, the truth has been known for some time now. Hoyle, in particular, has not been backward in publicizing its existence. It is to be hoped that McKellar will soon be given the credit he surely deserves.

It might usefully be noted that, long before Gamow and others began to espouse the Big Bang theory, several notable scientists had followed the lead of Guillaume and Eddington and proposed estimates of the temperature of intergalactic space. Following initial work by Millikan and Cameron in which it was deduced that the total energy of cosmic rays at the top of the atmosphere was a tenth of that due to the heat and light emitted by the fixed stars, Regener eventually concluded that both energy fluxes should possess more or less the same value. In an article of 1933, he used this as a basis for deducing a value of 2.8°K as the temperature characteristic of intergalactic space. This work was discussed favorably by no less a person than Walther Nernst who, by 1912 had developed the notion of a stationary state universe. By 1937, he had further developed this and actually proposed a 'tired light' explanation for the cosmological redshift; that is, he suggested that absorption of radiation by an aether which decreased the energy and frequency of galactic light. Whether one accepts or rejects these ideas now, it should be noted that, in all these separate pieces of work, as well as in subsequent examinations by such as Max Born, Stefan-Boltzmann's law, which is characteristic of black body radiation, is of paramount importance. Also, in none of this work, nor in that of McKellar, is any reference made to the big bang theory; it is simply not necessary to introduce it in order to achieve the results cited!

However, nowadays it is the papers by Gamow and by Alpher and Herman, dating from 1948[7], which tend to hold pride of place where discussion of the background temperature is concerned. They pointed out that, if helium was synthesized in the early universe, then, in present times, there should exist a radiation field with a temperature of approximately 5°K. Gamow offered another prediction of the temperature of the background radiation in his 1952 book, The Creation of the Universe, but this time the estimate, which was claimed to be "in reasonable agreement with the actual temperature of interstellar space", was roughly 50 °ak. Nevertheless, it is frequently claimed that Gamow and his collaborators predicted the 2.7 °ak temperature (even though their lowest estimate was in fact 5 °ak) before the 'discovery' of Penzias and Wilson, whereas the steady state theory did not. This was, and is still, hailed as one of the strongest arguments in favor of the big bang model. However, it must not be forgotten that the original steady state theory did not rule out the existence of a background radiation and, as is pointed out in Hoyle's last book, some unpublished calculations by Hoyle, Bondi and Gold, dating from about 1955, indicated a temperature associated with that radiation of 2.78°K. Obviously, revealing this at the time of the supposed discovery of the cosmic background radiation would have produced totally the wrong reaction. However, it is important to note that, from the outset, the steady state theory never ruled out the possibility of there being a background radiation in existence. Therefore, it is obviously totally incorrect to use the existence of this background radiation as a major reason for attempting to discount that theory.

4. The Synthesis of Helium

When introducing the articles by Gamow and by Alpher and Herman above, it was noted that they made reference to the synthesis of helium in the early universe. They were using this to support their claim that, if this were so, a radiation field pervading the whole of space should exist now. This, of course, raises the entire question of the process behind the synthesis of helium and the other light elements. It is of interest to realize that, once again, the papers referred to here were not the earliest attempts to raise this problem. Actually, the earliest article by Gamow appeared in 1946[8]. In it, he argued that, in the early universe, the chemical elements were synthesized by neutron addition. Hoyle also produced his first article on stellar nucleosynthesis in 1946[9] and, interestingly, his view was the direct opposite of that proffered by Gamow. In fact, it is quite widely accepted now that the originator of the theory behind the synthesis of the light elements was Hoyle and a great many people are still puzzled by the fact that he received no part of the Nobel Prize awarded for that work. While the overall thesis of this present work is concerned with the place of accepted 'conventional wisdom' in the scientific world, this treatment of Hoyle inevitably raises the spectre of 'politics' within the scientific establishment. However, now return to the articles by Gamow and by Alpher and Herman. After one or two early hiccups, Gamow and his collaborators produced a theory whose key point was the essential requirement that an amount of helium be synthesized in agreement with the observed value of approximately 0.25 by mass when compared with hydrogen. It might be noted immediately that this fraction is not thought to be constant in time and that alone raises questions. Although it is known that helium is produced from hydrogen in the interior of stars, it was always felt, and still is, that stellar synthesis would make only a negligible contribution to this observed fraction. As has been pointed out by Hoyle and his collaborators, it would take of the order of 1011 years to increase the value of this fraction from zero to 0.25. Around 1950, when these initial calculations were instigated, the Hubble constant was believed to hold a value leading to the age of the universe being only of the order of 109 years. Since this figure was so much less than the time apparently required for the mass fraction of helium to be explainable from astrophysical processes, it was decided that it needed to be explained via primordial synthesis in the very early universe. The first crucial realization to follow this decision was that it could be true only if the energy density of radiation in the early universe was large compared with the rest mass energy of matter. Accepting this was a major change in thinking for many since, up to that point, the opposite had been assumed true. An immediate consequence was that the radiation temperature had to be inversely proportional to the square root of the time. Up to this point, the argument was not unreasonable given the initial assumptions but what followed was a completely ad hoc step and it should be noted that it remains ad hoc today. The mass density of stable non-relativistic particles – neutrons and protons – decreases with the expansion of the universe and Alpher and Herman denoted this by ρ and took

$$\rho = 1.7\times 10^{-2} t^{-\frac{3}{2}} gcm^{-3} \tag{1}$$

Here it is the choice of the coefficient of proportionality as 1.7×10^{-2} which is the ad hoc step. There is absolutely nothing in the theory of the big bang which actually fixes the value of this coefficient. It is a choice made quite freely but a choice which has the enormous, but to many acceptable, effect of ensuring that the calculated value for the mass fraction of helium is indeed 0.25, in accordance with the observed value. This must mean, however, that as the value of the said mass fraction changes, as it surely must over an extended period of time, the value of this constant of proportionality must change also. The obvious question to follow then is, does the big bang theory, therefore, actually predict the correct value for the mass fraction of helium? The answer has to be an emphatic 'No'!

5. The Common Perception

It is, unfortunately, true to note that often, at the end of their undergraduate days, many students emerge totally convinced that the big bang theory correctly describes the beginnings of our universe and also many of its subsequently developed properties. They believe it to be the only theory which explains the cosmic microwave background radiation; they believe it to be the only theory to explain the mass fraction of helium. This, and much more, has all been learnt in undergraduate courses as being absolutely sacrosanct. Further, these beliefs are vigorously supported by so many popular science books, such as Simon Singh's Big Bang, and by many popular science lectures. Young people with impressionable minds leave such talks totally convinced that they have just been exposed to an enunciation of the complete truth regarding the birth of our universe. But have they? They will have been told, amongst other things, that the cosmic background radiation was discovered by Penzias and Wilson in 1965. McKellar's work will have been ignored. The steady state theory will have been dismissed totally with hardly a glance in its direction and no mention will have been made of the newer modified theory. The constant need to add to, and modify, the original Big Bang theory with entities such as dark matter and dark energy will have been glossed over completely. Herein lies a very real danger. The scientists of tomorrow are not being trained to have open questioning minds. Rather they are having their minds programmed to be closed to all thoughts which might possibly conflict with 'conventional wisdom'. The message often appears to be delivered with what amounts to an almost religious fervor, – what might be termed scientific evangelism.

It must be remembered that the steady state theory is still summarily dismissed as a serious attempt to explain the universe in which we exist. However, at this point in time, it should be noted that, even without the latest modifications to the theory, the advocates of steady state had answered many of the criticisms of that theory quite convincingly. The whole history of what Hoyle and his associates term 'the war of the source counts' provides a classic example of this. The details of this controversy are well documented, by those deeply involved on one side of the argument, in 'A Different Approach to Cosmology' by Hoyle, Burbidge and Narlikar [10]. Here it is discussed in detail how, initially, it appeared that the steady state theory indicated incorrect results when it came to examining radio sources and their distribution. Essentially, it seemed that the data collected allowed either of two possible conclusions to be drawn. Ryle and his collaborators at Cambridge took one view; Hoyle subscribed to the alternative. This meant that Ryle and his supporters viewed the data in a way which opposed the validity of the steady state theory. The argument certainly raged fast and furious for many years but, in the end, following queries raised by Robert Hanbury-Brown at the Paris Symposium as early as 1958, the truth finally emerged following work published in 1988. In truth, some objections to the original steady state theory were destroyed at this point. However, this occurred some thirty years after the queries first erupted onto the scientific scene. Too much time had elapsed; too many opinions had been irrevocably formed; there was little or no chance that any change in popular scientific opinion would be accomplished. The modified theory, presented so eloquently in the above-mentioned book, is also not likely to create a revolution in scientific thought on this matter, - at least not immediately. Positions are far too entrenched; too much 'face' – and, possibly more importantly, too many positions of power and influence – would be lost if any senior scientist completed a volte-face on this issue. It is also sad to realize that many have been deterred from studying the steady state theory because it is felt by so many to have been disproved by observations and, therefore, merits no further study. On the face of it, this is a not unreasonable stand-point, but no-one can claim seriously that there is a single undisputed theory describing all aspects of our universe and its origins. True the big bang theory seems, in some ways, the most successful theory so far but, at best, that is all that it is, - the most successful theory so far. In all aspects of science, practitioners should remain open-minded and, in this particular area, more so probably than in others. It is incumbent on all – amateur as well as professional – to keep all options open and that means remaining fully up-to-date and conversant with all of the modified steady state theory, as well as the present version of the big bang.

6. Some Alternative Ideas

Another problem associated with the Big Bang concerns the apparent lack of antimatter in the present Universe. The question of whether or not there is actual predominance of matter over antimatter is not necessarily a trivial one. In the middle of the last century, Hannes Alfvén and Oskar Klein suggested cosmological models which start with perfect symmetry between matter and antimatter. Subsequently in the theories these two components which comprise the Universe separate into matter-dominated and antimatter-dominated regions. Several objections were raised concerning this theory but an important one involved the manner of separation of the regions of matter and antimatter, since it was understood that even intergalactic space contains a small amount of matter and so galaxies could not be completely separate from antigalaxies. Alfvén[11] did propose a possible mechanism for achieving the required separation but most astrophysicists remained skeptical.

The mechanism proposed by Alfvén was effectively a generalization of a phenomenon investigated in the 19th century by a German physician, Johann Leidenfrost. It was noted that, if a drop of water is placed on a surface whose temperature is in the region of 100oC, it will evaporate almost immediately. However, if the surface temperature is several hundred degrees, the drop does not boil off immediately; rather it becomes smaller gradually before disappearing completely. The explanation is that, at the higher temperature, as the drop evaporates, a layer of steam forms between the drop and the surface and this layer acts to insulate the drop from the surface so that heat is conveyed from the surface to the drop more slowly. Alfvén's idea was that a similar situation might exist in some circumstances between matter and antimatter.

Another model introduced just a little later in the 1970's by Omnès, Stecker and others had as an initial state a mixture of matter and antimatter separated by a Jordan surface, which is a simple closed curve separating two different components, each of which is fully connected. This state was referred to as an 'emulsion'. However, before too long, these efforts were abandoned because it emerged that separation on the scale of clusters of galaxies was needed to satisfy the then current observations but the model was found unable to demonstrate that coalescence could continue long enough for the accumulation of matter and of antimatter to grow even to the size of galaxies, let alone clusters of galaxies, before separation occurred. The problem of an initial baryon, anti-baryon asymmetry, necessary in today's dominant model to ensure the apparent dominance of matter in the Universe as it is today, remains. The fact is that the existence of an initial imbalance between baryons and anti-baryons is a purely ad hoc assumption. That being so, people have continued to speculate on the presence of antimatter in our Universe, even though the models of Alfvén, Omnès and others have long since been discarded. However, it is possibly of interest to note that, although, as mentioned, Omnès and his co-workers referred to a state as an 'emulsion', at no time did they utilize the properties exhibited by an actual emulsion in their deliberations. It is worth noting these particular properties and contemplating the effects of incorporating them into the model.

An emulsion is a mixture of two substances which normally wouldn't mix; that is, a mixture of two immiscible substances. One, referred to as the dispersed phase, is dispersed throughout the other, referred to as the continuous phase. Again, emulsions fall into two categories; colloidal emulsions which are stable so that one phase will remain dispersed throughout the other over a period of time, and non-colloidal emulsions which are unstable and in which the two components tend to separate out. On occasions, substances known as emulsifiers may be added to stabilize an emulsion. A very typical example of an unstable emulsion is provided by salad dressing. In this example, as is well known to all, the emulsion will separate out very quickly unless shaken very vigorously. However, for present purposes, this common example is worth bearing in mind as it is an example of an emulsion which illustrates very clearly what an emulsion is, how it looks and how it behaves.

In the original Omnès model, although the term 'emulsion' was used, the situation envisaged was more a mixture of individual blobs of matter and antimatter; there seemed no notion of one phase being dispersed throughout a second phase which remains fully connected. Normally, the two substances forming an emulsion will separate out over time if left undisturbed but the situation in the early universe described by Omnès was certainly not undisturbed, more akin in fact to the situation of a violently shaken salad dressing. However, simply introducing the notion of a genuine emulsion into the discussion cannot, of itself, help in the resolution of the problem of the missing antimatter since no conglomerations of antimatter have been identified in the Universe. Recently, an ingenious suggestion[12] has been advanced in an attempt to rectify this and that suggestion is that what might be termed the cores of black holes are all, both primordial and supermassive black holes, composed of antimatter. With the popular modern notion of a black hole, such a suggestion would mean all the antimatter being hidden from view inside the event horizon of the black hole. Also, considering the sizes of the postulated supermassive black holes, it is relatively easy to see how an equivalence of content of matter and antimatter in the Universe could be achieved; indeed, in the above mentioned article12 some rough figures are included to support the plausibility of this assertion.

However, what if matter manages to cross the event horizon and come into contact with the antimatter? Obviously, any matter/antimatter contact will result in the annihilation of both but, in the model, the annihilation rate would be slowed down tremendously due to the antimatter being condensed into an extremely small body. Also, this annihilation would occur inside the event horizon and so there need not be any observation of resulting radiation. Further, it is suggested that such annihilation might not proceed too rapidly if a Leidenfrost layer, such as suggested by Alfvén, were to exist inside the event horizon. One further point occasioned by this idea is that such matter/antimatter annihilation could help the gradual evaporation of the black hole without recourse to the possible phenomenon of Hawking radiation, if such evaporation does, in fact, occur as speculated.

In the discussion so far, the role of the event horizon has been simply to prevent evidence of any possible matter/antimatter annihilation being viewed by observers; apart from that possibility, it appears to play no significant part in the model. Event horizons, though, are only part of the notion of a black hole which seems to emerge from the theory of general relativity. In the simplest case of an uncharged, non-rotating black hole, the starting point for discussion of the model is the Schwarzschild solution to the Einstein field equations but, as has been pointed out on numerous occasions [13], the popular version of that solution on which this deduction is based is not actually Schwarzschild's original solution, as is easily verified by referring to his original article and comparing it with the popular version which appears in so many textbooks. Schwarzschild's original solution does not possess the singularity which leads to the idea of a black hole. Hence, serious question marks hang over the modern notion of a black hole, added to which, as again has been pointed out on numerous occasions13, so far no black hole candidate has satisfied the fundamental inequality to be satisfied by the ratio of its mass to its radius; that is, the inequality;

$$M/R \geq c^2/2G = 6.7\times10^{26}\ kgm^{-1} \tag{2}$$

However, even if the modern notion of a black hole has problems, theoretically the idea put forward by Michell in 1784[14]and based on purely Newtonian principles is sound. Michell investigated the problem of a body with an escape speed greater than, or equal to, the speed of light. He found that the mass and radius of such a body would satisfy the same inequality as that mentioned above for a black hole as derived from the principles of general relativity. Since the event horizon plays so small a part in the above mentioned model of a balanced matter/antimatter Universe, it would not seem too much of a problem to substitute a Michell dark body instead of a black hole in that model. The term 'dark body' is used more correctly to describe the Michell idea since, as was pointed out by McVittie [15], if such a body exists, it would be simply a very dense body which could be approached and, in fact, viewed from a suitable distance, unlike the modern notion of a black hole. Obviously, this latter comment is in accordance with the usual meaning of a so-called 'escape speed'. It follows that the ideas advanced in the mentioned recent article[15] would hold if the bodies referred to were Michell type dark bodies of the appropriate size rather than conventional black holes since, although such objects wouldn't be hidden behind an event horizon, they would be effectively hidden from view by the very fact that even light would be unable to escape completely from them. Also, as with the suggestion based on black holes advocated in reference 12, any annihilation occurring would be slowed down to a great degree by the antimatter being condensed into an extremely small compact body. Of course, with no event horizon, if the dark body was composed of antimatter, any annihilation with nearby matter could only be prevented, or the effects slowed down, by the Leidenfrost layer solution as advocated originally by Alfvén. That in itself is no drawback to this modified suggestion since it is such a Leidenfrost layer which proves so important in the model suggested. It might be commented also that, in the case of a Michell dark body, the visibility referred to above would not mean that photons would reveal the presence of annihilation reactions since such photons would be degraded in energy and would not be what would be expected from annihilation. Of course, all of this particular discussion of the matter/antimatter problem is basically dependent on the big bang model being accepted as fundamentally correct. If it is not, then no immediate argument springs to mind to suggest the existence of antimatter in the Universe, at least not in quantities comparable with the amount of matter actually observed. Of course, consideration of this suggested model for the possible existence of comparable quantities of matter and antimatter in the Universe offers yet another possibility for examining the validity of the big bang model. As always, it should be remembered that the big bang is simply a theoretical model of how the Universe originated and developed and, as such, it must be open to observational and experimental checks in an attempt to establish how accurate a model it is or, in fact, if it is valid at all.

However, to return to the actual big bang theory, a further problem faced by the adherents to the theory is the seemingly constant need to add to the basic theory in order to overcome problems. Obvious examples of this are the introduction of the ideas of inflation, dark matter and even dark energy. It is, however, the first of these additions to which attention must be turned. The big bang model was faced with the 'horizon' and 'flatness' problems. The first of these arises from the prediction that the Universe is both homogeneous and isotropic, which implies that, in the early Universe, disconnected regions would have had to have been in nearly the same state to achieve the present-day homogeneity. The lack of contact makes such a scenario extremely unlikely. The second paradox concerns the extrapolation of the present value of the ratio of the energy density of the Universe to the critical energy density back to the big bang. When this is done, the extremely unlikely value of nearly unity is found. In 1981, Guth [16] attempted to address these by releasing the assumption of the adiabaticity of the early expansion of the Universe. This resulted in the so-called inflationary scenario, which supposes that a supercooling of the material of the Universe led to a period of exponential growth involving the release of the latent heat of the phase transition and an increase in the entropy of the Universe. Modifications to this basic model were introduced by Linde [17] and Hawking and Moss[18] to attempt to overcome the fact that it would produce large inhomogeneities which are incompatible with observation. The exponential dependence of the scale factor on the time is certainly a solution of the equations of general relativity, but the association of the release of a latent heat is not. This central objection went unnoticed until relatively recently[19].

An expanding Universe, as suggested by Hubble's observation of galactic expansion, will involve progressively increasing compression in the past. All that the inflation hypothesis was designed to do would be achieved by a speed of light which increases with increasing temperature. Of course, this alternative description of the past is not compatible with the universal application of the principle of general relativity which requires a universal speed for light.

It is not without interest to realize that additions to the big bang theory are accepted unerringly. Seemingly, no questions are raised when these new notions such as inflation, dark matter and dark energy are introduced in attempts to preserve this theory as the only acceptable explanation for our universe as we see it. However, there doesn't appear to have been any significant upsurge in interest in the steady state theory since the publication of all the material – both strictly academic and semi-popular – advocating modifications to that theory. Many will claim this due to the fact that the theory is quite simply incorrect, but the facts don't support this view. Neither do they support the view that the big bang theory is true beyond all reasonable doubt. In reality, the truth must lie either somewhere between these two extremes or possibly completely outside these two interesting attempts buried in some, as yet, totally unknown theory. We really truly understand very little, however great mankind's scientific achievements may be thought to be. When we understand in detail what is meant by terms such as 'force' and 'mass', then we will be on the way to a complete understanding of our universe and all that exists in it but, until that time, it seems sensible to retain all options and that must include both the big bang model and the steady state theory, together with any other thoughts, as possible explanations. Prominent among these other thoughts must be the so-called 'tired light' theory. So much in our presently accepted theories depends on the interpretation of the red-shift phenomenon. It is commonly accepted, as has been mentioned already, that this red-shift is brought about by the Doppler shift of light due to the recession of distant galaxies. However, at least theoretically, other explanations are feasible. A brief outline of the worries expressed by Halton Arp has been discussed earlier. However, another possible explanation for the existence of the observed red-shifts is provided by the notion of 'tired light'. Here the basic idea is that quanta of light could actually lose energy during their journey through space from distant galaxies to us. The suggested decrease in photon energy would result in an increase in wavelength that would be proportional to the distance travelled. This would, of course, be viewed as a reddening. Another contributory factor to this reddening of light could be provided by scattering by particles of intergalactic dust. Probably the effect of scattering by dust particles may be discounted at this stage, though not entirely forgotten, because such scattering would be expected to result in a broadening of the spectral lines and that is not observed. However, the general notion of 'tired light', while dismissed almost out of hand by most workers in the field, cannot be totally abandoned as yet. Firstly, it is a theory which has a long history and which has never gone away completely. It has been advanced and supported by a powerful array of physicists from Max Born to Jean-Pierre Vigier. This, in itself, is not sufficient to make the theory acceptable, but it is surely a good enough reason for it to be taken seriously. Some wish to dismiss it on the grounds that only in Big Bang cosmology is there a satisfactory explanation provided for the origin of the cosmic background radiation and for the abundance of the light elements. However, as has been seen already, this is simply not true. The case of the steady state theory proves this beyond reasonable doubt. Whether one believes or disbelieves the steady state theory or, for that matter, the big bang theory, it is certainly true to say that, in attempting to destroy the steady state theory, the truth was not to the fore. It is disturbing to realize that this explanation is the one advanced for dismissing so many suggestions and it is no more true today than it was when first put forward and agreed. 'Tired light' may not be a true explanation for any of the questions arising in cosmology but, like anything else, it deserves to be viewed with a completely open mind before a decision is reached. Once again it is seen that this is the true problem facing cosmology as a whole and the big bang theory in particular – both must be viewed and assessed with completely open minds. Personal preferences and prejudices have no place in the evaluation of a scientific theory. The task must be accomplished purely by using the accepted methods of science and known scientific knowledge - always realizing, of course, that any conclusion will be subject to limitations placed on its validity by the extent of such knowledge at any one time.

Yet another major problem facing this area is associated with the advance of knowledge. In this colossal area, knowledge advances through careful, painstaking observation of the cosmos. All the observations made must then be processed most carefully. This again is something which is not quite so straightforward as might appear at first. Quite frequently, data has to be analyzed statistically and it is crucial that this is done completely honestly. There must never ever be even a suspicion that an effect is claimed which might be simply due to the statistical package used for the analysis. Hence, this again is something which must be undertaken by truly open-minded people and making use of professional statisticians to analyses data – rather than it being done by those who might be thought to have a vested interest in the end result – could be a sensible way forward in this area. Too often the impression is left that the conclusion announced is merely confirmation of the result 'expected' before the experiment or observation was begun. In a way, this brings a return to the case of Halton Arp. As has been noted earlier, many astronomers are said to doubt Arp's interpretation of the photographs he has taken and usually their skepticism is said to be based on some aspect of the statistical analysis of his data. It has been claimed, though, that if a continuous change in red-shift values could be measured along an apparently material link between a low red-shift galaxy and a high red-shift quasar, then Arp's view would be vindicated. However, it seems that no such effect has been found as yet, although strenuous efforts are said to have been made to establish the presence, or absence, of such an effect. This again raises the question of whether or not observers are finding what they want to find rather than the truth. Some ask at this point, 'What is truth?' No doubt a deep philosophical discussion could ensue here. However, suffice it to note that the Oxford Dictionary states that one meaning of the word 'true' is "in accordance with fact or reality, not false or erroneous". It goes on to state that 'truth' is the "quality, state, of being true". These elementary definitions of the two words give a clear everyday meaning of what they mean in the present circumstances. Indulging in philosophical discussions surrounding the meanings of words doesn't necessarily help anyone; it frequently serves simply to divert attention from the question at issue, - in this case that of the major problems facing science today. As with so many of the major controversies in science, positions have become entrenched, 'conventional wisdom' has become almost indelibly etched into the folklore surrounding the subject. Young scientists are, all too often, taught established truth as if it were religious dogma. They are not trained to really think; only to think along well-established lines – lines drawn by the 'Gods' of 'conventional wisdom'. This probably sounds harsh and seemingly linking science with religion again will undoubtedly offend some who feel the two separated by an infinite chasm. Unfortunately, the truth often does hurt and, in reality, young scientists are all too often indoctrinated with supposed 'facts', rather than educated to have open, enquiring minds. If the result of raising these unpleasant aspects of present day world science is to reintroduce an open questioning attitude into science, then the imagined hurt will have been more than worthwhile.

As an addendum to this discussion of the big bang model, it might be noted that an entire edition of the well-known and well-respected British Broadcasting Corporation's television science program, Horizon, was devoted to the present-day search for dark matter [20]. The program title was 'Most of our Universe is Missing'; an eye-catching title guaranteed to attract viewers. It pointed out that some scientists feel it not known from what much of our universe is made; others argue that some presently accepted theories, such as Newton's law of gravitation, may be wrong – or, at least, only apply locally rather than globally. The program itself contained much of genuine, but not probably general, interest. However, one worrying aspect in the present context was the fervor exhibited by several contributors in support of the big bang as explaining the origins of the universe. Only one really drew back to point out that the big bang is a theory, and only a theory! As was asked in a recent letter to The Observatory [21], "When will the Cosmological Establishment stop calling their theory the truth, the whole truth, and nothing but the truth?" Anyone who questions it is said to belong to a minority. Apparently, most cosmologists would offer strong odds on there having been a 'big bang', feeling that "everything in our observable universe started as a compressed fireball, far hotter than the centre of the Sun". The idea that this scenario is questioned by a minority only would seem true, but largely because so many in science possibly feel it in their own best personal interests to conform to the imposed dictates of 'conventional wisdom'. Of course, in these terms, that 'minority' might really be a 'silent majority'. As for those outside professional scientific circles, those who in the final analysis pay the bills, they have been subject to so much publicity, via all media forms, in favor of this theory to the exclusion of all else, that it is no wonder they believe it to be an unassailable truth, not simply a mere theory. However, as another contributor to The Observatory pointed out[22] because the steady state theory appears to provide precise predictions, it seems to have suffered in comparison with other theories, such as the Big Bang, which allow scope for empirical adjustment. This writer felt it precisely this which made the steady state theory a good theory and seemed to feel it likely that that theory would return eventually in some form. Be that as it may, it is undoubtedly of interest to speculate on what the future holds in this field, but one thing is absolutely certain, for real progress to be made, investigators must retain open minds, very little should be totally discarded at this juncture. In the present atmosphere that seems a lot to ask, but it is absolutely essential if science is to advance positively!

7. Plasma Cosmology/Electric Universe

Of course, all of the discussion up to this point has been based on a theory dependent solely on possible effects due to the force of gravity. It might be noted that, following the introduction of Newton's mechanical ideas, work still proceeded apace investigating electromagnetic phenomena and this continued at least into the earlier years of the twentieth century, as is evidenced by the contents of J. J. Thomson's book Electricity and Matter[23] (Archibald Constable & Co. Ltd, Westminster, 1904). However, this book provides but one example to illustrate the very real emphasis on work involving the effects of the electric and magnetic fields. However, after those early years of the century, the emphasis seems to have shifted to explanations of phenomena purely in terms of gravitational effects as far as most mainline research was concerned. Considering that it is accepted that much of the matter in the Universe is in the form of plasma, this seems a retrograde step and this view is surely strengthened when the work of such as Kristian Birkeland and Hannes Alfvén is considered. One may only speculate as to why the emphasis of much scientific research changed in this way. However, thanks to people like Birkeland, Alfvén and (more recently) Peratt, work in the areas of electromagnetism and plasma physics did continue and it should be noted that much of the work on plasmas has been via laboratory experiments, so hard experimental evidence is available to support any claims made.

The work on plasmas and other electromagnetic phenomena has inspired the examination of astronomical phenomena in these terms and has resulted in the so-called Electric Universe idea as expounded, for example, in the books The Electric Universe by Wallace Thornhill and David Talbott [24] and The Electric Sky by Donald Scott[25] Reading through this material makes one aware that, while like orthodox accepted theory, the electric universe ideas are supported by much computer modeling, it can also draw on parallels in astronomy with plasma phenomena observed in the laboratory. Admittedly, drawing such parallels involves scaling up tremendously but assuming this possible is little different from assuming that laws seemingly applicable here on the Earth are also applicable in the Solar System and, indeed, throughout the Universe. However, at least visually, some of the phenomena observed in the laboratory are very like what is observed by some of the most powerful of telescopes; - electric currents in plasma naturally form filaments due to the so-called 'pinch effect' of the induced magnetic field. Electromagnetic interactions cause these filaments to rotate about one another to form a helical 'Birkeland Current' filament pair and this is very much the structure seen in the Double Helix nebula near the galactic centre; again, the Hubble image of the planetary nebula NGC6751 looks remarkably like the view down the barrel of a plasma focus device. Examples such as these prove nothing but should awaken people to the possibility of alternative explanations for astronomical phenomena.

Much of the laboratory work originated with the work of Kristian Birkeland more than one hundred years ago. It was during his Arctic expeditions at the end of the 19th century that the first magnetic field measurements were made of the Earth's polar regions. His findings also indicated the likelihood that the auroras were produced by charged particles originating in the Sun and guided by the Earth's magnetic field. Birkeland, though, was an experimentalist and is still known for his Terrella experiments carried out in a near vacuum and in which he used a magnetized metallic sphere to represent the Sun or a planet and subjected it to electrical discharges. By this means, he was able to produce scaled down auroral-type displays as well as analogues of other astronomical phenomena. These claims, however, were only vindicated finally by satellite measurements in the 1960's and 70's. To that point in time, his experimental and observational achievements had tended to be overshadowed by the purely theoretical predictions and explanations of the geophysicist, Sydney Chapman. Once again, powerful mathematics seems to have held sway over the more expected techniques of physics – experimentation and observation, with mathematics a mere tool to be used when necessary. This is not to decry Chapman's work but to emphasize the overwhelming importance of the physics when investigating natural phenomena.

Birkeland also showed experimentally that electric currents tend to flow along filaments shaped by current induced magnetic fields. Of course, this confirmed observations of Ampère that indicated that two parallel currents flowing in wires experience a long range attractive magnetic force that brings them closer together. However, as plasma currents come closer together, they are free to rotate about each other. Such action generates a short range repulsive magnetic force which keeps the filaments separated so that they are, in effect, insulated from each other and able to maintain their separate identities. The end effect is for them to appear like a twisted rope and it is this configuration which is termed a 'Birkeland current', as was mentioned earlier when the Double Helix nebula was noted as a possible example. Satellites orbiting above the auroras in the 60's and 70's were able to detect a movement of ions, indicating that electric currents were present. Later missions found quasi-steady electric fields above the auroras following the magnetic field lines, thus lending some credence to Birkeland's claim of the existence of an electric circuit between the earth and the Sun. Some may be skeptical of this latter interpretation but it is undoubtedly true that much of the material in the Universe is in the form of plasma and there is certainly electric and magnetic activity occurring in abundance. This means there are numerous very good reasons for considering the effects of the electromagnetic force in the Universe, one of which could be the resolution of the problem of the missing mass.

However, precisely what is the Electric Universe? In truth, it is really simply an hypothesis, a new way of interpreting known data by utilizing both new and well-established knowledge relating to electricity and plasma. It should be emphasized immediately that, in this new interpretation, gravity still has a role to play but it is a secondary one since the electric force is so much more powerful. A major point to be stressed from the outset is that, in this interpretation of astronomical phenomena, scientists are able to call on evidence from laboratory based experiments to help form and support suggested explanations for a wide variety of phenomena. It has been found that, as explained in more detail in the above-mentioned books, an electrified plasma in a laboratory is a good model for providing possible explanations for many recently observed astronomical phenomena which, in several cases, have caused puzzlement for astronomers seeking explanations via more orthodox gravitationally based theories. This is not to say that gravity is ignored and regarded as irrelevant; rather, the possible effects of the electromagnetic force on astronomical phenomena are investigated while still recognizing the importance of gravitational effects. In the electric universe, the gravitational systems of galaxies, stars, moons and planets are felt to have their origins in the proven ability of electricity to generate both structure and rotation in plasma. It is felt further that the force of gravity assumes importance only as the electromagnetic forces approach equilibrium. As has been noted already, great consternation has been caused in astronomical circles by the realization that gravity, as presently understood, cannot explain much that is observed if the amount of mass available is as now felt to be present. Hence, instead of positing the existence of 'dark matter' or following the path of modifying Newton's well-tried law of gravitation, it is suggested here that the possible effects of the electromagnetic force be examined to see if, in conjunction with orthodox ideas on gravity, these puzzling observations can be explained.

A point which is often relegated to the background when discussing the solution of problems through the introduction of dark matter is the fact that the missing mass, if there really is any missing mass, is not absent homogeneously throughout the Universe; it is missing only in specific places - for example, in the outer regions of galaxies. Hence, possible solutions, such as the idea that neutrinos possess mass, which are essentially homogeneous in nature cannot be acceptable. It should be mentioned at this point though that, in the Electric Universe model, neutrinos do possess mass and are extremely important. They respond only weakly to massive objects such as stars and galaxies but form an extended atmosphere which, for example, refracts light around the Sun from distant stars and this offers an alternative explanation for the so-called gravitational bending of light. On the other hand, in this model, neutrinos are not required to explain galactic rotation although they must contribute to the masses of both stars and galaxies. Again, having some mass, neutrinos will not be distributed homogeneously.

However, returning to the realization that much of the matter permeating the Universe is in the form of plasma, it might be remembered that these clouds of plasma respond to the well-known laws of Maxwell. Also, as pointed out by Scott in his book, another law, formulated by Lorentz, does help explain the galactic speeds alluded to earlier. This law states that

a moving charged particle's momentum (speed or direction) can be changed by application of either an electric field or a magnetic field or both.

This seems a highly likely contributory factor, at least, causing galaxies to rotate as they are perceived to do but would indicate, contrary to the accepted view, that gravity has less to do with things than has been thought. However, it should be emphasized that nowhere is it being suggested that Newton's law of gravitation is wrong or in need of modification; it is simply being suggested that, in deep space where everything swims in a sea of plasma, the Maxwell – Lorentz electromagnetic forces dominate over those of gravity.

It might be remembered also that the Lorentz force alluded to here changes a charged particle's momentum and that change is directly proportional to the strength of the magnetic field through which the particle is moving. Further, the strength of a magnetic field produced by an electric current is inversely proportional to the distance from the current but the gravitational force between stars is inversely proportional to the square of the distance. This well-known difference between the two forces could lie at the heart of the problem of the galactic rotation curves; certainly it seems an avenue worth exploring further, especially considering the fact that more and more space missions are indicating that electromagnetic forces are distributed more widely throughout space and are, of course, many orders of magnitude stronger than gravitational forces.

Much time, effort and money is spent worldwide on producing elaborate computer programs which purport to support the prevailing belief in the big bang model as being the correct theory explaining how the Universe originated. However, as well as a great many laboratory experiments being performed to help establish plasma properties, it has been shown also, using the Maxwell and Lorentz equations, that streams of charged particles, such as are found in the intergalactic plasma, will evolve into the familiar galactic shapes under the influence of electromagnetic forces. The results fit extremely well with the observed velocity profiles in the galaxies and all this with no recourse to missing mass. Much of this simulation work has been carried out by Anthony Peratt and is reported in various issues of the IEEE Transactions on Plasma Science, a highly prestigious journal. It does seem, therefore, that the case for the existence of dark matter is questionable and hence, yet another query is raised concerning the validity of the big bang model.

8. Conclusion

Possibly that last word – 'model' – is the most significant one used for that is all the big bang is, a model! Admittedly, it has been successful to some extent in attempting to explain the origin and subsequent development of our Universe, but much of that success has really been apparent due to the fact that so much information and so many alternatives have been kept hidden from public scrutiny. Here the ideas of the so-called Electric Universe have been used mainly in relation to an examination of the question of the existence, or not, of so-called dark matter, that mystical entity introduced to shore up the wobbling framework of the big bang model. However, it has been shown that this is another means of attempting to explain at least some aspects of the behavior of our Universe. It seems immediately apparent though that this theory simply must play at least some part in the explanation since the electromagnetic force is so much more powerful than that of gravity. Again, so many more modern space missions are indicating more and more effects of both electric and magnetic fields in the space above us that it cannot be long before it is acknowledged that the effects of these can be ignored no longer.

There is little doubt that, just as he has done throughout the ages, man will continue to search for as much knowledge as possible about the Universe in which we live. Again there is little doubt that at least part of that search will involve seeking the answer to the age old question 'How did it all begin?' Whether or not man's intellect is capable of discovering the true answer to this question is another matter; he will continue to search. What really matters is that that search should be carried out scientifically; workers must be scrupulously honest in all their work and in the reporting of that work whether in academic journals or in more popular publications designed to keep the public, which eventually pays for all this endeavor, informed of progress. The cult of 'conventional wisdom' must, therefore, be eradicated and scientific research must be conducted in a completely open minded manner with no single theory or model being allowed to dominate purely to preserve the status of powerful individuals. As Hannes Alfvén said in his Nobel Lecture of December 1970, "The centre of gravity of the physical sciences is always moving. Every new discovery displaces the interest and the emphasis." Maybe those working in the fields of astronomy/astrophysics and cosmology especially should take note of these words of wisdom uttered by an acknowledged scientific thinker and open their minds to other possibilities when attempting to solve problems in their preferred scientific domains.

References

[1] Lemaître, G. (1927) *Ann. de la Societe Scientifique de Bruxelles*, 47, 49.
[2] Gamow, G. (1946) Phys. Rev., 70, 572.
[3] Penzias, A. & Wilson, R. (1965) Ap. J., 142, 419.
[4] Hoyle, F. Burbidge, G & Narlikar, J.V. (2000) A Different Approach to Cosmology, Cambridge: Cambridge Univ. Press.
[5] Bondi, H. & Gold, T. (1948) M. N. R. A. S., 108, 252; Hoyle, F. (1948) M. N. R. A. S., 108, 372.
[6] Eddington, A. (1988) The Internal Constitution of the Stars, Cambridge: Cambridge Univ. Press.
[7] Gamow, G. (1946) Phys. Rev., 70, 572.
[8] Gamow, G. (1948) Phys. Rev., 74, 505; Alpher, R.A. & Herman, R. (1948) Phys. Rev., 75, 1084.
[9] Hoyle, F. (1946) M.N.R.A.S., 106, 343.
[10] Hoyle, F. Burbidge, G & Narlikar, J.V. (2000) A Different Approach to Cosmology, Cambridge: Cambridge Univ. Press.
[11] Alfvén, H. (1966) Worlds-Antiworlds, San Francisco: Freeman & Co.
[12] May, H.D. (2002) Another Look at the Cosmological Model of Omnès, ArXiv: 1011.0017.
[13] Dunning-Davies, J. (2007) Exploding a Myth, Chichester : Horwood Publishing Ltd.
[14] Michell, J. (1784) Philos. Trans. R. Soc. 74, 35.
[15] McVittie, G.C. (1978) The Observatory, 98, 272.
[16] Guth, A. (1981) Phys. Rev. D, 23, 347.
[17] Linde, A.D. (1982) Phys. Lett. B, 108, 389.
[18] Hawking, S.W. Moss, I.G. (1982) Phys. Lett. B, 110, 35.
[19] Lavenda, B.H. & Dunning-Davies, J. (1992) Found. Phys. Lett. 5, 191.
[20] Horizon, B.B.C. (2006) 9th February.
[21] Welch, A. (2006) The Observatory, 126, no. 1190, 51.
[22] Fellgett, P. (2006) The Observatory, 126, no. 1190, 51.
[23] Thomson, J.J. (1904) Electricity and Matter, Westminster: Archibald Constable & Co. Ltd.
[24] Thornhill, W. & Talbott, D. (2002) The Electric Universe, Portland: Mikamar Publishing.
[25] Scott, D.E. (2006) The Electric Sky, Portland: Likamar Publishing.

The Theological Basis of Big Bang Cosmology and the Failure of General Relativity

STEPHEN J. CROTHERS

PO Box 1546 Sunshine Plaza, 4558, QLD, Australia
thenarmis@gmail.com

It is shown in this paper that the Big Bang Cosmology has its basis in theology, not in science, that it pertains to a Universe entirely filled by a single spherically symmetric continuous indivisible homogeneous body and therefore models nothing, that it violates the physical principles of General Relativity, that it violates the conservation of energy, and that General Relativity itself violates the usual conservation of energy and momentum and is therefore in conflict with experiment on a deep level, rendering Einstein's conception of the physical Universe and the gravitational field invalid.

Keywords: Creationism; Spherical symmetry; Conservation of energy and momentum.

1. Introduction

Big Bang Cosmology is invalid by the invalidity of General Relativity because it is easily proven that General Relativity violates the usual conservation of energy and momentum. But it is nonetheless worth noting that cosmology is not a science at all; it is theology. The relativistic cosmologists generally fail to mention to the public and their students the fact that the Big Bang Cosmology was first conjured up by the Belgian mathematician and priest Georges Lemaître. Tolman [1] remarks that

non-static homogeneous models ... were first theoretically investigated by Friedmann, and first considered in connexion with the phenomena of the actual universe by Lemaître.

Lemaître introduced a creation event into the equations of General Relativity and hence infused physics with the notion of God and His creation of the Universe. The Big Bang theory has been ratified by the Vatican owing to Lemaître's creationism. All Lemaître did was substitute one creation event with another creation event. Indeed, Lemaître admitted to the Swedish Nobel Prize winner in physics, Hannes Alfvén, that he came up with the idea to make physics accord with Catholicism and the teachings of St. Thomas Aquinas. Alfvén was not impressed. Here is what Alfvén [2] reported: "*I was there when Abbe Georges Lemaître first proposed this theory*," he recalled. Lemaître was, at the time, both a member of the Catholic hierarchy and an accomplished scientist. He said in private that this theory was a way to reconcile science with St. Thomas Aquinas' theological dictum of *creatio ex nihilo* or creation out of nothing.

"*There is no rational reason to doubt that the universe has existed indefinitely, for an infinite time,*" Alfvén explained. "*It is only myth that attempts to say how the universe came to be, either four thousand or twenty billion years ago. Since religion intrinsically rejects empirical methods, there should never be any attempt to reconcile scientific theories with religion. An infinitely old universe, always evolving, may not be compatible with the Book of Genesis. However, religions such as Buddhism get along without having any explicit creation mythology and are in no way contradicted by a universe without a beginning or end. Creatio ex nihilo, even as religious doctrine, only dates to around AD 200. The key is not to confuse myth and empirical results, or religion and science.*"

Furthermore, in January 1933, Georges Lemaître travelled with Albert Einstein to California for a series of seminars. After the Lemaître detailed his 'primordial atom' or 'primaeval egg' expanding Universe theory, now called the Big Bang theory, Einstein [3] stood up, applauded, and said, "*This is the most beautiful and satisfactory explanation of creation to which I have ever listened.*" So evidently Einstein was actually a creationist, revealing thereby that he too was actually theological in his real disposition, despite his often overt cryptic claims that he was not. Now Big Bang Cosmology only has the façade of science because it is couched in the mantle of complicated, but meaningless, mathematics, in terms of General Relativity which is an invalid theory because it violates the usual conserve-ation of energy and momentum and is therefore in conflict with experiment on a very deep level. Before the Big Bang theory there was actually no alleged 'scientific' basis attached to the question of the creation of the Universe; only theology dealt with this question then. It still does.

The point of this is that it does not matter whether one is a theist or atheist, Republican or Democrat, Marxist or Capitalist; one's ideology, which is subjective, must not influence the outcome of a scientific inquiry. Contrary to scientific method Lemaître used his Catholic ideology to *predetermine* the outcome of a scientific investigation. It is therefore not surprising that his creationist cosmology has no scientific basis. What is surprising is that Einstein and his followers embraced this theological notion and present it to the world as science. As Alfvén rightly pointed out, science and theology are two entirely different thought processes and employ entirely different methodologies and so the one cannot be misconstrued as or confounded with the other. The engineer does not allow his ideology, religion or philosophy of life to predetermine the technical outcome of an engineering project or to influence his methods for designing and building a bridge. No scientist is permitted to do otherwise either when conducting a scientific investigation be it by means of experiment or mathematical analysis because if he or she does he or she is not doing science at all.

2. Big Bang Artificial Symmetry and Matter

We will require a little terminology at this stage in relation to tensors since we will be looking at a few tensor expressions. You should not be alarmed by this because you will not be required to know how to carry out any tensor calculations in what follows or to even know what a tensor is. Almost all that you will be required to know is what some given tensor physically or geometrically represents in some tensor equation.

This is essentially no more difficult for instance than considering the equation for a straight line; $y = mx + b$. You will recall from high school that m is the slope of the line, b is the intercept of the line on the y-axis, and if m and b are known, given some value of x you can calculate the value of y, if required. Similarly, consider the famous equation, $E = mc^2$. There is nothing frightening in this equation either; we merely identify E as energy, m as mass, and c as the speed of light in vacuum, and given the values of m and c you can easily calculate E if required. And in identifying the components in either equation, one geometrical and one related to physics, there is nothing difficult at all.

The very same situation arises in the tensor expressions we will examine, except for a special case we will later investigate, where I will provide you with all additional information anyhow. In this way the frightening mystery of tensors to those not familiar with them will disappear and so they will never again be intimidated by tensors.

The *order* or *rank* of a tensor is simply the total number of suffixes attached to it. These suffixes may be all subscripts or all superscripts or a combination of subscripts and superscripts. If the suffixes of a tensor are all subscripts the tensor is said to be *covariant*. If the suffixes are all superscripts the tensor is said to be *contravariant*. If there are both superscripts and subscripts the tensor is said to be *mixed*. We will be looking at only some 2nd-order tensors. That's all we need to know about tensors for the time being, and there's nothing complicated in this.

Now Einstein's field equations [4] "... *couple the gravitational field (contained in the curvature of spacetime) with its sources*." Qualitatively Einstein's field equations are:
Spacetime geometry = $-\kappa \times$ *causative matter (i.e. sources)* where κ is merely a coupling constant. So although matter is still the cause of a gravitational field in General Relativity, the gravitational field is no longer a force of attraction between two or more bodies as it is in Newton's theory but instead a curvature in the geometry of spacetime induced by the presence of the material sources. Thus spacetime and matter are *causally* linked in General Relativity. Carroll and Ostlie [5] say that, "*Mass acts on spacetime, telling it how to curve. Spacetime in turn acts on mass, telling it how to move.*"

Einstein's field equations take the mathematical form

$$G_{\mu\nu} = -\kappa T_{\mu\nu} \tag{1}$$

where $G_{\mu\nu}$ is the Einstein tensor describing the curvature of spacetime (the geometry), $T_{\mu\nu}$ the energy-momentum tensor describing the material sources of the 'gravitational field' that induce the curvature in spacetime, and κ is a coupling constant. There is no shape inherent in these equations. Shape is imposed upon the field equations as an entirely arbitrary mathematical device in order to facilitate an analytic solution. Two primary shapes have been used to achieve this: cylindrical symmetry and spherical symmetry, and in the latter case usually to conform to the spherical symmetry used for Minkowski spacetime, which Minkowski developed in relation to Special Relativity. Spherical symmetry is that which is used most and is used in the case of the Big Bang 'solution' by assuming spatial homogeneity. Since this imposition is entirely arbitrary due to certain assumptions there is no *a priori* reason to suppose that any one shape of the Universe is somehow more 'real' than any other shape or that the Universe even has any shape at all! With 'shape' or 'symmetry' comes the equally arbitrary notion of 'boundary' and so there is no *a priori* reason to suppose that the Universe has an associated boundary. Indeed, Einstein [6] has remarked,

"*Given certain field variables and a system of field equations for them, the latter will not in general determine the field completely. There still remain certain free data for a solution of the field equations. The smaller the number of free data consistent with the field equations the 'stronger' is the system. It is clear that in the absence of any other viewpoint from which to select the equations, one will prefer a 'stronger' system to a lesser one.*"

The Big Bang creationism suffers from being subject to a large system of arbitrarily adjustable parameters that result in various models from which one is merely selected in order to satisfy Lemaître's theistic *creatio ex nihilo* with its associated expansion of the Universe, with the latter being justified by an *ad hoc* reinterpretation of the Hubble-Humason red-shift with distance relation to a red-shift with recessional-velocity relation (i.e. Doppler effect on light). The earlier redshift and blue-shift observations made by Vespo Slipher rarely even get a mention.

It is also a fact that there are no known solutions to Einstein's field equations for two or more masses and a fact that there is no existence theorem by which it can even be asserted that his field equations contain latent solutions for two or more masses. That is why it is totally false, for example, to talk about multiple black holes, black holes interacting with one another and other matter, being components of binary systems, swallowing surrounding matter, merging or colliding. Nonetheless S. W. Hawking [7] says,

"*Also, suppose two black holes collided and merged together to form a single black hole. Then the area of the event horizon of the final black hole would be greater than the sum of the areas of the event horizons of the original black holes.*"

According to B. Schutz [8],

"*... Hawking's area theorem: in any physical process involving a horizon, the area of the horizon cannot decrease in time. ... This fundamental theorem has the result that, while two black holes can collide and coalesce, a single black hole can never bifurcate spontaneously into two smaller ones.*"

"*Black holes produced by supernovae would be much harder to observe unless they were part of a binary system which survived the explosion and in which the other star was not so highly evolved.*"

It is however often asserted in the literature that numerical methods have resulted in solutions for multiple masses. This is not correct at all and is only an abuse of the term 'numerical methods' combined with wishful thinking, because without an existence theorem for multiple mass solutions to Einstein's field equations one cannot say that multiple mass solutions exist. This is compounded by the fact that Einstein's field equations are *highly nonlinear* and so the Principle of Superposition does not apply. In relation to the popular but erroneous method of linearisation of Einstein's field equations, even the ardent relativist Wald [9] admits that,

"*The existence of exact solutions corresponding to a solution to the linearised equations must be investigated before perturbation analysis can be applied with any reliability.*"

In relation to solutions without 'singularities' Einstein [6] remarks,

"*Approximation methods are of no avail since one never knows whether or not there exists to a particular approximate solution an exact solution free of singularities.*"

The same can be said for 'solutions' that contain singularities, such as those for the alleged black hole. After all, all alleged black hole solutions to Einstein's field equations pertain to a Universe that by mathematical construction contains no matter and so it is impossible to use numerical methods to generate multiple black holes in violation of the very mathematical definition of all black hole solutions being generated by solutions for a Universe that by mathematical construction contains no matter. In fact, since all black holes are obtained from a spacetime that by mathematical construction contains no matter, there is in fact no such thing as a black hole because the alleged black hole has mass and so it cannot appear in a spacetime that by mathematical construction contains no matter. It is very easy in fact to prove that General Relativity does not predict the black hole [10-14] at all and that Newton's [10,13,15] theory too does not predict the black hole. Furthermore, as already mentioned, the Principle of Superposition does not apply in General Relativity. Mathematically this means that if $\boldsymbol{X}$ is a solution to Einstein's field equations and $\boldsymbol{Y}$ is another solution then the linear combination $a\boldsymbol{X} + b\boldsymbol{Y}$, where a and b are scalars, is *not* a solution.

Physically this means that for some given solution (which we know pertains to either an empty Universe or a Universe containing only *one* body because there are no known solutions to Einstein's field equations for two or more bodies) one cannot simply pile up (i.e. superimpose) matter into that solution to get multiple masses, charges, photons, electromagnetic fields, etc. as one might desire. It is for this reason that all claims for the discovery of multiple black holes are patently false. Multiple black holes and indeed multiple masses cannot be accounted for by General Relativity at all. Indeed, General Relativity cannot account for the simple experimental fact that two suspended fixed masses will approach one another upon release.

Thus, *all* known solutions to Einstein's field equations pertain to a Universe that contains no matter or allegedly either contains only one mass floating around in an infinite spacetime (e.g. the black hole) or is entirely filled by a single continuous indivisible homogeneous mass of some supposed macroscopic density and pressure either constant or a function of time. The Big Bang model allegedly consists of the latter case. Both models do not reflect the actual structure of the Universe we observe and so neither has any physical meaning. So how do the astrophysical scientists get multiple masses (including black holes), galaxies, stars, charges, electromagnetic fields, etc. into their General Relativistic models of the Universe? - they do so by simply applying the Principle of Superposition via a false analogy with Newton's theory of gravitation where the Principle of Superposition applies, in violation of the fact that the Principle of Superposition does not apply in General Relativity and in violation of the fact that there are no known solutions for two or more masses to Einstein's field equations or an existence theorem for two or more masses, and in violation of the additional physical principles of General Relativity and of experimental physics as we shall soon see.

3. Conservation of Energy and Mysticism

The conservation of energy is a very well established experimental principle in physics and it simply states that *energy can be neither created nor destroyed; only transformed from one type into another type*. Now according to the Big Bang *creatio ex nihilo* there was initially nothing; no spacetime, no mass, no energy, no photons, just nothingness. Then, from this nothingness, there allegedly appeared an instant presence of a huge amount of energy that expanded and formed fundamental particles, larger masses such as stars and galaxies, and ultimately all that now exists in the Universe. This obviously violates the experimentally well determined conservation of energy. To argue that physical principles themselves did not exist before the Big Bang *creatio ex nihilo* does not constitute a scientific argument by the very nature of the scientific method and so such an argument involves mysticism and myth, not the empirical methods of science. Oddly too is that in the literature it is sometimes asserted that the Big Bang *creatio ex nihilo* was caused by a quantum fluctuation, which would mean that some strange kind of unsubstantiated quantum principle existed before the Big Bang *creatio ex nihilo* anyhow. But what does a quantum fluctuation in nothingness mean? Such an assertion also smacks of a linguistic vacuum. It is therefore quite meaningless too and so explains nothing for want of scientific validity.

It is well known to anthropologists that all human societies, prehistoric, ancient and modern (civilized or tribal) have all developed some kind of creation myth to account for the existence of the Universe or at least the immediate world around us, and the related fearful questions of life and death. It has been well established by anthropologists that the human condition craves for a meaning to and explanation of existence and associates this with the fundamental notion of cause and effect that is observed all around us in our everyday lives. It is from this basic inclination that mythology, superstition, sympathetic magic and theology have their etiology. The Big Bang *creatio ex nihilo* is no exception, but it is couched in such pseudo-scientific jargon and elaborate mathematics all in violation of actual physical science in both facts and methods, and an abuse of mathematical methods that confounds thereby physical entities with mathematical entities to give the façade of true scientific inquiry. Heaviside[16] made a penetrating quip in this regard:

"It was once told as a good joke upon a mathematician that the poor man went mad and mistook his symbols for realities; as M for the moon and S for the sun."

When an engineer designs and builds a bridge he does not confound his design equations with his physical bridge. Astrophysical scientists however tend to confound their mathematical symbols and equations with physical objects; for example, infinitely dense point-mass singularities. In this way anything can be and has been concocted and falsely presented as legitimate astrophysical science. It is rather ironic that many astrophysical cosmologists oppose theological notions of creationism in science but themselves resort to a creationism by means of Lemaître's theological Big Bang *creatio ex nihilo* of St. Thomas Aquinas to rebuke creationism.

4. Additional Physical Principles of General Relativity

Einstein asserted that his Principle of Equivalence and his laws of Special Relativity must hold in sufficiently small finite regions of his gravitational field, and that these regions can be located anywhere in his gravitational field. Here is what Einstein [6] says,

"Let now K be an inertial system. Masses which are sufficiently far from each other and from other bodies are then with respect to K, free from acceleration. We shall also refer these masses to a system of co-ordinates K', uniformly accelerated with respect to K. Relatively to K' all the masses have equal and parallel accelerations; with respect to K' they behave just as if a gravitational field were present and K' were unaccelerated. Overlooking for the present the question as to the 'cause' of such a gravitational field, which will occupy us later, there is nothing to prevent our conceiving this gravitational field as real, that is, the conception that K' is 'at rest' and a gravitational field is present we may consider as equivalent to the

conception that only K is an 'allowable' system of co-ordinates and no gravitational field is present. The assumption of the complete physical equivalence of the systems of coordinates, K and K', we call the 'principle of equivalence'; this principle is evidently intimately connected with the law of the equality between the inert and the gravitational mass, and signifies an extension of the principle of relativity to co-ordinate systems which are in non-uniform motion relatively to each other. In fact, through this conception we arrive at the unity of the nature of inertia and gravitation. For, according to our way of looking at it, the same masses may appear to be either under the action of inertia alone (with respect to K) or under the combined action of inertia and gravitation (with respect to K').

Stated more exactly, there are finite regions, where, with respect to a suitably chosen space of reference, material particles move freely without acceleration, and in which the laws of special relativity, which have been developed above, hold with remarkable accuracy."

In their textbook, Foster and Nightingale [4] succinctly state the Principle of Equivalence thus,

"*We may incorporate these ideas into the principle of equivalence, which is this: In a freely falling (nonrotating) laboratory occupying a small region of spacetime, the laws of physics are the laws of special relativity.*"

Of the Principle of Equivalence Pauli [17] says,

"*We can think of the physical realization of the local coordinate system* K_0 *in terms of a freely floating, sufficiently small, box which is not subjected to any external forces apart from gravity, and which is falling under the influence of the latter. ... It is evidently natural to assume that the special theory of relativity should remain valid in* K_0."

In relation to the Principle of Equivalence Taylor and Wheeler[18] state in their book, "*General Relativity requires more than one free-float frame.*"

Carroll and Ostlie [5] write,

"***The Principle of Equivalence****: All local, freely falling, nonrotating laboratories are fully equivalent for the performance of all physical experiments. ... Note that special relativity is incorporated into the principle of equivalence. ... Thus general relativity is in fact an extension of the theory of special relativity.*"

For the Principle of Equivalence; the Dictionary of Geophysics, Astrophysics and Astronomy [19] says, "*Near every event in spacetime, in a sufficiently small neighborhood, in every freely falling reference frame all phenomena (including gravitational ones) are exactly as they are in the absence of external gravitational sources.*"

Note that both the Principle of Equivalence and Special Relativity are defined in terms of the *a priori* presence of multiple arbitrarily large finite masses and photons. Therefore neither the Principle of Equivalence nor Special Relativity can manifest in a spacetime that by mathematical construction either contains no matter or by mathematical construction contains only one mass. But all known solutions to Einstein's field equations pertain to a Universe that consists either no matter or only one mass. The Big Bang *creatio ex nihilo* consists of a Universe that is entirely filled by a continuous indivisible distribution of mass with a monotonically decreasing macroscopic density and pressure or a finite averaged macroscopic density and zero pressure in terms of the energy-momentum tensor for a perfect fluid. It therefore violates the Principle of Equivalence and Special Relativity as required by Einstein for his gravitational field.

Big Bang *creatio ex nihilo*, owing to its mathematical structure of a single continuous indivisible mass distribution throughout the entire Universe, cannot account for the presence of multiple masses, such as stars, black holes, and galaxies, bearing in mind that the Principle of Superposition does not apply in General Relativity as explained in Section 2 above. So once again it is a physically meaningless model. So how do the astrophysical scientists get multiple masses such as stars and galaxies and black holes (primordial or otherwise), photons, electromagnetic fields, nebulae, etc. into the Big Bang *creatio ex nihilo* Universe? Very simple; they do so by applying the Principle of Superposition in violation of the fact that in General Relativity the Principle of Superposition does not apply.

5. The Big Bang Equation

We shall now consider the so-called FLRW (Friedmann-Lemaître-Robertson-Walker) metric or line-element i.e. distance formula, given by

$$ds^2 = dt^2 - (R(t))^2\left(\frac{dr^2}{(1-kr^2)} + r^2\left(d\theta^2 + \sin^2\theta\, d\phi^2\right)\right) \quad (2)$$

where k, which is related to the spatial curvature, can take only the values -1, 0, 1, and the speed of light in vacuum c is set to unity so that the coefficient of dt^2 is 1 rather than c^2. The term $R(t)$ is a dimensionless scale factor that causes the spatial part of the metric to expand or contract, depending upon its form. There are a number of important things to note about this equation. It is obtained without any hypothesis about the presence of matter. The only requirements (see for example Tolman [1]) in its derivation are that the Universe is homogeneous and isotropic (an empty Universe is certainly homogeneous and isotropic) and on the assumption of homogeneity the metric is supposed to have spherical symmetry in the form,

$$ds^2 = dt^2 + g_{ij}dx^i dx^j \quad (i, j = 1,2,3).$$

The quantity g_{ij} is called the metric tensor. The dx^i and dx^j are not tensors; merely differential elements of the variables x^i and x^j.

The spatial geometry of the FLRW metric is,

$$d\sigma^2 = \left(\frac{dr^2}{(1-kr^2)} + r^2\left(d\theta^2 + \sin^2\theta \, d\phi^2\right) \right). \quad (3)$$

Metric (3) "*... is clearly flat if k = 0, but for k = ±1 it is curved. For k = 1 it is a space of constant positive curvature, the three-dimensional counterpart of a sphere, and the space is closed in the sense that it has a finite volume. For k = -1 it is a space of constant negative curvature, and it is open in the sense that its volume is infinite.*" [4]

In the derivation of metric (2) Tolman [1] remarks: "*... we have made no hypothesis as to the nature of the material filling the model ...*" "*We may now, however, introduce a more specific hypothesis by assuming the material filling the model can be treated as a perfect fluid.*"

"*... it will be noticed that the pressure and density are functions of the time t alone, and at a given value of t would be independent of position in the universe, in agreement with the spatial homogeneity of the model which we have already discussed.*"

McMahon [20] says,

"*To model the large-scale behavior of the universe such that Einstein's equations are satisfied, we begin by modeling the matter and energy in the universe by a perfect fluid. The particles in the fluid are galaxy clusters and the fluid is described by an average density ρ and pressure P.*"

Notice that McMahon has falsely 'slipped in' the notion of multiple masses by the words "*The particles in the fluid are galaxy clusters...*" despite introducing "*average density ρ and pressure P*" by which the model is mathematically constructed as a Universe entirely filled by a single continuous indivisible homogeneous mass in terms of a perfect fluid. McMahon effectively applies the Principle of Superposition where it does not apply in order to obtain multiple galaxies in what is a one body model by means of a perfect fluid with averaged density and pressure. This is the standard method by which the astrophysical scientists insert multiple masses (including the fictitious black holes) and photons and electromagnetic fields and everything else in the real Universe into what is in actual fact a mathematically constructed model that is either an empty Universe or a one-body Universe, since General Relativity cannot accommodate two or more masses. Tolman [1] however remarks,

In interpreting the expressions for density ρ_{oo} and pressure p_o ... it must be remembered that these quantities apply to the idealized fluid in the model, which we have substituted in place of the matter and radiation actually present in the real universe.

Now the Big Bang *creatio ex nihilo* with its associated expansion of the Universe is obtained by means of selecting an appropriate *ad hoc* value of k, since $k = 0$ and $k = -1$ produce continuously expanding universes when taken in conjunction with an *ad hoc* formulation of the scale factor $R(t)$ such that it produces expansion. In the case of $k = 0$ there is no upper limit on the quantity r in the spatial metric (2) and hence in the metric (3) and so it is effectively an infinite spherically symmetric flat Universe without the presence of $R(t)$ and so there is no meaning to expansion of an already infinite Universe. Taking $k = 1$ produces a universe that expands to a maximum value and then contracts back down to zero. In addition, $R(t)$ is conveniently formulated by means of a whole host of *ad hoc* assumptions, both quantitative and qualitative, which include re-interpreting the Hubble-Humason relation as a red-shift with recessional-velocity relation (i.e. Doppler effect on light) instead of the red-shift with distance relation originally proposed by Hubble and Humason, and making $R(t)$ such that $R(0) = 0$; for instance arranging things so that $R(t) = t^{2/3}$ or $R(t) = t^{1/2}$. All this is accomplished by a convoluted mathematical process in developing the Friedmann equations and playing around with them by adjusting various parameters in order to get the desired result – *creatio ex nihilo* and expansion of the Universe in the fashion of Lemaître's creationism.

There is no point in investigating the mathematical complexities associated with the so-called Friedmann equations because that would plunge us into the pointless drudgery of playing the physically meaningless mathematical games of Big Bang *creatio ex nihilo* and its associated falsity of expansion of the Universe. The fact that a model that treats the Universe as a single continuous indivisible distribution of mass has no physical meaning is sufficient to invalidate the model, especially when we recall that such a model cannot satisfy Einstein's requirement that his Principle of Equivalence and his Special Relativity must manifest in sufficiently small finite regions of his gravitational field and that both are defined in terms of the *a priori* presence of multiple arbitrarily large finite masses and photons; and that the Principle of Superposition does not apply in General Relativity; and that there are no known solutions to Einstein's field equations for two or more masses and no existence theorem by which it can even be asserted that his field equations contain latent solutions for two or more masses; and that it is easily proven that General Relativity is invalid because it violates the usual conservation of energy and momentum, placing it in conflict with experiment on a very deep level.

6. Invalidity of General Relativity

To satisfy the requirement of the usual conservation of energy and momentum Einstein introduced his 'pseudo-tensor', denoted by the symbol, t_{ν}^{μ}. We note that the pseudo-tensor, as we shall soon see, is *not* a tensor owing to its definition, and so it is not in keeping with Einstein's requirement that all equations of physics must be tensorial so that the laws of physics are the same for all observers independent of their motion. This is a serious problem. The notation for the pseudo-tensor requires explanation. We see that there are two suffixes, one superscript and one subscript. Thus it is 2nd-order and mixed: its covariant order is one and its contravariant order is one and so its total order is two. We can write the energy-momentum tensor as a mixed tensor as well, thus: T_{ν}^{μ} and also the Einstein tensor too, G_{ν}^{μ}. We can also write the metric tensor as g_{ν}^{μ}. Similarly we can write these tensors in contravariant form: $T^{\mu\nu}$, $G^{\mu\nu}$ and $g^{\mu\nu}$ merely for mathematical convenience when necessary. How this is achieved is of no consequence for our discussion.

Now, according to Paul [17] the components of Einstein's pseudo-tensor are, '*the "energy components" of the gravitational field*'.

According to Eddington [21] t_{μ}^{σ} denotes "*...potential energy, momentum and stress*", and he also says, "*We call t_{μ}^{σ} the pseudo-tensor-density of potential energy.*"

Einstein [22] asserts that the sum of the energy and momentum of his gravitational field and its sources is given by,

$$\mathbf{E} = \left(t_{\mu}^{\sigma} + T_{\mu}^{\sigma}\right). \qquad (4)$$

Note that this is *not* a tensor sum since t_{μ}^{σ} is not a tensor. For energy and momentum to be conserved the divergence of the expression for the total energy and momentum of the gravitational field and the sources thereof must be zero. But the divergence of Einstein's expression for the conservation of energy and momentum is an *ordinary divergence*, not a *tensor divergence*, contrary to his requirement that all the equations of physics be tensorial. Einstein [22] gives the *ordinary divergence* of his energy-momentum expression thus;

$$\frac{\partial\left(t_{\mu}^{\sigma} + T_{\mu}^{\sigma}\right)}{\partial x_{\sigma}} = 0 \qquad (5)$$

Einstein [22] says of equation (5),

"*Thus it results from our field equations of gravitation that the laws of conservation of momentum and energy are satisfied.*"

"*... we have to introduce the totality of the energy components of matter and gravitational field.*"

Now Einstein's allegation that by equation (5) "*... the laws of conservation of momentum and energy are satisfied*" is completely false because Einstein's pseudo-tensor is a meaningless concoction of mathematical symbols and so it *cannot* be used to make any calculations or to represent any physical entity or to model any physical phenomena. Thus, Einstein's expression (4) for the total energy and momentum of his gravitational field and the ordinary divergence (5) of it are nonsense. Here is the proof. Einstein's pseudo-tensor is defined as [23]:

$$\sqrt{-g}\, t_{\nu}^{\mu} = \frac{1}{2}\left[\delta_{\nu}^{\mu} L - \left(\frac{\partial L}{\partial g_{,\mu}^{\sigma\beta}}\right) g_{,\nu}^{\sigma\beta}\right]$$

where

$$L = -g^{\alpha\beta}\left(\Gamma_{\alpha\kappa}^{\gamma}\Gamma_{\beta\gamma}^{\kappa} - \Gamma_{\alpha\beta}^{\gamma}\Gamma_{\gamma\kappa}^{\kappa}\right)$$

Quite often in the literature $\partial g_{ab}/\partial x^{c}$ is written in

$$\Gamma_{bc}^{a} = \frac{1}{2} g^{ad}\left(\frac{\partial g_{dc}}{\partial x^{b}} + \frac{\partial g_{bd}}{\partial x^{c}} - \frac{\partial g_{bc}}{\partial x^{d}}\right)$$

the simplified form $g_{ab,c}$ and $\partial g^{ab}/\partial x^{c}$ as $g_{,c}^{ab}$. In tensor analysis we can perform a simple operation called *contraction* on a mixed tensor, which reduces the order of the tensor by two. We *contract* a mixed tensor (and Einstein's pseudo-tensor) by setting $\nu = \mu$. So contracting Einstein's pseudo-tensor gives an *invariant* t, thus [23]:

$$\sqrt{-g}\, t_{\mu}^{\mu} = \frac{1}{2}\left[4L - \left(\frac{\partial L}{\partial g_{,\mu}^{\sigma\beta}}\right) g_{,\mu}^{\sigma\beta}\right] \qquad t_{\mu}^{\mu} = t$$

Performing the calculation of the second part inside the brackets (the details are not important for our purposes) gives:

$$\left(\frac{\partial L}{\partial g_{,\mu}^{\sigma\beta}}\right) g_{,\mu}^{\sigma\beta} = 2L$$

Substituting this result into the expression above and rearranging gives the invariant:

$$t = \frac{L}{\sqrt{-g}}$$

where g is the determinant of the metric tensor and hence is composed of the components of the metric tensor. By the definitions of L, Γ^a_{bc} and g, we see that t is an invariant that is composed *solely* of the components of the metric tensor and their first derivatives[23]. Tolman[1] also remarks:

"*... it will be noted that the value of* t^{ν}_{μ} *at any point will be determined by the values of the components of the metrical tensor* $g_{\alpha\beta}$ *and their first derivatives* $\partial g_{\alpha\beta}/\partial x^{\gamma}$ *at that point.*"

Now the pure mathematicians G. Ricci-Curbastro and T. Levi-Civita [24], inventors of the tensor calculus, proved in 1900 that invariants that are composed *solely* of the components of the metric tensor and their first derivatives *do not exist!* Thus Einstein's pseudo-tensor is totally meaningless and hence his formulation of the usual conservation of energy and momentum totally invalid.

The upshot of this is that Einstein's field equations *must* take the following form [23,25]

$$\frac{G_{\mu\nu}}{\kappa} + T_{\mu\nu} = 0 \tag{6}$$

The $G_{\mu\nu}/\kappa$ are the components of a gravitational energy tensor. This expression is not only the necessary form of Einstein's field equations, it is also an expression for the total sum of the energy and momentum of his gravitational field (compare with equation (4) above). Now the tensor divergence of both sides of this equation is zero so energy and momentum are conserved *but* the total energy of Einstein's gravitational field *is always zero*; the $G_{\mu\nu}/\kappa$ and the $T_{\mu\nu}$ *must vanish identically* (so that when $T_{\mu\nu} = 0$ there is no spacetime and hence no gravitational field); there is *no possibility* for the localization of gravitational energy (i.e. *there are no Einstein gravitational waves*). This also means that Einstein's gravitational field *violates* the experimentally well-established usual conservation of energy and momentum making them inconsistent with experiment on a deep level and hence invalid. According to Pauli [17] Einstein

"*... raised the objection that, with this definition of the gravitational energy, the total energy of a closed system would always be zero, and the maintenance of this value of the energy does not require the continued existence of the system of one form or other. The usual kind of conclusions could not then be drawn from the conservation laws.*"

But Einstein's objections are futile on account of the failure of his formulation of the usual conservation of energy and momentum. Thus, General Relativity is invalid.

We can of course rewrite equation (6) in mixed tensor form so that it more closely resembles expression (4), thus

$$\frac{G^{\mu}_{\nu}}{\kappa} + T^{\mu}_{\nu} = 0.$$

In either case the result is necessarily the very same.

7. The Cosmic Microwave Background

In view of the foregoing discussion it is quite clear that the so-called Cosmic Microwave Background Radiation (CMBR) is not the afterglow of a Big Bang *creatio ex nihilo* since the Big Bang Cosmology has no valid basis whatsoever in any scientific theory, and especially since General Relativity itself is invalid.

What then is the true nature of the CMBR? In 1965 Penzias and Wilson [26] reported the detection of an isotropic signal that they interpreted as of cosmic origin. However, their observations were from the ground and in all directions they pointed their antenna they detected this signal. There was no valid reason for them to have supposed that this signal is of cosmic origin. It did not occur to them and the astrophysical scientists of the time that the signal they detected could have been quite possibly due to emissions from the Earth that are scattered by the atmosphere thereby resulting in an isotropic signal from an Earth-bound anisotropic source. Contemporary astrophysical scientists still cling to the alleged cosmic origin of the CMBR as a remnant of the Big Bang *creatio ex nihilo* and rely heavily on this claim to validate Big Bang creationism.

In recent years it has been alleged by the scientific teams of the COBE and WMAP satellites that the CMBR has been detected as a cosmic source. However, these claims do not stand up to scientific analysis by any stretch of the imagination. Professor Pierre-Marie Robitaille of Ohio State University, an foremost expert in imaging science, has carried out very detailed analyses of the COBE [27] and WMAP [28] satellites and the reports of their scientific teams and has revealed thereby that both COBE and WMAP have contributed nothing of any value to science other than confirming the presence of a dipole signal, already detected previously by the Russian Relikt-1 satellites [29] and experiments with balloons. Robitaille has shown without any doubt that both COBE and WMAP are so riddled with design faults and inappropriate and substandard signal processing methods that neither satellite has produced anything useful to science. For instance, the imaging instruments of WMAP have an effective signal to noise ratio barely greater than unity, at best, and so WMAP is incapable of distinguishing signal from noise, yet its scientific team claims to have successfully mapped the Galaxy. Robitaille [28] remarks:

WMAP is unable to confirm that the 'anisotropic signal' observed at any given point is not noise. The act of attributing signal characteristics to noise does not in itself create signal.

The requirement that the signals of interest are frequency independent cannot be met, and has certainly never been proven.

There is no single map of the anisotropy, since all maps are equally valid, provided coefficients sum to 1.

There is no unique solution and therefore each map is indistinguishable from noise. There are no findings relative to anisotropy, since there are no features in the maps which could guide astrophysics relative to the true solution.

WMAP images do not meet accepted standards in medical imaging research.

The radiation shield of the FIRAS instrument on COBE is defective and so signal from the sky has been diffracted over the shield into the FIRAS horn. Furthermore, the scientific teams of both the COBE and WMAP satellites claim to have extracted a signal that is ~1000 times weaker than the galactic foreground (i.e. the noise) when they have no *a priori* knowledge of the nature of the signal and no ability to manipulate the signal source. They have therefore claimed to have achieved a feat with obsolete equipment that laboratory experience has shown to be impossible, even with the most sophisticated equipment in the best radiological laboratories in the world today.

George Smoot [30], the principal investigator for the COBE Differential Microwave Radiometers (DMR), relates that to extract the weak multipoles by data processing, which Smoot calls "*wrinkles in the fabric of time*", required first the removal of the dipole, galactic foreground, and the quadrupole signals. Smoot puzzled over why the multipoles did not appear until the quadrupole was finally removed by data processing methods, since the raw data contained no systematic signal variations. Robitaille's [27] answer is simple: "*when Smoot and his colleagues imposed a systematic removal of signal, they produced a systematic remnant. In essence, the act of removing the quadrupole created the multipoles and the associated systematic anisotropies.*" Smoot's "*wrinkles in the fabric of time*" are nothing more than consistent residual ghost signals produced by his data processing. The appearance of such systematic ghost signals throughout an image when processing large contaminating signals is very well known in medical radiology. Robitaille [27,28 advises that "*Apparent anisotropy must not be generated by processing.*"

Owing to the very many defects in the COBE satellite and the inappropriate and substandard signal processing methods applied by the COBE-FIRAS team, the claim that COBE determined the *most perfect blackbody spectrum ever measured* is patently false.

Robitaille [27-29,31,32] has cogently argued that the CMBR is actually due to emissions from the oceans of the Earth, scattered by the atmosphere, thereby producing an isotropic signal from an anisotropic source. He has pointed out that we know from submarines at sea and from microwave ovens in the home that water is a powerful absorber, and hence emitter, in the far-infrared and microwave bands. Robitaille [33] has also analysed the emissions from hydrogen bonds in water and proven that Kirchhoff's Law of Thermal Emission [34-36] is not universal, contrary to Kirchhoff's claims and adopted by physicists ever since. Crothers [37] remarks,

The COBE and WMAP teams model the Earth as a blackbody source of emission at ~ 280 K. But Robitaille points out that "since the oceans are not enclosed" they do not satisfy the requirements for application of Kirchhoff's law of thermal emission, and so the emission profiles of the oceans "do not necessarily correspond to their true temperatures." By means of scattering in steady-state conditions, Robitaille argues: "Consequently, a mechanism for creating isotropy from an anisotropic ocean signal is indeed present for the oceanic ~3 K Earth Microwave Background."

Even in the unlikely event that the so-called CMBR is not ultimately reassigned to the Earth's oceans this does not alter the fact that neither COBE nor WMAP have contributed anything of value to science, or the fact that the CMBR is not the afterglow of a Big Bang *creatio ex nihilo*, since Big Bang Cosmology is fiction.

Dedication

I dedicate this paper to my late brother: Paul Raymond Crothers 12th May 1968 – 25th December 2008.

References

[1] Tolman, R.C. (1987) Relativity Thermodynamics and Cosmology, New York: Dover Publications Inc.
[2] Alfvén, H.
http://public.lanl.gov/alp/plasma/people/alfven.html
[3] Einstein, A.
www.catholiceducation.org/articles/science/sc0022.html
[4] Foster J. & Nightingale, J.D. (1995) A Short Course in General Relativity, New York: Springer-Verlag.
[5] Carroll, B. W. & Ostlie D. A. (1996) An Introduction to Modern Astrophysics, Reading: Addison-Wesley.
[6] Einstein, A. (1967) The Meaning of Relativity, New York: Science Paperbacks and Methuen & Co. Ltd.
[7] Hawking, S.W. (2002) The Theory of Everything, The Origin and Fate of the Universe; Beverly Hills: New Millennium Press.
[8] Schutz, B.F. (1990) A First Course In General Relativity, Cambridge: Cambridge University Press.
[9] Wald, R.M. (1984) General Relativity, Chicago: Univ. Chicago Press.
[10] Crothers, S.J. (2010) The Black Hole, the Big Bang: a Cosmology in Crisis, 2010, http://vixra.org/abs/1103.0047

[11] Crothers, S.J. (2012) Proof of no "black hole" binary in Nova Scorpii, Global Journal of Science Frontier Research Physics and Space Science, Volume 12, Issue 4 Version 1.0, http://vixra.org/pdf/1206.0080v2.pdf
[12] Crothers, S.J. (2012) General relativity – A theory in crisis, Global Journal of Science Frontier Research Physics and Space Science, Volume 12, Issue 4 Version 1.0.
[13] Crothers, S.J. (2009) The Schwarzschild solution and its implications for gravitational waves, mathematics, physics and philosophy, in the Interpretations of Relativity Theory, Budapest, 4–6 September. www.sjcrothers.plasmaresources.com/Budapest09-b.pdf
[14] Crothers, S.J. (2005) On the general solution to Einstein's vacuum field and its implications for relativistic degeneracy, Progress in Physics, Vol. 1, http://vixra.org/abs/1012.0018
[15] McVittie, G.C. (1978) Laplace's alleged "black hole", The Observatory, v.98, 272, www.sjcrothers.plasmaresources.com/McVittie.pdf
[16] Heaviside, O. (1893) Electromagnetic Theory, 1,133.
[17] Pauli, W. (1981) The Theory of Relativity, New York: Dover Publications.
[18] Taylor, E.F. & Wheeler, J.A. (2000) Exploring Black Holes – Introduction to General Relativity, Reading: Addison Wesley Long- man (in draft).
[19] Matzner, R.A. (ed.) (2001) Dictionary of Geophysics, Astrophysics, and Astronomy, Boca Raton: CRC Press LLC.
[20] McMahon, D. (2006) Relativity Demystified, New York: McGraw-Hill.
[21] Eddington, A.S. (1960) The mathematical theory of relativity, 2nd edition, Cambridge: Cambridge Univ. Press,
[22] Einstein, A. (1916) The foundation of the general theory of relativity, *Annalen der Physik*, 49.
[23] Levi-Civita, T. (1917) Mechanics. - On the analytical expression that must be given to the gravitational tensor in Einstein's theory, *Rendiconti della Reale Accadmeia dei Lincei,* 26, 381, http://arxiv.org/pdf/physics/9906004.
[24] Ricci-Curbastro G. & Levi-Civita, T. (1900) *Méthodes de calcul différentiel absolu et leurs applications, Matematische, Annalen*, B. 54, p.162.
[25] Lorentz, H.A. (1916) *Versl. Gewone Vergad. Akad.*, 24, 1389 and 1759; 25, 468 and 1380.
[26] Penzias, A.A. & Wilson, R.W. (1965) A measurement of excess antenna temperature at 4080 Mc/s. Astrophys. J., v.1, 419-421.
[27] Robitaille, P-M (2009) COBE: A radiological analysis, Progress in Phys., v.4, 17-42, www.ptep-online.com/index_files/2009/PP-19-03.PDF
[28] Robitaille, P-M (2007) WMAP: A radiological analysis, Progress in Physics, v.1, 3-18, www.ptep-online.com/index_files/2007/PP-08-01.PDF
[29] Robitaille, P-M (2007) On the origins of the CMB: insight from the COBE, WMAP, and Relikt-1 Satellites, Progress in Physics, www.ptep-online.com/index_files/2007/PP-08-02.PDF
[30] Smoot, G & Davidson, K. (1993) Wrinkles in time: witness to the birth of the Universe, New York: Harper Perennial.
[31] Robitaille, P-M (2007) On the earth microwave background: absorption & scattering by the atmosphere, Prog. Phys, www.ptep-online.com/index_files/2007/PP-10-01.PDF
[32] Robitaille, P-M (2008) The earth microwave background (emb), atmospheric scattering and the generation of isotropy, Progress in Physics, www.ptep-online.com/index_files/2008/PP-13-24.PDF
[33] Robitaille, P-M (2009), Water, hydrogen bonding, and the microwave background, Prog. Phys, Vol.2, www.ptep-online.com/index_files/2009/PP-17-L2.PDF
[34] Robitaille, P-M (2009) Kirchhoff's law of thermal emission: 150 years, Prog Phys. www.ptep-online.com/index_files/2009/PP-19-01.PDF
[35] Robitaille, P-M (2009), A critical analysis of universality and kirchhoff's law: a return to stewart's law of thermal emission, Progress in Physics. *www.ptep-online.com/index_files/2008/PP-14-06.PDF*
[36] Robitaille, P-M (2009) Blackbody radiation and the loss of universality: implications for planck's formulation and boltzman's constant, www.ptep-online.com/index_files/2009/PP-19-02.PDF
[37] Crothers, S.J. (2010) COBE and WMAP: Signal analysis by fact or fiction? Electronics World, http://vixra.org/abs/1101.0009, http://viewer.zmags.com/htmlCat/index.php?mid=rpdgfh&pageid=25

The Neurogenetic Correlates of Consciousness

JOHN K. GRANDY

Whitestone Consulting LLC- Department of Defense Fort Drum NY, USA
North Country Urgent Care, Watertown New York, USA
khyber_john@yahoo.com

The neurogenetic correlates of consciousness (NgCC) is a new field of consciousness studies that focuses on genes that have an effect on or are involved in the continuum of neuron-based consciousness. A framework of consciousness based on the neural correlates of consciousness (NCC) has already been established by Francis Crick and Christof Kock. In this work I propose that there are NgCC underlying the NCC which are both active during the conscious experience. So how are genes involved? There are two significant connections between DNA and neurons that are involved in the conscious experience. First, any brain system can be adversely affected by underlying genetic abnormalities which can be expressed in an individual at birth, in adulthood, or later in life. Second, the DNA molecule does not lay dormant while the neuron runs on autopilot. DNA is active in translating and transcribing RNA and protein products that are utilized during neuron functioning. Without these products being continuously produced by the DNA during a conscious experience the neurons would cease to function correctly and be rendered unable to provide a continuum of human consciousness. Consequently, in addition to NCC, NgCC must be factored in when appreciating a conscious event. In this work I will discuss and explain some NgCC citing several examples.

Keywords: Neurogenetic correlates, Consciousness, DNA consciousness, Interaction-based model of consciousness, Neurogenetic phases of consciousness.

1. Introduction

The notion of consciousness continues to baffle and mystify man. What has always been problematic in consciousness studies, as well as physics, is the concept of the observer. The phenomenon of the observer has also made it clear that a *quantum enigma* exists, which is where physics meets consciousness [1]. Often when we speak of human consciousness we are referring to, at least partially, the functioning of the brain. Consequently, we could also view the quantum enigma as being where physics meets the neuron. However, we must ask how do neurons and ultimately the brain arise and manifest consciousness? This could shed light on how the phenomenon of the quantum enigma arises.

Currently, global networking models that involve many brain regions that are dynamically interacting have been proposed. The materialist and neuro-biological models of consciousness maintain that human consciousness is a manifestation of the brain and the neurons. In 2003, a framework of consciousness that was based on the neural correlates of consciousness (NCC) was proposed by Francis Crick and Christof Koch [2]. In this proposal brain systems are active in tandem with the conscious experience. More recently specific areas of the brain have been the focus of research. Some of these brain regions were discussed and summarized at the *Towards a Science of Consciousness 2012 Conference* [3]. Here is a brief summary of some of the brain regions that appear to be NCC:

- Areas affected by anesthesia e.g. the frontal cortex integration to the posterior parietal cortex.
- Decreases in brain connectivity and cerebral integration seen in PET scans and fMRI studies in patients in vegetative states (unresponsive wakefulness syndrome) and minimally conscious states.
- Frontoparietal connections that provide a global workspace. Two examples of these connections are: 1) lateral prefrontal and parietal cortices that function to provide *external sensory awareness* and 2) precuneal and mesiofrontal midline activity functions to provide an *internal awareness*.
- Critically emergent properties of collective wide-spread connectivity of consciousness found in the thalamo-cortical regions.

With these four examples above we can appreciate specific brain regions that can extend globally in order to interact during the process of human consciousness. Each brain region is composed of specialized neurons that give that brain region its unique character. Therefore, the proper functioning of the neurons is pivotal to the process of human consciousness as well.

However, is it possible that there is something tangible beneath the neuron?

In previous works I have proposed that the DNA molecule is just as important to human consciousness as the neurons and the brain. In an attempt to establish a genetic account of consciousness I have proposed the neurogenetic correlates of consciousness (NgCC). In this work I will identify several NgCC and discuss their importance to consciousness. However, consciousness is interpreted and defined very differently by scholars and scientists. Before I discuss any of the NgCC I will first establish what I mean by consciousness and how these three neurogenetic phases were derived.

2. The Interaction-Based Theory of Consciousness

There are many different theories of consciousness. When I discuss consciousness I implement the *interaction-based theory of consciousness*. This theory defines consciousness as the interaction of *things* (be it an organism, DNA molecule, or atom) with other *things*, the external environment, different forms of energy, and forces. Essentially this theory maintains that consciousness *is* interactions. By accepting this premise everything in the universe has a *degree* of consciousness ranging from quarks to molecules to cells to brains as they all interact in various degrees. Another way of depicting this is to imagine a universe with no interactions. In this depiction would consciousness exist or would consciousness appear frozen?

If we look at the smaller end of the scale, for example molecules, and observe those interactions we will notice that as the interactions increase in any given system there is a resulting increase in complexity. So the smaller molecules interact and become larger molecules which become more complex and can ultimately give rise to cells. At the cellular level there is a larger degree of consciousness than that of the simple molecules. This is based on the cell's ability to interact on a more complex level than the individual molecule. Therefore, as *things* interact they become more complex.

In this model we also observe that as complexity increases so does the degree of consciousness. For example if an individual cell has a degree of consciousness (sometimes referred to as cellular consciousness or polyzoism) and if we compare this degree of consciousness to a brain we will notice an obvious difference between those two degrees of consciousness. The degree of consciousness seen in the brain is higher due to its biological complexity. Another way of saying this is that the complexity of a system is directly related to the complexity of the consciousness experience or that degree of consciousness.

If we go back to the beginning of this idea of the interaction-based model of consciousness we will see this conceptual relationship emerge- things interact (which is consciousness) and become more complex and as complexity increases so does this degree of consciousness. In this simplistic version the concept of interaction-complexity-consciousness (ICC) can be appreciated. In this concept of the ICC there is a perpetuation that goes in an ascending fashion from the level of quarks to molecules and from cells to multicellular organisms. It is possible that this schematic may reflect a charge-mass aggregation that accounts for this scale dependency. Similar concepts have been proposed by Wolfgang Baer [4]. In the Baer proposal there is a force (Fc) that physically balances mass and charge forces. This involves a cognitive process loop in where a description of a phenomenological experience is converted (in a quantum physics sense) into a description of the model of a physical world that we believe- when applied to human consciousness. The description is then converted back and thus completing the cognitive process loop.

So what does the interaction-based model of consciousness and the ICC have to do with DNA or the neurogenetic correlates of consciousness? Next, allow me to describe the theory of DNA consciousness.

3. DNA Consciousness

While observing the trajectory of the ICC that proceeds in an ascending fashion from low complexity (lower degree of consciousness) to higher complexity (higher degree of consciousness) we can appreciate that something special seems to happen in the molecular realm when nucleotides begin to assemble and give rise to RNA and DNA. The evolution of networks of RNA and DNA allows the emergence of higher degrees of consciousness at the microscopic level with cells and then eventually at the macroscopic scale with the evolution of multicellular organisms. At this point we have DNA and RNA at a particular degree of consciousness and at the same time possessing the ability to give rise to higher degrees of consciousness. This phenomenon is known as *DNA consciousness*.

In essence the theory of DNA consciousness maintains two main ideas. First, that DNA has a degree of consciousness within the ICC where everything has a degree of consciousness. Second, that DNA can give rise to higher degrees of consciousness. However, why does DNA receive all the credit? Is not RNA just as important to the assembly of the cell? The reason why DNA is given precedence over RNA is due to the fact that RNA by itself cannot give rise to macroscopic organisms and the subsequent higher degrees of consciousness. Although some RNA viruses exist in nature no macroscopic RNA-based organisms have been identified to date.

The theory of DNA consciousness forces us to view DNA as not merely a genetic storage unit. DNA is dynamic, autopoietic, and a form of consciousness [5]. In addition, DNA is required for the transformation of macroscopic scales of consciousness and without DNA consciousness higher degrees of consciousness cannot emerge from the molecular degree of consciousness.

The importance of DNA consciousness in the ICC and development of higher degrees of consciousness lay in the ability of the DNA molecule to create and store genetic characteristics. For example the development of novel genes allows cells to become neurons. Another way of stating this is that in order for neurons to emerge new genes are required to evolve. I will briefly discuss two of these gene families to serve as examples of this statement.

One gene family that plays a significant role in the emergence of neurons is the synapsins gene family [6]. Synapsins I, II, and III are not expressed in other cells, only neurons, and these genes allow the synapses of the neuron to change [7]. This ability of the neuron's synapse to change is known as neuron plasticity. A second gene that is important to the development of vertebrate neurons is the *myelin transcription factor gene* which allows the production of myelin [8]. Myelin is important because it allows faster transduction of nerve impulses. However, the myelin transcription factor gene is only found in the vertebrates.

At this point multicellular organisms begin to evolve that utilize neurons to interact with the environment in a more complex fashion. Once again the development of new genes allows the increase in degrees of consciousness which is the connection that DNA consciousness provides. Another example is when the Hox and Pax gene families emerge and will eventually pave the way for the appearance of the primate brain- some of these genes will be discussed in further sections. These two gene families allow for the centralization and cephalization of the central nervous system. The human brain emerges from this schematic and gives rise to human consciousness. I have proposed that this happens in three neurogenetic phases:

- Phase One: The emergence of neuron-based consciousness
- Phase Two: The continuum of neuron-based consciousness
- Phase Three: Neurodegeneration

Now that I have established what I mean by consciousness and how we have arrived at the three neurogenetic phases of human consciousness- first the interaction-based theory of consciousness gives rise to the concept of the ICC, second within the ICC DNA consciousness emerges, and third DNA consciousness will eventually give rise to neurogenetic correlates of human consciousness that are categorized in three phases. As I stated earlier the NgCC are genes that have an effect on or are involved in the process of human consciousness. Next, I will discuss a few genes in each of the three phases. I will not go into great molecular detail but rather only a brief summary as I discuss these genes in more detail in two other forthcoming works [9, 10].

4. The Neurogenetic Correlates of Consciousness — Phase One

In phase one the focus is on the emergence of neuron-based consciousness. The neurons and brain regions that develop will be the machinery that allows the degree of human consciousness to function. In this section I will briefly discuss four genes that are involved in this phase to use as examples while fully recognizing that many more are active during this process- Pax3, Pax6, Otx1, and Oxt2.

The Pax3 gene produces the Pax3 protein which is a transcription factor. This is active during the formation of neural tube. It is also serves a mandatory function in the closure of the neural tube and this is accomplished by p53 ubiquitination which is the inactivation of a protein by attaching the regulatory protein ubiquitin [11]. The p53 protein is signification in the cell cycle as it delays the transition from G1 phase to G0 phase. During neural tube development cells divide undifferentiated in G1. In G0 mitosis suppressor proteins activate tissue specific proteins. Closure of the neural tube needs to be complete prior to this is and Pax3 is vital to this process.

The function of the Pax3 gene is significant to the proper formation of the neural tube in the early embryonic stages of development. However, it also acts as a 'master gene' that controls the action of other genes involved in brain formation. For example it has been demonstrated that Pax3 regulates the Hes1 and Neurog2 genes, which is done by acetylation (the addition of a $COOH_3$ group) [12]. These two genes have very important roles in the development of the brain. Hes1 is involved in the maintenance of neural stem cells that determine what type of neurons will develop in the nervous system and Neurog2 plays a role in the specification of neuronal subtypes, sensory neurogenesis, and neural crest cell neurogenesis.

Another example of how Pax3 acts as a master gene is demonstrated when it interacts with Pax7 to regulate Meis2 which is required for the development of a region of the brain called the tectum [13]. The tectum is located in the midbrain and is made up of the inferior colliculi and superior colliculi which perform an auditory and visual function, respectively. This is a more specific example of how Pax3 is involve in the control of another gene (Meis2) that gives rise to a region of the brain that is involved in human consciousness.

Pax6 is a highly conserved gene in the animal kingdom and is involved in eye development [14]. It also is involved in the regulation of Spag5 which is associated with spindle machinery and regulation of kinetochore microtubule dynamics during cell division [15], influences the fates of the glial promoter cells of the forebrain [16], and regulates the neurogenesis or Neuro-Glia2 progenitor cells in the hippocampus [17]. Additionally, Pax6 is also involved in the proper development of the thalamocortical neurons [18] which are extremely significant to human consciousness. Recall that critically emergent properties of collective widespread connectivity of consciousness that is located in the thalamocortical regions was one of the NCC mentioned earlier in the introduction.

Two homeobox genes (Hox) Otx1 and Otx2 are required for the formation of a region in the embryonic brain called the zona limitans intrathalamica (ZLI) [19]. The ZLI is an important developmental boundary that is crucial for the development of the thalamus. Otx1 and Otx2 are also important for the development of the midbrain in the anteroposterior direction and dorsoventral direction [20]. This is accomplished by a dose-dependent antagonism of Shh gene in the dorsoventral direction and the Fgf8 gene in the anteroposterior direction.

These two genes have additional roles in the development of other brain regions. The Otx1 gene is critical for the development of the neocortex [21] and also affects the overall size of the cerebral cortex [22]. The Otx2 gene is critical in the development of the diencephalon, mesencephalon, choroid plexus, basal telencephalon, and hippocampal anlage [21] and regulates the neuronal progenitor domains of the ventral midbrain (tegmentum) [20].

What does Pax3, Pax6, Otx1, and Otx2 tell us about the emergence of neuron-based consciousness? The functions that I have enumerated demonstrate that the emergence of human consciousness is an orchestrated event that depends on the interactions of many genes, with some acting as master genes. This is not a random event that serendipitously evolved. It is a degree of consciousness- DNA consciousness which humans cannot perceive but can appreciate the end result.

5. The Neurogenetic Correlates of Consciousness — Phase Two

In phase two the focus is on the continuum of neuron-based consciousness. This continuum is composed of the continuous series of neuronal events active during the continuous seamless process of human consciousness. In phase one we saw that genes acting under a degree of consciousness- DNA consciousness give rise to the parts of the brain that will give rise to human consciousness. After the parts of the brain are formed DNA needs to dynamically and continuously be active in tandem with the conscious experience. In this section I will briefly discuss seven genes (some are transcription factors that are altered versions of the original gene product)- PTCHD1, three schizophrenia-associated genes (PDE$_4$B, DISC$_1$, and ZNF804a transcription factor), BDNF, FGF2, and the ΔFosB transcription factor. I will demonstrate that when there is a malfunction of some of these genes e.g. a mutation, deletion, or a single nucleotide polymorphism (SNP) that there is an adverse effect on the degree of human consciousness.

Disruptions in the gene locus of the PTCHD1 gene are strongly related to autism-spectrum disorders and intellectual disability [23]. This is significant in understanding how a gene locus disruption can have an effect on the continuum of human consciousness as autism is a well characterized clinically. In autism there is impairments in the abilities to communicate and socialize; particularly with reciprocal interactions. These patients appear introverted and out of touch for the most part with the external world.

Schizophrenia (SZ) is a mental illness in where there is a noticeable disturbance in the sensorium, improper processing of information in the brain, and the perception of stimuli that is not actually there e.g. visual and auditory hallucinations. Other symptoms accompanying SZ can be paranoia and delusions in where the patient misinterprets reality. Recently several SZ-associated genes have been under investigation. Here I will discuss three of them.

Two gene mutations have shown a strong correlation with SZ. The PDE4B and DISC1 genes produce the two proteins respectively that form the PDE4B-DISC1 complex. Mutations in either of these two genes which would result in a dysfunction of this complex have been strongly implicated in many of the molecular mechanisms underlying SZ [24, 25, & 26]. Although PDE4B and DISC1 interact extensively they also have individual functions. PDE4B regulates cAMP signaling cascades in neurons and is involved in the synaptic plasticity during learning and memory. DISC1 serves as a part of the microtubule organizing center called the centrosome, as well as being involved in cytoskeletal processes in neuronal migration.

ZNF804a is a transcription factor that regulates the expression of PDE4B and three other genes- PRSS16, COMT, and DRD2, which are all considered SZ-associated genes. So once again we have an example of a gene that regulates other genes that can have an effect on the continuum of consciousness. Recently ZNF804a expression has been proposed as a candidate mechanism that confirms SZ risk [27]. This is due to the fact that it influences the expression of these other SZ-associated genes.

Neuron plasticity is the ability of neurons in the brain to change and make new connections with other neurons or dissolve weakened connections. This happens every day and gives the brain the ability to change in response to new information. Neuron plasticity has been associated with the BDNF gene, FGF2 gene, and ΔFosB transcription factor.

BDNF and FGF2 genes produce proteins that are well known neurotrophic factors that play vital roles in neuron plasticity which underlies processes e.g. learning and memory [28, 29]. Both genes are also known to have decreased activity in disorders e.g. major depressive disorder and Alzheimer's disease [30, 31]. However, I will not discuss these genes in detail as I have done this in other works [9, 10]. At this juncture it is important to note that these two genes are associated with neuron plasticity and when either of their activities are decreased there is an association with decreases in degrees of consciousness that are observable in disorders like Alzheimer's disease and major depressive disorder.

The ΔFosB transcription factor, which is a truncated product of the FosB gene, is involved in neuron plasticity in regions of the brain e.g. the amygdala. The amygdala is involved in many processes that are important to consciousness i.e. in pain perception and rating the emotional response to the environment. The expression of the ΔFosB transcription factor can be affected by opioids which can result in the decrease volume of the amygdala seen on brain imaging studies [32]. These changes in the amygdala also have demonstrated correlations to clinical symptoms associated with addiction e.g. opioid-induced hyperalgesia (an increase in pain sensitivity secondary to opioid use) and hyperkatifeia (a hypersensitivity to negative emotional states and/or increased intensity of emotional distress secondary to opioid use) [33]. The ΔFosB transcription factor has been shown to affect the expression of several genes; for example it induces the expression of GluR2, Cdk5, and NFκB genes, whereas it represses the expression of the dynorphin gene [34]. Addiction is also interesting in relation to consciousness as addictive behaviors tend to be counter intuitive to self-preservation. These changes in gene expression cause profound changes in the central nervous system which contribute to drug-seeking behavior.

So what do the genes that were discussed in the second phase tell us? They demonstrate that disruptions or mutations in certain genes can have an affect on modalities of human consciousness, which can be manifested as disorders like autism and SZ. Secondly, drugs can affect genes that have downstream affects on brain regions that can influence the continuum of consciousness. Lastly, one gene product can influence several other genes that can affect multiple brain regions. This was seen with the ΔFosB and ZNF804a transcription factors.

Overall there is a collective genetic system that possesses a degree of consciousness which in turn gives rise to the continuum of human consciousness. If there is any damage or abnormal expression in certain genes an observable change in the continuum of human consciousness can be manifested which can be studied objectively.

6. The Neurogenetic Correlates of Consciousness — Phase Three

In phase three the focus is on neurodegeneration or the breakdown of the neurons and ultimately brain regions. The breakdown of certain brain regions due to pathological conditions which are involved in the cognitive process illustrates the direct relationship of human consciousness to the brain. In some of these pathological conditions genes are involve and a good example of this is seen in Alzheimer's disease (AD). In this section I will briefly discuss four genes- APP, PSEN1, PSEN2, and APOE-ε4 (and their protein products). These four genes are all strongly associated with AD [35]- in fact these genes are being explored to diagnose AD long before symptoms emerge later in life [36].

One of the main pathological features of AD is the accumulation of beta-amyloid protein in the brain, which eventually causes damage to the neurons. Mutations in any of the four AD genes- APP, PSEN1, PSEN2, and APOE-ε4 are associated with AD as these mutations produce abnormalities which contribute to the accumulation of beta-amyloid.

AD is commonly associated with memory loss, decline in cognitive functions, and behavioral abnormalities. These signs and symptoms are a result of prolonged damage to the neurons. The loss of neurons is what is referred to as neurodegeneration. As already mentioned AD-associated genes can accelerate this process. As the process of neurodegeneration progresses, as seen in AD, modalities of consciousness decline and are eventually lost e.g. the ability to form new memories. Recently the clinical criteria for diagnosing AD have been updated and AD is now recognized as a disease process that begins years to decades prior to the manifestation of signs and symptoms [37].

So what do APP, PSEN1, PSEN2, and APOE-ε4 genes tell us about the third neurogenetic phase of consciousness? These genes demonstrate a direct relationship to the maintenance of cognitive processes involved in human consciousness [38]. When there are abnormalities in any or several of these genes the molecular effects ensue (in the case of AD beta-amyloid accumulation) and cause damage to neurons and ultimately are manifested as a loss of gross brain mass. This damage translates into decreases in the modalities of human consciousness or more specifically in the degree of consciousness.

7. Conclusion

In this article many NgCC have been discussed and categorized into three neurogenetic phases. It is obvious that these three phases demonstrate a link between DNA and human consciousness: 1) during neurogenesis a genetic cascade is initiated that eventually constructs the human brain and central nervous system which will serve as the machinery of consciousness, 2) during the continuum of neuron-based consciousness the neurons rely on protein and RNA products in order to function properly, and these are continuously produced by DNA, and 3) during neurodegeneration there is a loss of neurons and resulting brain mass which can be cause by genetic mutations. This demonstrates that genes do influence degrees of human consciousness. Finally, by observing the many NgCC enumerated into three phases a connection between DNA consciousness and human consciousness can be appreciated.

In the future more NgCC shall be indentified and a more complete picture will be constructed. In addition, quantum physics may assist in understanding the interactions of the DNA-RNA-protein networks that manifest human consciousness.

References

[1] Rosenblum, B. & Kuttner, F. (2011) Quantum Enigma: Physics Encounters Consciousness, 2nd edition, New York: Oxford University Press.

[2] Crick, F. & Koch C. (2003) A framework for consciousness, Nature Neuroscience, 6:2, 119-126.

[3] Grandy, J. (2012) The Explosion of Consciousness: TSC Conference Tucson Arizona, Journal of Consciousness Exploration and Research; 3: 4, 432-445. http://jcer.com/index.php/jcj/article/view/217/234

[4] Baer W., Pereira A. & Bernroider G. (2012) The cognitive force in the hierarchy of the quantum brain, http://sbs.arizona.edu/project/consciousness/report_poster_detail.php?abs=1278

[5] Grandy, J. (2011) The DNA molecule is autopoietic, dynamic, evolving, and a form of consciousness, The International Journal of Arts and Sciences, 4: 20, 7-28.

[6] Fornasiero, E.F. et al. (2010) The role of synapsins in neuronal development, Cell Mol Life Sci., 67: 9, 1386-96.

[7] Cesca, F. et al. (2011) The synapsins: key actors of synapse function and plasticity, Prog. Neurobiol, 93: 4, 313-348.

[8] Emery, B. et al. (2009) Myelin Gene Regulatory Factor is a Critical Transcriptional Regulator Required for CNS Myelination, Cell, 138, 172-185.

[9] Grandy, J. (2014) DNA consciousness: From theory to science, in I. Frederickson (ed.) Aspects of Consciousness II.

[10] Grandy, J. The three neurogenetic phases of human consciousness, Forthcoming.

[11] Wang, X.D., et al. (2011) Pax3 Stimulates p53 Ubiquitination and Degradation Independent of Transcription. PLos ONE, 6: 12, e29379.

[12] Shunsuke, I. et al. (2011) Role of Pax3 acetylation in the regulation of Hes1 and Neurog2, Molecular Biology of the Cell, 22, 503-512.

[13] Agoston, Z. et al. (2012) Genetic and physical interaction of Meis2, Pax3, and Pax7 during dorsal midbrain development, BMC Developmental Biology, 12, 10.

[14] Callaerts, P. et al. (1997) PAX-6 in development and evolution, Annu Rev Neurosci, 20: 483-532.

[15] Asami, M., et al. (2011) The role of Pax6 in regulating the orientation and mode of cell division of progenitors in the mouse cerebral cortex, Development, 138: 5067-5078.

[16] Jang, E.S. & Goldman, J.E. (2011) Pax6 expression is sufficient to induce a neurogenic fate in glial progenitors of the neonatal subventricular zone, PloS ONE 2011, 6: 6, e20894.

[17] Klempin, F, et al. (2012) Modification of pax6 and olig2 expression in adult hippocampal neurogenesis selectively induces stem cell fate and alters both neuronal and glial populations, Stem Cells, 30: 3, 500-509.

[18] Wang, L. et al. (2011) Basal progenitor cells in the embryonic mouse thalamus- their molecular characterization and the role of neurogenins and Pax6, Neural Development, 6:35.

[19] Scholpp, S. et al. (2007) Otx1l, Otx2, and Irx1b establish and position the ZLI in the diencephalon. Development, 134: 3167-3176.

[20] Puelles, E. et al. (2004) Otx2 regulates the extent, identity and fate of neuronal progenitor domains in the ventral midbrain. Development, 131: 9, 2037-2048.

[21] Larsen, K. et al. (2010) Expression of the homeobox genes otx2 and otx1 in the early developing human brain, Journal of Histochemistry and Cytochemistry, 58: 7, 669-678.

[22] Ando, K, et al. (2008) Establishment of framework of the cortical area is influenced by Otx1, Neurosci Res, 60: 4, 457-459.

[23] Noor, A. et al. (2010) Disruption at the PTCHD1 locus on Xp22.11 in autism spectrum disorder and intellectual disability, Sci Transl Med., 2: 2, 49ra68.

[24] Guan, F, et al. (2012) Association of PDE4B polymorphisms and schizophrenia in Northwestern Han Chinese, Human Genetics, 131: 7, 1047-1056.

[25] Bradshaw, N. & Porteous, D.J. (2012) DISC1-binding proteins in neural development, signaling and schizophrenia.

Neuropharmacology, 62: 1230-1241.
[26] Millar, J. K. et al. (2007) Disrupted in schizophrenia 1 and phosphodiesterase 4B: towards an understanding of psychiatric illness, J. Physiol, 584: 2, 401-405.
[27] Girgenti, M.J, et al. (2012) ZNF804a regulates expression of the Schizophrenia-associated genes PRSS16, COMT, PDE4B, and DRD2, PLoS ONE, 7: 2, e32404.
[28] Zechel, S. et al. (2010) Expression and functions of fibroblast growth factor 2 (FGF-2) in hippocampal formation, Neuroscientist, 16: 4, 357-73.
[29] Binder, D.K. & Scharfman, H. (2004) Brain-derived Neurotrophic Factor, Growth Factors, 22: 3, 123-131.
[30] Gaughran, F, et al. (2006) Hippocampal FGF-2 and FGFR1 mRNA expression in major depression, schizophrenia and bipolar disorder. Brain Research Bulletin, 70: 3, 221-227.
[31] Gatz M, Reynolds, CA, & Fratiglionin L. (2006) Role of genes and environments for explaining Alzheimer disease. Arch Gen Psychiarty, 63: 2, 168-174.
[32] Upadhyay, J. et al. (2010) Alterations in brain structure and functional connectivity in prescription opioid-dependent patients, Brain, 133: 7, 2098-2114.
[33] Grandy, J. (2012) A clinical correlation made between opioid-induced hyperalgesia and hyperkatifeia with brain alterations induced by long-term prescription opioid use, Research & Reviews: A Journal of Neuroscience, 2: 2, 1-11.
[34] Grandy, J. (Forthcoming) Molecular Genetic Mechanisms of Addiction Involving the ΔFosB Pathways, New York: NOVA Science.
[35] Querfurth, HW & LaFerla, F. (2010) Mechanisms of disease: Alzheimer's Disease. N Engl J Med, 362: 329-344.
[36] Grandy, J. (2011) What's new in Alzheimer's disease: Biomarkers and gene mutations as aids for detecting AD early, Journal of the American Academy of Physician Assistants, 24: 6, 56-57.
[37] Grandy, J. (2012) Updated guidelines for the diagnosis of Alzheimer disease: A clinical review. Journal of the American Academy of Physician Assistants April 2012 25 (4): 50-55.
[38] Grandy, J. (2012) Alzheimer disease and DNA consciousness, Academic Journal of Science, 1: 3, 169-184.

Are Life, Consciousness, and Intelligence Cosmic Phenomena?

LUDWIK KOSTRO

Department for Logic, Methodology and Philosophy of Science, University of Gdansk, ul .Bazynskiego 4, 80-952 Gdansk, Poland
and
Ateneum Higher School, ul Waly Piastowskie 1.80-855 Gdansk, Poland
fizlk@univ.gda.pl

First of all the scientific reasons of astrophysics, astronomy and modern astrobiology in favor of the existence of life, consciousness and intelligence in the Universe will be presented and estimated (e.g. the Nobel Laureate Christian de Duve's arguments). The part played in this kind of scientific debate by the Copernicus principle will be stressed from the scientific and philosophical point of view. Since there are also philosophers and theologians who argue in favor of the existence of life in the Universe, their arguments will be shortly presented and estimated as well.

Keywords: Astrobiology, Bio-cosmology, Exobiology

1. Some Preliminary Remarks

We have not yet any sure direct or indirect observational evidence that in our galaxy and in others there are extraterrestrial life, consciousness and intelligence. But as has shown the history of science human mind reaches often truths earlier than human senses, sometimes many centuries before any observational evidence although the human mind alone without observations and experiments can mislead.

In arriving at scientific truths by means of human reasoning specially mathematics plays an important part. In order to make mathematical previsions in our subject new mathematical (algebraic, geometrical and statistical) methods are looked for and some of them were recently invented.

The methods of modern empirical sciences are operational. When we observe and make experiments we perform measuring operations and operations of mathematical calculus as well. We use these operational methods to reach an exact scientific knowledge. Mathematical calculations made on some empirical and theoretical basis enable us to know the truth before the main observational and experimental operations are done. Here is a recent example. In 1964, Peter Higgs predicted the existence of a field and of bosons that are excitations of this field. On 4th July 2012, the CMS and the ATLAS experimental teams at the Large Hadron Collider independently announced that they each confirmed the formal discovery of a previously unknown boson of mass between 125-127 GeV/c2. However, further data and analysis are needed before positively identifying the new boson beyond doubt as being a Higgs boson of some type [1]. As we can see, in such a way, we approach step by step to experimental evidence. As concerns exo-life, - consciousness and - intelligence we are approaching to the empirical evidence in a similar way.

Nowadays we are also in such favorable circumstances that we have not to wait long centuries for experimental evidences because our civilization recently became a technological one. Our observational technologies become more and more precise and effective.

2. Scientific Data

Twenty years ago we had no sure observational evidence that there are extrasolar planets. Now we have already sure indirect and direct observational evidence that our galaxy is a great producer of exo-planets [2] Until now astronomers and astrophysicists have already detected more than 850 extrasolar planets and about 3.500 candidates [3]. Such a sample is sufficient to conclude that not only our galaxy but also many others are great producers of enormous numbers of planets.

As it is well known there are five methods to detect exo-planets Until 5 October 2012:
(1) 716 exo-planets were detected by radial velocity or astrometry
(2) 239 by monitoring transiting planets as they cross the disc of their host star
(3) 16 by gravitational microlensing
(4) 31 by direct imaging
(5) 16 by pulsar timing

Until 19 November 2012 the total number of detected exoplanets is 851, of detected planetary systems 670 and of detected multiple planet systems 126 [3].

These extrasolar planetary systems were detected within a radius of 22.000 ly. The nearest ex-oplanet is at the distance of 10,5 ly and the most distant at the distance 22.000 ly. Among the 851 exoplanets there are Earth-like planets. A rocky Earth –like exoplanet is called exo-Earth when its mass is equal to the Earth-mass or Earth-twin when not only its mass, but also its radius, temperature and atmosphere are the same. An Earth-like exoplanet the mass of which is greater than Earth-mass is called super-Earth. When it is smaller it is called sub-Earth.

The radial velocity or astrometry and the pulsar timing methods were used by the Polish astronomer Aleksander Wolszczan from the Copernican University in Torun The pulsar timing method is so sensitive that it is capable of detecting planets far smaller than any other method can.

The method of monitoring transiting planets allows also studies of their atmospheres that are not possible for non-transiting planets. This method is a potential tool for habitability screening of Earth-like planets. When the ingredients of these atmospheres are the same as on our Earth, they can be considered as biomarkers. The transiting planets method was used by the Canadian astronomer David Charbonneau now working at Harvard University The CoRot and Kepler satellite missions which monitor thousands of stars are just looking for the transiting planets. The American of Polish origin William J. Borucki is the Principal Investigator of Keppler Mission

David Bennett from the Notre Dame University (USA) is the pioneer of the technique for finding distant planets called gravitational microlensing.

31 exoplanets were detected by direct imaging (e.g. Gliese 581d in a habitable zone with terrestrial mass and Gliese 581 g). Other examples of solar systems detected by direct imaging are HR 8799 a,b,c,d,e, HR 8799c and CFBDSIR J 1458+1013AB. The last one was obtained with the Keck II Telescope in Hawaii. (2011)

J. Kasting, J. Lunine, J. Baross are of the opinion that in 20 or at least 30 years the method of direct imaging will be so performed that we will be able to observe life on some of the exoplanets. J Kasting is the author of the book *How to find a habitable planet?* [4].

3. Steps of the Chemical Evolution of the Universe

(1) The formation of chemical elements was the first step. The simplest chemical elements (hydrogen, helium and to a certain extend lit) appeared before the formation of galaxies. The other elements (until iron) were formed inside the stars of different generations and the following ones during the explosions of supernovae. The formation of chemical elements lasted billions of years. On our Earth there are 92 natural chemical elements. The other ones until 118 were formed in an artificial way. The number of chemical elements is limited. It cannot be greater than 137 because otherwise the electrons moving around the atomic nucleus had to move with the velocity greater than the speed of light.

(2) The formation of chemical compounds is the second step of the chemical evolution. The continuation of the chemical evolution of the universe takes place almost exclusively inside the planets and on them. Chemical compounds especially the complicated ones cannot appear inside the stars because the temperature is there much too high. They appear almost exclusively inside the planets and on them. As can be shown by spectroscopy only very simple chemical compounds (e.g. oxide of titan) appear on the surface of stars whose temperature of surface is lower than 3.500 K^{o}

The number of possible chemical compounds is also finite but enormous. We do not know it yet. But we know already that the nature allows such complicated molecules of compounds like DNA, RNA. Thus they can appear on exo-planets if there are favorable conditions and circumstances.

(3) The chemical evolution of the bio-universe is the third step of the chemical evolution of our universe. Life, consciousness and intelligence on our Earth can serve as an example of biochemical evolution in the universe. Although we have not yet observational evidence of the existence of exo-life, many scientists are convinced that life, consciousness and intelligence are natural cosmic phenomena. There exists already a new branch of science called modern astrobiology (called also bio-astronomy, exo-biology or sometimes bio-cosmology). This new interdisciplinary science has already two international scientific journals and there are organized every year several international conferences on the subject. These conferences are interdisciplinary meetings of astronomers, astro-physicists, chemists, biologists and biochemists and sometimes historian of science (especially of astronomy, astrophysics, biology and suchlike) philosophers and theologians as well.

Modern bio cosmology is looking for biomarkers on the earth-like transiting planets investigating their atmospheres. If such an atmosphere is composed of the same chemical elements and compounds like on our earth there is an imputation that on the planet there is life.

In USA a Space Telescope is scheduled for launch in 2014 to search just for biomarkers. In such a projects is involved also NASA.

The modern bio-cosmology has already introduced the notion of habitable zones in galaxies and around the stars with favorable conditions for life. The firsts are called "galactic habitable zones" the seconds "solar habitable zones" around sun-like stars, hotter stars and cooler stars. It's obvious that the habitable zones of hotter stars are of farer distance from the star than in the case of sun-like star and in the case of cooler stars they are nearer the star [5]. Let's add that an exo-planet that exists in its galactic and solar habitable zone to be able to become really habitable must be characterized still by the so-called greenhouse effect. It means e.g. that the daily received and lost energy must be in equilibrium. Otherwise it becomes cooler and cooler and changes in a snow ball earth or becomes hotter and hotter and does not enable life [4].

Lisa Kaltenegger (EN Research Group leader at MPIA, Lecturer at Harvard, Research Associate at SAO) created a team to investigate the atmospheres of transiting Earth-like planets. They are looking for biomarkers in their atmospheres: i.e. for water, oxygen, ozone, nitrous oxide, carbon dioxide and methane [6].

4. Some Empirical Data and Statistical Calculations

Although we have not yet any sure observational evidence that life, consciousness and intelligence are cosmic phenomena we are, to a certain extent, in a situation similar to that which exists after political elections but before the computing of the votes. Thanks to good samples of observational data and statistical calculations we know already in the evening which political party won the elections though the computing of the votes has not yet begun. Christian de Duve, cytologist, biochemist, Nobel Prize Laureate, Member of Pontifical Academy of Science, professor of the Catholic University in Louvain Belgium and of the Rockefeller University in New York, indicates many observational data and uses the probability calculus to show that we can already state that "Life is a cosmic imperative". He presents the empirical data and his statistical calculations in his books: (1) *Vital Dust. The Origin and Evolution of Life on Earth*, [7]. (2) *Life Evolving: Molecules, Mind, and Meaning*, [8] (3) *Singularities. Landmarks on the Pathways of Life*, [9]. As regards the empirical data he indicates that 20% of the cosmic dust (which constitutes 0,5% of the mass of our galaxy) is composed of organic compounds [7]. According to him, our universe is a great melting pot of organic synthesis because the amino acids can be found in meteorites and are found, using spectroscopy, in the comet tails [7-9]. Therefore one of his books has the title *Vital Dust*. Note that peptides and proteins are composed of amino acids. All these compounds are shares of life.

De Duve opposes also to the Einstein's statement: "God doesn't play dice with the world" his own statement: "God plays dice with the universe because he is sure to win".

He proves the truth of his statement using the following equation of probability calculus

$$P_n = 1 - (1 - P)^n$$

where

n = number of opportunities (trials)
P = probability
P_n = probability after n opportunities (trials) [8,9].

According to the modern science in the process of the physical, chemical biological evolution there play part three constituents

1. Chance (spontaneity)
2. Necessity (regularities, laws)
3. Opportunities (number of trials)

Many people when hearing the words "chance" and "probability" exclude automatically "inevitability" and "certainty". According to de Duve that's a great mistake because chance does not exclude inevitability and probability does not exclude certainty. If the number n of opportunities is very great and the process of evolution lasts many billions years then inevitably P_n becomes 1 with full certainty. According to de Duve, in the process of evolution the rule of congruence plays an important part. The chemical elements fit to each other better or worse. When they fit well they form chemical compounds. For instance hydrogen and oxygen fit to each other and in favorable conditions form the chemical compound called water. The chemical element carbon fits to several other chemical elements and in favorable circumstances form organic compounds. According to de Duve there exists a certain chemical determinism in which the congruence rule plays its role. Nature step by step arrives to more and more complicated compounds, amino acids, proteins, RNA, DNA and so on. For him the appearance of life is an natural chemical process.

Therefore de Duve concludes: If on an exo-earth there are favorable conditions and the necessary key events have happened then life and consciousness will appear and evolve [7-9]. Therefore, according to de Duve: Life is a Cosmic imperative.

5. Impact of the Scientific Debate Concerning Exolife on Philosophy of Science and Logic

Under the influence of the mentioned above debate philosophy of science and modern logic provide their own arguments in the favor of the cosmic nature of life, consciousness and intelligence.

They use for this purpose especially the philosophical principle of homogeneity of Nature (in science called often Copernicus principle) that constitutes, in their opinion, the basis of modern science. According to this philosophical principle "The same physical, chemical and biological phenomena natural or statistical in the same natural or statistical circumstances, run naturally or statistically the same way", or in other words: "What the laws of Nature (described by physics, chemistry and biology) admit necessarily happens when there are favorable conditions" (George R. Price) [10].

On the basis of this principle philosophy, using logic concludes: If on an exo-earth there are required, adequate and favorable conditions (included the key events in Latin called *conditiones sine qua non*) life and consciousness will spontaneously appear and evolve.

The philosophical principle of homogeneity of Nature has found, especially in the XX century, its expression in all branches of natural sciences. Therefore it must be stressed that this principle became a natural basic component of all natural sciences. It has stopped to be only a philosophical principle used by philosophers. It entered in the internal structure of modern science. Therefore we must say that it belongs to the modern science as its cognitive backbone (spinal column). In physics, for instance, the principle is expressed in a mathematical way thanks to Emmy Noether in her theorem of the universality of physical laws. According to her, the laws of physics are invariant (1) with respect to the displacement in space what finds its expression especially in the law of momentum conservation, (2) with respect to the displacement in time what finds its expression especially in the law of energy conservation (3) and with respect to rotation what finds its expression especially in the law of angular momentum conservation. In such a way, thanks to Emmy Noether, the principle of the homogeneity of Nature entered in the mathematical structure of the classical physics.

The known physicist Richard Feynman doubted whether the considered principle can be used in Quantum Mechanics because the Noether's theorem concerns just the physical quantities that appear in the Heisenberg uncertainty relations. As it is well known the Heisenberg uncertainty relations are products (1) of uncertainty of space location and uncertainty of momentum, (2) of uncertainty of time location and uncertainty of energy and (3) of uncertainty of angle displacement and uncertainty of angular momentum.

However as it was shown in subsequent researches there exists a proper invariance of quantum mechanical laws with respect to the displacement in space and time and with respect to rotation. (see e.g. R. Shankar, *Principles of Quantum Mechanics,* chapters 11 and 12)[11]. Thus the principle of the homogeneity of Nature finds its expression not only in the mathematical structure of the classical physics but also, in a proper way, in the mathematical structure of Quantum Mechanics and in this way in the structure of Quantum Chemistry, Quantum Biochemistry and so on. It means that the principle under consideration finds its expression in the biological sciences as well.

In Italy Prof. Philippo Selvaggi philosopher and physicist, who was conferring my doctor degree thesis, was one of the defenders of the principle of homogeneity of Nature indicating its scientific nature. In his book *Filosofia delle Scienze* [12] he shows that in natural sciences the principle under consideration causes that our scientific reasoning is often nearly unfailing.

In Poland e.g. the logician Leopold Regner indicates the role of the principle in science as follows: *the conviction that what* happens *in determined circumstances W, will happen again exactly and unfailingly everywhere and always where and when there will be circumstances totally similar to W, is based on the recognition of the principle of homogeneity of Nature. The homogeneity of Nature consists on this that the course of a phenomenon does not depend on the circumstances of place (where) and time (when)*[13].

Regner investigated the logical status of this principle in natural sciences. He considers this principle as a logical preamble of induction, as a postulate or presumption concerning the properties of the universe.

Concluding let's repeat that the principle of homogeneity entered, and in my opinion forever, in the mathematical structure of natural sciences.

According to many scientists and philosophers the considered principle expressed also in the affirmation: "What is allowed by the laws of Nature has necessarily to be happen if there are adequate and favorable circumstances" has to be considered as one of the basic affirmations of modern Science and provides the basis to the affirmation that life and consciousness are cosmic phenomena. We can even say that the natural appearance of life and consciousness in the universe is a law of Nature. But we have not yet any sure observational evidence of this fact

6. Impact of Exoplanet Astronomy and Modern Cosmo-biology on Theology

1) On the one hand some theologians and Christian scientists provide theological arguments in favor of the affirmation that life, consciousness and intelligence are cosmic phenomena

Some authors (e.g. Pierre Theihard de Chardin, Karl Rahner) argue that if the universe was deprived of life and consciousness, the act of creation could not be called and regarded as God's gift granting sensible existence, but only as His toy to play with, totally deprived of sensibility. A feeling being needs to be at least in a minimal conscious state. It must have an elementary sensitiveness. Only beings that enjoy sensibility and consciousness can experience existence that can be considered as a gift. The beings that are deprived totally of sensibility and consciousness do not know that they exist and therefore they cannot enjoy their existence. They are beings for others and not for themselves. If in the whole universe only on our Earth there was sensible and conscious existence then the rest i.e. the enormously great Universe would be merely a toy for the Creator to play with.

2) On the other hand modern bio-cosmology poses problems for Christian theology that is still geocentric, if not provincial and parochial. According to the Christian beliefs, Heaven regarded as the place of eternal felicity, is governed only by the representatives of our Earth: Jesus of Nazareth who is sitting at the right-hand side of his Father, His Mother Mary who is considered by Catholicism to be the Queen of Heaven, and Saint Peter who holds the keys in the Kingdom of God. All of them lived on our Earth. If in our and other galaxies there are habitable planets, then Christian theology must change, must take into account the existence of many exo-civilizations. But that is not a scientific problem but a theological problem of different Christian Churches. For example the notion of multi-incarnation proposed already by the French Jesuit, philosopher and scientist P. Teilhard de Chardin is propagated in France in the book "*God, Church and the Extraterrestrials*"[14]. According to the notion of "multi-incarnation", the same Person, the second Person of the Holy Trinity, The Son of God incarnated on our Earth in Jesus from Nazareth and on other exo-Earths in other intelligent beings.

The Polish cosmologist and theologian Prof. Michał Heller (Templeton Prize Laureate) is of the opinion that in Christianity the notion of incarnation is too absolute with respect to its uniqueness [15]. So he follows the opinion of those who admit the multi-incarnation.

Other theologians try to resolve the problem returning to the ancient adoptive Christology. According to the adoptive Christology, there is not incarnation of God but an adoption of intelligent beings on every Earth-like planet. According to this theological opinion, the Creator has adopted, on every habitable planet, an intelligent cosmic being as his Son. The idea of adoptive Christology, based on the evolution of Christology of Jesus from Nazareth, in the New Testament, is proposed e.g by Alfons Weiser [16] and by Michael L. Cook S. J. [17].

Let's add that Jose G, Funes, Jesuit, director of the Vatican Astronomical Observatories in an interview published in "*Osservatore Romano*" said "The extraterrestrial is my brother" [18]. According to him, Christians have, however, to respect the tradition according to which the act of redemption is unique and took place on our Earth. He seems to follow the opinion of the French theologian Yves Congar who was of the opinion that the incarnation and redemption took place only on our Earth, but the Holy Spirit will illuminate the minds of the habitants of exo-planets that Jesus, born on our Earth, is the Savior of the whole universe [19].

All these theological opinions do not belong to science they appeared in the minds of theologians and scientists who believe that Jesus from Nazareth is the Savior of the whole universe. As we can see the results of modern bio-cosmology can provoke a new profound overturn of people's religious outlook on the whole world.

7. Conclusions

1. On the basis of scientific and philosophical reasoning and data we already have the courage to claim that in the universe there was a second great Big Bang i.e. the explosion of life and consciousness everywhere there were and are favorable conditions and where the necessary key events happened. With such enthusiastic convictions many astronomers, astrophysicists and bio-cosmologists are working and making discoveries.

2. The appearance and existence of life and consciousness in the universe has to be considered as a law of Nature, i.e. as a natural, normal thing although we have not yet any sure observational evidence that it is so.

3. As we could see the modern bio-cosmology being concerned with life and consciousness as cosmic phenomena (e.g. looking for biomarkers in the atmospheres of transiting Earth-like exo-planets) implicates philosophical and theological ideas that can provoke a new profound overturn of people's philosophical and religious outlook on the whole world. Such phenomenon will happen when the Science arrives at the sure observational evidence that we are not alone.

References

[1] Randall, L. (2012) Higgs Dicovery, The Power of Empty Space, London: The Bodley Head.

[2] Cassen, P. et al. (2006) Extrasolar Planets, Springer.

[3] The Extrasolar Planets Encyclopedia, http://exoplanet.eu/.

[4] Kasting, J. (2010) How to Find a Habitable Planet..., Princeton: Princeton Univ. Press.

[5] http:winteryknight.files.wordpress.com

[6] https://www.cfa.harvard.edu/~lkaltenegger/Home.html
[7] de Duve, Ch (1994) Vital Dust, The Origin and Evolution of Life on Earth, New York: Basic Books.
[8] de Duve, Ch (2002) Life Evolving: Molecules, Mind, and Meaning, Oxford: Oxford University Press.
[9] de Duve, Ch (2005) Singularities, Landmarks on the Pathways of Life, Cambridge: Cambridge University Press.
[10] Price, G.R. http://www.gougle.pl/search?q=george+r+price+equation
[11] Shankar, R. (2004) Principles of Quantum Mechanics, Dordrecht: Kluwer Academic.
[12] Selvaggi, F. (1953) *Filosofia delle scienze* Edizioni, Roma: *La Civilta Cattolica.*
[13] Regner, R. (1973) *Logika,* PTT, Cracow.
[14] Vigne, A. (2000) (red.), *Dieu, l'èglise et les extraterrstres,* Paris : Albin Michel.
[15] Heller, M. (2009) *Czy kosmita może być zbawiony?* Tygodnik Powszechny, 23, 3074.
[16] Weiser, A. (1974) *Jesus - Gottes Sohn? Antwort auf eine Herausforderung,*Stuttgart: KBW.
[17] Cook, M.L. (1993) Responses to 101 Questions about Jesus, Mahwah: Paulist Press.
[18] Funes, J. (2008) The extraterrestrial is my brother, in *L'Osservatore Romano.*
[19] Życzyński, J. (2011) *Kosmiczni bracia w człowieczeństwie?* in *Aniołowie i kosmici. Kośció wobec cywilizacji pozaziemskich,* Cracow: Wyd. "M".

A Holoinformational Model of the Physical Observer

FRANCISCO DI BIASE

Dept. of Neurosurgery-Neurology Santa Casa Hospital, Barra do Piraí, Rio de Janeiro, Brazil
Dept. of Electroencephalography and Brain Mapping, Clínica Di Biase, Barra do Piraí, Brazil
dibiase@terra.com.br

The author proposes a holoinformational view of the observer based, on the holonomic theory of brain/mind function and quantum brain dynamics developed by Karl Pribram, Sir John Eccles, R.L. Amoroso, Hameroff, Jibu and Yasue, and in the quantum-holographic and holomovement theory of David Bohm. This conceptual framework is integrated with nonlocal information properties of the Quantum Field Theory of Umesawa, with the concept of negentropy, order, and organization developed by Shannon, Wiener, Szilard and Brillouin, and to the theories of self-organization and complexity of Prigogine, Atlan, Jantsch and Kauffman. Wheeler's "it from bit" concept of a *participatory* universe, and the developments of the physics of information made by Zureck and others with the concepts of statistical entropy and algorithmic entropy, related to the number of bits being processed in the mind of the observer are also considered. This new synthesis gives a self-organizing quantum nonlocal informational basis for a new model of awareness in a participatory universe. In this synthesis, awareness is conceived as meaningful quantum nonlocal information interconnecting the brain and the cosmos, by a holoinformational unified field (integrating nonlocal holistic (quantum) and local (Newtonian). We propose that the cosmology of the physical observer is this unified nonlocal quantum-holographic cosmos manifesting itself through awareness, interconnected in a participatory holistic and indivisible way the human mind-brain to all levels of the self-organizing holographic anthropic multiverse.

Keywords: Holoinformational awareness, Quantum information, Nonlocality, Self-Organization, Medicine, Transpersonal psychology.

The Tao obscures when we only see small parts of existence - Chuang-Tzu

1. Introduction

Models that try to explain the nature of awareness, generally share the Cartesian-Newtonian paradigm by insisting on an approach exclusively reductionist. This reductionism has been impairing the grasp of the true essence of what awareness is since the 17th century. Hameroff [35] believes that such dispute *"may potentially be resolved by views which contend that awareness has a distinct quality, but one which emerges from brain processes which can be accounted for by natural science."* As a solution he proposes an awareness model based upon the emergence of quantum coherence in neural microtubules, developed with Penrose [36]. These models use a traditional interpretation of quantum mechanics, and as Clarke [16] shows, *"start from a basically quantum-mechanical position but then impose modifications of the quantum formalism so as to ensure that the net result is basically Newtonian ... Strong emphasis is placed on the wave function as the fundamental object of quantum theory and a "collapse" is invoked to pass to a Newtonian picture. As a result, they are very firmly bound to a spatial picture."* By transforming the quantum logic into a Newtonian logic, they leave aside the nonlocality function that is quantum logic's essence, a fundamental property of the universe - and, as we shall see, also a fundamental property of awareness linking the observer to the universe.

Wilber [68] considers that an integral theory of awareness should embody all the essential characteristics of the main schools that study awareness, *"not as an eclecticism but rather a tightly integrated approach that follows intrinsically from the holonic nature of the Cosmos"*. Such a holonic nature of the Cosmos is based upon the self-organizing holoarchy described by Jantsch [40] that correlates the co-evolutionary interactions amongst the micro-evolution of the holons described by Koestler [42], to the macroevolution of its collective/social forms.

Wilber's theory, however, leaves open what we consider the key point in the understanding of awareness, that is, the way by which information, order, and negentropy, are transmitted amongst the infinite levels of self-organization of the cosmic-brain holoarchy, giving meaning to them. This common ground capable of integrating awareness and cosmos in an ordered and indivisible whole, can only be fulfilled by a holoinformational (Newtonian local information + quantum nonlocal unified field information), model of awareness that connects the universe's nonlocal quantum-holographic anthropic informational structure with the brain's nonlocal informational quantum-holographic fields and also with the classic local neural networks of the brain, as already described elsewhere [21,24-26,70-73] . Also this holoinformational model to operate in a deterministic cosmos has to be compatible with Einstein's theory of relativity.

Wheeler [66] realized how important information is in such context. With his genius, Wheeler describes an elegant information-participatory universe that is the most brilliant and fundamental model of interaction brain-mind and cosmos ever described in the science of awareness. With his famous "the it from bit" concept he unite quantum information theory to awareness and physics: *"...every it - every particle, every field of force, even the space-time continuum itself - derives its function, its very existence, entirely - even if in some contexts, indirectly - from the apparatus-elicited answers to yes-or-no questions, binary choices, bits". "It from bit symbolizes the idea that every item of the physical world has at bottom – at a very deep bottom, in most instances – an immaterial source and explanation; that which we call reality arises in the last analysis from the posing of yes-no question and the registering of equipment-evoked responses, in short, that all things physical are information-theoretic in origin and this is a participatory universe".*

In the same paper Wheeler [66] gives the example of a photon being detected by a photodetector under watch, when we ask the yes-or-no question: *"Did the counter register a click during the specified second?" If yes, we often say "a photon did it." We know perfectly well that the photon existed neither before the emission nor after the detection. However, we also have to recognize that any talk of the photon "existing" during the intermediate period is only a blown-up version of the raw fact, a count. The yes or no that is recorded constitutes an unsplitable bit of information".*

2. Order Information Self-organization Negentropy

A more wide conception of order, organization, information and negentropy that goes beyond the classical works of Wiener [67], Shannon [60], Szilard and Brillouin [11] is essential for the development of our holoinformational model capable of integrating awareness to nature. Leon Brillouin, in his famous theorem, showed the equivalence between information and negentropy. Norbert Wiener put this identity on the very conceptual basis of cybernetics stating that *"information represents negative entropy"*, and prophetically for the first time in the history of science emphasizing that *"information is information, not matter or energy"*. Chalmers [14] states that information is an essential property of reality, as matter and energy, and that *"conscious experience must be considered a fundamental feature, irreducible to anything more basic"*. Bateson [5] defines information as *"the difference that makes a difference"*, a conception that Chalmers [15] retakes stating that this is *"the natural way to make the connection between physical systems and informational states."*

The equivalence/identity between order, negentropy and information, is the way that allows us to build upon and understand the whole irreducible and natural flow of order transmission in the universe, organized in a meaningful and intelligent informational mode. In the classical thermodynamic theory, the definition of order is probabilistic and dependent on the entropy concept, which measures the degree of disorder of a system, reducing to uncertainty the immense dimension of natural meanings.

For Atlan [2,3], as well as for Di Biase [17,20,22], *"entropy shouldn't be understood as a disorder measure, but much more as a measure of complexity."* In order to do this, it is necessary to consider that the notion of information implies a certain ambiguity, meaning the bit capacity of a physical system as Shannon put it, or the semantic content (meaning) conducted by the bits during a communication. In the information theory, the organization, the order expressed by the amount of information in the system (Shannon's H function) is the information measure that is missing to us, the uncertainty about the system.

Relating this uncertainty, this ambiguity to the variety and the non-homogeneity of the system, Atlan could solve certain logical paradoxes of self-organization and complexity, widening Shannon's theory and defining organization in a quantitatively formal mode. Atlan showed that the system's order corresponds to a commitment between the maximum informational content (i.e., the maximum variety) and the maximum redundancy, and showed also that the ambiguity can be described as a noise function, or even a time one, if we consider the time effects as related to the random factors accumulated by the environment's action.

Such ambiguity, peculiar to self organizing systems, can be manifested in a negative way (*"destructive ambiguity"*) with the classical meaning of disorganizing effect, or in a positive way (*"autonomy producer ambiguity"*) that acts by increasing the relative autonomy of a part of the system in relation to the others, that is reducing the system's natural redundancy and increasing its informational content. Atlan developed this self-organizing theory of complexity for biological systems. Jantsch [40] studying the evolution of the universe, showed that cosmological evolution is also a self-organizing process, with the microevolution of the individual systems (holons) co-evolving towards macrosystemic collective structures better organized, with a big reduction in the amount of these collective systems. This whole self-organizing process represents, actually, an universal expression of a bigger acquisition of variety or informational content that, as Atlan demonstrated, is a consequence of a reduction of redundancy in the totality of the system.

3. Information and Dissipative Structures

Ilya Prigogine [54,55] Nobel Prize winner, developed an extension of thermodynamics that shows how the second law can also allow the emergence of novel structures, and indicates the ways in which order can emerge from chaos. This type of self-organization generates dissipative structures that are created and maintained through the energy's exchanges with the environment in non-equilibrium conditions. These dissipative structures are dependent upon a new order, called by Prigogine *"order from fluctuations"*, which corresponds to a *"giant fluctuation"* stabilized by the exchanges with the environment. In these self-organizing processes the structure is maintained through an energy-information dissipation, that displaces itself, simultaneously generating (*in-formatting*) the structure through a continuous process.

The more complex the dissipative structure, the more information is needed to keep its interconnections, making it consequently more vulnerable to the internal fluctuations, which means a higher instability potential and higher reorganization possibilities. If fluctuations are small, the system accommodates them and does not change its organizational structure. If the fluctuations reach a critical size, however, they cause disequilibrium in the system, generating new intra-systemic interactions and reorganization. *"The old patterns interact between themselves in new ways, and establish new connections. The parts reorganize themselves in a new whole. The system reaches a higher order"* [55].

4. Awareness Self-organization and Information

Seager [59] states that awareness, self-organization and information are connected at the level of semantic significance, not at the level of "bit capacity", and that *"as the classical theory of information is situated at the level of "bit capacity" it would seem unable to provide the proper connection to awareness"*...and *"we can begin to move towards a more radical view of the fundamental nature of awareness with a move towards a more radical view of information"*. Seager still reminds us that in the famous two-slit experiment, and in the quantum eraser experiment, what is at stake is not the bit capacity, but the semantically significant correlation of *information laden* distinct physical systems, in a non-causal mode.

Chalmers [14] argues that each informational state has two different aspects, one as conscious experience, and the other as a physical process in the brain, that is, one internal/intentional and the other external/physical. This view finds support in the present developments of the so-called "information physics", developed by the physician Wojciech Zureck [69] and others, that propose that the physical entropy would be a combination of two magnitudes that compensate each other: the observer's ignorance, measured by Shannon's statistical entropy, and the disorder degree of the observed system, measured by the algorithmic entropy which is the smallest number of bits needed to register it in the memory. During the measurement, the observer's ignorance is reduced, as a result of the increase in bit numbers in its memory, remaining, however, constant the sum of these two magnitudes, that is, the physical entropy.

In this informational view of the universe, the observer remains included as part of the system, and the quantum universe changes because the observer's mind unleashed a transfer of information at a subatomic level. From this all results a **Law of Conservation of Information**, as well as or more fundamental than the law of conservation of energy.

Also Stonier [61,62] identifies information with the structure and organization of the universe, arguing that information is the *cosmic organizational principle* with a "status" equal to matter and energy.

In this holoinformational view I propose that what self-organizes significantly the cosmic evolution is the relationship between the physical entropy and the universe's quantum-holographic informational content, through a process in which the complexity using the pre-existing informational content reaches each time higher organizational levels and variety. In this way, complexity in the universe grows gradually, from gravity and nuclear powers, intensifies with the emergence of the self-organizing macromolecular systems of the biosphere, and reaches an almost infinite anti-entropic state of complexity, variety and informational content with the emergence of the noosphere with its quantum-holographic neural networks conscience. As we shall see soon ahead, there is a quantum holographic physical theory of the universe that has implicit in its conceptual framework, besides the mechanistic local interactions, a nonlocal quantum informational unfolding, that self-organizes matter, life and awareness in a meaningful way.

5. Transpersonal Awareness

Awareness as nonlocal information and something essential, primary and irreducible as we are proposing, is also found in the awareness maps obtained from thousands of very consistent and converging psychotherapeutics experiences reports, observed by several researchers of the medical and psychological areas like Di Biase [18,19,22,23], Jung [41], Grof [34], Moody Jr.[46], Ring [56], Sabom [57], Kubler-Ross [43], Weiss [65]. These researchers work with persons submitted to altered states of awareness, through various methods, like hypnosis, relaxation, meditation, holotropic breathing, near-to-death experiences, etc.

Surprisingly, such maps reveals *"an ontology and a cosmology in which awareness cannot originate from, or be explained in terms of any other thing. It is a primordial factor of existence and from it emerges everything that exist"* as Grof states, in Capra [12]. The replication of these numerous clinical observations by researchers of notorious scientific reputation are extremely important data, many times despised. It consistently prove an irreducibility of awareness, being one of the few non-philosophical, non-religious and non-physical ways that allow us to investigate and understand directly, *in toto*, the awareness phenomenon, in a controlled clinical scientific way.

Presently, there are available a series of psycho-technologies that are usually ignored and/or marginalized by the academic community, which allow us to use the human mind as a reliable system of investigation and elucidation of the nature of awareness, that are possibilities of replication an corroboration. I also emphasize here the numerous philosophical psycho-spiritual systems that during the history of humanity have been exploring through meditation and others psycho-technologies the nature of awareness, having described a vast and systematized *cannon* of experiential knowledge and wisdom about awareness (see for instance the interface between Buddhism and Neuroscience that is being studied by western neuroscientists with the Dalai Lama and Buddhist *scholars* at the Mind and Life Institute).

6. Nature, Information, and Awareness

We understand like Weil [64] that *"intelligence's nature is nature's intelligence"* and like Atkins [1] that *"awareness is emergent information itself at the moment of its generation, ongoing, self organizing change in a self/world model"*. It follows that only a holoinformational and self-organizing theory, capable of integrating intelligence and awareness to the nonlocal quantum-informational tessitura of the universe, can solve the question of awareness nature.

Fortunately, there is a quantum theory of the universe that integrates awareness as an irreducible dimension of nature in its conceptual framework. Nevertheless, this theory has been inexplicably considered in an insufficient way by the scientific world, going unnoticed its revolutionary implications about the awareness-universe interaction. It is the quantum-holographic theory of the holomovement developed by the physicist David Bohm [7] that mathematically demonstrates the existence of a hidden, spectral, implicit order in the universe, that is a primary reality. Matter, life and awareness (the explicate order), would originate from this common ground (the implicate order) by means of a continuous movement of unfolding and enfolding of the cosmos called holomovement.

Bohm [8] states that *"in the implicate order everything is folded into everything. But it's important to note here that the whole universe, is in principle enfolded into each part actively through the holomovement, as well as the parts. Now this means that the dynamic activity-internal and external- which is fundamental for what each part is, is based on its enfoldement of all the rest, including the whole universe. But of course, each part may unfold others in different degrees and ways. That is, they are not all enfolded equally in each part. But the basic principle of enfoldment in the whole, is not thereby denied. Therefore enfoldment is not merely superficial or passive but, I emphasize again, that each part is in a fundamental sense internally related in its basic activities to the whole and to all the other parts. The mechanistic idea of external relation as fundamental, is therefore denied. Of course such relationships are still considered to be real, but of secondary significance.*

That is, we can get approximations to a mechanistic behavior out of this. That is to say, the order of the world, as a structure of things that is basically external to each other, comes as secondary and emerges from the deeper implicate order". So, we can say that we live in a quantum-holographic universe in which reality is essentially nonlocal, and the classical Newtonian world with its external local interactions, emerges as a special case from this deeper quantum order.

According to Bohm [9], the analogy with the hologram in which each part of the system is an image of the total object, even if it is a static image that does not transmit the ever dynamic nature of the infinite unfolding and enfolding which at each moment create our universe, is a functional metaphor, because *"the mathematical laws of the quantum theory that apply to these waves, and therefore all matter, can be seen to describe just such a movement in which there is a continual enfoldment of the whole into each region, along with the unfoldment of each region into the whole again. Although this may take many particular forms - some known, and others not yet known - this movement is universal as far as we know"*.

This universal movement of enfoldment and unfoldment is Bohm's "holomovement". Bohm states - **and this is of supreme importance** - that these laws are also capable of being compatible with the theory of relativity, and therefore the implicate order is able to have support from the two most fundamental theories of modern physics, the theory of relativity and the quantum theory. In a posterior development, Bohm [9] postulated the existence of a superimplicit order, a still more subtle dimension of the universe's organization. In this model, a quantum superinformation field of the totality of the universe would organize the implicate first level in multiple wave-like structures which would unfold in the explicate order. According to Bohm [see Weber, 63] *"there is a physical model developed by De Broglie that proposes a new type of field, which activity is dependent upon the information content that is conducted to the whole experimental field, which, if extended to the quantum mechanics, results in the superimplicit order"*. This De Broglie-Bohm *information field* is the nonlocal informational field that can interact with the brain's nonlocal quantum-holographic fields. As we will see latter, these nonlocal fields can, through quantum coherence and superradiance in microtubules and synapses, and trough Bose-Einstein condensates and Frölich Effect in water molecules of the CSF, generate local fields in brain's neural network. That is the basis of our holoinformational field concept.

7. Awareness and Nonlocality

Adding to its equations a Quantum Potential that satisfies Schrödinger's equation, that depends on the form but not on the amplitude of the wave function, Bohm [9] developed a model in which the quantum potential, carries *"active information"* that *"guides"* the particle along its way. The quantum potential has inedited characteristics unknown up to then, because differently from the other nature's forces, it is subtle in its form, and does not decay with the distance. Such interpretation allows communication between this "pilot wave" and the particle, to be processed in a higher speed than the light, unveiling the quantum paradox of nonlocality, i.e., of the instantaneous causality, fundamental in the holoinformational view of awareness. In 1982, Alain Aspect and col. experimentally proved the existence of nonlocal actions, and more recently, in July 1997 Nicolas Gisin and col (cf. Science, vol. 277, pg. 481) proved the existence of this nonlocal quantum informational instantaneous action in large scale.

For Bohm, differently from Bohr, the elementary particles do not have dual nature wave/particle but are particles all the time, and not only when observed. Actually, the particle originates from a global quantum field fluctuation, being its behavior determined by the quantum potential *"that carries information about the environment of the quantum particle and thus informs and effects its motion. Since the information in the potential is very detailed, the resulting trajectory is so extremely complex that it appears chaotic or indeterminist"* [49]. Any attempt of measuring particles properties, changes the quantum potential, destroying its information. Actually, according to Bohm, Bohr had interpreted the uncertainty principle as meaning "not that there is uncertainty, but that there is an inherent ambiguity" in a quantum system [39]. As John Bell [6] observed, *"De Broglie-Bohm's idea seems... so natural and simple, to resolve the wave-particle, dilemma in such a clear and ordinary way, that it is a great mystery... that it was so generally ignored"*.

In the quantum-holographic theory, as Bohm [63] put it, *no field organized the implicate order, and it was consequently linear and difficult to unfold. The implicate order is a wave function, and the superimplicate order or superior informational field, is a function of the wave function, i.e., a superwave function that makes the implicate order non-linear organizing it in complex and relatively stable structures. Besides that, the holographic model as a way of organization of the implicate order was dependent upon the quantum informational potential field, that did not have capacity for self-organization and transmission of the information, essential for the understanding of the genesis and development of matter, life and awareness. The superimplicate order fills this need, allowing the understanding of awareness energy and matter as expression varieties of a same informational order. As a result awareness would already have been present since the beginnings of creation in the various levels of nature's unfolding and enfolding.*

"Even a stone is in some ways alive" states Bohm.

8. Towards a Holoinformational Model of Mind

The quantum potential guides by means of active information the particle alongside its course. As any elementary particle is united to the whole cosmos by means of a nonlocal quantum potential capable of change the structure of the universe, information then starts to be understood as nature's fundamental process. This active nonlocal information that organizes the particle's world reveal that the whole nature is informational, organized in a meaningful way.

In the brain, this meaning informational process is at the same time nonlocal quantum holistic and local classical Newtonian and mechanistic i.e. holoinformational. This is crucial to understand the holoinformational nature of awareness and intelligence in the universe. Matter, life and awareness are meaningful activities, intelligent quantum-informational processes, order transmitted through the cosmic evolution, originated from a generating informational field beyond our perception limits. A universe full of quantum potential and meaning activity, is an intelligent universe *functioning like a mind*, as Sir James Jeans already had observed. So, as awareness has always been present in all nature's levels of organization, matter, life and awareness cannot be considered as separated entities, capable of being analyzed under a fragmentary Cartesian-Newtonian framework. Actually, it must be considered as a indivisible unity, with quantum informational processes interacting by means of nonlocal holistic relationships, and simultaneously by local Newtonian mechanistic relationships, generating self-organization, complexity intelligence and evolution. Such view of a holoinformational intelligent "continuum", a fundamental generating order with a quantum-holographic informational creative flow permeating the whole cosmos, permits to understand the basic nature of the universe as an intelligent self-organizing unbroken wholeness, i.e., a *cosmic awareness*.

A kind of universal awareness unfolding in an infinite holoarchy. As a quantum-holographic system this universal awareness is distributed in every part of the cosmosphere, and each part of this holosphere contains the information of the whole cosmos in a holistic indivisible way. Quantum-informational fluctuations generated from this universal awareness through the holomovement (a nonlocal holoflux) self-organizes the universe's basic informational levels:

The **Cosmosphere** with the **Nuclear Code** that organizes energy and matter. It is the physical level and information is stored in atomic structures.

1) The **Biosphere** with the **Genetic Code** that organizes life. It is the biological level and information is stored in the DNA molecules.

2) The **Noosphere** with the **Neural Code** that organizes the brain and mind. It is the physical-social level and information is stored in neural networks

3) The **Technosphere** with **Artificial Intelligence Codes.** It is the technology level and information is stored in hardware and software designs

4) The **Holosphere** with the **Quantum-Holographic Code.** It is the awareness level that organizes the interconnectivity between the mind and the universe.

The spiritual level. Information is stored in quantum-holographic networks of the brain and the cosmos. This relegation (Latin *religare* and English religion) between us and the universe connects us with our primordial source, and has been described in a symbolic way in all humanity religious metaphors like "*the father is within us*"; "*as above so below*";" *as in earth so in heavens*", or the beautiful Buddhist metaphor of the Indra network. Such informational codes, this order that is transmitted in a meaningful and intelligent way through all levels of complexity of the universe, is the negentropic self-organization nature of the information awareness, a irreducible physical dimension of the cosmos as energy and matter.

9. Quantum Brain Dynamics

Experimental research developed by Pribram and other awareness researchers like Hameroff [35] and Penrose [36], Jibu and Yasua [37], and Ho [38] confirm the existence of a Quantum Brain Dynamics in neural microtubules, in synapses and in the molecular organization of the cerebrospinal fluid. This Quantum Brain Dynamics can generate Bose-Einstein condensates and the Fröhlich Effect. Bose-Einstein condensates consist of atomic particles, or in the case of the Fröhlich Effect biological molecules, that can assume a high level of coherent alignment, functioning as a highly ordered and unified informational state, as seen in lasers and superconductivity. Eccles psychons operate on synapses by way of quantum coherence processes.

These quantum dynamics show us that the interaction process between dendrons and psychons are not limited to the synaptic cleft, as stated by Eccles, but have a much wider embodiment throughout the whole brain, and as some researchers (see Popp), are saying, also throughout the whole body. Ho [38] have been demonstrating that "*Highly polarized multiple layers of liquid crystalline water molecules form dynamically coherent units with the macromolecules, enabling them to function as quantum molecular energy machines that transform and transfer energy with close to 100 percent efficiency. This liquid crystalline continuum of intimately associated polarized water and macromolecules extends throughout the extracellular matrix into the interior of every single cell, enabling each cell, ultimately each molecule, to intercommunicate with every other*".

Pribram [52,53] demonstrates good evidence that Eccles' dendrons make up receptive fields in cortical sensory units. Dendrons are composed of pre-synaptic teledendrons, synapses and post-synaptic dendrites, and they compose the fine fiber structure wherein brain processing occurs. As Pribram states [54], "*as sensory generated receptive fields they can be mapped in terms of wavelets, or wavelet-like patterns such as Gabor Elementary Functions. Dennis Gabor (1946) called these units Quanta of Information. The reason for this name is that Gabor used the same mathematics to describe his units as had Heisenberg in describing the units of quantum microphysics. Here they define the unit structure of processes occurring in the material brain*". I see this quantum holographic interactions not as a contradiction but as a natural extension of Eccles [22,23,26,35] ideas of an interactionism between dendrons and psychons.

Dejan Rakovic, in his site shows us how *the* ***Quantum-Holographic and Classically-Reduced Neural Networks can model Psychosomatic Functions***: *"The prevailing scientific paradigm considers information processing within the central nervous system as occurring through hierarchically organized and interconnected neural networks. However, it seems that this hierarchy of biological neural networks is going down sub-cellular cytoskeleton level, being according to some scientists a kind of interface between neural and quantum levels. At the same time it appeared, within the Feynman propagator version of the Schrödinger equation, that the quantum level is described by analogous mathematical formalism as Hopfield-like quantum-holographic associative neural network. The mentioned analogy opens additional fundamental question as to how the quantum parallel processing level gives rise to classical parallel processing level, which is a general problem of the relationship between quantum and classical levels within the quantum decoherence theory as well. The same question is closely related to the fundamental nature of awareness, whose in-deterministic manifestations of free will and other holistic manifestations of awareness like transitional states of awareness, altered states of awareness, and awareness pervading body – necessarily imply that some manifestations of awareness must have deeper quantum origin, with significant psychosomatic implications"* (see www.dejanrakovic.com).

This posits the necessity to explain how a self-organizing quantum mind can overlap quantum decoherence and maintain a persistent coherent state for a long time, at room temperature. I already cited Hameroff and Penrose's theory of quantum coherence in microtubules, and beyond this it is very well established experimentally today that the molecules of chlorophyll responsible for the photosynthesis process that transforms light photons in chemical energy in plants, can do this with extraordinary efficiency, for about 750 femtoseconds, compared with 1 to 1.5 femtoseconds frequency of chemical-bond vibrations. This is due to the action of a protein called antenna protein that holds the chlorophyll molecule sustaining the quantum coherence state and suppressing decoherence, by reinducing coherence in decohering parts of the chlorophyll molecule .

This show us that the capability to suppress decoherence at environment temperature is a common process in nature. So the capability to suppress quantum decoherence must be seen as a natural process in the wet brain, and we must work with the possibility that neurons and glia can sustain a quantum coherent state for milliseconds in the organized cellular complex molecular system full of proteins macromolecules, small molecules, ions and water. It is well known that in the vicinity of these macromolecules there is ordered water, and that proteins with a cavity in its 3 D structure can hold one or a few water molecules by means of hydrogen bonds. Quantum chemical computation shows that these ordered water molecules within and between two proteins separated by 12 to 16 Ångstroms permits the occurrence of quantum coherent electron transfer. This quantum coherence can propagate through nonlocal information transfer by quantum entanglement and superradiance. As self-organized systems these molecular systems have a huge structural and functional redundance that facilitates the interconnection of the neuronal quantum-holographic computer networks dynamics with the classic neuronal linear computer network.

I expanded my conjecture that the interconnectedness between brain and cosmos is an instantaneous holistic nonlocal connection and proposed the concept of a holoinformational flux, from which both mind and matter are in-formed, that resembles Bohm's holomovement. But in this new concept, quantum holographic brain dynamic patterns are conceived as an active part of the universal quantum-holographic informational field, and capable of generating an informational interconnection that is simultaneously nonlocal quantum-holistic(mind-cosmos holographic connection), and local Newtonian-mechanistic (brain-mind neural networks connections), i.e., holoinformational. Taking yet in consideration the basic mathematical property of holographic systems in which the information of the whole system is distributed in each part of the system, plus Bohm's holographic quantum physics data, and the experimental data of the holonomic theory of Pribram, we propose that this universal interconnectedness [24-26] could permit us to access all the information codified in the wave interference patterns existing in all the universe since its origin [18-21]. This quantum-holoinformational nature of the universe interconnects each part, each brain-mind- awareness, with all the quantum information stored in the holographic patterns distributed in the whole cosmos, in an indivisible irreducible informational cosmic unity [7,9,10,24-26].

10. Awareness and the Human Mind

The cybernetic networks of cyclical hierarchical relations through which we try to characterize life and awareness, interrelate themselves in a multilevel dynamic of *"hypercycles"* [32], organizing in *"self-catalytic"* cycles [54,41] in the *"edge of chaos"* [45]. Self-catalytic cycles can organize in higher levels, by means of catalytic hypercycles, (e.g. a virus) capable of evolving to more complex and more efficient structures, until the *"emergence of sets, of sets of... of sets of neurons"* [58]. In this way, the network generate *"creative loops"* [37] and *"hyperstructures"* [4] integrating themselves in systems with patterns of connectivity distributed and parallels, as the *"Global Workspace"* [47] and the "Extended Reticular-Thalamic Activation System"- ERTAS [48]

Dynamic non-linear systems like the human brain, with all these "neural correlates of awareness" are generated not only by such complexifications of matter's mechanistic external relations, but as we already saw, also as a prime reason by an harmonic unfolding of an indivisible universal holoinformational field. This quantum-holographic intelligent self-organizing field, is self-referred and continuously creates (unfolds) and recreates itself as a holographic distributed medium, and goes on experiencing continuously new possibilities of existence and non-existence, in an eternal and ever new unfolding-enfolding cycle. The *"self-consistent non big bang cosmology"* of Prigogine-Geheniau et al., describes the main features of this multi-cyclic learning scenario in which the cosmic evolution is the result of an interaction between the quantum vacuum (a *plenum*) and the particles of matter that are synthesized in it. Laszlo [44] adds to this scenario *"the postulate according to which the quantum vacuum is the fifth universal field interacting with matter"* stating that *"the field acts as a holographic medium, registering and conserving the scalar wave-transform of the 3n-dimensional configuration spaces assumed by matter in space"* (pg. 204).

This universal fifth field is not inferred from space-time interactions like the gravitational, the electromagnetic, the strong and the weak nuclear forces. In this new type of field, space and time become implicate, enfolded, as described mathematically by Bohm, in a spectral and holographic organized medium, made of the energy present in the interference patterns of the waveforms. The transformations from space-time order to this spectrum dimension are described by holographic mathematical formulations. This type of formulations was first described by Leibniz that created the conception of monads. Dennis Gabor in our century described the mathematical principles of holography and defined a quantum of information he named *logon*, a channel which can carry a unity of communication with the least amount of uncertainty.

11. Quantum-Holographic Neural Network Fields

Pribram in his Holonomic Brain Theory [50-54] proposes that there is a holographic process of information treatment, he named *multiplex neural hologram* distributed by the whole brain's cortex, dependent of neurons of local circuit that do not show long fibers and do not transmit the common nervous impulses. *"They are neurons that work in an undulatory way, and are overall responsible for the horizontal connections of the neural tissue's layers, connections in which holographic interference patterns can be built"* [52]. He describes a *"neural wave equation"* [53] resulting from the workings of the brain's neural networks, similar to the Schrödinger wave equation of the quantum theory.

Pribram has also demonstrated that hyper-stimulation of the fronto-limbic brain allows primates including humans to operate in a holistic, holographic-like mode. The electric excitation of these brain areas relaxes the Gaussian constraints, as Laszlo put it. *"While during ordinary levels of excitation of the frontolimbic system the signal processing creates the usual narrative awareness, when the excitation of this system exceeds a certain threshold, conscious experience is dominated by unconstrained holographic processes. The result is timeless, spaceless, causeless, 'oceanic' sensation"*. In these states the nervous system becomes, *'attuned to the holographic aspects of - the holograph-like order- in the universe'* [44]. Electroencephalographic and brain mapping studies made during altered states of awareness as meditation, prayer, and others brain-mind relaxation techniques, shows a high synchronization of the brain waves as if all the neurons of all brain centers were all playing the same symphony In this highly synchronized states of awareness the holographic brain treatment of information is optimized facilitating the interaction of the quantum-holographic brain network with the quantum holographic cosmic network [18,24].

Pribram have demonstrated that the receptor field of the cortical neurons reacts selectively to multiples sensorial modes making the harmony curves of adjacent receptors fields to mix as in a piano. In this way the harmony field of the cortex originates a resonance as a string instrument. The mathematical formulations that describes the resulting harmony curve are the Fourier transformations that Gabor applied in the creation of the hologram enriching these transformations with a model that can be reconstructed by the application of the inverse process. That holographic organization is what Bohm calls implicate order, a model that includes space and time in its structure as an enfolded dimension. Functioning in this holograph mode our brain "mathematically builds the objective reality" interpreting frequencies originating from a spectral dimension, a fundamental order, an informational field located beyond time and space.

12. Eccles Interactive Dualism — Pribram's Monism

Sir John Eccles described in the brain fine fibers structures he called dendrons composed of pre-synaptic teledendrons, synapses and post-synaptic dendrites connections, that he postulated could interact with the mind side of the interaction by way of units he called psychons. He proposed that these psychons could operate on synapses through quantum processes, and with Beck developed a beautiful and logical quantum interpretation of the synaptic function. Pribram [53] demonstrated that Eccles' dendrons make up receptive fields in cortical sensory units, that *"as sensory receptive fields they can be mapped in terms of wavelets, or wavelet-like patterns such as Gabor Elementary Functions. Dennis Gabor (1946) called these units Quanta of Information. The reason for this name is that Gabor used the same mathematics to describe his units as had Heisenberg in describing the units of quantum microphysics. Here they define the unity structure of processes occurring in the material brain. However, Gabor invented his function, not to describe brain processes, but to find the maximum compressibility of a telephone message that could be sent over the Atlantic Cable without destroying its intelligibility. The Gabor function thus describes both a unit of brain processing and a unit of communication. Brain is material, communication is mental. The same mathematical formulation describes both. The elementary structure of processing in Eccles' material dendron is identical to the elementary structure of processing of a mental (communication) psychon. There is a structural identity to the dual interactive process".*

To summarize: The structural identity between a material brain process and a mental communication process is provided by the Gabor "wavelet". The wavelet instantiates and accounts for the dual interactive process that Eccles and Popper are promoting. Eccles places the interaction within the synapse. This is not contradicted by the emphasis on the receptive field properties of the fine fibered pre and post-synaptic arbors except that the interaction is not limited to the synaptic cleft. Fine fibered membranes are also involved (Jibu and Yasue 1995). This extension overcomes two problems: There is no need to have ineffable, undefined mental processes acting on the synapse; and the energetics involved are brought into the realm of ordinary science.

Pribram proposes a monistic basis for Eccles dualism, showing that *there is an interactive mind/matter duality that is a "ground" from which both matter and mind are "formed" and the "dual" emerges*. That ground functions as a potential reality similar to Heisenberg potential world. *"This flux provides the ontological roots from which our experience regarding matter as well as mind (psychological processing) itself become actualized in spacetime"*. To illuminate this claim, Pribram relates the following story: *"Once, Eugene Wigner remarked that in quantum physics we no longer have observables (invariants) but only observations. Tongue in cheek I asked whether that meant that quantum physics is really psychology, expecting a gruff reply to my sassiness. Instead, Wigner beamed a happy smile of understanding and replied, "yes, yes, that's exactly correct". If indeed one wants to take the reductive path, one ends up with psychology, not particles. In fact, it is a psychological process, mathematics, that describes the relationships that organize matter. In a non-trivial sense current physics is rooted in both matter and mind. Communication depends on being embodied, instantiated in some sort of material medium. This convergence of matter on mind, and of mind on matter, gives credence to their common ontological root. My claim is that this root, though constrained by measures in spacetime, needs a more fundamental order, a potential order that underlies and transcends spacetime. The spectral basis of both matter and of communication portrayed by the Fourier relationship delineate this claim.*

As the brain has the capacity of function in the holographic nonlocal mode as in the space-temporal local mode, we think that we are dealing here with Bohr's concept of complementarity in the quantum functioning of the central nervous system.

The holonomic brain theory of Pribram, and the holographic quantum theory of Bohm, added with Laszlo's fifth field contribution quoted above, shows us that we are part of something much greater and vast than ours individual mind. Our mind is a subsystem of a universal hologram, accessing and interpreting this holographic universe. We are interactive resonances and harmonics systems with this unbroken self-organizing wholeness. We are this holoinformational field of awareness, and not observers external to it. The external observer's perspective made us lose the sense and the feeling of unity or supreme identity, generating the immense difficulties we have in understanding that we are one with the whole and not part of it. In this holoinformational model of awareness the nonlocal quantum-informational flow in a continuous holomovement of expansion and enfoldment, between the brain and the superimplicate order, is the universal awareness self-organizing itself as human mind. We are this very nonlocal quantum-holographic cosmos manifesting itself through our awareness, seeing itself through our eyes, and interconnecting in a participatory holistic and indivisible way the human brain to all levels of the self-organizing multiverse.

13. Qualia and the Hard Problem

Also, the essential characteristic of quantum nonlocal information of this dynamic process makes the question about the phenomenal quality (qualia) of conscious experience raised by Chalmers[13-15], multicontextual, multidimensional, relative not only to the observer, but also to the observation process and to what one observes, that is, to the holographic distributed information of the whole in question. The level of this informational quality is capable of increasing or reducing, in a phase transition, depending on the amount of information contained in the part of the universal hologram in focus and on the relations referential, whether external (Newtonian mechanistic) or internal (quantum holistic). The hard problem of awareness proposed by Chalmers is only difficult and problematic in a mechanistic and reductionist Cartesian-Newtonian-incomplete Copenhagen quantum context in which awareness and universe are considered separated entities. In a holoinformational context of internal relations, indivisible and nonlocal, it ceases to exist, because the self-organizing sublevels of the universe that get structured in a mechanistic-local way are understood as secondary manifestations of the harmonic, holistic and nonlocal nature of the universal holoinformational "continuum". Matter, life and awareness, are expressions of this holoinformational field, with fundamental nonlocal quantum relations, unfolding in myriads of possibilities [70,71].

Theoretically this sends us also to the question of the unconscious, which could be hypothetically understood this way, as a part of the universal holographic awareness unfolded in the brain/mind that gets "out of focus", gets "obscured", when it self-organizes as human awareness, as in a hologram, in which little parts contains the whole in a less clear way. Holoinformational awareness when structured (embodied) in the human brain, reduces the quality (qualia) of the perception of the unity/totality (holos) of nature, making these aspects remain usually unconscious, restricting the being's conscious field, and limiting it mentally and symbolically.

Matter, life and awareness, will never be understood by means of a fragmented and reductionist emergent view that only considers the external and mechanistic relationships. This is a perceptual error, already pointed out by oriental traditions, thousands of years ago, under the name of *"maya"*. As symbolical beings we can better understand this process going through the flower and fruit metaphor. We can say the fruit comes from the flower. However, the fruit is already implicit in the seed, making it impossible for us to state that it only and essentially comes from the flower. This would be a reductionism, a perceptive fragmentation of reality. Actually, not even the seed originates the fruit. The fruit comes from an indivisible totality, clearly intelligent and holo-related: sun, rain, earth, air, wind, cosmic rays, seasons, weather, microorganisms, insects, birds, seed, sap, steam, leaves, life principle...*ad infinitum*, in an irreducible holoinformational order.

14. Noetic Field Theory — The Quantization of Minds

Any form of Cartesian-Eccles interactive dualism is eminently unpopular in the face of the currently dominant view of the mind-body problem - The Cognitive approach which claims mind equals brain; and the related AI approach that professes 'if we knew the correct program/algorithm, we could duplicate all of human intelligence on a contemporary computer'. AI proponents also say a thermostat is intelligent because it has two states; to us this must mean that a smoke alarm is a genius because it can also speak! Cognitive theorists have asked 'What processes in the brain give rise to awareness?'. This has created a 'Hard Problem' too difficult to research [13-15]. Chalmers has also stated the problem of consciousness is tantamount (the same) as the problem of qualia.

Amoroso, on the other hand by simply asking 'What processes give rise to awareness' has managed to solve the mind-body problem in a comprehensive and empirically testable manner [70-73]. One proof of concept is that while 'qualia' has remained a philosophical construct in cognitive theory; the tenets of Amoroso's Noetic Field Theory: The Quantization of Mind (NFT) has actually physicalized the basis of qualia breaking down the 1st person-3rd person barrier. He has also filed patents [73] to develop new technologies based on the quantification of mind. He says that, 'as we have developed the periodic table of chemical elements over the past hundreds of years, so in the coming age of mind we will map the noumenon of awareness such that qualia can be extracellularly stored and transferred to 'other minds' by a special class of sensory-bypass enabled quantum computers [72,73].

Amoroso claims to have designed 13 empirical tests (at time of this writing) to falsify NFT; the simplest and least expensive experimental protocol is outlined in this volume [75]. Being able to physically quantify qualia has also led to a formal model of the Eccles' Psychon. Recall the unit of measure called 'The Einstein' signifying a mole or Avogadro's number of photons used in photosynthesis; in a similar fashion Amoroso has created a unit of measure called 'The Psychon' in honor of Nobelist Sir John C. Eccles that quantifies the energy of qualia or measures the energy of awareness [76].

We will summarize briefly here 'Discovery of the Mind' in terms of NFT, a true unified field theory of mind-body interaction. Firstly Descartes was unnecessarily criticized for claiming mind stuff - *res cogitans* is nonphysical. Even today hundreds of years later a salient definition of 'nonphysical' is spiritual which we assume is how Descartes used the term. In NFT a life principle based on the unified field of physics (or spirit of God) is a physically real aspect of the unified field. This removes the main stigma of Cartesian dualism that *res cogitans* violates the laws of thermodynamics and the conservation of energy. The ordering principle of the Unified Field is not a 5th fundamental force of physics; rather it is a 'force of coherence' applied ontologically (rather than phenomenologically which requires the exchange of energy by quanta transfer) by what is called topological switching [70-73].

Secondly in the very simplest terms the Planck scale is not the 'basement of reality' but just another stopping point in our understanding of the universe. Thus there are three regimes to reality: Classical-Quantum and Unified. It is in the new 3rd regime that access to the principles of the mind resides. Just as Quantum Mechanics was invisible to the tools of Newtonian Mechanics so until now has the regime of the unified field been invisible to the tools of quantum mechanics [71].

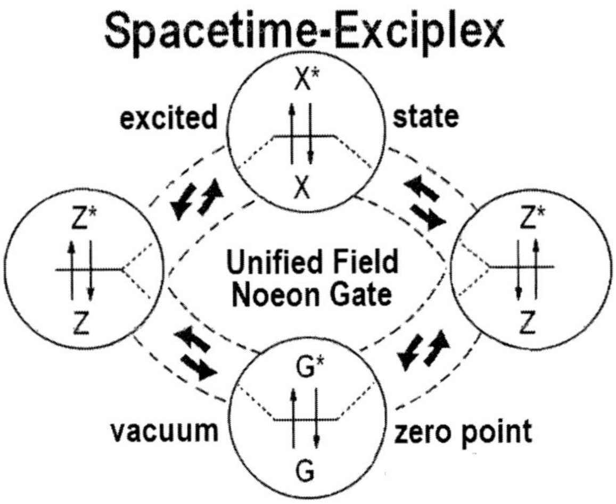

Figure 1. The four circles represent Least Cosmological Units (LCU) (See Fig. 2) forming an exciplex complex tiling the spacetime backcloth of HAM cosmology. The array of LCUs acts as a gating mechanism for ontological entry of the exchange quanta of the U_F into every point in Minkowski 4-spacetime and all matter. Figure courtesy R.L. Amoroso.

To summarize briefly: we now have a 3rd physical regime wherein lies a 'life principle' that interacts with the brain/body forming a self-organized living system (SOLS). To Amoroso this is most profoundly stated theologically: *"And the light which shineth, which giveth you light, is through him who enlighteneth your eyes, which is the same light that quickeneth your understandings; Which light proceedeth forth from the presence of God to fill the immensity of space—The light which is in all things, which giveth life to all things, which is the law by which all things are governed,..."* [74].

What is left to describe is the basic gating mechanism bringing the life principle into living systems. Developing NFT required a whole new anthropic cosmology to introduce this essential component absent from naturalistic Big Bang cosmology. Essentially NFT's description of the 'mind gate' requires violation of the quantum uncertainty principle [72]. Uncertainty being the shield 'hiding' the 3rd regime. Related to the uncertainty principle is the zero-point field (ZPF) where virtual quantum particles wink in and out of existence momentarily for a duration of the Planck time (as governed by the uncertainty principle). The 1st component of the gate he developed is called an 'exciplex', short for excited complex - meaning it stays excited and never returns to zero as the ZPF does in terms of the exclusion principle of the Copenhagen Interpretation of quantum theory.

Operation of the exciplex gate requires Large-Scale Extra Dimensions (LSXD). This includes an oscillating form of Planck's constant, $\hbar$ fluctuating from the continuous-state asymptotic virtual Planck scale (never reached) of the usual theory to the Larmor radius of the hydrogen atom. This is part of the process in which the exciplex gating mechanism violates the quantum uncertainty principle [70] utilizing LSXD in a continuous-state process (local alternative to expansion of the universe) such that the gate is only periodically open - cycling like a holophote or lighthouse beacon into each point and atom in spacetime.

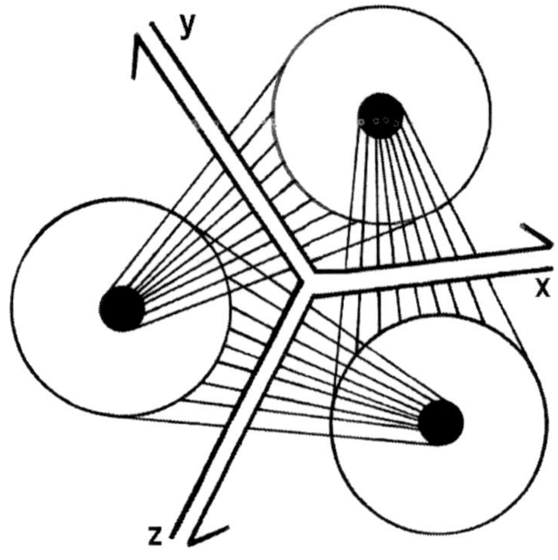

Figure 2. Triune structure of a conceptualized solitary LCU that like an isolated quark cannot exist free in nature. The central parallel lines are a Witten string vertex with properties of a complex Riemann sphere able to continuously rotate from zero to infinity. The field lines represent the 'super quantum potential' of the unified field, U_F. Figure courtesy R.L. Amoroso.

The Unified Field, U_F is not a 5th force per se, and is also not phenomenological (forces are mediated by the exchange of energy transferred by quanta, i.e. the em field is mediated by the photon). The U_F does imply force, however it is an ontological or energyless 'force of coherence'. Information is transferred by a process called 'topological switching'; this is what occurs when staring at a Necker cube and the vertices change position.

The caption in Fig. 2 mentions a piloted 'super quantum potential'. This arises from NFTs use of and LSXD extension of the de Broglie-Bohm interpretation of quantum theory. Recall their theory that quantum evolution is continuous and guided. Thus in terms of NFT one would say that the quantum potential/pilot wave are a subset of the action of the unified field.

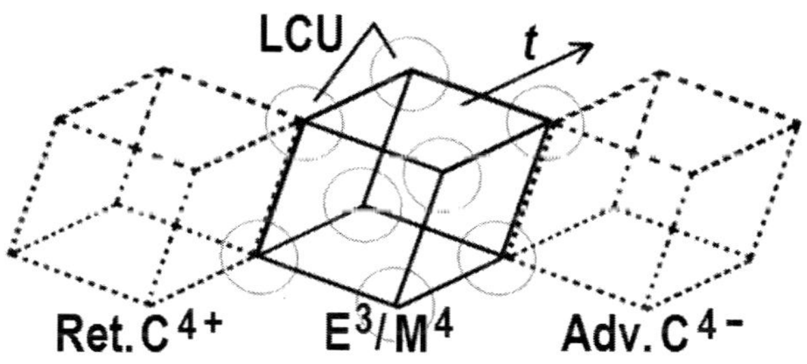

Figure 3. How observed (virtual) 3D reality (center) arises from the infinite potentia of 12D space. The 'standing-wave-like' (retarded-advanced future-past) mirror symmetric elements, C^{4+} / C^{4-} (where C^4 signifies an 8D potentia of complex continuous-state space is distinguished from the locally realized visible 3D spacetime) produce the observed Euclidian, E_3, Minkowski, M_4 space (center) as a closed resultant. Figure courtesy R.L. Amoroso.

In Fig. 3 Least Cosmological Units (LCU) governing evolution of the 'points' of 3D reality are represented by circles. The Advanced-Retarded future-past cubes in HD space guide the evolution of the central cube (our virtual reality) that emerges from potentia elements of HD U_F space.

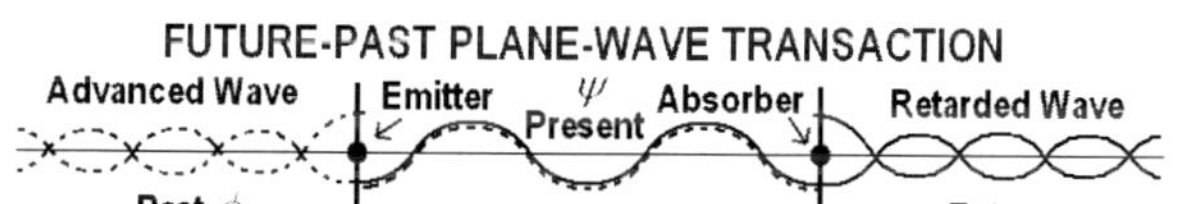

Figure 4. Conceptualized structure of a transaction (present quantum state or TI event) where the present (simplistically) is a standing-wave of future-past elements. A point is not a rigid singularity as in the classical sense, but has a complex structure like a mini-wormhole where R_1 & R_2 (like the frets holding the wire of a stringed instrument) represent opposite ends of its vibrational diameter. However a NFT singularity is only cyclically discrete as shown in the text below. Figure courtesy R.L. Amoroso.

This is also illustrated in Fig. 4 in terms of NFTs LSXD extension Cramer's transactional interpretation of quantum theory in a more dynamic sense; meaning that the three cubes in Fig. 3 act like a standing-wave of the future-past similar in principle to how a standing-wave operates on 1 1D stringed instrument or 2D drum. Realize of course that the standing-wave of reality is hyperdimensional.

Figure 5 represents a unique M-Theoretic model of 'Continuous-State' U_F dynamics as it relates to NFT and its putative exchange quanta of the U_F called the noeon. What Amoroso proposes as an essential part of this continuous-state anthropic multiverse cosmology is that our observed reality is closed and finite in time as a 'Poincare-Dodecahedral Space' at the cosmological scale and virtual Euclidean cube at the microscopic.

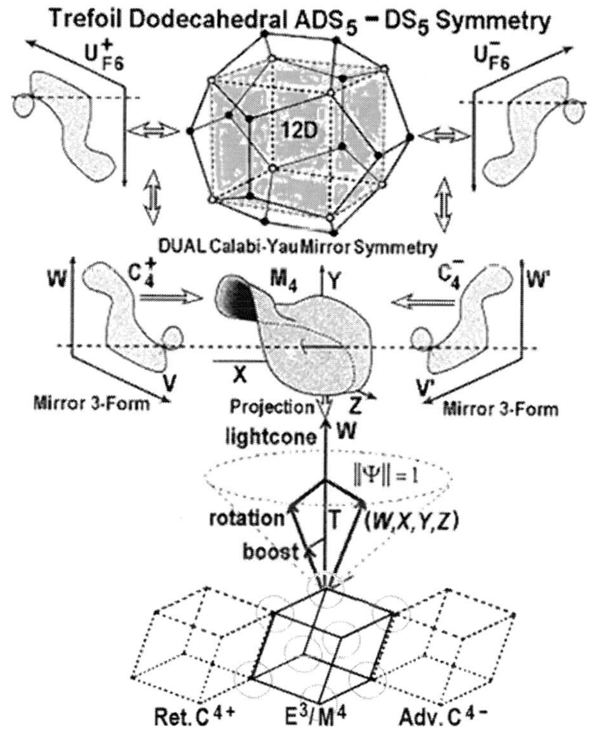

Figure 5. Complete LSXD regime of NFT cosmology illustrating the hierarchy of its geometric topology. Dodecahedral involute properties, as well as the continuous-state exciplex 'hysteresis loop' of noeon injection (not shown) of 'spirit' into every point (and thus atom) in spacetime animating SOLS and mediating qualia. Figure courtesy R.L. Amoroso.

15. Final Considerations

Besides outlining the fundamentals of a nonlocal, self-organizing and holoinformational conception of awareness, this approach also gives the directions for understand information as the unifying principle, capable of connecting awareness to the universe and to the totality of space and time (Amoroso [73-76,78,79]). It also allows a better understanding of phenomena and theories related to awareness which up to now we could not explain or understand adequately, such as synchronicities, archetypes, collective unconscious

(Jung), unconscious complexes (Freud), near-death experiences (Moody Jr.), premonitory dreams, psychokinesia and telepathy (Rhine), morphogenetic fields and morphic resonance (Sheldrake), out-cerebral memory (Stevenson), memories of previous existences (Weiss), amongst others.

Brian Josephson, Nobel Prize in physics, believes that Bohm's theory of the implicate order can even lead someday to the inclusion of God in the science network (see forward this volume). We believe that the holoinformational view of awareness which has in Bohm's quantum theory one of its very foundations, implies the inclusion in science's framework of a Cosmic awareness. A Universal Intelligence that originates, permeates, maintains and transforms the universe, life and mind, through the holoinformational process [70-72].

Finally, we would like to state that in the Cartesian-Newtonian reductionist paradigm, the question about the nature of awareness is unanswerable. It can be useful to unfold new knowledge and generate new questions and answers. However, the inherent fragmentation of this perspective, obscures more and more our understanding of what reality and awareness are. This missing piece is supplied in Sect. 14 by the addition of the heretofore 3rd regime: Newtonian-Quantum- Unified.

We did not come to this world: we came from it, like the leaves of a tree. Like the ocean produces waves, the universe produces people. Each individual is an expression of the whole kingdom of nature, a single action of the total universe. Rarely this is, if at any time it is at all, felt by the majority of the individuals. - Allan Watts

Acknowledgement

Heartfelt thanks expressed to Richard L. Amoroso for input in Sect. 14 and the figures utilized therein.

References

[1] Atkin, A. (1992) "On awareness: what is the role of emergence?", Medical Hypothesis, 38 pp. 311-14.
[2] Atlan, H. (1972) *L'Organization Biologique et la Théorie de L'Information*, Hermann, Paris.
[3] Atlan, H. (1979) *Entre le Cristal et la Fumée, essai sur l'organization du vivant, Seuil*, Paris.
[4] Baars, B.J. (1997) In the Theater of awareness: The Workspace of the Mind, Oxford University Press.
[5] Bateson, G. (1979) Mind and Nature: a necessary unity, Dutton, New York.
[6] Bell, J. (1987) Speakable and Unspeakable in Quantum Mechanics, Cambridge University Press.
[7] Bohm, D. (1983) Wholeness and the Implicate Order, Routledge, New York.
[8] Bohm, D. (1987) Unfolding Meaning, a weekend of dialogue with David Bohm. ARK Paperbacks, Routledge& Kegan Paul Ltd.
[9] Bohm, D. , Hiley, B. J. (1993) The Undivided Universe. Routledge, London.
[10] Bohm, D., and Peat, F. D. (1987) Science Order, and Creativity. A dramatic new look at the creative roots of science and life, Bantam Books, New York.
[11] Brillouin, L. (1959) Vie Matière et Observation. Editions Albin Michel.
[12] Capra, F. (1988) Uncommon Wisdom. Simon & Schuster, New York.
[13] Chalmers, D. J. (1995) "Facing Up To The Problem Of Consciousness", J Awareness Studies, 2, no 3, pp. 200-19
[14] Chalmers, D.J.(1995a)"The Puzzle of Conscious Experience", Scientific American, December.
[15] Chalmers, D.J. (1996) The Conscious Mind. In Search of a Fundamental Theory, Oxford. Univ. Press, New York.
[16] Clarke, C.J.S. (1995) "The Nonlocality of Mind", Journal of awareness Studies,2, N.3, pp.231-240
[17] Di Biase, F. (1981) *Auto-organização nos sistemas biológicos". Ciência e Cult.*,33(9): 1155-1159, *Sociedade Brasileira para o Progresso da Ciência,* Brazil.
[18] Di Biase, F. (1995) *O Homem Holístico, a Unidade Mente-Natureza*, Editora Vozes, Rio de Janeiro, Brazil.
[19] Di Biase, F. , Rocha, M.S.F.(1998) *Caminhos da Cura* , Editora Vozes, Rio de Janeiro, Brazil
[20] Di Biase, F., Rocha, M.S.F. (1999) Information Self-organization and awareness World Futures- the Journal of General Evolution vol. 53, pp.309-327, UNESCO – Ed. Erwin Laszlo. The Gordon and Breach Group, U.K.
[21] Di Biase, F.,& Rocha, M.S. (2000) Information Self-Organization and awareness: Toward a Holoinformational Theory of awareness, in Amoroso R.L. et al. (eds.) Science and the Primacy of Consciousness: Intimation of a 21st Century Revolution, Oakland: Noetic Press; also in The Noetic Journal, Vol. 2 No. 3, july (1999) Noetic Press
[22] Di Biase, F. & Rocha, M.S. (2004) *Ciência Espiritualidade e Cura- PsicologiaTranspessoal e Ciências Holísticas*, Rio: Editora Qualitymark.
[23] Di Biase, F., Amoroso, R.L (eds.) (2005) *A Revolução da Consciência. Novas Descobertas sobre a Mente no Século* XXI. Editora Vozes, Rio, Brasil.
[24] Di Biase, F.,Amoroso, R.L. (2008) Holoinformational awareness: An extension of interactive dualism with anticipatory parameters. Intl. J Computing Anticipatory Systems, Vol. 22, D.M. Dubois (ed.) Liège, Belgium.
[25] Di Biase, F. (2009),A Holoinformational Model of awareness, Quantum Biosystems, 3, 207-220,Italy.
[26] Di Biase, F.(2009), Quantum-holographic informational awareness , NeuroQuantology , Vol 7 , Issue 4. pp 657-664
[27] Eccles, J.C. (1952) The Neurophysiological Basis of Mind, Oxford University Press, Oxford.
[28] Eccles, J.C. (1989) A unitary hypothesis of mind-brain interaction in the cerebral cortex, Proc. R. Soc. Lond. B 240, pp. 433-451.
[29] Eccles, J.C. (1993) Evolution of complexity of the brain with the emergence of awareness, In Pribram, K. (ed.) Rethinking Neural Networks: Quantum Fields and Biological Data, Manwah: Lawrence Erlbaum.
[30] Eccles, J.C. (1994) *Evolution du Cerveau et Création de la Conscience, ch. 8.8 Une nouvelle hypothèse sur*

l'interaction esprit/cerveau à partir de la physique quantique: l'hypothèse des micro-sites, Paris : Flammarion.

[31] Eccles, J.C. (1998) Do mental events cause neural events analogously to the probability fields of quantum mechanics? Proc R Soc Lond [Biol] 227:411-28.

[32] Eigen, M., Schuster, P. (1979) The Hypercycle. A principle of natural self-organization.Springer-Verlag, Berlin.

[33] Gisin, N. et al. (1997) Science, vol. 277, pg 481.

[34] Grof, S. (1985) Beyond the Brain: Birth, Death, and Transcendence in Psychotherapy. State University of New York Press, Albany.

[35] Hameroff, S.R.(1994) "Quantum Coherence In Microtubules: A Neural Basis For Emergent awareness?', J Awareness Studies, 1, N. 1, Summer 1994, pp.91-118.

[36] Hameroff, SR., and Penrose R. (1996) Orchestrated reduction of quantum coherence in brain microtubules: A model for awareness, in Toward a Science of Awareness: The First Tucson Discussions and Debates, Stuart R. Hameroff, Alfred W. Kaszniak, & Alwyn C. Scott (eds.) Cambridge: MIT Press.

[37] Harth, E. (1993) The Creative Loop. How the brain makes a mind. Reading: Addison-Wesley.

[38] Ho, Mae-Wan, (2008) The Rainbow and the Worm, The Physics of Organisms, 3rd edition, World Scientific, London.

[39] Horgan, J. (1996) The End of Science, Helix Books, Addison-Wesley Publishing Company.

[40] Jantsch, E. (1980) The Self-Organizing Universe. Pergamon Press, New York.

[41] Kauffman, S. (1995), At Home in the Universe, The Search for the Laws of Self-Organization and Complexity. Oxford University Press, New York.

[42] Koestler, A. (1967), The Ghost in the Machine, Hutchinson & Co. (Publishers) Ltd., London.

[43] Kübler-Ross, E. (1983), On Children and Death, Mac Millan, New York.

[44] Laszlo, E. (1993), The Creative Cosmos, A unified science of matter, life and mind, Floris Books, Edinburgh, Great Britain.

[45] Lewin,R.(1992) Complexity: Life on the Edge of Chaos,.MacMillan, New York.

[46] Moody, R.A.(1976) Life after Life, Bantam Books, NY.

[47] Newman, J. (1997) Putting thePuzzle Together. Part I: Towards a General Theory of the Neural Correlates of awareness, Journal of awareness Studies, 4 (1), pp. 47-66

[48] Newman, J. & Baars, B.J. (1993) 'A neural attentional mode access to awareness: A Global Workspace perspective, Concepts in Neuroscience,4(2), pp.255-90.

[49] Peat, SD. (1987) Synchronicity, the bridge between matter and mind, Bantam Books, New York

[50] Pribram, K. (1969) "The Neurophysiology of Remembering", Scientific American 220, Jan. pp.75.

[51] Pribram, K. (1977) Languages of the Brain, Monterey, California, Wadsworth Publishing.

[52] Pribram, K. (1980) '*Esprit, cerveau et conscience*', in *Science et Conscience, les deux lectures de l'univers.* Éditions Stock et France-Culture, Paris.

[53] Pribram, K. (1991) Brain and Perception: Holonomy and Structure in Figural Processing. Erlbaum, Hilsdale, NJ.

[54] Pribram, K. (1997a) In Memoriam: Nobel Laureate Sir John Eccles, Noetic J, Vol. 1, pp 2-5. Noetic Press, Orinda.

[55] Pribram, K. Ed. (1993) Rethinking Neural Networks: Quantum Fields and Biological Data, Hillsdale: L. Erlbaum.

[56] Prigogine, I., Stengers, I. (1979) *La Nouvelle Alliance*, Editions Gallimard, Paris, France.

[57] Prigogine, I., Stengers, I., (1988), *Entre le Temps et L'Eternité, Fayard,* Paris, France.

[58] Ring, K. (1980), Life at Death. Quil, New York.

[59] Sabom, M. B. (1982) Recollections of Death, New York: Harper and Row.

[60] Scott, A. (1995) Stairway to the Mind. The Controversial New Science of awareness, Copernicus, New York: Springer-Verlag.

[61] Seager, W. (1995) Conciousness, Information and Panpsychism. J Awareness Studies, 2, No 3, pp. 272-88.

[62] Shannon, C. E., W. Weaver (1949) The Mathematical Theory of Communication.Univ of Illinois Press, Urbana.

[63] Stonier, T., (1990) Information and the Internal Structure of the Universe. Springer Verlag, New Addison-Wesley, Reading, Mass.

[64] Stonier, T.(1997) Information and Meaning. An Evolutionary Perspective. Springer, Great Britain.

[65] Umezawa, H (1993) Advanced Field Theory, New York: AIP Press

[66] Weber, R. (1982) The enfolding unfolding universe: A conversation with David Bohm, In The Holographic Paradigm, Ken Wilber (ed.) Boulder: New Science Library.

[67] Weil, P. (1993) *Axiomática transdisciplinar para um novo paradigma holístico*, in *Rumo À Nova Transdisciplinaridade: Sistemas Abertos De Conhecimento*; P. Weil, *Ubiratan D'Ambrosio, R. Crema, Summus,* São Paulo, Brasil.

[68] Weiss, B. (1996) Muitas Vidas Muitos Mestres. Editora Salamandra, Rio de Janeiro, Brazil.

[69] Wheeler, J. (1990) Information, physics, quantum: the search for links, in Complexity, Entropy and the Physics of Inform., Wojciech H. Zurek (ed.) Reading: Addison-Wesley.

[70] Wiener, N. (1948) Cybernetics, or Control and Communication in the Animal and Machine, The Technology Press & John Wiley & Sons, New York.

[71] Wilber, K. (1997) An Integral Theory of awareness, Journal of awareness Studies, 4, No 1, pp. 71-92.

[72] Zurek, W. H., (ed.) (1990) Complexity, entropy and the physics of information, Santa Fé Institute, Studies in the Science of Complexity, Vol. 8. Reading: Addison- Wesley.

[73] Amoroso, R.L., & Rauscher, E. (2009). The Holographic Anthropic Multiverse, Singapore, World Scientific.

[74] Amoroso, R.L. (ed) (2010) Complementarity of mind and body: Realizing the dream of Descartes, Einstein and Eccles, New York: Nova Science.

[75] Amoroso, R.L. (2014) Through the Looking Glass: The Physical Basis of the Observer, Hackensack: World Sci.

[76] Amoroso, R.L. (2012) US Patent: Spacetime energy resonator: a transistor of complex Dirac polarized vacuum topology, https://docs.google.com/a/google.com/viewer?url=www.google.com/patents/US20120075682.pdf.

[77] The Doctrine & Covenants (1999) 88: 11-13, Salt Lake City: The Church of Jesus Christ of Latter-day Saints.

[78] Amoroso, R.L. (2013) "Shut The Front Door!": Obviating the Challenge of Large-Scale Extra Dimensions and Psychophysical Bridging, in R.L. Amoroso, L.H. Kauffman, & P. Rowlands (eds.) The Physics of Reality: Space, Time, Matter, Cosmos, Singapore: World Scientific.

[79] Amoroso, R.L. (2013) Crossing the Psycho-Physical Bridge: Elucidating the Objective Character of Experience.

Time?

RICHARD L. AMOROSO

Noetic Advanced Studies Institute, Escalante Desert, UT 84714-5104 USA
amoroso@noeticadvancedstudies.us, www.noeticadvancedstudies.us

The concept of time in the 'clockwork' Newtonian world was irrelevant; and has generally been ignored until recently by several generations of physicists since the implementation of quantum mechanics. We will set aside the utility of time as a property relating to physical calculations of events relating to a metrics line element or as an aspect of the transformation of a particles motion/interaction in a coordinate system or in relation to thermodynamics etc., i.e. we will discard all the usual uses of time as a concept used to circularly define physical parameters in terms of other physical parameters; concentrating instead on time as an aspect of the fundamental cosmic topology of our virtual reality especially as it inseparably relates to the nature and role of the observer in natural science.

Keywords: M-Theory, Multiverse, Observer, Quantum theory, Reality, Time, Unified field mechanics

"*The distinction between past, present and future is an illusion...*" - Einstein.

1. Introduction

This is a brief introductory exploration into the nature and place of time in in what is cosmologically perceived as a virtual anthropic reality relative to 3D Euclidean space and the related matter of the role of the observer in natural science. A more rigorous development is being attempted in a joint monograph [1]. Time in physical science is one of seven fundamental quantities in the International System of Units used to define aspects of other quantities thus giving physical time a circularity keeping its fundamental basis elusive. There are two philosophical viewpoints on the nature of time:

- Time is a fundamental component or dimension of the universe in which events occur. This is called the realistic view and ascribed to Newtonian Mechanics.
- Time is not a 'bucket' or domain processing events or objects. Time is rather a fundamental property of the mind of the observer like number or space; time is not a thing that exists and can therefore not be measured or traveled. This position is taken by philosophers like Kant, Descartes or Leibniz.

These premises are also conjoined with arguments whether space is relational or absolute and the issue of whether space is finite and closed or open and infinite [2-5].

Our position seems aligned with the latter point of view above; however the work we propose to do is to lift the philosophical perspective into a form of anthropic physical cosmology with sufficient rigor to systematize the basis of the observer to a degree that is theoretically sound and thus empirically testable. In Newtonian mechanics time was considered irrelevant because of its mechanical nature and could thus be ignored. Einstein systematized a fundamental entity called spacetime with relativistic effects related to motion and matter and the quantization of energy such as the photon. Still time itself, if there was such a thing with real physical properties, remained a mystery essentially ignored for centuries. Recently however interest in the physics of time has increased because of its relevance to astrophysics, quantum field theory and especially the nature and role of the observer in the philosophy of measurement theory [6-8].

Here we attempt to make a definitive delineation of the concept of time in terms of Unified Field Theory, U_F. Our initial question is whether this can be done without taking into account not only the role of the observer but the nature of awareness and observer's very essence? It is well known that at the microscopic level the laws of physics are reversible; but at the local level (the level of the observer) an unexplained apparent arrow of time enters the world picture throwing causality and the nature of consciousness or awareness into the mix.

We know if by no other reason than the operation of quantum theory, that the 3D reality of the observer is virtual. This makes the nature of reality and the fundamental basis and essence of the observer within that system a key or essential element to the degree that it seems a mandatory prerequisite to solve both the nature of reality and the observers place within it before one can consider a proper exploration of the 'essence' of time. That is the required context. Any attempt otherwise would be doomed to failure, and indeed this seems to be the salient reason why such an attempt has not occurred until now.

The literature has discussed five features of times arrow [5]:

- Thermodynamic time or entropic flow
- Cosmological time - Perceived expansion of a Big Bang universe
- Psychological time
- Clock time or electromagnetic arrow of time
- Kaon decay - Temporal asymmetry of the neutral kaon and its role in CP violation, generating the observed matter-antimatter asymmetry of the universe.

But we will ignore all these aspects here and continue with our discourse in a different arena.

2. Crux of the Matter — Anthropic Reality ↔ Observer

Time is not ultimately phenomenological or a 'thing in itself' (noumenon) but rather an aspect of anthropic cosmology - the geometric topology of the observer's reality operationally organized (a complex system) to:

1) Partially isolate the observer's frame of awareness (in an anthropic multiverse) from the totality of God's eternity.
2) The geometric topology (of 3D virtual reality) operates so that 'time' 'surfs' on the face of that eternity (regime of the unified field, U_F).
3) Anthropic cosmology provides a gating mechanism for the teleological coherent action of the U_F to operate as a life principle.
4) Hides the function and operation of God's '*omnies*' (omniscience, omnipotence, omni-presence) from the observer's 'normal' awareness. The mechanistic properties of the gating mechanism are profoundly efficient and elegant.
5) A form of continuous-state 'subtractive interfer-ommetry' isolates the observer's awareness from the HD whole.
6) The additional dimensionality postulated by string theory is not invisible because it is curled up at the Planck scale. The unified field regime of large-scale additional dimensions (LSXD) is invisible by a process of subtractive interferometry relative to the observer.

The continuous-state cyclical operation of the observer's reality is a form of hypercycle. We will discuss this in more detail later. But for now this is the Continuous-State mantra of Holographic Anthropic Multiverse (HAM) Cosmology [5-8]: A 'continuous-state spin-exchange dimensional reduction compact-ification process. In the conventional wisdom the Planck scale is the basement of reality. We can apply reductionism from 4D Minkowski space or 11D M-Theory and reduce to a 0D rigid barrier, stochastic quantum foam or singularity.

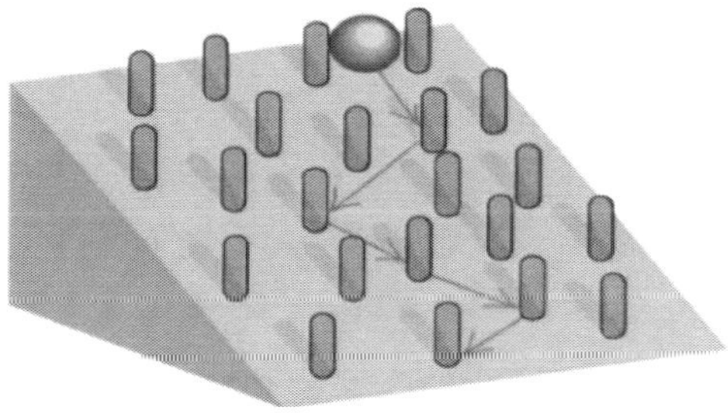

Figure 1. Conceptualizing the gravitational 'free-fall' flow property of the continuous-state process as the path of a fermion singularity (ball) embedded in a spacetime raster (not as shown but instead embedded in the Calabi-Yau topology of a 6D Riemann hypersphere) continuously rotating relativistically such that the fermion remains centralized like it was a stationary or standing wave.

Simplistically following a LSXD extension of Cramer's Transactional Interpretation, a continuous-state present event is like a 'standing-wave of the future-past', not like a 1D string or a 2D drum but a 12D hyperstructure. This standing-wave process also operates in conjunction with an HD extension of the de Broglie-Bohm interpretation where 'piloted' spacetime and matter are created-annihilated and recreated as a wave-particle duality. In fact in terms of the continuous-state process we elevate wave-particle duality to a principle of anthropic cosmology.

Now finally in this simplistic rendering imagine a 3-sphere - a reduced least cosmological unit (LCU) bouncing down a pin raster but where the raster is continuously rotated such that the 3 sphere never reaches ground. This conceptualizes primitively the continuous-state process. This is quite detailed in both its parameters and derivation; such that the reader would have to invest sufficient time delving into it in order to sort it out [6-8]. But the salient point we wish to make here is that the 'dimensional reduction compactification' Calabi-Yau mirror symmetry process on the way down from 12D to asymptotically approached 0D is a smooth 'cascade' of odd even dimensional jumps. Calabi-Yau manifolds are Kahler forms of Riemann manifolds. The Riemann sphere forms an extended complex plane or set of extended complex numbers consisting of the usual complex numbers, C together with infinity. We write the extended complex numbers as, $\hat{\mathbb{C}} = \mathbb{C} \cup \{\infty\}$.

A directed infinity in our anthropic dimensional reduction process in the extended complex plane has a defined cascade angle, θ with an infinite absolute value, r. For the limit $1/x$ where x a positive numbered dimensionality approaching zero it is a directed infinity with angle 0; however, 1/0 is not a directed infinity but a complex Riemann infinity. A directed infinity in direction z is an infinite numerical quantity that is a positive real multiple of the complex number z.

But when the continuously transforming topology of the Riemann sphere asymptotically approaches 0D, a catastrophic 'jump' back to infinity or in this case 12D occurs (actually 6D or 9D, as the other dimensions or degrees of freedom act as control factors. This is a property of the Hubble sphere being a self-organized complex system, closed and finite in 'time' but nested in an open and infinite LSXD. This 'cascade-jump' will be of prime importance later in discussing the subtractive interferometry within the observer's reality.

Thus we begin to see the Holographic Anthropic Multiverse Cosmology (HAM), and its inherent continuous-state mirror symmetric synchronization backbone [6] wonderfully restricts the framework for delineating the nature of the concept of time. If it helps in any way; we have been working on this cosmological paradigm for about 18 years, and it is only now that we understand it sufficiently well to attempt a description of the nature of time. What we are saying is that the restrictions seem to give us no other choice in the path to take. Not that we don't like it; rather why it might seem so 'weird' to the HAM novitiate.

The conundrum of the philosophical nomenclature is difficult. We mentioned earlier that we sit on the side of the fence where time is not considered a 'thing' but an aspect of the observer's mind. It is probably best to let the sleeping dogs of categorization lie and instead proceed with a delineation of the framework.

We will begin with the unpleasantries of theology. The rigid and myopic ecumenisms of church authorities at the time of Copernicus and Galileo is no worse than the current disdain by for 'metaphysical truth' by physicists wielding an umbrella of rigorous pragmatism as an exclusionary defense shield against the slightest possibility of transcendent inroads [9-11]. It seems an obvious misinterpretation of Bible passages to insist that 1) for example - 'The earth is the center of the universe, and 2) 'The orbit of celestial spheres must be perfectly circular'. And to have Copernicus hide his 1543 *On the Revolutions of the Heavenly Spheres*, "*after being buried among my papers and lying concealed not merely until the ninth year but by now the fourth period of nine years*" or bring Galileo to the brink of execution for a contrary opinion. This myopic 'force' was so strong that Hipparchus, 2000 years prior to Copernicus (who discovered the same heliocentric calculations) abandoned his work because of the perfect circularity issue. We make the trouble of mentioning this scenario because epistemologically an 'empirical metaphysics' is not only finally possible but we feel has now been demonstrated [10-12]. But worse for the tethered myopic the seemingly critical aspects of the observer and nature of reality are steeped in the teleology of an anthropic multiverse cosmology!

The continuous-state flux is not a temporal property but an aspect of the wave nature of reality itself. There are significant cosmological differences between Big Bang cosmology where protons where created and fixed from the original singularity as described by the Copenhagen interpretation of quantum theory and HAM cosmology. In HAM cosmology there is no wavefunction collapse but rather a continuous evolution guided by the pilot wave/quantum potential where the continuum of spacetime (duality of discrete-continuous by the exiplex gate of the LCU) is a wave-particle duality-like property (wave-particle duality elevated from quanta to a principle of reality itself) where the locus of spacetime is cyclically created - annihilated and recreated. Wave function collapse is phenomenological; the annihilation-creation locus locus of reality is different - it is an ontological or energyless exchange allowing continuous evolution. Within this continuous-state 'harmonic oscillation' a subtractive interferometry occurs (Figs. 6-8) that subtracts out the LSXD form the perceptual apparatus of the observer [6-8]. The cascade-jump of Riemann sphere rotational subtractive interferometry offers an explanation of why there is more matter than antimatter in the Hubble universe.

All quantum force interactions are mediated phenomenologically by the exchange of a particle. Electromagnetism is mediated by the photon, the strong nuclear force by eight gluons and the weak nuclear force by the W^+ and W^- boson and the Z boson. The graviton has not yet been detected, we believe as Feynman suggests [6] that gravity is not quantized. That gravity is not mediated phenomenologically by ontologically instead by 'topological charge' in HD brane dynamics. Our case is bolstered by the fact that one cannot easily bring easily bring another particle into the symmetry of the standard model without upsetting the logical implications, delicate balance or mathematics. The symmetry of the standard model qualifies it as a geometric model. The symmetry is all about spin. The *spin exchange dimensional reduction compactification process* contains a spin-spin coupling that has an inherent 'deficit angle' in the gravitational Bianchi identities that allow an energyless topological switching 1D above (a topological cover) the regime of gravitational interaction, i.e. matter resides on the brane and gravity is free to pass between them' [6]. This is like the Dirac spherical rotation of the Klein bottle (electron 360-720 spinor rotation) where the 'baton passing' (would be energy exchange) occurs outside the 4D regime of quantization!

3. Subtractive Interferometry — Limiting the Local Realm of Awareness

We have postulated a 12D HAM cosmology with the observer's awareness limited to a 3D virtual subspace. Since the noumenon of the observer's awareness is 12D, subtractive interferometry provides the mechanism for maintaining this limit. This scenario is a very rich complex topology [6-8]. The first new element beyond current thinking is the nature of the singularity of least cosmological unit (LCU) tessellating spacetime. Instead of a 0D or 3D vertex, it is 12D and provides a gating mechanism for control by and entry of the U_F.

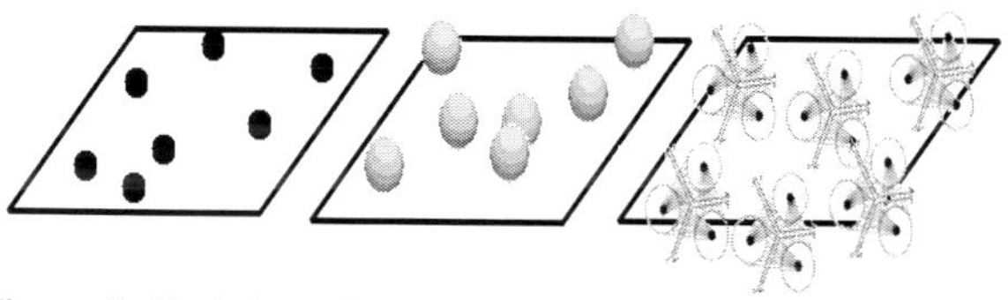

Figure 2. Evolution of the physical understanding of a point or singularity in spacetime. Each point in HAM cosmology is actually a multidimensional hyperstructure. While a) a 0D fermionic singularity and b) a 6D dual Calabi-Yau 3-torus are fixed, c) contains a central continuous flux Witten vertex giving it properties of a mini 'wormhole' able to gate information from higher to lower dimensions.

Spacetime is tessellated by a raster of LCUs. The stringy vertex is a quaternionic singularity representing the 'tip of the iceberg' in 3 space as it cycles through the LSXD of 12-space. The hidden or nonlocal HD portion requiring a topological raster of LCUs operates as a gating mechanism not only mediating the U_F but also supplying the mechanism for applying subtractive interferometry.

As we have said the observers 3D reality is virtual. Describing it in terms of Cramer's transactional interpretation it (each point) is like a standing-wave of the future past. Reality beyond the awareness of the observer is a 12D nilpotent standing wave present where the dual complex nilpotent future-past, $\pm = 0$ or the Minkowski virtual present, $\mathbb{C}_4^- \leftrightarrow \hat{M}_4^0 \leftrightarrow \mathbb{C}_4^+$ is also nilpotent and also sums to zero totality.

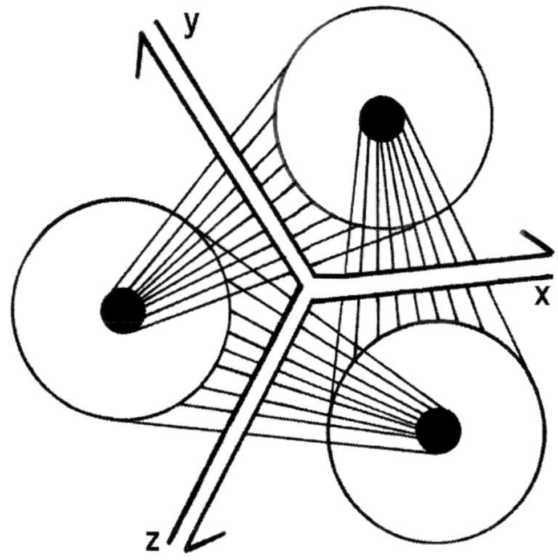

Figure 3. Topological geometry of a LCU tessellating spacetime revealing the triune structure of a conceptualized solitary LCU that like an isolated quark does not exist in nature. The central parallel lines are a 3D Witten string vertex with properties of a complex Riemann sphere able to continuously rotate from asymptotic 0D to infinity (12D). The field lines represent the 'super quantum potential' of the unified field, U_F. This LCU structure is the HAM extension of the Rowlands' fermionic nilpotent singularity.

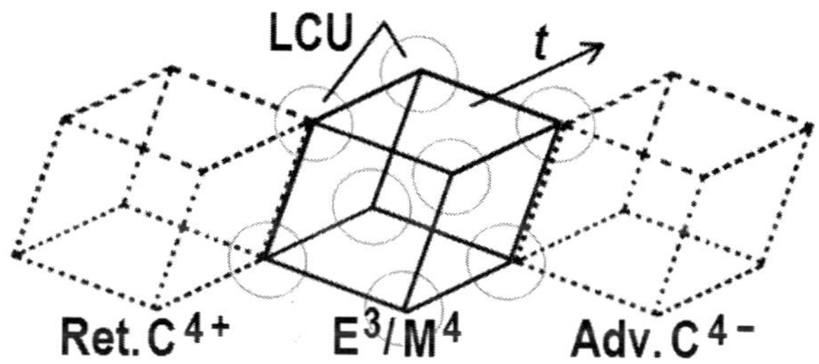

Figure 4. How observed (virtual) 3D reality (center) arises from the infinite potentia of 12D space (like a CI transaction). The 'standing-wave-like' (retarded-advanced future-past) mirror symmetric elements, C^{4+} / C^{4-} (where C^4 signifies an 8D potentia of complex continuous-state space is distinguished from the locally realized visible 3D spacetime) produce the observed Euclidian, E_3, Minkowski, M_4 space (center) as a closed resultant. Least Cosmological Units (LCU) governing evolution of the 'points' of 3D reality are represented by circles. The Advanced-Retarded future-past cubes in HD space guide the evolution of the central cube (our virtual reality) that emerges from potentia elements of HD U_F space.

In singularity form this is represented as a transaction.

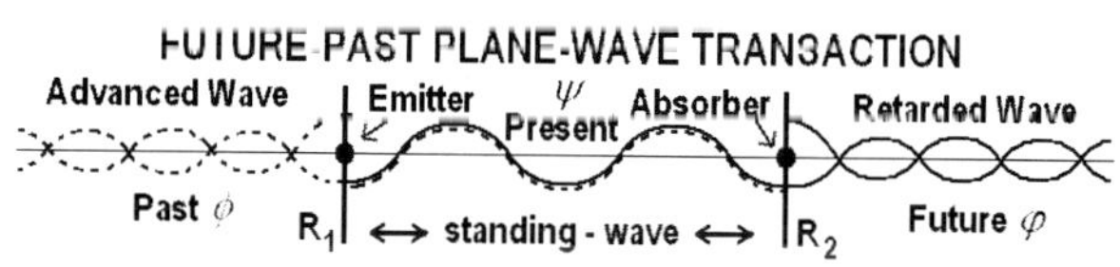

Figure 5. Conceptualized structure of a transaction (present quantum state or TI event) where the present (simplistically) is a standing-wave of future-past elements. A point is not a rigid singularity as in the classical sense, but has a complex structure like a mini-wormhole where R_1 & R_2 (like the frets holding the wire of a stringed instrument) represent opposite ends of its vibrational diameter.

However a NFT HAM singularity is only cyclically discrete as shown in the text below.

The tessellation of space by LCUs is complex.

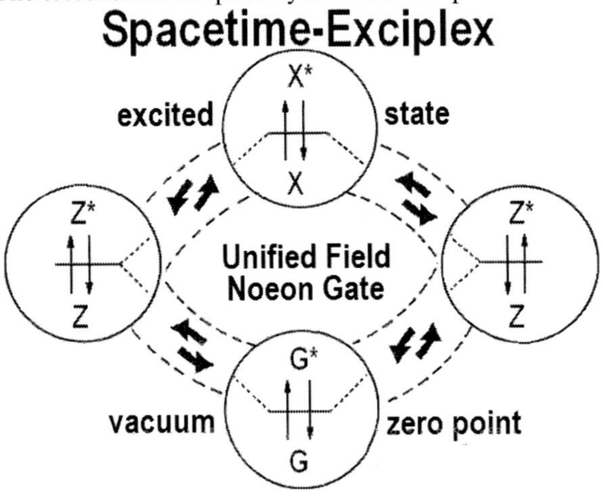

Figure 6. Least cosmological unit (LCU) exciplex complex tiling the spacetime backcloth of HAM cosmology. The array of LCUs acts as a gating mechanism for entry of the U_F into every point in Minkowski 4-space and all matter.

In this LCU continuous-state singularity context

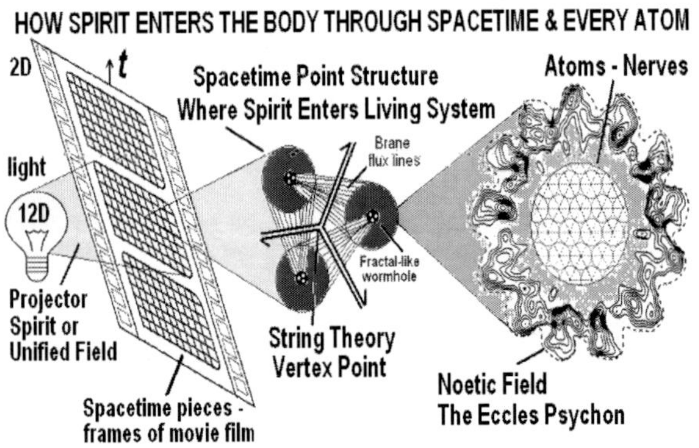

Figure 7. Microscopic details of transduction of the U_F through the complex exciplex spacetime raster LCU gate into every point, atom and thus molecule of self-organized living systems (SOLS) the propagation of which also produces a locus of spacetime points associated with the arrow of time because it is part of the structure and content of the observers mind.

When macroscopy is added to the microscopic details we can conceptually understand how subtractive interferometry occurs by modeling the process like a movie theater. Discrete 'frames of film'. i.e. the tessellation of LCUs pass through the projector - the standing-wave evanescing into 3D reality. In a movie projector the discrete frames of film pass over the lens at a few centimeters per second which since that is too fast for the observers perception allows the image to appear smooth or continuous on the screen. In reality LCUs are created and annihilated relativistically and the observer is made out of and embedded in spacetime.

The Cartesian dualistic form of the observer is HD and triune. Descartes dualism postulated a mind stuff '*res cogitans*' and a body stuff '*res extensa*' as the basis for consciousness. That is an excellent beginning, but as a self-organized living system (SOLS) an additional component of 'elemental intelligence' is required to define the basis for the observer. Elemental intelligence is a boundary condition demarking the individuality of the observer and is 'co-eternal with God. This is why the observer's basis is 12D; and why subtractive interferometry may destructively interfere or cancel out the HD portion of the observer's reality or awareness.

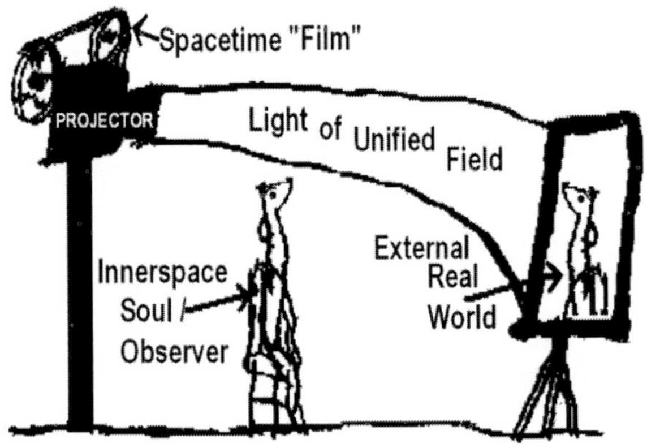

Figure 8. Macroscopic movie theatre model of the observers 3D virtual reality (like a hyperdimensional hologram) and the observer's place in the theatre. This is a 12D continuous-state structure but the HD are removed from the observer's awareness by 'subtractive interferometry'.

4. On the Cosmology of Mind - The Shape of Space

It seems reasonable to accept the postulate put forward by several authors that there is a 'wave function of the universe' satisfying the Wheeler-Dewitt equation which is a gravitational form of the Schrödinger equation taking the general unexpanded form, $H\Psi = 0$. HAM cosmology is willing to accept this premise if it is applied only to the Hubble, H_R portion of the universe which is perceived as a self-organized complex system closed and finite in time and not to the more extant multiverse open and infinite in eternity which does not have a quantization as suggested by the Wheeler-Dewitt equation.

The operation of reality depends on the geometric topology of space which can be radically different from local observation. WMAP satellite data has given preliminary evidence that we live inside a dodecahedron [3]. HAM cosmology utilizes a mix of three oscillating topologies: Dodecahedral - positively curved, (if one looks 'far out' one will see the back of one's head etc.), Hyperbolic - negatively curved (like a saddle) and flat. Flat is interesting and locally observed. If one folds the opposite edges of a square seamlessly into a cylinder and then a torus (a torus looks curved because we embed it in 3-space). But here is the next point. Take a cube with opposite faces wrapped around which forms a 3D flat torus [3,4]. To denizens living in such a space it appears flat in every direction, which is what we see!

Astrophysicists doing this 'geometric cosmology' do so in a Big Bang universe ending at an impenetrable wall. Since we are doing Cartesian dualism, the 'wiggle room' (oscillating backcloth) utilizing an oscillatory mix of the three topologies provides a hole (cyclic) (into large scale extra dimensionality - infinity) for "God" to squirt a little spirit in to give life, excite mind, steer gravity etc. And this Multiverse is open into other Hubble Spheres, 'worlds without number have I created, like grains of sand at the seashore".

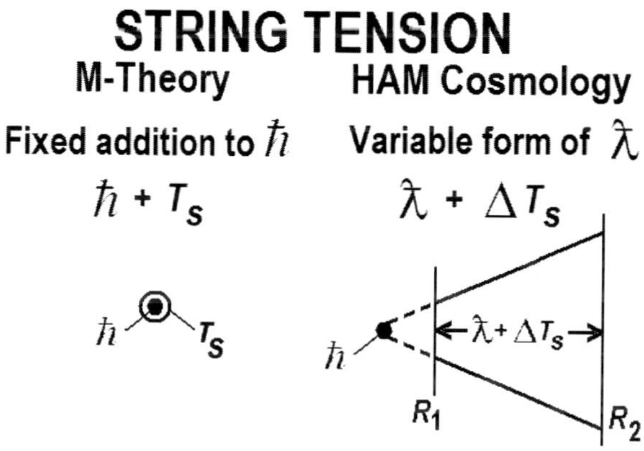

Figure 9. (a) Fixed String Tension, T_S as defined in M-Theory as a small addition or correction to $\hbar$. (b) In HAM cosmology (as in the original Hadronic form of String Theory, T_S is variable and based on the original Stoney, $\lambda\!\!\!^{-}$ (precursor to $\hbar$) that oscillates from virtual $\hbar$ (asymptote never reached in the continuous-state cycle) to the Larmor radius of the hydrogen atom.

The oscillation of T_S at the microscopic level operates (couples) in conjunction with a fluctuating, Λ at the cosmological level of H_R. This conformal scale-invariant duality is part of the gating mechanism of the U_F. It occurs because the Hubble sphere, H_R is a self-organized complex system closed and finite in time but nested in the multiverse that is open and infinite. It is the hierarchical and incursive properties of complex systems that allow subtractive interferometry to operate efficiently. Dark energy arises from the rest of the multiverse beyond our Hubble sphere.

We have left out the gravitational aspects of U_F dynamics as beyond the scope of this paper. Suffice it to mention here that parallel transport around the topology of the propagating LCU continuous-state array produces a deficit angle. This means in the compactification process the rotation does not return precisely to the original starting point. This so-called deficit angle provides a 'grease' for subtractive interferometry, i.e. instead of coupling to the HD topology or remaining coupled, the Witten vertices may rotate or cascade down the ladder of dimensionality thus removing the 'vision of eternity' from the observer's awareness.

References and Notes

[1] Amoroso, R. L., Kauffman, L.H., Rowlands, P. & Rauscher, E. A., The Basis of Unified Field Mechanics: Introduction to the 3rd Regime of Reality, in process.

[2] Sklar, L. (1985). Philosophy and spacetime Physics, Berkeley: Univ of California Press.

[3] Luminet, J-P (2008) The Wraparound Universe, Wellesley: AK Peters.

[4] Stuart, I. (2001) Flatterland, Cambridge: Persius.

[5] Amoroso, R. L. (2000)The parameters of temporal correspondence in a continuous-state conscious universe, in R. Buccheri, V. Di Gesu & M. Saniga (eds.) Studies on the Structure of Time: From Physics to Psycho(patho)logy, Dordrecht: Kluwer Academic.

[6] Amoroso, R. L., & Rauscher, E. A. (2009) The Holographic Anthropic Multiverse: Formalizing the Geometry of Ultimate Reality, Singapore: World Scientific.

[7] Rauscher, E. A. & Amoroso, R. L. (2011) Orbiting the Moons of Pluto, Hackensack: World Scienitfic.

[8] Amoroso, R.L. (2010) (ed.) The Complementarity of Mind and Body: Realizing the Dream of Descartes, Einstein and Eccles, New York: Nova Science Publishers.

[9] See Foreword, this volume for discussion on Transcendence.

[10] Amoroso, R.L. (2009) In R.L. Amoroso (ed.) Complementarity of Mind and Body: Realizing the Dream of Descartes, Einstein and Eccles, New York: Nova Science.

[11] Amoroso, R. L. (1995) Transcendent revelation or noetic Platonic insight (Voice heard), Oakland, CA "*Quantum mechanical uncertainty is the mystery, even the mystery of God*". Meaning 'behind' the virtual stochastic 4D barrier imposed by the Copenhagen interpretation of quantum theory, the closed and finite Hubble universe opens into an infinite multiverse which in certain higher dimensional modalities 'the hand of God can be observed to operate'.

'Shut The Front Door!': Obviating the Challenge of Large-Scale Extra Dimensions and Psychophysical Bridging

RICHARD L. AMOROSO

Noetic Advanced Studies Institute, Escalante Desert, UT 84714-5104 USA
amoroso@noeticadvancedstudies.us, www.noeticadvancedstudies.us

Physics has been slowly and reluctantly beginning to address the role and fundamental basis of the 'observer' which has until now also been considered metaphysical and beyond the mandate of empirical rigor. It is suggested that the fundamental premise of the currently dominant view of 'Cognitive Theory' - "Mind Equals Brain" is erroneous; and the associated belief that the 'Planck scale, 'the so-called basement level of reality', as an appropriate arena from which to model psycho-physical bridging is also in error. In this paper we delineate a simple, inexpensive experimental design to 'crack the so-called cosmic egg' thereby opening the door to large-scale extra dimensions (LSXD) tantamount to the regime of the unified field and thus awareness. The methodology surmounts the quantum uncertainty principle in a manner violating Quantum Electrodynamics, (QED), a cornerstone of modern theoretical physics, by spectrographic analysis of newly theorized Tight-Bound State (TBS) Bohr orbits in 'continuous-state' transition frequencies of atomic hydrogen. If one wonders why QED violation in the spectra of atomic hydrogen relates to solving the mind-body (observer) problem; consider this a 1st wrench in a forthcoming tool box of Unified Field Mechanics, U_F that will soon enough in retrospect cause the current tools of Classical and Quantum Mechanics to appear as stone axes. Max Planck is credited as the founder of quantum mechanics with his 1900 quantum hypothesis that energy is radiated and absorbed discretely by the formulation, $E = h\nu$. Empirically implementing this next paradigm shift utilizing parameters of the long sought associated 'new physics' of the 3rd regime (classical-quantum-unified) allows access to LSXD of space; thus pragmatically opening the domain of mental action for the 1st time in history. This rendering constitutes a massive paradigm shift to Unified Field Theory creating a challenge for both the writer and the reader!

Keywords: Bohr orbit, Psycho-physical bridging, QED, Quantum theory, Spectroscopy, Tight-bound states, Uncertainty principle.

1. Introduction — Threshold of a Third Regime

"Shut The Front Door!", a slang interjection or statement representing a moment of disbelief in US vernacular [1], is an appropriate 'reverse psychological' insult for this paper's experimental challenge to the myopic presumptions of both Cognitive Neuroscience and contemporary physics that has thwarted progress in solving the mind-body problem (nature of the observer) like the most honored Mullah Nasrudin who losing his keys in his backyard looked for them instead at night under the street light in front of his house because it was easier to look there with his neighbors helping.

Quantum Electrodynamics (QED) sacrosanct for the past 50 years is now summarily being violated [2-4]. QED, the relativistic quantum field theory (RQFT) of electrodynamics describes the interaction of light and matter; and was the first theory to mesh full agreement between quantum mechanics and special relativity. Feynman, a QED founder called it "the jewel of physics" for its extreme accuracy in predictions such as the anomalous magnetic moment of the electron, and the Lamb shift in energy levels of hydrogen.

It is interesting that new TBS energy levels of hydrogen will now lead to QED being violated. Rowlands in terms of current thinking in contemporary physics hints at the dilemma that we sachet past here:

> Physics at the fundamental level can be effectively reduced to an explanation of the structures and interactions of fermions. Fermions appear to be singularities rather than extended objects, but there is no obvious way of creating such structures within the 3-dimensional space of observation. However, the algebra associated with the Dirac equation appears to suggest that the fermion requires a double, rather than a single, vector space, and this would seem to be confirmed by the double rotation required by spin ½ objects, and the associated effects of *zitterbewegung* and Berry phase shift. Further investigation of the second 'space' reveals that it is, in effect, an 'antispace', which contains the same information as real space but in a less accessible form. The two spaces effectively cancel to produce a norm 0 (nilpotent) object which has exactly the mathematical structure required to be a fermionic singularity [5].

Although the mathematics of Rowlands *avant-garde* nilpotent physics works brilliantly in extending our understanding of the nature of a fermionic singularity (as yet an open question in physics), for Noetic Field Theory (NFT): The quantization of Mind, for which Rowland's model has recently even become a basis for solidifying NFT's foundation [6,7]; elegant math is not necessarily tantamount to physics. The nature of the so-called fermionic singularity is profoundly unique in NFT; and even though the protocol introduced here would not work without that noetic nature being physically realistic, its technical description is beyond the scope of this paper [8-10] within which we outline the experiment and only delineate a minimum of associated parameters to help explain its radical theoretical foundations.

2. Physics, New and Old — Protocol Philosophy

A wrinkle in atomic spectra may not initially appear as an appropriate crack in the 'cosmic egg' - meaning a door to empirically demonstrating the 1st indicia of the existence of Cartesian Interactive Dualism. To clarify, NFT [7-13] now has sufficiently firm theoretical grounds to declare that the Planck scale is 'not the basement of reality', albeit it is an oasis between the 1st and 3rd regimes of reality - a nilpotent boundary of the infinite potentia [5-7] hidden behind it or domain wall forming the virtual reality perceived by the Euclidean/ Minkowski observer. Just as the quantum regime was until recently invisible to the empirical tools of Newtonian or Classical Mechanics, so until now Unified Field Mechanics, U_F has remained invisible to the tools of quantum mechanics primarily because of the barrier associated with the uncertainty principle.

The uncertainty principle has been empirically demonstrated by the Stern-Gerlach experiment [14] where a continuous non-uniform em-field arbitrarily projected along the z-axis produces 'space-quantization' and by definition the quantum uncertainty principle under the auspices of the Copenhagen Interpretation of Quantum Theory. The simple solution to surmount uncertainty is to 'do something else' [6,8,15] which will only be given brief conceptual delineation here. The experiment to be presented revolves around an as yet obscure work by Vigier called 'tight bound states' (TBS) in the hydrogen atom [9,17]. Hydrogen is chosen for the experiment because of its atomic simplicity; it is surmised that a more complex atom might have a tendency to mask the anticipated effect making the putative result more difficult to observe.

The experiment if successful would demonstrate a violation of QED. The timing for theoretical acceptance is good because QED has recently been violated in Titanium atoms to a discernible degree [3]. Such preliminary experiments in the history of physics often start as tiny cracks in a theory that have to be repeated and extended many times to produce sufficiently salient effects. The TBS experiment introduced below promises to 'Open the Front Door' wide because it leaps beyond the standard model on a faster track by utilizing aspects of U_F mechanics; i.e. the several new QED violation experiments are standard 4D quantum mechanical and therefore unable to access the 'hidden' large-scale extra dimensional (LSXD) U_F regime.

History has incessantly demonstrated that it is human nature to resist new ideas. Thus it would be easier if one could merely describe the mechanical details of the TBS experiment without delving into a description of the theory behind it because it is currently extremely radical and such a perambulation most certainly will initially also appear sufficiently metaphysical (especially since it purports discovery of Cartesian dualism) to current thinking in mainstream physics to mow down the delicate bloom of open-minded consideration…

The spectra of hydrogen and associated Zeeman and Stark effects have been rigorously investigated over the last 100 years. What we propose requires no repeat or manipulation of that data which is used only as a comparative benchmark with the new TBS experimental data. We propose three new lines in the atomic transition frequency of hydrogen inconsistent with QED standards suggesting the existence of new TBS below the lowest traditional Bohr orbit (heretofore hidden behind the uncertainty principle) caused by a cyclical or holophote-like periodicity of LSXD resonant transitions. Imagine for the sake of simplistic illustration that the 1st Bohr orbit, *s* (without descriptive units) has a spectral energy volume of 10 and the 2nd Bohr orbit, *p* a volume of 20 units. Before we get more deeply into the TBS scenario let's further explain the 10-20 energy volume metaphor. Imagine being in a symmetrically organized orchard, auditorium or grave yard where an observer will notice alternating paths extending to infinity or local barriers blocking the view in the near field. Coincidently this appears like the interference fringes of a Moiré pattern. In HAM cosmologies Calabi-Yau-Dodecahedral AdS_5-DS_5 12D symmetry the 'mirror image of the mirror image' (conformal scale invariant copies) of the 3D 'particle in a box' is causally free of the 4D nilpotent Minkowski space resultant. 12D appears to be the minimal dimensionality for this condition to occur. This inherent structure allows us to set up the required rf-pulsed resonance hierarchy.

Figure 13e shows the possibility of three spectrographic results - spectral energy volumes of 12, 14 or 16 depending on whether the standing-wave structure of the resonance hierarchy cycle is fully parameterized or not. Thus if successful the TBS protocol will be distinct from the usual 10-20 QED spectrum of the hydrogen atom. This result would not only violate QED but also demonstrate the existence of LSXD which for our purposes opens the door to a regime of new physics able to cross the psycho-physical bridge leading to the 1st pragmatic model of 1st person 3rd person entanglement [8,11].

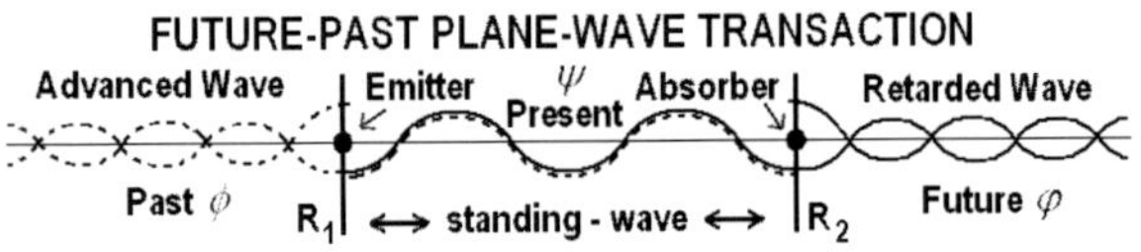

Figure 1. Conceptualized structure of a transaction (present quantum state or TI event) where the present (simplistically) is a standing-wave of future-past elements [22]. A point is not a rigid singularity as in the classical sense, but has a complex structure like a mini-wormhole where R_1 & R_2 (like the frets holding the wire of a stringed instrument) represent opposite ends of its vibrational diameter. However a NFT HAM singularity is only cyclically discrete as shown in the text below.

Under the auspices of the Copenhagen Interpretation protons are said to be created near the time of the Big Bang and their half-life has been confirmed to this order of temporal magnitude [18,19]. But this condition is independent of what one might say about de Broglie matter-waves applied to the de Broglie-Bohm Causal Stochastic Interpretation of quantum theory [20,21] well known to include a quantum potential or pilot wave that NFT additionally interprets to suggest that fermions embedded in spacetime are cyclically annihilated and recreated as an inherent ontological basis for the generation of observed reality (Figs. 2 and 3); which NFT further postulates in terms of Cramer's Transactional Interpretation (TI) [22], acts like a hyperspherical standing-wave of the future-past (Fig. 1).

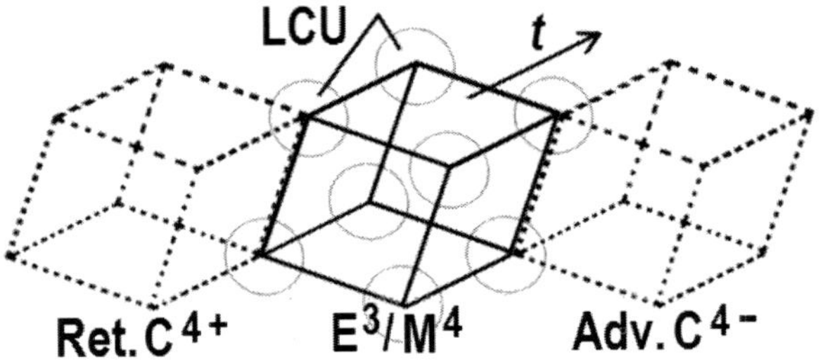

Figure 2. How observed (virtual) 3D reality (center) arises from the infinite potentia of 12D space (like a CI transaction). The 'standing-wave-like' (retarded-advanced future-past) mirror symmetric elements, C^{4+} / C^{4-} (where C^4 signifies an 8D potentia of complex continuous-state space is distinguished from the locally realized visible 3D spacetime) produce the observed Euclidian, E_3, Minkowski, M_4 space (center) as a closed resultant. Least Cosmological Units (LCU) governing evolution of the 'points' of 3D reality are represented by circles. The Advanced-Retarded future-past cubes in HD space guide the evolution of the central cube (our virtual reality) that emerges from potentia elements of HD U_F space.

Cramer derived this basis for his interpretation by correspondence to the Wheeler-Feynman radiation law [23]. NFT takes this scenario a significant additional step by elevating both Cramer's TI and wave-particle duality to Principles of Cosmology. One gets from this concatenation an application of the Dirac equation where the annihilation-creation vectors applied to matter embedded in spacetime act as a locus of resultant LCU points able to not only model the arrow of time with large-scale continuous-state extra dimensions (LSXD); but also a description of the exciplex gating mechanism mediating the U_F. (See Figs. 2, 3 & 10)

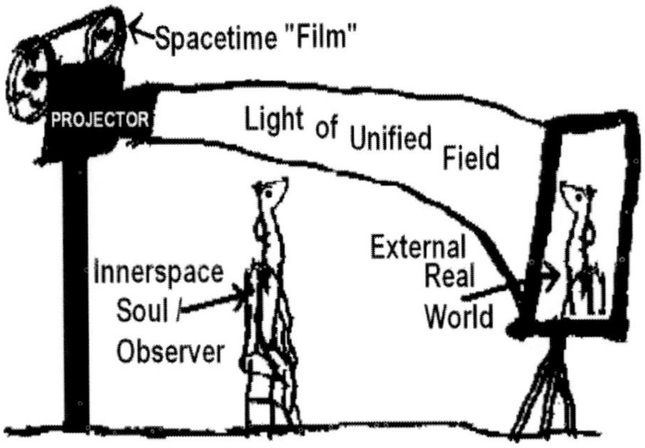

Figure 3(a). Macroscopic movie theatre model of virtual reality (like a hyperdimensional hologram) and the observer's place in the theatre.

Putting all this together (more is left out for brevity here) one gets instead of an inflationary Doppler Big Bang universe, a 'continuous-state' anthropic multiverse [8] where the energy of inflation is internalized locally as a form of gravitational free-fall - giving the Euclidean observer his façade of virtual reality (Figs. 2 & 3). If this sounds like a horrendous concatenation violating Occam's razor, be reminded that if one is inclined to accept the need for a 'life principle' - even the founders of the current interpretation of quantum theory maintained that it was incapable of describing biological systems; and the ever so popular Big Bang cosmology is Darwinian and naturalistic - thus void of an anthropic principle. Therefore to this author's thinking there is little choice other than to follow the path presented here especially since additional degrees of freedom are required to go beyond the mechanistic fallacy of cognitive neuro-science.

The Continuous-State property of the LCU space-spacetime raster forms the open-closed structure of its 'singularity'. Leibnitz in an argument with Newton over whether spacetime was relational or absolute called this conundrum an antinomy - that space was an open-closed duality of the two. In HAM cosmology [8,11] the Hubble sphere, H_R is a self-organized complex system - closed and finite in time, but open and infinite in eternity which is a conformal scale-invariant property symbolic especially of the continuous-state LCU structure modulated by the coherence force of the U_F.

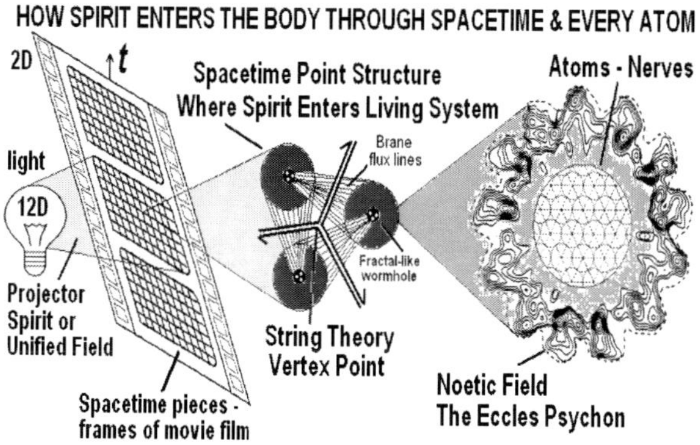

Figure 3(b). Microscopic details of transduction of the U_F through the complex exciplex spacetime raster LCU gate (Fig. 10) into every point, atom and thus molecule of self-organized living systems (SOLS) the propagation of which also produces a locus of spacetime points associated with the arrow of time because it is part of the structure and content of the observers mind.

3. Devil in the Details — Continuous-State, LCUs, and the Synchronization Backbone

One of the most amazing realizations during development of HAM cosmology was the inherent synchronization backbone associated with the nonlocal Continuous-State cycle. Even the author himself does not yet fully comprehend the structure of an LCU or its gating mechanism so additional profundity looms. The idea of a spacetime synchronization backbone arose with Feynman's suggestion of its requirement in the development of quantum computers (QC) [8,24].

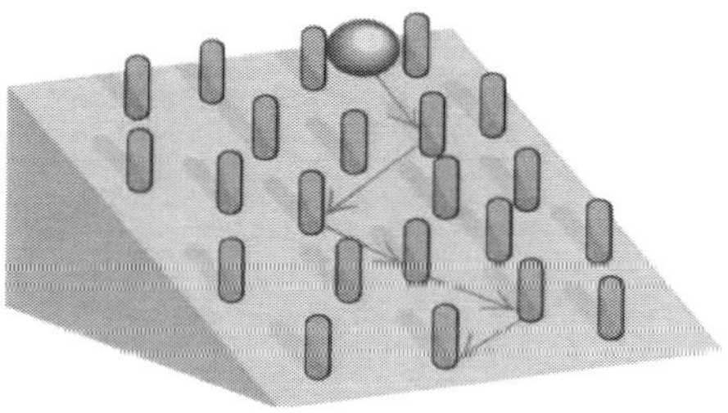

Figure 4. The Continuous-State process can be thought of as the path of a fermion (ball) embedded in a spacetime raster (not as shown but instead embedded in a Riemann hypersphere) that is continuously rotating relativistically such that the fermion remains centralized as if in a form of gravitational free-fall

But researchers in the QC field gave up the idea as intractable not realizing that this was the case only in terms of the 4D Copenhagen Interpretation of quantum theory. In 12D it is miraculous and like getting half the QC for free! What researchers attempted could best be described as a form of 'bi-locality' not a breaking through uncertainty to the essential nonlocal LSXD criteria for a synchronization backbone. However Feynman's comment was brief leaving that restriction unclear to Copenhagen regime philosophy and tools.

The chaotic motion of a bouncing ball on a harmonically forced oscillator like the motion of a ball rolling down an inclined plane and bouncing off pegs in its path is analogous to the continuous-state process of the LCU raster of spacetime in the presence of the U_F. The synchronization backbone forms the geometric topology or domain wall of continuous-state Calabi-Yau mirror symmetry. There is a sort of mantra associated with its structural-phenomenology: A '*Continuous-state mirror symmetric spin exchange dimensional reduction compactification process*'.

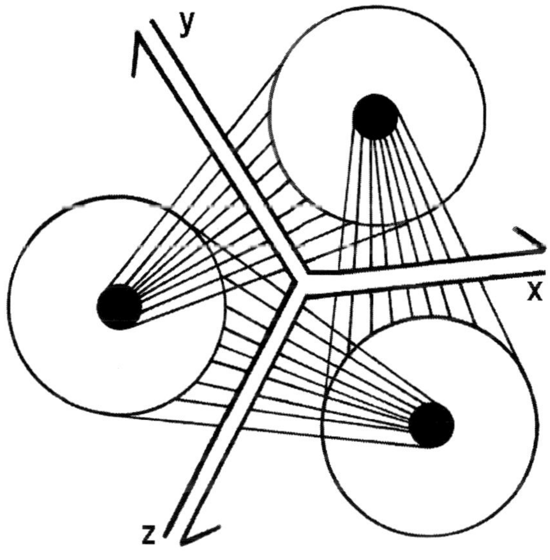

Figure 5. Triune structure of a conceptualized solitary LCU that like an isolated quark does not exist in nature. The central parallel lines are a Witten string vertex [32] with properties of a complex Riemann sphere able to continuously rotate from zero to infinity. The field lines represent the 'super quantum potential' of the unified field, U_F. This LCU structure is the NFT extension of the Rowland fermionic singularity [5].

As a portion of U_F dynamics the topology of the continuous-state process also includes gravitational elements called 'parallel transport' [8,31] which creates what physicists call a 'deficit angle'. What this means is that a fermion following a geodesic path around the edge of a hypercube does not return to the precise fermionic vertex it originated from at the beginning of the cycle but incurs a deficit angle or gap between the original point and new final point. This gap in the Calabi-Yau mirror symmetric spin-exchange process is like a relay race baton passing where the electron for example 'leapfrogs' across the deficit angle gap (and recouples) in the standing wave structure of the Minkowski space resultant maintaining the properties of the uncertainty principle while in the nonlocal Calab-Yau mirror symmetric background the continuous-state dimensional reduction compacti-

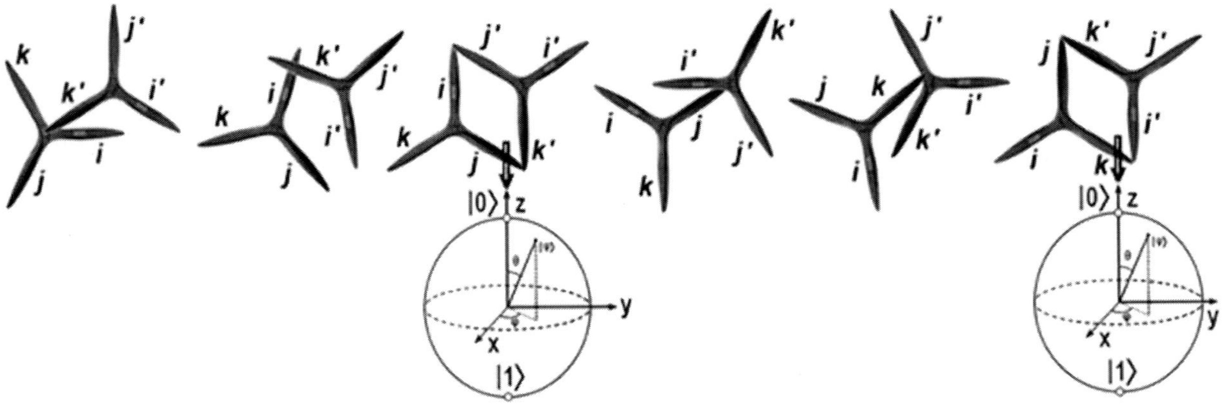

Figure 6. Segment of proximal HD Calabi-Yau mirror symmetric 3-folds represented as dual quaternion trefoils undergoing continuous-state rotation periodically coupling into a Euclidean 3-space resultant.

fication process runs smoothly through its 12D to 0D cycle - a stereographic property of a Riemann sphere rotation from zero to infinity.

This horrendous concatenation occurs as the basis for the arrow of time because our temporal virtual reality surfs as it were of the face of the LSXD eternal realm hidden behind it. (The discrete frames of spacetime film in Fig. 3 producing the continuous virtual image of reality relativistically on the screen) These parameters are essential to the LCU as an exciplex gating mechanism - This is how each point (LCU open-closed singularity - like a rotating light house beacon) is created in the temporal locus and allows the U_F's 'force of coherence' to modulate complex self-organized living systems (SOLS) [11] as the organizing principle itself and likewise mediate the physical basis of qualia [10-12]. We will do our best to clarify this scenario in the text and figures in sections following.

We are not yet finished outlining the required battery of new physical parameters; recall that NFT represents a whole paradigm shift. (why it hasn't been easy this past 1,000 years) WE MUST also utilize the parameters of another well established and generally ignored aspect of contemporary physics called Extended Electromagnetic Theory [25-27] in conjunction with a covariant Dirac polarized vacuum (ignored also) [28]; both because physicists erroneously believe they conflict with Gauge Theory which has been eminently successful for decades. Now we are finally be set up with enough parameters to putatively manipulate the spacetime backcloth (Einstein energy dependent spacetime metric) [29,30].

Metaphorically if one throws a stone in water concentric ripples occur. If one throws two stones regions of destructive and constructive interference occur. We will utilize an M-Theoretic Calabi-Yau symmetric version of this model to set up an rf-pulsed spacetime resonance hierarchy to access the 'hidden' regime of the U_F [8]. In the next series of several figures (Figs. 5-8) we will attempt to clarify the continuous-state structure of LSXD as it applies to HAM cosmology.

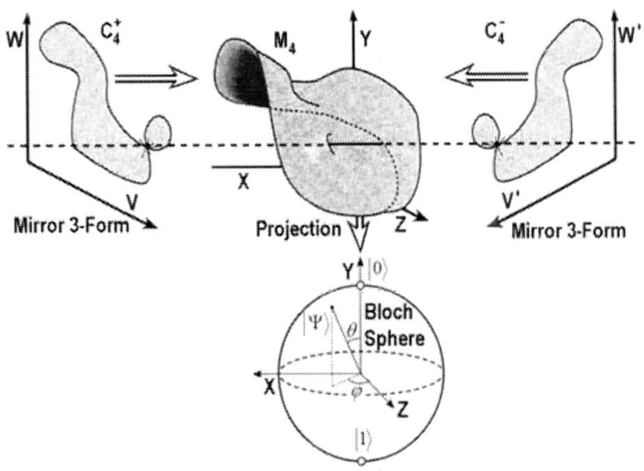

Figure 7. Complex Calabi-Yau mirror symmetric 3-form potentia, C_4 become an embedded quantum resultant in Minkowski space, M_4. This resultant projection entails a continuous quantum state evolution represented as a Bloch 2-Sphere. Representing the lower portion that embeds in local spacetime; there is an additional 5D dodecahedral duality above this projection embedded in the infinite potentia of the U_F from which it arises.

The noeon mediating the U_F does not imply the usual phenomenological exchange of energy as in a standard field interaction such as the photon of the electromagnetic field; but constitutes an ontological exchange (without energy transfer). This is achieved by a process called 'topological switching' and implies instead a 'force of coherence' inherent in the action of the U_F. This process also allows the quantum uncertainty principle defined by the Copenhagen Interpretation of quantum theory to be surmounted [8]. Figure 8 illustrates a nilpotent continuous-state regime

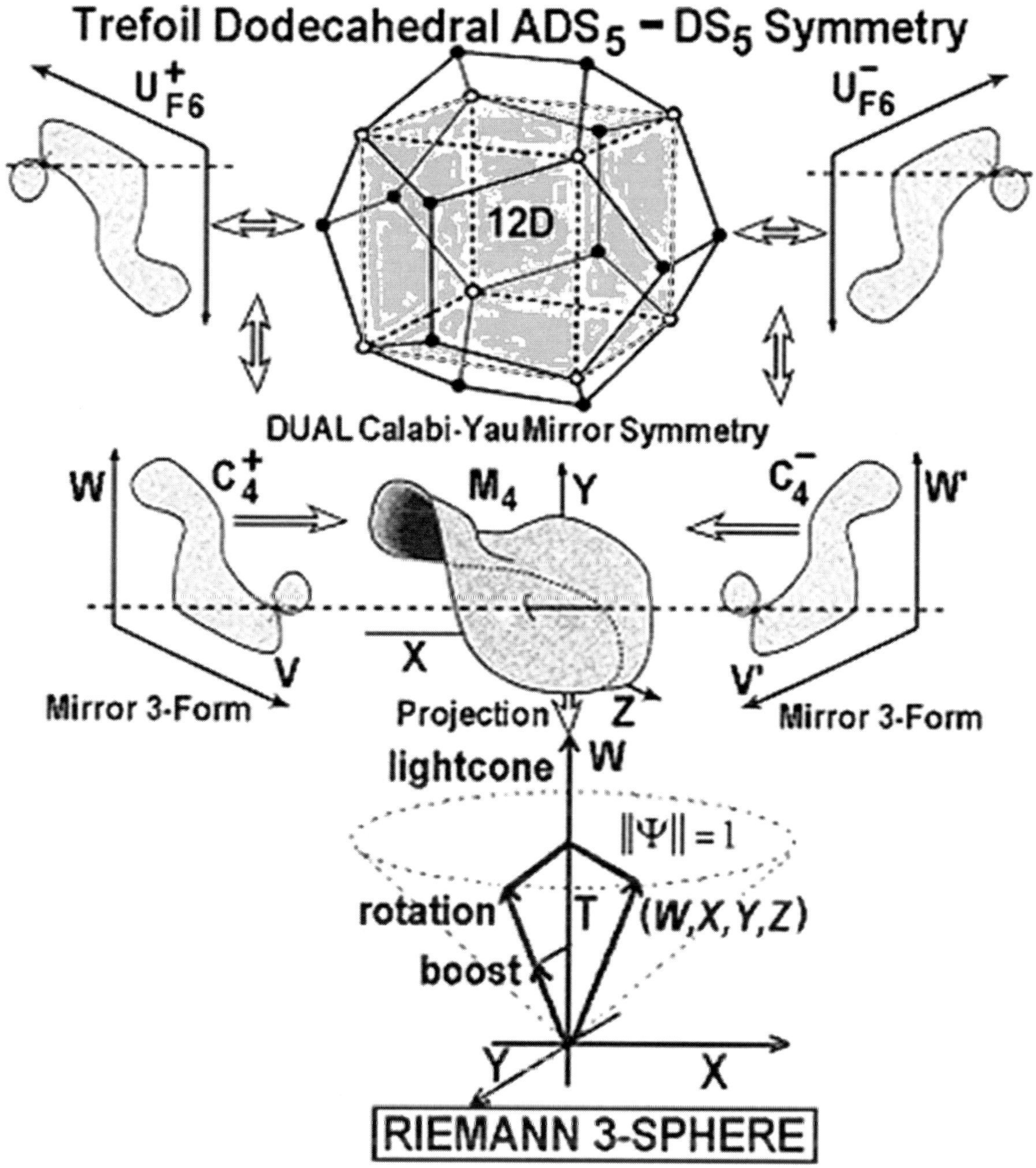

Figure 8. Complete LSXD regime of HAM cosmology illustrating the hierarchy of its geometric topology. Dodecahedral involute properties, as well as the continuous-state exciplex 'hysteresis loop' of noeon injection not shown. It represents a unique M-Theoretic model of 'Continuous-State' U_F dynamics as it relates to NFT and its putative exchange quanta of the U_F called the noeon.

cycling from a 12D dodecahedral de Sitter space [38] through an intermediate Calabi-Yau mirror symmetry to a virtual 'standing-wave' nilpotent 3D Euclidean space resultant.

We extend Vigier's original model of TBS in hydrogen [17] to include a unique 12D M-Theoretic perspective [8] with a Calabi-Yau: Dodecahedral involute mirror symmetry [6] in Continuous-State HD space [8,11] elevating Cramer's Transactional model [22] and wave-particle duality to principles of cosmology. The 4-Space nilpotent resultant of the E3/M4 virtual present which is a standing-wave of the future-past is shown in Figs. 2-9. M_4 being 'locked' into place cyclically by the uncertainty principle [8].

The HD or XD are not curled up at the Planck scale because they are invisible; but they are Large-scale XD

(LSXD) [8] because of subtractive interferometry as it were of the C_4^+ - M_4 - C_4^- standing-wave modes [8] that operates like a movie theatre where discrete frames of film moving through the projector at a few cm/sec appear continuous on the screen. For our virtual reality - Exchange quanta of the U_F is relativistically ‘pumped’ through discrete holographic-like LCU tiling the raster of spacetime producing the virtual image of the Minkowski space present. So behind the virtual veil is a continuous-state cycle from $0 \Leftrightarrow \infty$ as shown in Fig. 3.

4. More LSXD Experimental Theory

The Born-Oppenheimer (BO) approximation [33] which is based on the fact that within molecular systems fast-moving electrons can be distinguished from slow-moving nuclei allows the wavefunction of a molecule to be broken into its electronic and nuclear (vibrational, rotational) $\Psi_{\text{Total}} = \psi_{\text{Electronic}} \times \psi_{\text{Nuclear}}$ components for easier calculation.

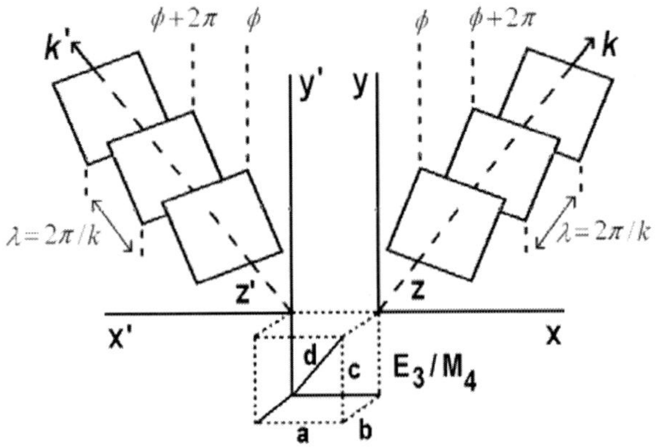

Figure 9. Calabi-Yau future-past mirror symmetry potentia illustrated as tiered surfaces (additional dimensionality suppressed) of constant phase, in this case to represent cyclic components of evenly spaced orthogonal standing reality waves with the E_3/M_4 cubic resultant localized at the bottom. The resultant locked in 4D by the uncertainty principle.

The assumption is made that if nonadiabatic coupling terms are negligibly small then the upper electronic surfaces have no effect on the nuclear wave function of the lower surface. This assumption is not considered dependent on the systems energy. However, the ordinary BO approximation was also employed for cases where these coupling terms are not necessarily small, assuming that the energy can be made as low as required. The justification for applying the approximation in such a case is that for a low enough energy the upper adiabatic surfaces are classically forbidden, implying that the components of the total wave function related to these states are negligibly small. As a result the terms that contain the product of these components with the nonadiabatic coupling terms are also small, and will have a minor effect on the dynamical process.

The protocol tests for both the existence of TBS and also for LSXD. HAM Cosmology predicts novel periodic HD cavities in the brane topology of Calabi-Yau : Dodecahedral AdS_5-DS_5 mirror symmetry. Simplistically a tunable NMR device acts on a vial of hydrogen over a range of de Broglie wavelengths set for specific Cavity-QED resonances to probe the lowest Bohr orbit for TBS conditions. If our cosmological model is correct there will be novel resonances that cannot correspond to either classical wave mechanics or Copenhagen QED modes. One might suspect C-QED to detect nodes in the Dirac spherical rotation of the electron (cyclical pattern of Klein bottle open-closed modes). Critics might say this is just a 4D Dirac effect of the putative Klein bottle symmetries in the electron’s dual spinor rotation. But our LSXD cosmology predicts a much richer Calabi-Yau mirror symmetry within the higher 9D brane topology so there "should" be a cycle of novel TBS resonances in the Calabi-Yau symmetry. Likewise these resonance nodes would have de Broglie wavelengths different than any higher Bohr orbit excitation in Hydrogen. It may be possible to predict the de Broglie wavelengths in the resonance hierarchy if the topology can be theoretically determined or if a clear C-QED resonance hierarchy appears, the topological structure of higher dimensions may be revealed. Vigier discussed using deuterium; it is an open question if that would make a qualitative difference in success or results in such an experiment. It will be easy enough to test both cases.

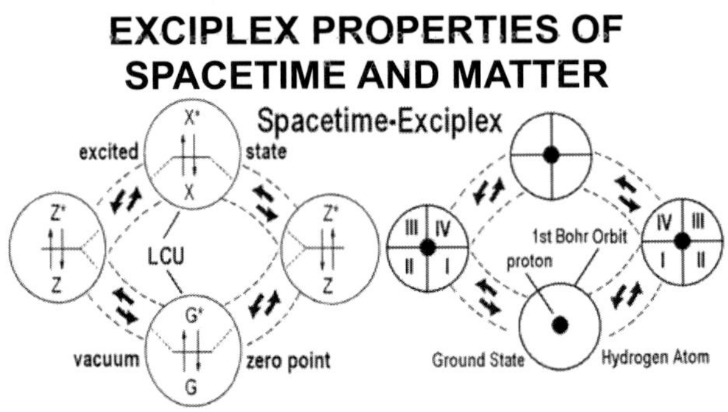

Figure 10. (a) Least cosmological unit (LCU) exciplex complex tiling the spacetime backcloth of HAM cosmology. The array of LCUs acts as a gating mechanism for entry of the U_F into every point in Minkowski 4-space and all matter [8]. (b) Conformal scale-invariant properties related to TBS in the hydrogen atom as it pertains to LSXD and the oscillating form of Planck’s constant fluctuating from the continuous-state asymptotic virtual Planck scale to the Larmor radius of the hydrogen atom 9,17].

Another way of looking at the experimental design set up is as a special case for manipulating the Lorentz Transformation [34,35] - Aspects of a spacetime exciplex model [8] in terms of restrictions imposed by Cramer's Transactional Interpretation [22] on mirror symmetry can be used for the putative detection of virtual tachyon-tardyon interactions in *zitterbewegun* [5]. We have designed a tachyon measurement experiment [36] by initially considering Bohr's starting point for the development of quantum theory, i.e. the emission of photons by atoms from quantum jumps between stable Bohr orbits. We do this from the point of view of the de Broglie-Bohm causal stochastic interpretation in order to take into consideration new laser experimental results by Kowalski [34,35].

Re-quoting Rowlands from pg. 1: "Fermions appear to be singularities rather than extended objects... requires a double, rather than a single, vector space, and this would seem to be confirmed by the double rotation required by spin ½ objects," [5]. This represents the crux of the nut to be cracked. As we already stated Rowlands' addition of an antispace is a brilliant extension of how to understand the nature of a fermionic vertex especially as we feel it represents a clear conceptualization of the lock-down of HD potentia components of the local quantum state 'resultant' that produces the uncertainty principle. The absolute nature of the fermionic singularity requires the structure of the LCU.

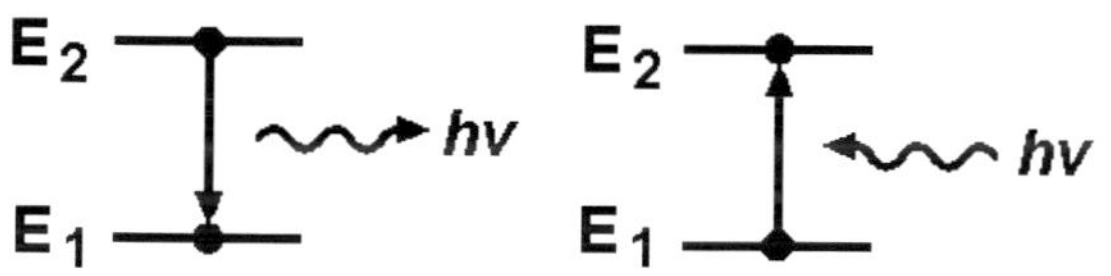

Figure 11. (a) Decrease in energy level E2 to E1 resulting in photon emission (squiggly arrow) with quantized energy, $\underline{h}\nu$. (b) An increase in energy level from E1 to E2 resulting from absorption of a photon (squiggly arrow) also with energy, $h\nu$. This is the simplistic basis for spectrographic analysis of the quanta in a particular atomic orbit. The TBS experiment anticipates 3 new spectral lines in atomic hydrogen under the novel conditions set up by the apparatus.

This is conceptually illustrated in the usual 2D illustration of a 3D space immersion of a 4D Klein bottle. Dissecting a Klein bottle into halves along its plane of symmetry results in two mirror image Möbius strips, one that has left-handed half-twist symmetry and the other with a right-handed half-twist symmetry. The Klein bottle unlike the Möbius strip is a closed 2D manifold without boundary; but unlike the Möbius strip which can be embedded in 3D Euclidean space, the Klein bottle can only be embed in a 4D or higher dimensional space. By adding a fourth dimension to the 3D space, the self-intersection can be eliminated. This is the property we wish to extend beyond the Rowlands singularity.

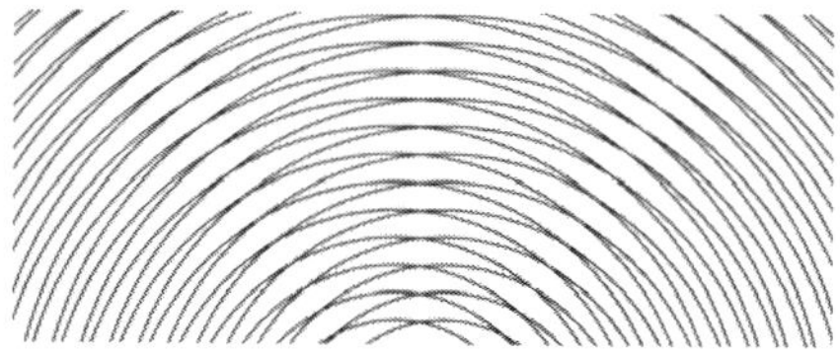

Figure 12. Periodic Moiré-like properties of the Continuous-state backcloth allow access to the LSXD regime by utilizing an rf-pulsed resonance hierarchy.

Just like Classical Mechanics had no access beyond 3D Euclidean space, so Quantum Mechanics or the Standard Model has no access beyond 4D Einstein-Minkowski-Riemann space. Thus the U_F Mechanics of NFT introduces an additional complex 8D space - A 12D total comprised of nine spatial and three temporal; but because of the 4D Minkowski resultant being virtual and the involute properties of the topology operationally the HD space is 6D, with three temporal dimensions and three 'energetic' dimensional control factors of the U_F acting like a 'super quantum potential' [8,21]. When this 12D structure is put into the HAM Continuous-State cycle with properties like a Moiré pattern added to the periodicity of the LCU gating mechanism; we are finally set up to understand the topological geometry of TBS parameters.

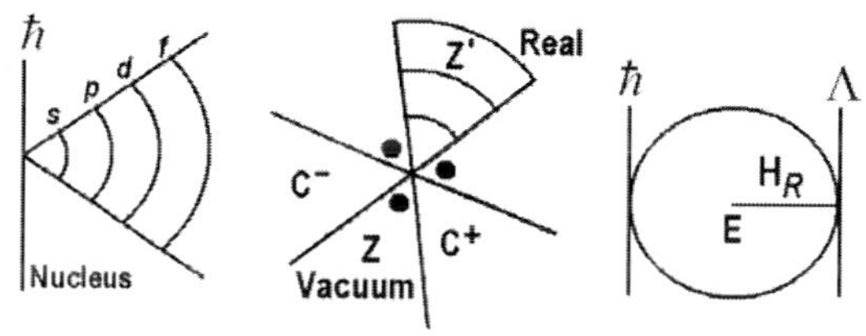

Figure 13(a).

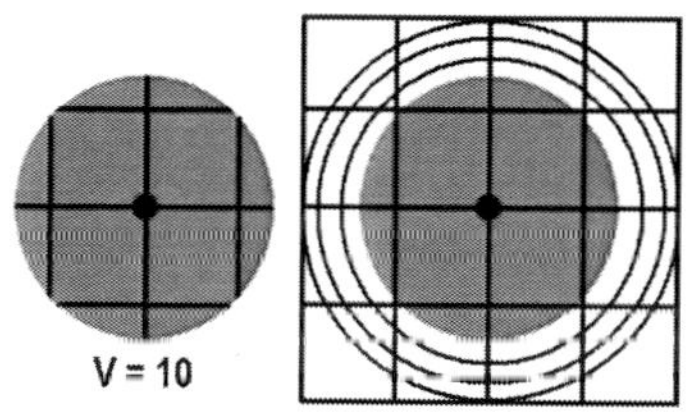

Figure 13(b). Historical evolution of atomic concepts. (a) 2D representation of 3D reduction cone of standard Bohr orbits. (b) *Z'* Real is same as (a), *Z* Vacuum is the Rowland's nilpotent doubling, C^+ & C^- complex Calabi-Yau mirror symmetric generators & the 3 dots represent continuous-state control elements. (c) Wave function domain of the Hubble Universe, $H = \Psi$ with a 1st and 2nd quantization, $\hbar$, Λ respectively of the Hubble radius, H_R with the Earth observer, E in center. (d) Conceptualized area, V =10; see text. (central dot is proton) for the usual representation of the *s* Bohr orbital in a hydrogen atom. (e) Same as (d) but with additional LSXD complex TBS space area represented. In reality the square would be folded up into a 3-torus or HD Klein bottle. The additional three concentric circles beyond the V = 10 volume of the gray circle representing the hydrogen atom *s* orbital are meant to represent the added TBS periodic LSXD 12-14-16 energy volume cycles of the continuous-state matter-wave annihilation-creation process.

As one knows light emitted from atoms during transitions of electrons from higher to lower energy states takes the form of photon quanta carrying energy and angular momentum. Any causal description of such a process implies that one adds to the restoring force of the harmonic oscillator an additional radiation (decelerating) resistance associated (derived from) with the electromagnetic (force) field of the emitted photon by the action-equal-reaction law. Any new causal condition thus implies that one must add a new force to the Coulomb force acting at random and which we suggest is related to ZPF vacuum resonant coupling and motions of the polarized Dirac aether. In this respect we prefer an extension of the model proposed by Rowlands [5] as mentioned above. We assume that the wave and particle aspects of electrons and photons are built with real extended spacetime structures containing internal oscillations of point-like electromagnetic topological charges, $e^{\pm}$ within an extended form of the causal stochastic interpretation of quantum mechanics.

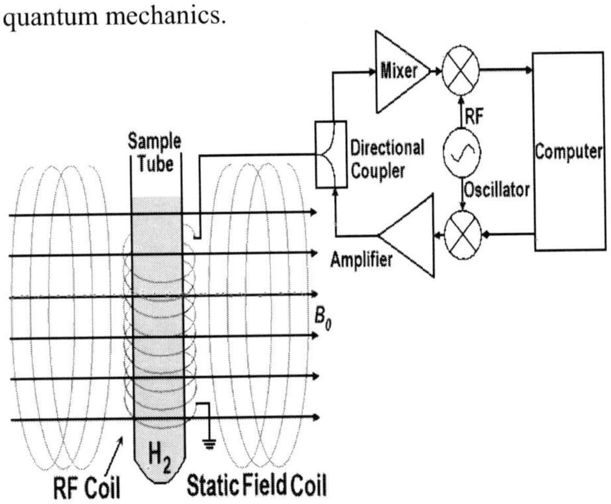

Figure 14. NMR apparatus designed to manipulate TBS in Hydrogen. The Fig. only shows possible details for rf-modulating TBS QED resonance, not the spectrographic recording and analysis components.

Kowalski's interpretation [34,35] drawn from recent laser experiments showing that emission and absorption between Bohr atomic states take place within a time interval equal to one period of the emitted-absorbed photon wave, the corresponding transition time is the time needed for the orbiting electron to travel one full orbit around the nucleus. Kowalski describes the additions to the usual theory of atomic structure. In the noetic theory the Kowalski additions, specifically the 'Kowalski time interval' are probably a key element in setting up the HD symmetry conditions of the TBS resonance hierarchy which can be further manipulated in terms of new parameters related to LSHD to test for U_F action and new cyclical continuous-state quantum cavity dynamics.

5. Experimental Design — Testing for and Manipulating TBS

Vigier et al. [17] has proposed TBS below the 1st Bohr orbit in the Hydrogen atom. Utilizing tenets of the original hadronic form of string theory such as a variable string tension, T_S where the Planck constant, $\hbar$ is replaced with a version of the original Stoney, λ where $\hbar$ is an asymptote never reached and instead oscillates from a virtual Planck asymptote to the Larmor radius of the hydrogen atom, i.e. the so-called Planck scale is a mathematical restriction (not physical) imposed by the limitations of the Copenhagen Interpretation and is not a fundamental physical barrier other than for Minkowski space [8].

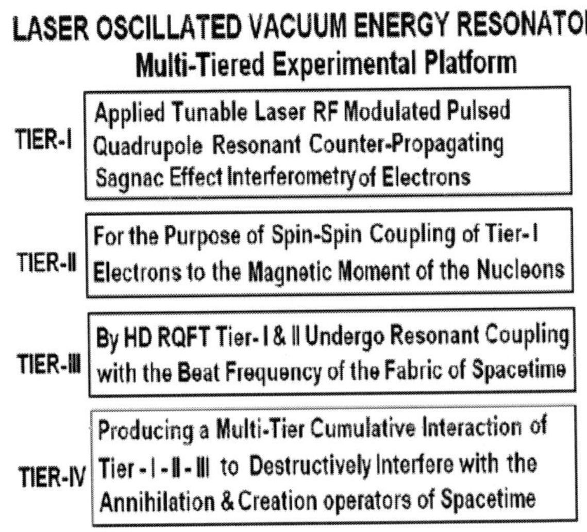

Figure 15(a).

Hierarchical Harmonic Oscillator Parameters	
classical	$X = A\cos(\omega t)$
quantum	$\frac{\hbar^2}{2m}\frac{d^2\psi}{dx^2} + \left(E - \frac{kx^2}{2}\right)\psi = 0$
annihilation creation	$x(t) = x_0\left[a\exp(-i\omega t) + a^{\dagger}\exp(i\omega t)\right]$
future-past retarded-advanced	$F_1 = F_0 e^{-ikx} e^{-2\pi i f t},\ F_2 = F_0 e^{ikx} e^{-2\pi i f t},$ $F_3 = F_0 e^{-ikx} e^{2\pi i f t},\ F_4 = F_0 e^{ikx} e^{2\pi i f t}$
incursive	$\frac{dx(t+\Delta t)}{dt} - v(t) = 0,\ \frac{dv(t+\Delta t)}{dt} + \omega^2 = 0$

Figure 15(b). (a) Top down list of the rf-resonance hierarchy for accessing HD space. (b) Basic conceptual mathematical components of the rf-pulsed applied harmonic oscillator resonance hierarchy in descending order: classical, quantum, relativistic, transactional and incursive are all required in order to achieve coherent control of the inherent conformal, scale-invariant properties of the HD Dirac polarized vacuum.

LSXD exist behind it. Key to operation of the experiment is what we have termed a 'couple-punch' (Table 1). Utilizing relativistic quantum field theory (RQFT) at the moment of spin-spin coupling an rf-pulse is 'kicked' at various phases of a Bessel function coordinate harmonically set to coincide with putative phases in the Continuous-State cycle of TBS [8-10]. This protocol is considered simplest and least expensive of the 12 empirical tests so far derived by the author [10-13].

Some experimental evidence has been found to support this view showing the possibility that the interaction of these extended structures in space involve real physical vacuum couplings by resonance with the subquantum Dirac ether. Because of photon mass the CSI model, any causal description implies that for photons carrying energy and momentum one must add to the restoring force of the harmonic oscillator an additional radiation (decelerating) resistance derived from the EM (force) field of the emitted photon by the action-equal-reaction law.

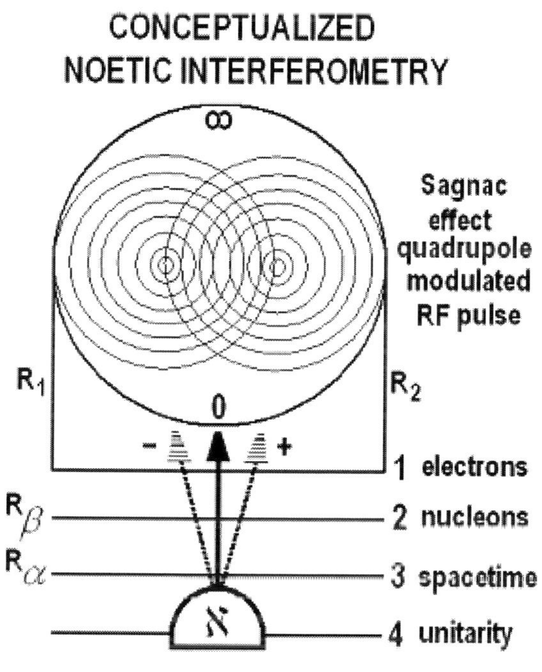

Figure 16. The spheroid is a 2D representation of a HD complex Riemann sphere able to spin-flip HD brane coupling dynamics from zero to infinity and act as a holophote or 'light house' exciplex pumping mechanism to mediate the action of the U_F into spacetime.

Kowalski has shown that emission and absorption between atomic states take place within a time interval equal to one period of the emitted or absorbed photon wave. The corresponding transition time corresponds to the time required to travel one full orbit around the nucleus. Individual photons are extended spacetime structures containing two opposite point-like charges rotating at a velocity near *c, at the opposite sides of a rotating diameter with a mass,* $m = 10^{-65}$ *g* and with an internal oscillation $E = m^2 = h\nu$. *Thus a new causal description implies the addition of a new component to the* Coulomb force acting randomly and may be related to quantum fluctuations. We believe this new relationship has some significance for our model of vacuum C-QED blackbody absorption/emission equilibrium.

6. Experimental Procedure

If for the sake of illustration we arbitrarily assume the *s* orbital of a hydrogen atom has an energy volume of 10 and the *p* orbital an energy volume of 20. To discover TBS we will investigate the possibility of heretofore unknown volume possibilities arising from cyclical fluctuations in LSXD Calabi-Yau mirror symmetry dynamics. This is in addition to the Vigier TBS model.

As in the perspective of rows of seats in an auditorium, rows of trees in an orchard or rows of headstones in a cemetery, from certain positions the line of sight is either open to infinity or blocked. This is the assumption we make about the continuous-state cyclicality of HD space.

Then if the theory has a basis in physical reality and we are able to measure it, we propose that at certain nodes in the cycle we would discover cavity volumes of say 12, 14, and 16. We propose the possibility of three LSXD cavity modes like 'phase locked loops' depending on the cycle position - maximal, intermediate and minimal.

The mechanical engineering property of perfect rolling motion or perfect rolling contact is another metaphor for illustrating how the continuous-state cycle operates. A logarithmic spiral can be sliced into curved segments; and ellipsoids of various radii can be constructed from the curved segments. If the spheroids are made to touch and roll without slippage during some rotational cycles the points of contact will return to the original position, i.e. forming a pattern of open and closed moments like orchard rows.

TABLE I

KICK	PULL	P + K	K + P
R	R	R + A	R + A
A	A	A + R	A + R
R + A	R + A	A = R	
A + R	A + R		
A = R	A = R		

Table I. Possible TBS experimental Kick-Pull (*K-P*) coupling mode parameters. *R* signifies a retarded *K* or *K* coupling, *A* respectively of advanced *K* coupling. The plus (+) sign signifies sequential order and the equals sign (=) means simultaneous action for a total of 15 experimental *KP* coupling options. Table 1 shows 15 ways to test Calab-Yau symmetric - CQED TBS parameters and LSXD.

7. Nine Empirical Tests of Noetic-Field-Theory Cosmology

A partial list of the 15 current putative empirical tests relating to both demonstrating aspects of noetic cosmology as well as mind-body parameters is listed. If experiments prove viable we anticipate a new class of research platform for studying properties of the spacetime vacuum, the structure of matter with bench top apparatus instead large accelerators and fundamental aspects of living systems. We propose nine derivatives of the experimental protocol to test the HD continuous-state hypotheses:

1. Basic Experiment - Fundamental test that the concatenation of principles is theoretically sound. A laser oscillated rf-pulsed Sagnac Effect resonance hierarchy set to interfere with the periodic conformal scale-invariant structure of the inherent 'beat frequency' of spacetime in a covariant Dirac polarized vacuum to detect the new action principle associated with a cyclical entry of the Unified Field, U_F into 4-space.
2. Bulk Quantum Computing - Utilizing protocol (1) Bulk Universal QC can be achieved by superseding the quantum uncertainty principle. (see [6] for details) Programming and data I/O are performed without decoherence by utilizing the inherent conformal scale-invariant mirror symmetry properties that act like a 'synchronization backbone' [6,8] whereby the local quantum state is causally free (measureable without decoherence) at a specific HD node in the continuous-state conformal symmetry cycle.
3. Protein Conformation - Utilizing aspects of protocols (1 & 2) dual Hadamard quantum logic gates are set as a Cavity-QED spacetime cellular automata experiment to facilitate conformational propagation in the prion protein from normal to the pathological form [39,40].
4. Manipulating a special case of the Lorentz Transformation [9] - Aspects of a spacetime exciplex model [8,9] in terms of restrictions imposed by Cramer's Transactional Interpretation on mirror symmetry can be used for the putative detection of virtual tachyon-tardyon interactions in *Zitterbewegung* [40]. We design our tachyon measurement experiment by initially considering Bohr's starting point for the development of quantum theory, i.e. the emission of photons by atoms from quantum jumps between stable Bohr orbits. We do this from the point of view of the de Broglie-Bohm causal stochastic interpretation in order to take into consideration new laser experimental results described by Kowalski [42]. As one knows light emitted from atoms during transitions of electrons from higher to lower energy states takes the form of photon quanta carrying energy and angular momentum. Any causal description of such a process implies that one adds to the restoring force of the harmonic oscillator an additional radiation (decelerating) resistance associated (derived from) with the electromagnetic (force) field of the emitted photon by the action equal reaction law. Any new causal condition thus implies that one must add a new force to the Coulomb force acting at random and which we suggest is related to ZPF vacuum resonant coupling and motions of the polarized Dirac aether. We assume that the wave and particle aspects of electrons and photons are built with real extended spacetime structures containing internal oscillations of point-like electromagnetic topological charges, e^{+-} within an extended form of the causal stochastic interpretation of quantum mechanics. Kowalski's interpretation drawn from recent laser experiments showing that emission and absorption between Bohr atomic states take place within a time interval equal to one period of the emitted-absorbed photon wave, the corresponding transition time is the time needed for the orbiting electron to travel one full orbit around the nucleus. Kowalski describes the additions to the usual theory of atomic structure. In the noetic theory the Kowalski additions can be further manipulated in terms of new parameters related to large scale HD to test for U_F action and new quantum cavity dynamics.
5. Extended Quantum Theory - Test of causal properties of de Broglie-Bohm-Vigier quantum theory by utility of the U_F holophote effect (protocol 1) as a "super" quantum potential to summate by constructive interference the density of de Broglie matter waves.
6. Coherent Control of Quantum Phase - Additional test of de Broglie-Bohm for existence of a nonlocal 'pilot- quantum potential' to manipulating the phase 'space quantization' in the double slit experiment by controlling which slit quanta passes through.
7. Manipulating Spacetime Structure - (similar to protocol 6) Test of conformal scale-invariant properties of the putative Dirac conformal polarized vacuum, a possible 'continuous-state' property related to an arrow of time (Similar to basic experiment, but more advanced)
8. Testing for and Manipulating Tight Bound States (TBS) [9] - (similar to protocol 4) Vigier [17] has proposed TBS below the 1st Bohr orbit in the Hydrogen atom. Utilizing tenets of the original hadronic form of string theory such as a variable string tension, T_S where the Planck constant, $\hbar$ is replaced with a version of the original Stoney, λ where $\hbar$ is an asymptote never reached and instead oscillates from virtual Planck to the Larmor radius of

the hydrogen atom, i.e. the so-called Planck scale is a restriction imposed by the limitations of the Copenhagen Interpretation and is not a fundamental physical barrier. Large scale XD exist behind it. Key to operation of the experiment is what we have termed a 'couple-punch'. Utilizing relativistic quantum field theory (RQFT) at the moment of spin-spin coupling an rf-pulse is kicked at various phases of a Bessel function harmonically set to coincide with putative phases in the cycle of TBS. [9]. Considered simplest and least expensive empirical test.

9. Test of the Unique String Vacuum - Until now the structure of matter has been explored by building ever bigger supercolliders like the LHC. If the model described here for accessing HD space in terms of a Dirac covariant polarized vacuum proves to be correct utilizing the inherent conformal scale-invariant mirror symmetry properties of de Broglie matter waves will allow examining the various cross sections in the structure of matter in symmetry interaction during the cyclic continuous-state future-past annihilation-creation modes of matter in the spacetime metric.

Experimental access to vacuum structure or for surmounting the uncertainty principle can be done by two similar methods. One is to utilize an atomic resonance hierarchy and the other an incursive spacetime resonance hierarchy.

8. Conclusions

Utilizing a postulates of the recently formulated paradigm shift in cosmology called the Holographic Anthropic Multiverse [8,11] that employs extensions of quantum theory, gravitation and electromagnetism, one is able to design experiments that putatively open the door to the 3rd regime of reality (classical-quantum-unified). If the model proves correct it will be the beginning of empirical access to tenets of unified field mechanics opening the door to a battery of new technologies including bulk universal quantum computing [8].

The key aspect is the cyclical nature of the continuous-state hypothesis that provides a simple method for surmounting quantum uncertainty, suggested to be a limiting factor of the Copenhagen Interpretation of quantum theory; 'behind' which the door opens to large-scale extra dimensions wherein lies the regime of action of the unified field and the teleology of mind-body interaction.

We have proposed a simple inexpensive experiment that suggests three new spectral lines in the hydrogen atom. The protocol operates by utility of a simple rf-pulsed resonance hierarchy timed to coincide with specific parameters of the continuous-state process when periodic openings occur into the higher dimensional regime.

With a final whimsical sortie through the entelechies of your imagination we mention a theory about the magnetic sense of birds that assumes an as yet undiscovered protein in the retina of the eye is changed due to the Zeeman effect. The door is now open and cannot be shut again. We are curious to see who will be the experimentalist. Will it be you...?

References and Notes

[1] The slang phrase 'Shut the Front Door' was originally coined by Buckethead on 101.1 WJRR Radio, Orlando, Fl, USA, www.wjrr.com.

[2] Chantler, C.T. et al. (2012) Testing three-body quantum electrodynamics with trapped Ti^{20+} ions: Evidence for a Z-dependent divergence between experiment and calculation, PRL 109, 153001.

[3] Chantler, C. T. (2004). Discrepancies in quantum electrodynamics, Radiation Physics and Chemistry, 71(3), 611-617.

[4] Pohl, R., Antognini, A., Nez, F. et al. (2010) The size of the proton, Nature (London) 466; 7303: 213-216.

[5] Rowlands, P. (2013) Space and Antispace, in RL Amoroso, LH Kauffman & P. Rowlands (eds.) The Physics of Reality: Space, Time, Matter, Cosmos, Singapore: World Scientific Publishers.

[6] Amoroso, R. L., Kauffman, L. H. & Giandinoto, S. (2013) Universal Quantum Computing; 3rd Gen Prototyping Utilizing Relativistic 'Trivector' Qubit Modeling Surmounting Uncertainty, in Amoroso, R. L., Kauffman, L. H. & Rowlands, P. (eds.) The Physics of Reality: Space, Time, Matter, Cosmos, Singapore: World Scientific.

[7] Amoroso, R. L. & Kauffman, L. H. & Rowlands, P. (2013) Exploring Novel Cyclical Extensions of Hamilton's Dual-Quaternion Algebra, in Amoroso, R. L., Kauffman, L. H. & Rowlands, P. (eds.) The Physics of Reality: Space, Time, Matter, Cosmos, Singapore: World Scientific.

[8] Amoroso, R. L., & Rauscher, E. A. (2009) The Holographic Anthropic Multiverse: Formalizing the Geometry of Ultimate Reality, Singapore: World Scientific.

[9] Amoroso, R.L. & Vigier, J-P (2013) Evidencing 'tight bound states' in the hydrogen atom: Empirical manipulation of large-scale XD in violation of QED, in RL Amoroso, LH Kauffman & P. Rowlands (eds.), The Physics of Reality: Space, Time, Matter, Cosmos, Singapore: World Scientific.

[10] Amoroso, R.L. (2013) Empirical protocol for mediating long-range coherence in biological systems, in press.

[11] Amoroso, R.L. (2010) (ed.) The Complementarity of Mind and Body: Realizing the Dream of Descartes, Einstein and Eccles, New York: Nova Science Publishers.

[12] Amoroso, R. L. (2013) Crossing the Psycho-Physical Bridge: Elucidating the objective character of experience.

[13] Amoroso, R. L. (2012) Physical Parameters of Mind-Body Interaction: Breaking the 1st Person 3rd Person Barrier, Journal of Nonlocality, 1(1).

[14] Gerlach, W. & Stern, O. (1922). *Das magnetische Moment des Silberatoms. Zeitschrift für Physik*, 9: 353–355.

[15] Amoroso, R. L. (2013) A Simple Method for Surmounting Quantum Uncertainty, in Press.
[16] Osoroma, D.S. (2013) programmable cellular automata properties of covariant dirac polarized spacetime, in RL Amoroso, LH Kauffman & P. Rowlands (eds.) Physics of Reality: Space, Time, Matter, Cosmos, London: World Sci.
[17] Dragic´, M., Maric´, A.Z. & Vigier, J-P (2000) New quantum mechanical tight bound states and 'cold fusion' experiments, Physics Letters A: 265;163–167.
[18] Nishino, H. et al. (2009) Super-K Collaboration Search for proton decay via $p^+ \rightarrow e^+\pi^0$ & $p^+ \rightarrow \mu+\pi^0$ in a Large Water Cherenkov Detector, Phys Rev Let 102: 14; 141801.
[19] Sreekantan, B. V. (1984) Searches for proton decay and superheavy magnetic monopoles, Journal of Astrophysics and Astronomy 5 (3): 251–271.
[20] Bohm, D. & Vigier, J-P (1954) Model of the causal interpretation of quantum theory in terms of a fluid with irregular fluctuations, Phys. Rev. 96:1; 208-217.
[21] Holland, P.R. (1995) The Quantum Theory of Motion: An Account of the de Broglie-Bohm Causal Interpretation of Quantum Mechanics, Cambridge: Cambridge Univ. Press.
[22] Cramer, J. (1986) The transactional interpretation of quantum mechanics, Rev Mod Phys 58, 647-687.
[23] Wheeler, J.A. & Feynman, R. P. (1945) Interaction with the absorber as the mechanism of radiation, Rev. Mod. Phys. 17, 1578.
[24] Feynman, R. P. (1985) QED: The Strange Theory of Light and Matter, Princeton: Princeton Univ. Press.
[25] Lehnert, B. (1994) Basic concepts of an extended electromagnetic field theory, Speculations in Science and Technology-Complete Edition, *17*(4), 259-266.
[26] Lehnert, B. (2006) An Extended Electromagnetic Theory, *Physica Scripta*,1999 (T82), 89.
[27] Lehnert, B., & Scheffel, J. (2006). On the minimum elementary charge of an extended electromagnetic theory. *Physica Scripta*, 65(3), 200.
[28] Hainzl, C., Lewin, M., & Séré, É. (2005). Existence of a stable polarized vacuum in the Bogoliubov-Dirac-Fock approximation. Comm. in Math. Physics, *257*(3), 515-562.
[29] Cardone, F., & Mignani, R. (1999). Energy-dependent metric for gravitation from clock-rate experiments. International Journal of Modern Physics A, 14(24), 3799-3811.
[30] Cardone, F., Francaviglia, M., & Mignani, R. (1999). Energy-Dependent Phenomenological Metrics & Five-Dimensional Einstein Equations.Foundations of Physics Letters, 12(3), 281-289.
[31] Misner, Thorne, K. & Wheeler, J.A. (1990) Gravitation, San Francisco: AOS Press.
[32] Witten, E. (1996) Reflections on the fate of spacetime, Physics Today (April), pp. 24-30.
[33] Born, Max; Oppenheimer, J. Robert (1927). *Zur Quantentheorie der Molekeln* (On the Quantum Theory of Molecules) *Annalen der Physik* 389 (20): 457–484.
[34] Kowalski, M. (2003). The Process of photon emission from atomic hydrogen, in RL Amoroso et al (eds) Gravitation and Cosmology: From the Hubble Radius to the Planck Scale, 207-222, Dordrecht: Kluwer Academic.
[35] Kowalski, M. (1999) Classical description of photon emission from atomic hydrogen, Phys. Essays, 12:2, 312-331.
[36] Amoroso, R.L., Kauffman, L. H., Rauscher, E. A., Rowlands, P., & Vigier, J. P. (2010). "Hidden" parameters describing internal motion within extended particle elements, in R.L. Amoroso, P. Rowlands &S. Jeffers (eds.) Search for Fundamental Theory: The VII International Symposium Honoring French Mathematical Physicist Jean-Pierre Vigier, AIP Conf. Proc., Vol. 1316, pp. 1-20.
[37] Amoroso, R. L., & Rauscher, E. A. (2010, December). Empirical Protocol for Measuring Virtual Tachyon⁄ Tardon Interactions in a Dirac Vacuum. In AIP Conference Proceedings (Vol. 1316, p. 199-222).
[38] Luminet, J-P (2003) Dodecahedral space topology as an explanation for weak wide-angle temperature correlations in the cosmic microwave background, arXiv:astro-ph/0310253v1.
[39] Chu, M-Y.J. & Amoroso, R.L. (2008) Empirical mediation of the primary mechanism initiating protein conformation in prion propagation, in D. Dubois (ed.) Partial Proceedings of CASYS07, IJCAS, Vol. 22, Univ. Liege Belgium.
[40] Varela, F.G., Maturana, H.R. & Uribe, R. (1974) Autopoiesis: The organization of living systems, its characterization and a model, BioSystems, 5, 187-196.
[41] Amoroso, R.L. & Rauscher, E.A. (2011) in Richard L. Amoroso, Peter Rowlands, Stanley Jeffers (eds.) Search for Fundamental Theory: The VII International Symposium Honoring French Mathematical Physicist Jean-Pierre Vigier, AIP conference proceedings, no. 1316.
[42] Kowalski, M, (2002) The process of photon emission from atomic hydrogen, in R.L. Amoroso et al (eds.), Gravitation and Cosmology: From the Hubble Radius to the Planck Scale, 207-222, Kluwer Academic Publishers.

Index